INTERNATIONAL UNION OF CRYSTALLOGRAPHY BOOK SERIES

This volume forms part of a series of books sponsored by the International Union of Crystallography (IUCr) and published by Oxford University Press. There are three IUCr series: IUCr Monographs on Crystallography, which are in-depth expositions of specialized topics in crystallography; IUCr Texts on Crystallography, which are more general works intended to make crystallographic insights available to a wider audience than the community of crystallographers themselves; and IUCr Crystallographic Symposia, which are essentially the edited proceedings of workshops or similar meetings supported by the IUCr.

IUCr Texts on Crystallography

1 *The solid state: From superconductors to superalloys*
A. Guinier and R. Jullien, *translated by* W. J. Duffin

2 *Fundamentals of crystallography*
C. Giacovazzo, *editor*

IUCr Monographs on Crystallography

1 *Accurate molecular structures: Their determination and importance*
A. Domenicano and I. Hargittai, *editors*

2 *P. P. Ewald and his dynamical theory of X-ray diffraction*
D. W. J. Cruickshank, H. J. Juretschke, and N. Kato, *editors*

3 *Electron diffraction techniques, Volume 1*
J. M. Cowley, *editor*

4 *Electron diffraction techniques, Volume 2*
J. M. Cowley, *editor*

5 *The Rietveld method*
R. A. Young, *editor*

IUCr Crystallographic Symposia

1 *Patterson and Pattersons: Fifty years of the Patterson function*
J. P. Glusker, B. K. Patterson, and M. Rossi, *editors*

2 *Molecular structure: Chemical reactivity and biological activity*
J. J. Stezowski, J. Huang, and M. Shao, *editors*

3 *Crystallographic computing 4: Techniques and new technologies*
N. W. Isaacs and M. R. Taylor, *editors*

4 *Organic crystal chemistry*
J. Garbarczyk and D. W. Jones, *editors*

5 *Crystallographic computing 5: From chemistry to biology*
D. Moras, A. D. Podjarny, and J. C. Thierry, *editors*

Fundamentals of Crystallography

C. GIACOVAZZO, H. L. MONACO, D. VITERBO
F. SCORDARI, G. GILLI, G. ZANOTTI, M. CATTI

Edited by

C. GIACOVAZZO

Dipartimento Geomineralogico, University of Bari, Italy
and Istituto di Ricerca per lo Sviluppo delle Metodologie Cristallografiche,
CNR, Bari, Italy

INTERNATIONAL UNION OF CRYSTALLOGRAPHY
OXFORD UNIVERSITY PRESS
1992

Oxford University Press, Walton Street, Oxford OX2 6DP

Oxford New York Toronto
Delhi Bombay Calcutta Madras Karachi
Petaling Jaya Singapore Hong Kong Tokyo
Nairobi Dar es Salaam Cape Town
Melbourne Auckland

and associated companies in
Berlin Ibadan

Oxford is a trade mark of Oxford University Press

Published in the United States
by Oxford University Press, New York

Parts of this work first appeared as Introduzione alla cristallografia moderna by M. Bolognesi, A. Coda, C. Giacovazzo, G. Gilli, F. Scordari, and D. Viterbo, edited by C. Giacovazzo, and published by Edizioni Fratelli Laterza, © 1985

A catalogue record for this book is available from the British Library

Library of Congress Cataloging in Publication Data
Fundamentals of crystallography/C. Giacovazzo . . . [et al.]; edited by C. Giacovazzo. (International Union of Crystallography texts on crystallography; 2). Includes bibliographical references.
1. Crystallography. I. Giacovazzo, Carmelo. II. Series. QD905.2.F86 1992 548—dc20 91-27271

ISBN 0 19 855579 2 (h/b)
ISBN 0 19 855578 4 (p/b)

Typeset by The Universities Press (Belfast) Ltd
Printed and bound in Great Britain by
The Bath Press, Avon

Preface

Crystallography, the science concerned with the study of crystals, is a very old subject. However, only in this century has it developed into a modern science, after the discovery of X-rays and their diffraction by crystals. In recent years crystallography has assumed an increasingly important role in the modern sciences because of its interdisciplinary nature, which has acted as a bridge between, and often as a stimulus for, various rapidly evolving disciplines. Indeed, Chemistry, Physics, Earth Sciences, Biology, Mathematics, and Materials Science have all provided stimuli to the development of new crystallographic interests and techniques. In turn, crystallography has significantly contributed to the advancement of these sciences. Thus, while on the one hand crystallography has been enriched, on the other hand writing a textbook describing all of its aspects has been made more difficult.

Recently, the demand for a compact book that gives a comprehensive account of the modern crystallographic subjects has increased. This volume should therefore be a useful and handy textbook for university courses that cover crystallography, fully or only partially. It should also be useful at the more advanced level required for doctorate studies as well as for experienced researchers.

It was with these ideas in mind that I first set out to co-ordinate the publication, in 1985, of a textbook in Italian (*Introduzione alla cristallografia moderna,* Edizioni Fratelli Laterza, Bari) of which *Fundamentals of crystallography* is not only an English translation, but a completely revised and updated version with a new chapter on crystal physics. It was clear to me that (a) the book had to be written by several authors in order to take advantage of their specific expertise; (b) the different chapters had to be carefully harmonized in order to conform them to a unified plan.

It seems to me that these two requirements are even more valid today and their achievement is entirely due to the creative co-operation of the co-authors of this book.

Two of the co-authors of the Italian textbook, M. Bolognesi and A. Coda, were unable to carry out the translation and revision of their chapters. I wish to express my thanks for their valuable contribution to the previous edition. In this book their topics are treated by G. Zanotti and H. L. Monaco. An additional chapter on crystal physics has been written by M. Catti. I thank the three new authors for entering our team and all the authors for their enthusiastic participation in this project.

Bari, Italy C.G.
August 1991

Acknowledgements

We have the pleasant task of expressing our gratitude to Dr C. K. Prout and to Dr D. J. Watkin for encouraging us to write this book and for their useful advice. We have also to thank many friends and colleagues who have commented on the manuscript or given technical support. The cover picture was kindly provided by Professor A. Zecchina and Dr S. Bordiga, who used the Chem-X software of Chemical Design Ltd, Oxford, UK. Thanks are also due from C. Giacovazzo to Mr D. Trione for his accurate and rapid typing of the manuscript, to Mr G. Pellegrino for his aid with the drawings and to Ms C. Chiarella for her technical support; from H. L. Monaco to Mr M. Tognolin for the figures, to Mr R. Pavan for the photographs, and to Mrs A. Migliazza for typing the manuscript; from F. Scordari to Professor F. Liebau for the critical reading of the silicate section and for helpful suggestions, Mr Trione is also acknowledged for the typing of the manuscript; from G. Gilli to Professor V. Bertolasi, Dr V. Ferretti, and Dr P. Gilli for their help in the critical reading of the manuscript; from M. Catti to Mr M. Bandera for his aid with the drawings.

Contents

Nec possunt oculi naturam noscere rerum: proinde animi vitium hoc oculis adfingere noli.

Lucretius, *De Rerum Natura,* IV, 386–7.

Contributors

Carmelo Giacovazzo, Dipartimento Geomineralogico, Università di Bari, Campus Universitario, 70124 Bari, Italy; Istituto di Ricerca per lo Sviluppo delle Metodologie Cristallografiche, CNR, 70124 Bari, Italy.

Hugo L. Monaco, Dipartimento di Genetica e Microbiologia, Sezione di Biologia Molecolare e Biofisica, Università di Pavia, Via Abbiategrosso 207 27100, Pavia, Italy.

Davide Viterbo, Dipartimento di Chimica, Inorganica, Chimica Fisica e Chimica dei Materiali, Università di Torino, Via P. Giuria 7, 10125 Torino, Italy.

Fernando Scordari, Dipartimento Geomineralogico, Università di Bari, Campus Universitario, 70124 Bari, Italy.

Gastone Gilli, Dipartimento di Chimica and Centro di Strutturistica Diffrattometrica, Università di Ferrara, Via L. Borsari 46, 44100 Ferrara, Italy.

Giuseppe Zanotti, Dipartimento di Chimica Organica, Università di Padova, Via Marzolo 1, 35126 Padova, Italy.

Michele Catti, Dipartimento di Chimica Fisica ed Elettrochimica, Università di Milano, Via Golgi 19, 20133 Milano, Italy.

Symmetry in crystals

1

CARMELO GIACOVAZZO

The crystalline state and isometric operations

Matter is usually classified into three states: gaseous, liquid, and solid. Gases are composed of almost isolated particles, except for occasional collisions; they tend to occupy all the available volume, which is subject to variation following changes in pressure. In liquids the attraction between nearest-neighbour particles is high enough to keep the particles almost in contact. As a consequence liquids can only be slightly compressed. The thermal motion has sufficient energy to move the molecules away from the attractive field of their neighbours; the particles are not linked together permanently, thus allowing liquids to flow.

If we reduce the thermal motion of a liquid, the links between molecules will become more stable. The molecules will then cluster together to form what is macroscopically observed as a rigid body. They can assume a random disposition, but an ordered pattern is more likely because it corresponds to a lower energy state. This ordered disposition of molecules is called the **crystalline state**. As a consequence of our increased understanding of the structure of matter, it has become more convenient to classify matter into the three states: gaseous, liquid, and crystalline.

Can we then conclude that all solid materials are crystalline? For instance, can common glass and calcite (calcium carbonate present in nature) both be considered as crystalline? Even though both materials have high hardness and are transparent to light, glass, but not calcite, breaks in a completely irregular way. This is due to the fact that glass is formed by long, randomly disposed macromolecules of silicon dioxide. When it is formed from the molten state (glass does not possess a definite melting point, but becomes progressively less fluid) the thermal energy which remains as the material is cooled does not allow the polymers to assume a regular pattern. This disordered disposition, characteristic of the liquid state, is therefore retained when the cooling is completed. Usually glasses are referred to as **overcooled liquids**, while non-fluid materials with a very high degree of disorder are known as **amorphous solids**.

A distinctive property of the crystalline state is a regular repetition in the three-dimensional space of an object (as postulated as early as the end of the eighteenth century by R. J. Haüy), made of molecules or groups of molecules, extending over a distance corresponding to thousands of molecular dimensions. However, a crystal necessarily has a number of defects at non-zero temperature and/or may contain impurities without losing its order. Furthermore:

1. Some crystals do not show three-dimensional periodicity because the

basic crystal periodicity is modulated by periodic distortions incommensurated with the basic periods (i.e. in incommensurately modulated structures, IMS). It has, however, been shown (p. 171 and Appendix 3.E) that IMSs are periodic in a suitable (3 + d)-dimensional space.

2. Some polymers only show a bi-dimensional order and most fibrous materials are ordered only along the fiber axis.

3. Some organic crystals, when conveniently heated, assume a state intermediate between solid and liquid, which is called the **mesomorphic** or **liquid crystal** state.

These examples indicate that periodicity can be observed to a lesser or greater extent in crystals, depending on their nature and on the thermodynamic conditions of their formation. It is therefore useful to introduce the concept of a **real crystal** to stress the differences from an ideal crystal with perfect periodicity. Although non-ideality may sometimes be a problem, more often it is the cause of favourable properties which are widely used in materials science and in solid state physics.

In this chapter the symmetry rules determining the formation of an ideal crystalline state are considered (the reader will find a deeper account in some papers devoted to the subject, or some exhaustive books,[1–5] or in the theoretical sections of the *International Tables for Crystallography*).[6]

In order to understand the periodic and ordered nature of crystals it is necessary to know the operations by which the repetition of the basic molecular motif is obtained. An important step is achieved by answering the following question: given two identical objects, placed in random positions and orientations, which operations should be performed to superpose one object onto the other?

The well known coexistence of **enantiomeric** molecules demands a second question: given two **enantiomorphous** (the term enantiomeric will only be used for molecules) objects, which are the operations required to superpose the two objects?

An exhaustive answer to the two questions is given by the theory of **isometric transformations**, the basic concepts of which are described in Appendix 1.A, while here only its most useful results will be considered.

Two objects are said to be **congruent** if to each point of one object corresponds a point of the other and if the distance between two points of one object is equal to the distance between the corresponding points of the other. As a consequence, the corresponding angles will also be equal in absolute value. In mathematics such a correspondence is called **isometric**.

The congruence may either be **direct** or **opposite**, according to whether the corresponding angles have the same or opposite signs. If the congruence is direct, one object can be brought to coincide with the other by a convenient **movement** during which it behaves as a rigid body. The movement may be:

(1) a **translation**, when all points of the object undergo an equal displacement in the same direction;

(2) a **rotation** around an axis; all points on the axis will not change their position;

(3) a **rototranslation** or **screw** movement, which may be considered as the combination (product) of a rotation around the axis and a translation along the axial direction (the order of the two operations may be exchanged).

If the congruence is opposite, then one object will be said to be enantiomorphous with respect to the other. The two objects may be brought to coincidence by the following operations:

(1) a symmetry operation with respect to a point, known as **inversion**;

(2) a symmetry operation with respect to a plane, known as **reflection**;

(3) the product of a rotation around an axis by an inversion with respect to a point on the axis; the operation is called **rotoinversion**;

(4) the product of a reflection by a translation parallel to the reflection plane; the plane is then called a **glide plane**.

(5) the product of a rotation by a reflection with respect to a plane perpendicular to the axis; the operation is called **rotoreflection**.

Symmetry elements

Suppose that the isometric operations described in the preceding section, not only bring to coincidence a couple of congruent objects, but act on the entire space. If all the properties of the space remain unchanged after a given operation has been carried out, the operation will be a **symmetry operation. Symmetry elements** are points, axes, or planes with respect to which symmetry operations are performed.

In the following these elements will be considered in more detail, while the description of translation operators will be treated in subsequent sections.

Axes of rotational symmetry

If all the properties of the space remain unchanged after a rotation of $2\pi/n$ around an axis, this will be called a symmetry axis of order n; its written symbol is n. We will be mainly interested (cf. p. 9) in the axes 1, 2, 3, 4, 6. Axis 1 is trivial, since, after a rotation of 360° around whatever direction the space properties will always remain the same. The graphic symbols for the 2, 3, 4, 6 axes (called two-, three-, four-, sixfold axes) are shown in Table 1.1. In the first column of Fig. 1.1 their effects on the space are illustrated. In keeping with international notation, an object is represented by a circle, with a + or − sign next to it indicating whether it is above or below the page plane. There is no graphic symbol for the 1 axis. Note that a 4 axis is at the same time a 2 axis, and a 6 axis is at the same time a 2 and a 3 axis.

Table 1.1. Graphical symbols for symmetry elements: (a) axes normal to the plane of projection; (b) axes 2 and 2_1 parallel to the plane of projection; (c) axes parallel or inclined to the plane of projection; (d) symmetry planes normal to the plane of projection; (e) symmetry planes parallel to the plane of projection

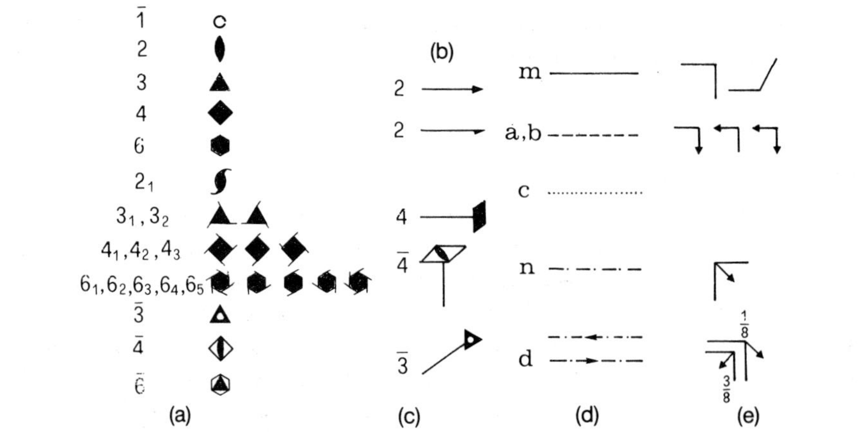

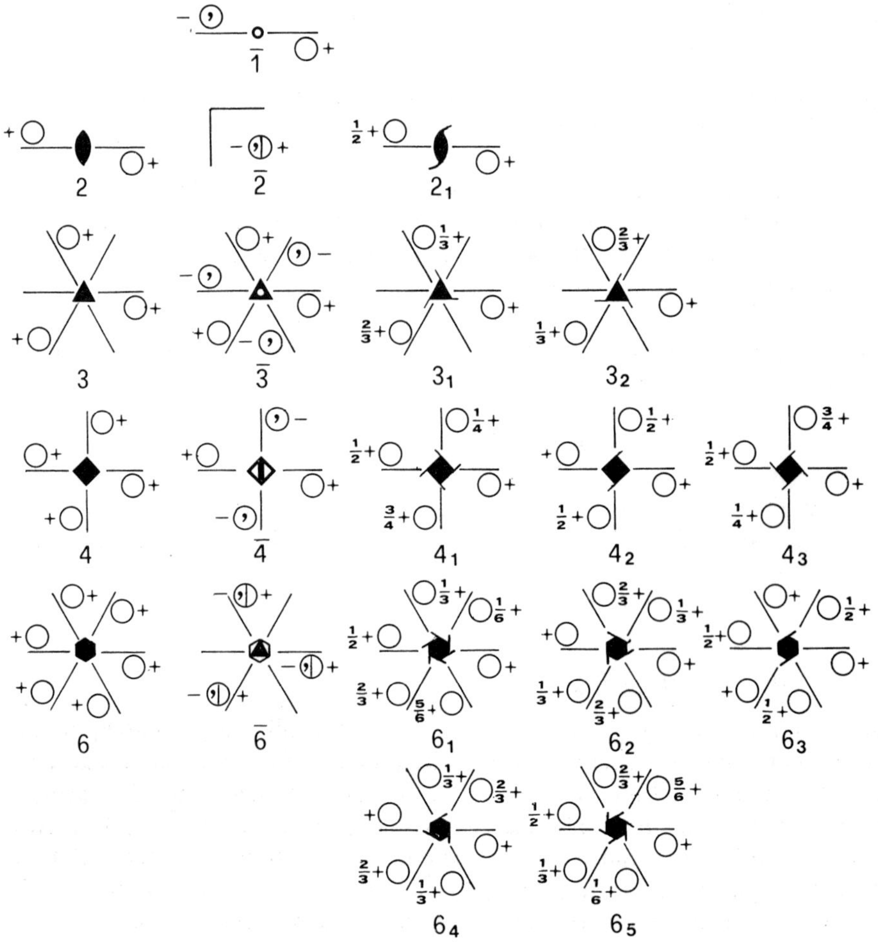

Fig. 1.1. Arrangements of symmetry-equivalent objects as an effect of rotation, inversion, and screw axes.

Axes of rototranslation or screw axes

A rototranslational symmetry axis will have an order n and a translational component $\boldsymbol{t}$, if all the properties of the space remain unchanged after a $2\pi/n$ rotation around the axis and the translation by $\boldsymbol{t}$ along the axis. On p. 10 we will see that in crystals only screw axes of order 1, 2, 3, 4, 6 can exist with appropriate translational components.

Axes of inversion

An inversion axis of order n is present when all the properties of the space remain unchanged after performing the product of a $2\pi/n$ rotation around the axis by an inversion with respect to a point located on the same axis. The written symbol is $\bar{n}$ (read 'minus n' or 'bar n'). As we shall see on p. 9 we will be mainly interested in $\bar{1}$, $\bar{2}$, $\bar{3}$, $\bar{4}$, $\bar{6}$ axes, and their graphic symbols are given in Table 1.1, while their effects on the space are represented in the second column of Fig. 1.1. According to international notation, if an object is represented by a circle, its enantiomorph is depicted by a circle with a comma inside. When the two enantiomorphous objects fall one on top of the other in the projection plane of the picture, they are represented by a single circle divided into two halves, one of which contains a comma. To each half the appropriate + or − sign is assigned.

We may note that:

(1) the direction of the $\bar{1}$ axis is irrelevant, since the operation coincides with an inversion with respect to a point;

(2) the $\bar{2}$ axis is equivalent to a reflection plane perpendicular to it; the properties of the half-space on one side of the plane are identical to those of the other half-space after the reflection operation. The written symbol of this plane is m;

(3) the $\bar{3}$ axis is equivalent to the product of a threefold rotation by an inversion: i.e. $\bar{3} = 3\bar{1}$;

(4) the $\bar{4}$ axis is also a 2 axis;

(5) the $\bar{6}$ axis is equivalent to the product of a threefold rotation by a reflection with respect to a plane normal to it; this will be indicated by $\bar{6} = 3/\mathrm{m}$.

Axes of rotoreflection

A rotoreflection axis of order n is present when all the properties of the space do not change after performing the product of a $2\pi/n$ rotation around an axis by a reflection with respect to a plane normal to it. The written symbol of this axis is $\tilde{n}$. The effects on the space of the $\tilde{1}$, $\tilde{2}$, $\tilde{3}$, $\tilde{4}$, $\tilde{6}$ axes coincide with those caused by an inversion axis (generally of a different order). In particular: $\tilde{1} = \mathrm{m}$, $\tilde{2} = \bar{1}$, $\tilde{3} = \bar{6}$, $\tilde{4} = \bar{4}$, $\tilde{6} = \bar{3}$. From now on we will no longer consider the $\tilde{n}$ axes but their equivalent inversion axes.

Reflection planes with translational component (glide planes)

A glide plane operator is present if the properties of the half-space on one side of the plane are identical to those of the other half-space after the product of a reflection with respect to the plane by a translation parallel to the plane. On p. 11 we shall see which are the glide planes found in crystals.

Symmetry operations relating objects referred by direct congruence are called **proper** (we will also refer to **proper symmetry axes**) while those relating objects referred by opposite congruence are called **improper** (we will also refer to **improper axes**).

Lattices

Translational periodicity in crystals can be conveniently studied by considering the geometry of the repetion rather than the properties of the motif which is repeated. If the motif is periodically repeated at intervals a, b, and c along three non-coplanar directions, the repetition geometry can be fully described by a periodic sequence of points, separated by intervals a, b, c along the same three directions. This collection of points will be called a **lattice**. We will speak of line, plane, and space lattices, depending on whether the periodicity is observed in one direction, in a plane, or in a three-dimensional space. An example is illustrated in Fig. 1.2(a), where HOCl is a geometrical motif repeated at intervals a and b. If we replace the molecule with a point positioned at its centre of gravity, we obtain the lattice of Fig. 1.2(b). Note that, if instead of placing the lattice point at the centre of gravity, we locate it on the oxygen atom or on any other point of the motif, the lattice does not change. Therefore the position of the lattice with respect to the motif is completely arbitrary.

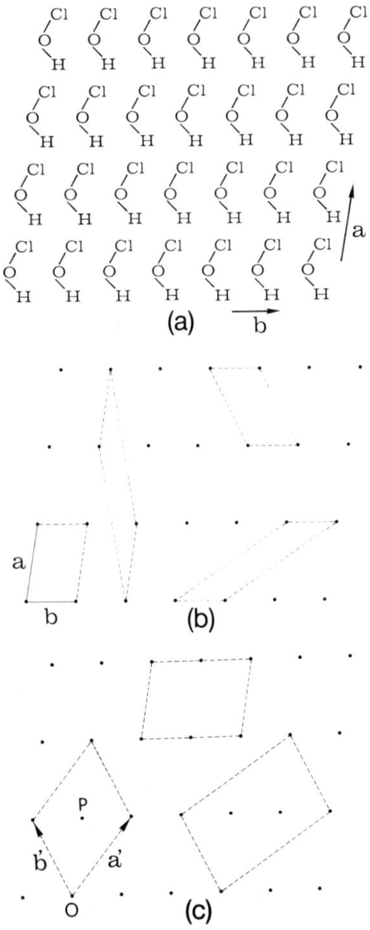

Fig. 1.2. (a) Repetition of a graphical motif as an example of a two-dimensional crystal; (b) corresponding lattice with some examples of primitive cells; (c) corresponding lattice with some examples of multiple cells.

If any lattice point is chosen as the origin of the lattice, the position of any other point in Fig. 1.2(b) is uniquely defined by the vector

$$\boldsymbol{Q}_{u,v} = u\boldsymbol{a} + v\boldsymbol{b} \tag{1.1}$$

where u and v are positive or negative integers. The vectors $\boldsymbol{a}$ and $\boldsymbol{b}$ define a parallelogram which is called the *unit cell*: $\boldsymbol{a}$ and $\boldsymbol{b}$ are the basis vectors of the cell. The choice of the vectors $\boldsymbol{a}$ and $\boldsymbol{b}$ is rather arbitrary. In Fig. 1.2(b) four possible choices are shown; they are all characterized by the property that each lattice point satisfies relation (1.1) with integer u and v.

Nevertheless we are allowed to choose different types of unit cells, such as those shown in Fig. 1.2(c), having double or triple area with respect to those selected in Fig. 1.2(b). In this case each lattice point will still satisfy (1.1) but u and v are no longer restricted to integer values. For instance, the point P is related to the origin O and to the basis vectors $\boldsymbol{a}'$ and $\boldsymbol{b}'$ through $(u, v) = (1/2, 1/2)$.

The different types of unit cells are better characterized by determining the number of lattice points belonging to them, taking into account that the points on sides and on corners are only partially shared by the given cell.

The cells shown in Fig. 1.2(b) contain only one lattice point, since the four points at the corners of each cell belong to it for only 1/4. These cells are called **primitive**. The cells in Fig. 1.2(c) contain either two or three points and are called **multiple** or **centred cells**. Several kinds of multiple cells are possible: i.e. double cells, triple cells, etc., depending on whether they contain two, three, etc. lattice points.

The above considerations can be easily extended to linear and space lattices. For the latter in particular, given an origin O and three basis vectors $\boldsymbol{a}$, $\boldsymbol{b}$, and $\boldsymbol{c}$, each node is uniquely defined by the vector

$$\boldsymbol{Q}_{u,v,w} = u\boldsymbol{a} + v\boldsymbol{b} + w\boldsymbol{c}. \tag{1.2}$$

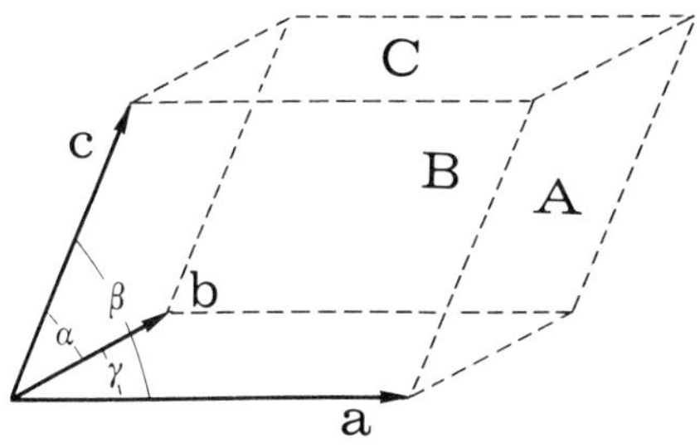

Fig. 1.3. Notation for a unit cell.

The three basis vectors define a parallelepiped, called again a **unit cell**. The directions specified by the vectors $\boldsymbol{a}$, $\boldsymbol{b}$, and $\boldsymbol{c}$ are the X, Y, Z **crystallographic axes**, respectively, while the angles between them are indicated by α, β, and γ, with α opposing $\boldsymbol{a}$, β opposing $\boldsymbol{b}$, and γ opposing $\boldsymbol{c}$ (cf. Fig. 1.3). The volume of the unit cell is given by

$$V = \boldsymbol{a} \cdot \boldsymbol{b} \wedge \boldsymbol{c}$$

where the symbol '·' indicates the scalar product and the symbol '$\wedge$' the vector product. The orientation of the three crystallographic axes is usually chosen in such a way that an observer located along the positive direction of $\boldsymbol{c}$ sees $\boldsymbol{a}$ moving towards $\boldsymbol{b}$ by an anti-clockwise rotation. The faces of the unit cell facing $\boldsymbol{a}$, $\boldsymbol{b}$, and $\boldsymbol{c}$ are indicated by A, B, C, respectively. If the chosen cell is primitive, then the values of u, v, w in (1.2) are bound to be integer for all the lattice points. If the cell is multiple then u, v, w will have rational values. To characterize the cell we must recall that a lattice point at vertex belongs to it only for 1/8th, a point on a edge for 1/4, and one on a face for 1/2.

The rational properties of lattices

Since a lattice point can always be characterized by rational numbers, the lattice properties related to them are called **rational**. Directions defined by two lattice points will be called rational directions, and planes defined by three lattice points rational planes. Directions and planes of this type are also called **crystallographic directions** and **planes**.

Crystallographic directions

Since crystals are anisotropic, it is necessary to specify in a simple way directions (or planes) in which specific physical properties are observed.

Two lattice points define a **lattice row**. In a lattice there are an infinite number of parallel rows (see Fig. 1.4): they are identical under lattice translation and in particular they have the same translation period.

A lattice row defines a crystallographic direction. Suppose we have chosen a primitive unit cell. The two lattice vectors $\boldsymbol{Q}_{u,v,w}$ and $\boldsymbol{Q}_{nu,nv,nw}$, with u, v, w, and n integer, define two different lattice points, but only one direction. This property may be used to characterize a direction in a unique way. For instance, the direction associated with the vector $\boldsymbol{Q}_{9,3,6}$ can be uniquely defined by the vector $\boldsymbol{Q}_{3,1,2}$ with no common factor among the indices. This direction will be indicated by the symbol [3 1 2], to be read as 'three, one, two' and not 'three hundred and twelve'.

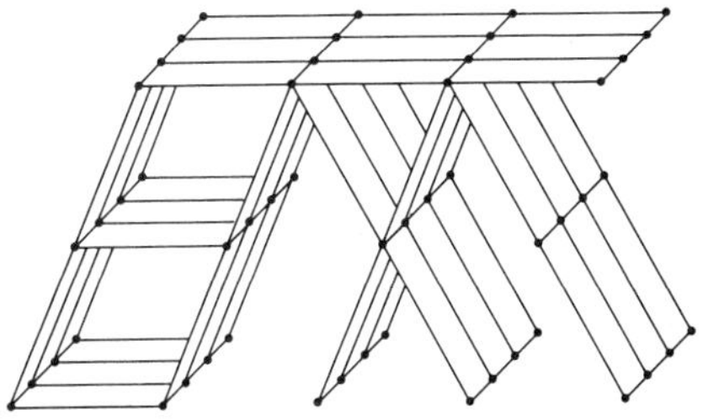

Fig. 1.4. Lattice rows and planes.

When the cell is not primitive u, v, w, and n will be rational numbers. Thus $\boldsymbol{Q}_{1/2,3/2,-1/3}$ and $\boldsymbol{Q}_{5/2,15/2,-5/3}$ define the same direction. The indices of the former may therefore be factorized to obtain a common denominator and no common factor among the numerators: $\boldsymbol{Q}_{1/2,3/2,-1/3} = \boldsymbol{Q}_{3/6,9/6,-2/6} \rightarrow$ [3 9 −2] to be read 'three', nine, minus two'.

Crystallographic planes

Three lattice points define a crystallographic plane. Suppose it intersects the three crystallographic axes X, Y, and Z at the three lattice points $(p, 0, 0)$, $(0, q, 0)$ and $(0, 0, r)$ with integer p, q, r (see Fig. 1.5). Suppose that the largest common integer factor of p, q, r is 1 and that $m = pqr$ is the least common multiple of p, q, r. Then the equation of the plane is

$$x'/pa + y'/qb + z'/rc = 1.$$

If we introduce the fractional coordinates $x = x'/a$, $y = y'/b$, $z = z'/c$, the equation of the plane becomes

$$x/p + y/q + z/r = 1. \tag{1.3}$$

Multiplying both sides by pqr we obtain

$$qrx + pry + pqz = pqr$$

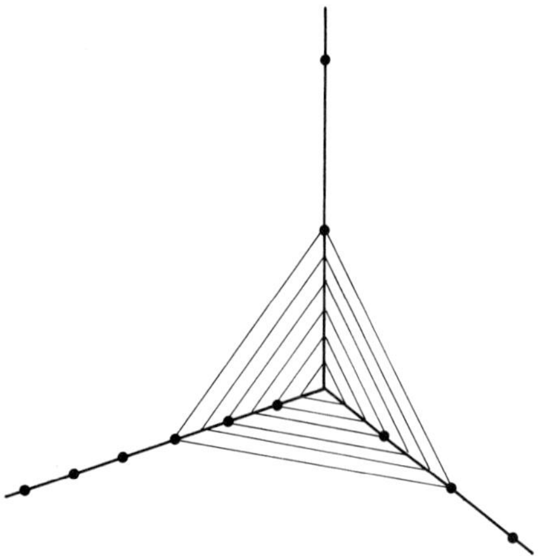

Fig. 1.5. Some lattice planes of the set (236).

which can be rewritten as

$$hx + ky + lz = m \tag{1.4}$$

where $h = qr$, $k = pr$, $l = pq$ and $m = pqr$. As for p, q, and r, also for h, k, and l the largest common integer factor will be 1.

We can therefore construct a family of planes parallel to the plane (1.4), by varying m over all integer numbers from $-\infty$ to $+\infty$. These will also be crystallographic planes since each of them is bound to pass through at least three lattice points.

The rational properties of all points being the same, there will be a plane of the family passing through each lattice point. For the same reason each lattice plane is identical to any other within the family through a lattice translation.

Let us now show that (1.4) represents a plane at a distance from the origin m times the distance of the plane

$$hx + ky + lz = 1. \tag{1.5}$$

The intercepts of the plane (1.5) on X, Y, Z will be $1/h$, $1/k$ and $1/l$ respectively and those of (1.4) m/h, m/k, m/l. It is then clear that the distance of plane (1.4) from the origin is m times that of plane (1.5). Since m is the least common multiple of h, k, and l, plane (1.4) is also the first plane of the family intersecting the axes X, Y, and Z at three lattice points. We can therefore conclude that eqn (1.4) defines, as m is varied, a family of identical and equally spaced crystallographic planes. The three indices h, k, and l define the family and are its **Miller indices**. To indicate that a family of lattice planes is defined by a sequence of three integers, these are included within braces: $(h\ k\ l)$. A simple interpretation of the three indices h, k, and

l can be deduced from (1.4) and (1.5). In fact they indicate that the planes of the family divide a in h parts, b in k parts, and c in l parts.

Crystallographic planes parallel to one of the three axes X, Y, or Z are defined by indices of type $(0kl)$, $(h0l)$, or $(hk0)$ respectively. Planes parallel to faces A, B, and C of the unit cell are of type $(h00)$, $(0k0)$, and $(00l)$ respectively. Some examples of crystallographic planes are illustrated in Fig. 1.6.

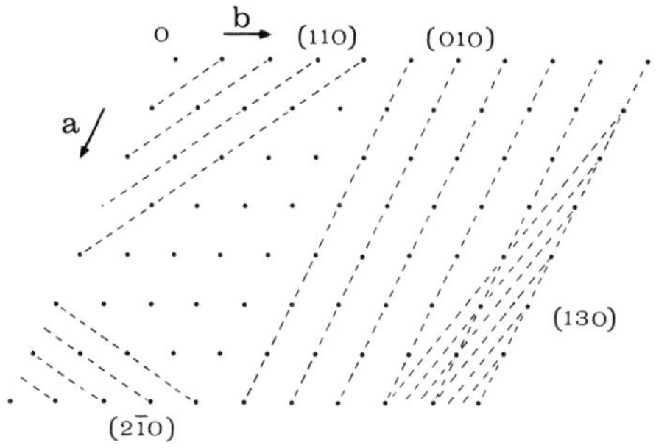

Fig. 1.6. Miller indices for some crystallographic planes parallel to Z (Z is supposed to be normal to the page).

Suppose now that the largest common integer factor of p, q, and r is >1. Let us consider for instance the plane

$$x/9 + y/6 + z/15 = 1 \tag{1.6}$$

which can be written as

$$10x + 15y + 6z = 90. \tag{1.7}$$

The first plane of the family with integer intersections on the three axes will be the 30th (30 being the least common multiple of 10, 15, and 6) and all the planes of the family can be obtained from the equation $10x + 15y + 6z = m$, by varying m over all integers from $-\infty$ to $+\infty$. If we divide p, q, and r in eqn (1.6) by their common integer factor we obtain $x/3 + y/2 + z/5 = 1$, from which

$$10x + 15y + 6z = 30. \tag{1.8}$$

Planes (1.7) and (1.8) belong to the same family. We can therefore conclude that a family of crystallographic planes is always uniquely defined by three indices h, k, and l having the largest common integer factor equal to unity.

Symmetry restrictions due to the lattice periodicity and vice versa

Suppose that the disposition of the molecules in a crystal is compatible with an n axis. As a consequence the disposition of lattice points must also be compatible with the same axis. Without losing generality, we will assume that n passes through the origin O of the lattice. Since each lattice point has identical rational properties, there will be an n axis passing through each and every lattice point, parallel to that passing through the origin. In particular each symmetry axis will lie along a row and will be perpendicular to a crystallographic plane.

Let $\boldsymbol{T}$ be the period vector of a row passing through O and normal to n. We will then have lattice points (see Fig. 1.7(a)) at $\boldsymbol{T}$, $-\boldsymbol{T}$, $\boldsymbol{T}'$, and $\boldsymbol{T}''$. The vector $\boldsymbol{T}' - \boldsymbol{T}''$ must also be a lattice vector and, being parallel to $\boldsymbol{T}$, we will have $\boldsymbol{T}' - \boldsymbol{T}'' = m\boldsymbol{T}$ where m is an integer value: in a scalar form

$$2\cos(2\pi/n) = m \qquad (m \text{ integer}). \tag{1.9}$$

Equation (1.9) is only verified for $n = 1, 2, 3, 4, 6$. It is noteworthy that a 5 axis is not allowed, this being the reason why it is impossible to pave a room only with pentagonal tiles (see Fig. 1.7(b).

A unit cell, and therefore a lattice, compatible with an n axis will also be compatible with an $\bar{n}$ axis and vice versa. Thus axes $\bar{1}$, $\bar{2}$, $\bar{3}$, $\bar{4}$, $\bar{6}$ will also be allowed.

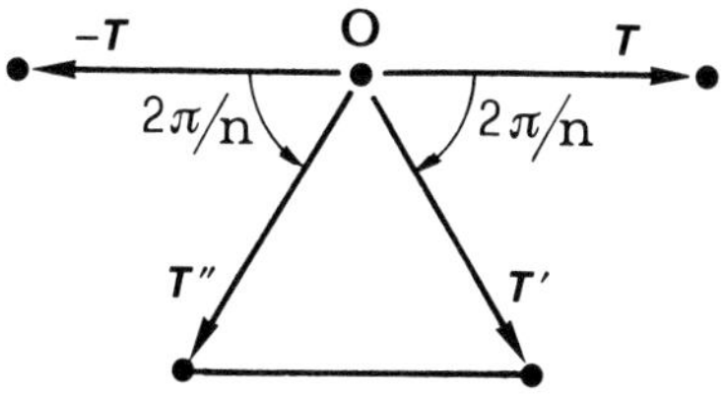

(a)

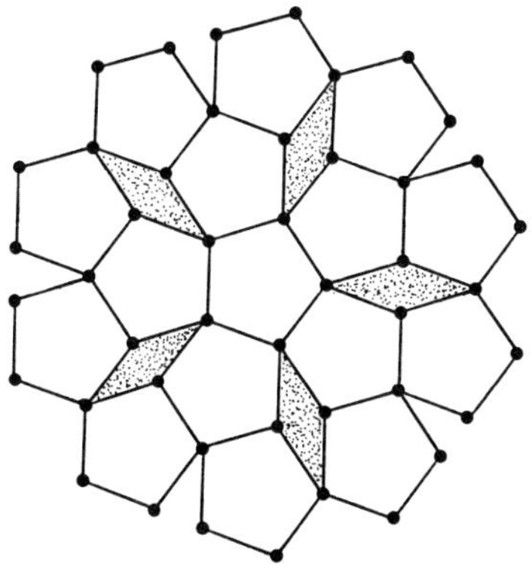
(b)

Fig. 1.7.(a) Lattice points in a plane normal to the symmetry axis n passing through O. (b) Regular pentagons cannot fill planar space.

Let us now consider the restrictions imposed by the periodic nature of crystals on the translational components $\boldsymbol{t}$ of a screw axis. Suppose that this lies along a row with period vector $\boldsymbol{T}$. Its rotational component must correspond to $n = 1, 2, 3, 4, 6$. If we apply the translational component n times the resulting displacement will be $n\boldsymbol{t}$. In order to maintain the periodicity of the crystal we must have $n\boldsymbol{t} = p\boldsymbol{T}$, with integer p, or

$$\boldsymbol{t} = (p/n)\boldsymbol{T}. \tag{1.10}$$

For instance, for a screw axis of order 4 the allowed translational components will be $(0/4)\boldsymbol{T}$, $(1/4)\boldsymbol{T}$, $(2/4)\boldsymbol{T}$, $(3/4)\boldsymbol{T}$, $(4/4)\boldsymbol{T}$, $(5/4)\boldsymbol{T}$, . . .; of these only the first four will be distinct. It follows that:

(1) in (1.10) p can be restricted within $0 \leqslant p < n$;

(2) the n-fold axis may be thought as a special screw with $\boldsymbol{t} = \boldsymbol{0}$. The nature of a screw axis is completely defined by the symbol n_p. The graphic symbols are shown in Table 1.1: the effects of screw axes on the surrounding space are represented in Fig. 1.8. Note that:

1. If we draw a helicoidal trajectory joining the centres of all the objects related by a 3_1 and by a 3_2 axis, we will obtain, in the first case a right-handed helix and in the second a left-handed one (the two helices are enantiomorphous). The same applies to the pairs 4_1 and 4_3, 6_1 and 6_5, and 6_2 and 6_4.

2. 4_2 is also a 2 axis, 6_2 is also a 2 and a 3_2, 6_4 is also a 2 and a 3_1, and 6_3 is also a 3 and a 2_1.

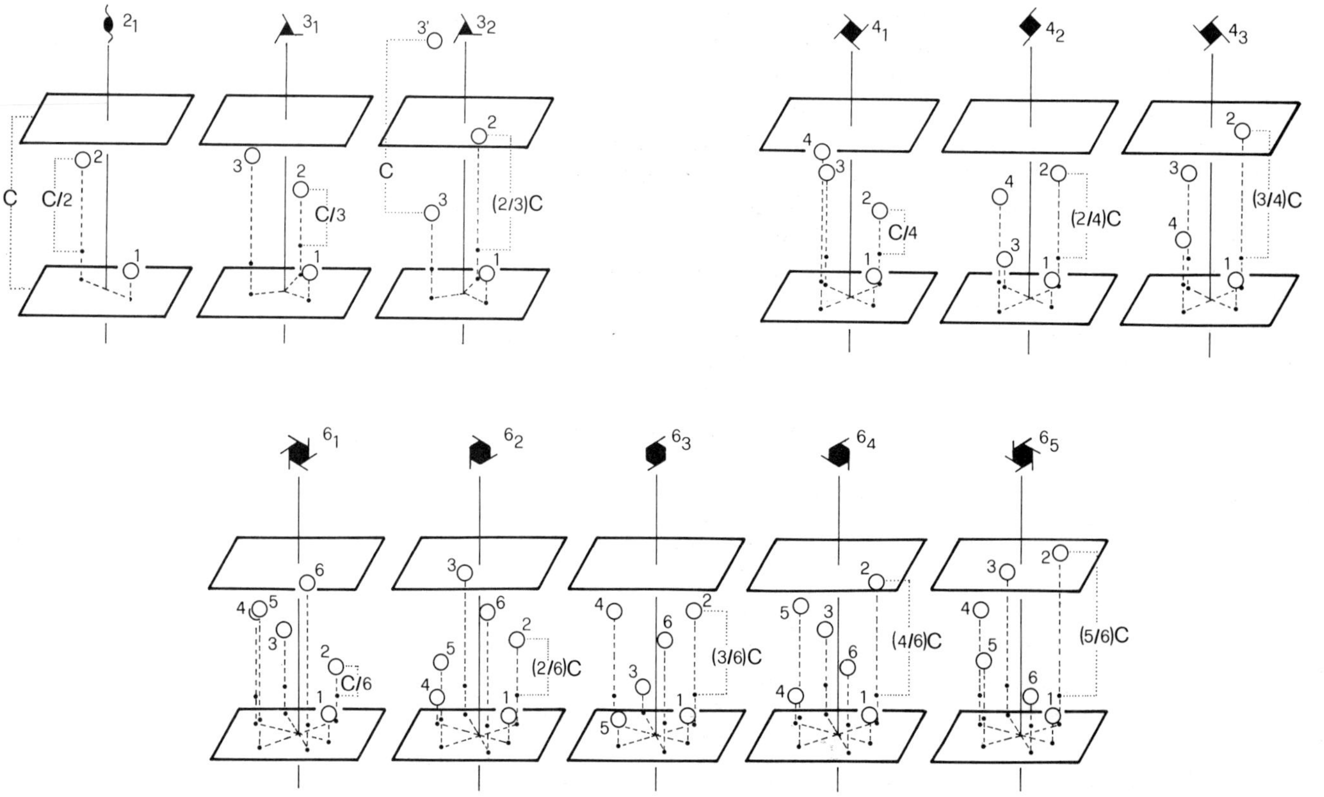

Fig. 1.8. Screw axes: arrangement of symmetry-equivalent objects.

We will now consider the restrictions imposed by the periodicity on the translation component $\boldsymbol{t}$ of a glide plane. If we apply this operation twice, the resulting movement must correspond to a translation equal to $p\boldsymbol{T}$, where p may be any integer and $\boldsymbol{T}$ any lattice vector on the crystallographic plane on which the glide lies. Therefore $2\boldsymbol{t} = p\boldsymbol{T}$, i.e. $\boldsymbol{t} = (p/2)\boldsymbol{T}$. As p varies over all integer values, the following translations are obtained $0\boldsymbol{T}$, $(1/2)\boldsymbol{T}$, $(2/2)\boldsymbol{T}$, $(3/2)\boldsymbol{T}, \ldots$ of which only the first two are distinct. For $p = 0$ the glide plane reduces to a mirror m. We will indicate by a, b, c axial glides with translational components equal to $\boldsymbol{a}/2$, $\boldsymbol{b}/2$, $\boldsymbol{c}/2$ respectively, by n the **diagonal** glides with translational components $(\boldsymbol{a}+\boldsymbol{b})/2$ or $(\boldsymbol{a}+\boldsymbol{c})/2$ or $(\boldsymbol{b}+\boldsymbol{c})/2$ or $(\boldsymbol{a}+\boldsymbol{b}+\boldsymbol{c})/2$.

In a non-primitive cell the condition $2\boldsymbol{t} = p\boldsymbol{T}$ still holds, but now $\boldsymbol{T}$ is a lattice vector with rational components indicated by the symbol d. The graphic symbols for glide planes are given in Table 1.1.

Point groups and symmetry classes

In crystals more symmetry axes, both proper and improper, with or without translational components, may coexist. We will consider here only those combinations of operators which do not imply translations, i.e. the combinations of proper and improper axes intersecting in a point. These are called **point groups**, since the operators form a mathematical **group** and leave one point fixed. The set of crystals having the same point group is called **crystal class** and its symbol is that of the point group. Often point group and crystal class are used as synonyms, even if that is not correct in principle. The total number of crystallographic point groups (for three-dimensional crystals) is 32, and they were first listed by Hessel in 1830.

The simplest combinations of symmetry operators are those characterized by the presence of only one axis, which can be a proper axis or an inversion one. Also, a proper and an inversion axis may be simultaneously present. The 13 independent combinations of this type are described in Table 1.2. When along the same axis a proper axis and an inversion axis are simultaneously present, the symbol $n/\bar{n}$ is used. Classes coinciding with other classes already quoted in the table are enclosed in brackets.

The problem of the coexistence of more than one axis all passing by a common point was first solved by Euler and is illustrated, with a different approach, in Appendix 1.B. Here we only give the essential results. Let us suppose that there are two proper axes l_1 and l_2 intersecting in O (see Fig. 1.9). The l_1 axis will repeat in Q an object originally in P, while l_2 will

Table 1.2. Single-axis crystallographic point groups

Proper axis		Improper axis		Proper and improper axis	
1		$\bar{1}$		$(1/\bar{1} = \bar{1})$	
2		$\bar{2} \equiv m$		$2/\bar{2} = 2/m$	
3		$\bar{3} \equiv 3\bar{1}$		$(3/\bar{3} = \bar{3})$	
4		$\bar{4}$		$4/\bar{4} = 4/m$	
6		$\bar{6} \equiv 3/m$		$6/\bar{6} = 6/m$	
5	+	5	+	3	= 13

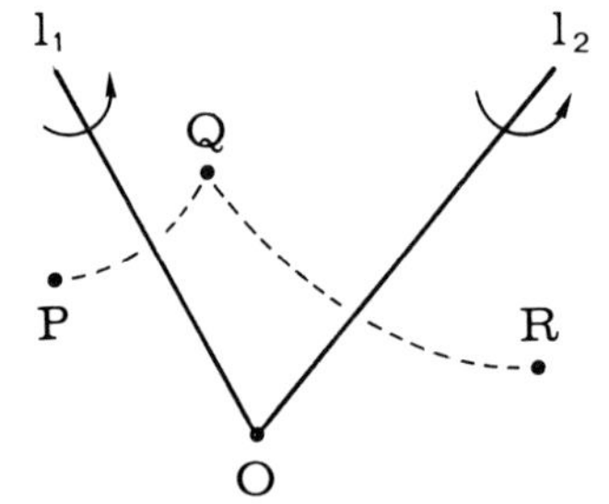

Fig. 1.9. Arrangement of equivalent objects around two intersecting symmetry axes.

Table 1.3. For each combination of symmetry axes the minimum angles between axes are given. For each angle the types of symmetry axes are quoted in parentheses

Combination of symmetry axes	α (deg)	β (deg)	γ (deg)
2 2 2	90 (2 2)	90 (2 2)	90 (2 2)
3 2 2	90 (2 3)	90 (2 3)	60 (2 2)
4 2 2	90 (2 4)	90 (2 4)	45 (2 2)
6 2 2	90 (2 6)	90 (2 6)	30 (2 2)
2 3 3	54 44′08″ (2 3)	54 44′08″ (2 3)	70 31′44″ (3 3)
4 3 2	35 15′52″ (2 3)	45 (2 4)	54 44′08″ (4 3)

repeat in R the object in Q. P and Q are therefore directly congruent and this implies the existence of another proper operator which repeats the object in P directly in R. The only allowed combinations are *n*22, 233, 432, 532 which in crystals reduce to 222, 322, 422, 622, 233, 432. For these combinations the smallest angles between the axes are listed in Table 1.3, while their disposition in the space is shown in Fig. 1.10. Note that the combination 233 is also consistent with a tetrahedral symmetry and 432 with a cubic and octahedral symmetry.

Suppose now that in Fig. 1.9 l_1 is a proper axis while l_2 is an inversion one. Then the objects in P and in Q will be directly congruent, while the object in R is enantiomorphic with respect to them. Therefore the third operator relating R to P will be an inversion axis. We may conclude that if one of the three symmetry operators is an inversion axis also another must be an inversion one. In Table 1.4 are listed all the point groups characterized by combinations of type PPP, PII, IPI, IIP (P = proper, I = improper), while in Table 1.5 the classes with axes at the same time proper and improper are given. In the two tables the combinations coinciding with previously considered ones are closed within brackets. The

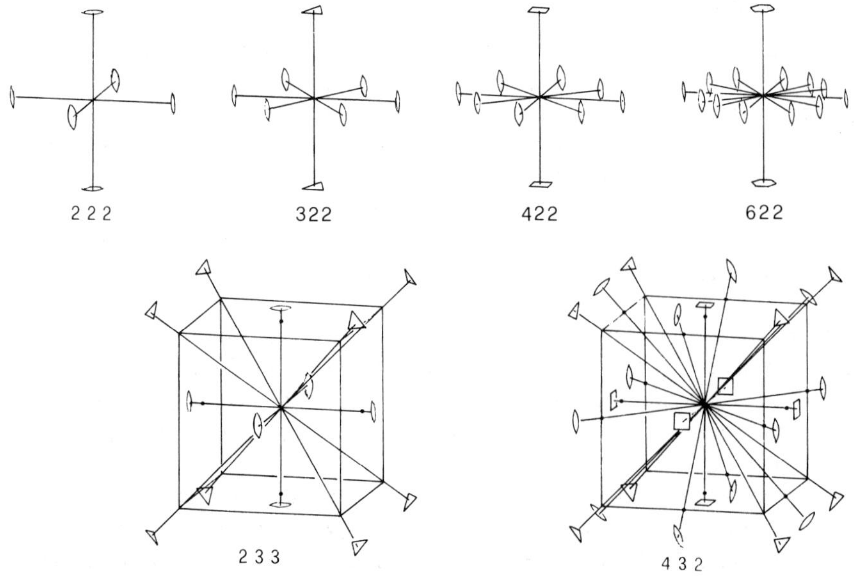

Fig. 1.10. Arrangement of proper symmetry axes for six point groups.

Table 1.4. Crystallographic point groups with more than one axis

P P P	**P I I**	**I P I**	**I I P**	
2 2 2	2 m m			
3 2 2	3 m m	$\bar{3}\ 2\ m$		
4 2 2	4 m m	$\bar{4}\ 2\ m$		
6 2 2	6 m m	$\bar{6}\ 2\ \bar{2} \rightarrow \bar{6}\,2\,m$		
2 3 3	$2\ \bar{3}\ \bar{3} \rightarrow \frac{2}{m}\bar{3}\ \bar{3}$			
4 3 2	$4\ \bar{3}\ \bar{2} \rightarrow \bar{4}\frac{3}{\bar{1}}\frac{1}{m} \rightarrow \frac{4}{m}\bar{3}\frac{2}{m} \rightarrow m\ \bar{3}\ m$	$\bar{4}\ 3\ m$	$\left(\bar{4}\ \bar{3}\ 2 \rightarrow \frac{\bar{4}}{m}\frac{3}{\bar{1}}\frac{2}{m} \rightarrow \frac{4}{m}\bar{3}\frac{2}{m}\right)$	
6	+6	+4	+0	= 16

Table 1.5. Crystallographic point groups with more than one axis, each axis being proper and improper simultaneously

$\frac{2}{\bar{2}}\frac{2}{\bar{2}}\frac{2}{\bar{2}}=\frac{2}{m}\frac{2}{m}\frac{2}{m}$;	$\left(\frac{3}{\bar{3}}\frac{2}{\bar{2}}\frac{2}{\bar{2}}=\bar{3}\frac{2}{m}\frac{2}{m}\right)$;	$\frac{4}{\bar{4}}\frac{2}{\bar{2}}\frac{2}{\bar{2}}=\frac{4}{m}\frac{2}{m}\frac{2}{m}$;
$\frac{6}{\bar{6}}\frac{2}{\bar{2}}\frac{2}{\bar{2}}=\frac{6}{m}\frac{2}{m}\frac{2}{m}$;	$\left(\frac{2}{\bar{2}}\frac{3}{\bar{3}}\frac{3}{\bar{3}}=\frac{2}{m}\bar{3}\ \bar{3}\right)$;	$\left(\frac{4}{\bar{4}}\frac{3}{\bar{3}}\frac{2}{\bar{2}}=\frac{4}{m}\bar{3}\frac{2}{m}\right)$

results so far described can be easily derived by recalling that:

1. If two of the three axes are symmetry equivalent, they can not be one proper and one improper; for example, the threefold axes in 233 are symmetry referred by twofold axes, while binary axes in 422 differing by 45° are not symmetry equivalent.
2. If an even-order axis and a $\bar{1}$ axis (or an m plane) coexist, there will also be an m plane (or a $\bar{1}$ axis) normal to the axis and passing through the intersection point. Conversely, if m and $\bar{1}$ coexist, there will also be a 2 axis passing through $\bar{1}$ and normal to m.

In Tables 1.2, 1.4, and 1.5 the symbol of each point group does not reveal all the symmetry elements present: for instance, the complete list of symmetry elements in the class $2/m\bar{3}\bar{3}$ is $2/m\ 2/m\ 2/m\bar{1}3333$. On the other hand, the symbol $2/m\bar{3}\bar{3}$ is too extensive, since only two symmetry operators are independent. In Table 1.6 are listed the conventional symbols used for the 32 symmetry classes. It may be noted that crystals with inversion symmetry operators have an equal number of 'left' and 'right' moieties; these parts, when considered separately, are one the enantiomorph of the other.

The conclusions reached so far do not exclude the possibility of crystallizing molecules with a molecular symmetry different from that of the 32 point groups (for instance with a 5 axis). In any case the symmetry of the crystal will belong to one of them. To help the reader, some molecules and their point symmetry are shown in Fig. 1.11.

It is very important to understand how the symmetry of the physical properties of a crystal relates to its point group (this subject is more extensively described in Chapter 9). Of basic relevance to this is a postulate

Table 1.6. List of the 32 point groups

Crystal systems	Point groups: Non-centro-symmetric		Point groups: Centro-symmetric	Laue classes	Lattice point groups
Triclinic	1		$\bar{1}$	$\bar{1}$	$\bar{1}$
Monoclinic	2	m	2/m	2/m	2/m
Orthorhombic	222	mm2	mmm	mmm	mmm
Tetragonal	4	$\bar{4}$	4/m	4/m	4/mmm
	422	4mm, $\bar{4}$2m	4/mmm	4/mmm	
Trigonal	3		$\bar{3}$	$\bar{3}$	$\bar{3}$m
	32	3m	$\bar{3}$m	$\bar{3}$m	
Hexagonal	6	$\bar{6}$	6/m	6/m	6/mmm
	622	6mm, $\bar{6}$2m	6/mmm	6/mmm	
Cubic	23		m$\bar{3}$	m$\bar{3}$	m$\bar{3}$m
	432	$\bar{4}$3m	m$\bar{3}$m	m$\bar{3}$m	

of crystal physics, known as the **Neumann principle**: 'the symmetry elements of any physical property must include the symmetry elements of the crystal point group'. In keeping with this principle, the physical properties may present a higher, but not a lower, symmetry than the point group. For instance:

1. **Cubic** crystals (see later for their definition) are optically isotropic: in this case the physical property has a symmetry higher than that of the point group.

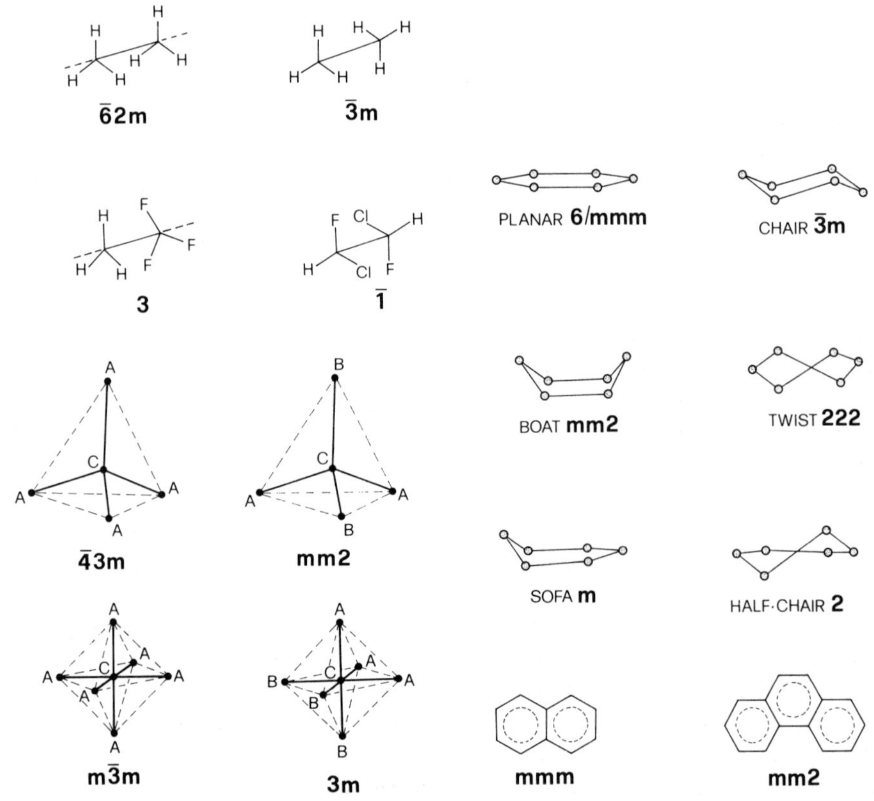

Fig. 1.11. Molecular examples of some point groups.

2. The variation of the refractive index of the crystal with the vibration direction of a plane-polarized light wave is represented by the **optical indicatrix** (see p. 607). This is in general a three-axis **ellipsoid**: thus the lowest symmetry of the property 'refraction' is 2/m 2/m 2/m, the point group of the ellipsoid. In crystal classes belonging to tetragonal, trigonal, or hexagonal systems (see Table 1.6) the shape of the indicatrix is a rotational ellipsoid (the axis is parallel to the main symmetry axis), and in symmetry classes belonging to the cubic system the shape of the indicatrix is a sphere. For example, in the case of tourmaline, with point group 3m, the ellipsoid is a revolution around the threefold axis, showing a symmetry higher than that of the point group.

We shall now see how it is possible to guess about the point group of a crystal through some of its physical properties:

1. The morphology of a crystal tends to conform to its point group symmetry. From a morphological point of view, a crystal is a solid body bounded by plane natural surfaces, the **faces**. The set of symmetry-equivalent phases constitutes a **form**: the form is *open* if it does not enclose space, otherwise it is *closed*. A crystal form is named according to the number of its faces and to their nature. Thus a **pedion** is a single face, a **pinacoid** is a pair of parallel faces, a **sphenoid** is a pair of faces related by a diad axis, a **prism** a set of equivalent faces parallel to a common axis, a **pyramid** is a set of planes with equal angles of inclination to a common axis, etc. The morphology of different samples of the same compound can show different types of face, with different extensions, and different numbers of edges, the external form depending not only on the structure but also on the chemical and physical properties of the environment. For instance, galena crystals (PbS, point group m$\bar{3}$m) tend to assume a cubic, cube-octahedral, or octahedral habit (Fig. 1.12(a)). Sodium chloride grows as cubic crystals from neutral aqueous solution and as octahedral from active solutions (in the latter case cations and anions play a different energetic role). But at the same temperature crystals will all have constant dihedral angles between corresponding faces (J. B. L. Rome' de l'Ile, 1736–1790). This property, the observation of which dates back to N. Steno (1669) and D. Guglielmini (1688), can be explained easily, following R. J. Haüy (1743–1822), by considering that faces coincide with lattice planes and edges with lattice rows. Accordingly, Miller indices can be used as form symbols, enclosed in braces: $\{hkl\}$. The indices of well-developed faces on natural crystals tend to have small values of h, k, l, (integers greater than six are rarely involved). Such faces correspond to lattice planes with a high density of lattice points per unit area, or equivalently, with large intercepts a/h, b/k, c/l on the reference axes (**Bravais' law**). An important extension of this law is obtained if space group symmetry (see p. 22) is taken into account: screw axes and glide planes normal to a given crystal face reduce its importance (Donnay–Harker principle).

The origin within the crystal is usually chosen so that faces (hkl) and $(\bar{h}\bar{k}\bar{l})$ are parallel faces an opposite sides of the crystal. In Fig. 1.13 some idealized crystal forms are shown.

The orientation of the faces is more important than their extension. The orientations can be represented by the set of unit vectors normal to them. This set will tend to assume the point-group symmetry of the given crystal

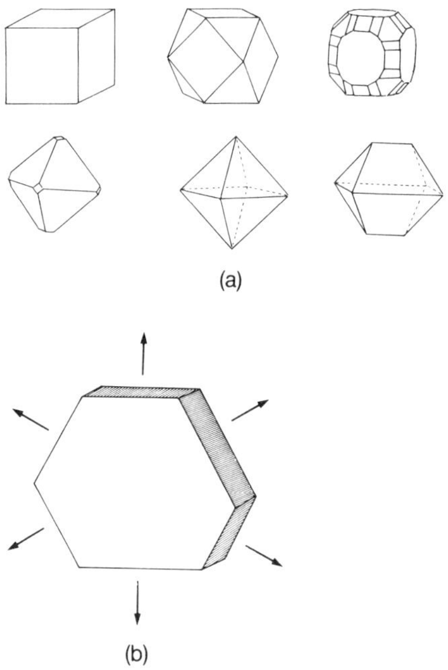

Fig. 1.12. (a) Crystals showing cubic or cube-octahedral or octahedral habitus, (b) crystal with a sixfold symmetry axis.

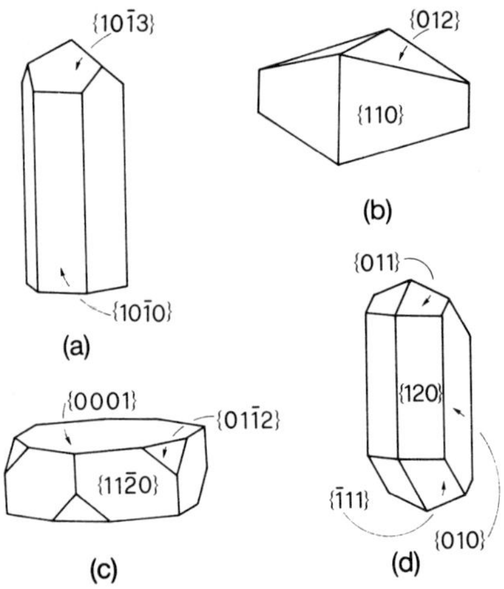

Fig. 1.13. Some simple crystal forms: (a) cinnabar, HgS, class 32; (b) arsenopyrite, FeAsS, class mmm; (c) ilmenite, $FeTiO_3$, class $\bar{3}$; (d) gypsum, $CaSO_4$, class 2/m.

independently of the morphological aspect of the samples. Thus, each sample of Fig. 1.12(a) shows an m$\bar{3}$m symmetry, and the sample in Fig. 1.12(b) shows a sixfold symmetry if the normals to the faces are considered instead of their extensions. The morphological analysis of a crystalline sample may be used to get some, although not conclusive, indication, of its point-group symmetry.

2. Electrical charges of opposite signs may appear at the two hands of a **polar axis** of a crystal subject to compression, because of the **piezoelectric** effect (see p. 619). A polar axis is a rational direction which is not symmetry equivalent to its opposite direction. It then follows that a polar direction can only exist in the 21 non-centrosymmetric point groups (the only exception is the 432 class, where piezoelectricity can not occur). In these groups not all directions are polar: in particular a direction normal to an even-order axis or to a mirror plane will never be polar. For instance, in quartz crystals (SiO_2, class 32), charges of opposite sign may appear at the opposite hands of the twofold axes, but not at those of the threefold axis.

3. A point group is said to be **polar** if a polar direction, with no other symmetry equivalent directions, is allowed. Along this direction a permanent electric dipole may be measured, which varies with temperature (**pyroelectric** effect, see p. 606). The ten polar classes are: 1, 2, m, mm2, 4, 4mm, 6, 6mm, 3, 3m. Piezo- and pyroelectricity tests are often used to exclude the presence of an inversion centre. Nevertheless when these effects are not detectable, no definitive conclusion may be drawn.

4. Ferroelectric crystals show a permanent dipole moment which can be changed by application of an electric field. Thus they can only belong to one of the ten polar classes.

5. The symmetry of a crystal containing only one enantiomer of an optically active molecule must belong to one of the 11 point groups which do not contain inversion axes.

6. Because of non-linear optical susceptibility, light waves passing through non-centrosymmetric crystals induce additional waves of frequency twice the incident frequency. This phenomenon is described by a third-rank tensor, as the piezoelectric tensor (see p. 608): it occurs in all non-centrosymmetric groups except 432, and is very efficient[7] for testing the absence of an inversion centre.

7. Etch figures produced on the crystal faces by chemical attack reveal the face symmetry (one of the following 10 two-dimensional point groups).

Point groups in one and two dimensions

The derivation of the crystallographic point groups in a two-dimensional space is much easier than in three dimensions. In fact the reflection with respect to a plane is substituted by a reflection with respect to a line (the same letter m will also indicate this operation); and $\bar{n}$ axes are not used. The total number of point groups in the plane is 10, and these are indicated by the symbols: 1, 2, 3, 4, 6, m, 2mm, 3m, 4mm, 6mm.

The number of crystallographic point groups in one dimension is 2: they are 1 and m $\equiv$ ($\bar{1}$).

The Laue classes

In agreement with Neumann's principle, physical experiments do not normally reveal the true symmetry of the crystal: some of them, for example diffraction, show the symmetry one would obtain by adding an inversion centre to the symmetry elements actually present. In particular this happens when the measured quantities do not depend on the atomic positions, but rather on the interatomic vectors, which indeed form a centrosymmetric set. Point groups differing only by the presence of an inversion centre will not be differentiated by these experiments. When these groups are collected in classes they form the 11 **Laue classes** listed in Table 1.6.

The seven crystal systems

If the crystal periodicity is only compatible with rotation or inversion axes of order 1, 2, 3, 4, 6, the presence of one of these axes will impose some restrictions on the geometry of the lattice. It is therefore convenient to group together the symmetry classes with common features in such a way that crystals belonging to these classes can be described by unit cells of the same type. In turn, the cells will be chosen in the most suitable way to show the symmetry actually present.

Point groups 1 and $\bar{1}$ have no symmetry axes and therefore no constraint axes for the unit cell; the ratios a:b:c and the angles α, β, γ can assume any value. Classes 1 and $\bar{1}$ are said to belong to the **triclinic** system.

Groups 2, m, and 2/m all present a 2 axis. If we assume that this axis coincides with the b axis of the unit cell, $\boldsymbol{a}$ and $\boldsymbol{c}$ can be chosen on the lattice plane normal to $\boldsymbol{b}$. We will then have $\alpha = \gamma = 90°$ and β unrestricted and the ratios a:b:c also unrestricted. Crystals with symmetry 2, m, and 2/m belong to the **monoclinic** system.

Classes 222, mm2, mmm are characterized by the presence of three mutually orthogonal twofold rotation or inversion axes. If we assume these as reference axes, we will obtain a unit cell with angles $\alpha = \beta = \gamma = 90°$ and with unrestricted a:b:c ratios. These classes belong to the **orthorhombic** system.

For the seven groups with only one fourfold axis [4, $\bar{4}$, 4/m, 422, 4mm, $\bar{4}$2m, 4/mmm] the $\boldsymbol{c}$ axis is chosen as the direction of the fourfold axis and the $\boldsymbol{a}$ and $\boldsymbol{b}$ axes will be symmetry equivalent, on the lattice plane normal to $\boldsymbol{c}$. The cell angles will be $\alpha = \beta = \gamma = 90°$ and the ratios a:b:c = 1:1:c. These crystals belong to the **tetragonal** system.

For the crystals with only one threefold or sixfold axis [3, $\bar{3}$, 32, 3m, $\bar{3}$m, 6, $\bar{6}$, 6/m, 622, 6mm, $\bar{6}$2m, 6/mm] the $\boldsymbol{c}$ axis is assumed along the three- or sixfold axis, while $\boldsymbol{a}$ and $\boldsymbol{b}$ are symmetry equivalent on the plane perpendicular to $\boldsymbol{c}$. These point groups are collected together in the **trigonal** and **hexagonal** systems, respectively, both characterized by a unit cell with angles $\alpha = \beta = 90°$ and $\gamma = 120°$, and ratios a:b:c = 1:1:c.

Crystals with four threefold axes [23, m$\bar{3}$, 432, $\bar{4}$3m, m$\bar{3}$m] distributed as the diagonals of a cube can be referred to orthogonal unit axes coinciding with the cube edges. The presence of the threefold axes ensures that these directions are symmetry equivalent. The chosen unit cell will have $\alpha = \beta = \gamma = 90°$ and ratios a:b:c = 1:1:1. This is called the **cubic** system.

The Bravais lattices

In the previous section to each crystal system we have associated a primitive cell compatible with the point groups belonging to the system. Each of these primitive cells defines a lattice type. There are also other types of lattices, based on non-primitive cells, which can not be related to the previous ones. In particular we will consider as different two lattice types which can not be described by the same unit-cell type.

In this section we shall describe the five possible plane lattices and fourteen possible space lattices based both on primitive and non-primitive cells. These are called Bravais lattices, after Auguste Bravais who first listed them in 1850.

Plane lattices

An oblique cell (see Fig. 1.14(a)) is compatible with the presence of axes 1 or 2 normal to the cell. This cell is primitive and has point group 2.

If the row indicated by m in Fig. 1.14(b) is a reflection line, the cell must be rectangular. Note that the unit cell is primitive and compatible with the point groups m and 2mm. Also the lattice illustrated in Fig. 1.14(c) with $a = b$ and $\gamma \neq 90°$ is compatible with m. This plane lattice has an oblique primitive cell. Nevertheless, each of the lattice points has a 2mm symmetry and therefore the lattice must be compatible with a rectangular system. This can be seen by choosing the rectangular centred cell defined by the unit vectors $\boldsymbol{a}'$ and $\boldsymbol{b}'$. This orthogonal cell is more convenient because a simpler coordinate system is allowed. It is worth noting that the two lattices shown in Figs. 1.14(b) and 1.14(c) are of different type even though they are compatible with the same point groups.

In Fig. 1.14(d) a plane lattice is represented compatible with the presence of a fourfold axis. The cell is primitive and compatible with the point groups 4 and 4mm.

In Fig. 1.14(e) a plane lattice compatible with the presence of a three- or a sixfold axis is shown. A unit cell with a rhombus shape and angles of 60° and 120° (also called hexagonal) may be chosen. A centred rectangular cell can also be selected, but such a cell is seldom chosen.

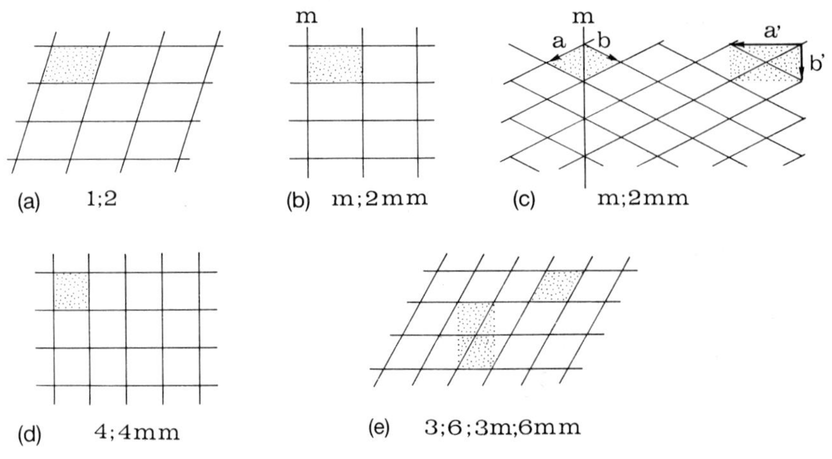

Fig. 1.14. The five plane lattices and the corresponding two-dimensional point groups.

Table 1.7. The five plane lattices

Cell	**Type of cell**	**Point group of the net**	**Lattice parameters**
Oblique	p	2	a, b, γ
Rectangular	p, c	2mm	$a, b, \gamma = 90°$
Square	p	4mm	$a = b, \gamma = 90°$
Hexagonal	p	6mm	$a = b, \gamma = 120°$

The basic features of the five lattices are listed in Table 1.7.

Space lattices

In Table 1.8 the most useful types of cells are described. Their fairly limited number can be explained by the following (or similar) observations:

1. A cell with two centred faces must be of type F. In fact a cell which is at the same time A and B, must have lattice points at $(0, 1/2, 1/2)$ and $(1/2, 0, 1/2)$. When these two lattice translations are applied one after the other they will generate a lattice point also at $(1/2, 1/2, 0)$;
2. A cell which is at the same time body and face centred can always be reduced to a conventional centred cell. For instance an I and A cell will have lattice points at positions $(1/2, 1/2, 1/2)$ and $(0, 1/2, 1/2)$: a lattice point at $(1/2, 0, 0)$ will then also be present. The lattice can then be described by a new A cell with axes $\boldsymbol{a}' = \boldsymbol{a}/2$, $\boldsymbol{b}' = \boldsymbol{b}$, and $\boldsymbol{c}' = \boldsymbol{c}$ (Fig. 1.15).

It is worth noting that the positions of the additional lattice points in Table 1.8 define the minimal translational components which will move an object into an equivalent one. For instance, in an A-type cell, an object at (x, y, z) is repeated by translation into $(x, y + m/2, z + n/2)$ with m and n integers: the shortest translation will be $(0, 1/2, 1/2)$.

Let us now examine the different types of three-dimensional lattices grouped in the appropriate crystal systems.

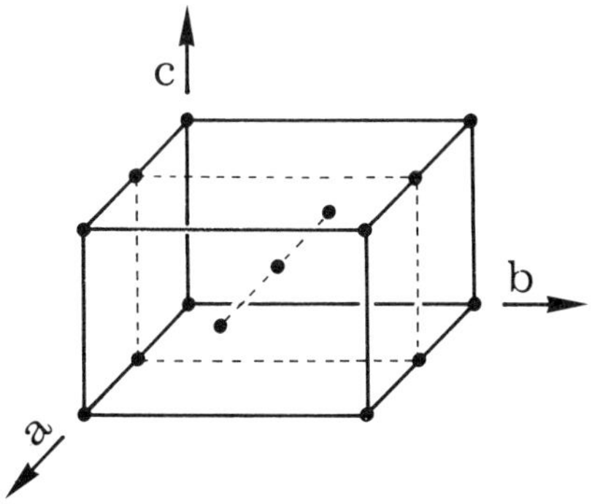

Fig. 1.15. Reduction of an I- and A-centred cell to an A-centred cell.

Table 1.8. The conventional types of unit cell

Symbol	**Type**	**Positions of additional lattice points**	**Number of lattice points per cell**
P	primitive	—	1
I	body centred	(1/2, 1/2, 1, 2)	2
A	A-face centred	(0, 1/2, 1/2)	2
B	B-face centred	(1/2, 0, 1/2)	2
C	C-face centred	(1/2, 1/2, 0)	2
F	All faces centred	(1/2, 1/2, 0), (1/2, 0, 1/2) (0, 1/2, 1/2)	2 4
R	Rhombohedrally centred (de scription with 'hexagonal axes')	(1/3, 2/3, 2/3), (2/3, 1/3, 1/3)	3

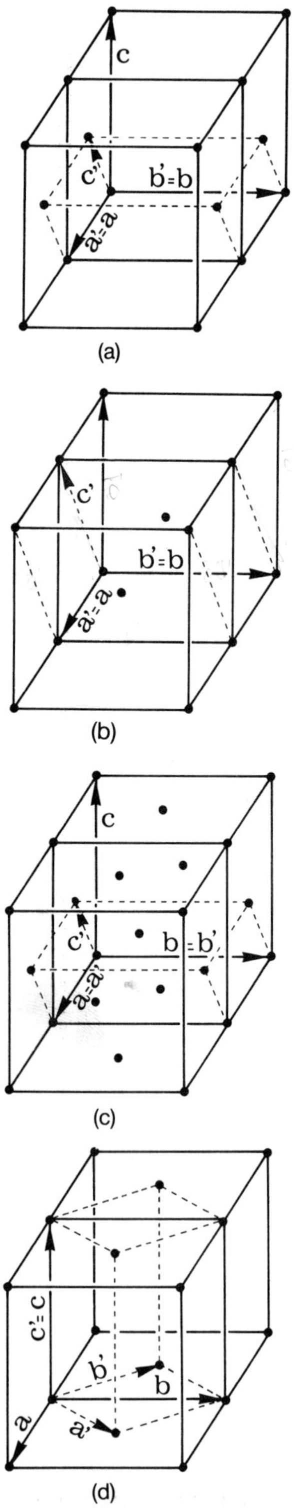

Fig. 1.16. Monoclinic lattices: (a) reduction of a B-centred cell to a P cell; (b) reduction of an I-centred to an A-centred cell; (c) reduction of an F-centred to a C-centred cell; (d) reduction of a C-centred to a P non-monoclinic cell.

Triclinic lattices

Even though non-primitive cells can always be chosen, the absence of axes with order greater than one suggests the choice of a conventional primitive cell with unrestricted α, β, γ angles and $a{:}b{:}c$ ratios. In fact, any triclinic lattice can always be referred to such a cell.

Monoclinic lattices

The conventional monoclinic cell has the twofold axis parallel to $\boldsymbol{b}$, angles $\alpha = \gamma = 90°$, unrestricted β and $a{:}b{:}c$ ratios. A B-centred monoclinic cell with unit vectors $\boldsymbol{a}$, $\boldsymbol{b}$, $\boldsymbol{c}$ is shown in Fig. 1.16(a). If we choose $\boldsymbol{a}' = \boldsymbol{a}$, $\boldsymbol{b}' = \boldsymbol{b}$, $\boldsymbol{c}' = (\boldsymbol{a} + \boldsymbol{c})/2$ a primitive cell is obtained. Since $\boldsymbol{c}'$ lies on the $(\boldsymbol{a}, \boldsymbol{c})$ plane, the new cell will still be monoclinic. Therefore a lattice with a B-type monoclinic cell can always be reduced to a lattice with a P monoclinic cell.

An I cell with axes $\boldsymbol{a}$, $\boldsymbol{b}$, $\boldsymbol{c}$ is illustrated in Fig. 1.16(b). If we choose $\boldsymbol{a}' = \boldsymbol{a}$, $\boldsymbol{b}' = \boldsymbol{b}$, $\boldsymbol{c}' = \boldsymbol{a} + \boldsymbol{c}$, the corresponding cell becomes an A monoclinic cell. Therefore a lattice with an I monoclinic cell may always be described by an A monoclinic cell. Furthermore, since the $\boldsymbol{a}$ and $\boldsymbol{c}$ axes can always be interchanged, an A cell can be always reduced to a C cell.

An F cell with axes $\boldsymbol{a}$, $\boldsymbol{b}$, $\boldsymbol{c}$ is shown in Fig. 1.16(c). When choosing $\boldsymbol{a}' = \boldsymbol{a}$, $\boldsymbol{b}' = \boldsymbol{b}$, $\boldsymbol{c}' = (\boldsymbol{a} + \boldsymbol{c})/2$ a type-C monoclinic cell is obtained. Therefore, also, a lattice described by an F monoclinic cell can always be described by a C monoclinic cell.

We will now show that there is a lattice with a C monoclinic cell which is not amenable to a lattice having a P monoclinic cell. In Fig. 1.16(d) a C cell with axes $\boldsymbol{a}$, $\boldsymbol{b}$, $\boldsymbol{c}$ is illustrated. A primitive cell is obtained by assuming $\boldsymbol{a}' = (\boldsymbol{a} + \boldsymbol{b})/2$, $\boldsymbol{b}' = (-\boldsymbol{a} + \boldsymbol{b})/2$, $\boldsymbol{c}' = \boldsymbol{c}$, but this no longer shows the features of a monoclinic cell, since $\gamma' \neq 90°$, $\boldsymbol{a}' = \boldsymbol{b}' \neq \boldsymbol{c}'$, and the 2 axis lies along the diagonal of a face. It can then be concluded that there are two distinct monoclinic lattices, described by P and C cells, and not amenable one to the other.

Orthorhombic lattices

In the conventional orthorhombic cell the three proper or inversion axes are parallel to the unit vectors $\boldsymbol{a}$, $\boldsymbol{b}$, $\boldsymbol{c}$, with angles $\alpha = \beta = \gamma = 90°$ and general $a{:}b{:}c$ ratios. With arguments similar to those used for monoclinic lattices, the reader can easily verify that there are four types of orthorhombic lattices, P, C, I, and F.

Tetragonal lattices

In the conventional tetragonal cell the fourfold axis is chosen along $\boldsymbol{c}$ with $\alpha = \beta = \gamma = 90°$, $a = b$, and unrestricted c value. It can be easily verified that because of the fourfold symmetry an A cell will always be at the same time a B cell and therefore an F cell. The latter is then amenable to a tetragonal I cell. A C cell is always amenable to another tetragonal P cell. Thus only two different tetragonal lattices, P and I, are found.

Cubic lattices

In the conventional cubic cell the four threefold axes are chosen to be parallel to the principal diagonals of a cube, while the unit vectors $\boldsymbol{a}$, $\boldsymbol{b}$, $\boldsymbol{c}$ are parallel to the cube edges. Because of symmetry a type-A (or B or C) cell is also an F cell. There are three cubic lattices, P, I, and F which are not amenable one to the other.

Hexagonal lattices

In the conventional hexagonal cell the sixfold axis is parallel to $\boldsymbol{c}$, with $\boldsymbol{a} = \boldsymbol{b}$, unrestricted $\boldsymbol{c}$, $\alpha = \beta = 90°$, and $\gamma = 120°$. P is the only type of hexagonal Bravais lattice.

Trigonal lattices

As for the hexagonal cell, in the conventional trigonal cell the threefold axis is chosen parallel to $\boldsymbol{c}$, with $\boldsymbol{a} = \boldsymbol{b}$, unrestricted $\boldsymbol{c}$, $\alpha = \beta = 90°$, and $\gamma = 120°$. Centred cells are easily amenable to the conventional P trigonal cell.

Because of the presence of a threefold axis some lattices can exist which may be described via a P cell of rhombohedral shape, with unit vectors $\boldsymbol{a}_R$, $\boldsymbol{b}_R$, $\boldsymbol{c}_R$ such that $a_R = b_R = c_R$, $\alpha_R = \beta_R = \gamma_R$, and the threefold axis along the $\boldsymbol{a}_R + \boldsymbol{b}_R + \boldsymbol{c}_R$ direction (see Fig. 1.17). Such lattices may also be described by three triple hexagonal cells with basis vectors $\boldsymbol{a}_H$, $\boldsymbol{b}_H$, $\boldsymbol{c}_H$ defined according to[6]

$$\boldsymbol{a}_H = \boldsymbol{a}_R - \boldsymbol{b}_R, \qquad \boldsymbol{b}_H = \boldsymbol{b}_R - \boldsymbol{c}_R, \qquad \boldsymbol{c}_H = \boldsymbol{a}_R + \boldsymbol{b}_R + \boldsymbol{c}_R$$

or

$$\boldsymbol{a}_H = \boldsymbol{b}_R - \boldsymbol{c}_R, \qquad \boldsymbol{b}_H = \boldsymbol{c}_R - \boldsymbol{a}_R, \qquad \boldsymbol{c}_H = \boldsymbol{a}_R + \boldsymbol{b}_R + \boldsymbol{c}_R$$

or

$$\boldsymbol{a}_H = \boldsymbol{c}_R - \boldsymbol{a}_R, \qquad \boldsymbol{b}_H = \boldsymbol{a}_R - \boldsymbol{b}_R, \qquad \boldsymbol{c}_H = \boldsymbol{a}_R + \boldsymbol{b}_R + \boldsymbol{c}_R.$$

These hexagonal cells are said to be in **obverse** setting. Three further triple hexagonal cells, said to be in **reverse** setting, can be obtained by changing $\boldsymbol{a}_H$ and $\boldsymbol{b}_H$ to $-\boldsymbol{a}_H$ and $-\boldsymbol{b}_H$. The hexagonal cells in obverse setting have centring points (see again Fig. 1.17)) at

$$(0, 0, 0), \qquad (2/3, 1/3, 1/3), \qquad (1/3, 2/3, 2/3)$$

while for reverse setting centring points are at

$$(0, 0, 0), \qquad (1/3, 2/3, 1/3),, \quad (2/3, 1/3, 2/3).$$

It is worth noting that a rhombohedral description of a hexagonal P lattice is always possible. Six triple rhombohedral cells with basis vectors $\boldsymbol{a}'_R$, $\boldsymbol{b}'_R$,

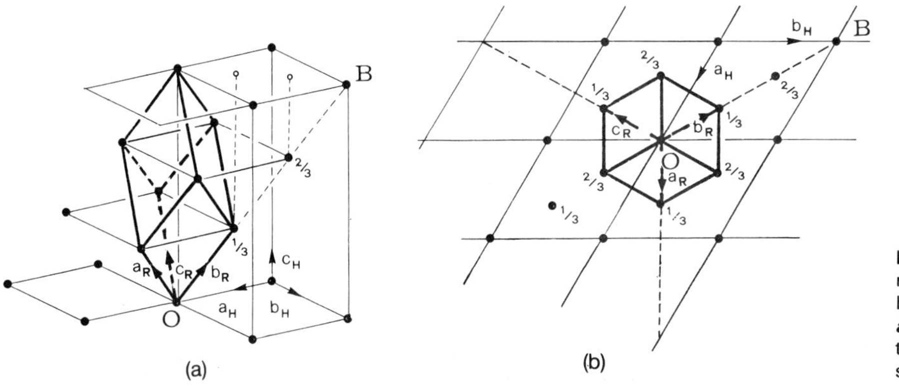

Fig. 1.17. Rhombohedral lattice. The basis of the rhombohedral cell is labelled $\boldsymbol{a}_R$, $\boldsymbol{b}_R$, $\boldsymbol{c}_R$, the basis of the hexagonal centred cell is labelled $\boldsymbol{a}_h$, $\boldsymbol{b}_h$, $\boldsymbol{c}_h$ (numerical fractions are calculated in terms of the $\boldsymbol{c}_h$ axis). (a) Obverse setting; (b) the same figure as in (a) projected along $\boldsymbol{c}_h$.

c'_R can be obtained from a_H, b_H, c_H by choosing:

$$a'_R = a_H + c_H, \qquad b'_R = b_H + c_H, \qquad c'_R = -(a_H + b_H) + c_H$$

$$a'_R = -a_H + c_H, \qquad b'_R = -b_H + c_H, \qquad c'_R = a_H + b_H + c_H$$

and cyclic permutations of a'_R, b'_R, c'_R. Each triple rhomobohedral cell will have centring points at (0, 0, 0), (1/3, 1/3, 1/3), (2/3, 2/3, 2/3).

In conclusion, some trigonal lattices may be described by a hexagonal P cell, others by a triple hexagonal cell. In the first case the nodes lying on the different planes normal to the threefold axis will lie exactly one on top of the other, in the second case lattice planes are translated one with respect to the other in such a way that the nth plane will superpose on the $(n + 3)$th plane (this explains why a rhombohedral lattice is not compatible with a sixfold axis).

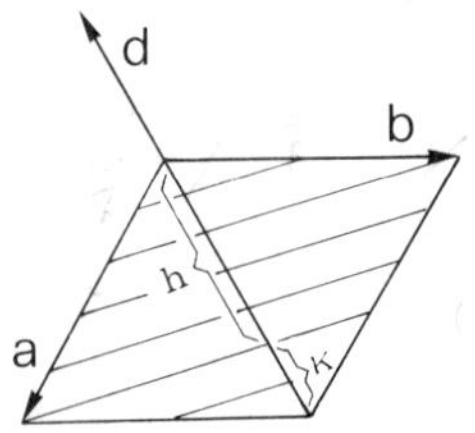

Fig. 1.18. Intersections of the set of crystallographic planes ($h\,k\,l$) with the three symmetry-equivalent ***a***, ***b***, ***d*** axes in trigonal and hexagonal systems.

When, for crystals belonging to the hexagonal or trigonal systems, a hexagonal cell is chosen, then on the plane defined by $\boldsymbol{a}$ and $\boldsymbol{b}$ there will be a third axis equivalent to them. The family of planes (hkl) (see Fig. 1.18) divides the positive side of $\boldsymbol{a}$ in h parts and the positive side of $\boldsymbol{b}$ in k parts. If the third axis (say $\boldsymbol{d}$) on the ($\boldsymbol{a}$, $\boldsymbol{b}$) plane is divided in i parts we can introduce an extra index in the symbol of the family, i.e. ($hkil$). From the same figure it can be seen that the negative side of $\boldsymbol{d}$ is divided in $h + k$ parts, and then $i = -(h + k)$. For instance (1 2 −3 5), (3 −5 2 1), (−2 0 2 3) represent three plane families in the new notation. The four-index symbol is useful to display the symmetry, since ($hkil$), ($kihl$), and ($ihkl$) are symmetry equivalent planes.

Also, lattice directions can be indicated by the four-index notation. Following pp. 7–8, a direction in the ($\boldsymbol{a}$, $\boldsymbol{b}$) plane is defined by a vector $(\mathrm{P} - \mathrm{O}) = m\mathbf{a} + n\boldsymbol{b}$. If we introduce the third axis $\boldsymbol{d}$ in the plane, we can write $(\mathrm{P} - \mathrm{O}) = m\boldsymbol{a} + n\boldsymbol{b} + 0\boldsymbol{d}$. Since a decrease (or increase) of the three coordinates by the same amount j does not change the point P, this may be represented by the coordinates: $u = m - j$, $v = n - j$, $i = -j$.

If we choose $j = (m + n)/3$, then $u = (2m - n)/3$, $v = (2n - m)/3$, $i = -(m + n)/3$. In conclusion the direction $[mnw]$ may be represented in the new notation as $[uviw]$, with $i = -(u + v)$. On the contrary, if a direction is already represented in the four-index notation $[uviw]$, to pass to the three-index one, $-i$ should be added to the first three indices in order to bring to zero the third index, i.e. $[u - i \;\; v - i \;\; w]$.

A last remark concerns the point symmetry of a lattice. There are seven three-dimensional lattice point groups, they are called **holohedries** and are listed in Table 1.6 (note that $\bar{3}$m is the point symmetry of the rhombohedral lattice). In two dmensions four holohedries exist: 2, 2mm, 4mm, 6mm.

The 14 Bravais lattices are illustrated in Fig. 1.19 by means of their conventional unit cells (see Appendix 1.C for a different type of cell). A detailed description of the metric properties of crystal lattices will be given in Chapter 2.

The space groups

A crystallographic space group is the set of geometrical symmetry operations that take a three-dimensional periodic object (say a crystal) into itself.

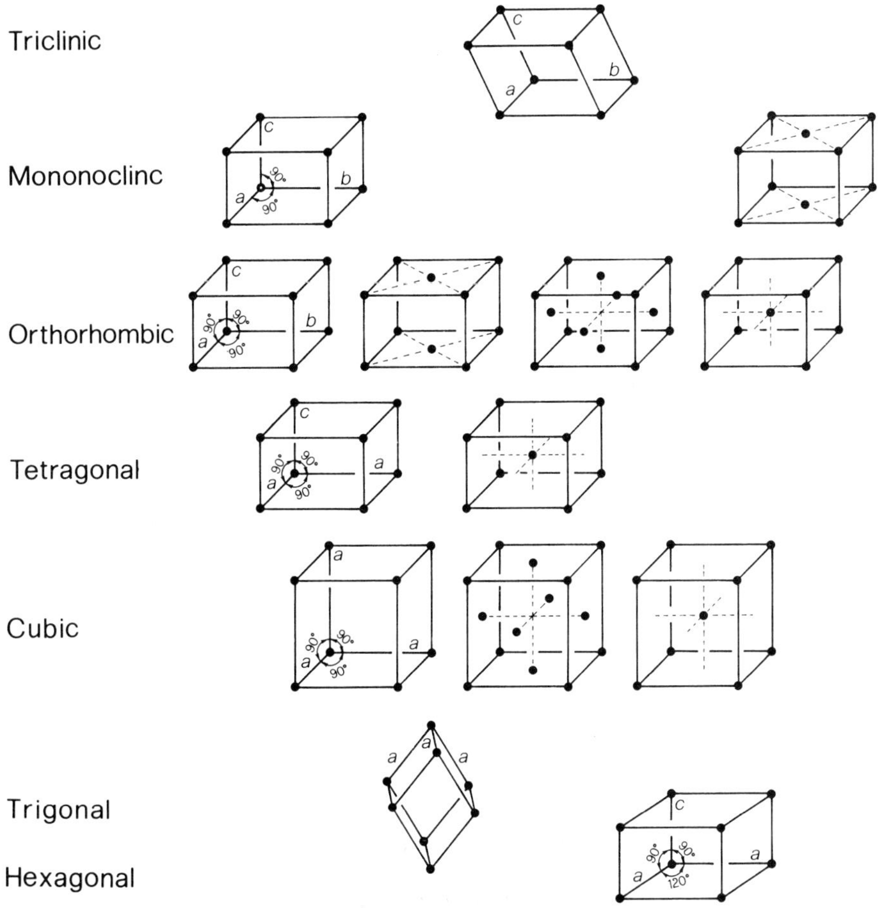

Fig. 1.19. The 14 three-dimensional Bravais lattices.

The total number of crystallographic space groups is 230. They were first derived at the end of the last century by the mathematicians Fedorov (1890) and Schoenflies (1891) and are listed in Table 1.9.

In Fedorov's mathematical treatment each space group is represented by a set of three equations: such an approach enabled Fedorov to list all the space groups (he rejected, however, five space groups as impossible: *Fdd*2, *Fddd*, I$\bar{4}$3*d*, P$4_3$32, P$4_1$32). The Schoenflies approach was most practical and is described briefly in the following.

On pp. 11–16 we saw that 32 combinations of either simple rotation or inversion axes are compatible with the periodic nature of crystals. By combining the 32 point groups with the 14 Bravais lattices (i.e. P, I, F, . . .) one obtains only 73 (symmorphic) space groups. The others may be obtained by introducing a further variation: the proper or improper symmetry axes are replaced by screw axes of the same order and mirror planes by glide planes. Note, however, that when such combinations have more than one axis, the restriction that all symmetry elements must intersect in a point no longer applies (cf. Appendix 1.B). As a consequence of the presence of symmetry elements, several symmetry-equivalent objects will coexist within the unit cell. We will call the smallest part of the unit cell which will generate the whole cell when applying to it the symmetry

Table 1.9. The 230 three-dimensional space groups arranged by crystal systems and point groups. Space groups (and enantiomorphous pairs) that are uniquely determinable from the symmetry of the diffraction pattern and from systematic absences (see p. 159) are shown in bold-type. Point groups without inversion centres or mirror planes are emphasized by boxes

Crystal system	Point group	Space groups
Triclinic	[1]	P1
	$\bar{1}$	$P\bar{1}$
Monoclinic	[2]	P2, P2$_1$, C2
	m	Pm, P*c*, Cm, C*c*
	2/m	P2/m, P2$_1$/m, C2/m, P2/*c*, **P2$_1$/*c***, C2/*c*
Orthorhombic	[222]	P222, **P222$_1$**, **P2$_1$2$_1$2**, **P2$_1$2$_1$2$_1$**, **C222$_1$**, C222, F222, I222, I2$_1$2$_1$2$_1$
	mm2	Pmm2, Pm*c*2$_1$, P*cc*2, Pm*a*2$_1$, P*ca*2$_1$, P*nc*2$_1$, Pm*n*2$_1$, P*ba*2, P*na*2$_1$, P*nn*2, Cmm2, Cm*c*2$_1$, C*cc*2, Amm2, A*b*m2, Am*a*2, A*ba*2, Fmm2, ***Fdd*2**, Imm2, I*ba*2, Im*a*2
	mmm	Pmmm, **P*nnn***, P*cc*m, **P*ban***, Pmm*a*, **P*nna***, Pm*na*, **P*cca***, P*ba*m, **P*ccn***, P*bc*m, P*nn*m, Pmm*n*, **P*bcn***, **P*bca***, P*n*m*a*, Cm*c*m, Cm*ca*, Cmmm, C*cc*m, Cmm*a*, **C*cca***, Fmmm, **F*ddd***, Immm, I*ba*m, **I*bca***, Imm*a*
Tetragonal	[4]	P4, **P4$_1$**, P4$_2$, **P4$_3$**, I4, **I4$_1$**
	$\bar{4}$	$P\bar{4}$, $I\bar{4}$
	4/m	P4/m, P4$_2$/m, **P4/*n***, **P4$_2$/*n***, I4/m, **I4$_1$/*a***
	[422]	P422, **P42$_1$2**, P4$_1$22, **P4$_1$2$_1$2**, **P4$_2$22**, **P4$_2$2$_1$2**, **P4$_3$22**, **P4$_3$2$_1$2**, I422, **I4$_1$22**
	4mm	P4mm, P4*b*m, P4$_2$*c*m, P4$_2$*n*m, P4*cc*, P4*nc*, P4$_2$m*c*, P4$_2$*bc*, I4mm, I4*c*m, I4$_1$m*d*, **I4$_1$*cd***
	$\bar{4}$m	$P\bar{4}$2m, $P\bar{4}$2*c*, $P\bar{4}$2$_1$m, **$P\bar{4}$2$_1$*c***, $P\bar{4}$m2, $P\bar{4}$*c*2, $P\bar{4}$*b*2, $P\bar{4}$*n*2, $I\bar{4}$m2, $I\bar{4}$*c*2, $I\bar{4}$2m, $I\bar{4}$2*d*
	4/mmm	P4/mmm, P4/m*cc*, **P4/*nb*m**, **P4/*nnc***, P4/m*b*m, P4/m*nc*, **P4/*n*mm**, **P4/*ncc***, P4$_2$/mm*c*, P4$_2$/m*c*m, **P4$_2$/*nbc***, **P4$_2$/*nn*m**, P4$_2$/m*bc*, P4$_2$m*n*m, **P4$_2$/*n*m*c***, **P4$_2$/*nc*m**, I4/mmm, I4/m*c*m, **I4$_1$/*a*m*d***, **I4$_1$/*acd***
Trigonal–hexagonal	[3]	P3, P3$_1$, P3$_2$, R3
	$\bar{3}$	$P\bar{3}$, $R\bar{3}$
	[32]	P312, P321, **P3$_1$12**, **P3$_1$21**, **P3$_2$12**, **P3$_2$21**, R32
	3m	P3m1, P31m, P3*c*1, P31*c*, R3m, R3*c*
	$\bar{3}$m	$P\bar{3}$1m, $P\bar{3}$1*c*, $P\bar{3}$m1, $P\bar{3}$*c*1, $R\bar{3}$m, $R\bar{3}$*c*
	[6]	P6, **P6$_1$**, **P6$_5$**, P6$_3$, **P6$_2$**, **P6$_4$**,
	$\bar{6}$	$P\bar{6}$
	6/m	P6/m, P6$_3$/m
	[622]	P622, **P6$_1$22**, **P6$_5$22**, **P6$_2$22**, **P6$_4$22**, **P6$_3$22**
	6mm	P6mm, P6*cc*, P6$_3$*c*m, P6$_3$m*c*
	$\bar{6}$m	$P\bar{6}$m2, $P\bar{6}$*c*2, $P\bar{6}$2m, $P\bar{6}$2*c*
	6/mmm	P6/mmm, P6/m*cc*, P6$_3$/m*c*m, P6$_3$/mm*c*
Cubic	[23]	P23, F23, I23, **P2$_1$3**, I2$_1$3
	m$\bar{3}$	Pm$\bar{3}$, **P*n*$\bar{3}$**, Fm$\bar{3}$, **F*d*$\bar{3}$**, Im$\bar{3}$, **P*a*$\bar{3}$**, **I*a*$\bar{3}$**
	[432]	P432, **P4$_2$32**, F432, **F4$_1$32**, I432, **P4$_3$32**, **P4$_1$32**, I4$_1$32
	$\bar{4}$3m	$P\bar{4}$3m, $F\bar{4}$3m, $I\bar{4}$3m, $P\bar{4}$3*n*, $F\bar{4}$3*c*, **$I\bar{4}$3*d***
	m$\bar{3}$m	Pm$\bar{3}$m, **P*n*$\bar{3}$*n***, Pm$\bar{3}$n, **P*n*$\bar{3}$m**, Fm$\bar{3}$m, Fm$\bar{3}$*c*, **F*d*$\bar{3}$m**, **F*d*$\bar{3}$*c***, Im$\bar{3}$m, **I*a*$\bar{3}$*d***

operations an **asymmetric unit**. The asymmetric unit is not usually uniquely defined and can be chosen with some degree of freedom. It is nevertheless obvious that when rotation or inversion axes are present, they must lie at the borders of the asymmetric unit.

According to the international (Hermann–Mauguin) notation, the space-group symbol consists of a letter indicating the centring type of the conventional cell, followed by a set of characters indicating the symmetry elements. Such a set is organized according to the following rules:

1. For triclinic groups: no symmetry directions are needed. Only two space groups exist: P1 and P$\bar{1}$.
2. For monoclinic groups: only one symbol is needed, giving the nature of the unique dyad axis (proper and/or inversion). Two settings are used: y-axis unique, z-axis unique.
3. For orthorhombic groups: dyads (proper and/or of inversion) are given along x, y, and z axis in the order. Thus P$ca2_1$ means: primitive cell, glide plane of type c normal to x-axis, glide plane of type a normal to the y-axis, twofold screw axis along z.
4. For tetragonal groups: first the tetrad (proper and/or of inversion) axis along z is specified, then the dyad (proper and/or of inversion) along x is given, and after that the dyad along [110] is specified. For example, P$4_2/nbc$ denotes a space group with primitive cell, a 4 sub 2 screw axis along z to which a diagonal glide plane is perpendicular, an axial glide plane b normal to the x axis, an axial glide plane c normal to [110]. Because of the tetragonal symmetry, there is no need to specify symmetry along the y-axis.
5. For trigonal and hexagonal groups: the triad or hexad (proper and/or of inversion) along the z-axis is first given, then the dyad (proper and/or of inversion) along x and after that the dyad (proper and/or of inversion) along [1$\bar{1}$0] is specified. For example, P6_3mc has primitive cell, a sixfold screw axis 6 sub 3 along z, a reflection plane normal to x and an axial glide plane c normal to [1$\bar{1}$0].
6. For cubic groups: dyads or tetrads (proper and/or of inversion) along x, followed by triads (proper and/or of inversion) along [111] and dyads (proper and/or of inversion) along [110].

We note that:

1. The combination of the Bravais lattices with symmetry elements with no translational components yields the 73 so-called **symmorphic** space groups. Examples are: P222, Cmm2, F23, etc.
2. The 230 space groups include 11 enantiomorphous pairs: P3_1 (P3_2), P$3_1$12 (P$3_2$12), P$3_1$21 (P$3_2$21), P4_1 (P4_3), P$4_1$22 (P$4_3$22), P$4_12_1$2 (P$4_32_1$2), P6_1 (P6_5), P6_2 (P6_4), P$6_1$22 (P$6_5$22), P$6_2$22 (P$6_4$22), P$4_1$32 (P$4_3$32). The (+) isomer of an optically active molecule crystallizes in one of the two enantiomorphous space groups, the (−) isomer will crystallize in the other.
3. Biological molecules are enantiomorphous and will then crystallize in space groups with no inversion centres or mirror planes; there are 65 groups of this type (see Table 1.9).
4. The point group to which the space group belongs is easily obtained from the space-group symbol by omitting the lattice symbol and by replacing

the screw axes and the glide planes with their corresponding symmorphic symmetry elements. For instance, the space groups $P4_2/mmc$, $P4/ncc$, $I4_1/acd$, all belong to the point group 4/mmm.

5. The frequency of the different space groups is not uniform. Organic compounds tend to crystallize in the space groups that permit close packing of triaxial ellipsoids.[8] According to this view, rotation axes and reflection planes can be considered as rigid scaffolding which make more difficult the comfortable accommodation of molecules, while screw axes and glide planes, when present, make it easier because they shift the molecules away from each other.

Mighell and Rodgers [9] examined 21 051 organic compounds of known crystal structure; 95% of them had a symmetry not higher than orthorhombic. In particular 35% belonged to the space group $P2_1/c$, 13.3% to $P\bar{1}$, 12.4% to $P2_12_12_1$, 7.6% to $P2_1$ and 6.9% to $C2/c$. A more recent study by Wilson,[10] based on a survey of the 54 599 substances stored in the Cambridge Structural Database (in January 1987), confirmed Mighell and Rodgers' results and suggested a possible model to estimate the number N_{sg} of structures in each space group of a given crystal class:

$$N_{sg} = A_{cc} \exp \{-B_{cc}[2]_{sg} - C_{cc}[m]_{sg})$$

where A_{cc} is the total number of structures in the crystal class, $[2]_{sg}$ is the number of twofold axes, $[m]_{sg}$ the number of reflexion planes in the cell, B_{cc} and C_{cc} are parameters characteristic of the crystal class in question. The same results cannot be applied to inorganic compounds, where ionic bonds are usually present. Indeed most of the 11 641 inorganic compounds considered by Mighell and Rodgers crystallize in space groups with orthorhombic or higher symmetry. In order of decreasing frequency we have: $Fm3m$, $Fd3m$, $P6_3/mmc$, $P2_1/c$, $Pm\bar{3}m$, $R\bar{3}m$, $C2/m$, $C2/c$, ...

The standard compilation of the plane and of the three-dimensional space groups is contained in volume A of the *International Tables for Crystallography.* For each space groups the *Tables* include (see Figs 1.20 and 1.21).

1. At the first line: the short international (Hermann–Mauguin) and the Schoenflies symbols for the space groups, the point group symbol, the crystal system.

2. At the second line: the sequential number of the plane or space group, the full international (Hermann–Mauguin) symbol, the Patterson symmetry (see Chapter 5, p. 327). Short and full symbols differ only for the monoclinic space groups and for space groups with point group mmm, 4/mmm, $\bar{3}$m, 6/mmm, m$\bar{3}$, m$\bar{3}$m. While in the short symbols symmetry planes are suppressed as much as possible, in the full symbols axes and planes are listed for each direction.

3. Two types of space group diagrams (as orthogonal projections along a cell axis) are given: one shows the position of a set of symmetrically equivalent points, the other illustrates the arrangement of the symmetry elements. Close to the graphical symbols of a symmetry plane or axis parallel to the projection plane the 'height' h (as a fraction of the shortest lattice translation normal to the projection plane) is printed. If $h = 0$ the height is omitted. Symmetry elements at h also occur at height $h + 1/2$.

4. Information is given about: setting (if necessary), origin, asymmetric unit, symmetry operations, symmetry generators (see Appendix 1.E) selected to generate all symmetrical equivalent points described in block 'Positions'. The origin of the cell for centrosymmetric space groups is usually chosen on an inversion centre. A second description is given if points of high site symmetry not coincident with the inversion centre occur. For example, for $Pn\bar{3}n$ two descriptions are available, the first with origin at 432, and the second with origin at $\bar{3}$. For non-centrosymmetric space groups the origin is chosen at a point of highest symmetry (e.g. the origin for $P\bar{4}2c$ is chosen at $\bar{4}1c$) or at a point which is conveniently placed with respect to the symmetry elements. For example, on the screw axis in $P2_1$, on the glide plane in Pc, at $1a2_1$ in $Pca2_1$, at a point which is surrounded symmetrically by the three 2_1 axis in $P2_12_12_1$.

5. The block **positions** (called also Wyckoff positions) contains the **general position** (a set of symmetrically equivalent points, each point of which is left invariant only by application of an identity operation) and a list of **special positions** (a set of symmetrically equivalent points is in special position if each point is left invariant by at least two symmetry operations of the space group). The first three block columns give information about **multiplicity** (number of equivalent points per unit cell), **Wyckoff letter** (a code scheme starting with *a* at the bottom position and continuing upwards in alphabetical order), **site symmetry** (the group of symmetry operations which leaves invariant the site). The symbol adopted[9] for describing the site symmetry displays the same sequence of symmetry directions as the space group symbol. A dot marks those directions which do not contribute any element to the site symmetry. To each Wyckoff position a **reflection condition**, limiting possible reflections, may be associated. The condition may be general (it is obeyed irrespective of which Wyckoff positions are occupied by atoms (see Chapter 3, p. 159) or special (it limits the contribution to the structure factor of the atoms located at that Wyckoff position).

6. Symmetry of special projections. Three orthogonal projections for each space group are listed: for each of them the projection direction, the Hermann–Mauguin symbol of the resulting plane group, and the relation between the basis vectors of the plane group and the basis vectors of the space group, are given, together with the location of the plane group with respect to the unit cell of the space group.

7. Information about maximal subgroups and minimal supergroups (see Appendix 1.E) is given.

In Figs. 1.20 and 1.21 descriptions of the space groups $Pbcn$ and $P4_222$ are respectively given as compiled in the *International Tables for Crystallography*. In order to obtain space group diagrams the reader should perform the following operations:

1. Some or all the symmetry elements are traced as indicated in the space-group symbol. This is often a trivial task, but in certain cases special care must be taken. For example, the three twofold screw axes do not intersect each other in $P2_12_12_1$, but two of them do in $P2_12_12$ (see Appendix 1.B).

Pbcn D_{2h}^{14} *mmm* Orthorhombic

No. 60 $P\,2_1/b\,2/c\,2_1/n$ Patterson symmetry *Pmmm*

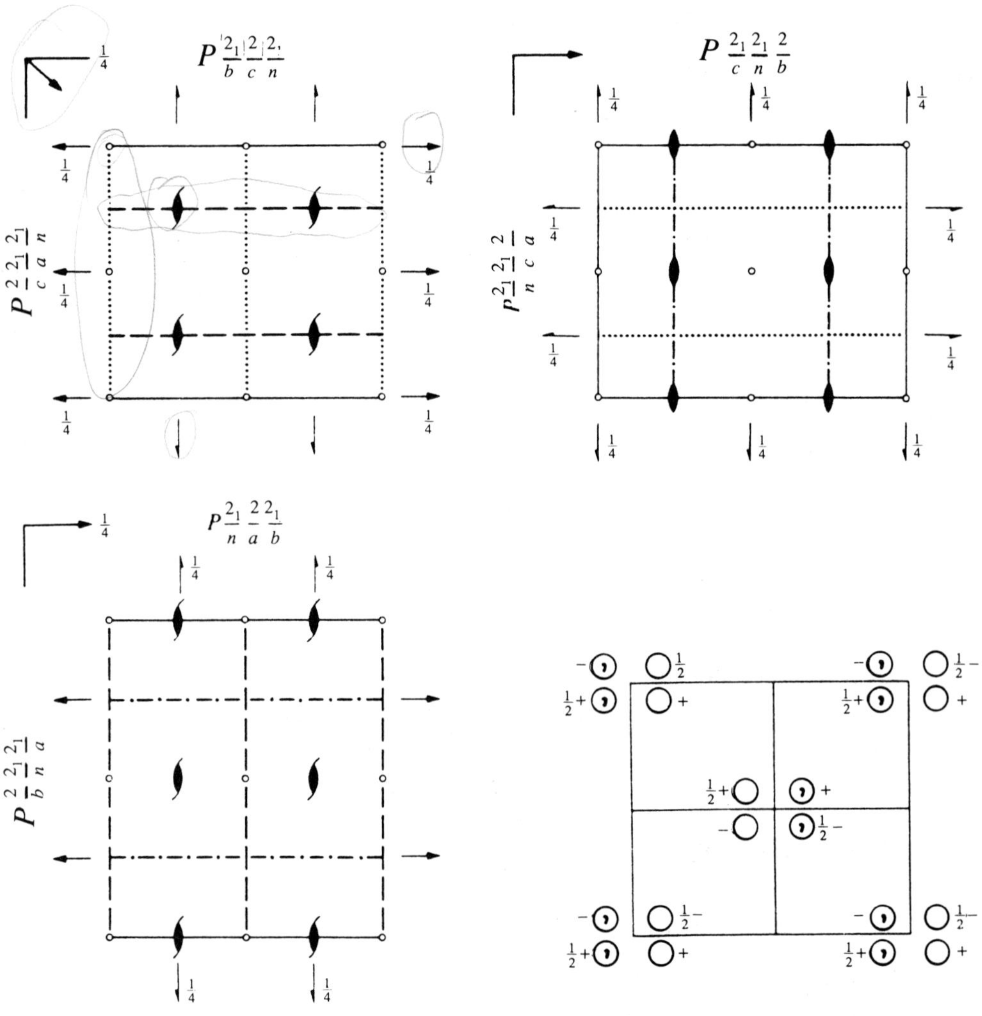

Origin at $\bar{1}$ on $1\,c\,1$

Asymmetric unit $0\leq x\leq \frac{1}{2};\quad 0\leq y\leq \frac{1}{2};\quad 0\leq z\leq \frac{1}{2}$

Symmetry operations

(1) 1	(2) $2(0,0,\frac{1}{2})\ \ \frac{1}{4},\frac{1}{4},z$	(3) $2\ \ 0,y,\frac{1}{4}$	(4) $2(\frac{1}{2},0,0)\ \ x,\frac{1}{4},0$
(5) $\bar{1}\ \ 0,0,0$	(6) $n(\frac{1}{2},\frac{1}{2},0)\ \ x,y,\frac{1}{4}$	(7) $c\ \ x,0,z$	(8) $b\ \ \frac{1}{4},y,z$

Fig. 1.20. Representation of the group P*bcn* (as in *International Tables for Crystallography*).

CONTINUED No. 60 ***Pbcn***

Generators selected (1); $t(1,0,0)$; $t(0,1,0)$; $t(0,0,1)$; (2); (3); (5)

Positions

Multiplicity, Wyckoff letter, Site symmetry			Coordinates				Reflection conditions
							General:
8	*d*	1	(1) x,y,z	(2) $\bar{x}+\frac{1}{2},\bar{y}+\frac{1}{2},z+\frac{1}{2}$	(3) $\bar{x},y,\bar{z}+\frac{1}{2}$	(4) $x+\frac{1}{2},\bar{y}+\frac{1}{2},\bar{z}$	$0kl$: $k=2n$
			(5) $\bar{x},\bar{y},\bar{z}$	(6) $x+\frac{1}{2},y+\frac{1}{2},\bar{z}+\frac{1}{2}$	(7) $x,\bar{y},z+\frac{1}{2}$	(8) $\bar{x}+\frac{1}{2},y+\frac{1}{2},z$	$h0l$: $l=2n$
							$hk0$: $h+k=2n$
							$h00$: $h=2n$
							$0k0$: $k=2n$
							$00l$: $l=2n$
							Special: as above, plus
4	*c*	. 2 .	$0,y,\frac{1}{4}$	$\frac{1}{2},\bar{y}+\frac{1}{2},\frac{3}{4}$	$0,\bar{y},\frac{3}{4}$	$\frac{1}{2},y+\frac{1}{2},\frac{1}{4}$	hkl: $h+k=2n$
4	*b*	$\bar{1}$	$0,\frac{1}{2},0$	$\frac{1}{2},0,\frac{1}{2}$	$0,\frac{1}{2},\frac{1}{2}$	$\frac{1}{2},0,0$	hkl: $h+k,l=2n$
4	*a*	$\bar{1}$	$0,0,0$	$\frac{1}{2},\frac{1}{2},\frac{1}{2}$	$0,0,\frac{1}{2}$	$\frac{1}{2},\frac{1}{2},0$	hkl: $h+k,l=2n$

Symmetry of special projections

Along [001] $c2mm$
$\mathbf{a}'=\mathbf{a}$ $\mathbf{b}'=\mathbf{b}$
Origin at $0,0,z$

Along [100] $p2gm$
$\mathbf{a}'=\frac{1}{2}\mathbf{b}$ $\mathbf{b}'=\mathbf{c}$
Origin at $x,0,0$

Along [010] $p2gm$
$\mathbf{a}'=\frac{1}{2}\mathbf{c}$ $\mathbf{b}'=\mathbf{a}$
Origin at $0,y,0$

Maximal non-isomorphic subgroups

I	$[2]P2_122_1\,(P2_12_12)$	1; 2; 3; 4
	$[2]P112_1/n\,(P2_1/c)$	1; 2; 5; 6
	$[2]P12/c1\,(P2/c)$	1; 3; 5; 7
	$[2]P2_1/b11\,(P2_1/c)$	1; 4; 5; 8
	$[2]Pbc2_1\,(Pca2_1)$	1; 2; 7; 8
	$[2]Pb2n\,(Pnc2)$	1; 3; 6; 8
	$[2]P2_1cn\,(Pna2_1)$	1; 4; 6; 7

IIa none

IIb none

Maximal isomorphic subgroups of lowest index

IIc $[3]Pbcn\,(\mathbf{a}'=3\mathbf{a})$; $[3]Pbcn\,(\mathbf{b}'=3\mathbf{b})$; $[3]Pbcn\,(\mathbf{c}'=3\mathbf{c})$

Minimal non-isomorphic supergroups

I none

II $[2]Abma\,(Cmca)$; $[2]Bbab\,(Ccca)$; $[2]Cmcm$; $[2]Ibam$; $[2]Pbcb\,(2\mathbf{a}'=\mathbf{a})\,(Pcca)$;
$[2]Pmca\,(2\mathbf{b}'=\mathbf{b})\,(Pbcm)$; $[2]Pbmn\,(2\mathbf{c}'=\mathbf{c})\,(Pmna)$

$P4_222$ D_4^5 422 Tetragonal

No. 93 $P4_222$ Patterson symmetry $P4/mmm$

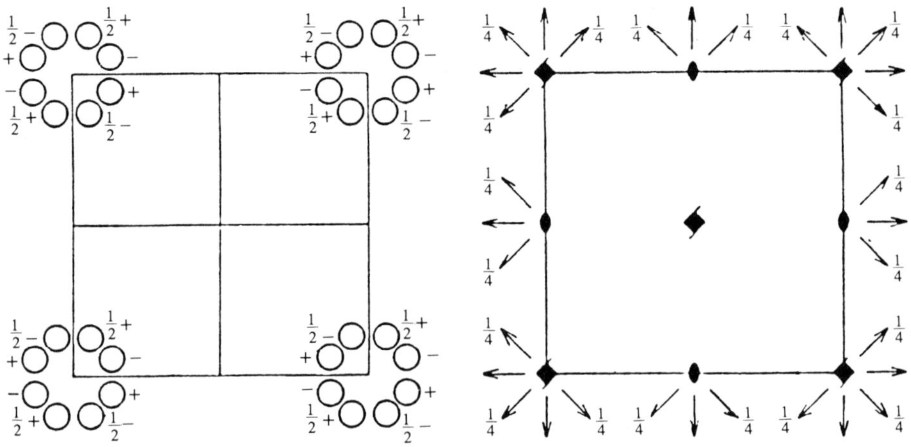

Origin at 222 at 4_221

Asymmetric unit $0 \leq x \leq \frac{1}{2}$; $0 \leq y \leq 1$; $0 \leq z \leq \frac{1}{4}$

Symmetry operations

(1) 1	(2) 2 0,0,z	(3) $4^+(0,0,\frac{1}{2})$ 0,0,z	(4) $4^-(0,0,\frac{1}{2})$ 0,0,z
(5) 2 0,y,0	(6) 2 x,0,0	(7) 2 $x,x,\frac{1}{4}$	(8) 2 $x,\bar{x},\frac{1}{4}$

Fig. 1.21. Representation of the group $P4_222$ (as in *International Tables for Crystallography*).

2. Once conveniently located, the symmetry operators are applied to a point P in order to obtain the **symmetry equivalent** points P′, P″, ... If P′, P″, . . . , fall outside the unit cell, they should be moved inside by means of appropriate lattice translations. The first type of diagram is so obtained.

3. New symmetry elements are then placed in the unit cell so producing the second type of diagram.

   Some space group diagrams are collected in Fig. 1.22. Two simple crystal structures are shown in Figs 1.23 and 1.24: symmetry elements are also located for convenience.

The plane and line groups

There are 17 plane groups, which are listed in Table 1.10. In the symbol g stays for a glide plane. Any space group in projection will conform to one of these plane groups. There are two line groups: p1 and pm.

A periodic decoration of the plane according to the 17 plane groups is shown in Fig. 1.25.

CONTINUED No. 93 $P4_222$

Generators selected (1); $t(1,0,0)$; $t(0,1,0)$; $t(0,0,1)$; (2); (3); (5)

Positions

Multiplicity, Wyckoff letter, Site symmetry			Coordinates				Reflection conditions
							General:
8	*p*	1	(1) x,y,z (5) $\bar{x},y,\bar{z}$	(2) $\bar{x},\bar{y},z$ (6) $x,\bar{y},\bar{z}$	(3) $\bar{y},x,z+\frac{1}{2}$ (7) $y,x,\bar{z}+\frac{1}{2}$	(4) $y,\bar{x},z+\frac{1}{2}$ (8) $\bar{y},\bar{x},\bar{z}+\frac{1}{2}$	$00l: l=2n$
							Special: as above, plus
4	*o*	..2	$x,x,\frac{3}{4}$	$\bar{x},\bar{x},\frac{3}{4}$	$\bar{x},x,\frac{1}{4}$	$x,\bar{x},\frac{1}{4}$	$0kl: l=2n$
4	*n*	..2	$x,x,\frac{1}{4}$	$\bar{x},\bar{x},\frac{1}{4}$	$\bar{x},x,\frac{3}{4}$	$x,\bar{x},\frac{3}{4}$	$0kl: l=2n$
4	*m*	.2.	$x,\frac{1}{2},0$	$\bar{x},\frac{1}{2},0$	$\frac{1}{2},x,\frac{1}{2}$	$\frac{1}{2},\bar{x},\frac{1}{2}$	$hhl: l=2n$
4	*l*	.2.	$x,0,\frac{1}{2}$	$\bar{x},0,\frac{1}{2}$	$0,x,0$	$0,\bar{x},0$	$hhl: l=2n$
4	*k*	.2.	$x,\frac{1}{2},\frac{1}{2}$	$\bar{x},\frac{1}{2},\frac{1}{2}$	$\frac{1}{2},x,0$	$\frac{1}{2},\bar{x},0$	$hhl: l=2n$
4	*j*	.2.	$x,0,0$	$\bar{x},0,0$	$0,x,\frac{1}{2}$	$0,\bar{x},\frac{1}{2}$	$hhl: l=2n$
4	*i*	2..	$0,\frac{1}{2},z$	$\frac{1}{2},0,z+\frac{1}{2}$	$0,\frac{1}{2},\bar{z}$	$\frac{1}{2},0,\bar{z}+\frac{1}{2}$	$hkl: h+k+l=2n$
4	*h*	2..	$\frac{1}{2},\frac{1}{2},z$	$\frac{1}{2},\frac{1}{2},z+\frac{1}{2}$	$\frac{1}{2},\frac{1}{2},\bar{z}$	$\frac{1}{2},\frac{1}{2},\bar{z}+\frac{1}{2}$	$hkl: l=2n$
4	*g*	2..	$0,0,z$	$0,0,z+\frac{1}{2}$	$0,0,\bar{z}$	$0,0,\bar{z}+\frac{1}{2}$	$hkl: l=2n$
2	*f*	2.22	$\frac{1}{2},\frac{1}{2},\frac{1}{4}$	$\frac{1}{2},\frac{1}{2},\frac{3}{4}$			$hkl: l=2n$
2	*e*	2.22	$0,0,\frac{1}{4}$	$0,0,\frac{3}{4}$			$hkl: l=2n$
2	*d*	222.	$0,\frac{1}{2},\frac{1}{2}$	$\frac{1}{2},0,0$			$hkl: h+k+l=2n$
2	*c*	222.	$0,\frac{1}{2},0$	$\frac{1}{2},0,\frac{1}{2}$			$hkl: h+k+l=2n$
2	*b*	222.	$\frac{1}{2},\frac{1}{2},0$	$\frac{1}{2},\frac{1}{2},\frac{1}{2}$			$hkl: l=2n$
2	*a*	222.	$0,0,0$	$0,0,\frac{1}{2}$			$hkl: l=2n$

Symmetry of special projections

Along [001] $p4mm$
$\mathbf{a}'=\mathbf{a}$ $\mathbf{b}'=\mathbf{b}$
Origin at $0,0,z$

Along [100] $p2mm$
$\mathbf{a}'=\mathbf{b}$ $\mathbf{b}'=\mathbf{c}$
Origin at $x,0,0$

Along [110] $p2mm$
$\mathbf{a}'=\frac{1}{2}(-\mathbf{a}+\mathbf{b})$ $\mathbf{b}'=\mathbf{c}$
Origin at $x,x,\frac{1}{4}$

Maximal non-isomorphic subgroups

I [2]$P4_211$ ($P4_2$) 1; 2; 3; 4
[2]$P221$ ($P222$) 1; 2; 5; 6
[2]$P212$ ($C222$) 1; 2; 7; 8

IIa none

IIb [2]$P4_122$ ($\mathbf{c}'=2\mathbf{c}$); [2]$P4_322$ ($\mathbf{c}'=2\mathbf{c}$); [2]$C4_222_1$ ($\mathbf{a}'=2\mathbf{a}, \mathbf{b}'=2\mathbf{b}$)($P4_22_12$);
[2]$F4_122$ ($\mathbf{a}'=2\mathbf{a}, \mathbf{b}'=2\mathbf{b}, \mathbf{c}'=2\mathbf{c}$)($I4_122$)

Maximal isomorphic subgroups of lowest index

IIc [3]$P4_222$ ($\mathbf{c}'=3\mathbf{c}$); [2]$C4_222$ ($\mathbf{a}'=2\mathbf{a}, \mathbf{b}'=2\mathbf{b}$)($P4_222$)

Minimal non-isomorphic supergroups

I [2]$P4_2/mmc$; [2]$P4_2/mcm$; [2]$P4_2/nbc$; [2]$P4_2/nnm$; [3]$P4_232$

II [2]$I422$; [2]$P422$ ($2\mathbf{c}'=\mathbf{c}$)

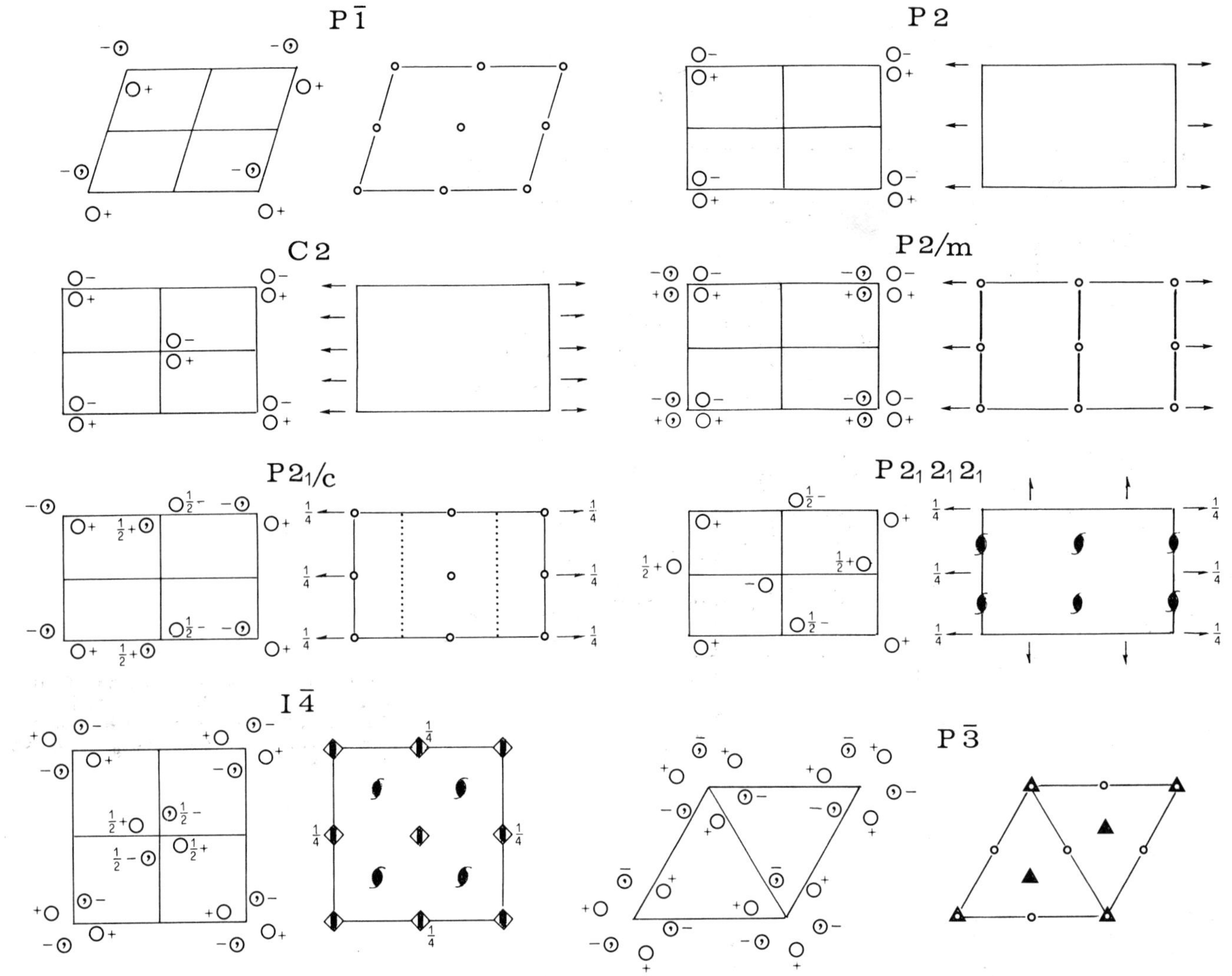

Fig. 1.22. Some space group diagrams.

On the matrix representation of symmetry operators

A symmetry operation acts on the fractional coordinates x, y, z of a point P to obtain the coordinates (x', y', z') of a symmetry-equivalent point P′:

$$\mathbf{X}' = \begin{pmatrix} x' \\ y' \\ z' \end{pmatrix} = \begin{pmatrix} R_{11} & R_{12} & R_{13} \\ R_{21} & R_{22} & R_{23} \\ R_{31} & R_{32} & R_{33} \end{pmatrix} \begin{pmatrix} x \\ y \\ z \end{pmatrix} + \begin{pmatrix} T_1 \\ T_2 \\ T_3 \end{pmatrix} = \mathbf{CX} = \mathbf{RX} + \mathbf{T}. \quad (1.11)$$

The **R** matrix is the **rotational component** (proper or improper) of the symmetry operation. As we shall see in Chapter 2 its elements may be 0, +1, −1 and its determinant is ±1. **T** is the matrix of the translational component of the operation. A list of all the rotation matrices needed to conventionally describe the 230 space groups are given in Appendix 1.D.

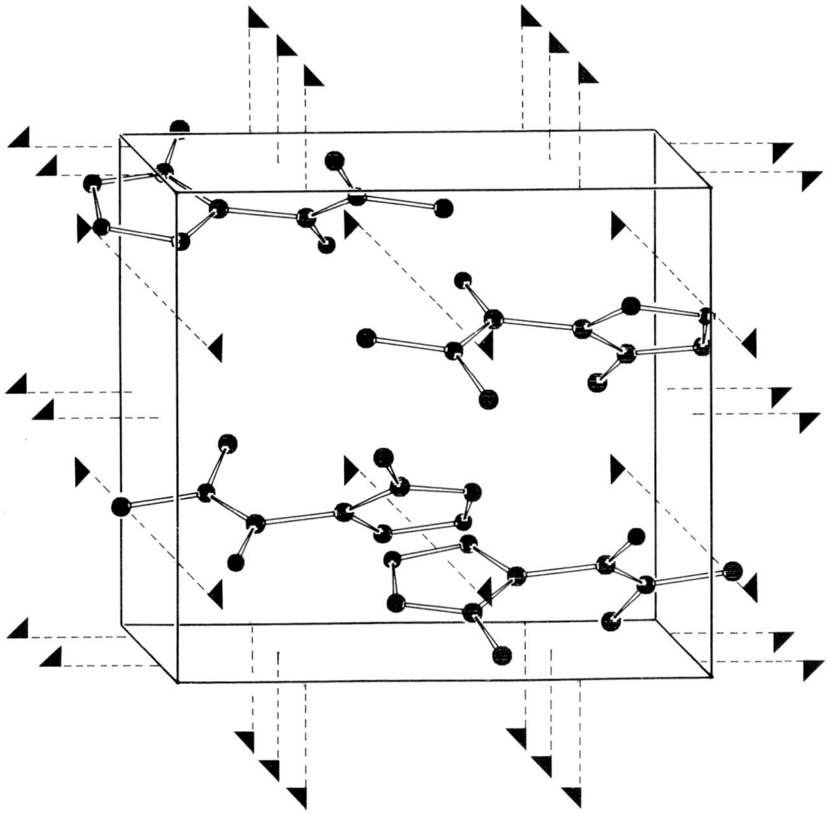

Fig. 1.23. A $P2_12_12_1$ crystal structure (G. Chiari, D. Viterbo, A. Gaetani Manfredotti, and C. Guastini (1975). *Cryst. Struct. Commun.*, 4,561) and its symmetry elements (hydrogen atoms are not drawn).

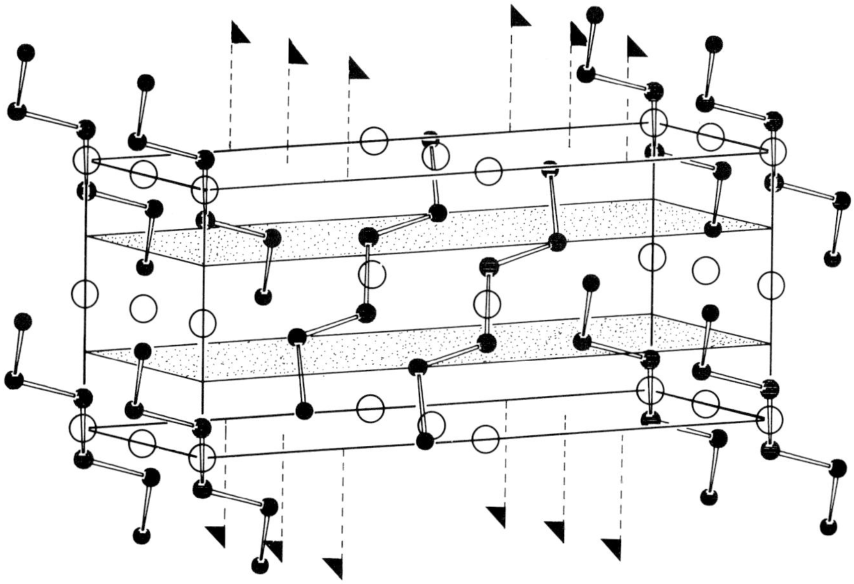

Fig. 1.24. A $P2_1/c$ crystal structure (M. Calleri, G. Ferraris, and D. Viterbo (1966). *Acta Cryst.*, **20,** 73) and its symmetry elements (hydrogen atoms are not drawn). Glide planes are emphasized by the shading.

Table 1.10. The 17 plane groups

Oblique cell	p1, p2
Rectangular cell	pm, pg, cm, p2mm, p2mg, p2gg, c2mm
Square cell	p4, p4mm, p4gm
Hexagonal cell	p3, p3m1, p31m, p6, p6mm

When applying the symmetry operator $\mathbf{C}_1 \equiv (\mathbf{R}_1, \mathbf{T}_1)$ to a point at the end of a vector $\boldsymbol{r}$, we obtain $\mathbf{X}' = \mathbf{C}_1\mathbf{X} = \mathbf{R}_1\mathbf{X} + \mathbf{T}_1$. If we then apply to $\boldsymbol{r}'$ the symmetry operator $\mathbf{C}_2$, we obtain

$$\mathbf{X}'' = \mathbf{C}_2\mathbf{X}' = \mathbf{R}_2(\mathbf{R}_1\mathbf{X} + \mathbf{T}_1) + \mathbf{T}_2 = \mathbf{R}_2\mathbf{R}_1\mathbf{X} + \mathbf{R}_2\mathbf{T}_1 + \mathbf{T}_2.$$

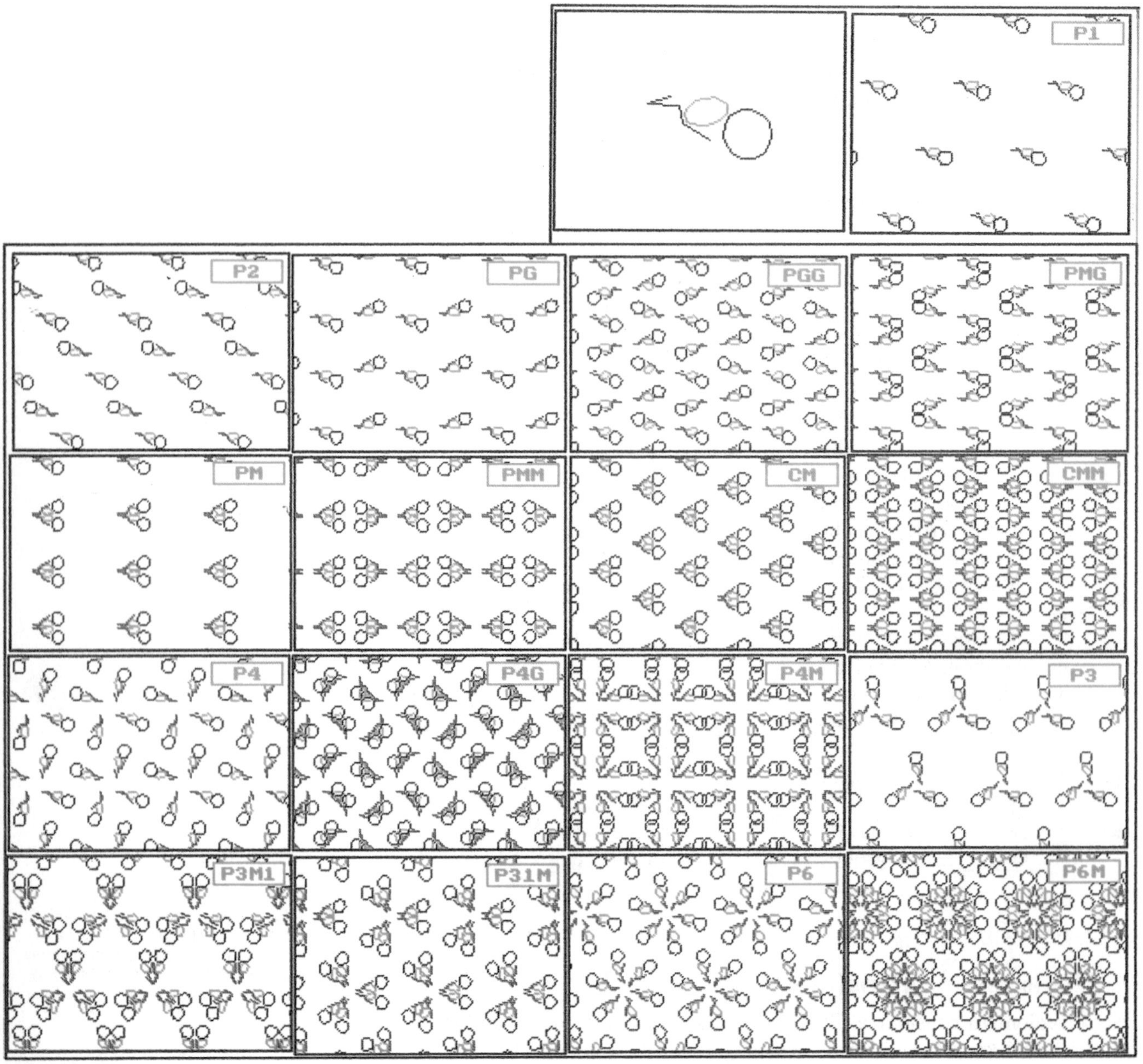

Fig. 1.25. A periodic decoration of the plane according to the 17 crystallographic plane groups (drawing by SYMPATI, a computer program by L. Loreto and M. Tonetti, *Pixel*, **9**, 9–20; Nov 1990).

Since the symmetry operators form a mathematical group, a third symmetry operator must be present (see also pp. 11–12),

$$\mathbf{C}_3 = \mathbf{C}_2\mathbf{C}_1 = (\mathbf{R}_2\mathbf{R}_1, \mathbf{R}_2\mathbf{T}_1 + \mathbf{T}_2), \tag{1.12}$$

where $\mathbf{R}_2\mathbf{R}_1$ is the rotational component of $\mathbf{C}_3$ and $(\mathbf{R}_2\mathbf{T}_1 + \mathbf{T}_2)$ is its translational component. In particular the operator $\mathbf{C}^2 = \mathbf{CC}$ will be present and in general also the $\mathbf{C}^j$ operator. Because of (1.12)

$$\mathbf{C}^j = [\mathbf{R}^j, (\mathbf{R}^{j-1} + \ldots + \mathbf{R} + \mathbf{I})\mathbf{T}]. \tag{1.13}$$

Let us now apply this result to the space group P6_1. Once we have defined the **R** and **T** matrices corresponding to an anti-clockwise rototranslation of 60° around z, we obtain all the six points equivalent to a point $\boldsymbol{r}$ by applying to it the operators $\mathbf{C}^j$ with j going from 1 to 6. Obviously $\mathbf{C}^6 = \mathbf{I}$ and $\mathbf{C}^{6+j} = \mathbf{C}^j$. For this reason we will say that the 6_1 operator is of **order** six (similarly 2 and m are of order two).

If $\boldsymbol{r}$ is transferred to $\boldsymbol{r}'$ by $\mathbf{C} = (\mathbf{R}, \mathbf{T})$ there will also be an inverse operator $\mathbf{C}^{-1} = (\mathbf{R}', \mathbf{T}')$ which will bring $\boldsymbol{r}'$ back to $\boldsymbol{r}$. Since we must have $\mathbf{C}^{-1}\mathbf{C} = \mathbf{I}$, because of (1.12) we will also have $\mathbf{R}'\mathbf{R} = \mathbf{I}$ and $\mathbf{R}'\mathbf{T} + \mathbf{T}' = \mathbf{0}$, and therefore

$$\mathbf{C}^{-1} = (\mathbf{R}^{-1}, -\mathbf{R}^{-1}\mathbf{T}) \tag{1.14}$$

where $\mathbf{R}^{-1}$ is the inverse matrix of **R**. In the P6_1 example, $\mathbf{C}^{-1} = \mathbf{C}^5$. When all the operators of the group can be generated from only one operator (indicated as the **generator of the group**) we will say that the group is **cyclic**.

All symmetry operators of a group can be generated from at most three generators. For instance, the generators of the space group P$6_1$22 are 6_1 and one twofold axis. Each of the 12 different operators of the group may be obtained as $\mathbf{C}_1^j$, $j = 1, 2, \ldots, 6$, say the powers of 6_1, or as $\mathbf{C}_2$, the twofold axis operator, or as their product. We can then represent the symmetry operators of P$6_1$22 as the product $\{\mathbf{C}_1\}\{\mathbf{C}_2\}$, where $\{\mathbf{C}\}$ indicates the set of distinct operators obtained as powers of **C**. Similarly there are two generators of the group P222 but three of the group P$\bar{4}$3m. In general all the operations of a space group may be represented by the product $\{\mathbf{C}_1\}\{\mathbf{C}_2\}\{\mathbf{C}_3\}$. If only two generators are sufficient, we will set $\mathbf{C}_3 = \mathbf{I}$, and if only one is sufficient, then $\mathbf{C}_2 = \mathbf{C}_3 = \mathbf{I}$. The list of the generators of all point groups is given in Appendix 1.E.

So far we have deliberately excluded from our considerations the translation operations defined by the Bravais lattice type. When we take them into account, all the space-group operations may be written in a very simple way. In fact the set of operations which will transfer a point $\boldsymbol{r}$ in a given cell into its equivalent points in any cell are:

$$\{\mathbf{T}\}\{\mathbf{C}_1\}\{\mathbf{C}_2\}\{\mathbf{C}_3\} \tag{1.15}$$

where $\mathbf{T} = m_1\boldsymbol{a} + m_2\boldsymbol{b} + m_3\boldsymbol{c}$ is the set of lattice translations.

The theory of symmetry groups will be outlined in Appendix 1.E.

Appendices

1.A The isometric transformations

It is convenient to consider a Cartesian basis $(\boldsymbol{e}_1, \boldsymbol{e}_2, \boldsymbol{e}_3)$. Any transformation which will keep the distances unchanged will be called an isometry or

an isometric mapping or a **movement C**. It will be a linear transformation, in the sense that a point P defined by the positional vector $\boldsymbol{r} = x\boldsymbol{e}_1 + y\boldsymbol{e}_2 + z\boldsymbol{e}_3$ is related to a point P′, with positional vector $\boldsymbol{r}' = x'\boldsymbol{e}_1 + y'\boldsymbol{e}_2 + z'\boldsymbol{e}_3$ by the relation

$$\mathbf{X}' = \begin{pmatrix} x' \\ y' \\ z' \end{pmatrix} = \begin{pmatrix} R_{11} & R_{12} & R_{13} \\ R_{21} & R_{22} & R_{23} \\ R_{31} & R_{32} & R_{33} \end{pmatrix} \begin{pmatrix} x \\ y \\ z \end{pmatrix} + \begin{pmatrix} T_1 \\ T_2 \\ T_3 \end{pmatrix} = \mathbf{CX} = \mathbf{RX} + \mathbf{T} \quad (1.A.1)$$

with the extra condition

$$\bar{\mathbf{R}}\mathbf{R} = \mathbf{I} \quad \text{or} \quad \mathbf{R} = \bar{\mathbf{R}}^{-1}. \quad (1.A.2)$$

$\bar{\mathbf{R}}$ indicates the transpose of the matrix $\mathbf{R}$ and $\mathbf{I}$ is the identity matrix.

We note that $\mathbf{X}$ and $\mathbf{X}'$ are the matrices of the components of the vectors $\boldsymbol{r}$ and $\boldsymbol{r}'$ respectively, while $\mathbf{T}$ is the matrix of the components of the translation vector $\boldsymbol{t} = T_1\boldsymbol{e}_1 + T_2\boldsymbol{e}_2 + T_3\boldsymbol{e}_3$.

A movement, leaving the distances unchanged, will also maintain the angles fixed in absolute value. Since the determinant of the product of two matrices is equal to the product of the two determinants, from (1.A.2) we have $(\det \mathbf{R})^2 = 1$, and then $\det \mathbf{R} = \pm 1$. We will refer to **direct** or **opposite movements** and to **direct** or **opposite congruence** relating an object and its transform, depending on whether $\det \mathbf{R}$ is $+1$ or -1.

Direct movements

Let us separate (1.A.1) into two movements:

$$\mathbf{X}' = \mathbf{X}_O + \mathbf{T} \quad (1.A.3)$$

$$\mathbf{X}_O = \mathbf{RX}. \quad (1.A.4)$$

(1.A.3) adds to each position vector a fixed vector and corresponds therefore to a translation movement. (1.A.4) leaves the origin point invariant. In order to find the other points left invariant we have to set $\mathbf{X}_O = \mathbf{X}$ and obtain

$$(\mathbf{R} - \mathbf{I})\mathbf{X} = \mathrm{O} \quad (1.A.5a)$$

(1.A.5a) will have solutions for $\mathbf{X} \neq 0$ only if $\det(\mathbf{R} - \mathbf{I}) = 0$. Since $\det(\mathbf{R} - \mathbf{I}) = \det(\mathbf{R} - \bar{\mathbf{R}}\mathbf{R}) = \det[(\mathbf{I} - \bar{\mathbf{R}})\mathbf{R}] = \det(\mathbf{I} - \bar{\mathbf{R}}) \det \mathbf{R} = \det(\mathbf{I} - \bar{\mathbf{R}}) = -\det(\mathbf{R} - \mathbf{I})$, then this condition is satisfied. Therefore one of the three equations represented by (1.A.4) must be a linear combination of the other two. The two independent equations will define a line, which is the locus of the invariants points; the movement described by (1.A.4) is therefore a rotation. In conclusion, a direct movement can be considered as the combination (or, more properly, the product) of a translation with a rotation around an axis.

If in eqn (1.A.1) is $\mathbf{R} \equiv \mathbf{I}$ then the movement is a pure translation, if $\mathbf{T} = \mathrm{O}$ the movement is a pure rotation. When the translation is parallel to the rotation axis the movement will be indicated as **rototranslation**. An example of direct movement is the transformation undergone by the points of a rigid body when it is moved. Another example is the anti-clockwise rotation around the z axis of an angle θ; this will move $\boldsymbol{r}(x, y, z)$ into

$\mathbf{r}'(x', y', z')$ through the transformation

$$x' = x\cos\theta - y\sin\theta$$
$$y' = x\sin\theta - y\cos\theta$$
$$z' = z$$

which in matrix notation becomes $\mathbf{X}' = \mathbf{RX}$, with

$$\mathbf{R} = \begin{pmatrix} \cos\theta & -\sin\theta & 0 \\ \sin\theta & -\cos\theta & 0 \\ 0 & 0 & 1 \end{pmatrix}. \qquad (1.A.5b)$$

$\mathbf{R}^{-1}$ can be obtained by substituting θ with $-\theta$ and it can be immediately seen that, in agreement with (1.A.2), $\mathbf{R} = \bar{\mathbf{R}}^{-1}$.

We will now show that any direct movement can be carried out by means of a translation or a rotation or a rototranslation. Let us suppose that an isometric transformation relates the three non-collinear points A, B, C with the points A′, B′, C′ respectively. If $\boldsymbol{a}$ is the translation bringing A on A′, then, if also B and C superpose on B′ and C′, the movement is a translation; if not, then the complete superposition can be achieved by a rotation $\mathbf{R}$ around an axis $\boldsymbol{l}$ passing through A′. If $\boldsymbol{l}$ is perpendicular to $\boldsymbol{a}$, the movement resulting from the combination of the rotation and translation operations is still a pure rotation around an axis parallel to $\boldsymbol{l}$ (see Appendix 1.B). If $\boldsymbol{a}$ is not perpendicular to $\boldsymbol{l}$, then it can be decomposed into two translational components $\boldsymbol{a}_1$ and $\boldsymbol{a}_2$, one perpendicular and the other parallel to $\boldsymbol{l}$. The product of $\mathbf{R}$ times $\boldsymbol{a}_1$ is a pure rotation around an axis parallel to $\boldsymbol{l}$, which, when composed with $\boldsymbol{a}_2$, results in a rototranslation movement.

Opposite movements

An opposite movement can be obtained from a direct one by changing the sign to one or three rows of the $\mathbf{R}$ matrix. For instance, when changing the sign of the third row, we substitute the vector (x', y', z') with $(x', y', -z')$, i.e. the point P′ with its symmetry related with respect to a plane at $z = 0$. This operation is called a **reflection** with respect to the plane at $z = 0$. Changing the signs of all three rows of the $\mathbf{R}$ matrix implies the substitution of the vector (x', y', z') with $(-x', -y', -z')$, i.e. of the point P with its symmetry related with respect to the origin of the coordinate system. This operation is called **inversion** with respect to a point.

We may conclude that each direct movement, followed by a reflection with respect to a plane or by an inversion with respect to a point yields an opposite movement. On the other hand an opposite movement may be obtained as the product of a direct movement by a reflection with respect to a plane or by an inversion with respect to a point.

1.B. Some combinations of movements

Only those combinations of movements explicitly mentioned in this book will be considered (for further insight, the reader is referred to the splendid book by Lockwood and MacMillan[4]). The stated laws may be interpreted

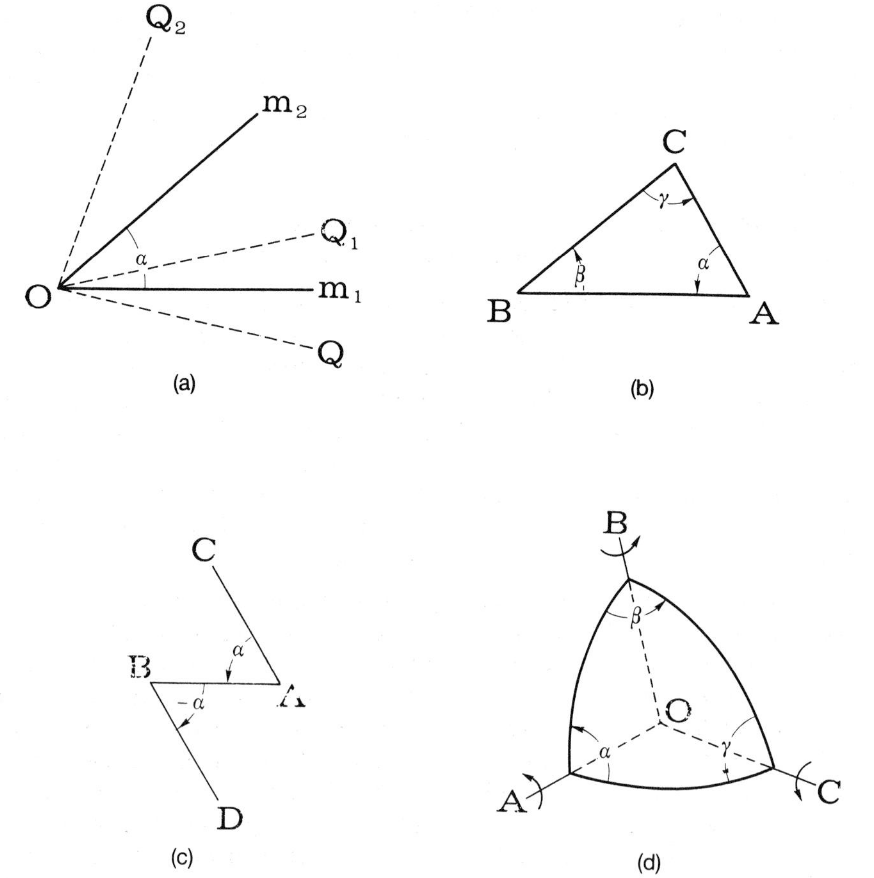

Fig. 1.B.1. (a) Composition of two reflections; (b) composition of two rotations about parallel axes; (c) composition of two rotations, the first through α and the second through $-\alpha$, about parallel axes; (d) composition of two rotations about axes passing through a point.

in terms of combinations of symmetry operations if all the space is invariant with respect to the movements.

1. *Composition of two reflections.* In Fig. 1.B.1(a) the two reflection planes m_1 and m_2 are at a dihedral angle α and intersect along a line, the trace of which is in O. The image of OQ with respect to m_1 is OQ_1 and the image of OQ_1 with respect to m_2 is OQ_2. It is possible to superpose OQ to OQ_2 by a rotation of 2α around the axis through O. We may conclude that the product of the two reflections is a rotation of 2α around O; in symbols $m_1m_2 = R$. The product is not commutative ($m_1m_2 \neq m_2m_1$): in fact by first reflecting OQ with respect to m_2 and then reflecting the image with respect to m_1 we obtain a 2α rotation in the opposite direction.

2. *The Silvester theorem relative to three parallel rotation axes.* The traces of the three axes on the plane of Fig. 1.B.1(b) are A, B, C. The Silvester theorem states that consecutive anti-clockwise rotations of 2α, 2β, 2γ around A, B, C respectively produce the identity. In fact, because of point 1, the rotations are equivalent to reflection operations with respect to the three pairs of planes AC–AB., BA–BC, CB–CA respectively and all these reflections cancel each other out. Since $\alpha + \beta + \gamma = \pi$, two successive rotations of 2α and 2β around A and B respectively, will be equivalent to a rotation of $+(2\alpha + 2\beta)$ around C. When $\alpha = -\beta$, the third axis goes to infinity and the resulting movement is a translation.

In Fig. 1.B.1(c) AC moves to AB by a rotation of α around A; by a $-\alpha$ rotation around B BA goes to BD. The resulting movement brings AC to BD and can be achieved by a translation of $DA = 2AB \sin(\alpha/2)$ perpendicular to the direction of the rotation axes and at an angle of $(\pi - \alpha)/2$ with respect to AB. In symbols: $\mathbf{R}_\alpha \mathbf{R}'_{-\alpha} = \mathbf{T}$. We can then deduce the following point 3.

3. A rotation and a translation perpendicular to the rotation axis combine in a resulting rotation movement around an axis parallel to the original axis.

4. *The Silvester theorem relative to three rotation axes passing through a point.* In Fig. 1.B.1(d) ABC is a spherical triangle with angles α, β, γ in A, B, C respectively. If A, B, C are in a clockwise order, rotations of 2α, 2β, 2γ around A, B, C leave the figure unchanged. In fact, because of point 1, the three rotations correspond to the products of the reflections with respect to the pairs of planes AOC–AOB, AOB–BOC, BOC–COA respectively, and these reflections cancel each other out. Since $\alpha + \beta + \gamma > \pi$, then consecutive rotations of 2α around A and of 2β around B are equivalent to a rotation of 2γ around C, with $2\gamma \neq (2\alpha + 2\beta)$.

5. *Coexistence of rotation axes passing through a point*: the *Euler theorem*. We will study this problem using the Silvester theorem treated in point 4. In Fig. 1.B.1(d) let OA and OB be two symmetry rotation axes of order m and n respectively. The angles α and β are chosen in such a way that $2\alpha = 2\pi/m$ and $2\beta = 2\pi/n$. Because of the Silvester theorem, anti-clockwise rotations of 2α and 2β around OA and OB are equivalent to a 2γ anti-clockwise rotation around OC. C is therefore a symmetry axis of order $p = 2\pi/2\gamma = \pi/\gamma$. The angles of the spherical triangle are then π/m, π/n and π/p. Since the sum of the angles must be greater than π, the inequality $1/m + 1/n + 1/p > 1$ follows. The possible solutions of this inequality are: l, 2, 2 with l integer >1; 2, 3, 3; 4, 3, 2; 5, 3, 2. We can now consider the different solutions, keeping in mind that the surface of a sphere of radius r is $4\pi r^2$ and that of a spherical triangle is $(\alpha + \beta + \gamma - \pi)r^2$:

(a) Solution l, 2, 2. α, β, and γ are equal to $\pi/2$, $\pi/2$, π/l respectively and A and B may be chosen on an equatorial circle with C as a pole. The binary axes are therefore always at 90° with respect to the l axis. Values of l different from 2, 3, 4, 6 correspond to non-crystallographic groups which occur as possible symmetries of molecules or as approximate local site symmetries in crystals.

(b) Solution 2, 3, 3. α, β, and γ are equal to $\pi/2$, $\pi/3$, $\pi/3$ respectively and the area of the spherical triangle is $\pi r^2/6$. On the sphere there will be 24 such triangles. The 24 $\pi/2$ angles meet four at a time at six vertices and the 48 $\pi/3$ angles meet six at a time at eight vertices. This implies the presence of three twofold axes and of four threefold axes. The three twofold axes will be perpendicular one to the other and can be assumed as the axes of a reference system. The four threefold axes run from the centre to the points $(1, 1, 1)$, $(1, -1, -1)$, $(-1, 1, -1)$, $(-1, -1, 1)$. The group of rotations is consistent with the symmetry of a tetrahedron.

(c) Solution 4, 3, 2. α, β, and γ are equal to $\pi/2$, $\pi/3$, $\pi/4$ respectively and the area of the spherical triangle is $\pi r^2/12$. On the sphere there will

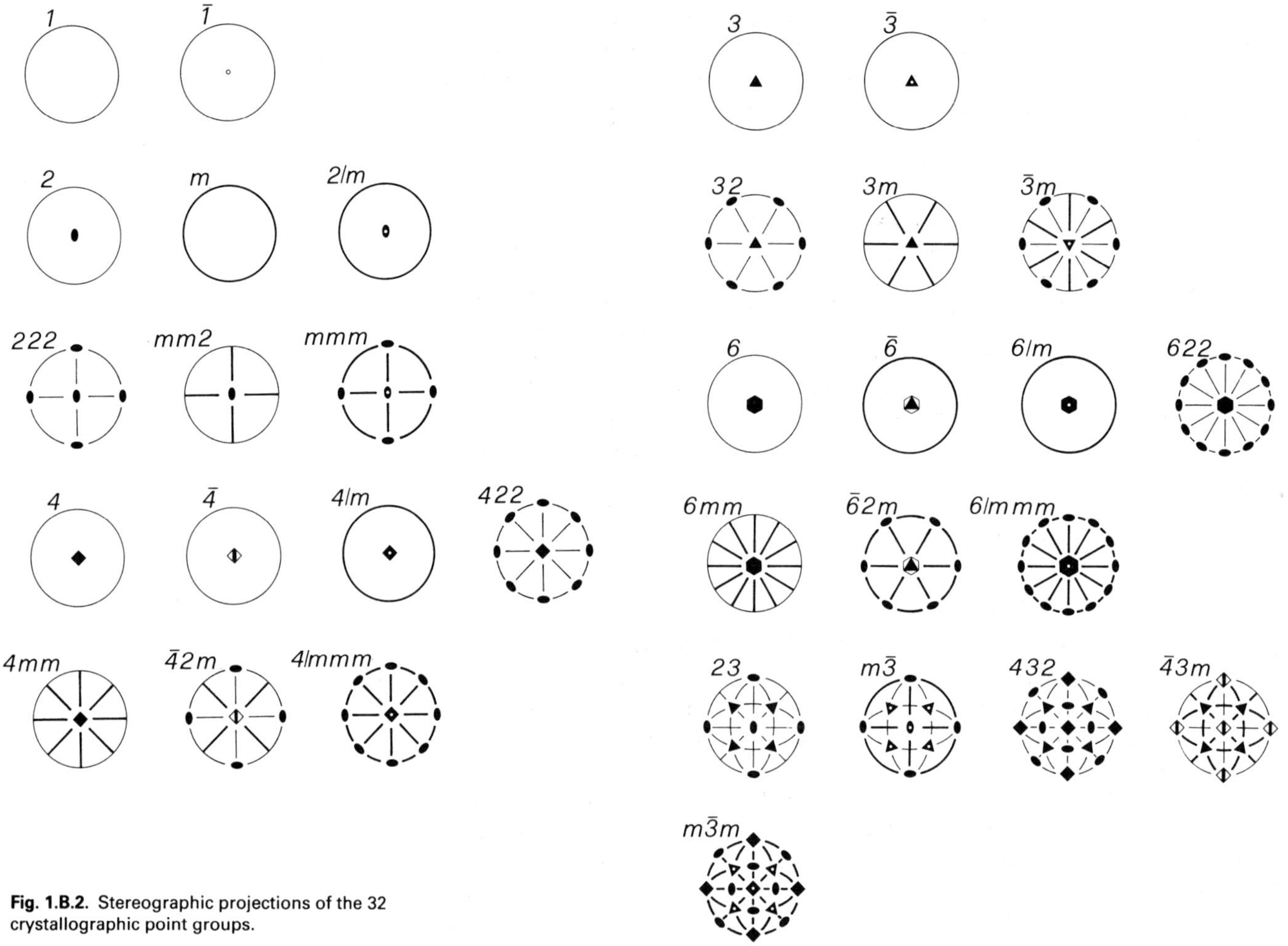

Fig. 1.B.2. Stereographic projections of the 32 crystallographic point groups.

be 48 such triangles. The 48 $\pi/2$ angles meet four at a time at twelve vertices, the 48 $\pi/3$ angles meet six at a time at eight vertices and the 48 $\pi/4$ angles meet eight at a time at six vertices. A total of six twofold axes, of four threefold axes, and of three fourfold axes will be present. The three fourfold axes will be perpendicular one to the other and can be assumed as the axes of a reference system. The threefold axes are located along the $[1, 1, 1]$, $[1, -1, -1]$, $[-1, 1, -1]$, $[-1, -1, 1]$ directions, while the twofold axes are on the bisecting lines of the angles between the fourfold axes. The group of rotations is consistent with the symmetry of a cube and of an octahedron (see Fig. 1.10).

The mutual disposition of the symmetry elements in the 32 crystallographic point groups is illustrated in Fig. 1.B.2, where the so called 'stereographic projections' are shown. The *c* axis is normal to the plane, the *a* axis points down the page, and the *b* axis runs horizontally in the page from left ro right. A stereographic projection is defined as follows (see Fig. 1.B.3(a)). A unit sphere is described around the crystal in C. A point P (terminal of some symmetry axis) in the $+z$ hemisphere is defined in the (x, y) plane as intersection P′ of that plane with the line connecting the point with the south pole of the unit sphere. If the point to be projected is in the $-z$ hemisphere then the north pole is used.

In Fig. 1.B.3(b) parts of the stereographic projections for m$\bar{3}$m are magnified in order to make clearer the statements made in the text.

(d) Solution 5, 3, 2. This solution, which is compatible with the symmetry of the regular icosahedron (20 faces, 12 vertices) and its dual, the regular pentagon–dodecahedron, (12 faces, 20 vertices), but not with the periodicity property of crystals, will not be examined.

It is however of particular importance in Crystallography as symmetry of viruses molecules and in quasi-crystals.

6. *Composition of two glide planes.* In Fig. 1.B.4 let S and S′ be the traces of two glide planes forming an angle α and O be the trace of their intersection line. The translational components OA and OB are chosen to lie on the plane of the drawing and Q is the meeting point of the axes of the OA and OB segments. X, Y, and Q′ are the reflection images of Q with respect to S, S′, and to the point O, respectively. The product S′S moves Q to Q′ and then back to Q. Since S′S is a direct movement it leaves Q unchanged and corresponds to a rotation around an axis normal to the plane of the figure and passing through Q. Since S′S moves first A to O and then to B, the rotation angle AQB $= 2\alpha$. Note that the two glides are equivalent to a rotation around an axis not passing along the intersection line of S and S′.

7. *Composition of two twofold axes, with and without translational component.* From point 5 we know that the coexistence of two orthogonal twofold axes passing by O, implies a third binary axis perpendicular to them and also passing through O (see Fig. 1.B.5(a)). The reader can easily verify the following conclusions:

(a) if one of the two axes is 2_1 (Fig. 1.B.5(b)), then another 2 axis, at 1/4 from O and intersecting orthogonally the screw axis, will exist;

(b) if two 2_1 intersect in O (Fig. 1.B.5(c)), then another 2 axis perpendicular to them and passing at (1/4, 1/4) from O will be present;

(c) if a pair of mutually perpendicular 2 axes is separated by 1/4 of a period (Fig. 1.B.5(d)), then a 2_1 axis orthogonally intersecting both axes will exist;

(d) if a 2 and a 2_1 axis are separated by 1/4 of a period (Fig. 1.B.5(e)) there will then be a new 2_1 axis normal to both of them and intersecting the first 2_1 axis at 1/4 from the 2 axis;

(e) if two orthogonal screws are separated by 1/4 of a period (Fig. 1.B.5(f)), then a third screw axis normal to them and passing at (1/4, 1/4) from them will be present.

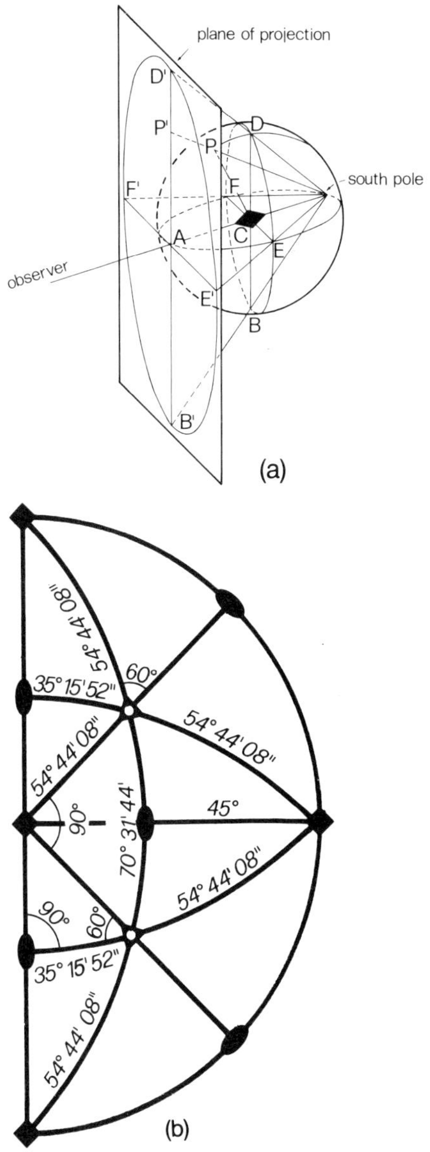

Fig. 1.B.3. (a) Geometry of the stereographic projection. (b) Angular values occurring in m$\bar{3}$m stereographic projection.

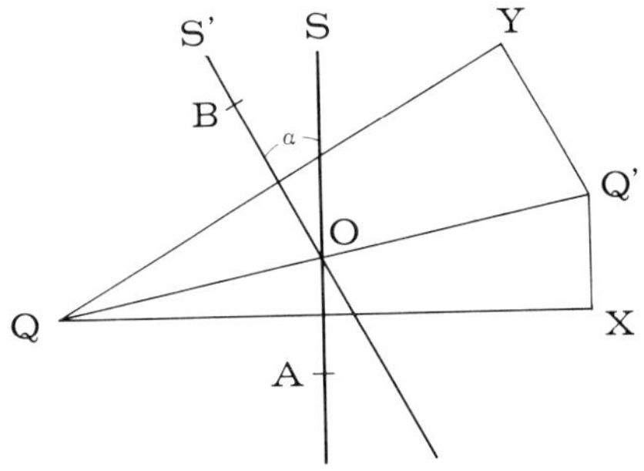

Fig. 1.B.4. Composition of two glide planes.

1.C. Wigner–Seitz cells

The 14 Bravais lattices are compatible with cells which are different from those conventionally associated to them. A conventional cell is a parallelepiped: as such it may be considered as a particular type of polyhedron. There are several families of polyhedra with which we can fill up the space

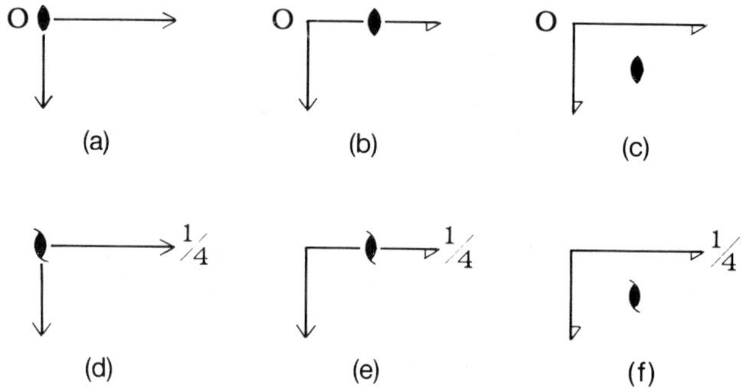

Fig. 1.B.5. Various arrangements of three orthogonal binary axes. The length of the graphical symbol for screw axes in the plane of the page corresponds to half repetition period.

by translation. A very important type is the one obtained through the **Dirichlet construction**. Each lattice point is connected with a line to its nearest neighbours. We then trace through the mid-points of the segments the planes perpendicular to them. These intersecting planes will delimit a region of the space which is called called the **Dirichlet region** or **Wigner–Seitz cell**. An example in two dimensions is given in Fig. 1.C.1(a) and two three-dimensional examples are illustrated in Fig. 1.C.1(b, c). The Wigner–Seitz cell is always primitive and coincides with the Bravais cell if this is rectangular and primitive. A construction identical to the Wigner–Seitz cell delimits in the reciprocal space (cf. Chapters 2 and 3) a cell conventionally known as the **first Brillouin zone**. There will be 14 first Brillouin zones corresponding to the 14 Bravais lattices. We recall here that to a lattice I in direct space corresponds an F lattice in reciprocal space and vice versa (see Appendix 2.D); then, the first Brillouin zone of an I lattice will look like a Wigner–Seitz cell of an F lattice and vice versa. The Brillouin zones are very important in the study of lattice dynamics and in electronic band theory.

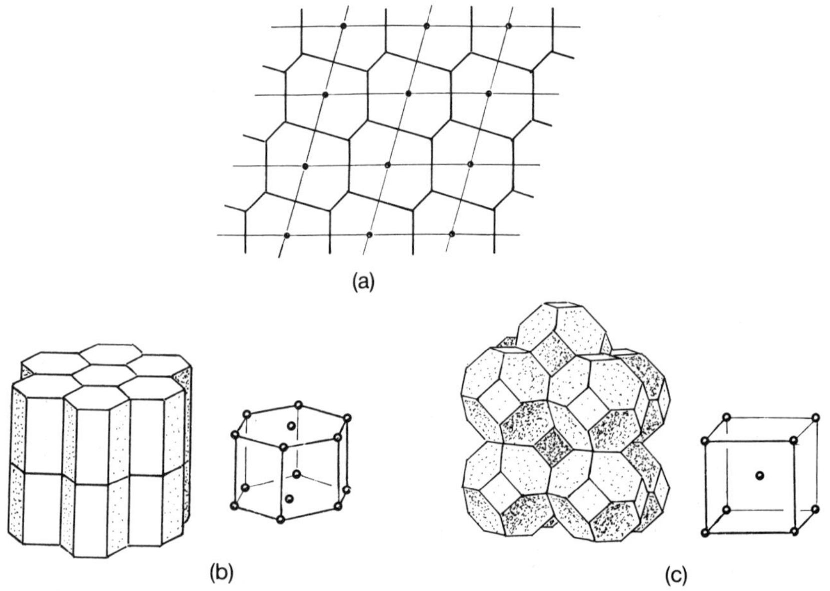

Fig. 1.C.1. Examples of Wigner–Seitz cells.

1.D. The space-group matrices

All the rotation matrices **R** needed to conventionally describe all 230 space groups will be listed. They operate according to the relation $\mathbf{X}' = \mathbf{RX}$. The matrices are grouped following the directions along which they operate, and for the hexagonal and trigonal systems they are preceded by the letter H. They may be constructed using the following practical criterion: in the first, second, and third columns are the coordinates of the points symmetry related to points $(1, 0, 0)$, $(0, 1, 0)$, and $(0, 0, 1)$ respectively.

The determinant of a matrix will have a $+1$ or a -1 value depending on whether the represented operation is of rotation or of inversion type. This number, together with the trace (sum of the diagonal elements) of a matrix, are characteristic of the symmetry element:

element	1	2	3	4	6	$\bar{1}$	$\bar{2}$	$\bar{3}$	$\bar{4}$	$\bar{6}$,
trace	3	$\bar{1}$	0	1	2	$\bar{3}$	1	0	$\bar{1}$	$\bar{2}$,
determinant	1	1	1	1	1	$\bar{1}$	$\bar{1}$	$\bar{1}$	$\bar{1}$	$\bar{1}$.

The matrix of the $\bar{n}$ operator may be simply obtained from that of the n operator by multiplying by the matrix corresponding to the $\bar{1}$ operator:

$$\bar{1} = \begin{pmatrix} -1 & 0 & 0 \\ 0 & -1 & 0 \\ 0 & 0 & -1 \end{pmatrix}.$$

This corresponds to changing the sign of all the elements of the original matrix. Therefore in the following list we will not give all the 64 matrices necessary to describe the space groups, but only the 32 matrices corresponding to proper symmetry elements.

Direction [0 0 0]

$$1 \equiv \begin{pmatrix} 1 & 0 & 0 \\ 0 & 1 & 0 \\ 0 & 0 & 1 \end{pmatrix}$$

Direction [1 0 0]

$$2 \equiv \begin{pmatrix} 1 & 0 & 0 \\ 0 & \bar{1} & 0 \\ 0 & 0 & \bar{1} \end{pmatrix}; \ \mathrm{H}2 \equiv \begin{pmatrix} 1 & \bar{1} & 0 \\ 0 & \bar{1} & 0 \\ 0 & 0 & \bar{1} \end{pmatrix}; \ 4 \equiv \begin{pmatrix} 1 & 0 & 0 \\ 0 & 0 & 1 \\ 0 & \bar{1} & 0 \end{pmatrix}; \ 4^3 \equiv \begin{pmatrix} 1 & 0 & 0 \\ 0 & 0 & \bar{1} \\ 0 & 1 & 0 \end{pmatrix}.$$

Direction [0 1 0]

$$2 \equiv \begin{pmatrix} \bar{1} & 0 & 0 \\ 0 & 1 & 0 \\ 0 & 0 & \bar{1} \end{pmatrix}; \ \mathrm{H}2 \equiv \begin{pmatrix} \bar{1} & 0 & 0 \\ \bar{1} & 1 & 0 \\ 0 & 0 & \bar{1} \end{pmatrix}; \ 3 \equiv \begin{pmatrix} 0 & 0 & \bar{1} \\ 0 & 1 & 0 \\ 1 & 0 & 0 \end{pmatrix}; \ 4^3 \equiv \begin{pmatrix} 0 & 0 & 1 \\ 0 & 1 & 0 \\ \bar{1} & 0 & 0 \end{pmatrix}.$$

Direction [0 0 1]

$$2 \equiv \begin{pmatrix} \bar{1} & 0 & 0 \\ 0 & \bar{1} & 0 \\ 0 & 0 & 1 \end{pmatrix}; \ \mathrm{H}3 \equiv \begin{pmatrix} 0 & \bar{1} & 0 \\ 1 & \bar{1} & 0 \\ 0 & 0 & 1 \end{pmatrix}; \ \mathrm{H}3^2 \equiv \begin{pmatrix} \bar{1} & 1 & 0 \\ \bar{1} & 0 & 0 \\ 0 & 0 & 1 \end{pmatrix}; \ 4 \equiv \begin{pmatrix} 0 & \bar{1} & 0 \\ 1 & 0 & 0 \\ 0 & 0 & 1 \end{pmatrix}$$

$$4^3 \equiv \begin{pmatrix} 0 & 1 & 0 \\ \bar{1} & 0 & 0 \\ 0 & 0 & 1 \end{pmatrix}; \ \mathrm{H}6 \equiv \begin{pmatrix} 1 & \bar{1} & 0 \\ 1 & 0 & 0 \\ 0 & 0 & 1 \end{pmatrix}; \ \mathrm{H}6^5 \equiv \begin{pmatrix} 0 & 1 & 0 \\ \bar{1} & 1 & 0 \\ 0 & 0 & 1 \end{pmatrix}.$$

Direction [1 1 0]

$$2 \equiv \begin{pmatrix} 0 & 1 & 0 \\ 1 & 0 & 0 \\ 0 & 0 & \bar{1} \end{pmatrix}.$$

Direction [1 0 1]

$$2 \equiv \begin{pmatrix} 0 & 0 & 1 \\ 0 & \bar{1} & 0 \\ 1 & 0 & 0 \end{pmatrix}.$$

Direction [0 1 1]

$$2 \equiv \begin{pmatrix} \bar{1} & 0 & 0 \\ 0 & 0 & 1 \\ 0 & 1 & 0 \end{pmatrix}.$$

Direction [1 $\bar{1}$ 0]

$$2 \equiv \begin{pmatrix} 0 & \bar{1} & 0 \\ \bar{1} & 0 & 0 \\ 0 & 0 & \bar{1} \end{pmatrix}.$$

Direction [$\bar{1}$ 0 1]

$$2 \equiv \begin{pmatrix} 0 & 0 & \bar{1} \\ 0 & \bar{1} & 0 \\ \bar{1} & 0 & 0 \end{pmatrix}.$$

Direction [0 1 $\bar{1}$]

$$2 \equiv \begin{pmatrix} \bar{1} & 0 & 0 \\ 0 & 0 & \bar{1} \\ 0 & \bar{1} & 0 \end{pmatrix}.$$

Direction [1 1 1]

$$3 \equiv \begin{pmatrix} 0 & 0 & 1 \\ 1 & 0 & 0 \\ 0 & 1 & 0 \end{pmatrix}; \; 3^2 \equiv \begin{pmatrix} 0 & 1 & 0 \\ 0 & 0 & 1 \\ 1 & 0 & 0 \end{pmatrix}.$$

Direction [$\bar{1}$ 1 1]

$$3 \equiv \begin{pmatrix} 0 & \bar{1} & 0 \\ 0 & 0 & 1 \\ \bar{1} & 0 & 0 \end{pmatrix}; \; 3^2 \equiv \begin{pmatrix} 0 & 0 & \bar{1} \\ \bar{1} & 0 & 0 \\ 0 & 1 & 0 \end{pmatrix}.$$

Direction [1 $\bar{1}$ 1]

$$3 \equiv \begin{pmatrix} 0 & \bar{1} & 0 \\ 0 & 0 & \bar{1} \\ 1 & 0 & 0 \end{pmatrix}; \; 3^2 \equiv \begin{pmatrix} 0 & 0 & 1 \\ \bar{1} & 0 & 0 \\ 0 & \bar{1} & 0 \end{pmatrix}.$$

Direction $[1\ 1\ \bar{1}]$

$$3 \equiv \begin{pmatrix} 0 & 1 & 0 \\ 0 & 0 & \bar{1} \\ \bar{1} & 0 & 0 \end{pmatrix}; \; 3^2 \equiv \begin{pmatrix} 0 & 0 & \bar{1} \\ 1 & 0 & 0 \\ 0 & \bar{1} & 0 \end{pmatrix}.$$

Direction $[2\ 1\ 0]$

$$\mathrm{H}2 \equiv \begin{pmatrix} 1 & 0 & 0 \\ 1 & \bar{1} & 0 \\ 0 & 0 & \bar{1} \end{pmatrix}.$$

Direction $[1\ 2\ 0]$.

$$\mathrm{H}2 \equiv \begin{pmatrix} \bar{1} & 1 & 0 \\ 0 & 1 & 0 \\ 0 & 0 & \bar{1} \end{pmatrix}.$$

1.E. Symmetry groups

A **group** G is a set of elements $g_1, g_2, \ldots g_j, \ldots$ for which a combination law is defined, with the following four properties: **closure**, the combination of two elements of the group is an element of the group $g_i g_j = g_k$; **associativity**, the associative law $(g_i g_j) g_k = g_i (g_j g_k)$ is valid; **identity**, there is only one element e in the group such that $eg = ge = g$; **inversion**, each element g in the group has one and only one inverse element g^{-1} such that $g^{-1}g = gg^{-1} = e$.

Examples of groups are:

(1) the set of all integer numbers (positive, negative, and zero), when the combination law is the sum. In this case $e = 0$, $g^{-1} = -g$;

(2) the rational numbers, excluding zero, when the law is the product: $e = 1$, $g^{-1} = 1/g$;

(3) the set of all $n \times n$ son-singular matrices under the product law: e is the diagonal matrix with $a_{ii} = 1$;

(4) the set of all lattice vectors $\boldsymbol{r}_{u,v,w} = u\boldsymbol{a} + v\boldsymbol{b} + w\boldsymbol{c}$ with u, v, w positive, negative, or null integers, when the combination law is the vector sum: then $e = \boldsymbol{r}_{0,0,0}$, $g^{-1} = -g$.

The number of different elements of a group is called the order of the group and can be finite or infinite. If a group also possesses the commutative property $g_i g_j = g_j g_i$ for any i and j, then G is said to be **Abelian**. With reference to the examples given above, (1), (2), (4) are infinite and Abelian groups, (3) is infinite and non-Abelian.

If g is an element of G, all powers of g must be contained in G. An integer m may exist for which

$$g^n = e; \tag{1.E.1}$$

then $g^{n+1} = g$, $g^{n+2} = g^2, \ldots$ If n is the smallest integer for which (1.E.1) is satisfied, there will only be n distinct powers of g. Since $g^j g^{n-j} = g^{n-j} g^j = e$, then g^{n-j} is the inverse of g^j. The element g is then said to be of order n and

Table 1.E.1. List of generators for non-cyclic point groups. There are 21 proper generators in all

Point group	Generators	Point group	Generators
2/m	$2_{[010]}, \bar{2}_{[010]}$	$\bar{3}$m	$\bar{3}_{[001]}, \bar{2}_{[100]}$
222	$2_{[100]}, 2_{[010]}$	$\bar{6}$2m	$\bar{2}_{[100]}, 2_{[1\bar{1}0]}$
mm2	$\bar{2}_{[100]}, \bar{2}_{[001]}$	622	$2_{[100]}, 2_{[1\bar{1}0]}$
mmm	$\bar{2}_{[100]}, \bar{2}_{[010]}, \bar{2}_{[001]}$	6/m	$6_{[001]}, \bar{2}_{[001]}$
422	$2_{[100]}, 2_{[1\bar{1}0]}$	6mm	$\bar{2}_{[100]}, \bar{2}_{[1\bar{1}0]}$
4/m	$4_{[001]}, \bar{2}_{[001]}$	6/mmm	$\bar{2}_{[100]}, \bar{2}_{[1\bar{1}0]}, \bar{2}_{[001]}$
4mm	$\bar{2}_{[100]}, \bar{2}_{[1\bar{1}0]}$	23	$3_{[111]}, 2_{[001]}$
$\bar{4}$2m	$2_{[100]}, \bar{2}_{[1\bar{1}0]}$	432	$3_{[111]}, 2_{[110]}$
4/mmm	$\bar{2}_{[100]}, \bar{2}_{[1\bar{1}0]}, \bar{2}_{[001]}$	m$\bar{3}$	$3_{[111]}, \bar{2}_{[001]}$
32	$3_{[001]}, 2_{[100]}$	$\bar{4}$3m	$3_{[111]}, \bar{2}_{[110]}$
3m	$3_{[001]}, \bar{2}_{[100]}$	m$\bar{3}$m	$3_{[111]}, \bar{2}_{[1\bar{1}0]}, \bar{2}_{[001]}$

the set of all powers of g is a group of order n:

$$G_1 = (g, g^2, \ldots, g^{n-1}, g^n = e). \tag{1.E.2}$$

A group, such as (1.E.2), in which all the elements are powers of a single generating element, is called **cyclic**. All cyclic groups are Abelian, but the converse is not true. An example of cyclic group of order n is the set of rotations around a given axis, which are multiple of an angle $\alpha = 2\pi/n$.

Any point group can be represented as the product of powers of at most three elements, which are the **generators** of the group. In Table 1.E.1 the list of the generators of the non-cyclic groups is reported. We also note that the definition of the generators is not unique. For instance in the class 222 we may chose the generators $2_{[100]}$ and $2_{[010]}$, or $2_{[100]}$ and $2_{[001]}$, or $2_{[010]}$ and $2_{[001]}$.

When the physical properties of the group elements are not specified the group is said to be **abstract**. From a mathematical point of view all its properties are determined by its **multiplication table**. For a group of n elements this table has the form:

$$\begin{array}{c|cccc} & g_1 & g_2 & g_n & \ldots \\ \hline g_1 & g_1^2 & g_1 g_2 & g_1 g_n & \ldots \\ g_2 & g_2 g_1 & g_2^2 & g_2 g_n & \ldots \\ \ldots & \ldots & \ldots & \ldots & \ldots \\ g_n & g_n g_1 & g_n g_2 & g_n^2. & \ldots \end{array} \tag{1.E.3}$$

We note that

1. Each element appears once and only once in a given row (or column) of the table. In order to demonstrate this statement let us consider the ith row of the table and suppose that there are two different elements g_j and g_k, for which $g_i g_j = g_i g_k = g_p$. Then g_p would appear twice in the row, but by multiplying the two equations by g_i^{-1} we obtain $g_j = g_k$, in contrast with the hypothesis.

2. Each row (column) is different from any other row (column); this property follows immediately from property 1.

3. For abelian groups the table is symmetric with respect to the diagonal.

Table 1.E.2. The 18 abstract groups corresponding to the 32 crystallographic point groups

Point group	Order of the group	Characteristic relations
1	1	$g = e$
$\bar{1}$, 2, m	2	$g^2 = e$
3	3	$g^3 = e$
4, $\bar{4}$	4	$g^4 = e$
2/m, mm2, 222	4	$g_1^2 = g_2^2 = (g_1 g_2)^2 = e$
6, $\bar{6}$, $\bar{3}$	6	$g^6 = e$
32, 3m	6	$g_1^3 = g_2^2 = (g_1 g_2)^2 = e$
mmm	8	$g_1^2 = g_2^2 = g_3^2 = (g_1 g_2)^2 = (g_1 g_3)^2 = (g_2 g_3)^2 = e$
4/m	8	$g_1^4 = g_2^2 = g_1 g_2 g_1^3 g_2 = e$
4mm, 422, $\bar{4}$2m	8	$g_1^4 = g_2^2 = (g_1 g_2)^2 = e$
6/m	12	$g_1^6 = g_2^2 = g_1 g_2 g_1^5 g_2 = e$
$\bar{3}$m, $\bar{6}$2m, 6mm, 622	12	$g_1^6 = g_2^2 = (g_1 g_2)^2 = e$
23	12	$g_1^3 = g_2^2 = (g_1 g_2)^3 = e$
4/mmm	16	$g_1^2 = g_2^2 = g_3^2 = (g_1 g_2)^2 = (g_1 g_3)^2 = (g_2 g_3)^4 = e$
432, $\bar{4}$3m	24	$g_1^4 = g_2^2 = (g_1 g_2)^3 = e$
m$\bar{3}$	24	$g_1^3 = g_2^3 = (g_1^2 g_2 g_1 g_2)^2 = e$
6/mmm	24	$g_1^2 = g_2^2 = g_3^2 = (g_1 g_2)^2 = (g_1 g_3)^2 = (g_2 g_3)^6 = e$
m$\bar{3}$m	48	$g_1^4 = g_2^6 = (g_1 g_2)^2 = e$

Groups having the same multiplication table, even though their elements might have different physical meaning, are called **isomorphous**. They must have the same order and may be considered as generated from the same abstract group. For instance the three point groups 222, 2/m and mm2 are isomorphous. To show this let us choose g_1, g_2, g_3, g_4 in the following way.

group 222: 1, 2, 2, 2;
group 2/m; 1, 2, m, $\bar{1}$;
group mm2; 1, m, m, 2.

The multiplication table of the abstract group is

	1	2	3	4
1	1	2	3	4
2	2	1	4	3
3	3	4	1	2
4	4	3	2	1

In Table 1.E.2 the 32 crystallographic point groups are grouped into 18 abstract groups and for each of them the defining relationships are listed. We note that all cyclic groups of the same order are isomorphous.

Subgroups

A set H of elements of the group G satisfying the group conditions is called a **subgroup** of G. The subgroup H is **proper** if there are symmetry operations of G not contained in H. Examples of subgroups are:

(1) the set of even integers (including zero) under the sum law is a subgroup of the group of all integers;

(2) the point group 32 has elements $g_1 = 1$, $g_2 = 3_{[001]}$, $g_3 = 3^2 = 3^{-1}$, $g_4 = 2_{[100]}$, $g_5 = 2_{[010]}$, $g_6 = 2_{[\bar{1}\bar{1}0]}$; $\mathrm{H} = (g_1, g_2, g_3)$ is a subgroup of G;

(3) the point groups 1, 2, $\bar{1}$, m are subgroups of the point group 2/m;

(4) 222 is a subgroup of 422;

(5) the set T of all primitive lattice translations is a subgroup of the space group G.

Conversely G may be considered a **supergroup** of H.

Subgroup–group–supergroup relationships are very important in the study of phase transitions and of order–disorder problems.

Cosets

Let $H=(h_1, h_2, \ldots)$ be a subgroup of G and g_i an element of G not contained in H. Then the products

$$g_iH=(g_ih_1, g_ih_2, \ldots) \text{ and } Hg_i=(h_1g_i, h_2g_i, \ldots)$$

form a left and a right coset of H respectively. In general they will not be identical.

Furthermore H can not have any common element with g_iH or Hg_i. In fact, if for instance, we had $g_ih_j=h_k$, it would follow that $g_i=h_kh_j^{-1}$, i.e. contradicting the hypothesis, g_i would belong to H.

It can be shown that two right (or left) cosets, either have no common element or are identical one to the other. This allows us to decompose G with respect to H in the following way:

$$\mathrm{G}=\mathrm{H} \cup g_i\mathrm{H} \cup g_q\mathrm{H} \cup \ldots \tag{1.E.4a}$$

or

$$\mathrm{G}=\mathrm{H} \cup \mathrm{H}g_i \cup \mathrm{H}g_q \cup \ldots. \tag{1.E.4b}$$

It follows that the order of a subgroup is a divisor of the order of the group and if this is a prime number, the only subgroup of G is e and G must also be cyclic.

The decomposition of the group 2/m into separate left cosets with respect to the subgroup 2 is:

$$2/\mathrm{m}=1(1,2) \cup \bar{1}(1,2)=(1,2) \cup (\bar{1}, \mathrm{m})$$

The number of distinct cosets obtained from the decomposition of a group with respect to a subgroup is called **index of the subgroup**. In the previous example, the index of the subgroup 2 of the group 2/m is two.

A coset is never a group because it does not contain the element e.

Conjugate classes

An element g_i is said to be conjugate to an element g_j of G if G contains an element g_k such that

$$g_i=g_k^{-1}g_jg_k. \tag{1.E.5}$$

If g_j is fixed and g_k varies within G, then the set of elements g_i forms a **class of conjugate elements**.

In agreement with relation (1.E.5) the element e forms a class on its own. Since each element of G can not belong to two different classes, it is possible to decompose G into the factorized set $G=e \cup \mathrm{T}_1 \cup \mathrm{T}_2 \cup \ldots$.

A physical or geometrical meaning may be attributed to the classes. In the transformation (1.E.5), let the element g_k be a coordinate transformation due to a symmetry operator and the element g_j a matrix operator related to another symmetry operation. Since g_j is transformed by (1.E.5) (see eqn (2.E.7)), the operators belonging to the same class are changed one onto the other by coordinate transformations represented by the elements of the group. For instance, for the point group 32 three classes may be set up:

$$(e),\ (3, 3^{-1}),\ (2_{[100]}, 2_{[010]}, 2_{[\bar{1}\bar{1}0]}.)$$

As we will see later, the character is identical for the matrix representation of all the elements of the same class.

The following rules may be useful to set up the point-group classes:

(1) the classes are formed by one element only for the point groups up to orthorhombic; this means that each symmetry operator commutes with all the others;

(2) in an Abelian group the classes are formed by only one element;

(3) the operators identity, inversion, and reflection with respect to a mirror plane perpendicular to the principal symmetry axis (the axis with the highest order), are each in a separate class.

Conjugate subgroups

Let H be a subgroup of G and g an element of G not in H. Then all the elements $g^{-1}Hg$, form a group. H and $g^{-1}Hg$ are conjugate subgroups.

Normal subgroups and factor groups

If the right and left cosets of the subgroup H are the same, i.e. $Hg_i = g_iH$ for every i, then $g_i^{-1}Hg_i = H$. This relation is still valid when g_i is in H since both g_iH and g_i belong to H. Conversely, if a subgroup is transformed into itself when applying all the elements of the group, the corresponding left and right cosets must be equal.

Subgroups which are transformed into themselves by applying all the elements of the group, are called **invariant** or **normal**. They must contain complete classes.

For instance the subgroup 2 of the point group 2/m is normal since $\bar{1}(1, 2) = (1, 2)\bar{1}$, $m(1, 2) = (1, 2)m$. Furthermore, in the group 32, the subgroup $(1, 3, 3^{-1})$ is normal, while $(1, 2_{[100]})$ is not; the subgroup T of all the primitive lattice translations is an invariant subgroup of the space group G.

Let H be a normal subgroup of G with index p, while n is the order of G. Because of (1.E.4) the order of H is n/p. We observe that for H and all its distinct cosets, the following multiplication law may be established:

$$(g_iHg_jH) = g_iHHg_j = g_iHg_j = g_ig_jH$$

Besides, it is:

$$(g_iH)^{-1}(g_iH) = H^{-1}g_i^{-1}g_iH = e,$$

$$H(g_iH) = g_iHH = g_iH.$$

We can now define a new type of group of order p, called a **factor group** or **quotient group**, indicated by the symbol G/H: its elements are cosets of

H. The following multiplication table is for the quotient group (we assume $g_1 = e$).

	g_1H	g_2H	g_pH
g_1H	g_1H	g_2H	g_pH
g_2H	g_2H	g_2^2H	$g_2 g_p$H
...	....	....	....
g_pH	g_pH	$g_p g_2$H	g_p^2H.

(1.E.6)

For instance, for the quotient group (2/m)/2, (1.E.6) becomes

	$(1, 2)$	$(\bar{1}, \mathrm{m})$
$(1, 2)$	$(1, 2)$	$(\bar{1}, \mathrm{m})$
$(\bar{1}, \mathrm{m})$	$(\bar{1}, \mathrm{m})$	$(1, 2)$

As a further example let us consider the elements of the point group 4mm: e, 4, 4^3, $4^2 = 2$, $\bar{2}_{[100]}$, $\bar{2}_{[010]}$, $\bar{2}_{[110]}$, $\bar{2}_{[1\bar{1}0]}$. Five classes may be formed: (e), (2), $(4, 4^3)$, $(\bar{2}_{[100]}, \bar{2}_{[010]})$, $(\bar{2}_{[110]}, \bar{2}_{[1\bar{1}0]})$. The subgroup $\mathrm{H} = (e, 2)$ is invariant and the factor group with respect to H may be written as: H, 4H, $\bar{2}_{[100]}$H, $\bar{2}_{[110]}$H.

The relation between G and G/H is a $n/p \to 1$ correspondence, i.e. G $\to$ H. In detail $g_1\mathrm{H} = \mathrm{H} \to e$, $g_2\mathrm{H} \to f_2, \ldots,$ $g_p\mathrm{H} \to f_p$. A correspondence of this type (many $\to$ one) is called **homomorphism** and G/H is said to be homomorphic with G.

Isomorphism is then a special case of homomorphism for which there is a one to one correspondence. A homomorphic correspondence allows us to reduce the study of the multiplication laws of the group G to that of the multiplication laws of the smaller G/H group.

To display many of the abstract definitions given so far, we consider as major example, the point group G $\equiv$ 23: say

$$\mathrm{G} \equiv \{1, 3_{[111]}, 3^2_{[111]}, 3_{[\bar{1}11]}, 3^2_{[\bar{1}11]}, 3_{[1\bar{1}1]}, 3^2_{[1\bar{1}1]},$$
$$3_{[11\bar{1}]}, 3^2_{[11\bar{1}]}, 2_{[100]}, 2_{[010]}, 2_{[001]}\}$$

Readers will find the four classes:

$$(1), (3_{[111]}, 3_{[\bar{1},1,1]}, 3_{[1\bar{1}1]}, 3_{[11\bar{1}]}),$$
$$(3^2_{[111]}, 3^2_{[\bar{1}11]}, 3^2_{[1\bar{1}1]}, 3^2_{[11\bar{1}]}),$$
$$(2_{[100]}, 2_{[010]}, 2_{[001]}),$$

and the ten subgroups

$$\{1\}, \{1, 2_{[100]}\}, \{1, 2_{[010]}\}, \{1, 2_{[001]}\}$$
$$\{1, 3_{[111]}, 3^2_{[111]}\}, \{1, 3_{[\bar{1}11]}, 3^2_{[\bar{1}11]}\}, \{1, 3_{[1\bar{1}1]}, 3^2_{[1\bar{1}1]}\}$$
$$\{1, 3_{[11\bar{1}]}, 3^2_{[11\bar{1}]}\}, \{1, 2_{[100]}, 2_{[010]}, 2_{[001]}\}, \mathrm{G},$$

of which only $\{1\}$, $\{1, 2_{[100]}, 2_{[010]}, 2_{[001]}\}$, and G are invariant subgroups.

Calculate now the factored group of $\mathrm{H} \equiv \{1, 2_{[100]}, 2_{[010]}, 2_{[001]})$: we multiply it by an element not in H, say $3_{[111]}$, and obtain

$$3_{[111]}\mathrm{H} = \{3_{[111]}, 3_{[\bar{1}11]}, 3_{[1\bar{1}1]}, 3_{[11\bar{1}]}\}.$$

We take now an element not in H or in $3_{[111]}$H, say $3^2_{[111]}$, and we get

$$3^2_{[111]}\mathrm{H} = \{3^2_{[111]}, 3^2_{[\bar{1}11]}, 3^2_{[1\bar{1}1]}, 3^2_{[11\bar{1}]}\}.$$

The group is now exhausted and we may write

$$G = H \cup (3_{[111]}H) \cup \{3^2_{[111]}H\}.$$

On assuming $g_1 = e$, then $\{e, 3_{[111]}H, 3^2_{[111]}H\}$ form a group under the following multiplication table

	e	$3_{[111]}H$	$3^2_{[111]}H$
e	e	$3_{[111]}H$	$3^2_{[111]}H$
$3_{[111]}H$	$3_{[111]}H$	$3^2_{[111]}H$	e
$3^2_{[111]}H$	$3^2_{[111]}H$	e	$3_{[111]}H$.

This group is homomorphic with G itself if we associate the elements in H to e, the elements in 3[111]H with the 'element' $3_{[111]}H$, the elements in $3^2_{[111]}H$ with the 'element' $3^2_{[111]}H$.

Maximal subgroups and minimal supergroups

The number of subgroups of a space group is always infinite. They may be the site-symmetry groups (groups without a lattice), or line groups, ribbon groups, rod groups, plane groups, or space groups. For example, the set T of all the primitive lattice translations is a subgroup (invariant) of the space group. Further subgroups may be found by considering the set of all the translations defined by a superlattice. We are here interested only in subgroups which are space groups themselves.

Let us first recall the concept of **proper subgroup**. A subgroup H is called proper subgroup of a group G if there are symmetry operations of G not contained in H. Now we define maximal subgroups: a subgroup H of a space group G is called a maximal subgroup of G if there is no proper subgroup M of G such that H is a proper subgroup of M. For example, $P112_1$, $P12_11$, $P2_111$ are maximal subgroups of $P2_12_12_1$, while P1 is not maximal.

For every H, according to (1.E.4), a right coset decomposition of G relative to H may be made. The index of the decomposition determines the degree of 'dilution' of the symmetry in H with respect to that in G. Such a dilution may be obtained in three different ways:

1. By eliminating some symmetry operators (e.g., from $G = P2_12_12_1$ to $H = P2_111$). Subgroups of this kind, are called **Translationengleiche** or **t subgroups**. Since the point group β of G is finite, the number of subgroups, and therefore, of maximal subgroups, is finite. All the maximal subgroups of type t for any given G are listed in the *International Tables* as **type *I* maximal non-isomorphous subgroups** (see Figs 1.20 and 1.21).

2. By loss of translational symmetry, i.e. by thinning out the lattice. Such subgroups are called **Klassengleiche** or ***k* subgroups** and are classified as **type *II***. A subset of k subgroups are those belonging to the same space group G or to its enantiomorphic: their number is infinite and they are called **maximal isomorphous subgroups**. Those of lowest index are listed as *IIc* in the *International Tables*. For example, if G = C222, maximal isomorphous subgroups of lowest index are C222 with $\boldsymbol{a}' = 3\boldsymbol{a}$ or $\boldsymbol{b}' = 3\boldsymbol{b}$ and C222 with $\boldsymbol{c}' = 2\boldsymbol{c}$.

Maximal non-isomorphous subgroups of C222 are also P222, $P2_12_12$, $P2_122$, $P22_12$, $P222_1$, which have the same conventional cell: for practical reasons they are labelled as subgroups of type *IIa*. Space groups with primitive cells have no entry in the block *IIa*. Some further subgroups of C222 are C222$_1$ (with $\mathbf{c}' = 2\mathbf{c}$), I222 (with $\mathbf{c}' = 2\mathbf{c}$) and I2$_1$2$_1$2$_1$ (with $\mathbf{c}' = 2\mathbf{c}$). These subgroups have conventional cells larger than that of C222 and are denoted as subgroups of type *IIb*. For *k* subgroups the point group β of G is unchanged.

3. By combination of 1 and 2. In this case both the translation group T and the point group β of G are changed.

A theorem by Hermann states that a maximal subgroup of G is either a *t* subgroup or a *k* subgroup. Thus in the *International Tables* only *I, IIa, IIb,* and *IIc* subgroups are listed.

Sometimes we are interested to the possible space groups G′ of which a given space group G is a subgroup. G′ is called a minimal supergroup of the group G if G is a maximal subgroup of G′. Of course we will have a minimal *t*, or a minimal non-isomorphous *k*, or a minimal isomorphous *k* supergroup G′ of G according to whether G is a maximal *t*, or a maximal non-isomorphous *k*, or a maximal isomorphous *k* subgroup of G′. The minimal non-isomorphous supergroups of C222 are:

of type *t*: Cmmm, C*cc*m, Cmm*a*, C*ca*, P422, P42$_1$2, P4$_2$22, P42$_1$2, P$\bar{4}$m2, P$\bar{4}c$2, P$\bar{4}b$2, P$\bar{4}n$2, P622, P6$_2$22, P6$_4$22,

of type *k*: F222, P222 (with $\mathbf{a}' = \mathbf{a}/2$, $\mathbf{b}' = \mathbf{b}/2$).

Maximal subgroups and minimal supergroups for three-dimensional crystallographic point groups

With trivial changes the definitions of maximal subgroup and minimal supergroup given above for space groups may be applied to three-dimensional crystallographic point groups. For example, it will be easily seen that Laue symmetry is always a minimal supergroup of index 2 of a non-centrosymmetric point group.

A scheme showing the subgroup and supergroup relationships for point groups is illustrated in Fig. 1.E.1. Maximal invariant subgroups are indicated by full lines: if two or three maximal invariant subgroups exist with the same symbol then double or triple full lines are used.

A set of maximal conjugate subgroups is referred to by a broken line. For example, from 3m three conjugate subgroups of type m can be formed. Thus a dashed line refers 3m to m. Furthermore, from 422 two invariant subgroups of type 222 with index 2 can be formed (no symmetry operation of 422 refers one subgroup to the other). Thus in Fig. 1.E.1 a double solid line refers 422 to 222.

Limiting groups in two and three dimensions

In two dimensions there are two types of point groups:

1, 2, 3, 4, 5, 6, . . .
m, 2mm, 3m, 4mm, 5m, 6mm, ...

For very large values of the order of the rotation axis the two types approach ∞ and ∞m respectively. From the geometrical point of view ∞ and ∞m are identical, and our standard notation will be ∞m (the situation

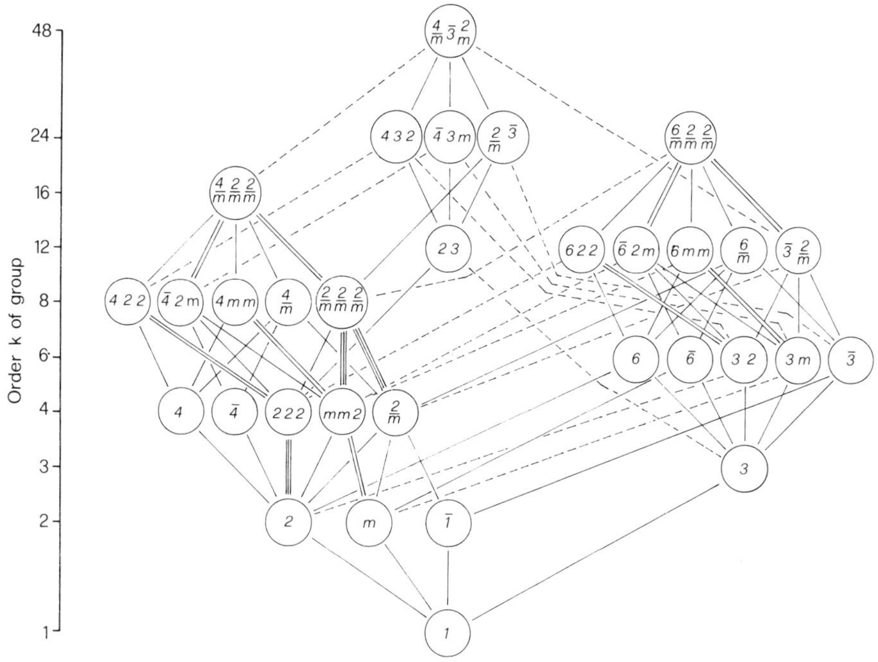

Fig. 1.E.1. Group, subgroup, supergroup relationships for point groups (from *International Tables for Crystallography*).

$\infty \neq \infty$m occurs when the rotation direction is taken into account: i.e. for a magnetic field round a disc).

In three dimensions a point group can include continuous rotations about one or about all axes (this is a consequence of the Euler theorem applied to such a limiting case). In the first case two groups can be identified, ∞m and $\frac{\infty}{\mathrm{m}}\frac{2}{\mathrm{m}}$, according to whether there is or not a mirror perpendicular to ∞ axis (the symmetry of the two groups can be represented by a circular cone and by a circular cylinder respectively). The symbol $\frac{\infty}{\mathrm{m}}\frac{\infty}{\mathrm{m}}$ represents the case in which continuous rotation about any axis is allowed (the symmetry is represented by a sphere).

Representation of a group

If a square matrix **d** can be associated to each $g \in \mathrm{G}$, in such a way that when $g_i g_j = g_k$ also $\mathbf{d}_i \mathbf{d}_j = \mathbf{d}_k$, then the matrices form a group D isomorphous with G. These matrices form an **isomorphous or exact representation** of the group: the order n of the matrices is the **dimension** of the representation. In accordance with this point of view, in Chapter 1 we have represented the symmetry groups through square matrices of order 3. Different representations of G may be obtained through a transformation of the type

$$\mathbf{d}_j' = \mathbf{q}^{-1}\mathbf{d}_j\mathbf{q}. \qquad (1.\mathrm{E}.7)$$

When condition (1.E.7) is verified for the two representations $\Gamma^1(\mathbf{d}_1, \mathbf{d}_2, \ldots)$ and $\Gamma^2(\mathbf{d}_1', \mathbf{d}_2', \ldots)$ then the two representations are said to be equivalent, since **q** can be interpreted as a change of coordinate system. It is often possible to find a new coordinate system for which each matrix **d** is transformed into

$$\left(\begin{array}{c|c} \mathbf{d}^1 & 0 \\ \hline 0 & \mathbf{d}^2 \end{array}\right) \qquad (1.A.13)$$

with $\mathbf{d}^1$ of order $m < n$ and $\mathbf{d}^2$ of order $(n - m)$. If this can not be obtained by any transformation, then the representation is called **irreducible**; otherwise it is called **reducible**. Sometimes $\mathbf{d}^1$ and $\mathbf{d}^2$ can be further reduced, and at the end of the process each matrix $\mathbf{d}_j$ will be transformed into

$$\mathbf{q}^{-1}\mathbf{d}_j\mathbf{q} = \text{diag}\,[\mathbf{d}_j^{(1)}, \mathbf{d}_j^{(2)}, \ldots, \mathbf{d}_j^{(s)}] = \mathbf{d}_j'$$

where $\mathbf{d}_j^{(i)}$ are themselves matrices.

The matrices $\mathbf{d}_1^{(1)}, \mathbf{d}_2^{(1)}, \mathbf{d}_3^{(1)}, \ldots$ all have the same dimension. Similarly $\mathbf{d}_1^{(2)}, \mathbf{d}_2^{(2)}, \mathbf{d}_3^{(2)}, \ldots$ have the same dimension. From the rule of the product of blocked matrices it follows that $(\mathbf{d}_1^{(1)}, \mathbf{d}_2^{(1)}, \mathbf{d}_3^{(1)}, \ldots)$ form a representation of the group, as well as $(\mathbf{d}_1^{(2)}, \mathbf{d}_2^{(2)}, \mathbf{d}_3^{(2)}, \ldots)$, etc.

It can be shown that for finite groups the number of irreducible representations is equal to the number of classes. For instance an isomorphous (reducible) representation of the point group 32 is

$$g_1 = 1 \Leftrightarrow \begin{pmatrix} 1 & 0 & 0 \\ 0 & 1 & 0 \\ 0 & 0 & 1 \end{pmatrix}; \quad g_2 = 3_{[001]} \Leftrightarrow \begin{pmatrix} 0 & \bar{1} & 0 \\ 1 & 1 & 0 \\ 0 & 0 & 1 \end{pmatrix};$$

$$g_3 = 3^2_{[001]} \Leftrightarrow \begin{pmatrix} \bar{1} & 1 & 0 \\ 1 & 0 & 0 \\ 0 & 0 & 1 \end{pmatrix}; \quad g_4 = 2_{[100]} \Leftrightarrow \begin{pmatrix} 1 & \bar{1} & 0 \\ 0 & \bar{1} & 0 \\ 0 & 0 & \bar{1} \end{pmatrix};$$

$$g_5 = 2_{[010]} \Leftrightarrow \begin{pmatrix} \bar{1} & 0 & 0 \\ \bar{1} & 1 & 0 \\ 0 & 0 & \bar{1} \end{pmatrix}; \quad g_6 = 2_{[110]} \Leftrightarrow \begin{pmatrix} 0 & 1 & 0 \\ 1 & 0 & 0 \\ 0 & 0 & \bar{1} \end{pmatrix}.$$

Also the two-dimensional irreducible representation

$$g_1 \Leftrightarrow \begin{pmatrix} 1 & 0 \\ 0 & 1 \end{pmatrix}; \quad g_2 \Leftrightarrow \begin{pmatrix} 0 & 1 \\ \bar{1} & \bar{1} \end{pmatrix}; \quad g_3 \Leftrightarrow \begin{pmatrix} \bar{1} & \bar{1} \\ 1 & 0 \end{pmatrix};$$

$$g_4 \Leftrightarrow \begin{pmatrix} 1 & 0 \\ \bar{1} & \bar{1} \end{pmatrix}; \quad g_5 \Leftrightarrow \begin{pmatrix} \bar{1} & \bar{1} \\ 0 & 1 \end{pmatrix}; \quad g_6 \Leftrightarrow \begin{pmatrix} 0 & 1 \\ 1 & 0 \end{pmatrix}$$

and the one-dimensional non-exact (homomorphic) representations exist:

$$g_1 \to 1;\ g_2 \to 1;\ g_3 \to 1;\ g_4 \to \bar{1};\ g_5 \to \bar{1};\ g_6 \to \bar{1}$$

$$g_1 \to 1;\ g_2 \to 1;\ g_3 \to 1;\ g_4 \to 1;\ g_5 \to 1;\ g_6 \to 1.$$

Character tables

The sum of the diagonal elements of a matrix, elsewhere called **trace**, in group theory is called character and is indicated by $\chi(g)$. It is obvious that $\chi(g_1)$ defines the dimensionality of the representation. The complete set of characters for a given representation is called the **character of the**

representation. Since the traces of two matrices related by a coordinate transformation are identical, the characters of two equivalent representations will be identical (the converse is also true). Several properties of a point group may be deduced from the characters of its irreducible representations alone. It is therefore convenient to set them up in tables called **character tables**. Each row of the table refers to a particular irreducible representation and each column to a given class.

1.F. Symmetry generalization

Only a few intuitive elements of this subject are given, since a full treatment would exceed the limits of the present book. The reader is referred to specific texts or papers.[1–3]

The symmetry groups G_n^m

A space may or may not be periodic in all its m dimensions. The corresponding symmetry groups are indicated by G_n^m, with $m \geqslant n$, where n is the number of dimensions of the subspace in which the group is periodic. In this space only symmetry operations transforming the space into itself will be allowed. For instance in the G_1^3 groups, describing objects periodic in one direction and finite in the other two, at least one line will remain invariant with respect to all symmetry operations.

The G^1 groups

1. G_0^1 groups. In a one-dimensional space (a line), which is non-periodic, only two symmetry operators are conceivable: 1 and $\bar{1}$ (which is the reflection operator m). The only two (point) groups are therefore 1 and $\bar{1}$.

2. G_1^1 groups. Besides the 1 and $\bar{1}$ operators, they contain the translation operator. Only two groups of type G_1^1 are then possible.

The G^2 groups

1. G_0^2 groups. In a two-dimensional space (a plane), which is non-periodic, the only conceivable operators are those of rotation around an axis perpendicular to the plane and of reflection with respect to a line on the plane. The number of (point) groups is infinite, but there are only ten crystallographic groups (see p. 16).

2. G_1^2 (**border**) groups. In a two-dimensional space, periodic in one dimension, only the symmetry operators (and their combinations), which transform that direction into itself, are allowed. We may therefore consider reflection planes parallel or perpendicular to the invariant direction, glides with translational component parallel to it and two-fold axes. There are seven G_1^2 groups (the symmetries of **linear decorations**) which are represented in Fig. 1.F.1.

3. G_2^2 groups. There are the 17 plane groups described on the pages 30 and 34.

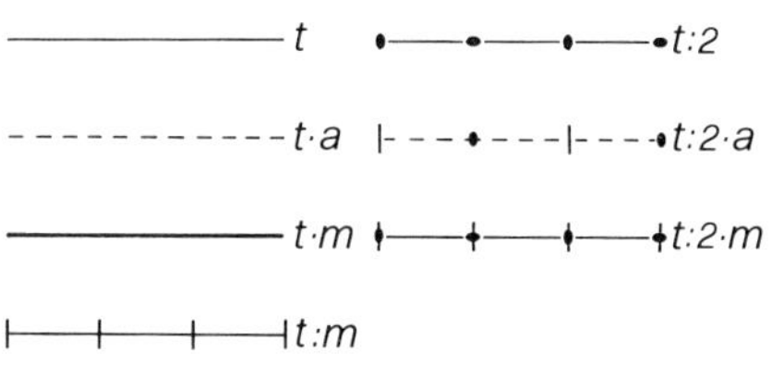

Fig. 1.F.1. The seven border groups.

The G^3 groups

1. G_0^3 groups. There describe non-periodic spaces in three dimensions. The number of (point) groups is infinite (see Appendix 1.B), but there are only 32 crystallographic point groups (see pp. 11–16).

2. G_1^3 (**rod**) groups. Rod groups may be considered as arising from the combination of one-dimensional translation groups with point groups G_0^3. They describe three-dimensional objects which are periodic in only one direction (say z). This must remain invariant with respect to all symmetry operations. The only allowed operations are therefore n and $\bar{n}$ axes coinciding with z, 2 and $\bar{2}$ axes perpendicular to it, screw axes and glide planes with a translational component parallel to the invariant direction.

There are 75 G_1^3 crystallographic groups. In Table 1.F.1 the rod group symbols are shown alongside the point groups from which they are derived. The first position in the symbol indicates the axis (n or $\bar{n}$) along z, the

Table 1.F.1. The 75 rod groups

Point group	**Rod groups**					
1	1					
2	2	2_1				
3	3	3_1	3_2			
4	4	4_1	4_3	4_2		
6	6	6_1	6_5	6_2	6_4	6_3
1m	1m	1c				
2mm	2mm	2_1mc	2cc			
3m	3m	3c				
4mm	4mm	4_2mc	4cc			
6mm	6mm	6_3mc	6cc			
m	m					
2/m	2/m	2_1/m				
4/m	4/m	4_2/m				
6/m	6/m	6_3/m				
m2m	m2m	m2c				
$\frac{2}{m}\frac{2}{m}\frac{2}{m}$	$\frac{2}{m}\frac{2}{m}\frac{2}{m}$	$\frac{2_1}{m}\frac{2}{m}\frac{2}{c}$	$\frac{2}{m}\frac{2}{c}\frac{2}{c}$			
$\frac{4}{m}\frac{2}{m}\frac{2}{m}$	$\frac{4}{m}\frac{2}{m}\frac{2}{m}$	$\frac{4_2}{m}\frac{2}{m}\frac{2}{c}$	$\frac{4}{m}\frac{2}{c}\frac{2}{c}$			
$\frac{6}{m}\frac{2}{m}\frac{2}{m}$	$\frac{6}{m}\frac{2}{m}\frac{2}{m}$	$\frac{6_3}{m}\frac{2}{m}\frac{2}{c}$	$\frac{6}{m}\frac{2}{c}\frac{2}{c}$			
12	12					
222	222	$2_1$22				
32	32	$3_1$2	$3_2$2			
422	422	$4_1$22	$4_3$22	$4_2$22		
622	622	$6_1$22	$6_5$22	$6_2$22	$6_4$22	$6_3$22
$\bar{1}$	$\bar{1}$					
$\bar{3}$	$\bar{3}$					
$\bar{4}$	$\bar{4}$					
$\bar{6}$	$\bar{6}$					
$\bar{1}\frac{2}{m}$	$\bar{1}\frac{2}{m}$	$\bar{1}\frac{2}{c}$				
$\bar{3}\frac{2}{m}$	$\bar{3}\frac{2}{m}$	$\bar{3}\frac{2}{c}$				
$\bar{4}$m2	$\bar{4}$m2	$\bar{4}c$2				
$\bar{6}$m2	$\bar{6}$m2	$\bar{6}c$2				

second position refers to axes (n or $\bar{n}$) in the plane (x, y) normal to the invariant direction, the third position for axes in the (x, y) plane bisecting the previous ones. Glides in the z direction are denoted by c.

3. G_2^3 groups. There are called **layer** groups and describe[4,5,12,13] the symmetries of doubly periodic three-dimensional objects. They are useful in describing the patterns of walls, panels, and, at an atomic level, in describing structures with layer molecular units.

Let us denote by z the direction perpendicular to the layer plane. It will be the vertical axis from now on: directions in the plane will be called horizontal. Rotations can only occur about a vertical axis, and if twofold, also about a horizontal one. No more than one horizontal reflection plane can exist, otherwise translational symmetry should also occur along the z axis. The layer groups can be enumerated according to the five nets quoted in Table 1.7 for the plane lattices. In Table 1.F.2 the 80 layer groups are divided in blocks, each block divided by the subsequent by a double line: each block refers to a specific net (in sequence, parallelogram, rectangular, centred rectangular, square, hexagonal) and contains the point groups compatible with the net and the corresponding layer groups.

The number of point groups which can be used is 31: the 32 three-dimensional point groups minus the five incompatible groups 23, m$\bar{3}$, 432, $\bar{4}$3m, m$\bar{3}$m plus four second settings (this time the z axis is distinguishable from x and y: so 2 is different from 12, or 2/m from $\bar{1}$2/m, . . .).

The first position in the layer group symbol gives the type of cell, the second refers to the z direction. The third and fourth positions refer:

(a) for rectangular nets, to x and y respectively;

(b) for a square net, to x (and therefore to y) and to a diagonal direction;

(c) for a hexagonal net, to x (and therefore symmetry related axes) and to diagonals directions.

As subgroups of the G_2^3 groups we may consider the G_2^2 groups, which are obtained by projecting the G_2^3 groups along the axis normal to the singular plane.

4. G_3^3 groups. These are the 230 space groups (see pp. 22–30).

The G_n^4 groups

The three-dimensional Euclidean space may be insufficient to describe the symmetries of some physical objects. We can therefore introduce one or more additional continuous variables (e.g. the time, the phase of a wave function, etc.), thus passing from a three-dimensional space into a space with dimensions $m > 3$. In a four-dimensional Euclidean space the symmetry groups G_n^4 may be constructed from their three-dimensional projections G_n^3, which are all well known. Thus there are 227 point groups G_0^4 and 4895 groups G_4^4.

The groups of colour symmetry

Groups in which three variables have a geometrical meaning while the fourth has a different physical meaning and is not continuous, are

Table 1.F.2. The layer groups

Point group	Plane groups			
1	p1			
2	p2			
m	pm	p*b*		
2/m	p2/m	p2/*b*		
$\bar{1}$	p$\bar{1}$			
1m	p1m	p1*b*		
2mm	p2mm	p2*b*m	p2*ba*	
m2m	pm2m	$\mathrm{pm2_1}a$	$\mathrm{p}a\mathrm{2_1m}$	p*a*2*a*
	$\mathrm{p}n\mathrm{2_1m}$	p*n*2*a*	p*b*2m	$\mathrm{p}b\mathrm{2_1}a$
$\frac{2}{\mathrm{m}}\frac{2}{\mathrm{m}}\frac{2}{\mathrm{m}}$	$\mathrm{p}\frac{2}{\mathrm{m}}\frac{2}{\mathrm{m}}\frac{2}{\mathrm{m}}$	$\mathrm{p}\frac{2}{\mathrm{m}}\frac{2}{b}\frac{2_1}{\mathrm{m}}$	$\mathrm{p}\frac{2}{\mathrm{m}}\frac{2_1}{b}\frac{2_1}{a}$	
	$\mathrm{p}\frac{2}{b}\frac{2}{\mathrm{m}}\frac{2_1}{\mathrm{m}}$	$\mathrm{p}\frac{2}{b}\frac{2_1}{\mathrm{m}}\frac{2_1}{a}$	$\mathrm{p}\frac{2}{b}\frac{2}{b}\frac{2}{\mathrm{m}}$	$\mathrm{p}\frac{2}{b}\frac{2_1}{b}\frac{2}{a}$
	$\mathrm{p}\frac{2}{n}\frac{2_1}{\mathrm{m}}\frac{2_1}{\mathrm{m}}$	$\mathrm{p}\frac{2}{n}\frac{2_1}{b}\frac{2}{\mathrm{m}}$	$\mathrm{p}\frac{2}{n}\frac{2}{b}\frac{2}{a}$	
12	p12	$\mathrm{p12_1}$		
222	p222	$\mathrm{p222_1}$	$\mathrm{p22_12_1}$	
$\bar{1}\frac{2}{\mathrm{m}}$	$\mathrm{p}\bar{1}\frac{2}{\mathrm{m}}$	$\mathrm{p}\bar{1}\frac{2}{b}$	$\mathrm{p}\bar{1}\frac{2_1}{\mathrm{m}}$	$\mathrm{p}\bar{1}\frac{2_1}{b}$
1m	*c*1m			
2mm	$c_2\mathrm{mm}$			
m2m	*c*m2m	*cg*2m		
$\frac{2}{\mathrm{m}}\frac{2}{\mathrm{m}}\frac{2}{\mathrm{m}}$	$c\frac{2}{\mathrm{m}}\frac{2}{\mathrm{m}}\frac{2}{\mathrm{m}}$	$c\frac{2}{g}\frac{2}{\mathrm{m}}\frac{2}{\mathrm{m}}$		
12	c12			
222	*c*222			
$\bar{1}\frac{2}{\mathrm{m}}$	$c\bar{1}\frac{2}{\mathrm{m}}$			
4	p4			
4mm	p4mm	p4*g*m		
4/m	p4m/m	p4/*n*		
$\frac{4}{\mathrm{m}}\frac{2}{\mathrm{m}}\frac{2}{\mathrm{m}}$	$\mathrm{p}\frac{4}{\mathrm{m}}\frac{2}{\mathrm{m}}\frac{2}{\mathrm{m}}$	$\mathrm{p}\frac{4}{\mathrm{m}}\frac{2_1}{g}\frac{2}{\mathrm{m}}$	$\mathrm{p}\frac{4}{\mathrm{m}}\frac{2_1}{g}\frac{2}{m}$	$\mathrm{p}\frac{4}{n}\frac{2}{g}\frac{2}{\mathrm{m}}$
$\bar{4}$	p$\bar{4}$			
422	p22	$\mathrm{p42_12}$		
$\bar{4}$m2	p$\bar{4}$m2	p$\bar{4}$g2	p$\bar{4}$2m	$\mathrm{p\bar{4}2_1m}$
3	p3			
3m	p3m1	p31m		
$\bar{3}$m	p$\bar{3}$			
32	p321	p312		
$\bar{3}\frac{2}{\mathrm{m}}$	$\mathrm{p}\bar{3}\frac{2}{\mathrm{m}}1$	$\mathrm{p}\bar{3}1\frac{2}{\mathrm{m}}$		
6	p6			
6mmm	p6mm			
6/m	p6/m			
$\frac{6}{\mathrm{m}}\frac{2}{\mathrm{m}}\frac{2}{\mathrm{m}}$	$\mathrm{p}\frac{6}{\mathrm{m}}\frac{2}{\mathrm{m}}\frac{2}{\mathrm{m}}$			
$\bar{6}$	p$\bar{6}$			
622	p622			
$\bar{6}$m2	p$\bar{6}$m2	p$\bar{6}$2m		

particularly important in crystallography. For instance:

1. The position and orientation of the magnetic moments of the cobalt atoms in the $CoAl_2O_4$ structure (space group $Fd\bar{3}m$) may be described by means of a two-colour group (see Fig. 1.F.2(a)) in which each colour corresponds to a given polarity of the magnetic moment. Groups of this type are also called 'groups with antisymmetry' or 'black-white' symmetry. More complicated cases require more colours, and the term **colour symmetry** is used. Classical groups involve only neutral points.

   As a further example let us consider the case of NiO, a material used in the ceramic and electronic industries. At room temperature, NiO is rhombohedral with edge length $a_R \simeq 2.952$ Å and $\alpha_R \simeq 60°4'$: α_R approaches 60° with increasing temperature, and, above 250°C, NiO is cubic with $a_C \simeq 4.177$ Å. The relation between the two cells is shown in Fig. 1.F.2(b): the same set of lattice points is described by the primitive rhombohedral unit cell and by the face-centred cubic cell provided that $\alpha_R = 60°$ exactly and $a_R = a/\sqrt{2}$. If the cube is compressed (or extended) along one of the four threefold axes of the cubic unit cell then symmetry reduces from cubic to rhombohedral (the only threefold axis is the compression axis). The polymorphism of NiO is due to its magnetic properties. Each Ni^{2+} ion has two unpaired spins (the $[Ar]3d^8$ electronic configuration). At room temperature the spins in NiO form an ordered antiferromagnetic array: layers of Ni^{2+} with net spin magnetic moments all in the same direction alternate with layers of Ni^{2+} with magnetic moments all in the opposite direction, as in Fig. 1.F.2(c). In these conditions the threefold axis is unique and the structure is rhombohedral. Above 250 °C the antiferromagnetic ordering is lost: the rhombohedral → cubic transition occurs and NiO displays ordinary paramagnetism.

2. If we project a G_3^3 group, in which a 6_1 axis is present, on a plane perpendicular to the axis, we obtain a G_2^2 group. But, if we assign a different colour to each of the six atoms related by the 6_1 axis, we will obtain a colour group $G_2^{2,(6)}$ with a clear meaning of the symbols.

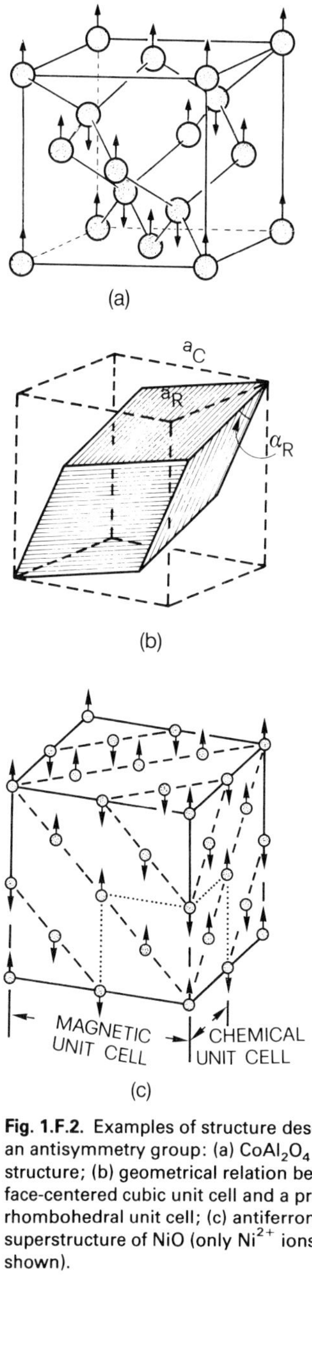

Fig. 1.F.2. Examples of structure described by an antisymmetry group: (a) $CoAl_2O_4$ magnetic structure; (b) geometrical relation between a face-centered cubic unit cell and a primitive rhombohedral unit cell; (c) antiferromagnetic superstructure of NiO (only Ni^{2+} ions are shown).

In the groups with antisymmetry there will be four types of equivalence between geometrically related objects: identity, identity after an inversion operation, anti-identity (the two objects differ only in the colour), identity after both an inversion operation and a change in colour. A general rotation matrix may be written in the form

$$\mathbf{R} = \begin{pmatrix} R_{11} & R_{12} & R_{13} & 0 \\ R_{21} & R_{22} & R_{23} & 0 \\ R_{31} & R_{32} & R_{33} & 0 \\ 0 & 0 & 0 & R_{44} \end{pmatrix}$$

where $R_{44} = -1$ or $+1$ depending on whether or not the operation changes the colour.

For the three-dimensional groups with antisymmetry we observe that, because of the existence of the anti-identity operation 1′ (only the colour is changed), the anti-translation operation t′ = t1′ will also exist. New types of Bravais lattices, such as those given in Fig. 1.F.3, will come out. As an example, in Fig. 1.F.2(c) the quasi-cubic magnetic unit cell of NiO has an

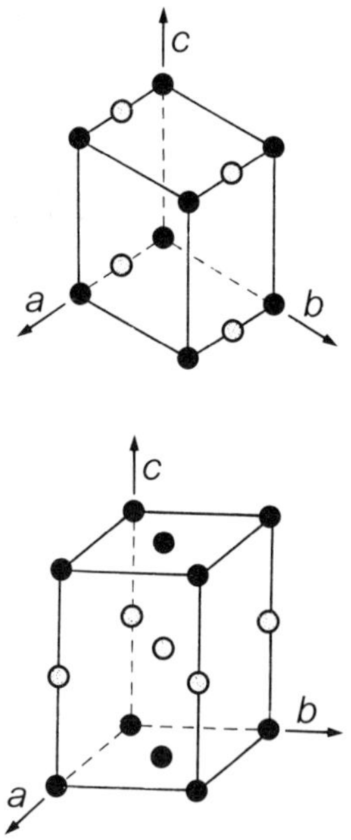

Fig. 1.F.3. Two antisymmetrical Bravais lattices.

edge length twice that of the chemical unit cell. It may be seen[14] that if the five Bravais lattices are centred by black and white lattice points (in equal percentage) then five new plane lattices are obtained. In three dimensions there are 36 black and white Bravais lattices, including the traditional uncoloured lattices.

References

1. Shubnikov, A. V. (1960). *Krystallografiya,* **5,** 489.
2. Shubnikov, A. V. and Belov, N. V. (1964). *Coloured symmetry.* Pergamon, Oxford.
3. Bradley, C. J. and Cracknell, A. P. (1972). The mathematical theory of symmetry in solids. *Representation theory for point groups and space groups.* Clarendon Press, Oxford.
4. Lockwood, E. H. and MacMillan, R. H. (1978). *Geometric symmetry.* Cambridge University Press.
5. Vainshtein, B. K. (1981). *Modern crystallography I: Symmetry of crystals, methods of structural crystallography.* Springer, Berlin.
6. (1983). *International tables for crystallography,* Vol. A, Space group symmetry. Reidel, Dordrecht.
7. Dougherty, J. P. and Kurtz, S. K. (1976). *Journal of Applied Crystallography,* **9,** 145.
8. Kitaigorodskij, A. I. (1955). *Organic crystallochemistry.* Moscow.
9. Mighell, A. and Rodgers, J. R. (1980). *Acta Crystallographica,* **A36,** 321.
10. Wilson, A. J. C. (1988). *Acta Crystallographica,* **A44,** 715.
11. Fischer, W., Burzlaff, H, Hellner, E., and Donnay, J. D. M. (1973). *Space groups and lattice complexes,* NBS Monograph 134. National Bureau of Standards, Washington, D.C.
12. Alexander, E. and Herrmann, K. (1929). *Zeitschrift für Kristallographie,* **70,** 328.
13. Alexander, E. (1929). *Zeitschrift für Kristallographie,* **70,** 367.
14. Mackay, A. L. (1957). *Acta Cryst.,* **10,** 543.

Crystallographic computing

2

CARMELO GIACOVAZZO

Introduction

In this chapter elements of crystallographic computing are described. Material is treated in order to answer day-to-day questions and to provide a basis for reference. Among the various topics, those which are of more frequent use have been selected: axis transformations, geometric calculations (bond angles and distances, torsion angles, principal axes of the quadratic forms, metric considerations on the lattices, structure factors, Fourier calculations, . . .). The method of least squares and its main crystallographic applications are treated in greater detail. For practical reasons some calculations useful in characterization of thermal ellipsoids are developed in Appendix 3.B.

The following notation is adopted: $\boldsymbol{r}_1 \cdot \boldsymbol{r}_2$ denotes the scalar product between the two vectors $\boldsymbol{r}_1$ and $\boldsymbol{r}_2$, $\boldsymbol{r}_1 \wedge \boldsymbol{r}_2$ is their cross product, r will be the modulus of $\boldsymbol{r}$. $\mathbf{S}_1\mathbf{S}_2$ is the (row by columns) product of two matrices $\mathbf{S}_1$ and $\mathbf{S}_2$: $\bar{\mathbf{S}}$ is the transposed matrix of $\mathbf{S}$, and S is the determinant of the square matrix $\mathbf{S}$.

We will also distinguish between **coordinate matrices** and **vectors**. For example, with respect to a coordinate system $[0, \boldsymbol{a}, \boldsymbol{b}, \boldsymbol{c}]$ the vector $\boldsymbol{r}$ will be written as

$$\boldsymbol{r} = x\boldsymbol{a} + y\boldsymbol{b} + z\boldsymbol{c} = (\boldsymbol{abc})\begin{pmatrix} x \\ y \\ z \end{pmatrix} = \bar{\mathbf{A}}\mathbf{X},$$

where $\mathbf{X}$ is the coordinate matrix and $\mathbf{A}$ is the matrix which represents the basis vectors of the rectilinear coordinate system.

The metric matrix

Let $[0, \boldsymbol{a}, \boldsymbol{b}, \boldsymbol{c}]$ be our coordinate system. The scalar product of $\boldsymbol{r}_1$ and $\boldsymbol{r}_2$ is

$$\begin{aligned} \boldsymbol{r}_1 \cdot \boldsymbol{r}_2 &= (x_1\boldsymbol{a} + y_1\boldsymbol{b} + z_1\boldsymbol{c}) \cdot (x_2\boldsymbol{a} + y_2\boldsymbol{b} + z_2\boldsymbol{c}) \\ &= x_1x_2a^2 + y_1y_2b^2 + z_1z_2c^2 + (x_1y_2 + x_2y_1)ab\cos\gamma \\ &\quad + (x_1z_2 + x_2z_1)ac\cos\beta + (y_1z_2 + y_2z_1)bc\cos\alpha \end{aligned}$$

or, in matrix notation,

$$\boldsymbol{r}_1 \cdot \boldsymbol{r}_2 = (x_1 y_1 z_1)\begin{pmatrix} \boldsymbol{a}\cdot\boldsymbol{a} & \boldsymbol{a}\cdot\boldsymbol{b} & \boldsymbol{a}\cdot\boldsymbol{c} \\ \boldsymbol{b}\cdot\boldsymbol{a} & \boldsymbol{b}\cdot\boldsymbol{b} & \boldsymbol{b}\cdot\boldsymbol{c} \\ \boldsymbol{c}\cdot\boldsymbol{a} & \boldsymbol{c}\cdot\boldsymbol{b} & \boldsymbol{c}\cdot\boldsymbol{c} \end{pmatrix}\begin{pmatrix} x_2 \\ y_2 \\ z_2 \end{pmatrix} = \bar{\mathbf{X}}_1\mathbf{G}\mathbf{X}_2. \tag{2.1}$$

G is the metric matrix, also called the metric tensor: its elements define both the moduli of $\boldsymbol{a}$, $\boldsymbol{b}$, $\boldsymbol{c}$ and the angles between them. The value of its determinant is

$$\mathbf{G} = a^2b^2c^2(1 - \cos^2\alpha - \cos^2\beta - \cos^2\gamma + 2\cos\alpha\cos\beta\cos\gamma) \tag{2.2}$$

which (see Table 2.1 and p. 69) is equal to V^2 (square of the volume of the unit cell). If $\boldsymbol{r}_1 = \boldsymbol{r}_2 = \boldsymbol{r}$ then (2.1) becomes

$$r^2 = \bar{\mathbf{X}}\mathbf{G}\mathbf{X} = x^2a^2 + y^2b^2 + z^2c^2 + 2xyab\cos\gamma$$
$$+ 2xzac\cos\beta + 2yzbc\cos\alpha \tag{2.3a}$$

which gives the modulus square of a vector. We can now calculate the following:

1. The interatomic distance d between two atoms positioned in (x_1, y_1, z_1) and $(x_2 y_2 z_2)$. Denoting

$$\Delta_1 = a(x_1 - x_2), \qquad \Delta_2 = b(y_1 - y_2), \qquad \Delta_3 = c(z_1 - z_2)$$

gives

$$d^2 = \Delta_1^2 + \Delta_2^2 + \Delta_3^2 + 2\Delta_1\Delta_2\cos\gamma + 2\Delta_1\Delta_3\cos\beta + 2\Delta_2\Delta_3\cos\alpha. \tag{2.3b}$$

2. The angle θ between two vectors

$$\cos\theta = \bar{\mathbf{X}}_1\mathbf{G}\mathbf{X}_2/(r_1r_2). \tag{2.4}$$

3. The cross product $\boldsymbol{r}_2 \wedge \boldsymbol{r}_3$:

$$\boldsymbol{r}_2 \wedge \boldsymbol{r}_3 = (x_2\boldsymbol{a} + y_2\boldsymbol{b} + z_2\boldsymbol{c}) \wedge (x_3\boldsymbol{a} + y_3\boldsymbol{b} + z_3\boldsymbol{c})$$
$$= (x_2y_3 - x_3y_2)\boldsymbol{a}\wedge\boldsymbol{b} + (y_2z_3 - y_3z_2)\boldsymbol{b}\wedge\boldsymbol{c}$$
$$+ (z_2x_3 - z_3x_2)\boldsymbol{c}\wedge\boldsymbol{a}.$$

4. The scalar triple product $\boldsymbol{r}_1 \cdot \boldsymbol{r}_2 \wedge \boldsymbol{r}_3$:

$$\boldsymbol{r}_1 \cdot \boldsymbol{r}_2 \wedge \boldsymbol{r}_3 = V \det\begin{pmatrix} x_1 & y_1 & z_1 \\ x_2 & y_2 & z_2 \\ x_3 & y_3 & z_3 \end{pmatrix} \tag{2.5}$$

where

$$V = \boldsymbol{a}\cdot\boldsymbol{b}\wedge\boldsymbol{c} = \boldsymbol{b}\cdot\boldsymbol{c}\wedge\boldsymbol{a} = \boldsymbol{c}\cdot\boldsymbol{a}\wedge\boldsymbol{b}$$

is the volume of the unit cell.

The following formulae of vector algebra are also recalled for future convenience:

$$\boldsymbol{u}\wedge(\boldsymbol{v}\wedge\boldsymbol{w}) = (\boldsymbol{u}\cdot\boldsymbol{w})\boldsymbol{v} - (\boldsymbol{u}\cdot\boldsymbol{v})\boldsymbol{w}, \tag{2.6}$$

$$(\boldsymbol{u}\wedge\boldsymbol{v})\cdot(\boldsymbol{w}\wedge\boldsymbol{z}) = (\boldsymbol{u}\cdot\boldsymbol{w})(\boldsymbol{v}\cdot\boldsymbol{z}) - (\boldsymbol{u}\cdot\boldsymbol{z})(\boldsymbol{v}\cdot\boldsymbol{w}), \tag{2.7}$$

$$(\boldsymbol{u}\wedge\boldsymbol{v})\wedge(\boldsymbol{w}\wedge\boldsymbol{z}) = (\boldsymbol{u}\cdot\boldsymbol{v}\wedge\boldsymbol{z})\boldsymbol{w} - (\boldsymbol{u}\cdot\boldsymbol{v}\wedge\boldsymbol{w})\boldsymbol{z}. \tag{2.8}$$

Some properties of the rotation component **R** of the symmetry operators can now be proved:

1. If $\boldsymbol{a}$, $\boldsymbol{b}$, $\boldsymbol{c}$ define a primitive cell, the elements of $\mathbf{R}$ are integers. Indeed the relation $\mathbf{X}' = \mathbf{RX}$ transforms lattice vectors $\boldsymbol{r}$ into lattice vectors $\boldsymbol{r}'$. Both $\boldsymbol{r}$ and $\boldsymbol{r}'$ have integer components, consequently also the elements of $\mathbf{R}$ must be integers.
2. A symmetry operator does not change the moduli of the vectors and the angles between vectors: i.e. $\boldsymbol{r}_1' \cdot \boldsymbol{r}_2' = \boldsymbol{r}_1 \cdot \boldsymbol{r}_2$. Therefore, according to (2.1), $\bar{\mathbf{X}}_1'\mathbf{G}\mathbf{X}_2' = \bar{\mathbf{X}}_1\mathbf{RGRX}_2 = \bar{\mathbf{X}}_1\mathbf{G}\mathbf{X}_2$, from which
$$\mathbf{G} = \bar{\mathbf{R}}\mathbf{G}\mathbf{R} \quad \text{or} \quad \mathbf{G} = \bar{\mathbf{R}}^{-1}\mathbf{G}\mathbf{R}^{-1}. \tag{2.9}$$
Only the matrices $\mathbf{R}$ which satisfy (2.9) can be symmetry rotation matrices in the coordinates system defined by $\mathbf{G}$.
3. Multiplying the second eqn (2.9) by $\mathbf{R}$ gives $\mathbf{GR} = \bar{\mathbf{R}}^{-1}\mathbf{G}$. If only determinants are taken into consideration then $R = \pm 1$ arises. Properties 1, 2, 3 can be verified for the rotation matrices quoted in Appendix 1.D.

The reciprocal lattice

Very useful in metric calculations (as well as in diffraction geometry), the reciprocal lattice was introduced by P. Ewald in 1921. Let $\boldsymbol{a}$, $\boldsymbol{b}$, $\boldsymbol{c}$ be the elementary translations of a space lattice (called here a **direct lattice**). A second lattice, **reciprocal** to the first one, is defined by translations $\boldsymbol{a}^*$, $\boldsymbol{b}^*$, $\boldsymbol{c}^*$, which satisfy the following two conditions:

$$\boldsymbol{a}^* \cdot \boldsymbol{b} = \boldsymbol{a}^* \cdot \boldsymbol{c} = \boldsymbol{b}^* \cdot \boldsymbol{a} = \boldsymbol{b}^* \cdot \boldsymbol{c} = \boldsymbol{c}^* \cdot \boldsymbol{a} = \boldsymbol{c}^* \cdot \boldsymbol{b} = 0, \tag{2.10a}$$

$$\boldsymbol{a}^* \cdot \boldsymbol{a} = \boldsymbol{b}^* \cdot \boldsymbol{b} = \boldsymbol{c}^* \cdot \boldsymbol{c} = 1. \tag{2.10b}$$

Equation (2.10a) suggests that $\boldsymbol{a}^*$ is normal to the plane $(\boldsymbol{b}, \boldsymbol{c})$, $\boldsymbol{b}^*$ to the plane $(\boldsymbol{a}, \boldsymbol{c})$ and $\boldsymbol{c}^*$ to the plane $(\boldsymbol{a}, \boldsymbol{b})$. The modulus and sense of $\boldsymbol{a}^*$, $\boldsymbol{b}^*$, $\boldsymbol{c}^*$ are fixed by (2.10b).

According to (2.10a) $\boldsymbol{a}^*$ may be written as

$$\boldsymbol{a}^* = p(\boldsymbol{b} \wedge \boldsymbol{c}) \tag{2.11}$$

where p is a constant. The value of p is obtained if the scalar product of both the sides of (2.11) by $\boldsymbol{a}$ is taken:

$$\boldsymbol{a}^* \cdot \boldsymbol{a} = 1 = p(\boldsymbol{b} \wedge \boldsymbol{c} \cdot \boldsymbol{a}) = pV$$

from which $p = 1/V$. Equation (2.11) and its analogue may then be written as

$$\boldsymbol{a}^* = \frac{1}{V}(\boldsymbol{b} \wedge \boldsymbol{c}), \qquad \boldsymbol{b}^* = \frac{1}{V}(\boldsymbol{c} \wedge \boldsymbol{a}), \qquad \boldsymbol{c}^* = \frac{1}{V}(\boldsymbol{a} \wedge \boldsymbol{b}), \tag{2.12a}$$

or, in terms of moduli,

$$a^* = \frac{1}{V} bc \sin \alpha, \qquad b^* = \frac{1}{V} ca \sin \beta, \qquad c^* = \frac{1}{V} ab \sin \gamma. \tag{2.12b}$$

Equation (2.10) also suggest that the roles of direct and reciprocal space may be interchanged: i.e. the reciprocal of the reciprocal lattice is the direct lattice. Therefore

$$\boldsymbol{a} = \frac{1}{V^*}(\boldsymbol{b}^* \wedge \boldsymbol{c}^*), \qquad \boldsymbol{b} = \frac{1}{V^*}(\boldsymbol{c}^* \wedge \boldsymbol{a}^*), \qquad \boldsymbol{c} = \frac{1}{V^*}(\boldsymbol{a}^* \wedge \boldsymbol{b}^*). \quad (2.13)$$

It may be easily verified that the reciprocals of triclinic, monoclinic, . . . lattices are triclinic, monoclinic, . . . themselves. However, the reciprocal of an F lattice is an I lattice and vice versa (see Appendix 2.D). In detail:

1. In monoclinic lattices $\boldsymbol{b}^* \parallel \boldsymbol{b}$ while $\boldsymbol{a}^*$ and $\boldsymbol{c}^*$ are in the plane $(\boldsymbol{a}, \boldsymbol{c})$: then

$$a^* = 1/(a \sin\beta), \qquad b = 1/b, \qquad c^* = 1/(c \sin\beta),$$
$$\alpha^* = \gamma^* = \pi/2, \qquad \beta^* = \pi - \beta.$$

2. In rhombic, tetragonal, and cubic lattices $\boldsymbol{a}^* \parallel \boldsymbol{a}$, $\boldsymbol{b}^* \parallel \boldsymbol{b}$, $\boldsymbol{c}^* \parallel \boldsymbol{c}$, and

$$a^* = 1/a, \quad b^* = 1/b, \quad c^* = 1/c, \quad \alpha^* = \beta^* = \gamma^* = \pi/2.$$

3. In trigonal and hexagonal lattices $\boldsymbol{c}^* \parallel \boldsymbol{c}$ while a^* and b^* are in the plane $(\boldsymbol{a}, \boldsymbol{b})$:

$$a^* = b^* = 2/(a\sqrt{3}), \quad c^* = 1/c, \quad \alpha^* = \beta^* = \pi/2, \quad \gamma^* = \pi/3.$$

For the rhombohedrical basis

$$a^* = b^* = c^* = \sin\alpha/[a(1 - 3\cos^2\alpha + 2\cos^3\alpha)^{1/2}],$$
$$\alpha^* = \beta^* = \gamma^*, \quad \cos\alpha^* = -\cos\alpha/(1 + \cos\alpha).$$

Some relations between direct and reciprocal bases are discussed in Appendix 2.A: a summary is given in Table 2.1.

Some important properties of the reciprocal lattice are:

1. The scalar product of two vectors, the first defined with respect to the reciprocal basis and the second to the direct basis, assumes a very simple expression:

$$\boldsymbol{r}_1 \cdot \boldsymbol{r}_2^* = (x_1\boldsymbol{a} + y_1\boldsymbol{b} + z_1\boldsymbol{c}) \cdot (x_2^*\boldsymbol{a}^* + y_2^*\boldsymbol{b}^* + z_2^*\boldsymbol{c}^*)$$
$$= x_2^* x_1 + y_2^* y_1 + z_2^* z_1 = \tilde{\mathbf{X}}_1 \mathbf{X}_2^*.$$

Table 2.1. Relationships among direct and reciprocal lattice parameters. Inverse relationships are obtained by interchanging the starred by unstarred parameters

$a^* = \dfrac{bc\sin\alpha}{V}$; $\quad b^* = \dfrac{ac\sin\beta}{V}$;	$c^* = \dfrac{ab\sin\gamma}{V}$
$\sin\alpha^* = \dfrac{V}{abc\sin\beta\sin\gamma}$;	$\cos\alpha^* = \dfrac{\cos\beta\cos\gamma - \cos\alpha}{\sin\beta\sin\gamma}$
$\sin\beta^* = \dfrac{V}{abc\sin\alpha\sin\gamma}$;	$\cos\beta^* = \dfrac{\cos\alpha\cos\gamma - \cos\beta}{\sin\alpha\sin\gamma}$
$\sin\gamma^* = \dfrac{V}{abc\sin\alpha\sin\beta}$;	$\cos\gamma^* = \dfrac{\cos\alpha\cos\beta - \cos\gamma}{\sin\alpha\sin\beta}$

$$V = abc(1 - \cos^2\alpha - \cos^2\beta - \cos^2\gamma + 2\cos\alpha\cos\beta\cos\gamma)^{1/2}$$
$$= abc\sin\alpha\sin\beta\sin\gamma^* = abc\sin\alpha\sin\beta^*\sin\gamma$$
$$= abc\sin\alpha^*\sin\beta\sin\gamma$$
$$V^* = 1/V$$

2. The vector $\boldsymbol{r}_{\mathbf{H}}^* = h\boldsymbol{a}^* + k\boldsymbol{b}^* + l\boldsymbol{c}^*$ is normal to the family of lattice planes (hkl). In Fig. 2.1 the plane of the family nearest to the origin is drawn. In accordance with p. 8 the vectors (A–O), (B–O) and (C–O) are equal to $\boldsymbol{a}/h$, $\boldsymbol{b}/k$ and $\boldsymbol{c}/l$ respectively. Consequently

$$(\text{B–A}) = \boldsymbol{b}/k - \boldsymbol{a}/h, \qquad (\text{C–A}) = \boldsymbol{c}/l - \boldsymbol{a}/h, \qquad (\text{C–B}) = \boldsymbol{c}/l - \boldsymbol{b}/k.$$

Therefore

$$\boldsymbol{r}_{\mathbf{H}}^* \cdot (\text{B–A}) = (h\boldsymbol{a}^* + k\boldsymbol{b}^* + l\boldsymbol{c}^*) \cdot (\boldsymbol{b}/k - \boldsymbol{a}/h) = 0,$$
$$\boldsymbol{r}_{\mathbf{H}}^* \cdot (\text{C–A}) = \boldsymbol{r}_{\mathbf{H}}^* \cdot (\text{C–B}) = 0.$$

Since $\boldsymbol{r}_{\mathbf{H}}^*$ is perpendicular to two lines in the (hkl) plane, it is normal to the plane.

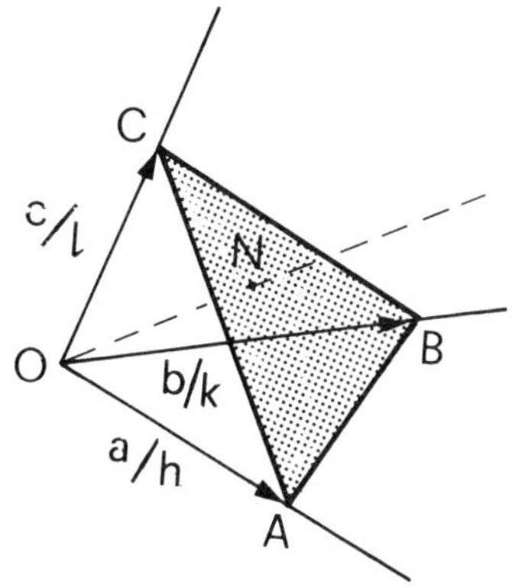

Fig. 2.1. The plane of the family (hkl) nearest to the origin.

3. If h, k, l have no common factor, then

$$r_{\mathbf{H}}^* = 1/d_{\mathbf{H}} \tag{2.14}$$

where $d_{\mathbf{H}}$ is the spacing of the planes (hkl) in the direct lattice. This may be proved by observing that $d_{\mathbf{H}}$ is equal to the length of the normal ON to the plane ABC from the origin O. Since $\boldsymbol{r}_{\mathbf{H}}^*$ has the same direction as ON,

$$d_{\mathbf{H}} = (\boldsymbol{a}/h) \cdot \frac{\boldsymbol{r}_{\mathbf{H}}^*}{r_{\mathbf{H}}^*} = \frac{1}{r_{\mathbf{H}}^*}.$$

4. As well as for direct lattice, a metric matrix $\mathbf{G}^*$ may be defined for the reciprocal lattice:

$$\mathbf{G}^* = \begin{pmatrix} \boldsymbol{a}^* \cdot \boldsymbol{a}^* & \boldsymbol{a}^* \cdot \boldsymbol{b}^* & \boldsymbol{a}^* \cdot \boldsymbol{c}^* \\ \boldsymbol{b}^* \cdot \boldsymbol{a}^* & \boldsymbol{b}^* \cdot \boldsymbol{b}^* & \boldsymbol{b}^* \cdot \boldsymbol{c}^* \\ \boldsymbol{c}^* \cdot \boldsymbol{a}^* & \boldsymbol{c}^* \cdot \boldsymbol{b}^* & \boldsymbol{c}^* \cdot \boldsymbol{c}^* \end{pmatrix}. \tag{2.15}$$

It may be easily verified that

$$\mathbf{G}^* = \mathbf{G}^{-1}, \qquad G^* = 1/G. \tag{2.16}$$

Consequently (see (2.1) and (2.3) for the direct lattice)

$$\boldsymbol{r}_1^* \cdot \boldsymbol{r}_2^* = \tilde{\mathbf{X}}_1^* \mathbf{G}^* \mathbf{X}_2^* \qquad \boldsymbol{r}^{*2} = \tilde{\mathbf{X}} \mathbf{G}^* \mathbf{X}^* \tag{2.17a}$$

from which

$$d_{\mathbf{H}} = (h^2 a^{*2} + k^2 b^{*2} + l^2 c^{*2} + 2hka^*b^* \cos\gamma^* + 2hla^*c^* \cos\beta^* + 2klb^*c^* \cos\alpha^*)^{-1/2} \tag{2.17b}$$

is easily obtained.

Specific expressions of $d_{\mathbf{H}}$ for the various crystal systems are given in Table 2.2.

Basis transformations

In three-dimensional space the coordinate system defined by the base vectors $\boldsymbol{a}'$, $\boldsymbol{b}'$, $\boldsymbol{c}'$ may be defined in terms of the base vectors $\boldsymbol{a}$, $\boldsymbol{b}$, $\boldsymbol{c}$ by

Table 2.2. The algebraic expressions of d_H for the various crystal systems

System	$1/d^2_{hkl}$
Cubic	$(h^2+k^2+l^2)/a^2$
Tetragonal	$\frac{h^2+k^2}{a^2}+\frac{l^2}{c^2}$
Orthorhombic	$\frac{h^2}{a^2}+\frac{k^2}{b^2}+\frac{l^2}{c^2}$
Hexagonal and trigonal (P)	$\frac{4}{3a^2}(h^2+k^2+hk)+\frac{l^2}{c^2}$
Trigonal (R)	$\frac{1}{a^2}\left(\frac{(h^2+k^2+l^2)\sin^2\alpha+2(hk+hl+kl)(\cos^2\alpha\text{-}\cos\alpha)}{1+2\cos^3\alpha-3\cos^2\alpha}\right)$
Monoclinic	$\frac{h^2}{a^2\sin^2\beta}+\frac{k^2}{b^2}+\frac{l^2}{c^2\sin^2\beta}-\frac{2hl\cos\beta}{ac\sin^2\beta}$
Triclinic	$(1-\cos^2\alpha-\cos^2\beta-\cos^2\gamma+2\cos\alpha\cos\beta\cos\gamma)^{-1}\Big(\frac{h^2}{a^2}\sin^2\alpha$ $+\frac{k^2}{b^2}\sin^2\beta+\frac{l^2}{c^2}\sin^2\gamma+\frac{2kl}{bc}(\cos\beta\cos\gamma-\cos\alpha)$ $+\frac{2lh}{ca}(\cos\gamma\cos\alpha-\cos\beta)+\frac{2hk}{ab}(\cos\alpha\cos\beta-\cos\gamma)\Big)$

three equations (suppose the transformation leaves the origin invariant)

$$\begin{aligned}\boldsymbol{a}' &= m_{11}\boldsymbol{a}+m_{12}\boldsymbol{b}+m_{13}\boldsymbol{c}\\ \boldsymbol{b}' &= m_{21}\boldsymbol{a}+m_{22}\boldsymbol{b}+m_{23}\boldsymbol{c}\\ \boldsymbol{c}' &= m_{31}\boldsymbol{a}+m_{32}\boldsymbol{b}+m_{33}\boldsymbol{c}\end{aligned} \tag{2.18}$$

where m_{ij} are any real numbers. In matrix notation (2.18) is written as

$$\mathbf{A}'=\begin{pmatrix}\boldsymbol{a}'\\ \boldsymbol{b}'\\ \boldsymbol{c}'\end{pmatrix}=\begin{pmatrix}m_{11}&m_{12}&m_{13}\\ m_{21}&m_{22}&m_{23}\\ m_{31}&m_{32}&m_{33}\end{pmatrix}\begin{pmatrix}\boldsymbol{a}\\ \boldsymbol{b}\\ \boldsymbol{c}\end{pmatrix}=\mathbf{MA}. \tag{2.19}$$

The reverse transformation will be $\mathbf{A}=\mathbf{M}^{-1}\mathbf{A}'$. The vector $\boldsymbol{r}=x\boldsymbol{a}+y\boldsymbol{b}+z\boldsymbol{c}$ may be written in the new coordinate system $\mathbf{A}'$ as $\boldsymbol{r}'\equiv\boldsymbol{r}=x'\boldsymbol{a}'+y'\boldsymbol{b}'+z'\boldsymbol{c}'$. Introducing (2.18) into the $\boldsymbol{r}'$ expression leads to

$$\mathbf{X}=\begin{pmatrix}x\\ y\\ z\end{pmatrix}=\begin{pmatrix}m_{11}x'+m_{21}y'+m_{31}z'\\ m_{12}x'+m_{22}y'+m_{32}z'\\ m_{13}x'+m_{23}y'+m_{33}z'\end{pmatrix}=\begin{pmatrix}m_{11}&m_{21}&m_{31}\\ m_{12}&m_{22}&m_{32}\\ m_{13}&m_{23}&m_{33}\end{pmatrix}\begin{pmatrix}x'\\ y'\\ z'\end{pmatrix}=\bar{\mathbf{M}}\mathbf{X}'. \tag{2.20}$$

The reverse transformation is $\mathbf{X}'=(\bar{\mathbf{M}})^{-1}\mathbf{X}$. It should be noted that (2.20) provides the transformation rule for the components of the vector $\boldsymbol{r}$ while the vector itself is unaffected ($\boldsymbol{r}'\equiv\boldsymbol{r}$) by a change of axes.

The metric matrix $\mathbf{G}'$ in $\mathbf{A}'$ may be calculated by substituting into (2.1) $\boldsymbol{a}'$, $\boldsymbol{b}'$, $\boldsymbol{c}'$ for $\boldsymbol{a}$, $\boldsymbol{b}$, $\boldsymbol{c}$ and then using (2.18). The result is

$$\mathbf{G}'=\mathbf{MG}\bar{\mathbf{M}} \quad \text{or} \quad \mathbf{G}=\mathbf{M}^{-1}\mathbf{G}'(\bar{\mathbf{M}})^{-1}. \tag{2.21}$$

The unit cell volume defined by $\mathbf{A}'$ is $V'=\boldsymbol{a}'\wedge\boldsymbol{b}'\cdot\boldsymbol{c}'$ therefore, according to (2.19) and (2.5), $V' = VM$. Thus, for any transformation of axes the unit cell volume is multiplied by the determinant of the transformation matrix.

The set of matrices conventionally used to pass from centred cells to primitive ones and vice versa is shown in Table 2.C.1 (however, transformation matrices are not unique).

Let us now apply (2.20) to derive the transformation rules of a quadratic form. Let

$$\bar{\mathbf{X}}\mathbf{Q}\mathbf{X} = q_{11}^2 x^2 + q_{22}^2 y^2 + q_{33}^2 z^2 + 2q_{12}xy + 2q_{13}xz + 2q_{23}yz,$$

be a quadratic form defined in the coordinate system $\mathbf{A}$. In accordance with (2.20) $\bar{\mathbf{X}}\mathbf{Q}\mathbf{X} = \bar{\mathbf{X}}'\mathbf{M}\mathbf{Q}\bar{\mathbf{M}}\mathbf{X}'$ from which

$$\mathbf{Q}' = \mathbf{M}\mathbf{Q}\bar{\mathbf{M}} \quad \text{or} \quad \mathbf{Q} = \mathbf{M}^{-1}\mathbf{Q}'(\bar{\mathbf{M}})^{-1}. \tag{2.22}$$

A special linear transformation is that which relates $\boldsymbol{a}' = \boldsymbol{a}^*$, $\boldsymbol{b}' = \boldsymbol{b}^*$, $\boldsymbol{c}' = \boldsymbol{c}^*$ to $\boldsymbol{a}$, $\boldsymbol{b}$, $\boldsymbol{c}$ (in this case $\mathbf{A}'$ will be replaced by $\mathbf{A}^*$). Let us show that in this case $\mathbf{M} = \mathbf{G}^{-1} = \mathbf{G}^*$. Write the vector $\boldsymbol{r} = x\boldsymbol{a} + y\boldsymbol{b} + z\boldsymbol{c}$ as

$$\boldsymbol{r} = (\boldsymbol{r} \cdot \boldsymbol{a}^*)\boldsymbol{a} + (\boldsymbol{r} \cdot \boldsymbol{b}^*)\boldsymbol{b} + (\boldsymbol{r} \cdot \boldsymbol{c}^*)\boldsymbol{c}, \tag{2.23}$$

(in terms of reciprocal axes

$$\boldsymbol{r}' \equiv \boldsymbol{r}^* = x'\boldsymbol{a}^* + y'\boldsymbol{b}^* + z'\boldsymbol{c}^* = (\boldsymbol{r}' \cdot \boldsymbol{a})\boldsymbol{a}^* + (\boldsymbol{r}' \cdot \boldsymbol{b})\boldsymbol{b}^* + (\boldsymbol{r}' \cdot \boldsymbol{c})\boldsymbol{c}^*). \tag{2.24}$$

On assuming in (2.23) $\boldsymbol{r} = \boldsymbol{a}^*, \boldsymbol{b}^*, \boldsymbol{c}^*$, the following relations

$$\begin{aligned}
\boldsymbol{a}^* &= (\boldsymbol{a}^* \cdot \boldsymbol{a}^*)\boldsymbol{a} + (\boldsymbol{a}^* \cdot \boldsymbol{b}^*)\boldsymbol{b} + (\boldsymbol{a}^* \cdot \boldsymbol{c}^*)\boldsymbol{c} \\
\boldsymbol{b}^* &= (\boldsymbol{b}^* \cdot \boldsymbol{a}^*)\boldsymbol{a} + (\boldsymbol{b}^* \cdot \boldsymbol{b}^*)\boldsymbol{b} + (\boldsymbol{b}^* \cdot \boldsymbol{c}^*)\boldsymbol{c} \\
\boldsymbol{c}^* &= (\boldsymbol{c}^* \cdot \boldsymbol{a}^*)\boldsymbol{a} + (\boldsymbol{c}^* \cdot \boldsymbol{b}^*)\boldsymbol{b} + (\boldsymbol{c}^* \cdot \boldsymbol{c}^*)\boldsymbol{c}
\end{aligned}$$

are respectively obtained, which in matrix notation reduce to

$$\mathbf{A}^* = \mathbf{G}^*\mathbf{A} \quad \text{or} \quad \mathbf{A} = \mathbf{G}\mathbf{A}^*. \tag{2.25}$$

It is easily seen that relations (2.25) are special cases of (2.19). Similarly:

1. Relations (2.26) are special cases of (2.20)

$$\mathbf{X} = \mathbf{G}^*\mathbf{X}^* \quad \text{or} \quad \mathbf{X}^* = \mathbf{G}\mathbf{X}. \tag{2.26}$$

   They transform components defined in direct space into components defined in reciprocal space and vice versa. In accordance with (2.1) and (2.26) we have

$$\boldsymbol{r}_1 \cdot \boldsymbol{r}_2 = \bar{\mathbf{X}}_1\mathbf{G}\mathbf{X}_2 = \bar{\mathbf{X}}_1\mathbf{X}_2^* = \bar{\mathbf{X}}_2\mathbf{X}_1^*. \tag{2.27}$$

   In particular $r^2 \equiv r^{*2} = \bar{\mathbf{X}}\mathbf{X}^*$ is the square modulus of a vector.

2. Equations (2.28) are special cases of (2.22):

$$\mathbf{Q}^* = \mathbf{G}^*\mathbf{Q}\mathbf{G}^* \quad \text{or} \quad \mathbf{Q} = \mathbf{G}\mathbf{Q}^*\mathbf{G}. \tag{2.28}$$

   Additional transformation rules are given in Appendix 2.E. The reader is referred to the splendid book by Sands[1] for further insight.

Transformation from triclinic to orthonormal axes

Geometrical calculations are often more easily made in an orthonormal than in a crystallographic frame.

Let

$$\mathbf{A} = \begin{pmatrix} \boldsymbol{a} \\ \boldsymbol{b} \\ \boldsymbol{c} \end{pmatrix} \quad \text{and} \quad \mathbf{E} = \begin{pmatrix} \boldsymbol{e}_1 \\ \boldsymbol{e}_2 \\ \boldsymbol{e}_3 \end{pmatrix}$$

be the crystallographic and the orthonormal bases respectively. Then, according to (2.18), $\mathbf{E} = \mathbf{MA}$, and inversely $\mathbf{A} = \mathbf{M}^{-1}\mathbf{E}$.

If we choose (see Fig. 2.2) $\boldsymbol{e}_1$ along $\boldsymbol{a}$, $\boldsymbol{e}_2$ normal to $\boldsymbol{a}$ in the $(\boldsymbol{a}, \boldsymbol{b})$ plane, $\boldsymbol{e}_3$ normal to $\boldsymbol{e}_1$ and $\boldsymbol{e}_2$ (and therefore parallel to $\boldsymbol{c}^*$), then the unit vectors $\boldsymbol{a}/a$, $\boldsymbol{b}/b$, $\boldsymbol{c}/c$ are referred to $\boldsymbol{e}_1$, $\boldsymbol{e}_2$, $\boldsymbol{e}_3$ by means of

$$\begin{pmatrix} \boldsymbol{a}/a \\ \boldsymbol{b}/b \\ \boldsymbol{c}/c \end{pmatrix} = \begin{pmatrix} l_1 & l_2 & l_3 \\ m_1 & m_2 & m_3 \\ n_1 & n_2 & n_3 \end{pmatrix} \begin{pmatrix} \boldsymbol{e}_1 \\ \boldsymbol{e}_2 \\ \boldsymbol{e}_3 \end{pmatrix} \tag{2.29}$$

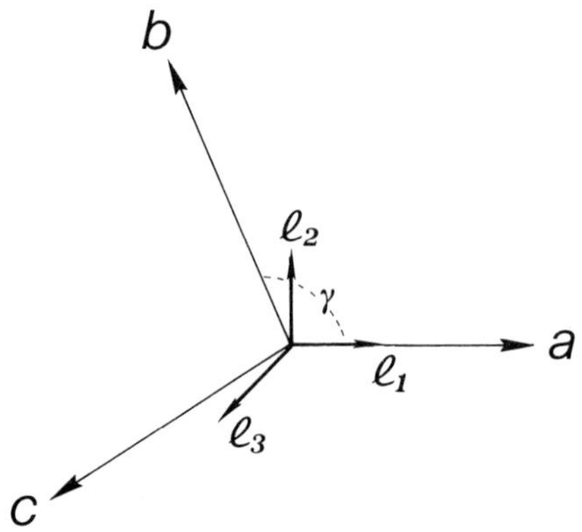

Fig. 2.2. Orthonormalization of crystallographic bases.

where (l_1, l_2, l_3), (m_1, m_2, m_3), (n_1, n_2, n_3) are direction cosines of the unit vectors $\boldsymbol{a}/a$, $\boldsymbol{b}/b$, $\boldsymbol{c}/c$ in $\mathbf{E}$. Therefore

$$\sum_i l_i^2 = \sum_i m_i^2 = \sum_i n_i^2 = 1.$$

From Fig. 2.2 it may be deduced that

$$l_1 = 1, \quad l_2 = 0, \quad l_3 = 0, \quad m_1 = \cos\gamma,$$
$$m_2 = \sin\gamma, \quad m_3 = 0, \quad n_1 = \cos\beta.$$

Since

$$\cos\alpha = \sum_i m_i n_i = \cos\gamma\cos\beta + \sin\gamma\sin n_2$$

we obtain

$$n_2 = (\cos\alpha - \cos\beta\cos\gamma)/\sin\gamma = -\sin\beta\cos\alpha^*.$$

Furthermore, from the relation $\sum_i n_i^2 = 1$

$$n_3^2 = \sin\beta\sin\alpha^* = 1/(cc^*)$$

is easily obtained. Finally

$$\begin{pmatrix} \boldsymbol{a}/a \\ \boldsymbol{b}/b \\ \boldsymbol{c}/c \end{pmatrix} = \begin{pmatrix} 1 & 0 & 0 \\ \cos\gamma & \sin\gamma & 0 \\ \cos\beta & -\sin\beta\cos\alpha^* & 1/(c^*c) \end{pmatrix} \begin{pmatrix} \boldsymbol{e}_1 \\ \boldsymbol{e}_2 \\ \boldsymbol{e}_3 \end{pmatrix}$$

from which

$$\mathbf{A} = \begin{pmatrix} \boldsymbol{a} \\ \boldsymbol{b} \\ \boldsymbol{c} \end{pmatrix} = \begin{pmatrix} a & 0 & 0 \\ b\cos\gamma & b\sin\gamma & 0 \\ c\cos\beta & -c\sin\beta\cos\alpha^* & 1/c^* \end{pmatrix} \begin{pmatrix} \boldsymbol{e}_1 \\ \boldsymbol{e}_2 \\ \boldsymbol{e}_3 \end{pmatrix} = \mathbf{M}^{-1}\mathbf{E}. \tag{2.30}$$

The matrix $\mathbf{M}$ to use in $\mathbf{E} = \mathbf{MA}$ is therefore

$$\mathbf{M} = \begin{pmatrix} 1/a & 0 & 0 \\ -\cos\gamma/(a\sin\gamma) & 1/(b\sin\gamma) & 0 \\ a^*\cos\beta^* & b^*\cos\alpha^* & c^* \end{pmatrix}. \tag{2.31a}$$

A crystallographic frame may be orthonormalized in infinite ways. For example, $\boldsymbol{e}_1$ may be chosen along $\boldsymbol{a}^*$ and $\boldsymbol{e}_2$ in the plane $(\boldsymbol{a}^*, \boldsymbol{b}^*)$: $\boldsymbol{e}_3$ is then along $\boldsymbol{c}$. In this case

$$\mathbf{M} = \begin{pmatrix} a^* & b^* \cos\gamma^* & c^* \cos\beta^* \\ 0 & b^* \sin\gamma^* & -c^* \sin\beta^* \cos\alpha \\ 0 & 0 & 1/c \end{pmatrix} \tag{2.31b}$$

and

$$\mathbf{M}^{-1} = \begin{pmatrix} 1/a^* & -\cot\gamma^*/a^* & a\cos\beta \\ 0 & 1/(b^* \sin\gamma^*) & b\cos\alpha \\ 0 & 0 & c \end{pmatrix}.$$

Also, we may choose $\boldsymbol{e}_1$ along $\boldsymbol{a}^*$ and $\boldsymbol{e}_2$ along $\boldsymbol{b}$ (then $\boldsymbol{e}_3$ is in the plane $(\boldsymbol{b}, \boldsymbol{c})$). We obtain

$$\mathbf{M} = \begin{pmatrix} 1/(a\,\mathrm{s}\gamma\,\mathrm{s}\beta^*) & 1/(b\,\mathrm{t}\alpha\,\mathrm{t}\beta^*) - 1/(b\,\mathrm{t}\gamma\,\mathrm{s}\beta^*) & -1/(c\,\mathrm{s}\alpha\,\mathrm{t}\beta^*) \\ 0 & 1/b & 0 \\ 0 & -1/(b\,\mathrm{t}\alpha) & 1/(c\,\mathrm{s}\alpha) \end{pmatrix} \tag{2.31c}$$

and

$$\mathbf{M}^{-1} = \begin{pmatrix} \alpha\,\mathrm{s}\gamma\,\mathrm{s}\beta^* & a\,\mathrm{c}\gamma & a\,\mathrm{s}\gamma\,\mathrm{c}\beta^* \\ 0 & b & 0 \\ 0 & c\,\mathrm{c}\alpha & c\,\mathrm{s}\alpha \end{pmatrix}$$

where sγ, cγ, tγ stand for sin γ, cos γ, and tan γ, etc. . . .

The family of all the possible transformations **M** which orthonormalize a given frame **A** according to $\mathbf{E} = \mathbf{MA}$ may be obtained from the following decomposition of the metric matrix **G** of **A**:

$$\mathbf{M}^{-1}(\bar{\mathbf{M}})^{-1} = \mathbf{G}.$$

Such a relation is obtained by requiring that in (2.21) the condition $\mathbf{G}' = \mathbf{I}$ is satisfied.

The volume of the unit cell defined by $\boldsymbol{a}$, $\boldsymbol{b}$, $\boldsymbol{c}$ may be easily obtained if **M** is known. Indeed, if we express $\boldsymbol{a}$, $\boldsymbol{b}$, $\boldsymbol{c}$ in terms of $\boldsymbol{e}_1$, $\boldsymbol{e}_2$, $\boldsymbol{e}_3$ according to (2.30), and if (2.5) is then used for calculating $\boldsymbol{a} \cdot \boldsymbol{b} \wedge \boldsymbol{c}$, then the result

$$V = \boldsymbol{a} \cdot \boldsymbol{b} \wedge \boldsymbol{c} = \det(\mathbf{M}^{-1})$$

is obtained. Furthermore, since $\det(\mathbf{M}^{-1}\bar{\mathbf{M}}^{-1}) = V^2 = G$, the assumption made on p. 62 according to which G is the square of the volume of the unit cell is also proved.

Rotations in Cartesian systems

In a right-handed Cartesian coordinate system $[0, \boldsymbol{e}_1, \boldsymbol{e}_2, \boldsymbol{e}_3]$ the anti-clockwise rotation of a vector $\boldsymbol{r}$ through angle α_1 about $\boldsymbol{e}_1$ or through α_2 about $\boldsymbol{e}_2$ or through α_3 about $\boldsymbol{e}_3$ produces a vector $\boldsymbol{r}' = \mathbf{R}_s\boldsymbol{r}$ where $s = x, y, z$

and

$$\mathbf{R}_x(\alpha_1) = \begin{pmatrix} 1 & 0 & 0 \\ 0 & \cos\alpha_1 & -\sin\alpha_1 \\ 0 & \sin\alpha_1 & \cos\alpha_1 \end{pmatrix}$$

$$\mathbf{R}_y(\alpha_2) = \begin{pmatrix} \cos\alpha_2 & 0 & \sin\alpha_2 \\ 0 & 1 & 0 \\ -\sin\alpha_2 & 0 & \cos\alpha_2 \end{pmatrix} \qquad (2.32a)$$

$$\mathbf{R}_z(\alpha_3) = \begin{pmatrix} \cos\alpha_3 & -\sin\alpha_3 & 0 \\ \sin\alpha_3 & \cos\alpha_3 & 0 \\ 0 & 0 & 1 \end{pmatrix}.$$

Since the matrices are orthogonal, the following relations hold:

$$\mathbf{R}(\alpha) = \bar{\mathbf{R}}^{-1}(\alpha) = \mathbf{R}^{-1}(-\alpha) = \bar{\mathbf{R}}(-\alpha).$$

Corresponding clockwise rotations are obtained by changing α_i into $-\alpha_i$. Matrices corresponding to rotatory–reflection operations about $\boldsymbol{e}_1$, $\boldsymbol{e}_2$, $\boldsymbol{e}_3$ are obtained by replacing in (2.32) the integer 1 by -1. Matrices corresponding to rotatory–inversion operations about $\boldsymbol{e}_1$, $\boldsymbol{e}_2$, $\boldsymbol{e}_3$ are obtained by changing the signs of all the elements. The traces of the matrices that represent proper rotations, rotatory–reflection and rotatory–inversion operations are $1 + 2\cos\alpha$, $2\cos\alpha - 1$, $-2\cos\alpha - 1$ respectively.

It should be noted that rotating $\boldsymbol{r}$ about $\boldsymbol{e}_1$, $\boldsymbol{e}_2$, $\boldsymbol{e}_3$ in anti-clockwise mode is equivalent to rotating $\boldsymbol{e}_1$, $\boldsymbol{e}_2$, $\boldsymbol{e}_3$ in clockwise mode. For example if the framework [O, $\boldsymbol{e}_1$, $\boldsymbol{e}_2$, $\boldsymbol{e}_3$] may be superimposed to the new framework [O, $\boldsymbol{e}'_1$, $\boldsymbol{e}'_2$, $\boldsymbol{e}'_3$] by a clockwise rotation through α_3 about $\boldsymbol{e}_3$, then $\mathbf{E}' = \mathbf{R}_z(\alpha_3)\mathbf{E}$. According to (2.20), $\mathbf{X}' = \bar{\mathbf{R}}_z^{-1}(\alpha_3)\mathbf{X} = \mathbf{R}_z(\alpha_3)\mathbf{X}$, which corresponds to an anti-clockwise rotation of a vector $\boldsymbol{r}$ in $\mathbf{E}$.

See now some useful applications:

1. Rotation about the unitary vector $\boldsymbol{l}$ in a rectilinear coordinate system.

Given the crystal base $\bar{\mathbf{A}} \equiv (\boldsymbol{a}, \boldsymbol{b}, \boldsymbol{c})$, an orthonormal base $\bar{\mathbf{E}} \equiv (\boldsymbol{e}_1, \boldsymbol{e}_2, \boldsymbol{e}_3)$ may be chosen ($\mathbf{E} = \mathbf{MA}$) such that $\boldsymbol{e}_1$ coincides with $\boldsymbol{l}$. In accordance with (2.32) a rotation about $\boldsymbol{l}$ in $\mathbf{E}$ is represented by $\boldsymbol{R}_x$, and, according to Table 2.E.1, the same rotation in $\mathbf{A}$ is represented by $\mathbf{R} = \bar{\mathbf{M}}\mathbf{R}_x(\bar{\mathbf{M}})^{-1}$. As an example we calculate in the hexagonal system the matrix corresponding to an anti-clockwise rotation through χ about $\boldsymbol{a}$. According to (2.30) and (2.31) the matrices $\mathbf{M}$ and $\mathbf{M}^{-1}$ are

$$\mathbf{M} = \begin{pmatrix} 1/a & 0 & 0 \\ 1/(a\sqrt{3}) & 2/(a\sqrt{3}) & 0 \\ 0 & 0 & 1/c \end{pmatrix}, \qquad \mathbf{M}^{-1} = \begin{pmatrix} a & 0 & 0 \\ -a/2 & a\sqrt{3}/2 & 0 \\ 0 & 0 & c \end{pmatrix}$$

from which

$$\mathbf{R} = \bar{\mathbf{M}}\mathbf{R}_x(\bar{\mathbf{M}})^{-1} = \begin{pmatrix} 1 & (\cos\chi - 1)/2 & -c\sin\chi/(a\sqrt{3}) \\ 0 & \cos\chi & -2c\sin\chi/(a\sqrt{3}) \\ 0 & a\sqrt{3}\sin\chi/(2c) & \cos\chi \end{pmatrix}.$$

Incidentally, it may be noted that the ratio a/c is unconstrained. Thus $\mathbf{R}$ (which is an integer matrix) may correspond to a symmetry axis along $\boldsymbol{a}$ in the hexagonal system only if $\chi = 0,\ \pi$ (onefold or twofold proper or improper axis).

The reader will easily find that the general rotation matrix about $\boldsymbol{b}$ in the monoclinic system is

$$\mathbf{R} = \begin{pmatrix} \cos\chi + \cot\beta/\sin\chi & 0 & c\sin\chi/(a\sin\beta) \\ 0 & 1 & 0 \\ -a\sin\chi/(c\sin\beta) & 0 & \cos\chi - \cot\beta\sin\chi \end{pmatrix}.$$

$\mathbf{R}$ will correspond to a symmetry operator if $\chi = 0,\ \pi$.

2. Explicit expression for the rotation matrix $\mathbf{R}$ about the unit vector $\boldsymbol{l}$ in an orthonormal system. Let l_1, l_2, l_3 be the direction cosines of $\boldsymbol{l}$ in the orthogonal system $\mathbf{A}$. A new orthonormal system $\bar{\mathbf{E}} \equiv (\boldsymbol{e}_1, \boldsymbol{e}_2, \boldsymbol{e}_3)$ is chosen so that $\boldsymbol{l}$ coincides with $\boldsymbol{e}_1$: then, according to (2.29),

$$\mathbf{M} = \begin{pmatrix} l_1 & l_2 & l_3 \\ m_{21} & m_{22} & m_{23} \\ m_{31} & m_{32} & m_{33} \end{pmatrix}$$

where m_{ij} are suitable parameters. The matrix corresponding to a rotation about $\boldsymbol{l}$ in $\mathbf{A}$ is then

$$\mathbf{R} = \bar{\mathbf{M}}\mathbf{R}_x(\bar{\mathbf{M}})^{-1} = \bar{\mathbf{M}}\mathbf{R}_x\bar{\mathbf{M}}$$

$$= \begin{pmatrix} c\chi + l_1^2(1 - c\chi) & l_3 s\chi + l_1 l_2(1 - c\chi) & -l_2 s\chi + l_1 l_3(1 - c\chi) \\ -l_3 s\chi + l_1 l_2(1 - c\chi) & c\chi + l_2^2(1 - c\chi) & l_1 s\chi + l_2 l_3(1 - c\chi) \\ l_2 s\chi + l_1 l_3(1 - c\chi) & -l_1 s\chi + l_2 l_3(1 - c\chi) & c\chi + l_3^2(1 - c\chi) \end{pmatrix}$$

$$= -c\chi\begin{pmatrix} 1 & 0 & 0 \\ 0 & 1 & 0 \\ 0 & 0 & 1 \end{pmatrix} - s\chi\begin{pmatrix} 0 & -l_3 & l_2 \\ l_3 & 0 & -l_1 \\ -l_2 & l_1 & 0 \end{pmatrix} + (1 - c\chi)\begin{pmatrix} l_1^2 & l_1 l_2 & l_1 l_3 \\ l_1 l_2 & l_2 & l_2 l_3 \\ l_1 l_3 & l_2 l_3 & l_3 \end{pmatrix} \tag{2.32b}$$

where $s\chi$ and $c\chi$ stand for $\sin\chi$ and $\cos\chi$.

3. Sequence of rotations about the same axis or about different axes.

The overall effect of a sequence of three successive rotations $\mathbf{R}_z(\omega)$, $\mathbf{R}_y(\psi)$, $\mathbf{R}_x(\varphi)$ on the vector $\boldsymbol{r}$ may be so described:

$$\mathbf{X}' = \mathbf{R}_z\mathbf{X}, \qquad \mathbf{X}'' = \mathbf{R}_y\mathbf{X}', \qquad \mathbf{X}''' = \mathbf{R}_x\mathbf{X}''$$

from which

$$\mathbf{X}''' = \mathbf{R}_x\mathbf{R}_y\mathbf{R}_z\mathbf{X} = \mathbf{R}\mathbf{X}$$

where

$$\mathbf{R} = \begin{pmatrix} c\psi c\omega & -c\psi s\omega & s\psi \\ c\varphi s\omega + s\varphi s\psi c\omega & c\varphi c\omega - s\varphi s\psi s\omega & -s\varphi c\psi \\ s\varphi s\omega - c\varphi s\psi c\omega & s\varphi c\omega + c\varphi s\psi s\omega & c\varphi c\psi \end{pmatrix}$$

and $c\psi$ stands for $\cos\psi$, $s\psi$ for $\sin\psi$, etc. . . .

If the three rotations are made in different order, the overall effect changes. As an example, the reader will easily verify that rotating $\boldsymbol{r}$ through 120° about $\boldsymbol{e}_3$ and then through 90° about $\boldsymbol{e}_1$ leaves $\boldsymbol{r}$ in $\boldsymbol{r}''$ where

$$\mathbf{X}'' = \mathbf{R}_x(90°)\mathbf{R}_z(120°)\mathbf{X} = \mathbf{R}\mathbf{X}$$

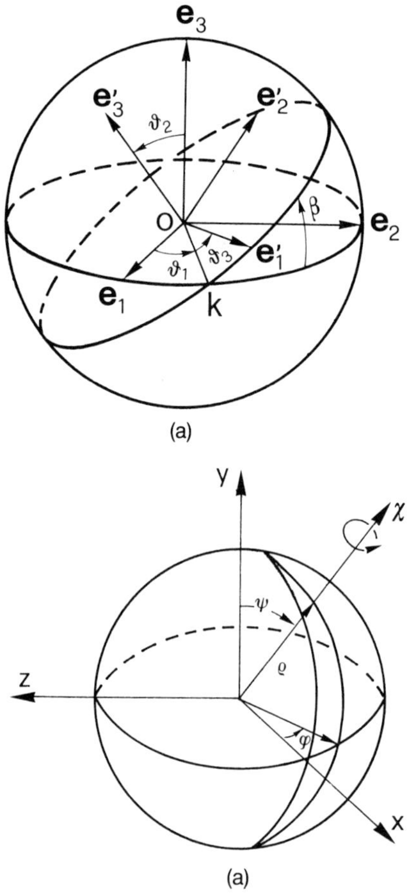

Fig. 2.3. (a) Eulerian angles. (b) Spherical polar coordinates.

and

$$\mathbf{R}=\begin{pmatrix}-1/2 & -\sqrt{3}/2 & 0\\ 0 & 0 & -1\\ \sqrt{3}/2 & -1/2 & 0\end{pmatrix}.$$

4. Eulerian angles. In Fig. 2.3(a) two orthonormal frameworks $[\mathrm{O}, \boldsymbol{e}_1, \boldsymbol{e}_2, \boldsymbol{e}_3]$ and $[\mathrm{O}, \boldsymbol{e}'_1, \boldsymbol{e}'_2, \boldsymbol{e}'_3]$ are shown. The axis OK, called the line of nodes, is the intersection of the $(\boldsymbol{e}_1, \boldsymbol{e}_2)$ and $(\boldsymbol{e}'_1, \boldsymbol{e}'_2)$ planes and is perpendicular both to $\boldsymbol{e}_3$ and $\boldsymbol{e}'_3$. $\mathbf{E}$ may be superimposed to $\mathbf{E}'$ by three anti-clockwise rotations in the following order: (a) rotate about $\boldsymbol{e}_3$ by the angle θ_1 (OK and $\boldsymbol{e}_1$ are now identical); (b) rotate through θ_2 about OK which will bring $\boldsymbol{e}_3$ into coincidence with $\boldsymbol{e}'_3$; (c) rotate about $\boldsymbol{e}'_3$ by θ_3 which then brings $\boldsymbol{e}_1$ to $\boldsymbol{e}'_1$ and $\boldsymbol{e}_2$ to $\boldsymbol{e}'_2$.

Suppose now that $\boldsymbol{r}$ is a vector with coordinates $\mathbf{X}$ in the space-fixed system and coordinates $\mathbf{X}'$ in a body-fixed system: then $\mathbf{X}$ and $\mathbf{X}'$ are connected by

$$\mathbf{X}'=\mathbf{R}_{\mathrm{Eu}}\mathbf{X}$$

where

$$\begin{aligned}\mathbf{R}_{\mathrm{Eu}}&=\mathbf{R}_z(\theta_3)\mathbf{R}_x(\theta_2)\mathbf{R}_z(\theta_1)\\ &=\begin{pmatrix}\mathrm{c}\theta_3 & \mathrm{s}\theta_3 & 0\\ -\mathrm{s}\theta_3 & \mathrm{c}\theta_3 & 0\\ 0 & 0 & 1\end{pmatrix}\begin{pmatrix}1 & 0 & 0\\ 0 & \mathrm{c}\theta_2 & \mathrm{s}\theta_2\\ 0 & -\mathrm{s}\theta_2 & \mathrm{c}\theta_2\end{pmatrix}\begin{pmatrix}\mathrm{c}\theta_1 & \mathrm{s}\theta_1 & 0\\ -\mathrm{s}\theta_1 & \mathrm{c}\theta_1 & 0\\ 0 & 0 & 1\end{pmatrix}\\ &=\begin{pmatrix}\mathrm{c}\theta_1\mathrm{c}\theta_3-\mathrm{s}\theta_1\mathrm{c}\theta_2\mathrm{s}\theta_3 & \mathrm{s}\theta_1\mathrm{c}\theta_3+\mathrm{c}\theta_1\mathrm{c}\theta_2\mathrm{s}\theta_3 & \mathrm{s}\theta_2\mathrm{s}\theta_3\\ -\mathrm{s}\theta_1\mathrm{c}\theta_2\mathrm{c}\theta_3-\mathrm{c}\theta_1\mathrm{s}\theta_3 & -\mathrm{s}\theta_1\mathrm{s}\theta_3+\mathrm{c}\theta_1\mathrm{c}\theta_2\mathrm{c}\theta_3 & \mathrm{s}\theta_2\mathrm{c}\theta_3\\ \mathrm{s}\theta_1\mathrm{s}\theta_2 & -\mathrm{c}\theta_1\mathrm{s}\theta_2 & \mathrm{c}\theta_2\end{pmatrix}\end{aligned}$$

where $\mathrm{c}\theta_i$ and $\mathrm{s}\theta_i$ stand for $\cos\theta_i$ and $\sin\theta_i$ respectively.

The Eulerian angles can also be used in order to calculate, in any crystallographic system $\mathbf{A}$, the rotation function corresponding to any desired rotation. The simplest procedure could be:

(a) Transform coordinates in $\mathbf{A}$ (say $\mathbf{X}_A$) into coordinates in $\mathbf{E}$ (say $\mathbf{X}_E$). If $\mathbf{E}=\mathbf{MA}$, then according to (2.20), $\mathbf{X}_E=(\bar{\mathbf{M}})^{-1}\mathbf{X}_A$. For example, $\mathbf{M}$ may be (2.31a) or (2.31b) or (2.31c).

(b) Transform the Cartesian coordinates into a rotated set of axes. Then $\mathbf{X}_E$ transforms in $\mathbf{X}'_E=\mathbf{R}_{\mathrm{Eu}}\mathbf{X}_E$.

(c) Return these coordinates into the system $\mathbf{A}$. Then the inverse operation described in (a) has to be made.

The final coordinates are $\mathbf{X}'_A=\bar{\mathbf{M}}\mathbf{X}'_E=\bar{\mathbf{M}}\mathbf{R}_{\mathrm{Eu}}(\bar{\mathbf{M}})^{-1}\mathbf{X}_A$ so that the desired rotation function is

$$\mathbf{R}_A=\bar{\mathbf{M}}\mathbf{R}_{\mathrm{Eu}}(\bar{\mathbf{M}})^{-1}.$$

Since

$$\begin{aligned}\mathbf{R}_{\mathrm{Eu}}(\theta_1, \theta_2, \theta_3)&=\mathbf{R}_{\mathrm{Eu}}(\theta_1+2n_1\pi, \theta_2+2n_2\pi, \theta_3+2n_3\pi)\\ &=\mathbf{R}_{\mathrm{Eu}}(\theta_1+\pi, -\theta_2, \theta_3+\pi)\end{aligned}$$

the full range of rotation operations is

$$0\leqslant\theta_1<\pi,\qquad 0\leqslant\theta_2<2\pi,\qquad 0\leqslant\theta_3<2\pi.$$

Since Eulerian angles are difficult to visualize it is preferred to specify a rotation in terms of spherical polar coordinates (see Fig. 2.3(b)). The rotation about a given axis is specified by the angle χ and the polar coordinates ψ and φ define the direction of the rotation axis. According to Fig. 2.3(b) the orthonormal coordinates x, y, z are given by

$$x = \rho \sin \psi \cos \varphi, \qquad y = \rho \cos \psi, \qquad z = -\rho \sin \psi \sin \varphi.$$

Since (2.32b) represents the anti-clockwise rotation matrix about the unit vector $\boldsymbol{l}$ in an orthonormal frame we can replace the direction cosines l_1, l_2, l_3 of the rotation axis $\boldsymbol{l}$ by

$$l_1 = \sin \psi \cos \varphi, \qquad l_2 = \cos \psi, \qquad l_3 = -\sin \psi \sin \varphi$$

and so obtain the expression of the rotation matrix $\mathbf{R}_{sp}$ in terms of the rotation angle χ and the spherical polar coordinates φ and ψ:

$$\mathbf{R}_{sp} = \begin{pmatrix} c\chi + (1-c\chi)s^2\psi c^2\varphi & -s\psi s\varphi s\chi + (1-c\chi)c\psi s\psi c\varphi & -c\psi s\chi - (1-c\chi)s^2\psi c\varphi s\varphi \\ s\psi s\varphi s\chi + (1-c\chi)c\psi s\psi c\varphi & c\chi + (1-c\chi)c^2\psi & s\psi c\varphi s\chi - (1-c\chi)c\psi s\psi s\varphi \\ c\psi s\chi - (1-c\chi)s^2\psi c\varphi s\varphi & -s\psi c\varphi s\chi - (1-c\chi)c\psi s\psi s\varphi & c\chi + (1-c\chi)s^2\psi s^2\varphi \end{pmatrix}.$$

Since

$$\begin{aligned} \mathbf{R}_{sp}(\chi, \psi, \varphi) &\equiv \mathbf{R}_{sp}(\chi + 2n_1\pi, \psi + 2n_2\pi, \varphi + 2n_3\pi) \\ &= \mathbf{R}_{sp}(\chi, 2\pi - \psi, \varphi + \pi) = \mathbf{R}_{sp}(-\chi, \pi - \psi, \pi + \varphi) \end{aligned}$$

all rotation operations are included in

$$0 \leq x < 2\pi, \qquad 0 \leq \psi \leq \pi, \qquad 0 \leq \varphi < \pi.$$

Some simple crystallographic calculations

The reader is referred to Appendix 2.B for calculations concerning crystallographic directions and planes. Here we limit ourselves to the description of some calculations which occur very frequently in crystal structure analysis.

Torsion angles

For a sequence of four atoms A, B, C, D, the torsion angle ω(ABCD) is defined as the angle between the normals to the planes ABC and BCD (see Fig. 2.4). By convention[2] ω is positive if the sense of rotation from BA to CD, viewed down BC, is clockwise, otherwise it is negative. Note that ω(ABCD) and ω(DCBA) have the same sign; furthermore, the sign of a torsion angle does not change by rotation or translation, and is reversed by reflection or inversion. According to the definition (see again Fig. 2.4)

$$\cos \omega = \frac{(\boldsymbol{a} \wedge \boldsymbol{b}) \cdot (\boldsymbol{b} \wedge \boldsymbol{c})}{ab^2c \sin \alpha \sin \gamma}, \qquad \frac{\boldsymbol{b}}{b} \sin \omega = \frac{(\boldsymbol{a} \wedge \boldsymbol{b}) \wedge (\boldsymbol{b} \wedge \boldsymbol{c})}{ab^2c \sin \alpha \sin \gamma}$$

which, owing to (2.7) and (2.8), become

$$\cos \omega = \frac{\cos \alpha \cos \gamma - \sin \beta}{\sin \alpha \sin \beta}, \qquad \sin \omega = \frac{Vb}{ab^2c \sin \alpha \sin \gamma}.$$

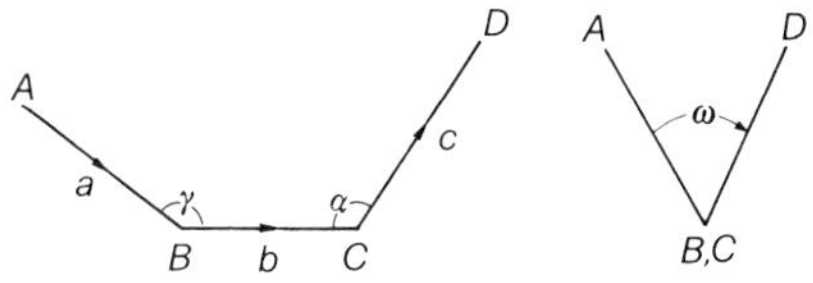

Fig. 2.4. Definition of the torsion angle ω.

If the vectors $\boldsymbol{a}^*$, $\boldsymbol{b}^*$, $\boldsymbol{c}^*$ reciprocal to $\boldsymbol{a}$, $\boldsymbol{b}$, $\boldsymbol{c}$ are considered, it may easily be seen from Table 2.1 that ω coincides with β^*.

Best plane through a set of points

Consider a set of p atoms at positions $\boldsymbol{r}_1, \boldsymbol{r}_2, \ldots, \boldsymbol{r}_p$ where $\boldsymbol{r}_j = \bar{\mathbf{A}}\mathbf{X}_j$. The best plane through them is that for which the sum of the squares of the distances of the atoms from the plane, multiplied by the weights of the atomic positions, is minimum. Such a plane is characterized by the minimum of the function (see eqns (2.B.5) and (2.B.7))

$$Q = \sum_j w_j(\bar{\mathbf{N}}^*\mathbf{X}_j - d)^2, \qquad (2.33)$$

where d is the distance of the plane from the origin of the coordinate system, $\boldsymbol{n} = n_1\boldsymbol{a} + n_2\boldsymbol{b} + n_3\boldsymbol{c} = \bar{\mathbf{A}}\mathbf{N} = n_1^*\boldsymbol{a}^* + n_2^*\boldsymbol{b}^* + n_3^*\boldsymbol{c}^* = \bar{\mathbf{A}}^*\mathbf{N}^*$ is the normal to the plane. The weights w_j should be taken as being inversely proportional to the variances of the atomic positions in the direction normal to the desired plane, but they are often assumed to be unitary.

If the atoms are considered as point masses of weight w_j, the least squares plane coincides with the principal plane of least inertia.

The minimum of Q will be searched with respect to d and n_1^*, n_2^*, n_3^* under the condition that $\boldsymbol{n}$ is a unit vector. This kind of problem is best solved by the method of Lagrange multipliers. The function to minimize is then[3]

$$Q' = \sum_j w_j(\bar{\mathbf{N}}^*\mathbf{X}_j - d)^2 - \lambda(\bar{\mathbf{N}}^*\mathbf{G}^*\mathbf{N}^* - 1). \qquad (2.34)$$

The partial derivative of (2.34) with respect to d gives

$$\sum_j w_j(\bar{\mathbf{N}}^*\mathbf{X}_j - d) = 0$$

from which

$$d = \bar{\mathbf{N}}^*\left[\left(\sum_j w_j\mathbf{X}_j\right)\left(\sum_j w_j\right)^{-1}\right] = \bar{\mathbf{N}}^*\mathbf{X}_0. \qquad (2.35)$$

Equation (2.35) states that the plane passes through the centroid $\boldsymbol{r}_0 = \bar{\mathbf{A}}\mathbf{X}_0$. Owing to (2.35), eqn (2.34) becomes

$$\begin{aligned} Q' &= \sum_j w_j(\bar{\mathbf{N}}^*\mathbf{X}_j')^2 - \lambda(\bar{\mathbf{N}}^*\mathbf{G}^*\mathbf{N}^* - 1) \\ &= \bar{\mathbf{N}}^*\mathbf{S}\mathbf{N}^* - \lambda(\bar{\mathbf{N}}^*\mathbf{G}^*\mathbf{N}^* - 1) \end{aligned} \qquad (2.36)$$

where $\boldsymbol{r}_j' = \boldsymbol{r}_j - \boldsymbol{r}_0$ and

$$\mathbf{S} = \begin{pmatrix} \sum_j w_j x_j'^2 & \sum_j w_j x_j' y_j' & \sum_j w_j x_j' z_j' \\ \sum_j w_j x_j' y_j' & \sum_j w_j y_j'^2 & \sum_j w_j y_j' z_j' \\ \sum_j w_j x_j' z_j' & \sum_j w_j y_j' z_j' & \sum_j w_j z_j'^2 \end{pmatrix}.$$

Note that $\bar{\mathbf{N}}^*\mathbf{S}\mathbf{N}^*$ is the weighted sum of the squares of the distances of the atoms from the plane. Setting to zero the derivative of (2.36) with respect to

$\mathbf{N}^*$ (in practice with respect to the components n_1^*, n_2^*, n_3^*) gives $\mathbf{SN}^* - \lambda\mathbf{G}^*\mathbf{N}^* = 0$, which may be also written as

$$(\mathbf{A} - \lambda\mathbf{I})\mathbf{N} = 0 \tag{2.37}$$

where $\mathbf{A} = \mathbf{SG}$ and $\mathbf{N} = \mathbf{G}^*\mathbf{N}^*$ (see Table 2.E.1). Writing (2.37) as $\mathbf{AN} = \lambda\mathbf{N}$ and multiplying both sides for $\mathbf{N}^*$ gives

$$\bar{\mathbf{N}}^*\mathbf{AN} = \bar{\mathbf{N}}^*\mathbf{SGN} = \bar{\mathbf{N}}^*\mathbf{SN}^* = \lambda\bar{\mathbf{N}}^*\mathbf{N} = \lambda.$$

The eigenvalue λ is therefore the weighted sum of the squares of the distances of the atoms from the plane, $\mathbf{N}$ is the corresponding eigenvector. There are three solutions of eqn (2.37), which in general correspond to three different λ values, say $\lambda_\alpha \geqslant \lambda_\beta \geqslant \lambda_\gamma$. Each eigenvalue of λ gives a stationary value of Q': the three corresponding eigenvectors $\mathbf{N}_\alpha$, $\mathbf{N}_\beta$, $\mathbf{N}_\gamma$ define the principal axis of inertia of the system of atoms when considered as points of masses $w_1, \ldots, w_p$. The best plane corresponds to the eigenvalue λ_γ and coincides with the plane passing through $\boldsymbol{r}_0$ and normal to $\mathbf{N}_\gamma$ (see eqn (2.B.6)), while λ_α and $\mathbf{N}_\alpha$ define the worst plane. Furthermore, $\bar{\mathbf{N}}_\gamma^*\mathbf{X}_i'$ is the distance of the ith atom from the best plane and $\bar{\mathbf{N}}_\alpha^*\mathbf{X}_i'$ and $\bar{\mathbf{N}}_\beta^*\mathbf{X}_i'$ are useful to prepare a diagram of the projection of the system of atoms on to the best plane.

The search of the best plane is remarkably simplified if Cartesian systems are used (then $\mathbf{G} = \mathbf{G}^* = \mathbf{I}$).

Best line through a set of points

This problem is strictly connected to results described in the above subsection. It may be shown that the least squares line passes through the centroid $\boldsymbol{r}_0$ of the atoms and is normal to the plane corresponding to the eigenvalue λ_1.

Principal axes of a quadratic form

Let

$$q = \bar{\mathbf{X}}\mathbf{QX} = q_{11}x^2 + 2q_{12}xy + \ldots + q_{33}z^2$$

be the quadratic form. Finding its principal axes is equivalent to finding the directions $\boldsymbol{n}$ in which q is stationary. As in calculating the best plane through a set of points, the problem may be solved via the Lagrange multipliers by minimizing

$$q' = \bar{\mathbf{N}}\mathbf{QN} - \lambda(\bar{\mathbf{N}}\mathbf{GN} - 1).$$

The derivative of q' with respect to $\mathbf{N}$ brings to

$$\mathbf{QN} - \lambda\mathbf{GN} = 0$$

from which

$$(\mathbf{A} - \lambda\mathbf{I})\mathbf{N}^* = 0 \tag{2.38a}$$

where $\mathbf{A} = \mathbf{QG}^*$, and $\mathbf{N}^* = \mathbf{GN}$ is the general eigenvector the components of which are referred to the reciprocal axis. The eigenvalue λ gives the value of q in the $\boldsymbol{n}$ direction. Indeed, if (2.38) is written as $\mathbf{AN}^* = \lambda\mathbf{N}^*$ and both sides are multiplied by $\bar{\mathbf{N}}$ we obtain

$$\bar{\mathbf{N}}\mathbf{A}\mathbf{N}^* = \bar{\mathbf{N}}\mathbf{Q}\mathbf{G}^*\mathbf{N}^* = \bar{\mathbf{N}}\mathbf{Q}\mathbf{N} = \bar{\mathbf{N}}\lambda\mathbf{N}^* = \lambda.$$

Substituting the three eigenvalues λ_1, λ_2, λ_3 into (2.38a) provides the three eigenvectors $\mathbf{N}^*_\alpha$, $\mathbf{N}^*_\beta$, $\mathbf{N}^*_\gamma$ which represent the principal axes of q.

If the quadratic form is referred to the reciprocal basis (i.e. $q^* = \bar{\mathbf{H}}\boldsymbol{\beta}\mathbf{H} = \beta_{11}h_1^2 + 2\beta_{12}hk + \ldots + \beta_{33}l^2$) the problem may be solved in the same way on condition that $\mathbf{G}^*$ and $\mathbf{N}^*$ replace $\mathbf{G}$ and $\mathbf{N}$ respectively. As an example let us determine the principal axes of an atomic thermal ellipsoid for which $\beta_{11} = 0.00906$, $\beta_{12} = -0.00049$, $\beta_{13} = -0.00102$, $\beta_{22} = 0.00401$, $\beta_{23} = 0.00038$, $\beta_{33} = 0.01424$. Let the orthohombic unit cell parameters be $a = 8.475$, $b = 10.742$, $c = 5.899$Å. The function to minimize is

$$\bar{\mathbf{N}}^*\boldsymbol{\beta}\mathbf{N}^* - \lambda(\bar{\mathbf{N}}^*\mathbf{G}^*\mathbf{N}^* - 1)$$

which, derived with respect to $\mathbf{N}^*$, leads to the condition

$$\boldsymbol{\beta}\mathbf{N}^* - \lambda\mathbf{G}^*\mathbf{N}^* = 0.$$

Since $\mathbf{N} = \mathbf{G}^*\mathbf{N}^*$, that condition may be written

$$(\boldsymbol{\beta}\mathbf{G} - \lambda\mathbf{I})\mathbf{N} = 0.$$

The corresponding secular equation is

$$\left(\begin{pmatrix} 0.00906 & -0.00049 & -0.00102 \\ -0.00049 & 0.00401 & 0.00038 \\ -0.00102 & 0.00038 & 0.01424 \end{pmatrix}\begin{pmatrix} 71.826 & 0 & 0 \\ 0 & 115.391 & 0 \\ 0 & 0 & 34.798 \end{pmatrix} - \lambda\begin{pmatrix} 1 & 0 & 0 \\ 0 & 1 & 0 \\ 0 & 0 & 1 \end{pmatrix}\right) = 0.$$

Expansion of the determinant produces the cubic equation

$$0.1468 - 0.8528\lambda + 1.6089\lambda^2 - \lambda^3 = 0$$

which has solutions $\lambda_\alpha = 0.483$, $\lambda_\beta = 0.448$, $\lambda_\gamma = 0.677$. Using the first root gives

$$\begin{pmatrix} 0.1677 & -0.0565 & -0.0354 \\ -0.0352 & -0.0203 & 0.0132 \\ -0.0733 & 0.0438 & 0.0125 \end{pmatrix}\begin{pmatrix} n_{1\alpha} \\ n_{2\alpha} \\ n_{3\alpha} \end{pmatrix} = \begin{pmatrix} 0 \\ 0 \\ 0 \end{pmatrix}.$$

Since the three equations are linearly dependent $n_{1\alpha}$ and $n_{2\alpha}$ can be found in terms of $n_{3\alpha}$: $n_{1\alpha} = +0.2592n_{3\alpha}$, $n_{2\alpha} = 0.185n_{3\alpha}$. The eigenvector $\mathbf{N}_\alpha$ will have unitary modulus (remember that $n_{1\alpha}$, $n_{2\alpha}$, $n_{3\alpha}$ are the components of $\mathbf{N}_\alpha$ in $\mathbf{A}$) if $n_{3\alpha} = 0.1515$. Therefore $\bar{\mathbf{N}}_\alpha = -[0.0393, 0.0280, 0.1515]$. In an analogous way $\bar{\mathbf{N}}_\beta = [-0.0139, -0.0866, 0.0587]$ and $\bar{\mathbf{N}}_\gamma = [-0.1097, 0.0210, 0.0492]$. Since

$$\bar{\mathbf{N}}^*\boldsymbol{\beta}\mathbf{N}^* = \bar{\mathbf{N}}^*\boldsymbol{\beta}\mathbf{G}\mathbf{N} = \bar{\mathbf{N}}^*\lambda\mathbf{N} = \lambda\bar{\mathbf{N}}^*\mathbf{N} = \lambda$$

each eigenvalue λ fixes the value of q^* along the corresponding eigenvector. If this conclusion is referred to the tensor $\mathbf{U}^* = \boldsymbol{\beta}/(2\pi^2)$, (see eqn (3.B.6) the following relations follow:

$$\langle u^2\rangle_1 = \frac{\lambda_1}{2\pi^2}, \qquad \langle u^2\rangle_2 = \frac{\lambda_2}{2\pi^2}, \qquad \langle u^2\rangle_3 = \frac{\lambda_3}{2\pi^2}, \tag{2.38b}$$

where $\langle u^2\rangle_i$ is the mean-square displacement along the ith principal axis. The reader will easily find, for the example above, that the root-mean-square displacements along the principal axes are 0.156 Å, 0.151 Å and 0.185 Å respectively.

It would be worthwhile remembering two basic properties of the eigenvectors:

(1) eigenvectors corresponding to different eigenvalues are orthogonal to each other;

(2) the matrix $\mathbf{M} = (\mathbf{GV})^{-1}$, where $\mathbf{V}$ is the eigenvector matrix

$$\mathbf{V} = \begin{pmatrix} n_{1\alpha} & n_{1\beta} & n_{1\gamma} \\ n_{2\alpha} & n_{2\beta} & n_{2\gamma} \\ n_{3\alpha} & n_{3\beta} & n_{3\gamma} \end{pmatrix},$$

transforms the basis $\mathbf{A}$ into a Cartesian coordinate system $\mathbf{A}'$ in which the axes are the eigenvectors of $\boldsymbol{\beta}$. Indeed, according to (2.E.8), $\boldsymbol{\beta}$ transforms into $\boldsymbol{\beta}' = \bar{\mathbf{V}}\mathbf{G}\boldsymbol{\beta}\mathbf{G}\mathbf{V}$. Because of (2.38b)

$$\boldsymbol{\beta}\mathbf{GV} = \mathbf{V}\begin{pmatrix} \lambda_1 & 0 & 0 \\ 0 & \lambda_2 & 0 \\ 0 & 0 & \lambda_3 \end{pmatrix}$$

so that

$$\boldsymbol{\beta} = \bar{\mathbf{V}}\mathbf{GV}\begin{pmatrix} \lambda_1 & 0 & 0 \\ 0 & \lambda_2 & 0 \\ 0 & 0 & \lambda_3 \end{pmatrix} = \begin{pmatrix} \lambda_1 & 0 & 0 \\ 0 & \lambda_2 & 0 \\ 0 & 0 & \lambda_3 \end{pmatrix}.$$

Metric considerations on the lattices

The results obtained so far can be used to characterize lattices and their properties.

Niggli reduced cell

A unit cell defines the lattice: conversely any lattice may be described by means of several types of cell. However, a special cell exists, called the Niggli reduced cell, which uniquely describes the lattice:[4] it is primitive and is built on the shortest three non-coplanar lattice translations (the Delaunay[5] reduction procedure is also a suitable tool for the identification of a crystal lattice: readers will find careful description of it in some recent papers[6–8]).

A unit cell characterized by the three shortest non-coplanar translations is called a Buerger cell.[9,10] Several algorithms can be used to obtain it. The easiest is: lattice vectors of magnitudes $r_{u,v,w}$ are calculated by (2.3) where u, v, w vary over the smallest integers set (usually between 3 and -3). The smallest three non-coplanar translations will be the Buerger cell edges. This cell is, however, not unique: if it is, then it coincides with the Niggli cell. For 7 of the 14 Bravais lattices a unique Buerger cell exists,[11] while in a face centred cubic lattice (see later) two Buerger cells can be found. In other lattice types up to five types of Buerger cell can be found (the values 4 and 5 occur only in triclinic lattices) according to whether some conditions on the parameters of the conventional cell are satisfied or not. For example, the triclinic lattice described by a Buerger cell with

$$a = 2\ \text{Å}, \quad b = 4\ \text{Å}, \quad c = 4\ \text{Å}, \quad \alpha = 60°, \quad \beta = 79°12', \quad \gamma = 75°31'$$

may be described by means of four other Buerger cells having the same a, b, c values, but with

$$\begin{aligned} &\alpha = 60°00', && \beta = 86°24', && \gamma = 75°31'; \\ &\alpha = 120°00', && \beta = 93°36', && \gamma = 100°48'; \\ &\alpha = 117°57', && \beta = 93°36', && \gamma = 104°28'; \\ &\alpha = 113°58', && \beta = 100°48', && \gamma = 104°28'. \end{aligned}$$

It will be shown later that only the first of the five cells is the Niggli cell.

If g_{ij} are the elements of the metric matrix, the Niggli cell is defined by the following conditions:[12,13]

1. Positive reduced cell (all the angles <90°). Main conditions:

$$g_{11} \leqslant g_{22} \leqslant g_{33}; \qquad g_{23} \leqslant 1/2 g_{22}; \qquad g_{13} \leqslant 1/2 g_{11}; \qquad g_{12} \leqslant 1/2 g_{11}.$$

Special conditions:

(a) if $g_{11} = g_{22}$ then $g_{23} \leqslant g_{13}$; if $g_{22} = g_{33}$ then $g_{13} \leqslant g_{12}$;

(b) if $g_{23} = 1/2 g_{22}$ then $g_{12} \leqslant 2 g_{13}$; if $g_{13} = 1/2 g_{11}$ then $g_{12} \leqslant 2 g_{23}$; if $g_{12} = 1/2 g_{11}$ then $g_{13} \leqslant 2 g_{23}$.

2. Negative reduced cell (all the angles ⩾90°). Main conditions:

$$g_{11} \leqslant g_{22} \leqslant g_{33}; \qquad |g_{23}| \leqslant 1/2 g_{22}; \qquad |g_{13}| \leqslant 1/2 g_{11}; \qquad |g_{12}| \leqslant 1/2 g_{11};$$
$$(|g_{23}| + |g_{13}| + |g_{12}|) \leqslant 1/2 (g_{11} + g_{22}).$$

Special conditions:

(a) if $g_{11} = g_{22}$ then $|g_{23}| \leqslant |g_{13}|$; if $g_{22} = g_{33}$ then $|g_{13}| \leqslant |g_{12}|$

(b) if $|g_{23}| = 1/2 g_{22}$ then $g_{12} = 0$; if $|g_{13}| = 1/2 g_{11}$ then $g_{12} = 0$; if $|g_{12}| = 1/2 g_{11}$ then $g_{13} = 0$; if $(|g_{23}| + |g_{13}| + |g_{12}|) = 1/2 (g_{11} + g_{22})$ then $g_{11} \leqslant 2\,|g_{13}| + |g_{12}|$.

The main conditions define a cell based on the three shortest non-coplanar vectors. Conditions (a) break down ambiguity when two cell edges are equal, conditions (b) define the Niggli cell when there is more than one symmetrically independent Buerger cell.

As an example of systematic ambiguity let us consider the face-centred cubic lattice with cubic edge a. If we move to the primitive unit cell by means of the appropriate matrix quoted in Table 2.C.1 we get

$$g'_{11} = g'_{22} = g'_{33} = a^2/2; \qquad g'_{12} = g'_{13} = g'_{23} = a^2/4.$$

If we move to the primitive cell by means of the transformation matrix

$\|1/2\ 1/2\ 0\|\ \overline{1/2}\ 1/2\ 0\|\ 0\ \overline{1/2}\ 1/2\|$ then we get

$$g''_{11} = g''_{22} = g''_{33} = a^2/2; \qquad g''_{12} = 0; \qquad g''_{13} = g''_{23} = -a^2/4.$$

Both the primitive cells are Buerger cells but the second violates the conditions (a): thus the first is the Niggli reduced cell.

Matrices which derive Niggli cells from Buerger cells are given by Santoro and Mighell.[12] A very efficient algorithm to derive the Niggli cell from any primitive cell is described by Krivy and Gruber.[13]

For any Bravais lattice Niggli[4] defined the algebraic relations that the g_{ij}s of the reduced cell must satisfy. The type of Bravais lattice may be thus derived from the Niggli cell just by comparing the found with the expected relations. For example, in a face-centred cubic lattice the g_{ij} of the Niggli cell must satisfy $g_{11} = g_{22} = g_{33}$, $g_{12} = g_{13} = g_{23} = g_{11}/2$.

The use of automatic procedures devoted to identify the Niggli cell may yield incorrect conclusions as a consequence of errors in the cell parameters or of rounding errors in the calculations. Some auxiliary procedures recently suggested by different authors[14–16] are less sensitive to these error sources.

The final steps from the Niggli cell to the conventional cell may be performed by means of suitable transformation matrices.[17] It would be worthwhile recalling that the lattice symmetry determined via the Niggli cell is only of metric nature, and that may be equal to or larger than the symmetry of the crystal structure.

Reduced cells may be used:

1. As a useful step for the correct definition of the space group (see also Chapter 3). An advisable sequence may the following:[18] from the conventional cell to a primitive cell, and then to the Niggli cell; analysis of the latttice symmetry, analysis of Laue symmetry and of systematic extinctions; space group choice.

2. As an effective tool for the identification and characterization of crystalline materials[19] (as an alternative to powder methods in which the identification is based on matching diffraction positions and intensities). An advisable sequence may be: a unit cell is determined, the reduced cell is derived together with derivative supercells and subcells (derivative cells are calculated to overcome possible errors made by the experimentalist). These cells are checked against a suitable file containing as complete as possible a file containing crystallographic data (the *NBS Crystal Data File* handles data of more than 60 000 materials).

It could be asked now if Niggli cell expresses some geometrical property. Gruber[20] has shown that a cell is a Niggli cell if and only if the following conditions are fulfilled:

(1) $a + b + c$ is a minimum when calculated for all primitive cells of the lattice;

(2) $$\left.\begin{array}{l} |\pi/2 - \alpha| + |\pi/2 - \beta| + |\pi/2 - \gamma| \\ |\cos\alpha| + |\cos\beta| + |\cos\gamma| \\ |\cos\alpha\cos\beta\cos\gamma| \end{array}\right\} = \begin{array}{l}\text{max. for the cells}\\ \text{defined in (1).}\end{array}$$

As an example let us consider the triclinic lattice defined by the primitive cell

$$a = 9.562\ \text{Å}, \quad b = 12.487\ \text{Å}, \quad c = 8.070\ \text{Å},$$
$$\alpha = 95.77°, \quad \beta = 110.03°, \quad \gamma = 110.79°.$$

The Niggli cell is obtained from the previous one by application of the matrix 001/100/111:

$$a = 8.070\ \text{Å}, \quad b = 9.562\ \text{Å}, \quad c = 12.434\ \text{Å},$$
$$\alpha = 100.97°, \quad \beta = 106.54°, \quad \gamma = 110.03°.$$

This cell satisfies the geometrical properties suggested by Gruber.

Sublattices and superlattices

The term superlattice is commonly used for a structure closely related to a parent structure and having cell dimensions superior to that of the parent structure. The superlattice then belongs to a subgroup of the parent structure: indeed increasing cell dimensions causes a loss of translational symmetry. Let

$$\mathbf{A} = \begin{pmatrix} \boldsymbol{a} \\ \boldsymbol{b} \\ \boldsymbol{c} \end{pmatrix} \quad \text{and} \quad \mathbf{A}' = \begin{pmatrix} \boldsymbol{a}' \\ \boldsymbol{b}' \\ \boldsymbol{c}' \end{pmatrix}$$

be two triplets of non-coplanar vectors defining two primitive cells in two different lattices L and L′ respectively. The cell in L′ is related to the cell in L by

$$\mathbf{A}' = \mathbf{MA}. \tag{2.39}$$

Consider four important cases:

1. The m_{ij} elements are integers and $M = 1$. Then L and L′ will coincide.
2. The m_{ij}s are integers and $M > 1$. Then $\mathbf{A}'$ defines a new lattice L′ called a superlattice of L whose elementary cell is M times larger than the primitive cell in L.
3. $\mathbf{M} = \mathbf{Q}^{-1}$ where $\mathbf{Q}$ is a matrix with integer elements for which $Q > 1$. In this case $\mathbf{A} = \mathbf{QA}'$ and L is a superlattice of L′, or, with equivalent terminology, L′ is a sublattice of L. It should be noted that if L′ is a sublattice of L then L′* is a superlattice of L*.
4. $\mathbf{M}$ is a rational matrix. This case is described in the following subsection.

Super- and sublattices are frequently related to important properties of the crystals. For example,

(1) twinning by merohedry takes place only if a superlattice exists with symmetry higher than that of the crystal lattice

(2) crystallographic phase transitions often take place between the structures for which one lattice is a superlattice of the other. The knowledge of the possible superlattices of a given lattice limits the set of possible structures of a new phase. That is of particular usefulness when the phases are simultaneously present in a polycrystalline sample and

Table 2.3.

	$M = 2$	
200/010/001	100/020/001	100/010/002
200/110/001	200/010/101	100/011/002
110/011/101		
	$M = 3$	
300/010/001	100/030/001	100/010/003
$1\bar{1}0/210/001$	$110/\bar{2}10/001$	$\bar{1}01/201/010$
$101/20\bar{1}/010$	$01\bar{1}/021/100$	$011/0\bar{2}1/100$
211/110/021	$121/\bar{1}\bar{1}0/201$	112/101/210
111/120/021		

diffraction peaks having identical positions are generated. In particular, if an order–disorder transformation occurs, the ordered phase is characterized by a cell larger than that of the disordered phase;

(3) magnetic structures are based on cells often larger than those of the corresponding conventional chemical structure (see Figs. 1.F.2 and 1.F.3).

For any value of M there is a finite number of matrices **M** that produce distinct superlattices.[21] This number quickly gets large with M: in Table 2.3 the unique matrices **M** are given generating superlattices for $M = 2$, 3. The unique matrices generating sublattices for $M = 1/2$, $1/3$ are obtained by applying the same matrices to the reciprocal lattice and then by calculating the lattices that are reciprocal to the resulting superlattices.

Coincidence-site lattices

Most materials of technological interest are used in their polycrystalline form. Their mechanical and chemical properties are controlled to a large extent by the boundary between crystallites. The energy of a polycrystal is higher than that of a single crystal with the same mass: the additional energy is stored in the grain boundary areas, and depends on the orientations of the neighbouring grains. Thus modern treatment of these materials tend to optimize the size of the grains and quality of the grain boundaries.

The mathematical model of the crystalline interfaces is today based on the properties of coincidence-site lattice (CSL) and related lattices. Consider two lattices L and L′ with bases **A** and **A**′. Without loss of generality it will be assumed that the two lattices have one lattice point in common, taken as the origin of the coordinate systems. Let **N** and **N**′ be two matrices with integer elements. The lattices L and L′ will have a common superlattice if a lattice point of L (defined by **NA**) can be found which is also a lattice point of L′ (defined by **N′A′**): then $\mathbf{NA} = \mathbf{N'A'}$, or also

$$\mathbf{A'} = \mathbf{X_c A}$$

where $\mathbf{X}_c = \mathbf{N'}^{-1}\mathbf{N}$ is a matrix with rational elements. In this case the CSL is defined as that superlattice (at 1 or 2 or 3 dimensions) of L and L′ which contains all (and only) the lattice points in common to L and L′. Note that several other lattices could be defined having points in common with L and L′ but all of them will be superlattices of the CSL.

To determine the CSL one has to find[22,23] a factorization of $\mathbf{X}_c$ of the type $\mathbf{X}_c = \mathbf{N}'^{-1}\mathbf{N}$ with the smallest possible values of N and N'. If $\mathbf{N}_0$ and $\mathbf{N}'_0$ satisfy this condition then N_0 and N'_0 indicate the reciprocal fraction of coincidence points (**degree of coincidence**) in lattices L and L′ respectively, and the CSL basis will be $\mathbf{N}_0\mathbf{A} = \mathbf{N}'_0\mathbf{A}'$. If N or N' are sufficiently small (a large fraction of points of one of the two lattices consists of coincidence sites) and if the boundary coincides with a dense net plane of CSL, then the boundary energy per unit area will be a minimum.

Analogously, two lattices L and L′ will have a common sublattice if two matrices $\mathbf{N}$ and $\mathbf{N}'$ (with integer elements) can be found such that $\mathbf{N}^{-1}\mathbf{A} = \mathbf{N}'^{-1}\mathbf{A}'$ or also

$$\mathbf{A}' = \mathbf{X}_d\mathbf{A},$$

where $\mathbf{X}_d = \mathbf{N}'\mathbf{N}^{-1}$ is a matrix with rational elements. In this case the **displacement-shift-complete lattice** (DSC) is defined as the sublattice with the largest volume of the primitive cell. All lattices which are sublattices of both L and L′ will be sublattices of DSC.

The DSC may be determined by means of a factorization similar to that used for $\mathbf{X}_c$: again we will look for the smallest possible values of N and N'. If $\mathbf{N}_0$ and $\mathbf{N}'_0$ satisfy such a factorization process then the DSC basis will be $\mathbf{N}_0^{-1}\mathbf{A} = \mathbf{N}_0'^{-1}\mathbf{A}'$, which defines a cell with volume $1/N_0$ times the volumes of the cell defined by $\mathbf{A}$ and $1/N'_0$ times the volume defined by $\mathbf{A}'$.

It may be also shown that:

1. The CSL (DSC) of the reciprocal lattices is the reciprocal lattice of the DSC (CSL) of the two lattices.[24]

2. The coarsest lattice which contains all vectors of the form $\boldsymbol{u} + \boldsymbol{u}'$, where $\boldsymbol{u}$ and $\boldsymbol{u}'$ are vectors of L and L′ respectively, is the DSC lattice.[25] With respect to the energy of the grain boundaries DSC lattices have the same importance as CSL: indeed translations by DSC vectors do not destroy the coincidence sites. Such vectors are the geometrically possible Burgers vectors (energy considerations will dictate the most probable of them) of dislocations in grain boundaries.

If the two lattices L and L′ are congruent then one can be transformed into the other by means of a rotation: this is called **coincidence rotation** if L and L′ have a CSL in common. The ratio between the volume of the CSL unit cell and the volume V of the crystal unit cell is called the multiplicity of the CSL and is denoted by Σ (the analogous ratio for the DSC cell will be $1/\Sigma$). The determination of all possible coincidence orientations with low values of Σ is an important premise for the understanding of grain boundaries. If $\mathbf{A}' = \mathbf{X}\mathbf{A}$ defines one of the required orientations, then, owing to (2.21),

$$\mathbf{G}' = \mathbf{X}\mathbf{G}\bar{\mathbf{X}}. \tag{2.40}$$

Several attempts have been made to find the general solution of (2.40). Special methods for the solution of this problem were developed for cubic,[26,27] hexagonal,[28,29] and rhombohedral[30] lattices. The problem may be so stated: determine all the rotation angles θ about a given lattice axis $[uvw]$ which generate CSLs. It may be shown that in cubic lattices a CSL is obtained by a rotation θ about an axis $[uvw]$ coincident with a lattice direction if

$$\tan(\theta/2) = (u^2 + v^2 + w^2)^{1/2}/m$$

where m is an integer. A unitary greatest common divisor among m, u, v, w may be chosen without restricting generality. Denoting by N the number of odd integers among m, u, v, w, the multiplicity Σ is given by

$$\Sigma = \frac{m^2 + u^2 + v^2 + w^2}{\alpha}$$

where $\alpha = 1$ if $N = 1, 3$, $\alpha = 2$ if $N = 2$, $\alpha = 4$ if $N = 4$.

An example is illustrated in Fig. 2.5(a). The lattice direction [001] has been chosen as the rotation axis: if we choose $m = 3$ then $N = 2$, $\alpha = 2$, $\Sigma = 5$, $\theta = 36°52'$. It may easily be seen from the figure that the volume of the CSL unit cell is formed by the vectors $[2\bar{1}0]$, [120], [001].

In Fig. 2.5(b) the CSL with $\Sigma = 7$ is shown for the cubic (111) plane ($\theta = 38.21°$). The black and the cross-like points belong to lattices 1 and 2 respectively. The CSL points are those on which cross-like and black points overlap. Black and cross-like points are both part of the DSC lattice, whose units are evidently smaller than crystal units. If lattice 1 is shifted by a DSC unit with respect to lattice 2 the pattern shown in the figure is reproduced in another position.

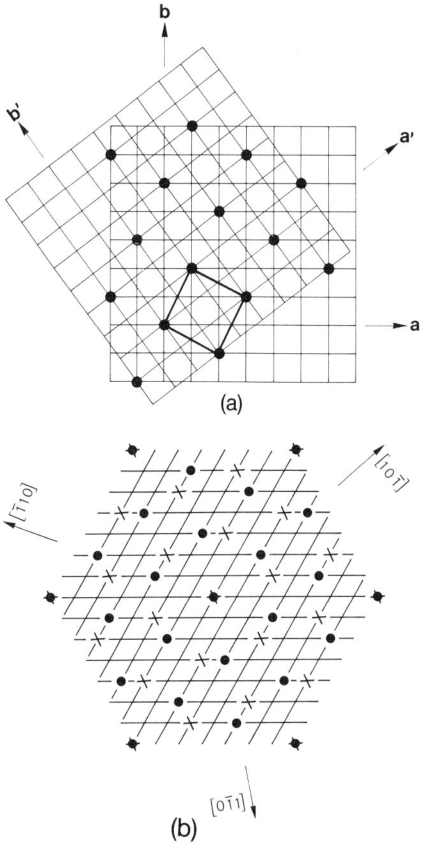

Fig. 2.5. (a): CSL lattice and CSL unit cell for a rotation of a cubic lattice about [001]. (b): CSL and DSC lattices for the cubic (111) plane.

Twins

Twins are regular aggregates consisting of individual crystals of the same species joined together in some definite mutual orientation. There are three principal types of twin: **growth twins** (produced by accident as the crystal grows from its initial nucleus), **deformation twins** (considered as a means of relieving the strain induced by some applied stress), and **transformation twins** (the product of a polymorphic transformation, i.e., when a higher symmetry crystal is cooled and converts to a lower symmetry structure).

From a geometrical point of view a twin is characterized by the symmetry operations which relate one individual to the other individuals in the composite crystal. The operation is very frequently a rotation through π about a zone axis (in this case the axis is the **twin axis** and the twin is a **rotation twin**), or a reflection in a lattice plane called the **twin plane** (the twin is then a **reflection twin**). Rotation twins of $\pi/3$, $\pi/2$, $2\pi/3$ also occur but are less common.

Obviously diad, tetrad, or hexad axes cannot be considered twin axes (at least for rotation through π). If a triad is a twin axis the twinning operation may be equivalently described as a $\pi/3$, π, or $4\pi/3$ rotation about the axis: conventionally the π rotation is preferred.

A twin is called a **contact twin** if the two components are joined in a plane (known as the **composition plane**). In the case of a rotation twin the composition plane is parallel to the twin axis, in reflection twins the composition plane is parallel to the twin plane. In **interpenetrating twins** the twin components intergrow so as to generate an irregular interface between components.

Multiple twins consist of three or more components. If the twinning operations relating adjacent components are all identical then the twins are known as **lamellar** or **polysynthetic twins** (the components have a lamellar form parallel to the composition plane). Polysynthetic twins may be on a microscopic or macroscopic scale.

Supplementary information on the most common types of twin and some morphological details are given in Appendix 2.J. Here, we are interested in the characterization of the twins in terms of lattice geometry. Twins are usually classified in accordance with the (pre-X-ray) theory developed by the French crystallographic school (Bravais, Mallard, Friedel). A recent classification[31,32] divides the twin kingdom into two species, TLQS and TLS. TLQS (twin-lattice quasi-symmetry) twins are characterized by two or more reciprocal lattices differently oriented, giving rise to double or multiple diffraction spots: they are generally recognizable simply by optical observations. TLS (twin-lattice symmetry) twins show a single orientation of the reciprocal lattice so that they give rise to single diffraction spots: they are optically indistinguishable, and their presence may be indicated by a high irreducible residual R in the attempted crystal structure determination.

The twin lattice is the lattice with the smallest cell that is common to both (or to all) individuals of the twin. Such a lattice will show perfect continuity at the composition surface of the twin in case of TLS, and will suffer some deviation for TLQS twins. The slight deviation, expressed in degrees, is referred to as the twin obliquity ω, and is usually less than 6°. Occurrence of TLQS mostly depends on special relationships between cell parameters while the existence of TLS depend on the symmetry properties of the lattice.

A further criterion for subdividing twins is the twin index, n:

$$n = (\text{vol. per node in twin lattice})/(\text{vol. per node in crystal lattice}).$$

For $n > 1$ the twin lattice is a superlattice of the crystal lattice (see CSL in previous subsection); for $n = 1$ the twin lattice coincides with the crystal lattice. TLS twins with $n > 1$ are often found in highly symmetrical lattices, for example, in minerals.

The twin law specifies the mutual orientation of the two twinned crystals: it is usually expressed in terms of the rotation necessary to bring one of the lattices into coincidence with the other. According to Friedel[33] twinning may occur because of the accidental presence in a lattice of a net and a row, sufficiently dense, which are exactly or approximately perpendicular.

Friedel's necessary (but not sufficient) condition for twinning in triclinic lattices (specialized conditions can be found for the other lattice types) is that

$$g_{11}:g_{22}:g_{33}:g_{12}:g_{13}:g_{23}$$

are, or approach rational numbers. Friedel's rule may be obtained by requiring $\mathbf{G}' \equiv \mathbf{G}$ in (2.40), and may be so expressed:[34] 'a crystal may twin if the condition

$$\mathbf{G} = \mathbf{M}\mathbf{G}\bar{\mathbf{M}} \tag{2.41}$$

is satisfied where $\mathbf{M}$ is a matrix with rational coefficients'.

Obviously only matrices with $M = 1$ have to be considered: furthermore symmetry operators $\mathbf{M}$ must also be excluded since they are trivial solutions of (2.41). If $\mathbf{M}$ satisfies (2.41) $-\mathbf{M} = (-\mathbf{I})\mathbf{M}$ also satisfies (2.41): therefore, for any twin rotation through α about the direction $[uvw]$ there should exist a twin operation which is the combination of $\mathbf{M}$ and of the inversion of the lattice. For example, if $\alpha = 180°$, $(-\mathbf{I})\mathbf{M}$ will correspond to a reflection in the plane (hkl) normal to $[uvw]$.

Twins of special interest are TLS twins with $n = 1$. They were called by Friedel twins by merohedry since the crystal symmetry is merohedry of order n (subgroup of order n) of the symmetry of its lattice. Accordingly, merohedrical twins have one or more symmetry operations which are present in the lattice and not in the crystal. In order to explain their diffraction behaviour, they may be divided into two classes:[35]

1. Twins in class I show the same crystal Laue symmetry as the lattice symmetry. Then the twin operation belongs to the Laue symmetry of the crystal: in these conditions the set of intensities collected from the twin coincides, except for anomalous scattering, with that which would be measured on a single crystal. Structure determination is therefore not hindered but the determination of the absolute configurations (using the methods described on p. 97) is impossible.

2. Twins in class II are characterized from the fact that the Laue symmetry of the crystal is lower than the crystal lattice symmetry. Then at least one of the twin operations belongs to the lattice symmetry but not to the Laue symmetry of the crystal. Twins by hemiedry, tetartohedry, and ogdohedry can be found: they are made by two, four, and eight crystals respectively.

A scheme for twin classification is drawn in Fig. 2.6. In Fig. 2.7 some examples of twinning are collected.[36]

In Fig. 2.7(a) the projection along the $\boldsymbol{b}$ axis of a monoclinic lattice with $\beta \cong 90°$ is shown, together with its twinned lattice (the assumed twin operation is the mirror plane m perpendicular to $\boldsymbol{a}$, but we could also choose the mirror plane perpendicular to $\boldsymbol{c}$). The ω misfit is intentionally exaggerated.

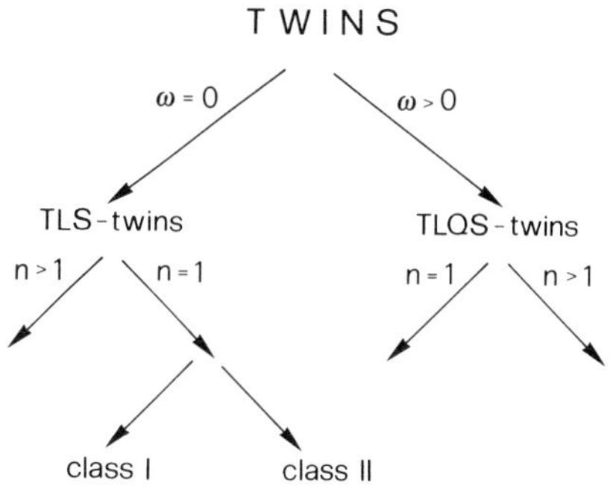

Fig. 2.6. A scheme for the classification of twins.

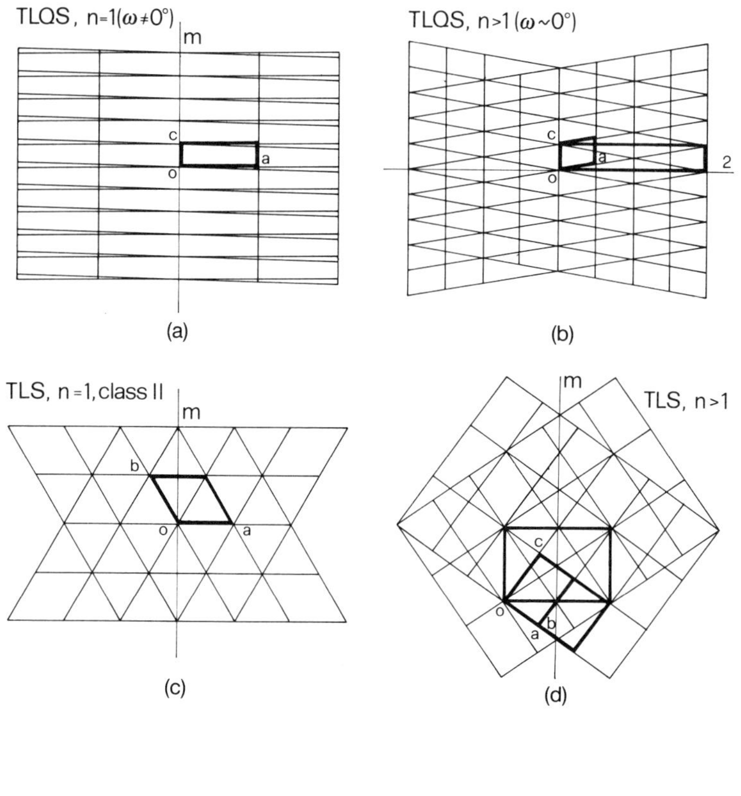

Fig. 2.7. Examples of twins (from reference [36]).

In Fig. 2.7(b) the projection along the **b** axis of the monoclinic lattice of l-aspartic acid with $a = 7.617$, $b = 6.982$, $c = 5.142$, $\beta = 99.84°$ is shown: the lattice is also shown after a two-fold rotation about the $\boldsymbol{a}^*$ axis. It is easily seen that $2\boldsymbol{a}$ of the original lattice nearly coincides with $2\boldsymbol{a}-\boldsymbol{c}$ of the rotated lattice, while $\boldsymbol{a}^*$ is the common reciprocal lattice of the two lattices. The twin lattice unit cell is defined by $\boldsymbol{a}' = 2\boldsymbol{a}$, $\boldsymbol{c}' = \boldsymbol{c}$, $\boldsymbol{b}' = \boldsymbol{b}$, $\beta' = \beta$ (but also a B-centred orthohombic cell may be chosen, four times larger than the original cell).

In Fig. 2.7(c) the projection of a hexagonal lattice along the **c** axis is shown. If the space group of the crystal is supposed to be $R\bar{3}$, a TLS twin of class II may be generated by reflection with respect to the plane m drawn in the figure. The diffraction pattern will show then $R\bar{3}m$ symmetry.

In Fig. 2.7(d) the classical penetration twin of fluorite (CaF_2) is described. Two cubic lattices, referred by the twin operation (111) mirror plane, are viewed along the direction $[\bar{1}10]$. The twin lattice has a volume three times the volume of the original cell.

An elegant derivation of twin laws by merohedry has been recently proposed.[37] Denote by H and G the point-group symmetry of the crystal and the point group of its lattice respectively (G may be obtained by the process of cell reduction). Since H is a subgroup of G, the coset decomposition of G with respect to H may be made (see Appendix 1.E). Any system of g_i operations ($g_i \in \mathrm{G}$, $g_i \notin \mathrm{H}$) used for the (left) coset decomposition will lead to the superposition of the lattice onto itself, and therefore will contain the possible merohedral twin laws for a crystal of point symmetry H in a lattice of point symmetry G.

For example, α-quartz crystallizes in $P3_121$ with $a = 4.913$ and $c = 5.404$ Å. The crystal point group is 321 and the metric symmetry is 6/mmm: thus

$$\mathrm{H} \equiv \{1;\ 2_{[110]};\ 2_{[100]};\ 2_{[010]};\ 3_{[001]};\ 3^2_{[001]}\}$$

$$\begin{aligned}\mathrm{G} \equiv \{&\mathrm{H};\ 2_{[001]};\ 2_{[1\bar{1}0]};\ 2_{[120]};\ 2_{[210]};\ 6_{[001]};\ 6^5_{[001]};\\ &\bar{1};\ \bar{2}_{[001]};\ \bar{2}_{[1\bar{1}0]};\ \bar{2}_{[120]};\ \bar{2}_{[210]}:\ \bar{3}_{[001]};\\ &\bar{3}^5_{[001]};\ \bar{6}_{[001]};\ \bar{6}^5_{[001]};\ \bar{2}_{[110]};\ \bar{2}_{[100]};\ \bar{2}_{[010]}\}.\end{aligned}$$

The coset decomposition is therefore

$$\mathrm{G} = \mathrm{H} \cup (2_{[001]}\mathrm{H}) \cup (\bar{1}\mathrm{H}) \cup (\bar{2}_{[001]}\mathrm{H}).$$

It may be seen that $2_{[001]}$, $\bar{1}$, and $\bar{2}_{[001]}$ correspond to the classical twin laws for Dauphiné, Brazil, and combined twinning respectively. The twin-related reflections are therefore (hkl), $(\bar{h}\bar{k}l)$, $(\bar{h}\bar{k}\bar{l})$, $(hk\bar{l})$.

The same procedure, applied to a crystal with point group H = 4/m, will decompose the metric symmetry group G = 4/mmm into

$$\mathrm{G} = \mathrm{H} \cup (2_{[010]}\mathrm{H}).$$

Hence the twin-related reflections are (hkl) and $(\bar{h}k\bar{l})$.

In the case of hemiedry (twins in class II, two individuals) two reflections which are not equivalent by Laue symmetry contribute to a twin reflection:[38] then

$$|F_{t\boldsymbol{H}}|^2 = \alpha\,|F_{\boldsymbol{H}}|^2 + (1-\alpha)\,|F_{\boldsymbol{K}}|^2$$
$$|F_{t\boldsymbol{K}}|^2 = (1-\alpha)\,|F_{\boldsymbol{H}}|^2 + \alpha\,|F_{\boldsymbol{K}}|^2$$

where $|F_{tH}|^2$ and $|F_{tK}|^2$ are twin reflection intensities, α is the volume fraction of crystal 1, and $|F_H|^2 \neq |F_K|^2$ are the intensities of the two overlapping reflections. The algebraic relation between $\boldsymbol{H}$ and $\boldsymbol{K}$ is fixed by the twinning law. A trial estimate of α may be obtained via suitable statistical tests on diffraction data:[38–41] α may then be refined as an extra parameter in adapted crystallographic least-squares programs.

An alternative procedure may be that of 'detwinning' reflection data: i.e. $|F_H|^2$ and $|F_K|^2$ can be calculated from observed $|F_{tH}|^2$ and $|F_{tK}|^2$ according to

$$|F_H|^2 = |F_{tH}|^2 + \frac{\alpha}{1-2\alpha}\{|F_{tH}|^2 - |F_{tK}|^2\}$$

$$|F_K|^2 = |F_{tK}|^2 - \frac{\alpha}{1-2\alpha}\{|F_{tH}|^2 - |F_{tK}|^2\}.$$

Previous knowledge of α is a necessary condition for the application of the above formulae: these cannot be applied if $\alpha \cong 0.5$. However, alternative methods are available for such a specific case (see [36] and references quoted therein).

Calculation of the structure factor

Let m be the order (i.e. the total number of symmetry operators) of the space group and t the number of symmetry-independent atoms. The structure factor may be then written as (see notation on pp. 152–4)

$$F_H = A_H + \mathrm{i}B_H$$

where

$$A_H = \sum_{j=1}^{t} n_j f_{0j}(\boldsymbol{H}) \sum_{s=1}^{m} \exp(-\bar{\mathbf{H}}\boldsymbol{\beta}_{js}\mathbf{H}) \cos 2\pi\, \bar{\mathbf{H}}(\mathbf{R}_s\mathbf{X}_j + \mathbf{T}_s) = \sum_{j=1}^{t} A_j \quad (2.42a)$$

$$B_H = \sum_{j=1}^{t} n_j f_{0j}(\boldsymbol{H}) \sum_{s=1}^{m} \exp(-\bar{\mathbf{H}}\boldsymbol{\beta}_{js}\mathbf{H}) \sin 2\pi\, \bar{\mathbf{H}}(\mathbf{R}_s\mathbf{X}_j + \mathbf{T}_s) = \sum_{j=1}^{t} B_j. \quad (2.42b)$$

A_j and B_j are the contributions of the jth atom and of its symmetry equivalents to A_H and B_H respectively, $\boldsymbol{\beta}_{js}$ is the 3×3 temperature factor matrix for the atom j in symmetry position s, n_j is the occupation number of atom j, defined as m_j/m, where m_j is the number of different atomic positions which are symmetry equivalent to the jth atom. Accordingly, $n_j = 1$ for an atom in a general position, $n_j < 1$ for atoms in special positions (the use of n_j allows that summation over s is always extended from 1 to m, independently of the atomic site type). If the jth site is only partially occupied because of some statistical disorder then n_j will be proportionally reduced.

The calculation of (2.42) will be simpler if, for a given $\mathbf{H}$ the symmetry equivalent indices $\mathbf{H}_s = \bar{\mathbf{H}}\mathbf{R}_s$, $s = 1, \ldots, m$ are calculated. In this case (see eqn (3.36)) $\bar{\mathbf{H}}\mathbf{R}_s\mathbf{X}_j$ may be replaced by $\bar{\mathbf{H}}_s\mathbf{X}_j$ and (see eqn (2.E.8)) $\bar{\mathbf{H}}\boldsymbol{\beta}_{js}\mathbf{H}$ by $\bar{\mathbf{H}}_s\boldsymbol{\beta}_j\mathbf{H}_s$.

We will then write eqns (2.42) as

$$A_{\boldsymbol{H}} = \sum_{j=1}^{t} \left[\sum_{s=1}^{m} n_j f_{0j}(\boldsymbol{H}) \exp(-\bar{\mathbf{H}}_s \boldsymbol{\beta}_j \mathbf{H}_s) \cos 2\pi (\bar{\mathbf{H}}_s \mathbf{X}_j + \bar{\mathbf{H}} \mathbf{T}_s) \right]$$

$$= \sum_{j=1}^{t} \sum_{s=1}^{m} u_{js} \qquad (2.43a)$$

$$B_{\boldsymbol{H}} = \sum_{j=1}^{t} \left[\sum_{s=1}^{m} n_j f_{0j}(\boldsymbol{H}) \exp(-\bar{\mathbf{H}}_s \boldsymbol{\beta}_j \mathbf{H}_s) \sin 2\pi (\bar{\mathbf{H}}_s \mathbf{X}_j + \bar{\mathbf{H}} \mathbf{T}_s) \right]$$

$$= \sum_{j=1}^{t} \sum_{s=1}^{m} v_{js}. \qquad (2.43b)$$

For each j the maximum number of u_{js} (and v_{js}) to calculate is 24. Indeed, if the space group is centrosymmetric (origin on a centre of symmetry), s may vary only over the symmetry matrices not referred by the inversion centre; $A_{\boldsymbol{H}}$ is then multiplied by 2 and $B_{\boldsymbol{H}}$ is settled to zero (origin on a centre of symmetry). For space groups with centred unit cell s may vary only over the matrices not referred by non-primitive lattice translations: $A_{\boldsymbol{H}}$ and $B_{\boldsymbol{H}}$ are then multiplied by the centring order of the cell.

Scattering factors f_{0j} have been tabulated[42] for all elements: their accuracy depends on the wave functions and on the numerical methods used. The values of f_{0j} at the actual $\sin\theta/\lambda$ value may be obtained from the tables by interpolation. A more usual procedure is to approximate scattering factors by the sum of one or more Gaussian functions: for accurate structure factor calculations four Gaussians are used according to [43]

$$f_0(\theta) = \sum_{i=1}^{4} a_i \exp[-b_i \sin^2\theta/\lambda^2] + c.$$

It should be noted that only nine parameters have to be stored for each element.

Calculation of the electron density function

According to (3.46) we have to calculate

$$\rho(x, y, z) = 2/V \sum_{h=0}^{+\infty} \sum_{k=-\infty}^{+\infty}{}' \sum_{l=-\infty}^{+\infty}{}' [A_{\boldsymbol{H}} \cos 2\pi \bar{\mathbf{H}}\mathbf{X} + B_{\boldsymbol{H}} \sin 2\pi \bar{\mathbf{H}}\mathbf{X}], \qquad (2.44)$$

where the prime to the summation implies that only half of the reflections $(0, k, l)$ have to be considered.

The calculations may be performed in a trivial fashion starting from the list of symmetry independent F_{hkl}, generating symmetry equivalents, and evaluating the sum in (2.44) for every **X**. The crystal symmetry may be more conveniently exploited by combining in advance the terms containing the symmetrical structure factors, thus obtaining an expression valid for that given symmetry. The summations in (2.44) are then limited to the set of independent F_{hkl} values. For example, in Pmmm

$$F_{hkl} = F_{h\bar{k}\bar{l}} = F_{h\bar{k}l} = F_{hk\bar{l}}.$$

then the right-hand side of (2.44) reduces to

$$\frac{8}{V}\sum F_{hkl}\cos 2\pi hx \cos 2\pi ky \cos 2\pi lz \quad (h, k, l \geqslant 0).$$

A drastic reduction of the number of operations ('operation' means a complex product followed by a complex sum) is attained by using the Beevers–Lipson factorization procedure. Since

$$\begin{aligned}\cos 2\pi\, \bar{\mathbf{H}}\mathbf{X} &= \cos 2\pi hx \cos 2\pi ky \cos 2\pi lz - \sin 2\pi hx \sin 2\pi ky \cos 2\pi lz \\ &\quad -\sin 2\pi hx \cos 2\pi ky \sin 2\pi lz - \cos 2\pi hx \sin 2\pi ky \sin 2\pi lz \\ &= \mathrm{ccc} - \mathrm{ssc} - \mathrm{scs} - \mathrm{css}, \\ \sin 2\pi \bar{\mathbf{H}}\mathbf{X} &= \mathrm{scc} + \mathrm{csc} + \mathrm{ccs} - \mathrm{sss},\end{aligned}$$

the right-hand side of (2.44) will be written as

$$\frac{2}{V}\sum_{h=0}^{\infty}\sum_{k=-\infty}^{+\infty}{}'\sum_{l=-\infty}^{+\infty}{}' A_{hkl}(\mathrm{ccc} - \mathrm{ssc} - \mathrm{scs} - \mathrm{css}) + B_{hkl}(\mathrm{scc} + \mathrm{csc} + \mathrm{ccs} - \mathrm{sss}). \tag{2.45}$$

On assuming

$$S_1(hkz) = \sum_{l=-\infty}^{+\infty}{}' (A_{hkl}\cos 2\pi lz + B_{hkl}\sin 2\pi lz),$$

$$S_2(hkz) = \sum_{l=-\infty}^{+\infty}{}' (-A_{hkl}\sin 2\pi lz + B_{hkl}\cos 2\pi lz),$$

$$S_3(hkz) = \sum_{l=-\infty}^{+\infty}{}' (-A_{hkl}\sin 2\pi lz + B_{hkl}\cos 2\pi lz),$$

$$S_4(hkz) = \sum_{l=-\infty}^{+\infty}{}' (-A_{hkl}\cos 2\pi lz + B_{hkl}\sin 2\pi lz),$$

we should write $\rho(x, y, z) = 2/V \times$

$$\sum_{h=0}^{+\infty}\sum_{k=-\infty}^{+\infty}{}' \{\cos 2\pi hx \cos 2\pi ky S_1(hkz) + \cos 2\pi hx \sin 2\pi ky S_2(hkz) + \sin 2\pi hx \cos 2\pi ky S_3(hkz) + \sin 2\pi hx \sin 2\pi ky S_4(hkz)\}. \tag{2.46}$$

Note that the S_i terms are nothing else but the components of a one-dimensional Fourier transform (for fixed h and k value):

$$S(hkz) = \sum_l F_{hkl}\exp(-2\pi i lz).$$

Equation (2.46) may be factored on assuming

$$T_1(hyz) = \sum_{k=-\infty}^{+\infty}{}' \cos 2\pi ky S_1(hkz) + \sin 2\pi ky S_2(hkz)$$

$$T_2(hyz) = \sum_{k=-\infty}^{+\infty}{}' \cos 2\pi ky S_3(hkz) + \sin 2\pi ky S_4(hkz).$$

Note that the T_i terms are nothing else but the components of a one-dimensional Fourier transform (for fixed h and z):

$$T(hyz) = \sum_h S(hkz)\exp(-2\pi i ky).$$

Then

$$\rho(x, y, z) = \frac{2}{V} \sum_{h=0}^{\infty} [\cos 2\pi hx T_1(hkz) + \sin 2\pi hx T_2(hyz)]$$

which is again a one-dimensional transform for fixed y and z:

$$R(xyz) = \sum_h T(hyz) \exp(-2\pi i hx).$$

Why is this procedure preferable to the direct use of (2.44)? If ρ is calculated directly from (2.44) the work to do is of the order n^6 (say $n_h n_k n_l n_x n_y n_z$, where n_h is written for the number of different h values, etc. . . .); if the Beevers–Lipson technique is used the work for the three transforms is of the order n^4 (say $n_h n_k n_l n_z + n_h n_k n_y n_z + n_h n_x n_y n_z$). Thus for large values of n the Beevers–Lipson technique makes a considerable saving. Furthermore, within every transform a single line of F_{hkl} is involved and, after its completion, the line is no longer necessary: thus $S(hkz)$ can replace the corresponding set of F_{hkl} values after the first transform, $T(hyz)$ can replace $S(hkz)$ after the second transform, and $R(xyz)$ can replace $T(hyz)$ at the end of the process.

A further reduction in the number of operations is obtained if the fast Fourier transform[44] (FFT) algorithm is used. The technique factorizes one-dimensional into two-dimensional transforms by a procedure quite similar to that of Beevers and Lipson (see Appendix 2.I). The saving of the computer time may be so described: a one-dimensional transform with n values of h and x requires n^2 operations, while the FFT technique reduces this to $2n \log_2 n$. The saving is not impressive for $n < 16$ but becomes large for larger values of n (as in protein crystallography).

Electron density maps are usually not printed: frequently peak search routines are used to locate electron density maxima and to print their list in order of peak intensity. A typical process for locating maxima involves two stages:

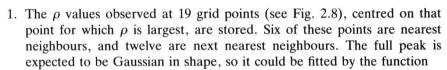

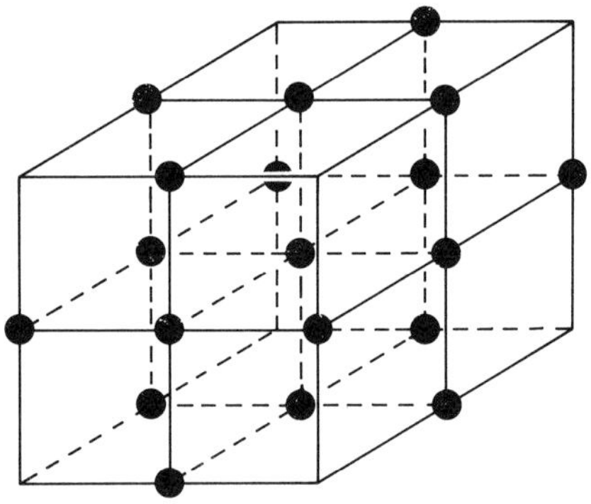

Fig. 2.8. The 19 grid points used for peak search routines.

1. The ρ values observed at 19 grid points (see Fig. 2.8), centred on that point for which ρ is largest, are stored. Six of these points are nearest neighbours, and twelve are next nearest neighbours. The full peak is expected to be Gaussian in shape, so it could be fitted by the function
$$\rho(xyz) = \exp(a + bx + cy + dz + ex^2 + fy^2 + gz^2 + hyz + kzx + lxy).$$
2. By least squares the ten values $a, \ldots, l$ are obtained which minimize $\sum_{19} (\ln \rho_{\text{obs}} - \ln \rho_{\text{calc}})^2$ (nine degrees of freedom are retained). Then the position of the maximum is obtained from the conditions $\delta\rho/\delta x = \delta\rho/\delta y = \delta\rho/\delta z = 0$.

The method of least squares

Linear least squares

A very common problem found in physical sciences is the following: given a set of experimental observations $\{f_i\}$ and a theoretical model which, from trial values of some parameters, generates a set of calculated f values, find the values of parameters which give the best fit to the data, estimate their accuracy, and comment on the adequacy of the assumed model. There are several approaches for answering this problem. Because of its wide use in crystallography we will mostly be interested in the method of least squares. Alternative approaches are briefly mentioned on pp. 108–9.

Suppose that a set of n experimental observations

$$\bar{\mathbf{F}} \equiv (f_1, f_2, \ldots, f_n)$$

is available for which:

(1) f_i is subjected to some random error e_i due to the finite precision of the measurement process;

(2) f_i is known to linearly depend on a set of $m \leqslant n$ parameters $\bar{\mathbf{X}} \equiv (x_1, x_2, \ldots, x_m)$.

Then the observational equations may be written[45] as

$$\mathbf{F} = \mathbf{AX} + \mathbf{E} \tag{2.47}$$

where $\mathbf{A} = \{a_{ij}\}$ is the $n \times m$ **design matrix** (n rows, m columns) of rank m which is assumed to be known, and $\bar{\mathbf{E}} \equiv (e_1, e_2, \ldots, e_n)$. The condition $n > m$ emphasizes the fact that the parameters are overdetermined. Our problem is to obtain satisfactory estimates $\hat{\mathbf{X}}$ of the parameters $\mathbf{X}$ and, at the same time, estimates of the variances of such estimates. We will assume that errors e_i have joint distributions with zero means and finite second moments: i.e.,

$$\langle \mathbf{F} \rangle \equiv \mathbf{F}^0 = \mathbf{AX} \tag{2.48}$$

$$\mathbf{M}_f = \begin{pmatrix} \mathrm{var}\,(f_1) & \mathrm{cov}\,(f_1, f_2) & \ldots\ldots & \mathrm{cov}\,(f_1, f_n) \\ \ldots\ldots & \ldots\ldots\ldots & \ldots\ldots & \ldots\ldots\ldots \\ \mathrm{cov}\,(f_n, f_1) & \mathrm{cov}\,(f_n, f_2) & \ldots\ldots & \mathrm{var}\,(f_n) \end{pmatrix}$$

where $\mathbf{M}_f = \langle (\mathbf{F} - \mathbf{F}^0)(\overline{\mathbf{F} - \mathbf{F}^0}) \rangle$ is the **variance–covariance matrix** of rank n, and

$$\mathrm{var}\,(f_i) = \sigma_i^2 = \langle e_i^2 \rangle, \qquad \mathrm{cov}\,(f_i, f_j) = \sigma_i \sigma_j \rho_{ij} = \langle e_i e_j \rangle,$$

where ρ_{ij} are the correlation coefficients.

The case $\mathbf{M}_f = \mathbf{I}$ (uncorrelated errors, equal variances) was treated by Gauss who showed that, among the various $\mathbf{X}$, the most satisfactory was that (say $\hat{\mathbf{X}}$) for which the residual

$$S = \sum_{i=1}^{n} v_i^2 = \bar{\mathbf{V}}\mathbf{V} = \text{minimum} \tag{2.49}$$

where $\bar{\mathbf{V}} \equiv (v_1, v_2, \ldots, v_n)$ is given by

$$\mathbf{V} = \mathbf{F} - \mathbf{AX}. \tag{2.50}$$

This result suggests that, when $\mathbf{M}_f$ is a diagonal matrix with elements σ_i^2 (in this case errors on f_i and f_j are statistically independent and $\rho_{ij} = 0$ for $i \neq j$), the most satisfactory estimate $\hat{\mathbf{X}}$ is that for which the weighted deviance

$$S = \sum_{i=1}^{n} w_i v_i^2 = \sum_{i=1}^{n} w_i (f_i - f_i^0)^2 = \bar{\mathbf{V}}\mathbf{W}\mathbf{V} = \bar{\mathbf{V}}\mathbf{M}_f^{-1}\mathbf{V} = \text{minimum} \tag{2.51a}$$

where $w_i = 1/\sigma_i^2$.

If $\mathbf{M}_f$ is not diagonal then

$$S = \sum_{i,j=1}^{n} w_{ij} v_i v_j = \sum_{i,j=1}^{n} w_{ij}(f_i - f_i^0)(f_j - f_j^0)$$
$$= \tilde{\mathbf{V}}\mathbf{W}\mathbf{V} = \tilde{\mathbf{V}}\mathbf{M}_f^{-1}\mathbf{V} = \text{minimum}. \qquad (2.51b)$$

In most applications $\mathbf{M}_f$ is taken to be diagonal.

Relations (2.51) suggest that, for a general variance–covariance matrix, the most satisfactory solution $\hat{\mathbf{X}}$ is given by

$$S = \tilde{\mathbf{V}}\mathbf{M}_f^{-1}\mathbf{V} = \text{minimum}. \qquad (2.52)$$

In accordance with Appendix 2.F, the relation (2.52) leads to the so called **normal equations**

$$\mathbf{B}\hat{\mathbf{X}} = \mathbf{D} \qquad (2.53)$$

where

$$\mathbf{B} = \tilde{\mathbf{A}}\mathbf{M}_f^{-1}\mathbf{A}, \qquad \mathbf{D} = \tilde{\mathbf{A}}\mathbf{M}_f^{-1}\mathbf{F}. \qquad (2.54)$$

$\mathbf{B}$ is a $m \times m$ symmetrical ($b_{ij} = b_{ji}$) square matrix. From (2.53) the least-squares estimate of $\mathbf{X}$ is obtained as

$$\hat{\mathbf{X}} = \mathbf{B}^{-1}\mathbf{D} \qquad (2.55)$$

It may be easily seen that $\hat{\mathbf{X}}$ is independent of the scale of the covariance matrix. Indeed suppose that $\mathbf{M}_f = K_v \mathbf{N}_f$, where $\mathbf{N}_f$ (the working matrix) is known and K_v unknown. Substituting $K_v \mathbf{N}_f$ for $\mathbf{M}_f$ in (2.55) eliminates K_v.

Reliability of the parameter estimates

It may be shown (see Appendix 2.G) that the variance–covariance matrix $\mathbf{M}_x$ relative to the unbiased estimate $\hat{\mathbf{X}}$ is given by

$$\mathbf{M}_x = \langle(\hat{\mathbf{X}} - \mathbf{X})\overline{(\hat{\mathbf{X}} - \mathbf{X})}\rangle = \mathbf{B}^{-1}. \qquad (2.56)$$

If $\mathbf{M}_f$ is known within the scale factor K_v ($\mathbf{M}_f = K_v \mathbf{N}_f$) then

$$\mathbf{M}_x = (\tilde{\mathbf{A}}\mathbf{M}_f^{-1}\mathbf{A})^{-1} = \left(\frac{\tilde{\mathbf{A}}\mathbf{N}_f^{-1}\mathbf{A}}{K_v}\right)^{-1} = K_v \mathbf{B}_v^{-1} \qquad (2.57)$$

where $\mathbf{B}_v$ is the working matrix calculated via the known $\mathbf{N}_f$ matrix. According to (2.57) $\mathbf{M}_x$ is completely determined only if K_v is known: luckily (see Appendix 2.H) an unbiased estimate of K_v is available from the least-square treatment:

$$\hat{K}_v = \frac{\tilde{\mathbf{V}}\mathbf{N}_f^{-1}\mathbf{V}}{n - m} = \frac{\hat{S}}{\langle\hat{S}\rangle} \qquad (2.58)$$

so that the unbiased estimate of $\mathbf{M}_x$ is given by

$$\mathbf{M}_x = \hat{K}_v \mathbf{B}_v^{-1}. \qquad (2.59)$$

$\hat{K}_v$ is often called goodness of fit and denoted by GofF.

Linear least squares with constraints

Assume that the parameters x_i are not independent but are constrained to satisfy the set of b linear equations

$$\mathbf{Q}\mathbf{X} = \mathbf{Z} \qquad (2.60)$$

where $\mathbf{Q}$ is a $b \cdot m$ matrix of rank b, and $\mathbf{Z}$ is a column vector with b components. In this case the unbiased estimate of $\mathbf{X}$ may be obtained by considering the variation function

$$S_c = \bar{\mathbf{V}}\mathbf{M}_f^{-1}\mathbf{V} - 2\bar{\boldsymbol{\lambda}}(\mathbf{QX} - \mathbf{Z})$$

where $\boldsymbol{\lambda}$ is a column vector with b components (the Lagrange multipliers). The condition $\delta S_c = 0$ brings to

$$2\bar{\mathbf{V}}\mathbf{M}_f^{-1}\delta\mathbf{V} - 2\bar{\boldsymbol{\lambda}}(\mathbf{QX} - \mathbf{Z}) = 2\overline{(\mathbf{F} - \mathbf{AX})}\mathbf{M}_f^{-1}\delta\mathbf{V} - 2\bar{\boldsymbol{\lambda}}\mathbf{Q}\delta\mathbf{X} = 0.$$

Since $\delta\mathbf{V} = -\mathbf{A}\delta\mathbf{X}$, we obtain

$$\bar{\mathbf{F}}\mathbf{M}_f^{-1}\mathbf{A} - \bar{\mathbf{X}}\bar{\mathbf{A}}\mathbf{M}_f^{-1}\mathbf{A} + \bar{\boldsymbol{\lambda}}\mathbf{Q} = 0.$$

Calling the solution $\mathbf{X}_c$ gives

$$\bar{\boldsymbol{\lambda}}\mathbf{Q} = \bar{\mathbf{X}}_c\mathbf{B} - \bar{\mathbf{F}}\mathbf{M}_f^{-1}\mathbf{A} \tag{2.61}$$

which, according to (2.55), may be written as

$$\bar{\boldsymbol{\lambda}}\mathbf{Q} = \overline{(\mathbf{X}_c - \hat{\mathbf{X}})}\mathbf{B}. \tag{2.62}$$

The value of $\boldsymbol{\lambda}$ may be obtained by post-multiplying both sides of (2.62) by $\mathbf{B}^{-1}\bar{\mathbf{Q}}$ and by requiring that (2.60) is satisfied:

$$\bar{\boldsymbol{\lambda}} = \overline{(Z - \mathbf{Q}\hat{\mathbf{X}})}(\mathbf{QB}^{-1}\bar{\mathbf{Q}})^{-1}.$$

Eliminating $\boldsymbol{\lambda}$ in (2.62) gives

$$\mathbf{X}_c = \hat{\mathbf{X}} + \mathbf{B}^{-1}\bar{\mathbf{Q}}(\mathbf{QB}^{-1}\bar{\mathbf{Q}})^{-1}(\mathbf{Z} - \mathbf{Q}\hat{\mathbf{X}}) \tag{2.63}$$

The variance–covariance matrix for $\mathbf{X}_c$ is ($\mathbf{B}^{-1}$ is the variance–covariance for $\mathbf{X}$)

$$\mathbf{M}_x = \mathbf{B}^{-1} - \mathbf{B}^{-1}\bar{\mathbf{Q}}(\mathbf{QB}^{-1}\bar{\mathbf{Q}})^{-1}\mathbf{QB}^{-1}. \tag{2.64}$$

By a procedure analogous to that described in Appendix 2.H it may be shown that the expected value of the weighted deviance S_c is given by

$$\langle \hat{S}_c \rangle = \langle \bar{\mathbf{V}}\mathbf{N}^{-1}\mathbf{V} \rangle / K_v = (n - m + b)/K_v$$

which is larger (see Appendix 2.H) than the expected value of the unconstrained residual

$$\langle \hat{S} \rangle = \langle \bar{\mathbf{V}}\mathbf{N}^{-1}\mathbf{V} \rangle = (n - m)/K_v.$$

Non-linear (unconstrained) least squares

Let us suppose that the observations $\bar{\mathbf{F}} \equiv (f_1, \ldots, f_n)$ do not linearly depend on the $\mathbf{X}$ parameters. Then the residual S will have several local minima (see Fig. 2.9) and condition (2.51) will not usually provide a satisfactory estimate $\hat{\mathbf{X}}$ for the parameters, unless a good approximation $\bar{\mathbf{X}}^0 \equiv (x_1^0, x_2^0, \ldots, x_n^0)$ of $\mathbf{X}$ is available. To this aim each f_j may be expanded in a Taylor series about the point $\mathbf{X}^0$:

$$f_i(\mathbf{X}) \cong f_i(\mathbf{X}^0) + \sum_{j=1}^{m} \left(\frac{\delta f_i}{\delta x_j}\right)^0 \delta x_j + \frac{1}{2}\sum_{j,p=1}^{m} \left(\frac{\delta^2 f_i}{\delta x_j \delta x_p}\right)^0 \delta x_j \delta x_p + \ldots. \tag{2.65}$$

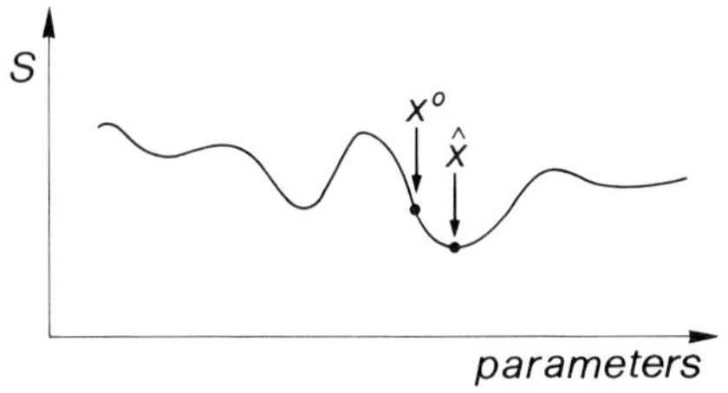

Fig. 2.9. Examples of local minima for the residual S.

If $\mathbf{X}$ is sufficiently close to $\mathbf{X}^0$ second and upper derivatives can be dropped in (2.65) (Gauss–Newton approximation) to get

$$\Delta f_i \simeq \sum_{j=1}^{m} \left(\frac{\delta f_i}{\delta x_j}\right)^0 \Delta x_j + e_i$$

where derivatives are calculated in $\mathbf{X}^0$, and

$$\Delta f_i = f_i(\mathbf{X}) - f_i(\mathbf{X}^0), \qquad \Delta x_j = x_j - x_j^0.$$

In matrix notation

$$\Delta \mathbf{F} = \mathbf{A}\,\Delta \mathbf{X} + \mathbf{E} \tag{2.66}$$

where

$$\mathbf{A} = \{a_{ij}\} = \{\delta f_i / \delta x_j\}$$

is the **design matrix** (compare (2.66) with (2.47)).

If the Taylor expansion is valid, the problem has been reduced to a linear one: thus an estimate for $\Delta\mathbf{X}$, say $\Delta\hat{\mathbf{X}}$. will be obtained in the same way as described in the previous paragraphs, from which more satisfactory values $\mathbf{X} = \mathbf{X}^0 + \Delta\hat{\mathbf{X}}$ may be derived. $\mathbf{M}_f$ can still be considered as the variance–covariance matrix of $\Delta\mathbf{F}$ if $\mathbf{X}^0$ is a good starting model.

Since in (2.66) second and upper derivatives have been ignored, $\Delta\hat{\mathbf{X}}$ will not be an unbiased estimate of $\Delta\mathbf{X}$. Therefore for non-linear problems an iterative procedure must be followed, according to which the new parameters $\mathbf{X} = \mathbf{X}^0 + \Delta\hat{\mathbf{X}}$ are used as a new starting model for the application of (2.66): in each cycle the derivatives will have changed so that the design matrix has to be recalculated. The iterative process will continue until the changes $\Delta\hat{\mathbf{X}}$ are very small or zero: then it may be concluded that the least-squares procedure has **converged**, and that the **refinement** of the parameters has been completed.

Least-squares refinement of crystal structures

Least-squares techniques are widely applied in crystallography. Examples are: refinement of unit cell parameters from diffraction angles (see p. 300); calculation of least squares planes for molecular fragments (p. 74), thermal motion analysis (pp. 117–20), calculation of the Wilson plot (p. 322), and profile analysis in powder methods (pp. 109–17). In the following we will describe the application of least-squares techniques to crystal structure refinement, by taking into account two different situations:

1. The Rietveld technique for the analysis of powder diffraction data. The entire diffraction profile is calculated and compared with the observed step profile, point by point: the model parameters are then adjusted by the least-squares method. We will describe this technique on pp. 109–17.

2. A large number of moduli $|F_{\mathbf{H}}|_{\mathrm{o}}$ are measured. Parameters are then refined in order to minimize the difference between the $|F_{\mathbf{H}}|_{\mathrm{c}}$s (structure factors moduli calculated from the structural model) and the $|F_{\mathbf{H}}|_{\mathrm{o}}$s. This technique is of general use for single-crystal data and of remarkable usefulness also for powder data; it is described in the following.

To carry out least-squares refinement of a crystal structure let us associate a weight $w_{\mathbf{H}}$ to each $|F_{\mathbf{H}}|_0$. Then, in accordance with (2.51), we want to make minimum the quantity

$$S = \sum_{\mathbf{H}} w_{\mathbf{H}}(|F_{\mathbf{H}}|_0 - |F_{\mathbf{H}}|_c)^2. \tag{2.67}$$

According to the previous subsection if a satisfactory model $\mathbf{X}^0$ is available, we expand $|F_{\mathbf{H}}|_c$ in a Taylor series: then

$$S = \sum_{\mathrm{H}} w_{\mathbf{H}}\left(\Delta F_{\mathbf{H}} - \sum_k \frac{\delta\,|F_{\mathbf{H}}|_c}{\delta x_k}\,\Delta x_k\right)^2$$

which, according to (2.51b), may be written as

$$S = (\overline{\Delta\mathbf{F} - \mathbf{A}\,\Delta\mathbf{X}})\mathbf{W}(\Delta\mathbf{F} - \mathbf{A}\,\Delta\mathbf{X}) = \bar{\mathbf{V}}\mathbf{W}\mathbf{V} = \bar{\mathbf{V}}\mathbf{M}_f^{-1}\mathbf{V}.$$

We have denoted

$$\mathbf{A} = \{a_{ik}\} = \left\{\frac{\delta\,|F_{\mathbf{H}_i}|_c}{\delta x_k}\right\},$$

$$\Delta\mathbf{F} = \{\Delta F_{\mathbf{H}_i}\} = \{|F_{\mathbf{H}_i}|_0 - |F_{\mathbf{H}_i}|_c\}.$$

$|F_{\mathbf{H}_i}|_c$ is the modulus of the ith structure factor calculated in $\mathbf{X}^0$; its derivatives are also calculated in $\mathbf{X}^0$. Thus eqn (2.66) is again obtained.

The normal equations may be obtained by settling to zero the derivatives of S with respect to $\Delta\mathbf{X}$:

$$\sum_{\mathbf{H}} w_{\mathbf{H}}\left(\Delta F_{\mathbf{H}} - \sum_k \frac{\delta\,|F_{\mathbf{H}}|_c}{\delta x_k}\,\Delta x_k\right)\frac{\delta\,|F_{\mathbf{H}}|_c}{\delta x_j} = 0 \qquad \text{for } j = 1, \ldots, m.$$

In matrix form (see eqns (2.53) and (2.54))

$$\mathbf{B}\,\Delta\hat{\mathbf{X}} = \mathbf{D} \tag{2.68}$$

or, more explicitly,

$$\begin{pmatrix} b_{11} & b_{12}\ldots\ldots b_{1m} \\ b_{21} & b_{22}\ldots\ldots b_{2m} \\ \ldots\ldots & \ldots\ldots\ldots\ldots \\ b_{m1} & b_{m2}\ldots\ldots b_{mm} \end{pmatrix}\begin{pmatrix} \Delta x_1 \\ \Delta x_2 \\ \ldots \\ \Delta x_m \end{pmatrix} = \begin{pmatrix} d_1 \\ d_2 \\ \ldots \\ d_m \end{pmatrix},$$

where

$$\mathbf{B} = \bar{\mathbf{A}}\mathbf{M}_f^{-1}\mathbf{A} = \mathbf{M}_f^{-1}\bar{\mathbf{A}}\mathbf{A} = \{b_{jk}\} = \sum_{\mathbf{H}} w_{\mathbf{H}}\left\{\frac{\delta\,|F_{\mathbf{H}}|_c}{\delta x_j}\,\frac{\delta\,|F_{\mathbf{H}}|_c}{\delta x_k}\right\}$$

$$\mathbf{D} = \bar{\mathbf{A}}\mathbf{M}_f^{-1}\,\Delta\mathbf{F} = \mathbf{M}_f^{-1}\bar{\mathbf{A}}\,\Delta\mathbf{F} = \{d_j\} = \left\{\sum_{\mathbf{H}} w_{\mathbf{H}}[|F_{\mathbf{H}}|_0 - |F_{\mathbf{H}}|_c]\,\frac{\delta\,|F_{\mathbf{H}}|_c}{\delta x_j}\right\}.$$

Then (compare with (2.55))

$$\Delta\hat{\mathbf{X}} = \mathbf{B}^{-1}\mathbf{D}$$

provides the required solution.

Parameters usually refined are: are overall scale factor (observed intensities are usually on an arbitrary scale); a parameter defining chirality in non-centrosymmetric space groups; for each atom, up to three coordinates (fewer if the atom is on a special position); thermal parameters (1 for isotropic motion, up to 6 for anisotropic); the site occupancy if the atomic position is statistically occupied because of some structural disorder.

Let us now calculate the explicit expressions of the derivatives for the various parameters. If x_{ji} is the ith parameter of the jth atom (i.e. in increasing order of i we will refer to n_j, x_j, y_j, z_j, B_j or n_j, x_j, y_j, z_j, β_{j11}, β_{j22}, β_{j33}, β_{j12}, β_{j13}, β_{j23}), then

$$\frac{\delta\,|F_{\mathbf{H}}|_c}{\delta x_{ji}}=\frac{\delta}{\delta x_{ji}}[A_{\mathbf{H}}^2+B_{\mathbf{H}}^2]^{1/2}=\frac{1}{2\,|F_{\mathbf{H}}|_c}\left[2A_{\mathbf{H}}\frac{\delta A_{\mathbf{H}}}{\delta x_{ji}}+2B_{\mathbf{H}}\frac{\delta B_{\mathbf{H}}}{\delta x_{ji}}\right]$$

$$=\cos\varphi\frac{\delta A_{\mathbf{H}}}{\delta x_{ji}}+\sin\varphi\frac{\delta B_{\mathbf{H}}}{\delta x_{ji}}$$

$$=\cos\varphi\frac{\delta A_j}{\delta x_{ji}}+\sin\varphi\frac{\delta B_j}{\delta x_{ji}}$$

where A_j and B_j are the contributions of the jth atom to $A_{\mathbf{H}}$ and $B_{\mathbf{H}}$ respectively.

Consider the various cases:

1. x_{ji} is an atomic coordinate: then

$$\frac{\delta A_j}{\delta x_{ji}}=-2\pi\sum_{s=1}^{m}h_{is}v_{is},\qquad \frac{\delta B_j}{\delta x_{ji}}=2\pi\sum_{s=1}^{m}h_{is}u_{is}$$

where h_{is}, $i=1, 2, 3$ are the components of $\mathbf{H}_s$ and u, v are defined by eqns (2.43).

2. x_{ji} is the component U^*_{jpq} of the vibrational tensor $\boldsymbol{U}^*_j$ (see eqn (3.20)). Then

$$\frac{\delta A_j}{\delta U^*_{jpq}}=-2\pi^2\sum_{s=1}^{m}h_{ps}h_{qs}u_{js},\qquad \frac{\delta B_j}{\delta U^*_{jpq}}=-2\pi^2\sum_{s=1}^{m}h_{ps}h_{qs}v_{js}.$$

For isotropic motion (Q denotes here the thermal factor, in order to avoid confusion with the normal matrix $\mathbf{B}$)

$$\frac{\delta A_j}{\delta Q_j}=-\frac{\sin^2\theta}{\lambda^2}\sum_{s=1}^{m}u_{js},\qquad \frac{\delta B_j}{\delta Q_j}=-\frac{\sin^2\theta}{\lambda^2}\sum_{s=1}^{m}v_{js}.$$

3. x_{ji} is the overall scale factor K. Even if $|F_{\mathbf{H}}|_c$ are on the absolute scale and $|F_{\mathbf{H}}|_0$ on a relative one, during a least-squares refinement it is the structural model which has to be refined and not vice versa. Therefore

$$S=\sum_{\mathbf{H}}w_{\mathbf{H}}(|F_{\mathbf{H}}|_0-K\,|F_{\mathbf{H}}|_c)^2$$

from which the following estimate for K is obtained

$$K=\left(\sum_{\mathbf{H}}w_{\mathbf{H}}\,|F_{\mathbf{H}}|_0\,|F_{\mathbf{H}}|_c\right)\left(\sum_{\mathbf{H}}w_{\mathbf{H}}\,|F_{\mathbf{H}}|_c^2\right)^{-1}.$$

Then the $|F_{\mathbf{H}}|_0$s will be multiplied by $1/K$ in order to lead them on the absolute scale. The $|F_{\mathbf{H}}|_0$ so rescaled will work as observations in subsequent cycles of least squares.

Since $\delta(K\,|F_{\mathbf{H}}|_c)/\delta K=|F_{\mathbf{H}}|_c$, the second derivative of S with respect to K is zero: therefore K is the parameter which converges faster. It should be noted that, if we erroneously had chosen to minimize $S'=\sum_{\mathbf{H}}w_{\mathbf{H}}(K\,|F_{\mathbf{H}}|_0-$

$|F_{\mathbf{H}}|_c)^2$, the minimum could be found at $K=0$ and with an extremely large thermal motion.

4. x_{ji} is the atomic occupancy factor. Then

$$\frac{\delta A_j}{\delta n_j}=\frac{1}{n_j}\sum_{s=1}^{m} u_{js}, \qquad \frac{\delta B_j}{\delta n_j}=\frac{1}{n_j}\sum_{s=1}^{m} v_{js}.$$

5. x_{ji} is a parameter defining chirality in non-centrosymmetry space groups. An efficient way for determining the absolute configuration is based on measurement of the intensities of the Bijvoet pairs of reflections (see pp. 165–9) when dispersion effects occur. In this case the symmetry of $|F|_0$ in reciprocal space degrades from Laue to the true point-group symmetry; indeed $(h\ k\ l)$ and its symmetry equivalents have an intensity value different from that of $(\bar{h}\ \bar{k}\ \bar{l})$ reflection and its equivalents. Thus Bijvoet pairs $(h\ k\ l)$ and $(\bar{h}\ \bar{k}\ \bar{l})$ can be used for the assignment of chirality.

More often only the asymmetric set of reflections is collected: then chirality may be fixed by the following procedure: (a) the structure is solved and refined by using real atomic scattering factors (and preferably absorption-corrected data); (b) the final values of scale factor, coordinates, and thermal factors are kept intact to calculate structure factors. By merely reversing the signs of all the if_j'' two R values are obtained (see eqn (5.88) for the meaning of R), say R_w^+ and R_w^-; (c) the ratio $T=R_w^-/R_w^+$ is used in the Hamilton test (the unweighted ratio R^-/R^+ is usually regarded as an acceptable approximation). For a given value of T one looks up Hamilton's tables $R\ (1,\ N,\ \alpha)$ to estimate α, where N is the number of degrees of freedom (see p. 103).

Reasons for considering the ratio test over-optimistic were given by Rogers[46] who suggested a better procedure: a single least-squares variable η is introduced into every dispersion term ($i\eta f_j''$ instead of if_j'') that contributes to the structure factor. η is kept zero in the early stages of refinement: later it should converge to values close to 1 or -1, with a readily computable precision $\sigma(\eta)/\eta$.

A very efficient alternative procedure[47] considers the crystal as an inversion twin:

$$|F_{\mathbf{H}}(x)|^2=(1-x)\,|F_{\mathbf{H}}|^2+x\,|F_{-\mathbf{H}}|^2$$

where $(1-x)$ and x are the fractions of the structure and its inverse in the macroscopic sample. The parameter x is refined: usually it converges in a few cycles to the final value.

6. x_{ij} is the secondary extinction parameter g (see p. 164). While primary extinction is often negligible in single-crystal studies, the effects of secondary extinction are usually corrected by means of the equation

$$(I_0)_{\text{corr}}=I_0(1-gI_0)^{-1}.$$

Since this equation corrects observed quantities, it cannot be directly used in least-squares procedures, where it is replaced by[48–50]

$$(I_c)_{\text{corr}}=I_c(1+gI_c)^{-1}.$$

In this case

$$|F|_{0\,\text{corr}}=K\,|F|_c(1+gLP\,|F|_c^2)^{-1/2},$$

where L is the Lorentz factor, P a suitable polarization factor, and K the scale factor.

Suppose now that $\mathbf{N}_f$ is known instead of $\mathbf{M}_f$ $(=K_v\mathbf{N}_f)$: then $\Delta\hat{\mathbf{X}}$ will be insensitive to the scale factor K_v, while the matrix $\mathbf{M}_x$ for the estimated parameters will strongly depend on it. According to eqn (2.59)

$$\hat{\mathbf{M}}_x = \hat{K}_v\mathbf{B}_v^{-1} = \frac{\hat{S}}{\langle\hat{S}\rangle}\mathbf{B}_v^{-1} = \frac{\sum_{\mathbf{H}} w_{\mathbf{H}}(\Delta F_{\mathbf{H}})^2}{n-m}\mathbf{B}_v^{-1},$$

where $\mathbf{B}_v = \bar{\mathbf{A}}\mathbf{N}_f^{-1}\mathbf{A}$, from which variance and covariance values for the parameters may be calculated. In particular the variance of the estimated parameters is given by

$$\sigma_{ii} = (b_{ii})^{-1}\frac{\sum_{\mathbf{H}} w_{\mathbf{H}}(\Delta F_{\mathbf{H}})^2}{n-m}$$

where b_{ii}^{-1} is the ith diagonal element of $\mathbf{B}_v^{-1}$. It is easily seen that increasing the number of observations reduces the variance of the parameters ($(b_{ii})^{-1}$ and n are mutually related in an inverse fashion so that the standard deviation $\sqrt{\sigma_{ii}}$ varies as $\sqrt{1/n}$). If the quality of the data is improved then $w_{\mathbf{H}} = 1/\sigma_{\mathbf{H}}^2$ will increase: since $(b_{ii})^{-1}$ varies inversely with w it follows that standard deviations of the parameters are proportional to $\sigma_{\mathbf{H}}$.

The standard deviations of the parameters are a useful guide during the refinement process. If they are sufficiently small, the refinement process may be considered complete if the ratios (calculated shifts/standard deviations) are sufficiently small (e.g. less than 0.2–0.3).

Correlation coefficients between two parameters are given by

$$\rho_{ij} = b_{ij}/(b_{ii}b_{jj})^{1/2}.$$

ρ_{ij} can range from 0 to ± 1: as a rule, $\rho_{ij} \leqslant 0.2$–0.3 are frequent, $\rho_{ij} = \pm 1$ refers to two completely dependent parameters, one of which has to be eliminated.

As already stressed, owing to the presence of systematic errors very often the working matrix $\mathbf{N}_f$ experimentally available is referred to $\mathbf{M}_f$ by an unpredictable relation much more complex than a scaling factor. In these cases the technique of multiplying the working variance–covariance matrix $\mathbf{B}_v^{-1}$ by $\hat{S}/\langle\hat{S}\rangle$ in order to obtain $\mathbf{M}_x$ may be highly questionable. A report[51] of the International Union of Crystallography Subcommittee suggests that besides indices R or R_w as given in eqns (5.3) and (5.85) the goodness of fit ratio $\hat{S}/\langle\hat{S}\rangle$ (see also eqn (5.86)) should be also reported in publications as a global measure of fit.

Practical considerations on crystallographic least squares

Crystallographic least-squares refinement may not be a trivial task. Suppose that, for a structure of modest complexity (40 atoms in the asymmetric unit), $n = 4000$ reflections have been measured. Generally speaking, an overall scale factor, 120 positional, and 240 anisotropic thermal parameters should be refined, for a total of 361 parameters. Then a square matrix $\mathbf{B}$ of order 361, with $m = 361 \times 360/2 = 64\,980$ distinct elements will arise, each one constituted by a summation on thousants of terms. The computing time

(Cpu) and the storage (St) needed for a 'structure factor calculation–least-squares refinement' cycle will comply with the following table:

Step	Cpu	St
Calculate structure factors	nm	n
Calculate derivatives	nm	m
Calculate normal matrix	nm^2	$m(m+1)/2$
Invert matrix	m^3	
Calculate shifts $\Delta\mathbf{X}$	m	

In accordance with the above table, computing time and storage rapidly increase with the complexity of the structure, so that the task soon becomes prohibitive even for large and fast computers when large-scale problems (thousands of parameters) are treated. A useful suggestion arises by observing that the elements on the principal diagonal of **B** are sums of squares, so that they are always positive and rather large. On the contrary, the off-diagonal elements are sums of products which may be positive or negative; therefore they are generally expected to be smaller than diagonal elements. Accordingly computer storage and computing time may be reduced by setting all off-diagonal elements to zero (**diagonal-least-squares approximation**). That is equivalent to assuming a complete statistical independence among the parameters, but this is often unrealistic. For example:

(1) errors in the thermal parameters generate a systematic error (larger to high $\sin\theta/\lambda$) on $|F|_c$, which on its turn produces a bias into the estimate of the scale factor K;

(2) in oblique coordinate systems, non-negligible correlations between some coordinates of the same atom may be found. Indeed an error on a coordinate is compensated by errors on some other coordinates (see Fig. 2.10);

(3) a high correlation will be found among the site occupancy and temperature factors if they are contemporaneously refined.

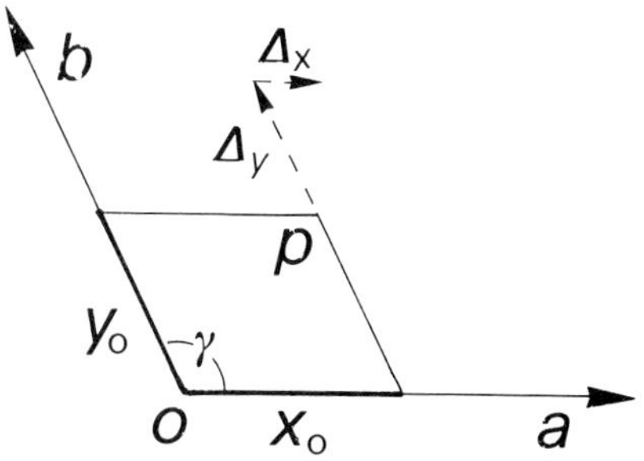

Fig. 2.10. P(x_0, y_0) is the true atomic position. If an error Δ_y is introduced, the 'best' value for x is obtained by minimizing, along the line $y_0 + \Delta y = \text{const}$, the distance of the atom from the true position P. That produces $\Delta_x = -\Delta_y \cos\gamma$.

An alternative to the diagonal approximation is the **block-diagonal approximation**; a first block may involve the correlation between the scale and the overall thermal parameter; the other blocks, one for each atom, are 9×9 matrices comprising positional and anisotropic temperature factors (a 4×4 matrix for an isotropic atom). The matrix **B** will then appear as in Fig. 2.11. Larger blocks are sometimes used; e.g. all the atoms in the same molecule could belong to the same block; or, also, a block for all positional parameters and a block for all vibrational parameters with the overall scale factor; or, ...

The storage requirement for the block diagonal approximation is certainly smaller than for full matrix methods. Each refinement cycle is faster, but convergence is slower: so the complete refinement process requires more cycles and almost the same computing time. Thus the major advantage of the method is the lessened storage requirements for problems too large to be treated by a full matrix.

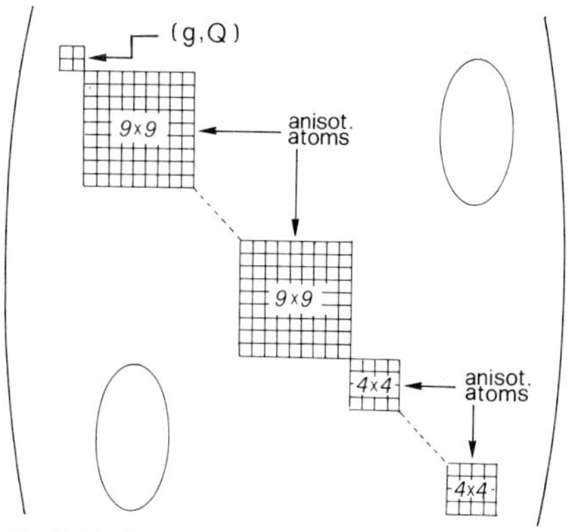

Fig. 2.11. A scheme for block-diagonal approximation.

A very large saving of computer time has been recently achieved by application of the fast Fourier transform (FFT) algorithm. FFT based programs allow full-matrix refinement of protein structures at a reasonable cost (see Appendix 2.I for some details).

A critical point in least squares is the occurrence of large correlation coefficients among parameters: this can give rise to ill-conditioned (determinant close to zero) **B** matrices and therefore to serious problems for the calculation of $\mathbf{B}^{-1}$ and for the accuracy of $\Delta\mathbf{X}$. That occurs when the following conditions hold.

1. Two atomic sites nearly overlap because of some structure disorder (or some error).

2. The symmetry of the structural model is higher than the space-group symmetry in which refinement is made. As an example, let us suppose that refinement of two symmetry independent molecules has been unsatisfactorily carried out in P2/m. The question arises as to whether or not the space group is really centrosymmetric. Refinement in P2 is then carried out on four symmetry independent molecules, the second two obtained from the first two by application of the inversion centre. Then the new **B** matrix will be singular: indeed the imaginary parts of the structure factors are vanishing, the derivatives of $|F|_c$ with respect to pairs of positional parameters referred by the inversion centre have equal moduli and opposite signs, the derivatives with respect to corresponding pairs of thermal parameters have equal moduli and signs. Since the elements of **B** are sums of products of derivatives, pairs of lines will be identical or opposite, so that **B** is singular. To overcome this difficulty small random shifts are applied to parameters when refinement moves from the first to the second space group. However, the matrix still remains ill-conditioned and further unpredictable shifts can be generated. A more efficient procedure is that which modifies the normal matrix to limit the parameter shifts (not equivalent to dumping shifts after the matrix solution).

As a further example, a singularity may occur when two atoms with the same x and y coordinates but differing by a 50 along the x axis are refined without using reflections with h odd.

3. Symmetry constraints are not introduced. For example if an atom lies on $4_{[001]}$, the following constraints will hold: $\beta_{11}=\beta_{22}$, $\beta_{23}=\beta_{13}=\beta_{12}=0$. If β_{11} and β_{22} are erroneously refined as if they were independent, the matrix **B** will have two rows and two columns equal and will be singular.

4. In some non-centrosymmetric space groups the origin may float in some directions. The normal equations matrix will become singular if the atomic coordinates of one atom along the free directions are not fixed (e.g. z in P$ca2_1$, or x and z in Pm).

Such a practice was frequently used in the primitive least squares programs but provides incorrect estimated standard deviations unless the full variance–covariance matrix is used. The problem may be solved by constraining (or restraining, see p. 107) the sum of the free coordinates to a constant value.

An efficient least-squares program should:

1. Provide easy ways for applying symmetry constraints to parameters of atoms in special positions. These are handled by not refining some

parameters or by equivalencing atomic parameter shifts. For example, in a hexagonal space group for an atom in special position at $(x, 2x, z)$ the only independent positional variables are x and z, so that

$$\delta\,|F|_c/\delta x = \delta\,|F|_c/\delta x + 2\delta\,|F|_c/\delta y.$$

Analogous problems occur for thermal parameters: symmetry constraints on them are described in Appendix 3.B.

2. Allow for a variety of weighting schemes (see p. 369). The weights play a critical role in any refinement process. In the initial stages of refinement artificial weights may be used to accelerate convergence (e.g. $w_{\mathbf{H}} = 1$ for all reflections is commonly used when a partial structure is under refinement). In the final stages the weights should reflect the accuracy of the individual observations ($w_{\mathbf{H}} \cong 1/\sigma^2(|F|_0)$ should be deduced from an analysis of the experiment). Then observations which are believed to be unreliable get very small weights.

Unfortunately it is not trivial to assess the accuracy of observations and the correctness of the assumed model. A long list of errors[52] can affect the values of the deviates $(|F_{\mathbf{H}}|_0 - |F_{\mathbf{H}}|_c)$ and therefore the results of a crystal structure analysis: experimental accidents (fluctuating power in the incident beam, crystals moving on their mounts, etc.); statistical fluctuations in quantum counts; crystal damage by the incident beam; errors in corrections for absorption, background radiation, thermal diffuse scattering, extinction, anomalous dispersion, anisotropic and/or anharmonic vibration, disorder, multiple reflections; deviations from the isolated atom approximation; incorrect symmetry assignment; correlation among observations.

Different types of errors can have similar effects and may be compensated by others. For example, absorption and extinction corrections tend to reduce intensities at low $\sin\theta/\lambda$ values more than at high values: thus errors in correcting these effects produce, as a partial compensation, errors in the thermal tensors.

In most cases the effects of each type of error cannot be estimated: then the standard deviation of individual intensity measurements is assumed as the standard deviation calculated from counting values plus an additional term which allows for errors of an undefined nature. It should be stressed that estimated variances of observations, and therefore weights, should not be based on counting statistics alone. Indeed variances estimated in this way are proportional to the intensities themselves: the consequence is that low intensities will systematically have lower variances than high intensities. That results in a bias in the estimate variances, and consequently (the Gauss–Markov theorem states that only if errors are random and uncorrelated it may be expected that the least-squares methods give the best unbiased estimate of the parameters) in a bias in the parameter estimates which may be of some relevance in high-precision structure determinations. In these cases it is highly recommended that all symmetry equivalent reflections are measured, that the averaged intensity is used in least squares, and that the variance based on counting statistics is combined with that arising from differences among symmetry-equivalent data.[51]

The possible presence of errors may be checked by observing that S is expected to be distributed like χ^2 with $n - m$ degrees of freedom. For large $n - m$ values $\chi^2/(n - m)$ is expected to be close to unity: so, if $w_{\mathbf{H}} = 1/\sigma^2_{\mathbf{H}}$, the expectation (see Appendix 2.H) for $\hat{S}$ is $\langle \hat{S} \rangle \cong n - m$. Accordingly the set of observations may be divided in large subsets (any significant criterion may be used for obtaining subsets: intensity ranges, or intervals of Bragg angles, or parity of Miller indices, etc.): then the weighted sum of discrepancies for each subset may be analysed. Each jth sum should be nearly $(n-m)(p_j/n)$, where p_j is the number of terms in the jth subset. If some unexpected behaviour appears then one may guess that some source of error has not been taken into account, and weight should be suitably changed.

A practical weighting criterion which takes into account all random experimental errors for which precise estimates are not available from experiment may be so described: the $|F|_0$ range is subdivided into a number of intervals each containing approximately the same number of reflections. For each ith interval the average $\langle |F|_0 \rangle_i$ and the difference $\langle \Delta F \rangle_c = (\langle |F|_0 - |F|_c \rangle)_i$ are calculated. Plotting $\langle \Delta F \rangle_c$ versus $\langle |F_0| \rangle$ (see Fig. 2.12) will show a distribution of points which can be approximated by an analytical function such as

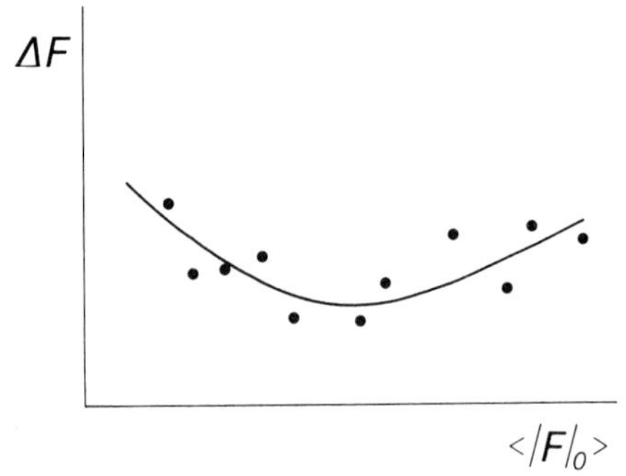

Fig. 2.12.

$$p(|F|_0) = a_0 + a_1 |F|_0 + a_2 |F|_0^2 + a_3 |F|_0^3.$$

The weighting scheme will be designed in order to make differences $\langle \Delta F_c \rangle$ constant; thus

$$\omega_{\mathbf{H}} = 1/p(|F_{\mathbf{H}}|_0).$$

More efficient empirical weights which make $\langle \Delta F \rangle_c^2$ constant as a function of $|F|_0$ can be obtained through truncated Chebyshev polinomials.[53]

But the above practices are not able to eliminate systematic errors or correlation among errors. If a careful check does not suggest any new strategy the only way for lessening the effects of systematic errors is based on a weighting scheme based on the deviates $(|F_0| - |F_c|)$. Wang and Robertson[54] modified least-squares weights by adding to the calculated variance a term which is a function of the disagreement between the experimental and the expected distribution of the deviates.

Robust-resistant techniques[55] also use weights based on the deviates. Due to some random or systematic errors, or to imperfections in modelling parameters, or to many other reasons it may occasionally occur that a small number of data, called **outliers**, show a very large discrepancy with respect to the corresponding calculated values. Least squares are very sensitive to the presence of large residuals which, having a high leverage, may cause non-negligible distorsions in results. A naive approach, often used in practice, may be to discard outliers by setting to zero the corresponding weights. The robust-resistant techniques decrease weights when disagreement increases, so avoiding any discontinuity in the process.

3. Allow for a proper choice of the number of reflections to be included in the refinement. In order to save computing time, or to avoid many low-quality data reducing the efficiency of the process, reflections for which $|F| < q\sigma(|F|)$ are often excluded from the refinement (q depends also on the number of available data: $q = 2$, 3 is a frequent choice. Higher values of q could bring to a unsatisfactory ratio number of observations/number of parameters).

However, the omission of observations is equivalent to assigning zero weights to them. If $\sigma(|F|)$ depends on $|F|$ such a practice preferentially discards low-intensity reflections so introducing a bias in the process. Often, that has little effect on the results,[56] but sometimes it discards important information (very unreliable results can be obtained when a structure with superstructure effects is refined without using weak reflections).

The problem of identifying meaningful changes in R produced when the model is altered has been faced by Hamilton.[45, 57] He considered the weighted residual (5.88) instead of the conventional residual (5.85), and proposed the test (known now as **Hamilton's test**) based on the ratio $T = R_w(1)/R_w(2)$, where $R_w(1)$ and $R_w(2)$ (with $R_w(1) > R_w(2)$] are obtained from the two different models. In the practice $T = R(1)/R(2)$ is often used because usually $R_w(1)/R_w(2) \simeq R(1)/R(2)$. The test compares T with the function

$$R_{m,n-m,\alpha} = \left(\frac{m}{n-m} F_{m,n-m,\alpha} + 1\right)^{1/2}.$$

If the computed T satisfies $T > R_{m,n-m,\alpha}$ the hypothesis can be rejected at the $100\alpha\%$ significance level (a significance level of $\alpha = 0.05$ is commonly used). Tables for $F_{m,n-m,\alpha}$ were computed from Hamilton. For large values of $n - m$ (very frequent in crystallography).

$$F_{m,n-m} \simeq \chi^2_m/m$$

where χ^2 is the well known chi-square function: in these cases

$$R_{m,n-m,\alpha} = \left(\frac{\chi^2_{n-m}}{n-m} + 1\right)^{1/2}.$$

As an example, let us suppose that two least-squares refinements based on the same number of parameters m and on the same number of observations n are carried out by two different investigators. The refinements resulted in two different R factors, say $R(1) > R(2)$. The hypothesis 'the m parameters obtained by investigator 1 are consistent with the experimental data' may be rejected at the α significance level if $T > R_{m,n-m,\alpha}$.

Sometimes the two models used for refinement involve a different number of parameters. For example one refinement is carried out where b parameters are held fixed. Then Hamilton's test may be so modified: the value of T given by

$$T = \frac{R\ (b \text{ fixed param., } m-b \text{ varied param.})}{R\ (\text{all } m \text{ param. varied})}$$

may be tested against values of

$$R_{b,n-m,\alpha} = \frac{b}{n-m}(F_{b,n-m,\alpha} + 1)^{1/2}$$

For example, a least-squares refinement was first carried out up to the $R(1)$ value on assuming that all the atoms have isotropic temperature factors (m_1 is the number of parameters). A subsequent refinement was carried out up to a residual $R(2) < R(1)$ on assuming anisotropic atoms ($m_2 > m_1$ is the number of parameters). The hypothesis 'all the atoms vibrate isotropically' may be rejected at the α significance level if

$$T = R(1)/R(2) > R_{m_2-m_1,n-m_2,\alpha}.$$

As a further example, let us suppose that refinement in a centrosymmetric space group (say P2/m), stops with the residual $R(1)$, while refinement in a non-centrosymmetric space group (say P2 or Pm) results in the residual $R(2) < R(1)$. If m_1 and m_2 are the numbers of parameters for the two refinements, the hypothesis that the correct space group is centrosymmetric may be rejected at the α significance level if

$$T = R(1)/R(2) > R_{m_2-m_1,n-m_2,\alpha}.$$

A last remark on the crystallographic least-squares technique concerns which function to minimize. In place of the function S defined by eqn (2.67) the function

$$S' = \sum_{\mathbf{H}} w'_{\mathbf{H}}(|F_{\mathbf{H}}|_0^2 - |F_{\mathbf{H}}|_c^2)^2$$

is sometime minimized. The normal equations (2.68) still hold if $w_{\mathbf{H}}$ is replaced by $w'_{\mathbf{H}} = w_{\mathbf{H}}/(4\,|F_{\mathbf{H}}|_0^2)$, $\delta\,|F_{\mathbf{H}}|_c/\delta x_j$ by $2\,|F_{\mathbf{H}}|_c\,(\delta\,|F_{\mathbf{H}}|_c/\delta x_j)$ and, in $\mathbf{D}$, $[|F_{\mathbf{H}}|_0 - |F_{\mathbf{H}}|_c]$ by $[|F_{\mathbf{H}}|_0^2 - |F_{\mathbf{H}}|_c^2]$.

There is still a strong disagreement[51] over the question of whether S or S' should be used in refinement. In favour of $|F|$-refinement it may be observed[58] that $\delta\,|F|^2/\delta x_j$ is small if $|F|^2$ is small: thus weak reflections have too weak a leverage in $|F|^2$-refinement. Arguments in favour of $|F|^2$-refinement are based on the observation[59, 60] that $|F|$ are obtained from intensity data via a square root. This is a non-linear operation which may introduce a bias proportional to the variance of $|F|^2$. Furthermore the formula $\sigma(|F|) = \sigma(|F|^2)/(2\,|F|)$ so widely used in practice is not appropriate for zero or close to zero $|F|$ values, for which $\sigma(|F|)$ become infinite or extremely large (we must, however, mention that methods[61, 62] for obtaining more reliable $|F|$ and $\sigma(|F|)$ from $|F|^2$ and $\sigma(|F|^2)$ values are now available).

Constraints and restraints in crystallographic least squares

The accuracy of the derived parameters is generally small (e.g. it would be found that C–C distances in a phenyl group were in the range 1.2–1.3 Å, which is highly improbable) when at least one of the following conditions occurs:

1. Low ratio between the number of the observations and the number of parameters to refine. The ratio is particularly unfavourable for biological macromolecules. Indeed, owing to their intrinsic flexibility, the amplitudes of the thermal vibrations from the average positions are rather large (usually between 0.2 and 0.7 Å), so limiting the diffraction data to a resolution of about 2 Å. While to a resolution of 0.86 Å (typical limit

for small molecules) there are about 30 observations for each coordinate being refined, at 2.7 Å the number of parameters is nearly equal to the number of the observations.

2. Atoms with very large atomic number and very light atoms coexist in the unit cell. Then modest errors on the heavy-atom parameters cause strong errors on the light-atoms parameters.

3. Too high thermal motion, presence of structural disorder, etc. Only poor and scarce data are then available.

If, however, some prior stereochemical information is available on parts of the structure its use in the least-squares procedures may increase the degree of overdetermination of the system and improve the accuracy of the results. Three methods will be briefly recalled here.

Rigid body refinement[63]

If the structure contains rather rigid molecular fragments of well established geometry (e.g. benzene rings), the number of parameters to refine may be considerably reduced. The atoms concerned are initially regularized to some ideal geometry which is preserved during the refinement.

For each rigid group of n atoms the number of independent positional parameters reduces from $3n$ to 6 ((x_0, y_0, z_0) parameters defining the position of some reference point and ω, ψ, ϕ for the group orientation). Thermal parameters are usually considered to be isotropic, and, to keep the number of parameters small, a single overall thermal parameter may be assumed for the group.

Least-squares calculations could be performed according to the following scheme.[64] For the jth atom of the group let $\mathbf{X}_j$ be the coordinates with respect to the system $\mathbf{A}$ defined by the conventional unit cell, $\mathbf{X}'_j$ the coordinates (in Å) with respect to a set of orthogonal internal axes $\mathbf{A}'$, $\mathbf{X}''_j$ the coordinates with respect to an orthogonal cell of unit dimensions (say $\mathbf{A}''$) oriented in some way with respect to $\mathbf{A}$. Assume $\mathbf{A}'' = \mathbf{MA}$ where $\mathbf{M}$ is given by (2.31c) or a similar matrix. A rotation matrix $\mathbf{R}$ may be then defined which aligns $\bar{\mathbf{A}}'' \equiv (\boldsymbol{e}''_1, \boldsymbol{e}''_2, \boldsymbol{e}''_3)$ on to $\mathbf{A}'$ (say $\mathbf{A}' = \mathbf{RA}''$): according to the discussion on p. 72 the matrix $\mathbf{R}$ may be expressed as the product of three successive anti-clockwise rotations, say by ω about $\boldsymbol{e}_3$, by ψ about the new $\boldsymbol{e}_2$, and by ϕ about the new $\boldsymbol{e}_1$. Then $\mathbf{X}''_j = \mathbf{RX}'_j$ (see Table 2.E.1) and

$$\mathbf{X}_j = \mathbf{X}_0 + \bar{\mathbf{M}}\mathbf{X}''_j = \mathbf{X}_0 + \bar{\mathbf{M}}\bar{\mathbf{R}}\mathbf{X}'_j.$$

The atomic coordinates so obtained may be used in the usual manner for calculating structure factors. The problem is now to calculate the appropriate contribution for the rigid parameters x_0, y_0, z_0, ω, ψ, ϕ to the matrix of normal equations.

Derivatives with respect to such parameters are calculated from those for the atomic parameters $\mathbf{X}_j$ using the chain rule: i.e.

$$\frac{\delta\,|F_{\mathbf{H}}|_c}{\delta\omega} = \sum_{j=1}^{n}\left(\frac{\delta\,|F_{\mathbf{H}}|_c}{\delta x_j}\frac{\delta x_j}{\delta\omega} + \frac{\delta\,|F_{\mathbf{H}}|_c}{\delta y_j}\frac{\delta y_j}{\delta\omega} + \frac{\delta\,|F_{\mathbf{H}}|_c}{\delta z_j}\frac{\delta z_j}{\delta\omega}\right),$$

$$\frac{\delta\,|F_{\mathbf{H}}|_c}{\delta x_0} = \sum_{j=1}^{n}\frac{\delta\,|F_{\mathbf{H}}|_c}{\delta x_j},$$

and, for a group temperature factor B_g,

$$\frac{\delta\,|F_{\mathbf{H}}|_c}{\delta B_g}=\sum_{j=1}^{n}\frac{\delta\,|F_{\mathbf{H}}|_c}{\delta B_j}.$$

The size of the normal matrix is therefore reduced, convergence is accelerated but there is no remarkable saving in computing time (derivatives with respect to each atomic parameter have to be calculated for each structure factor and chain rule implemented).

The success of this method relies on the accuracy of the assumed geometry: if this is wrong systematic errors on other parameters will be introduced (e.g. the symmetry of phenyl groups sometime deviates remarkably from the sixfold symmetry[65]).

A technique related to, but not coincident with, rigid body refinement is that used for putting hydrogen atoms into structures (X-ray data). Sometimes H atoms are geometrically positioned at the end of each round of LSQ, but more frequently H parameters ride upon those of a reference atom (riding refinement). In this last case the normal matrix has to be modified in order to take into account all physical parameters (the common practice fixes the thermal parameter of H, which is set equal to that of the reference atom). At the end of calculations H and the reference atom will have the same positional estimated standard deviations, while the bond distance between them must have an estimated standard deviation equal to zero (if the full variance–covariance matrix is used in the calculation).

Constraints via Lagrangian multipliers

In the absence of complete information on a fragment's geometry one may wish to fix some molecular parameters (e.g. some bond distances and angles are fixed to given values, a group of atoms is held in coplanar arrangement, etc.). In such cases Lagrangian multipliers can be used: the function to minimize is

$$S=\sum_{\mathbf{H}}w_{\mathbf{H}}(|F_{\mathbf{H}}|_o-|F_{\mathbf{H}}|_c)^2+\sum_{i=1}^{b}\lambda_i G_i$$

where the (generally non-linear equations)

$$G_i(\mathbf{X})=0,\qquad i=1,2,\ldots,b, \tag{2.69}$$

represent the fixed constraints. Usually the available model $\mathbf{X}^0$ will not exactly satisfy (2.69). The problem will be linearized by expanding both the $|F_{\mathbf{H}}|_c$s and the G_is in Taylor series:

$$S=\sum_{\mathbf{H}}w_{\mathbf{H}}\left(\Delta F_{\mathbf{H}}-\sum_k\frac{\delta\,|F_{\mathbf{H}}|_c}{\delta x_k}\Delta x_k\right)^2+\sum_{q=1}^{b}\lambda_q\left(G_q^0+\sum_k\frac{\delta G_q^0}{\delta x_k}\Delta x_k\right)$$

where $G_q^0=G_q(\mathbf{X}^0)$ and $\delta G^0/\delta x_k$ is the derivative of G calculated in $\mathbf{X}^0$. The normal equations can be derived by settling to zero the derivatives (with respect to Δx and λ) of S:

$$\sum_{\mathbf{H}}w_{\mathbf{H}}\left(\Delta F_{\mathbf{H}}-\sum_k\frac{\delta\,|F_{\mathbf{H}}|_c}{\delta x_k}\Delta x_k\right)\frac{\delta\,|F_{\mathbf{H}}|_c}{\delta x_j}-\frac{1}{2}\sum_{q=1}^{b}\lambda_q\frac{\delta G_q^0}{\delta x_j}=0,$$

$$G_q^0+\sum_j\frac{\delta G_q^0}{\delta x_j}\Delta x_j=0.$$

The augmented normal equations, expressed in terms of partitioned matrices, are

$$\begin{pmatrix} b_{jq} & \vdots & m_{jq} \\ \cdots & & \cdots \\ m_{qj} & \vdots & 0 \end{pmatrix} \begin{pmatrix} \Delta x_k \\ \cdots \\ \lambda_q \end{pmatrix} = \begin{pmatrix} d_j \\ \cdots \\ G_q^0/2 \end{pmatrix}$$

where $m_{jq} = \frac{1}{2}(\delta G_q^0/\delta x_j)$.

Even if a large variety of stereochemical constraints can be taken into account by the method, the approach is not very attractive for large-size problems: indeed the order of the normal matrix is increased.

Use of restraints

Soft, flexible constraints (say restraints) may be imposed to some functions of the parameters in order to permit only realistic deviations of their values from fixed standard ones. These functions are used as supplementary observations, so that the order of the normal equation matrix is neither increased nor reduced. In this situation the function to minimize[66,67] is

$$S = \sum_{\mathbf{H}} w_{\mathbf{H}}(|F_{\mathbf{H}}|_0 - |F_{\mathbf{H}}|_c)^2 + \sum_q w_q(g_{eq} - g_q)^2$$

where g_q is the function describing the gth restraint, g_{eq} is its standard (or optimal) value, w_q is the weight to associate to the qth restraint. The normal equations are obtained by expanding S in Taylor series and equalling to zero its derivatives (with respect to $\Delta\mathbf{X}$):

$$\left(\sum_{\mathbf{H}} w_{\mathbf{H}} \sum_k \frac{\delta\,|F_{\mathbf{H}}|_c}{\delta x_k}\frac{\delta\,|F_{\mathbf{H}}|_c}{\delta x_j} + \sum_q w_q \sum_k \frac{\delta g_q}{\delta x_k}\frac{\delta g_q}{\delta x_j}\right)\Delta x_k$$
$$= \sum_{\mathbf{H}} w_{\mathbf{H}} \frac{\delta\,|F_{\mathbf{H}}|_c}{\delta x_j}\Delta F_{\mathbf{H}} + \sum_q w_q \frac{\delta g_q}{\delta x_j}(g_{eq} - g_q).$$

The effect of restraints is therefore to add contributions to the elements of the normal equation matrix. Thus parameters highly correlated can be found, but this is expected since parameters are deliberately correlated by restraints.

For example, a restraint may be to set an interatomic distance d (bonded or non-bonded) to a given value taken from the literature. If we recall, according to p. 62, that

$$d^2 = \Delta\bar{\mathbf{X}}\mathbf{G}\,\Delta\mathbf{X}$$

where

$$\Delta\bar{\mathbf{X}} = (\Delta_1, \Delta_2, \Delta_3),$$
$$\Delta_1 = (x_1 - x_2), \qquad \Delta_2 = (y_1 - y_2), \qquad \Delta_3 = (z_1 - z_2),$$

then

$$\delta d/\delta x_1 = (g_{11}\,\Delta_1 + g_{12}\,\Delta_2 + g_{13}\,\Delta_3)/d = -\delta d/\delta x_2$$

and so on.

Another restraint may be: the sum of all the atomic coordinates along a polar direction can be fixed to its current value in order to keep fixed the centre of gravity of the molecule and so determine the origin. Several other types of restraints can be imposed (see Chapter 8, p. 564) and they concern van der Waals distances, planarity of groups, chirality (a restraint on the chiral volume about an asymmetric carbon atom may maintain the conformation in the correct hand), bond and torsion angles, thermal motion, positional and thermal parameters related by non-crystallographic symmetry, potential energy functions, etc. Also, the sum of site occupancies of two or more atoms can be restrained to some fixed value when disorder problems occur. When the number of restraints is sufficiently large the equations are overdetermined and refinement will also converge with low-resolution data.

The approximate nature of the restraints well agrees with our approximate knowledge of the chemistry involved. A basic difficulty of their use is to assess an optimal weighting scheme: indeed the first component of S is in electrons squared while the second may be considered an energy quantity, where w_q plays the role of an effective force constant. A convenient mechanism is to indicate the relative precisions (i.e. the expected standard deviations) of the restraints and scale them with respect to the observations.

As an example, let us suppose that the bond distance d_j is to be restrained to a set value. If the weights of X-ray observations are (as usually) sufficiently accurate but on an arbitrary scale, the normalization factor $\hat{K}_v = \sum_{\mathbf{H}} w_{\mathbf{H}}(\Delta F_{\mathbf{H}})^2/(n-m)$ may be obtained. Then the restraint equation involving d_j with estimated standard deviation $\sigma(d_j)$ is given the weight

$$w_j = \hat{K}_v/\sigma^2(d_j).$$

Very small standard deviations make restraints equivalent to constraints.

If refinement provides a bond distance differing from d_j by several times $\sigma(d_j)$ it should be concluded that experimental data are in conflict with the fixed restraint.

Applications of constraints and/or restraints in least-squares processes usually lead to larger residuals than by unconstrained refinement. That does not necessarily mean that the constrained solution fits less well to the data. A sensitive criterion may be to compare estimated standard deviations obtained by constrained and unconstrained refinements. When constraints are applied both S and $n-m$ increase but σ_{ii} (see p. 98) may be larger or smaller according to circumstances. An alternative criterion may be the use of the Hamilton's test (see p. 103).

Alternatives to the method of least squares

In the above sections some deficiencies of least squares have emerged, deficiencies which invite the application of alternative methods. In statistical sciences the main alternative to least squares are maximum-likelihood methods. Let us suppose that the errors e defined by (2.47) are distributed according to some probability density function $P(e)$. The likelihood function L is then given by

$$L = \prod_{i=1}^{n} P(e).$$

Since P is everywhere greater than or equal to zero, its logarithm is a monotomically increasing function of the argument. Thus the maximum of L is attained when $\ln L$ is a maximum. If the error function is Gaussian then

$$P(e_i) = (2\pi\sigma_i^2)^{-1/2} \exp\left[-e_i^2/(2\sigma_i^2)\right]$$

where σ is the variance of the ith observation. Then

$$\ln L = -\frac{1}{2}\sum_{i=1}^{n}\frac{e_i^2}{\sigma_i^2} - \sum_{i=1}^{n}\ln\sigma_i - \frac{n}{2}\ln 2\pi.$$

Since the second and third terms are constant, $\ln L$ has its maximum when

$$S = \sum_{i=1}^{n}\frac{e_i^2}{\sigma_i^2} = \sum_{i=1}^{n} w_i e_i^2 = \min.$$

where $w_i = 1/\sigma_i^2$. We have so shown that, if the distribution of the errors is Gaussian, and observations are weighted by the reciprocal of the variance, the least-squares and the maximum-likelihood methods provide equivalent results. Conversely, if errors are not distributed according to a Gaussian model, the method of least squares does not lead to a maximum-likelihood estimate.

The above results also explain why least squares are practiclly the only ones used in crystallographic refinement: they only depend on the variance of the errors while maximum-likelihood methods require the knowledge of the distribution of the errors. When this is available maximum-likelihood methods provide better estimates of the parameters, since such methods can exploit more information. However, such a situation is very infrequent in crystallographic refinement, where only rather imperfect estimates of the variances are available.

Parameters may also be estimated by maximum entropy methods, which, based on Jaynes' extension of Shannon's theory, require the maximization of the entropy. But such methods, as well as maximum-likelihood methods, are far from being competitive with least squares in most of crystallographic applications.

Rietveld refinement

The basis of the technique

Powder diffraction patterns (see p. 293) may be collected in a step scan mode: intensity is measured for a given interval of time, and the theta and two-theta axes are then stepped to the next position. The pattern is then indexed: i.e. appropriate Miller indices are associated with observed reflections and simultaneously accurate unit-cell dimensions are calculated. Because of unavailable experimental errors in the estimates of the diffraction angles and because of the frequent overlapping of peak intensities, indexing is a rather difficult task for relatively large cell volumes and/or low-symmetry crystals. Several approaches are today available for this aim: implemented in computer programs, they often provide more than one solution, so that proper figures of merit[68–70] can be used to distinguish between bad and good solutions. The reader will find general remarks on the various approaches, and tests on their efficiency, in a paper by Shirley[71] and in some more recent papers.[72–76] This stage of analysis is followed by the examination of possible systematic absences to suggest a space group.

If a (even imperfect) structural model is available then the intensity y_{io} observed at the ith step may be compared with the corresponding intensity

y_{ic} calculated via the model. According to Rietveld,[77] the model may be refined by minimizing by a least-squares process the residual

$$S = \sum w_i \, |y_{io} - y_{ic}|^2 \tag{2.70}$$

where w_i, given by

$$(w_i)^{-1} = \sigma_i^2 = \sigma_{ip}^2 + \bar{\sigma}_{ib}^2,$$

is a suitable weight. σ_{ip} is the standard deviation associated with the peak (usually based on counting statistics) and σ_{ib} is that associated with the background intensity y_{ib}.

y_{ic} is the sum of the contributions from neighbouring Bragg reflections and from the background:

$$y_{ic} = s \sum_k m_k L_k \, |F_k|^2 \, G(\Delta\theta_{ik}) + y_{ib}$$

where s is a scale factor, L_k is the Lorentz-polarization factor for the reflection k, F_k is the structure factor, m_k is the multiplicity factor, $\Delta\theta_{ik} = 2\theta_i - 2\theta_k$ where $2\theta_k$ is the calculated positon of the Bragg peak corrected for the zero-point shift of the detector, and $G(\Delta\theta_{ik})$ is the reflection profile function.

The parameters to adjust by refinement include unit cell, atomic positional and thermal parameters, and parameters defining the functions G and y_{ib}.

The determination of an accurate model for the profile function $G(\Delta\theta_{ik})$ is one of the fundamental problems in both single-peak and in Rietveld analysis. This is particularly true today when high-resolution neutron and X-ray spectra are available: deficiencies in the profile model are no longer lost in relatively large instrumental diffraction profiles. The shape of a diffraction peak depends on several parameters: the radiation source, the wavelength distribution in the primary beam (e.g. possibly selected by a monochromator crystal with its specific mosaic spread), the beam characteristics (as influenced by slits and collimator arrangements between the primary radiation source and monochromator, between monochromator and sample, and between sample and detector), the detector system. Accordingly, there are many choices of analytical peak-shape functions. We quote:

$$\frac{C_0^{1/2}}{\sqrt{\pi}\, H_k} \exp(-C_0 X_{ik}^2) \qquad \text{(Gaussian);}$$

$$\frac{C_1^{1/2}}{\pi H_k} (1 + C_1 X_{ik}^2)^{-1} \qquad \text{(Lorentzian);}$$

$$\frac{2C_2^{1/2}}{\pi H_k} (1 + C_2 X_{ik}^2)^{-2} \qquad \text{(modif. 1 Lorentzian);}$$

$$\frac{C_3^{1/2}}{2H_k} (1 + C_3 X_{ik}^2)^{-1.5} \qquad \text{(modif. 2 Lorentzian);}$$

$$\frac{\eta C_1^{1/2}}{\pi H_k} (1 + C_1 X_{ik}^2)^{-1} + (1-\eta) \frac{C_0^{1/2}}{\pi^{1/2} H_k} \exp(-C_0 X_{ik}^2)$$

$$\text{with } 0 \leqslant \eta \leqslant 1 \quad \text{(pseudo-Voigt);}$$

$$\frac{\Gamma(\beta)}{\Gamma(\beta-0.5)}\frac{C_4}{\pi}\frac{2}{H_k}(1+4C_4X_{ik}^2)^{-\beta} \qquad \text{(Pearson VII);}$$

where $C_0 = 4 \ln 2$, $C_1 = 4$, $C_2 = 4(\sqrt{2} - 1)$, $C_3 = 4\ (2^{2/3} - 1)$, $C_4 = 2^{1/\beta} - 1$, $X_{ik} = \Delta\theta_{ik}/H_k$. H_k is the full-width at half-maximum (FWHM) of the kth Bragg reflection, and Γ is the gamma function.

It is easily seen that the pseudo-Voigt function presents the mixing parameter η which gives the per cent Lorentzian character of the profile. When $\beta = 1, 2, \infty$, Person VII becomes a Lorentzian, modified Lorentzian, and Gaussian function respectively. Of some use also is the pure Voigt function which is the convolution of Gaussian and Lorentzian forms.

The FWHM is usually considered to vary with scattering angle according to

$$(\text{FWHM})_{\text{G}} = (U \tan^2 \theta + V \tan \theta - W)^{1/2} \qquad (2.71)$$

for the Gaussian component,[78] according to

$$(\text{FWHM})_{\text{L}} = X \tan \theta + Y/\cos \theta \qquad (2.72)$$

for the Lorentzian component.[79] U, V, W, and/or X, Y are variable parameters in the profile refinement.

Besides analytical functions non-analytical functions arising from an analysis of resolved peaks may also be used[80] to describe peak shape (in the Rietveld method the peak shape is not the end but a tool of the method).

There is no well established approach to the background. It is mainly due to insufficient shielding, to diffuse scattering, to incoherent scattering (rather high for neutrons), to electronic noise of the detector system. The background and its variation with angle is usually defined by refinement of the coefficients of a power series in 2θ:

$$y_{i\text{b}} = \sum_n b_n(2\theta_i)^n$$

where the b_n terms are refinable parameters.

All the above described parameters are introduced in the Rietveld refinement process according to (2.70). The agreement between the observations and the model is estimated by the following indicators:[81]

(1) The profile $R_{\text{p}} = \Sigma\,|y_{i\text{o}} - y_{i\text{c}}|/(\Sigma\, y_{i\text{o}})$.

(2) The weighted profile $R_{\text{wp}} = [\Sigma\, w_i(y_{i\text{o}} - y_{i\text{c}})^2/\Sigma\, w_i y_{i\text{o}}^2]^{1/2}$.

(3) The Bragg $R_{\text{B}} = \Sigma\,|I_{k\text{o}} - I_{k\text{c}}|/(\Sigma\, I_{k\text{o}})$. The values $I_{k\text{o}}$ are obtained by partitioning the row data in accordance with the $I_{k\text{c}}$ values of the component peaks.

(4) The expected $R_{\text{E}} = [(N - P)/(\Sigma\, w_i y_{i\text{o}}^2)]^{1/2}$, where N and P are the number of profile points and refined parameters respectively.

(5) The goodness of fit $\text{GofF} = \Sigma\, w_i(y_{i\text{o}} - y_{i\text{c}})^2/(N - P) = (R_{\text{wp}}/R_{\text{E}})^2$, which should approach the ideal value of unity.

The most meaningful indices for the progress of refinement are R_{wp} and GofF since they show in the numerator the quantity being minimized . Also R_{B} is of considerable use since it depends on the fit of structural parameters more than on the profile parameters.

The use of restraints in the Rietveld method has proved to be of particular usefulness for complex structures.[80, 82, 83]

Some practical aspects of Rietveld refinement

The generation of instrument profile calibration curves usually precedes Rietveld refinement. Such an analysis allows one to set up the angular dependence of the peak shape. A peak profile may be expressed as

$$y(2\theta) = b(2\theta) + [w(2\theta) * g(2\theta)] * f(2\theta), \tag{2.73}$$

where $b(2\theta)$ is the background function, $f(2\theta)$ is a specimen related function, and $w(2\theta) * g(2\theta)$ is the so-called instrument function. This may be considered as the convolution of the function $g(2\theta)$ representing the diffractometer's optics and a second function $w(\theta)$ representing the wavelength distribution of the incident radiation.

As a first step, diffraction profiles from a standard specimen (e.g. silicon, corundum, or quartz standard materials) devoid of crystallite size or strain broadening effects are collected: in this case $f(2\theta)$ may be represented by a delta function and (2.73) becomes

$$y(2\theta) = b(2\theta) + [w(2\theta) * g(2\theta)]. \tag{2.74}$$

If a single wavelength is used (i.e. in synchrotron or neutron experiments) a single peak may be represented by one or by a combination of the functions described in the previous subsection. If a Cu X-ray source is used and α_2 and α_3 are not eliminated by some experimental device (i.e. a germanium monochromator) a compound profile consisting of three split functions has to be chosen, one for each spectral component. The number of variable parameters in the profile function may be reduced by fixing α_2 and α_3 intensities relative to α_1 intensity via a previous profile-fitting experiment. Analogously, the angles for α_2 and α_3 lines may be fixed according to their wavelengths, and their FWHM can be set equal to that of α_1.

The correct profile function is in general obtained by trial-and-error methods from single well behaved profiles of the standard material. The goodness of the profile fit is measured by

$$R = \left(\sum_{i=1}^{n} w_i(y_{io} - y_{ic})^2\right)\left(\sum_{i=1}^{n} w_i y_{io}^2\right)^{-1}$$

where y_{io} and y_{ic} are the observed and calculated values at the position $2\theta_i$. For good experimental data the R value is expected to be around one per cent.

The values of the profile parameters may be refined for each single peak collected from the standard: each peak is separated and analysed in a region with enough points to allow a good sampling of background on each side of the peak (a parameter for background may also be refined). In particular, the values of FWHM obtained from refinement are used for a first estimate of the U, V, W values in eqn (2.71) and/or of X, Y values in eqn (2.72). It is worthwhile mentioning that standard specimen data are also used for the determination of wavelength λ (if necessary) and for the zero-point calibration $2\theta_0$ of the detector scale.

Analysis of the specimen profile f usually follows instrument profile calibration. Specimen broadening of the profiles is often present: it is traditionally assumed to be Lorentzian if profile broadening is present.

The profile analysis of the full diffraction pattern is often used to refine[84–86] peak positions and peak heights for each reflection: from them lattice parameters are calculated and refined in an iterative way. Peaks generated by overlapping of more reflections are deconvoluted and integrated intensities of individual *hkl* are calculated, together with their standard deviations and their correlations (see Appendix 3.A, p. 184).

A specific feature of Rietveld refinement is that the number of observations N can be chosen arbitrarily large simply by decreasing the step interval in 2θ. The increase of N does not remarkably affect the conventional Rietveld agreement but decreases the parameter estimated standard deviations approximately in proportion to $N^{1/2}$. Such an improvement is often fictitious. For example, a single Bragg reflection may be measured using 10, 100, or 1000 steps. The observed intensities are not statistically independent[87] and, for single peaks, yield structural information nearly equivalent to a single integrated intensity count (even if the precision obtained by 1000 steps count may be higher than by 10 or 100 counts). As a consequence the correlation among observations becomes higher with $N = 10$, 100, 1000 and least-squares residuals $\Delta_i = y_{io} - y_{ic}$ are correlated themselves, i.e. they tend to assume the same sign. In such a situation Durbin–Watson d-statistics[88] may be used to assess the nature of correlation. It is based on the parameter

$$d = \sum_{i=2}^{N} \left(\frac{\Delta_i}{\sigma_i} - \frac{\Delta_{i-1}}{\sigma_{i-1}} \right)^2 \left[\sum_{i=1}^{N} \left(\frac{\Delta_i}{\sigma_i} \right)^2 \right]^{-1}.$$

For no serial correlation a d value close to 2 is expected. For positive serial correlations adjacent Δ tend to have the same sign and d will be smaller than 2, whereas for negative serial correlation (alternating signs of adjacent Δ) d is expected to be larger than 2 (and smaller than 4).

The parameter d may be tested against the 0.1 per cent significance point via the formula[89]

$$Q = 2[(N-1)(N-P) - 3.0902/(N+2)^{1/2}]$$

where P is the number of least-squares parameters involved in the calculations. $Q < d < 4 - Q$ if consecutive terms are uncorrelated, if $d < Q$ or $d > 4 - Q$ consecutive terms tend to have positive or negative serial correlation. d-statistics suggest that an optimum value of the step width is[90] between one-fifth and one-half of the minimum FWHM of well resolved peaks.

The success of the Rietveld method is based on some fundamental requirements:

1. A starting model for atomic positional and thermal parameters should be available in advance. That can be provided by the following procedure:[91] (a) deconvolution techniques and profile-fitting methods are used to derive intensities I_k of individual reflections and to resolve overlapping peaks; (b) integrated intensities are calculated and structure solved by direct or Patterson methods. The same intensities are then used in a full-matrix least-squares refinement as for single-crystal structure determination. Because of peak overlapping, this time error margins for each

observation and correlation among reflections intensities could be processed to calculate a non-diagonal matrix $\mathbf{M}_f$ (see pp. 91–3).

2. The crystallites should be randomly distributed. That is not easy to achieve in practice.[92,93] Traditionally preferred orientations are minimized by loose packing of the powder: however, sufficient compression improves homogeneity and reduces surface roughness effects. If a specimen spinner is used in the experimental set-up the preferred orientation can be characterized by a few parameters only, thus it may be introduced in a rather simple way in the fitting schemes and in Rietveld refinement. If the preferred orientation is not taken into account, errors in the relative intensities may occur which limit the accuracy of the refinement process. Some geometries are less sensitive to this problem (i.e. Guinier cameras). The problem is also minimized in neutron diffraction because of the large transmission specimen, but it is severe when synchrotron radiation is used because the number of crystallites correctly oriented to reflect the parallel beam is much smaller.[94]

The most simple way to correct the intensities calculated from the model structure is to use an empirical correction function depending on ϕ, the acute angle between the preferred orientation plane and the diffracting plane k:

$$I_k(\text{corr}) = I_k P_k(\phi).$$

P_k may be Gaussian,[77, 95]

$$P_k(\phi) = \exp(-p\phi^2),$$

or

$$P_k(\phi) = \exp[p(\pi/2 - \phi^2)],$$

or of trigonometric type[96]

$$P_k(\phi) = (p^2 \cos^2 \phi + \sin^2 \phi/p)^{-3/2}.$$

p is a parameter to refine in the refinement process.

3. The crystallites should be of equal size. Particles much larger than 10 μm should be removed from the sample, as well as very small particles (<1 μm), which are responsible for profile broadening. Diffraction patterns of materials with single composition but two different distinct distributions of crystallite size (or of microstrain) could be characterized by profile-broadening effects with bimodal distribution.[97] In this case unimodal distributions are inadequate to describe profile functions, and unavoidable profile misfit errors will still appear at the end of refinement when observed and calculated patterns are compared.

4. Correction for instrumental aberrations[98–100] should be made.

The instrument function g in eqn (2.73) may be considered as the result of six specific functions:

$$g = g_1 * g_2 * \ldots. * g_6.$$

The function g_1 depends on the projected focal spot profile, g_2 is due to the varying displacements of the various parts of the flat specimen surface from the focusing circle, g_3 is due to the axial divergence (as regulated by Soller slit collimators), g_4 arises from specimen transparency (i.e. from the penetration by the beam of a sample with finite absorption), g_5 is defined by the receiving slit, g_6 is due to the possible misalignment of the experimental set-up. As may be seen from Fig. 2.13, only g_1, g_5, and g_6 broaden the profile symmetrically, while g_2, g_3, g_4 broaden it asymmetrically and shift the peak from the theoretical position. The deleterious effects of the aberrations can be minimized by the use of appropriate analytical functions in Rietveld refinement or by a careful design and/or alignment of the instrument.

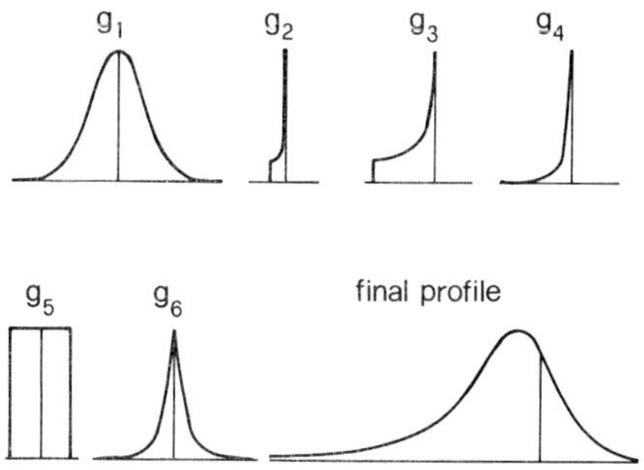

Fig. 2.13. Functions defining the instrument function *g*.

5. Resolution of the pattern may be improved by mathematical techniques:[101–103] i.e. by deconvolution of the pure from the observed profile, this last containing the effect of instrumental broadening (see Appendix 3.A, p. 185).

The resolution of the pattern as well as the signal-to-noise ratio can be appreciably improved with carefully designed diffraction experiments. When a non-conventional diffraction apparatus is used one of the first choices to be made is fixing the wavelength. Complex structures will produce peaks closely spaced and frequently overlapping: a long wavelength improves separation between peaks but reduces the number of accessible reflections which cannot then offset the large number of structural parameters to refine. A short wavelength increases the number of accessible peaks but produces severe overlap of them. Thus a useful compromise has to be chosen between the two requirements.[104] When higher resolution is needed for conventional X-ray sources it is advisable to use the K_β doublet (but the total experiment time will increase).

In order to compare the performances of different experimental arrangements, in Fig. 2.14 typical FWHM (instrument-only contribution) are plotted against the diffraction angle for modern neutron diffractometer (N), for a conventional Bragg–Brentano with (XCS) and without (XCW) diffracted beam Soller slit (Cu K_α radiation and a diffracted-beam curved graphite monochromator) limiting vertical beam divergence to less than the standard value of 5°, and for a synchrotron powder diffractometer (S). The use of incident beam monochromators can further improve performances of conventional divergent beam diffractometry.[105] Indeed, the K_{α_2} component may be removed, so halving the number of lines in the pattern. As a consequence, the resolution of the remaining lines is improved, the

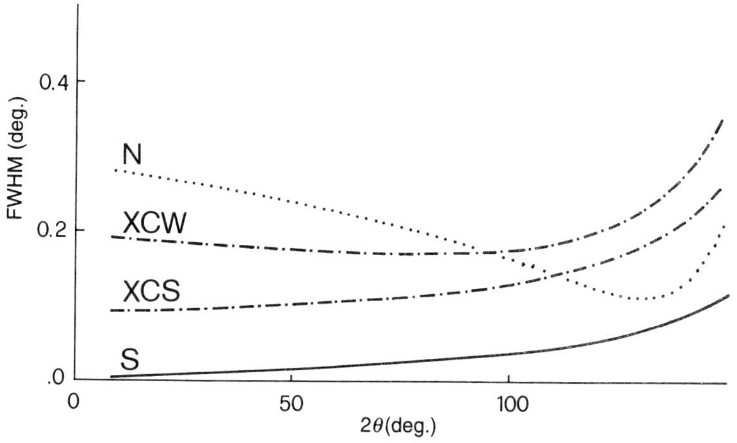

Fig. 2.14. FWHM for different experimental arrangements.

background is kept essentially constant (indeed K_β lines are removed at source, and also structures arising from the filter absorption edge are eliminated).

The best resolution is attained by synchrotron radiation: the probability of measuring single peaks is larger than for any other radiation source. It is not infrequent[106] that synchrotron data present so large a percentage of unique $I_{\mathbf{H}}$ values that structure determination may be carried on by techniques identical to those used for single-crystal data (for rather large structures specialized solution techniques have still to be applied[107]).

Neutrons have smaller resolution, but a better signal-to-noise ratio at high $\sin\theta/\lambda$ values. Because of the narrow range of neutron scattering lengths, refinement is facilitated when some heavy atoms are present; however, the structure solution may be difficult. This situation is just complementary to that occurring for X-ray data, where location of some dominant atoms may be easy but subsequent accurate location of very light atoms may be difficult. It is therefore not infrequent that the Rietveld method benefits by a combined use of neutron and X-ray data. When some heavy atoms are present in the structure or high-angle X-ray data are not attainable, refinement with neutron data may be used to confirm results obtained with X-ray data[79, 108, 109] (however, improvement in precision may be obtained if refinement is carried out on the combined X-ray and neutron data sets[110]). As an example in Fig. 2.15(a,b) observed (points), calculated (bold line), and difference profiles generated by powder samples of α-$CrPO_4$ are shown.[79] Synchrotron intensity data show a very good instrumental resolution (FWHM $= 0.07°$ at $2\theta = 50°$) but a remarkable intensity decay with $\sin\theta/\lambda$: the attained maximum $\sin\theta/\lambda$ value is $0.35\ \text{Å}^{-1}$, corresponding to 97 independent reflections. Neutron data have smaller resolution (FWHM $= 0.6°$ at $2\theta = 50°$) but no serious intensity decay with $\sin\theta/\lambda$: the maximum $\sin\theta/\lambda$ value is $0.40\ \text{Å}^{-1}$, corresponding to 127 independent reflections. Since the neutron scattering amplitudes for Cr, P, and O are 0.352, 0.53, and 0.577 respectively (see Appendix 3B, p. 198) neutron refinement is expected to be more sensitive than X-ray to oxygen positions.

The Rietveld procedure has been applied to a very wide range of inorganic and organic materials[111] and scientific interest in this area is increasing (the present annual publication rate is approximately 100). The limits of the method depend on progress in experimental aspects (sample preparation, higher resolution, better treatment of profiles, etc.) and on the available prior information which may be introduced into refinement. Such information is often generated by complementary experimental techniques. For example, intensities of overlapping reflections of $LaMo_5O_8$ were broken up by considering electron diffraction patterns:[109] 14 symmetry-independent atoms were then positioned and their isotropic thermal motion determined.

As a further example we recall the application of the Rietveld method to polymers.[82,112,113] Electron diffraction is a generally applied technique when chain-folded crystals are of minimal thickness and crystal deformation negligible. If such conditions are not satisfied quantitative kinematic treatment leads to models of low accuracy and the Rietveld method may be applied in order to discriminate unsatisfactory structural models. Owing to the smallness of crystallites and to the greater disorder present in the

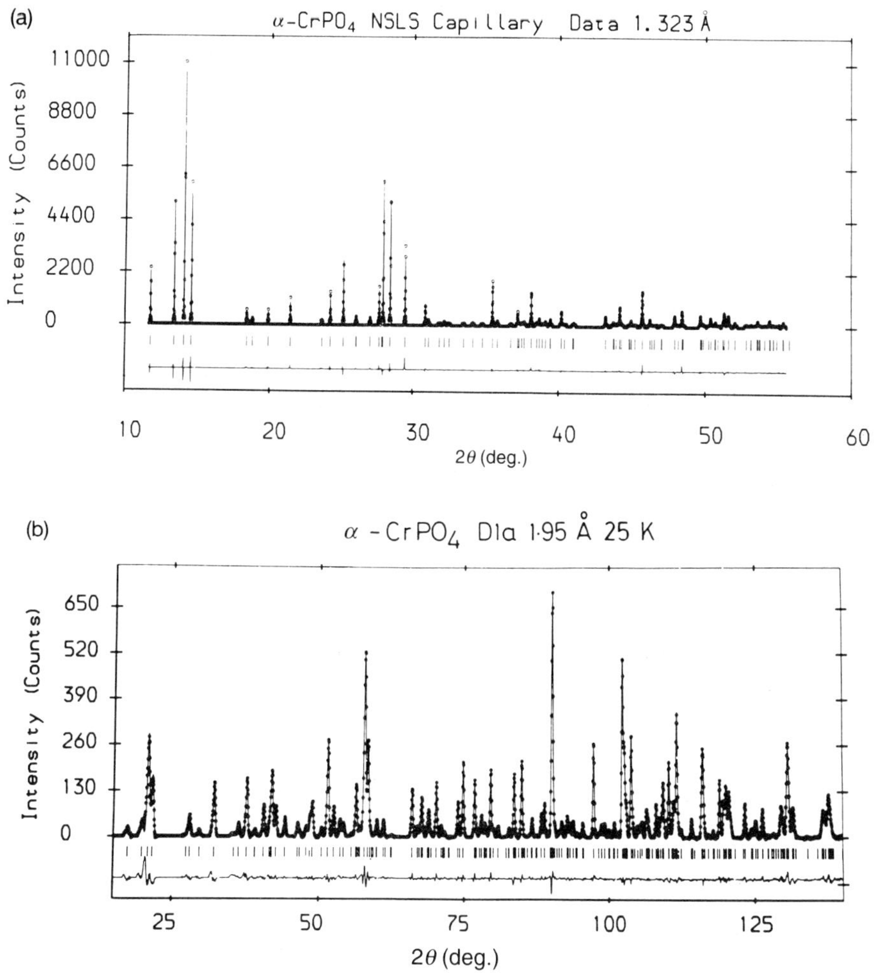

Fig. 2.15. Powder diffraction profiles for α-$CrPO_4$ from: (a) the synchrotron X-ray experiment; (b) neutron experiment. Reflection positions are marked.

polymers the width of peaks is rather large for polymers so that overlapping events in their pattern are more frequent. In spite of that, discrimination among various structural models may often be accomplished provided prior information (size and shape of rigid molecular fragments) is adequately introduced into refinement.

Analysis of thermal motion

In biological macromolecules the root-mean-square displacements of atoms from their mean positions are of the order of tenths of an angstrom. On the other hand the allowed variations for covalently bonded distances are no greater than some hundredths of an angstrom. The apparent contradiction may be overcome by assuming a high correlation among the thermal motions of bonded atoms. In other words, if an atom at a certain instant is out from its equilibrium position, some other atoms coherently move in such a way that bond distances are chemically meaningful.

The above conclusion is valid also for small molecules. Indeed more or less rigid groups of atoms can be found (e.g. rings, condensed rings, etc.)

for which the internal molecular motion may be negligible and the thermal motion may be described in terms of a rigid body model.[114–116]

In the crystallographic least-squares procedures (see p. 94) any correlation among $\boldsymbol{\beta}$ tensors of different atoms is usually ignored. However, if thermal ellipsoids correctly represent the thermal motion some a posteriori correlations among them can be found. For example, since bond stretching vibrations have a much smaller amplitude than other sources (i.e. bond bending or torsional vibrations) the mean square displacement of pairs of bonded atoms should be approximately equal in the bond direction. Or also, in long-chain molecules the thermal motion normal to the chain should be greater than at right angles to it; or, terminal atoms such as the O atoms in carbonyl groups or H atoms in methyl groups have generally greater thermal motion than the atoms to which they are bonded. Thus if the crystal contains more-or-less rigid groups of atoms, it makes sense to analyse thermal motion in terms of translational and librational oscillations of these groups. In the following, such an analysis will be described on assuming a Cartesian coordinate system.

In accordance with Appendix 1.A the most general motion of a rigid body is a screw rotation. If the axis of rotation is correctly oriented but incorrectly positioned, the rotation and the translation component parallel to the axis do not vary, but additional translation components perpendicular to the rotation axis are introduced. Thus the most general motion of a rigid body may be considered as the combination of a rotation with a suitable translation. These operations do not commute in general; luckily an adequate treatment of anisotropic thermal motion, accurate to the level of quadratic approximation, can be used upon infinitesimal rotations which do commute.

To illustrate this representation of rotation, consider an atom at $\bar{\mathbf{X}} = (x_1, x_2, x_3)$ in a Cartesian coordinate system $\mathbf{E}$. According to p. 71, in such a system, a rotation through χ about the unitary vector $\boldsymbol{l}$ (l_1, l_2, l_3 will be its direction cosines), is represented by relation (2.32b). If χ is sufficiently small (2.32b) may be written as

$$\mathbf{X}' = \mathbf{RX}, \quad \text{where} \quad \mathbf{R} = \mathbf{I} + \chi\mathbf{P} \quad \text{and} \quad \mathbf{P} = \begin{pmatrix} 0 & -l_3 & l_2 \\ l_3 & 0 & -l_1 \\ -l_2 & l_1 & 0 \end{pmatrix}. \tag{2.75}$$

The change of variable $\boldsymbol{q} = \chi\boldsymbol{l}$ brings to

$$\mathbf{X}' = (\mathbf{I} + \mathbf{K})\mathbf{X} = \mathbf{X} + \mathbf{KX} = \mathbf{X} + \mathbf{AQ}$$

where

$$\mathbf{K} = \begin{pmatrix} 0 & -q_3 & q_2 \\ q_3 & 0 & -q_1 \\ -q_2 & q_1 & 0 \end{pmatrix}, \qquad \mathbf{A} = \begin{pmatrix} 0 & x_3 & -x_2 \\ -x_3 & 0 & x_1 \\ x_2 & -x_1 & 0 \end{pmatrix}, \qquad \mathbf{Q} = \begin{pmatrix} q_1 \\ q_2 \\ q_3 \end{pmatrix}.$$

It may be concluded that the most general infinitesimal displacement of an atom in a molecular crystal from its equilibrium position is

$$\boldsymbol{u} = \boldsymbol{t} + \boldsymbol{q} \wedge \boldsymbol{r} = \boldsymbol{t} + (\boldsymbol{r}' - \boldsymbol{r}) = \boldsymbol{t} + \mathbf{A}\boldsymbol{q}. \tag{2.76}$$

The variance–covariance matrix for the displacement $\boldsymbol{u}$ is

$$\begin{aligned} \mathbf{U} = \langle \boldsymbol{u}\bar{\boldsymbol{u}} \rangle &= \langle \boldsymbol{t}\bar{\boldsymbol{t}} \rangle + \langle \mathbf{A}(\boldsymbol{q}\bar{\boldsymbol{q}})\bar{\mathbf{A}} \rangle + \langle \mathbf{A}(\boldsymbol{q}\bar{\boldsymbol{t}}) \rangle + \langle (\boldsymbol{t}\bar{\boldsymbol{q}})\bar{\mathbf{A}} \rangle \\ &= \mathbf{T} + \mathbf{AL\bar{A}} + \mathbf{AS} + \mathbf{\bar{S}\bar{A}} \end{aligned} \tag{2.77}$$

where

$$\mathbf{U}=\begin{pmatrix}\langle u_1^2\rangle & \langle u_1u_2\rangle & \langle u_1u_3\rangle\\ \langle u_2u_1\rangle & \langle u_2^2\rangle & \langle u_2u_3\rangle\\ \langle u_3u_1\rangle & \langle u_3u_2\rangle & \langle u_3^2\rangle\end{pmatrix},\qquad \mathbf{L}=\begin{pmatrix}\langle q_1^2\rangle & \langle q_1q_2\rangle & \langle q_1q_3\rangle\\ \langle q_2q_1\rangle & \langle q_2^2\rangle & \langle q_2q_3\rangle\\ \langle q_3q_1\rangle & \langle q_3q_2\rangle & \langle q_3^2\rangle\end{pmatrix}$$

$$\mathbf{S}=\begin{pmatrix}\langle q_1t_1\rangle & \langle q_1t_2\rangle & \langle q_1t_3\rangle\\ \langle q_2t_1\rangle & \langle q_2t_2\rangle & \langle q_2t_3\rangle\\ \langle q_3t_1\rangle & \langle q_3t_2\rangle & \langle q_3t_3\rangle\end{pmatrix},\qquad \mathbf{T}=\begin{pmatrix}\langle t_1^2\rangle & \langle t_1t_2\rangle & \langle t_1t_3\rangle\\ \langle t_2t_1\rangle & \langle t_2^2\rangle & \langle t_2t_3\rangle\\ \langle t_3t_1\rangle & \langle t_3t_2\rangle & \langle t_3^2\rangle\end{pmatrix}.$$

U, **T**, and **L** are symmetrical matrices while **S** is not.

T, the translation matrix, describes the translational motion of the molecule: it defines an ellipsoid and $\bar{\boldsymbol{l}}\mathbf{T}\boldsymbol{l}$ is the mean-square amplitude of translational vibration along the direction $\boldsymbol{l}$.

L, the libration matrix, describes the librational motion of the molecule: $\bar{\boldsymbol{l}}\mathbf{L}\boldsymbol{l}$ provides the mean-square amplitude of libration of the molecule about the direction $\boldsymbol{l}$. The corresponding ellipsoid defines the preferred directions for the libration of the rigid body. Thus diagonalization of **T** and **L** yields the magnitudes and directions of their principal axes. The eigenvectors of **L** represent the three principal libration axes, its eigenvalues the mean-square librational amplitudes about them.

S is the called translation–libration matrix or screw correlation matrix, even if only its diagonal terms correspond to screwlike motions. The off-diagonal elements indeed correspond to rotations around axes that do not pass through the origin. $\bar{\boldsymbol{l}}\mathbf{S}\boldsymbol{l}$ gives the mean correlation between the libration about $\boldsymbol{l}$ and the translation parallel to that axis. Since $\langle q_it_j\rangle \neq \langle q_jt_i\rangle$ the **S** matrix is not symmetric.

Equation (2.77) may be expressed as a partitioned matrix equation

$$\mathbf{U}=[\mathbf{I}\cdot\mathbf{A}]\begin{pmatrix}\mathbf{T} & \bar{\mathbf{S}}\\ \mathbf{S} & \mathbf{L}\end{pmatrix}\begin{pmatrix}\mathbf{I}\\ \bar{\mathbf{A}}\end{pmatrix}=\mathbf{E}_{(3\times 6)}\mathbf{C}_{(6\times 6)}\bar{\mathbf{E}}_{(6\times 3)}. \qquad (2.78)$$

The symmetric matrix **C** may be interpreted as the product $\langle \boldsymbol{c}\bar{\boldsymbol{c}}\rangle$ of the six-dimensional vector $\bar{\boldsymbol{c}}=[t_1, t_2, t_3, q_1, q_2, q_3]$ in the configurational space of rigid-body rotations and translations. At any instant the position of the rigid body is specified by a point of such a space.

Shifting the origin by the vector $\boldsymbol{r}_0$ converts $\boldsymbol{r}$ into $\boldsymbol{r}-\boldsymbol{r}_0$, but does not modify the rotational vector $\boldsymbol{q}$ and the displacement $\boldsymbol{u}$. Thus (2.76) becomes

$$\boldsymbol{u}'=\boldsymbol{u}=\boldsymbol{t}'+\boldsymbol{q}\wedge(\boldsymbol{r}-\boldsymbol{r}_0)=(\boldsymbol{t}'-\boldsymbol{q}\wedge\boldsymbol{r}_0)+\boldsymbol{q}\wedge\boldsymbol{r}=\boldsymbol{t}+\boldsymbol{q}\wedge\boldsymbol{r}$$

from which $\boldsymbol{t}'=\boldsymbol{t}+\boldsymbol{q}\wedge\boldsymbol{r}_0$. It may be concluded that the tensor **L** is independent on the chosen origin while **T** and **S** vary with it.

Once the anisotropic vibrational parameters U_{ij} for the individual atoms in the molecule have been determined by a careful least-squares process, the elements of **T**, **L**, **S** can be found by a least-squares fit. Let the rigid fragment contain n symmetry-independent atoms in general position: then $6n$ observational equations can be stated. While **T** and **L** are symmetric (six independent components each), **S** is unsymmetric and has nine components. However, the total number of unknowns in the least squares process is 20 rather than 21 owing to the fact that the system of equations is degenerate. Redundancy arises because relation (2.77) is satisfied both by **S** and by

$\mathbf{S} + \mathbf{D}$ provided

$$\mathbf{AD} + \bar{\mathbf{D}}\bar{\mathbf{A}} = 0 \tag{2.79}$$

is satisfied. Indeed, if $\mathbf{S}$ in (2.77) is replaced by $\mathbf{S} + \mathbf{D}$

$$\mathbf{U} = \mathbf{T} + \mathbf{AL}\bar{\mathbf{A}} + \mathbf{AS} + \bar{\mathbf{S}}\bar{\mathbf{A}} + \mathbf{AD} + \bar{\mathbf{D}}\bar{\mathbf{A}}$$

is obtained from which (2.79) arises. Since $\bar{\mathbf{A}} = -\mathbf{A}$ we will have $\mathbf{D} = k\mathbf{I}$ where k is an arbitrary constant. It may be concluded that the values of U_{ij} calculated from $\mathbf{T}$, $\mathbf{L}$, $\mathbf{S}$ do not vary if $\mathbf{D}$ is added to $\mathbf{S}$. Vice versa, only the differences $S_{11} - S_{22}$, $S_{22} - S_{33}$, $S_{33} - S_{11}$ may be obtained from the observed U_{ij}. Since their sum is zero the number of determinable components of $\mathbf{S}$ is eight: usually it is preferred to have the trace of $\mathbf{S}$ equal to zero.

Often thermal motion analysis produces an excellent agreement between observed and calculated U_{ij} (e.g. $\langle(\Delta U_{ij})^2\rangle^{1/2} \simeq (1\text{–}4) \times 10^{-3}$ Å^2): sometimes worse agreement occurs (e.g. of the order of 10^{-2} Å^2 or more). Disagreement may be due to internal molecular motion of the atoms as well as to errors in the observed U_{ij}, or also to the inadequacy of the approximation (2.75).

The effect of thermal motion on bond lengths and angles

From a crystal structure analysis atomic coordinates are obtained which represent the centroid of a distribution of electron density arising from the combined effect of the atomic structure and of the thermal vibration. Usually interatomic distances are calculated as separations between a pair of atomic positions (distances from mean positions), but it has become clear now[117] that, if the thermal motion is not negligible, the most suitable measure of an interatomic distance is the mean separation (usually longer than inter-mean distances), the average being calculated by taking into account the joint distribution of the thermal motion of the two atoms. Such complex information is seldom available, but simplified models of the vibrating system do provide useful estimates of the mean separation.

Let $\boldsymbol{d}_0$ represent the separation of the mean positions of two atoms A and B, and $\boldsymbol{d} = \boldsymbol{d}_0 + \boldsymbol{s}$ be the instantaneous separation. For convenience consider a cylindrical coordinate system with axis z in the direction of $\boldsymbol{d}_0$. If $\boldsymbol{v}$ and $\boldsymbol{w}$ are the axial and radial vectorial components of $\boldsymbol{s}$ respectively then $\boldsymbol{d} = \boldsymbol{d}_0 + \boldsymbol{v} + \boldsymbol{w}$ from which $d = [(d_0 + v)^2 + w^2]^{1/2}$.

Taylor–Maclaurin expansion of $d(v, w)$ yields

$$d = d_0 + v + w^2/(2d_0)$$

from which

$$\langle d\rangle \simeq d_0 + \langle w^2\rangle/(2d_0). \tag{2.80}$$

Express now the value of $\langle w^2\rangle$ in terms of quantities which are experimentally accessible for three simplified models of thermal motion. It may be noted that $\boldsymbol{w}$ is the difference between the projected instantaneous displacements $\boldsymbol{w}_{\mathrm{B}}$ and $\boldsymbol{w}_{\mathrm{A}}$ of the two atoms B and A ($\boldsymbol{w} = \boldsymbol{w}_{\mathrm{B}} - \boldsymbol{w}_{\mathrm{A}}$) so that

$$\langle w^2\rangle = \langle w_{\mathrm{B}}^2\rangle - 2\langle w_{\mathrm{B}} w_{\mathrm{A}}\rangle + \langle w_{\mathrm{A}}^2\rangle.$$

We will examine three simplified models:

1. The motion of the two atoms is not correlated (as a first approximation, that occurs for non-bonded atoms). In this case $\langle w^2 \rangle = \langle w_A^2 \rangle + \langle w_B^2 \rangle$; therefore

$$\langle d \rangle = d_0 + (\langle w_A^2 \rangle + \langle w_B^2 \rangle)/(2d_0). \tag{2.81}$$

2. The atom B has all of the translational motion of the atom A plus an additional motion uncorrelated with the instantaneous position of A (**riding motion**). Such a model well fits the case in which the riding atom B is strongly linked only to the atom A and the mass of B is considerably smaller than the mass of A (eventually summed to neighbouring atoms strongly linked to A, as occurs for example if A belongs to a rigid group). Then

$$\langle d \rangle = d_0 + (\langle w_B^2 \rangle - \langle w_A^2 \rangle)/(2d_0). \tag{2.82}$$

Express now $\langle w_A^2 \rangle$ and $\langle w_B^2 \rangle$ in terms of the vibrational parameters usually obtainable from least-squares refinements: the corresponding expressions could be introduced into (2.81) and (2.82) in order to obtain $\langle d \rangle$.

If the thermal motion of the atom A is assumed to be isotropic then, according to (3.19), $\langle w_A^2 \rangle = 2B_A/8\pi^2$, $\langle w_B^2 \rangle = 2B_B/8\pi^2$, where B_A and B_B are the thermal factors estimates arising from least-squares calculations for A and B atoms respectively. Then (2.81) and (2.82) become

$$\langle d \rangle = d_0 + (B_A + B_B)/(8\pi^2 d_0)$$

and

$$\langle d \rangle = d_0 + (B_B - B_A)/(8\pi^2 d_0),$$

respectively. If the thermal motion is assumed to be anisotropic, then a suitable coordinate system in the plane normal to $\boldsymbol{d}_0$ may be chosen, and (see eqn (2.38b)) we can write

$$\langle w_A^2 \rangle = \langle w_{Ax}^2 \rangle + \langle w_{Ay}^2 \rangle = (\lambda_{1A} + \lambda_{2A})/(2\pi^2).$$

According to (3.B.10a)

$$\langle w_A^2 \rangle = 3\langle u_A^2 \rangle_{\text{aniso}} - \frac{\lambda_{3A}}{2\pi^2}$$

where $\lambda_{3A}/(2\pi^2)$ is the mean-square displacement of the atom A along $\boldsymbol{d}_0$. On introducing (3.B.10c) and (3.B.9a) we obtain

$$\langle w_A^2 \rangle = \frac{1}{2\pi^2}\left(\text{Tr}\,(\boldsymbol{\beta}_A \mathbf{G}) - \frac{\tilde{\mathbf{X}}_0 \mathbf{G} \boldsymbol{\beta}_A \mathbf{G} \mathbf{X}_0}{\tilde{\mathbf{X}}_0 \mathbf{G} \mathbf{X}_0}\right)$$

where $\mathbf{X}_0$ represents now the direct lattice components of the vector $\boldsymbol{d}_0$.

3. The rigid body model. Let $\boldsymbol{n}$ be the unit vector about which a group of atoms oscillates with mean square amplitude $\langle \omega^2 \rangle$. A small rotation $d\omega$ about $\boldsymbol{n}$ will produce over the interatomic vector $\boldsymbol{d}$ the variation $\delta \boldsymbol{d} = (\boldsymbol{n} \wedge \boldsymbol{d})\, d\omega$.

Since $\delta \mathbf{d}$ is perpendicular to $\boldsymbol{d}$, we can write

$$\langle w^2 \rangle = d^2 \langle \omega^2 \rangle \sin^2 \psi \simeq d_0^2 \langle \omega^2 \rangle \sin^2 \psi$$

where ψ is the angle between $\boldsymbol{d}_0$ and $\boldsymbol{n}$. In accordance with (2.80)

$$\langle d \rangle \simeq d_0 + (d_0 \omega^2 \sin^2 \psi)/2 = d_0(1 + \omega^2 \sin^2 \psi/2).$$

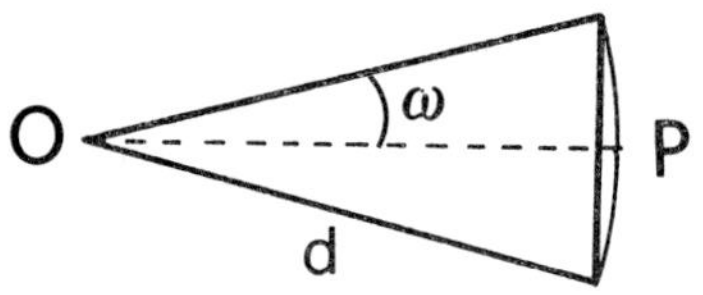

Fig. 2.16. Shortening of interatomic distances due to libration motion.

Shortening of interatomic distances will be more clearly deduced from Fig. 2.16. The libration axis $\boldsymbol{n}$ is normal to the page through O, d is the distance from P to the axis. The radial displacement is approximately given by $d - d\cos\omega \simeq d(1-\cos\omega) \simeq d(\omega^2/2)$.

As an example, a librational mean-square amplitude of 10′ may lead to shortening of interatomic distances of up to 0.025 Å.

Thermal motion is expected to produce distortion also in apparent angles. To study this type of effect the joint distribution of three correlated thermal motions has to be taken into account.[116]

About the accuracy of the calculated parameters

At the end of a structure analysis both positional and vibrational parameters are available, together with variances of the parameters and covariances among them. This supplementary information is of primary importance in assessing the accuracy of the parameters themselves. The variance of the parameter f in terms of variances of and covariances among other directly determined parameters p is

$$\sigma^2(f) = \sum_{i,j} \left(\frac{\delta f}{\delta p_i}\right)\left(\frac{\delta f}{\delta p_j}\right) \operatorname{cov}(p_i, p_j). \tag{2.83}$$

If the p_js are uncorrelated then (2.83) reduces to

$$\sigma^2(f) = \sum_i \left(\frac{\delta f}{\delta p_i}\right)^2 \sigma^2(p_i). \tag{2.84}$$

As an example, let us calculate the standard deviations of the unit cell volume V and of the angle α^* when the standard deviations of a, b, c, α, β, γ are known. On applying (2.84) to the expression of V in Table 2.1 we obtain

$$\sigma^2(V) = V^2\Big(\sum_i A_i^2\Big) + \frac{abc}{V^2}\Big(\sum_i B_i^2\Big)$$

where

$$\begin{aligned} A_1 &= \sigma(a)/a, & B_1 &= \sin\alpha(\cos\alpha - \cos\beta\cos\gamma)\sigma(\alpha), \\ A_2 &= \sigma(b)/b, & B_2 &= \sin\beta(\cos\beta - \cos\alpha\cos\gamma)\sigma(\beta), \\ A_3 &= \sigma(c)/c, & B_3 &= \sin\gamma(\cos\gamma - \cos\alpha\cos\beta)\sigma(\gamma). \end{aligned}$$

On applying (2.84) to the expression of $\cos\alpha^*$ in Table 2.1 we obtain

$$\begin{aligned} \sigma^2(\alpha^*) = \{&\sigma^2(\alpha)\sin^2\alpha + \sigma^2(\beta)(\sin\beta\cos\gamma + \cos\alpha^*\cos\beta\sin\gamma)^2 \\ &+ \sigma^2(\gamma)(\cos\beta\sin\gamma + \cos\alpha^*\sin\beta\cos\gamma)^2\}/(\sin\alpha^*\sin\beta\sin\gamma)^2. \end{aligned}$$

Apply now (2.84) to the relation (2.3b) in order to derive $\sigma^2(d^2)$, where d is the distance between two uncorrelated atoms positioned in (x_1, y_1, z_1) and (x_2, y_2, z_2) respectively. Then

$$\begin{aligned} \sigma^2(d) = \frac{1}{d^2}\{&(\Delta_1 + \Delta_2\cos\gamma + \Delta_3\cos\beta)^2[\Delta_1^2A_1^2 + a^2(\sigma^2(x_1) + \sigma^2(x_2))] \\ &+ (\Delta_1\cos\gamma + \Delta_2 + \Delta_3\cos\alpha)^2[\Delta_2^2A_2^2 + b^2(\sigma^2(y_1) + \sigma^2(y_2))] \\ &+ (\Delta_1\cos\beta + \Delta_2\cos\alpha + \Delta_3)^2[\Delta_3^2A_3^2 + c^2(\sigma^2(z_1) + \sigma^2(z_2))] \\ &+ (\Delta_1\Delta_2\sigma(\gamma)\sin\gamma)^2 + (\Delta_1\Delta_3\sigma(\beta)\sin\beta)^2 + (\Delta_2\Delta_3\sigma(\alpha)\sin\alpha)^2\}. \end{aligned} \tag{2.85a}$$

(2.85a) is simpler if the errors on the unit cell parameters can be neglected. An additional simplification is obtained if the errors are isotropic (i.e. $a^2\sigma_{x_1}^2 \simeq b^2\sigma_{y_1}^2 \simeq c^2\sigma_{z_1}^2 \simeq \sigma_1^2$ and $a^2\sigma_{x_2}^2 \simeq b^2\sigma_{y_2}^2 \simeq c^2\sigma_{z_2}^2 \simeq \sigma_2^2$) and the axes are orthogonal. Then (2.85a) reduces to

$$\sigma^2(d) = \sigma_1^2 + \sigma_2^2. \tag{2.85b}$$

For pairs of symmetry-related atoms (2.85a) cannot be applied. Thus, for a pair of atoms referred by an inversion centre, any error in one position is completely correlated with the error in the other and

$$\sigma(d) = 2\sigma,$$

where σ is measured in the direction of a line joining the atomic centres. It may be observed that 2σ (standard deviation for the distance of two completely correlated atoms) is larger than $\sqrt{2}\sigma$ (standard deviation for the distance of two uncorrelated atoms: see eqn (2.85b) for $\sigma_1 = \sigma_2$).

In general any pair of atoms in a structure are neither completely independent nor completely correlated, and the general expression (2.83) has to be invoked. An estimate of the covariance between the positional parameters p_i and p_j is usually obtained from the structure least-squares refinement (as the ijth element of the matrix $\mathbf{M}_x$ defined on p. 92). The covariances among the unit cell parameters are usually available from the least-squares refinement of the unit cell parameters.

By an analogous technique the standard deviations[118–121] of bond angles, torsion angles, and least-squares planes may be obtained. To assist the reader the simplified expressions of the standard deviations for bond angles θ and torsion angles ω are quoted:

$$\sigma_\theta^2 = \frac{\sigma_A^2}{d_{AB}^2} + \frac{\sigma_B^2 d_{AC}^2}{d_{AB}^2 d_{BC}^2} + \frac{\sigma_C^2}{d_{BC}^2}$$

$$\begin{aligned}\sigma_{(ABCD)}^2 = {} & \frac{\sigma_A^2}{d_{AB}^2 \sin^2(ABC)} + \frac{\sigma_D^2}{d_{CD}^2 \sin^2(BCD)} \\ & + \frac{\sigma_B^2}{d_{BC}^2}\left[\cot^2(BCD) + \left(\frac{d_{BC} - d_{AB}\cos(ABC)}{d_{AB}\sin(ABC)}\right)^2\right. \\ & \left. - 2\cos\omega\cot(BCD)\left(\frac{d_{BC} - d_{AB}\cos(ABC)}{d_{AB}\sin(ABC)}\right)\right] \\ & + \frac{\sigma_C^2}{d_{BC}^2}\left[\cot^2(ABC) + \left(\frac{d_{BC} - d_{CD}\cos(BCD)}{d_{CD}\sin(BCD)}\right)^2\right. \\ & \left. - 2\cos\omega\cot(ABC)\left(\frac{d_{BC} - d_{CD}\cos(BCD)}{d_{CD}\sin(BCD)}\right)\right].\end{aligned}$$

If in the first expression $d_{AB} \simeq d_{BC}$ and a common σ^2 is assumed then

$$\sigma_\theta^2 = 2\sigma^2(2 - \cos\theta)/d^2.$$

If in the second expression $d_{AB} \simeq d_{BC} \simeq d_{CD}$, equal bond angles (ABC = BCD = φ), and a common σ^2 are assumed then

$$\sigma^2(\omega) = \frac{4\sigma^2}{d^2\sin^2\varphi}[1 - \cos\varphi(1 - \cos\varphi)(1 - \cos\omega)].$$

Two simple examples on the use of the variances of the parameters in crystal structure analysis may be described:

1. It is often required to take a decision whether a bond length or angle differ significantly from some others. For example, suppose that two independent C–C bonds d_1 and d_2 are found which differ by 0.02 Å, while the isotropic standard deviations for the atomic positions are $\sigma_1 = \sigma_2 = \sigma_3 = \sigma_4 = 0.003$ Å. According to (2.85b) $\sigma(d) = 0.0042$ Å while the standard deviaton associated with $\delta = d_1 - d_2$ is
$$\sigma_\delta = [\sigma^2(d_1) + \sigma^2(d_2)]^{1/2} = 0.0059 \text{ Å}.$$
It is therefore found that the observed value of δ is $\lambda = 3.39$ times larger than σ_δ. If δ is assumed to be distributed around zero according to a normal distribution with standard deviation σ_δ, then a statistical criterion may be used to decide if d_1 and d_2 are significantly different. The conclusive inference is that the two bonds differ.

2. We are required to decide if a sample of n observed bond distances d_i is drawn from a common population or not. If the d_i are assumed to be normally distributed about the mean $\langle d \rangle$, the quantity
$$\chi^2 = \sum_{i=1}^{n} (d_i - \langle d \rangle)^2 / \sigma_i^2$$
is evaluated and compared with the tables of the $\chi^2_{n,p}$ distribution.

Appendices

2.A Some metric relations between direct and reciprocal lattices

Let us prove that $V^* = 1/V$. By definition
$$V^* = \boldsymbol{a}^* \cdot \boldsymbol{b}^* \wedge \boldsymbol{c}^* = \frac{1}{V^3} (\boldsymbol{b} \wedge \boldsymbol{c}) \cdot [(\boldsymbol{c} \wedge \boldsymbol{a}) \wedge (\boldsymbol{a} \wedge \boldsymbol{b})].$$

Owing to (2.8)
$$V^* = \frac{1}{V^3} (\boldsymbol{b} \wedge \boldsymbol{c}) \cdot [(\boldsymbol{c} \cdot \boldsymbol{a} \wedge \boldsymbol{b})\boldsymbol{a}] = \frac{1}{V}.$$

Derive now the values of α^*, β^*, γ^* from direct lattice parameters. According to the first eqn (2.13)
$$\sin \alpha^* = \frac{V^* a}{b^* c^*} = \frac{1}{V} \frac{a}{b^* c^*},$$
from which, owing to second and third relations (2.12b),
$$\sin \alpha^* = \frac{V}{abc \sin \beta \sin \gamma}$$
is obtained. Expressions for $\sin \beta^*$ and $\sin \gamma^*$ quoted in Table 2.1 are obtained by cyclic permutation of the parameters. Derive now the expres-

sion of $\cos\alpha^*$. By definition

$$\cos\alpha^* = \frac{\boldsymbol{b}^* \cdot \boldsymbol{c}^*}{b^* c^*} = \frac{1}{b^* c^*} \cdot \frac{1}{V^2} (\boldsymbol{c} \wedge \boldsymbol{a}) \cdot (\boldsymbol{a} \wedge \boldsymbol{b}).$$

Using (2.7) and the above derived expression of $\sin\alpha^*$ gives

$$\cos\alpha^* = (\cos\beta \cos\gamma - \cos\alpha)/(\sin\beta \sin\gamma).$$

The expressions for $\cos\beta^*$ and $\cos\gamma^*$ quoted in Table 2.1 may be obtained by cyclic permutation of the parameters.

2.B Some geometrical calculations concerning directions and planes

All crystallographic planes parallel to a given direction are said to belong to the same zone. According to p. 7 any vector $\boldsymbol{r}_U = u\boldsymbol{a} + v\boldsymbol{b} + w\boldsymbol{c}$ defines the crystallographic direction $[uvw]$ which is also the symbol of the zone. We derive now the following results:

1. Angle ϕ between two planes $\mathbf{H}_1 = (h_1 k_1 l_1)$ and $\mathbf{H}_2 = (h_2 k_2 l_2)$:

$$\cos\phi = \frac{\boldsymbol{r}^*_{\mathbf{H}_1} \cdot \boldsymbol{r}^*_{\mathbf{H}_2}}{r^*_{\mathbf{H}_1} r^*_{\mathbf{H}_2}} = d_{\mathbf{H}_1} d_{\mathbf{H}_2} \boldsymbol{r}^*_{\mathbf{H}_1} \cdot \boldsymbol{r}^*_{\mathbf{H}_2}$$

where $\boldsymbol{r}^*_{\mathbf{H}_1} \cdot \boldsymbol{r}^*_{\mathbf{H}_2}$ may be calculated by means of eqn (2.17a).

2. The family of planes $\mathbf{H} \equiv (hkl)$ belongs to the zone $[uvw]$ if

$$\boldsymbol{r}^*_{\mathbf{H}} \cdot \boldsymbol{r}_U = hu + kv + lw = 0. \qquad (2.B.1)$$

Indeed if (2.B.1) is verified then $\boldsymbol{r}^*_{\mathbf{H}}$ is normal to $\boldsymbol{r}_U$ so that the planes (hkl) are parallel to $\boldsymbol{r}_U$.

3. Symbol of the zone $[uvw]$ defined by two planes $\mathbf{H}_1$ and $\mathbf{H}_2$. Because of (2.B.1)

$$h_1 u + k_1 v + l_1 w = 0 \qquad h_2 u + k_2 v + l_2 w = 0$$

from which

$$u{:}v{:}w = (k_1 l_2 - k_2 l_1){:}(l_1 h_2 - l_2 h_1){:}(h_1 k_2 - h_2 k_1)$$

4. The planes $\mathbf{H}_1$, $\mathbf{H}_2$, $\mathbf{H}_3$ shall belong to the same zone if $\boldsymbol{r}^*_{\mathbf{H}_1}$, $\boldsymbol{r}^*_{\mathbf{H}_2}$, $\boldsymbol{r}^*_{\mathbf{H}_3}$ are coplanar. Then they shall define a cell in the reciprocal lattice whose volume is zero:

$$\begin{pmatrix} h_1 & k_1 & l_1 \\ h_2 & k_2 & l_2 \\ h_3 & k_3 & l_3 \end{pmatrix} = 0.$$

5. The condition that the point $P(x, y, z)$ lies in a plane (the nth of the set from the origin) of the family (hkl). We require that the projection of $\boldsymbol{r} = x\boldsymbol{a} + y\boldsymbol{b} + z\boldsymbol{c}$ onto the direction of $\boldsymbol{r}^*_{\mathbf{H}}$ be equal to n times the interplanar spacing $d_{\mathbf{H}}$:

$$\boldsymbol{r} \cdot \boldsymbol{r}^*_{\mathbf{H}} / r^*_{\mathbf{H}} = n d_{\mathbf{H}} = n / r^*_{\mathbf{H}}$$

from which

$$\boldsymbol{r} \cdot \boldsymbol{r}^*_{\mathbf{H}} = hx + ky + lz = \tilde{\mathbf{H}}\mathbf{X} = n. \qquad (2.B.2)$$

6. The condition that the point $P(h, k, l)$ of the reciprocal lattice lies in the nth (starting from the origin) plane of the set of planes (uvw) of the reciprocal lattice. It is the same problem as in 5 above, but transferred in the reciprocal lattice. The required condition is therefore

$$hu + kv + lw = n. \tag{2.B.3}$$

Note that the lattice vector $\boldsymbol{r}_U = u\boldsymbol{a} + v\boldsymbol{b} + w\boldsymbol{c}$ defines the direction $[uvw]$ orthogonal to the planes (uvw) of the reciprocal lattice. Therefore eqn (2.B.3) defines the reciprocal lattice points which lie on a plane normal to the direction $[uvw]$ defined in the direct space. As it will be seen on p. 247, these points give rise to the nth layer line in a rotation photograph when the crystal is rotated about the axis $[uvw]$.

7. Condition for the existence of a lattice plane perpendicular to the direction $[uvw]$. Such a plane will exist if a triplet of integer numbers h, k, l can be found for which

$$\boldsymbol{r}^*_{\mathbf{H}} \wedge \boldsymbol{r}_U = (h\boldsymbol{a}^* + k\boldsymbol{b}^* + l\boldsymbol{c}^*) \wedge (u\boldsymbol{a} + v\boldsymbol{b} + w\boldsymbol{c}) = 0.$$

In cubic lattices h, k, l coincide with u, v, w respectively (in other words the plane (uvw) is perpendicular to the direction $[uvw]$. Indeed $\boldsymbol{r}^*_U \wedge \boldsymbol{r}_U = 0$ owing to the fact that

$$\boldsymbol{a}^* \wedge \boldsymbol{a} = \boldsymbol{b}^* \wedge \boldsymbol{b} = \boldsymbol{c}^* \wedge \boldsymbol{c} = 0 \quad \text{and} \quad u\boldsymbol{a}^* \wedge v\boldsymbol{b} + v\boldsymbol{b}^* \wedge u\boldsymbol{a} = \ldots = 0.$$

8. Line determined by two points: $(\boldsymbol{r} - \boldsymbol{r}_1) \wedge (\boldsymbol{r} - \boldsymbol{r}_2) = 0$, where $\boldsymbol{r}_1$ and $\boldsymbol{r}_2$ are the positional vectors of the two points.

9. Line parallel to the direction $\boldsymbol{n}$ and going through the point $\boldsymbol{r}_0$:$(\boldsymbol{r} - \boldsymbol{r}_0) \wedge \boldsymbol{n} = 0$.

10. Line through the point P_0 and perpendicular to two unitary vectors $\boldsymbol{u}$ and $\boldsymbol{v}$:$(\boldsymbol{r} - \boldsymbol{r}_0) \wedge (\boldsymbol{u} \wedge \boldsymbol{v}) = 0$.

11. Projection of the vector $\boldsymbol{r}_1$ on to the unitary vector $\boldsymbol{n}$ (from now on we will denote by $\mathbf{N}$ and $\mathbf{N}^*$ the matrices of the components of $\boldsymbol{n}$ with respect to direct and reciprocal bases respectively):

$$P(\boldsymbol{r}_1/\boldsymbol{n}) = (\boldsymbol{r}_1 \cdot \boldsymbol{n})\boldsymbol{n} = (\bar{\mathbf{N}}^*\mathbf{X}_1)\boldsymbol{n}.$$

12. Plane determined by the three points P_1, P_2, P_3 whose positional vectors are $\boldsymbol{p}_1, \boldsymbol{p}_2, \boldsymbol{p}_3$:$(\boldsymbol{r} - \boldsymbol{p}_1) \cdot (\boldsymbol{p}_2 - \boldsymbol{p}_1) \wedge (\boldsymbol{p}_3 - \boldsymbol{p}_1) = 0$.
In terms of coordinates

$$\begin{pmatrix} x - x_1 & y - y_1 & z - z_1 \\ x_2 - x_1 & y_2 - y_1 & z_2 - z_1 \\ x_3 - x_1 & y_3 - y_1 & z_3 - z_1 \end{pmatrix} = 0. \tag{2.B.4}$$

Equation (2.B.4) derives from (2.5): indeed the unit cell volume defined by the vectors

$$\boldsymbol{r}_1 = \boldsymbol{r} - \boldsymbol{p}_1, \qquad \boldsymbol{r}_2 = \boldsymbol{p}_2 - \boldsymbol{p}_1, \qquad \boldsymbol{r}_3 = \boldsymbol{p}_3 - \boldsymbol{p}_1$$

is vanishing.

13. Plane normal to the unitary vector $\boldsymbol{n}$:$\boldsymbol{r} \cdot \boldsymbol{n} = d$, where d is the distance of the plane from the origin. It may also be written

$$\boldsymbol{r} \cdot \boldsymbol{n} = \bar{\mathbf{N}}^*\mathbf{X} = d. \tag{2.B.5}$$

14. Plane normal to the unitary vector $\boldsymbol{n}$ and through the point defined by $\boldsymbol{r}_0$:

$$(\boldsymbol{r} - \boldsymbol{r}_0) \cdot \boldsymbol{n} = \bar{\mathbf{N}}^*(\mathbf{X} - \mathbf{X}_0) = 0. \tag{2.B.6}$$

15. Distance from P_1 to the plane $\bar{\mathbf{N}}^*\mathbf{X} - d = 0$:

$$D = \bar{\mathbf{N}}^*\mathbf{X}_1 - d. \tag{2.B.7}$$

If the plane is defined by means of the equation $\boldsymbol{r} \cdot \boldsymbol{m} = d'$, where $\boldsymbol{m}$ is a general vector, before applying (2.B.7) $\boldsymbol{n} = \boldsymbol{m}/n$ and $d = d'/m$ have to be calculated.

16. Projection of the vector $\boldsymbol{r}_1$ on to the plane $\bar{\mathbf{N}}^*\mathbf{X} - d = 0$:

$$P(\boldsymbol{r}_1 \parallel \boldsymbol{n}^*) = \boldsymbol{r}_1 - (\bar{\mathbf{N}}^*\mathbf{X}_1)\boldsymbol{n}^*$$

17. Principal axes of the symmetry operator $\mathbf{R}$. A vector $\boldsymbol{r}$ along an axis of $\mathbf{R}$ should satisfy the eigenvalue equation $\mathbf{R}\boldsymbol{r} = \lambda \boldsymbol{r}$.

To give an example, let $\mathbf{R} = \mathbf{R}_z$ in a Cartesian system. The secular equation will be

$$\det(\mathbf{R}_z - \lambda \mathbf{I}) = \det \begin{pmatrix} \cos\theta - \lambda & -\sin\theta & 0 \\ \sin\theta & \cos\theta - \lambda & 0 \\ 0 & 0 & 1 - \lambda \end{pmatrix}$$
$$= (1 + \lambda^2 - 2\lambda\cos\theta)(1 - \lambda) = 0.$$

$\lambda = 1$ is one root. If $\sin\theta \neq 0$ the other eigenvalues are $\lambda_2 = e^{i\theta}$, $\lambda_3 = e^{-i\theta}$. The corresponding eigenvalues are $[0, 0, 1]$, $[1/\sqrt{2}, -i/\sqrt{2}, 0]$, $[1/\sqrt{2}, i/\sqrt{2}, 0]$ respectively.

The above result may be generalized to a rectilinear coordinate system: there will always be a real eigenvalue the eigenvector of which corresponds to the direction of the symmetry axis. For a proper rotation axis the eigenvalue is $+1$, for an improper axis it is -1. If $\sin\theta = 0$ all the eigenvalues are real. If two eigenvalues are equal, any linear combination of the corresponding eigenvectors is an eigenvector itself. For example, the matrix $|1\ \bar{1}\ 0/0\ \bar{1}\ 0/0\ 0\ \bar{1}|$ represents the twofold symmetry axis $2_{[100]}$ in a hexagonal framework. The eigenvectors corresponding to the doubly degenerate eigenvalues -1 are normal to [100] . Furthermore, the identity operator has three eigenvalues 1, 1, 1: the eigenvectors are any three unitary vectors orthogonal to each other.

2.C Some transformation matrices

Transformation matrices conventionally used to generate primitive from centred cells are shown in Table 2.C.1.

2.D Reciprocity of F and I lattices

Let $\bar{\mathbf{A}}_F \equiv (\boldsymbol{a}_F, \boldsymbol{b}_F, \boldsymbol{c}_F)$, the basis vectors of a face-centred lattice, and $\bar{\mathbf{A}}_P \equiv (\boldsymbol{a}_P, \boldsymbol{b}_P, \boldsymbol{c}_P)$, the basis vectors of a primitive unit cell for the same lattice. Then $\mathbf{A}_P = \mathbf{M}\mathbf{A}_F$, where $\mathbf{A}_P$ is one of matrices shown in Table 2.C.1: more explicitly

$$\boldsymbol{a}_P = (\boldsymbol{b}_F + \boldsymbol{c}_F)/2, \qquad \boldsymbol{b}_P = (\boldsymbol{a}_F + \boldsymbol{c}_F)/2, \qquad \boldsymbol{c}_P = (\boldsymbol{a}_F + \boldsymbol{b}_F)/2.$$

Table 2.C.1. Transformation matrices **M** conventionally used to generate centered from primitive lattices and vice versa, according to the relationships $\mathbf{A}' = \mathbf{MA}$

$\mathrm{I}\rightarrow\mathrm{P}$	$\mathrm{P}\rightarrow\mathrm{I}$	$\mathrm{R_h}\rightarrow\mathrm{R_{obv}}$	$\mathrm{R_{obv}}\rightarrow\mathrm{R_h}$
$\begin{pmatrix} -1/2 & 1/2 & 1/2 \\ 1/2 & -1/2 & 1/2 \\ 1/2 & 1/2 & -1/2 \end{pmatrix}$	$\begin{pmatrix} 0 & 1 & 1 \\ 1 & 0 & 1 \\ 1 & 1 & 0 \end{pmatrix}$	$\begin{pmatrix} 2/3 & 1/3 & 1/3 \\ -1/3 & 1/3 & 1/3 \\ -1/3 & -2/3 & 1/3 \end{pmatrix}$	$\begin{pmatrix} 1 & -1 & 0 \\ 0 & 1 & -1 \\ 1 & 1 & 1 \end{pmatrix}$
$\mathrm{R_h}\rightarrow\mathrm{R_{rev}}$	$\mathrm{R_{rev}}\rightarrow\mathrm{R_h}$	$\mathrm{F}\rightarrow\mathrm{P}$	$\mathrm{P}\rightarrow\mathrm{F}$
$\begin{pmatrix} 1/3 & -1/3 & 1/3 \\ 1/3 & 2/3 & 1/3 \\ -2/3 & -1/3 & 1/3 \end{pmatrix}$	$\begin{pmatrix} 1 & 0 & -1 \\ -1 & 1 & 0 \\ 1 & 1 & 1 \end{pmatrix}$	$\begin{pmatrix} 0 & 1/2 & 1/2 \\ 1/2 & 0 & 1/2 \\ 1/2 & 1/2 & 0 \end{pmatrix}$	$\begin{pmatrix} -1 & 1 & 1 \\ 1 & -1 & 1 \\ 1 & 1 & -1 \end{pmatrix}$
$\mathrm{A}\rightarrow\mathrm{P}$	$\mathrm{P}\rightarrow\mathrm{A}$	$\mathrm{B}\rightarrow\mathrm{P}$	$\mathrm{P}\rightarrow\mathrm{B}$
$\begin{pmatrix} -1 & 0 & 0 \\ 0 & -1/2 & 1/2 \\ 0 & 1/2 & 1/2 \end{pmatrix}$	$\begin{pmatrix} -1 & 0 & 0 \\ 0 & -1 & 1 \\ 0 & 1 & 1 \end{pmatrix}$	$\begin{pmatrix} -1/2 & 0 & 1/2 \\ 0 & -1 & 0 \\ 1/2 & 0 & 1/2 \end{pmatrix}$	$\begin{pmatrix} -1 & 0 & 1 \\ 0 & -1 & 0 \\ 1 & 0 & 1 \end{pmatrix}$
$\mathrm{C}\rightarrow\mathrm{P}$	$\mathrm{P}\rightarrow\mathrm{C}$		
$\begin{pmatrix} 1/2 & 1/2 & 0 \\ 1/2 & -1/2 & 0 \\ 0 & 0 & -1 \end{pmatrix}$	$\begin{pmatrix} 1 & 1 & 0 \\ 1 & -1 & 0 \\ 0 & 0 & -1 \end{pmatrix}$		

According to Table 2.E.1 $\mathbf{A}_\mathrm{P}^* = (\bar{\mathbf{M}})^{-1}\mathbf{A}_\mathrm{F}^*$, from which

$$\boldsymbol{a}_\mathrm{P}^* = -\boldsymbol{a}_\mathrm{F}^* + \boldsymbol{b}_\mathrm{F}^* + \boldsymbol{c}_\mathrm{F}^* = (-2\boldsymbol{a}_\mathrm{F}^* + 2\boldsymbol{b}_\mathrm{F}^* + 2\boldsymbol{c}_\mathrm{F}^*)/2, \ldots .$$

Denoting

$$\boldsymbol{a}_\mathrm{I}^* = 2\boldsymbol{a}_\mathrm{F}^*, \qquad \boldsymbol{b}_\mathrm{I}^* = 2\boldsymbol{b}_\mathrm{F}^*, \qquad \boldsymbol{c}_\mathrm{I}^* = 2\boldsymbol{c}_\mathrm{F}^*$$

gives

$$\boldsymbol{a}_\mathrm{P}^* = (-\boldsymbol{a}_\mathrm{I}^* + \boldsymbol{b}_\mathrm{I}^* + \boldsymbol{c}_\mathrm{I}^*)/2, \qquad \boldsymbol{b}_\mathrm{P}^* = (\boldsymbol{a}_\mathrm{I}^* - \boldsymbol{b}_\mathrm{I}^* + \boldsymbol{c}_\mathrm{I}^*)/2, \qquad \boldsymbol{c}_\mathrm{P}^* = (\boldsymbol{a}_\mathrm{I}^* + \boldsymbol{b}_\mathrm{I}^* - \boldsymbol{c}_\mathrm{I}^*)/2$$

or, in matrix notation, $\mathbf{A}_\mathrm{P}^* = \mathbf{Q}\mathbf{A}_\mathrm{I}^*$, where $\mathbf{Q}$ coincides (see Table 2.C.1) with the matrix which transforms an I cell into a P cell. It may be concluded that the reciprocal of an F lattice is an I lattice whose cell is defined by the vector $2\boldsymbol{a}_\mathrm{F}^*$, $2\boldsymbol{b}_\mathrm{F}^*$, $2\boldsymbol{c}_\mathrm{F}^*$.

If we index the reciprocal lattice with respect to $\mathbf{A}_\mathrm{I}^*$ and $\mathbf{A}_\mathrm{F}^*$ we obtain

$$\boldsymbol{r}^* = h_\mathrm{I}\boldsymbol{a}_\mathrm{I}^* + k_\mathrm{I}\boldsymbol{b}_\mathrm{I}^* + l_\mathrm{I}\boldsymbol{c}_\mathrm{I}^* = h_\mathrm{F}\boldsymbol{a}_\mathrm{F}^* + k_\mathrm{F}\boldsymbol{b}_\mathrm{F}^* + l_\mathrm{F}\boldsymbol{c}_\mathrm{F}^*$$

where all of $h_\mathrm{F} = 2h_\mathrm{I}$, $k_\mathrm{F} = 2k_\mathrm{I}$, $l_\mathrm{F} = 2l_\mathrm{I}$ will either assume even values (when h_I, k_I, l_I are integer numbers) or odd values (when h_I, k_I, l_I are of type $m/2$, $n/2$, $p/2$ with integer values of m, n, p). These are just the conditions for systematic extinctions in an F lattice.

2.E Transformations of crystallographic quantities in rectilinear spaces

In this paragraph we describe the transformation properties of crystallographic quantities not studied on pp. 65–7.

Let us first prove that the basis vector transformation (2.18) (from $\mathbf{A}$ to $\mathbf{A}'$) transforms the space group symmetry operator $\mathbf{C}_\mathrm{p} \equiv (\mathbf{R}_\mathrm{p}, \mathbf{T}_\mathrm{p})$ into $\mathbf{C}_\mathrm{p}' \equiv (\mathbf{R}_\mathrm{p}', \mathbf{T}_\mathrm{p}')$, where

$$\mathbf{R}_\mathrm{p}' = (\bar{\mathbf{M}})^{-1}\mathbf{R}_\mathrm{p}\bar{\mathbf{M}}, \qquad \mathbf{T}_\mathrm{p}' = (\bar{\mathbf{M}})^{-1}\mathbf{T}_\mathrm{p}. \qquad (2.E.1)$$

The relation $\mathbf{X}_\mathrm{p} = \mathbf{R}_\mathrm{p}\mathbf{X} + \mathbf{T}_\mathrm{p}$ is transformed, according to (2.20), into $\bar{\mathbf{M}}\mathbf{X}_\mathrm{p}' = \mathbf{R}_\mathrm{p}\bar{\mathbf{M}}\mathbf{X}' + \mathbf{T}_\mathrm{p}$ from which $\mathbf{X}_\mathrm{p}' = (\bar{\mathbf{M}})^{-1}\mathbf{R}_\mathrm{p}\bar{\mathbf{M}}\mathbf{X}' + (\bar{\mathbf{M}})^{-1}\mathbf{T}_\mathrm{p}$ which coincides with (2.E.1). Note that the trace of $\mathbf{R}$ is invariant under transformations such as (2.E.1).

A particular case of (2.E.1) occurs when $\mathbf{A}' \equiv \mathbf{A}^*$. Then

$$\mathbf{R}_p^* = \mathbf{G}\mathbf{R}_p\mathbf{G}^*, \qquad \mathbf{T}_p^* = \mathbf{G}\mathbf{T}_p. \tag{2.E.2}$$

If the second of the equations (2.9) is introduced into the first equation (2.E.2) the following result is obtained

$$\mathbf{R}_p^* = (\bar{\mathbf{R}}_p)^{-1}\mathbf{G}\mathbf{R}_p^{-1}\mathbf{R}_p\mathbf{G}^{-1} = (\bar{\mathbf{R}}_p)^{-1}. \tag{2.E.3}$$

In conclusion, if the operator $\mathbf{R}$ is a symmetry element in the direct space, $(\bar{\mathbf{R}})^{-1}$ is a symmetry element in the reciprocal space. The list of the symmetry operators in the reciprocal space may be obtained without inverting the various matrices. Indeed, if $\mathbf{R}$ is a symmetry element in the direct space, group properties guarantee that $\mathbf{R}^{-1}$ is also an element of the direct space symmetry group: therefore $\bar{\mathbf{R}}$ is a symmetry operator of the reciprocal space symmetry group. Consequently, if the set of matrices $\mathbf{R}$ operate in direct space, the set $\bar{\mathbf{R}}$ operate in reciprocal space (however, $\mathbf{R}_p$ and $\bar{\mathbf{R}}_p$ may pertain to different symmetry elements).

In an orthonormal system $\boldsymbol{a}^* \equiv \boldsymbol{a}$, $\boldsymbol{b}^* \equiv \boldsymbol{b}$, $\boldsymbol{c}^* \equiv \boldsymbol{c}$ and $\mathbf{G} = \mathbf{I}$. Therefore any symmetry operator $\mathbf{C} \equiv (\mathbf{R}, \mathbf{T})$ will have identical expression both in direct and in the reciprocal space. Furthermore $\mathbf{R} = \mathbf{R}^{-1}$ holds, as already obtained in Appendix 1.A.

It is often useful to know the transformation rules valid in reciprocal space when the basis vector transformation $\mathbf{A}' = \mathbf{M}\mathbf{A}$ is performed in direct space. We describe some of them:

1. $\mathbf{A}^* \Leftrightarrow \mathbf{A}'^*$: according to (2.25), $\mathbf{A}' = \mathbf{G}'\mathbf{A}'^* = \mathbf{M}\mathbf{G}\mathbf{A}^*$. Owing to (2.21), that gives

$$\mathbf{A}'^* = (\bar{\mathbf{M}})^{-1}\mathbf{A}^* \quad \text{or} \quad \mathbf{A}^* = \bar{\mathbf{M}}\mathbf{A}'^*. \tag{2.E.4}$$

2. $\mathbf{X}^* \Leftrightarrow \mathbf{X}'^*$: by analogy to point (1) it may be obtained

$$\mathbf{X}'^* = \mathbf{M}\mathbf{X}^* \quad \text{or} \quad \mathbf{X}^* = \mathbf{M}^{-1}\mathbf{X}'^*. \tag{2.E.5}$$

Note that relations (2.E.5) are the transformation rules of the Miller indices (*hkl*).

3. $\mathbf{G}^* \Leftrightarrow \mathbf{G}'^*$: let us introduce the first eqn (2.16) in the first eqn (2.21), which thus becomes $\mathbf{G}'^{*-1} = \mathbf{M}\mathbf{G}^{*-1}\bar{\mathbf{M}}$. On post-multiplying both sides of this equation by $\mathbf{G}'^*$ we obtain $\mathbf{I} = \mathbf{M}\mathbf{G}^{*-1}\bar{\mathbf{M}}\mathbf{G}'^*$ from which

$$\mathbf{G}'^* = (\bar{\mathbf{M}})^{-1}\mathbf{G}^*\mathbf{M}^{-1} \quad \text{or} \quad \mathbf{G}^* = \bar{\mathbf{M}}\mathbf{G}'^*\mathbf{M} \tag{2.E.6}$$

4. $\mathbf{C}^* \Leftrightarrow \mathbf{C}'^*$: introduce the first eqn (2.E.2) in the first eqn (2.E.1). We obtain $\mathbf{G}'^{-1}\mathbf{R}_p'^*\mathbf{G}' = (\bar{\mathbf{M}})^{-1}\mathbf{G}^{-1}\mathbf{R}_p^*\mathbf{G}\bar{\mathbf{M}}$, which, introduced into (2.E.6) gives

$$\mathbf{R}_p'^* = \mathbf{M}\mathbf{R}_p^*\mathbf{M}^{-1} \quad \text{and} \quad \mathbf{T}_p'^* = \mathbf{M}\mathbf{T}_p^* \tag{2.E.7}$$

Note that (2.E.7) represents the transformation rules for the reciprocal space symmetry operators generated by a basis vector change in direct space.

5. $\mathbf{Q}^* \Leftrightarrow \mathbf{Q}'^*$. According to (2.E.5) it will be

$$\bar{\mathbf{X}}^*\mathbf{Q}^*\mathbf{X}^* = \bar{\mathbf{X}}'^*(\bar{\mathbf{M}})^{-1}\mathbf{Q}^*\mathbf{M}^{-1}\mathbf{X}'^*$$

Table 2.E.1. Transformation relationships. In the table $\mathbf{M}$ is the matrix transforming $\bar{\mathbf{A}} \equiv (\boldsymbol{a}, \boldsymbol{b}, \boldsymbol{c})$ into $\bar{\mathbf{A}}' = (\boldsymbol{a}', \boldsymbol{b}', \boldsymbol{c}')$. $\mathbf{G}$ and $\mathbf{G}'$ are the metric matrices of $\mathbf{A}$ and $\mathbf{A}'$ respectively, $\mathbf{G}^*$ and $\mathbf{G}'^*$ are the metric matrices of $\bar{\mathbf{A}}^* \equiv (\boldsymbol{a}^*, \boldsymbol{b}^*, \boldsymbol{c}^*)$ and $\bar{\mathbf{A}}'^* \equiv (\boldsymbol{a}'^*, \boldsymbol{b}'^*, \boldsymbol{c}'^*)$ respectively. $\mathbf{C} \equiv (\mathbf{R}, \mathbf{T})$ is a symmetry operator ($\mathbf{R}$ is its rotational part, $\mathbf{T}$ its translational part): $\mathbf{C}$, $\mathbf{C}'$, $\mathbf{C}^*$, $\mathbf{C}'^*$ are symmetry operators defined in $\mathbf{A}$, $\mathbf{A}'$, $\mathbf{A}^*$, $\mathbf{A}'^*$ respectively. $\mathbf{Q}$ and $\mathbf{Q}^*$ are the quadratic forms of $\mathbf{A}$ and $\mathbf{A}^*$.

$\mathbf{A}' = \mathbf{MA}$	$\mathbf{A} = \mathbf{M}^{-1}\mathbf{A}'$
$\mathbf{X}' = (\bar{\mathbf{M}})^{-1}\mathbf{X}$	$\mathbf{X} = \bar{\mathbf{M}}\mathbf{X}'$
$\mathbf{G}' = \mathbf{MG}\bar{\mathbf{M}}$	$\mathbf{G} = \mathbf{M}^{-1}\mathbf{G}'(\bar{\mathbf{M}})^{-1}$
$\mathbf{Q}' = \mathbf{MQ}\bar{\mathbf{M}}$	$\mathbf{Q} = \mathbf{M}^{-1}\mathbf{Q}'(\bar{\mathbf{M}})^{-1}$
$\mathbf{R}' = (\bar{\mathbf{M}})^{-1}\mathbf{R}\bar{\mathbf{M}},\ \mathbf{T}' = (\bar{\mathbf{M}})^{-1}\mathbf{T}$	$\mathbf{R} = \bar{\mathbf{M}}\mathbf{R}'(\bar{\mathbf{M}})^{-1},\ \mathbf{T} = \bar{\mathbf{M}}\mathbf{T}'$
$\mathbf{A}^* = \mathbf{G}^*\mathbf{A}$	$\mathbf{A} = \mathbf{GA}^*$
$\mathbf{X}^* = \mathbf{GX}$	$\mathbf{X} = \mathbf{G}^*\mathbf{X}^*$
$\mathbf{Q}^* = \mathbf{G}^*\mathbf{QG}^*$	$\mathbf{Q} = \mathbf{GQ}^*\mathbf{G}$
$\mathbf{R}^* = \mathbf{GRG}^* = (\bar{\mathbf{R}})^{-1}$	$\mathbf{R} = (\bar{\mathbf{R}}^*)^{-1}$
$\mathbf{A}'^* = (\bar{\mathbf{M}})^{-1}\mathbf{A}^*$	$\mathbf{A}^* = \bar{\mathbf{M}}\mathbf{A}'^*$
$\mathbf{X}'^* = \mathbf{MX}^*$	$\mathbf{X}^* = \mathbf{M}^{-1}\mathbf{X}'^*$
$\mathbf{G}'^* = (\bar{\mathbf{M}})^{-1}\mathbf{G}^*\mathbf{M}^{-1}$	$\mathbf{G}^* = \bar{\mathbf{M}}\mathbf{G}'^*\mathbf{M}$
$\mathbf{R}'^* = \mathbf{MR}^*\mathbf{M}^{-1},\ \mathbf{T}'^* = \mathbf{MT}^*$	$\mathbf{R}^* = \mathbf{M}^{-1}\mathbf{R}'^*\mathbf{M},\ \mathbf{T}^* = \mathbf{M}^{-1}\mathbf{T}'^*$
$\mathbf{Q}'^* = (\bar{\mathbf{M}})^{-1}\mathbf{Q}^*\mathbf{M}^{-1}$	$\mathbf{Q}^* = \bar{\mathbf{M}}\mathbf{Q}'^*\mathbf{M}$

from which

$$\mathbf{Q}'^* = (\bar{\mathbf{M}})^{-1}\mathbf{Q}^*\mathbf{M}^{-1}, \qquad \mathbf{Q}^* = \bar{\mathbf{M}}\mathbf{Q}'^*\mathbf{M}. \tag{2.E.8}$$

The transformation rules obtained in this paragraph and on pp. 65–7 are collected in Table 2.E.1.

A particular basis transformation is that corresponding to a symmetry operation. In accordance with eqns (2.19) and (2.20) a transformation $\mathbf{R}$ acting on the coordinates is equivalent to the transformation $\mathbf{M} = (\bar{\mathbf{R}})^{-1}$ acting on the basis vectors. In this case the metric matrix will not vary: indeed, according to (2.21), $\mathbf{G}' = (\bar{\mathbf{R}})^{1}\mathbf{G}\mathbf{R}^{-1}$ which is identical (because of (2.9)) to $\mathbf{G}$. Find now the relationships existing between the matrices $\boldsymbol{\beta}'$ and $\boldsymbol{\beta}$ defining the anisotropic temperature factors of the atoms related by a symmetry operator. In accordance with (2.E.8)

$$\boldsymbol{\beta}' = \mathbf{R}\boldsymbol{\beta}\bar{\mathbf{R}}. \tag{2.E.9}$$

As an example, the reader will easily verify that the relationships existing in the cubic system along the components of $\boldsymbol{\beta}$ of two atoms related by the symmetry axis $3_{[111]}$ (see the matrix given in Appendix 1.D) are

$$\beta'_{11} = \beta_{33}, \qquad \beta'_{12} = \beta_{13}, \qquad \beta'_{13} = \beta_{23},$$
$$\beta'_{22} = \beta_{11}, \qquad \beta'_{23} = \beta_{12}, \qquad \beta'_{33} = \beta_{22}.$$

2.F Derivation of the normal equations

Because of (2.50) the parameter S in (2.52) may be written as

$$S = \bar{\mathbf{V}}\mathbf{M}_f^{-1}\mathbf{V} = \bar{\mathbf{F}}\mathbf{M}_f^{-1}\mathbf{F} - \bar{\mathbf{X}}\bar{\mathbf{A}}\mathbf{M}_f^{-1}\mathbf{F} - \bar{\mathbf{F}}\mathbf{M}_f^{-1}\mathbf{AX} + \bar{\mathbf{X}}\bar{\mathbf{A}}\mathbf{M}_f^{-1}\mathbf{AX}. \tag{2.F.1}$$

By applying to (2.F.1) the differential operator δ we obtain (note that $\delta\bar{\mathbf{X}} = \delta\mathbf{X}$)

$$\delta S = (\overline{\delta\mathbf{X}})\bar{\mathbf{A}}\mathbf{M}_f^{-1}\mathbf{AX} + \bar{\mathbf{X}}\bar{\mathbf{A}}\mathbf{M}_f^{-1}\mathbf{A}\,\delta\mathbf{X} - \bar{\mathbf{F}}\mathbf{M}_f^{-1}\mathbf{A}\,\delta\mathbf{X} - (\overline{\delta\mathbf{X}})\bar{\mathbf{A}}\mathbf{M}_f^{-1}\mathbf{F} = 0. \tag{2.F.2}$$

Since $\mathbf{X}$ and $\delta\mathbf{X}$ are vectors, the first two terms on the right-hand side of (2.F.2) are 1×1 matrices which are equal ($\mathbf{M}_f^{-1}=\bar{\mathbf{M}}_f^{-1}$). For the same reason the third and the fourth terms are equal too. Therefore

$$\delta S = 2(\overline{\delta\mathbf{X}})(\bar{\mathbf{A}}\mathbf{M}_f^{-1}\mathbf{A}\mathbf{X}-\bar{\mathbf{A}}\mathbf{M}_f^{-1}\mathbf{F})=0$$

from which

$$(\bar{\mathbf{A}}\mathbf{M}_f^{-1}\mathbf{A})\hat{\mathbf{X}}=\bar{\mathbf{A}}\mathbf{M}_f^{-1}\mathbf{F},$$

which coincides with (2.53).

2.G Derivation of the variance–covariance matrix $\mathbf{M}_x$

Because of (2.55), eqn (2.56) may be written as

$$\mathbf{M}_x=\langle(\mathbf{B}^{-1}\bar{\mathbf{A}}\mathbf{M}_f^{-1}\mathbf{F}-\mathbf{A}^{-1}\mathbf{F}^0)(\overline{\mathbf{B}^{-1}\bar{\mathbf{A}}\mathbf{M}_f^{-1}\mathbf{F}-\mathbf{A}^{-1}\mathbf{F}^0})\rangle. \qquad (2.G.1)$$

According to (2.54) $\mathbf{A}^{-1}=\mathbf{B}^{-1}\bar{\mathbf{A}}\mathbf{M}_f^{-1}$ so that (2.G.1) becomes

$$\mathbf{M}_x=\mathbf{B}^{-1}\bar{\mathbf{A}}\mathbf{M}_f^{-1}\langle(\mathbf{F}-\mathbf{F}^0)(\overline{\mathbf{F}-\mathbf{F}^0})\rangle\mathbf{M}_f^{-1}\mathbf{A}\mathbf{B}^{-1}=\mathbf{B}^{-1}\bar{\mathbf{A}}\mathbf{M}_f^{-1}\mathbf{M}_f\mathbf{M}_f^{-1}\mathbf{A}\mathbf{B}^{-1}=\mathbf{B}^{-1}.$$

2.H Derivation of the unbiased estimate of $\mathbf{M}_x$

Let us calculate (2.F.1) for $\mathbf{X}=\hat{\mathbf{X}}$: then

$$\hat{S}=\bar{\mathbf{F}}\mathbf{M}_f^{-1}\mathbf{F}-\hat{\bar{\mathbf{X}}}\bar{\mathbf{A}}\mathbf{M}_f^{-1}\mathbf{F}-(\bar{\mathbf{F}}\mathbf{M}_f^{-1}\mathbf{A}-\hat{\bar{\mathbf{X}}}\bar{\mathbf{A}}\mathbf{M}_f^{-1}\mathbf{A})\hat{\mathbf{X}}. \qquad (2.H.1)$$

Because of (2.54) and (2.55) the eqn (2.H.1) reduces to

$$\hat{S}=\bar{\mathbf{F}}\mathbf{M}_f^{-1}\mathbf{F}-\hat{\bar{\mathbf{X}}}\mathbf{B}\mathbf{X}=(\overline{\mathbf{F}-\mathbf{F}^0})\mathbf{M}_f^{-1}(\mathbf{F}-\mathbf{F}^0)-(\overline{\hat{\mathbf{X}}-\mathbf{X}})\mathbf{B}(\hat{\mathbf{X}}-\mathbf{X})$$

where $\mathbf{X}=\mathbf{A}^{-1}\mathbf{F}^0$. The expected value of $\hat{S}$ will then be

$$\langle\hat{S}\rangle=\langle(\overline{\mathbf{F}-\mathbf{F}^0})\mathbf{M}_f^{-1}(\mathbf{F}-\mathbf{F}^0)\rangle-\langle(\overline{\hat{\mathbf{X}}-\mathbf{X}})\mathbf{B}(\hat{\mathbf{X}}-\mathbf{X})\rangle.$$

$(\mathbf{F}-\mathbf{F}^0)$ and $(\hat{\mathbf{X}}-\mathbf{X})$ are random variables with zero means and finite variances, and with variance–covariance matrices of rank n and m respectively. In this condition it is possible to show[45] that

$$\langle(\overline{\mathbf{F}-\mathbf{F}^0})\mathbf{M}_f^{-1}(\mathbf{F}-\mathbf{F}^0)\rangle=n, \qquad \langle(\overline{\hat{\mathbf{X}}-\mathbf{X}})\mathbf{M}_x^{-1}(\hat{\mathbf{X}}-\mathbf{X})\rangle=m.$$

Therefore

$$\langle\hat{S}\rangle=\langle\bar{\mathbf{V}}\mathbf{M}_f^{-1}\mathbf{V}\rangle=n-m.$$

If $\mathbf{M}_f$ is known within a scale factor ($\mathbf{M}_f=K_v\mathbf{N}_f$) then

$$\langle\hat{S}\rangle=\langle\bar{\mathbf{V}}\mathbf{N}_f^{-1}\mathbf{V}\rangle/k_v=n-m$$

which coincides with (2.58).

2.I The FFT algorithm and its crystallographic applications

Suppose that we want to calculate the one-dimensional transform

$$\rho(x)=\sum_{h'}F(h')\exp(2\pi i h'x) \qquad (2.I.1)$$

where $-N/2\leqslant h'<N/2$. Subdivide the x axis in N parts so that

$$x=j/N \qquad (j=0,1,\ldots,N-1).$$

Denoting $h' = h - N/2$ gives

$$\exp(2\pi i h' x) = \exp(2\pi i hj/N)\exp(-\pi i j) = (-1)^j \exp(2\pi i hj/N)$$

where $0 \leqslant h < N$. Then (2.A.21) becomes

$$\begin{aligned}\rho(j) &= (-1)^j \sum_{h=0}^{N-1} F(h)\exp(2\pi i hj/N)\\ &= (-1)^j \sum_{h=0}^{N-1} F(h) w_N^{hj} \quad (j = 0, 1, \ldots, N-1)\end{aligned}$$

where $w_N = \exp(2\pi i/N)$.

Because of the transformation $h' \rightarrow h$ the negative indices are removed: furthermore, the number of Fourier coefficients is chosen equal to the number of points in which ρ has to be calculated (the F_h array may eventually be completed with a number of zeros).

If N is not a prime number, say $N = r \cdot s$ (of particular interest and simplicity is the case $N = 2^m$), then the expression $h_1 s + h_0$ will generate all the integers h in the range $[0, (N-1)]$:

$$h = h_1 s + h_0 \qquad (h_1 = 0, 1, \ldots, r-1; \qquad h_0 = 0, 1, \ldots, s-1). \quad (2.\text{I}.2)$$

Similarly, the expression $j_1 r + j_0$ will generate all the integers j in the range $[0, (N-1)]$

$$j = j_1 r + j_0 \qquad (j_1 = 0, 1, \ldots, s-1; \qquad j_0 = 0, 1, \ldots, r-1).$$

Then

$$w_N^{hj} = w_s^{h_0 j_1} w_r^{h_1 j_0} w_N^{h_0 j_0}$$

and

$$\begin{aligned}\rho(j) \equiv \rho(j_0, j_1) &= (-1)^j \sum_{h_0=0}^{s-1} \sum_{h_1=0}^{r-1} F(h_1, h_0) w_s^{h_0 j_1} w_r^{h_1 j_0} w_N^{h_0 j_0}\\ &= (-1)^j \sum_{h_0=0}^{s-1} w_s^{h_0 j_1} w_N^{h_0 j_0} \sum_{h_1=0}^{r-1} F(h_1, h_0) w_r^{h_1 j_0} \qquad (2.\text{I}.3)\end{aligned}$$

where $F(h_1, h_0)$ represents the F_h factor obtained by combination of h_1 and h_0 in eqn (2.I.2).

As in the Beevers–Lipson procedure the expression (2.I.3) may be calculated in two steps:[122–124]

1. Step 1: calculate

$$T(j_0, h_0) = \sum_{h_1=0}^{r-1} F(h_1, h_0) w_r^{h_1 j_0}$$

for $j_0 = 0, 1, \ldots, r-1$ and $h_0 = 0, 1, \ldots, s-1$. $T(j_0, h_0)$ values can replace $F(h_1, h_0)$ values in the memory of a computer since these last are no longer necessary,

2. Step 2: calculate

$$R(j_0, j_1) = \sum_{h_0=0}^{s-1} [T(j_0, h_0) w_N^{h_0 j_0}] w_s^{h_0 j_1}$$

for $j_0 = 0, 1, \ldots, r-1$ and $j_1 = 0, 1, \ldots, s-1$.

While the evaluation of (2.I.3) by the classical method requires N^2

operations, the FFT method involves $r(s+r)$ operations in the first step and $s(s+r)$ operation in the second step. If N is sufficiently large that corresponds to a large saving of time.

Besides calculating electron density maps, the FFT method is also used for the calculation of the structure factors when the number of atoms is very large and computing time has to be saved. Calculations are organized into two steps:

1. Step 1: All the atoms (in the asymmetric unit) which contribute to the electron density are selected. For each of them the electron density is nothing else but the Fourier transform of the atom scattering curve corrected for thermal motion. Usually each atom is represented by a single Gaussian function (for low resolution data) or as the sum of two or three Gaussian components. The overall model electron density map is sampled on a grid not too fine (that would greatly increase cost and computer storage) and not too coarse (in order to avoid too rough an approximation of the electron density).
2. Step 2: The fourier inversion of the map is made, which provides structure factor magnitudes and phases.

The speed of steps 1 and 2 depends on the number of grid points in which the density has been sampled and on the number of terms in the Gaussian approximation to the atomic scattering curve. Too coarse a sampling grid would produce structure factors which are the sum of the desired ones and those with indices spaced away by multiples of N. Ten Eick[122] proved that the number of grids may be taken small by artificially increasing the B value of the atoms.

Besides structure factor calculations most of the steps involved in LSQ structure refinement can be performed by the FFT algorithm:[125,126] i.e. the calculation of the gradient vector **D** and of the normal matrix **B** (see p. 92). Indeed, in accordance with p. 145, the following properties of the Fourier transform hold:

Real space		Reciprocal space
$\rho(\mathbf{r})$	← FT →	$F(\mathbf{H})$
$\dfrac{\delta\rho(\mathbf{r})}{\delta x}$	← FT →	$(-2\pi ih)F(\mathbf{H})$
$\dfrac{\delta\rho(\mathbf{r})}{\delta y}$	← FT →	$(-2\pi ik)F(\mathbf{H})$
etc.		

These more recent developments have made possible the use of a full-matrix least-squares for macromolecule structure at a reasonable cost.

2.J Examples of twin laws

Some examples of twin laws are described here, with special references to minerals. From the figures the reader will conclude that morphology exhibited by twins often shows re-entrant angles, which are not displayed by single crystals.

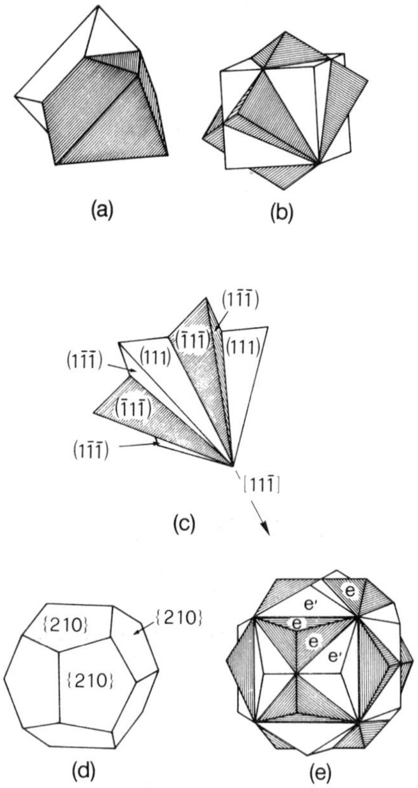

Fig. 2.J.1. Spinel law. (a) Twin plane (111); (b) penetration twin; (c) interpenetrant rotation twin; (d) classical pentagonal dodecahedron of pyrite; (e) iron cross law.

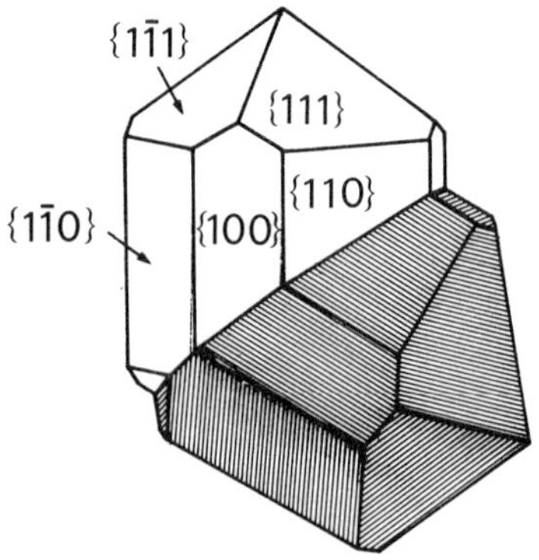

Fig. 2.J.2. Elbow twin.

Cubic system

Twinning according to the **spinel law** is commonly found in crystals of the class $4\bar{3}m$, which often exhibit forms {111}. Twinning occurs with {111} as a twin plane (see Fig. 2.J.1(a)). The same law applies to the penetration twin shown in Fig. 2.J.1(b), where the two cubes are rotated about [111] (classical mineral: **fluorite**). An interpenetrant rotation twin about $[11\bar{1}]$ is shown in Fig. 2.J.1(c).

Crystals of pyrite, point group $m\bar{3}$, frequently obey the **iron cross law**: they twin with {110} as a twin plane.

In Fig. 2.J.1(d) the classical pentagonal dodecahedron is shown, with form $e \equiv \{210\}$: twinned crystals display the form shown in Fig. 2.J.1(e).

Tetragonal system

Cassiterite (SnO_2, point group 4/mmm), presents the so called **elbow twin**, twin plane {101}, shown in Fig. 2.J.2. The elbow twin is also found in rutile (TiO_2) and zircon ($ZrSiO_4$), which are often polysynthetic.

Calcopyrite crystals ($CuFeS_2$, point group $\bar{4}2m$) are commonly tetrahedral, with {112} as the dominant form. Twinning occurs on {112}, contact, lamellar, or penetration.

Hexagonal and trigonal systems

Twinning is rare in minerals of the hexagonal system and frequent in minerals of the trigonal system.

The twin plane (and the composition plane) in calcite ($CaCO_3$, point group $\bar{3}m$) is often {0001} (see Fig. 2.J.3(a)), more frequently $\{1\bar{1}02\}$ (see Fig. 2.J.3(b)). This last twin is readily provoked by mechanical deformation (for example, the pressure of a razor blade on a rhombohedron of calcite). The twinning is due to the ability of structural layers parallel to $\{1\bar{1}02\}$ to slide past each other.

Quartz (SiO_2, point group 32) twins in several ways. The most important are listed below.

1. **Brazil twins**, with $\{11\bar{2}0\}$ as the twin plane. This combines right- and left-handed crystals in a penetration twin (often with plane composition surface). A left-handed single crystal is shown in Fig. 2.J.4(a), a right-handed crystal is shown in Fig. 2.J.4(b), and the Brazil twin is displayed in Fig. 2.J.4(c). Such types of twin are useless for electrical work, and may be detected in polarized light (the plane of polarization is rotated in opposite directions by the two parts of the twin).

2. **Dauphiné twins**, with $\boldsymbol{c}$ as twin axis. Two right- or two left-handed individuals are combined, separated by a very irregular surface. Then the faces $\{10\bar{1}1\}$ of one individual and the faces $\{01\bar{1}1\}$ of the other will coincide (see Fig. 2.J.4(d)).

3. **Japan twins**, contact twins with $\{11\bar{2}2\}$ as twin and composition plane (see Fig. 2.J.4(e)). The $\boldsymbol{c}$ axes intersect at 84°33′, one pair of faces $\{10\bar{1}0\}$ is common to both pairs of twins.

Orthorhombic system

Aragonite ($CaCO_3$, point group mmm, polymorphous with calcite) often twins on {110} faces. In Fig. 2.J.5(a) a single crystal of aragonite in shown, in Fig. 2.J.5(b) a twinned crystal is displayed, in Fig. 2.J.5(c) the structural scheme of the twin is given.

Twinning is often repeated and produces interpenetrant triplets of pseudo-hexagonal shape (Fig. 2.J.5(d)).

Staurolite ($Al_4FeSi_2O_{10}(OH)_2$, point group mmm) forms cruciform penetration twins: if the twin plane is {032} then the two individuals form a nearly right-angled cross (Fig. 2.J.6(a)); an oblique cross (angle ≈60°) is formed by twinning on {232} (see Fig. 2.J.6(b)).

Monoclinic system

Orthoclase crystals ($KAlSi_3O_8$, point group 2/m) often twin according to the Carlsbad, Baveno, or Manebach law.

In the Carlsbad law ***c*** is the twin axis and {010} the composition plane (see Fig. 2.J.7(a)). The twin and composition plane is {021} in Baveno twins (see Fig. 2.J.7(b)), and {001} in Manebach twins (see Fig. 2.J.7(c)). While Carlsbad twins are penetrating twins, Baveno and Manebach are usually contact twins.

Triclinic system

In twins of plagioclase feldspars [(Ca, Na)(Al, Si)$AlSi_2O_8$, point group $\bar{1}$] the **albite** law is often satisfied: twin plane {010}, type of twinning usually multiple and polysynthetic (see Fig. 2.J.8(a,b,c)). Also common is the

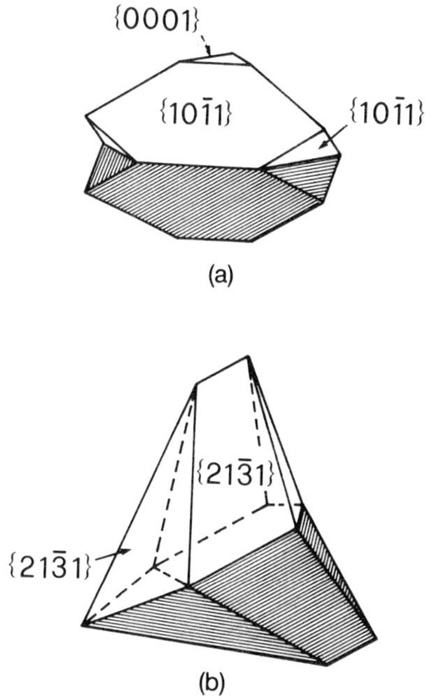

Fig. 2.J.3. Some examples of calcite twins. (a) Twin plane {0001}; (b) twin plane {1$\bar{1}$02}.

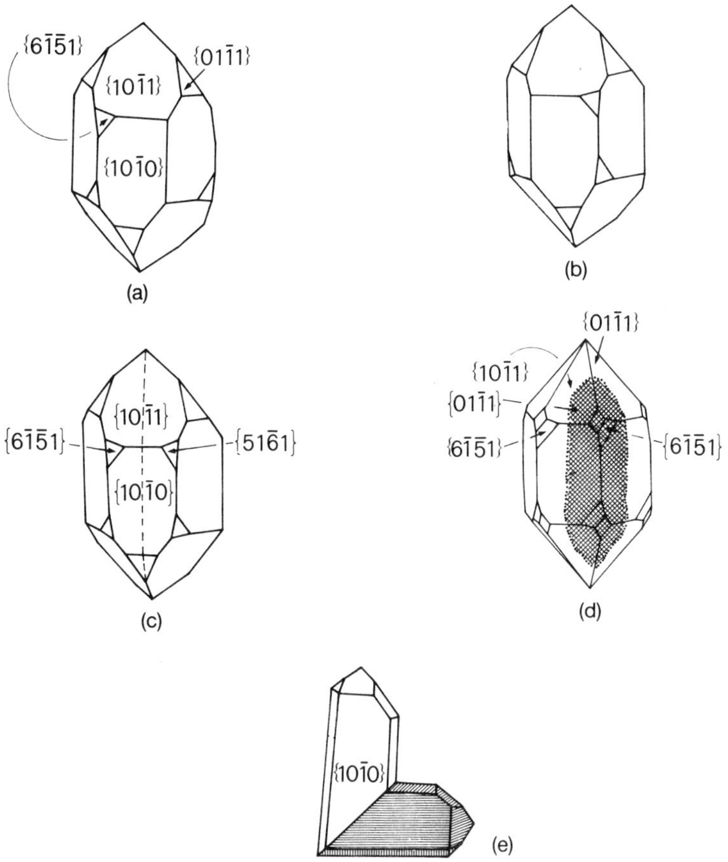

Fig. 2.J.4. Quartz twins. (a) Left-handed quartz; (b) right-handed quartz; (c) Brazil twin: the twinning plane is indicated by a broken line; (d) Dauphiné twin: interpenetration of two left-hand crystals.

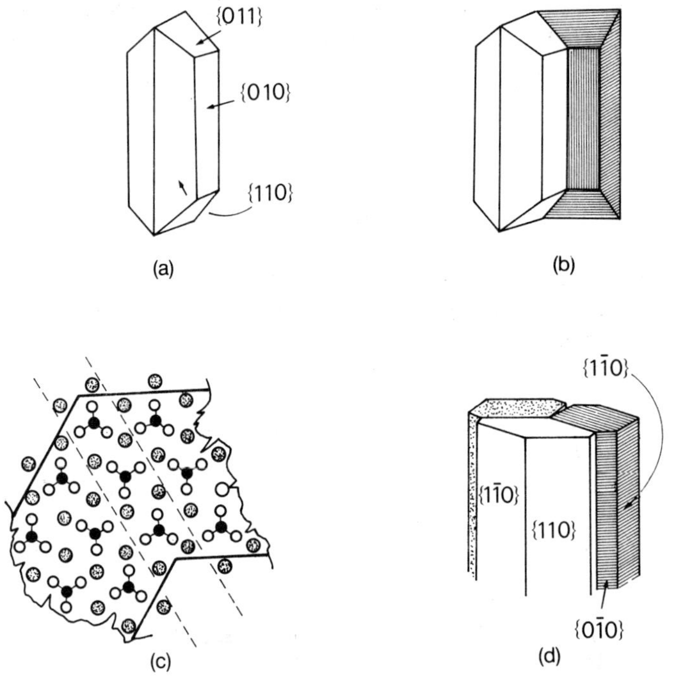

Fig. 2.J.5. Aragonite twinning. (a) Single crystal; (b) twinned crystal; (c) structural scheme of the twin; (d) triplet.

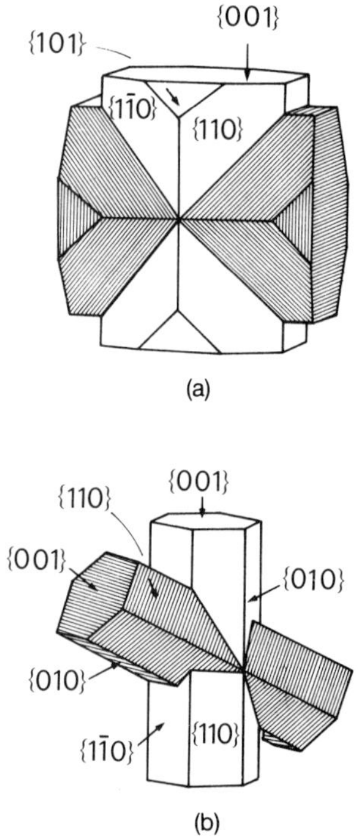

Fig. 2.J.6. Staurolite twinning. (a) Twin after {032}; (b) twin after {232}.

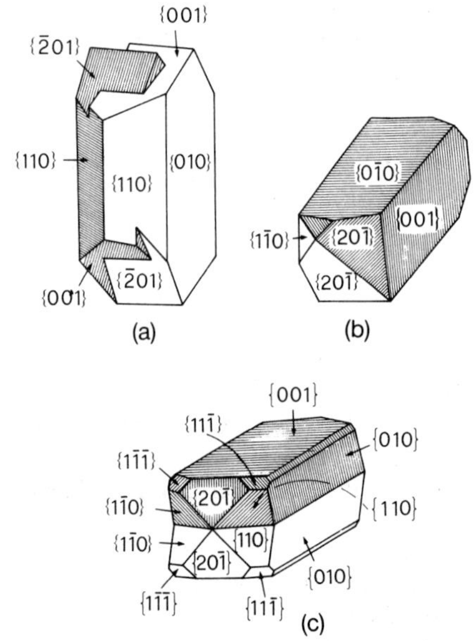

Fig. 2.J.7. Orthoclase twins. (a) Carlsbad twin; (b) Baveno twin; (c) Manebach twin.

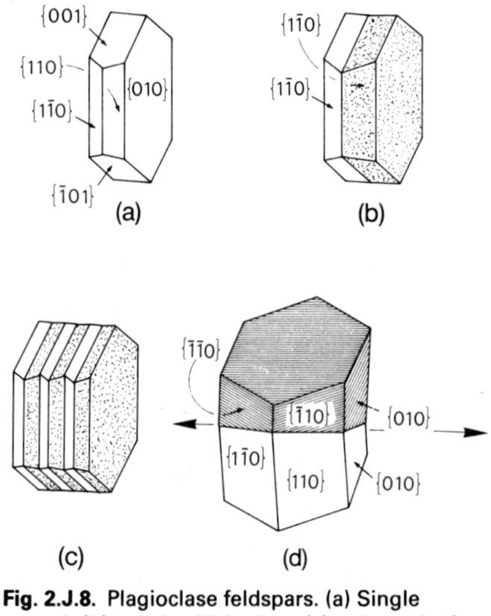

Fig. 2.J.8. Plagioclase feldspars. (a) Single crystal; (b) twin by Albite law; (c) polysynthetic albite twin; (d) twin by pericline law.

pericline law, with twin axis [010]. The composition plane is the so-called 'rhombic section', a plane parallel to ***b*** whose orientation does not correspond to rational indices; see Fig. 2.J.8(d).

References

1. Sands, D. E. (1982). *Vectors and tensors in crystallography*. Addison-Wesley, Reading.
2. Klyne, W. and Prelog, V. (1960). *Experientia,* **16,** 521.
3. Schomaker, V., Waser, J., Marsh, R. E., and Bergman, G. (1958). *Acta Crystallographica,* **8,** 600.
4. Niggli, P. (1982). *Handbuch der Experimentalphysik,* Vol. 7, Part 1. Academische Verlagsgesellschaft, Leipzig.
5. Delaunay, B. N. (1983). *Zeitschrift für Kristallographie,* **84,** 109.
6. Katayama, C. (1986). *Journal of Applied Crystallography*, **19,** 69.
7. Burzlaff, H. and Zimmermann, H. (1984). *Zeitschrift für Kristallographie*, **170,** 241.
8. Burzlaff, H. and Zimmermann, H. (1984). *Zeitschrift für Kristallographie,* **170,** 247.
9. Buerger, M. J. (1957). *Zeitschrift für Kristallographie*, **109,** 42.
10. Buerger, M. J. (1960). *Zeitschrift für Kristallographie*, **113,** 52.
11. Gruber, B. (1973). *Acta Crystallographica,* **A29,** 433.
12. Santoro, A. and Mighell, A. D. (1970). *Acta Crystallographica,* **A26,** 124.
13. Krivy, I. and Gruber, B. (1976). *Acta Crystallographica,* **A32,** 297.
14. Clegg, W. (1981). *Acta Crystallographica,* **A37,** 913.
15. Ferraris, G. and Ivaldi, G. (1983). *Acta Crystallographica,* **A39,** 595.
16. Andrews, L. C. and Bernstein, H. J. (1988). *Acta Crystallographica,* **A44,** 1009.
17. Azaroff, L. V. and Buerger, M. J. (1958). *The powder method in x-ray crystallography*. McGraw-Hill, New York.
18. Mighell, A. D. and Rodgers, J. R. (1980). *Acta Crystallographica,* **A36,** 321.
19. Mighell, A. D. and Himes, V. L. (1986). *Acta Crystallographica,* **A42,** 101.
20. Gruber, B. (1989). *Acta Crystallographica,* **A45,** 123.
21. Santoro, A. and Mighell, A. D. (1972). *Acta Crystallographica,* **A28,** 284.
22. Fortes, M. A. (1983). *Acta Crystallographica,* **A39,** 351.
23. Grimmer, H. (1976). *Acta Crystallographica,* **A32,** 783.
24. Grimmer, H. (1974). *Scripta Metallurgica,* **8,** 1221.
25. Bolmann, W. (1970). *Crystal defects and crystalline interfaces.* Springer, Berlin.
26. Ranganathan, S. (1966). *Acta Crystallographica,* **A21,** 197.
27. Grimmer, H. (1984). *Acta Crystallographica,* **A40,** 108.
28. Grimmer, H. (1989). *Acta Crystallographica,* **A45,** 320.
29. Bonnet, R., Cousineau, E., and Warrington, D. M. (1981). *Acta Crystallographica,* **A37,** 184.
30. Grimmer, H. (1989). *Acta Crystallographica,* **A45,** 505.
31. Donnay, G. and Donnay, J. D. H. (1974). *Canadian Mineralogist,* **12,** 422.
32. Le Page, Y., Donnay, J. D. H., and Donnay, G. (1984). *Acta Crystallographica,* **A40,** 679.
33. Friedel, G. (1926). *Leçons de cristallographie.* Paris: Berger-Levrault, Paris. (Reprinted 1964. Blanchard, Paris.)
34. Santoro, A. (1974). *Acta Crystallographica,* **A30,** 224.
35. Catti, M. and Ferraris, G. (1976). *Acta Crystallographica,* **A32,** 163.
36. Van der Sluis, P. (1989). Thesis. University of Utrecht.
37. Flack, H. D. (1987). *Acta Crystallographica,* **A43,** 564.
38. Britton, D. (1972). *Acta Crystallographica,* **A28,** 296.

39. Stanley, E. (1972). *Acta Crystallographica*, **5**, 191.
40. Rees, D. C. (1982). *Acta Crystallographica*, **A38**, 201.
41. Yeates, T. O. (1988). *Acta Crystallographica*, **A44**, 142.
42. *International tables for x-ray crystallography* (1962), vol. III. Kynock Press, Birmingham.
43. Cromer, D. T. and Mann, J. B. (1968). *Acta Crystallographica*, **A24**, 321.
44. Cooley, J. W. and Tukey, J. W. (1965). *Mathematics of Computation*, **19**, 297.
45. Hamilton, W. C. (1964). *Statistics in physical science*. Ronald Press Company, New York.
46. Rogers, D. (1981). *Acta Crystallographica*, **A37**, 734.
47. Berardinelli, G. and Flack, H. D. (1985). *Acta Crystallographica*, **A41**, 500.
48. Larson, A. C. (1970). In *Crystallographic computing*, pp. 291–4. Munksgaard, Copenhagen.
49. Zachariasen, W. H. (1963). *Acta Crystallographica*, **16**, 1139.
50. Zachariasen, W. H. (1967). *Acta Crystallographica*, **23**, 558.
51. Report of the International Union of Crystallography. Subcommittee on Statistical Descriptors (1989). *Acta Crystallographica*, **A45**, 63–75.
52. Rollett, J. S. (1988). Error analysis. In *Crystallographic computing 4. Techniques and new technologies* (ed. N. W. Isaacs and M. R. Taylor), pp. 149–66. Oxford University Press and International Union of Crystallography, Oxford.
53. Carruthers, J. R. and Watkin, D. J. (1979). *Acta Crystallographica*, **A35**, 698.
54. Wang, H. and Robertson, B. E. (1985). In *Structure and statistics in crystallography* (ed. A. J. C. Wilson), pp. 125–35. Adenine Press, Guiderland, NY.
55. Prince, E. (1982). *Mathematical techniques in crystallography and material sciences*. Springer, New York.
56. Seiler, P., Schweizer, W. B., and Dunitz, J. D. (1984). *Acta Crystallographica*, **B40**, 319.
57. Hamilton, W. C. (1965). *Acta Crystallographica*, **18**, 502.
58. Prince, E. and Nicholson, W. L. (1985). In *Structure and statistics in crystallography* (ed. A. J. C. Wilson), pp. 183–95. Adenine Press, Guiderland, NY.
59. Wilson, A. J. C. (1976). *Acta Crystallographica*, **A32**, 994.
60. Wilson, A. J. C. (1979). *Acta Crystallographica*, **A35**, 122.
61. French, S. and Wilson, K. (1978). *Acta Crystallographica*, **A34**, 517.
62. Gonschorek, W. (1985). *Acta Crystallographica*, **A41**, 189.
63. Scheringer, C. (1963). *Acta Crystallographica*, **16**, 546.
64. Doedens, R. J. (1970). In *Crystallographic computing* (ed. by F. R. Ahmed), pp. 198–204. Munksgaard, Copenhagen.
65. Domenicano, A. and Vaciago, A. (1975). *Acta Crystallographica*, **B31**, 2553.
66. Waser, J. (1963). *Acta Crystallographica*, **16**, 1091.
67. Watkin, D. (1988). In *Crystallographic computing 4. Techniques and new technologies* (ed. N. W. Isaacs and M. R. Taylor), pp. 111–29. Oxford Science Publications and International Union of Crystallography, Oxford.
68. Wolff, P. M. de (1968). *Journal of Applied Crystallography*, **1**, 108.
69. Taupin, D. (1988). *Journal of Applied Crystallography*, **21**, 485.
70. Wu, E. (1988). *Journal of Applied Crystallography*, **21**, 530.
71. Shirley, R. (1978) In *Computing in crystallography* (ed. H. Schenk, O. Olthof-Hazekamp, H. van Koningsveld, and G. C. Bassi), pp. 221–34. Delft University Press.
72. Pawley, G. S. (1981). *Journal of Applied Crystallography*, **14**, 357.
73. Louër, D. and Vargas, R. (1982). *Journal of Applied Crystallography*, **15**, 542.
74. Werner, P. E., Ericksson, L., and Westdahl, M. (1985). *Journal of Applied Crystallography*, **18**, 367.
75. Paszkowicz, W. (1987). *Journal of Applied Crystallography*, **20**, 166.

76. Taupin, D. (1989). *Journal of Applied Crystallography,* **22,** 455.
77. Rietveld, H. M. (1969). *Journal of Applied Crystallography,* **2,** 65.
78. Caglioti, G., Paoletti, A., and Ricci, F. P. (1958). *Nuclear Instruments and Methods,* **3,** 223.
79. Attfield, J. P., Cheetham, A. K., Cox, D. E., and Sleight, A. W. (1988). *Journal of Applied Crystallography,* **21,** 452.
80. Baerlocker, C. (1984). *Proceedings of the 6th International Zeolite Conference, Reno, USA, July 1983,* p. 823. Butterworth, Guildford.
81. Young, R. A. and Wiles, D. B. (1982). *Journal of Applied Crystallography,* **15,** 430.
82. Immirzi, A. (1980). *Acta Crystallographica,* **B36,** 2378.
83. Pawley, G. S. (1980). *Journal of Applied Crystallography,* **13,** 630.
84. Pawley, G. S. (1981). *Journal of Applied Crystallography,* **14,** 357.
85. Jansen, E., Schaefer, W., and Will, G. (1988). *Journal of Applied Crystallography,* **21,** 228.
86. Howard, S. A. and Snyder, R. L. (1989). *Journal of Applied Crystallography,* **22,** 238.
87. Sakata, M. and Cooper, M. J. (1979). *Acta Crystallographica,* **12,** 554.
88. Durbin, J. and Watson, G. S. (1971). *Biometrika,* **58,** 1.
89. Theil, H. and Nagar, A. L. (1961). *Journal of the American Statistical Association,* **56,** 793.
90. Hill, R. J. and Madsen, I. C. (1987). *Powder Diffraction,* **2,** 146.
91. Will, G., Jansen, E., and Schäfer, W. (1983). POWLS-80: a program for calculation and refinement of powder diffraction data. *Report Juel-1967.* KFA Jülich.
92. Ahtee, M., Nurmela, M., Suortti, P., and Järvinen, M. (1989). *Journal of Applied Crystallography,* **22,** 261.
93. Dahms, M. and Bunge, H. J. (1989). *Journal of Applied Crystallography,* **22,** 439.
94. Parrish, W., Hart, M., and Huang, T. C. (1986). *Journal of Applied Crystallography,* , **19,** 92.
95. Will, G., Parrish, W., and Huang, T. C. (1983). *Journal of Applied Crystallography,* **16,** 611.
96. Dollase, W. A. (1986). *Journal of Applied Crystallography,* **19,** 267.
97. Young, R. A. and Sakthivel, A. (1988). *Journal of Applied Crystallography,* **21,** 416.
98. Wilson, A. J. C. (1963). *Mathematical theory of x-ray powder diffractometry.* Centrex, Eindhoven.
99. Prince, E. (1983). *Journal of Applied Crystallography,* **16,** 508.
100. Klug, P. H. and Alexander, L. E. (1974). *X-ray diffraction procedures.* Wiley, New York.
101. Kauppinen, J. K., Moffatt, D. J., Mantsch, H. H., and Cameron, D. G. (1981). *Applied Spectroscopy,* **35,** 271.
102. Toraya, H. (1988). *Journal of Applied Crystallography,* **21,** 192.
103. Wiedemann, K. E., Unnam, J., and Clark, R. K. (1987). *Powder Diffraction* **2,** 3, 130.
104. Christensen, A. N., Lehmann, M. S., and Nielsen, M. (1985). *Australian Journal of Physics,* **38,** 497.
105. Louër, D. and Langford, J. I. (1988). *Journal of Applied Crystallography,* **21,** 430.
106. Cheetham, A. K. (1986). *High resolution powder diffraction,* Materials Science Forum, Vol. 9 (ed. C. R. A. Catlow), pp. 103–12. Transtech Publications Limited, Switzerland.
107. Cascarano, G., Favia, L., and Giacovazzo, C. (1990). *XV IUCr Congress, Bordeaux,* abs. 02.08.22, page C77.

108. Lehmann, M. S., Christensen, A. N., Fjellvåg, H., Feidenhans, R., and Nielsen, M. (1987). *Journal of Applied Crystallography,* **20,** 123.
109. Hibble, S. J., Cheetham, A. K., Bogle, A. R. L., Wakerley, H. R., and Cox, D. E. (1988). *Journal of the American Chemical Society,* **110,** 3295.
110. Maichle, J. K., Ihringer, J., and Prandl, W. (1988). *Journal of Applied Crystallography,* **21,** 22.
111. Cheetham, A. K. and Taylor, J. C. (1977). *Journal of Solid State Chemistry,* **21,** 253.
112. Meille, S. V., Brückner, S., and Lando, J. B. (1989). *Polymer,* **30,** 786.
113. Brückner, S., Meille, S. V., Porzio, W., and Ricci, G. (1988). *Makromolekulare Chemie,* **189,** 2145.
114. Cruickshank, D. W. J. (1956). *Acta Crystallographica,* **9,** 754.
115. Schomaker, V. and Trueblood, K. N. (1968). *Acta Crystallographica,* **B24,** 63.
116. Johnson, C. K. (1970). In *Crystallographic computing* (ed. F. R. Ahmed), pp. 207–19. Munksgaard, Copenhagen.
117. Busing, W. R. and Levy, H. A. (1964). *Acta Crystallographica,* **17,** 142.
118. Cruickshank, D. W. J. and Robertson, A. P. (1953). *Acta Crystallographica,* **6,** 698.
119. Stanford, R. H. Jr and Waser, J. (1972). *Acta Crystallographica,* **A28,** 213.
120. Nardelli, M. (1983). *Computers and Chemistry,* **7,** 3, 95.
121. Shmueli, V. (1974). *Acta Crystallographica* **A30,** 848.
122. Ten Eick, L. F. (1973). *Acta Crystallographica,* **A29,** 183.
123. Ten Eick, L. F. (1977). *Acta Crystallographica,* **A33,** 486.
124. Immirzi, A. (1976). In *Crystallographic computing techniques* (ed. F. R. Ahmed), pp. 399–412. Munksgaard, Copenhagen.
125. Agarwal, R. (1978). *Acta Crystallographica,* **A34,** 791.
126. Jack, A. and Levitt, M. (1978). *Acta Crystallographica,* **A34,** 931.

The diffraction of X-rays by crystals

3

CARMELO GIACOVAZZO

Introduction

Crystal structure analysis is usually based on diffraction phenomena caused by the interaction of matter with X-rays, electrons, or neutrons. Although the theory of diffraction is the same for all types of radiation, we shall consider X-ray scattering with particular interest: some references to electron and neutron scattering are made in Appendix 3.B pp. 195 and 198.

The most important properties of X-rays were described by Röntgen in 1896. However, with equipment in common use in optics at that time he could not measure any effect of interference, reflection, or refraction. Several years later Sommerfeld measured an X-ray wavelength of about 0.4 Å. In 1912 M. von Laue, starting from an article by Ewald, a student of Sommerfeld, suggested the use of crystals as natural lattices for diffraction. This experiment was successfully performed by Friedrich and Knipping, both students of Röntgen. In 1913 W. L. Bragg and M. von Laue used X-ray diffraction patterns for deducing the structure of NaCl, KCl, KBr, and KI. In such a way, in only few years, the electromagnetic nature of X-rays and their usefulness in the determination of crystal structure was indisputably demonstrated.

Let us recall some properties of electromagnetic radiation:

1. Electromagnetic radiation propagates in a vacuum with the velocity c equal to 300 000 km s^{-1}.

2. The vectors electric field $\boldsymbol{E}$ and magnetic field $\boldsymbol{H}$ are both orthogonal to the direction of wave propagation. In addition, they are mutually orthogonal and vary sinusoidally with time.

3. The range of X-ray wavelengths is placed between the ultraviolet region and the region of γ-rays emitted by radioactive substances. The interval of wavelengths of particular usefulness in crystallography ranges between 0.4 and 2.5 Å.

4. The refractive index of X-rays is very near to unity: for $\lambda = 2$ Å and for high-density substances the difference from unity is of the order 10^{-4}, being 10^{-5} for most cases. For this reason the X-rays cannot be focused by means of suitable lenses like ordinary light or electrons. Thus, if X-rays are used, we cannot talk about a direct observation of objects by means of instruments equivalent to optical or electron electron microscopes.

The interaction of X-rays with matter essentially occurs by means of two processes:

1. Some photons of the incident beam are deflected without a loss of energy. They constitute the **scattered radiation** with exactly the same wavelength as the incident radiation. Other photons are scattered with a small loss of energy; they constitute the **Compton radiation** (see pp. 185–6) with wavelength slightly greater than the wavelength of the incident radiation.

2. The photons can be absorbed by the atoms of a target and will increase its temperature. Discontinuities in curves of absorption are caused by the photoelectric effect. In this case the photon energy is used to remove one of inner-shell electrons of the absorber element. This atom can come back to its minimal energy state by emission of a photon X whose wavelength is characteristic of the atom (fluorescent radiation). Absorption phenomena will be discussed in Chapter 4.

The present chapter is mainly devoted to the so-called **kinematic theory of diffraction** (only few brief remarks on some aspects of **dynamic theory of diffraction** are given on pp. 162–5 and Appendix 3.B). Major emphasis is given to diffraction by perfect crystals, but basic elements of diffraction by gases, liquids, and amorphous solids are described in Appendix 3.C.

Diffraction theory is described by making use of Fourier transforms. Readers not acquainted with such a mathematical concept are urged to consult Appendix 3.A.

Thomson scattering

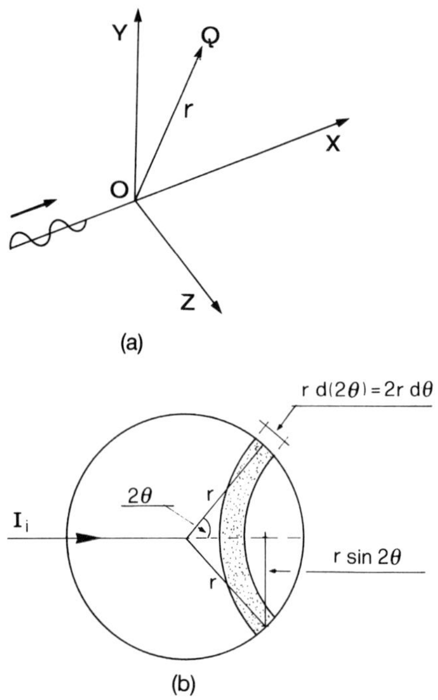

Fig. 3.1. (a) A free charged particle is in O: a plane monochromatic electromagnetic wave propagates along the *x* axis. (b) Surface element at scattering angle 2*θ*.

Let suppose that (see Fig. 3.1(a)) a free material particle with electric charge e and mass m is at the origin O of our coordinates system and that a plane monochromatic electromagnetic wave with frequency ν and electric vector $\boldsymbol{E}_i$ propagates along the x axis in positive direction. Its electric field is described by equation

$$\boldsymbol{E}_i = \boldsymbol{E}_{0i} \exp 2\pi i\nu(t - x/c)$$

where $\boldsymbol{E}_{oi}$ is the amplitude of the wave and $\boldsymbol{E}_i$ is the value of the field at position x at time t. The field exerts on the particle a periodic force $\boldsymbol{F} = e\boldsymbol{E}_i$ and therefore the particle will undergo oscillatory motion with acceleration $\boldsymbol{a} = \boldsymbol{F}/m = e\boldsymbol{E}_i/m$ and frequency ν. In accordance with classical theory of electromagnetism a charged particle in accelerated motion is a source of electromagnetic radiation: its field at $\boldsymbol{r}$ is proportional to acceleration and lies in the plane $(\boldsymbol{E}_i, \boldsymbol{r})$. Let us orient the axes y and z of our coordinates system in such a way that the observation point Q defined by vector $\boldsymbol{r}$ is in the plane (x, y). At the point Q we will measure the electric field $\boldsymbol{E}_d$ due to scattered radiation

$$\boldsymbol{E}_d = \boldsymbol{E}_{0d} \exp [2\pi i\nu(t - r/c) - i\alpha].$$

Thomson showed that (see also pp. 165–6)

$$\boldsymbol{E}_{0d} = \frac{1}{r} \boldsymbol{E}_{0i}(e^2/mc^2) \sin \varphi \tag{3.1}$$

where φ is the angle between the direction of acceleration of electron and the direction of our observation. The term $\sin\varphi$ is a polarization term: we 'see' only the component of vibration parallel to the observer and normal to the direction of propagation. α is the phase lag with which the charge re-emits the incident radiation. The decrease of $\boldsymbol{E}_d$ with $\boldsymbol{r}$ is caused by the scattering of radiation in all directions.

In terms of intensity eqn (3.1) becomes

$$I_{eTh} = I_i \frac{e^4}{m^2 r^2 c^4} \sin^2 \varphi \tag{3.2}$$

where I_{eTh} is the density of scattered radiation and I_i is the intensity of incident radiation. This simple result excludes neutrons from the category of X-ray scatterers because they do not have electric charge, and makes negligible the contribution to scattering by protons whose factor $(e/m)^2$ is about 1837^2 times less than that of electrons. Therefore, from now on and according to tradition, the symbol e will represent the electron charge.

If the primary beam is completely polarized: (α) with $\boldsymbol{E}_i$ along the z axis, then $I_{eTh} = I_i e^4/(m^2 r^2 c^4)$; (b) with $\boldsymbol{E}_i$ along the y axis, then $\boldsymbol{E}_{eTh} = \boldsymbol{I}_i e^4 \cos^2 2\theta/(m^2 r^2 c^4)$, where 2θ is the angle between the primary beam and the direction of observation. In general, the computation can be executed by decomposing the primary beam into two beams whose electric vectors are perpendicular and parallel respectively to the plane containing the primary beam and the scattered radiation being observed. If K_1 and K_2 are parts of these two beams in percentage we obtain

$$I_{eTh} = I_i \frac{e^4}{m^2 r^2 c^4} (K_1 + K_2 \cos^2 2\theta).$$

If the primary beam is not polarized, then $K_1 = K_2 = 1/2$ and

$$I_{eTh} = I_i \frac{e^4}{m^2 r^2 c^4} \frac{1 + \cos^2 2\theta}{2} \tag{3.3}$$

where $P = (1 + \cos^2 2\theta)/2$ is called the **polarization factor** (see also p. 303). It suggests that the radiation scattered in the direction of the incident beam is maximum while it is minimum in the direction perpendicular to the primary beam.

Equation (3.3) gives the intensity scattered into a unit solid angle at angle 2θ. If we want to obtain the total scattered power P we have to integrate (3.3) from 0 to π (see Fig. 3.1(b)).

$$P = I_i \frac{e^4}{m^2 r^2 c^4} \int_0^{\pi} \frac{1 + \cos^2 2\theta}{2} 2\pi r^2 \sin 2\theta \, d(2\theta)$$
$$= \frac{8\pi e^4}{3m^2 c^4} I_i$$

where $(2\pi r \sin 2\theta)\, r(d(2\theta))$ is the surface element at angle 2θ. The total scattering 'cross-section' P/I_i is equal to 6.7×10^{-25} cm^2/electron, which is a very small quantity. It may be calculated that the total fraction of incident radiation scattered by one 'crystal' composed only of free electrons and having dimensions less that 1 mm is less than 2 per cent.

The scattered radiation will be partially polarized even if the incident radiation is not. Thus, if the beam is scattered first by a crystal (monochromator) and then by the sample the polarization of the beam will be different. The scattering is coherent, according to Thomson, because there is a well defined phase relation between the incident radiation and the scattered one: for electrons $\alpha = \pi$.

Unfortunately it is very difficult to verify by experiment the Thomson formula since it is almost impossible to have a scatterer composed exclusively of free electrons. One could suppose that scatterers composed of light elements with electrons weakly bound to the nucleus is a good approximation to the ideal Thomson scatterer. But experiments with light elements have revealed a completely different effect, the Compton effect.

Compton scattering

The process can be described in terms of elastic collision between a photon and a free electron. The incident photon is deflected by a collision from its original direction and transfers a part of its energy to the electron. Consequently there is a difference in wavelength between the incident radiation and the scattered one which can be calculated by means of the relation (see also Appendix 3.B, p. 185)

$$\Delta\lambda\,(\text{Å}) = 0.024\,(1 - \cos 2\theta). \tag{3.4}$$

The following properties emerge from eqn (3.4): $\Delta\lambda$ does not depend on the wavelength of incident radiation; the maximum value of $\Delta\lambda$ ($\Delta\lambda = 0.048$) is reached for $2\theta = \pi$ (backscattering) which is small but significant for wavelengths of about 1 Å. Besides, $\Delta\lambda = 0$ for $2\theta = 0$.

Compton scattering is incoherent; it causes a variation in wavelength but does not involve a phase relation between the incident and the scattered radiation. It is impossible to calculate interference effects for Compton radiation.

Interference of scattered waves

Here we shall not be interested in wave propagation processes, but only in diffraction patterns produced by the interaction between waves and matter. These patterns are constant in time since they are produced by the system of atoms, which can be considered stationary. This fact permits us to omit the time from the wave equations.

In Fig. 3.2 two scattering centres are at O and at O′. If a plane wave excites them they become sources of secondary spherical waves which mutually interfere. Let $\boldsymbol{s}_0$ be the unit vector associated with the direction of propagation of the primary X-ray beam. The phase difference between the wave scattered by O′ in the direction defined by the unit vector $\boldsymbol{s}$ and that scattered by O in the same direction is

$$\delta = \frac{2\pi}{\lambda}(\boldsymbol{s} - \boldsymbol{s}_0)\cdot\boldsymbol{r} = 2\pi\boldsymbol{r}^*\cdot\boldsymbol{r}$$

where

$$\boldsymbol{r}^* = \lambda^{-1}(\boldsymbol{s} - \boldsymbol{s}_0). \tag{3.5}$$

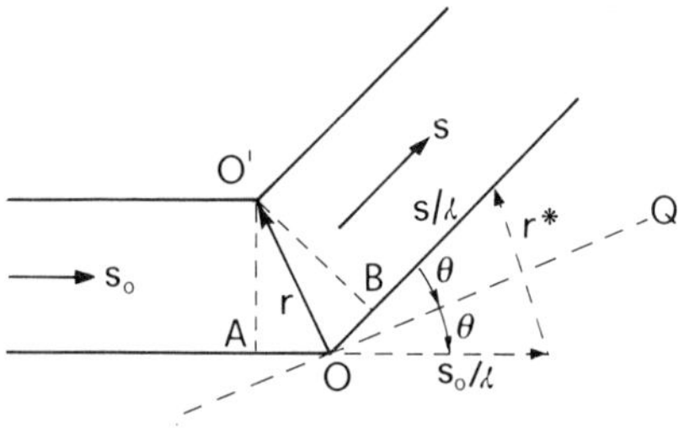

Fig. 3.2. Point scatterers are in O and O′, $\boldsymbol{s}_0$ and $\boldsymbol{s}$ are unit vectors. Therefore AO = $-\boldsymbol{r}\cdot\boldsymbol{s}_0$, BO = $\boldsymbol{r}\cdot\boldsymbol{s}$.

If λ is much greater than $\boldsymbol{r}$ there will be no phase difference between the scattered waves and consequently no appreciable interference phenomena will occur. Since interatomic bond distances lie between 1 and 4 Å no interference effect could be observed by using visible-light waves.

The modulus of $\boldsymbol{r}^*$ can be easily derived from Fig. 3.2:

$$r^* = 2\sin\theta/\lambda \tag{3.6}$$

where 2θ is the angle between the direction of incident X-rays and the direction of observation. If we outline two planes normal to $\boldsymbol{r}^*$ passing through O and O′ (OQ in Fig. 3.2 is the trace of the plane passing through O) we can consider interference as a consequence of specular reflection with respect to these planes.

If A_{O} is the amplitude of the wave scattered by the material point O (its phase is assumed to be zero) the wave scattered by O′ is described by $A_{\mathrm{O'}} \exp(2\pi i\boldsymbol{r}^* \cdot \boldsymbol{r})$. If there are N point scatterers along the path of the incident plane wave we have

$$F(\boldsymbol{r}^*) = \sum_{j=1}^{N} A_j \exp(2\pi i\boldsymbol{r}^* \cdot \boldsymbol{r}_j) \tag{3.7a}$$

where A_j is the amplitude of the wave scattered by the jth scatterer.

The Thomson formula plays an essential role in all calculations to obtain the absolute values of scattering. In our case it is more convenient to express the intensity I scattered by a given object (for example, an atom) in terms of intensity I_{eTh} scattered by a free electron. The ratio I/I_{eTh} is f^2, where f is the **scattering factor** of the object. Vice versa, for obtaining the observed experimental intensity it is sufficient to multiply f^2 by I_{eTh}. To give an example, let us imagine a certain number of electrons concentrated at O′ which undergo Thomson scattering. In this case $f_{\mathrm{O'}}$ expresses the number of electrons.

According to the convention stated above eqn (3.7a) becomes

$$F(\boldsymbol{r}^*) = \sum_{j=1}^{N} f_j \exp(2\pi i\boldsymbol{r}^* \cdot \boldsymbol{r}_j). \tag{3.7b}$$

If the scattering centres constitute a continuum, the element of volume $\mathrm{d}\boldsymbol{r}$ will contain a number of electrons equal to $\rho(\boldsymbol{r})\,\mathrm{d}\boldsymbol{r}$ where $\rho(\boldsymbol{r})$ is their density. The wave scattered on the element $\mathrm{d}\boldsymbol{r}$ is given, in amplitude and phase, by $\rho(\boldsymbol{r})\,\mathrm{d}\boldsymbol{r}\exp(2\pi i\boldsymbol{r}^* \cdot \boldsymbol{r})$ and the total amplitude of the scattered wave will be

$$F(\boldsymbol{r}^*) = \int_V \rho(\boldsymbol{r}) \exp(2\pi i\boldsymbol{r}^* \cdot \boldsymbol{r})\,\mathrm{d}\boldsymbol{r} = \mathrm{T}[\rho(\boldsymbol{r})] \tag{3.8}$$

where T represents the Fourier transform operator.

In crystallography the space of the $\boldsymbol{r}^*$ vectors is called **reciprocal space**. Equation (3.8) constitutes an important result: the amplitude of the scattered wave can be considered as the Fourier transform (see Appendix 3.A, p. 175) of the density of the elementary scatterers. If these are electrons, the amplitude of the scattered wave is the Fourier transform of the electron density. From the theory of Fourier transforms we also know that

$$\rho(\boldsymbol{r}) = \int_{V^*} F(\boldsymbol{r}^*) \exp(-2\pi i\boldsymbol{r}^* \cdot \boldsymbol{r})\,\mathrm{d}\boldsymbol{r}^* = \mathrm{T}^{-1}[F(\boldsymbol{r}^*)]. \tag{3.9}$$

Therefore, knowledge of the amplitudes of the scattered waves (in modulus and phase) unequivocally defines $\rho(\boldsymbol{r})$.

Scattering by atomic electrons

The processes of Thomson and Compton scattering are an example of wave–particle duality and they seem to be mutually incompatible.

In fact both processes are simultaneously present and they are precisely described by modern quantum mechanics. In common practice the scatterers are atomic electrons: they can occupy different energetic states corresponding to a discontinuous set of negative energies and to a continuous band of positive energies. If, after interaction with the radiation, the electron conserves its original state the photon conserves entirely its proper energy (conditions for coherent scattering). If the electron changes its state, a portion of the energy of the incident photon is converted into potential energy of an excited atom (conditions for incoherent scattering). Quantum-mechanical calculations indicate that the processes of coherent and incoherent scattering are simultaneously present and that $I_{\text{coe}} + I_{\text{incoe}} = I_{\text{eTh}}$.

The coherent intensity I_{coe} can be calculated on the basis of the following observations. An atomic electron can be represented by its distribution function $\rho_e(\boldsymbol{r}) = |\psi(\boldsymbol{r})|^2$, where $\psi(\boldsymbol{r})$ is the wave function which satisfies the Schrödinger equation. The volume dv contains $\rho_e\, dv$ electrons and scatters an elementary wave which will interfere with the others emitted from all the elements of volume constituting the electron cloud. In accordance with p. 145 the electron scattering factor will be

$$f_e(\boldsymbol{r}^*) = \int_S \rho_e(\boldsymbol{r}) \exp(2\pi i \boldsymbol{r}^* \cdot \boldsymbol{r})\, d\boldsymbol{r} \tag{3.10}$$

where S is the region of space in which the probability of finding the electron is different from zero. If we assume that $\rho_e(\boldsymbol{r})$ has spherical symmetry (what is in fact justifiable for s electrons, less so for p, d, etc. electrons) then eqn (3.10) can be written (see eqn (3.A.33)) as:

$$f_e(r^*) = \int_0^\infty U_e(r) \frac{\sin 2\pi r r^*}{2\pi r r^*}\, dr \tag{3.11}$$

where $U_e(r) = 4\pi r^2 \rho_e(r)$ is the radial distribution of the electron and $r^* = 2\sin\theta/\lambda$. For instance, there are two 1s electrons, two 2s electrons, and two 2p electrons in carbon atom. In radial approximation the 2s and 2p electrons have an equivalent distribution. For carbon the Slater formulae give

$$(\rho_e)_{1s} = \frac{c_1^3}{\pi} \exp(-2c_1 r); \qquad (\rho_e)_{2s} = \frac{c_2^5}{96\pi} r^2 \exp(-c_2 r) \tag{3.12}$$

with $c_1 = 10.77\ \text{Å}^{-1}$, $c_2 = 6.15\ \text{Å}^{-1}$. Then, eqn (3.11) gives

$$(f_e)_{1s} = \frac{c_1}{(c_1^2 + \pi^2 r^{*2})^2}, \qquad (f_e)_{2s} = \frac{c_2(c_2 - 4\pi^2 r^{*2})}{(c_2^2 + 4\pi^2 r^{*2})^4} \tag{3.13}$$

respectively.

Equation (3.12) are illustrated in Fig. 3.3(a) and eqns (3.13) in Fig. 3.3(b). In accordance with eqn (3.10) the electron scattering factor is equal to 1 when $r^* = 0$. Moreover, the scattering of 1s electrons, whose distribution is very sharp, is more efficient at higher values of r^*. If the distribution of 1s electrons could really be considered point-like their scattering factor would be constant with varying r^* (see Appendix 3.A, p. 177 for the transform of a Dirac delta function).

According to the premise of this section the intensity of the Compton radiation of an atomic electron will be

$$I_{\text{coe}} = I_{\text{eTh}}(1 - f_{\text{e}}^2)$$

where I_{eTh} is given by eqn (3.2) or eqn (3.3). The intensity of the Compton radiation has the same order of magnitude as the radiation scattered coherently.

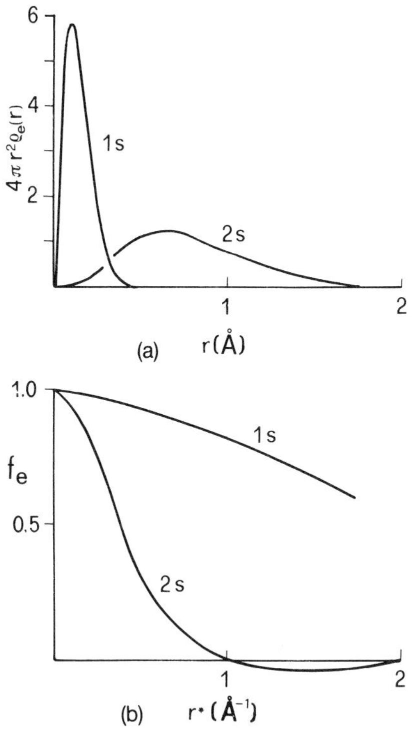

Fig. 3.3. (a) Radial distribution for 1s and 2s electrons of a C atom as defined by Slater functions. (b) ScatteRing factors for 1s and 2s electrons.

Scattering by atoms

Let $\psi_1(\boldsymbol{r}), \ldots, \psi_Z(\boldsymbol{r})$ be the wave functions of Z atomic electrons: then $\rho_{ej}\, dv = |\psi_j(\boldsymbol{r})|^2\, dv$ is the probability of finding the jth electron in the volume dv. If every function $\psi_j(\boldsymbol{r})$ can be considered independent of the others, then $\rho_a(\boldsymbol{r})\, dv = (\sum_{j=1}^{Z} \rho_{ej})\, dv$ is the probability of finding an electron in the volume dv. The Fourier transform of $\rho_a(\boldsymbol{r})$ is called the **atomic scattering factor** and will be denoted by f_a.

Generally the function $\rho_a(\boldsymbol{r})$ does not have spherical symmetry. In most crystallographic applications the deviations from it, for instance because of covalent bonds, are neglected in first approximation. If we assume that ρ_a is spherically symmetric and, without loss of generality, that the centre of the atom is at the origin, we will have

$$f_a(r^*) = \int_0^\infty U_a \frac{\sin(2\pi r r^*)}{2\pi r r^*}\, dr = \sum_{j=1}^{Z} f_{ej} \qquad (3.14)$$

where $U_a(r) = 4\pi r^2 \rho_a(r)$ is the radial distribution function for the atom. The ρ_a function is known with considerable accuracy for practically all neutral atoms and ions: for lighter atoms via Hartree–Fock methods, and for heavier atoms via the Thomas–Fermi approximation. In Fig. 3.4(a) the f_a functions for some atoms are shown. Each curve reaches its maximum value, equal to Z, at $\sin\theta/\lambda = 0$ and decreases with increasing $\sin\theta/\lambda$. According to the previous paragraph most of radiation scattered at high values of $\sin\theta/\lambda$ is due to electrons of inner shells of the electron cloud (core). Conversely scattering of valence electrons is efficient only at low $\sin\theta/\lambda$ values. f_a can thus be considered the sum of core and valence electron scattering:

$$f_a = f_{\text{core}} + f_{\text{valence}}.$$

In Fig. 3.4(b) f_{core} and f_{valence} of a nitrogen atom are shown as function of $\sin\theta/\lambda$.

As a consequence of eqn (3.14) the intensity of the radiation coherently scattered from an atom can be obtained by summing the amplitudes relative

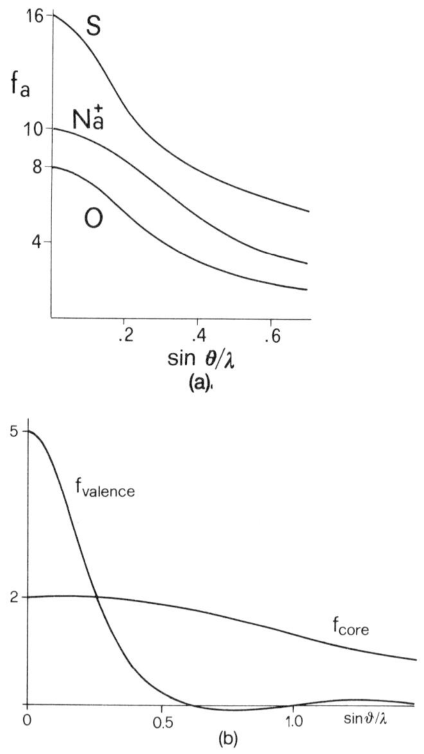

Fig. 3.4. (a) Scattering factors for S, Na^+, O. (b) core and valence scattering for nitrogen atom.

to the electrons taken individually:

$$I_{eTh} f_a^2 = I_{eTh} \left(\sum_{j=1}^{Z} f_{ej} \right)^2.$$

The Compton radiation scattered from an electron is incoherent with respect to that scattered from another electron: its intensity is obtained by summing the individual intensities relative to every single electron:

$$I_{eTh} \sum_{j=1}^{Z} [1 - (f_{ej})^2].$$

Since $f_e = 1$ for $\sin\theta/\lambda = 0$ there is no Compton radiation in the direction of the primary beam. Nevertheless it is appreciable at high values of $\sin\theta/\lambda$.

When we consider the diffraction phenomenon from one crystal the intensity coherently diffracted will be proportional to the square of the vectorial sum of the amplitudes scattered from the single atoms while the intensity of the Compton radiation will be once more the sum of the single intensities. As a consequence of the very high number of atoms which contribute to diffraction, Compton scattering can generally be ignored: its presence is detectable as background radiation, easily recognizable in crystals composed of light atoms.

The temperature factor

In a crystal structure an atom is bound to others by bond forces of various types. Their arrangement corresponds to an energy minimum. If the atoms are disturbed they will tend to return to the positions of minimal energy: they will oscillate around such positions gaining thermal energy.

The oscillations will modify the electron density function of each atom and consequently their capacity to scatter. Here we will suppose that the thermal motion of an atom is independent of that of the others. This is not completely true since the chemical bonds introduce strong correlations between the thermal motions of various atoms (see pp. 117–20 and Appendix 3.B, p. 186).

The time-scale of a scattering experiment is much longer than periods of thermal vibration of atoms. Therefore the description of thermal motion of an atom requires only the knowledge of the time-averged distribution of its position with respect to that of equilibrium. If we suppose that the position of equilibrium is at the origin, that $p(\boldsymbol{r}')$ is the probability of finding the centre of one atom at $\boldsymbol{r}'$, and that $\rho_a(\boldsymbol{r} - \boldsymbol{r}')$ is the electron density at $\boldsymbol{r}$ when the centre of the atom is at $\boldsymbol{r}'$, then we can write

$$\rho_{at}(\boldsymbol{r}) = \int_{S'} \rho_a(\boldsymbol{r} - \boldsymbol{r}') p(\boldsymbol{r}') \, d\boldsymbol{r}' = \rho_a(\boldsymbol{r}') * p(\boldsymbol{r}') \tag{3.15}$$

where $\rho_{at}(\boldsymbol{r})$ is the electron density corresponding to the thermically agitated atom. Notice that the rigid body vibration assumption has been made; i.e., the electron density is assumed to accompany the nucleus during thermal vibration.

In accordance with Appendix 3.A, p. 181), ρ_{at} is the convolution of two functions and its Fourier transform (see eqn (3.A.38)) is

$$f_{at}(\boldsymbol{r}^*) = f_a(\boldsymbol{r}^*) q(\boldsymbol{r}^*) \tag{3.16}$$

where

$$q(\boldsymbol{r}^*) = \int_{S'} p(\boldsymbol{r}') \exp(2\pi i \boldsymbol{r}^* \cdot \boldsymbol{r}') \, d\boldsymbol{r}' \tag{3.17}$$

the Fourier transform of $p(\boldsymbol{r}')$, is known as the Debye–Waller factor.

The function $p(\boldsymbol{r}')$ depends on few parameters; it is inversely dependent on atomic mass and on chemical bond forces, and directly dependent on temperature. $p(\boldsymbol{r}')$ is in general anisotropic. If assumed isotropic, the thermal motion of the atom will have spherical symmetry and could be described by a Gaussian function in any system of reference:

$$p(\boldsymbol{r}') = p(r') \simeq (2\pi)^{-1/2} U^{-1/2} \exp[-(r'^2/2U)] \tag{3.18}$$

where r' is measured in Å and $U = \langle r'^2 \rangle$ is the square mean shift of the atom with respect to the position of equilibrium. The corresponding Fourier transform is (see eqn (3.A.25))

$$q(r^*) = \exp(-2\pi^2 U r^{*2}) = \exp(-8\pi^2 U \sin^2\theta/\lambda^2)$$
$$= \exp(-B \sin^2\theta/\lambda^2) \tag{3.19}$$

where

$$B = 8\pi^2 U \ (\text{Å}^2).$$

The factor B is usually known in the literature as the **atomic temperature factor**.

The dependence of B on the absolute temperature T has been studied by Debye who obtained a formula valid for materials composed of only one chemical element. From X-ray diffraction structure analysis it is possible to conclude schematically that the order of value of $\sqrt{U}$ is in many inorganic crystals between 0.05 and 0.20 Å (B lying between 0.20 and 3.16 Å^2) but can also reach 0.5 Å ($B \simeq 20$ Å^2) for some organic crystals. The consequence of this is to make the electron density of the atom more diffuse and therefore to reduce the capacity for scattering with increasing values of $\sin\theta/\lambda$.

In general an atom will not be free to vibrate equally in all directions. If we assume that the probability $p(\boldsymbol{r}')$ has a three-dimensional Gaussian distribution the surfaces of equal probability will be ellipsoids called vibrational or thermal, centred on the mean position occupied by the atom.

Now eqn (3.19) will be substituted (see Appendix 3.B, pp. 186 and 188) by the anisotropic temperature factor (3.20) which represents a vibrational ellipsoid in reciprocal space defined by six parameters U_{11}^*, U_{22}^*, U_{33}^*, U_{12}^*, U_{13}^*, U_{23}^*:

$$q(\boldsymbol{r}^*) = \exp[-2\pi^2(U_{11}^* x_2^* + U_{22}^* y^{*2} + U_{33}^* z^{*2} + 2U_{12}^* x^* y^*$$
$$+ 2U_{13}^* x^* z^* + 2U_{23}^* y^* z^*)]. \tag{3.20}$$

The six parameters U_{ij}^* (five more than the unique parameter U necessary to characterize the isotropic thermal motion) define the orientation of the thermal ellipsoid with respect to the crystallographic axes and the lengths of the three ellipsoid axes. In order to describe graphically a crystal molecule

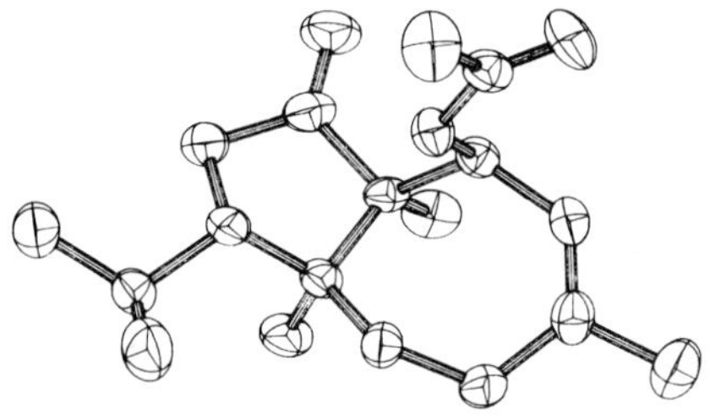

Fig. 3.5.

and its thermal motion each atom is usually represented by an ellipsoid, centred on the mean position of the atom, and surrounding the space within which the atomic displacement falls within the given ellipsoid with a probability of 0.5 (see Fig. 3.5).

Scattering by a molecule or by a unit cell

Let $\rho_j(\boldsymbol{r})$ be the electron density of the jth atom when it is thermically agitated, isolated, and localized at the origin. If the atom is at position $\boldsymbol{r}_j$ its electron density will be $\rho_j(\boldsymbol{r}-\boldsymbol{r}_j)$. If we neglect the effects of redistribution of the outer electrons because of chemical bonds, the electron density relative to an N-atom molecule or to one unit cell containing N atoms is

$$\rho_{\mathrm{M}}(\boldsymbol{r}) = \sum_{j=1}^{N} \rho_j(\boldsymbol{r}-\boldsymbol{r}_j). \tag{3.21}$$

The amplitude of the scattered wave is

$$\begin{aligned} F_{\mathrm{M}}(\boldsymbol{r}^*) &= \int_S \sum_{j=1}^{N} \rho_j(\boldsymbol{r}-\boldsymbol{r}_j) \exp(2\pi i \boldsymbol{r}^* \cdot \boldsymbol{r})\, \mathrm{d}\boldsymbol{r} \\ &= \sum_{j=1}^{N} \int_S \rho_j(\boldsymbol{R}_j) \exp[2\pi i \boldsymbol{r}^* \cdot (\boldsymbol{r}_j + \boldsymbol{R}_j)]\, \mathrm{d}\boldsymbol{R}_j \\ &= \sum_{j=1}^{N} f_j(\boldsymbol{r}^*) \exp(2\pi i \boldsymbol{r}^* \cdot \boldsymbol{r}_j), \end{aligned} \tag{3.22}$$

where $f_j(\boldsymbol{r}^*)$ is the atomic scattering factor of the jth atom (thermal motion included; in the previous section indicated by f_{at}). The fact that in eqn (3.21) we have neglected the redistribution of the outer electrons leads to negligible errors for $F_{\mathrm{M}}(\boldsymbol{r}^*)$, except in case of small r^* and for light atoms, where the number of outer electrons represents a consistent fraction of Z.

$\rho_{\mathrm{M}}(\boldsymbol{r})$, as defined by (3.21), is the electron density of a **promolecule**, or, in other words, of an assembly of spherically averaged free atoms thermically agitated and superimposed on the molecular geometry. Such a model is unsatisfactory if one is interested in the deformation of the electron density consequent to bond formation. In a real molecule the electron density is generated by superposition of molecular space orbitals ψ_i with occupation n_i:

$$\rho_{\text{molecule}} = \sum_i n_i\, |\psi_i|^2.$$

Since ρ_{molecule} can be decomposed into atomic fragments, a finite set of appropriately chosen basis functions can be used to represent each jth atomic fragment (see Appendix 3.D). Then

$$\rho_{\text{molecule}} = \rho_{\text{promolecule}} + \Delta\rho$$

where $\Delta\rho$ models the effects of bonding and of molecular environment (in particular, pseudoatoms may become aspherical and carry a net charge).

By Fourier transform of $\Delta\rho$ the deformation scattering is obtained:

$$\Delta F = F_{\text{molecule}} - F_{\text{promolecule}}.$$

Since the core deformation scattering is negligible ΔF practically coincides with deformation scattering of the valence shells.

Diffraction by a crystal

One three-dimensional infinite lattice can be represented (see Appendix 3.A, p. 174) by the lattice function

$$L(\boldsymbol{r}) = \sum_{u,v,w=-\infty}^{+\infty} \delta(\boldsymbol{r} - \boldsymbol{r}_{u,v,w})$$

where δ is the Dirac delta function and $\boldsymbol{r}_{u,v,w} = u\boldsymbol{a} + v\boldsymbol{b} + w\boldsymbol{c}$ (with u, v, w being integers) is the generic lattice vector. Let us suppose that $\rho_{\mathrm{M}}(\boldsymbol{r})$ describes the electron density in the unit cell of an infinite three-dimensional crystal. The electron density function for the whole crystal (see Appendix 3.A, p. 183) is the convolution of the $L(\boldsymbol{r})$ function with $\rho_{\mathrm{M}}(\boldsymbol{r})$:

$$\rho_{\infty}(\boldsymbol{r}) = \rho_{\mathrm{M}}(\boldsymbol{r}) * L(\boldsymbol{r}). \tag{3.23}$$

As a consequence of eqns (3.A.35), (3.A.30), and (3.22) the amplitude of the wave scattered by the whole crystal is

$$\begin{aligned} F_{\infty}(\boldsymbol{r}^*) &= \mathrm{T}[\rho_{\mathrm{M}}(\boldsymbol{r})] \cdot \mathrm{T}[L(\boldsymbol{r})] \\ &= F_{\mathrm{M}}(\boldsymbol{r}^*) \cdot \frac{1}{V} \sum_{h,k,l=-\infty}^{+\infty} \delta(\boldsymbol{r}^* - \boldsymbol{r}^*_{\mathbf{H}}) \\ &= \frac{1}{V} F_{\mathrm{M}}(\mathbf{H}) \sum_{h,k,l=-\infty}^{+\infty} \delta(\boldsymbol{r}^* - \boldsymbol{r}^*_{\mathbf{H}}) \end{aligned} \tag{3.24}$$

where V is the volume of the unit cell and $\boldsymbol{r}^*_{\mathbf{H}} = h\boldsymbol{a}^* + k\boldsymbol{b}^* + l\boldsymbol{c}^*$ is the generic lattice vector of the reciprocal lattice (see pp. 63–5).

If the scatterer object is non-periodic (atom, molecule, etc.) the amplitude of the scattered wave $F_{\mathrm{M}}(\boldsymbol{r}^*)$ can be non-zero for any value of $\boldsymbol{r}^*$. On the contrary, if the scatterer object is periodic (crystal) we observe a non-zero amplitude only when $\boldsymbol{r}^*$ coincides with a reciprocal lattice point:

$$\boldsymbol{r}^* = \boldsymbol{r}^*_{\mathbf{H}}. \tag{3.25}$$

The function $F_{\infty}(\boldsymbol{r}^*)$ can be represented by means of a pseudo-lattice: each of its points has the position coinciding with the corresponding point of the reciprocal lattice but has a specific 'weight' $F_{\mathrm{M}}(\mathbf{H})/V$. For a given node the diffraction intensity $I_{\mathbf{H}}$ will be function of the square of its weight.

Let us multiply eqn (3.25) scalarly by $\boldsymbol{a}$, $\boldsymbol{b}$, $\boldsymbol{c}$ and introduce the definition (3.5) of $\boldsymbol{r}^*$: we obtain

$$\boldsymbol{a} \cdot (\boldsymbol{s} - \boldsymbol{s}_0) = h\lambda \qquad \boldsymbol{b} \cdot (\boldsymbol{s} - \boldsymbol{s}_0) = k\lambda \qquad \boldsymbol{c} \cdot (\boldsymbol{s} - \boldsymbol{s}_0) = l\lambda. \tag{3.26}$$

The directions $\boldsymbol{s}$ which satisfy eqns (3.26) are called diffraction directions and relations (3.26) are the **Laue conditions**.

Finiteness of the crystal may be taken into account by introducing the form function $\Phi(\boldsymbol{r})$: $\Phi(\boldsymbol{r}) = 1$ inside the crystal, $\Phi(\boldsymbol{r}) = 0$ outside the crystal. In this case we can write

$$\rho_{\mathrm{cr}} = \rho_{\infty}(\boldsymbol{r})\Phi(\boldsymbol{r})$$

and, because of eqn (3.A.35), the amplitude of the diffracted wave is

$$F(\boldsymbol{r}^*) = \mathrm{T}[\rho_\infty(\boldsymbol{r})] * [\Phi(\boldsymbol{r})] = F_\infty(\boldsymbol{r}^*) * D(\boldsymbol{r}^*) \tag{3.27}$$

where

$$D(\boldsymbol{r}^*) = \int_S \Phi(\boldsymbol{r}) \exp(2\pi i \boldsymbol{r}^* \cdot \boldsymbol{r})\, \mathrm{d}\boldsymbol{r} = \int_\Omega \exp(2\pi i \boldsymbol{r}^* \cdot \boldsymbol{r})\, \mathrm{d}\boldsymbol{r}$$

and Ω is the volume of the crystal. Because of eqn (3.A.40) the relation (3.27) becomes

$$\begin{aligned} F(\boldsymbol{r}^*) &= \frac{1}{V} F_{\mathrm{M}}(\mathbf{H}) \sum_{h,k,l=-\infty}^{+\infty} \delta(\boldsymbol{r}^* - \boldsymbol{r}^*_{\mathbf{H}}) * D(\boldsymbol{r}^*) \\ &= \frac{1}{V} F_{\mathrm{M}}(\mathbf{H}) \sum_{h,k,l=-\infty}^{+\infty} D(\boldsymbol{r}^* - \boldsymbol{r}^*_{\mathbf{H}}). \end{aligned} \tag{3.28}$$

If we compare eqns (3.28) and (3.24) we notice that, going from an infinite crystal to a finite one, the point-like function corresponding to each node of the reciprocal lattice is substituted by the distribution function D which is non-zero in a domain whose form and dimensions depend on the form and dimensions of the crystal. The distribution D is identical for all nodes.

For example, let suppose that the crystal is a parallelepiped with faces A_1, A_2, A_3: then

$$D(\boldsymbol{r}^*) = \int_{-A_1/2}^{A_1/2} \int_{-A_2/2}^{A_2/2} \int_{-A_3/2}^{A_3/2} \exp[2\pi i(x^*x + y^*y + z^*z)]\, \mathrm{d}x\, \mathrm{d}y\, \mathrm{d}z.$$

If we integrate this function over separate variables, it becomes, in accordance with Appendix 3.A, p. 174

$$D(\boldsymbol{r}^*) = \frac{\sin(\pi A_1 x^*)}{\pi x^*} \frac{\sin(\pi A_2 y^*)}{\pi y^*} \frac{\sin(\pi A_3 z^*)}{\pi z^*}. \tag{3.29}$$

Each of the factors in eqn (3.29) is studied in Appendix 3.A and shown in Fig. 3.A.1 (p. 174). We deduce:

1. The maximum value of $D(\boldsymbol{r}^*)$ is equal to $A_1A_2A_3$, i.e. to the volume Ω of the crystal;

2. The width of a principal maximum in a certain direction is inversely proportional to the dimension of the crystal in that direction. Thus, because of the finiteness of the crystals each node of the reciprocal lattice is in practice a spatial domain with dimensions equal to A_i^{-1}. In Fig. 3.6 some examples of finite lattices with the corresponding reciprocal lattices are shown.

When we consider the diffraction by a crystal the function $F_{\mathrm{M}}(\mathbf{H})$ bears the name of **structure factor** of vectorial index $\mathbf{H}$ (or indexes h, k, l if we make reference to the components of $\boldsymbol{r}^*_{\mathbf{H}}$) and it is indicated as:

$$F_{\mathbf{H}} = \sum_{j=1}^{N} f_j \exp(2\pi i \boldsymbol{r}^*_{\mathbf{H}} \cdot \boldsymbol{r}_j)$$

where N is the number of atoms in the unit cell. In accordance with

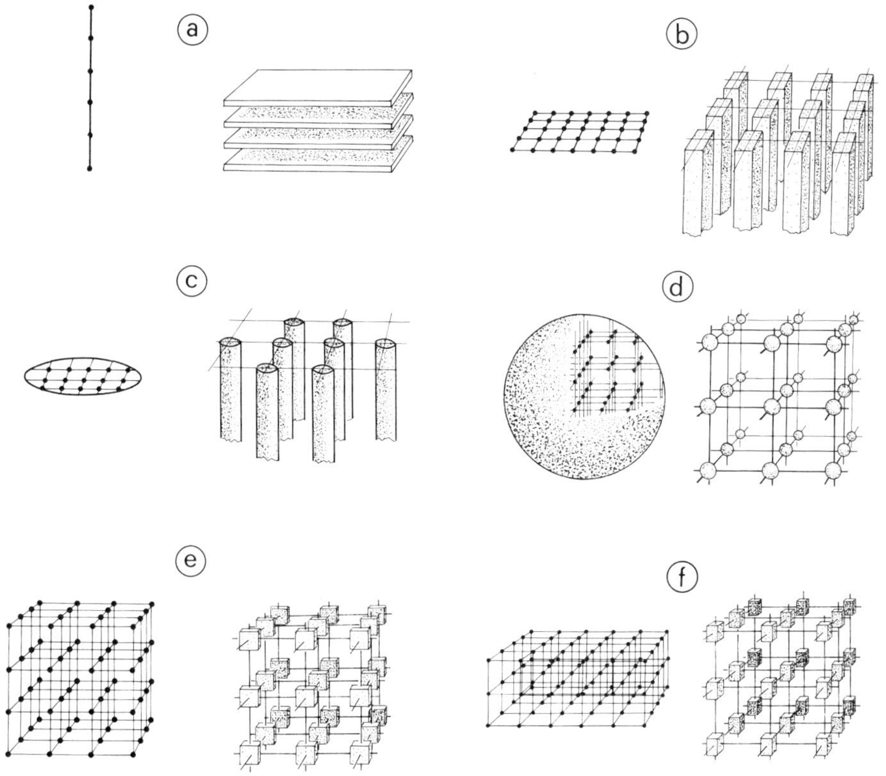

Fig. 3.6. Direct and reciprocal lattices for: (a) a one-dimensional lattice; (b) a two-dimensional lattice in the form of a rectangle; (c) a two-dimensional lattice in the form of a circle; (d) a cubic crystal in the form of a sphere; (e) a cubic crystal in the form of a cube; (f) a crystal in the form of a parallelepiped (from Kitaigorodskii, A. I. (1951). *The theory of crystal structure analysis*, Consultants Bureau, New York).

p. 64 we write

$$F_{\mathbf{H}} = \sum_{j=1}^{N} f_j \exp(2\pi i \bar{\mathbf{H}} \mathbf{X}_j) = A_{\mathbf{H}} + iB_{\mathbf{H}} \tag{3.30a}$$

where

$$A_{\mathbf{H}} = \sum_{j=1}^{N} f_j \cos 2\pi \bar{\mathbf{H}} \mathbf{X}_j, \qquad B_{\mathbf{H}} = \sum_{j=1}^{N} f_j \sin 2\pi \bar{\mathbf{H}} \mathbf{X}_j. \tag{3.30b}$$

According to the notation introduced in Chapter 2, we have indicated the vector as $\boldsymbol{r}^*_{\mathbf{H}}$ and the transpose matrix of its components with respect to the reciprocal coordinates system as $\bar{\mathbf{H}} = (hkl)$. In the same way $\boldsymbol{r}_j$ is the jth positional vector and the transpose matrix of its components with respect to the direct coordinates system is $\bar{\mathbf{X}}_j = [x_j\ y_j\ z_j]$. In a more explicit form (3.30a) may be written

$$F_{hkl} = \sum_{j=1}^{N} f_j \exp 2\pi i(hx_j + ky_j + lz_j).$$

In different notation (see Fig. 3.7)

$$F_{\mathbf{H}} = |F_{\mathbf{H}}| \exp(i\varphi_{\mathbf{H}}) \text{ where } \varphi_{\mathbf{H}} = \arctan(B_{\mathbf{H}}/A_{\mathbf{H}}). \tag{3.31}$$

$\varphi_{\mathbf{H}}$ is the **phase** of the structure factor $F_{\mathbf{H}}$.

If we want to point out in eqn (3.30a) the effect of thermal agitation of the atoms we write, in accordance with p. 149 and Appendix 3.B

$$F_{\mathbf{H}} = \sum_{j=1}^{N} f_{0j} \exp(2\pi i \bar{\mathbf{H}} \mathbf{X}_j - 8\pi^2 U_j \sin^2\theta/\lambda^2)$$

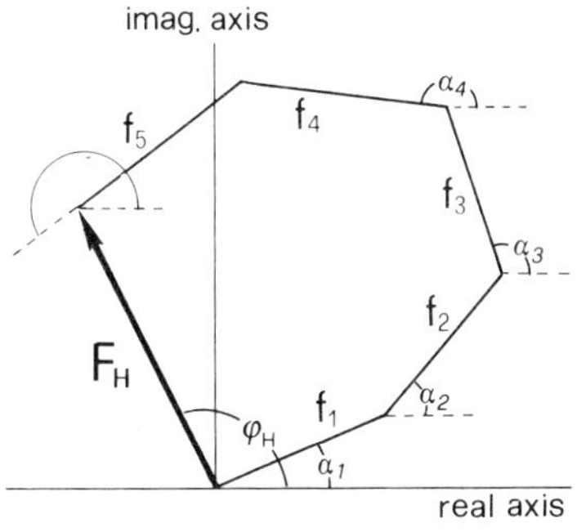

Fig. 3.7. $F_{\mathbf{H}}$ is represented in the Gauss plane for a crystal structure with $N = 5$. It is $\alpha_j = 2\pi\bar{\mathbf{H}}\mathbf{X}_j$.

or

$$F_{\mathbf{H}} = \sum_{j=1}^{N} f_{0j} \exp\,(2\pi i \bar{\mathbf{H}}\mathbf{X}_j - 2\pi^2 \bar{\mathbf{H}}\mathbf{U}_j^* \mathbf{H})$$

depending on the type of the thermal motion (isotropic or anisotropic) of the atoms. f_{0j} is the scattering factor of the jth atom considered at rest. Let us note explicitly that the value of $F_{\mathbf{H}}$, in modulus and phase, depends on the atomic positions i.e. on the crystal structure.

Details of the structure factors calculation from a known structural model are given on pp. 87–8 and Appendix 2.I.

Bragg's law

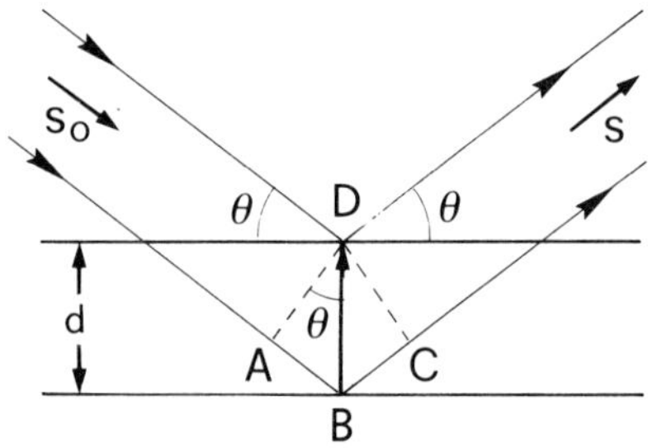

Fig. 3.8. Reflection of X-rays from two lattice planes belonging to the family $\mathbf{H} \equiv (h, k, l)$. d is the interplanar spacing.

A qualitatively simple method for obtaining the conditions for diffraction was described in 1912 by W. L. Bragg who considered the diffraction as the consequence of contemporaneous reflections of the X-ray beam by various lattice planes belonging to the same family (physically, from the atoms lying on these planes). Let θ be (see Fig. 3.8) the angle between the primary beam and the family of lattice planes with indices h, k, l (having no integer common factor larger than unity). The difference in 'path' between the waves scattered in D and B is equal to $AB + BC = 2d \sin\theta$. If it is multiple of λ then the two waves combine themselves with maximum positive interference:

$$2d_{\mathbf{H}} \sin\theta = n\lambda. \qquad (3.32)$$

Since the X-rays penetrate deeply in the crystal a large number of lattice planes will reflect the primary beam: the reflected waves will interfere destructively if eqn (3.32) is not verified. Equation (3.32) is the **Bragg equation** and the angle for which it is verified is the **Bragg angle**: for $n = 1, 2, \ldots$ we obtain reflections (or diffraction effects) of first order, second order, etc., relative to the same family of lattice planes **H**.

The point of view can be further simplified by observing that the family of fictitious lattice planes with indices $h' = nh$, $k' = nk$, $l' = nl$ has interplanar spacing $d_{\mathbf{H}'} = d_{\mathbf{H}}/n$. Now eqn (3.32) can be written as

$$2(d_{\mathbf{H}}/n) \sin\theta = 2d_{\mathbf{H}'} \sin\theta = \lambda \qquad (3.33)$$

where h', k', l' are no longer obliged to have only the unitary factor in common.

In practice, an effect of diffraction of nth order due to a reflection from lattice planes **H** can be interpreted as reflection of first order from the family of fictitious lattice planes $\mathbf{H}' = n\mathbf{H}$.

It is easy to see now that eqn (3.33) is equivalent to eqn (3.25). Indeed, if we consider only the moduli of eqn (3.25) we will have, because of eqns (2.14) and (3.6),

$$r^* = 2 \sin\theta/\lambda = 1/d_{\mathbf{H}}.$$

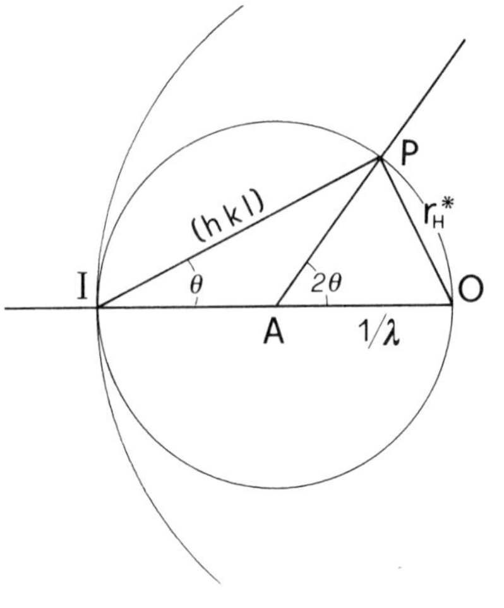

Fig. 3.9. Reflection and limiting spheres.

The reflection and the limiting spheres

Let us outline (see Fig. 3.9) a sphere of radius $1/\lambda$ in such a way that the primary beam passes along the diameter IO. Put the origin of the reciprocal lattice at O. When the vector $\boldsymbol{r}^*_{\mathbf{H}}$ is on the surface of the sphere then the corresponding direct lattice planes will lie parallely to IP and will make an angle θ with the primary beam. The relation

$$\mathrm{OP} = r^*_{\mathbf{H}} = 1/d_{\mathbf{H}} = \mathrm{IO} \sin \theta = 2 \sin \theta/\lambda$$

holds, which coincides with Bragg's equation. Therefore: the necessary and sufficient condition for the Bragg equation to be verified for the family of planes (hkl) is that the lattice point defined by the vector $\boldsymbol{r}^*_{\mathbf{H}}$ lies on the surface of the sphere called the **reflection** or **Ewald sphere**. AP is the direction of diffracted waves (it makes an angle of 2θ with the primary beam): therefore we can suppose that the crystal is at A.

For X-rays and neutrons $\lambda \simeq (0.5\text{–}2)$ Å, which is comparable with the dimensions of the unit cell ($\simeq 10$ Å): the sphere then has appreciable curvature with respect to the planes of the reciprocal lattice. If the primary beam is monochromatic and the crystal casually oriented, no point of the reciprocal lattice should be in contact with the surface of the Ewald sphere except the (000) point which represents scattering in the direction of the primary beam. It will be seen in Chapter 4 that the experimental techniques aim to bring as many nodes of the reciprocal lattice as possible into contact with the surface of the reflection sphere.

In electron diffraction $\lambda \simeq 0.05$ Å: therefore the curvature of the Ewald sphere is small with respect to the planes of the reciprocal lattice. A very high number of lattice points can simultaneously be in contact with the surface of the sphere: for instance, all the points belonging to a plane of the reciprocal lattice passing through O.

If $r^*_{\mathbf{H}} > 2/\lambda$ (then $d_{\mathbf{H}} < \lambda/2$) we will not be able to observe the reflection $\mathbf{H}$. This condition defines the so-called **limiting sphere**, with centre O and radius $2/\lambda$: only the lattice points inside the limiting sphere will be able to diffract. Vice versa if $\lambda > 2a_{\max}$, where $a_{\max}$ is the largest period of the unit cell, then the diameter of the Ewald sphere will be smaller than $r^*_{\min}$ (the smallest period of the reciprocal lattice). Under these conditions no node could intercept the surface of the reflection sphere. That is the reason why we can never obtain diffraction of visible light (wavelength $\simeq$5000 Å) from crystals.

The wavelength determines the amount of information available from an experiment. In ideal conditions the wavelength should be short enough to leave out of the limiting sphere only the lattice points with diffraction intensities close to zero due to the decrease of atomic scattering factors.

Symmetry in reciprocal space

There are some relationships among structure factors relative to different indexes $\mathbf{H}$ which are originated either from the physics of diffraction and from the crystal symmetry. Let us give the relevant rules.

Friedel law

In accordance with eqn (3.30) we write $F_{\mathbf{H}} = A_{\mathbf{H}} + \mathrm{i}B_{\mathbf{H}}$. Then it will also be: $F_{-\mathbf{H}} = A_{\mathbf{H}} - \mathrm{i}B_{\mathbf{H}}$ and consequently

$$\varphi_{-\mathbf{H}} = -\varphi_{\mathbf{H}} \tag{3.34}$$

The value of $\varphi_{-\mathbf{H}}$ is opposite to the value of $\varphi_{\mathbf{H}}$. Since the intensities $I_{\mathbf{H}}$ and $I_{-\mathbf{H}}$ depend on $|F_{\mathbf{H}}|^2$ and $|F_{-\mathbf{H}}|^2$ respectively, we have

$$I_{\mathbf{H}} \propto (A_{\mathbf{H}} + \mathrm{i}B_{\mathbf{H}})(A_{\mathbf{H}} - \mathrm{i}B_{\mathbf{H}}) = A_{\mathbf{H}}^2 + B_{\mathbf{H}}^2$$
$$I_{\mathbf{H}} \propto (A_{\mathbf{H}} - \mathrm{i}B_{\mathbf{H}})(A_{\mathbf{H}} + \mathrm{i}B_{\mathbf{H}}) = A_{\mathbf{H}}^2 + B_{\mathbf{H}}^2.$$

From that the Friedel law is deduced, according to which the diffraction intensities associated to the vectors $\boldsymbol{H}$ and $-\boldsymbol{H}$ of the reciprocal space are equal. Since these intensities appear to be related by a centre of symmetry, usually, although imperfectly, it is said that the diffraction by itself introduces a centre of symmetry.

Effects of symmetry operators in the reciprocal space

Let us suppose that the symmetry operator $\mathbf{C} = (\mathbf{R}, \mathbf{T})$ exists in direct space. Then the points $\boldsymbol{r}_j$ and $\boldsymbol{r}_j'$ whose coordinates are related by $\mathbf{X}_j' = \mathbf{R}\mathbf{X}_j + \mathbf{T}$ are symmetry equivalent. We wonder which kind of relationships brings in the reciprocal space the presence of the operator $\mathbf{C}$. Since

$$F_{\bar{\mathbf{H}}\mathbf{R}} \exp(2\pi\mathrm{i}\bar{\mathbf{H}}\mathbf{T}) = \sum_{j=1}^{N} f_j \exp(2\pi\mathrm{i}\bar{\mathbf{H}}\mathbf{R}\mathbf{X}_j) \cdot \exp(2\pi\mathrm{i}\bar{\mathbf{H}}\mathbf{T})$$
$$= \sum_{j=1}^{N} f_j \exp 2\pi\mathrm{i}\bar{\mathbf{H}}(\mathbf{R}\mathbf{X}_j + \mathbf{T}) = F_{\mathbf{H}}$$

we can write

$$F_{\bar{\mathbf{H}}\mathbf{R}} = F_{\mathbf{H}} \exp(-2\pi\mathrm{i}\bar{\mathbf{H}}\mathbf{T}). \tag{3.35}$$

Sometimes it is convenient to split eqn (3.35) into two relations:

$$|F_{\bar{\mathbf{H}}\mathbf{R}}| = |F_{\mathbf{H}}|, \tag{3.36}$$
$$\varphi_{\bar{\mathbf{H}}\mathbf{R}} = \varphi_{\mathbf{H}} - 2\pi\bar{\mathbf{H}}\mathbf{T}. \tag{3.37}$$

From (3.36) it is concluded that intensities $I_{\mathbf{H}}$ and $I_{\bar{\mathbf{H}}\mathbf{R}}$ are equal while their phases are related by eqn (3.37). The most relevant consequences of eqn (3.35) are described in the following.

Determination of the Laue class

The Laue class of a crystal (see p. 17) may be determined by means of the qualitative examination of the diffraction intensities.

Let P1[(x, y, z)] be the space group (symmetry equivalent positions are shown in square brackets). The only symmetry operator is the identity, and its use in eqn (3.36) does not give us any useful result. Anyhow, because of the Friedel law $|F_{-\mathbf{H}}| = |F_{\mathbf{H}}|$ holds. If the space group is P$\bar{1}$ [$(x, y, z), (\bar{x}, \bar{y}, \bar{z})$] we can use $\mathbf{R}_2$ in eqn (3.36) and consequently obtain $F_{-\mathbf{H}} = F_{\mathbf{H}}$. In this case the Friedel law does not add any additional observable relationship. In conclusion, the diffraction intensities from crystals with symmetry P1 and P$\bar{1}$ will both show a centre of symmetry or, in other words, they will show the symmetry of the Laue class $\bar{1}$.

As a further example let P2 [$(x, y, z), (\bar{x}, y, \bar{z})$] be the space group. By introducing $\mathbf{R}_2$ in eqn (3.36) and by applying the Friedel law we obtain the following relationship (**symmetry equivalent reflections**):

$$|F_{hkl}| = |F_{\bar{h}k\bar{l}}| = |F_{\bar{h}\bar{k}\bar{l}}| = |F_{h\bar{k}l}|. \tag{3.38}$$

If the space group was Pm $[(x, y, z), (x, \bar{y}, z)]$, by using $\mathbf{R}_2$ and by applying the Friedel law we would obtain eqn (3.38) again. If the space group was P2/m eqn (3.38) would be obtained again only by using matrices $\mathbf{R}_2$, $\mathbf{R}_3$, and $\mathbf{R}_4$ in eqn (3.36). This time the Friedel law does not add any additional relationship to those obtained from eqn (3.36). We can conclude that the symmetry of the diffraction intensities from crystals belonging to space groups P2, Pm and P2/m is that of the Laue class 2/m.

The reader will easily verify that the crystals belonging to groups P222, Pmm2, and P2/m 2/m 2/m show intensity symmetry of the 2/m 2/m 2/m Laue class:

$$\begin{aligned}|F_{hkl}| &= |F_{h\bar{k}\bar{l}}| = |F_{\bar{h}k\bar{l}}| = |F_{\bar{h}\bar{k}l}| \\ &= |F_{\bar{h}\bar{k}\bar{l}}| = |F_{\bar{h}kl}| = |F_{h\bar{k}l}| = |F_{hk\bar{l}}|\end{aligned}$$

and those belonging to space groups P4, P$\bar{4}$, P4/m show the intensity symmetry

$$\begin{aligned}|F_{hkl}| &= |F_{\bar{h}\bar{k}l}| = |F_{\bar{k}hl}| = |F_{k\bar{h}l}| \\ &= |F_{\bar{h}\bar{k}\bar{l}}| = |F_{hk\bar{l}}| = |F_{k\bar{h}\bar{l}}| = |F_{\bar{k}h\bar{l}}|\end{aligned}$$

of the Laue class 4/m.

Determination of reflections with restricted phase values

Let us suppose that for a given set of reflections the relationship $\bar{\mathbf{H}}\mathbf{R} = -\bar{\mathbf{H}}$ is satisfied. If we apply (3.37) to this set we will obtain $2\varphi_{\mathbf{H}} = 2\pi\bar{\mathbf{H}}\mathbf{T} + 2n\pi$ from which

$$\varphi_{\mathbf{H}} = \pi\bar{\mathbf{H}}\mathbf{T} + n\pi. \qquad (3.39)$$

Equation (3.39) restricts the phase $\varphi_{\mathbf{H}}$ to two values, $\pi\bar{\mathbf{H}}\mathbf{T}$ or $\pi(\bar{\mathbf{H}}\mathbf{T} + 1)$. These reflections are called reflections with restricted phase values, or less properly, 'centrosymmetric'.

If the space group is centrosymmetric the inversion operator

$$\mathbf{R} = \begin{pmatrix} \bar{1} & 0 & 0 \\ 0 & \bar{1} & 0 \\ 0 & 0 & \bar{1} \end{pmatrix}, \qquad \mathbf{T} = \begin{pmatrix} T_1 \\ T_2 \\ T_3 \end{pmatrix}$$

will exist. In this case every reflection is a restricted phase reflection and will assume the values $\pi\bar{\mathbf{H}}\mathbf{T}$ or $\pi(\bar{\mathbf{H}}\mathbf{T} + 1)$. If the origin is assumed on the centre of symmetry then $\mathbf{T} = \mathbf{0}$ and the permitted phase values are 0 and π. Then according to eqn (3.30b), $F_{\mathbf{H}}$ will be a real positive number for $\varphi_{\mathbf{H}}$ equal to 0, and a negative one for $\varphi_{\mathbf{H}}$ equal to π. For this reason we usually talk in centrosymmetric space groups about the sign of the structure factor instead of about the phase.

In Fig. 3.10 $F_{\mathbf{H}}$ is represented in the complex plane for a centrosymmetric structure of six atoms. Since for each atom at $\boldsymbol{r}_j$ another symmetry equivalent atom exists at $-\boldsymbol{r}_j$, the contribution of every couple to $F_{\mathbf{H}}$ will have to be real.

As an example of a non-centrosymmetric space group let us examine $P2_12_12_1$, $[(x, y, z), (\frac{1}{2} - x, \bar{y}, \frac{1}{2} + z), (\frac{1}{2} + x, \frac{1}{2} - y, \bar{z}), (\bar{x}, \frac{1}{2} + y, \frac{1}{2} - z)]$ where the reflections $(hk0)$, $(0kl)$, $(h0l)$ satisfy the relation $\bar{\mathbf{H}}\mathbf{R} = -\bar{\mathbf{H}}$ for $\mathbf{R} = \mathbf{R}_2$, $\mathbf{R}_3$, $\mathbf{R}_4$ respectively. By introducing $\mathbf{T} = \mathbf{T}_2$ in eqn (3.39) we obtain

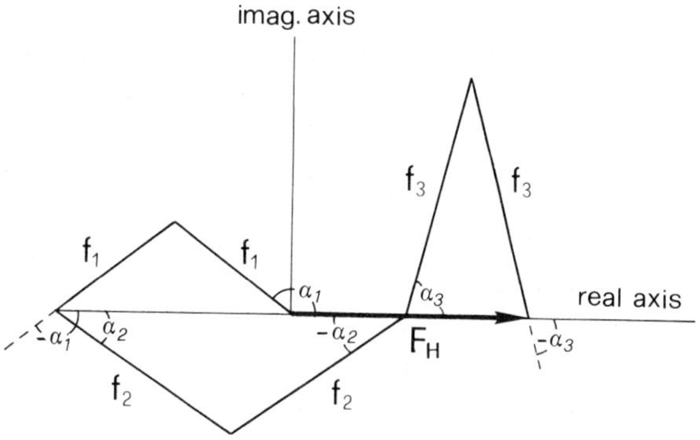

Fig. 3.10. F_H is represented in the Gauss plane for a centrosymmetric crystal structure with $N = 6$. It is $\alpha_j = 2\pi\bar{\mathbf{H}}\mathbf{X}_j$.

$\varphi_{hk0} = (\pi h/2) + n\pi$. Thus φ_{hk0} will have phase 0 or π if h is even, phase $\pm\pi/2$ if h is odd. By introducing $\mathbf{T} = \mathbf{T}_3$ in eqn (3.39) we obtain $\varphi_{0kl} = (\pi k/2) + n\pi$: i.e. φ_{0kl} will have phase 0 or π if k is even, $\pm\pi/2$ if k is odd. In the same way, by introducing $\mathbf{T} = \mathbf{T}_4$ in eqn (3.39) we obtain $\varphi_{h0l} = (\pi l/2) + n\pi$: i.e. φ_{h0l} will have phase 0 or π if l is even, $\pm\pi/2$ if l is odd.

Phase restrictions to $(\pi/4, 5\pi/4)$ or to $(3\pi/4, 7\pi/4)$ or to $(\pi/6, 7\pi/6)$ or . . . , can be found for some space groups. The allowed phase values depend

Table 3.1. Restricted phase reflections for the 32 crystal classes

Point group	Sets of restricted phase reflections
1	None
$\bar{1}$	All
m	$(0, k, 0)$
2	$(h, 0, l)$
2/m	All
mm2	$[(h, k, 0)$ masks $(h, 0, 0), (0, k, 0)]$
222	Three principal zones only
mmm	All
4	$(h, k, 0)$
$\bar{4}$	$(h, k, 0)$; $(0, 0, l)$
4/m	All
422	$(h, k, 0)$; $\{h, 0, l\}$; $\{h, h, l\}$
$\bar{4}$2m	$[(h, k, 0), \{h, h, 0\}]$; $[\{h, 0, l\}, (0, 0, l)]$
4mm	$[(h, k, 0), \{h, 0, 0\}, \{h, h, 0\}]$
4/mmm	All
3	None
$\bar{3}$	All
3m	$\{h, 0, \bar{h}, 0\}$
32	$\{h, 0, \bar{h}, l\}$
$\bar{3}$m	All
6	$(h, k, 0)$
$\bar{6}$	$(0, 0, l)$
6/m	All
$\bar{6}$m2	$[\{h, h, l\}, \{h, h, 0\}, (0, 0, l)]$
6mm	$[(h, k, 0), \{h, h, 0\}, \{h, 0, 0\}]$
62	$(h, k, 0)$; $(h, 0, l)$; (h, h, l)
6/mmm	All
23	$\{h, k, 0\}$
m$\bar{3}$	All
$\bar{4}$3m	$(\{h, k, 0\}, \{h, h, 0\}]$
432	$\{h, k, 0\}$; $\{h, h, l\}$
m$\bar{3}$m	All

on the translational component of the symmetry element and on its location with respect to the cell origin.

A different point of view may also be used: the existence of reflections with restricted phase is due to the presence of symmetry elements which at least in projection simulate the centre of inversion. For instance, the projection of a structure in $P2_12_12_1$ along the c axis is centrosymmetric: correspondingly, the reflections belonging to the zone $(hk0)$, being insensitive to the z coordinate, have restricted phase values. The reader will easily verify that zones orthogonal to axes 2, 4, $\bar{4}$, and 6 (and, of course, to the corresponding screw axis) have symmetry restricted phases, as well as zones parallel to the axes $\bar{2}$, $\bar{4}$, $\bar{6}$. In Table 3.1 the sets of restricted phase reflections are given for the 32 crystal classes.

Systematic absences

Let us look for the class of reflections for which $\bar{\mathbf{H}}\mathbf{R} = \bar{\mathbf{H}}$ and let us apply eqn (3.35). This relation would be violated for those reflections for which $\bar{\mathbf{H}}\mathbf{T}$ is not an integer number unless $|F_{\mathbf{H}}| = 0$. From this fact the rule follows: reflections for which $\bar{\mathbf{H}}\mathbf{R} = \bar{\mathbf{H}}$ and $\bar{\mathbf{H}}\mathbf{T}$ is not integer will have diffraction intensity zero or, as usually said, will be systematically absent or extinct. Let us give a few examples.

In the space group $P2_1$ $[(x, y, z), (\bar{x}, y + \frac{1}{2}, \bar{z})]$ the reflections $(0k0)$ satisfy the condition $\bar{\mathbf{H}}\mathbf{R}_2 = \bar{\mathbf{H}}$. If k is odd, $\bar{\mathbf{H}}\mathbf{T}_2$ is semi-integer. Thus, the reflections $(0k0)$ with $k \neq 2n$ are systematically absent.

In the space group $P4_1$ $[(x, y, z), (\bar{x}, \bar{y}, \frac{1}{2} + z), (\bar{y}, x, \frac{1}{4} + z), (y, \bar{x}, \frac{3}{4} + z)]$ only the reflections $(00l)$ satisfy the condition $\bar{\mathbf{H}}\mathbf{R}_j = \bar{\mathbf{H}}$ for $j = 2, 3, 4$. Since $\bar{\mathbf{H}}\mathbf{T}_2 = l/2$, $\bar{\mathbf{H}}\mathbf{T}_3 = l/4$, $\bar{\mathbf{H}}\mathbf{T}_4 = 3l/4$, the only condition for systematic absence is $l \neq 4n$, with n integer.

In the space group Pc $[(x, y, z), (x, \bar{y}, z + \frac{1}{2})]$ the reflections $(h0l)$ satisfy the condition $\bar{\mathbf{H}}\mathbf{R}_2 = \bar{\mathbf{H}}$. Since $\bar{\mathbf{H}}\mathbf{T}_2 = l/2$ the reflections $(h0l)$ with $l \neq 2n$ will be systematically absent.

Note that the presence of a glide plane imposes conditions for systematic absences to bidimensional reflections. In particular, glide planes opposite to $\boldsymbol{a}$, $\boldsymbol{b}$, and $\boldsymbol{c}$ impose conditions on classes $(0kl)$, $(h0l)$, and $(hk0)$ respectively. The conditions will be $h = 2n$, $k = 2n$, $l = 2n$ for glide planes of type a, b, or c respectively. The reader can easily check the data listed in Table 3.2.

Let us apply now the same considerations to the symmetry operators centring the cell. If the cell is of type A, B, C, I, symmetry operators will exist whose rotational matrix is always the identity while the translational matrix is:

$$\mathbf{T}_{\mathrm{A}} = \begin{pmatrix} 0 \\ \frac{1}{2} \\ \frac{1}{2} \end{pmatrix} \quad \mathbf{T}_{\mathrm{B}} = \begin{pmatrix} \frac{1}{2} \\ 0 \\ \frac{1}{2} \end{pmatrix} \quad \mathbf{T}_{\mathrm{C}} = \begin{pmatrix} \frac{1}{2} \\ \frac{1}{2} \\ 0 \end{pmatrix} \quad \mathbf{T}_{\mathrm{I}} = \begin{pmatrix} \frac{1}{2} \\ \frac{1}{2} \\ \frac{1}{2} \end{pmatrix} \qquad (3.40)$$

respectively. If we use these operators in eqn (3.35) we obtain: the relation $\bar{\mathbf{H}}\mathbf{R} = \bar{\mathbf{H}}$ is satisfied for any reflection; the systematic absences, of three-dimensional type, are those described in Table 3.2.

A cell of type F is simultaneously A-, B-, and C-centred, so the respective conditions for systematic absences must be simultaneously valid. Conse-

Table 3.2. Systematic absences

Symmetry elements		Set of reflections	Conditions
Lattice	P		none
	I		$h+k+l=2n$
	C		$h+k=2n$
	A		$k+l=2n$
	B	hkl	$h+l=2n$
	F		$\begin{cases} h+k=2n \\ k+l=2n \\ h+l=2n \end{cases}$
	R_{obv}		$-h+k+l=3n$
	R_{rev}		$h-k+l=3n$
Glide-plane ‖ (001)	a		$h=2n$
	b	$hk0$	$k=2n$
	n		$h+k=2n$
	d		$h+k=4n$
Glide-plane ‖ (100)	b		$k=2n$
	c	$0kl$	$l=2n$
	n		$k+l=2n$
	d		$k+l=4n$
Glide-plane ‖ (010)	a		$h=2n$
	c	$h0l$	$l=2n$
	n		$h+l=2n$
	d		$h+l=4n$
Glide-plane ‖ (110)	c		$l=2n$
	b	hhl	$h=2n$
	n		$h+l=2n$
	d		$2h+l=4n$
Screw-axis ‖ c	$2_1, 4_2, 6_3$		$l=2n$
	$3_1, 3_2, 6_2, 6_4$	$00l$	$l=3n$
	$4_1, 4_3$		$l=4n$
	$6_1, 6_5$		$l=6n$
Screw-axis ‖ a	$2_1, 4_2$		$h=2n$
	$4_1, 4_3$	$h00$	$h=4n$
Screw-axis ‖ b	$2_1, 4_2$		$k=2n$
	$4_1, 4_3$	$0k0$	$k=4n$
Screw-axis ‖ [110]	2_1	$hh0$	$h=2n$

quently only the reflections for which h, k, and l are all even or all odd will be present.

The same criteria lead us to establish the conditions for systematic absences ($-h+k+l \neq 3n$ for obverse setting and $h-k+l \neq 3n$ for reverse setting) for a hexagonal cell with rhombohedral lattice.

Rules for systematic absences may be also derived by using the explicit algebraic form of the structure factor. Suppose, for example, that the space group contains a c-glide plane perpendicular to the $\boldsymbol{b}$ axis (then (x, y, z) and $(x, \bar{y}, z+\frac{1}{2})$ will be symmetry equivalent points). The structure factor is then

$$F_{hkl} = \sum_{j=1}^{N/2} f_j \exp 2\pi i(hx_j + ky_j + lz_j) + \sum_{j=1}^{N/2} f_j \exp 2\pi i[hx_j - ky_j + l(z_j + \tfrac{1}{2})].$$

For l even

$$F_{hkl} = 2\sum_{j=1}^{N/2} f_j \exp\,[2\pi i(hx_j + lz_j)] \cos 2\pi k y_j;$$

for l odd

$$F_{hkl} = 2\sum_{j=1}^{N/2} f_j \exp\,[2\pi i(hx_j + lz_j)] \sin 2\pi k y_j.$$

It is easily seen that $F_{h0l} = 0$ for l odd, in accordance with our previous results.

Unequivocal determination of the space group

Determination of the space group is an important first step in any crystal structure analysis. From the qualitative examination of the diffraction intensities (which leads to the identification of the Laue class) and from the analysis of the systematic extinctions 58 space groups are unequivocally determinable (without making any distinction between enantiomorph couples). The symbols of these groups are printed in bold character in Table 1.9. As an example, suppose that relations (3.38) and systematic absences $(0k0)$ for $k = \text{odd}$ and $(h0l)$ for $l = \text{odd}$ are observed. Then the space group $P2_1/c$ is unequivocally deduced since:

(1) the cell has to be primitive (no systematic absences involving three-dimensional reflections are observed);

(2) the Laue-class is 2/m (from (3.38));

(3) a twofold screw axis along ***b*** and a c-glide plane normal to ***b*** do exist according to systematic absences.

Sometimes the identification of space group extinctions is disturbed by the presence of one or a few reflections seemingly violating one of the extinction rules (e.g. only one $(hk0)$ reflection with $h + k = \text{odd}$ has a meaningful non-zero intensity). One way to produce such forbidden reflections is the Renninger effect, described in Appendix 3.B, p. 191, which occurs when two or more reciprocal lattice points happen to intersect the reflection sphere simultaneously. One way to avoid the Renninger effect is to rotate the crystal about the reflecting plane's normal and repeat measurements.

Space groups which are not unequivocally determined could, with few exceptions, be unambiguously identified if their point group is known. The methods which can give us information about point groups have been schematically described in Chapter 1 (p. 15) (crystal habit, piezoelectricity, pyroelectricity, etc.). Let us add to them the anomalous dispersion effects (see p. 165) and the statistical methods (see Chapter 5, p. 322) based on the expectation that the distribution of the intensities is influenced by the presence of symmetry elements and, in particular, it is different for centrosymmetric and non-centrosymmetric space groups.

Diffraction intensities

The theory so far described is called kinematic: basically it calculates interference effects between the elementary waves scattered inside the volume of the crystal. However, it neglects two important phenomena: when the incident wave propagates inside the crystal its intensity decreases gradually because a part of its energy is transferred to the scattered beam or it is absorbed; the diffracted waves interfere with each other and with the incident beam.

The theory which takes into account all these phenomena and analyses the wave field set up as a whole is called the **dynamic theory of diffraction**. It was initiated by Ewald:[1] later on Laue showed that Ewald's theory is equivalent to analysing the propagation of any electromagnetic field through a medium having a periodically varying complex dielectric constant. The description of the dynamical theory is out of the scope of this book. The reader is referred to specialist books or review articles[2, 3] for exhaustive information. Only a dynamical effect of particular importance for crystal structure analysis, the Renninger effect, will be here described (in Appendix 3.B, p. 191) in any detail. Other dynamical interactions will be described on the basis of the kinematical theory properly modified by Darwin and other authors.

Dynamic effects develop gradually in a crystal: it may be shown that for sufficiently small thicknesses the incident beam is not weakened considerably, the diffracted waves are not yet so strong as to give rise to remarkable interference effects with the incident beam and the effects of absorption are negligible. Under these conditions (theoretically, thicknesses $<10^{-3}$–10^{-4} cm) the kinematic theory is a fairly accurate approximation to dynamic theory. However, in practice, corresponding equations proved to be valid even for crystals having dimensions of several tenths of a millimetre. This is due to the real crystal structures.

A simplified model of real crystal was proposed by Darwin: [4, 5] it can be ideally schematized like a **mosaic** of crystalline blocks with dimensions of about 10^{-5} cm, tilted very slightly to each other for angles of the order of fractions of one minute of arc: each block is separated by faults and cracks from other blocks. The interference between the waves only occurs inside every single block, whose dimensions satisfy the theoretical conditions of applicability of the kinematic theory. Because of the loss of coherence between the waves diffracted from different blocks, the diffracted intensity from the whole crystal is equal to the sum over the intensities diffracted from every single block.

Real crystals, however, differ by ideal ones also because they may contain a large variety of defects, which are convenient to classify into the following groups (see Chapter 9): **transient defects**, having lifetimes measured in microseconds (e.g. phonons, which are elastic waves propagating through the crystal and inducing atomic displacements); **point defects**, which can be missing (called vacancies), interstitial, or vicarious atoms; **line defects**, extending along straight or curved lines (e.g. dislocations); **plane defects**, extending along planes or curved surfaces (e.g. small angle boundaries, stacking faults); **volume defects**, extending throughout small volumes in the crystal (e.g. inclusions, precipitates, voids).

The importance of defects with respect to diffraction intensities depend on their nature and on their density. For example, a single point defect does not produce detectable effects on diffraction maxima but a large number of them, as in the case of order–disorder transitions, strongly affects diffraction. When agglomerated they can form voids or cracks in a crystal, or, clustering along certain planes, they form two-dimensional precipitates. Furthermore, individual dislocations have little effect on diffraction but they may array themselves to form small-angle boundaries separating two relatively perfect regions tilted relative to each other by about one minute of arc (mosaic blocks).

We can therefore expect that any type of defect which disturbs the crystal periodicity by lattice distortion or substitution or shift of atoms from the equilibrium position will produce some effect on diffracted intensities: among others, it will cause a more rapid statistical decrease of diffraction intensities with $\sin\theta/\lambda$. In particular, the lattice distortions cause variations in the unit cell dimensions and therefore modify the form and volume of the reciprocal lattice points (variations of the spacing $d_{\mathbf{H}}$ bring variations in the modulus $r^*_{\mathbf{H}}$, variations in the orientation of the planes **H** cause modifications in the orientation of $\boldsymbol{r}^*_{\mathbf{H}}$).

On the basis of these premises it is nonsense to affirm that a crystal is in the 'exact' Bragg position for a given family of lattice planes. Indeed, because of the finite size of the crystal, its mosaic structure, defects, and lattice distortions, etc., each node of the reciprocal lattice will have a finite volume and will be in contact with the surface of the reflection sphere for a finite angle interval. In addition the surface of the reflection sphere itself has in practice to be substituted by a solid domain. Indeed, the incident X-ray beam does have an inevitable divergence and an imperfect monochromaticity. As a consequence (see Fig. 3.11) the spherical surface of radius $1/\lambda$ is replaced by a family of spherical surfaces whose efficiency relative to diffraction depends on the distribution of the intensities as a function of the angle divergence and of the wavelength of the incident beam.

According to the above remarks quantity of practical interest is the integrated intensity and not the maximum intensity of the diffraction peak. Experimental arrangements normally used to measure diffraction intensities change the orientation of the crystal (see Chapter 4) so as to compel reciprocal lattice points to cross progressively the Ewald sphere while continuously recording the intensity of the diffracted beam. Thus the total diffracted energy during a fixed time is measured. Equivalently, the same total energy may be measured by integrating the diffracted intensity over a suitable angular range around the ideal Bragg angle. According to eqns (3.3) and (3.28) the integrated intensity is given by

$$I_{\mathbf{H}} = k_1 k_2 I_0 LPTE\,|F_{\mathbf{H}}|^2 \tag{3.41}$$

where I_0 is the intensity of the incident beam, $k_1 = e^4/(m^2c^4)$ takes into account the universal constants existing in eqn (3.3). $k_2 = \lambda^3\Omega/V^2$ is a

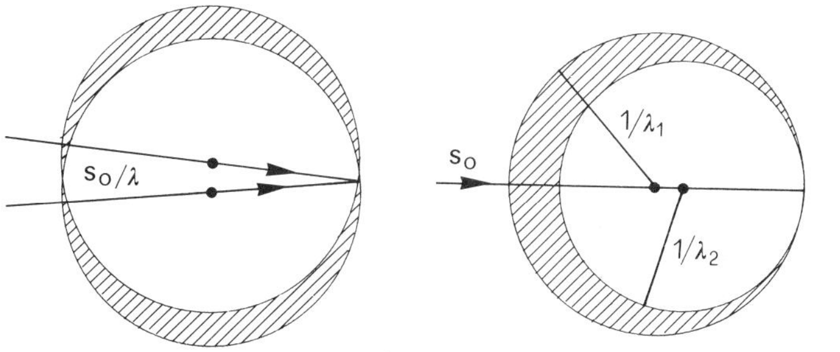

Fig. 3.11. (a) Incident radiation with non-vanishing divergence. (b) Non-monochromatic incident radiation.

constant for a given diffraction experiment (Ω is the volume of the crystal, V is the volume of the unit cell), P is the polarization factor, defined on p. 143, T is the transmission factor and depends on the capacity of the crystal to absorb the X-rays (see Chapter 4, p. 304), L is the Lorentz factor and depends on the diffraction technique (Chapter 4, p. 301). E is the extinction coefficient. It depends on the mosaic structure of the crystal and has two components. The most important one, called **secondary extinction**, takes into account the fact that the lattice planes first encountered by the primary beam will reflect a significant fraction of the primary intensity so that deeper planes receive less primary radiation. That causes a weakening of the diffracted intensity, mainly observable for high-intensity reflections at low $\sin\theta/\lambda$ values in sufficiently perfect crystals. If the mosaic blocks are misoriented (as they usually are) then they do not diffract together and shielding of deeper planes is consequently reduced. Secondary extinction is equivalent to an increase of the linear absorption coefficient: thus it is negligible for sufficiently small crystals. Reflections affected by secondary extinction can be recognized in the final stages of the crystal structure refinement when for some high-intensity reflection $|F_{obs}| < |F_{cal}|$. A method for inclusion of secondary extinction in least-squares methods is recalled in Chapter 2, p. 97.

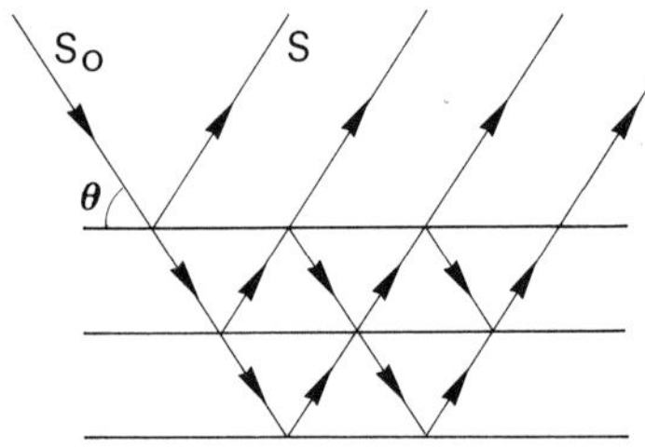

Fig. 3.12. Multiple reflections from a family of lattice planes.

The second component of the extinction coefficient, called **primary extinction**, takes into account the loss of intensity due to dynamic effects inside every single block. This phenomenon can be understood intuitively by means of Fig. 3.12. At the Bragg angle every incident wave can suffer multiple reflections from different lattice planes: after an odd number of reflections the direction will be the same as the diffracted beam: after an even number of reflections the direction will be the same as the primary beam. Each scattering causes a phase lag of $\lambda/4$. Thus, the unscattered radiation having direction $\mathbf{S}_0$ in Fig. 3.12 is joined by doubly scattered radiation (with much smaller intensity) with a phase lag of π: consequently destructive interference will result. The same consideration holds for waves propagating along the direction of the diffracted beam: the result is that both primary and diffracted beams are weakened because of dynamical effects.

A theory describing the mutual transfer of intensity between incident and diffracted beams was proposed by Zachariasen.[6] If absorption is neglected the intensity I_0 of the beam in the incident direction and the intensity I of the beam in the diffracted direction should be related by:

$$\frac{\delta I_0}{\delta t_0} = -\sigma I_0 + \sigma I$$

$$\frac{\delta I}{\delta t} = \sigma I_0 - \sigma I$$

where t_0 and t are lengths in the direction of the primary and diffracted beams respectively, σ is the diffracted power per unit distance and intensity. The equations have to be solved subject to the boundary conditions: I_0 should be equal to the intensity of the primary beam when $t_0 = 0$ and $I = 0$ when $t = 0$. The sum of the two equations is zero, which is the condition for the conservation of energy. Zachariasen's theory has been modified by other authors:[7–9] the introduction of an extinction correction parameter in least-squares analysis may or may not have, according to circumstances, an appreciable influence on the accuracy of the structural parameters included in the refinement. Indeed prior information on the mosaic structure is not usually available and therefore the corrections which have to be made are not easy to calculate a priori. An experimental (often efficient) way to reduce extinction consists of rapid cooling of the crystal by means of immersion in liquid air: this reduces the dimensions of the mosaic grains.

Anomalous dispersion

It is well known that electrons are bound to the nucleus by forces which depend on the atomic field strength and on the quantum state of the electron. Therefore they have to be considered as oscillators with natural frequencies. If the frequency of the primary beam is near to some of these natural frequencies resonance will take place. The scattering under these conditions is called anomalous and can be analytically expressed by substitution of the atomic scattering factor f_a defined earlier by a complex quantity

$$f = f_a + \Delta f' + \mathrm{i}f'' = f' + \mathrm{i}f''.$$

$\Delta f'$ and f'' are called the real and imaginary dispersion corrections. In order to have a simple insight into the problem (a rigorous quantum-mechanical treatment was carried out by Hönl) we recall that the classical differential equation describing the motion of a particle of mass m and charge e in an alternating field intensity $E_{0\mathrm{i}} \exp(\mathrm{i}\omega t)$ is

$$\frac{\mathrm{d}^2x}{\mathrm{d}t^2} + g\frac{\mathrm{d}x}{\mathrm{d}t} + \omega_0^2 x = E_{0\mathrm{i}} \exp(\mathrm{i}\omega t)$$

where $\omega/2\pi$ is the frequency of the incident wave, ω_0 is the natural angular frequency of the vibrating particle, $g\,\mathrm{d}x/\mathrm{d}t$ expresses a damping force proportional to the velocity. The steady-state solution of the above equations is

$$x(t) = X_0 \exp(\mathrm{i}\omega t)$$

where

$$X_0 = \frac{eE_{0\mathrm{i}}/m}{\omega_0^2 - \omega^2 + \mathrm{i}g\omega}.$$

If the displacement $x(t)$ is multiplied by e the polarizability moment $[ex(t)]$ of each dipole is obtained; the electrical susceptibility of a collection of Z uncoupled dipoles is then

$$\chi = \frac{ZeX_0}{E_{0\mathrm{i}}} = \frac{Ze^2}{m}\frac{1}{\omega_0^2 - \omega^2 + \mathrm{i}g\omega}$$

which is a complex function of the frequency of the incident radiation. The electric field produced by the dipole oscillator at a distance $r \gg X_0$ has a magnitude (we neglect the polarization factor and the phase shift due to the travelling in $\boldsymbol{r}$ of the scattered wave) which is $\omega^2/(rc^2)$ times its dipole moment:

$$E_d = E_{0d} \exp(i\omega t) = \frac{\omega^2 e x(t)}{rc^2} = \frac{\omega^2 e^2}{mrc^2} E_{0i} \frac{\exp(i\omega t)}{\omega_0^2 - \omega^2 + ig\omega}.$$

If the electron is unrestrained and undamped then $g = \omega_0 = 0$ and

$$E_d \equiv (E_d)_{Th} = \frac{-e^2}{mrc^2} E_{0i} \exp(i\omega t)$$
$$= \frac{e^2}{mrc^2} E_{0i} \exp[i(\omega t + \pi)]$$

which well agrees with eqn (3.1) suggested by Thomson: π is the phase lag between the scattered and the incident radiation.

Since $g \ll \omega$, when $\omega \gg \omega_0$ the expression of E_d is not very different from that of a free electron. Therefore Thomson scattering is only applicable when $\omega \gg \omega_0$.

We define now the scattering factor for an electron as the ratio

$$f_e = \frac{E_{0d}}{(E_{0d})_{Th}} = \frac{\omega^2(\omega^2 - \omega_0^2)}{(\omega^2 - \omega_0^2)^2 + g^2\omega^2} + i\frac{g\omega^3}{(\omega^2 - \omega_0^2)^2 + g^2\omega^2}$$
$$= f'_e + i f''_e.$$

While the imaginary term is always positive, the real term is negative when $\omega < \omega_0$ and positive when $\omega > \omega_0$. From the quantum-theory point of view, the frequency ω_0 coincides with that of a photon with just sufficient energy to eject the electron from the atom. Such an energy corresponds to the wavelength $\lambda_0 = 2\pi c/\omega_0$ corresponding to the absorption edge. Thus it may be expected that a remarkable deviation from Thomson scattering will arise when the primary beam wavelength is close to an absorption edge of the atom being considered.

An important question is whether $\Delta f'$ and f'' vary with diffraction angle. Existing theoretical treatments suggest changes of some per cent with $\sin\theta/\lambda$ but no rigorous experimental checks have been made so far: therefore in most of the routine applications $\Delta f'$ and f'' are considered to be constant.

For most substances at most X-ray wavelengths from conventional sources dispersion corrections are rather small. Calculated values for CrK_α ($\lambda = 2.291$ Å), CuK_α ($\lambda = 1.542$ Å), and MoK_α ($\lambda = 0.7107$ Å) are listed in the *International tables for x-ray crystallography*, Vol. III. In some special cases ordinary X-ray sources can also generate relevant dispersion effects. For example, holmium has the L_3 absorption edge ($\simeq$1.5368 Å) very close to CuK_α radiation: in this case the holmium scattering factor is not the same for K_{α_1} and K_{α_2} wavelengths. The following dispersion corrections are calculated:[10]

$$CuK_{\alpha_1}(\lambda = 1.5406\ \text{Å}): \quad \Delta f' \simeq -15.41 \quad f'' \simeq 3.70$$
$$CuK_{\alpha_2}(\lambda = 1.5444\ \text{Å}): \quad \Delta f' \simeq -14.09 \quad f'' \simeq 3.72$$

Furthermore, the holmium L_2 absorption edge ($\simeq$1.3905 Å) is very close to the CuK_β wavelength ($\lambda = 1.3922$ Å), so giving rise to

$$\Delta f' = -11.88, \qquad f' \simeq 8.75.$$

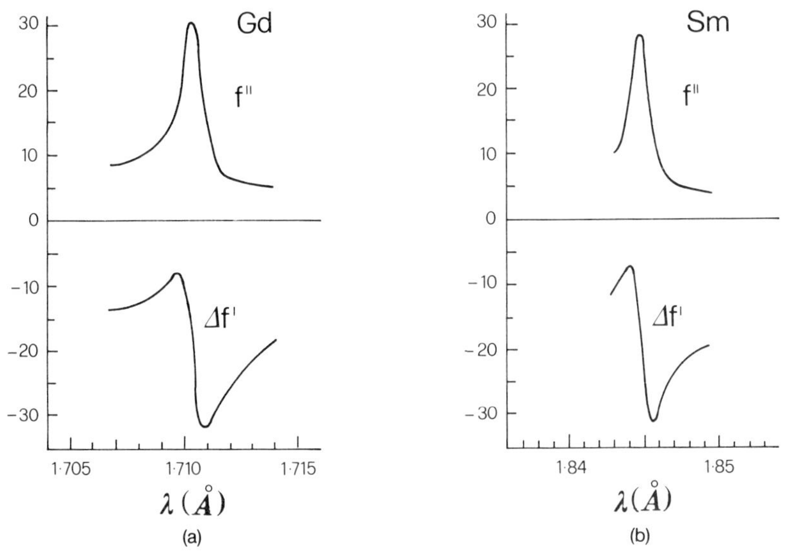

Fig. 3.13. Anomalous scattering terms $\Delta f'$ and f'' for: (a) gadolinium near the L_3 edge; (b) samarium near the L_3 edge.

If synchrotron radiation is used, its intense continuous spectrum may be chosen to high precision in order to provoke exceptionally large anomalous scattering. Very large effects have been measured[11] for rare-earth elements in the trivalent state near L_3 absorption edges (corresponding wavelengths are of crystallographic interest because edges span from 2.26 Å for lanthanum to 1.34 Å for lutetium). In Fig. 3.13 we show the anomalous scattering terms $\Delta f'$ and f'' for gadolinium and samarium near the L_3 absorption edge: spectacular effects as large as $\simeq 30$ electrons/atom could be measured.

Since anomalous dispersion may induce substantial variation of the diffracted intensities depending on the wavelength used, anomalous scattering is an important tool for solving crystal structures (see Chapter 8). Several recent works suggest an important role for multiple-wavelength methods: the power of such methods depends on the distances between the working points representing f in the complex plane.[12] As an example, we plot in Figs. 3.14 (a) and (b) the complex scattering factor $\Delta f' + \mathrm{i}f''$ near the L_3 edge for gadolinium and samarium respectively.

Now we will give only a few elementary ideas about the effects of anomalous dispersion: we postpone the methodological aspects for crystal structure analysis until Chapter 8. Let suppose that a non-centrosymmetric crystal contains N atoms in the unit cell from which P are anomalous scatterers and the remaining $Q = N - P$ atoms are normal scatterers. Then

$$\begin{aligned} F^{+} &= F_Q^{+} + F_P'^{+} + \mathrm{i}F_P''^{+} = F'^{+} + \mathrm{i}F_P''^{+} \\ F^{-} &= F_Q^{-} + F_P'^{-} + \mathrm{i}F_P''^{-} = F'^{-} + \mathrm{i}F_P''^{-} \end{aligned} \tag{3.42}$$

where + and − indicate that the magnitudes are calculated for the vectors $\boldsymbol{H}$ and $-\boldsymbol{H}$ respectively. The subscripts P and Q indicate that the structure factors are calculated only with the contribution of P or Q atoms

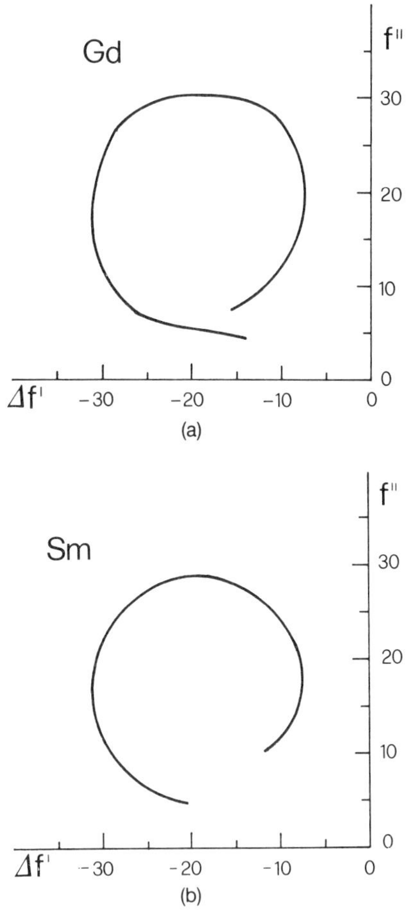

Fig. 3.14. Plot in the complex plane of $\Delta f' + \mathrm{i}f''$ near the L_3 edge for: (a) gadolinium near the L_3 edge; (b) samarium near the L_3 edge.

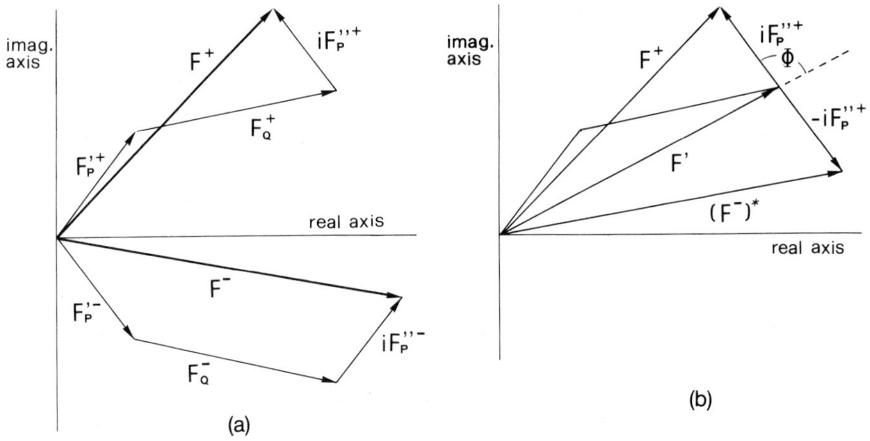

Fig. 3.15. (a) Relation between $F_{\mathbf{H}}$ and $F_{-\mathbf{H}}$ when anomalous dispersion is present; (b) relation between $F_{\mathbf{H}}$ and $F^*_{-\mathbf{H}}$ when anomalous dispersion is present.

respectively:

$$F_P'^{+} = \sum_{j=1}^{P} f_j' \exp 2\pi i \bar{\mathbf{H}}\mathbf{X}_j; \qquad F_P''^{+} = \sum_{j=1}^{P} f_j'' \exp 2\pi i \bar{\mathbf{H}}\mathbf{X}_j;$$

$$F_Q^{+} = \sum_{j=P+1}^{P+Q} f_j \exp 2\pi i \bar{\mathbf{H}}\mathbf{X}_j.$$

The vectors F^+ and F^- are described in Fig. 3.15(a). In Fig. 3.15(b) F^+ and $(F^-)^*$ are shown: the latter is the complex conjugate of F^- and they are symmetric with respect to the real axis:

$$(F^-)^* = F_Q^+ + F_P'^+ - iF_P''^+ = F'^+ - iF_P''^+. \qquad (3.43)$$

The difference $\Delta I = |F^+|^2 - |F^-|^2$ is known as the Bijvoet difference[13] and can be easily calculated by means of Fig. 3.15(b):

$$|F^+|^2 = |F'|^2 + |F_P''|^2 + 2\,|F'|\,|F_P''| \cos \varphi$$
$$|F^-|^2 = |F'|^2 + |F_P''|^2 - 2\,|F'|\,|F_P''| \cos \varphi$$

from which

$$\Delta I = 4\,|F'|\,|F_P''| \cos \varphi. \qquad (3.44)$$

Furthermore

$$\{|F^+|^2 + |F^-|^2\}/2 = |F'|^2 + |F_P''|^2.$$

In general, as we can see, $|F_{\mathbf{H}}| = |F_{-\mathbf{H}}|$ is no longer valid, i.e. the Friedel law is not satisfied in the presence of anomalous dispersion. The value of ΔI depends on the collinearity of F_P^+ and F_Q^+. If they are collinear then $|F^+| = |F^-|$: but this happens by mere chance. ΔI is a maximum when F_P and F_Q are approximately at the right angles.

The Friedel law is satisfied if: the structure is centrosymmetric—in this case $|F^+|$ and $|F^-|$ are always equal; the reflection is centrosymmetric even if the structure is non-centrosymmetric; the crystal is constituted of only one chemical element which is the anomalous scatterer.

As a last observation it should be mentioned that besides X-ray, neutron and gamma-ray anomalous dispersion are also very useful in crystal structure analysis. Neutron anomalous dispersion techniques employ nuclear isotopes with resonances in the range of thermal-neutron energies (see p. 198 for further details).

Gamma-rays are elastically scattered by the electrons of the atoms in the crystal. An elastic resonant scattering by the nucleus (Mössbauer effect) also occurs—the transition involves energies comparable with those employed in conventional X-ray diffraction. Since both the processes are coherent, scattering by resonant nuclei (there are no other nuclear scattering contributions) and by electrons can occur simultaneously and interfere with each other.

An ideal nucleus for gamma-ray resonance is ^{57}Fe; its 14.4 keV resonance corresponds to a wavelength of 0.86 Å. A widely used experimental set-up includes a radioactive source which emits the 14.4 keV radiation in the decay of the parent isotope ^{57}Co. This produces a resonance effect in ^{57}Fe (this atom may naturally be present in the crystal or implanted by techniques such as those used for isomorphous replacement) which is superimposed on the various Bragg reflections upon the gamma radiation elastically scattered by all the atoms in the crystal, iron atoms included. Since ^{57}Fe has a natural abundance of about 2.2 per cent, isotopic enrichment techniques must be applied in order to provide a sufficiently large resonant scattering.

The frequency of the incident radiation may be modified by moving the radiation source, at low velocity (some mm s^{-1}), towards or away from the crystal (linear Doppler effect).

The anomalous scattering amplitudes increase dramatically in going from X-ray to neutron to gamma-ray. Conversely, the intensities of the radiation sources decrease dramatically. This is the most severe drawback for the use of gamma rays: relatively large signals are produced by relatively very weak radiation sources.

The Fourier synthesis and the phase problem

If the structure factors are known in modulus and phase the atomic positions are unequivocally determinable. Indeed, according to eqn (3.A.16) the electron density is the inverse Fourier transform of $F(\boldsymbol{r}^*)$:

$$\rho(\boldsymbol{r}) = \int_{S^*} F(\boldsymbol{r}^*) \exp(-2\pi i \boldsymbol{r}^* \cdot \boldsymbol{r})\, d\boldsymbol{r}^*$$
$$= \frac{1}{V} \sum_{h,k,l=-\infty}^{+\infty} F_{hkl} \exp[-2\pi i(hx + ky + lz)]. \qquad (3.45)$$

$\bar{\mathbf{X}} \equiv [x, y, z]$ are the fractional coordinates of the point defined by the vector $\boldsymbol{r}$. The atomic positions will correspond to the maxima of $\rho(\boldsymbol{r})$.

If in eqn (3.45) we sum up the contributions of $\boldsymbol{H}$ and $-\boldsymbol{H}$ we will have

$$F_{\mathbf{H}} \exp(-2\pi i \bar{\mathbf{H}}\mathbf{X}) + F_{-\mathbf{H}} \exp(2\pi i \bar{\mathbf{H}}\mathbf{X})$$
$$= (A_{\mathbf{H}} + iB_{\mathbf{H}}) \exp(-2\pi i \mathbf{HX}) + (A_{\mathbf{H}} - iB_{\mathbf{H}}) \exp(2\pi i \bar{\mathbf{H}}\mathbf{X})$$
$$= 2[A_{\mathbf{H}} \cos 2\pi \bar{\mathbf{H}}\mathbf{X} + B_{\mathbf{H}} \sin 2\pi \bar{\mathbf{H}}\mathbf{X}]$$

from which

$$\rho(\boldsymbol{r}) = \frac{2}{V} \sum_{h=0}^{+\infty} \sum_{k=-\infty}^{+\infty} \sum_{l=-\infty}^{+\infty} \times [A_{hkl} \cos 2\pi(hx + ky + lz) + B_{hkl} \sin 2\pi(hx + ky + lz)] \quad (3.46)$$

is obtained. The right-hand side of (3.46) is explicitly real and is a sum over a half of the available reflections.

The mathematical operation represented by the synthesis (3.46) can be interpreted as the second step in the process of formation of an image in optics. The first step consists of the scattering of the incident radiation which gives rise to the diffracted rays with amplitudes $F_{\mathbf{H}}$. In the second step the diffracted beams are focused by means of lenses, and by interfering with each other they create the image of the object. There are no physical focusing lenses for X-rays but they can be substituted by a mathematical lens (exactly, by Fourier synthesis (3.46)).

Because of the decrease of the atomic scattering factors the diffraction intensities (and consequently $|F_{\mathbf{H}}|$) weaken 'on average' with the increase of $\sin\theta/\lambda$ and can be considered zero for values above a given $(\sin\theta/\lambda)_{\max} = 1/(2d_{\min})$. Since the reflections at high values of $\sin\theta/\lambda$ give the fine details of the structure (small variations of the atomic coordinates can produce big changes in high-angle structure factors) the quantity $d_{\min}$ is adopted as a measure of the natural resolution of the diffraction experiment. $d_{\min}$ depends on different factors such as: the chemical composition of the crystal (heavy atoms are good scatterers even at high values of $\sin\theta/\lambda$), the chemical stability under the experimental conditions of temperature and pressure, the radiation used (the resolution improves when we pass from electrons to X-rays and neutrons), the temperature of the experiment. Roughly speaking, for X-rays $d_{\min}$ can reach the limit of 0.5 Å in inorganic crystals, 0.7–1.5 Å in organic crystals, and 1.0–3 Å in protein crystals.

Because of the limit of natural resolution or of an artificially introduced limit (for instance in order to save time and calculations) the electron density function will be affected by errors of termination of series. The effect can be mathematically evaluated by calculating the function $\rho'(\boldsymbol{r})$ available via the function

$$F'(\boldsymbol{r}^*) = F(\boldsymbol{r}^*)\Phi(\boldsymbol{r}^*).$$

$\Phi(\boldsymbol{r}^*)$ is the form function: $\Phi(\boldsymbol{r}^*) = 1$ inside the available reflection sphere, $\Phi(\boldsymbol{r}^*) = 0$ outside this sphere. According to eqn (3.A.35) we have

$$\rho'(\boldsymbol{r}) = \mathrm{T}[F'(\boldsymbol{r}^*)] = \rho(\boldsymbol{r}) * \mathrm{T}[\Phi(\boldsymbol{r}^*)]. \quad (3.47)$$

If $\boldsymbol{r}^*$ is the radius of the available reflection sphere, according to Appendix 3.A, p. 181, $\mathrm{T}[\Phi(\boldsymbol{r}^*)]$ is a function with a maximum at $\boldsymbol{r} = 0$ and the subsidiary maxima of weight decreasing with $1/r^2$. The effect of the convolution (3.47) is qualitatively represented in Fig. 3.16. In particular, even if $\rho(\boldsymbol{r})$ is positive everywhere, $\rho'(\boldsymbol{r})$ can be negative in more or less extended regions: the atomic peaks can be broad and surrounded by a series of negative and positive ripples of gradually decreasing amplitude.

The number of reflections used in practice in eqn (3.46) varies from several tens or hundreds, for unit cells of small dimensions, to several tens of thousands for macromolecules.

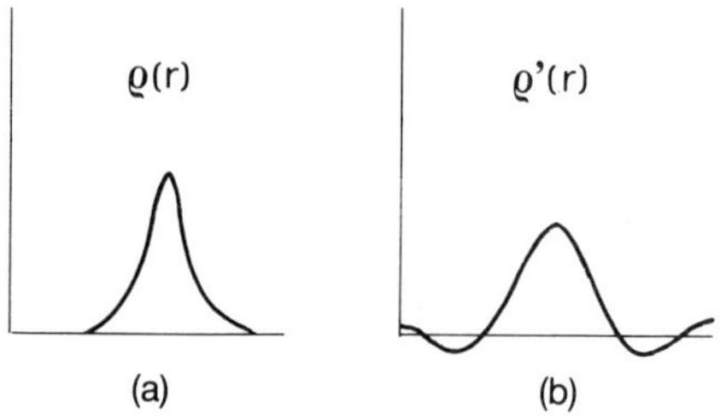

Fig. 3.16. (a) Electron density $\rho(\boldsymbol{r})$; (b) electron density obtained via a Fourier synthesis with series termination errors.

In order to reveal the atomic positions, $\rho'(\boldsymbol{r})$ is sampled upon a three-dimensional grid whose spacings along each of the unit-cell axes have to be fixed with some care. If the grid is too coarse the interpolation between grid points to find the maximum of the electron density may be uncertain, if the grid is too fine a great deal of computing may be unnecessary. In absence of symmetry, at a resolution $d_{\min}$ there are

$$N_r = \frac{4\pi}{3d_{\min}^3 V^*} = \frac{4\pi V}{3d_{\min}^3}$$

measurable (only $N_r/2$ independent) reflections and the number of grid points is $N_p \simeq V/\Delta^3$ where Δ is the grid spacing (say 0.2–0.4 Å in the three directions). In practical cases N_r and N_p are rather large: for instance, for $V \simeq 1000$ Å^3, $d_{\min} = 0.8$ Å, $\Delta = 0.25$ Å, we have $N_r \simeq 8180$ and $N_p \simeq 64\,000$.

If symmetry is present the amount of calculation is smaller. The number of independent reflections to be measured is roughly $N_r/(\tau m)$ where τ is the centring order of the cell and m the multiplicity factor of the Laue class (this is not strictly exact as the multiplicity factor refers to general reflections of type (hkl) and may be different and less than m for certain zones of reflections). Furthermore, it will be sufficient to sample ρ upon the grid points lying inside the asymmetric unit for reconstructing the whole content of the cell.

For instance, let P2/m be the space group with $a = 7.8$ Å, $b = 16.2$ Å and $c = 8.1$ Å and $\beta = 93^0$. If we divide a and c into 33 and b into 66 intervals the grid spacing will have a sufficient and almost identical resolution in all three directions. The number of grid points lying inside the asymmetric unit (1/4 of the unit cell) is now $33 \times 33 \times 17 = 18\,513$.

Very often the volume of the unit cell is much larger than 10^3 Å^3 ($V > 10^6$ Å^3 is not infrequent for macromolecules). Thus even with the use of high-speed computers, the calculation of ρ is a fairly arduous task involving time-consuming procedures. Different algorithms are used to make calculations faster. The most convenient are the Beevers–Lipson technique and the fast Fourier transform algorithm by Cooley and Tookey (see Chapter 2, pp. 88–90, and Appendix 2.I).

Unfortunately, it is not possible to apply eqn (3.47) only on the basis of information obtained directly from X-ray diffraction. Indeed, according to eqn (3.41), only the moduli $|F_{\mathbf{H}}|$ can be obtained from diffraction intensities because the corresponding phase information is lost. This is the so-called **crystallographic phase problem**: how to identify the atomic positions starting only from the moduli $|F_{\mathbf{H}}|$. A general solution to the problem has not been found, but there are methods we can successfully apply (see Chapters 5 and 8).

Modulated crystal structures

So far our attention has been devoted to condensed-matter systems with perfect three-dimensional space group symmetry. In recent years numerous systems have been found which can be considered as perfect crystals with periodic distortion (from some basic structure) of the atomic positions (displacive modulation) and/or of the occupation probability of atoms (density modulation). The presence of periodic distortions is revealed from the existence of the so-called **satellite reflections**, generally weak, which are regularly distributed around the so-called **main reflections**. If satellite reflections can be labelled by rational indices in terms of the reciprocal lattice L* of the basic structure, then the distortion periods are commensurate with the translation periods of the basic structure and a new lattice L′* can be introduced which contains L*. In such a case the main and the satellite reflections are called substructure and superstructure reflections respectively (see page 80).

For some crystals the positions of the satellite reflections vary continuously with respect to L*, depending on the temperature. In these cases we have to admit that satellite reflections can assume non-rational indices, or also, that the periodic distortions are incommensurable with the translation periods of the basic lattice. The result is the so-called **incommensurately modulated structure** (IMS).

In order to prove the existence of satellite reflections we briefly examine the simple case of a one-dimensionally displacively modulated structure when the displacement vector field is a harmonic function.[76]

In a perfect crystal the jth atom in the unit cell defined by the lattice vector $\boldsymbol{r_u} = u\boldsymbol{a} + v\boldsymbol{b} + w\boldsymbol{c}$ is located at $\boldsymbol{r}_{\mathbf{u},j} = \boldsymbol{r}_j^0 + \boldsymbol{r}_\mathbf{u}$. In a harmonically modulated crystal it will be

$$\boldsymbol{r}_{\mathbf{u},j} = \boldsymbol{r}_j^0 + \boldsymbol{r}_\mathbf{u} + \boldsymbol{g}_j \sin[2\pi\boldsymbol{K}\cdot(\boldsymbol{r}_j^0 + \boldsymbol{r}_\mathbf{u}) - \phi_j],$$

where $\boldsymbol{r}_j^0$ is now the average position of the jth atom, and $\boldsymbol{g}_j$ and ϕ_j are its displacement wave amplitude and phase respectively. $\boldsymbol{K}$ is the modulation vector, which may be expressed in the reciprocal space by $\boldsymbol{K} = k_1\boldsymbol{a}^* + k_2\boldsymbol{b}^* + k_3\boldsymbol{c}^*$. The contribution of the jth atom to the structure factor is now given by

$$\begin{aligned} &f_j(\boldsymbol{r}^*)\sum_\mathbf{u} \exp 2\pi i\boldsymbol{r}^*\cdot\{\boldsymbol{r}_j^0 + \boldsymbol{r}_\mathbf{u} + \boldsymbol{g}_j \sin[2\pi\boldsymbol{K}\cdot(\boldsymbol{r}_j^0 + \boldsymbol{r}_\mathbf{u}) - \phi_j]\} \\ &= f_j(\boldsymbol{r}^*)\exp(2\pi i\boldsymbol{r}^*\cdot\boldsymbol{r}_j^0)\sum_\mathbf{u}\exp(2\pi i\boldsymbol{r}^*\cdot\boldsymbol{r}_\mathbf{u}) \\ &\times \exp\{2\pi i\boldsymbol{r}^*\cdot\boldsymbol{g}_j \sin[2\pi\boldsymbol{K}\cdot(\boldsymbol{r}_j^0 + \boldsymbol{r}_\mathbf{u}) - \phi_j]\} \qquad (3.48) \end{aligned}$$

where $f_j(\boldsymbol{r}^*)$ is the usual atomic scattering factor. To the last exponential term in (3.48) the Jacobi expansion

$$\exp(iz\sin\alpha) = \sum_{m=-\infty}^{+\infty}\exp(-im\alpha)J_{-m}(z)$$

may be applied. J_m (see Fig. 3.17) is the Bessel function of the first kind of order m, satisfying $J_{-m}(z) = (-1)^m J_m(z)$. Then (3.48) reduces to

$$\begin{aligned} &f_j(\boldsymbol{r}^*)\exp(2\pi i\boldsymbol{r}^*\cdot\boldsymbol{r}_j^0)\sum_\mathbf{u}\exp(2\pi i\boldsymbol{r}^*\cdot\boldsymbol{r}_\mathbf{u}) \\ &\times\sum_m \exp\{im[\phi_j - 2\pi\boldsymbol{K}\cdot(\boldsymbol{r}_j^0 + \boldsymbol{r}_\mathbf{u})]\}J_{-m}(2\pi\boldsymbol{r}^*\cdot\boldsymbol{g}_j) \\ &= f_j(\boldsymbol{r}^*)\exp(2\pi i\boldsymbol{r}^*\cdot\boldsymbol{r}_j^0)\sum_m J_{-m}(2\pi\boldsymbol{r}^*\cdot\boldsymbol{g}_j) \\ &\times\exp\{im(\phi_j - 2\pi\boldsymbol{K}\cdot\boldsymbol{r}_j^0)\}\sum_\mathbf{u}\exp[2\pi i\boldsymbol{r}_\mathbf{u}\cdot(\boldsymbol{r}^* - m\boldsymbol{K})]. \end{aligned}$$

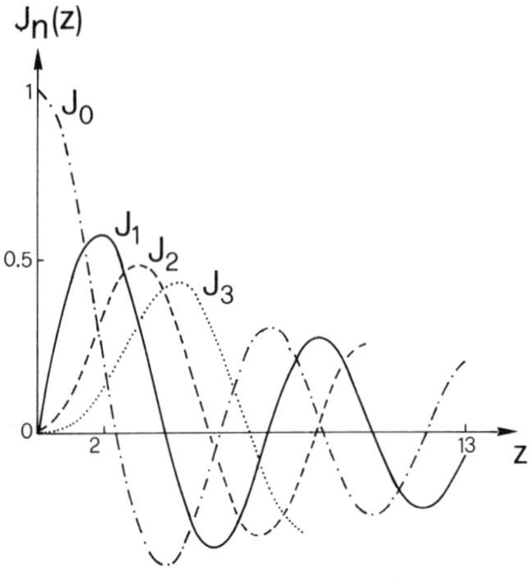

Fig. 3.17. The Bessel functions $J_n(z)$, $n = 0, 1, 2, 3$.

Provided the number of cells in the crystal is large enough the sum over $\boldsymbol{u}$ leads to $(1/V)\,\delta(\boldsymbol{H}-\boldsymbol{r}^*+m\boldsymbol{K})$ where $\boldsymbol{H}=h\boldsymbol{a}^*+k\boldsymbol{b}^*+l\boldsymbol{c}^*$. Consequently the reflections occur for $\boldsymbol{r}^*=\boldsymbol{H}'=\boldsymbol{H}+m\boldsymbol{K}$: for $m=0$ we have main reflections ($\boldsymbol{H}'=\boldsymbol{H}$), for $m\neq 0$ satellites are defined. We also see that four indices are now needed for the identification of a diffraction effect. The structure factor of the reflections $\bar{\boldsymbol{H}}'=(h, k, l, n)$ may then be written as

$$F_{\boldsymbol{H}'}=\sum_{j=1}^{N} f_j(\boldsymbol{H}')\exp(2\pi\mathrm{i}\boldsymbol{r}^*\cdot\boldsymbol{r}_j^0)J_m(2\pi\boldsymbol{H}'\cdot\boldsymbol{g}_j)(-1)^m \times\exp(-2\pi\mathrm{i}\boldsymbol{K}\cdot\boldsymbol{r}_j^0)\exp(\mathrm{i}m\phi_j). \quad (3.49)$$

According to Fig. 3.17 the average intensity of satellite reflections rapidly decreases with m.

The above formalism may be extended to one-dimensionally density modulated structures and also to multi-dimensional (harmonic or not) modulations. The above results also suggest that the reciprocal lattice of an IMS is aperiodic in the three-dimensional space, and that the symmetry group of an IMS cannot be a three-dimensional space group. We will show in Appendix 3.E (main references will also be given) that such a reciprocal lattice may be transformed into a periodic lattice provided a higher-dimensional space is taken into consideration. Thus the symmetry in the three-dimensional space can also be a poor residue of the full symmetry in the higher-dimensional space.

Appendices

3.A Mathematical background

The Dirac delta function

In a three-dimensional space the Dirac delta function $\delta(\boldsymbol{r}-\boldsymbol{r}_0)$ has the following properties

$$\delta=0 \quad \text{for} \quad \boldsymbol{r}\neq\boldsymbol{r}_0, \qquad \delta=\infty \quad \text{for} \quad \boldsymbol{r}=\boldsymbol{r}_0, \qquad \int_S \delta(\boldsymbol{r}-\boldsymbol{r}_0)\,\mathrm{d}\boldsymbol{r}=1 \quad (3.\text{A}.1)$$

where S indicates the integration space. Thus the delta function corresponds to an infinitely sharp line of unit weight located at $\boldsymbol{r}_0$. It is easily seen that, if $\boldsymbol{r}_0=x_0\boldsymbol{a}+y_0\boldsymbol{b}+z_0\boldsymbol{c}$, then

$$\delta(\boldsymbol{r}-\boldsymbol{r}_0)=\delta(x-x_0)\,\delta(y-y_0)\,\delta(z-z_0). \quad (3.\text{A}.2)$$

$\delta(x-x_0)$ may be considered as the limit of different analytical functions. For example, as the limit for $\sigma\to 0$ of the Gaussian function

$$N(\sigma, x_0)=\frac{1}{\sigma\sqrt{2\pi}}\exp\left(-\frac{(x-x_0)^2}{2\sigma^2}\right). \quad (3.\text{A}.3)$$

Of particular usefulness will be the relation

$$\delta(x-x_0)=\int_{-\infty}^{+\infty}\exp[2\pi\mathrm{i}x^*(x-x_0)]\,\mathrm{d}x^* \quad (3.\text{A}.4)$$

where x^* is a real variable. It easily seen that (3.A.4) satisfies the properties

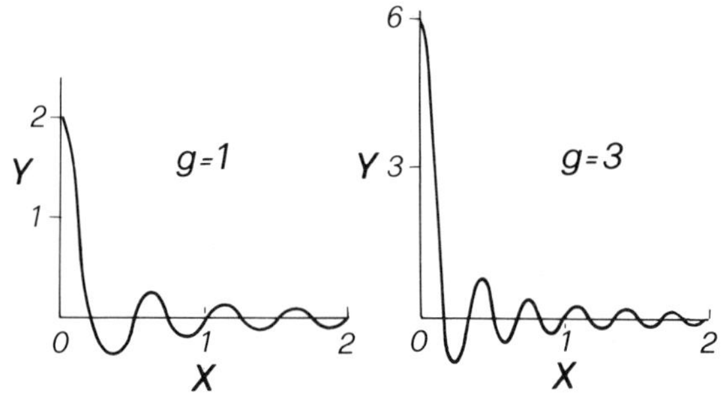

Fig. 3.A.1. The function $Y = (\sin 2\pi g x)/(\pi x)$ is plotted for $g = 1, 3$. Clearly $Y(-x) = Y(x)$.

(3.A.1): indeed its right-hand side may be written as

$$\lim_{g\to\infty}\int_{-g}^{+g} \exp\,[2\pi i x^*(x-x_0)]\,\mathrm{d}x^* = \lim_{g\to\infty}\frac{\sin\,[2\pi g(x-x_0)]}{\pi(x-x_0)}.$$

The function $\sin\,[2\pi g(x-x_0)]/[\pi(x-x_0)]$ takes the maximum value $2g$ at $x = x_0$ (see Fig. 3.A.1), oscillates with period $1/g$, and has decreasing subsidiary maxima with increasing x: the value of its integral from $-\infty$ to $+\infty$ is unitary for any value of g. Therefore the limit for $g\to\infty$ of $\sin\,[2\pi g(x-x_0)]/[\pi(x-x_0)]$ satisfies all the properties of a delta function. Consequently we can also write:

$$\delta(x-x_0) = \lim_{g\to\infty}\frac{\sin\,[2\pi g(x-x_0)]}{\pi(x-x_0)}. \tag{3.A.5}$$

In a three-dimensional space (3.A.4) becomes

$$\delta(\boldsymbol{r}-\boldsymbol{r}_0) = \int_{S^*} \exp\,[2\pi i \boldsymbol{r}^*\cdot(\boldsymbol{r}-\boldsymbol{r}_0)]\,\mathrm{d}\boldsymbol{r}^* \tag{3.A.6}$$

where S^* indicates the $\boldsymbol{r}^*$ space. Two important properties of the delta function are:

$$\begin{aligned}\delta(\boldsymbol{r}-\boldsymbol{r}_0) &= \delta(\boldsymbol{r}_0-\boldsymbol{r})\\ f(\boldsymbol{r})\,\delta(\boldsymbol{r}-\boldsymbol{r}_0) &\equiv f(\boldsymbol{r}_0)\,\delta(\boldsymbol{r}-\boldsymbol{r}_0).\end{aligned} \tag{3.A.7}$$

Indeed, for $\boldsymbol{r}\neq\boldsymbol{r}_0$, left- and right-hand members of (3.A.7) are both vanishing, for $\boldsymbol{r}=\boldsymbol{r}_0$ both are infinite. From (3.A.7)

$$\int_S f(\boldsymbol{r})\,\delta(\boldsymbol{r}-\boldsymbol{r}_0)\,\mathrm{d}\boldsymbol{r} = f(\boldsymbol{r}_0) \tag{3.A.8}$$

is derived. Consequently

$$\int_S \delta(\boldsymbol{r}-\boldsymbol{r}_2)\,\delta(\boldsymbol{r}-\boldsymbol{r}_1)\,\mathrm{d}\boldsymbol{r} = \delta(\mathbf{r}_2-\boldsymbol{r}_1). \tag{3.A.9}$$

The lattice function L

Delta functions can be used to represent lattice functions. For example, in a one-dimensional space a lattice with period a may be represented by

$$L(x) = \sum_{n=-\infty}^{+\infty} \delta(x-x_n) \tag{3.A.10}$$

where $x_n = na$ and n is an integer value. $L(x)$ vanishes everywhere except at the points na. Analogously a three-dimensional lattice defined by unit vectors $\boldsymbol{a}$, $\boldsymbol{b}$, $\boldsymbol{c}$ may be represented by

$$L(\boldsymbol{r}) = \sum_{u,v,w=-\infty}^{+\infty} \delta(\boldsymbol{r} - \boldsymbol{r}_{u,v,w}) \qquad (3.A.11)$$

where $\boldsymbol{r}_{u,v,w} = u\boldsymbol{a} + v\boldsymbol{b} + w\boldsymbol{c}$ and u, v, w are integer values.

Accordingly, in a three-dimensional space:

(1) a periodic array of points along the z axis with positions $z_n = nc$ may be represented as

$$P_1(\boldsymbol{r}) = \delta(x)\,\delta(y) \sum_{n=-\infty}^{+\infty} \delta(z - z_n); \qquad (3.A.12)$$

(2) a series of lines in the (x, z) plane, parallel to x and separated by c may be represented by

$$P_2(\boldsymbol{r}) = \delta(y) \sum_{n=-\infty}^{+\infty} \delta(z - z_n); \qquad (3.A.13)$$

(3) a series of planes parallel to the (x, y) plane and separated by c is represented by

$$P_3(\boldsymbol{r}) = \sum_{n=-\infty}^{+\infty} \delta(z - z_n). \qquad (3.A.14)$$

The Fourier transform

The Fourier transform of the function $\rho(\boldsymbol{r})$ is given (for practical reasons we follow the convention of including 2π in the exponent) by

$$F(\boldsymbol{r}^*) = \int_S \rho(\boldsymbol{r}) \exp(2\pi \mathrm{i}\boldsymbol{r}^* \cdot \boldsymbol{r})\, \mathrm{d}\boldsymbol{r}. \qquad (3.A.15)$$

The vector $\boldsymbol{r}^*$ may be considered as a vector in 'Fourier transform space', while we could conventionally say that $\boldsymbol{r}$ is a vector in 'direct space'.

We show now that

$$\rho(\boldsymbol{r}) = \int_{S^*} F(\boldsymbol{r}^*) \exp(-2\pi \mathrm{i}\boldsymbol{r}^* \cdot \boldsymbol{r})\, \mathrm{d}\boldsymbol{r}^*. \qquad (3.A.16)$$

Because of (3.A.15) the right-hand side of (3.A.16) becomes

$$\int_S \rho(\boldsymbol{r}')\left(\int_{S^*} \exp[2\pi i\boldsymbol{r}^* \cdot (\boldsymbol{r}' - \boldsymbol{r})]\, \mathrm{d}\boldsymbol{r}^*\right) \mathrm{d}\boldsymbol{r}',$$

which, in turn, because of (3.A.6), reduces to

$$\int_S \rho(\boldsymbol{r}')\, \delta(\boldsymbol{r}' - \boldsymbol{r})\, \mathrm{d}\boldsymbol{r}' = \rho(\boldsymbol{r}).$$

Relations (3.A.15) and (3.A.16) may be written as

$$F(\boldsymbol{r}^*) = \mathrm{T}[\rho(\boldsymbol{r})], \qquad (3.A.17)$$

$$\rho(\boldsymbol{r}) = \mathrm{T}^{-1}[F(\boldsymbol{r}^*)] \qquad (3.A.18)$$

respectively: we will also say that ρ is the inverse transform of F. Obviously

$$\mathrm{T}^{-1}\mathrm{T}[\rho(\boldsymbol{r})] = \rho(\boldsymbol{r})$$

but

$$\mathrm{TT}[\rho(\boldsymbol{r})] = \mathrm{T}[F(\boldsymbol{r}^*)] = \rho(-\boldsymbol{r}).$$

$F(\boldsymbol{r}^*)$ is a complex function: by denoting

$$A(\boldsymbol{r}^*) = \int_S \rho(\boldsymbol{r}) \cos(2\pi \mathrm{i} \boldsymbol{r}^* \cdot \boldsymbol{r})\, \mathrm{d}\boldsymbol{r} \qquad (3.A.19)$$

$$B(\boldsymbol{r}^*) = \int_S \rho(\boldsymbol{r}) \sin(2\pi \mathrm{i} \boldsymbol{r}^* \cdot \boldsymbol{r})\, \mathrm{d}\boldsymbol{r} \qquad (3.A.20)$$

then

$$F(\boldsymbol{r}^*) = A(\boldsymbol{r}^*) + \mathrm{i}B(\boldsymbol{r}^*).$$

We calculate now the Fourier transform of $\rho(-\boldsymbol{r})$. Since

$$\int_S \rho(-\boldsymbol{r}) \exp(2\pi \mathrm{i} \boldsymbol{r}^* \cdot \boldsymbol{r})\, \mathrm{d}\boldsymbol{r} = \int_S \rho(\boldsymbol{r}) \exp[2\pi \mathrm{i}(-\boldsymbol{r}^* \cdot \boldsymbol{r})]\, \mathrm{d}\boldsymbol{r} = A(\boldsymbol{r}^*) - \mathrm{i}B(\boldsymbol{r}^*),$$

it may be concluded that

$$\mathrm{T}[\rho(-\boldsymbol{r})] = F^-(\boldsymbol{r}^*) \qquad (3.A.21)$$

where $F^-(\boldsymbol{r}^*)$ is the complex conjugate of $F(\boldsymbol{r}^*)$.

If $\rho(\boldsymbol{r})$ is symmetric with respect to the origin, say $\rho(\boldsymbol{r}) = \rho(-\boldsymbol{r})$, because of (3.A.17) and (3.A.21) it will result $F(\boldsymbol{r}^*) = F^-(\boldsymbol{r}^*)$, or, in other words, $F(\boldsymbol{r}^*)$ will be a real function ($B(\boldsymbol{r}^*) \equiv 0$). Vice versa, if $F(\boldsymbol{r}^*) = F^-(\boldsymbol{r}^*)$ then $\rho(\boldsymbol{r})$ is symmetric with respect to the origin.

If $\rho(\boldsymbol{r}) = -\rho(-\boldsymbol{r})$ then $F(\boldsymbol{r}^*) = -F^-(\boldsymbol{r}^*)$, so that $F(\boldsymbol{r}^*)$ is pure imaginary (then $A(\boldsymbol{r}^*) = 0$).

Some examples of Fourier transform

In a one-dimensional space (3.A.15) becomes

$$F(x^*) = \int_{-\infty}^{+\infty} \rho(x) \exp(2\pi \mathrm{i} x^* x)\, \mathrm{d}x. \qquad (3.A.22)$$

Let us consider some examples for $\rho(x)$.

1. *Gaussian function*:

$$\rho(x) = N(\sigma, 0) = \frac{1}{\sigma\sqrt{2\pi}} \exp\left(\frac{x^2}{2\sigma^2}\right). \qquad (3.A.23)$$

Since

$$\int_{-\infty}^{+\infty} \exp[-\tfrac{1}{2}hu^2 + \mathrm{i}tu]\, \mathrm{d}u = \sqrt{\frac{2\pi}{h}} \exp\left(-\frac{t^2}{2h}\right), \qquad (3.A.24)$$

then

$$\mathrm{T}[\rho(x)] = F(x^*) = \exp(-2\pi^2\sigma^2 x^{*2}). \qquad (3.A.25)$$

It should be noted that the larger the 'width' of $\rho(x)$, the smaller is that of $F(x^*)$ (see Fig. 3.A.2(a))

2. *Exponential function*: $\rho(x) = \exp(-g\,|x|)$. Its Fourier transform is the Cauchy function (see Fig. 3.A.2(b))

$$F(x^*) = (2g)/(g^2 + 4\pi^2 x^{*2}).$$

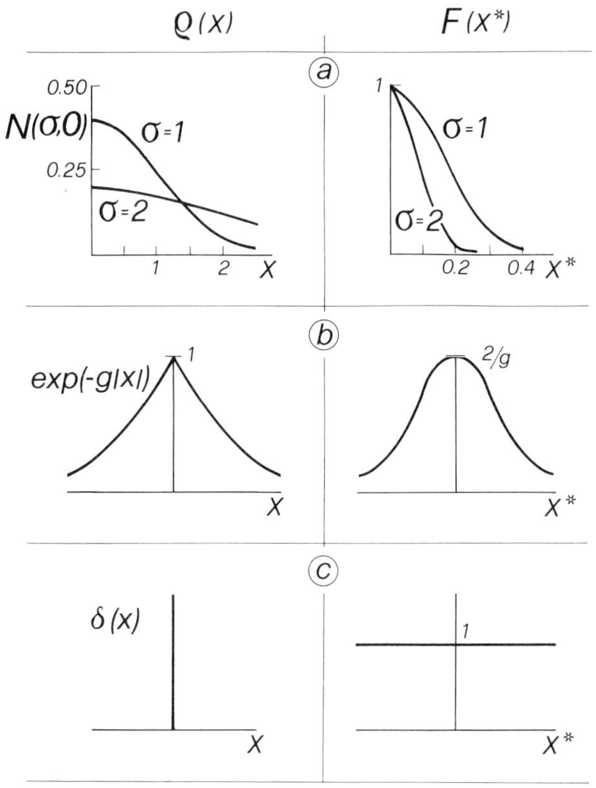

Fig. 3.A.2. Examples of Fourier transform.

3. *Rectangular aperture*:

$$\rho(x) = c \qquad \text{for } -g < x < g, \qquad \text{otherwise } \rho(x) = 0.$$

Then

$$F(x^*) = c\int_{-g}^{+g} \exp(2\pi i x^* x)\, dx = c\,\frac{\sin(2\pi g x^*)}{\pi x^*}$$

which is plotted in Fig. 3.A.1.

4. *Dirac delta function*: $\rho(x) = \delta(x)$. Then (see Fig. 3.A.2(c))

$$F(x^*) = \int_{-\infty}^{+\infty} \delta(x) \exp(2\pi i x^* x)\, dx = 1.$$

If $\rho(x) = \delta(x - a)$ then $F(x^*) = \exp(2\pi i a x^*)$.

5. *One-dimensional finite lattice*: $\rho_p = \sum_{n=-p}^{+p} \delta(x - na)$, where ρ_p represents a set of $N = 2p + 1$ equally spaced delta functions. Then

$$\begin{aligned}
T[\rho_p(x)] &= \sum_{n=-p}^{p} \exp(2\pi i n a x^*) = \sum_{n=-p}^{p} \cos(2\pi n a x^*) \\
&= \frac{1}{2\sin(\pi a x^*)} \sum_{n=-p}^{p} 2\cos(2\pi n a x^*)\sin(\pi a x^*) \\
&= \frac{1}{2\sin(\pi a x^*)} \sum_{n=-p}^{p} \sin[\pi(2n+1)a x^*] - \sin[\pi(2n-1)a x^*]
\end{aligned}$$

$$= \frac{1}{2\sin(\pi a x^*)}\{\sin[\pi(2p+1)ax^*] - \sin[\pi(2p-1)ax^*]$$
$$+ \sin[\pi(2p-1)ax^*] - \sin[\pi(2p-3)ax^*]$$
$$+ \ldots\ldots\ldots\ldots\ldots\ldots\ldots\ldots\ldots\ldots$$
$$+ \sin[\pi(-2p+1)ax^*] - \sin[\pi(-2p-1)ax^*]\}$$
$$= \frac{\sin[(2p+1)\pi a x^*]}{\sin \pi a x^*} = \frac{\sin N\pi a x^*}{\sin \pi a x^*}. \tag{3.A.26}$$

The function

$$f(y) = \frac{\sin N\pi y}{\sin \pi y} \tag{3.A.27}$$

is plotted in Fig. 3.A.3 for $N = 6, 7$. When $y = h$ (h is an integer value) numerator and denominator of the right-hand member of (3.A.27) both vanish. Then the value of $f(y)$ (as determined from its limit for $y \to h$) is equal to N if N is odd; if N is even, then $f(y)$ is equal to N if h is even, to $-N$ if h is odd. Between each pair of main maxima there are $N-2$ subsidiary peaks. Each main peak has width equal to $2/N$ ($1/N$ is that of the subsidiary peaks), therefore it becomes sharper as N increases. Furthermore the ratio between the amplitude of a main peak with respect that of a subsidiary one increases with N.

What we noted for $f(y)$ may be easily applied to (3.A.26), whose principal maxima are at $x^* = h/a$.

6. *One-dimensional infinite lattice*:

$$\rho(x) = L(x) = \sum_{n=-\infty}^{+\infty} \delta(x - na).$$

According to 5 above the Fourier transform of ρ will be

$$F(x^*) = \lim_{N\to\infty} \frac{\sin N\pi a x^*}{\sin \pi a x^*}.$$

The function $F(x^*)$ will present infinitely sharp lines at $x^* = h/a$ of weight $1/a$. Indeed

$$\int_{-\varepsilon}^{+\varepsilon} \frac{\sin N\pi a x^*}{\sin \pi a x^*}\,dx^* = \frac{1}{a}\lim_{N\to\infty}\int_{-\varepsilon}^{+\varepsilon} \frac{\sin N\pi y}{\sin \pi y}\,dy.$$

Whichever the value of ε', when $N \to \infty$ the value of the integral is unity.

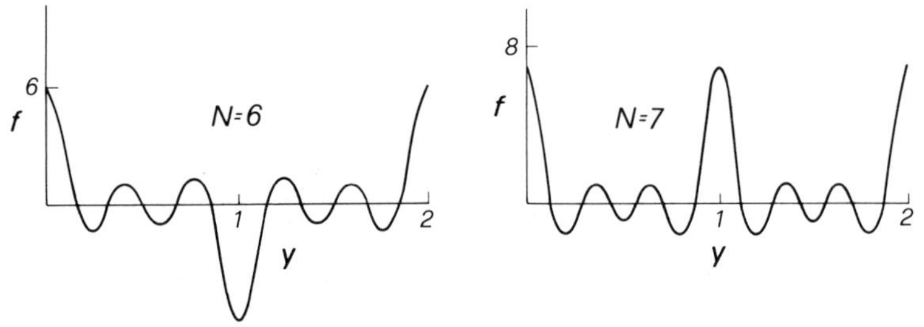

Fig. 3.A.3. The function $f(y) = \sin N\pi y/\sin \pi y$ for $N = 6, 7$.

Consequently we could write

$$F(x^*) = \frac{1}{a}\sum_{h=-\infty}^{+\infty} \delta\left(x^* - \frac{h}{a}\right) = \frac{1}{a}\sum_{h=-\infty}^{+\infty} \delta(ax^* - h) \qquad (3.A.28)$$

or, in words, the Fourier transform of a one-dimensional lattice with period a is a one-dimensional lattice with period $1/a$ represented by delta functions with weight $1/a$.

7. *Three-dimensional finite lattice*:

$$\rho(\boldsymbol{r}) = \sum_{u=-p_1}^{p_1} \sum_{v=-p_2}^{p_2} \sum_{w=-p_3}^{p_3} \delta(\boldsymbol{r} - \boldsymbol{r}_{u,v,w}).$$

Its Fourier transform is

$$\begin{aligned} F(\boldsymbol{r}^*) &= \sum_{u=-p_1}^{p_1} \sum_{v=-p_2}^{p_2} \sum_{w=-p_3}^{p_3} \exp(2\pi i \boldsymbol{r}^* \cdot \boldsymbol{r}_{u,v,w}) \\ &= \sum_{u=-p_1}^{p_1} \exp(2\pi i u \boldsymbol{r}^* \cdot \boldsymbol{a}) \cdot \sum_{v=-p_2}^{p_2} \exp(2\pi i v \boldsymbol{r}^* \cdot \boldsymbol{b}) \cdot \sum_{w=-p_3}^{p_3} \exp(2\pi i w \boldsymbol{r}^* \cdot \boldsymbol{c}) \\ &= \frac{\sin N_1 \pi \boldsymbol{a} \cdot \boldsymbol{r}^*}{\sin \pi \boldsymbol{a} \cdot \boldsymbol{r}^*} \cdot \frac{\sin N_2 \pi \boldsymbol{b} \cdot \boldsymbol{r}^*}{\sin \pi \boldsymbol{b} \cdot \boldsymbol{r}^*} \cdot \frac{\sin N_3 \pi \boldsymbol{c} \cdot \boldsymbol{r}^*}{\sin \pi \boldsymbol{c} \cdot \boldsymbol{r}^*}, \end{aligned} \qquad (3.A.29)$$

where $N_1 = (2p_1 + 1)$, $N_2 = (2p_2 + 1)$, $N_3 = (2p_3 + 1)$. In accordance with point 5, each of the terms in the right-hand side of (3.A.29) is maximum when $\boldsymbol{r}^*$ satisfies

$$\boldsymbol{a} \cdot \boldsymbol{r}^* = h, \qquad \boldsymbol{b} \cdot \boldsymbol{r}^* = k, \qquad \boldsymbol{c} \cdot \boldsymbol{r}^* = l$$

with integer values of h, k, l. It is easily seen that the solution of the above three equations is given by

$$\boldsymbol{r}^*_{\mathbf{H}} = h\boldsymbol{a}^* + k\boldsymbol{b}^* + l\boldsymbol{c}^*,$$

where

$$\boldsymbol{a}^* = \frac{\boldsymbol{b} \wedge \boldsymbol{c}}{V}, \qquad \boldsymbol{b}^* = \frac{\boldsymbol{c} \wedge \boldsymbol{a}}{V}, \qquad \boldsymbol{c}^* = \frac{\boldsymbol{a} \wedge \boldsymbol{b}}{V}$$

and $V = \boldsymbol{a} \cdot \boldsymbol{b} \wedge \boldsymbol{c}$. The vectors $\boldsymbol{a}^*$, $\boldsymbol{b}^*$, $\boldsymbol{c}^*$ are nothing else but the basic vectors of the reciprocal lattice defined in § 2.3. When N_1, N_2, N_3 are sufficiently large then $F(\boldsymbol{r}^*)$ has appreciable values only in the reciprocal lattice points defined by the triple of integers $\mathbf{H} \equiv (h, k, l)$.

8. *Three-dimensional infinite lattice*:

$$\rho(\boldsymbol{r}) = \sum_{u=-\infty}^{+\infty} \sum_{v=-\infty}^{+\infty} \sum_{w=-\infty}^{+\infty} \rho(\boldsymbol{r} - \boldsymbol{r}_{u,v,w}).$$

Its Fourier transform is the limit of (3.A.29) for N_1, N_2, N_3 tending to infinity:

$$F(\boldsymbol{r}^*) = \lim_{N_1, N_2, N_3 \to \infty} \frac{\sin N_1 \pi \boldsymbol{a} \cdot \boldsymbol{r}^*}{\sin \pi \boldsymbol{a} \cdot \boldsymbol{r}^*} \cdot \frac{\sin N_2 \pi \boldsymbol{b} \cdot \boldsymbol{r}^*}{\sin \pi \boldsymbol{b} \cdot \boldsymbol{r}^*} \cdot \frac{\sin N_3 \pi \boldsymbol{c} \cdot \boldsymbol{r}^*}{\sin \pi \boldsymbol{c} \cdot \boldsymbol{r}^*}.$$

According to points 6 and 7 $F(\boldsymbol{r}^*)$ represents a three-dimensional lattice by an array of delta functions the weight of which may be calculated by integrating $F(\boldsymbol{r}^*)$ on a domain $\mathrm{d}V^*$ about a specific lattice point **H**. Since

$$\mathrm{d}V^* = \mathrm{d}h\boldsymbol{a}^* \cdot \mathrm{d}k\boldsymbol{b}^* \wedge \mathrm{d}l\boldsymbol{c}^* = V^*\,\mathrm{d}h\,\mathrm{d}k\,\mathrm{d}l$$

because of the point 6

$$\int_{V^*} F(\boldsymbol{r}^*)\,\mathrm{d}V^* = V^* \lim_{N_1,N_2,N_3\to\infty} \int_{-\varepsilon_1}^{\varepsilon_1} \frac{\sin N_1\pi h}{\sin \pi h} \times \mathrm{d}h \int_{-\varepsilon_2}^{\varepsilon_2} \frac{\sin N_2\pi k}{\sin \pi k}\,\mathrm{d}k \int_{-\varepsilon_3}^{\varepsilon_3} \frac{\sin N_3\pi l}{\sin \pi l}\,\mathrm{d}l = V^* = \frac{1}{V}$$

arises. In conclusion, the Fourier transform of a lattice in direct space (represented by the function $L(\boldsymbol{r})$) is the function $L(\boldsymbol{r}^*)/V$:

$$\begin{aligned} \mathrm{T}[L(\boldsymbol{r})] &= \mathrm{T}\Big[\sum_{u,v,w=-\infty}^{+\infty} \delta(\boldsymbol{r} - \boldsymbol{r}_{u,v,w})\Big] \\ &= \frac{1}{V} L(\boldsymbol{r}^*) = \frac{1}{V} \sum_{h,k,l=-\infty}^{+\infty} \delta(\boldsymbol{a}\cdot\boldsymbol{r}^* - h) \\ &\quad \times \delta(\boldsymbol{b}\cdot\boldsymbol{r}^* - k)\,\delta(\boldsymbol{c}\cdot\boldsymbol{r}^* - l) \\ &= \frac{1}{V} \sum_{h,k,l=-\infty}^{+\infty} \delta(\boldsymbol{r}^* - \boldsymbol{r}^*_{\mathbf{H}}) \end{aligned} \qquad (3.A.30)$$

which represents a lattice again (called the **reciprocal lattice**) in the Fourier transform space.

9. *Fourier transform of a one-dimensional periodic array of points along the z axis*, as defined by (3.A.12). Then

$$F(\boldsymbol{r}^*) = \frac{1}{c} \sum_{l=-\infty}^{+\infty} \delta\left(z^* - \frac{1}{c}\right). \qquad (3.A.31)$$

10. *Fourier transform of a lattice plane* lying on the plane $z = 0$ and with translation constants a and b. Then

$$\rho(\boldsymbol{r}) = \delta(x - na)\,\delta(y - nb)\,\delta(z)$$

and

$$F(\boldsymbol{r}^*) = \frac{1}{a} \sum_{k=-\infty}^{+\infty} \delta\left(x^* - \frac{h}{a}\right) \frac{1}{b} \sum_{h=-\infty}^{+\infty} \delta\left(y^* - \frac{k}{b}\right). \qquad (3.A.32)$$

Fourier transform of spherically symmmetric functions

If $\rho(\boldsymbol{r})$ is spherically symmetric we can represent $\boldsymbol{r}$ and $\boldsymbol{r}^*$ in spherical polar coordinates (r, θ, φ) and $(r^*, \theta^*, \varphi^*)$ defined by (see Fig. 3.A.4)

$$x = r \sin\varphi \cos\theta \qquad y = r \sin\varphi \sin\theta \qquad z = r\cos\varphi$$

with $r > 0$, $0 < \varphi \leqslant \pi$, $0 \leqslant \theta < 2\pi$. Analogous transformations could be written for $\boldsymbol{r}^*$. Without loss of generality we can choose z along the $\boldsymbol{r}^*$ direction: then $\boldsymbol{r}\cdot\boldsymbol{r}^* = rr^*\cos\varphi$. Furthermore, for each point with coordinates (r, θ, φ) another point will exist, equivalent to the first, with coordinates $(r, \pi + \theta, \pi - \varphi)$. The contribution of both the points to the integral (3.A.15) will be $\exp(2\pi i rr^*\cos\varphi) + \exp[2\pi i rr^* \cos(\pi - \varphi)] = 2\cos(2\pi rr^*\cos\varphi)$. Thus (3.A.15) reduces to

$$F(r^*) = \int_0^\infty \int_0^\pi \int_0^{2\pi} \rho(r)\cos(2\pi rr^*\cos\varphi)\, r^2 \sin\varphi\, \mathrm{d}r\, \mathrm{d}\varphi\, \mathrm{d}\theta$$

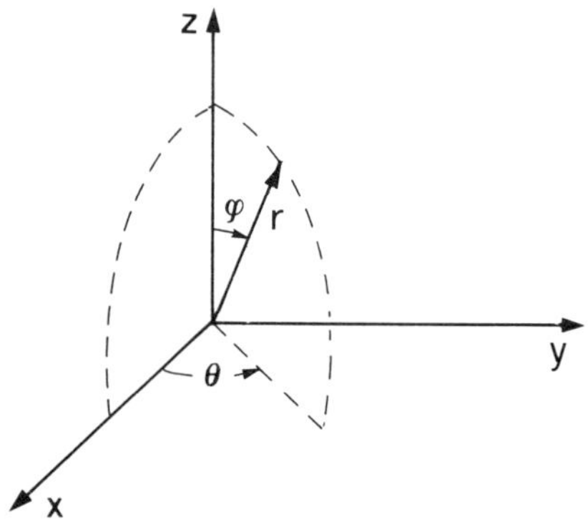

Fig. 3.A.4. Polar and Cartesian coordinates.

where $r^2 \sin \varphi$ is the Jacobian of the transformation (from Cartesian to polar coordinates). Integration over φ and θ gives

$$F(r^*) = \int_0^\infty 4\pi r^2 \rho(r) \frac{\sin 2\pi r r^*}{2\pi r r^*} \, \mathrm{d}r = \int_0^\infty U(r) \frac{\sin 2\pi r r^*}{2\pi r r^*} \, \mathrm{d}r \quad (3.A.33)$$

where $U(r) = 4\pi r^2 \rho(r)$ is the radial distribution function. Thus $F(r^*)$ is also spherically symmetric and its value at $r^* = 0$ is given by

$$F(0) = \int_0^\infty U(r) \, \mathrm{d}r.$$

As an example, let $\rho(r)$ be a spherically symmetric function equal to 1 for $r \leq R$ and vanishing elsewhere. Its Fourier transform

$$F(r^*) = \int_0^R 4\pi r^2 \frac{\sin 2\pi r r^*}{2\pi r r^*}$$

$$= \frac{2}{r^*} \int_0^R r \sin 2\pi r r^* \, \mathrm{d}r = \tfrac{4}{3}\pi R^3 \varphi(y)$$

where $\varphi(y) = 3(\sin y - y \cos y)/y^3$ and $y = 2\pi r^* R$. The function $\varphi(y)$ is plotted in Fig. 3.A.5. The main maximum occurs at $y = 0$. Intensities of the subsidiary maxima decrease when y increases. The distance between consecutive maxima is inversely related to R.

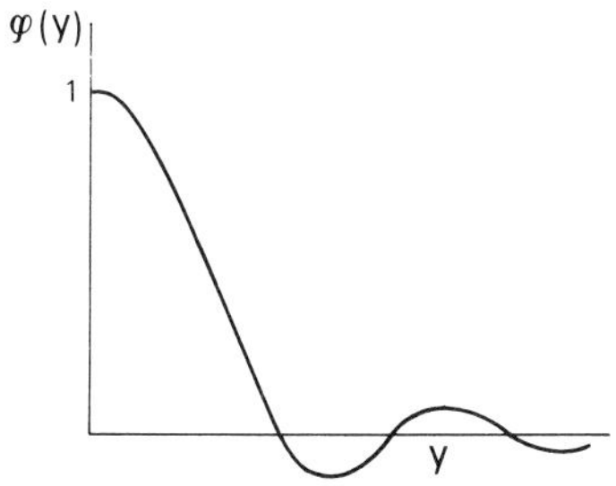

Fig. 3.A.5. The function $\varphi(y) = 3(\sin y - y \cos y)/y^3$ is plotted as a function of $y = 2\pi r^* R$.

Convolutions

The convolution (or folding) of two functions $\rho(\boldsymbol{r})$ and $g(\boldsymbol{r})$ (it will be denoted by $\rho(\boldsymbol{r}) * g(\boldsymbol{r})$) is defined by the integral

$$C(\boldsymbol{u}) = \rho(\boldsymbol{r}) * g(\boldsymbol{r}) = \int_S \rho(\boldsymbol{r}) g(\boldsymbol{u} - \boldsymbol{r}) \, \mathrm{d}\boldsymbol{r} \quad (3.A.34)$$

where S is the $\boldsymbol{r}$ space. Note that in (3.A.34) the integrand is a function of both $\boldsymbol{u}$ and $\boldsymbol{r}$ while the integral is only a function of $\boldsymbol{u}$. The relation between $\rho(\boldsymbol{r})$ and $g(\boldsymbol{r})$ is symmetrical in forming their convolution. That may be proved first by replacing in (3.A.34) $\boldsymbol{r}$ by $\boldsymbol{R} = \boldsymbol{u} - \boldsymbol{r}$ and after $\boldsymbol{R}$ by $\boldsymbol{r}$: then

$$\rho(\boldsymbol{r}) * g(\boldsymbol{r}) = g(\boldsymbol{r}) * \rho(\boldsymbol{r}).$$

Convoluting two functions very often has the effect of 'broadening' the one by the other. As an example, the convolution of two Gaussian functions $N(\sigma_1, a_1)$ and $N(\sigma_2, a_2)$ is the Gaussian function $N((\sigma_1^2 + \sigma_2^2)^{1/2}, a_1 + a_2)$.

The convolution operation appears in many scientific areas, and is involved in the interpretation of most experimental measurements. For example, when the intensity of a spectral line is measured by scanning it with a detector having a finite slit as input aperture, or when a beam of light passes through a ground-glass screen and is broadened out into a diffuse beam. Suppose in the second example that $\rho(\theta)$ is the angular distribution of the incident beam and $g(\theta)$ is the angular distribution which could be obtained if the incident beam was perfectly collimated. For any given $\rho(\theta)$ the angular distribution of the transmitted beam is given by:

$$C(\theta) = \int_{-\pi/2}^{\pi/2} \rho(\theta') g(\theta - \theta') \, \mathrm{d}\theta' = \rho(\theta) * g(\theta).$$

That may be explained by observing that the component of the transmitted beam emerging at angle θ due to the light component incident at angle θ' (and therefore deviated through the angle $\theta - \theta'$) has intensity $\rho(\theta')g(\theta - \theta')$. If interference effects are absent the total intensity in the direction θ is the integral of $\rho(\theta')g(\theta - \theta')$ over all values of θ'.

A very important theorem for crystallographers is the following:

$$\mathrm{T}[\rho(\boldsymbol{r}) * g(\boldsymbol{r})] = \mathrm{T}[\rho(\boldsymbol{r})] \cdot \mathrm{T}[g(\boldsymbol{r})]. \tag{3.A.35}$$

The left-hand side of (3.A.35) may be written

$$\int_{S''}\int_{S} \rho(\boldsymbol{r})g(\boldsymbol{u} - \boldsymbol{r}) \exp[2\pi i(\boldsymbol{u} \cdot \boldsymbol{r}^*)]\, \mathrm{d}\boldsymbol{r}\, \mathrm{d}\boldsymbol{u} \tag{3.A.36}$$

where $\boldsymbol{r}$ and $\boldsymbol{u}$ vary in S and S'' respectively. Denoting $\boldsymbol{u} = \boldsymbol{r} + \boldsymbol{r}'$ gives

$$\begin{aligned}\mathrm{T}[\rho(\boldsymbol{r}) * g(\boldsymbol{r})] &= \int_{S'}\int_{S} \rho(\boldsymbol{r})g(\boldsymbol{r}') \exp[2\pi i(\boldsymbol{r} + \boldsymbol{r}') \cdot \boldsymbol{r}^*]\, \mathrm{d}\boldsymbol{r}\, \mathrm{d}\boldsymbol{r}' \\ &= \int_{S} \rho(\boldsymbol{r}) \exp[2\pi i(\boldsymbol{r} \cdot \boldsymbol{r}^*)]\, \mathrm{d}\boldsymbol{r} \int_{S'} g(\boldsymbol{r}') \exp[2\pi i(\boldsymbol{r}' \cdot \boldsymbol{r})]\, \mathrm{d}\boldsymbol{r}' \\ &= \mathrm{T}[\rho(\boldsymbol{r})] \cdot \mathrm{T}[g(\boldsymbol{r})].\end{aligned}$$

In an analogous way the reader will prove that

$$\mathrm{T}[\rho(\boldsymbol{r}) \cdot g(\boldsymbol{r})] = \mathrm{T}[\rho(\boldsymbol{r})] * \mathrm{T}[g(\boldsymbol{r})]. \tag{3.A.37}$$

If $g(\boldsymbol{r}) = \rho(-\boldsymbol{r})$ (3.A.34) will represent the autoconvolution of $\rho(\boldsymbol{r})$ with itself inverted with respect to the origin: in crystallography it has a special significance, the 'Patterson function', and will be denoted by $P(\boldsymbol{u})$. It is

$$P(\boldsymbol{u}) = \rho(\boldsymbol{r}) * \rho(-\boldsymbol{r}) = \int_{S} \rho(\boldsymbol{r})\rho(\boldsymbol{u} + \boldsymbol{r})\, \mathrm{d}\boldsymbol{r} \tag{3.A.38}$$

the transform of which, according to (3.A.35), is given by

$$\mathrm{T}[P(\boldsymbol{u})] = [A(\boldsymbol{r}^*) + iB(\boldsymbol{r}^*)][A(\boldsymbol{r}^*) - iB(\boldsymbol{r}^*)] = |F(\boldsymbol{r}^*)|^2. \tag{3.A.39}$$

It is now easily seen that the Fourier transform of an autoconvolution is always a real function: therefore, in accordance with the conclusions of p. 176, $P(\boldsymbol{u})$ will always be centrosymmetric even if $\rho(\boldsymbol{r})$ is not.

Convolutions involving delta functions

Because of (3.A.8) and (3.A.34)

$$\delta(\boldsymbol{r} - \boldsymbol{r}_0) * \rho(\boldsymbol{r}) = \rho(\boldsymbol{u} - \boldsymbol{r}_0).$$

If $\boldsymbol{r}$ and $\boldsymbol{u}$ are assumed to belong to the same space, choosing the same coordinate system transforms the above relation onto

$$\delta(\boldsymbol{r} - \boldsymbol{r}_0) * \rho(\boldsymbol{r}) = \rho(\boldsymbol{r} - \boldsymbol{r}_0). \tag{3.A.40}$$

We see that the convolution of $\rho(\boldsymbol{r})$ with $\delta(\boldsymbol{r} - \boldsymbol{r}_0)$ is equivalent to a shift of the origin by $\boldsymbol{r}_0$ (see Fig. 3.A.6(a)).

Suppose now that $f(x)$ is a function defined between 0 and a. Because of (3.A.10)

$$L(x) * f(x) = \sum_{n=-\infty}^{+\infty} \delta(x - na) * f(x) = \sum_{n=-\infty}^{+\infty} f(x - na) = \rho(x)$$

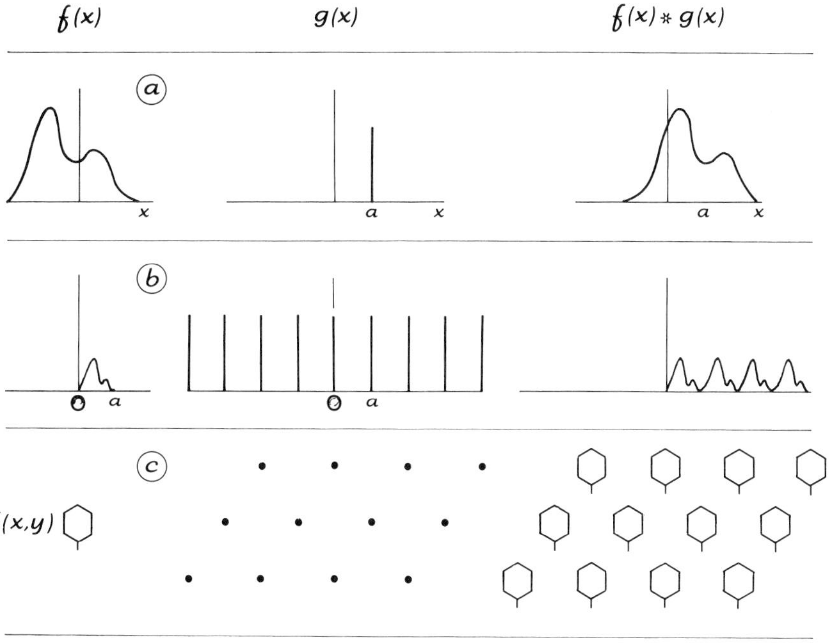

Fig. 3.A.6. Convolutions of the function f with: (a) the Dirac $\delta(x-a)$ delta function; (b) a one-dimensional lattice. In (c) the convolution of the function $f(x, y)$ with a two-dimensional lattice is shown.

where $\rho(x)$ is a periodic function defined between $-\infty$ and $+\infty$, equal to $f(x)$ for $0 \leqslant x \leqslant a$ and with period a. Indeed each term in the summation corresponds to a function equal to $f(x)$ but shifted by na (see Fig. 3.A.6(b)). It may be concluded that each function $\rho(x)$, periodic with period a, may be considered as the convolution of the function $f(x) = \rho(x)$ defined between 0 and a, with an array of delta functions located at the lattice point positions.

This result may be generalized to a three-dimensional space. Indeed

$$L(\boldsymbol{r}) * f(\boldsymbol{r}) = \sum_{u,v,w=-\infty}^{+\infty} \delta(\boldsymbol{r} - \boldsymbol{r}_{u,v,w}) * f(\boldsymbol{r}) = \sum_{u,v,w=-\infty}^{+\infty} f(\boldsymbol{r} - \boldsymbol{r}_{u,v,w}) = \rho(\boldsymbol{r}) \tag{3.A.41}$$

where: $f(\boldsymbol{r})$ is a function defined for $0 \leqslant x \leqslant a$, $0 \leqslant y \leqslant b$, $0 \leqslant z \leqslant c$; $\rho(\boldsymbol{r})$ is a function defined for x, y, z between $-\infty$ and $+\infty$, periodic with periods $\boldsymbol{a}$, $\boldsymbol{b}$, $\boldsymbol{c}$, and equal to $f(\boldsymbol{r})$ when $0 \leqslant x \leqslant a$, $0 \leqslant y \leqslant b$, $0 \leqslant z \leqslant c$.

In conclusion, a three-dimensional periodic function $\rho(\boldsymbol{r})$ may be considered as the convolution of a function defined in an elementary cell with a three-dimensional lattice, this last represented by the lattice function L.

Some properties of convolutions

Let $\rho(x)$ and $g(x)$ be two one-dimensional distributions defined in the interval $(-\infty, +\infty)$ and $C(u)$ their convolution. Consider the characteristic equation of $C(u)$:

$$\int_{-\infty}^{+\infty} C(u) \exp(\mathrm{i}tu)\, \mathrm{d}u = \int_{-\infty}^{+\infty} \left(\int_{-\infty}^{+\infty} \rho(x) g(u-x) \exp(\mathrm{i}tu)\, \mathrm{d}x \right) \mathrm{d}u.$$

Table 3.A.1. Properties of distributions under the convolution operation

$A_C = A_\rho A_g$
$m_C = m_\rho + m_g$
$\mu_2(C) = \mu_2(\rho) + \mu_2(g)$
$\mu_3(C) = \mu_3(\rho) + \mu_3(g)$
$\mu_4(C) = \mu_4(\rho) + 6\mu_2(\rho)\mu_2(g) + \mu_4(g)$
$\gamma_1(C) = (\gamma_1(\rho)\sigma^3(\rho) + \gamma_1(g)\sigma^3(g))/(\sigma^2(\rho) + \sigma^2(g))^{3/2}$
$\gamma_2(C) = (\gamma_2(\rho)\sigma^4(\rho) + \gamma_2(g)\sigma^4(g))/(\sigma^2(\rho) + \sigma^2(g))^2$

Changing the variable u into $q = u - x$ gives

$$\int_{-\infty}^{+\infty} C(u) \exp(\mathrm{i}tu)\, \mathrm{d}u = \int_{-\infty}^{+\infty} \rho(x) \exp(\mathrm{i}tx)\, \mathrm{d}x \int_{-\infty}^{+\infty} g(q) \exp(\mathrm{i}tq)\, \mathrm{d}q \tag{3.A.42}$$

which implies that the characteristic function of a distribution obtained by convolution is equal to the product of the characteristic functions of the constituent distributions. From (3.A.42) several properties of great import follow.

Setting $t = 0$ in (3.A.42) gives

$$A_C = \int_{-\infty}^{+\infty} C(u)\, \mathrm{d}u = \int_{-\infty}^{+\infty} \rho(x)\, \mathrm{d}x \int_{-\infty}^{+\infty} g(q)\, \mathrm{d}q = A_\rho A_g$$

where A represents the area under the distribution.

Taking the derivative of (3.A.42) with respect to t and setting $t = 0$ yields

$$\int_{-\infty}^{+\infty} C(u) u\, \mathrm{d}u = \int_{-\infty}^{+\infty} \rho(x) x\, \mathrm{d}x + \int_{-\infty}^{+\infty} g(q) q\, \mathrm{d}q$$

or

$$m_C = m_\rho + m_g.$$

Thus, the mean of the convolution is equal to the sum of the means of the constituent distributions.

By extending the procedure Table 3.A.1 may be obtained. The following notation has been used:

$$\mu_p(C) = \frac{1}{A_C} \int_{-\infty}^{+\infty} C(u)(u - m_C)^p\, \mathrm{d}u$$

are the central moments of order p for the convolution distribution (similar expression can be derived for the constituent distributions). Accordingly $\mu_0 = 1$, $\mu_1 = 0$, μ_2 coincides with the variance σ^2, while $\gamma_1 = \mu_3/\sigma^3$ and $\gamma_2 = [(\mu_4/\sigma_4) - 3]$ are the skewness and the excess parameters for any distribution.

Deconvolution of spectra

Often it occurs that an experimentally measured function C may be considered as the convolution of the functions ρ and g. If ρ is known in advance then it may be of some interest to obtain g. That frequently occurs in spectroscopy or in powder diffraction, where a spectrum is often constituted by overlapping peaks and it is wanted to deconvolute from such a spectrum a given lineshape function. Effects of such self-deconvolution are:[14–16] the component lines are more clearly distinguished, their location, area, etc. are more correctly defined, the signal to noise ratio is increased.

Let us consider C as the convolution of the lineshape function ρ and of the ideal spectrum g. Then ρ may be deconvoluted[17] from C by taking the Fourier transform of C,

$$\mathrm{T}[C]=\mathrm{T}[\rho]\cdot\mathrm{T}[g],$$

and by calculating

$$\mathrm{T}[g]=\mathrm{T}[C]/\mathrm{T}[\rho]. \tag{3.A.43}$$

g is finally obtained as inverse Fourier transform of the right-hand side of (3.A.43):

$$g\simeq g'=\mathrm{T}^{-1}\{\mathrm{T}[C]/\mathrm{T}[\rho]\}. \tag{3.A.44}$$

An example of what can be achieved is shown in Fig. 3.A.7. In (a) a diffractogram of quartz measured between 67° and 69° is shown, in (b) a possible Fourier self-deconvoluted spectrum is drawn. The widths of deconvoluted lines are narrower, peak centroids shift as a result of changes in degree of overlap: since integrated intensities of peaks are retained (see p. 184) more careful estimates of integrated intensities of single peaks may now be accomplished.

The practical use of (3.A.44) is not always straightforward.[18,19] The limited instrumental resolution and the random errors associated with experimental data are a difficult problem. In particular, random errors are usually amplified by the process so that the right-hand side of (3.A.44) could coincide with a function g' substantially different from g. Filtering operations or supplementary considerations are then introduced to reduce spurious features.

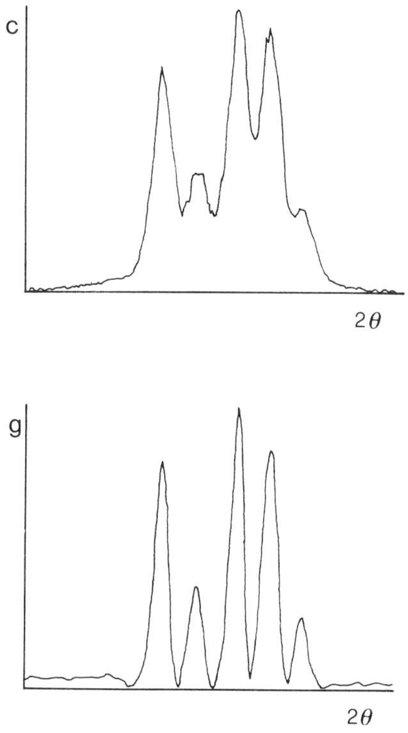

Fig. 3.A.7.

3.B Scattering and related topics

Compton scattering

A photon of energy $h\nu$ ($\nu=c/\lambda$) moving along the direction $\boldsymbol{s}_0$ is scattered by a completely free electron located in O, initially at rest. Let $\boldsymbol{s}$ be the direction of the scattered photon, 2θ the scattering angle, $h\nu'=hc/(\lambda+\mathrm{d}\lambda)$ its energy and $mv^2/2$ the recoil electron energy. Conservation of energy (neglecting any relativistic effect) requires

$$\frac{hc}{\lambda}=\frac{hc}{\lambda+\mathrm{d}\lambda}+\tfrac{1}{2}mv^2$$

which may be approximated to

$$\frac{hc}{\lambda^2}\mathrm{d}\lambda=\tfrac{1}{2}mv^2. \tag{3.B.1}$$

Conservation of momentum requires

$$\frac{h\nu}{c}\boldsymbol{s}_0=\frac{h}{\lambda}\boldsymbol{s}_0=\frac{h}{\lambda+\mathrm{d}\lambda}\boldsymbol{s}+m\boldsymbol{v} \tag{3.B.2}$$

which, on assuming $\lambda + d\lambda \simeq \lambda$, may be reduced to

$$\tfrac{1}{2}mv = \frac{h}{\lambda}\sin\theta. \tag{3.B.3}$$

Finally, v may be eliminated from (3.B.1) and (3.B.3) so as to obtain

$$d\lambda = \frac{h}{mc}(1 - \cos 2\theta). \tag{3.B.4}$$

The anisotropic temperature factor

Thermal agitation will cause the atoms to fluctuate about their equilibrium positions. Let us suppose that the force field around an atom located at the origin of the coordinate system (its equilibrium position) is anisotropic and harmonic. Then the mean-square vibrational amplitude is different for different directions and equal in the two ways of the same direction. In any rectilinear coordinate system defined by $\bar{\mathbf{A}} \equiv [\boldsymbol{a}, \boldsymbol{b}, \boldsymbol{c}]$ let $p(\boldsymbol{r}')$ be the probability of finding the atomic centre moved in $\boldsymbol{r}' = x'\boldsymbol{a} + y'\boldsymbol{b} + z'\boldsymbol{c} = \bar{\mathbf{X}}'\mathbf{A}$. In general

$$p(x', y', z') = (2\pi)^{-3/2}(\det \mathbf{U})^{-1/2} \exp[-\tfrac{1}{2}(\bar{\mathbf{X}}'\mathbf{U}^{-1}\mathbf{X}')], \tag{3.B.5}$$

where $\mathbf{U}$ is the variance–covariance matrix (it represents a symmetrical tensor)

$$\mathbf{U} = \begin{pmatrix} \langle x'^2 \rangle & \langle x'y' \rangle & \langle x'z' \rangle \\ \langle x'y' \rangle & \langle y'^2 \rangle & \langle y'z' \rangle \\ \langle x'z' \rangle & \langle y'z' \rangle & \langle z'^2 \rangle \end{pmatrix} = \langle \mathbf{X}'\bar{\mathbf{X}}' \rangle.$$

The Fourier transform of $p(\mathbf{X}')$ gives

$$\begin{aligned} q(\mathbf{X}^*) &= \exp(-2\pi^2\bar{\mathbf{X}}^*\mathbf{U}^*\mathbf{X}^*) = \exp[-2\pi^2(U_{11}^* x^{*2} + 2U_{12}^* x^* y^* + \ldots)] \\ &= \exp(-\bar{\mathbf{X}}^*\boldsymbol{\beta}\mathbf{X}^*) = \exp - (\beta_{11}x^{*2} + 2\beta_{12}x^*y^* + \ldots)] \end{aligned}$$

where $\mathbf{U}^*$ is the variance–covariance matrix expressed in reciprocal coordinates and

$$\boldsymbol{\beta} = 2\pi^2\mathbf{U}^*. \tag{3.B.6}$$

Usually crystallographic least-squares calculations (see p. 94) provide estimates of the tensor $\boldsymbol{\beta}$ from which $\mathbf{U}^*$ is easily derived via eqn (3.8.6). Then the atomic scattering factor, calculated for $\boldsymbol{r}^* = \boldsymbol{r}_{\mathbf{H}}^*$, and corrected for thermal motion is (see p. 153)

$$f_0 \exp(-\bar{\mathbf{H}}\boldsymbol{\beta}\mathbf{H}) = f_0 \exp[-2\pi^2(\bar{\mathbf{H}}\mathbf{U}^*\mathbf{H})]$$

where f_0 refers to the atom at rest.

In order to obtain $\mathbf{U}$ from $\mathbf{U}^*$ we only need to apply eqn (2.28). We obtain $\mathbf{U} = \mathbf{G}\mathbf{U}^*\mathbf{G}$ from which

$$\mathbf{U} = (\mathbf{G}\boldsymbol{\beta}\mathbf{G})/(2\pi^2). \tag{3.B.7}$$

In accordance with a well known property of tensors, the mean-squares vibrational amplitude $\langle u^2 \rangle$ in the direction defined by the unit vector $\boldsymbol{n} = \bar{\mathbf{N}}\mathbf{A} = \mathbf{N}^*\mathbf{A}^*$, is given by

$$\langle u^2 \rangle_{(\boldsymbol{n})} = \bar{\mathbf{N}}\mathbf{U}\mathbf{N} = \bar{\mathbf{N}}^*\mathbf{U}^*\mathbf{N}^*. \tag{3.B.8}$$

Let us now express $\boldsymbol{n}$ as a ratio between a vector $\boldsymbol{r} = \bar{\mathbf{X}}\mathbf{A} = \bar{\mathbf{X}}^*\mathbf{A}^*$ parallel to $\boldsymbol{n}$ and its modulus: $\boldsymbol{n} = \boldsymbol{r}/r$. Then eqn (3.B.8) may be written as

$$\langle u^2 \rangle_{(\boldsymbol{r})} = \frac{\bar{\mathbf{X}}\mathbf{U}\mathbf{X}}{\bar{\mathbf{X}}\mathbf{G}\mathbf{X}} = \frac{1}{2\pi^2}\frac{\bar{\mathbf{X}}\mathbf{G}\boldsymbol{\beta}\mathbf{G}\mathbf{X}}{\bar{\mathbf{X}}\mathbf{G}\mathbf{X}} \qquad (3.B.9a)$$

or also

$$\langle u^2 \rangle_{(\boldsymbol{r})} = \frac{\bar{\mathbf{X}}^*\mathbf{U}^*\mathbf{X}^*}{\bar{\mathbf{X}}^*\mathbf{G}^*\mathbf{X}^*} = \frac{1}{2\pi^2}\frac{\bar{\mathbf{X}}^*\boldsymbol{\beta}\mathbf{X}^*}{\bar{\mathbf{X}}^*\mathbf{G}^*\mathbf{X}^*}. \qquad (3.B.9b)$$

From (3.B.9) the following relations arise:

$$\langle u^2 \rangle_{(\boldsymbol{a})} = \frac{U_{11}}{g_{11}} = \frac{U_{11}}{a^2}, \qquad \langle u^2 \rangle_{(\boldsymbol{b})} = \frac{U_{22}}{b^2}, \qquad \langle u^2 \rangle_{(\boldsymbol{c})} = \frac{U_{33}}{c^2},$$

$$\langle u^2 \rangle_{(\boldsymbol{a}^*)} = \frac{U_{11}^*}{a^{*2}}, \qquad \langle u^2 \rangle_{(\boldsymbol{b}^*)} = \frac{U_{22}^*}{b^{*2}}, \qquad \langle u^2 \rangle_{(\boldsymbol{c}^*)} = \frac{U_{33}^*}{c^{*2}}.$$

The following exercises may be useful to clarify some aspects of the subject:

1. Calculate the β_{ij}s corresponding to the isotropic temperature factor B_{iso} defined by (3.19).

   If in (3.19) $\sin^2\theta/\lambda^2 = r^{*2}/4 = 1/(4d_{\mathbf{H}}^2)$ is replaced by (2.17b) then

   $$\boldsymbol{\beta} = B_{\text{iso}}\mathbf{G}^*/4$$

   is obtained.

2. If $\boldsymbol{\beta}$ is the anisotropic tensor in the coordinate system $\mathbf{A}$, calculate its expression in $\mathbf{A}' = \mathbf{MA}$.

   According to (2.A.15) $\boldsymbol{\beta}' = (\bar{\mathbf{M}})^{-1}\boldsymbol{\beta}\mathbf{M}^{-1}$.

3. If $\boldsymbol{\beta}$ is the thermal tensor of a given atom, calculate the tensor $\boldsymbol{\beta}'$ of the atom referred to the first one by the symmetry operation $\mathbf{C} = (\mathbf{R}, \mathbf{T})$.

   In accordance with (2.A.16) $\boldsymbol{\beta}' = \mathbf{R}\boldsymbol{\beta}\bar{\mathbf{R}}$.

4. Calculate the principal axes of the ellipsoid describing the anisotropic temperature factor $\boldsymbol{\beta}$ and their orientations with respect to the crystallographic axes.

   See p. 76.

5. Calculate the isotropic temperature factor **equivalent**[20] to a given anisotropic temperature factor (often the anisotropic parameters are deposited and equivalent isotropic are published). The equivalent isotropic motion is defined as that one which gives rise to the same value of $\langle u^2 \rangle$: in our case $\langle u^2 \rangle_{\text{aniso}} = \langle u^2 \rangle_{\text{iso}}$.

   Let us suppose that eigenvalues λ_1, λ_2, and λ_3 of the anisotropic tensor $\boldsymbol{\beta}$ have been calculated. Then, according to (2.38b) we can write

   $$\langle u^2 \rangle_{\text{aniso}} = \frac{1}{2\pi^2}\tfrac{1}{3}(\lambda_1 + \lambda_2 + \lambda_3) \qquad (3.B.10a)$$

   while, for an isotropic motion (see eqn (3.19))

   $$\langle u^2 \rangle_{\text{iso}} = B_{\text{iso}}/(8\pi^2).$$

Isotropic and anisotropic motion will be then equivalent if

$$B_{iso} = \tfrac{4}{3}(\lambda_1 + \lambda_2 + \lambda_3). \quad (3.B.10b)$$

Let us now observe that, under a transformation of axes, the matrix **βG** transforms, because of eqns (2.E.8) and (2.21), according to

$$\boldsymbol{\beta}'\mathbf{G}' = (\bar{\mathbf{M}})^{-1}\boldsymbol{\beta}\mathbf{M}^{-1}\mathbf{M}\mathbf{G}\bar{\mathbf{M}} = (\bar{\mathbf{M}})^{-1}(\boldsymbol{\beta}\mathbf{G})\bar{\mathbf{M}};$$

in other words, **βG** transforms in the same way as a rotation matrix of a symmetry operator (see eqn (2.E.1)). Therefore the trace of **βG** is invariant under the transformation and, according to p. 75, it is constantly equal to $\lambda_1 + \lambda_2 + \lambda_3$.

At the end we can replace eqn (3.B.10b) by the most simple equation

$$B_{equiv} = \tfrac{4}{3}\,\mathrm{Tr}\,(\boldsymbol{\beta}\mathbf{G}), \quad (3.B.10c)$$

or

$$\langle u^2 \rangle_{equiv} = \frac{1}{6\pi^2}\,\mathrm{Tr}\,(\boldsymbol{\beta}\mathbf{G}) = \frac{1}{6\pi^2}\sum_{i,j} \beta_{ij} g_{ij}. \quad (3.B.10d)$$

The reader will easily derive from (3.B.10d) specific formulae for specific crystallographic systems: e.g. for cubic, tetragonal, and orthorhombic systems

$$\langle u^2 \rangle_{equiv} = (\beta_{11}a^2 + \beta_{22}b^2 + \beta_{33}c^2)/(6\pi^2),$$

for hexagonal and trigonal systems (hexagonal setting)

$$\langle u^2 \rangle_{equiv} = (\beta_{11}a^2 + \beta_{22}b^2 + \beta_{33}c^2 - \beta_{12}ab)/(6\pi^2),$$

etc. Readers are referred to Schomaker and Marsh[21] for the estimated standard deviation of $\langle u^2 \rangle_{equiv}$.

In reports of crystal structure determination thermal motion is usually presented in direct space by means of a probability ellipsoid. Each ellipsoid is centred on the mean position of the atom: the surface has constant probability density obtained, in accordance with eqn (3.B.5), on assuming

$$\bar{\mathbf{X}}'\mathbf{U}^{-1}\mathbf{X}' = C^2$$

where C is a chosen constant. The probability P that a thermal displacement falls within the ellipsoids may be calculated as

$$P(C) = \int_{ellipsoid} p(\mathbf{X}')\,d\mathbf{X}'.$$

The most often used value is $C = 1.5282$: then[22] the ellipsoid encloses 50 per cent of the trivariate Gaussian probability density.

Symmetry restrictions on the anisotropic temperature factors

Coefficients β_{ij} as provided by crystallographic least-squares procedures (see p. 94) are affected by various error sources: for example, experimental errors on diffraction intensity measurements, inadequacy of the atomic scattering model, wrong or unapplied absorption correction (see Chapter 4), existence of non-harmonic force fields around the atoms, etc. Thus the thermal parameters estimated by least squares could be physically unrealistic; for example the variance–covariance matrix **U** should not result positive-definite. Necessary and sufficient conditions for positiveness are

$$(\det \mathbf{U}) > 0;\ U_{ii} > 0;\ \text{and}\ U_{ii}U_{ij} > 0\ \text{for}\ i, j = 1, 2, 3.$$

If the above conditions are violated then atomic vibrations could be described by imaginary ellipsoids or by paraboloids or by hyperboloids, with the obvious meaning that experimental data are not of sufficient accuracy to justify the use of anisotropic temperature factors.

If an atom lies on a special site of a space group the components of the thermal tensor are restricted according to the site symmetry. Indeed the thermal ellipsoids must remain invariant after application of the n symmetry operations which leave invariant the site. In accordance with point 3 in the previous section the relations

$$\boldsymbol{\beta}'_s = \mathbf{R}_s \boldsymbol{\beta} \bar{\mathbf{R}}_s = \boldsymbol{\beta} \qquad \text{for } s = 1, 2, \ldots, n$$

arise, from which the $\boldsymbol{\beta}$ restrictions can be derived. Such restrictions can also be obtained in an easier way by using Wigner's theorem according to which the symmetry of the $\boldsymbol{\beta}$ restrictions is displayed by the matrix

$$\boldsymbol{\beta}_{\text{inv}} = \sum_{s=1}^{n} \boldsymbol{\beta}'_s.$$

Since no atomic position is left invariant after the application of glide planes or screw axes, only point symmetries have to be taken into account. Furthermore, ellipsoids are centrosymmetric, thus no additional $\boldsymbol{\beta}$ restriction arises if an inversion centre is added to the site symmetry, or if a mirror plane is replaced by a twofold axis perpendicular to the plane. Thus it is not difficult to understand that ellipsoid symmetry is one of the following three cases:

(1) spherical symmetry, for atomic sites with symmetry 23 and its supergroups;

(2) ellipsoid of revolution, for site symmetries 3, 4, $\bar{4}$, and 'non-cubic' supergroups;

(3) general ellipsoid. For site symmetry 222 and its orthorhombic supergroups the orientations of the ellipsoid axes is fixed by symmetry; for site symmetry 2, m, or 2/m only the orientation of one ellipsoid axis is fixed by symmetry.

In Table 3.B.1 all the conventional special positions are collected in seven blocks.[22,23] Each block shows the possible point groups to which each site may belong, the 'minimal' site symmetry characterizing each position and a cross-reference number identifying the type of $\boldsymbol{\beta}$ restrictions described in Table 3.B.2. The letter h, when present, warns that an hexagonal setting is involved and the letter w suggests that the rule holds for hexagonal and non-hexagonal settings. To make Table 3.B.1 clearer we note that all the special positions in the first block belong to one of the point groups m$\bar{3}$m, $\bar{4}$3m, 432, m$\bar{3}$, 23. According to previous observations, we now neglect whether symmetry axes are proper or improper; thus the 'minimal' symmetry $3_{[111]}\,2_{[001]}$ is associated with the sites. Conversely, under the same convention, all sites with symmetry $2_{[001]}\,2_{[100]}\,\bar{1}$ or $2_{[001]}\,\bar{2}_{[100]}$ or $\bar{2}_{[001]}\,2_{[010]}$ or $\bar{2}_{[001]}\,2_{[100]}$ or $2_{[001]}\,2_{[100]}$ are all represented in block five of Table 3.B.1 by the site with 'minimal' symmetry $2_{[001]}\,2_{[100]}$.

To make Table 3.B.2 clearer we observe that case No. 23 may be rewritten as:

$$\beta_{11},\ \beta_{33} = \beta_{22},\ \beta_{12},\ \beta_{13} = -\beta_{12},\ \beta_{23}.$$

Table 3.B.1. Site symmetry table giving key for the 29 types of symmetry β-restrictions

	$m\bar{3}m$, $\bar{4}3m$, 432, $m\bar{3}$, 23			2/m, m, 2	
	$3_{[111]}$ $2_{[001]}$	1	w	$2_{[001]}$	16
				$2_{[010]}$	17
	6/mmm, $\bar{6}m2$, 6mm, 622, 6/m,			$2_{[100]}$	18
	$\bar{6}$, 6, $\bar{3}m$, 32, $\bar{3}$, 3		w	$2_{[110]}$	19
h	$3_{[001]}$	9	w	$2_{[1\bar{1}0]}$	20
				$2_{[101]}$	21
	4/mmm, $\bar{4}2m$, 4mm, 422			$2_{[10\bar{1}]}$	22
	4/m, $\bar{4}$, 4			$2_{[011]}$	23
	$4_{[001]}$	2		$2_{[01\bar{1}]}$	24
			h	$2_{[100]}$	25
	$4_{[010]}$	3	h	$2_{[210]}$	26
			h	$2_{[120]}$	27
	$4_{[100]}$	4	h	$2_{[010]}$	28
	$\bar{3}m$, 3m, 32, $\bar{3}$, 3				
	$3_{[111]}$	5			
	$3_{[11\bar{1}]}$	6		$\bar{1}$, 1	
	$3_{[1\bar{1}1]}$	7		1	29
	$3_{[\bar{1}11]}$	8			
	mmm, mm2, 222				
	$2_{[001]}$ $2_{[100]}$	10			
w	$2_{[001]}$ $2_{[110]}$	11			
	$2_{[010]}$ $2_{[101]}$	12			
	$2_{[100]}$ $2_{[011]}$	13			
h	$2_{[001]}$ $2_{[100]}$	14			
h	$2_{[001]}$ $2_{[010]}$	15			

Table 3.B.2. Symmetry β-restrictions.

Cross-reference No.	β_{11}	β_{22}	β_{33}	β_{12}	β_{13}	β_{23}	**Cross-reference No.**	β_{11}	β_{22}	β_{33}	β_{12}	β_{13}	β_{23}
1	A	A	A	0	0	0	16	A	B	C	D	0	0
2	A	A	C	0	0	0	17	A	B	C	0	E	0
3	A	B	A	0	0	0	18	A	B	C	0	0	F
4	A	B	B	0	0	0	19	A	A	C	D	E	−E
5	A	A	A	D	D	D	20	A	A	C	D	E	E
6	A	A	A	D	−D	−D	21	A	B	A	D	E	−D
7	A	A	A	D	−D	D	22	A	B	A	D	E	D
8	A	A	A	D	D	−D	23	A	B	B	D	−D	F
9	A	A	C	A/2	0	0	24	A	B	B	D	D	F
10	A	B	C	0	0	0	25	A	B	C	B/2	F/2	F
11	A	A	C	D	0	0	26	A	B	C	A/2	0	F
12	A	B	A	0	E	0	27	A	B	C	B/2	E	0
13	A	B	B	0	0	F	28	A	B	C	A/2	E	E/2
14	A	B	C	B/2	0	0	29	A	B	C	D	E	F
15	A	B	C	A/2	0	0							

The Renninger effect and experimental phase determination by means of multiple diffraction experiments

Let us suppose that the two nodes $\mathbf{H}_1$ and $\mathbf{H}_2$ are simultaneously at the surface of the reflection sphere: according to eqn (3.25) it will have to be

$$(\boldsymbol{s}_1 - \boldsymbol{s}_0)/\lambda = \boldsymbol{r}^*_{\mathbf{H}_1}, \qquad (\boldsymbol{s}_2 - \boldsymbol{s}_0)/\lambda = \boldsymbol{r}^*_{\mathbf{H}_2}. \tag{3.B.11}$$

By subtracting the second equation from the first we obtain

$$(\boldsymbol{s}_1 - \boldsymbol{s}_2)/\lambda = \boldsymbol{r}^*_{\mathbf{H}_1} - \boldsymbol{r}^*_{\mathbf{H}_2} = \boldsymbol{r}^*_{\mathbf{H}_1-\mathbf{H}_2}. \tag{3.B.12}$$

Equation (3.B.12) sets up that the diffracted beam $\mathbf{H}_2$ can act as an incident beam for the reflection $\mathbf{H}_1 - \mathbf{H}_2$ which is thus in the diffraction position (see Fig. 3.B.1(a)). Therefore, to the beam diffracted in direction $\boldsymbol{s}_1$ by lattice planes $\mathbf{H}_1$ a twice-reflected beam overlaps, first from the planes $\mathbf{H}_2$ and then from the planes $\mathbf{H}_1 - \mathbf{H}_2$ (see Fig. 3.B.1(b)). In the same way the beam diffracted in direction $\boldsymbol{s}_2$ from the planes $\mathbf{H}_2$ will overlap with a twice-reflected beam, first from the planes $\mathbf{H}_1$ and then from the planes $\mathbf{H}_2 - \mathbf{H}_1$.

In general, when diffraction data are collected for crystal structure solution, the multiple diffraction (Renninger) effect can be considered as a *disturbance*. As already noted, it can be the reason for apparent violation of the systematic extinctions (but the shape of their diffraction peaks is different from that of the normal reflections and therefore easily recognizable). On the other hand, the three-beam diffraction effect can be treated in accordance with dynamic diffraction theory and used for the experimental solution of the phase problem. This point of view is extremely promising, and a short account is given here.

First Lipscomb[24] investigated the possibility of using multiple diffraction intensities for phase determination, but conclusive results were not obtained. In spite of many subsequent investigations, only in more recent years[25–30] has it become clear that dynamical theory applied to three-beam diffraction is a powerful tool for the experimental solution of the phase problem even for common (say imperfect) crystals. General solutions for the three-beam case are now available which provide a satisfactory interpretation of experimental results. The reader is referred to a recent book[31] for a review of the various contributions and a description of the

Fig. 3.B.1. Renninger effect described in: (a) reciprocal space; (b) direct space.

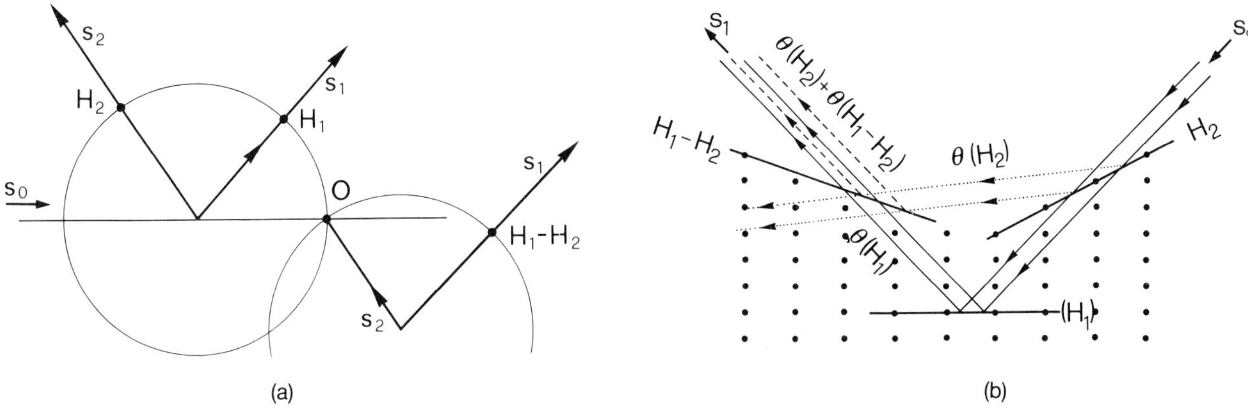

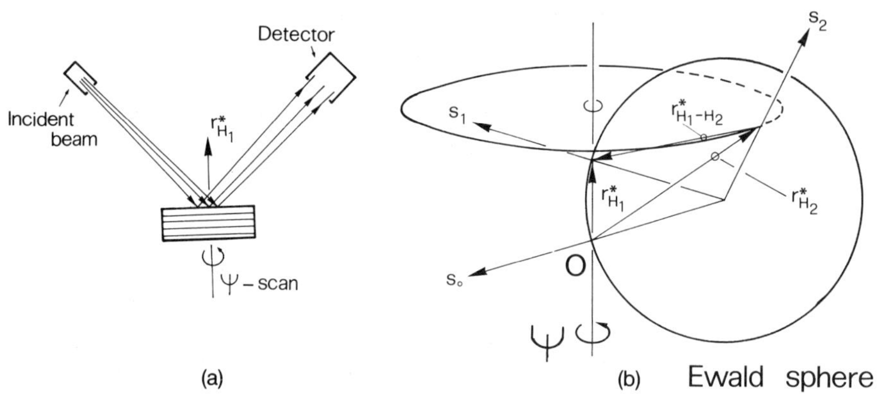

Fig. 3.B.2. ψ-scan experiment described in: (a) direct space; (b) reciprocal space.

theoretical background: here a short account of the results is given in accordance with Hummer and Billy.[32]

The interference between the wave directly diffracted from $\mathbf{H}_1$ planes (primary reflection) and the **'Renninger Umweg wave'** generated from $\mathbf{H}_2$ and $\mathbf{H}_1 - \mathbf{H}_2$ scatterings produces a wave whose intensity depends on the triplet invariant phase

$$\Phi_3 = -\varphi(\mathbf{H}_1) + \varphi(\mathbf{H}_2) + \varphi(\mathbf{H}_1 - \mathbf{H}_2).$$

Without anomalous dispersion effects it is equal to

$$\Phi_3 = \varphi(-\mathbf{H}_1) + \varphi(\mathbf{H}_2) + \varphi(\mathbf{H}_1 - \mathbf{H}_2).$$

Such an intensity can be measured by a ψ-scan experiment (see Fig. 3.B.2): the integrated intensity $I_{\mathbf{H}_1}(\psi)$ is monitored while the crystal is rotated about the vector $\boldsymbol{r}^*_{\mathbf{H}_1}$ and scanned through the three-beam position. The integrated ψ-scan profile may be approximately calculated (except for a range very close to the three-beam setting) according to (see Fig. 3.B.3)

$$I_{\mathbf{H}_1}(\psi) = |C_1 D^0_{\mathbf{H}_1} + C_2 D_u(\psi)|^2 \tag{3.B.13}$$

where the Ds are the amplitudes of the excited wavefields in the crystal (the intensities are given by $I \simeq |D|^2$). In eqn (3.B.13) $D^0_{\mathbf{H}_1}$ is the two-beam amplitude of the reflection $\mathbf{H}_1$, D_u is the amplitude of the Umweg wave defined by

$$D_u \cong R(\psi) F_{\mathbf{H}_2} F_{\mathbf{H}_1-\mathbf{H}_2}. \tag{3.B.14}$$

C_i are suitable parameters.

$|R(\psi)|$ and $\Delta(\psi)$ are the modulus and phase of the complex resonance[33] term $R(\psi) = |R(\psi)| \exp(\mathrm{i}\Delta(\psi)]$ which governs the amplitude and the resonance phase shift of the Umweg wave. Typical resonance curves for $|R(\psi)|$ and $\Delta(\psi)$ are given in Fig. 3.B.4. $|R(\psi)|$ is highest near the three-beam position. $\Delta(\psi)$ varies from zero to π when ψ is scanned through the three-beam position: it is less than $\pi/2$ when $\mathbf{H}_2$ is inside the Ewald sphere; it is between $\pi/2$ and π when it is outside.

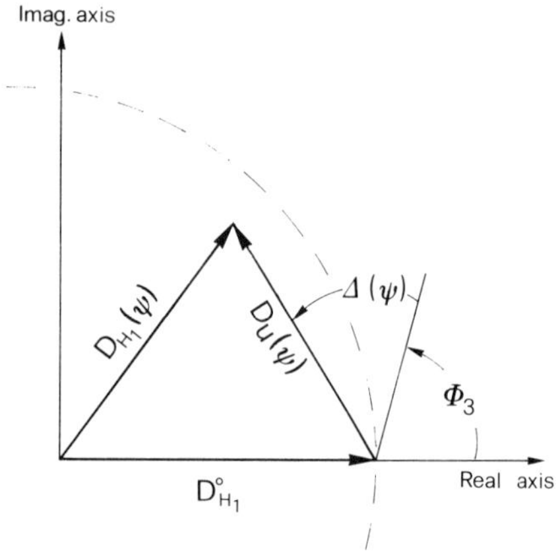

Fig. 3.B.3. Diagram of the interference between the unperturbed two-beam amplitude D_0 and the Umweg resonance term D_u.

Equation (3.B.13) and (3.B.14) may be used to interpret the main features of typical ψ-scan profiles (we assume that $\boldsymbol{r}^*_{\mathbf{H}_2}$ crosses the Ewald sphere from inside ($\psi < 0$) to outside ($\psi > 0$)).

Suppose $\Phi_3 \simeq 0$: for $\psi \ll 0$ it is $\Phi_3 + \Delta(\psi) \simeq 0$, interfering waves are essentially in phase, but $R(\psi) \simeq 0$. Then $I_{\mathbf{H}_1}(\psi) \simeq I^0_{\mathbf{H}_1}$. When we get near the three-beam position $|D_u(\psi)|$ rapidly increases while $\Delta(\psi)$ approaches $\pi/2$: then the intensity $I_{\mathbf{H}_1}(\psi) > I^0_{\mathbf{H}_1}$. Just after scanning through the three-beam position $|R(\psi)|$ is still large but $\cos(\Phi_3 + \Delta\psi)$ is negative: then $I_{\mathbf{H}_1}(\psi)$ drops below the two-beam value $I^0_{\mathbf{H}_1}$. For larger values of ψ $I_{\mathbf{H}_1}(\psi)$ will approach $I^0_{\mathbf{H}_1}$. The type of asymmetry is reversed for triplet phase $\Phi_3 = 180°$.

It is easy now to foresee that for triplet phases with $\Phi_3 \simeq \pi/2$ the ψ-scan profile will show a nearly symmetrical decrease of the intensity about $\psi \simeq 0$, where $\cos(\Phi_3 + \Delta(\psi)) \simeq -1$. If $\Phi_3 \simeq -\pi/2$ an increase in intensity will occur about $\psi \simeq 0$, where $\cos(\Phi_3 + \Delta(\psi)) \simeq 1$. According to this method, expected profiles for $\Phi_3 \simeq +45°$ (or $\Phi_3 \simeq -45°$) should present characteristics between those for $\Phi_3 \simeq 0$ and for $\Phi_3 \simeq 90°$ (or $\Phi_3 \simeq -90°$). Ideal ψ-scan profiles for $\Phi_3 \simeq 0$, π, $\pm\pi/2$, $\pm\pi/4$, have been recently secured[34] and are shown in Fig. 3.B.5. Profiles for other quadrants can be obtained by observing that, according to (3.B.14), $I_{\mathbf{H}_1}(\psi)$ for a given Φ_3 is equal to $I_{\mathbf{H}_1}(-\psi)$ for the phase $(180 - \Phi_3)$. Therefore ψ-scan profiles for Φ_3 and $(180 - \Phi_3)$ are related by a mirror line through $\psi = 0$.

Therefore one is able to define ideal profiles that satisfy the condition

$$\Delta I(\psi) = I^0_{\mathbf{H}_1}$$

where $\Delta I(\psi)$ is defined as

$$\Delta I(\psi) = \tfrac{1}{2}(I^+_{\mathbf{H}_1}(\psi) + I^-_{\mathbf{H}_1}(-\psi)).$$

$I^+_{\mathbf{H}_1}(\psi)$ and $I^-_{\mathbf{H}_1}(\psi)$ are the ψ-scan profiles for the positive $(+\Phi_3)$ and the negative triplet phase $(-\Phi_3)$ respectively. Thus the ideal profiles shown in Fig. 3.B.5 are marked by the condition $\Delta I(\psi)/I^0_{\mathbf{H}_1} \equiv 1$.

Ideal ψ-scan profiles can only be obtained if the dominant process in he three-beam interaction is due to the interference effect. That occurs when $|F_{\mathbf{H}_2}|$ and $|F_{\mathbf{H}_1-\mathbf{H}_2}|$ are about twice as strong as $|F_{\mathbf{H}_1}|$. If $|F_{\mathbf{H}_2}|$ is small and $|F_{\mathbf{H}_2-\mathbf{H}_1}|$ is large enough (or vice versa as the influence of $|F_{\mathbf{H}_2}|$ and $|F_{\mathbf{H}_2-\mathbf{H}_1}|$ is symmetric) then intensity is removed from the $\mathbf{H}_1$ reflection and coupled into the $\mathbf{H}_2$ reflection via the coupling $\mathbf{H}_2 - \mathbf{H}_1$. This loss of intensity is not compensated by the scattering power from $\mathbf{H}_2$ into $\mathbf{H}_1$ reflection. The result is the so-called **aufhellung effect**, that is a strong depletion of the $\mathbf{H}_1$ intensity. If $|F_{\mathbf{H}_1}|$ is very small and $|F_{\mathbf{H}_2}|$ is large then the intensity of $\mathbf{H}_1$ is increased at the cost of the $\mathbf{H}_2$ reflection intensity. This effect is called **umweganregung** by Renninger.[35]

Unweganregung and aufhellung effects can be evaluated[36] by comparing the ψ-scan profiles for Φ_3 and $-\Phi_3$. Then profiles can be separated into two parts: the symmetric ΔI curve, which represents the phase-independent umweganregung or aufhellung effect; the ideal ψ-scan profiles, which contain the phase information.

Some concluding remarks may be useful:

1. Exact ψ-scan can be difficult for conventional four-circle diffractometers.[37] A special six-circle diffractometer[36] proved to be useful: two circles (θ, ν) for the detector and four circles for the crystal motion (see Fig. 3.B.6). When the vector $\mathbf{H}_1$ is aligned into the ψ axis, the ψ scan is performed by rotating only about the ψ axis: the detector circles (θ, ν) may be moved to observe the reflections $\mathbf{H}_1$ or $\mathbf{H}_2$.

2. Since dynamical three-beam interaction has a very small angular

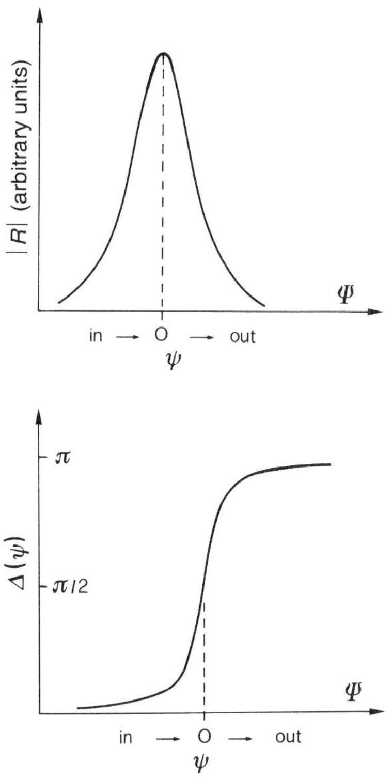

Fig. 3.B.4. Schematic drawings for the amplitude of the resonance term R and for its phase factor in function of ψ. ψ is assumed to be zero at the ideal three-beam position.

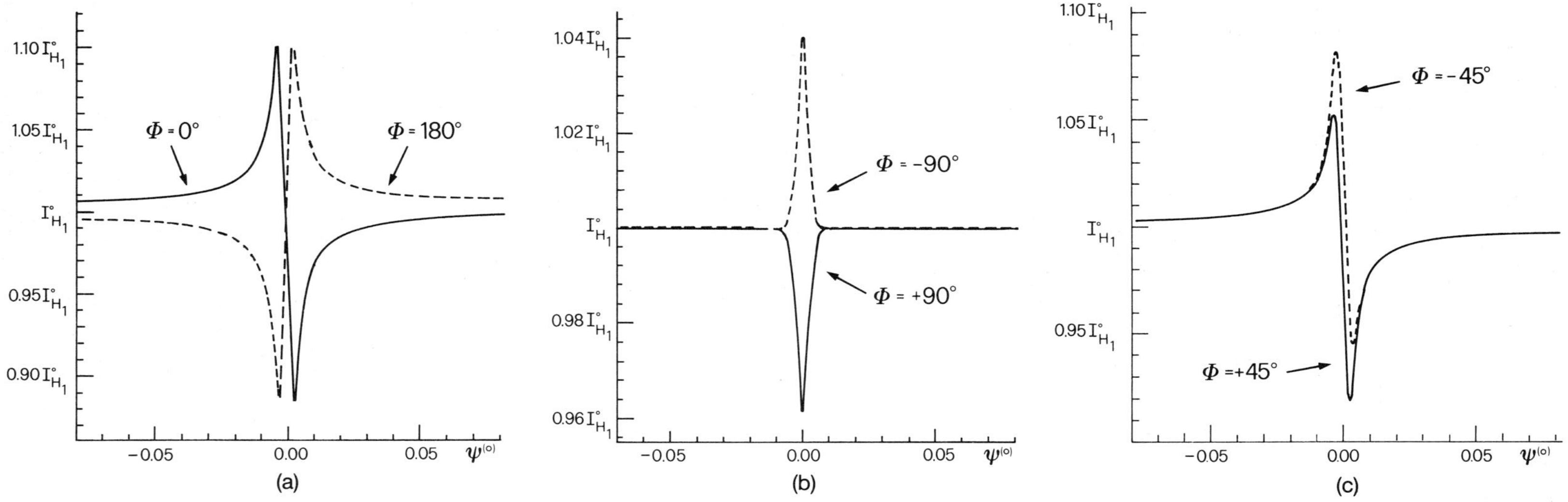

Fig. 3.B.5. Ideal ψ-scan profiles calculated with: (a) $\Phi = 0°$, 180°; (b) $\Phi = 90°$, $-90°$; (c) $\Phi = 45°$, $-45°$.

range, both the divergence and the spectral width of the primary beam should be small enough. Synchrotron radiation seems therefore the most suitable source for ψ-scan experiments. But also properly modified X-ray equipments based on rotating-anode generators can play an important role.

3. Measuring with the necessary statistical accuracy a single ψ-scan profile is a time-consuming process (say, half an hour when synchrotron radiation is used, some hours for less intense sources). In order to avoid the influence of long-range intensity variations, the ψ-scan profile is usually obtained as the sum of many fast scans.

4. Finding three beam points is not very easy for large structures (more than three reciprocal lattice points can simultaneously lie on the Ewald sphere). It is safe to calculate in advance the most suitable ψ angles for each reflection $\mathbf{H}_1$ and choose for measurements three-beam cases which are separated from all the others by an angular distance greater than 0.1°.

5. As a rule, more reliable phase estimates can be obtained if $F_{\mathbf{H}_1}$, $F_{\mathbf{H}_2}$, $F_{\mathbf{H}_1-\mathbf{H}_2}$ have comparable magnitudes. If polarized radiation is used, the polarization factor may be exploited to attain such a rule.

6. The absolute configuration of a non-centrosymmetric structure may be determined by accurately measuring one (or more) triplet phase, having a value near $\pm\pi/2$.

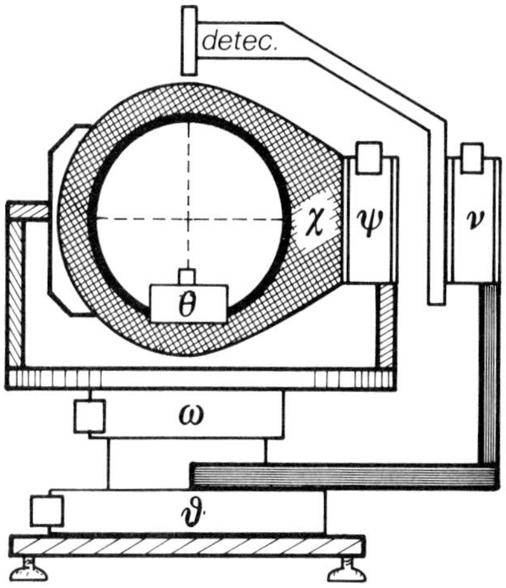

Fig. 3.B.6. Non-conventional six-circle diffractometer.

Electron diffraction

The diffraction of electrons (e-diffraction) was demonstrated by Davisson and Germer in 1927. The electron beam is produced in an electron gun by a 'hair-pin' filament with diameter of about 10 μm, or in a heated pointed filament of 1–2 μm size (see Chapter 9). Electromagnetic lenses restrict divergence to 10^{-3} or 10^{-4} rad, but also divergence of 10^{-6} rad may be achieved for special purposes. Electrons are accelerated through a potential difference of V volts. The following relation is valid:

$$\lambda = 12.3/(V + 10^{-6}\,V^2)^{1/2}.$$

Two energetical intervals are commonly used: we will talk about high-energy diffraction (HEED) when $V \simeq 50$–120 kV (with the advent of high-voltage electron microscopes this range needs to be extended to 1 MeV) and $\lambda \simeq 0.05$ Å, and low-energy diffraction (LEED) when $V \simeq 10$–300 V and $\lambda \simeq 4$–1 Å. Electrons are strongly absorbed by matter and therefore e-diffraction in transmission is applicable to very thin layers of matter (10^{-7}–10^{-5} cm). The scattering is caused by the interaction of the electrons with the electrostatic field $\varphi(\boldsymbol{r})$ of the atoms. $\varphi(\boldsymbol{r})$ is the sum of the field caused by the nucleus and the field caused by the electron cloud. Thus the interaction of electrons with matter may be divided into three processes: (a) no interaction—the electron passes straight through the specimen; (b) elastic scattering—the electrons are scattered by the Coulombic potential due to the nucleus. Since the proton mass is much larger than that of electron, no loss of energy occurs: such a scattering is the most important in electron microscopy. (c) Inelastic scattering—electrons of the primary beam interact with atomic electrons, and are scattered having suffered a loss of energy. In a microscope such electrons are focused at different positions and produce an effect called **chromatic aberration** (causing blurring of the image).

$\varphi(\boldsymbol{r})$ is related to the electron density by Poisson's equation

$$\nabla^2\varphi(\boldsymbol{r}) = -4\pi\{\rho_n(\boldsymbol{r}) - \rho_e(\boldsymbol{r})\}$$

where $\rho_n(\boldsymbol{r})$ is the charge density due to the atomic nucleus and ρ_e is the electron density function (as defined for X-ray scattering). While the nucleus with the charge Ze can be considered as a point, the electrons occupy a finite volume: therefore, as for X-rays (see p. 146), the e-scattering will have a geometric component which will take into account the distribution of the electrons around the nucleus. By supposing a spherical symmetry a simple relation may be obtained between the atomic scattering factor for X-rays, f_{ax}, and that for electrons, f_{ae}. Let us introduce into Poisson's equation the inverse Fourier transform

$$\nabla^2\left(\int f_{ae}(\boldsymbol{r}^*) \exp(-2\pi i\boldsymbol{r}^*\cdot\boldsymbol{r})\,d\boldsymbol{r}^*\right)$$

$$= 4\pi\int f_{ax}(\boldsymbol{r}^*) \exp(-2\pi i\boldsymbol{r}^*\cdot\boldsymbol{r})\,d\boldsymbol{r}^*$$

$$- 4\pi\int Z \exp(-2\pi i\boldsymbol{r}^*\cdot\boldsymbol{r})\,d\boldsymbol{r}^*.$$

Since the left-hand side is equal to

$$\int |-2\pi i\boldsymbol{r}^*|^2 f_{ae}(\boldsymbol{r}^*) \exp(-2\pi i\boldsymbol{r}^*\cdot\boldsymbol{r})\,d\boldsymbol{r}^*,$$

we obtain the Mott formula

$$f_{ae}(r^*) = \pi^{-1}(Z - f_{ax}(r^*))/r^{*2}.$$

For $\sin\theta/\lambda \Rightarrow 0$ the value of f_{ae} is indeterminate, but a sensible value may be obtained by the boundary condition

$$f_{ae}(0) = \int \varphi(\boldsymbol{r})\,d\boldsymbol{r}.$$

It is still preferred to tabulate (see *International tables for x-ray crystallography*, Vols. 3 and 4)

$$f^{B}_{ae}(r^*) = \frac{2\pi me}{h^2} f_{ae}(r^*) = \frac{0.0239[Z - f_{ax}(r^*)]}{\sin^2\theta/\lambda^2} \qquad (3.B.15)$$

where f_{ax} is in electrons and f^{B}_{ae} in Å (f^{B}_{ae} occurs in the first Born approximation for electron scattering by atoms).

If we compare numerically eqn (3.B.15) with the atomic scattering factor for the X-rays, we observe that:

1. e-scattering is much more efficient than X-scattering (see Fig. 3.B.7). Consequently diffraction effects are easily detected even from volumes much smaller than those required for X-rays (in practice, starting from thicknesses of 100 Å for simple structures constituted of heavy atoms).
2. The curves f_{ae} are less sensitive (see Table 3.B.3) to the atomic number Z than f_{ax} (on the average $f_{ae}(0) \propto Z^{1/3}$). Therefore the positions of the

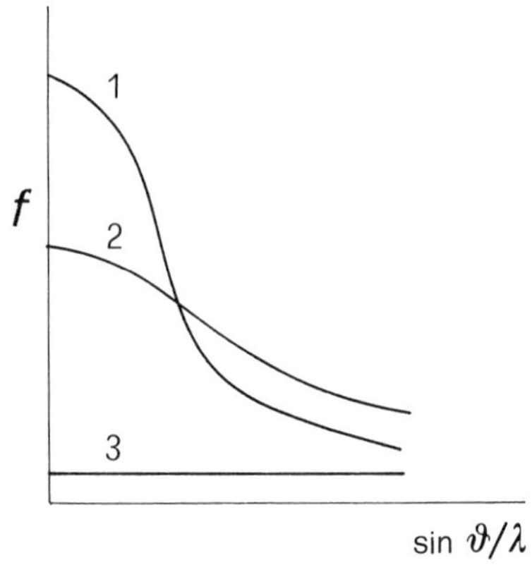

Fig. 3.B.7. Typical scattering curves for: (1) electrons; (2) X-rays; (3) neutrons.

Table 3.B.3. Some atomic scattering amplitudes for electrons at $\sin\theta/\lambda = 0$ based on the rest mass of the electron (for electrons of velocity v, multiply f_{ae} by $[1-(v/c)^2]^{-1/2}$) and nuclear scattering for neutrons for specific nuclei

Element	$f_{ae}^{B}(0)$ (10^{-8} cm)	Specific	b (10^{-12} cm)
H	0.529	^{1}H	−0.378
		^{2}H	0.65
Li	3.31	Li	−0.18
		^{6}Li	0.7
		^{7}Li	−0.25
Si	6.0		0.42
Cl	4.6		0.99
Ca	10.46		0.49
Fe	7.4	Fe	0.96
		^{54}Fe	0.42
		^{56}Fe	1.01
		^{57}Fe	0.23

light atoms in presence of heavy atoms (e.g. hydrogen atoms in an organic compound) may be more easily determined by e-diffraction.

3. As for X-rays, the atomic positions can be fixed by means of the three-dimensional Fourier synthesis whose maxima correspond to the positions of the atomic nuclei.

4. The electrons can exist in two spin states and therefore can be polarized. Usually internal fields of crystal structures are not strong enough to do that.

In HEED techniques diffraction angles are very small. There are a lot of reflections having 2θ between 0° and 2°. Under these conditions $\sin\theta$ in the Bragg law can be approximated by θ. Besides, the radius of the Ewald sphere is very large and its surface is in practice reduced to a plane: thus, all the nodes of a reciprocal plane can simultaneously be in Bragg position. Tens or hundreds diffracted beams can be simultaneously observed on a fluorescent screen, or collected on a photographic plate for measuring the intensities, or transformed into images by means of adequate electronic optics.

With $\lambda = 0.05$ Å, one might think it would be easy to obtain resolved images of the atoms (in some cases they *have* been effectively obtained). The aberrations of magnetic lenses, particularly the **spherical aberration** are the principal obstacle: electrons widely scattered are focused at positions different from those to which are focused electrons travelling close to the lens axis. The result is that a point object is spread over a length Δr in the image plane. The theoretical resolution of an electron microscope is approximately 2 Å at 100 kV and less than 2 Å at higher potentials. Because of these characteristics, HEED microscopy is particularly favourable for studying superstructures, order–disorder phenomena, grain boundaries, and generally the real structure of crystals.

LEED techniques are commonly used for studying crystal surfaces (electrons can penetrate into a crystal only for a very small thickness), in order to study processes of electronic and ionic emission, catalysis, nucleation of new phases, oxidation, etc. The function $\varphi(\boldsymbol{r})$ terminates at the crystal surface: thus in principle the structure of outer layers can be different from that of internal layers. Interpretation of the spectra is not easy because of multiple scattering of the electrons. Additional information is provided by means of Auger spectroscopy of scattered electrons.

Neutron scattering

A neutron is a heavy particle with spin $\frac{1}{2}$ and magnetic moment of 1.9132 nuclear magnetons. Its wave properties were shown in 1936 by Halban and Preiswerk and by Mitchell and Powers. Neutron diffraction experiments (n-diffraction) require high fluxes provided nowadays by modern reactors. They produce fast neutrons whose energy is reduced by collisions in a moderator of heavy water or graphite. So retarded neutrons are called **thermal** and their speed obeys a Maxwell distribution: the corresponding spectrum is white (see Fig. 3.B.8(a)). A monochromator (usually single crystals of Ge, Cu, Zn, Pb) selects the desired wavelength λ.

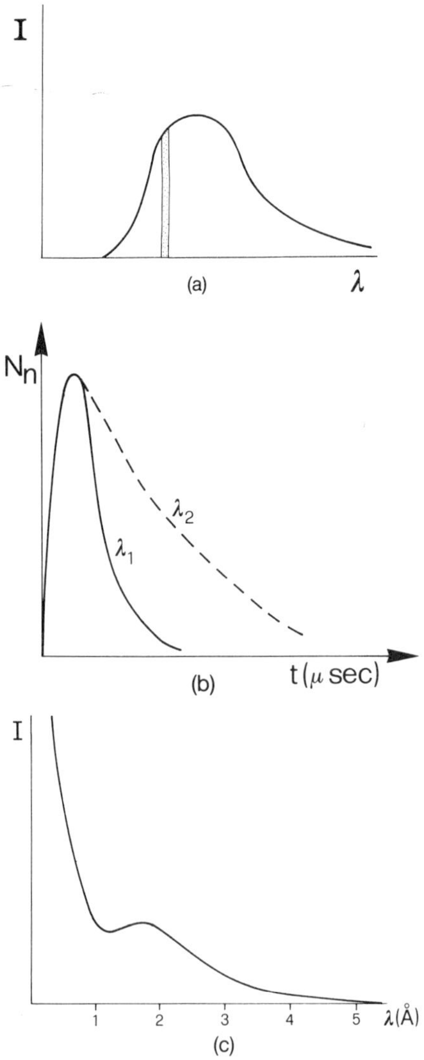

Fig. 3.B.8. (a) Spectrum from a nuclear reactor. The shaded wavelengths are selected by a monochromator. (b) A schematic time-dependence of two pulses of neutrons leaving the moderator with different energies ($\lambda_2 > \lambda_1$). A next pulse starts when the number of neutrons (N_n) is small enough. (c) Neutron flux distribution on a pulsed neutron machine.

Neutrons can also be produced in a pulsed manner by spallation, at a repetition rate between 24 Hz and 50 Hz. High-energy protons ($\simeq$1 GeV), in short pulses at the appropriate pulse frequency, strike a target such as uranium or tungsten releasing several tens of neutrons per proton ($\simeq$25 neutrons for ^{238}U). The pulsed neutron flux is only present for a very short time (the burst lasts around 0.4 μs): heat-removal is then easy and high fluxes are allowed (higher than those produced by reactors).

High-energy neutrons are slowed down to thermal energies by appropriate moderators. Water, polyethylene, liquid hydrogen, or liquid methane are frequently used: the choice is determined by the scattering experiment. During the thermalization process neutrons undergo a large number of collisions which cause pulse broadening. The width of the pulse leaving the moderator is roughly proportional to the wavelength (see Fig. 3.B.8(b)) so that the fractional wavelength resolution is nearly constant. A characteristic neutron spectrum is shown in Fig. 3.B.8(c): high intensities at short wavelength ($\lambda < 1$ Å) is a very significant characteristic of pulsed neutron sources.

The neutron–atom interaction comprises interaction with the nucleus and interaction of the magnetic momentum associated with the spin of the neutron with the magnetic momentum of the atom. This effect mainly occurs for atoms with incompletely occupied outer electron shells (for instance, transition elements).

The neutron–nucleus interaction is governed by very short range nuclear forces ($\sim 10^{-13}$ cm). Since the nuclear radius is of the order 10^{-15} cm, i.e. of several orders of magnitude less than the wavelength associated with the incident neutrons, the nucleus will behave like a point and its scattering factor b_0 will be isotropic and not dependent on $\sin\theta/\lambda$ (see Fig. 3.B.7). By convention, the scattering amplitude is assumed positive where there is a phase change of 180° between incident and scattered waves: it has the dimension of length and is measured in units of 10^{-12} cm. When the neutron is very close to the nucleus a metastable system, nucleus + neutron, is created which decays by re-emitting the neutron. For appropriate energy a resonance effect can occur: then the scattering factor assumes the form $b = b_0 - \Delta b'$. Since $\Delta b'$ can be greater than b, it is possible to have negative scattering factors for some nuclei (for instance, ^{1}H, ^{48}Ti, ^{62}Ni, ^{55}Mn): for them the scattering is out of phase by 180° with respect to the nuclei with $b > 0$. A few nuclei also present an imaginary wavelength-dependent term $\Delta b''$ (see pp. 165–9 for a more extensive treatment of anomalous scattering).

The angular momentum of the nucleus is another factor which influences the capacity of scattering. If I is the angular momentum, then it can be combined with the neutron spin in a parallel or in an antiparallel fashion yielding the total spin of either $J = I + \frac{1}{2}$ or $J = I - \frac{1}{2}$, and corresponding scattering factors b_+ and b_-. According to quantum mechanics, there are $2J + 1$ orientations in space compatible with one spin of value J. The compound nucleus will then have $[2(I + \frac{1}{2}) + 1] + [2(I - \frac{1}{2}) + 1] = 2(2I + 1)$ possible states. One fraction of these, $w_+ = [2(I + \frac{1}{2}) + 1]/[2(2I + 1)] = (I + 1)/(2I + 1)$ corresponds to states of parallel spins while the other one, $w_- = I/(2I + 1)$, corresponds to states of antiparallel spins. In conclusion, a single isotope with $I > 0$ can be described as a random mixture of nuclei with atomic fractions w_+ and w_- and scattering amplitudes b_+ and b_-. Then $b = w_+ b_+ + w_- b_-$ will be the coherent scattered amplitude. Incoherent scattering will also occur, with square amplitude given by $[(w_+ b_+^2 + w_- b_-^2) - (w_+ b_+ + w_- b_-)^2]$ which contributes to a uniform background with no diffraction effects. This is particularly important for hydrogen, for which $I = \frac{1}{2}$, $b_+ \sim 1.04 \times 10^{-12}$ cm, $b_- \sim -4.7 \times 10^{-12}$ cm. Then $w_+ \simeq 0.75$, $w_- \simeq 0.25$, $b \sim -0.39$, and most of the scattering is incoherent background.

In general, one chemical element exists in a crystal structure in different isotopic forms which are randomly distributed over all sites occupied by the element. The scattering factor of the element will be a weighted average over various states of spin of every isotope and over various isotopes. This disorder generates a pervasive coherent and incoherent background in the diffraction spectrum. If w_n is the relative abundance of the nth isotope then the effective scattering factor will be $b = \sum w_n b_n$ and the total diffuse background scattered intensity will be $\sum_n w_n b_n^2 - b^2$.

Atoms which possess a magnetic momentum because of the presence of unpaired electrons interact with the magnetic momentum of the neutron giving additional neutron scattering. Interaction occurs in a finite domain around the nucleus: therefore, as for X-rays and electrons, magnetic scattering will decline with $\sin\theta/\lambda$.

Let us consider the case of identical atoms with all spins parallel (ferromagnetic materials) and spins alternating parallel and antiparallel (antiferromagnetic materials). The magnetic scattering amplitude of each atom is given by

$$p = (e^2 \gamma S f_{\mathrm{mag}}/mc^2) \sim 0.54 S f_{\mathrm{mag}}\ (10^{-12}\ \mathrm{cm})$$

where γ is the neutron magnetic moment in nuclear magnetons, e and m are the charge and the mass of the electron, c the velocity of light, S the electron spin quantum of the scattering atom, f_{mag} the atomic scattering form factor given by the Fourier transform of the distribution of electrons having unpaired spin and normalized so as to be $f_{\mathrm{mag}} = 1$ for $\theta = 0$. p is of the same order as nuclear scattering.

Nuclear and magnetic scattering are additive for unpolarized neutron beams: then

$$|F|^2 = |F|^2_{\mathrm{nucl}} + \sin^2 \alpha\, |F_{\mathrm{mag}}|^2$$

where α is the angle between the unit vector in the direction of the spin orientation in the sample and the vector $\boldsymbol{r}^*$. For a sample containing a single magnetic domain $\sin^2 \alpha$ is equal for all atoms. For the multi-domain samples it will be necessary to average over the various spin orientations. For ferromagnetic and ferrimagnetic samples external fields for orienting the spins can be applied so as to obtain $\sin^2 \alpha = 0$ or $\sin^2 \alpha = 1$. Measurements of $|F|^2$ for both cases separate magnetic from nuclear scattering and allow the study of magnetic structures. Sometimes the magnetic cell coincides with the 'chemical' cell so that the magnetic contribution to the intensities is added to nuclear contribution. In other cases the magnetic structure has a cell which is a multiple of the chemical cell causing additional purely magnetic reflections. Magnetic symmetry can be described by means of space groups of antisymmetry or by means of colour groups (see Appendix 1.F and Fig. 1.F.2(b)) associated with classical three-dimensional space groups.

Monochromatic beams of polarized neutrons are easily obtained from unpolarized beams by using suitable single-crystal monochromators (for example, a Co–Fe alloy). For a given polarization, either constructive or destructive interference between nuclear and magnetic scattering amplitudes can occur. Thus the technique may detect a weak magnetic scattering even when it is accompanied by a strong nuclear scattering.

Among the most important characteristics and applications of neutron diffraction we quote:

1. The interaction of neutrons with the matter is weaker than that of X-rays or electrons (see Table 3.B.3). Generally speaking, scattering amplitudes are $f_{ax} \simeq (10^{-12}–10^{-11})$ cm for X-rays, $f_{ae} \sim 10^{-8}$ cm for electrons, $f_n \sim 10^{-12}$ cm for neutrons. Therefore high neutron fluxes and crystals with dimensions of several millimetres are needed for measuring appreciable scattered intensities.

2. The b values vary non-monotonically with the atomic number Z: isotopes of the same element can have very different values of b. This allows us to distinguish between atoms having very close values of Z (but very different values of b: e.g. $b_{\mathrm{Mn}} = -0.36$, $b_{\mathrm{Fe}} = 0.96$, $b_{\mathrm{Co}} = 0.25$) and to localize the positions of light atoms in the presence of heavy atoms. Neutrons are particularly useful for localizing hydrogen atoms. Usually they are partially or completely substituted by deuterium with a value $b > 0$ and with negligible incoherent scattering.

3. The three-dimensional Fourier synthesis with coefficients F_{nucl} gives the positions of the nuclei (they do not necessarily coincide with the atomic positions found by X-ray diffraction). The peak heights are proportional to the b values of the corresponding nuclei: if b is negative, the peak is negative too. Fourier syntheses with coefficients F_{mag} give the spin density distribution of the magnetic atoms.

4. Since b does not depend on $\sin \theta/\lambda$ nuclear scattering decreases only because of the temperature effect. Thus reflections with high values of $\sin \theta/\lambda$ can be collected giving atomic positions and thermal parameters with accuracy higher than from X-rays. In many cases X- and n-diffraction experiments are both performed in such a way that accurate maps of the electron density can be obtained.

5. The energy of a neutron with wavelength of 1 Å is of the order of 0.1 eV, which is comparable with the energy of the modes of thermal vibration of the crystal. This causes inelastic scattering but also allows for the possibility of studying energy changes with good accuracy.

6. Some additional information about neutron diffraction experiments with pulsed neutron sources should be useful. In diffraction techniques using monochromatic radiation, intensities are measured by moving a detector to different scattering angles $\theta_{\mathbf{H}}$ chosen according to Bragg's law $\lambda_0 = 2d_{\mathbf{H}} \sin \theta_{\mathbf{H}}$. At spallation neutron sources, where the neutron beam is pulsed, the discrimination among neutrons with different wavelengths may be accomplished by their time of arrival at the detector (time of flight): thus diffraction effects can be measured at a fixed scattering angle θ_0:

$$\lambda_{\mathbf{H}} = 2d_{\mathbf{H}} \sin \theta_0.$$

From de Broglie's relation $mv = h/\lambda$ (m and v are mass and velocity of the neutron respectively) the following relation arises:

$$mv = m(L/t) = h/\lambda,$$

where $L = L_1 + L_2$. Here L_1 is the flight path from moderator to sample and L_2 is that from sample to detector. Then

$$t_{\mathbf{H}} = \frac{m}{h} L\lambda_{\mathbf{H}} = 252.778\, L\lambda$$

or

$$t_{\mathbf{H}} = 505.555 L d_{\mathbf{H}} \sin \theta_0$$

where L is measured in meters, λ and $d_{\mathbf{H}}$ in Å, t in μs. Thus time of flight depends linearly on both the flight path and on the wavelength. In addition the equations suggest that resolution improves with increasing flight path.

Area detectors (see Chapter 4) are the best choice for time-of-flight techniques. Indeed, even if the source is on for brief pulses the full spectrum may be used (in monochromatic techniques coupled with a reactor the source is on all the time but a small portion of the spectrum is used).

3.C Scattering of X-ray by gases, liquids, and amorphous solids

Introduction

In this appendix we will describe the scattering of X-rays from gases, liquids, and amorphous solids. No attempt will be made to deal exhaustively with the great amount of work made in the field: only the basic principles of the phenomena will be described. For further information the reader is referred to specialist papers on the subject or to some enlightening books.[38–41]

First Debye[42] (but see also Ehrenfest[43]) realized that no arrangement of molecules of finite size could be considered completely random. The most random array which can be imagined is that of a monoatomic gas: but even in this case the finite size of the atoms prevents their approaching one another within a distance smaller than a given threshold. According to

Debye, two types of effects could be found on the diffraction patterns:

(1) the effects due to the more or less fixed spatial arrangements of the atoms in the molecules (even if molecules have random positions and orientations), called 'internal interference effects';

(2) the effects due to the mutual configuration of the molecules, which may be not arbitrary since it is conditioned by their dimensions and their packing, called 'external interference effects'.

The relative importance of the two types of effects is a matter of degree. Indeed, for monoatomic gases external interference effects are dominant, while for polyatomic gases internal interference effects are not negligible. Again in liquids (or amorphous solids) the existence of average interatomic distances produces well marked external interference effects: if the liquid is composed of groups of atoms the internal interference effects are again not neglibible.

In the following sections some formulae will be developed which refer to the coherent scattered radiation. However, in experiments the total, say both coherent and incoherent radiation, is usually collected; in the case of amorphous bodies the incoherent radiation may be an important fraction of the total radiation. According to the discussion on p. 147, the incoherent intensity relative to one atom is given by

$$\sum_k (1 - |f_k|^2) \qquad (3.C.1)$$

where k varies over the individual electrons of the atom. If we sum (3.C.1) over all the atoms in the assemblage the total incoherent scattering is obtained, which is a smooth function slowly increasing with $\sin\theta/\lambda$. Such a function has to be subtracted from the total scattering pattern: only after that can the coherently scattered intensity be used to deduce structural parameters.

Diffraction from a finite statistically homogeneous object

Usually the motion of atoms or molecules in gases and liquids is extremely rapid compared with the duration of a diffraction experiment. In these conditions the phase value of $F(\boldsymbol{r}^*)$ has no practical interest while its modulus can provide useful information on the statistical distribution of the atoms. Indeed the measured diffraction intensity is the weighted sum of contributions arising from the various configurations of the atoms. A similar conclusion holds for amorphous materials, for which trying to calculate the atomic positions has no practical meaning while it is possible to estimate interatomic distances (on which diffraction intensities depend).

Suppose that the object we want to investigate is finite and statistically homogeneous, of volume Ω. In accordance with eqn (3.27)

$$F(\boldsymbol{r}^*) = F_\infty(\boldsymbol{r}^*) * D(\boldsymbol{r}^*)$$

and the Patterson function becomes

$$P(\boldsymbol{u}) = \int_\Omega \rho_\infty(\boldsymbol{r})\rho_\infty(\boldsymbol{r}+\boldsymbol{u})\Phi(\boldsymbol{r})\Phi(\boldsymbol{r}+\boldsymbol{u})\,\mathrm{d}\boldsymbol{r}. \qquad (3.C.2)$$

The product $\Phi(\boldsymbol{r})\ \Phi(\boldsymbol{r}+\boldsymbol{u})$ is always vanishing except when both the points

$\boldsymbol{r}$ and $\boldsymbol{r}+\boldsymbol{u}$ are inside the object: in this case $\Phi(\boldsymbol{r})\Phi(\boldsymbol{r}+\boldsymbol{u})=1$. The space domain in which this condition is verified is a function of $\boldsymbol{u}$ and will be denoted by $\Omega v(\boldsymbol{u})$, where $0 \leqslant v(\boldsymbol{u}) \leqslant 1$. It is readily seen from Fig. 3.C.1, that $\Omega v(\boldsymbol{u})$ is the volume which belongs to the object and to the object shifted by $\boldsymbol{u}$. Furthermore, $v(\boldsymbol{u})$ is centrosymmetric ($v(\boldsymbol{u})=v(-\boldsymbol{u})$), it decreases with increasing $\boldsymbol{u}$ and takes its maximum for $\boldsymbol{u}=0$ ($v(0)=1$). Accordingly

$$P(\boldsymbol{u})=\int_{\Omega v(\boldsymbol{u})} \rho_\infty(\boldsymbol{r})\rho_\infty(\boldsymbol{r}+\boldsymbol{u})\,\mathrm{d}\boldsymbol{r} \tag{3.C.3}$$

where

$$\Omega v(\boldsymbol{u})=\int_V \Phi(\boldsymbol{r})\Phi(\boldsymbol{r}+\boldsymbol{u})\,\mathrm{d}\boldsymbol{r}=\Phi(\boldsymbol{r})*\Phi(-\boldsymbol{r}). \tag{3.C.4}$$

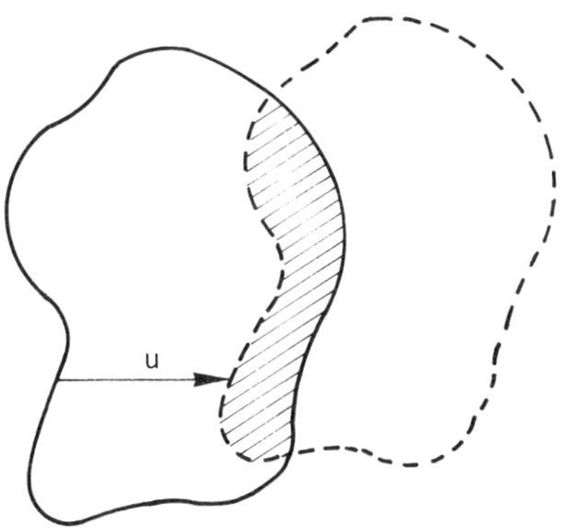

Fig. 3.C.1. The dashed area has volume $\Omega v(\mathbf{u})$.

For homogeneous objects (3.C.3) may be written as

$$P(\boldsymbol{u})=\Omega v(\boldsymbol{u})P_1(\boldsymbol{u}) \tag{3.C.5}$$

where $P_1(\boldsymbol{u})$ is the Patterson function for the unit volume.

According to (3.A.37)

$$|F(\boldsymbol{r}^*)|^2=\mathrm{T}[\Omega v(\boldsymbol{u})P_1(\boldsymbol{u})]=\mathrm{T}[\Omega v(\boldsymbol{u})]*\mathrm{T}[P_1(\boldsymbol{u})].$$

The Fourier transform of $\Omega v(\boldsymbol{u})$, according to eqns (3.C.4), (3.A.37), and (3.A.21) is

$$\mathrm{T}[\Omega v(\boldsymbol{u})]=\int \Omega v(\boldsymbol{u})\exp(2\pi \mathrm{i}\boldsymbol{r}^*\cdot\boldsymbol{u})\,\mathrm{d}\boldsymbol{u}=|D(\boldsymbol{r}^*)|^2 \tag{3.C.6}$$

so that

$$|F(\boldsymbol{r}^*)|^2=\mathrm{T}[P_1(\boldsymbol{u})]*|D(\boldsymbol{r}^*)|^2. \tag{3.C.7}$$

Since $\Phi(\boldsymbol{r})$ is real, $|D(\boldsymbol{r}^*)|^2$ is centrosymmetric: its maximum value is $|D(0)|^2$. According to (3.27) $D(0)=\Omega$ so that

$$|D(0)|^2=\Omega^2. \tag{3.C.8}$$

Furthermore, according to (3.C.6)

$$\int_{S^*}|D(\boldsymbol{r}^*)|^2\,\mathrm{d}\boldsymbol{r}^*=\int_S \Omega v(\boldsymbol{u})\,\mathrm{d}\boldsymbol{u}\int_{S^*}\exp(2\pi\mathrm{i}\boldsymbol{r}^*\cdot\boldsymbol{u})\,\mathrm{d}\boldsymbol{r}^*$$
$$=\int_S \Omega v(\boldsymbol{u})\,\delta(\boldsymbol{u})\,\mathrm{d}\boldsymbol{u}=\Omega v(0)=\Omega. \tag{3.C.9}$$

From (3.C.8) and (3.C.9) it is easily deduced that $|D(\boldsymbol{r}^*)|^2$ rapidly decreases with $|\boldsymbol{r}^*|$, and is appreciably different from zero in a domain near the origin of the reciprocal space (when the diffracting object is a crystal, $|D(\boldsymbol{r}^*)|^2$ repeats itself around each lattice point: the larger is Ω, the smaller the domain).

Replace in (3.C.9) the decreasing function $|D(\boldsymbol{r}^*)|^2$ by a constant which is equal to its maximum value Ω^2 in a volume w defined by $\Omega^2 w=\Omega$ (see Fig. 3.C.2 for a one-dimensional example). Then an approximate estimate of the domain volume is obtained, say $w=1/\Omega$, together with its approximate dimension $r_0^* \simeq (1/\Omega)^{1/3} \simeq 1/r_0$, where r_0 is the average dimension of the diffracting object.

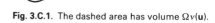

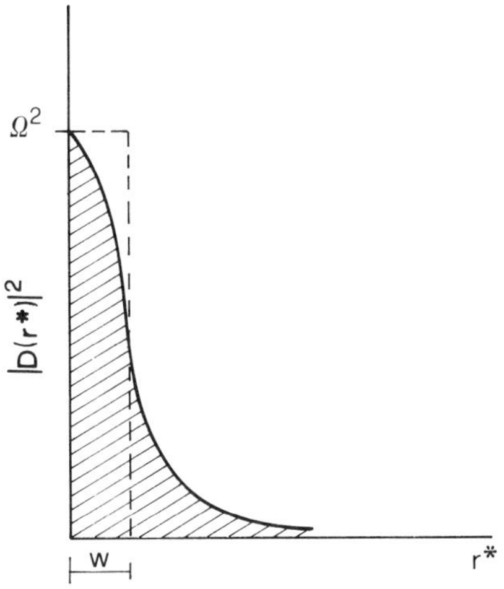

Fig. 3.C.2. The typical form of $|D(r^*)|^2$ as a function of r^*.

In terms of diffraction angles the width of the diffraction peak near the origin of the reciprocal space may be obtained from the Bragg relation $(2/r_0^*)\sin\theta_0 \simeq (2/r_0^*)\theta_0 \simeq 2r_0\theta_0 = \lambda$ from which

$$\theta_0 = \lambda/2r_0. \qquad (3.C.10)$$

For Cu K_α radiation and an object with $r_0 \simeq 0.1\ \mu$m it is $\theta_0 \simeq 4.4 \times 10^{-3}$. On assuming that diffraction around the central peak may be observed down to $\theta \simeq 10^{-3}$, 0.1 μm constitutes the upper limit for the dimension of the object.

Diffraction from a finite statistically homogeneous object with equal atoms

Suppose that the object is composed of N atoms of a single type, and N_1 is the number of atoms contained in the unit volume (of course fluctuations from this value are possible due to thermal motion or to structural features of amorphous solids). Then $v_1 = 1/N_1$ is the average volume available for each atom. Equations (3.C.5) and (3.C.7) may be written in terms of autocorrelation functions of **atomic density** rather than of electron density. Let us define

$$\rho(\boldsymbol{r}) = \rho_a(\boldsymbol{r}) * q_a(\boldsymbol{r})$$

where $\rho_a(\boldsymbol{r})$ is the electron density relative to one atom and $q_a(\boldsymbol{r})$ is the probability of finding an atom in $\boldsymbol{r}$. Then

$$P(\boldsymbol{u}) = \rho_a(\boldsymbol{r}) * \rho_a(-\boldsymbol{r}) * q_a(\boldsymbol{r}) * q_a(-\boldsymbol{r})$$

and

$$|F(\boldsymbol{r}^*)|^2 = \mathrm{T}[P(\boldsymbol{u})] = f^2(\boldsymbol{r}^*) \cdot \mathrm{T}[P_a(\boldsymbol{u})] \qquad (3.C.11)$$

where $P_a(\boldsymbol{u})$ is just the autocorrelation function of the atomic density.

Let $P_{a1}(\boldsymbol{u})$ be the function $P_a(\boldsymbol{u})$ calculated for the unit volume. Then, according to (3.C.5)

$$P_a(\boldsymbol{u}) = \Omega v(\boldsymbol{u}) P_{a1}(\boldsymbol{u})$$

and (3.C.11) becomes

$$|F(\boldsymbol{r}^*)|^2 = f^2(\boldsymbol{r}^*) \cdot \mathrm{T}[P_{a1}(\boldsymbol{u})] * |D(\boldsymbol{r}^*)|^2. \qquad (3.C.12)$$

Let us now express $P_{a1}(\boldsymbol{u})$ in a more convenient form. On assuming the origin on an atomic site, the probability of finding a second atom in the volume $\mathrm{d}v$ at the extremity of the vector $\boldsymbol{u}$ is $\mathrm{d}p(\boldsymbol{u}) = \mathrm{d}v/v_1$ only for a completely random atomic distribution. In general we could write

$$\mathrm{d}p(\boldsymbol{u}) = \frac{p(\boldsymbol{u})}{v_1}\,\mathrm{d}v \qquad (3.C.13)$$

where $p(\boldsymbol{u})$ is a distribution function defining the configuration of the atoms. In general $p(\boldsymbol{u})$ is not easily found. A very simple model for which $p(\boldsymbol{u})$ may be rigorously calculated was proposed by Zernicke and Prins:[44] it concerns the one-dimensional arrangement on a segment of length L of N rigid spherical particles of diameter d. Figure 3.C.3 shows $p(\boldsymbol{u})$ as a function of four values of the particle concentration d/l, where $l = L/N$ is the average length allotted to each particle. It is easily seen that always $p(\boldsymbol{u}) = 0$ when $u < d$ (the spheres are impenetrable). At very small concentrations (quasi-zero volume particles which do not exert any influence on each other) $p(\boldsymbol{u})$

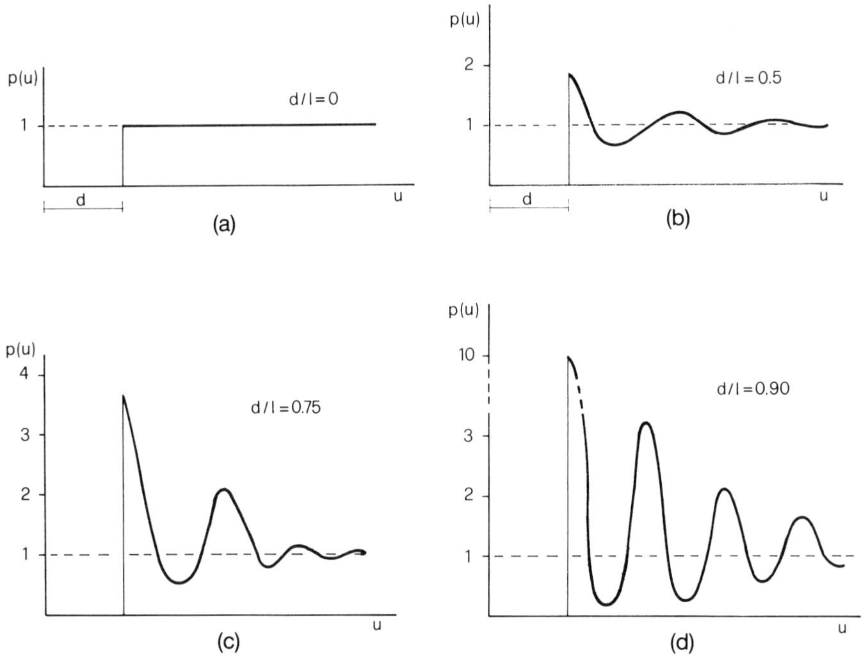

Fig. 3.C.3. The distribution $p(u)$ for one-dimensional arrangements of objects of length d as a function of the compactness.

is similar to the atomic distribution of the perfect gas. For higher concentrations the model represents distributions similar to those found in real gases, liquids, and amorphous solids. $p(\boldsymbol{u})$ will oscillate about unity at short distances from the origin, while $p(\boldsymbol{u}) \simeq 1$ at long distances. Oscillations of $p(\boldsymbol{u})$ are larger for higher concentrations. When d/l becomes maximum the close packing of the particles gives rise to a perfect lattice of period d: then very sharp maxima of $p(u)$ will occur at $u = nd$ (full short-range and long-range order).

Let us now denote by $z(\boldsymbol{u})$ the probability of finding an atom in the element of volume $\mathrm{d}v$ located at the extremity of the vector $\boldsymbol{u}$ from an atom located at the origin. Then

$$z(\boldsymbol{u}) = \delta(\boldsymbol{u}) + p(\boldsymbol{u})/v_1. \qquad (3.C.14)$$

It may be noted that $1/v_1$ is not only the asymptotic value of $z(\boldsymbol{u})$ but also its mean value. Indeed, integrating $z(\boldsymbol{u})$ on the unit volume gives

$$\int z(\boldsymbol{u})\,\mathrm{d}v = 1 + \int \frac{p(\boldsymbol{u})}{v_1}\,\mathrm{d}v = 1 + (N_1 - 1) = N_1 = 1/v_1. \qquad (3.C.15)$$

Since

$$P_{a1}(\boldsymbol{u}) = N_1 z(\boldsymbol{u})$$

from (3.C.12) the relation

$$|F(\boldsymbol{r}^*)|^2 = N_1 f^2(\boldsymbol{r}^*)\mathrm{T}[z(\boldsymbol{u})] * |D(\boldsymbol{r}^*)|^2 \qquad (3.C.16)$$

follows. Write now (3.C.14) by emphasizing the oscillating part of $p(\boldsymbol{u})$:

$$z(\boldsymbol{u}) = \frac{1}{v_1} + \delta(\boldsymbol{u}) + \frac{1}{v_1}(p(\boldsymbol{u}) - 1). \qquad (3.C.17)$$

Then the Fourier transform of $z(\boldsymbol{u})$ is

$$Z(\boldsymbol{r}^*) = \frac{1}{v_1}\,\delta(\boldsymbol{r}^*) + 1 + \frac{1}{v_1}\,\mathrm{T}[p(\boldsymbol{u}) - 1]$$

and (3.C.16) becomes $(N_1 = N/\Omega)$

$$|F(\boldsymbol{r}^*)|^2 = \frac{N}{\Omega} f^2(\boldsymbol{r}^*)\,\frac{\delta(\boldsymbol{r}^*)}{v_1} * |D(\boldsymbol{r}^*)|^2 + \frac{N}{\Omega} f^2(\boldsymbol{r}^*)\Big(1 + \frac{1}{v_1}\,\mathrm{T}[p(\boldsymbol{u}) - 1]\Big) * |D(\boldsymbol{r}^*)|^2.$$

The function $|D(\boldsymbol{r}^*)|^2$ can be considered very broad with respect to the Dirac delta function but very sharp with respect to $\mathrm{T}[p(\boldsymbol{u}) - 1]$ (which in disorder structures is a function slowly varying with $\boldsymbol{r}^*$). According to (3.C.9) the integral of the $|D(\boldsymbol{r}^*)|^2$ peak is Ω: therefore

$$|F(\boldsymbol{r})|^2 = \frac{N}{\Omega v_1} f^2(\boldsymbol{r}^*)\,|D(\boldsymbol{r}^*)|^2 + \frac{N}{\Omega} f^2(\boldsymbol{r}^*)\cdot\Big(\Omega + \frac{\Omega}{v_1}\int [p(\boldsymbol{u}) - 1]\exp(2\pi \mathrm{i}\boldsymbol{u}\cdot\boldsymbol{r}^*)\,\mathrm{d}\boldsymbol{u}\Big). \quad (3.C.18)$$

In terms of the normalized intensity $|E(\boldsymbol{r}^*)|^2$ eqn (3.C.18) becomes

$$|E(\boldsymbol{r}^*)|^2 = \frac{|F(\boldsymbol{r}^*)|^2}{Nf^2(\boldsymbol{r}^*)} = \frac{1}{\Omega v_1}|D(\boldsymbol{r}^*)|^2 + \Big(1 + \frac{1}{v_1}\int [p(\boldsymbol{u}) - 1]\exp(2\pi \mathrm{i}\boldsymbol{u}\cdot\boldsymbol{r}^*)\,\mathrm{d}\boldsymbol{u}\Big). \quad (3.C.19)$$

The first term in (3.C.19) corresponds to the peak at the origin of the reciprocal space. It is detectable only at very small angles and it is distinguishable from the primary beam (see the previous section) provided the diffracting object is very small (say, <1 μm). It does not depend on the internal structure of the object but only on its external shape. We will discuss such a peak on p. 213.

The second term depends exclusively on the statistical distribution of the atoms: if this is perfectly uniform $p(\boldsymbol{u}) = 1$ and $|E(\boldsymbol{r}^*)|^2 = 1$. Thus the variations of $|E(\boldsymbol{r}^*)|^2$ about its average contain information about the atomic distribution in the object. Such a distribution may be determined by inversion of (3.C.19):

$$p(\boldsymbol{u}) = 1 + v_1\int (|E(\boldsymbol{r}^*)|^2 - 1)\exp(2\pi \mathrm{i}\boldsymbol{r}^*\cdot\boldsymbol{u})\,\mathrm{d}\boldsymbol{r}^* \quad (3.C.20)$$

which is the most general formula describing the diffraction from a statistically homogeneous object composed of atoms of one kind.

Diffraction from an isotropic statistically homogeneous object

If the diffracting object is statistically homogeneous and isotropic (that occurs, for example, in gases, most liquids, and in finely dispersed crystal powders) the function $|E(\boldsymbol{r}^*)|^2$ and $p(\boldsymbol{u})$ will also be isotropic. Then using results in Appendix 3.A, p. 181, and neglecting the first term on the right-hand side of eqn (3.C.19) give

$$|E(r^*)|^2 = 1 + \frac{1}{v_1}\int_0^\infty 4\pi u^2(p(u)-1)\frac{\sin 2\pi r^* u}{2\pi r^* u}\,\mathrm{d}u$$
$$= 1 + \frac{2}{r^* v_1}\int_0^\infty u(p(u)-1)(\sin 2\pi r^* u)\,\mathrm{d}u \qquad (3.C.21)$$

where $4\pi u^2(p(u)-1)$ is the **radial atomic distribution function**: the product of such a distribution for du gives the number of atoms situated at a distance between u and $u+\mathrm{d}u$ from the atom assumed to be at the origin.

Inverting (3.C.21) gives

$$p(u) = 1 + \frac{2v_1}{u}\int_0^\infty r^*(|E(r^*)|^2 - 1)(\sin 2\pi r^* u)\,\mathrm{d}r^*. \qquad (3.C.22)$$

Equation (3.C.22) shows that from diffraction experiments the average number of neighbouring atoms can be deduced as a function of the distance from a given atom chosen to be at the origin.

Equations (3.C.21) and (3.C.22) also hold when the object is made by particles constituted by groups of atoms: then v_1 is the average volume available for each particle.

The Debye formula

A simple way to describe diffraction from an isotropic statistically homogeneous object composed by a collection of N identical groups of atoms with known geometry but random orientation and position was proposed by Debye. Let

$$F_n(\mathbf{r}^*) = \sum_{j=1}^{p} f_j(\mathbf{r}^*)\exp(2\pi i \mathbf{r}^* \cdot \mathbf{r}_j)$$

be the structure factor of the nth group of atoms (each group composed of p atoms). Then

$$F(\mathbf{r}^*) = \sum_{n=1}^{N} F_n(\mathbf{r}^*)$$

and

$$|F(\mathbf{r}^*)|^2 = \sum_{n,m=1}^{N} F_n(\mathbf{r}^*)F_m(-\mathbf{r}^*).$$

Since the observed intensity will be the average with respect to all the possible mutual configurations of the atomic groups

$$|F(\mathbf{r}^*)|^2 = \sum_{n,m=1}^{N}\langle F_n(\mathbf{r}^*)F_m(-\mathbf{r}^*)\rangle = \sum_{n=1}^{N}\langle |F_n(\mathbf{r}^*)|^2\rangle$$
$$= \sum_{n=1}^{N}\left(\sum_{j_1,j_2=1}^{p} f_{j_1}f_{j_2}\langle \exp(2\pi i \mathbf{r}^* \cdot \mathbf{u}_{j_1,j_2})\rangle\right)$$

where $\mathbf{u}_{j_1,j_2} = \mathbf{r}_{j_1} - \mathbf{r}_{j_2}$. In accordance with p. 180 we obtain

$$\langle |F(r^*)|^2\rangle = N\left(\sum_{i,j=1}^{p} f_i f_j \frac{\sin 2\pi r^* u_{ij}}{2\pi r^* u_{ij}}\right)$$
$$= N\left(\sum_{j=1}^{p} f_j^2 + \sum_{i\neq j=1}^{p} f_i f_j \frac{\sin 2\pi r^* u_{ij}}{2\pi r^* u_{ij}}\right). \qquad (3.C.23)$$

In terms of normalized intensity (3.C.23) becomes

$$\langle |E(r^*)|^2 \rangle = 1 + \sum_{i \neq j = 1}^{p} v_i v_j \frac{\sin 2\pi r^* u_{ij}}{2\pi r^* u_{ij}} \tag{3.C.24}$$

where $v = f/N(\sum_{j=1}^{p} (f_j^2))^{1/2}$. In order to give some examples, for a molecule composed of two atoms at distance l (3.C.23) becomes

$$\langle |F(r^*)|^2 \rangle = N\left(f_1^2 + f_2^2 + 2f_1 f_2 \frac{\sin 2\pi r^* l}{2\pi r^* l}\right). \tag{3.C.25}$$

For a tetrahedral molecule (i.e. $SiCl_4$) (3.C.23) becomes

$$\langle |F(r^*)|^2 \rangle = N\left(f_{Si}^2 + 4f_{Cl}^2 + 12 f_{Cl}^2 \frac{\sin 2\pi r^* l}{2\pi r^* l} + 8 f_{Si} f_{Cl} \frac{\sin 2\pi r^* d}{2\pi r^* d}\right) \tag{3.C.26}$$

where l is the edge of the tetrahedron and d its radius.

In favourable conditions estimates for interatomic distances can be readily obtained from the intensity diffraction pattern.[43,45] According to p. 174, the value of the interference term $\sin 2\pi r^* x/(2\pi r^* x)$ is unity for $r^* = 0$ and becomes zero for $r^* x = n/2$: subsidiary maxima occur when

$$2\pi r^* x = 2.459\pi,\ 4.477\pi,\ 6.484\pi, \ldots$$

or, in other terms, when

$$x(\sin\theta/\lambda) = 0.615,\ 1.119,\ 1.621, \ldots. \tag{3.C.27}$$

The position of the mth maximum for large values of m will approach more and more closely to

$$x \sin\theta/\lambda = (0.125 + 0.5m)$$

but maxima will become weak and ill defined.

It is easily seen that the maxima in the $|E(r^*)|^2$ curve will coincide with those given by (3.C.27). In particular, from the first maximum located in $(\sin\theta/\lambda)_1$ the interatomic distance

$$x = 0.615/(\sin\theta/\lambda)_1 \tag{3.C.28}$$

is obtained.

It is easily seen that Debye maxima will correspond to those provided by (3.C.21).

Diffraction by gases

For a perfect gas (consisting of identical atoms with negligible volume and exercising no action on each other) any atom can occupy any position with the same probability. Therefore $p(u) \equiv 1$, (3.C.14) reduces to

$$z(u) = \delta(u) + 1/v_1$$

and (3.C.21) reduces to

$$|E(r^*)|^2 = 1$$

or, in terms of $|F(r^*)|^2$, to

$$|F(r^*)|^2 = N f^2(r^*). \tag{3.C.29}$$

If the central peak is also taken into consideration, the observed intensity is of type described in Fig. 3.C.4. For $r^* = 0$ the value of $|F(r^*)|^2$ is N^2f^2 (replace $|D(r^*)|^2$ by $|D(0)|^2 = \Omega^2$ and v_1 by Ω/N in (3.C.18)). As soon as r^* is out of the peak at the origin the value of $|F(r^*)|^2$ is equal to Nf^2.

If the gas is a mixture of perfect monoatomic gases the scattering is practically given by

$$|F(r^*)|^2 = \sum_{j=1}^{N} f_j^2(r^*).$$

If the gas is composed of identical molecules formed by atoms with negligible volume and arranged as in a perfect gas then features will appear in the scattering curve which are influenced by the molecular structure. A typical diffraction pattern provided by chlorine diatomic molecules is shown in Fig. 3.C.5 (curve (a)). The maxima in the experimental curve are not well defined (they are represented by roughly horizontal portions of the curve) because of the continuous decay of f with $\sin\theta/\lambda$. The maxima should be better emphasized by plotting $|E(r^*)|^2$ (curve (b)).

For real gases the intrinsic volume of the atoms is no longer negligible. Since atoms are impenetrable, an atomic distribution function $p(u)$ such as that described in Fig. 3.C.3(a) may be chosen, where d is the atomic diameter. Then (3.C.21) becomes

$$\begin{aligned} |E(r^*)|^2 &= 1 - \frac{1}{v_1}\int_0^d 4\pi u^2 \frac{\sin 2\pi r^* u}{2\pi r^* u}\,du \\ &= 1 - \frac{1}{v_1}(\tfrac{4}{3}\pi d^3)\varphi(y) = 1 - 8\frac{v_0}{v_1}\varphi(y) \\ &= 1 - 8\frac{N v_0}{\Omega}\varphi(y) = 1 - 8C\varphi(y) \end{aligned}$$

where $v_0 = 4\pi(d/2)^3/3$ is the volume of one atom, C is the concentration factor (ratio between the effective volume occupied by the atoms and the gas volume) and (see p. 181) $\varphi(y) = 3(\sin y - y\cos y)/y^3$, where $y = 2\pi r^* d$.

Since $\varphi(0) = 1$ it is $|E(0)|^2 = 1 - 8C$: thus the asymptotic value of $|E(r^*)|^2$ for $r^* \Rightarrow 0$ is smaller than unity by a quantity proportional to C. For $C > 1/8$ (e.g. for high-pressure gases) $|E(0)|^2 < 0$, which is not acceptable: for such cases the diffraction by gases closely resembles that by liquids or amorphous solids, for which supplementary considerations are needed (see p. 212).

Since electron scattering is much more intense than X-ray scattering (see p. 195) electron diffraction is preferable for gases. Very small quantities of gas can be used with shorter exposure times, and peaks can be measured at greater values of $\sin\theta/\lambda$.

Gas electron diffraction apparatus[46,47] usually consists of an electron gun, a sample injection system, and the detector: the last is often a photographic plate. Typical electron accelerating voltages, highly stabilized, are between 30 and 80 kV, with exposure times of a few minutes. In order to compensate the fast decay of the scattered intensity with $\sin\theta/\lambda$ a rotating sector is introduced into the path of the scattered electrons: it is a metallic disc designed in such a way that the intensity reaching the photographic plate is larger at larger scattering angles. In these conditions the diffraction pattern will emphasize high-angle interference effects, whose intensity will now be

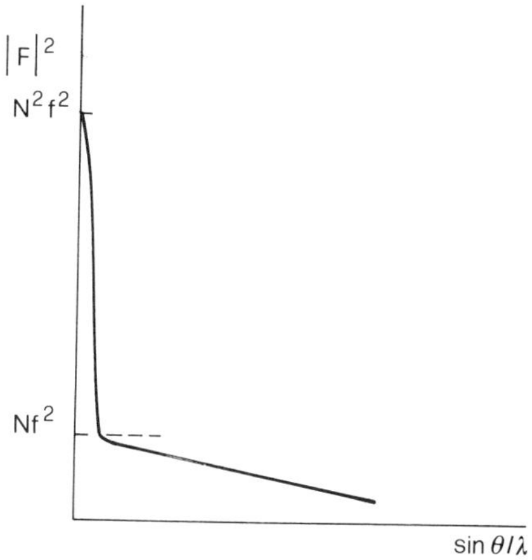

Fig. 3.C.4. Schematic diagram of the scattered intensity from a monoatomic gas.

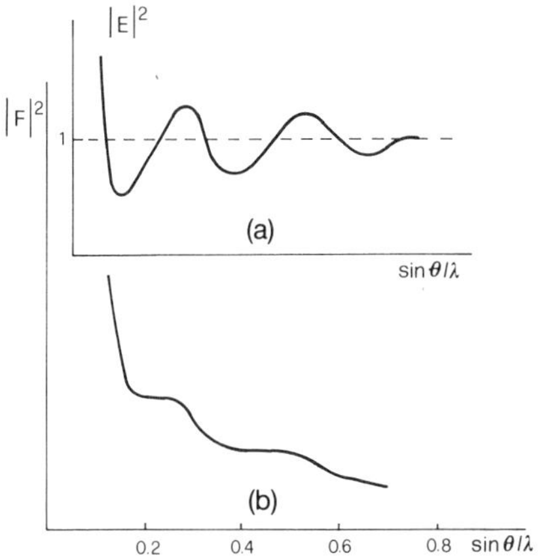

Fig. 3.C.5. Schematic diagram of the diffraction pattern for the diatomic molecule Cl_2.

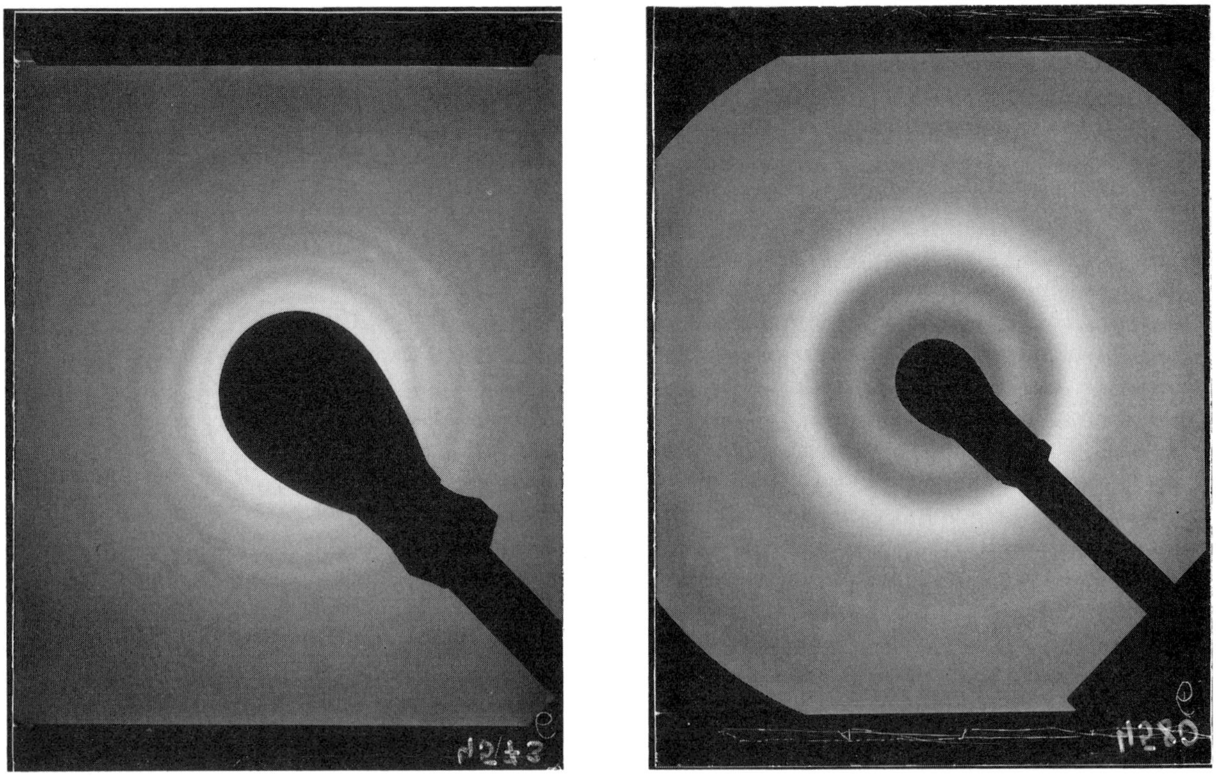

Fig. 3.C.6. Electron diffraction patterns from C_6H_5–NO_2 gas (working temperature about 353 K, two camera distances).[48]

sufficient to be measured by a microphotometer. After the background elimination, the scattering optical density distribution is converted into electron distribution.

In Fig. 3.C.6 two gas-phase electron diffraction patterns[48] of nitrobenzene C_6H_5–NO_2 are shown (corresponding to two different positions of the plate). The experiment was made in order to determine, in combination with *ab initio* molecular orbital calculations and X-ray single-crystal structure analysis of several derivatives, the molecular structure of C_6H_5–NO_2 in the planar and orthogonal conformation. From data in Fig. 3.C.6 the one-dimensional spectrum in Fig. 3.C.7 may be derived. In the figure the curve $sM(s)$ is given where $M(s) = (I_t - I_b)/I_b$, $s = 4\pi \sin\theta/\lambda$, I_t is the total observed radiation, I_b is the background intensity. The corresponding radial distribution curve $p(u)$ is shown in Fig. 3.C.8: the positions of the most important distances are marked by vertical bars whose height is proportional to the weight of the distances. The geometrical parameters of the theoretical model (the molecule as a whole was supposed to have a binary symmetry, while a local mm2 symmetry was assumed for the benzene ring and the nitro group), together with vibration amplitudes, were refined by a least-squares process applied to molecular intensities: the good final agreement may be deduced from the small differences between the *E* and *T* curves.

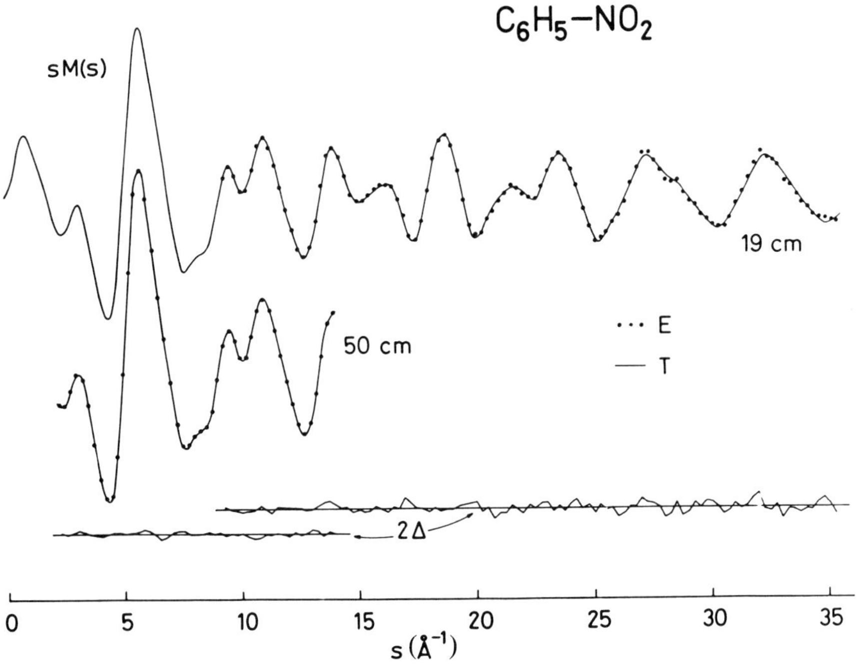

Fig. 3.C.7. Molecular intensity curves (E, experimental; T, theoretical) for $C_6H_5-NO_2$ (gas electron diffraction, two camera distances). Also shown are the difference curves.[48]

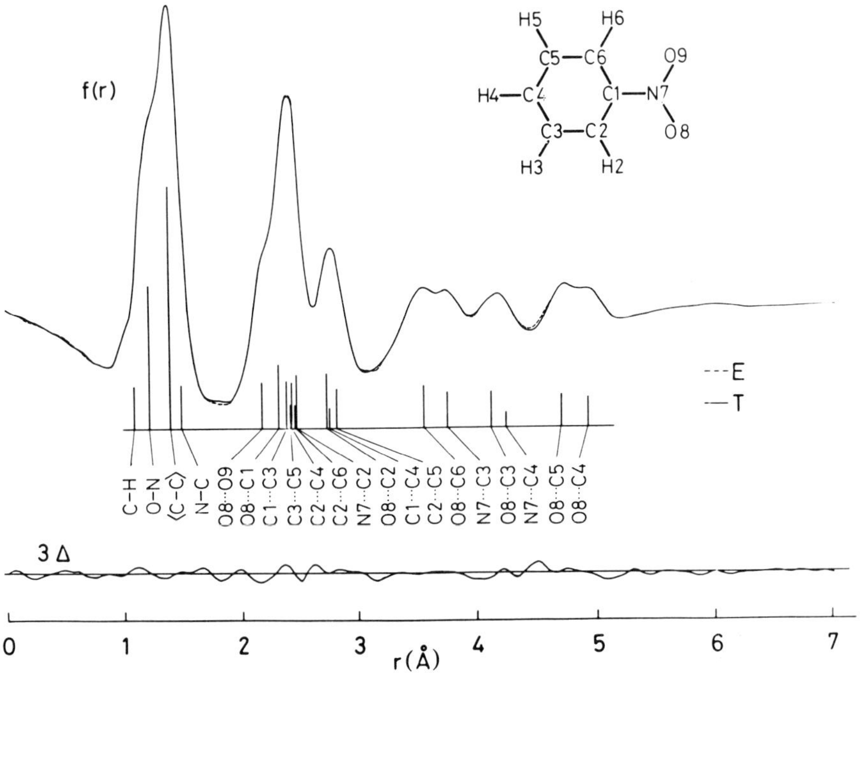

Fig. 3.C.8. Radial distribution curves for $C_6H_5-NO_2$ (E, experimental; T, theoretical). Also shown is the difference curve (E–T).[48]

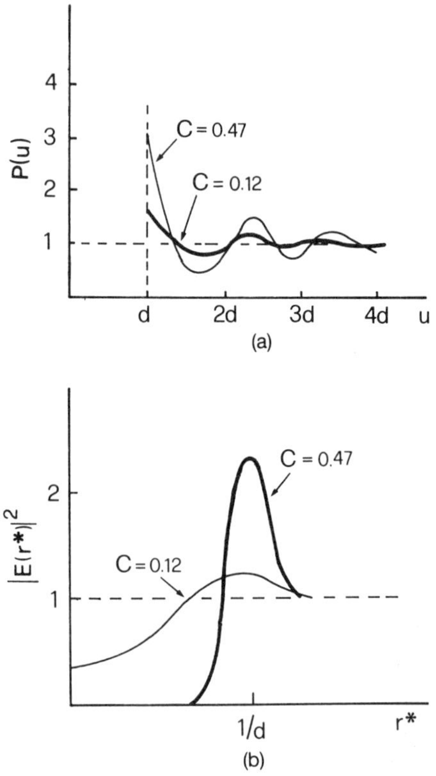

Fig. 3.C.9. (a) Distribution functions $P(u)$ for hard spheres of diameter d for two values of the concentration parameter C. (b) $|E(r^*)|^2$ curves corresponding to the two distribution shown in (a).

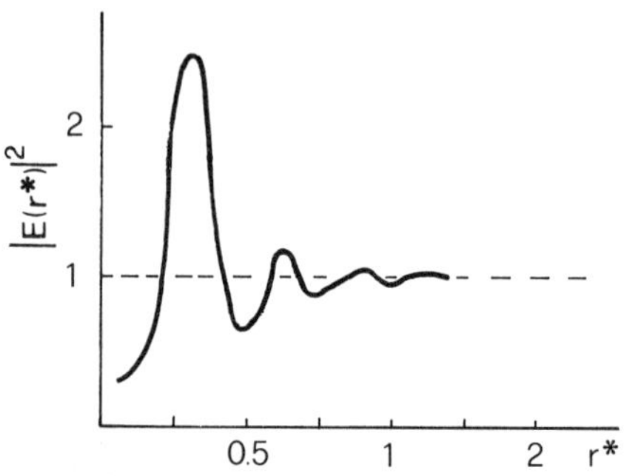

Fig. 3.C.10. $|E(r^*)|^2$ curves for liquid mercury.

Diffraction by liquids and amorphous bodies

In liquids or amorphous bodies each molecule may be considered in contact with a certain number of neighbours. Thus both internal and external interference effects will dominate the diffraction pattern. Let us first consider an atomic liquid: if atoms are assumed to be hard spheres which do not exert any force on each other the distribution $p(u)$ may be calculated.[49] In Fig. 3.C.9(a) $p(u)$ is plotted for two values of the concentration parameter C (ratio between the effective volume occupied by the spheres and the total volume of the liquid). The corresponding $|E(r^*)|^2$ functions may be derived by application of (3.C.21) and are shown in Fig. 3.C.9(b). Their main features are:

(1) for small values of r^* it is $|E(r^*)|^2 < 1$;

(2) a large maximum occurs at about $r^* \simeq 0.95/d$. Its intensity increases with C while its position does not vary;

(3) for lage values of r^* $|E(r^*)|^2$ tends to unity as for perfect gases.

In spite of the purely geometrical nature of the model the diffraction pattern of monoatomic liquids closely satisfies the model predictions. In Fig. 3.C.10 the $|E(r^*)|^2$ curve of liquid mercury is shown: its Fourier transform (eqn (3.C.22)) shows a maximum a little over 3 Å.

A further example concerns the diffraction pattern of water. As it is well known, the molecules are strongly polar and V-shaped with O–H distances of about 1 Å, and nearly tetrahedral HÔH angles (109°). For our purposes, they may be geometrically represented by spheres of about 2.8 Å diameter. The water scattering curves[50] at 1.5 °C and 83 °C are shown in Fig. 3.C.11(a); the corresponding radial distribution curves are given in Fig. 3.C.11(b). The main maximum for the radial distribution occurs at a radius of about 2.8 Å while a second broader maximum occurs at about 4.5 Å, nearly vanishing at high temperature. Such results suggested to Bernal and Fowler[51] that the arrangement of H_2O molecules in water may be described as a broken-down ice structure in which each molecule tries to bind four neighbours: since bonds are continually breaking and re-forming at any instant each molecule is bonded to fewer than four neighbours.

As a last example, examine now Fig. 3.C.12, where the diffraction patterns by vitreous silica, a cristobalite crystal powder, and silica gel are shown. The main maxima of all the three curves nearly overlap: but cristobalite shows numerous sharp maxima while only one maximum occurs in vitreous silica. Furthermore, vitreous silica intensity decreases with $\sin\theta/\lambda$ as we have just seen for liquids, while silica gel shows increasing intensity toward small $\sin\theta/\lambda$. The radial distribution function for vitreous silica[52] shows a first peak at $r = 1.62$ Å and a second one at 2.65 Å: that indicates that the tetrahedral coordination in the crystalline state persists in the vitreous state (but here the orientations of the tetrahedral groups are randomly distributed). The vitreous state is essentially homogeneous (diffraction intensity decreases at small $\sin\theta/\lambda$ as for liquids) while silica gel is made from very small discrete particles (10–100 Å) with voids among them (diffraction intensity increases at small scattering angles: see the next section).

Small-angle scattering

Small-angle scattering is a technique for studying structural features or inhomogeneities of colloidal dimensions.[40] For wavelengths of about 1 Å the typical angular domain of the technique ranges up to one or two degrees.

We have already seen (p. 204) that when diffraction occurs from a finite statistically homogeneous object of volume Ω a central peak in the intensity curve may be measured which becomes broader as the object size decreases, and does not depend on the internal structure of the object. Its intensity distribution varies according to (3.C.18):

$$|F(r^*)|^2 = \frac{N}{\Omega v_1} f^2(r^*)\,|D(r^*)|^2 = \frac{N^2}{\Omega^2} f^2(r^*)\,|D(r^*)|^2.$$

Since diffraction is considered at very small angles $f(r^*)$ will coincide with the number of electrons per atom. Thus the above equation reduces to

$$|F(r^*)|^2 = \rho^2\,|D(r^*)|^2. \qquad (3.C.30)$$

We consider now a system constituted by N particles with electron density ρ randomly dispersed in a homogeneous medium of electron density ρ_0 (solvent). Particles are supposed to be separated from each other widely enough, thus they will generate independent contributions to the diffracted intensity. The system may also be thought of as a medium of constant ρ_0 and macroscopic volume Ω to which N particles are summed with density $\Delta\rho = \rho - \rho_0$. If Ω is sufficiently large the central peak due to the homogeneous medium is totally unobservable, while the observed intensity distribution may be ascribed to electron inhomogeneties only. In this case the mean scattering power for particle is

$$|F_p(r^*)|^2 = (\rho - \rho_0)^2\,|D_p(r^*)|^2 \qquad (3.C.31)$$

where $|D_p(r^*)|^2$ is the average value of $|D_p(\boldsymbol{r}^*)|^2$ over all the possible particle orientations. Different shapes of particles will give rise to different shapes of the scattering function in reciprocal space: ideal scattering intensities can therefore be calculated for spherical, cylindrical, flat, ellipsoidal, etc., particles. The results are all rather similar, particularly in the central range, but remarkable differences occur at larger angles. It may then be expected that in the central part a universal approximation for all particle shapes must exist.[53]

Let us assume the origin O in the centre of gravity of a particle of volume

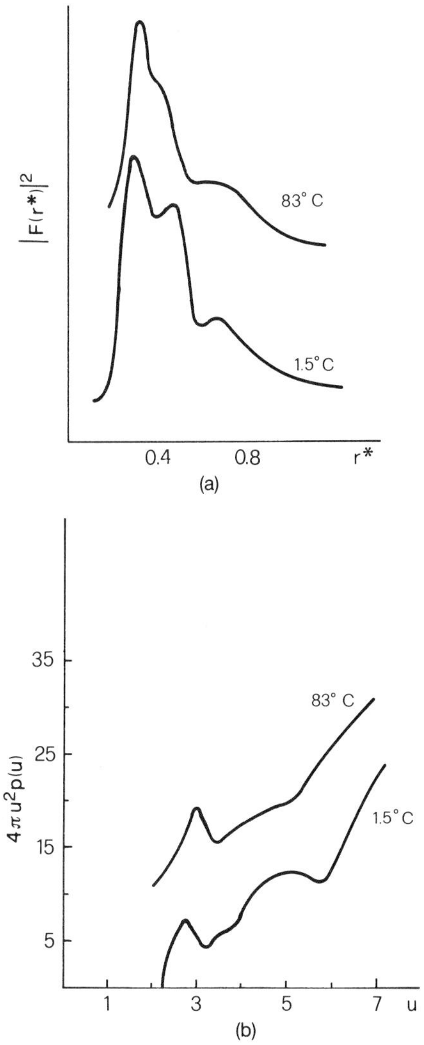

Fig. 3.C.11. (a) $|F(r^*)|^2$ water scattering curves at 1·5 °C and 83 °C; (b) the corresponding distributions $4\pi u^2 p(u)$.

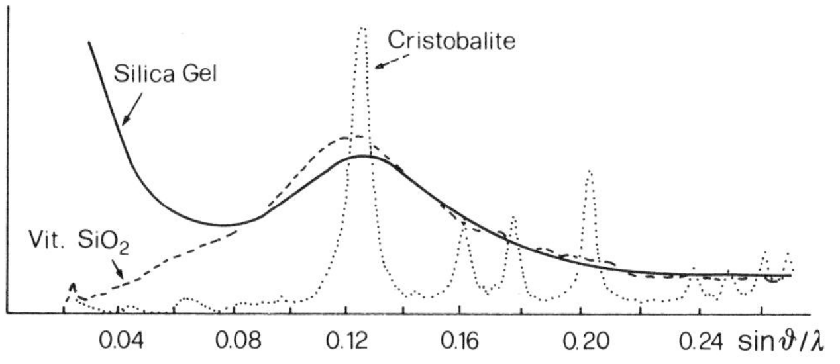

Fig. 3.C.12. Diffraction patterns for silica gel, vitreous SiO_2, and for cristobalite.

v_p and let us estimate

$$D_p(\boldsymbol{r}^*) = \int_{v_p} \exp(2\pi i \boldsymbol{r}^* \cdot \boldsymbol{r})\, dv. \tag{3.C.32}$$

At very small angles we can expand the exponential function in (3.C.32) in power series: it will suffice to expand $\cos 2\pi \boldsymbol{r}^* \cdot \boldsymbol{r}$ according to $\cos\theta \simeq 1 - \theta^2/2$. In Cartesian coordinates

$$\cos 2\pi \boldsymbol{r}^* \cdot \boldsymbol{r} \simeq 1 - 2\pi^2 r^{*2}(x^2\alpha^2 + y^2\beta^2 + z^2\gamma^2)^2,$$

where α, β, γ stand for the cosines between $\boldsymbol{r}^*$ and Cartesian axes. When integration is performed the contribution of mixed products vanishes by hypothesis (the centre of gravity has been located at the origin) and we have

$$D_p(\boldsymbol{r}^*) = v_p - 2\pi^2 r^{*2}\left(\alpha^2 \int x^2\, dx + \beta^2 \int y^2\, dy + \gamma^2 \int z^2\, dz\right).$$

Since we are interested to calculate $|D_p(r^*)|^2$, say the average of $|D_p(\boldsymbol{r}^*)|^2$ over all particle orientations, we have to rotate Cartesian axes about the origin. We can then replace α^2, β^2, γ^2 by $\langle\alpha^2\rangle = \langle\beta^2\rangle = \langle\gamma^2\rangle = \frac{1}{3}$ so as to obtain

$$\begin{aligned} |D_p(r^*)|^2 &= \left(v_p - \frac{2\pi^2}{3} r^{*2}\left\langle \int_{v_p} r^2\, dv \right\rangle\right)^2 \\ &= \left(v_p - \frac{2\pi^2}{3} r^{*2} R_g^2 v_p\right)^2 \\ &= v_p^2\left(1 - \frac{2\pi^2}{3} r^{*2} R_g^2\right)^2 \\ &= v_p^2\left(1 - \frac{4\pi^2}{3} r^{*2} R_g^2\right) \\ &\simeq v_p^2 \exp\left(-\frac{4\pi^2}{3} r^{*2} R_g^2\right) \end{aligned}$$

where r^2 stands for $(x^2 + y^2 + z^2)$ and

$$R_g^2 = (1/v_p)\left\langle \int_{v_p} r^2\, dv \right\rangle$$

is the **radius of gyration** ($\simeq$ the mean square distance) of the particle with respect to its centre of gravity.

According to (3.C.31) the scattering power per particle will be

$$|F_p(r^*)|^2 = n_e^2 \exp\left(-\frac{4\pi^2}{3} r^{*2} R_g^2\right) \tag{3.C.33}$$

where $n_e = (\rho - \rho_0)v_p$. By taking the logarithm to the base 10

$$\lg |F_p(r^*)|^2 \simeq \lg n_e^2 - \left(0.4343 \frac{4\pi^2}{3\lambda^2} R_g^2\right)(2\theta)^2$$

is obtained. If $\lg |F_p(r^*)|^2$ is plotted against $(2\theta)^2$ a curve will be obtained which, at small values of θ, is close to a straight line with slope $\alpha = -5.715 R_g^2/\lambda^2$, from which the gyration radius $R_g = 0.418\lambda\sqrt{-\alpha}$ is obtained.

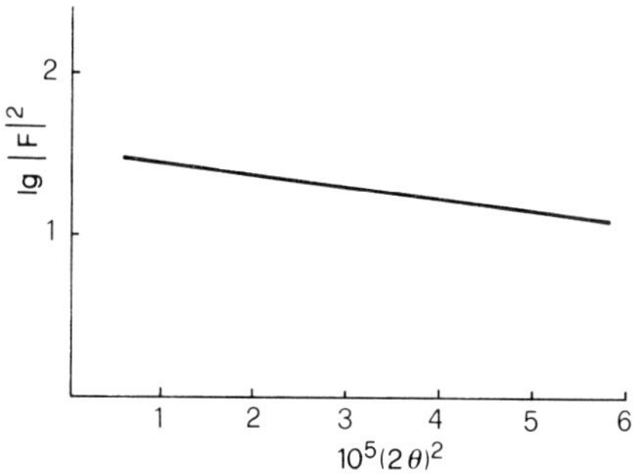

Fig. 3.C.13. A Guinier plot.

In Fig. 3.C.13 the Guinier plot for a low-concentration solution of haemocyanin from *Astacus leptodactylus* is shown.[54] The reader will easily derive from the figure (Cu K_α radiation used) a gyration radius of $R_g \simeq 62$ Å.

The radius of gyration is connected to the geometrical parameters of some simple homogeneous bodies in the following way:[55]

sphere of radius r:

$$R_g^2 = 3r^2/5;$$

hollow sphere (limiting radii r_1 and r_2):

$$R_g^2 = (3/5)(r_2^5 - r_1^5)/(r_2^3 - r_1^3);$$

ellipsoid (semi-axes a, b, c):

$$R_g^2 = (a^2 + b^2 + c^2)/5;$$

prism with edge lengths a, b, c:

$$R_g^2 = (a^2 + b^2 + c^2)/12;$$

elliptic cylinder (height h; semi-axes a, b):

$$R_g^2 = [(a^2 + b^2)/4] + [(h^2/12)];$$

hollow cylinder (height h, radii r_1, r_2):

$$R_g^2 = [(r_1^2 + r_2^2)/2] + [(h^2/12)].$$

Other important parameters can be evaluated directly from the scattering data: among others the molecular weight (from the scattering intensity at zero angles[56]), the total surface of the particles for unit mass (from the region where the small-angle scattering tends to zero[57,58]), and the volume of the particle. This latter may be obtained by integrating (3.C.31) over all the reciprocal space (by extrapolation of $r^{*2}\,|D_p(r^*)|^2$ at very small r^*):

$$Q = \int_0^\infty 4\pi r^{*2}\,|F_p(r^*)|^2\,dr^* = (\rho - \rho_0)^2 \int_0^\infty |D_p(r^*)|^2\,4\pi r^{*2}\,dr^*.$$

According to (3.C.9) the above relation reduces to

$$Q = (\rho - \rho_0)^2 v_p$$

from which v_p may be derived.

So far we have considered particles dispersed in a homogeneous solvent and widely separated from each other. If the concentration of the particles increases the mutual interference will enter into play and the scattered intensity is no longer the sum of the individual particle scatterings. As described in on p. 206 a probability function $p(u)$ can be introduced: then $4\pi u^2 p(u)\,du$ will represent the number of particles with distance u lying in the interval $(u, u + du)$ from a particle at the origin. Then the scattering power per particle will be (see eqn (3.C.21))

$$|F(r^*)|^2 = \langle |F_p(r^*)|^2 \rangle \left(1 + \frac{2}{r^* v_1} \int_0^\infty u[p(u) - 1] \sin(2\pi r^* u)\,du\right)$$

where $\langle |F_p(r^*)|^2 \rangle$ is the squared modulus of the structure factor of a particle averaged over all orientations. From the above equation the function $p(u)$ may be derived by inverse Fourier transform.

3.D About electron density mapping

The accurate determination of charge density distribution is of basic importance in several scientific areas. For example, for the study of

(1) electronic structure of metals and alloys;

(2) metal–ligand interaction;

(3) variation of solid state properties (i.e. conductivity) with temperature.

Spatial partitioning of charge density can also answer some specific questions, such as the location of lone-pair maxima, the net charge of a particular atom, the excess charge accumulated in a covalent bond, and so on.

Such results can be attained only if the selection of the compounds, the process of measuring diffraction data, and their subsequent treatment is carried out with great care. Among the various desiderata we mention the following.

1. Compounds with a small valence-to-core ratio are less suitable to charge density studies. If we assume that the number of valence electrons per unit volume is approximately constant for different compounds, the valence-to-core ratio is fixed by the number of core electrons per unit volume. Stevens and Coppens[59] introduced the criterion

$$S = V \Big/ \sum_{\text{core}} n_i^2$$

as a possible criterion to estimate the suitability of a system for charge density studies (V is the volume of the unit cell). S varies from 3–5 for the first-row-atom organic crystals to 0.1–0.3 for metals and alloys. Crystals with $S < 1$ require extremely precise diffraction data and careful work.

2. Centrosymmetrical crystals are often preferred (in acentric crystals the phases of the structure factors cannot be uniquely determined).

3. Collection of diffraction data has to be carried out at low temperature. Then thermal diffuse scattering can be neglected (*a posteriori* corrections are seldom accurate), high resolution data become available (then accurate structural parameters can be calculated), and electron charge density can be easily deconvoluted from the atomic thermal vibrations.

4. Sophisticated measuring programmes should be used, in order to perform optimal scanning mode and profile analysis, decide differentiated measuring times (in order to obtain a set of structure factors with equal variance), prevent multiple scattering, and perform careful correction for extinction, absorption, anomalous dispersion, etc.

5. When possible, neutron data should also be collected in order to obtain nuclear positional and thermal parameters which are unbiased by asphericity in the atomic charge density.

Each atomic fragment in which the molecular electron density may be subdivided has an invariant core and an external charge cloud deformable because of interatomic bonding and molecular environment (see p. 147). Accordingly, the bonded atom charge density may be represented as[60,61,62]

$$\rho_{\text{atom}}(\boldsymbol{r}) = \rho_{\text{core}}(\boldsymbol{r}) + \rho_{\text{valence}}(\boldsymbol{r}) + \rho_{\text{def}}(\boldsymbol{r}) \qquad (3.\text{D}.1)$$

where ρ_{core} and ρ_{valence} are spherically symmetric terms corresponding to the free-atom charge distribution, ρ_{def} accounts for non-spherical distortion due to bond formation. All the terms are constrained to obey local site symmetry[63]. The deformation term may be modelled by suitable nuclear-centred functions: in polar coordinates with origin at the nuclear position, $\Delta\rho$ may be expressed as

$$\rho_{\text{def}} = \sum_n R_n(r) \sum_{l,m} C_{nlm} Y_{l,m}(\theta, \phi) \qquad (3.\text{D}.2)$$

where n, l, m are integers ($m \leqslant l$), C_{nlm} is the deformation population parameter (to be determined from the analysis of experimental data). $R_n(r)$ is a one-electron radial function such a

$$R_n(r) = \frac{\alpha^{n+3}}{4\pi(n+2)!} r^n \exp(-\alpha r).$$

The reader can easily verify that $R_0(r)$ coincides with $(\rho_e)_{1s}$ in eqns (3.12) when $c_1 = \alpha/2$, and also, that $R_2(r)$ coincides with $(\rho_e)_{2s}$ when $c_2 = \alpha$. Common values for α are 4.69, 6.42, 7.37, and 8.50 Å^{-1} for H, C, N, and O respectively.[64]

For first-row atoms one can choose[65] $n = 2, 2, 2, 3, 4$ for $l = 0, 1, 2, 3, 4$ respectively.

$Y_{l,m}$ are the real spherical harmonics: the 'even' and the 'odd' terms can be written as

$$Y^{\text{e}}_{l,m}(\theta, \phi) = \cos(m\phi) P^m_l(\cos\theta)$$
$$Y^{\text{o}}_{l,m}(\theta, \phi) = \sin(m\phi) P^m_l(\cos\theta)$$

respectively.

$P^m_l(\cos\theta)$ are the un-normalized Legendre functions. The terms with $l = 0, 1, 2, 3, 4, \ldots$ of the multipole expansion are called monopole, dipole, quadrupole, octapole, hexadecapole, . . . , respectively.

Low-order $Y_{l,m}$ functions are:

$$Y^{\text{e}}_{0,0} = 1 \qquad \text{monopole.}$$

A positive value of the population parameter confers an excess electron density to the atom.

$$\left.\begin{aligned} Y^{\text{e}}_{1,1} &= \sin\theta\cos\phi = s_x \\ Y^{\text{o}}_{1,1} &= \sin\theta\sin\phi = s_y \\ Y^{\text{e}}_{1,0} &= \cos\theta = s_z \end{aligned}\right\} \quad \text{dipoles.}$$

$Y^{\text{e}}_{1,0}$, $Y^{\text{e}}_{1,1}$, $Y^{\text{o}}_{1,1}$ are distributed on hemispheres with opposite sign. A positive population parameter for $Y^{\text{o}}_{1,1}$ confers an excess electron density to one hemisphere and a deficiency to the other, but does not change the net charge.

$$\left.\begin{aligned}
Y^{e}_{2,2} &= 3\sin^2\theta\cos 2\phi = 3(s_x^2 - s_y^2)\\
Y^{o}_{2,2} &= 6\sin^2\theta\sin\phi\cos\phi = 6s_xs_y\\
Y^{e}_{2,1} &= 3\sin\theta\cos\theta\cos\phi = 3s_xs_z\\
Y^{o}_{2,1} &= 3\sin\theta\cos\theta\sin\phi = 3s_ys_z\\
Y^{o}_{2,0} &= 6\cos^2\theta - 2 = 6s_z^2 - 2
\end{aligned}\right\}\ \text{quadrupoles.}$$

$$\left.\begin{aligned}
Y^{e}_{3,3} &= 15\sin^3\theta\cos 3\phi = 15(s_x^2 - 3s_y^2)s_x\\
Y^{o}_{3,3} &= 15\sin^3\theta\sin 3\phi = (45s_x^2 - 15s_y^2)s_y\\
Y^{e}_{3,2} &= 15\sin^2\theta\cos\theta\cos 2\phi = 15(s_x^2 - s_y^2)s_z\\
Y^{o}_{3,2} &= 30\sin^2\theta\cos\theta\sin\phi\cos\phi = 30s_xs_ys_z\\
Y^{e}_{3,1} &= 1.5\sin\theta(5\cos^2\theta - 1)\cos\phi = 1.5(5s_z^2 - 1)s_x\\
Y^{o}_{3,1} &= 1.5\sin\theta(5\cos^2\theta - 1)\sin\phi = 1.5(5s_z^2 - 1)s_y\\
Y^{e}_{3,0} &= 10\cos^3\theta - 6\cos\theta = (10s_z^2 - 6)s_z
\end{aligned}\right\}\ \text{octapoles.}$$

In Fig. 3.D.1 some electron density plots in special sections of direct space are shown.[61]

In most organic molecules the expansion (3.D.2) may be stopped at octapoles: then the deformation charge density is described by a linear combination of 16 terms, for which 16 population parameters have to be estimated.

In some formulations[62] the free-atom valence shell is modified by an expansion–contraction radial parametrization in order to take into account the fact that an atom will expand or contract when it becomes more negative or more positive. Such a feature may be incorporated into a perturbed valence density $\rho'_{\text{valence}}(r)$ given by

$$\rho'_{\text{valence}}(r) = P_{\text{valence}}k^3\rho_{\text{valence}}(kr)$$

where ρ_{valence} is, as in eqn (3.D.1), the free-atom ground state density, P_{valence} is the valence shell population, and k^3 is a normalization factor. If $k > 1$ the atom is contracted relative to the free one [indeed $\rho'(r)$ is proportional to $\rho(kr)$ with $kr > r$].

The atomic scattering amplitude has the same type of multipole expansion as the atomic charge density. By Fourier transformation of (3.D.2) one obtains

$$f(\boldsymbol{r}^*) = \sum_{n,l,m} f_{nlm} = 4\pi \sum_{n,l,m} (\mathrm{i})^l f_{n,l}(\boldsymbol{r}^*) Y_{l,m}(\theta^*, \phi^*) \qquad (3.D.3)$$

where θ^*, ϕ^* are angular components of the Bragg vector,

$$f_{n,l}(\boldsymbol{r}^*) = \int_0^\infty R_n(r) j_l(r^*r) r^2\,\mathrm{d}r,$$

and $j_l(x)$ is the spherical Bessel function of order l.

From (3.D.3) the calculated structure factors F_{calc} are easily obtained. Then (full or block-diagonal) least-squares refinement can be carried out: as well as the usual positional and thermal parameters for each atom, the population parameters of the deformation terms may also be refined. Such a formalism should improve (with respect to traditional crystallographic refinements) the agreement between observed and calculated structure factors.

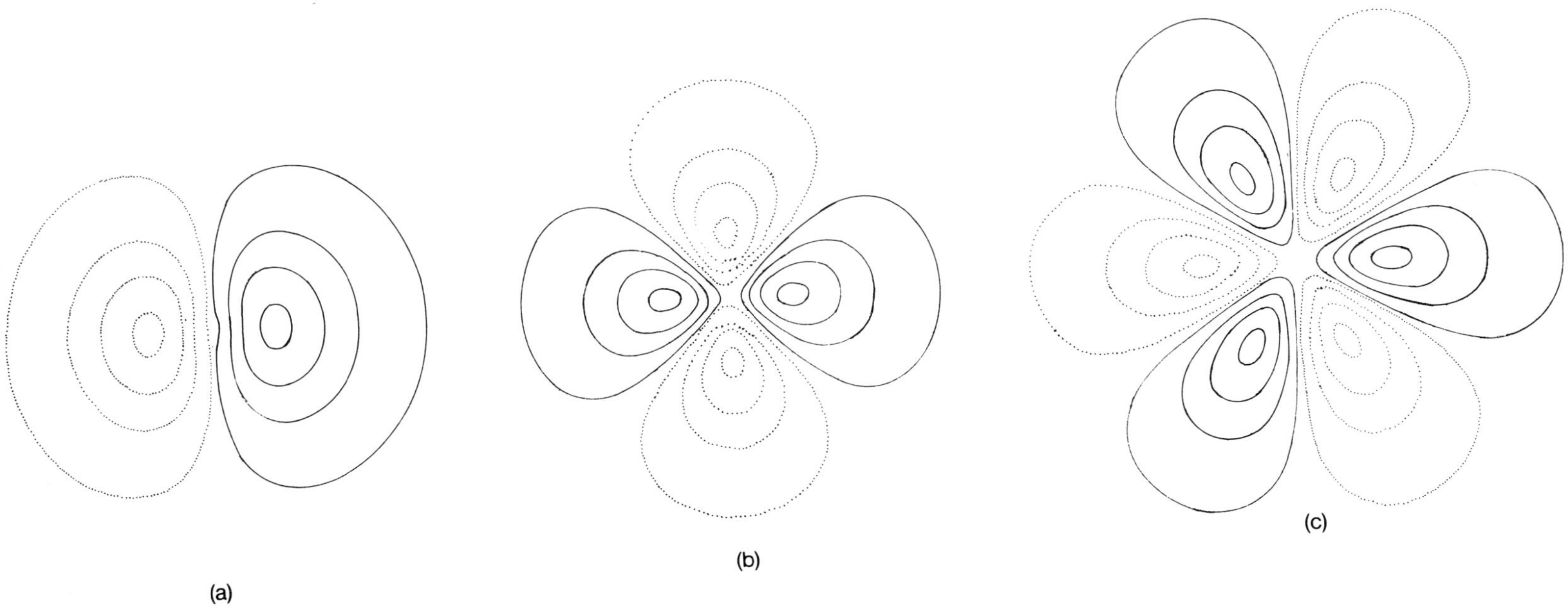

Fig. 3.D.1. Electron density plot for (a) $R_2(r)Y^{e}_{1,1}$ for $\cos\theta = 0$; (b) $R_2(r)Y^{e}_{2,2}$ for $\cos\theta = 0$; (c) $R_2(r)Y^{e}_{3,3}$ for $\cos\theta = 0$. In all figures, solid lines represent positive density, broken lines negative density.

Suitable Fourier syntheses can be calculated and analysed in order to reveal the effects of interatomic bonding on the charge clouds. Fourier maps of large usefulness are

$$\Delta\rho(\boldsymbol{r}) \simeq \frac{1}{V}\sum_{\boldsymbol{H}} (F_1(\boldsymbol{H}) - F_2(\boldsymbol{H})) \exp(-2\pi i \boldsymbol{H}\cdot\boldsymbol{r}) \qquad (3.D.4)$$

where

(1) $F_1 = F_{\text{calc, multipole}}$, $F_2 = F_{\text{calc, free atoms}}$. $\Delta\rho$ is then the deformation map, i.e. the difference between the atomic densities represented by spherical harmonics and those represented by free atoms;

(2) $F_1 = F_{\text{obs}}$, $F_2 = F_{\text{calc, multipole}}$. $\Delta\rho$ is then the **residual density**, e.g. the part of electron density not accounted for by the multipole model;

(3) $F_1 = F_{\text{obs}}$, $F_2 = F_{\text{calc, free atoms}}$, under the condition that F_2 is calculated with parameters from high-order (HO) refinement (high-order reflections are unaffected by chemical deformation of valence orbitals and locate atoms carefully). $\Delta\rho$ is then denoted as a $(X - X_{\text{HO}})$ deformation map;

(4) $F_1 = F_{\text{obs}}$, $F_2 = F_{\text{calc, free atoms}}$, with F_2 calculated as in (3) but for core electrons only. $\Delta\rho$ is the $(X - X_{\text{HO}})$ valence map;

(5) $F_1 = F_{\text{obs}}$, $F_2 = F_{\text{calc, free atoms}}$, with F_2 calculated with parameters from neutrons or a joint high-order X-ray and neutron data. Then $\Delta\rho$ is the $X - N$ or $X - (X_{\text{HO}} + N)$ deformation map;

(6) $F_1 = F_{\text{obs}}$, $F_2 = F_{\text{calc, free atoms}}$, with F_2 calculated as in (5) from core electrons only. $\Delta\rho$ is then a $X - (X_{\text{HO}} + N)$ valence map.

The estimation of the charge density model obtained at the end of calculations is different according to whether least squares or Fourier methods have been used[67]. However, direct and reciprocal fitting are nearly equivalent: indeed, if we assume in (3.D.4) that $F_1 = F_{\text{obs}}$, $F_2 = F_{\text{calc}}$, $\Delta F = F_{\text{obs}}(\boldsymbol{H}) - F_{\text{calc}}(\boldsymbol{H})$, and we calculated the difference Patterson

$$\int \Delta\rho(\boldsymbol{r})\Delta\rho(\boldsymbol{r}+\boldsymbol{u})\,\mathrm{d}\boldsymbol{r} = \frac{1}{V}\sum_{\boldsymbol{H}} |\Delta F|^2 \cos 2\pi \boldsymbol{H}\cdot\boldsymbol{u},$$

then for $\boldsymbol{u} = 0$ we obtained

$$\sum_{\boldsymbol{H}} |\Delta F|^2 = \frac{1}{V}\int |\Delta\rho(\boldsymbol{r})|^2\,\mathrm{d}\boldsymbol{r}.$$

Thus least-squares parameters which give the best fit between observed and calculated structure factors are expected to give rise to the lowest variance in $\Delta\rho$.

Direct evaluation of atomic or molecular charges, of dipoles and higher moments, etc, can be obtained from the estimated population parameters or directly from the electron density.[68,69] For example, the net charge on atom j is

$$q_j = -\int \rho(\boldsymbol{r})\,\mathrm{d}\boldsymbol{r}$$

when the integration is made over the (not always easily defined) atomic volume. Also, the dipole moment of a molecule may be calculated as

$$\int r\rho(r)\,dr$$

where the integration is over the molecular volume.

3.E Modulated structures and quasicrystals

Embedding of modulated structures in higher-dimensional space

In Chapter 3, pp. 171–3, it has been suggested that incommensurately modulated structures (IMS) can be better described in higher-dimensional spaces. From now on S_n and L_n will denote n-dimensional space and lattice respectively, while S_n^* and L_n^* will be their reciprocals. We will show that the reciprocal lattice L_3^* of an IMS, even if aperiodic in S_3^*, may be transformed into a $(3+d)$-dimensional periodic lattice (say L_{3+d}^*) by embedding the IMS in a metrical $(3+d)$-dimensional space S_{3+d}.

Let us suppose that the reciprocal vector H of a main or of a satellite reflection may be written in S_3^* as

$$H = h_1 a^* + h_2 b^* + h_3 c^* + \sum_{i=1}^{d} h_{3+i} K^i \qquad (3.E.1)$$

where

$$K^i = k_1^i a^* + k_2^i b^* + k_3^i c^*$$

and k_1^i, k_2^i, k_3^i are the a^*, b^*, c^* components of K^i. One of them should be irrational for incommensurate modulation. Since $(3+d)$ indices h_i are necessary to label a single diffraction effect, we will speak of d-dimensional periodic modulation or of d-dimensional modulated structure. In Fig. 3.E.1 a section of the three-dimensional diffraction pattern of a one-dimensional modulated structure is sketched. The main reflections are marked by the largest spots.

A L_{3+d}^* lattice may now be defined[70–74] with basis vectors

$$b_1 = a^*, \; b_2 = b^*, \; b_3 = c^*, \; b_{3+i} = K^i + e_i \qquad (i = 1, \ldots, d) \quad (3.E.2)$$

where the e_is are unit vectors perpendicular to S_3^*. Then in S_{3+d}^* a reciprocal vector may be written as

$$H' = \sum_{i=1}^{3+d} h_i b_i.$$

According to hypotheses, H is the projection of H' in S_3^*. Consequently the whole three-dimensional diffraction pattern L_3^* is the projection in S_3^* of a $(3+d)$-dimensional reciprocal lattice L_{3+d}^*. In Fig. 3.E.2 such a projection is shown schematically for a one-dimensional modulation: S_3^* is represented by a horizontal line, e_1 is a unit vector perpendicular to S_3^*, main and satellite reflections are labelled by M and W respectively.

In S_{3+d} the $(3+d)$-dimensional direct lattice L_{3+d} is defined by the set of

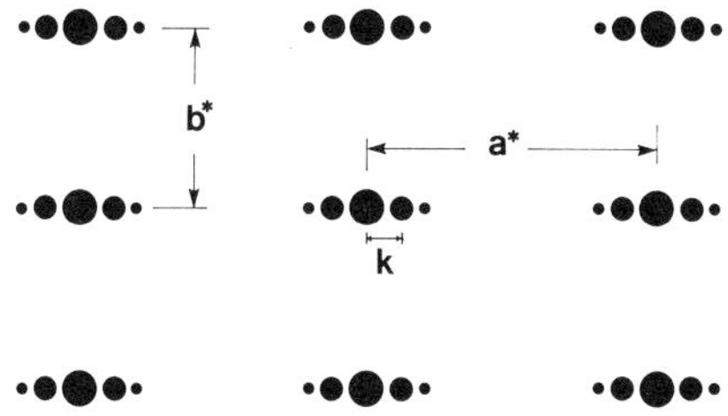

Fig. 3.E.1. One-dimensional modulated structure: sketch of a section of the three-dimensional diffraction pattern, showing main and satellite reflections.

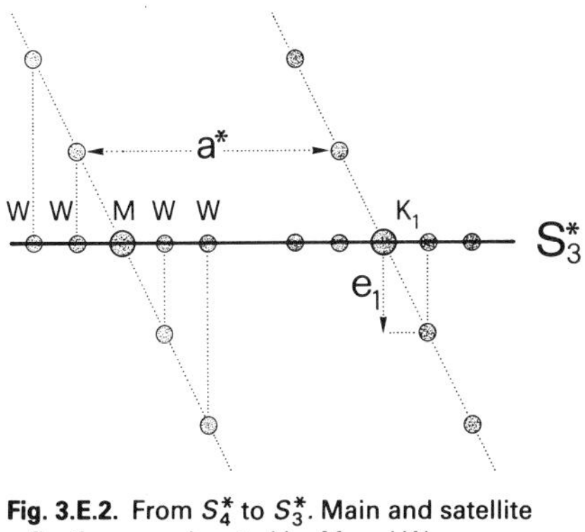

Fig. 3.E.2. From S_4^* to S_3^*. Main and satellite reflections are denoted by M and W respectively.

vectors $\boldsymbol{a}_i$ reciprocal to $\boldsymbol{b}_i$ according to the conditions $\boldsymbol{a}_i \cdot \boldsymbol{b}_j = \delta_{ij}$. We have

$$\begin{cases} \boldsymbol{a}_1 = \boldsymbol{a} - \sum_{i=1}^{d} k_1^i \boldsymbol{e}_i \\ \boldsymbol{a}_2 = \boldsymbol{b} - \sum_{i=1}^{d} k_2^i \boldsymbol{e}_i \\ \boldsymbol{a}_3 = \boldsymbol{c} - \sum_{i=1}^{d} k_3^i \boldsymbol{e}_i \\ \boldsymbol{a}_{3+i} = \boldsymbol{e}_i \qquad (i = 1, 2, \ldots, d) \end{cases} \tag{3.E.3}$$

where $\boldsymbol{a}$, $\boldsymbol{b}$, $\boldsymbol{c}$ are reciprocal to $\boldsymbol{a}^*$, $\boldsymbol{b}^*$, $\boldsymbol{c}^*$. The reciprocity condition is easily verified: for example,

$$\begin{aligned} \boldsymbol{a}_1 \cdot \boldsymbol{b}_{3+j} &= \left(\boldsymbol{a} - \sum_{i=1}^{d} k_1^i \boldsymbol{e}_i\right) \cdot (\boldsymbol{K}^j + \boldsymbol{e}_j) \\ &= \boldsymbol{a} \cdot (k_1^j \boldsymbol{a}^* + k_2^j \boldsymbol{b}^* + k_3^j \boldsymbol{c}^*) - k_1^j = k_1^j - k_1^j = 0 \\ \boldsymbol{a}_1 \cdot \boldsymbol{b}_3 &= \left(\boldsymbol{a} - \sum_{i=1}^{d} k_1^i \boldsymbol{e}_i\right) \cdot \boldsymbol{c}^* = 0 \\ \boldsymbol{a}_1 \cdot \boldsymbol{b}_1 &= \left(\boldsymbol{a} - \sum_{i=1}^{d} K_1^i \boldsymbol{e}_i\right) \cdot \boldsymbol{a}^* = 1. \end{aligned}$$

If a position is defined in S_3 by

$$\boldsymbol{r} = x\boldsymbol{a} + y\boldsymbol{b} + z\boldsymbol{c},$$

it will be defined in S_{3+d} by

$$\boldsymbol{r} = \sum_{i=1}^{3+d} x_i \boldsymbol{a}_i$$

where

$$x_{3+j} = k_1^j x_1 + k_2^j x_2 + k_3^j x_3. \tag{3.E.4}$$

Indeed, the first three coordinates in S_3 and in S_{3+d} coincide, owing to the fact that L_3 is the projection of L_{3+d} along $\boldsymbol{e}_i$; any point $\boldsymbol{r}$ in S_3 satisfies the relations

$$\boldsymbol{e}_j \cdot \boldsymbol{r} = \boldsymbol{e}_j \cdot \left(\sum_{i=1}^{3+d} x_i \boldsymbol{a}_i\right) = 0 \qquad \text{for } j = 1, \ldots, d.$$

In accordance with (3.E.3)

$$\begin{aligned} \boldsymbol{e}_j \cdot \sum_{i=1}^{3+d} x_i \boldsymbol{a}_i &\equiv \boldsymbol{e}_j \cdot \left\{\sum_{i=1}^{3} x_i \boldsymbol{a}_i + x_{3+j} \boldsymbol{e}_j\right\} \\ &= -k_1^j x_1 - k_2^j x_2 - k_3^j x_3 + x_{3+j} = 0 \end{aligned}$$

so that

$$x_{3+j} = k_1^j x_1 + k_2^j x_2 + k_3^j x_3$$

is obtained, which coincides with (3.E.4).

If in S_{3+d} the new coordinates

$$t_j = -k_1^j x_1 - k_2^j x_2 - k_3^j x_3 + x_{3+j} \tag{3.E.5}$$

are introduced, it may be concluded that S_3 is defined by

$$t_j = 0 \quad \text{for} \quad j = 1, \ldots, d.$$

The electron density ρ' in S_{3+d} is periodic in $(3+d)$-dimensions and may be calculated in the usual way:

$$\rho(x_1, \ldots, x_{3+d}) = \frac{1}{V} \sum_{h_1,\ldots,h_{3+d}} F_{\mathbf{H}'} \exp\left(2\pi i \sum_{j=1}^{3+d} h_j x_j\right),$$

where V is the volume of the unit cell in S_3.

According to reciprocity rule (the Fourier transform of a projection corresponds to a section and vice versa) the electron density distribution ρ in S_3 is the section of ρ' with S_3. Therefore in S_{3+d} the shape of the atoms is a hypersurface which extends in the extra dimensions. In Fig. 3.E.3, a schematic representation of ρ' for a one-dimensional modulation is described.

A scattering formalism for a large variety of modulated structures is now available:[75–77] substitutional and/or displacive modulations of each of the atoms in the crystal may be taken into account, together with the translational or rotational displacement of a molecule or one of its segments in molecular crystals.

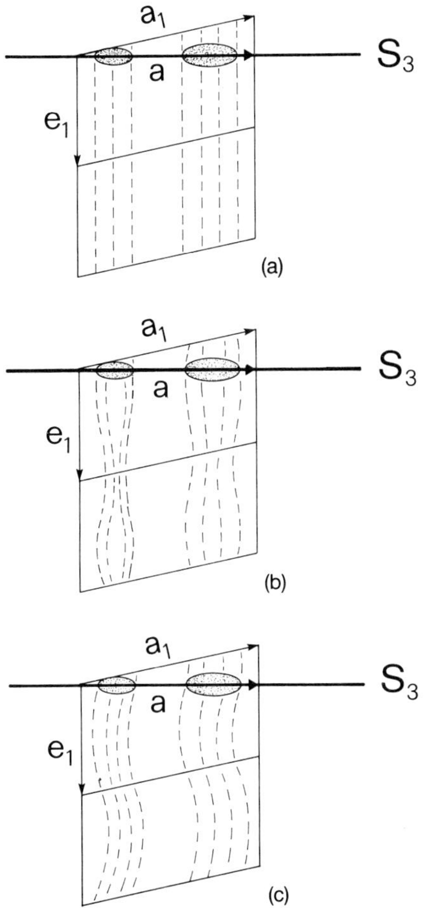

Fig. 3.E.3. Schematic representation of the density ρ' in S_4. S_3 is the horizontal line: its bulging parts represent real atoms. (a) Perfect crystal; (b) density modulation; (c) displacive modulation.

Quasicrystals

In a famous paper by Shechtman, Blech, Gratias, and Cahn[78] electron diffraction patterns of a rapidly solidified Al–Mn alloy were shown which suprisingly displayed five-fold symmetry. Sharpness of the spots suggested long-range translational order, but the presence of the five-fold symmetry violated the sacred rules of crystallography. In particular, by rotation of the specimen, five-fold axes (in six directions), three-fold axes (in 10 directions) and two-fold axes (in 15 directions) could be revealed: the subsequent ascertainment of the existence of an inversion centre fixed, for this Al–Mn phase, the icosahedral point group $m\bar{3}\bar{5}$. Somewhat later a large number of alloys with 'forbidden' symmetries were found: such kinds of materials (providing electron diffraction patterns displaying sharp peaks and forbidden symmetry such as icosahedral, octogonal, decagonal, dodecagonal, etc.) are called quasicrystals. A huge amount of theoretical and experimental publications are now available: for a more comprehensive treatment of the subject and for relevant literature the reader is referred to three excellent reviews.[79–81]

A useful premise to quasicrystals (as well as to the IMSs) is the definition of quasiperiodic functions. It is well known that a periodic function $f(x) = f(x+1)$ may be uniformly approximated by finite sums of functions $\exp(2\pi i n x)$. Accordingly, in a p-dimensional space

$$f'(x_1, \ldots, x_p) = \sum_{n_1,\ldots,n_p} q(n_1, \ldots, n_p) \exp(2\pi i(n_1 x_1 + \ldots + n_p x_p)$$

is a periodic function. However, its 'projection' on one-dimensional space

$$f(x) = \sum_{n_1,\ldots,n_p} q(n_1, \ldots, n_p) \exp(2\pi i(n_1 \nu_1 + \ldots + n_p \nu_p)x) \quad (3.E.6)$$

obtained by fixing in f' x_j to $\nu_j x$ is not periodic if some of the ν_js are irrational. Functions which may be uniformly approximated by finite sums of functions $\exp[2\pi i(n_1\nu_1 x + n_2\nu_2 x_2 + \ldots + n_p\nu_p x)]$ are quasiperiodic functions. We take three simple examples.

1. Let $f'(x_1, x_2) = A_1 \sin 2\pi x_1 + A_2 \sin 2\pi x_2$.
   If we assume $x_2 = \alpha x_1$ where α is an irrational number then
   $$f(x) = A_1 \sin 2\pi x_1 + A_2 \sin 2\pi\alpha x_1$$
   is not periodic.
2. $f(x)$ is the superposition of a periodic sequence of large open circles, schematically represented by $\sum_n \delta(x - na)$, and a periodic sequence of small solid circles represented by $\sum_m \delta(x - ma\tau/2)$. Here, $\tau = 2\cos(\pi/5) = (1+\sqrt{5})/2 = 1.618034\ldots$ is the golden mean, n and m are any integers. Since a and $a\tau/2$ are incommensurate numbers the structure is not periodic (see Fig. 3.E.4, first line) but the diffraction pattern will display delta peaks owing to the fact that the order is perfectly maintained at long distances. The existence of delta peaks may be demonstrated by examining periodic approximations of $f(x)$. Since
   $$\tau = 1 + 1/(1 + 1/(1 + 1/(1 + \ldots),$$
   successive approximations to τ are
   $$\tau_0 = 1,\ \tau_1 = 2,\ \tau_2 = 3/2,\ \tau_3 = 5/3, \ldots$$
   to which periodic structures of increasing periods $a/2$, a, $3a$, $5a$ can be associated. It is easily seen from Fig. 3.E.4 that better periodic approximations of the aperiodic sequence can be obtained by higher order approximants.
3. In Fig. 3.E.5(a), a square lattice with a unit cell with edge length $(1+\tau^2)^{1/2}$ is drawn. An irrational direction is drawn, at an angle of $\arctan(1/\tau)$ with the cell edge. All the lattice points contained in a band parallel to such a direction and with width $(1+\tau)$ are projected on it (strip-projection method), giving rise to a non-periodic pattern (the numerical sequence is 0, 1, 2.618, 3.618, 5.236, 6.854, 7.854, 9.472, 10.472, etc.). All intervals in the projection will have length either 1 or τ (L and S respectively), in the sequence *SLSLLSLSLLSLLSLS*... (Fibonacci sequence). The chain may also be written
   $$x_n = n(3-\tau) - (\tau - 1)\,\mathrm{frac}\,(n\tau),$$

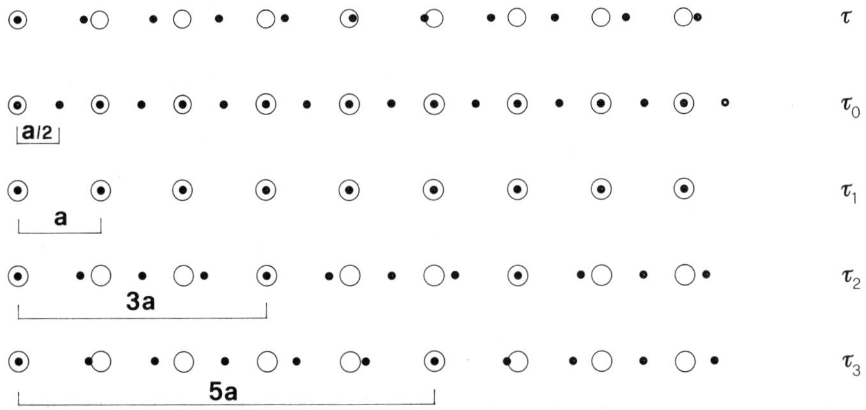

Fig. 3.E.4. An aperiodic long-range ordered structure and its periodic approximants.

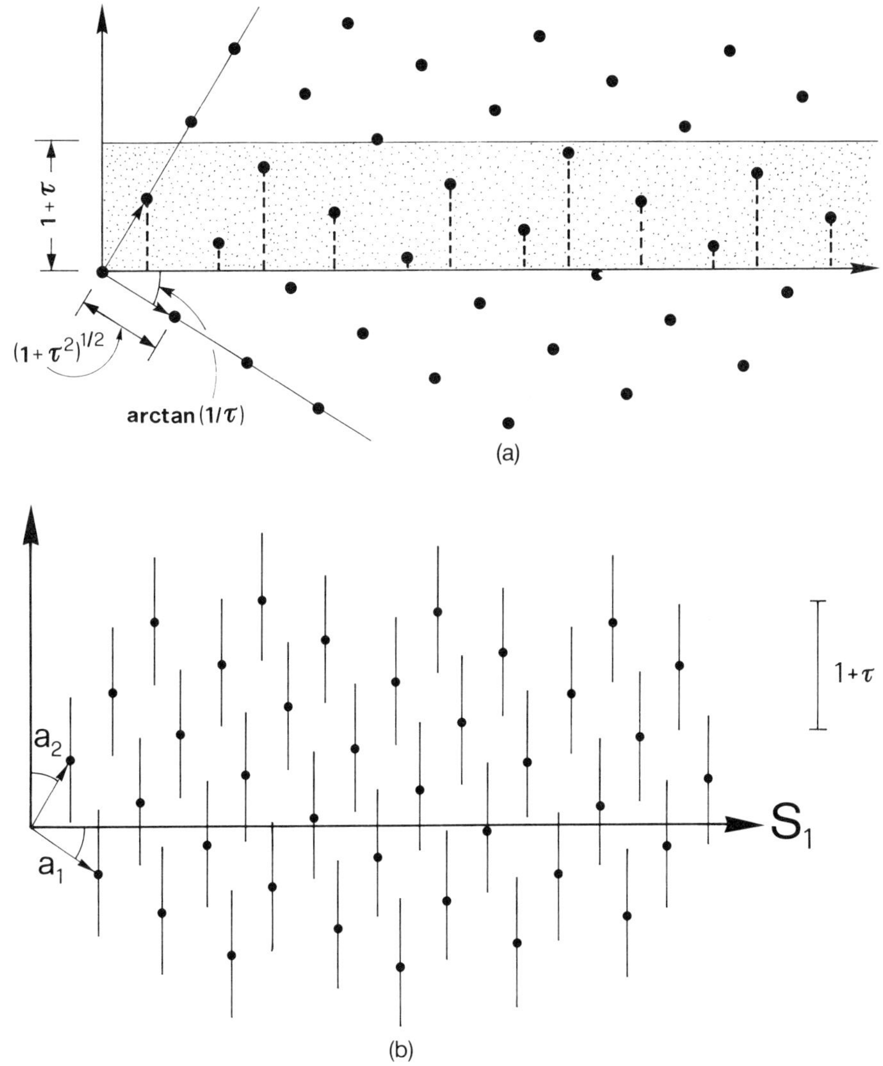

Fig. 3.E.5. (a) Projection of a band in a two-dimensional lattice to obtain a non-periodic ordered sequence (Fibonacci sequence). (b) The Fibonacci sequence results from a cut of the disconnected 'line atoms' of the two-dimensional crystal with the real space S_1.

where frac (x) is the fractional part of x. Since there is a weighted average lattice with constant $a = (3 - \tau)$ the sequence may be considered as a modulated structure with modulation factor $g(x) = (\tau - 1)\,\mathrm{frac}\,(x)$, which may be embedded in a two-dimensional space.

The above three examples suggest that aperiodic sequences in one dimension can be obtained by projection of periodic sequences in two dimensions. The procedure for embedding a one-dimensional aperiodic sequence in a two-dimensional space may be performed in accordance with the previous section. For the last example we can choose $\boldsymbol{a}_1 = (\tau, -1)$ and $\boldsymbol{a}_2 = (1, \tau)$ (see Fig. 3.E.5(b)): then the reciprocal average lattice is also square with basis vectors

$$\boldsymbol{a}_1^* = \frac{1}{(1+\tau^2)}(\tau, -1), \qquad \boldsymbol{a}_2^* = \frac{1}{(1+\tau^2)}(1, \tau).$$

In order to obtain the quasiperiodic one-dimensional sequence on S_1 (see

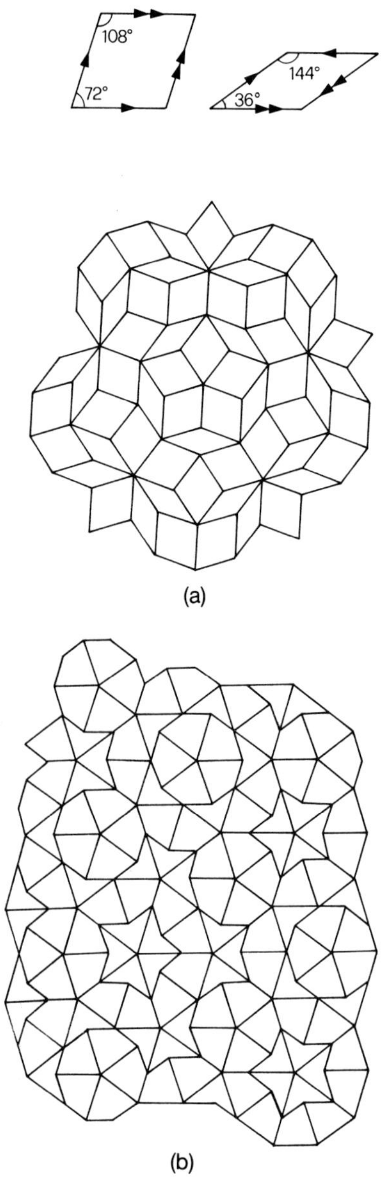

Fig. 3.E.6. (a) A two-dimensional quasilattice showing one five-fold rotation point (plane symmetry 5mm). Basic oblate and prolate rhombi with their matching rules are shown: similarly arrowed edges must fit. (b) Penrose tiling with kites and darts (it does not show five-fold symmetry).

again Fig. 3.E.5(b)) one has to attach parallel line elements in the lattice vertices. Thus the real one-dimensional quasicrystal results from a cut of the disconnected 'line atoms' of the two-dimensional crystal with the physical space S_1.

The procedure may be generalized to the n-dimensional case: the dimension of the space S_n in which a lattice with translational periodicity is obtained is determined by the number of n rationally independent reciprocal basis vectors which are necessary to index the diffraction pattern. The real aperiodic crystal is again the cut (in the direct space) of the n-dimensional crystal with the physical space.

Well known two-dimensional examples of quasiperiodic structures are Penrose tilings.[84,85] Two examples are shown in Fig. 3.E.6: in (a) tiling of the plane is achieved by putting together two rhombic units in accordance with some matching rules (without them the plane should be covered in a periodic way). The pattern shows a five-fold rotation point. In (b) 'kites' and 'darts' are used: no pentagonal symmetry is shown. Mackay[86] first showed that their Fourier transform satisfies a five-fold symmetry.

Octagonal, decagonal, and dodecagonal two-dimensional quasicrystals are also known; all of them can be embedded in a periodic five-dimensional space. The three-dimensional icosahedral lattice mentioned at the beginning of this section may be embedded in a six-dimensional space.

The characteristics of the quasicrystals do not coincide with those of the incommensurately modulated structures. While the latter show main and satellite reflections, an average structure, and crystallographic point symmetry, the quasicrystals show one kind of reflection only, no average structure, and non-crystallographic point symmetry.

References

1. Ewald, P. P. (1917). *Annalen der Physik,* **54,** 519.
2. Laue, M. von (1931). *Ergebnisse der exakten Naturwissenschaften,* **10,** 133.
3. Pinsker, Z. G. (1978). *Dynamical scattering of x-rays in crystals.* Springer, Berlin.
4. Darwin, C. G. (1914). *Philosophical Magazine,* **27,** 315.
5. Darwin, C. G. (1922). *Philosophical Magazine,* **43,** 800.
6. Zachariasen, W. H. (1967). *Acta Crystallographica,* **23,** 558.
7. Becker, P. J. and Coppens, P. (1974). *Acta Crystallographica,* **A30,** 129.
8. Kato, N. (1976). *Acta Crystallographica,* **A32,** 453.
9. Becker, P. J. (1982). In *Computational crystallography,* pp. 462–9. Clarendon, Oxford.
10. Chapuis, G., Templeton, D. H., and Templeton, L. K. (1985). *Acta Crystallographica,* **A41,** 274.
11. Templeton, L. K., Templeton, D. H., Phizackerley, R. P., and Hodgson, K. O. (1982). *Acta Crystallographica,* **A38,** 74.
12. Phillips, J. C. and Hodgson, K. O. (1980). *Acta Crystallographica,* **A36,** 856.
13. Bijvoet, J. M. (1949). *Proceedings of the K. Nederlandse Akademie van Wetenschappen,* **B52,** 313.
14. Phillips, D. L. (1962). *Journal of the Association for Computing Machinery,* **9,** 84.
15. Twomey, S. (1963). *Journal of the Association for Computing Machinery,* **10,** 97.
16. Kennett, T. J., Brewster, P. M., Prestwich, W. V., and Robertson, A. (1978). *Nuclear Instruments and Methods,* **153,** 125.

17. Kauppinen, J. K., Moffatt, D. J., Mantsch, H. H., and Cameron, D. G. (1981). *Applied Spectroscopy*, **35,** 3, 271.
18. Wiedemann, K. E., Unnam, J., and Clark, R. K. (1987). *Powder Diffraction*, **2,** 3, 130.
19. Toraya, H. (1988). *Journal of Applied Crystallography*, **21,** 192.
20. Hamilton, W. C. (1959). *Acta Crystallographica*, **12,** 609.
21. Schomaker, V. and Marsh, R. E. (1983). *Acta Crystallographica*, **A39,** 819.
22. Johnson, C. K. and Levy, H. A. (1974). In *International tables for x-ray crystallography*, Vol. IV, pp. 311–36. Kynock Press, Birmingham.
23. Peterse, W. J. A. M. and Palm, J. H. (1966). *Acta Crystallographica*, **20,** 147.
24. Lipscomb, W. N. (1949). *Acta Crystallographica*, **2,** 193.
25. Colella, R. (1974). *Acta Crystallographica*, **A30,** 413.
26. Post, B. (1977). *Physics Review Letters*, **39,** 760.
27. Chapman, L. D., Yoder, D. R., and Collella, R. (1981). *Physics Review Letters*, **46,** 1578.
28. Chang, S. L. (1982). *Acta Crystallographica*, **A38,** 516.
29. Hummer, K. and Billy, H. (1982). *Acta Crystallographica*, **A38,** 841.
30. Juretscke, H. J. (1984). *Acta Crystallographica*, **A42,** 127.
31. Chang, S. L. (1984). *Multiple diffraction of x-rays in crystals*. Springer, Berlin.
32. Hümmer, K. and Billy, H. (1986). *Acta Crystallographica*, **A42,** 127.
33. Ewald, P. P. (1917). *Annalen der Physik*, **54,** 519.
34. Weckert, E. and Hümmer, K. (1990). *Acta Crystallographica*, **A46,** 387.
35. Renninger, M. (1937). *Zeitschrift für Physik*, **106,** 141.
36. Hümmer, K., Weckert, E., and Bondza, H. (1989). *Acta Crystallographica*, **A45,** 182.
37. Mö, F., Hauback, B. C., and Thorkildsen, G. (1988). *Acta Chemica Scandinavica. Series A*, **42,** 130.
38. Guinier, A. (1963). *X-ray diffraction*. Freeman, San Francisco.
39. Hosemann, R. and Bacchi S. N. (1962). *Direct analysis of diffraction matter*. North-Holland, Amsterdam.
40. Glatter, O. and Kratky, O. (1982). *Small angle x-ray scattering*. Academic, London.
41. Guinier, A. and Fournet, G. (1955). *Small-angle scattering of x-rays*. Wiley, New York.
42. Debye, P. (1915). *Annalen der Physik*, **46,** 809.
43. Ehrenfest, P. (1915). *Proceedings of the Amsterdam Akademie*, **17,** 1132.
44. Zernicke, F. and Prins, J. A. (1927) *Zeitschrift für Physik*, **41,** 184.
45. Fournet, G. (1951). *Transactions of the Faraday Society*, **11,** 121.
46. Hildebrandt, R. L. (1975). *Molecular structure by diffraction methods*, The Chemical Society, special periodic report, No. 3 (ed. G. A. Sim and L. E. Sutton).
47. Oberhammer, H. (1976). *Molecular structure by diffraction methods*, The Chemical Society, special perodic report, No. 4 (ed. G. A. Sim and L. E. Sutton).
48. Domenicano, A., Schultz, G., Hargittay, I., Colapietro, M., Portalone, G., George, P., and Bock, C. W. (1989). *Structural Chemistry*, **1,** 107.
49. Kirkwood, J. G., Maun, E. K., and Alder, B. J. (1952). *Journal of Chemical Physics*, **85,** 777.
50. Morgan, I. and Warren, B. E. (1938). *Journal of Chemical Physics*, **6,** 666.
51. Bernal, I. D. and Fowler, R. H. (1933). *Journal of Chemical Physics*, **1,** 515.
52. Warren, B. E., Krutter, H., and Morningstar, O. (1936). *Journal of the American Ceramic Society*, **19,** 202.
53. Guinier, A. (1939). *Annales de Physique*, **12,** 161.
54. Pilz, I., Goral, K., Hoyaerts, M., Lontie, R., and Witters, R. (1980). *European Journal of Biochemistry*, **105,** 539.

55. Mittelbach, P. (1964). *Acta Physica Austriaca,* **19,** 53.
56. Kratky, O. and Porod, G. (1953). In *Die Physik der Hochpolymere,* Vol. II. (ed. H. A. Stuart). Springer, Berlin.
57. Porod, G. (1951). *Kolloid-Zeitschrift* **124,** 83.
58. Luzzati, V. (1960). *Acta Crystallographica,* **13,** 939.
59. Stevens, E. D. and Coppens, P. (1976). *Acta Crystallographica,* **A32,** 915.
60. Hirshfeld, F. L. (1971). *Acta Crystallographica,* **B27,** 769.
61. Stewart, R. F. (1976). *Acta Crystallographica,* **A32,** 565.
62. Hansen, N. K. and Coppens, P. (1978). *Acta Crystallographica,* **A34,** 909.
63. Kurki-Suonio, K. (1977). *Israel Journal of Chemistry,* **16,** 115.
64. Hehre, W. J., Stewart, R. F., and Pople, J. A. (1969). *Journal of Chemical Physics,* **51,** 2657.
65. Stewart, R. F. (1973). *Journal of Chemical Physics,* **58,** 4430.
66. Epstein, J., Ruble, J. R., and Craven, B. M. (1982). *Acta Crystallographica,* **B38,** 140–9.
67. Rees, B. (1977). *Israel Journal of Chemistry,* **16,** 180.
68. Hirshfeld, F. L. (1977). *Israel Journal of Chemistry,* **16,** 198.
69. Coppens, P. (1975). *Physical Review Letters,* **35,** 98.
70. de Wolff, P. M. (1974). *Acta Crystallographica,* **A30,** 777.
71. Janner, A. and Janssen, T. (1977). *Physical Review,* **B15,** 643.
72. Janner, A. and Janssen, T. (1980). *Acta Crystallographica,* **A36,** 399.
73. Janner, A. and Janssen, T. (1980). *Acta Crystallographica,* **A36,** 408.
74. de Wolff, P. M., Janssen, T., and Janner, A. (1981). *Acta Crystallographica,* **A37,** 625.
75. Yamamoto, A. (1982). *Acta Crystallographica,* **A38,** 87.
76. Petricek, V., Coppens, P., and Becker, P. (1985). *Acta Crystallographica,* **A41,** 478.
77. Petricek, V. and Coppens, P. (1988). *Acta Crystallographica,* **A44,** 1051.
78. Shechtman, D., Blech, I., Gratias, D., and Cahn, J. W. (1984). *Physical Review Letters,* **53,** 1951.
79. Janot, Ch. and Dubois, J. M. (1988). *Journal of Physics F: Metal Physics,* **18,** 2303.
80. Janssen, T. (1988). *Physics Reports,* **168,** 55.
81. Steurer, W. (1990). *Zeitschrift für Kristallographie,* **190,** 179.
82. Duneau, M. and Katz, A. (1985). *Physical Review Letters,* **54,** 2688.
83. Janssen, T. (1986). *Acta Crystallographica,* **A42,** 261.
84. Penrose, R. (1974). *Bulletin of the Institute of Applied Mathematics,* **10,** 266.
85. Gardner, M. (1977). *Scientific American,* **236,** 110.
86. Mackay, A. L. (1982). *Physica,* **A114,** 609.

Experimental methods in X-ray crystallography

4

HUGO L. MONACO

Introduction

This chapter discusses the experimental methods used to study the diffraction of X-rays by crystalline materials. Although, as seen in Appendix 3.B (pp. 195 and 198), electrons and neutrons are also diffracted by crystals, we will concentrate our attention on X-ray diffraction. We begin discussing how X-rays are produced and how one can define the beam of radiation that will interact with the crystalline sample. The specimens that will receive our attention are single crystals and polycrystalline materials, that is aggregates of a very large number of very small crystals. We discuss the methods used to record the diffraction pattern and to measure the intensities of the X-rays scattered by these two types of specimen in separate sections. The ultimate goal of extracting structure factor amplitudes from diffracted intensities requires the application of a series of correction factors. This process, called data reduction, is discussed in the final section of the chapter.

X-ray sources

Conventional generators

All the standard laboratory sources normally used for X-ray diffraction experiments generate radiation using the same physical principles but can vary substantially in their construction details. The two types of conventional generators that are used in conjunction with the data recording devices discussed on pp. 245 and 287 are sealed-tube and rotating-anode generators. Most of the techniques used for diffraction data collection require monochromatic radiation. Due to the way in which radiation is produced in the conventional generators, only a discrete number of possible wavelengths can be selected for experimental use. This limited choice and the difference in intensity are two of the major differences between this type of radiation and that generated by synchrotrons.

The origin of X-rays in the conventional sources

X-rays are produced when a beam of electrons, accelerated by a high voltage, strikes a metal target and is therefore rapidly decelerated by collision with the metal atoms. Most of the electrons do not lose their energy in a single collision but do it gradually through multiple events. The result is the production of a continuous spectrum of X-rays called white radiation. If all of the energy carried by an electron is transformed into radiation, the energy of an X-ray photon is

$$E_{max} = h\nu_{max} = eV$$

where h is Planck's constant, ν_{max} the photon frequency and the subscript max indicates that this is the maximum possible energy, e is the charge of the electron, and V the accelerating potential.

If we substitute the frequency in terms of the wavelength

$$h\nu_{max} = hc/\lambda_{min} = eV$$

and

$$\lambda_{min} = hc/eV = 12\,398/V \tag{4.1}$$

where the potential V has to be measured in volts and the minimum wavelength is obtained in angstroms. This equation shows that there is a minimum value for the wavelength of the X-rays that can be obtained by this process which is a function of the voltage accelerating the electrons.

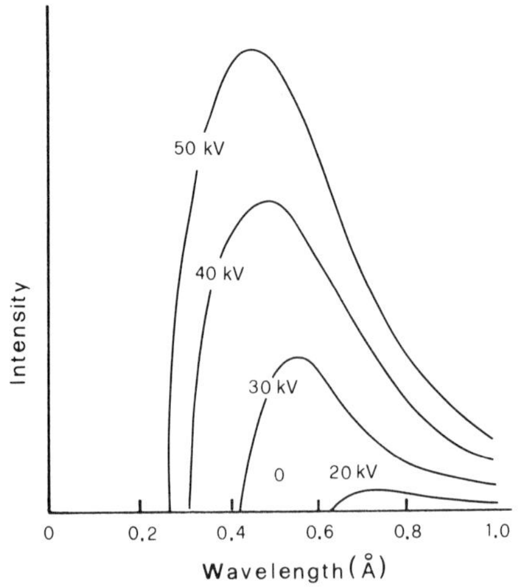

Fig. 4.1. X-ray white radiation spectra as a function of the accelerating voltage.

Total conversion of the electron energy into radiation is not a highly probable event and therefore the radiation of the highest intensity is obtained at a somewhat longer wavelength. Figure 4.1 shows some continuous X-ray spectra as a function of the accelerating voltage. Notice that as the voltage is increased both the minimum wavelength and the position of the intensity maximum shift to the left. The intensity maximum is found at a wavelength which is approximately 1.5 times λ_{min}.

The total X-ray intensity generated per second in this way, a quantity which is proportional to the areas under the curves, can be calculated using the equation

$$I_w = AiZV^n \tag{4.2}$$

where A is a proportionality constant, i the electrical current, a measure of the number of electrons which are generating X-rays, Z is the atomic number of the target, V the accelerating potential, and n a constant with a value of about 2.

When the energy of the electrons striking the target is higher than a certain threshold value, a second type of spectrum, discontinuous and with very sharp lines, appears superimposed on the white radiation curves just described. This second spectrum is called the characteristic spectrum because its peaks are found at precisely defined wavelengths which depend on the material constituting the target. The characteristic spectrum of copper is shown in Fig. 4.2. The electrons with an energy above the threshold potential are capable of ionizing the target atoms by ejecting an electron from one of the inner shells. When that happens another electron from a higher atomic energy level can move in to fill the vacancy created and, since the new level has a lower energy, emit the energy difference as a characteristic X-ray photon whose wavelength depends on the difference in energy between the two levels involved.

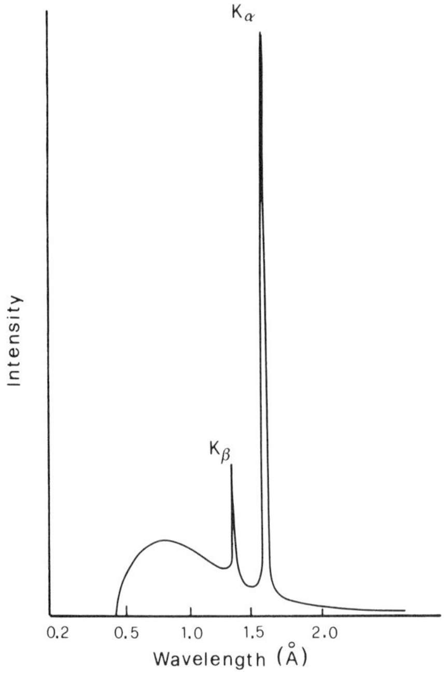

Fig. 4.2. Characteristic spectrum of copper superimposed on the white radiation spectrum. Notice the ratio of the relative intensities of the K_α and K_β lines.

The characteristic lines of this type of spectrum are called K, L, and M and correspond to transitions from higher energy orbitals to the K, L, and M orbitals, that is to the orbitals of principal quantum numbers $n = 1$, 2, and 3. When the two orbitals involved in the transition are adjacent the line is called α, if they are separated by another shell, the line is called β. Thus, the Cu K_α line is produced by a copper target in which the atoms lost an electron in the orbital of $n = 1$ and the vacancy was filled by an electron of the orbital $n = 2$. The X-ray photon energy is the difference between these two energy levels. Since for every principal quantum number n there are n energy levels corresponding to the possible values of the quantum number 1 (from 0 to $n - 1$), the α and β lines are actually split into multiple lines that are very close to one another because the difference between these energy levels is small. Still, X-ray radiation corresponding to all the possible energy differences is not observed because some energy transitions are forbidden by the selection rules. Thus, although Fig. 4.2 has a scale on the abscissa which cannot show it, the Cu K_α line is actually split into a doublet, the $K_{\alpha 1}$ and $K_{\alpha 2}$ lines, of very similar wavelengths and which are, for this reason, not easily separable.

The frequency of the characteristic line corresponding to a given transition is related to the atomic number of the element that gave rise to it, Z, by Moseley's law

$$\nu = C(Z - \sigma)^2 \tag{4.3}$$

where the constant C depends on the atomic energy levels involved in the transition and the constant σ takes into account the interactions with other electrons. Thus, in a plot of $\nu^{1/2}$ as a function of Z for a given transition the points corresponding to different target elements lie in a straight line and different lines are obtained for the K_{α_1}, K_{α_2}, K_{β_1}, etc., transitions. The characteristic frequency is higher the higher the atomic number and so the Mo K_α line ($Z = 42$) has a higher frequency and therefore a higher energy than the Cu K_α line ($Z = 29$). A full list of the wavelengths of the characteristic lines of the elements which are used in X-ray diffraction studies can be found in the *International tables for x-ray crystallography.*[1] Here, we will just point out that the two most frequently used lines are the Cu K_α line, $\lambda = 1.5418$ Å and the Mo K_α line, $\lambda = 0.7107$ Å. Both are doublets of slightly different wavelengths as pointed out before.

The intensity of a characteristic K line can be calculated using the equation:

$$I_K = Bi(V - V_K)^{1.5} \tag{4.4}$$

where B is a constant, i the electrical current, and V_K the excitation potential of the K series, a quantity which is proportional to the energy required to remove a K electron from the target atom. It can be shown[2] that the ratio I_K/I_W is a maximum if the accelerating potential is chosen to be $V = 4V_K$. If this condition is fulfilled, the K_α line is about 90 times more intense than the white radiation of equal wavelength (I_W). The K_{α_1} line is approximately twice as intense as the K_{α_2} line and the ratio K_α/K_β depends on Z but it averages 5 (see Rieck[1]). The data collection methods that use monochromatic radiation discussed later all use K_α radiation and therefore require the elimination of the K_β component of the spectrum which is always present. The methods used to achieve this are discussed on p. 241.

Sealed-tube and rotating-anode generators

A conventional generator consists of a high-voltage power supply with electronic controls, connected to either a sealed tube or to the cathode of a rotating-anode generator. In the second case there must also be a way of making, keeping, and monitoring the high vacuum required for X-ray generation. Just how high the voltage to be applied must be, can be estimated from the considerations of the previous paragraphs and taking an excitation voltage given by Rieck.[1] This value for the copper K energy level is 8.981 kV so if one wishes to apply a voltage of 4 times V_K about 36 kV have to be applied. For molybdenum even higher voltages are required.

The sketch of a sealed X-ray tube is shown in Fig. 4.3(a). It consists of a cathode with a filament that emits the electrons that are accelerated under vacuum by the high voltage applied and hit the fixed anode made of the metal whose characteristic spectrum has a K_α line of a wavelength which is adequate for the diffraction experiment to be performed. The high vacuum is necessary because the presence of gas molecules in the tube decreases the efficiency of the X-ray generating process by collisions with the electrons in the beam.

The process of X-ray production discussed in the previous section is highly inefficient: only about 0.1 per cent of the power applied is transformed into X-rays, the rest being dissipated as heat. In order to avoid melting of the target, it therefore becomes necessary to cool it, which is done by circulating cold water as shown in the figure. It is the efficiency of the cooling system that ultimately determines the maximum power that can be applied to the tube. The problem of heat dissipation also dictates the choice of the focal area of the beam of electrons on the anode. This area is always chosen to be rectangular with the sides in a ratio of at least six to one. In this way, when one looks at the focused electron beam on the target, which determines the surface that produces X-rays, in the direction of the longest side of the rectangle one can see, by choosing the appropriate angle, a small square focus with a side equal to the smallest rectangle side. At the same time, the real area that dissipates heat is several times that of a square of that size. The X-rays generated come out of the tube through four beryllium windows: the two that are shown in the figure, that are found in the direction of the longest axis of the rectangular focus, and another two in a direction perpendicular to it. The first two windows are used for single-crystal work, the others may be used when a linear focus is needed.

In rotating-anode generators, the target area seen by the electron beam is continuously renovated because the anode is rotated. In this way, much higher powers per unit of focal area can be applied to the unit and consequently higher X-ray intensities can be produced. Figure 4.3(b) is a sketch of a rotating anode chamber showing the two X-ray windows used for single-crystal work, the filament, placed with its longest axis parallel to the direction of the windows, the focusing cup, used to focus the electron beam on the target, and the anode which rotates about an axis parallel to the direction of the windows.

In a rotating-anode generator, the chamber has to be kept under vacuum for the reasons stated before, water has to be circulated to cool the anode as is done in the case of a sealed tube, and of course the anode has to be rotated at a certain speed. The system is therefore mechanically much more complicated than a tube and it requires a fairly frequent and demanding maintenance work. Figure 4.4 is a photograph of a modern rotating anode X-ray generator. Phillips[3] gives a detailed discussion of the advantages of

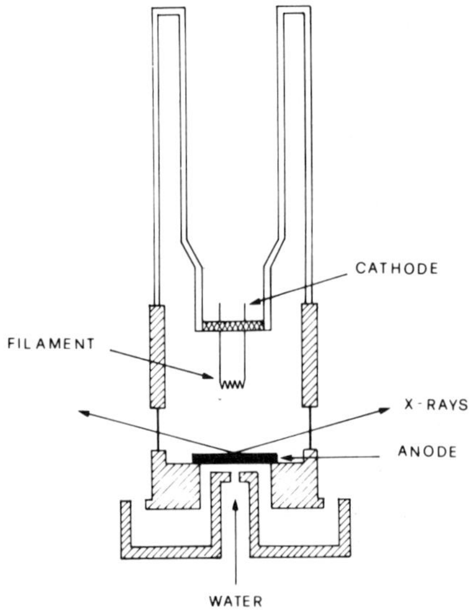

(a)

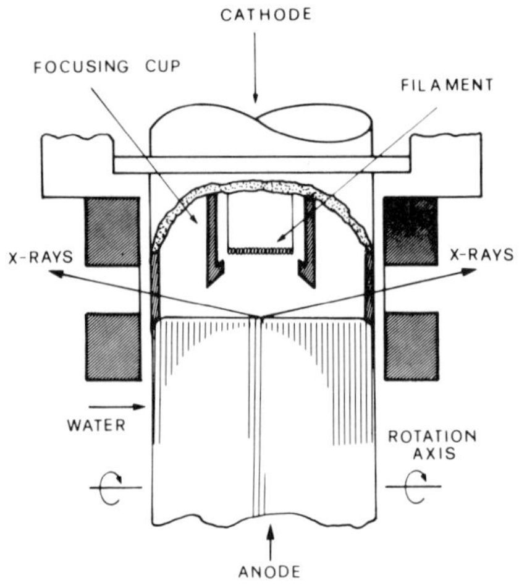

(b)

Fig. 4.3. (a) A sealed X-ray tube. (b) Sketch of a rotating-anode chamber. The path followed by the water that cools the anode is not shown in the figure. The anode is cylindrical and rotates about the axis shown.

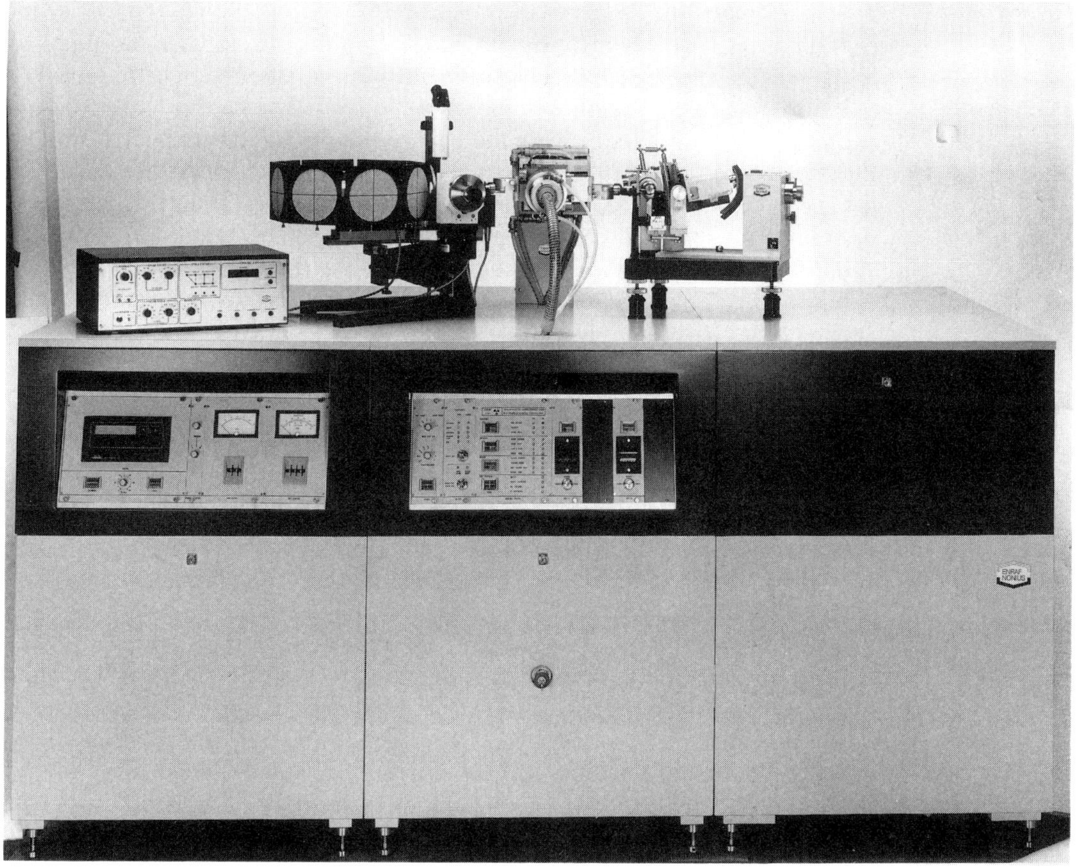

Fig. 4.4. A rotating-anode X-ray generator. (Photograph courtesy of Enraf Nonius.)

rotating-anode generators together with an exhaustive description of the practical problems encountered in their operation.

Choice of the type of radiation

As discussed above, the K_α lines are those of highest intensity and therefore they are the ones that are normally used for standard X-ray diffraction work. Thus, in practice, the choice of the wavelength of the X-rays generated by conventional equipment is limited to the values of the characteristic K_α lines of the metals commonly used as targets. Two of these metals are most frequently used: copper and molybdenum; their K_α transitions generate radiation with a wavelength of about 1.5 and 0.7 Å respectively. As will be seen later, for a given crystal-to-detector distance and unit cell, the diffracted beams will be more separated on the detector if a longer wavelength is used. Thus, copper is used for macromolecular work in which one usually was large unit cell parameters and for the structure determination of organic molecules, which do not contain atoms that absorb this radiation strongly. Absorption of the radiation by the sample is therefore also a primary consideration in the selection of the wavelength to be used.

Another consideration is the maximum resolution of the reflections that will be recorded. We saw on p. 155 that there is a limiting sphere of radius

$2/\lambda$ that limits the volume of reciprocal space accessible to diffraction experiments. This sphere obviously has a different size for the two types of radiation discussed and therefore molybdenum radiation may be required to record data to a resolution not accessible to copper radiation.

Finally we mention the detection efficiency of the method that will be used to record the diffracted intensities. Film is a better detector for copper than for molybdenum radiation and some area detectors have been optimized for use with copper radiation. On the other hand, diffractometer counters have a very high counting efficiency for the Mo K_α radiation which explains why this type of radiation is so widely used for small-molecule single-crystal X-ray diffraction work.

Synchrotron radiation

X-rays, as well as other types of electromagnetic radiation, can also be generated by sources known as synchrotron radiation facilities. In these installations either electrons or positrons are accelerated at relativistic velocities along orbits of very large radii, several metres or even hundreds of metres. These sources are, by necessity, very complex and those that produce suitable X-rays are limited, located mainly in the United States and Europe. However, since, as we will see, the X-rays they produce are in many ways much better than those generated by conventional sources, their use has grown steadily in the crystallographic community. As a result, more beam time has been made available to crystallographers and more synchrotron sources are planned for construction in different countries. Among those, the 6 GeV storage ring to be built in the USA[4] and the European Synchrotron Radiation Facility,[5] designed specifically to produce the best possible X-rays, promise to be of special importance is the development of this field.

From the extensive literature that exists in this ever expanding field we recommend two very elementary descriptions,[6,7] an introductory textbook,[8] and a more advanced treatise in several volumes.[9] In this last treatise chapters 1, 2, and 11 of volume 1 are specially relevant to our discussion; volume 3 of the series is totally devoted to X-ray methods.

Generation of X-rays in a synchrotron radiation source

It is well known that charged particles moving under the influence of an accelerating field emit electromagnetic radiation. The energy of this radiation is dependent on the velocity of the particle. If the velocity is like that of the electrons moving in an antenna, emission takes place in the radio-frequency range but when the charged particles, electrons, or positrons, move with a speed approaching that of light, the spectrum extends into higher energy regions covering the X-ray range. An additional consequence of relativistic effects is a distortion of the angular distribution of the radiation. When the charged particles move at velocities approaching that of light, radiation is emitted in a very narrow cone parallel to the instantaneous velocity. In a synchrotron source the charged particles are made to move in closed trajectories, often circular or elliptical, and so the radiation is generated in cones tangent to the path followed by the particles (see for example chapter 1 of Koch[9]).

Figure 4.5 shows the essential elements of a synchrotron radiation source.

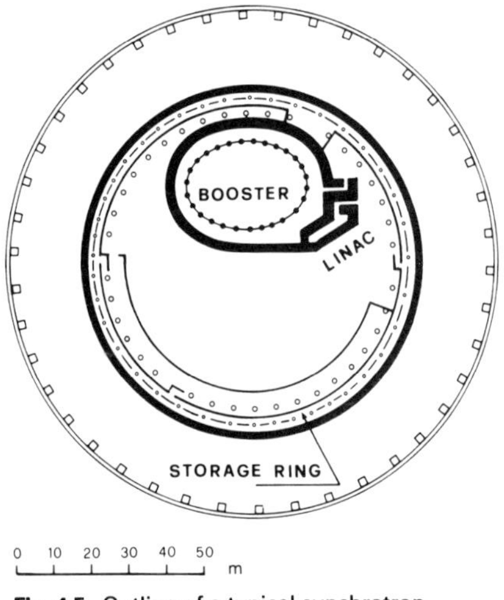

Fig. 4.5. Outline of a typical synchrotron radiation facility. The electrons or positrons are accelerated in the linear accelerator (linac) and then in the booster to be finally injected into the storage ring where they are kept circulating for periods of several hours. The beam lines, not shown in the figure, are tangential to the particle trajectory. Notice the scale at the bottom of the picture corresponding to 0–50 m.

Notice the scale at the bottom which gives an idea of the size of this type of facility. The basic element of the installation from which radiation is generated is the storage ring, a toroidal cavity in which the charged particles are kept circulating under vacuum. An extremely high vacuum is required or else the particles are lost by collision with the atoms present in the cavity. Prior to injection into the storage ring, the particles must be accelerated, for example, first by a linear accelerator and then by a booster as shown in the figure. The other two elements that are essential to the operation of the ring are the so-called lattice, that is the set of magnets which force the particles to follow a closed trajectory as well as performing other functions, and the radio-frequency cavity system which restores to the particles the energy they lose as synchrotron radiation. The beam lines, not shown in the figure, are tangential to the storage ring.

It can be shown that a charged particle moving along a circular orbit emits as electromagnetic radiation the following power[8,9]

$$P = \frac{2e^2cE^4}{3R^2(m_0c^2)^4} = \frac{2e^2c\gamma^4}{3R^2} \tag{4.5}$$

where P is the energy emitted per unit time, e is the particle charge, c the speed of light, E the energy of the particle, m_0 its mass at rest, and R the bending radius of the orbit. This equation explains why high-energy particles are required and also why only electrons or positrons are used. The power emitted by heavier particles such as protons is too low to be of significant importance.

The quantity γ, the ratio of the total energy to the rest energy of the particle, is of considerable importance since it is approximately related to the opening angle of the cone of radiation by

$$\Delta\psi \cong 1/\gamma \tag{4.6}$$

where $\Delta\psi$ is the emission angle in radians. A more exact relationship between these two parameters is given in reference [10].

The total power emitted by the ring is the power emitted by a particle in one revolution multiplied by the number of particles and divided by the time it takes them to complete a revolution.

The total power can be shown to be equal to[8,9]

$$P_{tot} = 26.6E^3Bi \tag{4.7}$$

where the energy of the particles is measured in GeV, the magnetic field B in tesla, the current i in amperes and the power is obtained in kW. The power is thus seen to be directly proportional to the current in the storage ring.

The radiation emitted at a modern storage ring comes from two sources: the bending magnets and the insertion devices. The first is the radiation we have discussed so far, the second is generated by special devices called wigglers and undulators which are briefly discussed below.

An important property of the radiation generated by bending magnets is its wide spectral distribution which can be described quantitatively in terms of the spectral flux N which is the number of photons emitted per unit time interval into a relative band width $\Delta\lambda/\lambda$ into an angle element $d\theta$ in the plane of the electron orbit and integrated in the vertical plane. It can be shown that the flux of radiation normalized to the ring current generated by bending magnets is equal to[8]

$$N(h\nu) = 1.256 \times 10^7 \, \gamma \, G_1(y) \text{ photons s}^{-1} \text{ mrad}^{-1} \text{ mA}^{-1} \quad (0.1\% \text{ band width}) \qquad (4.8)$$

where the factor $G_1(y)$ is an energy-dependent function that can be found tabulated for example in the book by Margaritondo.[8] The variable y is defined as follows

$$y = h\nu / E_c$$

where E_c is the critical energy associated with a magnetic field B which in keV is given by

$$E_c = 2.22 \, E^3 / R \qquad (4.9)$$

where E is measured in GeV and R, the radius of the bending magnet in metres. An alternative expression for E_c is

$$E_c = 0.665 \, E^2 B \qquad (4.10)$$

with B measured in tesla.

A widely used parameter related to the critical photon energy is the critical wavelength λ_c

$$E_c = h\nu_c = hc/\lambda_c = 12.4/\lambda_c \qquad (4.11)$$

where E_c is measured in keV and λ_c in Å.

The spectra of total emitted flux for several synchrotron radiation sources is shown in Fig. 4.6 taken from Materlik.[10] The critical wavelength, λ_c, is useful as an indicator of the suitability of a source for X-ray experiments. As can be seen in the figure although it does not correspond to the maximum flux it is close enough to give a good idea of the position of this maximum. It can be shown that ideally the wavelength used at a synchrotron source should fall in the range $0.25\lambda_c$ to $4.0\lambda_c$. Table 4.1, taken from the literature,[5] gives E_c, λ_c, and other relevant parameters for the

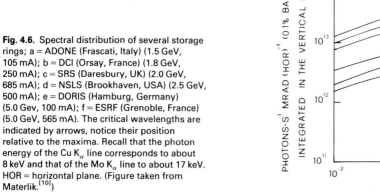

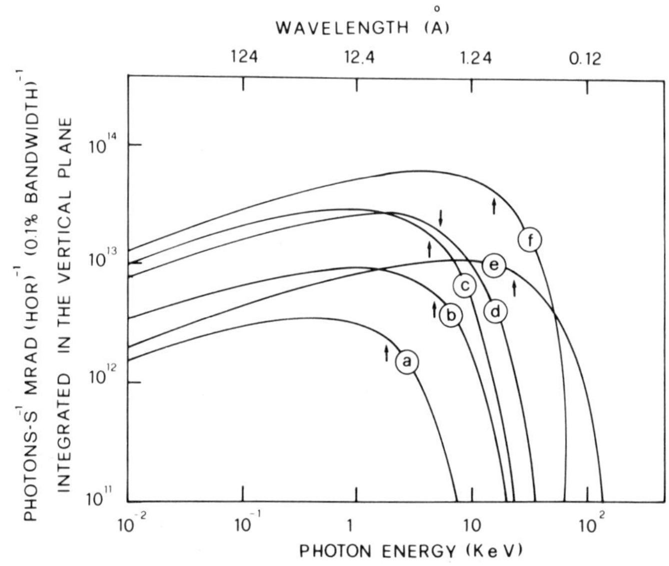

Fig. 4.6. Spectral distribution of several storage rings; a = ADONE (Frascati, Italy) (1.5 GeV, 105 mA); b = DCI (Orsay, France) (1.8 GeV, 250 mA); c = SRS (Daresbury, UK) (2.0 GeV, 685 mA); d = NSLS (Brookhaven, USA) (2.5 GeV, 500 mA); e = DORIS (Hamburg, Germany) (5.0 Gev, 100 mA); f = ESRF (Grenoble, France) (5.0 GeV, 565 mA). The critical wavelengths are indicated by arrows, notice their position relative to the maxima. Recall that the photon energy of the Cu K_α line corresponds to about 8 keV and that of the Mo K_α line to about 17 keV. HOR = horizontal plane. (Figure taken from Materlik.[10])

Table 4.1. Relevant parameters of the Synchrotron Radiation Sources in operation in 1987 (taken from reference 5)

Name	Location	E(GeV)	R(m)	i(mA)	E_c(keV)	λ_c(Å)	Emittance ($\times 10^8$ m rad)	Insertion (devices)
Group I, $E_c \leq 0.06$ keV								
N-100	Karkhov, USSR	0.10	0.5	50	0.004	3 100.0		
Surf II	Washington, USA	0.25	0.84	25	0.041	302.4	27	
Tantalus I	Wisconsin, USA	0.24	0.64	200	0.048	258.3	27	2
Group II, E_c 0.06–2 keV								
Sor ring	Tokyo, Japan	0.4	1.1	250	0.13	95.4	30	1
Siberia I	Moscow, USSR	0.45						
Max	Lund, Sweden	0.55	1.2	370	0.30	41.3	3	2
COSY	Berlin, Germany	0.56					250	
Teras	Tsukuba, Japan	0.6						
ACO	Orsay, France	0.7	1.1	100	0.32	38.8	15	1
NSLS	Brookhaven, USA	0.75	1.9	500	0.4	31.0	13	2
UV SOR	Okazaki, Japan	0.75	2.2	500	0.43	28.8	8.10	2
Fian C-60	Moscow, USSR	0.67	1.6	10	0.44	28.2		
Vepp-2M	Novosibirsk, USSR	0.80	1.22	100	0.54	23.0		2
Hesyrl	Hefei, China	0.8					9	3
Superaco	Orsay, France	0.8	1.75				3.8	6
SPRL	Stanford, USA	1	2.1	500	1.05	11.8	2	5
Bessy	Berlin, Germany	0.8	1.83	500	0.62	20.0	4	1/2
Aladdin	Wisconsin, USA	1.3	2.08	500	1.07	11.6	6	3
INS-ES	Tokyo, Japan	1.3	4.0	30	1.22	10.2		
Pakhara	Moscow, USSR	1.36	4.0	300	1.22	10.2		
Sirius	Tomsk, USSR	1.36	4.23	15	1.32	9.4		
Adone	Frascati, Italy	1.5	5.0	60	1.5	8.3	22.5	2–4
Group III, E_c 2–30 keV								
DCI	Orsay, France	1.8	4.0	300	3.63	3.4	1500	1
SRS	Daresbury, UK	2.0	5.55	500	3.2	3.9	150	1/4
Vepp-3	Novosibirsk, USSR	2.25	6.15	100	4.2	3.0	150	2
Photon Factory	Tsukuba, Japan	2.5	8.33	500	4.16	3.0	50/15	2/7
NSLS	Brookhaven, USA	2.5	8.17	500	4.2	3.0	8	4/5
Bonn	Bonn, Germany	2.0	7.6	50	2.3	5.4		
Siberia II	Moscow, USSR	2.5	5	300	6.9	1.8	8	3
BEPC	Beijing, China	2.8					66	8
Elsa	Bonn, Germany	3.5	10.1	50	9.3	1.3	50	1
Arus	Erevan, USSR	4.5	24.6	1.5	8.22	1.5		
Spear	Stanford, USA	4.0	12.7	100	11.1	1.1	45	4/12
Doris II	Hamburg, Germany	5.0	12.1	50	22.9	0.5	570	1/7
Tristan ACC	Tsukuba, Japan	6/8					48	3
Group IV, $E_c > 30$ keV								
CESR	Ithaca, USA	8.0	32.5	100	35.0	0.35	20	1/3
Vepp-4	Novosibirsk, USSR	7.0	16.5	10	46.1	0.27		2
Petra	Hamburg, Germany	18.0	192.0	18	67.4	0.18		
PEP	Stanford, USA	18.0	165.5	10	78.0	0.16	15	1
Tristan	Tsukuba, Japan	30.0					18	

synchrotron sources in operation in 1987. If the spectral flux is transformed into power, i.e. the number of photons is multiplied by the energy of a photon $h\nu$, it can be shown that half the total power is irradiated below and half above the critical value λ_c.

Other parameters of interest are the source size and the divergence of the irradiated beam. The charged particle beam in the storage ring has a Gaussian profile characterized by the parameter σ_x, in the horizontal and σ_z in the vertical plane. The full width at half maximum of the beam can then be calculated as 2.35σ and the source area F is estimated as

$$F = 2.35^2 \sigma_x \sigma_z.$$

Similarly the angular distribution is characterized by the parameters $\sigma_{x'}$ and $\sigma_{z'}$ and the solid angle of emission is then estimated as

$$\Omega = 2.35^2 \sigma_{x'} \sigma_{z'}$$

Although σ_x, $\sigma_{x'}$ and σ_z, $\sigma_{z'}$ vary along the orbit their variations are correlated and it is thus useful to define another parameter, the emittance, which at special symmetry positions is found to be

$$\begin{aligned} \varepsilon_x &= \sigma_x \sigma_{x'} \\ \varepsilon_z &= \sigma_z \sigma_{z'}. \end{aligned} \tag{4.12}$$

The emittance is instead a constant along the charged particle path and it is thus another important parameter characteristic of an installation. The emittances of the synchrotron sources in operation in 1987 are also shown in Table 4.1.

Another useful function, which is often used to compare the potential performance of two sources, is the spectral brightness, also called spectral brilliance,[8] defined as the number of photons emitted per unit area of the source at point x, z over a 0.1 per cent relative band width per unit solid angle $d\Omega$ and unit time in the direction defined by the angles ψ (defined by the instantaneous velocity of the charged particle and the projection of the direction of observation onto the vertical plane) and θ (defined by the projection onto the horizontal plane instead). It can be shown that the spectral brightness b is equal to (see Margaritondo[8] chapter 2)

$$b = N(2\pi)^{-3/2}(\sigma_x \sigma_z \sigma_\psi)^{-1} \exp[-(x/2\sigma_x)^2 - (z/2\sigma_z)^2 - (\psi/2\sigma_\psi)^2]$$

where N is the spectral flux. If one defines the central brightness b_c which is the brightness for $x = z = \psi = 0$ it is obvious that

$$b_c = N(2\pi)^{-3/2}(\sigma_x \sigma_z \sigma_\psi)^{-1}. \tag{4.13}$$

From this equation it can be seen that the brightness can be increased by increasing the flux or by decreasing the σs. Decreasing the σs can be accomplished by reducing the emittance of the storage ring (eqn (4.12)). The emittance of a ring is thus a fundamental parameter to be taken into account in comparing the expected performance from two different sources.

A radiation spectrum quite different from that produced by bending magnets can be obtained by the use of insertion devices. These are a series of periodically spaced magnets of alternating polarity which are inserted in a straight region of the ring and which do not alter the ideal closed orbit of the particles in the storage ring. Most insertion devices create a sinusoidal magnetic field which forces the charged particles to oscillate around the mean orbit. According to their characteristics they are called wigglers or undulators.

The parameter that has to be examined to determine whether an insertion device is a wiggler or an undulator is the K parameter

$$K = \alpha\gamma = \frac{eB_0\lambda_0}{2\pi m_0 c} = 0.934 B_0 \lambda_0 \tag{4.14}$$

where α is the maximum deflection angle of the electron or positron trajectory, B_0 is the oscillating magnetic field measured in tesla and λ_0 is the period of the magnetic array measured in centimetres. Recalling that $1/\gamma$ is approximately equal to the natural opening angle $\Delta\psi$, K becomes the ratio between the maximum deflection angle of the electron trajectory along the insertion device, α, and $\Delta\psi$.

When the parameter $K \gg 1$ the device is called a wiggler. The effect of a wiggler on the emitted spectrum is to shift the critical energy E_c to higher values and to increase the intensity of the radiation by a factor proportional to the number of periods of the magnetic array. No interference effects are observed and so the emitted spectrum is qualitatively very similar to that obtained from bending magnets.

If the parameter $K \ll 1$ the device is called an undulator and interference occurs between radiation which is emitted by the same electron at different points in its path through the device. As a result, radiation is emitted as a series of relatively sharp peaks whose wavelengths are given by the equation[10]

$$\lambda_i = \frac{\lambda_0}{2\gamma^2 j}\left(1 + \frac{\alpha^2\gamma^2}{2} + \gamma^2\theta^2\right) \tag{4.15}$$

where θ is the angle at which radiation is emitted and j is the harmonic number. As is evident from this equation, the wavelength of the peaks can be shifted by changing the parameters that appear in parentheses, θ the angle of observation and α which, as we have seen, is related to the magnetic field of the device B_0. The intensity which is produced by an undulator with N poles at $\theta = 0$ is amplified by a factor proportional to N^2.[10]

Comparison of synchrotron and conventionally generated X-rays

We have seen on p. 233 that in practice only the characteristic lines of copper and molybdenum are sufficiently intense for use in crystallographic X-ray work and thus with conventional sources one is limited to only a discrete number of possible wavelengths. Synchrotron radiation does not have this limitation as should be clear from Fig. 4.6. The spectrum emitted is very wide and in fact extends in both directions far beyond the range spanned by the Cu K_α and Mo K_α lines. The possibility of selecting the wavelength for X-ray diffraction work has not been fully exploited although it is clear that it has far-reaching consequences. It can be used for example to solve the phase problem in macromolecular work (see Chapter 8, p. 544) by measuring anomalous dispersion effects (see for example Hendrikson *et al.*[11]). Furthermore, crystal absorption and radiation damage (see pp. 304 and 308) are wavelength dependent and thus can be minimized by optimixing the X-ray wavelength used.[12]

Probably the best known property of synchrotron radiation is its high brightness, resulting from the small cross-section of the charged particle beam and the high degree of collimation of the radiation.

A detailed comparison of the brightness and other properties of X-rays produced by synchrotron sources and sealed-tube and rotating-anode generators can be found in Bonse.[13] In Fig. 4.7, taken from Eisenberger,[4]

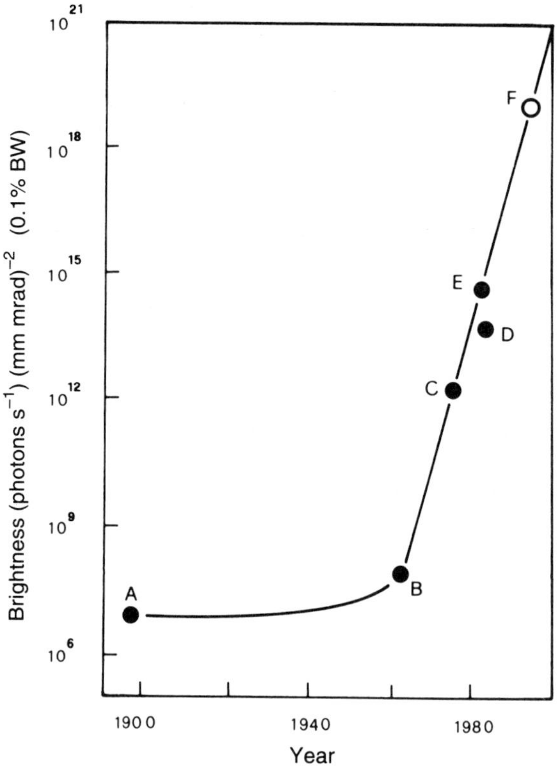

Fig. 4.7. The brightness of x-ray sources as a function of the year in which they became available. The points correspond to the following sources: A = X-ray tube; B = rotating-anode generator; C = Stanford Synchrotron Radiation Laboratory (SSRL) bending magnet; D = Brookhaven National Synchrotron Light Source (NSLS) bending magnet; E = SSRL 54 pole wiggler; F = 6 GeV source with undulator. (Figure taken from Eisenberger.[4])

the brightness of X-ray sources is plotted as a function of the year in which they became available. In the figure one can see that radiation from bending magnets, which, as we have seen, are the least intense of the possible synchrotron sources, can be 10^5 times more intense than that from a rotating-anode generator. As a result, data collection at a synchrotron source is faster, produces higher-resolution data, and permits single-crystal work with specimens that can be only a few μm^3 in volume.[14] A sample of such a small size can only be treated as a polycrystalline material with a conventional source.

An important property of synchrotron radiation is that it is completely linearly polarized in the plane of the orbit of the charged particles and elliptically above and below this plane.[9] X-rays generated from conventional sources are totally non-polarized. As a result diffracted intensities have to be corrected for this polarization effect in different ways as we will see later (p. 303).

We mentioned before that the energy emitted by the moving charged particles is replenished by a radio-frequency cavity. Only the particles with an adequate phase relation with respect to the radio-frequency field can keep a stable orbit and they end up gathered in bunches which have a length that is dependent on the radio frequency used. The number of bunches of charged particles circulating around the orbit is an integer and can be regulated by varying the parameters of operation. Thus, synchrotron radiation is, in fact, produced at a given point in the orbit as a flash when a bunch passes through that point. As a result, synchrotron radiation has a well defined time structure, that is, pulses of radiation are emitted at perfectly defined time intervals. For example the European Synchrotron Radiation Facility is planned to produce pulses lasting from 65 to 140 ps and the pulses will be separated by a minimum of 3 ns.[5] This property of synchrotron radiation, of clear importance in time-resolved studies, has received so far very little attention in X-ray diffraction work.

Monochromatization, collimation, and focusing of X-rays

We have seen that conventional X-ray sources generate the discrete lines of the characteristic spectrum of the anode superimposed on the white radiation continuum and that radiation is emitted in every possible direction, and that although synchrotron sources emit a spatially confined narrow beam, the wavelength of this radiation spans a very wide continuous spectrum. All of the data collection methods discussed later require that a narrow pencil of X-rays strike the specimen under examination and, in addition, most of them also require that the energy of the radiation be limited to a wavelength band as narrow as possible. Ideally the radiation should consist of photons of only a single wavelength. Here we discuss the methods used to select a narrow wavelength out of the spectrum generated by the source and how to define a narrow parallel beam of X-rays to be used for diffraction experiments in conjunction with the data collection devices discussed later.

Filters

One way to select a wavelength interval out of the spectrum generated by the source is by filtering the radiation through a material that selectively absorbs the unwanted radiation while letting through most of the photons of the wavelength that will be used for the diffraction experiment. The absorption of X-rays by a material follows Beer's law:

$$I/I_0 = e^{-\mu x} \tag{4.16}$$

where I is the transmitted intensity, I_0 the incident intensity, x the distance travelled by the X-rays in the material, i.e. the thickness of the filter, and μ is the linear absorption coefficient which depends on the substance, its density, and the wavelength of the X-rays. Since μ depends on the density of the material, the quantity that is usually tabulated is $\mu_m = \mu/\rho$, the mass absorption coefficient, a characteristic of the substance that depends only on the wavelength considered. Complete tables of μ_m as a function of the wavelength for different materials used as filters can be found in Koch and MacGillavry,[15] and a plot of μ_m versus λ for nickel is shown in Fig. 4.8 along with the radiation spectrum generated by a copper anode. In the figure it can be seen that the curve of μ_m versus λ shows two continuous branches separated by a sharp discontinuity, called the absorption edge. If the filter is a pure element, the continuous parts of the curve follow approximately the equation

$$\mu_m = kZ^3\lambda^3 \tag{4.17}$$

where k is a constant with different values for the two branches of the curve and Z is the atomic number of the element. This equation shows why harder X-rays, i.e. X-rays with a shorter wavelength, are absorbed less than those

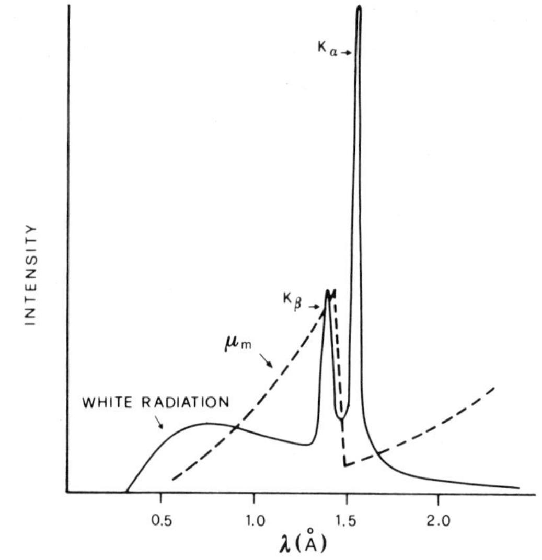

Fig. 4.8. The broken line represents the variation of the mass absorption coefficient μ_m as a function of the wavelength for nickel. The continuous line is the X-ray spectrum generated by a copper anode. Notice that the absorption edge of nickel falls in between the K_α and the K_β characteristic lines of copper.

with a longer wavelength. The presence of the absorption edge in the curve is explained by the fact that the photons at the edge have the wavelength corresponding to the energy necessary to eject an electron from the K orbital of the atoms of the filter. Thus, when this energy is reached massive absorption of radiation occurs with photoionization of the filter and production of fluorescent radiation.

Similarly to the displacement of the position of the characteristic lines with Z, absorption edges move to shorter wavelength as the atomic number of the element increases. A common single filter is chosen so that its absorption edge falls in between the K_α and the K_β peaks of the anode that has been used to generate the X-rays. In this way, the unwanted radiation of highest intensity, i.e. the unavoidable K_β peak, can be greatly attenuated without reducing too much the intensity of the K_α peak that will be used for the experiment. Figure 4.8 shows that a nickel filter has its absorption edge at the wavelength necessary to very strongly absorb the copper K_β peak. Since copper has $Z = 29$ and nickel $Z = 28$ there is a difference of one between the atomic numbers of the target anode used to generate the X-rays and the filter with an absorption edge falling in between its K_α and K_β peaks. This is generally true for every element with $Z \leq 70$ and for the elements of the second long row of the periodic table it is also true that both the elements of $Z - 1$ and $Z - 2$ can be used to absorb the K_β peak of the anode with atomic number Z. Thus both Nb ($Z = 41$) and Zr ($Z = 40$) can be used as filters for Mo ($Z = 42$) radiation.

The relative intensities of the K_α and the K_β peaks depend not only on the absorption coefficient of the filter but also on its thickness. Roberts and Parrish[16] give a table of the appropriate filter thicknesses necessary to produce K_α/K_β ratios of 100 and 500 for different elements used as targets and filters. The same table gives also the percentage of K_α peak lost by filtering which can vary between about 40 and 70 per cent.

A variation of the simple filter technique is the Ross balanced-filter method.[16] In this method two filters are used: one with its absorption edge at slightly shorter and the other at slightly longer wavelength than the K_α

peak selected. The thickness of the filters is chosen so that the radiation is absorbed to the same extent except in the interval in between the two absorption edges. With this technique two measurements are made with either one of the two filters in position and the measured intensity is then taken to be the difference between the two values obtained.

Crystal monochromators

An alternative and more selective way to produce a beam of X-rays with a narrow wavelength distribution is by using a single-crystal monochromator.

Bragg's equation (3.32) shows that when radiation of different wavelengths impinges upon a crystal, diffracted beams are observed at scattering angles θ that depend on the wavelength of the radiation λ. Thus, selecting a given diffraction angle θ is equivalent to choosing a particular wavelength out of the spectrum incident on the crystal.

The simplest type of crystal monochromator consists of a single crystal with one face parallel to a major set of crystal planes mounted so that its orientation with respect to the X-ray beam can be properly adjusted. The most important properties of a crystal monochromator are:

(1) the crystal used must be mechanically strong and should be stable in the X-ray beam;

(2) the interplanar distance should be in the appropriate range to allow the selection of the desired wavelength at a reasonable scattering angle;

(3) the presence of one or more strong diffracted intensities that can be chosen so that the intensity loss of the beam, which is always appreciable, may be reduced as much as possible; and

(4) the mosaicity of the crystal, which determines the divergence of the diffracted beam and the resolution of the crystal, should be small.[17,18]

The reflection chosen should also have a scattering angle as small as possible in order to minimize the loss of intensity due to the polarization factor (see p. 303). Roberts and Parrish[16] give a table with the important properties of crystals commonly used as monochromators.

In a variation of this simple type of flat monochromator, the crystal surface is cut so that it forms an angle with the set of planes that diffract the radiation. In this way, the diffracted beam has a smaller width and as a result more photons are concentrated in a smaller cross section of the beam.[16] By properly curving their surface, crystal monochromators can be used to focus the X-ray beam in a very small area.[19] The curvature of the surface can be produced by simply bending the crystal, in which case the diffracting planes should ideally be tangential to the curved surface. If the monochromator is bent in the shape of a cylinder of elliptical section with the source in one of the foci, the reflected radiation will concentrate on the other focus of the ellipse. A further variation consists of not only bending the crystal but also in grinding its surface so that the radius of curvature of the diffracting planes of the crystal is different from that of its surface. The advantage of this type of monochromator is that it does not suffer from some optical aberrations present in singly bent crystals.[16] Curved crystal monochromators are frequently used to select the wavelength of synchrotron radiation. In addition to the requirements stated before, the crystals should have in this case a very small thermal expansion and a large thermal conductivity because the power applied is much larger than in the case of conventional sources.[18]

Another type of monochromator of wide application in synchrotron sources is the double-crystal monochromator in which the incident X-ray beam is diffracted twice by two similar crystals. This type of monochromator can be constructed with different geometries designed to improve the resolution and/or to keep the X-ray beam in the original direction. A discussion of this type of crystal monochromator can be found in Margaritondo.[18] Crystal monochromators are more selective than filters and, in the case of conventional generators, are capable of resolving the K_{α_1} and K_{α_2} doublet which cannot be separated by any filtering method.

Collimators

The function of collimators is to define a narrow cylindrical beam of X-rays that ideally should be as parallel as possible.

A simple pinhole collimator is shown sketched in Fig. 4.9. It consists of a cylinder with two apertures defining the beam and a third guard aperture which does not affect the beam size defined by the other two but eliminates the radiation scattered by the defining aperture furthest from the X-ray source. These apertures are commonly circular, although slits can be used instead in which case square or rectangular beams can be defined. Cylindrical beam pinhole collimators are typically used with conventional sources to define a beam of radiation that is monochromatized by either a filter or a crystal monochromator. Such collimators never produce an ideally parallel X-ray beam but, in addition to the parallel X-rays, they also produce convergent and divergent X-rays as shown in the figure. A conventional X-ray source, when viewed at the appropriate take-off angle is seen as a square. If l is the distance between the two defining apertures S_1 and S_2 and d is the diameter of the collimator the maximum angle of divergence of the beam, γ, can be calculated as shown in the figure as

$$\tan \gamma/2 = \frac{d/2}{l/2} = d/l$$

and since the angle is very small

$$\tan \gamma/2 \simeq \sin \gamma/2 \simeq \gamma/2 = d/l \qquad \gamma = 2(d/l)$$

where γ is calculated in radians.

If we substitute in this equation d and l for two reasonable values: 50 and 0.5 mm we can get an estimate for the maximum divergence γ

$$\gamma = 2 \times 0.5/50 = 2 \times 10^{-2} \text{ radians.}$$

Huxley[20] and more recently Arndt and Sweet[17] have extensively discussed the conditions that can be varied in order to optimize collimation. The

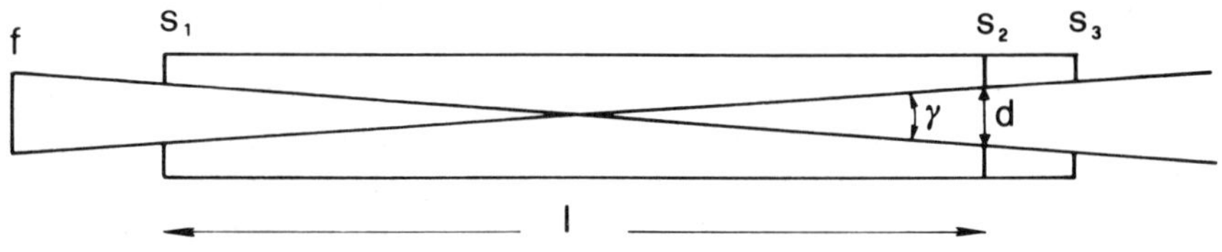

Fig. 4.9. Pinhole collimator showing the angle of greatest divergence γ, S_1 and S_2 are the two apertures defining the beam which are separated by the distance l, and have a diameter of d; S_3 is the guard aperture.

variables that can be adjusted are the crystal and X-ray focus size, the crystal to focus distance, and the crystal to detector distance. Depending also on the reflection to reflection resolution necessary for the experiment, different conditions are found which maximize the signal to noise ratio given the restrictions imposed by the experiment.

Mirrors

X-rays can be reflected by mirrors when the angle of incidence is smaller than a certain critical angle θ_c which is a function of the wavelength λ and can be calculated by the equation[17,19]

$$\theta_c = 2.32 \times 10^5 \left(\frac{Z\rho}{A}\right)^{1/2} \lambda \tag{4.18}$$

where Z is the atomic number, ρ the density, and A the atomic weight of the reflecting material. If the angle of incidence is chosen to be within ten per cent of θ_c for the Cu K_α radiation, the X-rays having a wavelength shorter than the corresponding λ, in particular the fairly intense K_β peak, will not be reflected and therefore they will be eliminated from the beam. Thus, by properly choosing the glancing angle of the X-rays on the mirror the radiation can be partially monochromatized. The values of the critical angles θ_c depend on the reflecting material as shown by the equation given above and they are, in general, very small: 14′ for glass and 23′ for nickel mirrors, calculated for a wavelength corresponding to the Cu K_α radiation. A table of this property and other parameters of interest of mirrors can be found in Witz.[19]

If the reflecting surface of the mirror is curved, ideally in the shape of an elliptical cylinder with the source in one focus, the reflected radiation will converge to the other focus of the ellipse and a very intense X-ray beam will be obtained at that point.

This principle is used in the design of a very powerful device that is used to focus and partially monochromatize an X-ray beam and which uses two curved mirrors with perpendicular axes of curvature.[21] This double-mirror system has been used for X-ray diffraction work on virus crystals which, having very large unit cell parameters, pose particularly serious problems for the spatial separation of the very close diffracted beams.[22] Mirrors are also very extensively used in the beamlines of synchrotrons. In addition to focusing and partial monochromatization they perform several other functions: splitting of a beam into two, magnification or demagnification of the source, and change in the polarization of the radiation.[18] The function to be performed determines their geometry and so their surface can be flat or curved and the mirror can be bent or segmented, that is constituted by several small flat pieces which are easier to produce than large curved single mirrors.

Data collection techniques for single crystals

We saw in Chapter 3 (p. 155) that the condition for X-ray diffraction to occur in a crystal is that the scattering vector $\boldsymbol{r}^*$ should coincide with a node of the reciprocal lattice associated to the crystalline lattice defined in Chapter 2 (p. 63). The sphere of reflection, also called Ewald sphere, was defined (on p. 155) and it was shown that diffraction is observed whenever a reciprocal lattice node lies on this sphere. The direction of the diffracted beam is determined by the vector joining the centre of the Ewald sphere and the reciprocal lattice node lying on its surface. We saw that the characteristics of the crystal and the conditions of the experiment cause these nodes of reciprocal space to have a volume. We will make use of all these concepts in our discussion of the data collection devices that will follow.

If the scattering experiment is performed with radiation of a single wavelength there will be a single Ewald sphere of radius $1/\lambda$ and the probability that a stationary reciprocal lattice node may by chance be on its surface will be fairly low. Furthermore, and as pointed out on p. 163, diffraction from only a cross-section of the node is not acceptable since it is the entire volume that should give rise to the diffracted intensity from which the structure factor amplitude is to be derived. In addition, in order to solve a structure, one needs all of the diffracted intensities that can be measured corresponding to the nodes found within a sphere of radius $D^* = 1/R_{max}$, the inverse of the resolution of the structure. Broadly speaking there are two ways to tackle these problems: the first is to use polychromatic radiation, i.e. to have a series of Ewald spheres corresponding to different wavelengths; the second method is to move the reciprocal lattice nodes, i.e. the crystal, so that all the nodes from which one wishes to measure the diffraction cross the Ewald sphere completely. The first method is historically the oldest, it is called the Laue method, and it was with it that diffraction from crystals was discovered.[23]

In Fig. 4.10, the shaded area represents the volume of reciprocal space containing all the nodes that will produce diffraction when the stationary specimen is hit with radiation of a wavelength in the interval between λ_{max} and λ_{min}. In the Laue method, different cross-sections of a node are excited by radiation of slightly different wavelengths and the result is an intensity integrated over the wavelength rather than over the volume sweeping through a single Ewald sphere. If the white radiation spectrum resulting from a conventional generator is used to produce Laue diffraction, the practical applications of the method are rather limited.[24,25] It is for this reason, and also because the diffraction pattern produced is rather difficult

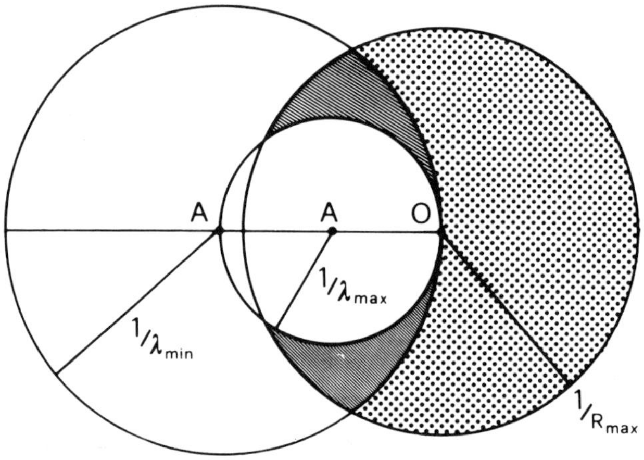

Fig. 4.10. The Laue method. A stationary specimen is irradiated with X-rays covering the wavelength interval $\lambda_{max} - \lambda_{min}$. The shaded area represents the volume of reciprocal space that produces diffraction.

to interpret, that the method fell into disuse until the advent of synchrotron radiation and the development of very powerful computers. The use of the Laue method with synchrotron radiation[26] provides an extremely fast and efficient method to record diffraction data. The applications are already important in the field of small-molecule[27,28] as well as macromolecular crystallography[29,30] where it has opened up the possibility of performing time-resolved studies in the crystalline state.[31]

All the other data collection methods discussed in this chapter use monochromatic radiation and therefore require a more or less complicated mechanism to move the crystal as the diffracted radiation is measured.

The Weissenberg camera

Although the Weissenberg camera,[32–34] introduced in 1924, has nowadays been largely superseded by other data collection methods, historically it has played a central role, since for many years it was the standard instrument used to quantitatively measure diffracted intensities. Currently, it is still used for space-group and unit cell determination of small-molecule crystals, that is for the preliminary characterization of crystals whose diffracted intensities are then measured with the diffractometer.

The cylindrical film rotation camera

The simplest way to produce a regular motion of the reciprocal lattice is to rotate the crystal about a direct lattice axis oriented normal to the incident beam direction: then, as seen on p. 126, there is a family of equidistant reciprocal lattice planes perpendicular to this axis of rotation. Whenever a reciprocal lattice point intersects the Ewald sphere a diffracted beam passes through the point as shown in Figs. 4.11(a) and (b). Such points all lie on equidistant circles of decreasing radii as we move away from the centre of the Ewald sphere. If a cylindrical cassette containing a piece of film is placed so that the cylinder axis coincides with the crystal rotation axis, as shown in Fig. 4.11(c), the circles will project undistorted onto the film and when the film is unrolled they will appear as straight lines (layer lines) that contain the diffraction spots.

Figure 4.12 shows a rotating-crystal photograph. From it one can calculate quite easily the spacing between the equidistant planes of reciprocal space and then use this value to compute one of the unit cell parameters of the crystal.

If the distance on the film between the layer line corresponding to the nth reciprocal lattice plane and the 0 layer line is d_f, r_f is the radius of the cylindrical cassette and $d_{\mathbf{H}}$ is the constant distance between two consecutive reciprocal lattice planes in the family then

$$\tan \alpha_n = d_f / r_f$$

and

$$\sin \alpha_n = \frac{n d_{\mathbf{H}}}{\lambda^{-1}}$$

as shown in Fig. 4.11(c). Recall that the radius of the Ewald sphere is λ^{-1}.

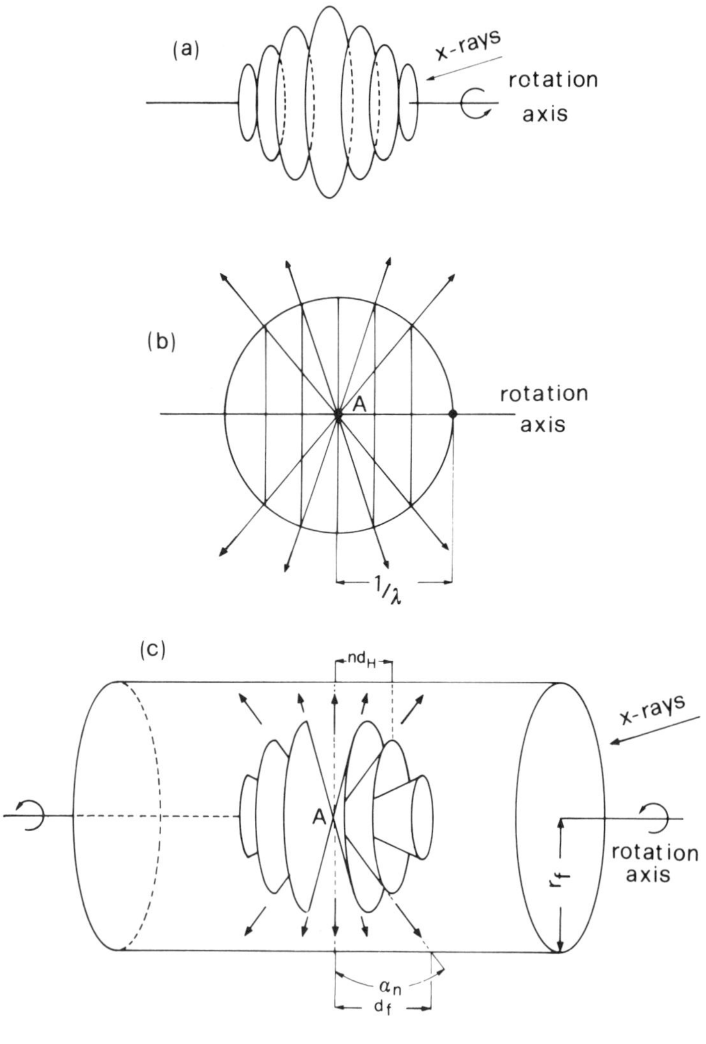

Fig. 4.11. The cylindrical film rotation method. (a) The intersection of a family of equidistant reciprocal lattice planes and the Ewald sphere. The points lying on the circles represented will produce diffraction. (b) Projection in the direction of the incident X-ray beam. The rotation axis is horizontal and is found in the plane of the figure. The circles shown in (a) are now projected as parallel vertical lines. (c) A cylindrical film is placed with its axis coincident with the rotation axis. The intersections of the reciprocal lattice planes and the Ewald sphere produce on the film a series of parallel lines.

Thus the interplanar spacing in reciprocal space is

$$d_{\mathbf{H}} = \frac{\sin \alpha_n}{n\lambda} = \frac{1}{n\lambda} \sin \tan^{-1} \frac{d_f}{r_f}. \tag{4.19}$$

Using this value, the unit cell parameter corresponding to the real axis coincident with the rotation axis can be calculated.

Incidentally, notice that because of this relationship equally spaced planes in reciprocal space do not produce equally spaced layer lines on the film.

Zero-level Weissenberg photographs

Although in a rotation photograph the index corresponding to each layer line is trivial to assign, the method cannot be used to easily map the reciprocal lattice. This is because all the reciprocal lattice points present in one plane are projected by this geometry on to one layer line and thus the information concerning the other two indices is not easy to extract.

This difficulty can be overcome by the use of the Weissenberg camera, which is essentially a cylindrical film rotation camera with two differences.

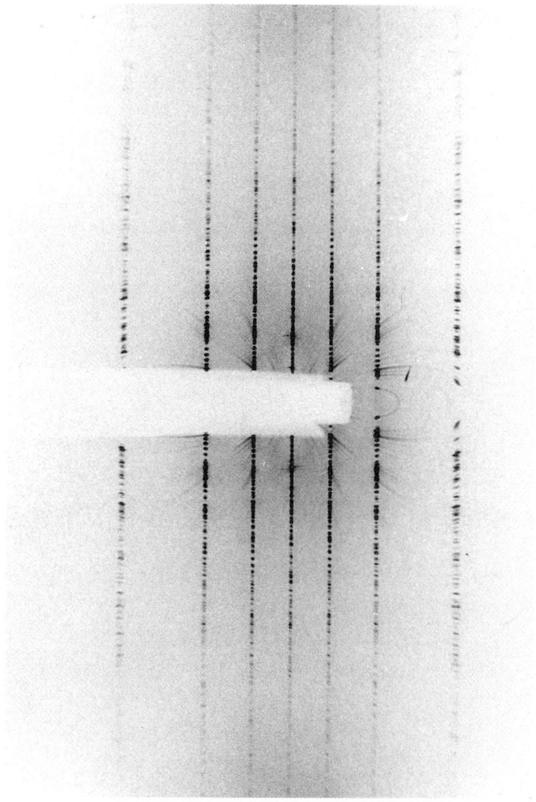

Fig. 4.12. A typical rotation photograph showing a family of parallel lines due to the set of parallel reciprocal lattice planes. Notice that the lines are not equally spaced.

The first is the use of a layer line screen that blocks all the diffracted radiation with the exception of that due to one selected reciprocal lattice plane at a time. The second difference is the coupling of the rotation motion to a displacement of the film along the cylinder axis. In this way, spots belonging to the same reciprocal lattice plane that cross the Ewald sphere at different times and which would end up recorded on the same layer line are recorded instead in different positions on the film. Thus, a single reciprocal lattice plane is mapped on to the film plane.

Figure 4.13(a) shows the Ewald sphere projected on to a reciprocal lattice plane that would produce one layer line in a normal cylindrical film rotation camera. The view is in the direction of the rotation axis and each of the lines represented corresponds to a reciprocal lattice row of a given index. In this representation the origin of reciprocal space is found at point O, the intersection of the incident X-ray beam with the Ewald sphere. The line tangent to this point has one index equal to zero and corresponds to one of the reciprocal lattice axes. In Fig. 4.13(b) the reciprocal lattice point P is crossing the Ewald sphere and therefore it is in diffracting position. Its coordinates, shown on the unrolled film on the right, are x and z; the first is proportional to the angle 2θ as can be seen in the same figure, the second is proportional to the rotation angle ω since, as already pointed out, crystal rotation and film translation along the cylinder axis are coupled.

If r_f is again the cylindrical cassette radius one can write

$$2\theta/360° = x/2\pi r_f$$

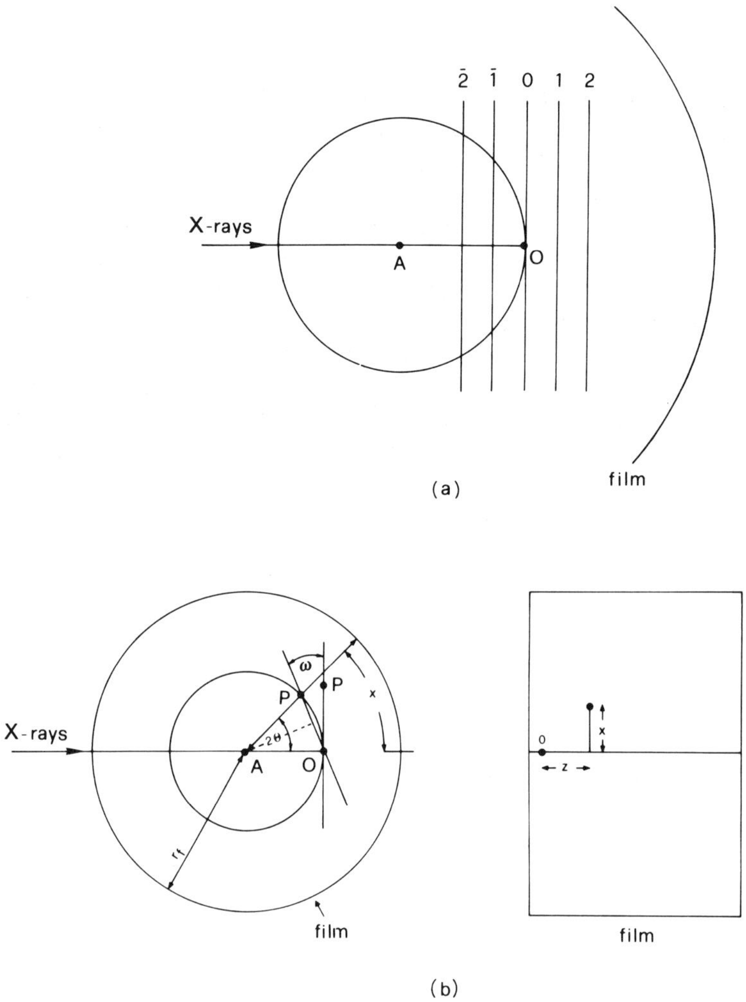

Fig. 4.13. Zero-level Weissenberg method. (a) When a layer line screen selects the zero-level plane, a series of equidistant lines will contain all the reciprocal lattice nodes that can produce diffraction on the film. (b) When the crystal is rotated an angle $\omega = \theta$ the point P belonging to one of the reciprocal lattice axes is in diffracting position. Its coordinates on the film are x and z.

and

$$2\theta = 360°x/2\pi r_f = C_1 x. \tag{4.20}$$

Normally r_f is chosen so that C_1 has the value of $2°\ \text{mm}^{-1}$. Thus measuring the x coordinate of a reflection in millimetres one can automatically calculate the corresponding value of 2θ.

The coupling parameters of the film movement to the crystal rotation are chosen so that the constant C_2 in

$$\omega = C_2 z$$

is also made equal to 2 and thus, the two angles ω and 2θ can be measured on the film on the same scale.

Figure 4.13(b) shows that a normal to the zero-level reciprocal lattice row passing through A bisects the angle 2θ. Thus, for this particular level, θ is equal to ω because the sides of the two angles are perpendicular and one can write

$$\theta = \tfrac{1}{2}C_1 x = \omega = C_2 z$$

and

$$x = 2C_2z/C_1 \tag{4.21}$$

which is the equation of a straight line of slope 2 if the two constants C_1 and C_2 are chosen to be equal. Thus the reciprocal lattice axis passing through O will produce on the film a series of spots that will be found on a straight line which will normally have a slope of 2.

When the angle ω reaches the value of 90° the zero-level line will be found in the direction of the X-ray beam and the line traced by the spots will have reached the point where the film is cut to let the X-rays through. Immediately thereafter the spots will be recorded on the other side of the film interruption, that is they will begin to be recorded on the bottom half of the film. When ω equals 180° the zero-level line is found again tangent to the Ewald sphere but it has been flipped over. The spots recorded after that will change the sign of the only index which is varying along the line. Figure 4.14 shows the Ewald construction and the film appearance at the beginning of the rotation cycle and after a rotation angle of slightly more than 180°.

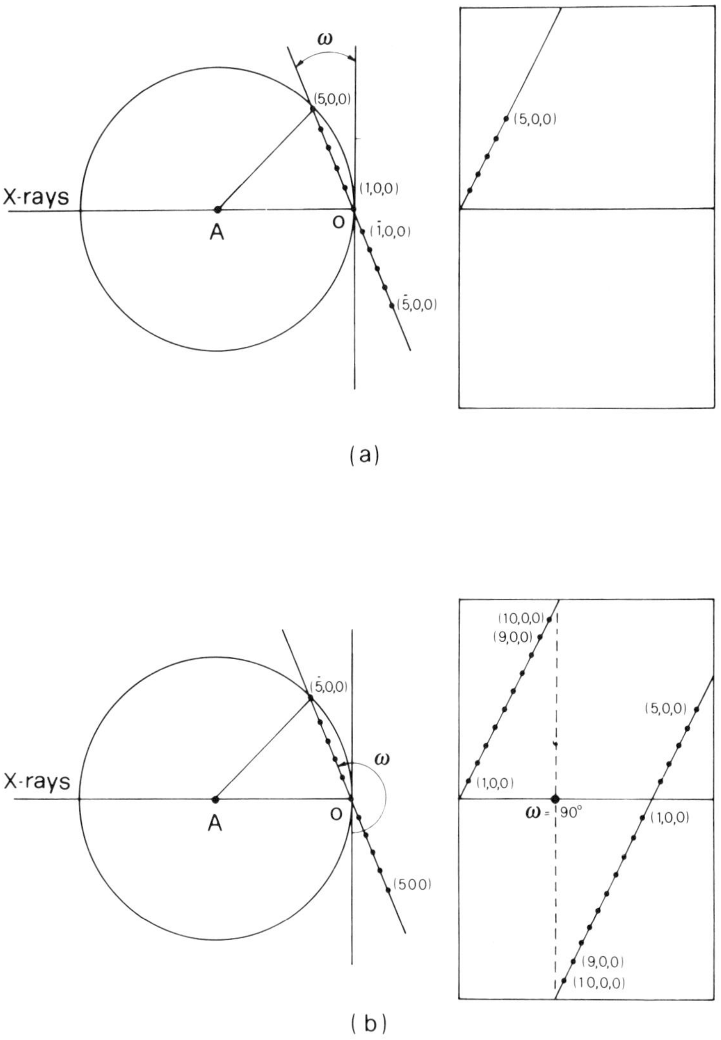

Fig. 4.14. (a) Five reciprocal lattice nodes have crossed the Ewald sphere and produced the five spots lying on a straight line on the film. (b) After a rotation of 180° plus the ω of (a) the film shows this pattern. Notice that after a 180° rotation the pattern is perfectly symmetrical about the point where $\omega = 90°$ on the projection of the rotation axis onto the film plane.

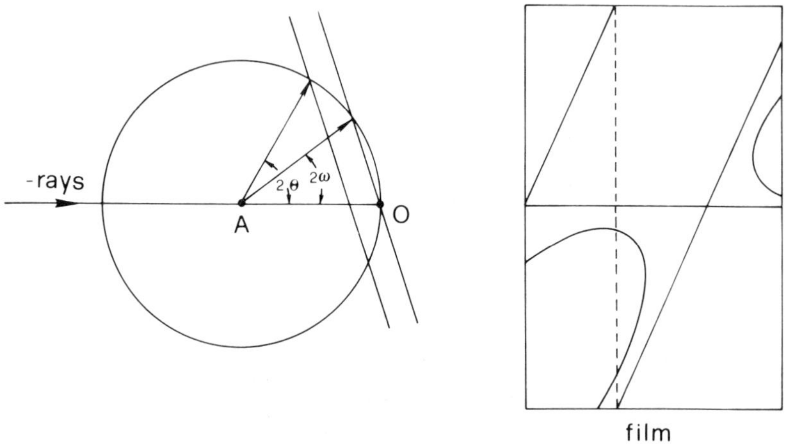

Fig. 4.15. For non-central reciprocal lattice lines 2θ is not equal to 2ω and the plot of x versus z is not a straight line but the curve shown in the figure.

If the second reciprocal lattice axis, which has to be found in the plane selected for recording, forms with this axis an angle of α^*, when ω equals α^* this second axis will be found tangent to the sphere of reflection and after that will begin to produce a second straight line parallel to that traced by the first axis and separated from it a distance $\omega = \alpha^*$. Thus, reciprocal lattice axes are identified in the photograph by the straight lines they produce and the angle between them is simply the ω value that separates the lines on the film.

Figure 4.15 shows that for reciprocal lattice lines not passing through the origin of reciprocal space, θ is not equal to ω and therefore the relationship between x and z is no longer the equation of a straight line. It is instead a curve of the type shown on the right-hand side of Fig. 4.15. Each of the layer lines that do not pass through O will produce a curve similar to the one shown in the figure and therefore the film will show a family of non-intersecting curves or festoons that will be found on both sides of the line crossing the centre of the film, at increasing distances from the centre. Each festoon corresponds to one reciprocal space line and therefore the reflections found on it will have one index in common. On every film there will be two festoon families, one for each of the two reciprocal lattice axes found in the plane that is being examined. All non-central reflections are found at the intersections of two festoons, one from each of the two families found on the film.

Figure 4.16 is a picture of the camera and Fig. 4.17 is a typical zero-level Weissenberg photograph. After the plane axes and the festoons have been identified, it is not difficult to index a photograph like this by inspection.

Upper-level Weissenberg photographs

Figure 4.18(a), a projection of the Ewald sphere on to the plane defined by the incident X-ray beam and a normal rotation axis, shows that for a non-zero-level photograph there is an area of the reciprocal lattice plane that cannot cross the Ewald sphere. This area becomes larger the further the plane is from the zero level.

Figure 4.18(b) shows the solution which is adopted to circumvent this problem. The rotation axis is no longer chosen to be normal to the X-ray

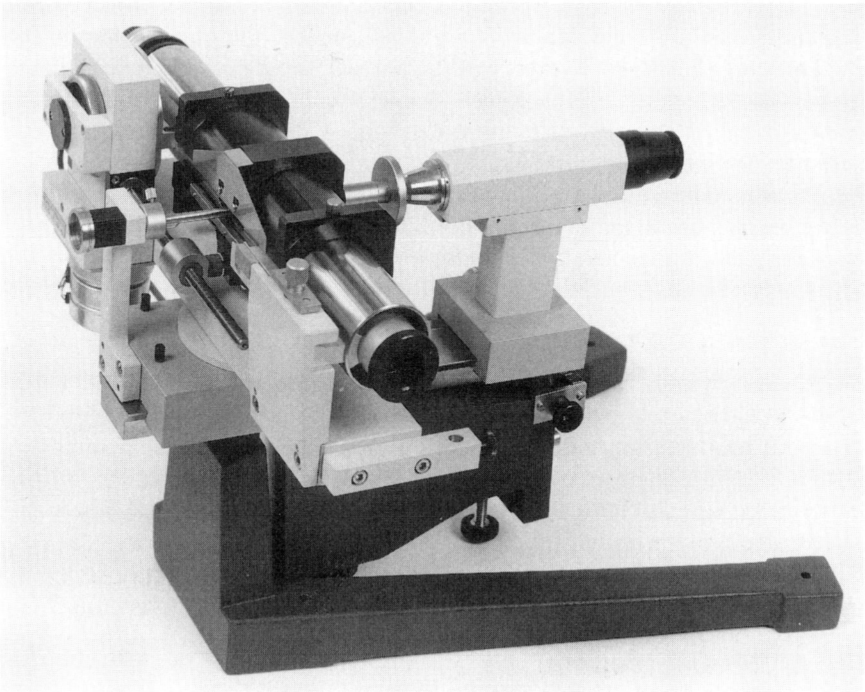

Fig. 4.16. A modern Weissenberg camera. (Photograph courtesy of Enraf Nonius.)

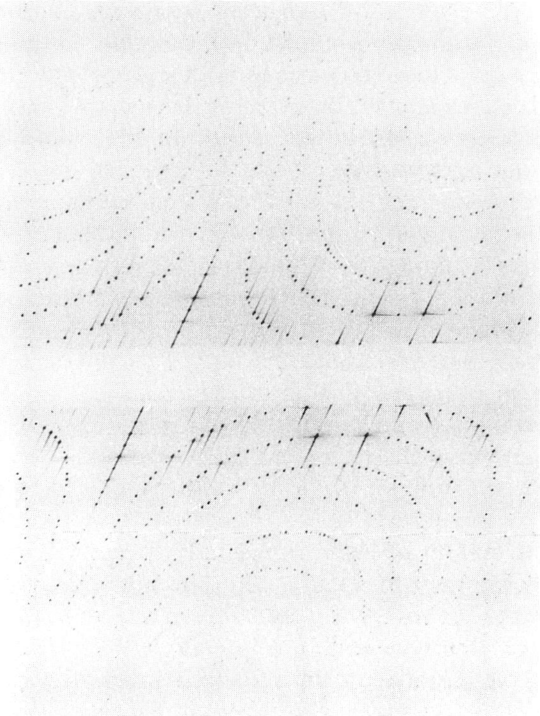

Fig. 4.17. Zero-level Weissenberg photograph. Notice the straight lines of points produced by the zero-level lines and the festoons due to the non-zero-level lines.

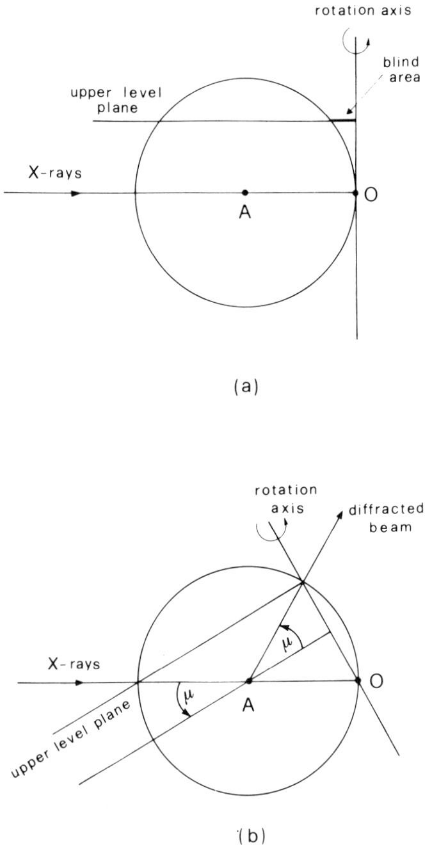

Fig. 4.18. Upper-level Weissenberg photographs. (a) If the incident X-ray beam is normal to the rotation axis, there will be a blind area for upper-level planes. (b) The equi-inclination method.

beam, it is instead tilted so that its intersection with the reciprocal lattice plane falls on the Ewald sphere. As a consequence, the angle made by the incident and diffracted X-rays with a normal to the rotation axis going through the centre of the Ewald sphere are equal (see Fig. 4.18(b)) and the method is called the equi-inclination method. A detailed description of this method is found in Buerger.[33]

Since the main uses of the Weissenberg camera are nowadays space-group and unit cell parameter determination and no longer quantitative intensity measurements, upper-level Weissenberg photographs are seldom recorded.

The precession camera

The precession camera, introduced by Martin Buerger in 1942, is still a very popular instrument, especially among macromolecular crystallographers, because it produces an undistorted picture of a reciprocal lattice plane.[35] A precession photograph is quite easily indexed by inspection, it shows very clearly the reciprocal lattice plane symmetry, and can be used to quickly and easily calculate the unit cell parameters. This is so because the distance between the spots on the film is simply a reciprocal lattice distance scaled by the X-ray wavelength and the crystal to film distance.

The precession motion

In the precession method a crystal is first oriented with one of its axes parallel to the X-ray beam, that is to say with a family of reciprocal lattice planes perpendicular to the beam as shown in Fig. 4.19(a). This figure shows the zero-level reciprocal lattice plane tangent to the Ewald sphere at the point O. In the precession camera the crystal is rotated in such a way that the normal to the plane family precesses about the X-ray beam, in other words it revolves about the beam keeping a constant angle μ, the precession angle. During this revolution the normal to the planes describes a cone whose axis is the X-ray beam. At the same time the film that will collect the reflections is made to move in the same way so that it is always parallel to the reciprocal lattice plane that will be photographed and keeps with it a constant distance.

When the precession angle μ is different from O, the zero-level reciprocal lattice plane is no longer tangent to the Ewald sphere; it intersects the sphere defining with it a circle. All the reciprocal lattice points which fall on this circle produce a diffracted beam that will be recorded on the film. This situation is represented in Fig. 4.19(b) which shows on the right-hand side the projection of this circle onto the film. This projection is also a circle because the film is always parallel to the reciprocal lattice plane and the distance between the two planes is constant.

When the normal to the reciprocal lattice planes has described half of its revolution movement about the X-ray beam we find the situation shown in Fig. 4.19(c).

After a full cycle has been completed we are back to the situation represented in Fig. 4.19(b). Figure 4.20 shows the film with the projections of the intersection of the reciprocal lattice plane and the Ewald sphere in four situations; two are those shown in Figs. 4.19(b) and (c), the other two correspond to the positions in which the projection of the normal on to the plane of the figure coincides with the X-ray beam. These situations are

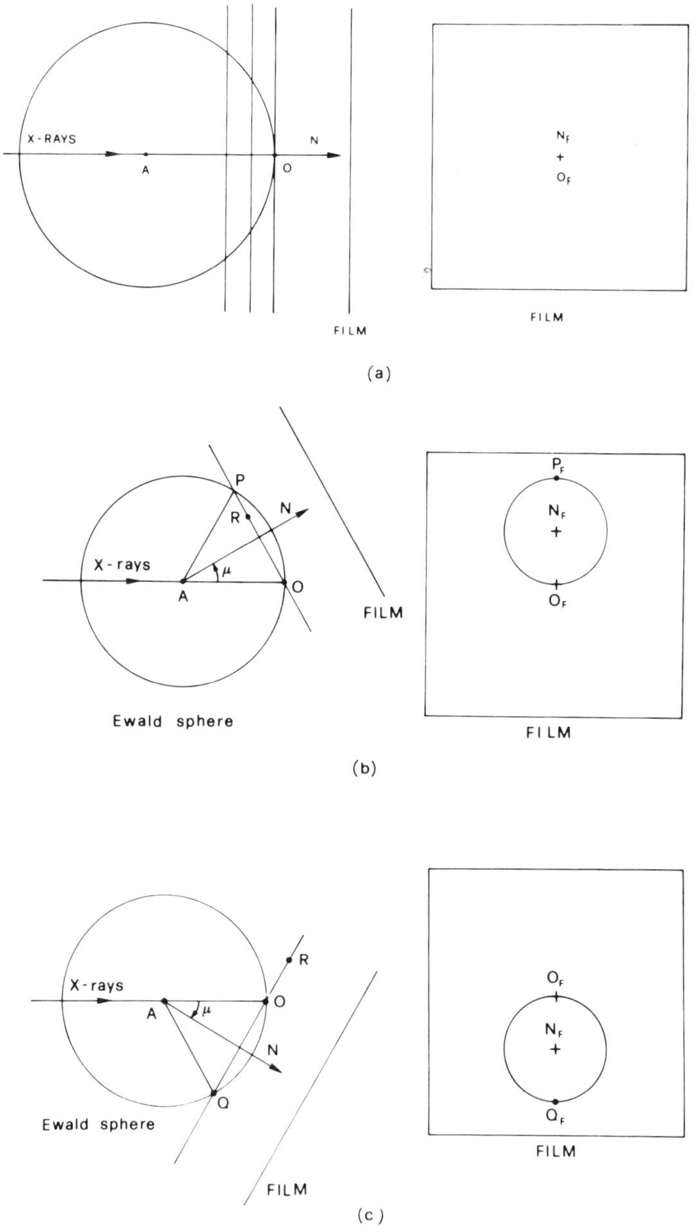

Fig. 4.19. Precession motion. (a) The crystal is first oriented with a family of reciprocal lattice planes perpendicular to the X-ray beam. On the right-hand side we view the film in the direction of the normal to the planes. (b) The normal to the reciprocal lattice planes forms an angle μ with the X-ray beam. The intersection of the zero-level plane with the Ewald sphere projects onto the parallel film as a circle centred on N_F and tangent to the film centere O_F. (c) After half a precession cycle we find the situation shown here. The circle on the film has now moved from above to below the film origin.

described by the same Figs 4.19(b) and (c) if we imagine that we are viewing the Ewald sphere not from a side but from the top or the bottom.

When a full revolution has been completed the intersection of the reciprocal lattice plane and the Ewald sphere has described a circle whose radius is thc diameter of the intersection and equal to

$$2 \sin \mu / \lambda. \qquad (4.22)$$

All the points in the reciprocal lattice plane which are found within this radius have passed through the Ewald sphere and therefore have produced a diffracted beam which has been recorded on the film. Since the precession

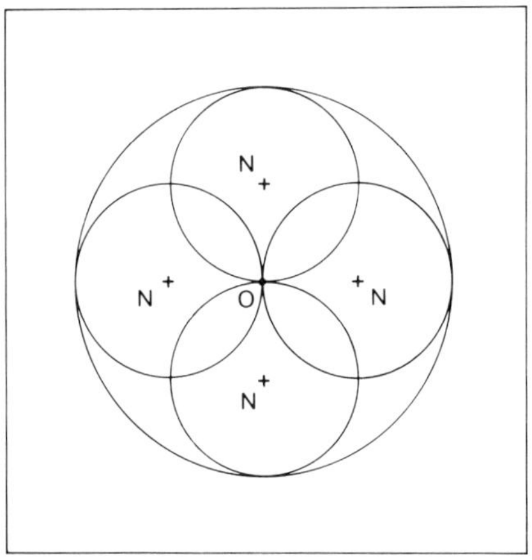

Fig. 4.20. Movement of the projection of the intersection of the reciprocal lattice plane and the Ewald sphere onto the film as one precession cycle is completed. Only four extreme situations are represented. The circle defined by the projection will contain all the reflections collected by the camera.

angle is usually not larger than 30°, the maximum radius of this circle is normally λ^{-1}.

In Fig. 4.19(b) the reciprocal lattice point R is inside the Ewald sphere, in Fig. 4.19(c) it is outside. When a full cycle has been completed and we are back to the situation shown in Fig. 4.19(b), the point is again inside the Ewald sphere. Thus in a full precession cycle, every reciprocal lattice point that will produce a signal crosses the Ewald sphere twice, moving each time in opposite directions.

Isolating a zero-level reciprocal lattice plane

In Figs. 4.19(b) and (c) we have shown the precession motion indicating only a zero-level reciprocal lattice plane. In fact, as can be easily understood reconsidering Fig. 4.19(a), it is a family of reciprocal lattice planes that always intersects the Ewald sphere generating a set of concentric circles of variable radii. In order to record only the zero-level plane, the diffracted beams originating from all the other planes have to be stopped using a layer line screen.

Figure 4.21 shows the cones generated by the zero- and first-level reciprocal lattice planes and the screen used to stop the beams diffracted from the first level. The screen is simply a metal plate with an annular aperture that will let through only the diffracted radiation coming from the selected reciprocal lattice plane. As the crystal and the film move, the screen also moves in such a way that it is always parallel to the film and therefore parallel to the reciprocal lattice planes. In the precession camera, the movement of crystal, film and screen are always coupled to satisfy this condition.

The position of the screen, that is the crystal to screen distance, depends on the radius of the annulus. If s is the crystal to screen distance and r_s the annular radius

$$\cot \mu = s/r_s$$

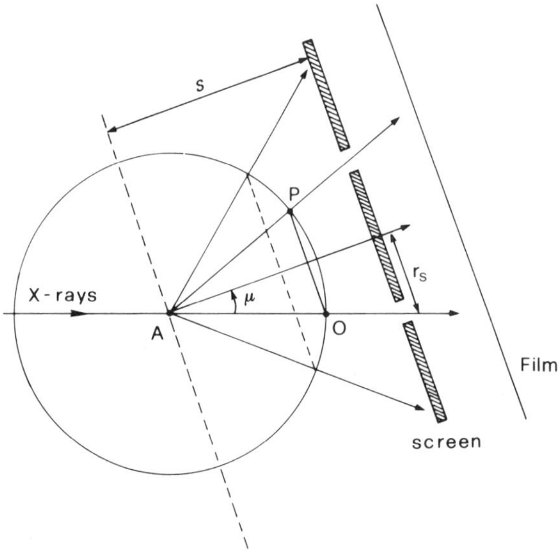

Fig. 4.21. Diffraction cones generated by the zero and first reciprocal lattice planes. The screen is set at a distance from the crystal s so that only the diffracted beams originated from the zero level are let through and therefore produce a signal on the film.

and

$$s = r_s \cot \mu \qquad (4.23)$$

as shown in Fig. 4.21.

Upper-level precession photographs

Let us consider an upper-level reciprocal lattice plane whose normal forms a precession angle μ with the direction of the X-ray beam as shown in Fig. 4.22.

The upper level origin O_u represented in the figure lies a distance nd^* from the intersection of the X-ray beam with the Ewald sphere. In order to produce an undistorted photograph of the reciprocal lattice plane, the film centre has to be moved forward from O_f^0 to O_f^u.

From the construction in Fig. 4.22 we can write

$$D/nd^* = f/\lambda^{-1}$$

where D is the distance the film has to be moved forward, nd^* the distance between the upper-level plane and the zero-level plane, f is the crystal to film distance, and λ^{-1} the radius of the Ewald sphere.

Thus we can calculate D

$$D = \lambda nd^* f \qquad (4.24)$$

for any upper-level plane and crystal to film distance.

Figure 4.23 is the construction used to calculate the crystal to screen distance for an upper-level precession photograph.

In this case we can write

$$\cot \nu = s/r_s$$

Also the distance between the upper-level layer and the zero-level plane is

$$nd^* = 1/\lambda(\cos \mu - \cos \nu).$$

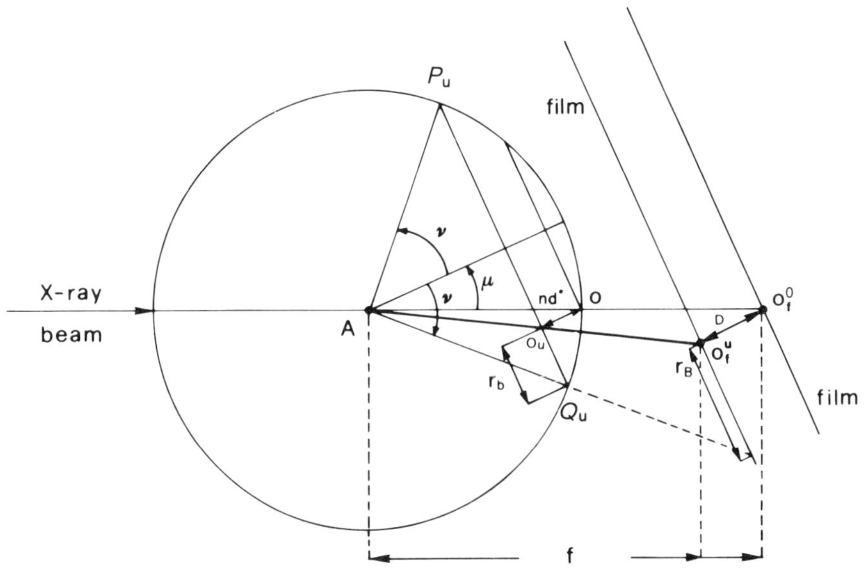

Fig. 4.22. Upper-level precession geometry. O_f^u is the new centre of the film that will permit the collection of an undistorted upper level; nd^* is the distance between the upper and the zero-level plane, f is the crystal to film distance.

Thus

$$\cos \nu = \cos \mu - nd^* \lambda$$

and finally

$$s = r_s \cot \cos^{-1}(\cos \mu - nd^* \lambda) \qquad (4.25)$$

which can be used to calculate the crystal to screen distance for any upper-level precession photograph. Incidentally, we notice that if n is made equal to 0 this equation reduces to (4.23), the expression derived for a zero-level precession photograph.

Re-examining Fig. 4.22, we notice that there is an area of the reciprocal lattice plane that will never pass through the Ewald sphere. The projection of this area onto the plane of the figure is the line that goes from the point O_u to Q_u. The blind region circle has a radius, r_b, that can be calculated from Fig. 4.22

$$r_b = 1/\lambda(\sin \nu - \sin \mu).$$

On the film the radius of the blind region, r_B, is

$$r_B / r_b = f / \lambda^{-1}$$

from which

$$r_B = f(\sin \nu - \sin \mu) \qquad (4.26)$$

where f is the crystal to film distance.

We can use Fig. 4.23 to calculate the radius of the limiting circle for any upper level. If r_M is the maximum radius and f the crystal to film distance

$$r_M / f = (\sin \nu + \sin \mu)(1/\lambda)/\lambda^{-1}$$

$$r_M = f(\sin \nu + \sin \mu). \qquad (4.27)$$

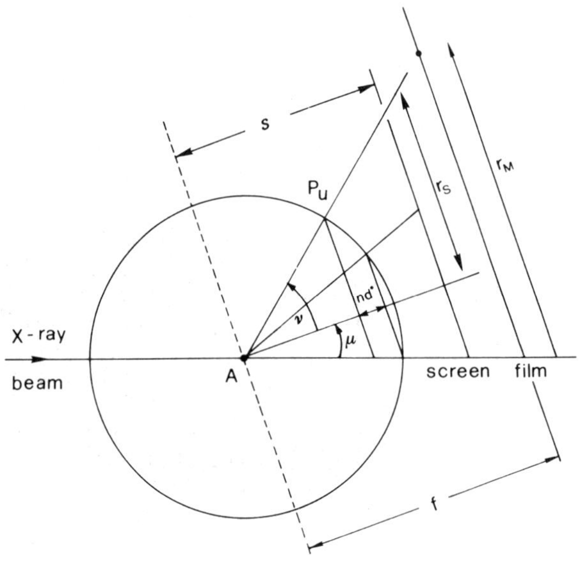

Fig. 4.23. Setting the layer line screen to select an upper-level plane.

Calculating unit cell parameters from a precession photograph

Figure 4.24 is a picture of a precession camera and Fig. 4.25 a zero-level precession photograph of a protein crystal that belongs to the space group $P3_221$. Figure 4.25 is an $h0l$ zone, i.e. the axes shown are $\boldsymbol{a}^*$ and $\boldsymbol{c}^*$. In order to determine the unit cell parameters one has to know the wavelength of the X-ray radiation, the crystal to film distance, and the distance between two consecutive reflections along the two axes. The precision of the parameter determination, can be increased if a larger distance is measured, i.e. the distance between several spots, if possible using one of the many optical devices available to improve the quality of the measurement. If df is the distance between two spots on the film, as shown in Fig. 4.26

$$a^*/\lambda^{-1} = \mathrm{d}f/f$$

and

$$a^* = \mathrm{d}f/\lambda f \qquad (4.28)$$

where f is the crystal to film distance and a^* the reciprocal lattice parameter from which the unit cell parameter can be easily calculated.

The rotation (oscillation) method in macromolecular crystallography

In our discussion of the Weissenberg and precession cameras we have seen that in both methods only one reciprocal lattice plane is recorded per film. As a result, indexing of the photographs is easy but the drawback in both cases is that most of the diffracted radiation is not recorded and therefore it

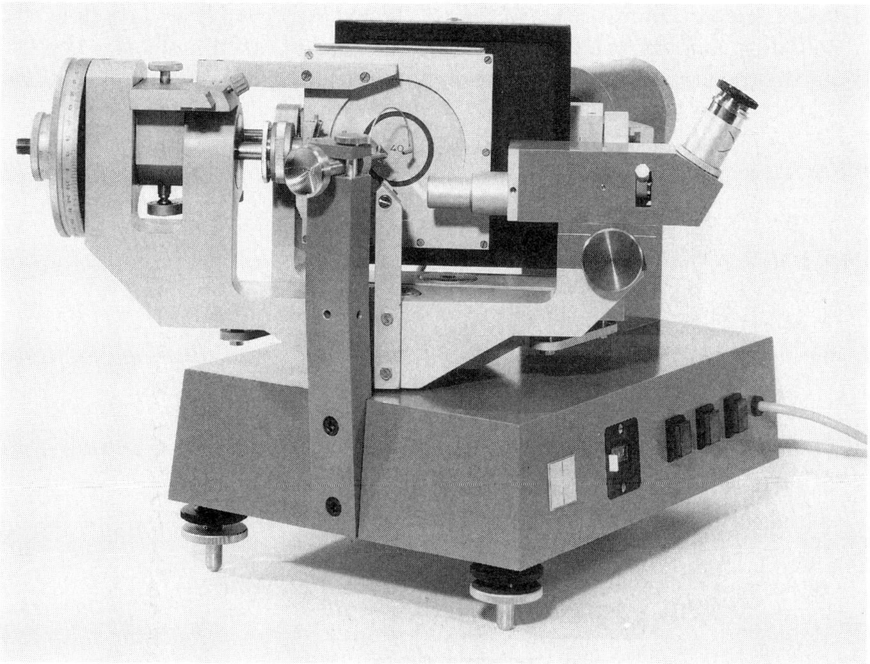

Fig. 4.24. The precession camera. (Photograph courtesy of Huber Diffraktionstechnik.)

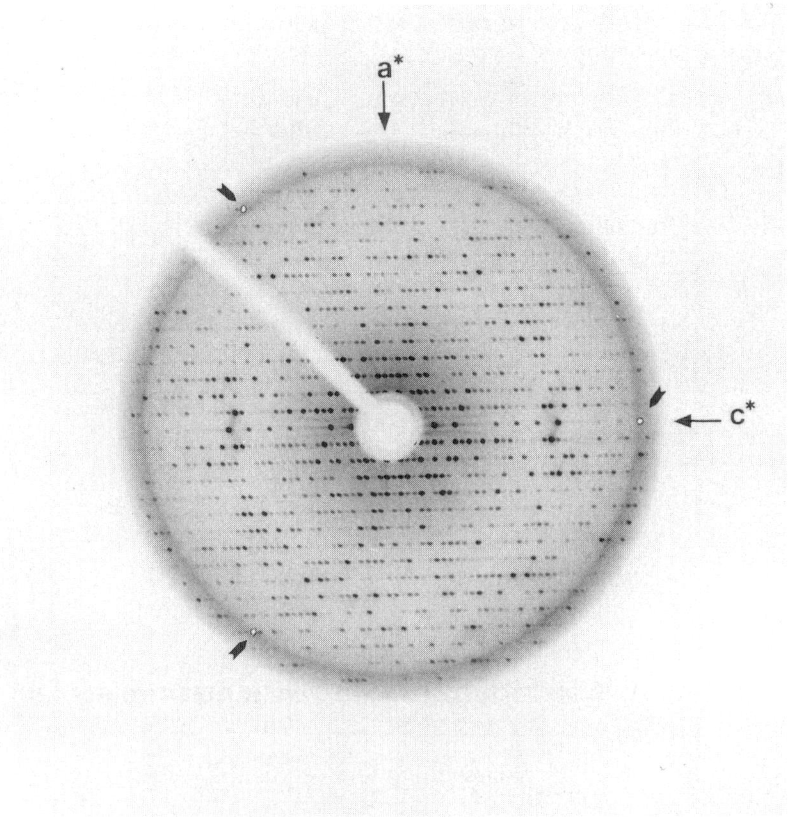

Fig. 4.25. Zero-level precession photograph of a protein crystal showing the two reciprocal lattice axes. From this picture it is quite easy to calculate the two unit cell parameters as described in the text. Notice the holes punched on the picture used to roughly define its centre during densitometry.

is wasted. If the crystals do not decay in time, that is if the intensity of the reflections does not change as the crystal is exposed to the X-ray, this is not too a serious a problem; with sufficient time available all the reciprocal lattice region of interest can, in theory, be explored one layer at a time. But if radiation damage is a problem, as is the case of macromolecular crystals, both the precession and Weissenberg methods are extremely inefficient. For

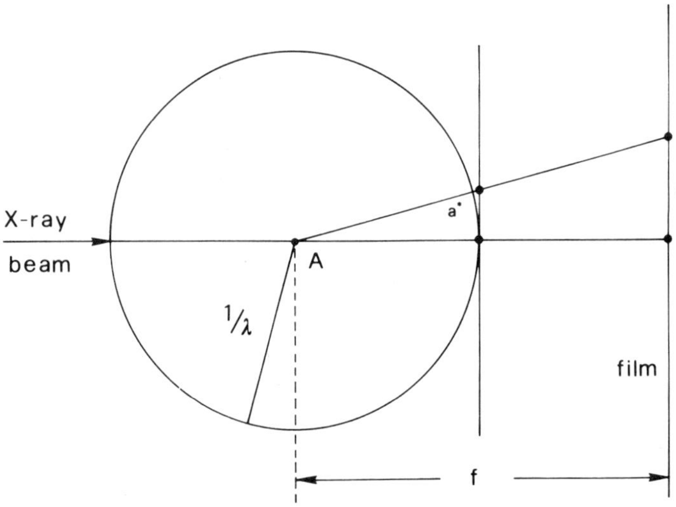

Fig. 4.26. Determining the unit cell parameters from a precession photograph.

example, the recording of a complete data set of a protein crystal with the precession camera usually requires one crystal per photograph, i.e. one crystal per reciprocal lattice plane. Although in the early days of protein crystallography this was the way that data were collected, the search for a more efficient way to record diffracted intensities led in the early 1970s to the reintroduction of the screenless flat-film rotation (oscillation) method for macromolecular data collection. The flat-film rotation method had been used since the very beginning of X-ray crystallography[36] but it had been abandoned due to the difficulties in indexing and quantitatively measuring the reflection intensities on the film. This situation was changed by the introduction of computer-controlled microdensitometers which assured that films could be conveniently scanned and reliable intensities could be extracted from them. A new type of flat-film rotation camera with eight cassettes that are used for successive exposures, the Arndt–Wonacott camera,[37,38] was built, and in a relatively short time the rotation method became one of the major, if not the major, method for macromolecular crystal data collection.

Geometry of the rotation method

In the rotation method a crystal is made to rotate or oscillate, that is it is moved back and forth, through a small angle about one axis perpendicular to the X-ray beam and the diffracted intensities are recorded on a flat film (that can also be V-shaped) contained in a casette perpendicular to the incident X-rays (see Fig. 4.27). Small rotation angles are required to avoid reflection overlap on the film. An asymmetric unit of reciprocal space is covered by successively exposing different cassettes.

When we discussed the Weissenberg camera, we saw that reciprocal lattice planes perpendicular to the rotation axis generate a series of layer lines on a cylindrical film. Let us now see what happens if we have a flat film perpendicular to the X-ray beam as shown in Fig. 4.27. In describing a crystal rotation, it is easier, and therefore often preferred, to leave the reciprocal lattice fixed and move instead the Ewald sphere. When the crystal is rotated a small angle $\Delta\Phi$, the two circles that result from the intersection of the Ewald sphere at the beginning and end of the rotation range and each reciprocal lattice plane, define two crescent-shaped lunes

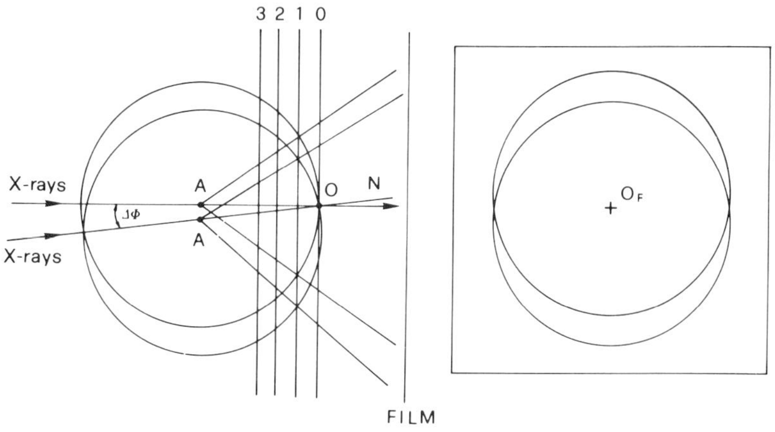

Fig. 4.27. The rotation motion. The motion is shown leaving the reciprocal lattice planes fixed and rotating the Ewald sphere about the point O. After the spindle has rotated an angle $\Delta\Phi$, the intersections of the Ewald sphere with each reciprocal lattice plane define two lunes per plane. A pair of lunes is shown on the right-hand side. The film is placed flat and normal to the X-ray beam. During the rotation of the crystal it is kept stationary.

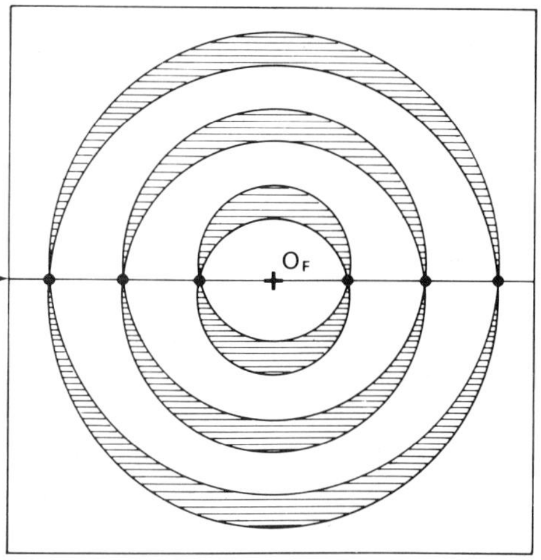

Fig. 4.28. Projection of the intersections of the Ewald sphere and three reciprocal lattice planes on the film. The innermost pair of lunes corresponds to the plane with index one because the plane with index zero is tangential to the Ewald sphere.

that contain all the reciprocal lattice points that will produce diffraction on the flat film perpendicular to the X-ray beam. Thus a rotation photograph contains reflections coming from all the reciprocal layers that intersect the Ewald sphere as shown in Fig. 4.28. The reflections contained in each of the lune pairs come from the same reciprocal lattice plane and therefore have one index in common.

Since nodes in reciprocal space have a volume, a reflection is not completely recorded on the film until the entire volume has passed through the Ewald sphere. Any reflection whose reciprocal lattice node has not completely passed through the Ewald sphere is called a partially recorded reflection or, more improperly, half spot, regardless of the percentage of the volume that has passed through the sphere of reflection. Since the rotation range in macromolecular crystallography is usually quite small, as we will see, a substantial number of reflections on a rotation film are partially recorded reflections. These have to be properly identified and dealt with during film processing. One of the reasons why the rotation method has been so successful is that it has been found that reflections partially recorded on different films can be added together[38] to yield a reliable value for the total diffracted intensity.

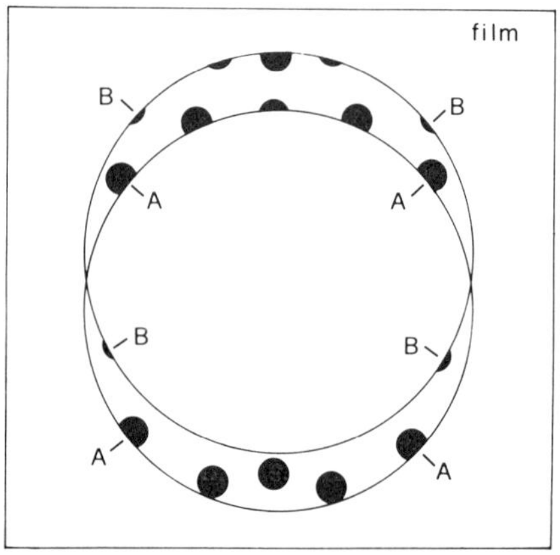

Fig. 4.29. The idealized shape of partially recorded reflections. The spots labelled A are recorded at one end of the rotation range and those labelled B at the other end. The missing parts of reflections A will be found on the previous rotation photograph and those of spots B on the following one.

Figure 4.29 shows the idealized shape of the partially recorded reflections, which is different if they are recorded at the beginning or the end of the rotation range. In one case the missing part of the reflection has been recorded in the previous film, in the other it will be recorded on the next. Re-examining Fig. 4.28 we notice that the area of reciprocal space that crosses the Ewald sphere becomes a series of points along the projection of the rotation axis. Thus, no matter how small their reflecting range, reflections found along this line will always be partially recorded.

Wonacott[39] has examined the factors that limit the maximum rotation range allowed without having reflection overlap from different reciprocal lattice planes on the film. The expression for the maximum rotation range is:

$$\Delta\Phi_{\max} \leqslant |\boldsymbol{r}^*|/R^*_{\max} - \Delta \tag{4.29}$$

where $\Delta\Phi_{\max}$ is the maximum allowed rotation range in radians, $R^*_{\max}$ the maximum resolution in reciprocal space of the data that will be collected, $\boldsymbol{r}^*$ is the pertinent reciprocal lattice vector, and Δ is the reflecting range of the crystal.

As an example let us calculate $\Delta\Phi_{max}$ for a maximum resolution of 2.5 Å when the relevant reciprocal lattice vector is 1/100 Å and Δ is 0.4°

$$\Delta\Phi_{max} \leqslant 2.5/100 \times 57.3 - 0.4 = 1°$$

where we have multiplied by 57.3 to convert radians into degrees.

In Fig. 4.30 the Ewald sphere is represented projected onto the plane determined by the X-ray beam and the crystal rotation axis. A limiting sphere of smaller radius represents the maximum resolution of the data that will be collected in the rotation experiment. The maximum rotation range of the crystal, represented by a 360° rotation of the Ewald sphere about the rotation axis, will cover the volume of reciprocal space limited by the toroid generated by the rotation of the sphere but will leave out the shaded volume shown projected onto the plane of the figure, a blind zone inaccessible to the rotation geometry. From the figure it should be apparent that the volume of the blind region increases with the resolution of the data to be collected. A plot of the percentage of reciprocal space which cannot be measured in a rotation experiment as a function of the resolution desired has been calculated by Wonacott.[39] For data of up to 5 Å resolution, it is only 0.72 per cent but it increases to 4.58 per cent for 2.0 Å resolution. In order to record the data generated by the reciprocal lattice nodes present in the blind region, the crystal has to be rotated about a different axis.

Figure 4.31 shows the total volume of reciprocal space swept out as the crystal is rotated a total angle of 30°, 90°, and 180°. From the figure it should be evident that even if the rotation axis has no symmetry it is never necessary to rotate a crystal much more than 180° in order to collect all of reciprocal space with the exception of the blind region, not represented in the figure. The volume which has not been covered by a 180° rotation has been dotted in Fig. 4.31(c) and its size depends on the resolution desired as shown in the same figure. From the figure it should also be clear that, in

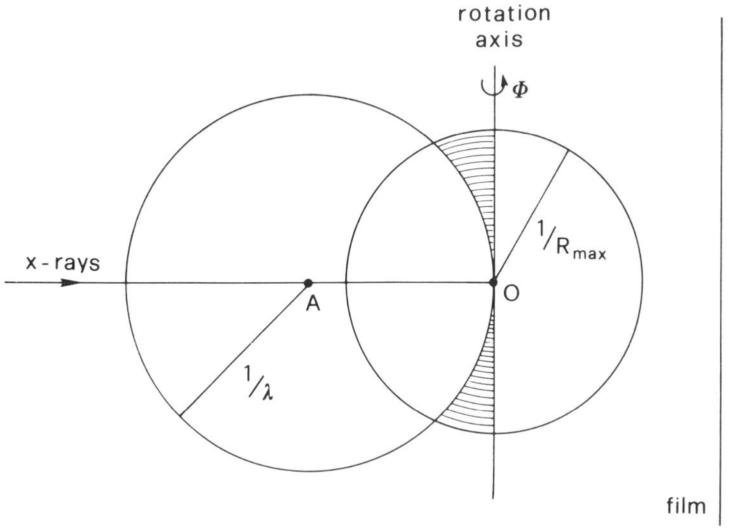

Fig. 4.30. Rotation geometry showing the Ewald sphere projected on to the plane defined by the X-ray beam and the rotation axis. A limiting sphere centred on O shows the resolution desired or attainable. The shaded region represents the volume of reciprocal space inaccessible to the rotation movement.

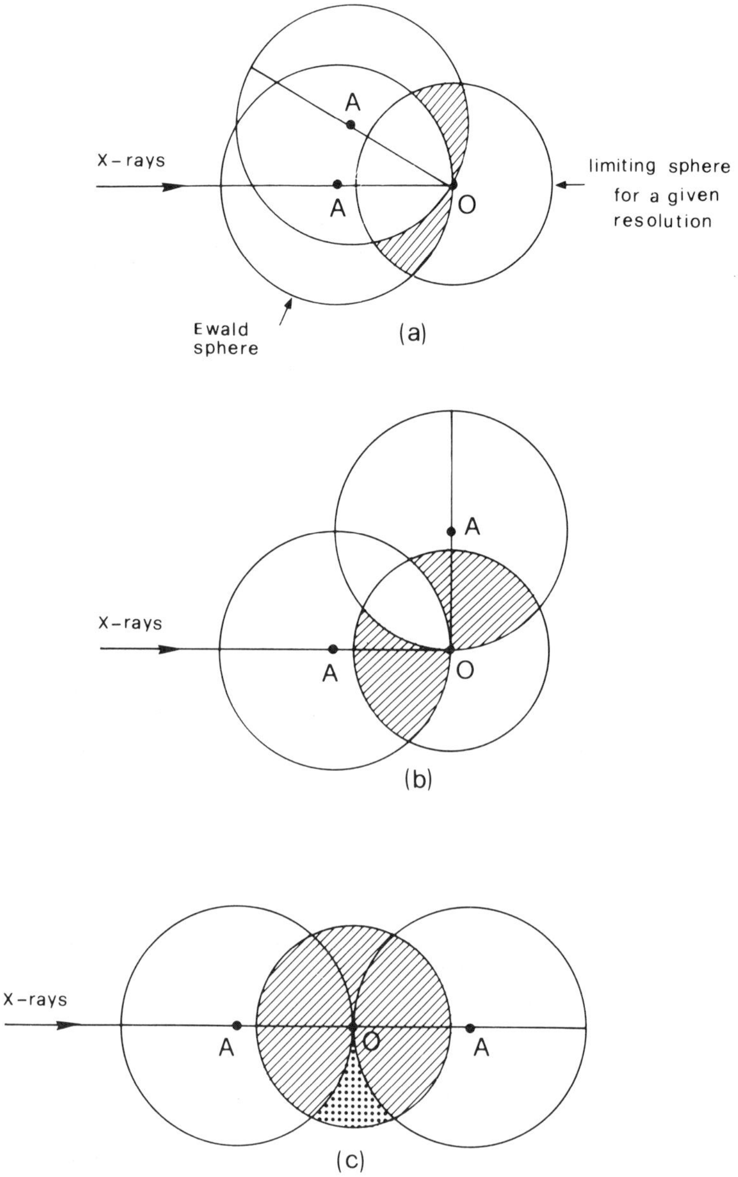

Fig. 4.31. Volume of reciprocal space explored by a total rotation range of 30° (a), 90° (b), and 180° (c). In each case a limiting sphere corresponding to a chosen resolution defines the volume of reciprocal space of interest.

order to minimize the total rotation range, it is desirable to rotate the crystal about the crystallographic axis of highest symmetry. Thus if a hexagonal crystal is rotated about the ***c*** axis, a 30° total rotation range will suffice to cover a reciprocal space asymmetric unit in which the only missing volume will be the blind region.

Symmetry is, however, not the only consideration in the choice of the rotation axis. Another factor, which is equally important, is the size of the unit cell parameters. We have seen that the maximum rotation interval per photograph is dependent on the unit cell parameters. Ideally, one would like to rotate about the axis with the largest unit cell parameter, which in this way does not become a limiting factor in the choice of the maximum

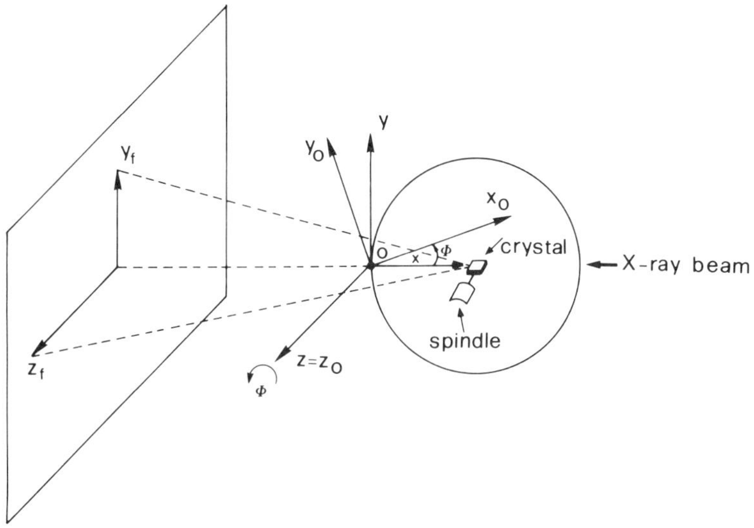

Fig. 4.32. The crystal, laboratory, and film coordinate systems. The origin of crystal and laboratory systems is at the intersection of the X-ray beam with the Ewald sphere. The origin of the film system at the intersection of the X-ray beam with the film.

rotation range per photograph. Still, there may be practical reasons that partially or totally limit the freedom of choice of the rotation axis. An example is crystal morphology. A crystal shaped as a very thin plate with its highest symmetry axis perpendicular to the plate cannot be easily mounted with that axis parallel to the spindle.

Calculation of the film coordinates for a reflection

In order to calculate the film coordinates for a reflection, four different sets of coordinate systems have to be introduced. The first is the crystallographic reciprocal lattice with its unit vectors $\boldsymbol{a}^*$, $\boldsymbol{b}^*$, and $\boldsymbol{c}^*$ and the angles α^*, β^*, and γ^*. In this coordinate system, a reciprocal lattice point is specified by the vector $\boldsymbol{r}^*$ (see p. 65).

$$\boldsymbol{r}^* = h\boldsymbol{a}^* + k\boldsymbol{b}^* + l\boldsymbol{c}^*.$$

The second coordinate system in an orthogonal system called the crystal coordinate system which is linked to the crystal, that is it rotates just as $\boldsymbol{a}^*$, $\boldsymbol{b}^*$, and $\boldsymbol{c}^*$ and differs from it in that it is always orthogonal (see Fig. 4.32).

In this new coordinate system, the coordinates of a reciprocal lattice point, (x_0, y_0, z_0) can be calculated as follows

$$\begin{aligned} x_0 &= ha_x^* + kb_x^* + lc_x^* \\ y_0 &= ha_y^* + kb_y^* + lc_y^* \\ z_0 &= ha_z^* + kb_z^* + lc_z^* \end{aligned} \tag{4.30}$$

where a_x^* is the projection of $\boldsymbol{a}^*$ onto the x_0 axis, etc.

The third coordinate system is the laboratory orthogonal coordinate system which remains fixed as the crystal is rotated. This system is defined so that z is parallel to the rotation axis and x is parallel to the X-ray beam but points in the opposite direction. If at the beginning of the rotation experiment the crystal and laboratory coordinate systems coincide, after the crystal has rotated a certain angle Φ, the coordinates of a point in the laboratory system will be

$$x = x_0 \cos \Phi + + y_0 \sin \Phi$$
$$y = y_0 \cos \Phi - x_0 \sin \Phi \tag{4.31}$$
$$z = z_0$$

as can be seen in Fig. 4.32. If the angle Φ at which a given reciprocal lattice point crosses the Ewald sphere is known, since x_0, y_0, and z_0 are only functions of the reflection indices and the unit cell parameters, x, y, and z, the reflection coordinates in the laboratory system, can be calculated.

The fourth coordinate system is the projection of the laboratory coordinate system on to the film plane. In order to convert from the laboratory to the film coordinate system we only need to know the crystal to film distance. Figure 4.32 shows the relationship between the crystal, laboratory, and film coordinate systems.

Thus we can calculate the film coordinates for a reflection if we know the angle Φ at which this reflection crossed the Ewald sphere.

Let us consider the reciprocal lattice point P which crosses the Ewald sphere at a rotation angle Φ. Figure 4.33 shows that P is on the Ewald sphere when the centre of the sphere has moved from A to A′ that is from $(\lambda^{-1}, 0, 0)$ to $(\lambda^{-1} \cos \Phi, \lambda^{-1} \sin \Phi, 0)$.

Since P lies on the Ewald sphere, its coordinates must satisfy the equation of a sphere

$$(x_0 - \lambda^{-1} \cos \Phi)^2 + (y_0 - \lambda^{-1} \sin \Phi)^2 + z_0^2 = \lambda^{-2}$$

and

$$x_0 \cos \Phi + y_0 \sin \Phi = \tfrac{1}{2}\lambda(x_0^2 + y_0^2 + z_0^2) \tag{4.32}$$

which can be solved to find the value of Φ.[39]

Once Φ for a reflection is known its coordinates in the laboratory system and its position on the film can be calculated in the way that has been outlined.

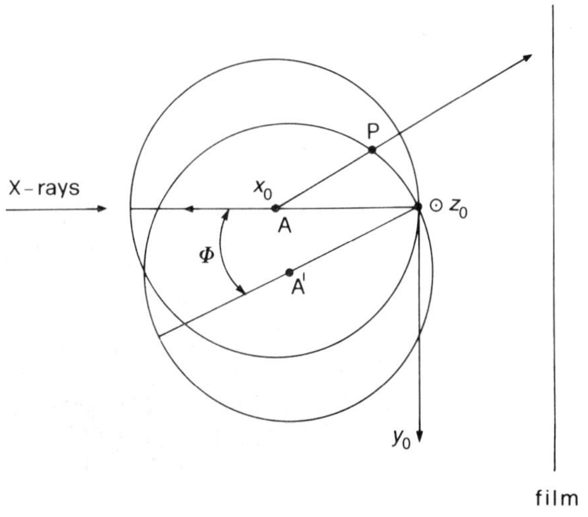

Fig. 4.33. When the crystal rotates an angle Φ point P lies on the surface of the Ewald sphere and the centre of the sphere has moved from A to A′. The crystal coordinate system is only shown before the rotation, that is when it is coincident with the laboratory system.

Indexing a rotation photograph

Figure 4.34 is a photograph of an Arndt–Wonacott rotation camera.[37] There are other commercial versions of rotation cameras, some of which, with the possibility to expose only one cassette at a time, which are very demanding on the operator unless very long exposure times per film are used. Figure 4.35 is a rotation photograph of a protein crystal. When data collection with the rotation method is started on a protein crystal, the space group has already been determined using the precession method and the unit cell parameters are known. Also precession photographs of the zero-level projections and some upper levels are usually available. The crystal that produced Fig. 4.35 is orthorhombic, space group $P2_12_12_1$, and it was mounted with the $\boldsymbol{c}^*$ axis on the spindle. Although a knowledge of the crystal setting and indexing of one or more rotation pictures is not essential to process the data,[40] we will briefly discuss how Fig. 4.35 can be indexed. The picture is the second of a series collected with the $\boldsymbol{c}^*$ axis mounted on the spindle. It was exposed at a crystal to film distance of 7.5 cm and $\Delta\Phi$ is in this case 3°, the unit cell parameters are $a = 46.1$, $b = 49.1$, and $\boldsymbol{c} = 76.1$ Å. Since in this case there is no higher symmetry axis, the choice of the rotation axis was dictated only by the size of the unit cell parameters. Partially recorded reflections can be identified by examining the spot shape (see Fig. 4.29) or, more easily when the spot shape is not very regular as in this case, by placing the photograph on top of the previous and the following one. Partially recorded reflections are those that appear recorded on more than one film.

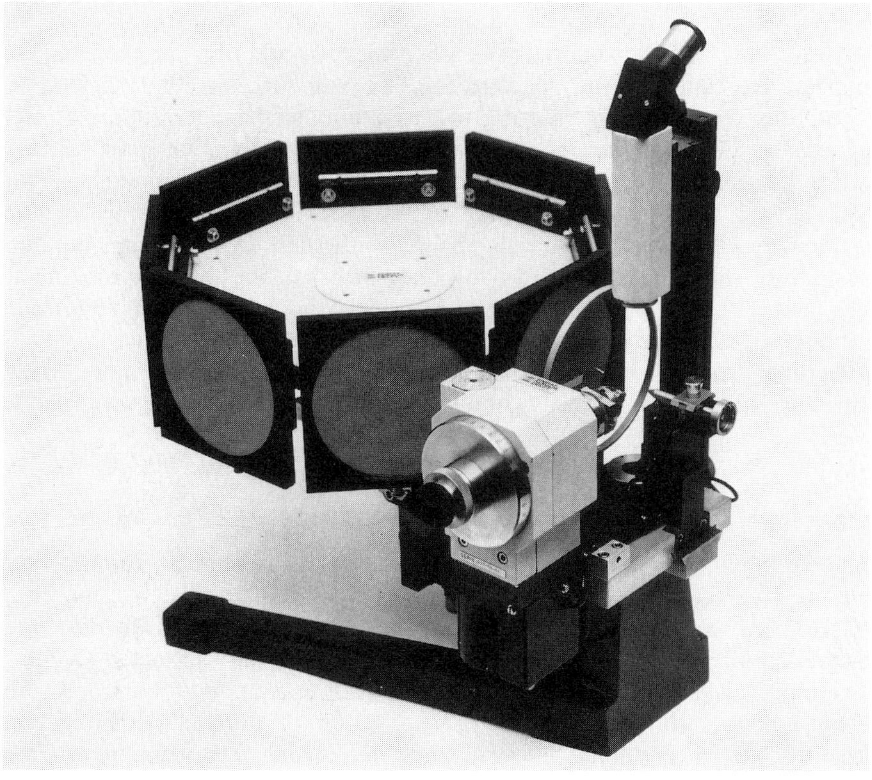

Fig. 4.34. The Arndt–Wonacott rotation camera. (Photograph courtesy of Enraf Nonius.)

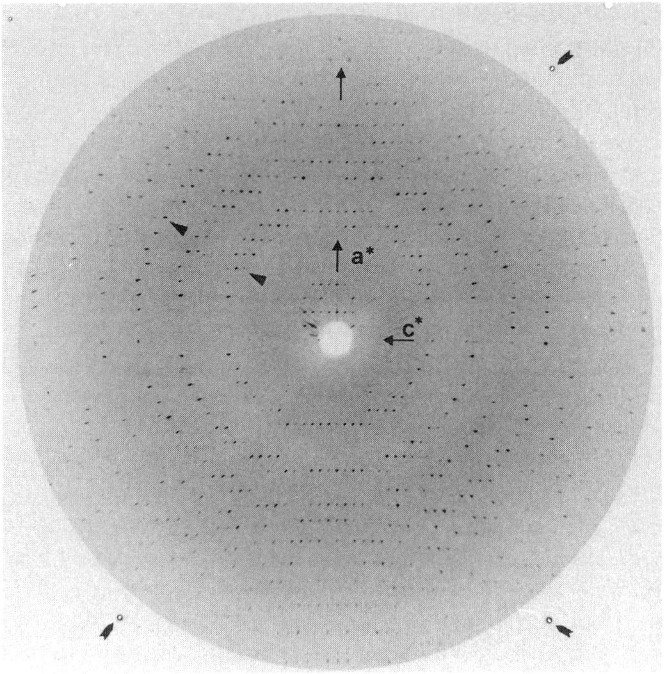

Fig. 4.35. Rotation photograph of a protein crystal. Notice the fiducial marks that are used to determine the centre of the film during densitometry. For a discussion of the indexing procedure see text.

The rotation picture recorded before the one shown in Fig. 4.35 was exposed with the X-rays in approximately the direction of the $\boldsymbol{b}$ axis. In fact in Fig. 4.35 we can see the small circle closer to the beam stop that corresponds to the $h0l$ plane.

The k index of the reflections is the easiest to identify; for the smallest circle it is 0 and for the concentric lunes it is successively 1, 2, 3, etc. (compare with Fig. 4.28). In order to find h and l we have to find the $\boldsymbol{c}^*$ and $\boldsymbol{a}^*$ axes on the film, then indexing is done simply by counting spots. The $\boldsymbol{c}^*$ axis is horizontal because it coincides with the rotation axis, its intersection with $\boldsymbol{a}^*$ can be found by locating $\boldsymbol{a}^*$, here we are aided by systematic extinctions. The first index h is the most difficult to determine but the position of the $\boldsymbol{c}^*$ axis, which corresponds to $h = 0$ can be found looking at the picture exposed before Fig. 4.35 which shows symmetry about the rotation axis. The axis of symmetry is the $\boldsymbol{c}^*$ axis.

Bernal[36] proposed the use of specially designed charts to more easily index rotation photographs. These charts are nowadays not very widely used.

Densitometry

The data collection cameras that we have seen in this chapter normally use film as a detector for the radiation diffracted by the crystals. In Chapter 3 (p. 161) we have seen that the quantity that is proportional to the structure factor amplitude is the integrated intensity of the diffracted X-rays. Therefore, in order to be able to quantitatively use the data recorded with these methods, the first step is the extraction of the relative integrated intensities from the film. This is currently done using an instrument called a densitometer or film scanner, hence the name densitometry applied to the procedure followed to measure the relative diffracted intensities on film.

Optical density and integrated intensity

The quantity that is measured by the densitometer is the optical density of the spots produced by the diffracted beams on the film. There is a relationship between optical density and number of X-ray photons striking the film that we will explore more carefully in the next section but which can be assumed, for the time being, to be linear up to X-ray optical densities of at least 1.0. The optical density of the spots is determined by shining visible light on the film and measuring the ratio of the intensity transmitted to that of incident light and it is defined as

$$D = \log I_0/I_t \tag{4.33}$$

where I_0 is the incident light intensity, I_t the transmitted light intensity, and D is the optical density.

Modern instruments are capable of measuring optical densities of very small areas, as small as $12.5 \times 12.5\ \mu m^2$ and are therefore called microdensitometers.

The total optical density of a spot on a film is the sum of the optical density of a certain number of squares, called pixels, covering the spot area minus the same number of terms with an optical density equal to that of the film background.

$$D = \sum_{\substack{\text{spot}\\\text{area}}} D_S - \sum D_B.$$

Figure 4.36 shows the computer output from a scanning program with a

```
010011111111110111101011000101110111001211211
01110111110111112110111111101110111122211111
101101111111011111110001201111211111111101111
101110111111110111121011101111111111112101011
101110110001111211111221110011111110112121101
102012111111121112110211111111101121111112111
101111011101111111111111111221111011111112111
111110001101111112122122211221121011111112121
121111011021213222323333322211122111110221201
10111111121122333544555444444323222111111101101
111111112223345577787776655532222211121111110
1011222123434688889999999877754333321111110000
10001122224568989999999999999998555433222111101
10101222345889999999999999998666443222221111
2110222334787999999999999999999875322122221111
110111335588899999999999999999886653332211121
21111132557899999999999999999987554321110111
101011244568999999999999999999964533321221010
1111122345789999999999999999987763322221 0011
101111245578999999999999999998886432221211111
011112135668999999999999999999988533221212121
111122245688999999999999999999888775322222211
101111135689999999999999999999987543221111101
112113134588999999999999999999987643222221111
111111235669999999999999999999976644221111111
110111245678999999999999999999999865532110112
110112234788999999999999999999997643312121211
011112335687899999999999999999997664332321211
111112244567999999999999999999998545422111111
000011134578899999999999999999995643322111111
111011124568899999999999999999998765232212012
001111113346678999999999999999998764322210111
000011134456878999999999999999998644322221001
011011124345798999999999999999996654322112110
100012223344578789999999999999988654432211211 0
111111223324679679999999999999997644222211110 1
111100111234466666689989999999977544322322211 11
111111122223344555567898999999754443422222222 10
011010022122344456566766787664432222111100121 1
000011010112223344544554555454322221210111111 1
011111100112113322224344444443422221112211112 1101
101111101111211222213223333422111010110011011 0
000111101001111111122222122222101011210101100 01
001021001001111111111111011111111111101111111 0012
001111011100210111111112111111101111001111110 110
010111011100111110111111110111111210212011001 1
101101110010110110011111211001111111011011101 1
```

Fig. 4.36. Computer output showing a reflection recorded on film and scanned with a 25 μm raster. The optical densities from 0 to 2 correspond to numbers that range from 0 to 9 in the figure.

diffraction spot in which the optical densities from 0 to 2 correspond to numbers that range from 0 to 9. This last interval was chosen to make the spot representation clearer, although normally these numbers do not range from 0 to 9 but from 0 to 255.

The sampling area has to be chosen bearing in mind that because of film graininess statistically more significant results are obtained with larger areas. On the other hand, the optical density should not vary significantly within the sampling area; otherwise one is averaging transmitted light rather than optical density and the final optical density will be in error because of the logarithmic relationship between the two quantities. This last effect is called the Wooster effect.[41]

Film as an X-ray detector

When an X-ray photon strikes a film, it excites a silver halide particle which, when the film is developed, will produce a black silver grain on the film. Here we will follow Vonk and Pijpers in their derivation of a relationship between the number of photons striking the film and the optical density they produce.[42]

If the film contains a number of excitable particles n_0 per unit area and n have been excited by the X-rays, as dE photons strike the film, the number of excited particles will increase by dn and

$$\mathrm{d}n = ma\frac{n_0 - n}{n_0}\,\mathrm{d}E$$

where $(n_0 - n)/n_0$ is the fraction of particles that have not been excited, a is the fraction of radiation absorbed by excited and unexcited particles, and m is the number of particles excited per photon, normally taken to be equal to 1. Solving the differential equation

$$n = n_0[1 - \exp(-maE/n_0)]$$

using as boundary conditions $n = 0$ for $E = 0$ and $n = n_0$ for very large values of E.

As more and more particles are excited on the film, the fraction of light transmitted, when the optical density is measured, will decrease. Assuming that the fraction of light transmitted by the film is proportional to the area that has not been excited one can obtain

$$\mathrm{d}I_t/I_0 = -fI_t\,\mathrm{d}n/I_0K$$

where I_t/I_0 is proportional to the unexcited area, K is the proportionality constant, f is the surface covered by a grain, and dn the increment in excited particles as before.

Solving the differential equation we obtain

$$I_t/I_0 = \exp(-fn/K),$$

$$2.3\log I_t/I_0 = -fn/K,$$

and

$$D = fn/2.3K.$$

The maximum optical density that can be measured, D_{max}, will correspond to $n = n_0$. If, in the expression of n as a function of E, we substitute

n and n_0 in terms of D and D_{max} we obtain

$$D = D_{max}[1 - \exp(-ES_0/D_{max})] \tag{4.34}$$

where $S_0 = maf/2.3K$.

This equation can be simplified[42] by two successive series expansions to

$$D/E = S_0(1 - D/2D_{max}). \tag{4.35}$$

From this equation we see that for large values of D_{max} and small optical densities the relationship between optical density and exposure is linear. The exposure E, that is the number of photons exciting the film, is proportional to exposure time when a film is exposed to successively higher X-ray doses by increasing the time that it is hit by a constant X-ray beam. The parameter D_{max} is characteristic of the film and can be determined plotting D/E as a function of D. Some typical values have been determined by Vonk and Pijpers[42] and more recently by Elder[43] for very widely used types of X-ray film. The value of D_{max}, representing the optical density of the film when all the silver halide granules on it have been excited, is usually well beyond the optical density range that can be measured with the densitometer.

The ratio D/E is called the film speed at density D, S_0 is the initial speed, and the function in parentheses shows how this value decreases as exposure progresses.

Arndt *et al.*[44] have derived a simple approximate expression for the fractional standard deviation of an optical density measurement on film. If σ_D is the standard deviation

$$\sigma_D/D = 1.35N_D^{-1/2} \tag{4.36}$$

where N_D is the total number of quanta producing the optical density D. This equation shows that the fractional standard deviation is not dependent on the film speed.

All radiation counters are assumed to be approximately Poissonian detectors, i.e. the statistics of counting rates can be analysed theoretically by a Poisson distribution. If film were an ideal Poisson detector the factor 1.35 would have to be 1.0 instead. Thus the fractional standard deviation of film is higher by a factor of 1.35 than the equivalent parameter in more ideal Poisson detectors, like for example those used for diffractometers.

Microdensitometers

There are currently three different types of microdensitometers in use: the rotating-drum, cathode ray tube and flat-bed microdensitometer.[45] The first kind is by far the most widely used by X-rays crystallographers and it is for that reason that it will be briefly described here.

Rotating-drum densitometers have been described by Abrahamsson[46] and Xuong[47] and their performance in the scanning of precession photographs has been analysed by Nockolds and Kretsinger[48] and by Matthews *et al.*[49]

In a rotating-drum scanner, the film is mounted on a cylindrical drum that has a rectangular aperture for the film and which rotates about the cylinder axis. The rotation speed is variable, and depending on the instrument can be as high as 12 revolutions per second. A beam of visible light is passed through the film and its intensity is measured by a detector. Source and detector are stationary during one revolution of the drum and are automatically stepped along the cylinder axis until the entire film is covered. The incident light intensity I_0, is measured as the beam goes through air.

The raster size is variable, it can be 12.5, 25, 50, 100, or 200 μm or more and the transmitted light is measured at intervals equal to these values. Thus if a raster size of say 100 μm has been chosen a strip 100 μm wide will be read in one revolution at 100 μm intervals. After that, source and detector will be advanced 100 μm and another strip will be read until the entire film is covered by 100×100 μm pixels. The instrument is interfaced to a computer and the process is totally computer controlled.

When the light beam goes through air, the instrument gives an optical density of 0. The maximum integer reading of $255(2^8 - 1)$ can be made to correspond to an optical density of either 2 or 3. The optical density values thus measured can be stored on magnetic tape or directly in the computer connected to the densitometer. There are two strategies used by computer programs in processing densitometer data.[48,49] In the first, the entire film is read as described above and the data are stored for subsequent computer processing to obtain the integrated intensities. In the second approach, integrated intensities are obtained on-line as scanning of the film proceeds. Both strategies have their advantages and disadvantages and have been used extensively to scan all types of diffraction films.

The equations relating optical density on the film to the total number of photons that have caused it ((4.34) and (4.35)) show that there is a maximum value of optical density that can be measured for a given type of film and that the optical density is a linear function of the exposure only for relatively low values of D. We will briefly discuss how these two limitations are handled experimentally.

The problem of a limited dynamic range of the film, i.e. of a limited optical density range that can be measured, can be solved by placing more than one film in the cassette that records the X-ray diffraction pattern. Since a substantial fraction of the radiation is absorbed by the X-ray film, those reflections which are too strong to be measured on the first film will normally fall within measuring range on the second or third. After the integrated intensities have been calculated, all the films in the pack can be scaled together.

The non-linearity of the film response can be handled by constructing experimentally a table that relates an optical density produced by the scanner to a given exposure. Since one is interested in relative integrated intensities, the table can be constructed by exposing the film to the same X-ray beam during different times and measuring the optical density of the spots under the conditions that will be used for data scanning. In this experiment the exposure times are proportional to the number of photons hitting the X-ray film.

An alternative approach proposed by Matthews *et al.*[49] is to assume a parabolic relationship between integrated intensity and optical density.

$$I_{\mathbf{H}} = AD_{\mathbf{H}} + BD_{\mathbf{H}}^2$$

where A and B are two constants; A is arbitrarily set to one and B is determined from the scaling factor between successive films in the pack. In this second approach the non-linearity of the film response is handled in the film scaling procedure.

Although the simplest approach to determine the integrated intensity is to simply subtract the background from the area covering the spot, using the technique called profile fitting, smaller estimated standard deviations can be obtained.[50] In this method, a model profile for the reflection is constructed, that is a model intensity distribution in two dimensions is determined by averaging the measured profiles of a certain number of reflections on the film. It is then assumed that all the reflections have this standard profile and the measured data are fitted to it.

The single-crystal diffractometer

An alternative to the data collection methods using film that we have described so far is the single-crystal diffractometer.[51,52] Nowadays, it is with this instrument, that almost all of the data required to solve small molecule structures are collected and the diffractometer can also be used in macromolecular crystallography. Its most important advantage is better precision since film has a much higher level of background. Its only disadvantage is that it measures one reflection at a time, a limitation that is not very important for non-demanding crystals but can become very severe if radiation damage of the sample is a problem.

Diffractometer geometry

A single-crystal diffractometer consists of an X-ray source, an X-ray detector, a goniostat that orients the crystal so that a chosen X-ray diffracted beam can be received by the detector, and a computer that controls goniostat and detector movements and performs the mathematical operations required to position the crystal and detector in the desired orientations.

The detector is usually of the scintillation counter type in which X-rays excite a fluorescent material and thus generate visible light which is measured in an appropriate way. Xenon filled proportional counter detectors are also used, particularly with Cu K_α radiation. This type of X-ray detector will be further discussed on p. 281.

Both the molecular excitation of the fluorescent material and the gas ionization in the counter detector are events that can be triggered by the arrival of only one photon. It can be shown experimentally[53] that if an X-ray beam is measured several times with a diffractometer detector, the different intensity values obtained follow a Poisson distribution. For such a distribution, the estimated standard deviation is

$$\sigma = N^{1/2}$$

where N is the number of counts. The fractional standard deviation is thus

$$\sigma_f = N^{1/2}/N = N^{-1/2}.$$

We saw on p. 271 that for film methods the fractional standard deviation is larger by a factor of 1.35. Thus, diffractometer measurements can be said to be intrinsically more precise than those obtained from film methods. Modern diffractometers use the equatorial geometry in which the diffracted beams are always measured in a horizontal plane defined by the incident

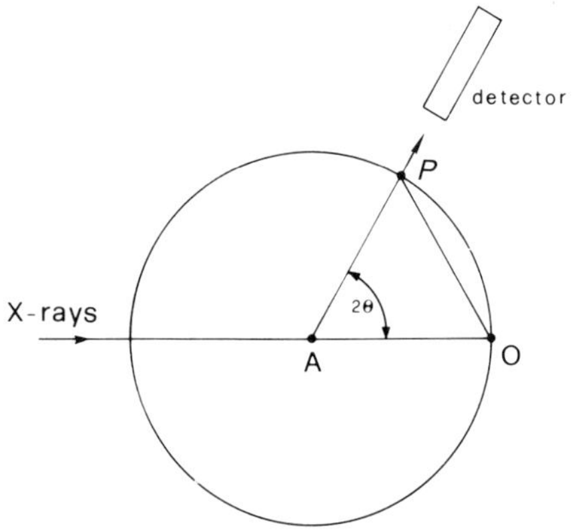

Fig. 4.37. The diffractometer equatorial geometry. The detector, rotating about the instrument main axis, defines a plane that contains the incident beam. Reflections will always be measured on this plane.

X-rays and the rotation of the detector about an axis passing through the crystal. The detector can only move on this plane and it forms an angle 2θ with the incident beam as shown in Fig. 4.37. In the figure, point P is in diffracting position because it is on the Ewald sphere and produces a scattered beam that can be detected because it is on the equatorial plane. In order to observe diffraction, all the reciprocal lattice nodes are brought in turn to some point on the circle defined by the intersection of the sphere of reflection and the equatorial plane. At the same time, the detector is moved to the appropriate 2θ angle so that it can receive the diffracted beam.

The most widely used type of goniostat uses the Eulerian cradle which gives rise to the four-circle diffractometer shown schematically in Fig. 4.38. The cradle is constituted by the χ circle which carries the goniometer head with the crystal. The instrument has a main axis that is normal to the equatorial plane and therefore to the incident and diffracted X-ray beams and passes through the crystal. Rotation of the cradle about the main axis defines the angle ω, rotation about the spindle axis of the goniometer head defines the angle Φ in exactly the same way as in the rotation camera. The angle χ is defined by the spindle of the goniometer head and the main instrument axis. The four circles of the diffractometer are thus the Φ and χ circles about which the crystal can be rotated, the ω circle, defined by the rotation of the cradle, and the 2θ circle described by the rotation of the detector about the main axis.

Both the 2θ and ω rotations are about the main axis but the first moves the detector and the second the cradle. In a three-circle diffractometer the degree of freedom that is missing is the rotation ω, in other words the cradle is fixed with its χ plane perpendicular to the incident X-ray beam. The angle 2θ is 0 when the detector is positioned in the direction of the incident X-ray beam, χ is 0 if the spindle axis is parallel to the main axis, ω is 0 when the χ circle is perpendicular to the incident X-ray beam, and the zero position of Φ is arbitrary and can be defined with respect to the crystal orientation. If the angle χ is 0 the ω and Φ rotations coincide.

In general, only two rotations are required to bring a reciprocal lattice node to the intersection of the Ewald sphere and the equatorial plane. If we

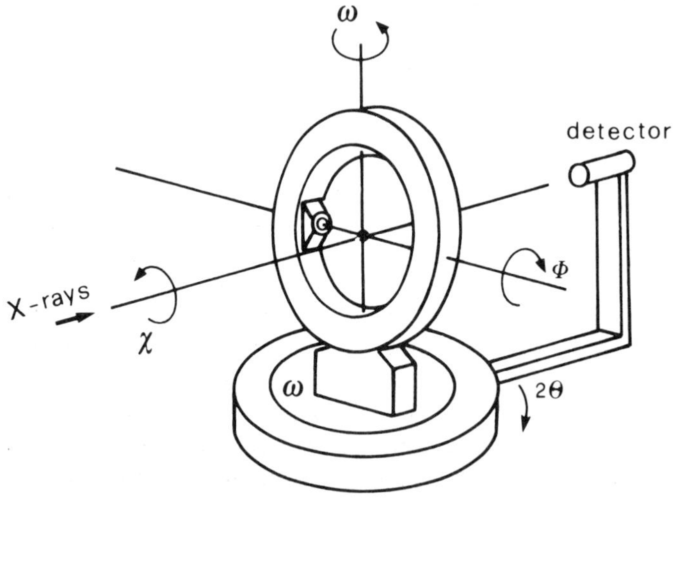

Fig. 4.38. The four-circle diffractometer. The χ circle carries the goniometer head with the crystal and together with it constitutes the Eulerian cradle. The cradle can rotate about the main instrument axis describing the angle ω. The detector moves independently and forms an angle 2θ with the incident X-ray beam.

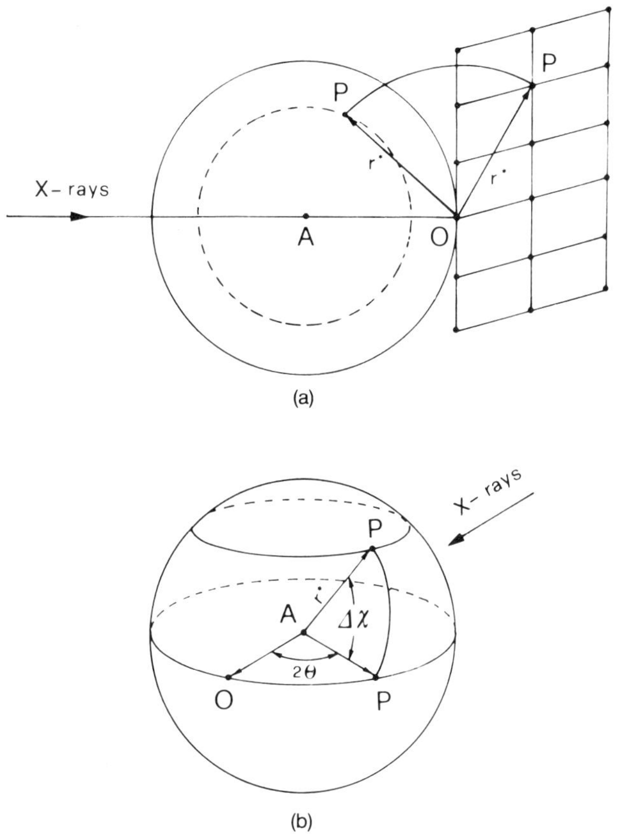

Fig. 4.39. Bringing a reciprocal lattice node to the intersection of the Ewald sphere and the equatorial plane. (a) The crystal can be first rotated about Φ until the node lies on the sphere. (b) A χ rotation brings the point on the sphere of reflection to the equatorial plane where it can be measured.

start with $\chi = 0$, a rotation about Φ or ω can bring the reciprocal lattice node on the Ewald sphere as shown in Fig. 4.39(a) which is a projection looking down the instrument main axis. Then, it can be brought to the equatorial plane by a rotation about the χ axis as shown in Fig. 4.39(b). Measurement of the scattered beam can then be accomplished by placing the detector at the appropriate 2θ angle. However, the direction of the scattered beam thus defined may be such that detection may be difficult or even impossible because of the interference of the χ circle, collimator, etc. It is for this reason that, although in principle rotation of the crystal about two axes plus the detector movement are sufficient to satisfy the diffraction condition, the extra degree of freedom provided by the ω circle can prove very valuable.

Quite often, the so-called bisecting geometry arrangement is chosen in which the χ plane is moved so that it is made to contain the vector $\boldsymbol{r}^*$. Under these conditions, the χ axis, perpendicular to the χ plane, bisects the angle formed by the incident and scattered X-rays as shown in Fig. 4.40.

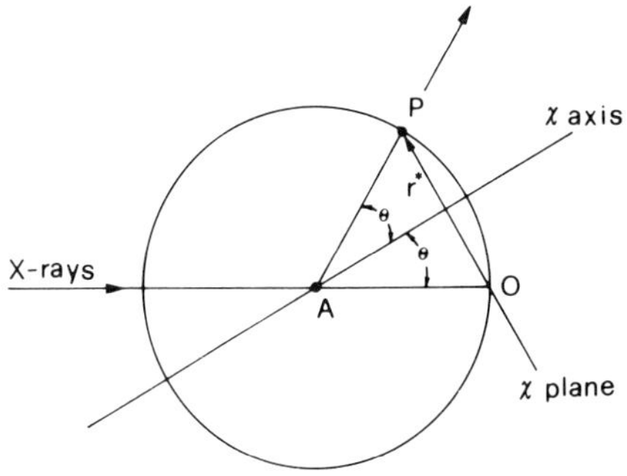

Fig. 4.40. The bisecting geometry arrangement. The reciprocal lattice vector $\boldsymbol{r}^*$ lies in the χ plane and the χ axis bisects the angle formed by the incident and diffracted beam.

An alternative to the Eulerian cradle, that is used in a widely diffused instrument, is the kappa goniostat. In this type of geometry the χ circle does not exist and the goniometer head carrying the crystal is mounted on an arm that can rotate about an axis, the κ axis, which forms an angle of 50° with the main instrument axis as shown in Fig. 4.41. The advantage of this system is that it is less cumbersome than the Eulerian cradle and thus

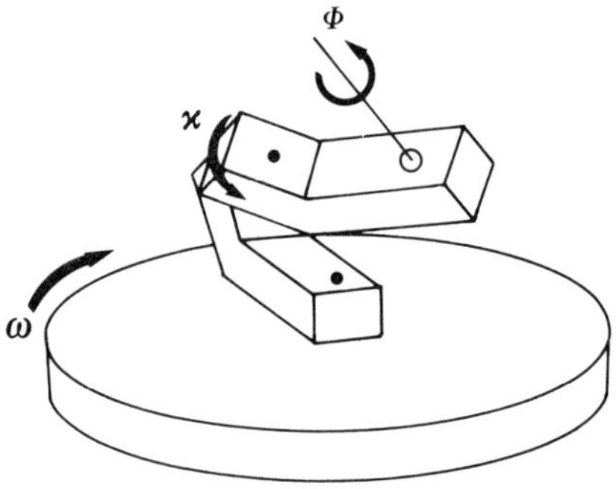

Fig. 4.41. The κ geometry goniostat.

rotations that would cause collisions or would produce diffracted beams that would be blocked by the χ circle in the conventional four-circle diffractometer are still practicable.

Determination of the crystal orientation and unit cell parameters: the orientation matrix

Before one can predict the values of the angles Φ, χ, ω, and 2θ for which scattering will be observed for a given reflection of a crystal, it is necessary to determine the crystal orientation with respect to the Eulerian cradle. In this section we follow the *International tables for x-ray crystallography*,[54] and define several coordinate systems similarly to what was done on p. 265 for the rotation method.

Let us define the diffraction orthonormal coordinate system, $\mathbf{A}_\mathrm{D}$, with basis vectors $\boldsymbol{a}_\mathrm{D}$, $\boldsymbol{b}_\mathrm{D}$, $\boldsymbol{c}_\mathrm{D}$ such that $\boldsymbol{a}_\mathrm{D}$ and $\boldsymbol{b}_\mathrm{D}$ are both in the equatorial plane; $\boldsymbol{a}_\mathrm{D}$ points towards the X-ray source for $2\theta = 0$, $\boldsymbol{b}_\mathrm{D}$ is parallel to the diffraction vector for positive values of 2θ and $\boldsymbol{c}_\mathrm{D}$ is parallel to the instrument main axis and has a direction such that the three unitary vectors $\boldsymbol{a}_\mathrm{D}$, $\boldsymbol{b}_\mathrm{D}$, and $\boldsymbol{c}_\mathrm{D}$ form a right-handed system. Accordingly, in Fig. 4.42, the vector $\boldsymbol{c}_\mathrm{D}$ is normal to the figure plane and points away from the observer. A second orthonormal coordinate system, $\mathbf{A}_\mathrm{G}$, the crystal coordinate system, is defined with respect to the goniometer head and therefore relative to the crystal. Its basis vectors are called $\boldsymbol{a}_\mathrm{G}$, $\boldsymbol{b}_\mathrm{G}$, and $\boldsymbol{c}_\mathrm{G}$ and it is coincident with the diffraction system when all the diffractometer angles are equal to 0. When one or more of the diffractometer angles differ from 0 the two coordinate systems are related by a rotation matrix $\mathbf{F}$ such that

$$\mathbf{A}_\mathrm{G} = \mathbf{F}\mathbf{A}_\mathrm{D}$$

where (see p. 72)

$$\mathbf{F} = \mathbf{R}_\Phi \mathbf{R}_\chi \mathbf{R}_\omega$$

$$\mathbf{F} = \begin{pmatrix} \cos\Phi\cos\omega - \sin\Phi\sin\omega\cos\chi & \cos\Phi\sin\omega + \sin\Phi\cos\omega\,\mathrm{cox}\,\chi & \sin\Phi\sin\chi \\ -\sin\Phi\cos\omega - \cos\Phi\sin\omega\cos\chi & -\sin\Phi\sin\omega + \cos\Phi\cos\omega\cos\chi & \cos\Phi\sin\chi \\ \sin\chi\sin\omega & -\sin\chi\cos\omega & \cos\chi \end{pmatrix}.$$

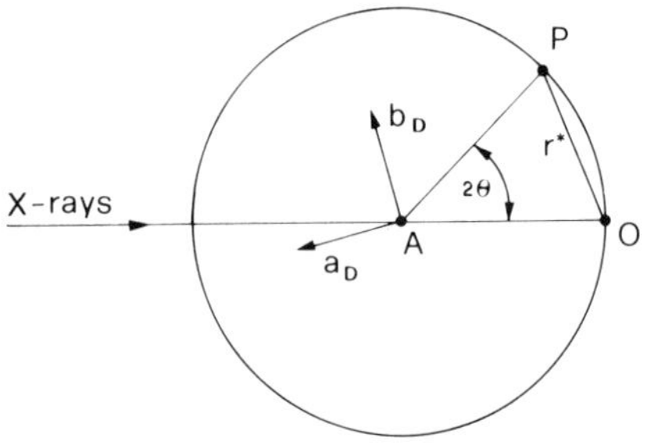

Fig. 4.42. The diffraction orthonormal system as seen in the projection on to the equatorial plane. $\boldsymbol{a}_\mathrm{D}$ and $\boldsymbol{b}_\mathrm{D}$ are the two basis vectors that are found on this plane.

Let

$$\mathbf{H} = \begin{pmatrix} h \\ k \\ l \end{pmatrix}$$

be the coordinate matrix of the vector $\boldsymbol{r}^*$ with respect to the reciprocal coordinate system

$$\mathbf{A} = \begin{pmatrix} \boldsymbol{a}^* \\ \boldsymbol{b}^* \\ \boldsymbol{c}^* \end{pmatrix}$$

and let

$$\mathbf{X}_G = \begin{pmatrix} x_G \\ y_G \\ z_G \end{pmatrix}, \qquad \mathbf{X}_D = \begin{pmatrix} x_D \\ y_D \\ z_D \end{pmatrix}$$

be the coordinate matrices of $\boldsymbol{r}^*$ with respect to the systems $\mathbf{A}_G$ and $\mathbf{A}_D$ respectively. Then, according to p. 66 and since $(\bar{\mathbf{F}})^{-1} = \mathbf{F}$

$$\mathbf{X}_G = \mathbf{F}\mathbf{X}_D.$$

If the angles ω, χ, and Φ are such that the diffraction condition is satisfied, remembering the way in which the diffraction coordinate system $\mathbf{A}_D$ was defined (see Fig. 4.42), then the vector coordinates in the diffraction system are

$$\mathbf{X}_D = \begin{pmatrix} 0 \\ |\boldsymbol{r}^*| \\ 0 \end{pmatrix}$$

and its coordinates in the crystal system $\mathbf{A}_G$

$$\mathbf{X}_G = |\boldsymbol{r}^*| \begin{pmatrix} \cos\Phi \sin\omega + \sin\Phi \cos\omega \cos\chi \\ -\sin\Phi \sin\omega + \cos\Phi \cos\omega \cos\chi \\ -\sin\chi \cos\omega \end{pmatrix}. \qquad (4.37)$$

If the bisecting geometry arrangement was chosen, then as was pointed out the diffraction vector $\boldsymbol{r}^*$ is in the χ plane; in other words ω is 0 and the vector coordinates in the crystal fixed system are:

$$\mathbf{X}_G = |\boldsymbol{r}^*| \begin{pmatrix} \sin\Phi \cos\chi \\ \cos\Phi \cos\chi \\ -\sin\chi \end{pmatrix}. \qquad (4.38)$$

Thus if we know the coordinates of a reciprocal lattice node in the fixed crystal system we can determine without any ambiguity from eqn (4.38) the values of the angles Φ, χ, and ω that will bring the node to a position that satisfies the diffraction condition. How do we then calculate the coordinates of a reciprocal lattice node in the fixed crystal system? In order to do this we need to introduce the orientation matrix $\mathbf{U}$ defined by

$$\mathbf{A}^* = \mathbf{U}\mathbf{A}_G \quad \text{or} \quad \mathbf{A}_G = \mathbf{U}^{-1}\mathbf{A}^* \qquad (4.39)$$

and (see. p. 66)

$$\mathbf{X}_G = \bar{\mathbf{U}}\mathbf{H} \quad \text{or} \quad \mathbf{H} = (\bar{\mathbf{U}})^{-1}\mathbf{X}_G \qquad (4.40)$$

where $\bar{\mathbf{U}}$ is the transposed orientation matrix.

Thus if we know the orientation matrix we can calculate the fixed crystal coordinates of any reciprocal lattice mode and therefore we can determine the Φ, χ, and ω values for which diffraction will be observed for the node. Let us now see how the orientation matrix can be determined in the general case in which the crystal unit cell parameters are unknown. If the angles χ, Φ, and ω corresponding to the diffraction condition of three general reflections, H_1, H_2, and H_3, are measured, the coordinates of the three reciprocal lattice points in the fixed crystal system, $\mathbf{X}_{G1}$, $\mathbf{X}_{G2}$, and $\mathbf{X}_{G3}$ can be calculated since they depend only on the angles and the magnitude of the vectors $|\boldsymbol{r}^*|$ (eqn (4.38)).

Multiplying both sides of the first eqn (4.39) by $\bar{\mathbf{H}}$ and introducing the second eqn (4.40) we obtain

$$\bar{\mathbf{H}}\mathbf{A}^* = \bar{\mathbf{H}}\mathbf{U}\mathbf{A}_G = \bar{\mathbf{X}}_G\mathbf{U}^{-1}\mathbf{U}\mathbf{A}_G = \bar{\mathbf{X}}_G\mathbf{A}_G. \tag{4.41}$$

Now we can write for the three reflections, 1, 2, and 3

$$\begin{aligned} \bar{\mathbf{H}}_1\mathbf{A}^* &= \bar{\mathbf{X}}_{G1}\mathbf{A}_G \\ \bar{\mathbf{H}}_2\mathbf{A}^* &= \bar{\mathbf{X}}_{G2}\mathbf{A}_G. \\ \bar{\mathbf{H}}_3\mathbf{A}^* &= \bar{\mathbf{X}}_{G3}\mathbf{A}_G. \end{aligned} \tag{4.42}$$

If one defines the matrices

$$\mathbf{H}_M = \begin{pmatrix} \bar{\mathbf{H}}_1 \\ \bar{\mathbf{H}}_2 \\ \bar{\mathbf{H}}_3 \end{pmatrix} \quad \text{and} \quad \mathbf{X}_M = \begin{pmatrix} \bar{\mathbf{X}}_{G1} \\ \bar{\mathbf{X}}_{G2} \\ \bar{\mathbf{X}}_{G3} \end{pmatrix}.$$

Equation (4.42) can be written in the more compact form

$$\mathbf{H}_M\mathbf{A}^* = \mathbf{X}_M\mathbf{A}_G \quad \text{or} \quad \mathbf{A}^* = \mathbf{H}_M^{-1}\mathbf{X}_M\mathbf{A}_G.$$

Recalling eqn (4.39)

$$\mathbf{U}\mathbf{A}_G = \mathbf{H}_M^{-1}\mathbf{X}_M\mathbf{A}_G$$

and finally,

$$\mathbf{U} = \mathbf{H}_M^{-1}\mathbf{X}_M. \tag{4.43}$$

Since $\mathbf{H}_M$ and $\mathbf{X}_M$ are known, one can calculate $\mathbf{U}$, the orientation matrix from (4.43).

If the unit cell parameters are known the angles corresponding to only two reflections are sufficient to calculate $\mathbf{U}$.[54]

In the more general case we have discussed, the orientation matrix yields also the unit cell parameters of the crystal

$$\mathbf{A}^* = \mathbf{U}\mathbf{A}_G$$

and

$$\mathbf{A}^*\bar{\mathbf{A}}^* = \mathbf{U}\mathbf{A}_G(\overline{\mathbf{U}\mathbf{A}_G}) = \mathbf{U}\mathbf{A}_G\bar{\mathbf{A}}_G\bar{\mathbf{U}} = \mathbf{U}\bar{\mathbf{U}}$$

since $\mathbf{A_G}$ is orthonormal (see Chapter 2, p. 68).

Now the matrix $\mathbf{A}^*\bar{\mathbf{A}}^*$ is nothing else than the metric matrix $\mathbf{G}^*$ of the reciprocal lattice defined by (2.15) and from it one can determine the unit cell parameters of the crystal.

Measurement of the integrated intensities

In order to obtain the integrated intensity of a reflection, the entire volume of the node must be made to sweep through the Ewald sphere while the detector counts the diffracted photons.

There are basically two scanning modes that are very widely used (but see also Wyckoff[52]) and they are called the ω scan and the 2θ or $\omega - 2\theta$ scan. In the ω scan the detector is left stationary while the crystal and thus the chosen reciprocal lattice node is made to cross the Ewald sphere by a rotation of $\Delta\omega$ about the main axis. This type of scanning mode is illustrated in Fig. 4.43(a). From the figure it can be seen that the section of the reciprocal lattice node sampled lies on an arc passing through the node and centred at the origin of the reciprocal lattice. In the 2θ or $\omega-2\theta$ scanning mode the crystal is moved in the same way but the detector follows the ω rotation at twice the angular speed of the crystal so that in the end $\Delta 2\theta = 2\Delta\omega$. This scanning mode is illustrated in Fig. 4.43(b). In this case, the section of the node sampled lies in the direction of the vector $\boldsymbol{r}^*$. The advantages and disadvantages of these two scanning modes are extensively discussed by Wyckoff.[52]

The simplest way to obtain the net integrated intensity is by using the

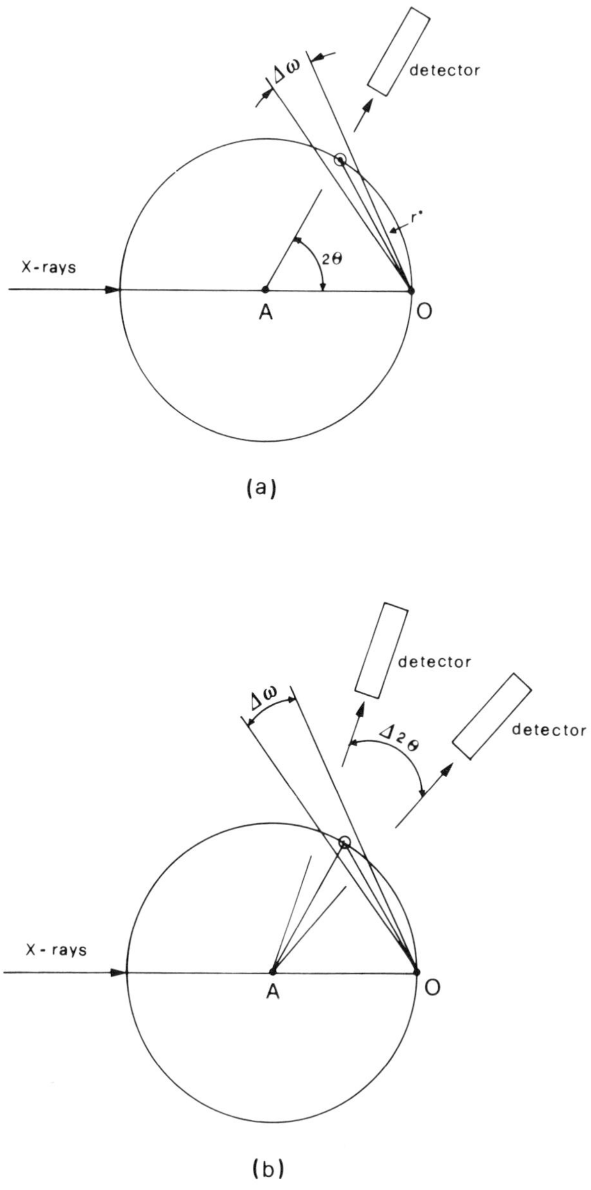

Fig. 4.43. Diffractometer scanning modes. (a) The ω scanning mode. (b) The ω-2θ scanning mode.

so-called background–peak–background (BPB) method in which the integrated intensity of the reflection is calculated by simply subtracting the background resulting from the average of measurements to the left and to the right of the peak from the total peak counts. This method requires an accurate definition of the scanning width $\Delta\omega$. In this simple case, the standard deviation of the measured intensity can be estimated by

$$\sigma(I) = (I_P + I_B)^{1/2}$$

where I_P is the intensity of the peak and I_B that of the background.

The recording of the complete reflection profile by carrying out step scans is clearly a more sophisticated alternative but it is much more demanding in terms of data storage space. If step scans are made, profile analysis of the peaks further improves the quality of the data.[55] In this method peak profiles are fitted to a shape that was previously determined and is periodically updated as data collection proceeds.

The problem of defining the peak width in a step-scan-measured reflection is discussed by Lehmann and Larsen.[56] The criterion used is the function σ/I, calculated for all possible peak widths, which becomes a minimum when the appropriate width is found.

Figure 4.44 is a photograph of a modern four-circle-diffractometer.

Once the appropriate λ and X-ray source operating conditions have been chosen, the main steps that have to be taken in order to record a data set

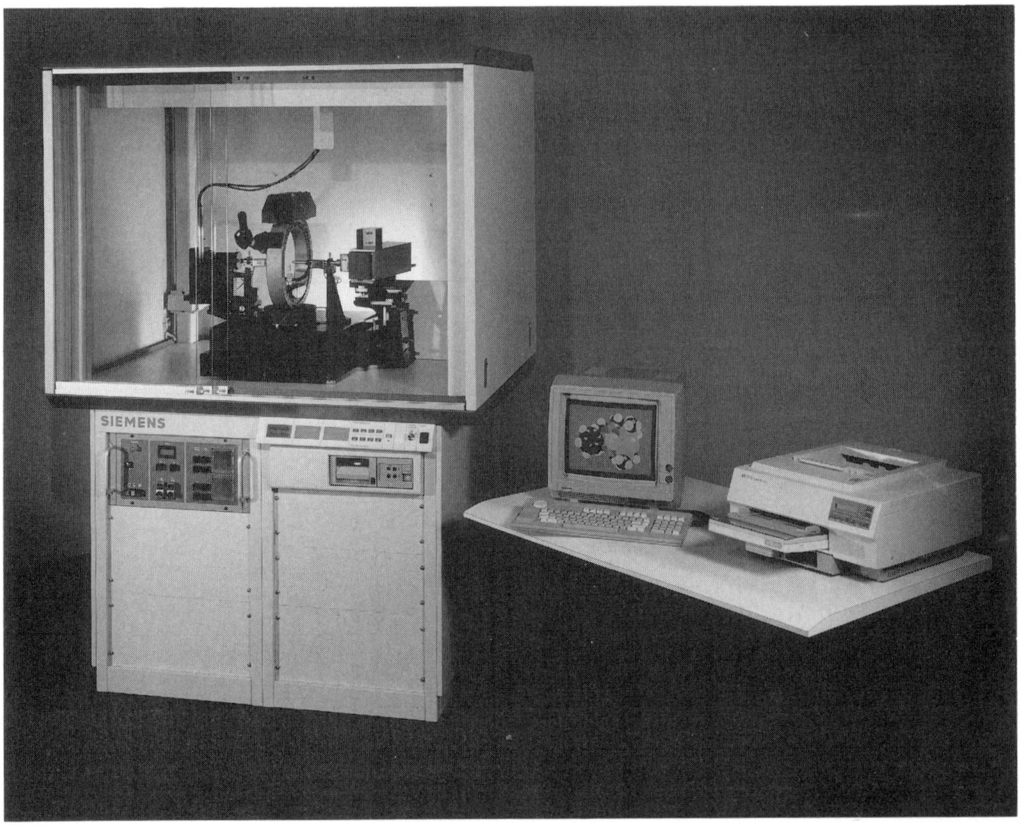

Fig. 4.44. A modern four-circle diffractometer. (Photograph courtesy of Siemens Analytical X-ray Instruments.)

from a crystal with a diffractometer can be summarized by the following check-list:

1. Mount the goniometer head with the specimen and centre the crystal optically.
2. Search for the reflections that will be used to determine the orientation matrix and unit cell parameters of the crystal. At this stage it is easier to look for reflections at fairly low 2θ values.
3. Control the quality of the crystal by examining the profile of some reflections. If at this point one suspects that a better crystal is available, it is wise to select it and repeat steps (1) through (3).
4. Once a satisfactory number of reflections has been found, determine a preliminary orientation matrix and the unit cell parameters.
5. Using the results of (4) find reflections at high 2θ values that will yield a better orientation matrix and more precise unit cell parameters. If required, at this point one is in a position to determine very accurate unit cell parameters by using an adequate number of Friedel pairs (see p. 155) at high 2θ values and least-square refinement methods (p. 90).
6. If possible, transform the unit cell that has been found into another of higher symmetry. Check for systematic absences.
7. Choose the scan mode, range, and speed by examining an adequate number of reflections carefully.
8. Select a subset of the reflections to periodically check the orientation of the crystal and another to monitor radiation damage (see p. 308).
9. Decide which reflections will be collected (one or more asymmetric units), in which order and 2θ range.
10. Start data collection.
11. When data collection is finished, before removing the crystal, remember that an absorption correction is essential, and most probably record the data necessary for an empirical absorption correction (see p. 306).

Area detectors

Due to its high precision, the diffractometer is the ideal data collection instrument for small-molecule crystals but it suffers, as we said in the previous section, from the drawback that it collects only one reflection at a time. When data have to be collected from macromolecular crystals which have very large unit cells and which therefore require the recording of many reflections and which in addition have, in general, a more or less serious radiation decay problem, the diffractometer is an inadequate data collection device. On the other hand, the rotation method described earlier (p. 259), that is with the reflections recorded on film and with a choice of the rotation range $\Delta\Phi$ made in order to minimize the number of films exposed and the fraction of partially recorded reflections, has an intrinsically lower precision. This is due to two main reasons; the first is that, as we have seen, film is a poorer detector than diffractometer counters, the second is that during the film exposure the signal is recorded by the film only during a fraction of the total exposure time. If the reflecting range of a reflection of the crystal is Δ and the rotation range selected $\Delta\Phi$ this fraction is $\Delta/\Delta\Phi$. Typically Δ is no more than a few tenths of a degree whereas $\Delta\Phi$ can be one degree or more. In other words, in the rotation method the signal is recorded on the film during a time equal to $t\Delta/\Delta\Phi$ where t is the total exposure time whereas the background is recorded instead during the total time t.[57,58] One could, in principle, improve this situation by simply reducing $\Delta\Phi$, so that Δ is spanned by several rotation photographs, and then measure the integrated intensities only on those films in which the reflection is found. There are many reasons why this is not done when working with film but this is instead perfectly feasible when the detection is done by the devices called area detectors or X-ray position-sensitive detectors.[59]

Area detectors were designed to combine the photon counting efficiency of the diffractometer with the ability to record a large fraction of the reflections which simultaneously cross the Ewald sphere, which is the main advantage of the rotation method. Area detectors are thus probably the best choice for the data collection of macromolecules, and although they have not yet found many important applications in small-molecule crystallography they will probably turn out to be very useful in cases when radiation damage is a problem. However, it should be pointed out that the devices that are currently available commercially have been optimized for the detection of copper radiation and are, in most cases, less efficient in the detection of the higher-energy molybdenum radiation which is very often used in small-molecule work.[60]

Principles of operation of area detectors

The most common area detectors types that are currently used in macromolecular work and that include the current commercially available instruments belong to two groups: multiwire proportional counters and television area detectors. To these two groups there has been a recent addition: the imaging plate, which is based on entirely different physical principles and, because it is more recent, has not yet been as thoroughly tested as the other two groups. We will briefly discuss the X-ray detection mechanisms of these three types of detector.

Multiwire proportional counters are gas filled chambers that contain three parallel planar electrodes; an anode sandwiched in between two cathodes. The anode and at least one of the cathodes are arrays of parallel wires which are perpendicular among them.[61,62] The gas filling the chamber is usually a xenon–carbon dioxide mixture (see also Mokulskaya *et al.*[63]). When an X-ray photon is absorbed by a Xe molecule an inner shell electron is emitted with a kinetic energy that is most of the energy of the absorbed photon and which is sufficient to produce the ionization of many more molecules. It has been calculated[62] that an X-ray photon of the Cu K_α wavelength has enough energy to induce on average the formation of 320 ion pairs. The free electrons and the positive ions move in opposite directions, the former in the direction of the anode where they produce an ionization avalanche with the formation of several orders of magnitude of new ion pairs.[62] These ion pairs move again in opposite directions under the influence of the electric field and in so doing generate the electrical signal which is measured in the detector and which is localized in the region where the initial photon hit the counter. The function of the carbon dioxide molecules is to absorb the ultraviolet photons which are generated in the avalanche process and which could produce the photoemission of electrons and thus start the whole process in another region of the counter.

In television area detectors the X-ray radiation is converted into visible light by a fluorescent phosphor. These visible photons, after suitable intensification, are detected by the photocathode of a standard high-sensitivity television camera tube which is linked to a computer.[64–67] The area detector phosphor, that is the fluorescent material that transforms X-rays into visible light, is either polycrystalline gadolinium oxysulphide or zinc sulphide and it produces between 250 and 500 visible photons per X-ray photon.[66] In spite of this gain in photon numbers, the sensitivity of the camera is not enough to measure them and so an increase in the signal is required to make it detectable. This enhancement is achieved by an image intensifier in which the photons produced by the first phosphor generate a certain number of electrons from a photocathode which are then accelerated and strike a second phosphor that is optically coupled to the camera. The photon gain of these intensifiers is of the order of either 100 or 1000.[66] The television camera tube consists of a photoemissive cathode with an intensifier that accelerates the electrons generated by the light producing a charge image which is scanned by an electron beam used to measure the signal. Since the photons arriving in 40 ms, which is the period necessary to scan the image, are not enough to give good counting statistics, a certain number of these images have to be added before the statistics become satisfactory.

The imaging plate is essentially a storage phosphor. This means that the X-ray photons produce on the plate a latent image that is then excited by stimulation with a He–Ne laser producing light at 633 nm. The light thus generated has a wavelength of 390 nm and is irradiated from the plate areas which were previously hit by the X-ray photons. This phenomenon is called photostimulated luminescence.[68] The radiation energy of the X-ray photons can be stored by the phosphor for fairly long periods; it has been found that the photostimulated luminescence is reduced to one half of its initial value after approximately ten hours. The photostimulable material covering the plate is Ba F Br:Eu^{2+} crystals. When the plate is hit by X-ray photons some of the Eu^{2+} ions are ionized to Eu^{3+} ions and the electrons that are freed are trapped in Br vacancies introduced in the crystal that are called F centres. Subsequent excitation of these centres by the laser liberates again these electrons that return to the Eu^{3+} ions which thus become excited Eu^{2+} ions. An electronic transition in these ions generates the luminescence with an intensity proportional to that of the original X-rays.[69] The storage phosphor is read by an image reader which releases the stored information by means of the laser and collects the emitted radiation and channels it into a photomultiplier tube which converts the radiation into an electrical signal. The plate can be used repeatedly, since exposure to visible radiation restores it to its initial condition. The two other elements of the detection system are an image processor and an image writer that can be used to imprint the plate image onto photographic film to produce a permanent record.[68] The characteristics of the image reader turn out to be crucial for the performance of the entire system and the best precision could not be obtained until an adequate read-out instrument was built.[70]

The performance of the three types of detector mentioned here has been analysed in several of the references given in this section and in particular by Arndt.[59] Multiwire proportional counters and television detectors are already fairly widely spread in many laboratories and the data they have produced have been used to solve several new protein structures.[71–73] The special characteristic of the imaging plate is its very wide dynamic range which appears to make it the ideal detector to be used with the very intense synchrotron radiation.

Data collection with an area detector

In addition to the area detector, the data collection system requires some means of orienting and moving the crystal. Multiwire proportional counters and television detectors use either a modified rotation (oscillation) camera with a vertical spindle, a Eulerian cradle, or another type of three-circle diffractometer or a κ goniostat. Figure 4.45 is a photograph of a television area detector which uses a κ goniostat. The detector is mounted on a horizontal arm and can be moved about an axis passing through the crystal in a way totally analogous to a standard diffractometer counter and, in addition, it can be moved back and forth providing another degree of freedom—the crystal to detector distance D. The angle formed by the X-ray beam and a normal to the centre of the detector is called θ_c and, as shown in Fig. 4.46, limits the resolution of the data to be collected. Following Xuong *et al.*[74] we can write Bragg's law as follows:

$$R_{max} = \lambda/2 \sin \theta_{max}$$
$$R_{min} = \lambda/2 \sin \theta_{min} \qquad (4.44)$$

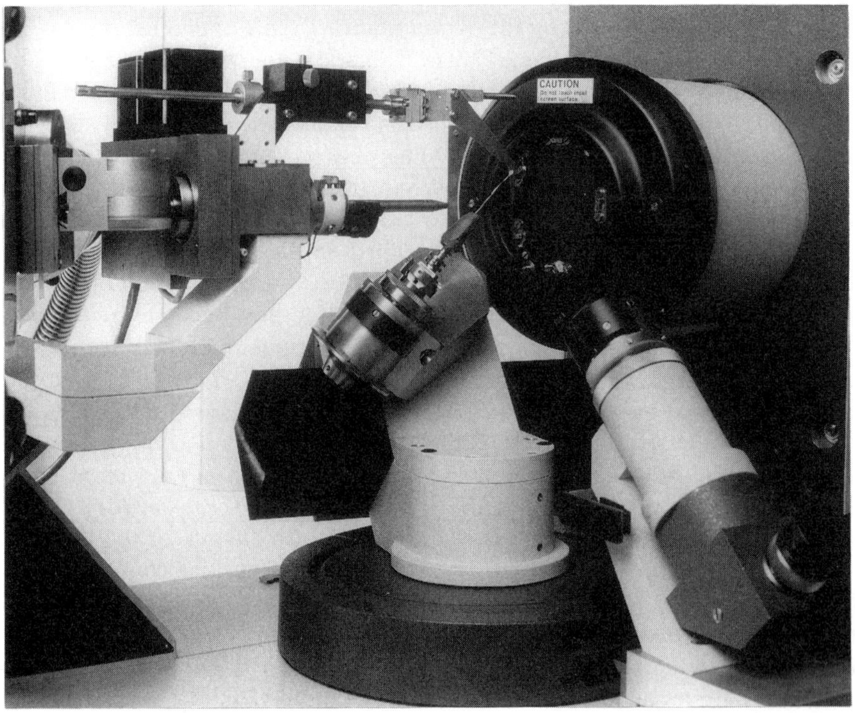

Fig. 4.45. A television area detector with a κ goniostat. (Photograph courtesy of Enraf Nonius).

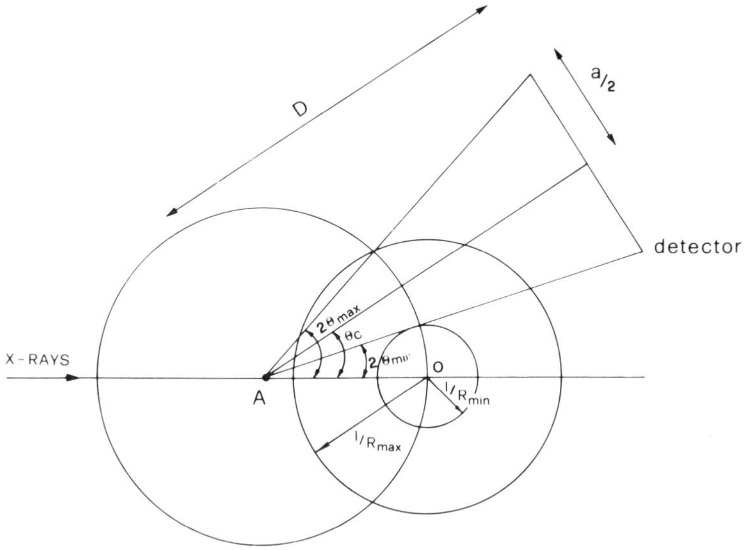

Fig. 4.46. Projection of the Ewald sphere and the area detector on to the plane defined by the incident X-ray beam and a normal to the detector passing through the crystal. The volume of reciprocal space accessible is limited by the two spheres of radii $1/R_{max}$ and $1/R_{min}$ which in turn are a function of D and θ_c.

where R_{max} is the maximum and R_{min} the minimum resolution of the reflections that can be measured by the detector in the position corresponding to the selected values of θ_c and D. The angles θ_{max} and θ_{min} are defined in the figure and are a function of the detector size, the angle θ_c and the crystal to detector distance D. From the figure it can be seen that

$$2\theta_{max} = \theta_c + \tan^{-1}(a/2D)$$
$$2\theta_{min} = \theta_c - \tan^{-1}(a/2D) \qquad (4.45)$$

where a is the detector width.

Equations (4.45) can be used to calculate the value of θ_c to be selected for data collection to a particular resolution once that D, the crystal to detector distance, has been chosen.

D must be selected according to the characteristics of the detector, the wavelength of the radiation, and the unit cell parameters of the crystal. The first two parameters do not normally change for different experiments performed at a given installation and therefore D can usually be calculated with a very simple formula in which the only variable is the maximum unit cell parameter of the crystal. For example in Howard *et al.*,[73] the crystal to detector distance for one type of multiwire proportional counter is calculated in centimetres, for Cu K_α radiation by the following equation:

$$D = a_{max}/8$$

where a_{max} is the longest unit cell parameter of the crystal measured in ångströms. The equivalent equation for another type of multiwire proportional counter[74] is:

$$D = a_{max}/1.93$$

Thus, data from the same crystal would have to be collected at very different Ds by the two detectors which, although based on the same general principles, differ in their construction details. Once D has been determined, knowing the detector width *a*, one can calculate the θ_c required to collect data to the resolution desired.

Each of the electronic pictures generated by the detector is called a frame and the individual elements of the picture are called pixels. The reflection size on the picture, the space between reflections, and, in general, the spatial resolution of the detector are expressed by the number and size of the pixels.

The camera or the goniostat and the detector are controlled by a computer which is, in general, connected to another computer which receives from it the frames that are then used to calculate the integrated intensities (see for example Blum *et al.*[75]).

Two methods have been proposed to make the reciprocal lattice nodes of the crystal cross the Ewald sphere: the rotation (oscillation) method and the stationary picture method. In both cases the detector does not move while data collection proceeds. In the first method the crystal is rotated about an axis which is often the vertical axis in pretty much the same way as when film is used. Many of the considerations discussed earlier (p. 259) are thus applicable to this technique in which data are collected in a series of consecutive rotation (oscillation) frames. In addition to the detection method there are basically two fundamental differences between the two techniques. The first is that the detector is not always perpendicular to the X-ray beam but can form with it an angle θ_c as pointed out before. Obviously this fact has to be taken into account in the prediction of the detector coordinates of the reflections collected. The second major difference is in the choice of $\Delta\Phi$, the rotation range, which in this case is selected so that each reflection appears in several frames.[73,75] As pointed out before this strategy improves the signal to noise ratio since reflections are integrated only in the frames in which they appear.

In the electronic stationary picture method,[76,77] the crystal is also rotated about an axis but the frame is recorded with the crystal held stationary. The reflection intensity is thus extracted from a series of still electronic pictures at slightly different values of Φ. The $\Delta\Phi$ between frames is of the order of 0.06° and subframes are sampled at distances of 0.01° in order to better scan reflections that in some cases can be very sharp.[77] This second data collection strategy is less widely used than the rotation (oscillation) method.

It is worth noticing that these strategies of data collection are the only ones described so far that truly sample the entire volume of a reciprocal lattice node. With film methods what one sees is a projection of the entire volume on to the film plane, whereas with the diffractometer one looks at a reflection profile on a single plane that can be chosen to cut the node volume in different ways as seen earlier (p. 278). Thus, area detector data are the only ones that can be profile fitted in three dimensions, a possibility that ought to further improve their quality.

In most cases the crystal is more or less accurately aligned before data collection can begin so that x_D and y_D, the coordinates of a reflection on the detector, and Φ, the rotation angle at which the node crosses the Ewald sphere for all the reflections to be collected can be predicted[76] (see also p. 265). However, a full data set collected with an area detector contains a very large fraction, if not all, of the reciprocal lattice nodes to a given resolution and, since the crystal orientation can be obtained automatically by efficient computer programs,[78] it is also possible not to orient the crystal before data collection begins and find the orientation afterwards, during frame processing.[73] A strategy that can be used to cover a section of reciprocal space with an area detector, which is obviously applicable when the crystal orientation is known before data collection begins, is discussed by Xuong *et al.*[74]

Data collection techniques for polycrystalline materials

X-ray diffraction of polycrystalline materials

An ideal polycrystalline material or powder is an ensemble of a very large number of randomly oriented crystallites. Figure 4.47 shows the effect that this random orientation has on the diffraction of a specimen assumed to contain only one reciprocal lattice node. The most remarkable difference with the single-crystal case is that we must now think of the scattering vectors not as lying on discrete nodes of reciprocal space but on the surfaces of spheres whose radii are the reciprocal lattice vectors $\boldsymbol{r}_{\mathrm{H}}^{*}$, the distances from the single-crystal reciprocal lattice nodes to the origin of reciprocal space. Thus, with these specimens, diffraction is observed when the scattering vectors lie at the intersection of the Ewald sphere and a series of concentric spheres centred at the reciprocal lattice origin. So, rather than having one point on the Ewald sphere, which together with the position of the sample A fixes the direction of the diffracted beam, we now have a series of circles. In strict analogy with the single-crystal case, these circles and the sample define a series of concentric cones with apex in A. The entire surface of these cones gives rise to diffraction.

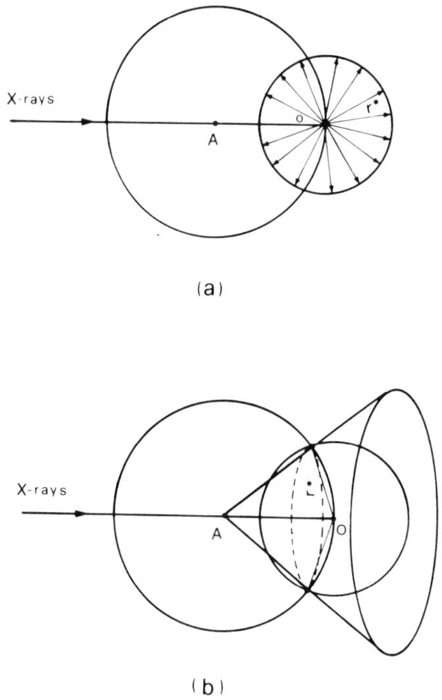

Fig. 4.47. (a) If the specimen is an aggregate of randomly oriented crystallites, the vector $\boldsymbol{r}^*$ is found in all the possible orientations with respect to the X-ray beam. These orientations define a sphere of radius $\boldsymbol{r}^*$. (b) The intersection of the sphere of radius $\boldsymbol{r}^*$ with the Ewald sphere is a circle that together with point A defines a diffraction cone of all the possible directions in which diffraction is observed.

A simple way to record the diffraction pattern of a polycrystalline material is by placing a film perpendicular to the incident X-ray beam. The diffraction cones will, in this case, give rise to a series of concentric rings. Alternatively, a narrow strip of film can be placed on a cylinder centred at the sample. In this case, the cones will generate concentric arcs, which are segments of the rings, on the strip. A final possibility is to reduce the strip to a line, that is to simply record the position and the intensity of the diffracted radiation on any plane that contains the incident X-ray beam. In this last case one only measures the radius of the cone and the diffracted intensity at a single position. If the sample can be considered perfectly isotropic this single measurement is sufficient to completely characterize the diffraction pattern. The parameters reported are 2θ, that is the angle made by any vector with origin in A and lying on the diffraction cone surface and the incident X-ray beam, and the relative intensity of the radiation along any direction on the cone. If the orientation of the crystallite in the specimen is not perfectly random, the pattern obtained will not be isotropic. It may even present spots corresponding to single reciprocal lattice nodes. In that case the powder sample can be appropriately rotated so that each crystallite adopts many different orientations in the course of data collection, thus generating a more homogeneous diffraction pattern. The result is equivalent to having a sample with many more possible crystallite orientations and therefore closer to isotropy. Rotation of the specimen is a standard practice in data recording of polycrystalline materials. Exceptions

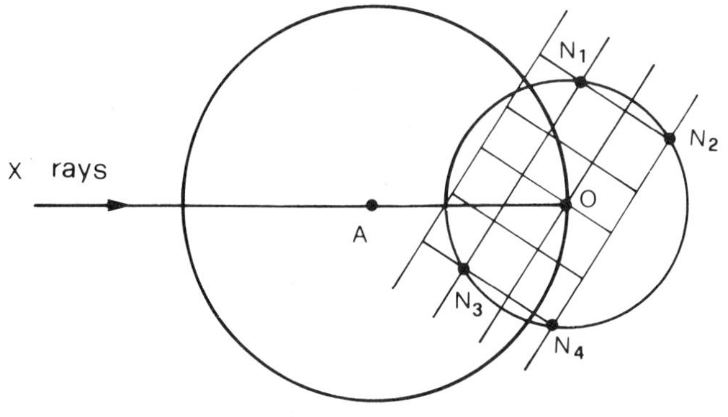

Fig. 4.48. The four reciprocal lattice nodes represented in the figure are found at the same distance from the origin of reciprocal space O and therefore they will all contribute to the intensity of the line of the cone corresponding to the distance $\boldsymbol{r}_i^*$.

are the cases in which the preferred orientations and other properties of the crystallites need to be studied.

Another important feature which distinguishes powder diffraction is that the intensity of the diffracted radiation on the cone surfaces can arise from the contributions of more than one single-crystal reciprocal lattice node. Figure 4.48 shows in projection that this can happen both as a result of chance and crystal symmetry. A powder diffraction maximum, measured along any direction on the cone surface, is thus said to have a certain multiplicity that will be higher the higher the symmetry of the crystallites under examination.

When the diffraction experiment is performed with monochromatic radiation, that is when there only a single Ewald sphere, there is only one diffraction cone corresponding to each sphere of a given radius r_i^* in reciprocal space. In other words, the angle $2\theta_1$ corresponds unambiguously to the sphere of radius r_1^*, $2\theta_2$ to that of radius r_2^*, etc., and we have only one possibility if we want to measure the diffraction that arises from the sphere of radius r_i^*: to have some means of detecting radiation at an angle $2\theta_i$ with the incident X-ray beam. It is, however, possible to shine on the specimen X-rays with a wavelength variable within a certain range. The experiment is exactly equivalent to the Laue method used for single crystals. In this case, there will be many Ewald spheres, one for each wavelength, and each will generate a diffraction cone with a given sphere of radius r_i^*. Figure 4.49 shows the Ewald spheres corresponding to the two values limiting the wavelength interval of the radiation used. In the figure it can be seen that the diffraction due to the sphere of radius r_i^* can be measured at many different values of the angle $2\theta_i$. For different acceptable choices of $2\theta_i$ there will be diffraction produced by radiation of different wavelengths. The methods which use polychromatic incident radiation and analyse the energy or wavelength of the scattered radiation at a fixed scattering angle are called energy dispersive methods in powder diffraction. They obviously require a detector that will discriminate the energy of the arriving scattered radiation and have some advantages that make them the best choice in certain situations.[79] Just like the Laue method they are best practised with a synchrotron source which can furnish, as we have seen, radiation of adequate intensity in a rather extended energy interval. For the remainder of this chapter we will assume that we are dealing with monochromatic X-rays. The methods which use them are the most widely diffused in

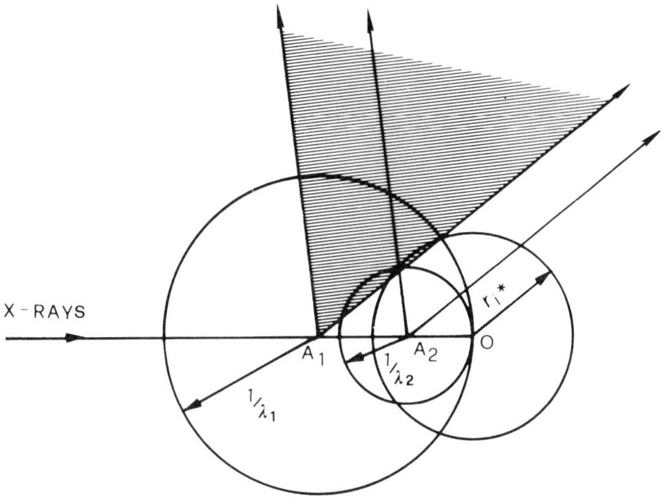

Fig. 4.49. The two Ewald spheres limiting the wavelength range of the polychromatic radiation used define with the sphere of radius r_i^* two limiting diffraction cones. All the cones in between correspond to r_i^* for different wavelengths. The shaded region in the figure shows the range of θ values that can be used to measure the diffraction in an energy dispersive experiment.

standard laboratories. From the rich literature that covers the diffraction of polycrystalline materials in depth we recommend two books.[80,81]

Cameras used for polycrystalline materials

We have already mentioned that a very simple way to record the diffraction pattern generated by a powder is by simply placing a film perpendicular to the X-ray beam, tacitly assuming that the specimen was positioned in between the source and the film. The cameras that use this geometry are called transmission cameras and in them the film is usually kept stationary during data collection. Although these cameras offer the advantage of recording the entire circle resulting from the projection of the diffraction cone onto the film plane, they suffer from a serious disadvantage: they are limited to maximum 2θ values which for the standard specimen to film distances and film sizes are about 45°.

If the powder diffraction at very high 2θ angles needs to be studied, the flat film can still be placed perpendicular to the X-ray beam but in between the source and the specimen instead. The cameras that use this geometry are called back-reflection cameras. In this case the problem is the opposite, i.e. only the diffraction at $2\theta_s$ higher than a minimum value can be recorded.

Figure 4.50 shows the sample and X-ray beam and the position of the film in the transmission and back-reflection cameras. It should be clear from our discussion that they are useful only in those cases in which the diffraction pattern of the polycrystalline material needs to be recorded at fairly low or rather high 2θ values.

A geometry which allows the recording of the diffraction pattern of a polycrystalline material at both high and low 2θ values is that used by the Debye–Scherrer camera,[82] probably the most popular diffraction camera for powders. In the Debye–Scherrer camera, the recording film is a strip which is placed on a cylindrical drum centred on the sample. Figure 4.51(a) is a photograph of the camera and Fig. 4.51(b) shows the appearance of the strip after recording a typical diffraction pattern. As seen in the figure, the

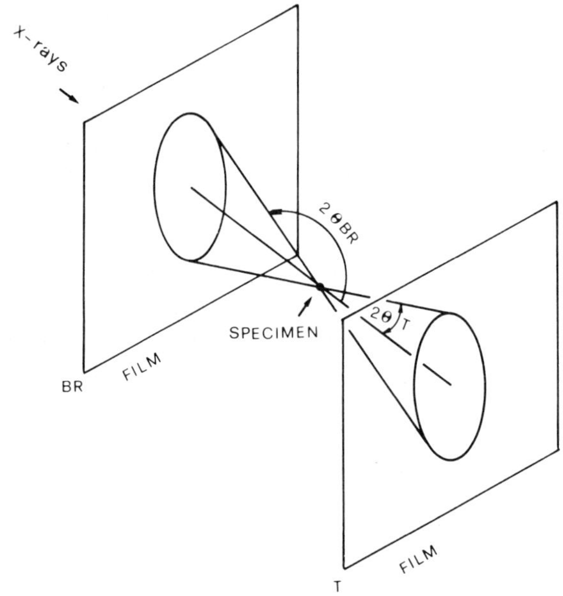

Fig. 4.50. Position of the specimen and the film in transmission (T) and back reflection (BR) cameras.

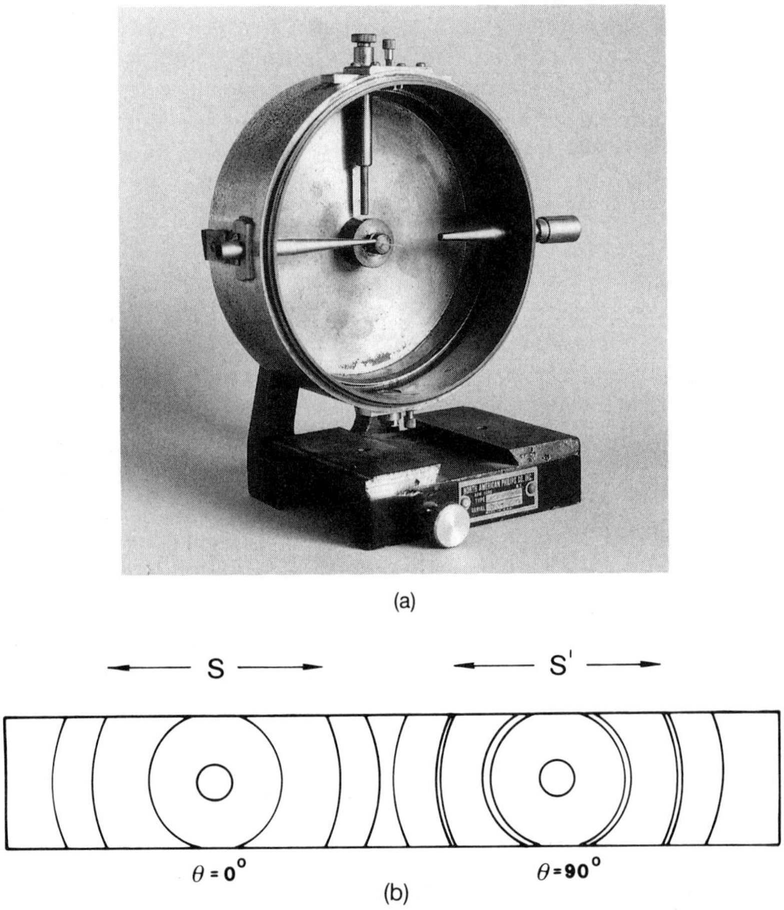

Fig. 4.51. (a) The Debye–Scherrer camera. (b) Sketch of a diffraction pattern recorded with the Debye–Scherrer camera using the Straumanis method of film mounting. Notice the doublets at high diffraction angles.

pattern shows a series of arches resulting from the projection of the diffraction cones on to the cylindrical surface. The big advantage of the camera is that it records the entire pattern generated at all possible values of 2θ; its main disadvantage is that it does not record the entire projection of the diffraction cone but only a segment. Since, as we have seen, in most cases the diffraction pattern is isotropic, and one is therefore only interested in the position of the arches and their relative intensities, this limitation is not very severe.

In addition to the cylinder that holds the film strip in place, the main body of the camera has a collimator that serves to define the incident X-ray beam and a beam trap that stops it after it has travelled through the specimen. Although one can place the film so that the cut in the cylindrical surface is made to coincide with the collimator or beam trap, punching a hole for the other, and both ways of mounting the film have been used, a third alternative is usually preferred. In the so-called Straumanis method of film mounting two holes are punched in the film strip positioned at about one quarter and three quarters of the total film length. One of the two holes is then used for the collimator and the other for the beam trap. The advantage of the Straumanis method of film mounting is that it provides accurate measurements for the positions of the arches that will then be translated into Bragg spacings, $d_{\mathbf{H}}$, for both high and low values of 2θ. As seen in Fig. 4.51(b), the arches centred on one of the two holes punched are present as doublets. They correspond to the K_{α_1} and K_{α_2} lines of conventional generators which are normally not resolved by X-ray filters but are clearly separated after diffraction by powder samples at high 2θ values. That the doublets correspond to high 2θ values can be seen by differentiating Bragg's law:

$$2d_{\mathbf{H}} \sin \theta = \lambda,$$

$$2d_{\mathbf{H}} \cos \theta \, d\theta = d\lambda, \qquad \lambda(\sin \theta)^{-1} \cos \theta \, \Delta\theta = \Delta\lambda,$$

and

$$\Delta\theta = \lambda^{-1} \tan \theta \, \Delta\lambda. \tag{4.46}$$

In the case of Cu K_α radiation, the doublet is separated by 0.0038 Å, if we take $\lambda = 1.5418$, $\Delta\theta = 0.0240°$ for $\theta = 10°$ and $\Delta\theta = 0.8009°$ for $\theta = 80°$ instead. Thus, the presence of double lines centred on one of the punched holes serves to unambiguously identify it as that corresponding to the collimator making it unnecessary to mark the strip. It is the diffraction pattern recorded that tells us which hole corresponds to $\theta = 90°$. An important advantage of this method of film mounting is that the positions $\theta = 0°$ and $\theta = 90°$ can be very precisely determined by taking the averages of the arch positions corresponding to several diffraction cones on the film.

From the position of the symmetrical arches, one can easily calculate the corresponding θ values since as seen in Fig. 4.52, if S is the distance between the arches due to a diffraction cone and R is the radius of the cylinder

$$S/2\pi R = 4\theta/360°$$

for the arches centred on the beam trap ($\theta = 0°$) and

$$S'/2\pi R = (360° - 4\theta)/360°$$

for those centred on the collimator ($\theta = 90°$).

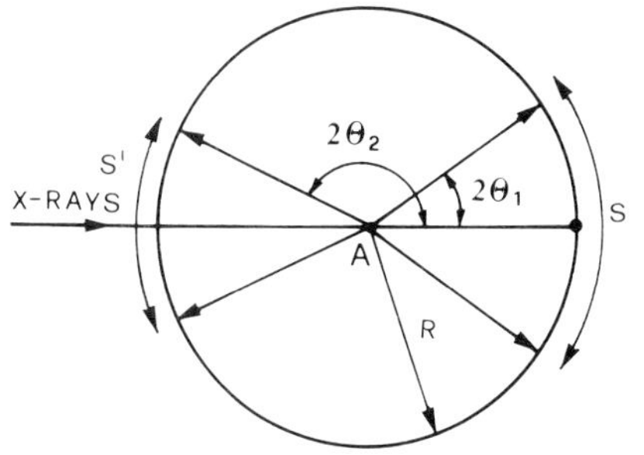

Fig. 4.52. Projection of the drum of the Debye–Scherrer camera on to its axis. The specimen is in the centre of the circle, S and S' are the distances between the symmetrical arches corresponding to one diffraction cone, and R is the radius of the camera.

From the first equation we find that

$$\theta = \frac{180s}{4\pi R}. \tag{4.47}$$

If the radius of the camera is chosen so that its value in millimetres is an exact multiple of $180/\pi$, a simple relationship will exist between the distance S measured in millimeters and the scattering angle θ measured in degrees. Two values of R that have been very extensively used are 57.26 mm and 114.6 mm. A camera that has a larger cylindrical radius will resolve the diffraction lines better since for a given $\Delta\theta$ there will be a larger corresponding ΔS. The price to be paid is longer exposure times and higher levels of background, since the diffracted X-rays must travel a longer path which is normally through air which with its scattering raises the background level.

After the θ values of a given diffraction cone have been determined, the corresponding interplanar spacings, $d_{\mathbf{H}}$, can be easily calculated by simple application of Bragg's law. In the Debye–Scherrer camera the sample is usually present in the form of a small cylinder placed in coincidence with the camera axis and it is slowly rotated about this axis during data collection in order to minimize the effect of the possible preferred orientation of the crystallites.

The Debye–Scherrer camera produces rather broad diffraction lines which for some applications may be a serious drawback since some of the lines on the pattern may be found overlapping on the film. If one attempts to separate the lines by using a collimator of smaller diameter or a camera of larger radius the exposure time may become unacceptably long. It is for those cases that the focusing or parafocusing cameras have been devised. Parafocusing cameras, of which there are many variants used for different applications, can be broadly classified into two groups: the Seemann–Bohlin[83,84] and the Guinier[85] types.

The principle on which parafocusing cameras is based is shown in Fig. 4.53(a) for the Seemann–Bohlin and in Fig. 4.53(b) for the Guinier geometry. In both cases the sample is ideally an arc of a circle which is irradiated by reflection by a divergent beam in the case of the Seemann–Bohlin camera and by transmission by a convergent beam in the case of the Guinier camera. Since an arc of a circle subtends equal angles at every point on the circle, the diffracted radiation will converge in both cases on the parafocusing circle of the camera thus producing a very narrow line on the film.

The convergent beam necessary for the Guinier camera can be produced by a bent or bent and ground crystal monochromator of the type we have discussed on p. 243.

As seen in the figure, both cameras restrict the range of 2θ that can be recorded. This range depends on the details of camera construction but one can roughly say, and from the figure it should be appreciated, that Seeman–Bohlin cameras record the diffraction lines only above a certain minimum value of 2θ whereas Guinier cameras do it up to a certain maximum value.

Parafocusing cameras offer the advantage of producing very sharp diffraction lines in times comparable with those required for standard exposures in the Debye–Scherrer camera.

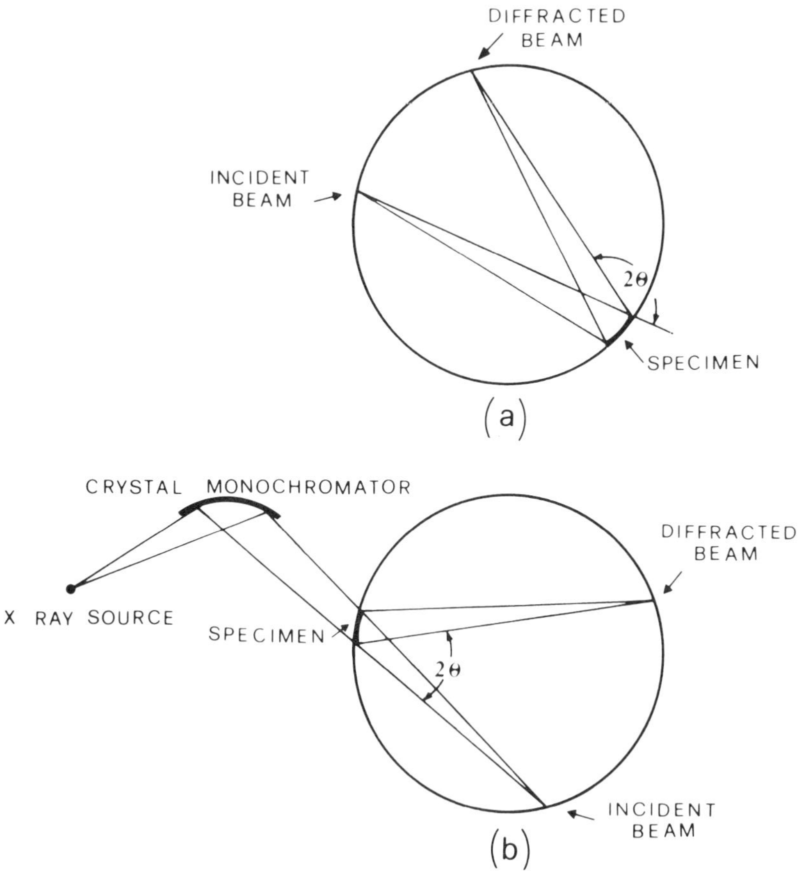

Fig. 4.53. (a) The Seeman–Bohlin parafocusing geometry. (b) The Guinier parafocusing geometry. Notice the different way in which the two geometries restrict the 2θ range available for measurements.

The final type of camera that we wish to mention here is a modification of the Debye–Scherrer camera that can be used to produce a powder pattern when the specimen is a single crystal. In the Gandolfi camera[86] the crystal is mounted on a spindle inclined at 45° with respect to the camera axis and is rotated about both this and the camera axis. If after a complete revolution about one of the axes there has not been an integral number of rotations about the other, the effect of the crystal motion will be that of randomizing its orientation as exposure proceeds. If necessary the crystal can be mounted in more than one orientation during a single exposure. The final result is a diffraction pattern that closely resembles that produced by a true polycrystalline material.

Diffractometers used for polycrystalline materials

The powder cameras described in the previous section all use film to record the intensity of the diffracted X-rays. Just as in the case of single crystals, an alternative to this measuring method is offered by quantum counter detectors. The instruments that use this type of detector to measure the position, that is Bragg angle, and relative intensity of the diffraction pattern produced by a polycrystalline material are called powder diffractometers.

Powder diffractometers are thus characterized by a counter which is no different from those used for single-crystal work: it is normally of the scintillation type, although gas ionization counters are also used. Whatever the detector, the measuring strategy is the same as that of the single-crystal diffractometer: the relative intensity and 2θ angle of each diffraction cone generated by the polycrystalline material are measured one at a time. As we have seen before, in this way of recording the data lies the only weakness of the diffractometer when compared with the diffraction cameras that simultaneously record all the reflections in the accessible 2θ range. For most applications, this limitation is not very severe in the case of polycrystalline materials and is more than counterbalanced by the greater precision offered by these instruments.

Since the 2θ angle and relative intensity of an isotropic diffraction cone can be measured in any plane that passes through the cone apex and bisects a projection of the cone, the powder diffractometer is simpler than the single-crystal diffractometers we have described. The detector simply moves, tracing a circle centred on the specimen, in a plane that contains the incident X-ray beam, and which is normally in the vertical position, to increasingly higher values of 2θ. When a diffraction line is found by the detector, its relative intensity is recorded as a function of 2θ, as measured in the plane on which the detector is moving. There is, in other words, only one way of scanning a reflection—by varying the angle 2θ that the detector makes with the incident X-ray beam. As in the case of the other recording methods described, the sample may also be rotated about an axis which is found in the plane in which the detector moves in those cases in which the preferred orientation of the crystallites in the specimen is a problem. Figure 4.54 is a photograph of a powder diffractometer and Fig. 4.55 shows a typical output record of a powder diffractometer in which the variable in ordinates is the number of counts measured plotted by the diffractometer as a function of the scattering angle 2θ. Each of the peaks in the figure fully describes one of the diffraction cones of the specimen.

Fig. 4.54. A powder diffractometer. (Photograph courtesy of Philips Analytical.)

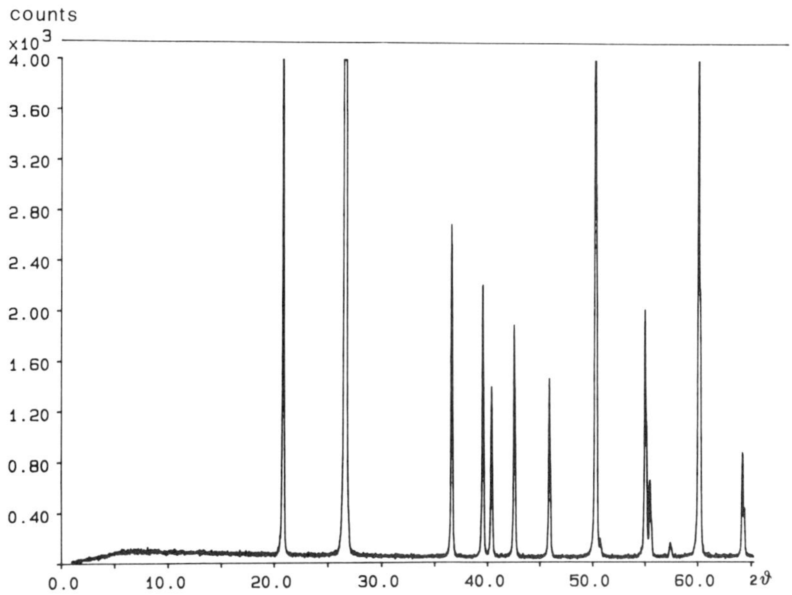

Fig. 4.55. Diffraction pattern of a polycrystalline material recorded with a diffractometer. Notice at higher 2θ values the separation of the peaks into doublets corresponding to the K_{α_1} and K_{α_2} lines.

A characteristic that distinguishes the modern powder diffractometer is the use of a parafocusing arrangement that improves the intensity and resolution of the diffraction maxima. The most widely used parafocusing geometry is that found in the instrument called the Bragg–Brentano powder diffractometer.[87] In the Bragg–Brentano diffractometer the specimen, which is flat, is irradiated by a divergent beam and so in a sense this geometry can be considered a modified version of the Seemann–Bohlin arrangement discussed before. As we have seen, if the source is at a distance r from the specimen, the parafocusing effect is produced on a circle of radius r to which the specimen is tangent (in the Seemann–Bohlin arrangement the specimen is an arc of the circle). Since in the Bragg–Brentano diffractometer we have the additional constraint that the detector moves around the sample keeping a constant distance R, it is obviously necessary to ensure that the parafocusing effect is also produced at a constant distance from the sample. This goal is achieved by moving the flat specimen so that it is tangent to focusing circles of variable radii for different values of 2θ but which produce the focusing effect at the constant distance from the specimen R.

Figure 4.56 shows the parafocusing geometry of the Bragg–Brentano diffractometer. As seen in the figure the triangle SOD is isosceles since two of its sides are the distance r. A normal drawn from O to the other side R defines with each of the sides r an angle equal to θ and we can then write

$$\sin\theta = \tfrac{1}{2}R/r \quad \text{and} \quad r = R/2\sin\theta.$$

This equation shows why if R, the radius of the goniometer circle, is a constant, r, the radius of the focusing circle must vary with the scattering angle of the diffracted radiation 2θ. In practice, the parafocusing geometry of the Bragg–Brentano diffractometer can be maintained by rotating the sample S in a range spanning from a few degrees to about 160°, the useful interval in which the instrument can be used.

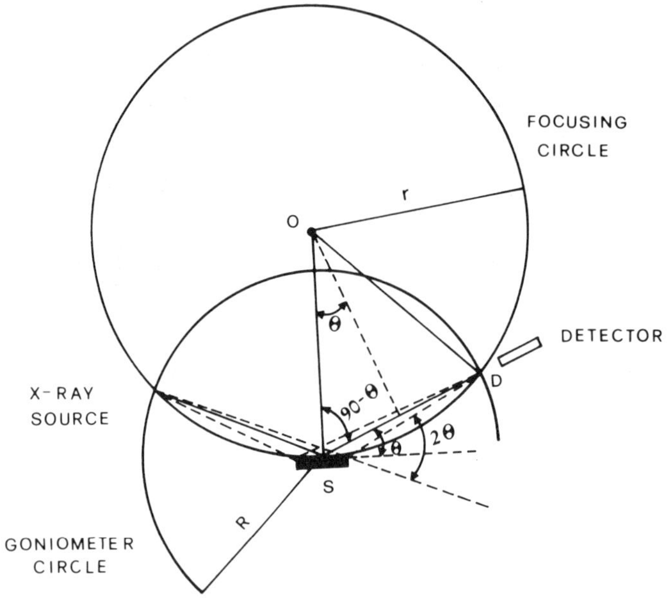

Fig. 4.56. Parafocusing geometry of the Bragg–Brentano diffractometer.

In our discussion of the parafocusing geometry, it has been tacitly assumed that divergence of the X-ray beam can only take place in the plane of the figures. In order to ensure that this is truly so and that no important divergence occurs out of that plane, X-rays are made to travel both before reaching the sample and after diffraction through a series of metal plates parallel to the plane where divergence is allowed. These plates are called Soller slits and they are an essential part of the optics of the powder diffractometer.

Although less widespread than the Bragg–Brentano diffractometers, instruments that use the Seemann–Bohlin parafocusing geometry have also been designed and marketed.[88,89] Their main advantage stems from the fact that in this arrangement all the reflections focus on a circle of fixed radius. This makes it possible to use more than one detector (or a curved position-sensitive detector) to simultaneously record several diffraction maxima. With this geometry the specimen does not have to be moved since it is assumed to be constantly an arc of the focusing circle but the specimen to detector distance requires adjusting as the scattering angle 2θ varies. Probably the most serious limitation of this geometry is that it excludes the value of $2\theta = 0°$ and so the instrument requires calibration using the known 2θ position of the maxima of samples used as standards.

Modern powder diffractometers can be used with equal ease in both the continuous and step-scanning modes. The first mode of operation is faster and, in general, adequate for most standard applications. In the step-scanning mode, the angle 2θ made by the detector and the incident X-ray beam is slowly incremented by small $\Delta 2\theta$ and the number of counts is measured and recorded. Although this operation mode requires more time and computer memory, it results, in the end, in a much more accurate diffraction line profile, which may be an essential requirement in some types of application, like for example in crystallite size or and strain measurements or if the data are going to be used for Rietveld refinement.

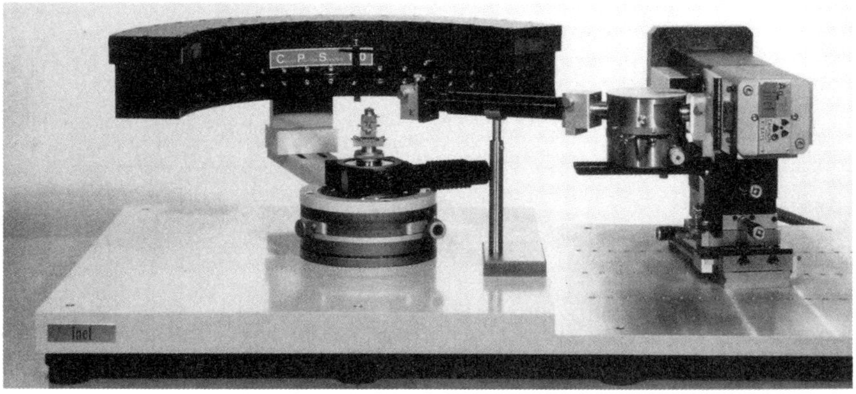

Fig. 4.57. A curved position-sensitive detector used for powder diffraction work. (Photograph courtesy of INEL.)

Curved position-sensitive detectors based on the same principles discussed on p. 282 have also been devised for use in powder diffraction work.[90–92] They offer the advantage of combining the high precision of the diffractometer with the possibility of simultaneously recording the entire diffraction pattern generated by a polycrystalline material.

These type of detector are curved gas filled chambers with a metal grid in the position of the cathode and either a curved wire or a blade which functions as the anode, in other words they are curved linear detectors. Thus, in a sense curved position-sensitive detectors are the equivalent of the area detectors used in single-crystal work adapted for use with polycrystalline materials. Figure 4.57 is a photograph of a curved position-sensitive detector used for powder diffraction work. The simultaneous recording of the entire diffraction pattern offers the possibility of following processes in which the relative intensities of the Bragg peaks varies in time as the experiment proceeds.

Uses of powder diffraction

Probably the best known and most widely used application of powder diffraction is as an analytical tool for both qualitative and quantitative analysis of crystalline materials. From our discussion it should be evident that the Bragg spacings $d_{\mathbf{H}}$ and relative intensities of the diffraction cones generated by a given polycrystalline material are a function not only of the substance under investigation but also of its crystalline form, which may be one of the several in which the material may crystallize.

Powder diffraction became a widely used technique for the identification of polycrystalline unknowns in the 1930s as a result of both the clear definition of the minimum number of parameters required to preliminarily identify a specimen as a member of a more or less restricted group of substances, and also of the compilation of a file containing the diffraction patterns of a number of known crystalline materials sufficiently large to give a reasonable probability that an unknown would be found among the standards present in the file.[93] This original file has since been expanded to such an extent that in 1986 it included data for more than 44 000 crystalline phases.[94] The file is currently known as the Powder Diffraction File (PDF) and it is updated and distributed by an international organization called the

25-58 25-59

d	3.55	2.22	2.34	4.96	$BaAlF_5$ (α-phase) ★
I/I_1	100	35	30	20	Barium Aluminum Fluoride

Rad. CuKα λ 1.5418 Filter Ni Dia.
Cut off I/I_1 Diffractometer I/I cor.
Ref. Schultz et al., Acta Chem. Scand., 26 2623-30 (1972)

Sys. Orthorhombic S.G.
a_0 5.156 b_0 7.575 c_0 19.64 A C
α β γ Z Dx
Ref. *

εα nωβ εγ Sign
2V D 4.53 mp Color
Ref. Ibid.

*Holter: Hovedfagsarbeide for den Matematisk Naturvitenskapelige Embetseksamen, Universitetet i Trondheim, Trondheim (1969)
α-phase is stable below 770°C.
Compound formed at 50 mol. % BaF_2 with AlF_3.

d Å	I/I_1	hkl	d Å	I/I_1	hkl
4.96	20	013	2.092	16	019
4.27	10	110	2.060	20	028
4.12	14	014	2.054	18	134
3.55	100	104	2.029	5	223
3.27	17	006	1.998	4	036
3.01	15	016	1.940	4	119
2.919	12	122	1.892	11	029
2.633	13	017			
2.584	25	200			
2.505	3	031			
2.478	7	026			
2.458	8	008			
2.403	5	203			
2.335	30	018			
2.284	30	213			
2.254	20	131			
2.230	16	126			
2.216	35	108			
2.176	17	009			
2.122	15	035			

FORM M-2
W

Fig. 4.58. Reproduction of a card of the J.C.P.D.S. powder diffraction data file.

Joint Committee for Powder Diffraction Standards (JCPDS), International Centre for Diffraction Data.

Figure 4.58 is a reproduction of a card in the PDF. The information contained in the card should be readily interpretable. Notice that the relative intensities are expressed as percentages of the strongest line which is arbitrarily assigned an intensity equal to 100. Not all of the cards contain all of the information shown in the figure. In particular it may not always be possible to unambiguously index the lines present in a pattern and therefore the Miller indices corresponding to a given Bragg spacing may not be available in the file.

Using the information contained in the PDF it is often possible to match the diffraction pattern of an unknown to that of one of the known substances present in the file. This task can be accomplished using both manual[95] and computer methods[96,97] with a current tendency in favour of the latter.

The simultaneous identification of more than one component in a sample is also possible using the method described above but clearly with a degree of difficulty that increases with the complexity of the diffraction pattern generated.

The quantitative analysis of the different crystalline phases present in an unknown is another important application of powder diffraction. Due to absorption effects of these specimens the assumption of a direct proportionality between the intensities measured and the amount of a given crystalline phase present in the sample is not possible. Alexander and Klug[98] have derived the equation that relates the intensity of a given diffraction line due to a component to its weight fraction in the sample for the case of a flat polycrystalline specimen. If the sample is a uniform mixture of n components and extinction and microabsorption effects can be neglected, it can be shown that[98]

$$I_{\mathbf{H}_i} = \frac{K_{\mathbf{H}_i} x_i}{\rho_i[x_i(\mu_i - \mu_M) + \mu_M]} \tag{4.48}$$

where $I_{\mathbf{H}_i}$ is the intensity of a given diffraction maximum **H** due to component i, $K_{\mathbf{H}_i}$ is a proportionality constant which depends on the component and the diffraction line, x_i is the weight fraction of component i in the sample, ρ_i its density, μ_i is its mass absorption coefficient (defined on p. 241), and μ_M the mass absorption coefficient of the matrix, that is of all the other components in the sample with the exclusion of component i. This equation can be specifically applied to three different cases.

In the first case the mass absorption coefficient of component i is identical to that of the matrix. This not very frequent case arises when the specimen is a mixture of different crystalline forms of the same compound. In this case eqn (4.48) reduces to

$$I_{\mathbf{H}_i} = \frac{K_{\mathbf{H}_i} x_i}{\rho_i \mu_M} = K_{x_i}$$

that is there is a direct proportionality between the intensity of a line **H** due to component i and the weight fraction of i in the unknown.

In the second case, μ_i is not equal to μ_M but there are only two components in the sample, 1 and 2, whose identity (and therefore their mass absorption coefficients) is known. In this case the intensity of line H due to component 1 is

$$I_{\mathbf{H}_1} = \frac{K_{\mathbf{H}_1} x_1}{\rho_1[x_1(\mu_1 - \mu_2) + \mu_2]}.$$

If $I^0_{\mathrm{H}_1}$ is the intensity of the same line for the pure component 1

$$I^0_{\mathbf{H}_1} = \frac{K_{\mathbf{H}_1}}{\rho_1 \mu_1}$$

and

$$\frac{I_{\mathbf{H}_1}}{I^0_{\mathbf{H}_1}} = \frac{x_1 \mu_1}{x_1(\mu_1 - \mu_2) + \mu_2}$$

so in this case $I_{\mathbf{H}_1}$ is not a linear function of x_1. Plots of the ratio $I_{\mathbf{H}_1}/I^0_{\mathbf{H}_1}$ as a function of x_1 can then be either calculated using the tabulated values of μ_1 and μ_2 or determined experimentally from the intensities measured from samples of known composition. These curves can then be used to determine x_1 for an unknown specimen.

In the general case in which μ_i is not equal to μ_M and there are more than two components, the determination of x_1 requires the addition of an internal standard. For this case it can be shown that[98]

$$x_i = K \frac{I_{\mathbf{H}_i}}{I_{\mathrm{KS}}}.$$

The linear plots, experimentally determined, of the ratio $I_{\mathbf{H}_i}/I_{KS}$ as a function of x_i can then be used to determine x_i for an unknown.

Alternative equations can be derived when the quantitative analysis is done in the presence of a so-called 'flushing agent'[99] which can be any pure crystalline compound added to the specimen. It can be shown that

$$x_i = x_f \frac{k_f}{k_i} \frac{I_i}{I_f} \tag{4.49}$$

where x_i and x_f are the weight fractions of the i component and the flushing agent, I_i and I_f their diffracted intensities and the two constants k_i and k_f their intensities relative to that of a reference substance, normally corundum (α-Al_2O_3). If the flushing agent is chosen to be corundum, this equation reduces to

$$x_i = \frac{x_c}{k_i} \frac{I_i}{I_c}$$

where the constant k_i is readily available for many substances.

Many other theoretical methods that try to avoid the use of an internal standard have been proposed and have had a more or less limited amount of success.[100]

Another common application of powder diffraction is in the determination of the unit cell parameters of a crystalline phase that can only be obtained as microcrystals. In this case the precision of the measurement depends on the θ angle of the diffraction maximum used in the determination as can be seen rewriting Bragg's law

$$2d_{\mathbf{H}} = \lambda \csc \theta$$

and differentiating

$$2dd_{\mathbf{H}} = -\lambda \csc \theta \cot \theta \, d\theta$$

whence

$$\Delta d_{\mathbf{H}}/d_{\mathbf{H}} = -\cot \theta \, \Delta\theta. \tag{4.50}$$

Therefore for a given precision in the angular measurement θ the relative error in the measurement of d_H decreases for larger values of θ.

A careful analysis of the profiles of the diffraction maxima of a polycrystalline material can give information on the distribution of sizes of the crystallites and on the presence of lattice strains. In order to carry out such an analysis, it is essential to previously extract the pure diffraction profile from the experimentally determined profiles that are influenced by the instrument used (see p. 112) and the experimental conditions selected for the experiment.

If the polycrystalline material has texture, that is if its crystallites do not easily adopt a random orientation, the intensity of a ring corresponding to a given diffraction cone will not be homogeneous. In this case, information about the preferred orientation of the sample constituents can be obtained by studying the entire diffraction ring in two dimensions.

The final application of powder diffraction that will be mentioned here is the Rietveld method used for the solution of a crystal structure with a limited number of parameters[101–103] and for quantitative phase analysis.[104]

We have already seen in Chapter 2 that in the Rietveld method of refinement the entire diffraction profile of a peak is calculated using a starting model for the unit cell content and the model is then refined by comparing point by point the calculated and measured diffraction profiles. This method of crystal structure determination has already permitted the solution of structures of moderate complexity and which could not be solved before by the conventional single-crystal methods because the substances yielded crystals of inadequate size.

Data reduction

In Chapter 3 we saw that in the relationship between integrated intensity and the square of the structure factor amplitude there are several factors that vary from reflection to reflection. In order to calculate the relative structure factor amplitudes to be used in the solution of the crystal structures as described in Chapter 5 one needs first to take these effects into account. The procedure followed to extract relative structure factor amplitudes from raw integrated intensities is called data reduction. In data reduction the different reflection dependent parameters present in eqn (3.41) are taken into account by multiplying the relative intensities by suitable correction factors. Here we will neglect E, the extinction coefficient which was discussed on pp. 97 and 164 and will concentrate on L, P, and T, the other three factors. The corrections applied are called, as we have seen, Lorentz, polarization, and absorption corrections respectively. In addition, we will also discuss the problem of radiation damage of the crystals which is usually handled before the other corrections are applied. Another problem that is often encountered, especially in the case of macromolecular crystal data sets, is that of scaling partial data sets originating from different crystals which when merged will produce the final total set of relative integrated intensities. We discuss this problem briefly at the end of the chapter.

Lorentz correction

We have seen that diffraction arises whenever reciprocal lattice nodes, that always have a non-negligible volume, cross the sphere of reflection. If a node is in diffracting position for a longer time, the intensity of the corresponding reflection will be proportionally higher. This factor would not be important if the method used to record the integrated intensities ensured that every reciprocal lattice node were in a diffracting position for exactly the same time, as it would affect every reflection in the same way and in the end it would simply scale all the intensities by the same factor. This, however, is not the case. Depending on the method used to record the reflection intensity and on the position of the reciprocal lattice node, the times required for different nodes to cross the Ewald sphere are different. The Lorentz correction simply takes this factor into account.

The time a node is in diffracting position is dependent on two factors: the position of the node and the velocity with which it sweeps through the sphere of reflection. We will derive the form of the Lorentz factor in a very simple case and then show the form it takes in a more complicated situation.

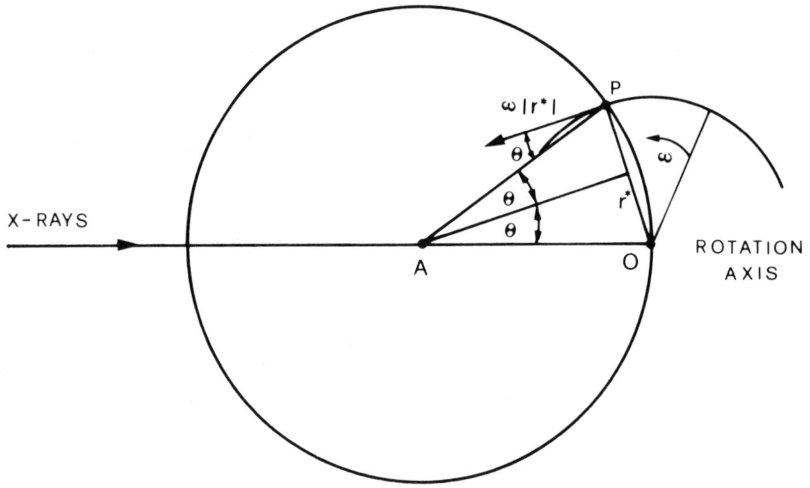

Fig. 4.59. Lorentz correction for a crystal rotated about an axis normal to the plane defined by the incident and scattered X-ray beams.

Figure 4.59 shows the Ewald sphere for a diffraction experiment in which the crystal is rotated about an axis which is normal to the plane defined by the incident and the diffracted beams. This is for example the case of a zero-level rotation or Weissenberg photograph or of the equatorial reflections measured with a diffractometer.

The crystal, and therefore the reciprocal lattice, is assumed to be rotated at a constant angular velocity ω; if V_n is the linear velocity component of the reciprocal lattice node along the radius of the sphere of reflection, the Lorentz factor can be defined as follows

$$L = \omega / V_n \lambda$$

which is indeed proportional to the time during which diffraction takes place for a given reciprocal lattice node.

The linear velocity of the point P is

$$V = |\boldsymbol{r}^*|\, \omega$$

and its component along the radius of the Ewald sphere

$$V_n = |\boldsymbol{r}^*|\, \omega \cos \theta$$

since θ is the angle formed by the linear velocity V and the radius of the sphere of reflection passing through the point P as shown in Fig. 4.59.

Substituting $|\boldsymbol{r}^*|$ in terms of Bragg's law

$$|\boldsymbol{r}^*| = 1/d = 2 \sin \theta / \lambda,$$

$$V_n = (\omega/\lambda) 2 \sin \theta \cos \theta,$$

and

$$L = (2 \sin \theta \cos \theta)^{-1} = (\sin 2\theta)^{-1}. \tag{4.51}$$

which is the simplest possible form that can be taken by the Lorentz factor.

In a more general case one has to compute V_n according to the geometry of the diffraction experiment and then apply the standard equation we have seen.

If the axis of rotation makes an angle $90°\text{-}\mu$ with the incident X-ray beam and the point P is not on a zero-level layer but rather on the nth layer with a diffraction cone with a semiangle equal to 90-v it can be shown that[105,106]

$$L = (\cos\mu \cos v \sin\gamma)^{-1} \tag{4.52}$$

where γ is the projection on to the zero layer of the angle 2θ between the incident and the diffracted beam.

If the rotation axis is normal to the X-ray beam and the reflection is on a zero-level $\mu = 0$ and $v = 0$. In this situation the projection onto the zero level of 2θ, i.e. γ, is identical to 2θ and the expression for L given by eqn (4.52) reduces to eqn (4.51).

Lipson[107] discusses the form of the L factor for the different experimental arrangements which are used in data collection and gives tables of the values of L as a function of the parameters which can be selected.

Polarization correction

The polarization correction depends on the state of polarization of the incident X-ray beam and on the scattering angle of the diffracted beam. In Chapter 3 we have seen that when a totally non-polarized beam is diffracted by a crystal, the diffracted intensity is affected by a factor, called the polarization factor, which in this simple case was shown to be equal to

$$P = \tfrac{1}{2}(1 + \cos^2 2\theta)$$

where θ is the Bragg angle of the reflection considered and the diffracting crystal was tacitly assumed to be ideally mosaic. This simple expression for the polarization correction can be applied whenever the incident X-rays are not polarized, that is when the radiation is produced by a conventional source and monochromatized using an appropriate filter. Notice that in theory this factor can have values ranging between 1.0 and 0.5 depending on the scattering angle, although in practice this variation is less substantial. For a data set collected with CuK_α radiation between 50 Å and 2 Å resolution it varies between $P_{50} = 0.9995$ and $P_2 = 0.7470$.

The more general form of the polarization correction for an incident beam monochromatized with a crystal is[17,108]

$$P = \frac{(\cos^2 2\theta \cos^2\rho + \sin^2\rho)\,|\cos^n 2\theta_M| + \cos^2 2\theta \sin^2\rho + \cos^2\rho}{1 + |\cos^n 2\theta_M|} \tag{4.53}$$

where θ is the Bragg angle of the reflection produced by the specimen and θ_M the angle of the reflection of the monochromator crystal which was used to select the wavelength. The angle ρ is the angle between the projection of the normal to the reflecting plane on to a plane perpendicular to the incident monochromatized X-rays and the plane of incidence.[108] When the original X-ray beam, the monochromated beam, and the scattered beam all lie in the same plane this angle is equal to 0 and the polarization factor takes the simpler form

$$P = \frac{\cos^2 2\theta\,|\cos^n 2\theta_M| + 1}{1 + |\cos^n 2\theta_M|}.$$

The exponent n depends on the characteristics of the monochromating crystal. If the crystal is assumed to be a perfectly mosaic crystal it is equal to

2, whereas if it is assumed to be an ideal crystal it is equal to 1. Real monochromator crystals are usually an intermediate between these two extreme cases and one should ideally examine the monochromator used to decide the appropriate form of the polarization correction that should be used in the specific case.[109–111]

If the radiation striking the crystal monochromator is the totally polarized synchrotron radiation, the form of the polarization correction changes again. An expression for the polarizaton correction to be used with synchrotron radiation has been derived by Kahn *et al.*[112]

$$P = P_0 - P' = \tfrac{1}{2}(1 + \cos^2 2\theta) - \tfrac{1}{2}\zeta' \cos 2\rho \sin^2 2\theta \qquad (4.54)$$

where the angle ρ is defined as above and

$$\zeta' = \frac{E'^2_\sigma - E'^2_\pi}{E'^2_\sigma + E'^2_\pi}.$$

Here E'_σ is the amplitude of the optical field in the plane of incidence of the X-rays and E'_π is the component perpendicular to it. In this expression for the polarization correction, the problem is to obtain an accurate value for the parameter ζ' which depends on the set-up of the facility used. This can be done in two ways; one is by measuring the polarization ratio of the beam that will strike the specimen. The second method is by calculating it theoretically on the basis of the characteristics of the source and of the crystal used to monochromatize the radiation.

The polarization correction is frequently grouped with the Lorentz correction in a single factor, the *LP* correction.

Absorption corrections

As pointed out in Chapter 3, the transmission factor T is related to the absorption of the incident and diffracted X-ray beams by the crystal. We have briefly discussed the absorption of X-rays on p. 241 where it was pointed out that according to Beer's law, absorption reduces the intensity of an X-ray beam travelling through a given material by an amount which depends on the material and the length of the path travelled by the radiation in it. Figure 4.60 shows that, for a given scattered beam, this path can be very different for different points in the crystal. The path lengths are dependent, as can be seen in the picture for points O, R, and T, on the location of the point scattering the X-rays, and on the incident and scattering angle, that is on the reflection considered.

The intensity of the diffracted X-rays is thus reduced, with respect to what it would be without absorption by the factor

$$I/I_0 = e^{-\mu x} \qquad (4.55)$$

which is valid for every point in the crystal. Here x is the total path length and μ is, as we have seen, the linear absorption coefficient, in this case, of the crystal.

Equation (4.55) can be used to calculate a very rough estimate of the optimum crystal size for a given compound of linear absorption coefficient μ. In eqn (3.41), the constant K_2 included Ω, the crystal volume, that we

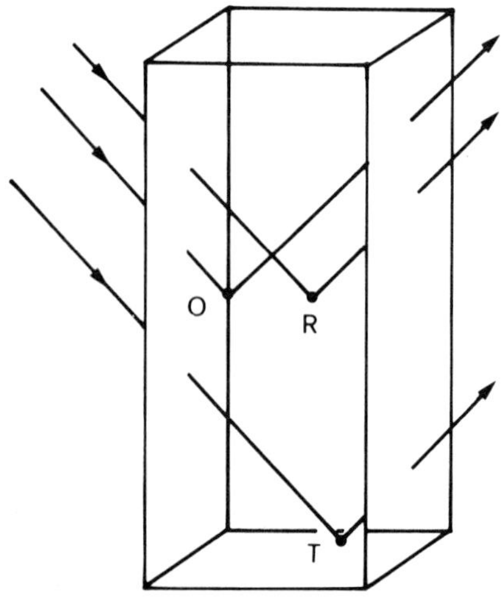

Fig. 4.60. For a given scattering angle 2θ, the path of the incident and scattered beams in the crystal depends on the position of the scattering point within the crystal.

will take to be proportional to x^3. Hence

$$I_0 \propto x^3 \quad \text{and} \quad I \propto x^3 e^{-\mu x}.$$

Since we want to maximize I, the diffracted intensity after absorption

$$\frac{dI}{dx} = 3x^2 e^{-\mu x} - x^3 \mu e^{-\mu x} = x^2 e^{-\mu x}(3 - \mu x) = 0$$

and $x \simeq 3/\mu$.

In order to get T, the transmission factor for an entire crystal, one simply has to integrate eqn (4.55) over the total crystal volume. If instead of writing x we decompose the path into p the incident or primary beam path and q the diffracted or secondary beam path the transmission factor T can be written as follows:

$$T = \frac{1}{V} \int_v e^{-\mu(p+q)}\, dv, \tag{4.56}$$

where the integration is over the entire crystal volume V.

The linear absorption coefficient for the crystal can be calculated from the mass absorption coefficients of the atoms present in the unit cell. No structural knowledge is required, only the values of the mass absorption coefficients of the elements which, as we have seen, can be found in the *International tables for x-ray crystallography*. From the values of μ_m for a given wavelength, μ can be calculated by the following equation:

$$\mu = \rho \sum_i g_i \mu_m^i. \tag{4.57}$$

where g_i is the mass fraction of element i present in the unit cell, μ_m^i is its mass absorption coefficient, and ρ is the crystal density. Recall that μ_m^i is a function of the atomic number of the element and of the wavelength of the radiation used: it is smaller for lower atomic numbers and for shorter wavelengths. This explains why absorption corrections become more important for heavy-element crystals and for radiation of longer wavelengths. Sometimes all it takes is a change from copper to molybdenum radiation to sufficiently reduce the absorption problem in a given crystal structure determination. In any case it is always instructive to calculate the value of μ for the crystal being examined in order to get an indication of the severity of the absorption problem.

An analytical evaluation of T according to eqn (4.56) would be, in theory, the ideal method to use in order to take care of the absorption correction. The result would depend on the beam path in the crystal which is a function of the reflection considered, i.e. one would get a different value of T for every reflection measured. The problem is that the integral in eqn (4.56) cannot be calculated analytically even in the case of the simplest crystal shapes. Numerical evaluations have been obtained in the case of spheres or cylinders, they can be found in Lipson[113] where T is given as a function of μR and θ, R being the sphere or cylinder radius and θ the scattering angle. Spheres or cylinders are, however, not very good approximations for the shape of most real crystals so that if one wants to assume that the specimen under study is a cylinder or a sphere, it is usually necessary to grind it into that shape. Mechanical devices exist that can be used to accomplish this.[114]

This approach is, however, not always possible since there are many crystals that will not survive the very harsh treatment required to shape them into an ideal form.

An analytical method that can be used to calculate T for any polyhedral crystal was proposed by de Meulenaer and Tompa.[115] This method divides an arbitrary crystal volume into smaller polyhedra that are ultimately subdivided into tetrahedra. The total transmission factor is then calculated as

$$T = \frac{\sum A_T}{\sum V_T}$$

where A_T is the contribution to the transmission factor of each tetrahedron and V_T its volume. The contribution A_T can be calculated analytically and it depends on the values of μx at each vertex of the tetrahedron and on V_T. In this method, the calculations are performed by a computer program which requires a precise knowledge of the crystal shape which is input as the equations of the planes defining the faces of the crystal.

In many cases, an accurate description of the crystal shape and size is very difficult, if not impossible, to obtain. Furthermore, if one is dealing with a protein crystal, it must be mounted inside a capillary and bathed in its mother liquor, both of which absorb the X-rays differently from the crystal. Thus, there are many cases in which analytical methods cannot be applied; it is for these situations that empirical absorption correction methods have been proposed. These methods, which attempt to measure T experimentally are currently the most widely used, mainly because they are quite easy to implement. The experimental absorption correction methods are reviewed by Bartels.[116] Here we will only briefly discuss one of them, the very widely used method of North *et al.*[117]

The method is designed to correct intensities measured with the diffractometer and the crystal is therefore assumed to be totally bathed by the incident X-ray beam. In Fig. 4.40 for example we have point P in the diffracting position, that is the point is in the equatorial plane and on the Ewald sphere. If the crystal is rotated about the vector $\boldsymbol{r}^*$ the point P will still remain in the diffracting position and the variations in intensity observed may be attributed to absorption effects. Such a rotation is called an azimuthal or ψ-scan. In practice what is done is to bring the Φ rotation axis to a direction coincident with that of vector $\boldsymbol{r}^*$ which requires the choice of a reflection with a χ value of 90° (see p. 274). Then the azimuthal angle ψ is equivalent to the diffractometer angle Φ and a Φ-scan produces a curve similar to that shown in Fig. 4.61.

The relative transmission factor for a given value of Φ is then given by

$$T = I(\Phi)/I_{max}(\Phi)$$

where $I_{max}(\Phi)$ is the maximum intensity observed as Φ is varied over the 360° range. Clearly, in this method, and as seen in Fig. 4.61, the transmission factor T at Φ and $\Phi + 180°$ must be identical since the only difference between these two positions is that the crystal has been flipped over. In the method of North *et al.* the transmission factor for a given reflection is given by the mean

$$T(\mathrm{H}) = \tfrac{1}{2}(T(\Phi_{inc}) + T(\Phi_{ref})) \tag{4.58}$$

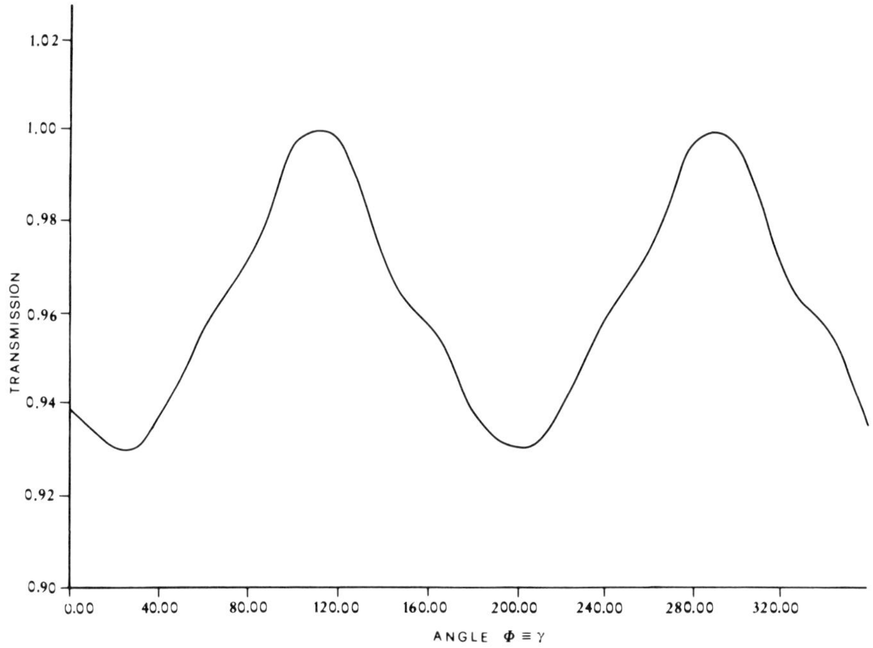

Fig. 4.61. Relative transmission factor plotted as a function of the scanning angle $\Phi \equiv \psi$ for a reflection chosen with a χ value close to 90°. The plot can be used to apply an empirical absorption correction as described in the text.

where Φ_{inc} and Φ_{ref} define the orientations of the crystal in which the incident and reflected beams for the H reflection coincide with or lie in the same plane as the incident X-ray beam.

An underlying assumption of this method is that the total transmission is the product of the transmission of the primary and the secondary beams. A discussion of the applicability of this assumption can be found in Kopfmann and Huber.[118] If the transmission curves are measured for symmetry-equivalent reflections this assumption is not necessary.[119] Other methods that calculate transmission curves based on the differences observed in symmetry-equivalent reflections have been proposed.[120,121] They have the disadvantage that they require several ψ-scans, i.e. much more experimental work to obtain a result that does not always justify it. An example of the way in which absorption correction effects can be taken into account, together with other factors, by examining the differences in intensity observed for symmetry equivalent reflections is presented by Takusagawa[122] who discusses together absorption and radiation damage corrections for data collected with an area detector. A final possibility is to take care of the absorption effects after a preliminary structure has been solved using the uncorrected data. From the differences between the calculated and measured structure factors it is then possible to calculate correction curves that can be applied to the raw data.[123] Whatever the absorption correction used, the redundancy in the data set can be used to check whether the correction has improved the data or not. This is normally done by calculating an R factor (see eqn (4.70)) that should obviously decrease after the absorption correction is applied.

Although, as we have seen, the theory of X-ray absorption by crystals is very well established, and in spite of the fact that so many different methods to determine T have been proposed, absorption still remains one of the most serious, if not the most serious, source of error in the experimental determination of relative integrated intensities.

Radiation damage corrections

So far we have tacitly assumed that as data collection proceeds the integrated intensities measured do not change with time, or at least that they do not change significantly. Unfortunately, this is very seldom the case since crystal decomposition triggered by exposure to X-rays is not at all an uncommon occurrence. Radiation damage can vary remarkably with the specimen considered, it is known to be substantial in the case of macromolecular crystals, but it is also present in many small-molecule crystals as a survey conducted in 1971 has shown.[124]

The causes that lead to the changes in intensity as a crystal is exposed to higher and higher X-ray doses have been listed by Abrahams.[124] If the specimen can be considered to be close to an ideal crystal, exposure to radiation can increase its mosaicity thereby reducing extinction effects. The result will be that, after exposure, some integrated intensities will be higher. More often, there is in real crystals a loss of short- and long-range order that produces instead a decrease in the measured intensities. Another effect is chemical damage, for example due to free radicals generated by the radiation, which can either increase or decrease the intensities. A combination of these factors is, of course, possible, which explains the difficulties encountered in proposing a satisfactory model for this phenomenon and the scarcity of research papers dealing with this subject.

Radiation damage is evidenced by monitoring the intensity, as a function of data collection time, of a certain number of reflections, ideally sampling as large as possible a region of reciprocal space. Figure 4.62 is a plot of the values measured for three reflections of a protein crystal that decayed, on the average, about 20 per cent during an exposure time of less than two days. This type of monitoring is usually done when data are collected with the diffractometer, although it is also possible for the rotation and the area-detector methods as well.

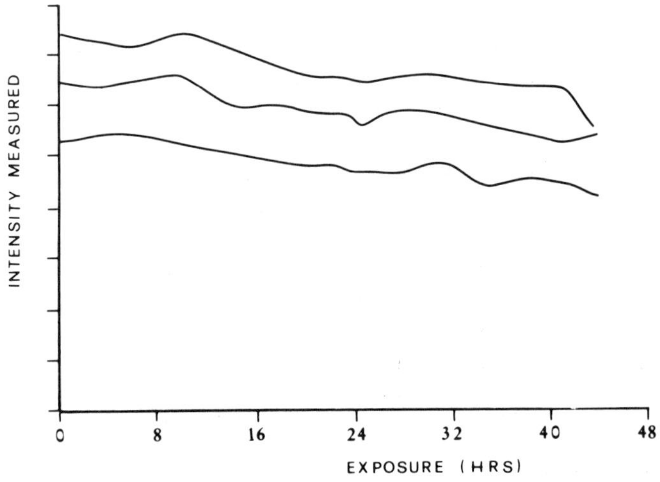

Fig. 4.62. Typical behaviour of the relative intensities of three reflections of a protein crystal monitored with a diffractometer at different times after exposure begins.

The radiation damage correction factor, $R(t, \theta)$, is defined as follows

$$R(t, \theta) = I(t)/I(0), \tag{4.59}$$

that is it is the inverse of the factor by which the intensity measured at time t has to be multiplied to yield the intensity corrected at time zero, the value that should be used to extract the relative structure factor amplitudes. This correction factor is a function, not only of the time after data collection started but also of the scattering angle of the reflection θ.

The value of R can be estimated without resorting to any particular model for the radiation damage process, for example by fitting the intensities of each monitored reflection to a polynomial of the form[125]

$$I(t) = I(0)\left(1 - \sum A_n t^n\right) \tag{4.60}$$

where t is the exposure time and n is a number that ranges from 1 to 7. The discrepancy index

$$R' = \frac{\sum |I - \langle I \rangle|}{\sum I} \tag{4.61}$$

is then calculated for each standard reflection as a function of n and used to decide the best form of the polynomial to be used, i.e. the value of n that best fits the data measured. If a general trend for the different standard reflections is identified, then the same form of the polynominal can be adopted and a weighted average of the coefficients A_n of the monitored reflections can be used to correct all the data collected. In the simple case in which the decay is found to be linear, i.e. if the best n turns out to be equal to 1, anisotropy can be readily taken into account.[125]

Sometimes, a correction based on just a few monitored reflections may not be acceptable. A scheme that attempts to overcome this difficulty has been proposed by Ibers.[126] If, after the structure has been solved, S_n is the scale factor for the nth reflection measured at time t that brings the calculated structure factor amplitude $|F_c|$ (see Chapter 5) in scale with the observed amplitude $|F_0|$, one can define S_n as the following function of exposure time:

$$S_n = S_0 + a_1 t_n + a_2 t_n^2 \tag{4.62}$$

and then determine the constants a_1 and a_2 by least-squares refinement along with the other structural parameters (see Chapter 5).

The problem of intensity variations due to radiation damage was perceived quite early on in protein crystallography. The first model that attempted to explain them in the case of myoglobin was proposed by Blake and Phillips in 1962.[127] The model describes the diffraction pattern from the irradiated crystal as due to three different components: an undamaged fraction A_1, a disordered fraction A_2 that scatters predominantly at low values of θ, and a totally amorphous fraction A_3 that no longer scatters coherently. The variation of the intensities as a function of time is then governed by the equation

$$I(t)/I(0) = A_1(t) + A_2(t)\exp(-DS^2) \tag{4.63}$$

where D is a disorder parameter, $S = \sin\theta/\lambda$, and the quantities $A_1(t)$ and

$A_2(t)$ can be calculated by making different simplifying assumptions for the rate processes involved. Hendrickson[128] has proposed a general method that considers all the possible paths between A_1, A_2, and A_3, and has also tested the validity of the different possible assumptions in the case of myoglobin. His conclusion is that for moderate radiation damage all the rate models work satisfactorily whereas for severe cases none do.

A refinement of the Blake and Phillips model has been recently proposed by Sygusch and Allaire.[129] In their model another state, A_1', with a dose-dependent rate of formation is introduced. The molecules in this state are thought to be only superficially perturbed. In addition, anisotropy is also taken into account.

The usefulness of all the models proposed is that they furnish a form for the variation of $R(t, \theta)$ as a function of a certain number of parameters that can be determined from the variation of $I(t)$ as a function of t for the monitored reflections. The parameters so determined can then be used to calculate a correction factor for each reflection present in the data set collected.

An efficient way to minimize radiation damage and therefore to make the application of a correction less significant is by collecting the data at low temperature. Even a modest decrease to say 4 °C has already important beneficial effects on most crystals but the best results are obtained when working at liquid nitrogen or, even better, liquid helium temperatures. Although the experimental problems posed in this case are more difficult to solve, low temperature data collection is a well established technique for small molecule crystals[130] and shows great promise in macromolecular work as well.[131,132] In addition to greatly reducing radiation damage, working at low temperature decreases the intensity losses due to thermal vibrations and thereby improves the resolution of the diffraction pattern and, by reducing the thermal diffuse scattering, that is the background radiation resulting from the correlated atomic displacements in the lattice, it improves the signal-to-noise ratio of the diffracted peaks. In Appendix 3.D we have already seen that precise electron density determinations of small molecule crystals require that the data be collected at low temperature.

Relative scaling

If the final total data set of relative integrated intensities has not been collected under strictly constant conditions but results from merging a certain number of subsets, each measured under more or less different conditions, before these sets can be merged together into a single one it is necessary to scale them by applying the appropriate relative scale factors. The subsets may be derived from different crystals, if radiation damage has made it necessary to stop data collection at a certain stage before the set was completed, or they may also come from the same crystal. For example data collection on films with the rotation method requires scaling as there is no way to ensure that all the data collection parameters, take for example film developing and fixing times, will be held strictly constant throughout the data collection process.

Relative scaling of partial data subsets is done on the basis of the reflections which these subsets have in common. In order to determine the relative scale factors to be applied to the subsets, it is first necessary to define the conditions on the data that the scale factors ought to satisfy. Among the different criteria proposed, that of Hamilton *et al.*[133] is currently the most widely used. In this method one defines

$$\varphi_{\mathbf{H}_i} = \sqrt{V_{\mathbf{H}_i}}\,(K_{l(i)} I_{\mathbf{H}_i} - I_{\mathbf{H}})$$

where $I_{\mathbf{H}_i}$ is the ith observation of reflection $\mathbf{H}$, l is the subset in which the ith reflection is present, K_l the relative scale factor to be applied and $V_{\mathbf{H}_i}$ the weight of the ith observation of reflection $\mathbf{H}$.

It can be shown that this condition can be stated in the equivalent form[133]

$$\varphi_{\mathbf{H}_i} = \sqrt{W_{\mathbf{H}_i}}\,(I_{\mathbf{H}_i} - G_{l(i)} I_{\mathbf{H}}). \tag{4.64}$$

Now if one defines

$$\psi = \sum_{\mathbf{H}} \sum_{i} \varphi^2_{\mathbf{H}_i}$$

the relative scale factors are chosen so that the quantity ψ is a minimum and therefore the condition from which the best value for the intensity of reflection $\mathbf{H}$, $I_{\mathbf{H}}$, is found is

$$\delta\psi/\delta I_{\mathbf{H}} = 0$$

and $I_{\mathbf{H}}$ is

$$I_{\mathbf{H}} = \frac{\sum_i W_{\mathbf{H}_i} G_{l(i)} I_{\mathbf{H}_i}}{\sum_i W_{\mathbf{H}_i} G^2_{l(i)}}. \tag{4.65}$$

Since in this formulation the residual is not linear in the G_ls, the best values for these parameters are determined using the iterative non-linear least-squares procedure described in Chapter 2 (p. 94), that is for each iteration $\varphi_{\mathbf{H}_i}$ is approximated by

$$\varphi_{\mathbf{H}_i} \simeq \sqrt{W_{\mathbf{H}i}}\,(I_{\mathbf{H}_i} - G_{l(i)} I_{\mathbf{H}}) + \sum_i \frac{\delta\varphi_{\mathbf{H}_i}}{\delta G_{l(K)}} \Delta G_{l(K)}. \tag{4.66}$$

Given that the $\Delta G_{l(i)}$s are not independent, one of them is arbitrarily set equal to zero, that is one G_l is made constant, and then the other G_ls are corrected until convergence is achieved.

An alternative way of solving the equations of Hamilton *et al.* was proposed by Fox and Holmes.[134] In their formulation one sets in turn all the derivatives of ψ with respect to the G_ls to zero, i.e.

$$\delta\psi/\delta G_l = 0 \quad \text{for } l = 1 \ldots L$$

and then approximates ψ by the Taylor expansion:

$$\psi = \psi_0 + \sum_l \frac{\mathrm{d}\psi}{\mathrm{d}G_l} \Delta G_l. \tag{4.67}$$

In the special simple case in which the weights can be written as the product of a term which depends on the reflection and another which depends on the subset an exact solution for this problem has been found.[134] It is useful in scaling the different films present in a pack which have in common many reflections and differ only in those which are outside the dynamic range of each film.

Another alternative procedure that avoids the use of iterations to determine the scale factors has been proposed by Rae[135] who defines the residual

$$\Delta'_{\mathbf{H}_{ij}} = \log K'_i I_{\mathbf{H}_i} - \log K'_j I_{\mathbf{H}_j} \tag{4.68}$$

and minimizes the quantity

$$\Delta = \sum_{\substack{\text{all pairs} \\ \mathbf{H}, i, j}} W_{ij} \Delta'^2_{\mathbf{H}_{ij}} \tag{4.69}$$

where the subscripts i and j refer to different subsets.

If the number of data points is large there is very little difference in the relative scale factors found by this method and the classical solution of Hamilton *et al.*

When merging together different subsets of data into a final data set, it is customary to calculate a reliability index for the set that can be defined either in terms of the structure factor amplitudes or of the intensities. A very widely used definition for this parameter in terms of the intensities is that given by eqn (4.61)

$$R = \frac{\sum_{\mathbf{H}} \sum_{i=1}^{N} \langle I(\mathbf{H}) \rangle - I(\mathbf{H})_i}{\sum_{\mathbf{H}} \sum_{i=1}^{N} I(\mathbf{H})_i} \tag{4.70}$$

where $I(\mathbf{H})_i$ is the ith measurement of reflection $\mathbf{H}$, $\langle I(\mathbf{H}) \rangle$ is its mean value and the summation extends over all the reflections measured more than once in the set.

In the cases in which R is calculated using independent reflections which ought to have equal intensities for symmetry reasons, the notation R_{sym} is used.

Appendix

4.A Determination of the number of molecules in the unit cell of a crystal

One of the first fundamental parameters required in a crystal structure determination is the number of molecules present in the unit cell of the crystal.

Small-molecule crystals do not usually contain important amounts of solvent and/or other substances whereas in macromolecular crystals a substantial fraction of the unit cell volume is occupied by solvent (see Chapter 8, p. 538). In both cases the number of molecules present in the unit cell is a function of the crystal density, the molecular weight of the substance under investigation, and the unit cell volume.

If we can assume that the only species present in the crystal is the compound under investigation, its mass in the unit cell is

$$m = n(M_s/N)$$

where n is the number of molecules in the unit cell, M_s the molecular weight of the substance, and N is Avogadro's number.

The density of the crystal is

$$\rho_c = m/V = nM_s/NV. \tag{4.A.1}$$

Equation (4.A.1) can also be written

$$n = 0.602(V\rho_c/M_s) \tag{4.A.2}$$

where the unit cell volume has to be measured in Å^3 and the density in g cm^{-3}.

Measuring the density of a small-molecule crystal usually poses no serious problem. There are several methods available and the measurements can be done with high precision.[136]

If the molecular weight of the compound is known, then, using (4.A.2), the number of molecules in the unit cell, n, can be easily calculated. Since n has to be an integer, if the density measurement is very reliable and the molecular weight is not, eqn (4.A.2) can be used to calculate a more accurate molecular weight. Alternatively, a precise molecular weight can be used to yield an accurate density for the crystal using the integer closest to the n determined experimentally.

In the case of protein crystals, the situation is not so simple and there are several alternative equations that are equivalent to (4.A.2) We will briefly discuss one of them.[137]

In a macromolecular crystal (see Chapter 8, p. 536) there is water which is eliminated when the crystal is dried and water which remains bound to the macromolecule and there is also salt dissolved in the solvent. If d is the fractional loss of mass when the crystal is dried, u is the fraction of liquid which remains in the crystal, s is the mass of salt per unit mass of solvent, and w is the solvent not accessible to the salt because it is strongly bound to the macromolecule, the total mass of the unit cell of the crystal is

$$m = m_p + dm + um_p + s(m - m_p - wm_p),$$

where m is the total mass of the unit cell and m_p the mass of the protein in the unit cell.

From this equation we can obtain

$$\frac{m_p}{m} = \frac{1 - d - s}{1 + u - s - sw}.$$

The mass of the unit cell is

$$m = V\rho_c$$

where V is the unit cell volume and ρ_c is the crystal density.

The mass of the protein in the unit cell is

$$m_p = nM_p/N$$

where n is the number of molecules in the unit cell, M_p the molecular weight of the protein, and N is Avogadro's number. Thus

$$n = \frac{NV\rho_c}{M_p}\frac{1-d-s}{1+u-s-sw}$$

or

$$n = 0.602\frac{V\rho_c}{M_p}\frac{1-d-s}{1+u-s-sw}. \tag{4.A.3}$$

Equation (4.A.3) can be used in much the same way as eqn (4.A.2) but in this case determining the crystal density is a much more serious experimental problem.[138] In addition, one needs to know u, w, s, and d; the first two parameters are usually not determined, they are instead estimated from the average of known protein crystals; s is the quantity most easily measured and d is quite difficult to determine and requires the use of several crystals for better precission.

An alternative to eqn (4.A.3) has been derived by Matthews.[139] Another approach to determine n for macromolecular crystals is discussed in Chapter 8 (p. 538).

References

1. Rieck, G. D. (1962). In *International tables for x-ray crystallography,* Vol. III, (ed. C. H. MacGillavry and G. D. Rieck), pp. 59–72. Kynock, Birmingham.
2. Luger, P. (1980). *Modern x-ray analysis on single crystals,* Ch. 2. Walter de Gruyter, Berlin.
3. Phillips, W. C. (1985). In *Methods in enzymology,* Vol. 114, (ed. H. W. Wyckoff, C. H. W. Hirs, and S. N. Timasheff), pp. 300–16. Academic, Orlando.
4. Eisenberger, P. (1986). *Science,* **231,** 687–93.
5. ESRF (1987). *Foundation Phase Report.* ESRF, Grenoble.
6. Bienenstock, A. and Winick, H. (1983). *Physics Today,* **36,** 48–58.
7. Winick, H. (1987). *Scientific American,* **257,** 72–81.
8. Margaritondo, G. (1988). *Introduction to synchrotron radiation.* Oxford University Press, New York.
9. Koch, E. E. (ed.) (1983). *Handbook on synchrotron radiation.* North Holland, Amsterdam.
10. Materlik, G. (1982). In *Uses of Synchrotron Radiation in Biology* (ed. H. B. Stuhrmann), pp. 1–21. Academic, New York.
11. Hendrickson, W. A., Smith, J. L., Phizackerley, R. P., and Merritt, E. A. (1988). *Proteins: Structure, Function and Genetics,* **4,** 77–88.
12. Arndt, U. W. (1984). *Journal of Applied Crystallography,* **17,** 118–19.
13. Bonse, U. (1980). In *Characterization of crystal defects by x-ray methods* (ed. B. K. Tanner and D. K. Bowen), pp. 298–319. Plenum, New York.
14. Rieck, W., Euler, H., and Schulz, H. (1988). *Acta Crystallographica* **A44,** 1099–101.
15. Koch, B. and MacGillavry, C. H. (1962). In *International tables for x-ray crystallography,* Vol. III (ed. C. H. MacGillavry and G. D. Rieck), pp. 157–200. Kynock, Birmingham.
16. Roberts, B. W. and Parrish, W. (1962). In *International tables for x-ray crystallography,* Vol. III, (ed. C. H. MacGillavry and G. D. Rieck), pp. 73–88. Kynock, Birmingham.
17. Arndt, U. W. and Sweet, R. M. (1977). In *The Rotation Method in*

Crystallography (ed. U. W. Arndt and A. J. Wonacott), pp. 43–63. North Holland, Amsterdam.
18. Margaritondo, G. (1988). *Introduction to synchrotron radiation,* pp. 50–81. Oxford University Press, New York.
19. Witz, J. (1969). *Acta Crystallographica,* **A25,** 30–42.
20. Huxley, H. E. (1953). *Acta Crystallographica,* **6,** 457–65.
21. Franks, A. (1955). *Proceedings of the Physical Society,* **B68,** 1054–64.
22. Harrison, S. C. (1968). *Journal of Applied Crystallography,* **1,** 84–90.
23. Friedrich, W., Knipping, P., and Laue, M. (1912). *Sitzungsberichte der mathematisch-physicalischen Klasse der Koeniglich Bayerischen Akademie der Wissenschaften zu Muenchen* (reprinted in *Naturwissenschaften* (1952) pp. 361–7).
24. Henry, N. F. M., Lipson, H., and Wooster, W. A. (1951). *The interpretation of x-ray diffraction photographs,* pp. 71–86. Macmillan, London.
25. Amorós, J. L., Buerger, M. J. and Canut de Amorós, M. (1975). *The Laue method.* Academic, New York.
26. Helliwell, J. R., Habash, J., Cruickshank, D. W. J., Harding, M. M., Greenhough, T. J., Campbell, J. W., Clifton, I. J., Elder, M., Machin, P. A., Papiz, M. Z., and Zurek, S. (1989). *Journal of Applied Crystallography,* **22,** 483–97.
27. Wood, I. G., Thompson, P., and Matthewman, J. C. (1983). *Acta Crystallographica,* **B39,** 543–7.
28. Gomez de Anderez, D., Helliwell, M., Habash, J., Dodson, E. J., Helliwell, J. R., Bailey, P. D., and Gammon, R. E. (1989). *Acta Crystallographica,* **B45,** 482–8.
29. Moffat, K., Szebenyi, D., and Bilderback, D. (1984). *Science,* **223,** 1423–5.
30. Hajdu, J., Machin, P. A., Campbell, J. W., Greenhough, T. J., Clifton, I. J., Zurek, S., Gover, S., Johnson, L. N., and Elder, M. (1987). *Nature,* **329,** 178–81.
31. Moffat, K. (1989). *Annual Reviews of Biophysics and Biophysical Chemistry,* **18,** 309–32.
32. Weissenberg, K. (1924). *Zeitschrift für Physik,* **23,** 229–38.
33. Buerger, M. J. (1942). *X-ray crystallography,* Ch. 12. Wiley, New York.
34. Stout, G. H. and Jensen, L. H. (1968). *X-ray structure determination,* Ch. 5. Macmillan, New York.
35. Buerger, M. J. (1964). *The precession method.* Wiley, New York.
36. Bernal, J. D. (1926). *Proceedings of the Royal Society of London,* **A113,** 117–64.
37. Arndt, U. W., Champness, J. N., Phizackerley, R. P., and Wonacott, A. J. (1973). *Journal of Applied Crystallography,* **6,** 457–63.
38. Arndt, U. W. and Wonacott, A. J. (ed.) (1977). *The rotation method in crystallography.* North Holland, Amsterdam.
39. Wonacott, A. J. (1977). In *The rotation method in crystallography* (ed. U. W. Arndt and A. J. Wonacott), pp. 75–117. North Holland, Amsterdam.
40. Rossmann, M. G. and Erickson, J. W. (1983). *Journal of Applied Crystallography,* **16,** 629–36.
41. Wooster, W. A. (1964). *Acta Crystallographica,* **17,** 878–82.
42. Vonk, C. G. and Pijpers, A. P. (1981). *Journal of Applied Crystallography,* **14,** 8–16.
43. Elder, M. (1985). In *Methods in enzymology,* Vol. 114, (ed. H. W. Wyckoff, C. H. W. Hirs, and S. N. Timasheff), pp. 199–211. Academic, Orlando.
44. Arndt, U. W., Gilmore, D. J., and Wonacott, A. J. (1977). In *The rotation method in crystallography* (ed. U. W. Arndt and A. J. Wonacott), pp. 207–18. North Holland, Amsterdam.
45. Wonacott, A. J. and Burnett, R. M. (1977). In *The rotation method in*

crystallography (ed. U. W. Arndt and A. J. Wonacott), pp. 119–38. North Holland, Amsterdam.
46. Abrahamsson, S. (1966). *Journal of Scientific Instruments,* **43,** 931–3.
47. Xuong, Ng. H. (1969). *Journal of Physics E: Scientific Instruments,* **2,** 485–9.
48. Nockolds, C. E. and Kretsinger, R. H. (1970). *Journal of Physics E: Scientific Instruments,* **3,** 842–6.
49. Matthews, B. W., Klopfenstein, C. E., and Colman, P. M. (1972). *Journal of Physics E: Scientific Instruments,* **5,** 353–9.
50. Ford, G. C. (1974). *Journal of Applied Crystallography,* **7,** 555–64.
51. Arndt, U. W. and Willis, B. T. M. (1966). *Single crystal diffractometry.* Cambridge University Press.
52. Wyckoff, H. W. (1985). In *Methods in enzymology,* Vol. 114, (ed. H. W. Wyckoff, C. H. W. Hirs, and S. N. Timasheff), pp. 330–86. Academic, Orlando.
53. Luger, P. (1980). *Modern x-ray analysis on single crystals,* Ch. 4. Walter de Gruyter, Berlin.
54. Hamilton, W. C. (1974). *International tables for x-ray crystallography,* Vol. IV, (ed. J. A. Ibers and W. C. Hamilton), pp. 275–81. Kynock, Birmingham.
55. Diamond, R. (1969). *Acta Crystallographica,* **A25,** 43–55.
56. Lehmann, M. S. and Larsen, F. K. (1974). *Acta Crystallographica,* **A30,** 580–4.
57. Arndt, U. W. and Faruqi, A. R. (1977). In *The rotation method in crystallography* (ed. U. W. Arndt and A. J. Wonacott), pp. 219–26. North Holland, Amsterdam.
58. Harrison, S. C. (1984). *Nature,* **309,** 408.
59. Arndt, U. W. (1986). *Journal of Applied Crystallography,* **19,** 145–63.
60. Hamlin, R., Cork, C., Howard, A., Nielsen, C., Vernon, W., Mathews, D., and Xuong, Ng. H. (1981). *Journal of Applied Crystallography,* **14,** 85–93.
61. Hamlin, R. (1985). In *Methods in Enzymology,* Vol. 114, (ed. H. W. Wyckoff, C. H. W. Hirs, and S. N. Timasheff), pp. 416–52. Academic, Orlando.
62. Faruqi, A. R. (1977). In *The rotation method in crystallography* (ed. U. W. Arndt and A. J. Wonacott), pp. 227–43. North Holland, Amsterdam.
63. Mokulskaya, T. D., Kuzev, S. V., Myshko, G. E., Khrenov, A. A., Mokulskii, M. A., Dobrokhotova, Z. D., Volodenkov, A. Y., Rubanov, V. P., Ryanzina, N. A., Shitikov, B. I., Baru, S. E., Khabakhpashev, A. G., and Sidorov, V. A. (1981). *Journal of Applied Crystallography,* **14,** 33–7.
64. Arndt, U. W. and Gilmore, D. J. (1979). *Journal of Applied Crystallography,* **12,** 1–9.
65. Arndt, U. W. (1982). *Nuclear Instrumentation and Methods,* **201,** 13–20.
66. Arndt, U. W. (1985). In *Methods in enzymology,* Vol. 114, (ed. H. W. Wyckoff, C. H. W. Hirs, and S. N. Timasheff), pp. 472–85. Academic, Orlando.
67. Arndt, U. W. and Int'Veld, G. A. (1988). *Advances in Electronics and Electron Physics* **74,** 285–96. Academic, Orlando.
68. Miyhara, J., Takahashi, K., Amemiya, Y., Kamiya, N., and Satow, Y. (1986). *Nuclear Instrumentation and Methods,* **A26,** 572–8.
69. Amemiya, Y., Kamiya, N., Satow, Y., Matsushita, T., Chikawa, J., Wakabayashi, K., Tanaka, H., and Miyahara, J. (1987). In *Biophysics and synchrotron radiation* (ed. A. Bianconi and A. Congiu Castellano), pp. 61–72. Springer, Berlin.
70. Amemiya, Y., Matushita, T., Nakagawa, A., Satow, Y., Miyahara, J., and Chikawa, J. (1988). *Nuclear Instrumentation and Methods,* **A266,** 645–53.
71. Xuong, Ng. H., Sullivan, D., Nielsen, C., and Hamlin, R. (1985). *Acta Crystallographica,* **B41,** 267–9.
72. Durbin, R. M., Burns, R., Moulai, J., Metcalf, P., Freymann, D., Blum, M.,

Anderson, J. E., Harrison, S. C., and Wiley, D. C. (1986). *Science*, **232**, 1127–32.
73. Howard, A. J., Gilliland, G. L., Finzel, B. C., Poulos, T. L., Ohlendorf, D. H., and Salemme, F. R. (1987). *Journal of Applied Crystallography*, **20**, 383–7.
74. Xuong, Ng. H., Nielsen, C., Hamlin, R. and Anderson, D. (1985). *Journal of Applied Crystallography*, **18**, 342–50.
75. Blum, M., Metcalf, P., Harrison, S. C., and Wiley, D. C. (1987). *Journal of Applied Crystallography*, **20**, 235–42.
76. Xuong, Ng. H., Freer, S. T., Hamlin, R., Nielsen, C., and Vernon, W. (1978). *Acta Crystallographica*, **A34**, 289–96.
77. Howard, A. J., Nielsen, C., and Xuong, Ng. H. (1985). In *Methods in enzymology*, Vol. 114, (ed. H. W. Wyckoff, C. H. W. Hirs, and S. N. Timasheff), pp. 452–72. Academic, Orlando.
78. Kabsch, W. (1988). *Journal of Applied Crystallography*, **21**, 916–24.
79. Giessen, B. C. and Gordon, G. E. (1968). *Science*, **159**, 973–75.
80. Klug, H. P. and Alexander, L. E. (1974). *X-ray diffraction procedures for polycrystalline and amorphous materials*. Wiley, New York.
81. Bisch, D. L. and Post, J. E. (ed.) (1989). *Modern powder diffraction*, Reviews in Mineralogy, Vol. 20. Mineralogical Society of America, Washington.
82. Debye, P. and Scherrer, P. (1917). *Physikalische Zeitschrift*, **18**, 291–301.
83. Seemann, H. (1919). *Annalen der Physik*, **59**, 455–64.
84. Bohlin, H. (1920). *Annalen der Physik*, **61**, 421–39.
85. Guinier, A. (1956). *Theorie et technique de la radiocristallographie*, pp. 185–202. Dunod, Paris.
86. Gandolfi, G. (1967). *Mineralogica Petrographica Acta*, **13**, 67–74.
87. Parrish, W. (ed.) (1962). *Advances in x-ray diffractometry and x-ray spectrography*. Centrex, Eindhoven.
88. Parrish, W. and Mack, M. (1967). *Acta Crystallographica*, **23**, 687–92.
89. Mack, M. and Parrish, W. (1967). *Acta Crystallographica*, **23**, 693–700.
90. Ortendahl, D., Perez-Mendez, V., Stoker, J., and Beyermann, W. (1978). *Nuclear Instrumentation and Methods*, **156**, 53–6.
91. Izumi, T. (1980). *Nuclear Instrumentation and Methods*, **177**, 405–9.
92. Ballon, J., Comparat, V., and Pouxe, J. (1983). *Nuclear Instrumentation and Methods*, **217**, 213–16.
93. Hanawalt, J. D. and Rinn, H. W. (1986). *Powder Diffraction*, **1**, 2–6.
94. McMurdie, H. F., Morris, M. C., Evans, E. H., Paretzkin, B., Wong-Ng, W., and Hubbart, C. R. (1986). *Powder Diffraction*, **1**, 40–3.
95. Hanawalt, J. D. (1986). *Powder Diffraction*, **1**, 7–13.
96. Frevel, L. K. (1965). *Analytical Chemistry*, **37**, 471–82.
97. Sai-Zhu, Z., Li-Jun, C., and Xin-Xing, C. (1983). *Journal of Applied Crystallography*, **16**, 150–4.
98. Alexander, L. and Klug, H. P. (1948). *Analytical Chemistry*, **20**, 886–9.
99. Chung, F. H. (1974). *Journal of Applied Crystallography*, **7**, 519–31.
100. Zevin, L. S., and Zevin, Sh. L. (1989). *Powder Diffraction*, **4**, 196–200.
101. Rietveld, H. M. (1969). *Journal of Applied Crystallography*, **2**, 65–71.
102. Albinati, A. and Willis, B. T. M. (1982). *Journal of Applied Crystallography*, **15**, 361–74.
103. Hill, R. J. and Madsen, I. C. (1987). *Powder Diffraction*, **2**, 146–62.
104. Hill, R. J. and Howard, C. J. (1987). *Journal of Applied Crystallography*, **20**, 467–74.
105. Buerger, M. J. (1960). *Crystal structure analysis*, Ch. 7. Wiley, New York.
106. Arndt, U. W. and Willis, B. T. M. (1966). *Single crystal diffractometry*, Ch. 11. Cambridge University Press.
107. Lipson, H. (1959). *International tables for x-ray crystallography*, Vol. II, (ed. J. S. Kasper and K. Lonsdale), pp. 237–90. Kynock, Birmingham.

108. Azaroff, L. V. (1955). *Acta Crystallographica,* **8,** 701–4.
109. Miyake, S., Togawa, S., and Hosoya, S. (1964). *Acta Crystallographica,* **17,** 1083–84.
110. Hope, H. (1971). *Acta Crystallographica,* **A27,** 392–3.
111. Kerr, K. A. and Ashmore, J. P. (1974). *Acta Crystallographica,* **A30,** 176–9.
112. Kahn, R., Fourme, R., Gadet, A., Janin, J., Dumas, C., and Andre, D. (1982). *Journal of Applied Crystallography,* **15,** 330–7.
113. Lipson, H. (1959). *International tables for x-ray crystallography,* Vol. II, (ed. J. S. Kasper and K. Lonsdale), pp. 291–315. Kynock, Birmingham.
114. Cordero-Borboa, A. E. (1985). *Journal of Physics E: Scientific Instruments,* **18,** 393–5.
115. De Meulenaer, J. and Tompa, M. (1965). *Acta Crystallographica,* **19,** 1014–18.
116. Bartels, K. (1977). In *The rotation method in crystallography* (ed. U. W. Arndt and A. J. Wonacott), pp. 153–71. North Holland, Amsterdam.
117. North, A. C. T., Phillips, D. C., and Mathews, F. S. (1968). *Acta Crystallographica,* **A24,** 351–9.
118. Kopfmann, G. and Huber, R. (1968). *Acta Crystallographica,* **A24,** 348–51.
119. Flack, H. D. (1974). *Acta Crystallographica,* **A30,** 569–73.
120. Katayama, C., Sakabe, N., and Sakabe, K. (1972). *Acta Crystallographica,* **A28,** 293–5.
121. Lee, B. and Ruble, J. R. (1977). *Acta Crystallographica,* **A33,** 629–41.
122. Takusagawa, F. (1987). *Journal of Applied Crystallography,* **20,** 243–5.
123. Walker, N. and Stuart, D. (1983). *Acta Crystallographica,* **A39,** 158–66.
124. Abrahams, S. C. (1973). *Acta Crystallographica,* **A29,** 111–16.
125. Abrahams, S. C. and Marsh, P. (1987). *Acta Crystallographica,* **A43,** 265–9.
126. Ibers, J. A. (1969). *Acta Crystallographica,* **B25,** 1667–8.
127. Blake, C. C. F. and Phillips, D. C. (1962). In *Biological effects of ionizing radiations at the molecular level,* pp. 183–91. IAEA, Vienna.
128. Hendrickson, W. A. (1976). *Journal of Molecular Biology,* **106,** 889–91.
129. Sygusch, J. and Allaire, M. (1988). *Acta Crystallographica,* **A44,** 443–8.
130. Rudman, R. (1976). *Low temperature x-ray diffraction.* Plenum, New York.
131. Hope, H. (1988). *Acta Crystallographica,* **B44,** 22–6.
132. Hope, H. (1990). *Annual Review of Biophysics and Biophysical Chemistry,* **19,** 107–26.
133. Hamilton, W. C., Rollett, J. S., and Sparks, R. A. (1965). *Acta Crystallographica,* **18,** 129–30.
134. Fox, G. C. and Holmes, K. C. (1966). *Acta Crystallographica,* **20,** 886–91.
135. Rae, A. D. (1965). *Acta Crystallographica,* **19,** 683–4.
136. Richards, F. M. and Berger, J. E. (1962). In *International tables for x-ray crystallography,* Vol. III (ed. C. H. MacGillavry and G. D. Rieck), pp. 17–20. Kynock, Birmingham.
137. Blundell, T. L. and Johnson, L. N. (1976). *Protein crystallography,* pp. 104–6. Academic, New York.
138. Westbrook, E. M. (1985). In *Methods in Enzymology,* Vol. 114, (ed. H. W. Wyckoff, C. H. W. Hirs, and S. N. Timasheff), pp. 187–96. Academic, Orlando.
139. Matthews, B. W. (1974). *Journal of Molecular Biology,* **82,** 513–26.

Solution and refinement of crystal structures

5

DAVIDE VITERBO

Introduction

The goal of a structural analysis is to obtain the distribution of atomic electron density in the unit cell (in practice the atomic positions) starting from the diffraction data. As already observed in Chapter 3 (p. 169) it is not possible to reach this goal in a unique and automatic way, because from the experimental data only the magnitudes, but not the phases, of the structure factors may be obtained. Therefore, in order to compute the electron density by means of eqn (3.45), we must somehow derive the missing information. In this chapter we shall analyse the most important methods commonly used to solve the **phase problem.**

The problem must in principle have a solution (even if not necessarily unique), since the measured intensities are proportional to the squares of the structure factors, which may be expressed as†

$$|F_{\boldsymbol{h}}|^2 = \sum_{j=1}^{N} f_j^2 + 2 \sum_{j>k=1}^{N} f_j f_k \cos 2\pi \boldsymbol{h} \cdot (\boldsymbol{r}_j - \boldsymbol{r}_k). \tag{5.1}$$

In these relationships, the number of which is equal to the number of observed reflections, the terms f and $\boldsymbol{h}$ are known quantities, while the atomic position vectors $\boldsymbol{r}$ are unknown. Unfortunately these unknowns appear as argument of trigonometric functions and the solution of systems of non-linear equations can not be obtained in any analytical way, even though the number of relationships greatly exceeds the number of unknowns (there may be up to 100 reflections per atom).

It is noteworthy that, as in the case of other physical experiments (cf. Chapter 1, p. 17), the intensities only depend on the interatomic vectors, which are independent of the arbitrarily chosen reference system.

It is theoretically possible that the solution of equations such as (5.1) is not unique. Indeed, more than one set of positional vectors $\boldsymbol{r}$ may correspond to the same set of intensities; these are called **homometric**

† Throughout this chapter the convention of using capital letters for the 1×3 matrix of the reciprocal lattice indices will not be followed, in order to conform with the notation generally used in the literature on Patterson and direct methods, where generally lower case letters indicate the general reciprocal vectors as well as the matrix of their components. With this notation no ambiguity should arise between, for instance, the scalar product of the reciprocal vectors $\boldsymbol{h}$ by the direct position vector $\boldsymbol{r}$, indicated by $\boldsymbol{h} \cdot \boldsymbol{r}$, and the product of the indices matrix $\boldsymbol{h}$ by the rotation matrix $\boldsymbol{R}$ of a symmetry operator, indicated by $\tilde{\boldsymbol{h}}\boldsymbol{R}$ (the transpose sign is usually omitted).

sets.[1] Nevertheless, in practice the constraint that the solution must obey stereochemical rules makes it extremely unlikely that more than one homometric set is chemically acceptable.

The possibility of solving a system of non-linear equations relies on that of obtaining a first approximate solution, constituting the so called **initial structural model**. This can then be refined until the best agreement with the experimental data is achieved.

Before considering the different methods employed to define an initial model, it is therefore necessary to establish the criteria which allow us to assess its correctness. From the M positional vectors of the model the structure factors

$$F_{\boldsymbol{h}}^{c} = \sum_{m=1}^{M} f_m \exp(2\pi i \boldsymbol{h} \cdot \boldsymbol{r}_m) \tag{5.2}$$

may be computed. A good agreement between the $|F_{\boldsymbol{h}}^{c}|$s and the observed moduli $|F_{\boldsymbol{h}}^{o}|$, obtained directly from the intensities, will indicate a correct model. The most common parameter used to express this agreement is the ***R* index** (also called agreement index or residual)

$$R = \frac{\sum_{\boldsymbol{h}} \left\| F_{\boldsymbol{h}}^{o}| - K\,|F_{\boldsymbol{h}}^{c}| \right|}{\sum_{\boldsymbol{h}} |F_{\boldsymbol{h}}^{o}|} = \frac{\sum_{\boldsymbol{h}} \Delta F_{\boldsymbol{h}}}{\sum_{\boldsymbol{h}} |F_{\boldsymbol{h}}^{o}|} \tag{5.3}$$

where K is a scale factor bringing $|F_{\boldsymbol{h}}^{c}|$ on the same scale of $|F_{\boldsymbol{h}}^{o}|$, obtained as $K = \sum_{\boldsymbol{h}} |F_{\boldsymbol{h}}^{o}| / \sum_{\boldsymbol{h}} |F_{\boldsymbol{h}}^{c}|$.

In the case of equal atom structures, the R value for totally random atomic positions has been statistically evaluated to be 0.83 for centrosymmetric structures and 0.59 for non-centrosymmetric structures.[2] Structural models yielding values of the R index lower than these extreme values may be considered as plausible initial guesses to start the refinement process. In general a model with $R \leq 0.5$ if centrosymmetric or 0.4 if non-centrosymmetric will be a good starting point. It may also happen that the postulated model contains errors which can not be corrected by the following refinement, and therefore it does not converge to the correct solution. A quite frequent case is represented by crystals containing one or more solvent molecules; if the presence of these molecules in the cell is overlooked, then the initial model will be incomplete and the index R will not decrease below 0.15–0.25, unless the positions of the solvent molecules are taken into account. We will consider the behaviour of the R index in some more detail in the paragraph on structure refinement.

Historically, the first crystal structures were solved by **trial and error** methods, consisting in a systematic trial of all structural hypotheses compatible with the known physical and chemical properties of the considered crystal. These methods require a great effort, ingenuity, and skill and can only be used with simple structures. They are seldom used today and for this reason they will not be treated (for a comprehensive account the reader is referred to Lipson and Cochran[3]). Only the methods based on the use of the Patterson function and the so-called direct methods will be considered here, while those using isomorphous replacement and anomalous dispersion, mainly used in solving biological macromolecular structures, will be dealt with in Chapter 8.

Before entering into the discussion of these methods, a short digression will be necessary to describe some important results derivable from a statistical analysis of the observed intensities.

Statistical analysis of structure factor amplitudes

The statistical analysis of the observed structure factor moduli gives very useful indications on the presence or not of those symmetry elements which do not give rise to systematic absences. It also allows us to obtain an estimate both of the scale factor by which one has to multiply the measured data to scale them to their absolute value, and of the temperature factor. Finally it is a basic step in the calculation of **unitary and normalized structure factors**, which will be seen to be useful quantities, especially in connection with the use of direct methods.

The derivation of the theoretical probability distributions of structure factor amplitudes is given in Appendix 5.A. They are derived[4,5] under the hypothesis that the atomic positions are random variables with uniform distribution throughout the unit cell (i.e. all points in the cell have the same probability of hosting an atom).

Their form depends on whether or not the crystal possesses an inversion center (**centric** distribution or **acentric** distribution respectively). For the two cases we have

$$P_{\bar{1}}(|F|) = \frac{\sqrt{2}}{\sqrt{\pi\Sigma}} \exp(-|F|^2/2\Sigma) \quad \text{(centric)}, \tag{5.4}$$

$$P_{1}(|F|) = \frac{2\,|F|}{\Sigma} \exp(-|F|^2/\Sigma) \quad \text{(acentric)}, \tag{5.5}$$

where

$$\Sigma = \sum_{j=1}^{N} f_j^2 = \frac{1}{\varepsilon} \langle |F|^2 \rangle,$$

and ε is a factor (defined in Appendix 5.A) depending on the specific symmetry of the weighted reciprocal lattice. Σ, being a function of the atomic scattering factors, will not have a constant value in all the reciprocal space, but it will decrease for increasing values of $\sin\theta/\lambda$. This is a problem in the practical use of distributions (5.4) and (5.5). In order to overcome this problem the **unitary structure factors**

$$|U_{\boldsymbol{h}}| = \frac{|F_{\boldsymbol{h}}|}{\sum_{j=1}^{N} f_j} \tag{5.6}$$

and the **normalized structure factors**

$$|E_{\boldsymbol{h}}| = \frac{|F_{\boldsymbol{h}}|}{\sqrt{\langle |F|^2 \rangle}} = \frac{|F_{\boldsymbol{h}}|}{\sqrt{\varepsilon\Sigma}} \tag{5.7}$$

are introduced. In the same way by which $\langle |F|^2 \rangle$ was defined, we have

$$\langle |U_{\boldsymbol{h}}|^2 \rangle = \varepsilon \sum_{j=1}^{N} \left[\frac{f_i}{\sum_{j=1}^{N} f_j} \right]^2 = \varepsilon \Sigma_U \tag{5.8}$$

and in the case of all equal atoms $\Sigma_U = 1/N$. From (5.7) we can immediately derive

$$\langle |E_{\boldsymbol{h}}|^2 \rangle = 1. \tag{5.9}$$

Both U and E are independent of the scattering angle θ and correspond to idealized point atom structures.

The centric and acentric distributions, when expressed in terms of normalized structure factors, become

$$P_{\bar{1}}(|E|) = \sqrt{\frac{2}{\pi}} \exp(-|E|^2/2) \tag{5.10}$$

$$P_1(|E|) = 2\,|E| \exp(-|E|^2) \tag{5.11}$$

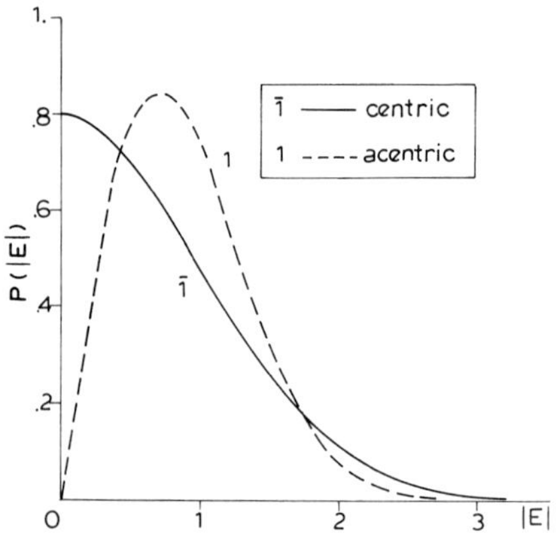

Fig. 5.1. Probability distributions of a normalized structure amplitude for centrosymmetric (*centric*) and for non-centrosymmetric (*acentric*) structures.

which are completely independent of the structure complexity. They are represented in the curves of Fig. 5.1. The two curves are quite different; the centric distribution foresees a higher percentage of reflections with extreme intensity (weak and strong reflections), with respect to the acentric one, showing a maximum corresponding to intermediate values of the intensities. The comparison of the distribution of the observed amplitudes with the theoretical distributions of Fig. 5.1 will allow us to establish the presence or not of the inversion centre in the crystal under consideration.

From the distributions it is possible to derive the theoretical mean values of several functions of $|E|$ or the theoretical percentage of $|E| > t$. Some of these values are reported in Table 5.1 and may be compared with the corresponding experimental values to verify whether the structure is centrosymmetric or not.

So far we have implicitly assumed that the observed structure factor moduli are on an absolute scale, but in general the values of $|F_{\boldsymbol{h}}|^2_{\text{obs}}$, obtained from the intensities $I_{\boldsymbol{h}}$ through eqn (3.41), are on a relative scale. Assuming that the thermal motion is isotropic and equal for all the atoms, we may write

$$|F_{\boldsymbol{h}}|^2_{\text{obs}} = K' \,|{}^{\circ}F_{\boldsymbol{h}}|^2 \exp(-2Bs^2) \tag{5.12}$$

where K' (reciprocal of the scale factor used in eqn (5.3)) is the scale factor, $|{}^{\circ}F_{\boldsymbol{h}}|$ is the structure amplitude in absolute scale for atoms at rest, B is the overall isotropic temperature factor, and $s = \sin\theta/\lambda$.

Table 5.1. Theoretical values of some functions of $|E|$ obtained from the centric (5.10) and acentric (5.11) distributions and their comparison with the corresponding experimental values for the AZOS structure

	Theoretical		
	Centrosymmetric	**Non-centrosymmetric**	**Experimental for AZOS**
$\langle \lvert E\rvert^2 \rangle$	1.000	1.000	1.000
$\langle \lvert E^2 - 1\rvert \rangle$	0.968	0.736	0.942
$\langle \lvert E\rvert \rangle$	0.798	0.886	0.797
% $\lvert E\rvert > 1.0$	31.7	36.8	32.7
% $\lvert E\rvert > 2.0$	4.6	1.8	3.8
% $\lvert E\rvert > 3.0$	0.3	0.01	0.4

Using the results of the statistical analysis of the intensities, Wilson[5] proposed a method to derive the values of K' and B.

Let us consider a set of observed intensities falling within a restricted range of s, such that within this range the decrease of the fs with s may be neglected. The average value of both sides of (5.12) will be

$$\langle |F_{\text{obs}}|^2 \rangle_s = K' \langle |{}^{\circ}F|^2 \rangle_s \exp(-2B\langle s^2 \rangle) = K' \Sigma_s \exp(-2B\langle s^2 \rangle) \quad (5.13)$$

from which

$$\ln\left(\frac{\langle |F_{\text{obs}}|^2 \rangle}{\Sigma_s}\right) = \ln K' - 2B\langle s^2 \rangle \quad (5.14)$$

where $\langle s^2 \rangle$ is the mean value of $\sin^2\theta/\lambda^2$ in the considered interval and

$$\Sigma_s = \sum_{j=1}^{N} {}^{\circ}f_j^2$$

is computed using the tabulated values of the atomic scattering factors for atoms at rest for $s = \sqrt{\langle s^2 \rangle}$. Dividing the reciprocal lattice into several intervals of s, (5.14) tells us that a linear relation will exist between $\ln(\langle |F_{\text{obs}}|^2 \rangle_s / \Sigma_s)$ and $\langle s^2 \rangle$ and that a plot of these values obtained from the experimental data can be interpolated by the best straight line passing through them. The intercept of the line on the vertical axis will give us $\ln K'$ and its slope the value of $2B$.

Figure 5.2 is an example of such **Wilson plot** for a typical small organic structure (p-carboxyphenylazoxycyanide-dimethyl sulphoxide,[6] AZOS hereinafter); the numerical values of the terms appearing in eqn (5.14), obtained from 1908 observed reflections, are given in Table 5.2. The main reason for the deviations of the experimental points from the straight line is the breakdown of the condition of equiprobability of all atomic positions, assumed in deriving (5.4) and (5.5); in fact, the presence of structural

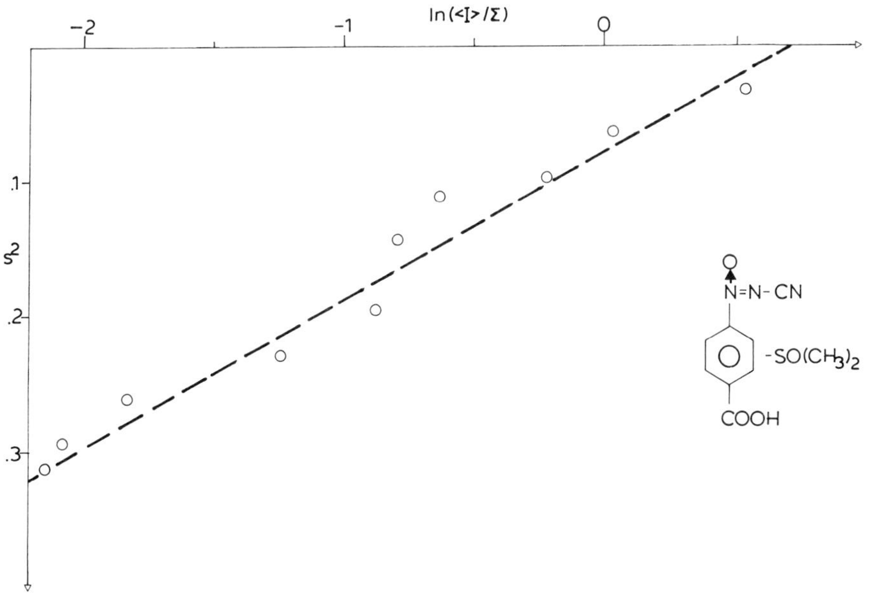

Fig. 5.2. Wilson plot of AZOS together with the formula of the compound.

Table 5.2. Numerical values of the different terms in eqn (5.14) employed to obtain the Wilson plot of AZOS shown in Fig. 5.2

Interval	Limits s^2	Reflection no.	$\langle s^2 \rangle$	$\langle \|F^2\| \rangle_s$	Σ_s	$\langle \|F\|^2 \rangle_s / \Sigma_s$	$\ln [\langle \|F\| \rangle_s^2 / \Sigma_s]$
1	0.0000–0.0655	82	0.0327	3736.8	2199.9	1.6986	0.5298
2	0.0327–0.0982	124	0.0655	1359.5	1322.9	1.0277	0.0273
3	0.0655–0.1309	156	0.0982	821.2	1020.9	0.8044	−0.2177
4	0.0982–0.1637	182	0.1309	420.2	791.3	0.5310	−0.6329
5	0.1309–0.1964	210	0.1637	292.8	646.5	0.4529	−0.7922
6	0.1637–0.2292	231	0.1964	234.4	562.7	0.4166	−0.8756
7	0.1964–0.2619	248	0.2292	146.2	507.8	0.2879	−1.2451
8	0.2292–0.2946	269	0.2619	73.0	450.9	0.1619	−1.8209
9	0.2619–0.3274	272	0.2946	52.0	410.4	0.1267	−2.0663
10	0.2946–0.3274	134	0.3110	46.3	392.7	0.1179	−2.1378

regularities, such as the phenyl hexagon in AZOS, is in contrast with this assumption. The broken line in Fig. 5.2, obtained by least-square fitting, has a slope of -9.20 and an intercept of 0.72. From these values we can derive that $B = 4.60$ Å^2 and $K' = 2.05$. The normalized structure factors can then be calculated by means of the relation

$$|E_{\boldsymbol{h}}| = \left(\frac{|F_{\boldsymbol{h}}|^2_{\text{obs}}}{K' \exp(-2Bs^2)\varepsilon \sum_{j=1}^{N} {}^{\circ}f_j^2} \right)^{1/2}. \tag{5.15}$$

The last column in Table 5.1 gives the experimental values of several statistical indicators based on the $|E|$ values of AZOS (space group $P2_1/a$), which confirm the presence of an inversion centre.

The Patterson function and its use

In Appendix 3.A (p. 182) the Patterson function was defined as the self-convolution of the electron density $\rho(\boldsymbol{r})$. According to (3.A.38) we have

$$P(\boldsymbol{u}) = \rho(\boldsymbol{r}) * \rho(-\boldsymbol{r}) = \int_v \rho(\boldsymbol{r})\rho(\boldsymbol{r}+\boldsymbol{u})\, \mathrm{d}\boldsymbol{r}. \tag{5.16}$$

We have shown (see (3.A.39)) that the Fourier transform of $P(\boldsymbol{u})$ is $|F(\boldsymbol{r}^*)|^2$ (in symbols $|F(\boldsymbol{r}^*)|^2 = \mathrm{T}[P(\boldsymbol{u})]$) and vice versa

$$P(\boldsymbol{u}) = \mathrm{T}^{-1}[|F(\boldsymbol{r}^*)|^2] = \int_{v^*} |F(\boldsymbol{r}^*)|^2 \exp(-2\pi i \boldsymbol{r}^* \cdot \boldsymbol{u})\, \mathrm{d}\boldsymbol{r}^*$$

$$= \frac{1}{V} \sum_{\boldsymbol{h}} |F_{\boldsymbol{h}}|^2 \exp(-2\pi i \boldsymbol{h} \cdot \boldsymbol{u}).$$

Since $|F_{\boldsymbol{h}}| = |F_{-\boldsymbol{h}}|$, we may write

$$P(\boldsymbol{u}) = \frac{1}{V} \sum_{\boldsymbol{h}} |F_{\boldsymbol{h}}|^2 \cos 2\pi \boldsymbol{h} \cdot \boldsymbol{u} \tag{5.17}$$

and then $P(\boldsymbol{u}) = P(-\boldsymbol{u})$, i.e. the Patterson function is always centrosymmetric even when $\rho(\boldsymbol{r})$ is not. This is in agreement with the deductions of p.

176 concerning the centrosymmetric nature of all functions with real Fourier transform. Since $|F_{\boldsymbol{h}}|^2$ depend on the interatomic vectors [cf. eqn (5.1)] we may expect that also $P(\boldsymbol{u})$ will contain information on these quantities. This can be verified starting from the definition (5.16). Let us, for simplicity, suppose we have an idealized structure made up of n point atoms with an associated weight equal to their atomic number (Fig. 5.3). The integral in (5.16) then becomes a summation over the n points, and

$$P(\boldsymbol{u}) = \sum_{j=1}^{n} \rho(\boldsymbol{r}_j)\rho(\boldsymbol{r}_j + \boldsymbol{u}). \tag{5.18}$$

In order to derive $P(\boldsymbol{u})$, all the atoms of the original structure $\rho(\boldsymbol{r}_j)$ are shifted by a fixed vector $\boldsymbol{u}$ to obtain the corresponding $\rho(\boldsymbol{r}_j + \boldsymbol{u})$, then the products $\rho(\boldsymbol{r}_j)\rho(\boldsymbol{r}_j + \boldsymbol{u})$ are performed and finally all these contributions are summed; the products will be non-zero only when $\rho(\boldsymbol{r}_j + \boldsymbol{u}) \neq 0$, that is when a point in the translated structure (broken lines in Fig. 5.3) coincides with an atom of the original structure. This condition is verified only when the vector $\boldsymbol{u}$ coincides with an interatomic vector (in Fig. 5.3(a) $\boldsymbol{u}$ coincides with the vector 2–4), while for a general $\boldsymbol{u}$ (Fig. 5.3(b)) all point of the translated image fall into regions where $\rho(\boldsymbol{r})$ is zero. In the first case the value of $P(\boldsymbol{u})$ will be proportional to the product of the weights of the two superposed atoms, in the second $P(\boldsymbol{u}) = 0$. In Fig. 5.3 a higher weight has been given to atom 1 (heavier atom) and in 5.3(c) the case of a vector $\boldsymbol{u}$ coinciding with the distance 1–4 is represented; it will be $P(\boldsymbol{u}_{1,4}) > P(\boldsymbol{u}_{2,4})$, both being single Patterson peaks as each corresponds to only one interatomic vector. Finally in 5.3(d) the case of a vector $\boldsymbol{u}$ coinciding with two parallel interatomic vectors of equal length is illustrated; two terms will contribute to the summation in (5.18) and $P(\boldsymbol{u})$ becomes twice as large as $P(\boldsymbol{u}_{2,4})$ and it is said to have a multiplicity of two. Let us, for instance, suppose that atom 1 is a sulphur and the others are carbons; we will then have: $P(\boldsymbol{u}_{2,4}) = 6 \times 6 = 36$, $P(\boldsymbol{u}_{1,4}) = 6 \times 16 = 96$ and, with reference to Fig. 5.3(d), $P(\boldsymbol{u}) = 2 \times (6 \times 6) = 72$.

From what we have seen so far it follows that the Patterson function will have maxima corresponding to all possible interatomic vectors within the unit cell; the height of each peak will be proportional to the product of the atomic numbers of the atoms connected by the vector $\boldsymbol{u}$, multiplied by the multiplicity of the same vector.

This concept can be further clarified by considering Fig. 5.4, where in 5.4(a) a set of $N = 5$ points is represented, while in 5.4(b) the corresponding distribution of interatomic vectors, and in 5.4(c) the same set of vectors, after they have been translated to a common origin, are shown; the last corresponds to the distribution of peaks in the Patterson function.

The Patterson function will have the same periodicity as the electron density and therefore the size of the unit cell will be identical. On the other hand, the number of peaks in the Patterson function is much greater than that in $\rho(\boldsymbol{r})$; given N atoms in the cell they will give rise to N^2 peaks in $P(\boldsymbol{u})$, N of which will superpose on a single peak at the origin (they correspond to the N zero distances of each atom with itself), while the remaining $N(N-1)$ are distributed over the cell. This higher density of peaks becomes a more serious problem for real structures with non-point-like atoms. In fact the Patterson peaks are wider than the maxima in an electron density map. As illustrated in the one-dimensional example of Fig.

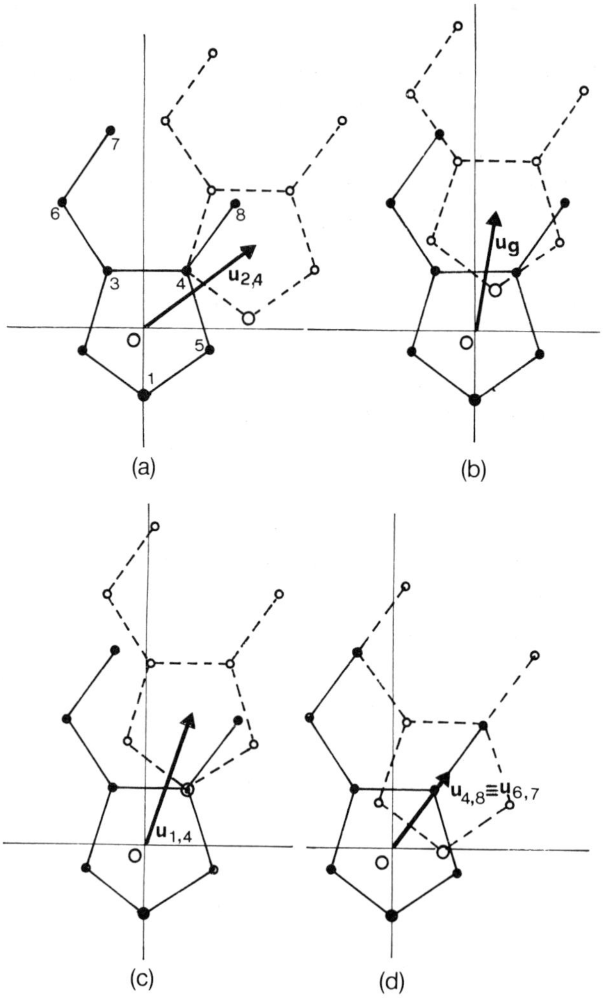

Fig. 5.3. Schemes for the construction of the Patterson function of a point-atom structure, in which each atom has a weight equal to its atomic number: (a) two images separated by a vector coinciding with the interatomic vector between atoms 2 and 4 of equal weight; (b) two images separated by a vector not coinciding with any interatomic vector; (c) two images separated by a vector coinciding with the interatomic vector between atoms 1 and 4 of different weight; (d) two images separated by a vector coinciding with the two parallel interatomic vectors of equal length between atoms 4, 8 and 6, 7.

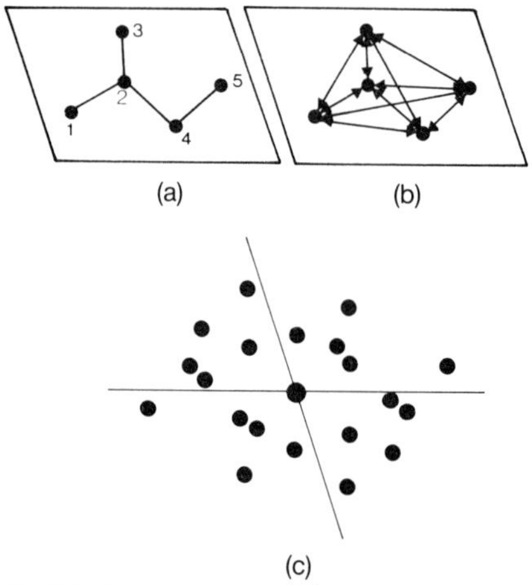

Fig. 5.4. (a) Scheme of a molecule formed by five point atoms; (b) corresponding representation of all possible interatomic vectors; (c) Patterson function obtained by translating all vectors in (b) to a common origin.

5.5, because of the non-zero width of the peaks in $\rho(x)$, the Patterson peaks will have a width twice as large.

For these reasons the Patterson map of a structure, with even a moderate number of atoms, may appear as an almost featureless distribution of vector density. To overcome this problem it is convenient to employ a **sharpening** procedure, consisting in computing the Patterson function with coefficients $|E_{\boldsymbol{h}}|^2$ or better $|F_{\boldsymbol{h}}E_{\boldsymbol{h}}|$. In fact the normalized structure factors correspond to a point-atom structure with no decrease of the atomic scattering factor with increasing $\sin\theta/\lambda$. The $|F_{\boldsymbol{h}}E_{\boldsymbol{h}}|$ coefficients are more convenient, because over-sharpening is sensitive to the series truncation errors (cf. Fig. 3.16) and may produce spurious peaks or a down-scaling of correct peaks in the map.

It is also possible to eliminate the origin peak, which may obscure some short vectors, by subtracting from the coefficients the terms corresponding to the interaction of each atom with itself (i.e. the first term in the right-hand side of (5.1)); the coefficient in the series (5.17) will be

$$|F'_{\boldsymbol{h}}|^2 = |F_{\boldsymbol{h}}|^2 - \sum_{j=1}^{N} f_j^2 \qquad (5.19)$$

where, of course, the $|F_{\mathbf{h}}|^2$s must be in an absolute scale. Figure 5.6 shows the comparison between a normal and a sharpened Patterson section.

From the previous considerations about the height of the Patterson peaks, we can infer that the map will have prominent maxima when:

1. The structure contains a limited number of *heavy* atoms (i.e. with an atomic number Z_p significantly higher than that of the other lighter atoms, Z_l), giving rise to three types of peaks:

Vectors	Height
heavy atom–heavy atom	Z_pZ_p very high
heavy atom–light atom	Z_pZ_l intermediate
light atom–light atom	Z_lZ_l very low

2. The molecular geometry is such that it will give rise to several interatomic vectors with almost the same length and direction (e.g. systems with condensed aromatic rings); these will correspond to multiple peaks.

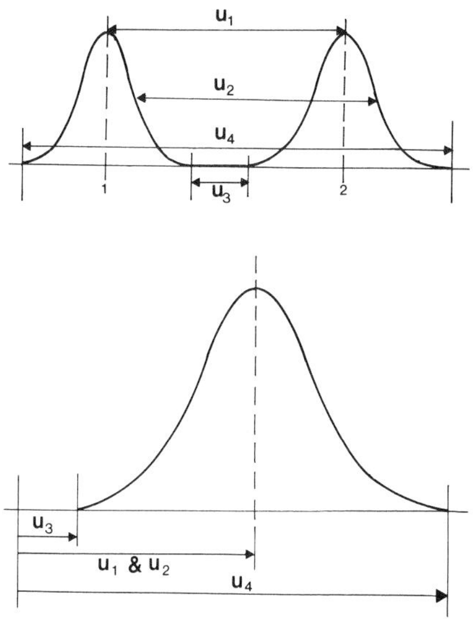

Fig. 5.5. Illustration of the widening of a Patterson maximum with respect to the electron density maxima contributing to it.

Before illustrating the methods for interpreting the Patterson function, let us first analyse how the **symmetry** of the crystal is reflected into the vector map. The following considerations apply:

1. Since all vectors are translated to a common origin, the symmetry elements in the crystal will be translated in the Patterson function so that they will all pass through the origin and at the same time they will lose their translational components. By applying these rules it can be seen that, of the 230 space groups, only 24 have a symmetry compatible with

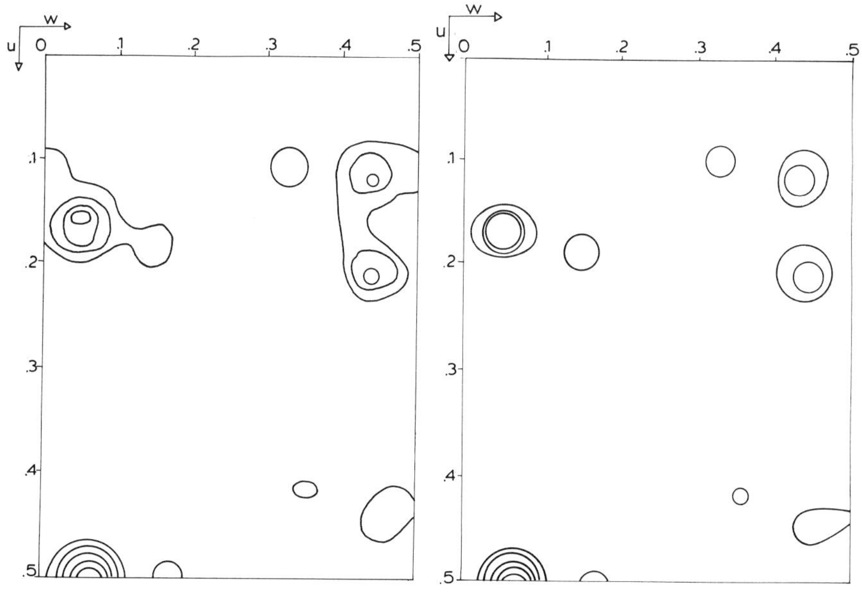

Fig. 5.6. Comparison between a normal (left) and a sharpened (right) Patterson section.

the Patterson function. Thus, for instance, we have:

Crystal space group	Patterson group
Triclinic (P1, P$\bar{1}$)	P$\bar{1}$
Primitive monoclinic (P2, P2_1, . . . , P$2_1/c$)	P2/m
Centred monoclinic (C2, C*c*, . . . , C2/*c*)	C2/m
Primitive orthorhombic (P$2_12_12_1$, . . . , P*na*2_1, . . . , P*bca* . . .)	Pmmm

2. The symmetry operators present in the crystal leave a trace in the Patterson function, consisting of particular clusterings of vector maxima on specific lines or planes of the map, called **Harker[7] lines and sections**. These are produced by vectors, with one or two constant components, between equivalent atoms related by symmetry elements other than the inversion centre. Thus, for instance, the space group P2_1, with equivalent positions: x, y, z; $\bar{x}, y+\frac{1}{2}, \bar{z}$, will have vectors between equivalent atoms located on a Harker section with coordinates $(2x, \frac{1}{2}, 2z)$. In fact

$$u = x - \bar{x} = 2x: \qquad v = y - y - \tfrac{1}{2} = -\tfrac{1}{2} \equiv \tfrac{1}{2}: \qquad w = z - \bar{z} = 2z.$$

In the space group P2, the equivalent positions x, y, z and $\bar{x}, y, \bar{z}$ give rise to maxima on the Harker section at $(2x, 0, 2z)$.

In a similar way it can be seen that the space groups Pm and Pc give rise to Harker lines with coordinates $(0, 2y, 0)$ and $(0, 2y, \frac{1}{2})$ respectively. The Harker lines and sections corresponding to some of the most common symmetry elements are listed in Table 5.3. The presence or lack of Harker lines or sections may also give valuable indications on the space group symmetry when this is not uniquely defined.

The heavy atom method

The procedures for the interpretation of the Patterson function are greatly simplified when the structure contains a limited number of **heavy atoms**. The peaks due to the vectors between these atoms will dominate the map and it is therefore usually quite an easy task to derive the positions of the heavy atoms. Then, if the atoms have a sufficiently high atomic number, the information about their coordinates represents a good initial model; in fact, the term $F_{\mathbf{h}}^{c}$ computed by relation (5.2) with the contribution of the M heavy atoms in the unit cell, generally represents the predominant contribution to the structure factor, as illustrated in Fig. 5.7.

One may then assume that the phase $\phi_{\mathbf{h}}^{c}$ of $|F_{\mathbf{h}}^{c}|$ is a good approximation of the true phase of $F_{\mathbf{h}}$ and compute an electron density map using as

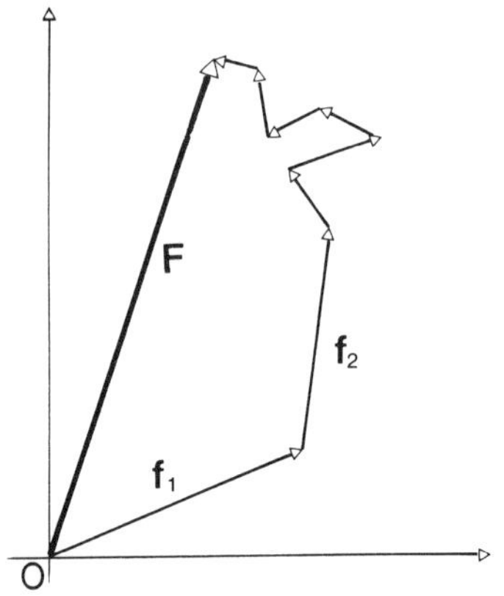

Fig. 5.7. Argand diagram in which two heavy atoms (with atomic scattering factors f_1 and f_2) and six light atoms contribute to the structure factor F; the resultant of the contributions of the two heavy atoms is quite close to F.

Table 5.3. Harker lines and sections relative to the most common symmetry elements

2 axis ∥ to **a**, **b**, **c**	$0, v, w;\ u, 0, w;\ u, v, 0$
2_1 axis ∥ to **a**, **b**, **c**	$\frac{1}{2}, v, w;\ u, \frac{1}{2}, w;\ u, v, \frac{1}{2}$
m mirror ⊥ to **a**, **b**, **c**	$u, 0, 0;\ 0, v, 0;\ 0, 0, w$
a glide ⊥ to **b**, **c**	$\frac{1}{2}, v, 0;\ \frac{1}{2}, 0, w$
b glide ⊥ to **a**, **c**	$u, \frac{1}{2}, 0;\ 0, \frac{1}{2}, w$
c glide ⊥ to **a**, **b**	$u, 0, \frac{1}{2};\ 0, v, \frac{1}{2}$

coefficients the observed amplitudes (to which all atoms in the structure will contribute) with the corresponding calculated phases ϕ_h^c. The map will not only reveal the heavy atoms but also other atoms of the structure. In the most favourable cases the structure may be completed from the first electron density map, but in general it is necessary to operate in more than one cycle by the so-called **method of Fourier synthesis recycling**. Each cycle requires the calculation of the structure factors from the coordinates of the known atoms; their phases will then be used to compute a new electron density map. If the initial model is correct, each cycle will reveal new atoms until the structure is completed.

From the previous considerations one may get the impression that it would be advantageous to have compounds containing atoms of high atomic number, but one has also to consider that their contribution to the diffracted amplitudes may became so dominant that the observed data will be almost unaffected by the contribution of the remaining light atoms. The definition of the final structure will then be rather inaccurate. It has been demonstrated empirically that the best ratio between heavy and light atoms is that for which

$$\frac{\sum Z_p^2}{\sum Z_l^2} = 1. \tag{5.20}$$

As the values of the ratio (5.20) become less than 1.0, then the interpretation of the Patterson function and the process of completing the structure become more and more difficult, but at the same time the accuracy of the refined positions of the light atoms will increase. As an example let us consider a hypothetical organic compound of formula $C_{30}H_{36}O_4X$, where X is a halogen considered as a heavy atom; then, supposing that there are two molecules in the cell, $\sum Z_l^2 = 2708$ and $\sum Z_p^2 = 578$, 2450, 5618 for X = Cl, Br, I respectively and the ratios (5.20) will be 0.21, 0.90, 2.07 respectively. Supposing that the data measured for the three derivatives are equally good, the chlorine compound will be difficult to solve, but the refined structure will be quite accurate; with bromine the solution should be quite easy with still a reasonable accuracy of the final structure, while with iodine it will be very easy to solve the structure but its accuracy will be further reduced.

In order to find the positions of the heavy atoms it is necessary to locate them with respect to the symmetry elements and to the conventional origin of the unit cell. Let as now consider some examples, assuming for the moment that the asymmetric unit only contains one heavy atom.

1. Space group $P\bar{1}$. The vector between equivalent atoms related by the inversion centre, has component $u = 2x$, $v = 2y$, $w = 2z$; once it has been localized on the map it will immediately give the heavy atom coordinates with respect to the origin chosen on the inversion centre.

2. Space group $P2_1$ (twofold screw axis parallel to b). As we have seen, the vector between equivalent heavy atoms gives rise to a peak on the Harker section at $2x, \frac{1}{2}, 2z$. From its position one may easily derive the x and z coordinates of the heavy atom; the y coordinate may be arbitrarily assigned in order to fix the origin along the twofold screw axis.† Let us

† This is correct in the process of finding a starting model formed by a heavy atom, but during the refinement a more robust way of fixing the origin should be used, as described in Chapter 2, p. 107.

assume $y = \frac{1}{4}$; we will then have two equivalent heavy atoms at positions $x, \frac{1}{4}, z$ and $\bar{x}, \frac{3}{4}, \bar{z}$, which are related by an inversion centre (this is true for any value of y, but the choice $y = \frac{1}{4}$ makes it more obvious). The structure factors computed with the contribution of the heavy atoms only will then be real quantities and the map obtained with the corresponding phases will be centrosymmetric, showing both the true structure and its mirror image (enantiomorph). It may be quite difficult to unravel one image from the other; a sufficient number of light atoms, all belonging to the same image, should be localized in order to break down the centrosymmetric nature of the subsequent Fourier syntheses. This may become easier when atomic groups of known stereochemistry or one or more atoms of intermediate atomic number (from Na to Cl) are present.

3. Space group $P2_1/c$. This very common space group will be illustrated by means of the structure of the lower-melting isomer of methyl-(phenylsulphonyl)furoxan, $C_9H_8N_2O_4S$[8]

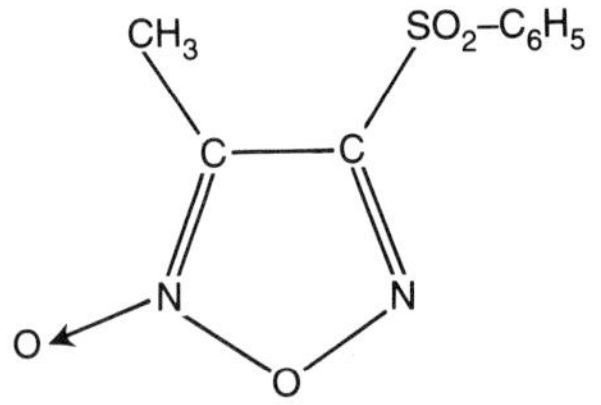

The heavy atom is the sulphur and the value of the ratio (5.20) is 0.37. The general equivalent positions are:

$$x, y, z; \quad \bar{x}, \tfrac{1}{2}+y, \tfrac{1}{2}-z; \quad \bar{x}, \bar{y}, \bar{z}; \quad x, \tfrac{1}{2}-y, \tfrac{1}{2}+z.$$

The twofold screw axis gives rise to a Harker section of type $\overline{2x}, \frac{1}{2}, \frac{1}{2}-2z$ (Fig. 5.8(a)), while the glide plane generates a Harker line at $0, \frac{1}{2}-2y, \frac{1}{2}$ (Fig. 5.8(b)). Finally the inversion centre will produce a vector between equivalent heavy atoms of components $2x$, $2y$, $2z$. In Fig. 5.8(a) the highest peak is at $u = -2x = 62.2/100$, $w = \frac{1}{2} - 2z = 26.9/100$ and then: $x = 0.189$, $z = 0.115$. In Fig. 5.8(b) there is a large peak centred at $v = \frac{1}{2} - 2y = 25.2/100$ and therefore $y = 0.124$. The coordinates of the S atom found in this way are confirmed by the presence of a fairly large peak at position $u = 2x = 0.384$, $v = 2y = 0.245$, $w = 2z = 0.230$. Despite the low value of the ratio (5.20), the electron density map, computed with the signs derived from the contribution of the S atoms only, allowed the localization of all the other non-hydrogen atoms. The final coordinates of the sulphur in the refined structure are: $x = 0.1880$, $y = 0.1249$, $z = 0.1143$ (shifted by only 0.097 Å from the original position).

4. Space group $P2_12_12_1$. The higher-melting isomer of methyl(phenyl-sulphonyl)furoxan[8]

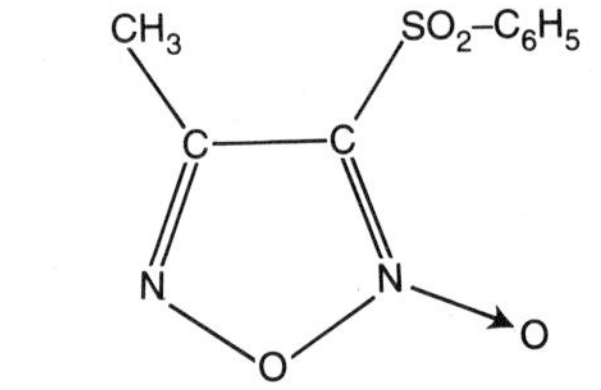

crystallizes in this common non-centrosymmetric space group.

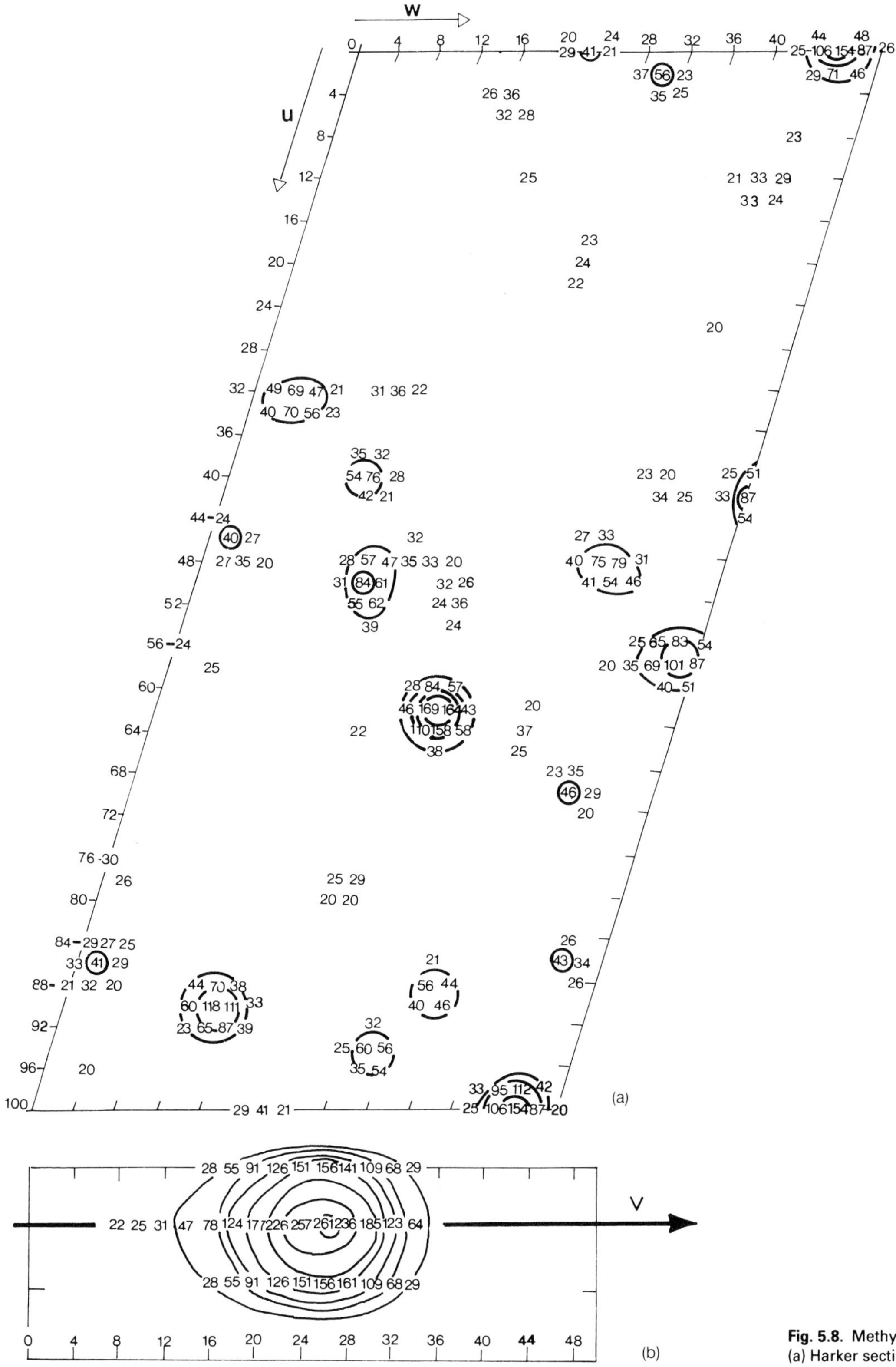

Fig. 5.8. Methyl(phenylsulphonyl)furoxan LMI: (a) Harker section $u, \frac{1}{2}, w$; (b) Harker line $0, v, \frac{1}{2}$.

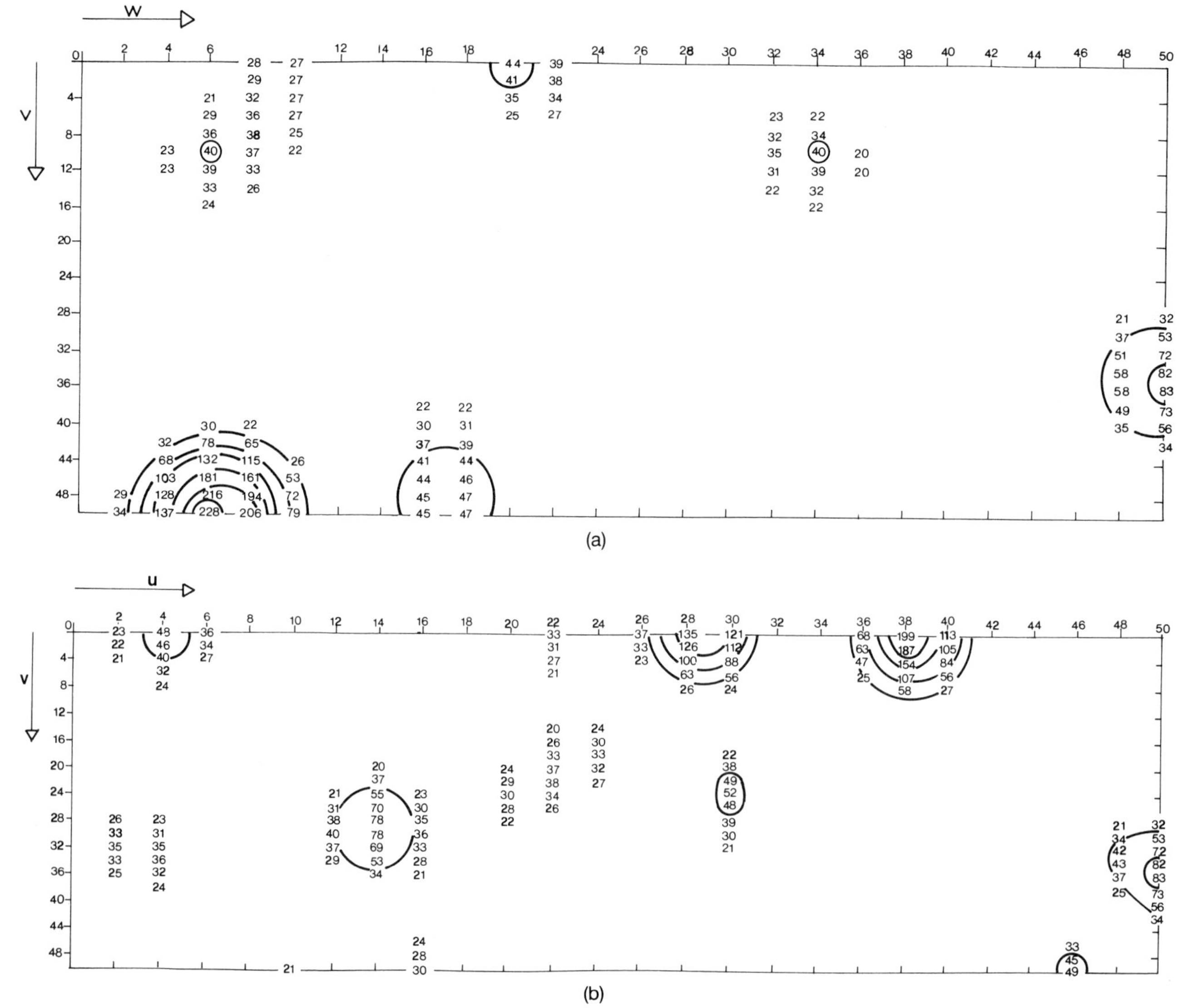

Fig. 5.9. Methyl(phenylsulphonyl)furoxan HMI: (a) Harker section $\frac{1}{2}$, *v*, *w*; (b) Harker section *u*, *v*, $\frac{1}{2}$; (c) Harker section *u*, $\frac{1}{2}$, *w*.

The general equivalent positions are

$$x,\ y,\ z;\qquad \tfrac{1}{2}-x,\ \bar{y},\ \tfrac{1}{2}+z;\qquad \tfrac{1}{2}+x,\ \tfrac{1}{2}-y,\ \bar{z};\qquad \bar{x},\ \tfrac{1}{2}+y,\ \tfrac{1}{2}-z.$$

The three mutually perpendicular twofold screw axes give rise to three Harker sections of the type $\frac{1}{2}$, $\frac{1}{2}-2y$, $2z$; $\frac{1}{2}-2x$, $2y$, $\frac{1}{2}$; $2x$, $\frac{1}{2}$, $\frac{1}{2}-2z$ (Fig. 5.9(a, b, c)). In the first the highest peak is positioned at $v=\frac{1}{2}-2y=50/100$, $w=2z=6.6/100$, giving $y=0.0$ and $z=0.033$. In the second the highest peak is at $u=\frac{1}{2}-2x=38.2/100$, $v=2y=0/100$ (in agreement with the first section) and then $x=0.059$ and $y=0.0$. In the third section the peak at $u=2x=11.8/100$ and $w=\frac{1}{2}-2z=43.2/100$, confirming the coordinates derived from the first two sections, is not the highest peak but the fourth highest. The largest peak at 50 and 6.6 is in common with the first section, because of the special value of $y=0$; the second and third peaks are

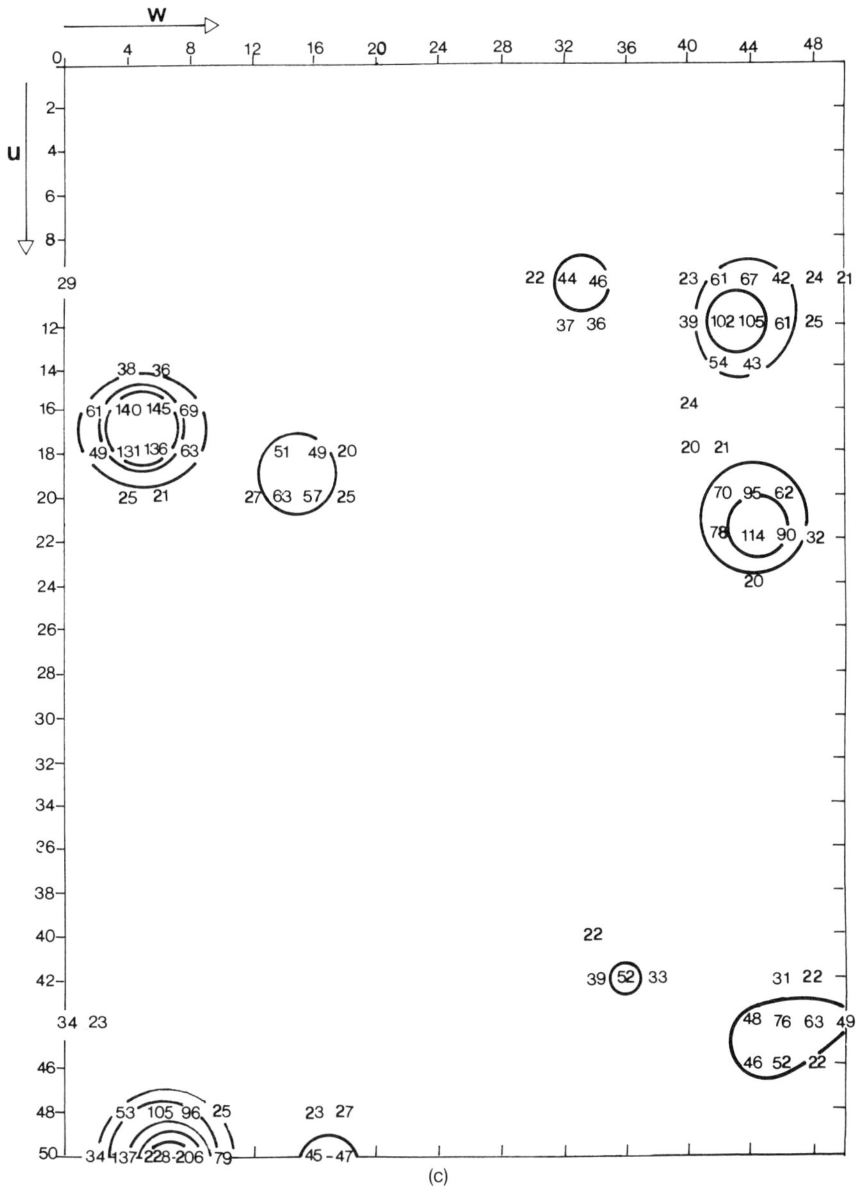

Fig. 5.9. (*Continued*)

probably due to an accidental superposition of several vectors connecting lighter atoms. Also in this case the Fourier map, computed with the phases due to the contribution of the S atoms only, revealed the whole molecule (excluding hydrogens) and the coordinates of the S in the final refined structure are $x = 0.0589$, $y = 0.0094$, $z = 0.0343$ (shifted by 0.0162 Å from the original position).

Let us now consider the case of a compound with more than one heavy atom, such as the complex $Fe_4(CO)_{11}(NOC_2H_5)(NC_2H_5)$,[9] crystallizing in the space group $P\bar{1}$ with $Z = 2$. In Table 5.4 the coordinates of the highest peaks in the Patterson map are listed in decreasing order of height. The values of the function are on an arbitrary scale and can be scaled to their absolute value by multiplying by the ratio $\sum_j Z_j^2/P(0\,0\,0) = 8240/364 = 22.6$; the height of a single Fe–Fe peak is $Z_{Fe}^2 = 676$ in absolute scale and about 30 in the relative scale. The vectors between equivalent atoms linked by an inversion centre have components $2x$, $2y$, $2z$ and correspond to single peaks; we may therefore assume that maxima 10, 11, 12, and 13 are of this type. If we choose peak 10 as Fe(1) − Fe(1) the coordinates of Fe(1) are simply obtained by dividing its components by two. Once this has been done, we are not allowed to simply divide by two the components of peaks 11, 12, 13 to obtain the coordinates of the other Fe atoms. In fact in this way we wouldn't be sure that the other three Fe atoms are referred to the same origin as the first. Indeed in the space group $P\bar{1}$ there are eight different positions of the inversion centre where the origin may be chosen. In order to resolve this ambiguity the peaks corresponding to vectors between non-equivalent Fe atoms should be used, also taking into account the periodic nature of the Patterson function. For instance, for peak 11, we must consider the eight combinations:

$$u_2, v_2, w_2; \quad 1+u_2, v_2, w_2; \quad u_2, 1+v_2, w_2; \ldots ; \quad 1+u_2, 1+v_2, 1+w_2;$$

giving for Fe(2) the coordinates referred to the eight possible origins:

$$x_2, y_2, z_2; \quad \tfrac{1}{2}+x_2, y_2, z_2; \quad x_2, \tfrac{1}{2}+y_2, z_2; \ldots ; \quad \tfrac{1}{2}+x_2, \tfrac{1}{2}+y_2, \tfrac{1}{2}+z_2.$$

Only one of these positions will account for the vectors Fe(1) − Fe(2) and Fe(1) + Fe(2) in the Patterson. In our case it will be the last position; in fact $u_2 = 1.380$, $v_2 = 1.300$, $w_2 = 1.645$ will give $x_2 = 0.690$, $y_2 = 0.650$, $z_2 = 0.822$. Peak 3 and peak 5 may then be interpreted as Fe(1) − Fe(2) and Fe(1) + Fe(2) respectively, after these Fe(2) coordinates have been translated by −1.0 along the three axes to obtain the values listed in the bottom

Table 5.4. List of the highest Patterson peaks of the complex $Fe_4(CO)_{11}(NOC_2H_5)(NC_2H_5)$ and their interpretation in terms of vectors between Fe atoms; Fe coordinates obtained from this interpretation

Peak	*u*	*v*	*w*	H	Interpretation
Origin	0	0	0	364	
1	0.080	0.260	0.110	144	Fe(1) + Fe(3) & Fe(2) + Fe(4)
2	0.020	0.000	0.710	138	Fe(1) − Fe(4) & Fe(2) − Fe(3)
3	0.490	0.560	0.470	120	Fe(3) − Fe(4) & Fe(1) − Fe(2)
4	0.070	0.275	0.400	72	Fe(3) + Fe(4)
5	0.110	0.245	0.820	65	Fe(1) + Fe(2)
6	0.485	0.455	0.810	65	Fe(1) − Fe(3)
7	0.420	0.180	0.420	63	Fe(1) + Fe(4)
8	0.405	0.290	0.360	59	Fe(2) + Fe(3)
9	0.480	0.570	0.760	55	Fe(2) − Fe(4)
10	0.400	0.195	0.700	38	Fe(1) − Fe(1)
11	0.380	0.300	0.645	33	Fe(2) − Fe(2)
12	0.430	0.285	0.075	30	Fe(3) − Fe(3)
13	0.450	0.165	0.135	27	Fe(4) − Fe(4)

	x	*y*	*z*
Fe(1)	0.200	0.097	0.350
Fe(2)	−0.310	−0.350	−0.178
Fe(3)	−0.290	−0.360	−0.463
Fe(4)	0.225	0.082	0.067

part of Table 5.4. With a similar procedure we may deduce the coordinates of the remaining two Fe atoms listed in the table.

Advanced Patterson methods

Procedures for the automatic determination of positions of heavy atoms from the Patterson function, have been developed both for small molecules[10–13] and for biological macromolecules.[14–16] The procedure implemented within the SHELXS-86 computer program[10] will be briefly described. The list of the highest peaks in the Patterson map is analysed in terms of Harker and cross peaks. Several possible solutions are given and the selection of the 'best' solution to be further processed (by the methods described on pp. 365–374) is done by means of two reliability indices (or figures of merit): an R-factor based on E values (RE) and R(Patt) which measures the agreement between the observed and predicted (from the atomic numbers) Patterson values.

We finally mention that there are other methods for interpreting the Patterson function, which may be applied also in the absence of heavy atoms. They are the so called **vector** and **superposition methods** which are described in some detail in Appendix 5.B.

Direct methods

Introduction

With the term **direct methods** are indicated those methods which try to derive the structure factor phases directly from the observed amplitudes through mathematical relationships. In general the phase and the amplitude of a wave are independent quantities and in order to understand how, in the case of X-ray diffraction, it is possible to relate these two quantities, two important properties of the electron density function should be considered:

(1) it is everywhere positive, i.e. $\rho(\boldsymbol{r}) \geqslant 0$ (**positivity**);

(2) it is composed of discrete atoms (**atomicity**).

The relation between positivity and phase values may be simply understood by just imagining the computation of $\rho(\boldsymbol{r})$ of a centrosymmetric structure as a Fourier series, first with all signs correct and then with all signs reversed: the first map will be everywhere positive or zero, while the second will be negative or zero and therefore physically unacceptable. Two pictorial examples of how positivity restricts the possible values of the phases are described in the Appendix 5.C, while a more formal explanation will be given later.

Historically, the first mathematical relationships capable of giving phase information were obtained, in the form of inequalities, by Harker and Kasper[17] in 1948 and then further developed by Karle and Hauptman[18] and by other authors. Because of their limited practical interest, they will not be treated here and the reader is referred to more specialized textbooks.[19] In 1953 Hauptman and Karle[20] established the basic concepts and the probabilistic foundations of direct methods; the great power of these methods in solving complex crystal structures had its highest recogni-

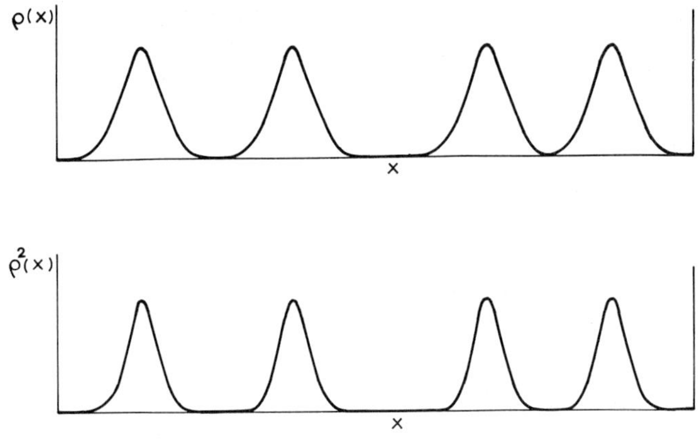

Fig. 5.10. Comparison between $\rho(x)$ and $\rho^2(x)$ for a one-dimensional structure with equal and well resolved atoms.

tion in the Nobel Prize for Chemistry conferred in 1985 on the mathematician H. Hauptman and the physicist J. Karle.

Also in 1953 Sayre,[21] using the atomicity condition, was able to derive a very important relation. He considered that for a structure formed by well resolved and almost equal atoms, the two functions $\rho(\boldsymbol{r})$ and $\rho^2(\boldsymbol{r})$ are quite similar and show maxima at the same positions. A one-dimensional example is illustrated in Fig. 5.10.

We have seen that the Fourier transform of $\rho(\boldsymbol{r})$ is $(1/V)F_{\boldsymbol{h}}$ and for the case of all equal atoms

$$F_{\boldsymbol{h}} = f_{\boldsymbol{h}} \sum_{j=1}^{N} \exp(2\pi i \boldsymbol{h} \cdot \boldsymbol{r}_j). \tag{5.21}$$

We can also define the structure factor corresponding to $\rho^2(\boldsymbol{r})$

$$G_{\boldsymbol{h}} = g_{\boldsymbol{h}} \sum_{j=1}^{N} \exp(2\pi i \boldsymbol{h} \cdot \boldsymbol{r}_j) \tag{5.22}$$

where $g_{\boldsymbol{h}}$ is the scattering factor of the 'squared' atom.

The Fourier transform of $\rho^2(\boldsymbol{r})$ is $(1/V)G_{\boldsymbol{h}}$ and, because of the convolution theorem, it will correspond to the convolution product $(1/V)F_{\boldsymbol{h}} * (1/V)F_{\boldsymbol{h}}$. Since $F_{\boldsymbol{h}}$ is a discrete function defined only at the nodes of the reciprocal lattice, the convolution integral (3.A.38) becomes a summation

$$G_{\boldsymbol{h}} = \frac{1}{V} \sum_{\boldsymbol{k}} F_{\boldsymbol{k}} F_{\boldsymbol{h}-\boldsymbol{k}}. \tag{5.23}$$

From the ratio of (5.21) and (5.22) we also have

$$F_{\boldsymbol{h}} = (f_{\boldsymbol{h}}/g_{\boldsymbol{h}})G_{\boldsymbol{h}} = \theta_{\boldsymbol{h}} G_{\boldsymbol{h}},$$

and (5.23) becomes

$$F_{\boldsymbol{h}} = \frac{\theta_{\boldsymbol{h}}}{V} \sum_{\boldsymbol{k}} F_{\boldsymbol{k}} F_{\boldsymbol{h}-\boldsymbol{k}}, \tag{5.24}$$

which is Sayre's equation, valid for both centrosymmetric and non-centrosymmetric structures. Multiplying both sides of (5.24) by $F_{-\boldsymbol{h}}$, we obtain

$$|F_{\boldsymbol{h}}|^2 = \frac{\theta_{\boldsymbol{h}}}{V} \sum_{\boldsymbol{k}} |F_{\boldsymbol{h}} F_{\boldsymbol{k}} F_{\boldsymbol{h}-\boldsymbol{k}}| \exp[i(\varphi_{-\boldsymbol{h}} + \varphi_{\boldsymbol{k}} + \varphi_{\boldsymbol{h}-\boldsymbol{k}})]. \tag{5.25}$$

For large values of $|F_{\mathbf{h}}|$ the left-hand side will be large, real, and positive. It is therefore likely that the largest terms in the sum on the right will also be real and positive. It follows that, if $|F_{\mathbf{k}}|$ and $|F_{\mathbf{h}-\mathbf{k}}|$ also have large values, it will be

$$\Phi_{\mathbf{hk}} = \varphi_{-\mathbf{h}} + \varphi_{\mathbf{k}} + \varphi_{\mathbf{h}-\mathbf{k}} \simeq 0 \tag{5.26}$$

which for centrosymmetric structures becomes

$$S(-\mathbf{h})S(\mathbf{k})S(\mathbf{h}-\mathbf{k}) \simeq + \tag{5.27}$$

where $S(\mathbf{h})$ stands for the sign of reflection $\mathbf{h}$ and the symbol $\simeq$ stays for 'probably equal'. We note that (5.27) coincides with the indication obtained in Appendix 5.C.

Relations (5.26) and (5.27) are expressed in a probabilistic form and indicate the necessity of applying probability methods to estimate their reliability. On the whole, the use of probability techniques to obtain relationships between phases and magnitudes, has proved to be the most important approach for the practical use of direct methods. We will therefore describe in more detail these methods and the procedures employed for their practical applications.

Structure invariants and semi-invariants

As already mentioned, the goal of direct methods is that of obtaining the phases directly from the observed amplitudes. These last quantities are independent of the chosen reference system, while in general the phases depend on it. From the observed amplitudes we can only obtain informations on single phases or linear combinations of phases which are independent of the choice of origin (the other features of the reference system being already defined by the choice of the space group). Since their value depends only on the structure, they are called **structure invariants** (s.i.).[20,22]

The most general s.i. is represented by the product

$$F_{\mathbf{h}_1}F_{\mathbf{h}_2}\ldots F_{\mathbf{h}_m} = |F_{\mathbf{h}_1}F_{\mathbf{h}_2}\ldots F_{\mathbf{h}_m}| \exp[\mathrm{i}(\varphi_{\mathbf{h}_1} + \varphi_{\mathbf{h}_2} + \ldots + \varphi_{\mathbf{h}_m})] \tag{5.28}$$

when

$$\mathbf{h}_1 + \mathbf{h}_2 + \ldots + \mathbf{h}_m = 0. \tag{5.29}$$

Let us show that its value does not change when the origin is moved by a general vector $\mathbf{r}_0$. The structure factor of index $\mathbf{h}$, referred to the new origin, will be

$$\begin{aligned} F'_{\mathbf{h}} &= \sum_{j=1}^{N} f_j \exp\{2\pi\mathrm{i}\mathbf{h}\cdot(\mathbf{r}_j - \mathbf{r}_0)\} \\ &= F_{\mathbf{h}} \exp\{-2\pi i\mathbf{h}\cdot\mathbf{r}_0\} \\ &= |F_{\mathbf{h}}| \exp[\mathrm{i}(\varphi_{\mathbf{h}} - 2\pi\mathbf{h}\cdot\mathbf{r}_0]; \end{aligned} \tag{5.30}$$

the modulus remains unchanged while the phase changes by $\Delta\varphi = 2\pi\mathbf{h}\cdot\mathbf{r}_0$.

The variation of the phase of the product (5.28), due to the same origin shift, will be

$$\Delta\varphi = 2\pi\mathbf{r}_0 \sum_{i=1}^{m} \mathbf{h}_i = 0 \tag{5.31}$$

which is zero because of condition (5.29).

The simplest structure invariants are:

1. $F_{000} = \sum_{j=1}^{N} Z_j$ giving the number of electrons in the unit cell; its phase is always zero.
2. $F_{\boldsymbol{h}}F_{-\boldsymbol{h}} = |F_{\boldsymbol{h}}|^2$, which does not contain any phase information.
3. $F_{-\boldsymbol{h}}F_{\boldsymbol{k}}F_{\boldsymbol{h}-\boldsymbol{k}}$ with phase $\varphi_{-\boldsymbol{h}} + \varphi_{\boldsymbol{k}} + \varphi_{\boldsymbol{h}-\boldsymbol{k}}$, called **triplet** invariant; we shall see that these invariants play a primary role in the procedures which will be described later.
4. $F_{-\boldsymbol{h}}F_{\boldsymbol{k}}F_{\boldsymbol{l}}F_{\boldsymbol{h}-\boldsymbol{k}-\boldsymbol{l}}$ with phase $\varphi_{-\boldsymbol{h}} + \varphi_{\boldsymbol{k}} + \varphi_{\boldsymbol{l}} + \varphi_{\boldsymbol{h}-\boldsymbol{k}-\boldsymbol{l}}$ called **quartet** invariant; its importance has been recognized in recent years.
5. With a simple extension it is possible to define **quintet, sextet**, etc., invariants. Considering that the practical use of these relations is still under discussion, we will just mention here that complex probabilistic formulae have been derived to allow the estimation of these higher invariants.

Structure semi-invariants (s.s.)[20,22] are single phases or linear combinations of phases which are invariant with respect to a shift of origin, provided that the position of the origin is restricted to those points in the cell which possess the same point symmetry (the so-called 'permissible origins'). A basic property of a s.s. is its capability of being transformed into a s.i. by adding one or more pairs of symmetry-equivalent phases. For instance, in a given space group possessing the symmetry operator $\mathbf{C} \equiv (\mathbf{R}, \mathbf{T})$, the phase $\varphi_{\boldsymbol{H}}$ is a s.s. if it is possible to find a reflection $\boldsymbol{h}$ such that

$$\psi = \varphi_{\boldsymbol{H}} - \varphi_{\boldsymbol{h}} + \varphi_{\boldsymbol{hR}} \tag{5.32}$$

is a s.i., i.e. $\boldsymbol{H} - \boldsymbol{h} + \boldsymbol{hR} = 0$.

Because of (3.37), relation (5.32) may be written as

$$\psi = \varphi_{\boldsymbol{H}} - 2\pi\boldsymbol{h}\cdot\boldsymbol{T}. \tag{5.33}$$

ψ is independent of any choice of origin, while $\varphi_{\boldsymbol{H}}$ and $\boldsymbol{T}$ will depend on it; however, the dependence is such that (5.33) holds and therefore the value of $\varphi_{\boldsymbol{H}}$ will not change if we move the origin within those points which maintain the vectors $\boldsymbol{T}$ unchanged (identical point symmetry). For instance in $P\bar{1}$, $\varphi_{2\boldsymbol{h}}$ is a s.s. since

$$\psi = \varphi_{2\boldsymbol{h}} - \varphi_{\boldsymbol{h}} - \varphi_{\boldsymbol{h}} \tag{5.34}$$

is a s.i. The origin is conveniently chosen on the inversion centres and then $\mathbf{T} = (0, 0, 0)$. The permissible origins are on the eight inversion centres present in the cell; any one of them, chosen as origin, will always give $\mathbf{T} = (0, 0, 0)$.

Alternatively, by applying (5.31) we can derive that the phase of $\varphi_{2\boldsymbol{h}}$ changes by $\Delta\varphi = 2\pi 2\boldsymbol{h}\cdot\boldsymbol{r}_0$ when the origin is shifted of a vector $\boldsymbol{r}_0$. It is easy to see that $\Delta\varphi = 0$ (or $2\pi m$) when the components of $\boldsymbol{r}_0$ are 0 or $\frac{1}{2}$ as happens for the eight inversion centres in $P\bar{1}$ (cf. Fig. 5.17). The qualitative example relative to the sign of $F_{2\boldsymbol{h}}$, described in Appendix 5.C, can now be related to a s.s.

As another example, let us consider, in space group $P2_1$, the phase $\varphi_{2h,0,2l}$, which is a s.s., because

$$\psi = \varphi_{2h,0,2l} - \varphi_{h,k,l} + \varphi_{\bar{h},k,\bar{l}} \tag{5.35}$$

is a s.i. for any value of k. When the origin is chosen on a twofold screw axis, then $\psi = \varphi_{2h,0,2l} - \pi k$. The permissible origins are located on any of the four screw axes present in the unit cell (cf. Fig. 5.18).

The two examples given above refer to single phases and are therefore one-phase semi-invariants. We can generalize what we have seen so far to the case of the linear combinations of more phases. Thus, the combination $\sum \phi_{\boldsymbol{h}}$ is a s.s. in $P\bar{1}$ if $\sum \boldsymbol{h} = 2\boldsymbol{H}$ (i.e. if the three components of the sum vector are all even); in $P2_1$ it is a s.s. if $\sum \boldsymbol{h} = (2H, 0, 2L)$. A compact way to indicate these two conditions is: $\sum \boldsymbol{h} \equiv 0 \bmod (2, 2, 2)$ (meaning that each of the three components of the vector $\sum \boldsymbol{h}$ gives a zero rest when divided by 2), and $\sum \boldsymbol{h} \equiv 0 \bmod (2, 0, 2)$ (with the same meaning as before for the first and third components, while the second must be zero). In general we may write $\sum \boldsymbol{h} \equiv 0 \bmod \boldsymbol{\omega}_s$, where $\boldsymbol{\omega}_s$ is a vector, called **semi-invariant modulus**, with integer components; the vector $\sum \boldsymbol{h}$ is called the **semi-invariant vector.**

In order to identify which phases or combinations of phases are s.s. one can refer to special tables (see, for example, Giacovazzo[19]) in which the space groups are classified in such a way that those belonging to the same class have the same permissible origins. The same process may be carried out automatically on a computer following an algebraic approach.[23]

Before considering probability methods, let us use the positivity property of the electron density function to obtain an indication on the value of triplet invariants. The development of the product $F_{-\boldsymbol{h}}F_{\boldsymbol{k}}F_{\boldsymbol{h}-\boldsymbol{k}}$ gives

$$\begin{aligned} F_{-\boldsymbol{h}}F_{\boldsymbol{k}}F_{\boldsymbol{h}-\boldsymbol{k}} &= \sum_{j_1} f_{j_1} \exp[-2\pi i \boldsymbol{h} \cdot \boldsymbol{r}_{j_1}] \sum_{j_2} f_{j_2} \exp[-2\pi i \boldsymbol{k} \cdot \boldsymbol{r}_{j_2}] \\ &\quad \times \sum_{j_3} f_{j_3} \exp[-2\pi i (\boldsymbol{h} - \boldsymbol{k}) \cdot \boldsymbol{r}_{j_3}] \\ &= \sum_{j_1, j_2, j_3} f_{j_1} f_{j_2} f_{j_3} \exp\{2\pi i [\boldsymbol{h} \cdot (\boldsymbol{r}_{j_3} - \boldsymbol{r}_{j_1}) \\ &\quad + [\boldsymbol{k} \cdot (\boldsymbol{r}_{j_2} - \boldsymbol{r}_{j_3})]\}. \end{aligned} \tag{5.36}$$

For simplicity let us assume that reflections $\boldsymbol{h}$, $\boldsymbol{k}$, and $\boldsymbol{h} - \boldsymbol{k}$ have similar $\sin \theta / \lambda$ values and therefore the variation of the atomic scattering factors with the indices can be overlooked. In (5.36) we can separate the terms:

(1) $j_1 = j_2 = j_3 = j$, $\quad \sum_j f_j^3$;

(2) $j_1 = j_2 \neq j_3$, $\quad \sum_{j_1, j_3} f_{j_1}^2 f_{j_3} \exp\{2\pi i [(\boldsymbol{h} - \boldsymbol{k}) \cdot (\boldsymbol{r}_{j_3} - \boldsymbol{r}_{j_1})]\}$,

which is proportional to (cf. (5.1)) $|F_{\boldsymbol{h}-\boldsymbol{k}}|^2 - \sum_j f_j^2$;

(3) $j_1 = j_3 \neq j_2$, $\quad \sum_{j_2, j_3} f_{j_3}^2 f_{j_2} \exp\{2\pi i [\boldsymbol{k} \cdot (\boldsymbol{r}_{j_2} - \boldsymbol{r}_{j_3})]\}$,

which is proportional to $|F_{\boldsymbol{k}}|^2 - \sum_j f_j^2$;

(4) $j_2 = j_3 \neq j_1$, $\quad \sum_{j_1, j_2} f_{j_2}^2 f_{j_1} \exp\{2\pi i[\boldsymbol{h} \cdot (\boldsymbol{r}_{j_2} - \boldsymbol{r}_{j_1})]\}$,

which is proportional to $|F_{\boldsymbol{h}}|^2 - \sum_j f_j^2$;

(5) $j_1 \neq j_2 \neq j_3$, $\quad \sum_{j_1, j_2, j_3} f_{j_1} f_{j_2} f_{j_3} \exp\{2\pi i[\boldsymbol{h} \cdot (\boldsymbol{r}_{j_3} - \boldsymbol{r}_{j_1})$

$$+ [\boldsymbol{k} \cdot (\boldsymbol{r}_{j_2} - \boldsymbol{r}_{j_3})]\} = R.$$

If the electron density is positive, the atomic scattering factors, which are the Fourier transform of the electron density around each atom, will also be positive and the summation in (1) will then be positive. Recalling equation (5.A.9) the terms (2), (3), and (4) are proportional to $|F_{\boldsymbol{h}}|^2 - \langle |F|^2 \rangle$, $|F_{\boldsymbol{k}}|^2 - \langle |F|^2 \rangle$, and $|F_{\boldsymbol{h}-\boldsymbol{k}}|^2 - \langle |F|^2 \rangle$ respectively and for large values of $|F|$ they will be large and positive. The last term R is the only complex term, but being a sum of positive and negative quantities, it will on average be small; R may be considered as a 'noise' term. For large values of the structure amplitudes we can then write

$$F_{-\boldsymbol{h}} F_{\boldsymbol{k}} F_{\boldsymbol{h}-\boldsymbol{k}} = \sum_j f_j^3 + K\{|F_{\boldsymbol{h}}|^2 + |F_{\boldsymbol{k}}|^2 + |F_{\boldsymbol{h}-\boldsymbol{k}}|^2 - 3\langle |F|^2 \rangle\} + R \geqslant 0. \quad (5.37)$$

In the case of a centrosymmetric structure this relation implies that, when $|F_{\boldsymbol{h}}|$, $|F_{\boldsymbol{k}}|$, and $|F_{\boldsymbol{h}-\boldsymbol{k}}|$ are large, then the product of the three signs of the structure factors, $s(\boldsymbol{h})s(\boldsymbol{k})s(\boldsymbol{h}-\boldsymbol{k}) = +$. This deduction is in agreement with (5.27), derived from Sayre's equation, and with the pictorial derivation given in Appendix 5.C.

Probability methods

For the same reasons considered on pp. 321–2, from now on we will use the normalized structure factors. The use of normalized amplitudes is also suggested by (5.37), which shows that the value of the triplet does not depend on the diffraction angle. Indeed we may note that at low $\sin\theta/\lambda$, where the scattering power is higher, both the positive terms and the noise term R are larger, and that they decrease by comparable amounts as the diffraction angle increases. It is also important to point out that the use of normalized structure factors, corresponding to a point-atom structure, implicitly corresponds to a sharpening of the atomicity assumption for the electron density.

In Appendix 5.D the probability formulae for triplet invariants are derived, here we will just consider the main results. For a non-centrosymmetric structure the distribution associated to (5.26), derived by Cochran,[24] is given by (cf. (5.D.17))

$$P(\Phi_{\boldsymbol{hk}}) = (1/L) \exp(G_{\boldsymbol{hk}} \cos \Phi_{\boldsymbol{hk}}) \quad (5.38)$$

where:

(1) for equal atoms $\quad G_{\boldsymbol{hk}} = (2/\sqrt{N})\, |E_{\boldsymbol{h}} E_{\boldsymbol{k}} E_{\boldsymbol{h}-\boldsymbol{k}}|; \quad (5.39a)$

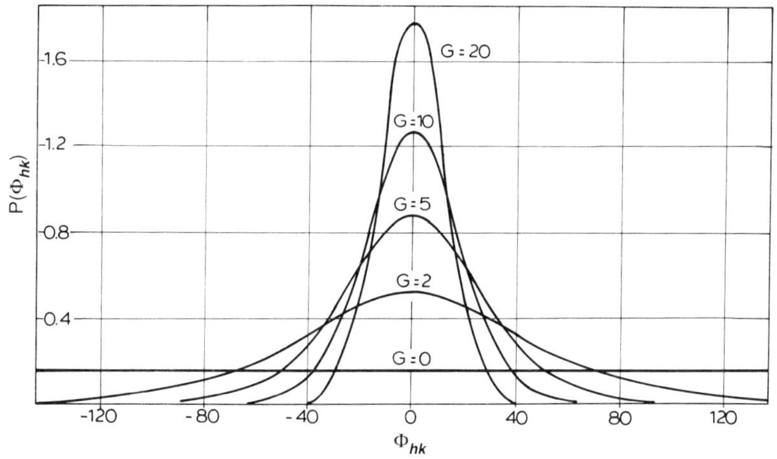

Fig. 5.11. Trends of the probability distributions (5.38) for different values of the parameter G_{hk} given by (5.39).

(2) for non-equal atoms

$$G_{\boldsymbol{hk}} = 2\sigma_3\sigma_2^{-3/2}\,|E_{\boldsymbol{h}}E_{\boldsymbol{k}}E_{\boldsymbol{h}-\boldsymbol{k}}|; \qquad (5.39b)$$

with $\sigma_n = \sum_{j=1}^{N} Z_j^n$, Z_j being the atomic number of the jth atom. In (5.38) L is a normalization term depending on $G_{\boldsymbol{hk}}$ only. $P(\Phi_{\boldsymbol{hk}})$ is a so-called **von Mises distribution**[25] and $G_{\boldsymbol{hk}}$ is its **concentration parameter**; its trend is very similar to that of a Gaussian, as it is shown in Fig. 5.11 for different values of $G_{\boldsymbol{hk}}$. All curves have a maximum at $\Phi_{\boldsymbol{hk}} = 0$, i.e. $\varphi_{\boldsymbol{h}} = \varphi_{\boldsymbol{k}} + \varphi_{\boldsymbol{h}-\boldsymbol{k}}$ and their width around the most probable value narrows as $G_{\boldsymbol{hk}}$ increases. The variance of the relation

$$\varphi_{\boldsymbol{h}} \simeq \varphi_{\boldsymbol{k}} + \varphi_{\boldsymbol{h}-\boldsymbol{k}} \qquad (5.40)$$

will therefore decrease as $G_{\boldsymbol{hk}}$ increases.

If more than one pair of phases $\varphi_{\boldsymbol{k}_j}$, $\varphi_{\boldsymbol{h}-\boldsymbol{k}_j}$ are known (with associated amplitudes $|E_{\boldsymbol{k}_j}|$ and $|E_{\boldsymbol{h}-\boldsymbol{k}_j}|$,), with $j = 1, 2, \ldots, r$, all defining the same phase $\varphi_{\boldsymbol{h}}$ through triplet relations such as (5.40), then the total probability distribution for $\varphi_{\boldsymbol{h}}$ is given by the product of the r distributions of the form (5.38), which are assumed to be independent of one another; i.e.

$$P(\varphi_{\boldsymbol{h}}) = \prod_{j=1}^{r} P_j(\varphi_{\boldsymbol{h}}) = A \exp\left(\sum_{j=1}^{r} G_{\boldsymbol{hk}_j} \cos\,(\varphi_{\boldsymbol{h}} - \varphi_{\boldsymbol{k}_j} - \varphi_{\boldsymbol{h}-\boldsymbol{k}_j})\right) \qquad (5.41)$$

where A is a normalization factor. The exponent of (5.41) may be developed as

$$\cos\varphi_{\boldsymbol{h}} \sum_{j=1}^{r} G_{\boldsymbol{hk}_j} \cos\,(\varphi_{\boldsymbol{k}_j} + \varphi_{\boldsymbol{h}-\boldsymbol{k}_j}) + \sin\varphi_{\boldsymbol{h}} \sum_{j=1}^{r} G_{\boldsymbol{hk}_j} \sin\,(\varphi_{\boldsymbol{k}_j} + \varphi_{\boldsymbol{h}-\boldsymbol{k}_j})$$
$$= \alpha_{\boldsymbol{h}} \cos\,(\varphi_{\boldsymbol{h}} - \beta_{\boldsymbol{h}}) \qquad (5.42)$$

where

$$\alpha_{\boldsymbol{h}} = \left[\left(\sum_{j=1}^{r} G_j \cos\omega_j\right)^2 + \left(\sum_{j=1}^{r} G_j \sin\omega_j\right)^2\right]^{1/2} \qquad (5.43)$$

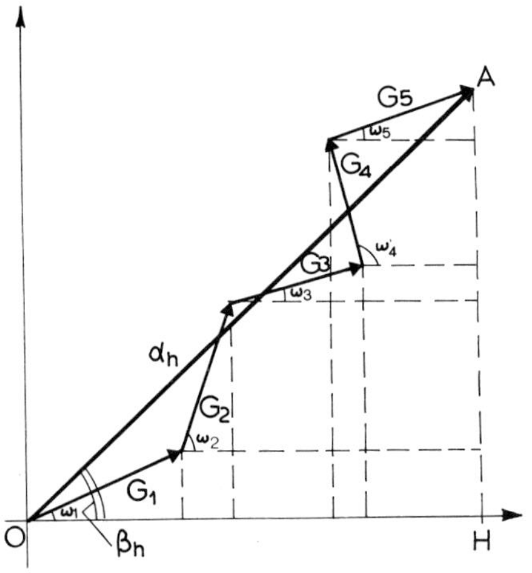

Fig. 5.12. Vector representation, in the complex plane, of the combination of five triplets of type (5.40) involving the same reflection $\boldsymbol{h}$.

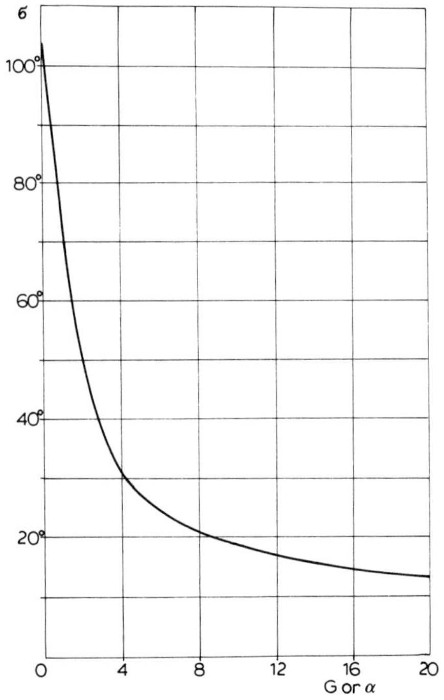

Fig. 5.13. Trend of the standard deviation, σ, of the distribution (5.45) as a function of $\alpha_{\boldsymbol{h}}$, defined by (5.43).

and

$$\tan \beta_{\boldsymbol{h}} = \frac{\left(\sum_{j=1}^{r} G_j \sin \omega_j\right)}{\left(\sum_{j=1}^{r} G_j \cos \omega_j\right)} \tag{5.44}$$

with $G_j = G_{\boldsymbol{hk}_j}$ and $\omega_j = \varphi_{\boldsymbol{k}_j} + \varphi_{\boldsymbol{h}-\boldsymbol{k}_j}$. Equations (5.42) to (5.44) are easily understood when the r relationships of type (5.40) are plotted on an Argand diagram as vectors of modulus G_j and phase ω_j; the angle between each vector and the real axis is an indication of a probable value of φ_h. In Fig. 5.12 the case with $r = 5$ is illustrated; it can be seen that

$$\mathrm{AH} = \alpha_{\boldsymbol{h}} \sin \beta_{\boldsymbol{h}} = \sum_{j=1}^{r} G_j \sin \omega_j, \qquad \mathrm{OH} = \alpha_{\boldsymbol{h}} \cos \beta_{\boldsymbol{h}} = \sum_{j=1}^{r} G_j \cos \omega_j$$

and relations (5.42), (5.43), (5.44) become immediately clear.

Finally eqn (5.41) becomes

$$P(\varphi_{\boldsymbol{h}}) = A \exp\left[\alpha_{\boldsymbol{h}} \cos(\varphi_{\boldsymbol{h}} - \beta_{\boldsymbol{h}})\right], \tag{5.45}$$

which still is a von Mises distribution with a maximum for $\varphi_{\boldsymbol{h}} = \beta_{\boldsymbol{h}}$ and a variance[26] depending on $\alpha_{\boldsymbol{h}}$ as shown in Fig. 5.13. For instance we may deduce that for $\alpha_{\boldsymbol{h}} = 2$, $\langle \varphi_{\boldsymbol{h}} \rangle = \beta_{\boldsymbol{h}} \pm 50°$, while for $\alpha_{\boldsymbol{h}} = 10$, $\langle \varphi_{\boldsymbol{h}} \rangle = \beta_{\boldsymbol{h}} \pm 19°$. Equation (5.44) gives the most probable value of $\varphi_{\boldsymbol{h}}$ and is known as the **tangent formula**;[27] as we shall see later, this formula plays an important role in the phase determination process.

In the case of centrosymmetric structures the probability that the sign relationship (5.27) is true, is given by[28] (Appendix 5.D)

$$P^+ = \tfrac{1}{2} + \tfrac{1}{2} \tanh\left[(\sigma_3 \sigma_2^{-3/2} \left|E_{\boldsymbol{h}} E_{\boldsymbol{k}} E_{\boldsymbol{h}-\boldsymbol{k}}\right|)\right]. \tag{5.46}$$

When several relations for the same $S(\boldsymbol{h})$ exist

$$S(\boldsymbol{h}) \simeq S(\boldsymbol{k}_j) S(\boldsymbol{h} - \boldsymbol{k}_j) \qquad j = 1, 2, \ldots, r \tag{5.47}$$

the product of the different probabilities $P(E_{\boldsymbol{h}})$ (given by (5.D.19)) should be considered. With a procedure similar to that described in Appendix 5.D, we obtain

$$P^+ = \tfrac{1}{2} + \tfrac{1}{2} \tanh\left(\sigma_3 \sigma_2^{-3/2} \left|E_{\boldsymbol{h}}\right| \sum_{j=1}^{r} E_{\boldsymbol{k}_j} E_{\boldsymbol{h}-\boldsymbol{k}_j}\right). \tag{5.48}$$

Figure 5.14 shows the trend of (5.48); it can be seen that, when several terms, all with the same sign, contribute to the summation, then the absolute value of the tanh argument may become rather large and P^+ approaches the extreme values 1 or 0.

In the past 10–15 years probability methods have seen new important developments. Not only it has been possible to improve the estimate of triplets, but also to derive reliable estimates of other phase relationships, both s.i. and s.s.

At the basis of the new approaches stands the following principle: '*It is possible to obtain a good estimate of s.i. or s.s. given "appropriate" sets of normalized structure factor moduli, which are statistically the most effective in determining the value of the given s.i. or s.s.*'

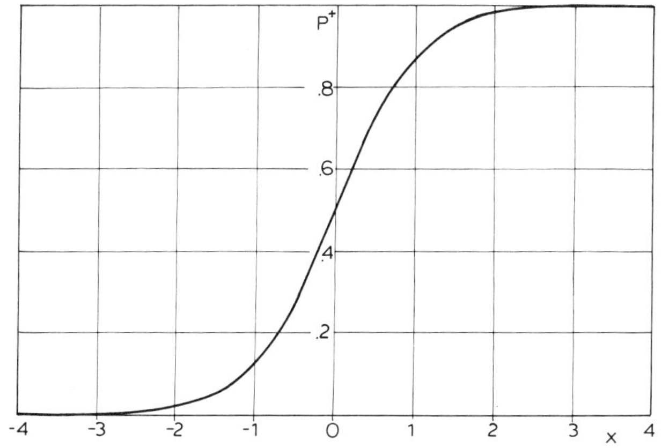

Fig. 5.14. Trend of $P^+ = \frac{1}{2} + \frac{1}{2}\tanh(x)$ (cf. (5.48)) as a function of x.

The first task will then be to identify these moduli, indicated as the **phasing magnitudes**, and rank them according to their effectiveness in estimating s.i. or s.s.. Given their set $\{|E|\}$, the second task will be to derive the probability distribution

$$P(\Phi \mid \{|E|\}) \tag{5.49}$$

where Φ is any phase relationship we want to estimate and the vertical bar after it stands for: 'given all magnitudes in $\{|E|\}$'.

Cochran's formula[24] for triplets, derived in Appendix 5.D, is a trivial example of such **conditional distribution**; in fact it may be seen as $P(\Phi_{\boldsymbol{hk}} \mid |E_{\boldsymbol{h}}|, |E_{\boldsymbol{k}}|, |E_{\boldsymbol{h-k}}|)$.

Schenk[29,30] extended this principle to the case of four-phase s.i. (quartets):

$$Q = \varphi_{\boldsymbol{h}} + \varphi_{\boldsymbol{k}} + \varphi_{\boldsymbol{l}} + \varphi_{-\boldsymbol{h}-\boldsymbol{k}-\boldsymbol{l}}. \tag{5.50}$$

Their distribution, given the four associated magnitudes, derived by Simerska,[31] indicates that $Q \approx 0$, with a variance depending on

$$B = (2/N)\,|E_{\boldsymbol{h}}E_{\boldsymbol{k}}E_{\boldsymbol{l}}E_{\boldsymbol{h}+\boldsymbol{k}+\boldsymbol{l}}|\,. \tag{5.51}$$

Because of the $1/N$ factor any reasonably sized structure will have very small B values, and for this reason quartets estimated in this way can not be used in practice. Schenk pointed out that quartet (5.50) could be considered as the sum of two triplets, such as $T_1 = \varphi_{\boldsymbol{h}} + \varphi_{\boldsymbol{k}} - \varphi_{\boldsymbol{h}+\boldsymbol{k}}$ and $T_2 = \varphi_{\boldsymbol{l}} + \varphi_{-\boldsymbol{h}-\boldsymbol{k}-\boldsymbol{l}} + \varphi_{\boldsymbol{h}+\boldsymbol{k}}$. Then, if also $|E_{\boldsymbol{h}+\boldsymbol{k}}|$ is large, we will have $T_1 \simeq 0$ and $T_2 \simeq 0$ and therefore $Q = T_1 + T_2 \simeq 0$ with a strengthened reliability with respect to that indicated by (5.51). In a similar way we can see that the same quartet can be written as the sum of two other pairs of triplets, and that Q also depends on $|E_{\boldsymbol{h}+\boldsymbol{l}}|$ and $|E_{\boldsymbol{k}+\boldsymbol{l}}|$. We can then say that the quartet not only depends on the four **basis** magnitudes $|E_{\boldsymbol{h}}|$, $|E_{\boldsymbol{k}}|$, $|E_{\boldsymbol{l}}|$, $|E_{\boldsymbol{h}+\boldsymbol{k}+\boldsymbol{l}}|$, but also on the three **cross** magnitudes $|E_{\boldsymbol{h}+\boldsymbol{k}}|$, $|E_{\boldsymbol{h}+\boldsymbol{l}}|$, $|E_{\boldsymbol{k}+\boldsymbol{l}}|$. If the last three moduli are also large, then the indication $Q \approx 0$ is strengthened. Empirically it was also found that, when the cross magnitudes have very small values, then $Q \approx \pi$ (since $\cos Q \approx -1$, these are called **negative quartets**).

Later Hauptman[32] derived the probability distribution of Q, given the seven basis and cross magnitudes, and confirmed Schenk's empirical findings. He then formulated the **neighbourhood principle**,[33] the concept of neighbourhoods being very close to that of phasing magnitudes. Later[34] he also gave rules for deriving the neighbourhoods of the main s.i. and of some s.s. in the most common space groups.

The **representation theory** proposed by Giacovazzo[35,36] is a generalization and a systematization of the previous ideas, as it gives precise general rules for identifying the phasing magnitudes, allowing, at the same time, a completely general use of the space-group symmetry.

The **first representation** of a s.i. Φ is formed by the invariant itself plus all symmetry equivalent s.i. differing from Φ by a constant angle, which is a function only of the symmetry operators. The set of all basis and cross magnitudes which appear in the first representation forms the first shell of phasing magnitudes (**first phasing shell**).

This definition will be clarified through some examples. The importance of taking the space-group symmetry into account is first illustrated by the case of a special triplet in the space group $P2_12_12_1$:

$$T = \varphi_{(0k_1l_1)} + \varphi_{(h_2\bar{k}_10)} + \varphi_{(\bar{h}_20\bar{l}_1)};$$

if the first reflection has k_1 even and l_1 odd and restricted phase $0, \pi$, the second has h_2 even and phase 0, π, and the third has consequently phase $\pm\pi/2$, then $T = \pm\pi/2$, in contrast with the indication $T \simeq 0$ given by (5.38) when the three normalized amplitudes are large. The symmetry operators in $P2_12_12_1$ are: $\mathbf{R}_1 = \mathbf{I}$, $\mathbf{T}_1 = (0, 0, 0)$; $\mathbf{R}_2 = (\bar{1}, \bar{1}, 1)$ (only the diagonal elements are indicated, all the others being zero), $\mathbf{T}_2 = (\frac{1}{2}, 0, \frac{1}{2})$; $\mathbf{R}_3 = (1, \bar{1}, \bar{1})$, $\mathbf{T}_3 = (\frac{1}{2}, \frac{1}{2}, 0)$; $\mathbf{R}_4 = (\bar{1}, 1, \bar{1})$, $\mathbf{T}_4 = (0, \frac{1}{2}, \frac{1}{2})$; and a triplet equivalent to T may be written as

$$T' = \varphi_{(0k_1l_1)\mathbf{R}_2} + \varphi_{(h_2\bar{k}_10)\mathbf{R}_3} + \varphi_{(\bar{h}_20\bar{l}_1)}$$
$$= \varphi_{(0\bar{k}_1l_1)} + \varphi_{(h_2k_10)} + \varphi_{(\bar{h}_20\bar{l}_1)} = T + 2\pi(l_1 - k_1)/2 \equiv T + \pi$$

because l_1 is odd and k_1 even. T and T', forming the first representation of the triplet, have the same G but opposite phase and their combination in a phase diagram (similar to that of Fig. 5.12) will give a null vector, which does not contradict the expected restricted value $T = \pm\pi/2$. The general use of the space-group symmetry in estimating triplets has been described by Giacovazzo,[37] who also showed that for non-primitive cells the $1/\sqrt{N}$ factor in (5.39a) should be replaced by $1/\sqrt{N_p}$, where N_p is the number of atoms in the primitive cell.

Let us now consider the example of a quartet such as (5.50). As we have seen, Q in general depends on seven magnitudes, and when no special symmetry conditions exist, the first representation is formed just by Q and the seven magnitudes form the first phasing shell. On the other hand, if, besides the identity, there is an other symmetry operator $\mathbf{C}_s = (\mathbf{R}_s, \mathbf{T}_s)$, which leaves some special reflections unchanged ($\boldsymbol{H} = \boldsymbol{H}\mathbf{R}_s$) and if one of the cross reflection (e.g. $\boldsymbol{h} + \boldsymbol{k}$) is of this special type, then the quartet

$$Q' = \varphi_{\boldsymbol{h}\mathbf{R}_s} + \varphi_{\boldsymbol{k}\mathbf{R}_s} + \varphi_{\boldsymbol{l}} + \varphi_{-\boldsymbol{h}-\boldsymbol{k}-\boldsymbol{l}} \tag{5.52}$$

is equivalent to Q $[Q' - Q = 2\pi(\boldsymbol{h} + \boldsymbol{k})\mathbf{T}_s]$, but with two new cross terms $|E_{\boldsymbol{h}\mathbf{R}_s+\boldsymbol{l}}|$ and $|E_{\boldsymbol{k}\mathbf{R}_s+\boldsymbol{l}}|$. In this case the first representation is formed by Q and Q' and the first phasing shell contains nine magnitudes. Since the larger the number of phasing magnitudes the better the estimate of the s.i., a significant advantage is obtained by considering also Q'.

A numerical example will further clarify the above ideas; let us consider the quartet

$$Q = \varphi_{153} + \varphi_{4\bar{5}1} + \varphi_{212} + \varphi_{\bar{7}\bar{1}\bar{6}} \qquad \text{(cross: 504, 365, 6}\bar{4}\text{3)}$$

in the space group $P2_1/c$, for which $\mathbf{R}_1 = \mathbf{I}$, $\mathbf{R}_2 = (\bar{1}, 1, \bar{1})$, $\mathbf{R}_3 = (\bar{1}, \bar{1}, \bar{1})$, $\mathbf{R}_4 = (1, \bar{1}, 1)$. The first cross reflection is such that $(504)\mathbf{R}_4 = (504)$, and an equivalent quartet may be set up

$$Q' = \varphi_{1\bar{5}3} + \varphi_{451} + \varphi_{212} + \varphi_{\bar{7}\bar{1}\bar{6}} \qquad \text{(cross: 504, 3}\bar{4}\text{5, 663)}$$

with two new cross terms.

As an example of **second representation** of a s.i. let us consider the case of a triplet T. Its second representation is defined as the set of all special quintets of type

$$C = T + \varphi_{\boldsymbol{H}} - \varphi_{\boldsymbol{H}} = \varphi_{\boldsymbol{h}} + \varphi_{\boldsymbol{k}} + \varphi_{-\boldsymbol{h}-\boldsymbol{k}} + \varphi_{\boldsymbol{H}} - \varphi_{\boldsymbol{H}} \tag{5.53}$$

and $\boldsymbol{H}$ is any reciprocal lattice vector for which $|E_{\boldsymbol{H}}|$ is large. Since the term added to T is null, then $C = T$, but the quintet will depend on four basis magnitudes $|E_{\boldsymbol{h}}|$, $|E_{\boldsymbol{k}}|$, $|E_{\boldsymbol{h}+\boldsymbol{k}}|$, $|E_{\boldsymbol{H}}|$ and on six cross terms $|E_{\boldsymbol{h}\pm\boldsymbol{H}}|$, $|E_{\boldsymbol{k}\pm\boldsymbol{H}}|$, $|E_{\boldsymbol{h}+\boldsymbol{k}\pm\boldsymbol{H}}|$. If M vectors $\boldsymbol{H}$ are selected (80–100 reflections with largest $|E_{\boldsymbol{H}}|$), then the second representation will be formed by M quintets and the second phasing shell will contain $10 \times M$ magnitudes.

Let us now consider a general n-phase s.s.

$$\Phi = \varphi_{\boldsymbol{h}_1} + \varphi_{\boldsymbol{h}_2} + \ldots + \varphi_{\boldsymbol{h}_n}. \tag{5.54}$$

In general (the few exceptions are beyond our scope) it is possible to find a phase $\varphi_{\boldsymbol{H}}$ and two symmetry operators $\mathbf{C}_i$ and $\mathbf{C}_j$ such that

$$\psi_1 = \varphi_{\boldsymbol{h}'_1} + \varphi_{\boldsymbol{h}'_2} + \ldots + \varphi_{\boldsymbol{h}'_n} + \varphi_{\boldsymbol{H}\mathbf{R}_i} - \varphi_{\boldsymbol{H}\mathbf{R}_j} \tag{5.55}$$

is a structure invariant; $\boldsymbol{h}'$ indicates a reflection equivalent to $\boldsymbol{h}$. Since ψ_1 differs from Φ by a constant term depending only on the symmetry operators, the estimate of ψ_1 gives at the same time the value of Φ. The collection of all s.i. ψ_1, obtained by varying $\boldsymbol{H}$, $\mathbf{C}_i$ and $\mathbf{C}_j$ forms the first representation of Φ and the first phasing shell will contain all the associated basis and cross magnitudes.

Let us now consider a few examples regarding one-phase s.s., for which (5.55) reduces to (5.32). As we have seen, in $P\bar{1}$, the s.s. $\varphi_{2\boldsymbol{h}}$ may be represented by the triplet (5.34), which alone will form its first representation with the corresponding phasing shell including the two magnitudes $|E_{2\boldsymbol{h}}|$ and $|E_{\boldsymbol{h}}|$. The probability that $E_{2\boldsymbol{h}}$ is positive was derived by Cochran and Woolfson[28] and is given by

$$P^+(E_{2\boldsymbol{h}}) = \tfrac{1}{2} + \tfrac{1}{2}\tanh\left[\tfrac{1}{2}\sigma_3\sigma_2^{-3/2}\,|E_{2\boldsymbol{h}}|\,(|E_{\boldsymbol{h}}|^2 - 1)\right], \tag{5.56}$$

which is greater than $\tfrac{1}{2}$ when $|E_{\boldsymbol{h}}| > 1$ and approaches 1 as $|E_{2\boldsymbol{h}}|$ and $|E_{\boldsymbol{h}}|$ increase.

Similarly in the space group $P2_1$ the first representation of the s.s. $\varphi_{2h,0,2l}$ is given by the set of triplets (5.35) and the probability formula will be a function of all the magnitudes in the first phasing shell: $|E_{2h,0,2l}|$ and $\{|E_{h,j,l}|\}$ with j assuming all possible values of the index k within the set of measured reflections. In $P2_1$ one-phase s.s. are special reflections with phase restricted to 0 or π and the probability formula will be of the tanh type; we then have

$$P^+(E_{2h,0,2l}) = \tfrac{1}{2} + \tfrac{1}{2}\tanh\left(\tfrac{1}{2}\sigma_3\sigma_2^{-3/2}\,|E_{2h,0,2l}|\sum_k (-1)^k(|E_{hkl}|^2-1)\right) \quad (5.57)$$

where the sum is over all values of k. The term $(-1)^k \equiv \exp(\pi k)$ corresponds to the constant angle πk and is obtained by applying (3.37) to express the phase relation between φ_{hkl} and $\varphi_{\bar{h}k\bar{l}}$ in (5.35). When all major terms in the sum have k even, then $P^+(E_{2h,0,2l}) > \frac{1}{2}$ and $S(2h,0,2l) \simeq +$, when they have k odd, then $P^+(E_{2h,0,2l}) < \frac{1}{2}$ and $S(2h,0,2l) \simeq +$.

Formulae (5.56) and (5.57) and those for some other space groups were derived long before the concept of representation was introduced, and they were known as the Σ_1 formulae[20].

The second representation of a s.s. will also be illustrated by means of an example. In space group $P2_12_12_1$ let us consider the s.s. $\Phi = \varphi_{406}$, for which the first representation (5.55) reduces to the set of triplets

$$\psi_1 = \varphi_{406} + \varphi_{\bar{2}k\bar{3}} - \varphi_{2k3} \quad (5.58)$$

with variable k index. With an extension similar to that used to derive (5.53), the second representation is defined as the set of quintets

$$\psi_2 = \varphi_{406} + \varphi_{2k3} - \varphi_{2k3} + \varphi_{\boldsymbol{q}} - \varphi_{\boldsymbol{q}} \quad (5.59)$$

with $\boldsymbol{q}$ varying over all reflections with $|E_{\boldsymbol{q}}|$ large. The second phasing shell will include $|E_{\boldsymbol{H}}| = |E_{406}|$, a term $|E_{\boldsymbol{h}}| = |E_{2k3}|$ for each triplet (5.58) and, for each quintet, $|E_{\boldsymbol{q}}|$ and the four cross terms $|E_{\boldsymbol{H}\pm\boldsymbol{q}}|$, $|E_{\boldsymbol{h}\mathbf{R}\pm\boldsymbol{q}}|$.

Once the phasing magnitudes have been identified it becomes possible to derive the appropriate probability distribution (5.49). So far formulae have been derived for estimating:

(1) one-phase s.s. by the second representation;[38–41]

(2) two-phase s.s. by the first representation;[42–44]

(3) triplets by the second representation (P10 formula);[45]

(4) quartets by the first representation.[46,47]

In most cases, the derived probability distributions for non-centrosymmetric structures have the form of a von Mises function as (5.38) and for centrosymmetric structures the tanh form of (5.47); the concentration parameter G or the tanh argument are now substituted by more complex functions of the considered phasing magnitudes, which will not be reported. Only their practical use will be illustrated in the following section.

Fixing the origin and the enantiomorph

In the previous sections we have seen how it is possible to derive the values of s.i. and s.s. from the observed magnitudes. In order to pass from these values to those of the individual phases it will be necessary to **fix the origin**, and, in non-centrosymmetric space groups, it will be necessary to choose between the two enantiomorphic forms compatible with the observed amplitudes; in fact, since all the phases change their sign in going from one form to the other, the same will apply to s.i. and s.s. For this reason all probability formulae, such as (5.38), only define the cosine (i.e.

the absolute value) of the s.i. and s.s. In order to obtain all phases referred to one form it will be necessary to **fix the enantiomorph**. The phase determination process can then be summarized as shown in the flow-diagram of Fig. 5.15.

The choice of the space group with the corresponding set of symmetry operators already imposes some restrictions on the points where the origin may be localized. Indeed, only the sites with the same point symmetry will be suitable and represent the so-called permissible origins.

We will then have to choose the origin among the permissible ones and this may be performed by fixing the value of a limited number of suitable phases. In order to show how this is possible, let us first consider a one-dimensional structure (Fig. 5.16(a)) formed by three atoms of different atomic numbers ($Z_1 = 3$, $Z_2 = 2$, $Z_3 = 1$ and $\sum Z_j = 6$). The unitary structure factor for this structure, referred to an origin at X_0, will be

$$U_h(X_0) = A_h(X_0) + iB_h(X_0) = (1/6) \sum_{j=1}^{3} Z_j \cos 2\pi h(x_j - X_0) + (i/6) \sum_{j=1}^{3} Z_j \sin 2\pi h(x_j - X_0). \quad (5.60)$$

When the origin is shifted so that X_0 varies from zero to the identity period a, the trigonometric terms in (5.60) will change and A_h, B_h, and therefore U_h, will follow these changes. The modulus $|U_h|$ does not depend on the position of the origin, and only the phase φ_h will depend on X_0. Figure 5.16(b) illustrates this variation in the case $h = 1$: we can see that at $X_0 = 0$, $\varphi_1 = 109°$, while at $X_0 = 0.7$, $\varphi_1 = 0°$. A change of enantiomorph is achieved by inverting the direction of the x axis; again $|U_h|$ will not change, while all phases will change their sign ($\varphi \rightarrow 360 - \varphi$).

Since there is only one value of X_0 for which $\varphi_1 = 0°$, then the origin is uniquely defined by fixing $\varphi_1 = 0°$. In Fig. 5.16(c) the variation of φ_2 with X_0 is also shown: it can be seen that φ_2 goes to zero at two points, $X_0 = 0.26$

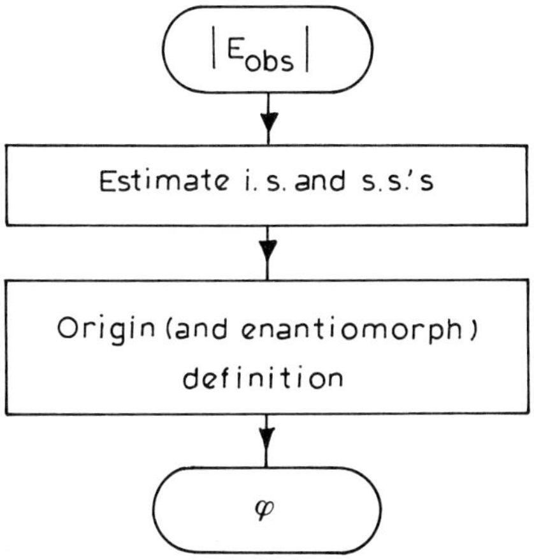

Fig. 5.15. Flow diagram of the phase determination process by direct methods.

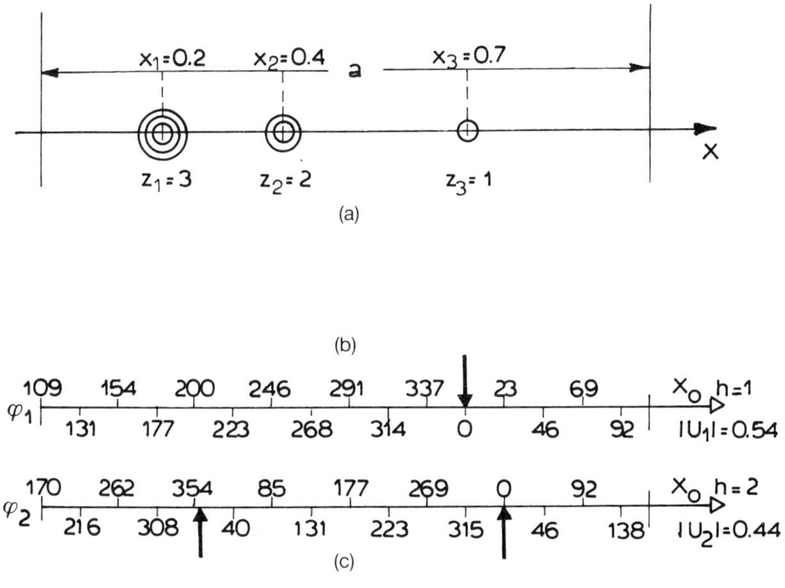

Fig. 5.16. (a) One-dimensional structure made of three non-equal atoms; (b) values of the phase φ_1 for different origin positions; (c) corresponding values of the phase φ_2.

and $X_0 = 0.76$, at a distance $a/2$ from each other. Fixing $\varphi_2 = 0°$ does not define the origin uniquely, but only restricts its possible position to two points; in general, by fixing $\varphi_h = 0°$, the origin is restricted to h possible positions. It should finally be observed that at $X_0 = 0.7$ (position for which $\varphi_1 = 0°$) $\varphi_2 = 315°$, and a change of enantiomorph will give $\varphi_2 = 45°$. It can then be seen that the enantiomorph can be chosen by restricting the value of φ_2 (or in general of another phase φ_h) within the interval 0–π (or, as in our case, π–2π).

Let us now generalize to three dimensions the procedure just described to fix the origin in one dimension. We will first consider the space group P1, in which any point of the unit cell is a permissible origin and therefore no semi-invariants which are not at the same time invariants exist. In analogy with the one-dimensional case, we can uniquely fix the origin along the x axis by fixing the phase of the (100) reflection and thus restricting the possible origins to lie on planes parallel to (100). Similarly the origin may be fixed in the other two directions by fixing the phases φ_{010} and φ_{001}. The three reflections (100), (010), and (001) define a primitive cell in the reciprocal lattice, but in a triclinic lattice there are infinite ways in which a primitive cell may be chosen. Indeed, any three non-coplanar vectors

$$\begin{aligned} \boldsymbol{H}_1 &= h_1\boldsymbol{a}^* + k_1\boldsymbol{b}^* + l_1\boldsymbol{c}^* \\ \boldsymbol{H}_2 &= h_2\boldsymbol{a}^* + k_2\boldsymbol{b}^* + l_2\boldsymbol{c}^* \\ \boldsymbol{H}_3 &= h_3\boldsymbol{a}^* + k_3\boldsymbol{b}^* + l_3\boldsymbol{c}^* \end{aligned} \tag{5.61}$$

will define a primitive cell if

$$\boldsymbol{H}_1 \cdot \boldsymbol{H}_2 \wedge \boldsymbol{H}_3 = V^* = \boldsymbol{a}^* \cdot \boldsymbol{b}^* \wedge \boldsymbol{c}^* = \text{volume of the reciprocal cell.}$$

Using (5.61) and developing the mixed product we can easily obtain

$$\boldsymbol{H}_1 \cdot \boldsymbol{H}_2 \wedge \boldsymbol{H}_3 = V^* \begin{pmatrix} h_1 & k_1 & l_1 \\ h_2 & k_2 & l_2 \\ h_3 & k_3 & l_3 \end{pmatrix}. \tag{5.62}$$

The three vectors will then form a primitive cell if

$$\Delta = \begin{pmatrix} h_1 & k_1 & l_1 \\ h_2 & k_2 & l_2 \\ h_3 & k_3 & l_3 \end{pmatrix} = \pm 1. \tag{5.63}$$

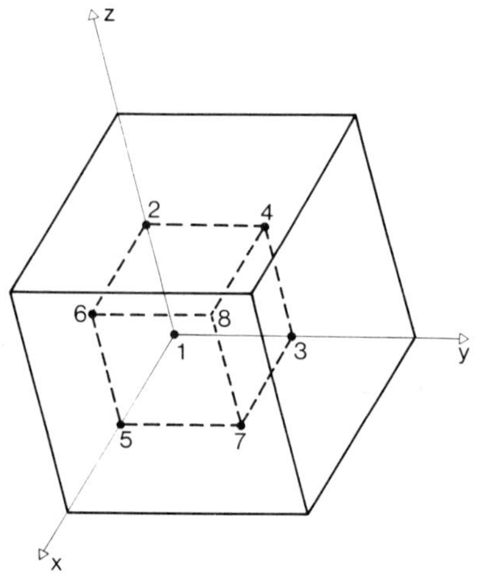

Fig. 5.17. Position of the eight distinct inversion centres in a P$\bar{1}$ unit cell, corresponding to the positions of the permissible origins, numbered as in Table 5.5.

By fixing $\varphi_{\boldsymbol{H}_1} = 0°$ we will restrict the possible origins to lie on planes parallel to the crystallographic planes of indices $\boldsymbol{H}_1$. If we also fix $\varphi_{\boldsymbol{H}_2} = 0°$, the possible origins will be at the same time restricted to lie on planes parallel to the planes $\boldsymbol{H}_2$, i.e. they will lie on the intersection lines between the two sets of planes. Finally, by fixing $\varphi_{\boldsymbol{H}_3} = 0°$ (with $\boldsymbol{H}_3 \neq m\boldsymbol{H}_1 + n\boldsymbol{H}_2$, m, n positive or negative integers) the origin will be further restricted to be at the intersection points of the above lines with the planes parallel to the $\boldsymbol{H}_3$ crystallographic planes. The number of such points within the unit cell is given by the value of Δ and only when the **primitivity condition** (5.63) is obeyed, the three phases will fix the origin in a unique way.

In the space group P$\bar{1}$ the permissible origins lie on the eight distinct inversion centres in the unit cell (Fig. 5.17). When the origin is shifted, for

Table 5.5. Sign variations for the reflections divided into parity groups, when the origin is placed at the different inversion centers in the $P\bar{1}$ cell of Fig. 5.17

Origin	Parity							
	ggg	ggu	gug	guu	ugg	ugu	uug	uuu
(1) 0, 0, 0	+	+	+	+	+	+	+	+
(2) 0, 0, $\frac{1}{2}$	+	−	+	−	+	−	+	−
(3) 0, $\frac{1}{2}$, 0	+	+	−	−	+	+	−	−
(4) 0, $\frac{1}{2}$, $\frac{1}{2}$	+	−	−	+	+	−	−	+
(5) $\frac{1}{2}$, 0, 0	+	+	+	+	−	−	−	−
(6) $\frac{1}{2}$, 0, $\frac{1}{2}$	+	−	+	−	−	+	−	+
(7) $\frac{1}{2}$, $\frac{1}{2}$, 0	+	+	−	−	−	−	+	+
(8) $\frac{1}{2}$, $\frac{1}{2}$, $\frac{1}{2}$	+	−	−	+	−	+	+	−

instance, from (0, 0, 0) to $(\frac{1}{2}, 0, 0)$, the phase $\varphi_{\boldsymbol{H}}$ of the reflection $\boldsymbol{H} \equiv (hkl)$ will change by $-\pi h$; this is equivalent to saying that the sign of the reflection will change or not depending on whether h is odd (u for ungerade in German) or even (g for gerade). Any change of origin among the eight permissible ones will have an effect on the sign of a reflection $\boldsymbol{H}$ which will depend on the parity of the three components h, k, or l (cf. Table 5.5). Reflections of type ggg are structure semi-invariants and may not be used to distinguish among the possible origins. If we consider a reflection of different parity, e.g. ggu, by imposing that its sign must be +, we restrict the possible origins to lie on four points (in Table 5.5 these are the origins 1, 3, 5, 7). In order to further reduce the ambiguity we will have to fix the sign of a reflection of different parity (not ggg), e.g. ugg; with reference to Table 5.5, by fixing the sign to be +, the possible origins are restricted to points 1 and 3. In order to fix uniquely the origin, we will have to fix the sign of a third different reflection. Its parity should not only be different from that of the two already chosen, but it must also be different from ugu (for which both origins 1 and 3 have a + sign); in fact the combination ggu + ugg + ugu = ggg is a s.s. We will, for instance, choose uug.

The above rules given for $P\bar{1}$ are also valid for all primitive centrosymmetric space groups with symmetry not higher than orthorhombic.

Similar procedures may be devised for the other space groups. If along a given direction the origin is restricted on points separated by $\frac{1}{2}$, then it can be fixed by fixing the phase of a reflection with an appropriate parity. When the origin can be shifted in a continuous way along an axis, it is possible to fix it by using a reflection the phase of which only takes unique values within the corresponding unit period.

Thus, in space group $P2_1$ (Fig. 5.18), with the b axis parallel to the twofold screw axis, the permissible origins are all the points on the four 2_1 axes at $(0, y, 0)$, $(\frac{1}{2}, y, 0)$, $(0, y, \frac{1}{2})$, $(\frac{1}{2}, y, \frac{1}{2})$. The choice of the twofold screw axis corresponds to that of one of the four inversion centers on the projection along y; the projection reflections $h0l$ have restricted phases 0, π. The phases φ_{g0g} are s.s. and indeed the (g0g) crystallographic planes pass through all permissible origins. On the other hand, a phase φ_{u0g} will have, for instance, a zero value if the origin is chosen at $(0, y, 0)$ or $(0, y, \frac{1}{2})$, and a π value if the origin is on the other two screw axes. By fixing $\varphi_{u0g} = 0°$ the

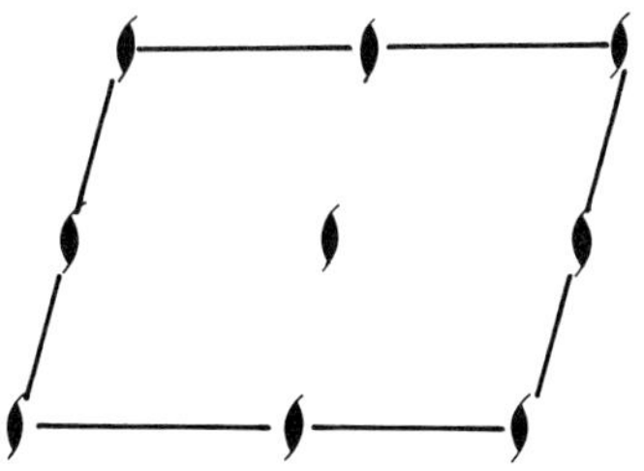

Fig. 5.18. Projection along the y axis of a $P2_1$ cell with the twofold screw axis parallel to b.

origins are restricted on the first two 2_1 axes. A second phase of type φ_{g0u} (or φ_{u0u}) will assume a zero value at $(0, y, 0)$ and a π value at $(0, y, \frac{1}{2})$, and by fixing $\varphi_{\mathrm{g0u}} = 0°$ the origins are restricted to lie on the first screw axis. In order to fix uniquely the origin along the y axis, we will have to fix a phase of type φ_{h1l}, because the $(h1l)$ planes intersect the y axis only once within the period b.

Let us finally consider the very common space group $P2_12_12_1$, in which the permissible origins are located at the eight points midway between the three orthogonal, non-intersecting 2_1 axes. The origin may be fixed in a simple way by using reflections belonging to the three principal zones $(0kl)$, $(h0l)$, and $(hk0)$. These reflections have phases restricted to two values: $0, \pi$ or $\pm\pi/2$ depending on whether the index following the zero, in a cyclic way, is even or odd (cf. Chapter 3, pp. 157–8). It is therefore possible to apply the rules derived for the space group $P\bar{1}$: the origin may be fixed by fixing the phases of three zone reflections belonging to three linearly independent parity groups (not ggg).

In the examples considered so far, we have always used three phases to fix the origin, and this is true for all primitive space groups up to the orthorhombic system. In the centred space groups some of the permissible origins are related by translational symmetry and are indistinguishable. For this reason the number of phases needed to fix the origin is reduced, as, at the same time, is reduced the number of allowed parity groups. Thus, for instance, in a C-centred lattice all ugg, ugu, gug, and guu reflections are systematically absent and the origin is fixed by fixing the phases or the signs of two reflections belonging to the other four parity groups.

Going from one enantiomorph form to the other will change the sign of all individual phases and, as a consequence, the sign of all linear combinations of phases. Therefore, the most general way of **fixing the enantiomorph** will be that of restricting a suitable s.i. or s.s. within the interval 0–π.

Let us for instance, consider the space group $P2_1$. Suppose that the two phases φ_{102} and φ_{213} have been assigned zero value to fix the origin; they form a triplet invariant with $\varphi_{\bar{3}\bar{1}1}$

$$\Phi = \varphi_{\bar{3}\bar{1}1} + \varphi_{102} + \varphi_{21\bar{3}}.$$

By restricting $0 \leqslant \varphi_{\bar{3}\bar{1}1} \leqslant \pi$ also $0 \leqslant \Phi \leqslant \pi$ and, if its value is sufficiently far from the extreme values zero or π, this choice will fix the enantiomorph. Unfortunately the most probable value of Φ is zero and in this space group it may be difficult to properly fix the enantiomorph by restricting the phase of one reflection. This problem does not arise in the space group $P2_12_12_1$ as will be shown by the following example. Suppose that the origin has been fixed using reflections of type: u0g (with phase $0, \pi$), gu0 $(0, \pi)$ and 0uu $(\pm\pi/2)$. Since all linear combinations of phases for which the sum of the indices is ggg, are s.s. in this space group, it is possible to set up s.s. with value restricted to $\pm\pi/2$; the choice of one of the two values allows an unambiguous choice of the enantiomorph. For instance

$$\varphi_{\mathrm{u0g}} + \varphi_{\mathrm{ug0}} = \pm\pi/2$$

is a two-phase s.s.; if φ_{u0g} is fixed to be zero, then the sign of the s.s. will depend on the value chosen for φ_{ug0}. By choosing $\varphi_{\mathrm{ug0}} = \pi/2$ we fix the enantiomorph.

For a complete and general theoretical treatment of the procedures for fixing the origin and the enantiomorph in all space groups, the reader is addressed to more specialized textbooks.[19,48]

A very effective way of defining the enantiomorph, and at the same time the absolute configuration (see later), is represented by the experimental measurement of some triplet phases with value close to $\pm\pi/2$, by multiple diffraction experiments (cf. Chapter 3, p. 191). In general the availability of a certain number of experimental triplet-phase values, which unfortunately are not easy to obtain, would be of great help in solving a structure by direct methods.

Phase determination procedures

We will first describe some of the steps which are common to all direct methods practical procedures.

Normalization

By the method described on p. 324 and using equation (5.15), the values of the normalized structure factors are first calculated. Most computer programs will supply a list of reflections sorted in decreasing order of $|E|$ and perform a statistical analysis of the normalized amplitudes as shown earlier. In some cases a more detailed statistical analysis is carried out in order to reveal the presence of pseudo-translational symmetry (cf. Appendix 5.E).[49–52]

Some of the most recent programs allow one to introduce the available a priori information, such as the existence of pseudo-translational symmetry or the coordinates of a previously located fragment;[53,54] examples of the use of this information will be given later.

Setting up phase relationships

We will focus our attention on triplets, the procedures for setting up other s.i. or s.s. being similar.

Since the most important relations are those estimated with the highest reliability, we will only set up triplets relating reflections with large $|E|$. The minimum $|E|$ value is chosen in such a way that the number of reflections employed is approximately equal to ten times the number of atoms in the asymmetric unit and usually lies between 1.2 and 1.6.

For each reflection $\boldsymbol{h}$ all pairs of reflections $\boldsymbol{k}$ and $\boldsymbol{h}-\boldsymbol{k}$ with large $|E|$ must be found; the set of all relationships obtained in this way is often called the Σ_2 **listing**. The triplet search must take into account the reciprocal lattice symmetry and the relations between the phases of equivalent reflections.

If all the indices are reconducted within the same independent portion of the reciprocal lattice, each reflection $\boldsymbol{h}$ will have to be combined with all vectors $\pm\boldsymbol{k}\mathbf{R}_i$, obtained by varying $\mathbf{R}_i$ over all rotation matrices of the space group not related by an inversion centre; if also the $|E|$ value of reflection $\boldsymbol{h}\pm\boldsymbol{k}\mathbf{R}_i$ is large then the triplet is retained. In general this third reflection will not be in the considered independent portion of reciprocal lattice, and it can be expressed as $\pm\boldsymbol{H}\mathbf{R}_j$. The phase relationship will then become

$$\varphi_{\boldsymbol{h}} \simeq \pm\varphi_{\boldsymbol{k}\mathbf{R}_i} \pm \varphi_{\boldsymbol{H}\mathbf{R}_i} = \pm\varphi_{\boldsymbol{k}} \pm \varphi_{\boldsymbol{H}} - 2\pi(\boldsymbol{k}\mathbf{T}_i + \boldsymbol{H}\mathbf{T}_j). \qquad (5.64)$$

Most computer programs store triplet relations in the form

$$n_{\boldsymbol{h}}, \quad \pm n_{\boldsymbol{k}}, \quad \pm n_{\boldsymbol{H}}, \quad \Delta\varphi, \quad G; \tag{5.65}$$

where $n_{\boldsymbol{h}}$, $n_{\boldsymbol{k}}$, $n_{\boldsymbol{H}}$ represent the sequential numbers of the corresponding reflections in the sorted list, with the sign indicated by (5.64), $\Delta\varphi = 2\pi(\boldsymbol{k}\mathbf{T}_i + \boldsymbol{H}\mathbf{T}_j)$ and G is either defined by (5.39) or it is the parameter obtained by means of the second representation. For centrosymmetric structures (5.64) reduces to

$$S(\boldsymbol{h}) \simeq S(\boldsymbol{k})S(\boldsymbol{H})(-1)^{2(\boldsymbol{k}\mathbf{T}_i + \boldsymbol{H}\mathbf{T}_j)} \tag{5.66}$$

and in (5.65) the signs are of no use, while G represents the tanh argument in (5.46) or the corresponding value from the second representation.

Definition of an optimum starting set of phases

From (5.64) or (5.66), given the two phases or signs in the right-hand side we can derive an estimate of the left-hand side phase, the accuracy of which will increase with increasing G. We have seen that the origin may be fixed by assigning an arbitrary value to up to three phases. Even if these are chosen among the reflections involved in large numbers of triplets, in general they will not be sufficient to start up the phase determination process. It is necessary to start with a larger number of phases, and the problem is not overcome even when a limited number of reliably estimated one-phase s.s. (usually fewer than five) is available. Two methods are used to define the values of the extra unknown phases necessary to start up the process. With the **symbolic addition**[55,26] method these phases are assigned a symbolic value and all other phases are determined as combinations of these symbols; it is more suitable for centrosymmetric structures as we shall see later through an example. With the **multisolution**[56,57] method, the unknown starting phases are directly assigned permuted numerical values. This method is easier to automate and can more readily incorporate some of the most recent developments in direct methods; for this reason it is more widely used and will be described in more detail.

Given the phases in the starting set, all other phases are determined one after the other in a chain process. The choice of the reflections in the starting set is therefore very important and the best starting set will be that allowing one to eliminate, as much as possible, the weak links in the chain, i.e. the poor determination of some phases by few low-reliability relationships. An optimal starting set of phases can be defined by means of the **convergence** procedure.[57] We have seen that $\alpha_{\boldsymbol{h}}$ as defined by (5.43) is a measure of the reliability with which a phase $\varphi_{\boldsymbol{h}}$ may be determined combining several phase relations. If the vector representing $\alpha_{\boldsymbol{h}}$ in Fig. 5.12 is rotated by an angle $\beta_{\boldsymbol{h}} = \langle \phi_{\boldsymbol{h}} \rangle$ in such a way that it falls on the real axis, Fig. 5.19 is obtained; it can be seen that

$$\alpha_{\boldsymbol{h}} = \sum_{j=1}^{r} G_j \cos \Phi_j \tag{5.67}$$

where $\Phi_j = \langle \varphi_{\boldsymbol{h}} \rangle - \varphi_{\boldsymbol{k}_j} - \varphi_{\boldsymbol{h}-\boldsymbol{k}_j} \approx \Phi_{\boldsymbol{hk}_j}$. Since at the beginning the phases are not known, $\alpha_{\boldsymbol{h}}$ can not be computed, but it can be estimated a priori by

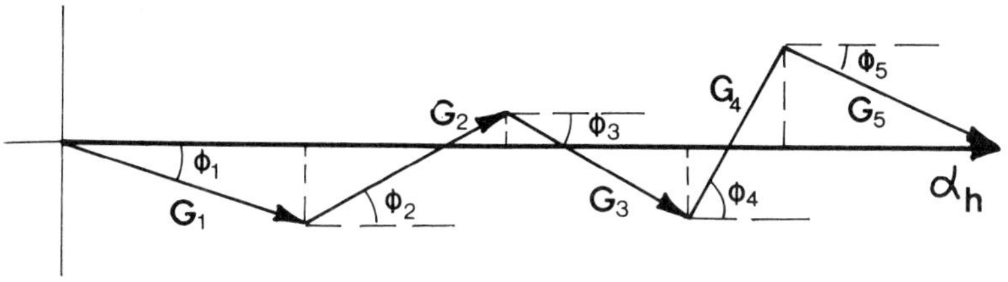

Fig. 5.19. Figure 5.12 rotated by an angle $\beta_{\boldsymbol{h}} = \langle \varphi_{\boldsymbol{h}} \rangle$ in order to bring the vector $\boldsymbol{\alpha}_{\boldsymbol{h}}$ on to the real axis.

substituting each $\cos \Phi_j$ in (5.67) with its expected value, defined as

$$\langle \cos \Phi_j \rangle = \frac{1}{L} \int_0^{2\pi} \cos \Phi_j \exp(G_j \cos \Phi_j) \, \mathrm{d}\Phi_j = \frac{I_1(G_j)}{I_0(G_j)} = D_1(G_j) \quad (5.68)$$

where $I_0(x)$ and $I_1(x)$ are modified Bessel functions of zero and first order. These functions and their ratio are readily available[58] and Fig. 5.20 illustrates the trend of (5.68) as a function of G; it may be seen that the ratio tends asymptotically to 1.0 as G increases.

The estimated value $\langle \alpha_{\boldsymbol{h}} \rangle$ is then given by

$$\langle \alpha_{\boldsymbol{h}} \rangle = \sum_{j=1}^{r} G_j D_1(G_j), \quad (5.69)$$

and it can be computed for each reflection, before any phase information has been obtained, using the G values only.

The convergence process is a step process in which, at each step, the reflection with minimum $\langle \alpha_{\boldsymbol{h}} \rangle$ is (temporarily) eliminated, provided it is not an accepted one-phase s.s. or another reflection already included in the starting set. When a reflection is eliminated, at the same time, all phase relationships contributing to it are eliminated and the $\langle \alpha_{\boldsymbol{h}} \rangle$ values of the other reflections involved in these relations are updated. Since, at each step, the reflection which is less related to the remaining reflections is eliminated, the process must converge towards the group of reflections which are most strongly interrelated and are therefore the most effective in starting the phase determination process.

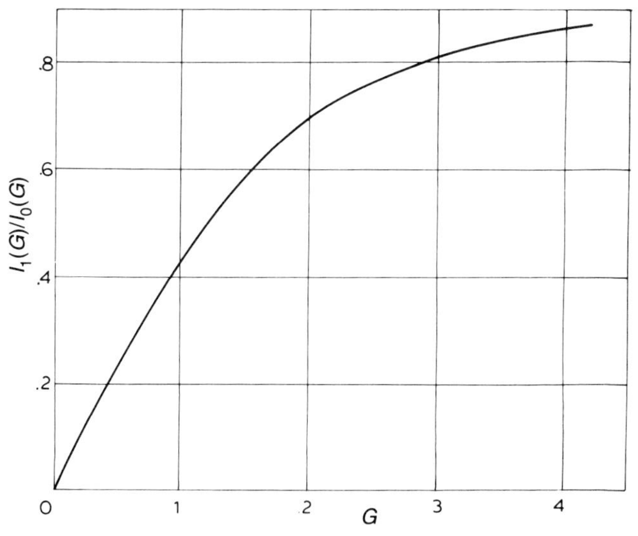

Fig. 5.20. Trend of the ratio, $I_1(G)/I_0(G)$, of the modified Bessel functions of first and zero order, as a function of G.

Before eliminating a reflection, the remaining ones are analysed to check if it is still possible to fix the origin; if not, the reflection is not removed but kept in the starting set as an origin-fixing reflection. If, in the final steps, a reflection is eliminated with an $\langle \alpha_{\mathbf{h}} \rangle$, which, at the elimination moment, is reduced to zero (or to a very low value), than this reflection is not (or is only poorly) connected with the remaining few reflections; on the other hand it is an essential link between these last reflections and those previously eliminated. By putting this reflection into the starting set, we assign a value to its phase and in this way we insert the missing link (or replace the weak link) in the chain process. The convergence process is repeated several times, each time adding a new reflection into the starting set, until the desired number of starting-set reflections has been reached. The flow diagram of Fig. 5.21 illustrates the convergence procedure. The procedure may be further improved by assigning to each contributor to the summation in (5.69) a weight depending on $\langle \alpha_{\mathbf{k}} \rangle$ and $\langle \alpha_{\mathbf{h-k}} \rangle$.[59]

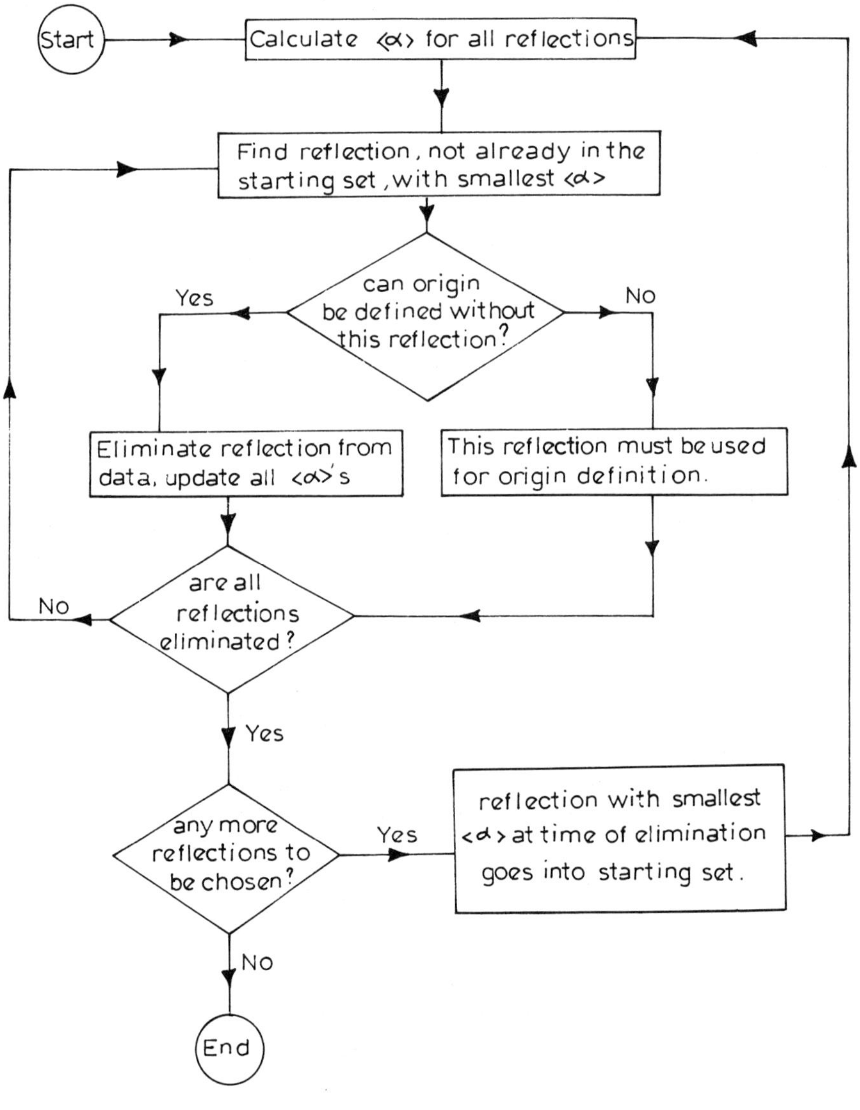

Fig. 5.21. Flow diagram of the convergence procedure.

Figures of merit

As we shall see, the phase determination process usually leads to more than one solution. Given several sets of phases it would be rather time consuming to compute and interpret all the corresponding electron density maps to see which yield the correct structure. It is instead easier to compute some appropriate functions, called **figures of merit** (fom), which allow an a priori estimate of the goodness of each phase set. Several functions have been proposed[60] and we will analyse those most commonly used.

1. **MABS** (absolute fom) represents a measure of the internal consistency of the employed triplet relationships in estimating the phases. It is defined as

$$\text{MABS} = \frac{\sum_{\boldsymbol{h}} \alpha_{\boldsymbol{h}}}{\sum_{\boldsymbol{h}} \langle \alpha_{\boldsymbol{h}} \rangle} = \frac{A}{A_{\mathrm{e}}}; \tag{5.70}$$

for a correct structure A should be close to the theoretically estimated A_{e} and $\text{MABS} \approx 1.0$. In practice it has been found that often for the correct set of phases $A > A_{\mathrm{e}}$ and MABS values between 0.9 and 1.3 indicate a promising phase set.

2. $\boldsymbol{R_\alpha}$ **fom**: this is a measure of how much triplets deviate from their expected statistical behavior and is defined as

$$R_\alpha = 100\left(\sum_{\boldsymbol{h}} |\alpha_{\boldsymbol{h}} - \langle \alpha_{\boldsymbol{h}} \rangle|\right) \Big/ A_{\mathrm{e}}; \tag{5.71}$$

it should be minimum for the correct set of phases.

3. $\boldsymbol{\psi_0}$ **fom**: this is defined as

$$\psi_0 = \frac{\sum_{\boldsymbol{l}} \left(\left| \sum_{\boldsymbol{k}} E_{\boldsymbol{k}} E_{\boldsymbol{l}-\boldsymbol{k}} \right| \right)}{\sum_{\boldsymbol{l}} \left(\sum_{\boldsymbol{k}} |E_{\boldsymbol{k}} E_{\boldsymbol{l}-\boldsymbol{k}}|^2 \right)^{1/2}} = \frac{\sum_{\boldsymbol{l}} \alpha'_{\boldsymbol{l}}}{\sum_{\boldsymbol{l}} v_{\boldsymbol{l}}^{1/2}}. \tag{5.72}$$

where the outer summations are over a certain number (50–150) of reflections with lowest $|E_{\boldsymbol{l}}|$ value. The triplet-generating routine will have to set up also the relationships (ψ_0 triplets) linking each of these $\boldsymbol{l}$ reflections with pairs of reflections $\boldsymbol{k}$ and $\boldsymbol{l} - \boldsymbol{k}$ with large $|E|$, the phases of which have been determined. The inner summation at the numerator corresponds to Sayre's equation (5.24) (written here in terms of normalized structure factors) relative to each reflection $\boldsymbol{l}$. Since $|E_{\boldsymbol{l}}| \approx 0$, the large terms in the summation must tend to cancel each other out and the correct set of phases will correspond to a minimum ψ_0 value. This function differs from the previous foms because it is not a self-consistency figure among the determined reflections only, but it establishes the coherence of the obtained phases with some weak reflections not used for their determination.

4. **SS1FOM, SS2FOM, NTREST, NQUEST:** these four functions have the same functional form:

$$\text{fom}_j = \frac{\sum_r G_{jr} \cos \Phi_{jr}}{\sum_r |G_{jr}|\, D_1(|G_{jr}|)} = \frac{T_j}{B_j} \tag{5.73}$$

with $j = 1$ for one-phase s.s., $j = 2$ for two-phase s.s., $j = 3$ for negative triplets, $j = 4$ for negative quartets respectively, and r running over all estimated relationships of each type; as shown in (5.68), $D_1(|G_{jr}|)$ is the expected value of $\cos \Phi_{jr}$. For instance in the case of negative quartets (5.73) becomes

$$\text{NQUEST} = \frac{\sum_r G_r \cos Q_r}{\sum_r |G_r| \, D_1(|G_r|)} = \frac{T_4}{B_4} \tag{5.74}$$

where the summation is extended to all quartets estimated to be negative ($G_r < 0.0$) by means of their first representation and Q_r are the values of the individual quartets computed with the determined phases. The correct set of phases should give rise to the largest number of negative $\cos Q_r$ values, thus maximizing NQUEST. Similarly for the other three functions, a good agreement between the estimated (through the G_j values) and the calculated $\cos \Phi_j$ values, maximizes their values. The four foms based on phase relationships, are finally combined to give

$$\text{CPHASE} = \frac{T_1 + T_2 + T_3 + T_4}{B_1 + B_2 + B_3 + B_4}. \tag{5.75}$$

A better understanding of the statistical meaning of the MABS, R_α and ψ_0 foms was recently achieved by a more rigorous probabilistic approach,[60] which also allowed the derivation of some new effective functions. At the basis of this approach is the use of the probability distributions $P(\alpha_{\boldsymbol{h}})$ of $\alpha_{\boldsymbol{h}}$ defined in (5.43) and $P(\alpha'_{\boldsymbol{h}})$ of $\alpha'_{\boldsymbol{h}}$ defined in (5.72), derived by Cascarano *et al.*[61] On p. 322 we saw how the expectation values in Table 5.1 could be obtained from the distributions (5.10) and (5.11). By a similar method, from $P(\alpha_{\boldsymbol{h}})$ and $P(\alpha'_{\boldsymbol{h}})$ it is possible to derive various theoretical moments, which may then be compared with the corresponding 'experimental' values computed using the different sets of phases; MABS and R_α are related to moments of $P(\alpha_{\boldsymbol{h}})$, while ψ_0 is related to a moment of $P(\alpha'_{\boldsymbol{h}})$, but the expected values of other appropriate functions of $\alpha_{\boldsymbol{h}}$ and $\alpha'_{\boldsymbol{h}}$ may also be derived and used as foms.

Once the different foms have been calculated for each phase set it is useful to derive a **combined figure of merit**, CFOM; its capability of discriminating the correct set of phases will in general be higher than that of the individual functions. Different ways of combining the individual foms have been used; a more objective procedure is described by Cascarano *et al.*[60] in which each function is introduced with a properly estimated weight and the resulting combined fom is normalized to fall between -1.0 and 1.0. A correct set of phases should lead to $\text{CFOM} \approx 1.0$, while $\text{CFOM} \ll 1.0$ should indicate a wrong solution.

The electron density map(s) corresponding to the most promising phase set(s) (with highest CFOM) will be calculated first.

Electron density maps (E-maps)

In the calculation of the electron density maps it is more convenient to use as coefficients of the Fourier series, E rather than F. In fact the reflections with large $|E|$, for which phases are determined, are mainly of two types

Table 5.6. AZOS: list of the 50 reflections with largest $|E|$ value

No.	*h*	*k*	*l*	*E*	No.	*h*	*k*	*l*	*E*
1	4	0	−2	4.09	2	16	0	4	3.35
3	0	1	2	3.24	4	6	4	7	3.24
5	19	2	0	3.23	6	14	4	−3	3.16
7	3	5	4	3.06	8	1	1	−2	3.02
9	20	1	0	2.84	10	2	4	6	2.83
11	16	3	−4	2.79	12	8	6	4	2.77
13	10	5	1	2.74	14	16	4	−2	2.70
15	10	4	5	2.67	16	10	4	−1	2.65
17	18	4	−5	2.64	18	4	1	−4	2.63
19	5	2	0	2.60	20	15	2	1	2.59
21	21	2	4	2.58	22	13	5	0	2.55
23	20	0	2	2.54	24	15	3	4	2.52
25	23	2	−2	2.52	26	18	0	3	2.51
27	3	6	2	2.47	28	19	1	1	2.47
29	19	3	2	2.45	30	20	2	−2	2.44
31	16	1	2	2.41	32	15	1	3	2.41
33	2	5	4	2.40	34	14	5	−1	2.39
35	12	6	2	2.37	36	20	4	−4	2.36
37	14	0	5	2.33	38	1	2	−4	2.33
39	4	7	0	2.31	40	8	6	0	2.28
41	7	5	2	2.27	42	18	1	2	2.27
43	5	6	5	2.26	44	9	6	3	2.25
45	7	1	2	2.24	46	8	7	−2	2.23
47	19	2	−1	2.23	48	5	1	−4	2.23
49	18	4	3	2.22	50	5	2	−6	2.21

(this may be verified in Table 5.6): some low-angle reflections (small indices) with very large $|F|$ and several high-angle (large indices) ones with relatively small $|F|$. The contribution of these small $|F|$ to the electron density would be obscured by that of the low-angle reflections, and by using the $|E|$ this problem is overcome and the resolution of the maps is improved. On the other hand, also for the limited number of terms in the Fourier summations, the E-maps often show spurious peaks.

The programs for the calculation of the electron density maps usually also perform an automatic peak search and supply a list of maxima sorted in decreasing order of height. This list may then be analysed in terms of distances and angles among the peaks; peaks related by a sensible chemical geometry may be selected and labelled as a possible molecular fragment. The peaks of each fragment, labelled by their sequential numbers, may then be represented in projection directly on the line-printer output, where it will be easy to reconstruct the image of the molecular fragment by connecting the peaks related by possible chemical bonds (cf. Figs. 5.22 and 5.23). Attempts have been made recently to establish an automatic link between the E-map output and the structure refinement programs;[62] to this aim the map interpretation procedure has been improved and the peaks are also assigned an atomic symbol. For simple structures it then becomes possible to go from the diffraction intensities to the refined molecular picture without human intervention.

Unfortunately the phase set with the highest CFOM does not always yield an E-map interpretable in terms of a chemically plausible structure. In these cases it will be necessary to compute some other E-maps, with the hope that another statistically less favoured phase set will yield the correct structure.

Besides, in the case of complex structures, the E-maps often only show a partial image of the structure, which will then have to be completed.

We may now illustrate the most common phase determination procedures.

Symbolic addition method

This method will be illustrated through an example. We will use the same AZOS structure employed earlier (p. 323); the 50 reflections with largest $|E|$ value are listed in Table 5.6. When using all sign relations of type (5.47) relating these reflections, the convergence procedure defines the following starting set of signs:

	No.	h	k	l	Sign	Parity
Origin-fixing reflections	3	0	1	2	+	gug
	8	1	1	2	+	uug
	16	10	4	1	+	ggu
Other reflections with symbolic sign	1	4	0	2	a	
	2	16	0	4	b	
	12	8	6	4	c	

In the unit cell there are four molecules of row formula $C_{10}H_{11}N_3O_4S$ and $\sigma_3\sigma_2^{-3/2} = 0.141$. The probability with which a sign is determined is given by (5.46) or (5.48) and it can be seen that it is never less than 0.95, the minimum tanh argument being $0.141 \times (2.21)^3 = 1.522$.

The sign expansion procedure may be followed in detail in Table 5.7. At step (17) a new symbol d must be introduced in order to define new signs; at step (20) we have two different, but not contradictory, symbolic indications for the sign of the same reflection and their comparison suggests that symbols b and c correspond to the same sign. Similarly at step (22) we have the indication that a and b represent opposite signs. Not all the signs of the first 50 reflections are determined; 10 signs are not defined with only four symbols. At this point it is not convenient to introduce new symbols, but rather to use other reflections with smaller $|E|$. The probability of the individual sign relations will decrease, but, with 40 signs already defined, we will have several multiple indications. For AZOS 200 reflections with $|E| \geqslant 1.61$ were used. As the sign determination is carried out, several other indications about the values of the symbols are obtained: the indications $b = c$ and $a = -b$ are confirmed several times and new indications that $a = -$ and $b = +$ (in agreement with $a = -b$) are obtained. No indication is obtained about symbol d. Out of the $2^4 = 16$ possible sign combinations, only two are consistent with the previous indications: $- + + +$ and $- + + -$. Also some contradictory indications are obtained; for instance, $S(165)$ has seven contributions: $-$, ab, ab, ab, $-$, $-$, $+$. The first six indications confirm the relation $a = -b$, while the seventh, with very low probability, is clearly contradictory with respect to the others and it will be neglected. If the determination of one or more signs turns out to have several contradictory indications of similar probability, it should be supposed that a wrong choice has been made at some previous step and the procedure should be reconsidered from the beginning. In our case this did not happen and it has been possible to define the signs of all reflections

Table 5.7. Step by step illustration of the symbolic addition procedure for the 50 reflections of Table 5.6

Step	Relation	Sign	Comments and indications
1	$S(23) \simeq S(1)\,S(2)$	$a\ b$	
2	$S(18) \simeq -S(1)\,S(3)$	$-a$	
3	$S(9) \simeq S(18)\,S(2)$	$-a\ b$	
	$-S(23)\,S(3)$	$-a\ b$	
4	$S(31) \simeq S(1)\,S(9)$	$-\ \ b$	$aa = +$
	$-S(2)\,S(3)$	$-\ \ b$	
5	$S(35) \simeq S(1)\,S(12)$	$a\ \ c$	
6	$S(40) \simeq S(2)\,S(12)$	$b\ c$	
	$S(23)\,S(35)$	$b\ c$	
7	$S(48) \simeq S(1)\,S(8)$	a	
8	$S(22) \simeq S(35)\,S(8)$	$a\ \ c$	
	$S(48)\,S(12)$	$a\ \ c$	
9	$S(7) \simeq S(22)\,S(2)$	$a\ b\ c$	
	$S(40)\,S(48)$	$a\ b\ c$	
10	$S(41) \simeq S(1)\,S(7)$	$b\ c$	
	$S(8)\,S(40)$	$b\ c$	
	$S(22)\,S(23)$	$b\ c$	
11	$S(50) \simeq S(8)\,S(18)$	$-a$	
	$-S(3)\,S(48)$	$-a$	
12	$S(38) \simeq -S(3)\,S(8)$	$-$	
	$S(50)\,S(1)$	$-$	symbol a cancels out
13	$S(46) \simeq -S(3)\,S(40)$	$-\ \ b\ c$	
	$S(50)\,S(7)$	$-\ \ b\ c$	
	$S(9)\,S(35)$	$-\ \ b\ c$	
	$S(12)\,S(31)$	$-\ \ b\ c$	
	$S(38)\,S(41)$	$-\ \ b\ c$	
14	$S(39) \simeq S(46)\,S(1)$	$-a\ b\ c$	
	$S(7)\,S(38)$	$-a\ b\ c$	
	$S(31)\,S(35)$	$-a\ b\ c$	
15	$S(27) \simeq -S(3)\,S(7)$	$-a\ b\ c$	
	$S(39)\,S(8)$	$-a\ b\ c$	
	$S(31)\,S(22)$	$-a\ b\ c$	
	$S(46)\,S(48)$	$-a\ b\ c$	
16	$S(10) \simeq S(7)\,S(8)$	$a\ b\ c$	
	$S(27)\,S(38)$	$a\ b\ c$	
	$S(41)\,S(48)$	$a\ b\ c$	
17	$S(14)$ new symbol	d	introduced because fewer than half the signs have been determined
18	$S(5) \simeq S(14)\,S(27)$	$-a\ b\ c\ d$	
19	$S(11) \simeq -S(5)\,S(7)$	d	
	$S(14)\,S(3)$	d	
20	$S(36) \simeq S(1)\,S(14)$	$a\ \ \ \ d$	Two different indications
	$S(38)\,S(5)$	$a\ b\ c\ d$	implying that $bc = +$ or
	$-S(39)\,S(11)$	$a\ b\ c\ d$	$b = c$; $abcd$ is chosen
21	$S(25) \simeq S(1)\,S(5)$	$-\ \ b\ c\ d$	
	$-S(41)\,S(11)$	$-\ \ b\ c\ d$	Confirm $b = c$
	$S(27)\,S(36)$	$-\ \ \ \ d$	
22	$S(13) \simeq S(16)\,S(3)$	$+$	Indication that $ab = -$
	$S(16)\,S(9)$	$-a\ b$	or $a = -b$
23	$S(6) \simeq S(16)\,S(1)$	a	
	$-S(13)\,S(18)$	a	
24	$S(34) \simeq S(1)\,S(13)$	a	
	$S(3)\,S(6)$	a	

Table 5.7. (*Continued*)

Step	Relation	Sign	Comments and indications
25	$S(42) = S(5)\,S(8)$	$-a\ b\ c\ d$	
	$-S(10)\,S(11)$	$-a\ b\ c\ d$	
	$S(48)\,S(25)$	$-a\ b\ c\ d$	
26	$S(33) \simeq -S(3)\,S(10)$	$-a\ b\ c$	
	$S(8)\,S(27)$	$-a\ b\ c$	
	$S(42)\,S(14)$	$-a\ b\ c$	
27	$S(17) \simeq S(1)\,S(6)$	$+$	
	$-S(18)\,S(34)$	$+$	
28	$S(44) \simeq S(8)\,S(13)$	$+$	
	$-S(50)\,S(6)$	$+$	
	$-S(16)\,S(38)$	$+$	
	$S(48)\,S(34)$	$+$	
29	$S(43) \simeq S(44)\,S(1)$	a	
	$S(48)\,S(13)$	a	
	$-S(50)\,S(16)$	a	
30	$S(45) \simeq S(11)\,S(25)$	$-\ b\ c$	
	$-S(27)\,S(39)$	$-$	Again $b = c$ confirmed
31	$S(29) \simeq S(3)\,S(5)$	$-a\ b\ c\ d$	
	$-S(18)\,S(25)$	$-a\ b\ c\ d$	
32	$S(24) \simeq S(29)\,S(1)$	$-\ b\ c\ d$	
	$-S(5)\,S(18)$	$-\ b\ c\ d$	
33	$S(30) \simeq -S(3)\,S(9)$	$a\ b$	
	$S(31)\,S(18)$	$a\ b$	

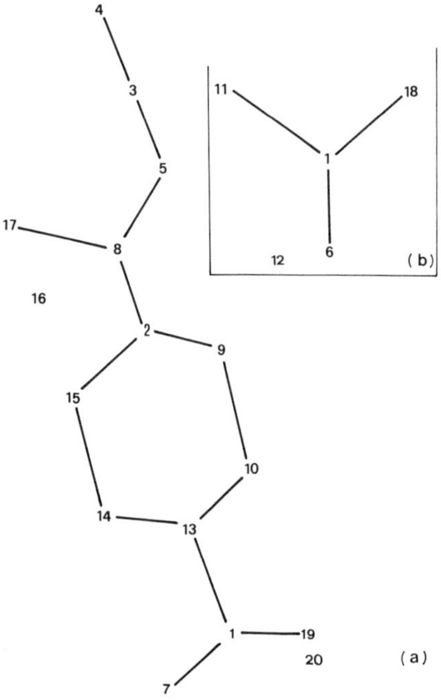

Fig. 5.22. AZOS: representation and interpretation of the E-map computed with the best sign set obtained by application of the symbolic addition procedure.

using four symbols. When the most probable values are substituted for the symbols, two sets of signs are obtained for AZOS and the corresponding electron density maps may be computed. The values of the MABS, R_α, and ψ_0 foms indicate that the set with $d = -$ is more reliable and, indeed, the corresponding E-map shows the two fragments interpreted in Fig. 5.22 in terms of a *p*-carboxyphenylazoxycyano group (5.22(a)) and a dimethylsulphoxide group (5.22(b)). The structure obtained in this way will then be refined with the procedures described in the next paragraph.

The symbolic addition method, both for centrosymmetric and for non-centrosymmetric structures, has also been automated into a number of computer programs, the most general of which is SIMPEL;[63,64] but, so far, the most striking results[65] have been obtained by I. and J. Karle using a careful manual procedure proposed in 1966.[26]

Multisolution methods

The basic idea of these methods is that of assigning approximate numerical values to the starting phases, instead of using symbols. It has been found empirically that initial errors of 40–50° usually do not spoil structure solutions. The starting set, defined by the convergence procedure, will include, besides the origin and, when needed, enantiomorph fixing reflections, also a limited number of other phases necessary to initiate the phase expansion process by means of the tangent formula (5.44). If these are general phases with values anywhere between 0 and 2π, we may tentatively give them the **four quadrant values**: $\pm\pi/4$, $\pm 3\pi/4$. One of these will be correct within 45°. All restricted phases are assigned the values defined by the space-group symmetry; e.g. 0, π, or $\pm\pi/2$.

If ng and ns are the numbers of general and special reflections in the starting set, the total number of combinations of numerical values to be developed by the tangent formula, will be

$$n = 4^{ng} \times 2^{ns}.$$

This number grows very rapidly with increasing ng and ns and only by limiting the number of starting-set reflections it is possible to maintain the computing time within reasonable limits. This limit can be greatly reduced by using the so called **magic integers** (described in the Appendix 5.F), which allow a considerable reduction of the number of combinations with a minimum increase in the phase error.

Most multisolution computer programs, such as MULTAN,[66] are mainly based on the use of triplets estimated by Cochran's formula, to which a few one-phase s.s. estimated by the Σ_1 formulae may be added. With these programs it is possible to solve structures with up to 60–70 atoms in the asymmetric unit. Although more complex structures have also been solved, it may also happen that simpler structures can not be solved. In fact, because of the rather crude probabilistic estimate of triplets, it may happen that at some stage of the 'chain' phase-expansion process, some triplets with an actual value quite far from zero are used; as a result the determined phases are completely wrong. In the recent years several new developments have been proposed in order to overcome these problems and to make direct methods capable of solving increasingly more complex structures. In Appendix 5.G the most promising developments of the multisolution techniques are outlined, while in the following we will describe, through a practical example, the use of the SIR (semi-invariant representation) program;[67] this program is based on the multisolution strategy strengthened by the use of all phase relationships for which a reliable estimate may be obtained by means of the representation method.

We will follow in detail the solution of the structure of the antibiotic 21-acetoxy-11-(R)-rifamicinol (RIFOL),[68] $C_{39}H_{49}NO_{13}.CH_3OH{\cdot}H_2O$, which crystallizes in the space group $P2_1$ with $Z = 2$. Its structural formula is

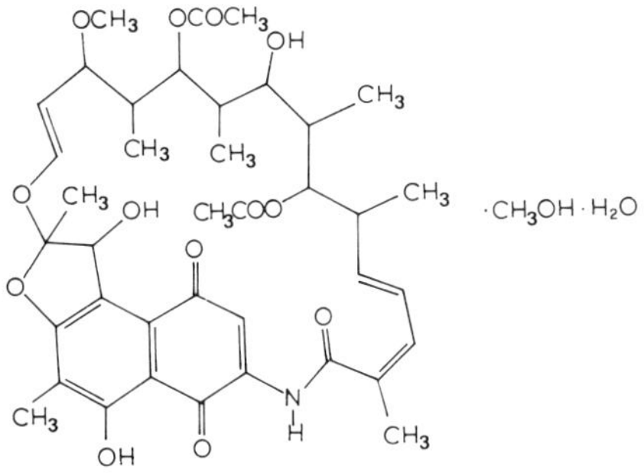

Normalized structure factors are computed by the normalization routine and 362 reflections with $|E| > 1.723$ selected for phase determination; the non-centrosymmetric space group is confirmed by the statistical analysis and no pseudo-translational symmetry (cf. Appendix 5.E) is detected.

None of the 17 one-phase s.s. estimated by means of the second representation has a sufficiently high tanh argument to be accepted and these indications are only used to compute SS1FOM. 326 two-phase s.s. are estimated using their first representation; all of them are used to compute SS2FOM, while 129 with $|G| > 0.6$ are actively used in the phase determination procedure.

Triplet invariants relating the 362 strongest reflections are set up and estimated by means of their second representation (P10 formula).[45] The concentration parameter of the von Mises distribution is given by

$$G' = G(1 + R) \tag{5.76}$$

where G is the Cochran parameter defined by (5.39) and R is a function of all the magnitudes in the second phasing shell. R may also be negative and for some triplets a negative G' value may be obtained, indicating that their expected value is π. The reliability of the negative estimates is in general too small for an active use of negative triplets in the phase determination process, but a passive use to compute an effective fom proved to be very useful. Besides, the capability of the P10 formula of sampling out most triplets with a value far from zero, allows their elimination from the phase determination process, which then becomes more stable. For RIFOL 2280 positive triplets estimated with $G' \geqslant 0.6$ are actively used, while 141 negatively estimated triplets are employed to compute the NTREST fom. At the same time 1399 ψ_0 triplets relating the 100 weakest reflections with the considered 362 strongest reflections are generated, to be used in the calculation of the corresponding fom. Finally also 1009 negative quartets are estimated by their first representation and the most reliable 500 are used to compute NQUEST.

The convergence procedure is carried out using positive triplets and two-phase s.s. When also some one-phase s.s. are accepted, these are treated as known phases and directly included in the starting set. Two-phase s.s. are treated in the same way as triplets. In fact for each reflection $\boldsymbol{h}$ forming a s.s. with $\boldsymbol{k}$, we have

$$\varphi_{\boldsymbol{h}} = \pm\varphi_{\boldsymbol{k}} + S2 \tag{5.77}$$

where $S2$ is the value of the s.s. estimated with a reliability depending on the concentration parameter G. The contribution of (5.77) to $\langle \alpha_{\boldsymbol{h}} \rangle$ will be given by $G \cdot D_1(G)$. For RIFOL the convergence procedure defines the starting set reported in Table 5.8. It may be noticed that the origin is fixed following the rules given earlier (p. 349) for the space group $P2_1$. All permuted phases are general and are expressed in terms of magic integers, giving rise to 24 phase permutations (cf. Appendix 5.F, enantiomorph fixed by the magic integer procedure).

Table 5.8. RIFOL: starting set of phases defined by the convergence procedure

Type	Code	*h*	*k*	*l*	*E*	Phase
Origin	5	5	0	−18	3.57	0
	15	2	0	−1	2.93	0
	4	6	1	−18	3.85	0
Permuted	18	6	4	22	2.92	MI
	72	10	4	−5	2.40	MI
	149	3	1	−17	2.09	MI
	25	1	2	−16	2.79	MI
	17	8	4	−4	2.92	MI

Total number of permutations 24

In the SIR program a weighted tangent formula is used, i.e.

$$\tan \varphi_{\boldsymbol{h}} = \frac{\sum_t G_t W_t \sin(\varphi_{\boldsymbol{k}_t} + \varphi_{\boldsymbol{h}-\boldsymbol{k}_t}) + \sum_s G_s W_s \sin(\varphi_{\boldsymbol{k}_s} + S2_s)}{\sum_t G_t W_t \cos(\varphi_{\boldsymbol{k}_t} + \varphi_{\boldsymbol{h}-\boldsymbol{k}_t}) + \sum_s G_s W_s \cos(\varphi_{\boldsymbol{k}_s} + S2_s)} = \frac{T_{\boldsymbol{h}}}{B_{\boldsymbol{h}}} \qquad (5.78)$$

where the summations over t extend to all triplets linking $\varphi_{\boldsymbol{h}}$ to two other known or previously determined phases, while the sums over s refer to two-phase s.s. The weight attributed to each relation is calculated from the $\alpha = (T^2 + B^2)^{1/2}$ values of the reflections contributing to it.[59] At the beginning the origin-fixing reflections and the restricted permuted phases are given $\alpha = 100$, one-phase s.s. (if any) are assigned their G values, while the general phases represented by magic integers are given an α value which depends on the root mean square deviation of the representation. Starting from the eight reflections in the starting set, the phase extension is carried out following the order indicated by the convergence procedure. As we have seen, the reflections eliminated at the end are those best related to the starting set; the phase determination process should therefore be carried out in an order inverse to that of elimination. In Table 5.9 the final part of the inverted convergence map (**divergence map**) of RIFOL is reported and with its aid we can follow the phase determination process.

Reflection 24 is the first to be determined by a single triplet relation (written in the form of (5.65)); it will be used to define the following reflections with its corresponding α value. The second reflection no. 99 forms a two-phase s.s. with 24 (written as a triplet but with an asterisk instead of the third code number) and is related by a triplet to the starting set; each summation in (5.78) will have one term. The remaining reflections are then determined by a similar chain process.

When a sufficiently large number (60–100) of phases has been determined, one should proceed to their refinement. It is in fact possible to redetermine the initial phases with a greater number of contributors to the sums in (5.78); for instance the phase of reflection 24 can now be determined using all relations involving 24 and two other reflections among the 60–100 determined ones. The tangent refinement process is repeated until self-consistency and then the remaining phases are determined and refined. At this point different foms are computed.

Table 5.9. RIFOL: divergence map illustrating the phase determination path starting from the set of reflections in Table 5.8

Code	*h*	*k*	*l*	$\langle\alpha\rangle$	Contributors							
24	4	1	−17	8.41	4	−15	0	8.94				
99	8	1	−19	11.01	24	*	0	6.00	4	15	0	6.09
40	2	5	14	9.50	4	17	π	8.10	72	99	π	5.72
188	4	5	13	13.75	4	72	π	6.35	17	24	π	6.15
					15	40	0	5.78				
31	3	1	−1	7.80	−24	25	π	6.69	−5	99	0	3.71
13	3	2	−17	10.46	15	25	0	7.29	4	31	π	6.02
135	7	6	12	8.60	13	72	0	4.88	17	25	0	4.42
					31	188	0	3.92				
7	7	0	−19	6.72	5	15	0	7.26				
44	9	5	−5	11.47	7	40	0	7.26	5	188	0	5.28
37	3	4	14	10.22	−5	17	0	8.37	−7	72	0	6.40

Table 5.10. RIFOL: list of the ten best sets of phases with their relative figures of merit

Set	Figures of merit				
	MABS	**ALCOMB**	**PSCOMB**	**CPHASE**	**CFOM**
1	1.093	0.920	0.925	0.885	0.911
2	1.090	0.843	0.921	0.878	0.885
3	1.016	0.388	0.683	0.479	0.554
4	1.056	0.457	0.624	0.440	0.539
5	1.032	0.348	0.601	0.495	0.517
6	1.052	0.422	0.554	0.431	0.502
7	1.051	0.432	0.538	0.435	0.501
8	1.049	0.401	0.566	0.431	0.500
9	1.044	0.383	0.567	0.431	0.496
10	1.054	0.429	0.533	0.426	0.495

The complete tangent procedure is repeated for each of the 24 permutations of the phases in the starting set. This step may become quite demanding of computer time when the number of phase permutations is increased. Table 5.10 gives the list of the 10 best sets output by SIR in decreasing order of CFOM. ALCOMB and PSCOMB are combinations of functions based on α and ψ_0 respectively. The first set with the highest

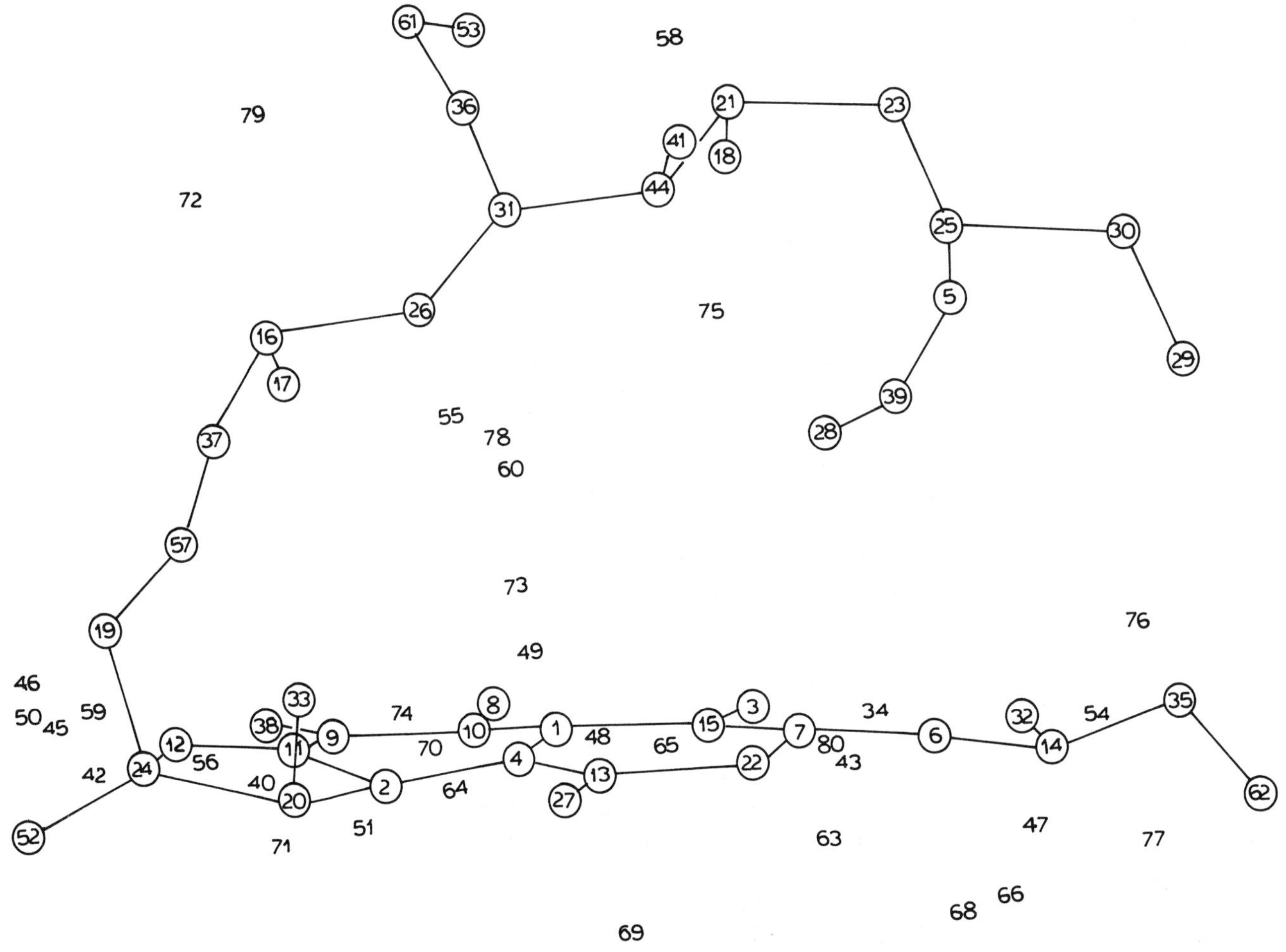

Fig. 5.23. RIFOL: representation and interpretation of the E-map computed with the first and best set of phases in Table 5.10; only part of the molecule can be located.

CFOM yields the E-map shown in Fig. 5.23, which may be interpreted in terms of the RIFOL structure as indicated by the connected circled maxima. Not all atoms are found in the map (solvent molecules plus eight atoms are missing, two to close the macrocycle and the others in the side chains), but the structure can be easily completed by the methods described in the next paragraph. The final refined structure is shown in Fig. 5.24.

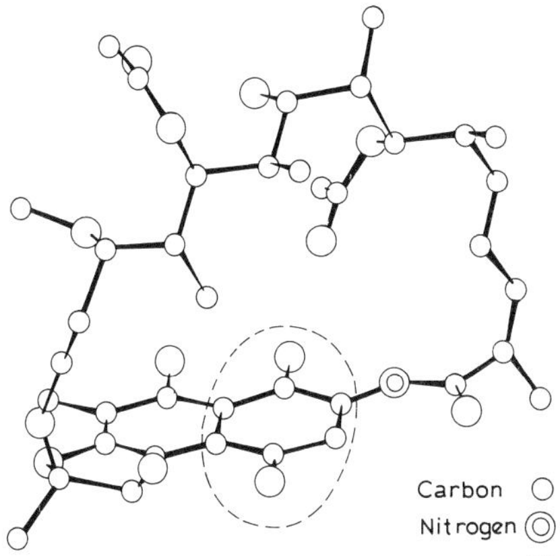

Fig. 5.24. RIFOL: final structure, without solvent molecules; the fragment outlined by the broken line can yield the complete molecule when input as a partial structure.

Partial structure recycling

It may happen that in the best E-map only a small molecular fragment can be recognized and in this case direct methods can be used to complete the structure. The first and simplest procedure was proposed by Karle[69] and is known as **Karle recycling**. With the coordinates of the fragment the corresponding structure factors are computed in modulus $|F_{\boldsymbol{h}}^{c}|$ and phase $\varphi_{\boldsymbol{h}}^{c}$, using (5.2). The most accurate phases, defined according to the criteria

$$|F_{\boldsymbol{h}}^{c}| \geqslant p\,|F_{\boldsymbol{h}}^{o}| \quad \text{and} \quad |E_{\boldsymbol{h}}^{o}| \geqslant 1.5 \quad (p = \Sigma_m Z_m / \Sigma_j Z_j)$$

are then used as starting set in a tangent-formula extension and refinement process. p represents the fraction of scattering matter in the fragment and is forced within the interval 0.25–0.65. If the fragment is a sufficiently good initial model the procedure should yield an E-map showing a larger fragment and further recycling should lead to the complete structure.

Other procedures for completing a partial structure have been proposed and among them we recall the DIRDIF procedure,[70] which is based on difference structure factors, i.e. structure factors corresponding to the unknown part of the structure; with initial phases obtained from the partial structure, the difference structure factors are used in a weighted tangent-formula procedure. In SIR a new probabilistic method[53,54] has been implemented. The structure factors for the known fragment are calculated and appropriate pseudo-normalized structure factors for the known and unknown part of the structure are obtained. These are then used in the new probabilistic formulae for the estimation of s.i.; on the basis of the estimated αs computed by these formulae, a convergence procedure will define the best starting set of phases, which will include both permuted phases and phases calculated from the partial structure. Phase expansion and refinement is performed by means of a new weighted tangent formula, and two specific foms, related to R_α and ψ_0, but computed with the pseudo-normalized structure factors, are used to select the correct solution. The procedure was also tested on the RIFOL structure; starting from the eight-atom fragment (11 per cent of the total number of electrons) surrounded by a broken line in Fig. 5.24 the complete structure was recovered.

Completing and refining the structure

The structural models obtained both by Patterson and direct methods are often incomplete (in the sense that not all atoms have been localized) and in all cases they only represent a crude first approximation of the real structure.

The method of Fourier recycling was outlined earlier (p. 329) in connection with the problem of locating the light atoms when the initial model is formed by one or more heavy atoms. The same method can be used to complete a molecular fragment when all atoms have approximately the same weight. If 50–60 per cent of the electron density has been located with sufficient accuracy it is quite easy to complete the structure. If the initial model only contains a smaller percentage of the electron density, the method can still be applied but the Fourier coefficients should be corrected by appropriate statistical weights, such as those proposed by Sim[71,72], taking into account the different contribution of the known atoms to the different structure factors. The derivation of these weights is given in Appendix 5.H, where some other procedures for completing a partial model will also be described. The Fourier cycles not only allow the location of the new atoms but also the improvement of the positions of the model atoms. For very small fragments the direct method procedures mentioned at the end of the last paragraph are usually easier to apply.

Difference Fourier method

Another convenient way of completing and refining a structural model is the **difference Fourier synthesis method**. A Fourier series having as coefficients the $|F^c_{\boldsymbol{h}}|$s computed using eqn (5.2)

$$\rho_c(\boldsymbol{r}) = \frac{1}{V}\sum_{\boldsymbol{h}} F^c_{\boldsymbol{h}} \exp(-2\pi i\boldsymbol{h}\cdot\boldsymbol{r}) \tag{5.79}$$

will show maxima at the positions of the atoms of the given model, while a series with coefficients $F^o_{\boldsymbol{h}} = |F^o_{\boldsymbol{h}}| \exp(i\varphi_{\text{true}})$

$$\rho_o(\boldsymbol{r}) = \frac{1}{V}\sum_{\boldsymbol{h}} F^o_{\boldsymbol{h}} \exp(-2\pi i\boldsymbol{h}\cdot r) \tag{5.80}$$

represents the true structure. In order to see how much the initial model deviates from the real structure, the difference series

$$\Delta\rho(\boldsymbol{r}) = \rho_o(\boldsymbol{r}) - \rho_c(\boldsymbol{r}) = \frac{1}{V}\sum_{\boldsymbol{h}} (F^o_{\boldsymbol{h}} - F^c_{\boldsymbol{h}}) \exp(-2\pi i\boldsymbol{h}\cdot\boldsymbol{r}) \tag{5.81}$$

should be computed. Unfortunately the values of φ_{true} are not known and we have to assume $\varphi_{\text{true}} \approx \varphi^c_{\boldsymbol{h}}$; this approximation, illustrated in the Argand diagram of Fig. 5.25, will hold better the better is the initial model. Equation (5.81) then becomes

$$\Delta\rho(\boldsymbol{r}) = \frac{1}{V}\sum_{\boldsymbol{h}} (|F^o_{\boldsymbol{h}}| - |F^c_{\boldsymbol{h}}|) \exp(-2\pi i\boldsymbol{h}\cdot\boldsymbol{r} + i\varphi^c_{\boldsymbol{h}}). \tag{5.82}$$

If in the model an atom is missing, then $\rho_c(\boldsymbol{r})$ will be zero at the corresponding position, while $\rho_o(\boldsymbol{r})$ will show a maximum. The difference synthesis will also show a peak at the same position but it will be almost zero at the positions of the model atoms (if these are correct) where $\rho_o(\boldsymbol{r}) \approx \rho_c(\boldsymbol{r})$.

An important property of the difference syntheses is that they are almost unaffected by series truncation errors. Indeed, because of the limited number of observations, the Fourier maps computed by means of (5.79) and (5.80) will show some ripples around each peak (see Fig. 3.16), the size of which increases with increasing peak height. As a consequence a light atom close to a heavy atom may be obscured by its ripples. Since the number of

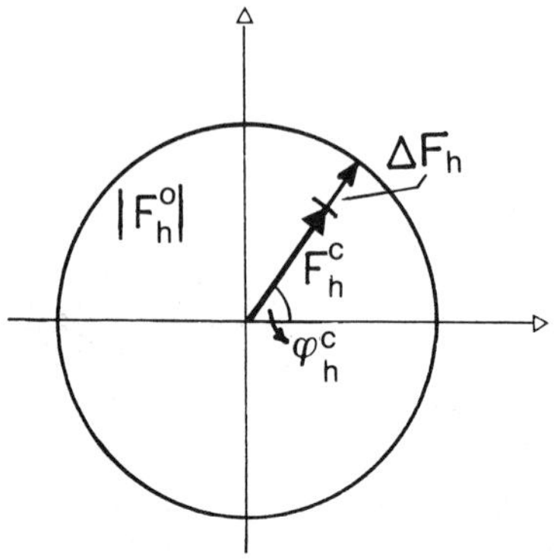

Fig. 5.25. Illustration of the approximation $\varphi_{\text{true}} \approx \varphi^c_{\boldsymbol{h}}$.

terms in the two series (5.79) and (5.80) is the same, the truncation errors will also be approximately the same and will cancel out in the difference (5.82).

Let us now see how the difference types of errors in the model are reflected in a difference synthesis.

1. *Missing atoms*. We have already seen that they appear as positive maxima, but, because of the approximation made for the phases, their height is usually smaller than that corresponding to the atomic number of the missing atom.

The lack of truncation errors allows the correct localization of light atoms even when the model contains much heavier atoms. For this reason, difference Fourier series, computed when the model has been corrected for most other important errors, allow the localization of hydrogen atoms (with only one electron they contribute very little to the X-ray diffraction) even in the presence of medium size atoms (cf. Fig. 5.28).

2. *Position errors*. Their effect is shown in Fig. 5.26. If ρ_o gives the correct position of the atom and ρ_c its wrong position in the model, in the $\Delta\rho$ map the latter will be close to a negative minimum, while the correct position will be towards the neighbouring positive maximum along the maximum gradient line. It is possible to have a quantitative estimate of the shifts to be applied, but in general, when the errors are not too large, it is easier to correct the position errors by the least-squares methods (cf. Chapter 2, pp. 90–108, and later in this section).

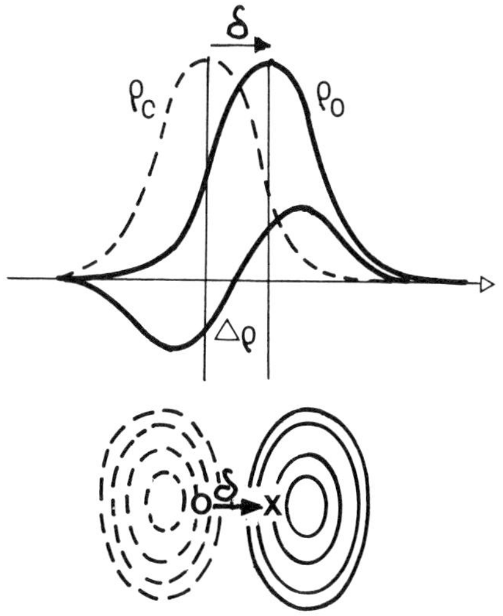

Fig. 5.26. Position error of a model atom (top) and corresponding difference synthesis (bottom).

3. *Errors in the thermal parameters*. As we have seen in Chapter 3 (p. 148), because of the thermal motion the electron density function around each atomic nucleus becomes wider. In Fig. 5.27(a) the case in which the thermal motion has been neglected or underestimated in the model is represented. The ρ_o density will therefore have a smaller and wider peak with respect to ρ_c, and in the difference synthesis a negative depression, surrounded by a positive ring, will appear. If, on the other hand, too high a thermal motion has been assumed for one atom of the model, then $\Delta\rho$ will show a small positive maximum surrounded by a negative ring.

Finally, the case in which an isotropic thermal motion has been assumed in the model (the ρ_c maximum has a spherical distribution) while the real motion is anisotropic (the ρ_o maximum has an ellipsoidal distribution), is illustrated in Fig. 5.27(b) together with the corresponding difference synthesis showing two positive maxima and two negative minima; the line joining the two positive lobes represents the direction of largest thermal motion. The qualitative indications obtained in this way allow one to recognize those atoms for which it is more important to carry out the least-squares refinement varying the six parameters of the anisotropic thermal motion (cf. Appendix 3.B).

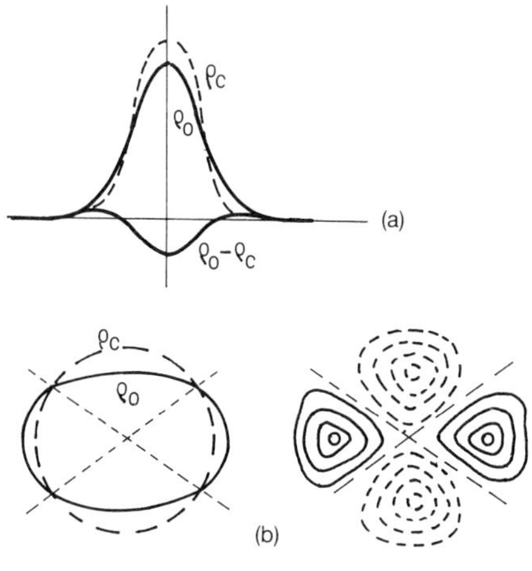

Fig. 5.27. Thermal parameter errors and their effect on the difference syntheses; for the atom of the model it is assumed: (a) too small an isotropic motion; (b) an isotropic motion when an anisotropic model should be assumed.

Least-squares method

By far the most widely used method of structure refinement is the **least-squares method.** The theory and the computing procedures have been described in Chapter 2, pp. 90–108. Several computer programs have been implemented to carry out the crystallographic least-squares refinement; usually they are integrated within complete crystallographic packages such

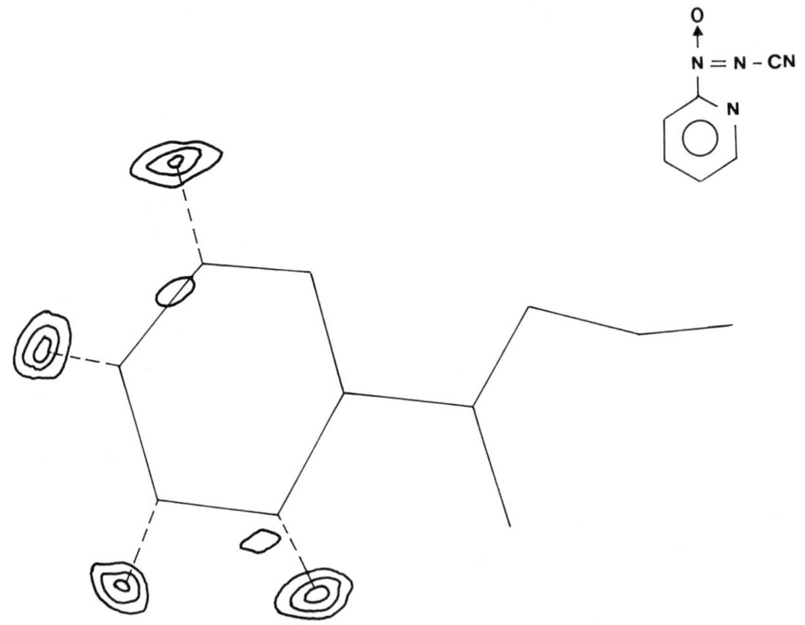

Fig. 5.28. AZOP: structural formula and projection on the mean plane of the molecule of the difference map computed when $R = 0.078$ (the scheme of the molecule is added to the map for reference).

as SHELX,[10] XTAL,[73] NRCVAX,[74] CRYSTALS[75] and those commercially available together with most single-crystal diffractometers. Here we will illustrate the practical application to the refinement of the fairly small structure of 2-azoxycyanopyridine[76] (AZOP, $C_6H_4N_4O$, space group $P\bar{1}$, $Z = 2$). The structure was solved using the SIR program;[67] from the best E-map the coordinates of all the 11 non-hydrogen atoms of the molecule (cf. Fig. 5.28) were obtained. The structure factors computed with this model, assuming an isotropic temperature factor equal for all atoms ($U = 0.05$ Å^2), yield a residual $R = 0.34$, a sufficiently low value to indicate the correctness of the model. As we have seen in Chapter 2 (p. 93) the refinement must be repeated in several cycles until convergence. For AZOP the refinement was performed using 1176 observed reflections by means of the system of programs SHELX;[10] the first four cycles were carried out varying the overall scale factor and the three coordinates plus the isotropic thermal parameter of each atom; the total number of varied parameters is $1 + 4 \times 11 = 45$ and the R factor reduces to 0.148. In the next step the refinement is carried out varying for each atom the six components of the anisotropic thermal parameter as described on pp. 186–8. Four anisotropic refinement cycles were performed with $1 + 9 \times 11 = 100$ varied parameters and at convergence $R = 0.078$.

At this point we may assume that the model formed by the heavier atoms is sufficiently well refined and we can compute a difference Fourier synthesis in order to localize the hydrogen atoms. Figure 5.28 shows a projection of the map onto the mean plane of the AZOP molecule; the known atomic positions are also shown linked by their chemical bonds, and the hydrogen atoms are clearly seen as maxima in the map. The coordinates of the four hydrogen atoms derived from the map are then introduced in the refinement process with appropriate isotropic temperature factors, but this usually can not be done in a straightforward way. In fact the contribution of the hydrogen atoms to the observed amplitudes is very small and rapidly decreasing with increasing $\sin\theta/\lambda$; the accuracy of their localization is therefore quite low and often, because of the errors in the experimental data, the refinement shifts them into stereochemically incompatible positions (e.g. bond distances which are too short or too long) with abnormal temperature factors (e.g. values which are too high and tend to cancel out the H atoms or too small for H linked to atom with significant thermal motion). As we have seen in Chapter 2 (p. 106) it is possible to impose some geometrical restraints during the least-squares refinement; in our example we have imposed that the C–H distances should remain within 0.98 ± 0.02 Å. In addition a common isotropic thermal parameter was assumed for all four H atoms, which was varied as a single parameter. Four cycles of refinement varying $100 + 3 \times 4 + 1 = 113$ parameters converged to $R = 0.055$. A frequently used alternative procedure for locating the H atoms consists in calculating their positions on the basis of the most common stereochemical rules. It is recommended that these positions are checked against those obtained from a difference Fourier synthesis. Their refinement may then be carried out as described above.

Up to this step all observations were given the same unitary weight, but following the indications of Chapter 2 (p. 101), in order to get meaningful values of the estimated standard deviations of the varied parameters, the observational equations should be properly weighted. In Chapter 2 we also saw how it is possible to express the weights in terms of a polynomial in $|F_o|$. Several weighting functions have been proposed and the first scheme suggested by Hughes[77] was

$$w = constant = 1/16F^2_{\min} \qquad \text{for} \quad |F_o| \leqslant 4F_{\min}$$

$$w = 16F^2_{\min}/|F_o|^2 \qquad \text{for} \quad |F_o| > 4F_{\min}$$

where $F_{\min}$ is close to the minimum observed amplitude. Later Cruickshank[78] proposed the scheme

$$w = 1/(a_0 + |F_o| + a_2\,|F_o|^2) \tag{5.83}$$

with $a_0 = 2F_{\min}$ and $a_2 = 2/F_{\max}$ as initial guesses.

For AZOP the intensities were measured with an automatic diffractometer using a scintillation counter; for each $|F_o|$ the corresponding estimated standard deviation $\sigma(F_o)$ due to counting errors may be derived (cf. Chapter 4, pp. 271 and 280). Since this is not the only error affecting our observations, it is necessary to modify the weights introducing an empirical dependence on $|F_o|$ itself. The weighting function available in the SHELX program is

$$w = \frac{1}{\sigma^2(F_o) + q\,|F_o|^2} \tag{5.84}$$

where q is an adjustable parameter chosen in such a way that the $\langle w\,\Delta F^2\rangle$ values remain constant when the reflections are grouped in different ways (e.g. in $|F_o|$ or $\sin\theta/\lambda$ intervals). The last four cycles with the weighting function (5.84) afforded a final R value of 0.052 with $q = 0.0007$. The weighted R factor

$$R_w = \frac{\sum_i w_i \Delta F_i}{\sum_i w_i\,|F_i^o|} \tag{5.85}$$

and the goodness-of-fit measure (also called standard deviation of an observation of unit weight)

$$\mathrm{GofF} = \frac{\sum_i w_i \, \Delta F_i^2}{n - m} \tag{5.86}$$

are also computed. Equation (5.86) is nothing but the quantity defined in (2.58) and GofF should be close to unity if the weights of the observations have been correctly assessed, the errors in the model are negligible in comparison with the errors in the data and there are no significant systematic errors. For AZOP $R_w = 0.054$ and GofF $= 1.504$.

In the final stages of the refinement one should check whether some intense low-angle reflections are affected by secondary extinction (cf. p. 97) and have $|F_o|$ systematically smaller than $|F_c|$; some least-squares programs allow an empirical correction of this effect but it is common practice to simply discard these few reflections from the refinement. In the case of AZOP three such reflections were eliminated.

It is also important to check for possible correlations between parameters during the least-squares refinement. In Chapter 2 we saw that the least-squares procedure gives the variance–covariance matrix of the derived parameters (eqn (2.56)). It is therefore straightforward to calculate, for each term in the matrix, the correlation coefficient

$$\rho_{ij} = \frac{\mathrm{cov}\,(p_i p_j)}{\sigma^2(p_i)\sigma^2(p_j)} \tag{5.87}$$

which is close to zero if the correlation is negligible and close to 1 for large correlation. Most programs compute the correlation matrix and output the off-diagonal elements which are greater than 0.5. No such correlation was found for AZOP.

How low should the final R factor be in order to have good confidence in the quality of the refined model? The answer to this question is not straightforward, because it depends on the type and complexity of the structure and on the quality of the experimental data. Nevertheless, for small or medium size structures giving good quality crystals (as it is usually the case for most organic or organometallic compounds), with intensities collected at room temperature on a diffractometer, we should expect R values in the range 0.03–0.07. For very accurate low-temperature measurements these values may be further reduced. The quality of the model should then also be assessed by looking at the standard deviations of the atomic parameters or at those of the derived geometrical quantities, such as bond distances, which should be less than 0.006 Å for non-hydrogen atoms. Usually as the complexity of the structure increases the quality of the crystals and of the diffraction data decreases, and, at convergence the R value may be in the range 0.08–0.15 for large molecules and up to 0.25–0.30 for macromolecules solved at low or medium resolution. In these cases the final model may still contain some fairly large errors, which may often be corrected by applying stereochemical or energy constraints (cf. Chapter 8, p. 564).

The final R value will also depend on the selection of reflections used during the refinement. It is in fact common practice to discard the very weak reflections, the intensities, I, of which are measured by diffractometer with large estimated standard deviations, $\sigma(I)$; reflections with $I \leqslant n\sigma(I)$, usually with $2 \leqslant n \leqslant 3$ (for AZOP $n = 2$ was used to select the 'observed' reflections) are omitted. Of course the higher is n the lower will be R, but too high a cut-off value will considerably reduce the number of observations and the least-squares refinement will yield higher standard deviations (cf. Chapter 2, pp. 102–4). There is no theoretical justification for discarding weak reflections, and a more rigorous procedure requires the use of all data with proper weights[79] and a prior statistical treatment of weak reflections to account also for negative measurements.[80] On the other hand, in most cases, only for very accurate structural analyses would the latter procedure reveal significant changes in the final model. When the choice between a centrosymmetric and a non-centrosymmetric space group is ambiguous or when the structure shows pseudotranslational symmetry, the inclusion of weak reflections becomes essential.[81]

In following the convergence of our refinement we have mainly used the R factor defined in (5.3); other agreement indices have been proposed, some of which are statistically more appropriate to assess the significance of small changes in the final refinement stages. Among them we recall the weighted residual proposed by Hamilton[82]

$$R'_{\mathrm{w}} = \frac{\sum_i w_i \, \Delta F_i^2}{\sum_i w_i \, |F_i^{\mathrm{o}}|^2} \tag{5.88}$$

already considered in Chapter 2 (p. 103), where it was shown that the ratio of the residuals $R'_{\mathrm{w}}(1)/R'_{\mathrm{w}}(2)$ obtained from two refinements of different models may be used for testing the significance of the difference between the two models. For instance this test may be used for determining the absolute configuration of optically active molecules from the effect of anomalous dispersion,[83] although, as pointed out by Rogers,[84] some caution is necessary.

Although the R index is the most widely used criterion to assess the goodness of a structural model, its indications should be regarded with some caution. As an example of the weakness of the R index, let us consider the case of a structure containing one heavy atom, such as Pt, Ir, Os, or Cd, and a certain number of lighter atoms (C, N, O). Since the R value is less sensitive to the position of the light atoms, a model with the heavy atom correctly placed and the light atoms quite inaccurately localized will give a lower R value than a model with a small error in the position of the heavy atom and an almost correct position of the light atoms. Nevertheless the second model, with a smaller average position error and a uniform distribution of errors over all atoms, is certainly a better one.

When the model is an incomplete representation of the molecule, R becomes a rather weak guide, as it measures the agreement between quantities which depend on largely different numbers of parameters ($|F_{\boldsymbol{h}}^{\mathrm{o}}|$ depends on all atomic parameters, while $|F_{\boldsymbol{h}}^{\mathrm{c}}|$ only depends on the limited number of parameters defining the incomplete model). The only valid check of the indication given by the R value is the convergence of the subsequent process of completing and refining the model.

As mentioned in the introduction of this chapter, the refinement of a structural model will not converge if the model is affected by large errors. This is a direct consequence of the non-linear nature of the problem, which, as observed in Chapter 2 (p. 93), implies the presence of several local minima in the function to be minimized. Two types of errors, which occur quite often, are particularly disturbing.

1. *Incorrectly positioned molecule.* It may happen that the structural model obtained by direct or Patterson methods is constituted by a correctly oriented molecule, but its position with respect to the symmetry elements is incorrect. In most cases the entire molecule is shifted by a vector corresponding to a large Patterson peak due to several superposing interatomic vectors. An example is given by the AZOP structure; the initial attempts at solving the structure by direct methods gave correctly oriented molecules, shifted from the correct position and their refinement would not converge to an R value lower than 0.25. The correct solution was only obtained after estimating the triplet invariants by their second representation. An alternative way, which is often successful when applied to structures belonging to the space group $P\bar{1}$, would be to try to solve the structure in the lower symmetry group P1. The solution will give two molecules related by an inversion centre in a general position; this can be located by inspection and then, after translating one of the two molecules in such a way that it is referred to the inversion centre at the origin (i.e. the general coordinates of the inversion centre are subtracted from all atomic coordinates), the structure can be refined in $P\bar{1}$. When these methods fail one can try to find the correct position by the Patterson method described in Appendix 5.B.

2. *Wrong space group.* As explained in Chapter 3 (p. 161), the determination of the space group from the diffraction data is not unique, and a wrong choice is more frequent than one might think. An interesting analysis of such errors is given by Marsh and Herbstein[85,86] who noted that a relatively large number of structures has been reported with space group of unnecessarily low symmetry. Two types of structure can be considered: (i) those for which the reported space group has a lower Laue symmetry than that of the actual structure; and (ii) those for which the reported space group lacks a centre of symmetry, which is instead present. For the first type the correct space group may often be found using geometrical considerations[87,88] such as those implemented in the computer programs NEWLAT[89] or LEPAGE.[90] For the second type, the refinement by least-squares should in theory yield a singular matrix (cf. Chapter 2, p. 100), as half of the refined parameters have a linear relation (the overlooked inversion operation) with the other half; in practice this relation may not be exact, but in any case the solution of the normal equations will be very unstable with large errors in the atomic parameters and large values of their estimated standard deviations. The most common errors occur for the following space groups: P1 instead of $P\bar{1}$, Pc instead of $P2_1/c$, C2 or Cm instead of C2/m, Cc instead of C2/c, $Pna2_1$ instead of Pnam, $Cmc2_1$ instead of Cmcm (several examples are given by Marsh and Herbstein[85,86]). They may be recognized because they usually lead to abnormal values of some bond distances and angles or of the thermal parameters, which are adjusted when the refinement is carried out in the correct space group.

A critical view (possibly a bit too pessimistic) of the accuracy of the reported structure determinations has been presented by Jones[91] and the reading of his paper is highly recommended.

Two important effects may cause a lack of convergence in the refinement process: **twinning** and **disorder**. As we saw on p. 83–7, the diffraction intensities of a twinned crystal must be carefully analysed in order to separate the contributions from the different individuals forming the twin. This will be more difficult when all reflections from the two individuals superpose exactly (merohedral twins, cf. p. 85). Often merohedral twinning is not recognized and then in most cases the structure fails to be solved; even when a solution is found, it is impossible to refine the model to an acceptable R value because of the systematic errors on the observed amplitudes.

Real crystals deviate from the ideal picture of a periodical perfect repetition of a unit cell. We saw on p. 148 that, because of the thermal motion, atoms oscillate around an equilibrium position and we may say that they are affected by a small dynamic disorder. But larger movements are sometimes possible in crystals, giving rise to a real **dynamic disorder.** Diffraction experiments will just show a time-averaged situation and if the movement is continuous the electron density will be smeared out. On the other hand if the movement is stepwise between two or more energy minima, the total electron density will be divided between the positions of the minima. An example of such dynamic disorder in the *cis* form of the complex $[(\eta^5C_5H_5)_2Fe_2(CO)_2(\mu_2\text{-}CO)_2]$ has been recently described;[92] the two independent cyclopentadienyl rings show different thermal motions, which may be interpreted in terms of different rates of stepwise rotation around their fivefold axes. This hypothesis was confirmed by a recent solid-state NMR study.[93]

Another type of disorder may occur in crystals; it is referred to as **static disorder** and in this case the average contents of the unit cell is obtained by averaging over the space, i.e. over the different unit cells. An important example of this type of disorder, which is common in minerals and alloys, is the **substitutional disorder** where the same site is occupied by different types of atoms in different cells.

The structure factors depend only on the time- and space-averaged contents of the unit cell and the result of a diffraction analysis is this averaged contents. No distinction can be made between dynamic and static disorder and the case of a molecular fragment oscillating between two positions in all unit cells is indistinguishable from that in which half of the cells have the fragment in one position and half in the other.

If a disordered structure is refined without taking into account the disorder, then only an average structure is obtained with abnormal thermal motion and the refinement will not converge. In the case of substitutional disorder the different atomic species are introduced with the same coordinates but with different occupation factors (cf. p. 87), depending on the fraction of each species. When the disorder is between, say, two positions in the cell, then the same atomic species are located in both possible positions with an occupation factor depending on the 'population' of each position (the term population will be associated with the fraction of time spent in the given position in the case of dynamic disorder and with the fraction of cells having atoms in that position in the case of static disorder); if this is not known one may initially assume an occupation factor of 0.5 and than refine the factor to convergence (the occupation factor is highly correlated to the temperature factor and some caution should be used in the least-squares refinement). Only when disorder is treated in a proper way will the refinement converge to an acceptably low R value.

Absolute configuration

In Chapter 3 (p. 168) it was shown that, when the anomalous dispersion effect is present, Friedel's law is no longer satisfied. We shall see in Chapter 7 (p. 489) that for compounds containing asymmetric carbon atoms, isomers of opposite chirality (enantiomers) are possible and that their solutions rotate the plane of polarized light in opposite directions. By simple chemical methods it is not possible to decide which of the two configurations corresponds to the isomer rotating the plane of the light to the right ((+)-rotamer). Fisher proposed to assume as reference that (+)-glyceraldehyde corresponds to the configuration 5a shown in the table on p. 487 and to establish the configuration of all other chiral molecules in a relative way with respect to this convention; thus (+)-tartaric acid corresponds to 7a. Since this is an arbitrary assumption there was only a 50 per cent chance of it being correct. Fortunately the first determination of the absolute configuration of NaRb-(+)-tartrate[94] by the method described below (using the anomalous scattering of the Zr K_α radiation by Rb), proved that the assumption was correct. Since then the anomalous dispersion method has been used to determine the absolute configuration of a large number of molecules and it will be shown in Chapter 8 (p. 545) that the method is also applied in macromolecular crystallography.

The two optical isomers differ in hand and may be related either by an inversion or by a reflection operation (inverse congruence).†

In a normal X-ray experiment, with no anomalous dispersion, because of Friedel's law, the two enantiomers are indistinguishable, as they give rise to the same diffraction pattern. When one or more anomalous scatterers are present, then both $|F_{\boldsymbol{h}}|$ and $|F_{-\boldsymbol{h}}|$ should be measured and compared with the computed values. Since the Bijvoet[83] differences $|F_{\boldsymbol{h}}|^2 - |F_{-\boldsymbol{h}}|^2$ are rather small, and anomalous dispersion is only large for heavier atoms near their absorption edges, it is essential to apply a proper absorption correction to the diffraction data.

In order to illustrate how the absolute configuration may be determined it is instructive to consider the procedure originally used by Bijvoet, Peerderman, and van Bommel,[94] which gives reliable results only when the anomalous dispersion effect is sufficiently large and for this reason it is no longer used. The ratios $|F^{\mathrm{o}}_{\boldsymbol{h}}|/|F^{\mathrm{o}}_{-\boldsymbol{h}}|$ and $|F^{\mathrm{c}}_{\boldsymbol{h}}|/|F^{\mathrm{c}}_{-\boldsymbol{h}}|$ are tabulated for a limited number of reflections with large Bijvoet differences; a one-by-one comparison of the two ratios indicates that the wrong configuration has been chosen if, when the first ratio is greater than 1, the second is less than 1 and

† We should note that the term 'absolute configuration' is not always correct in describing crystal structures, as it does not apply to the case of non-centrosymmetric but achiral space groups (polar groups with reflection symmetry operations, such as $Pna2_1$) or to that of achiral molecules (without asymmetric centres) crystallizing in non-centrosymmetric space groups, such as quartz. The term 'absolute structure' has been recently proposed and discussed.[95,96]

vice versa. If this is the case, then the correct enantiomer is obtained by changing the sign of all atomic coordinates.

When the anomalous dispersion effect is small all the reflections should be used and efficient methods have been proposed by Rogers[84] and Flack;[97,98] in the latter an absolute-structure or chirality parameter x is refined in a least-squares process (cf. p. 97) in which the structure factor is written as

$$|F(\boldsymbol{h}, x)|^2 = (1 - x)\,|F_{\boldsymbol{h}}|^2 + x\,|F_{-\boldsymbol{h}}|^2 \tag{5.89}$$

where x is close to zero when the model and the crystal are in the same chirality, and approaches 1 if they are inverted one with respect to another.

Appendices

5.A Structure factor probability distributions

We will use the **central limit theorem** stating that, given N independent random variables x_j $(j = 1, N)$, each with mean value $\langle x_j \rangle$ and variance α_j^2, their sum $X = \sum_{j=1}^{N} x_j$ has a mean value equal to the sum of the individual mean values and a variance equal to the sum of the variances, provided that N is sufficiently large, i.e.

$$\langle X \rangle = \sum_{j=1}^{N} \langle x_j \rangle \quad \text{and} \quad \sigma^2 = \sum_{j=1}^{N} \alpha_j^2. \tag{5.A.1}$$

Whatever the distributions of the variables x_j are (they may be all different), the sum X will have a normal probability distribution

$$P(X) = \frac{1}{\sqrt{2\pi\sigma^2}} \exp\left(- \frac{(x - \langle x \rangle)^2}{2\sigma^2} \right). \tag{5.A.2}$$

For a centrosymmetric structure the structure factor may be written as

$$F_{\boldsymbol{h}} = \sum_{j=1}^{N/2} 2f_j \cos 2\pi \boldsymbol{h} \cdot \boldsymbol{r}_j \tag{5.A.3}$$

where the sum is extended to the independent atoms only so that $F_{\boldsymbol{h}}$ can be considered as the sum of $N/2$ independent random variables $x_j = 2f_j \cos 2\pi \boldsymbol{h} \cdot \boldsymbol{r}_j$ with mean value

$$\langle x_j \rangle = 2f_j \langle \cos 2\pi \boldsymbol{h} \cdot \boldsymbol{r}_j \rangle. \tag{5.A.4}$$

Assuming a random distribution of the atomic coordinates, all $\boldsymbol{r}_j$ values will be equally probable and $\langle \cos 2\pi \boldsymbol{h} \cdot \boldsymbol{r}_j \rangle = 0$; hence

$$\langle x_j \rangle = 0. \tag{5.A.5}$$

The variance of x_j is defined as

$$\alpha_j^2 = \langle x_j^2 \rangle - \langle x_j \rangle^2. \tag{5.A.6}$$

Because of (5.A.5) the second term is zero and

$$\alpha_j^2 = 4f_j^2\langle \cos^2 2\pi \boldsymbol{h} \cdot \boldsymbol{r}_j \rangle = 2f_j^2 \tag{5.A.7}$$

given that, for a randomly variable angle θ, $\langle \cos^2\theta \rangle = \frac{1}{2}$.

Because of the central limit theorem, the mean value of $F_{\boldsymbol{h}}$ will be

$$\langle F \rangle = \sum_{j=1}^{N/2} \langle x_j \rangle = 0, \tag{5.A.8}$$

and its variance

$$\sigma^2 = \langle F^2 \rangle = \sum_{j=1}^{N/2} 2f_j^2 = \sum_{j=1}^{N} f_j^2 = \Sigma. \tag{5.A.9}$$

Finally the distribution function of F is given by

$$P_{\bar{1}}(F) = \frac{1}{\sqrt{2\pi\Sigma}} \exp(-F^2/2\Sigma). \tag{5.A.10}$$

F has the same probability of being positive or negative and the probability $P_{\bar{1}}(|F|)$ that its modulus has a given value is given by $P_{\bar{1}}(|F|) = 2P_{\bar{1}}(F)$; substitution of (5.A.10) gives eqn (5.4).

The approximate character of (5.4) is clearly indicated by the observation that the maximum value of $|F|$ is $\sum_{j=1}^{N} f_j$ while $P_{\bar{1}}(|F|)$ has a Gaussian shape and goes to zero only at $\pm\infty$.

For a non-centrosymmetric structure $F_{\boldsymbol{h}} = A_{\boldsymbol{h}} + \mathrm{i}B_{\boldsymbol{h}}$ and, in analogy with the derivation of (5.A.8) to (5.A.10), we have

$$\langle A \rangle = 0, \quad \sigma_A^2 = \tfrac{1}{2}\Sigma \qquad \langle B \rangle = 0, \quad \sigma_B^2 = \tfrac{1}{2}\Sigma \tag{5.A.11}$$

and

$$P(A) = \frac{1}{\sqrt{\pi\Sigma}} \exp(-A^2/\Sigma), \qquad P(B) = \frac{1}{\sqrt{\pi\Sigma}} \exp(-B^2/\Sigma). \tag{5.A.12}$$

The joint probability that A lies between A and $A + \mathrm{d}A$ and at the same time B lies between B and $B + \mathrm{d}B$ is given by the product of the two distributions (5.A.12)

$$P(A, B)\,\mathrm{d}A\,\mathrm{d}B = P(A)P(B)\,\mathrm{d}A\,\mathrm{d}B = \frac{1}{\pi\Sigma} \exp(-(\mathrm{A}^2 + \mathrm{B}^2)/\Sigma)\,\mathrm{d}A\,\mathrm{d}B. \tag{5.A.13}$$

Equation (5.A.13) represents the probability that the structure factor F lies within a region of the complex plane of area $\mathrm{d}S = \mathrm{d}A\,\mathrm{d}B$. If the structure factor is expressed in polar coordinates (i.e. in modulus and phase), then $\mathrm{d}S = |F|\,\mathrm{d}|F|\,\mathrm{d}\varphi$ and

$$P_1(|F|, \varphi)\,\mathrm{d}|F|\,\mathrm{d}\varphi = \frac{|F|}{\pi\Sigma} \exp(-|F|^2/\Sigma)\,\mathrm{d}|F|\,\mathrm{d}\varphi. \tag{5.A.14}$$

The distribution of the amplitude $|F|$ alone is obtained by integrating (5.A.14) over φ and thus (5.5) is immediately derived. From (5.A.11) we can see that also for non-centrosymmetric structures $\Sigma = \langle |F|^2 \rangle$. In deriving the value of $\langle |F|^2 \rangle$ we did not consider the effects due to the presence of

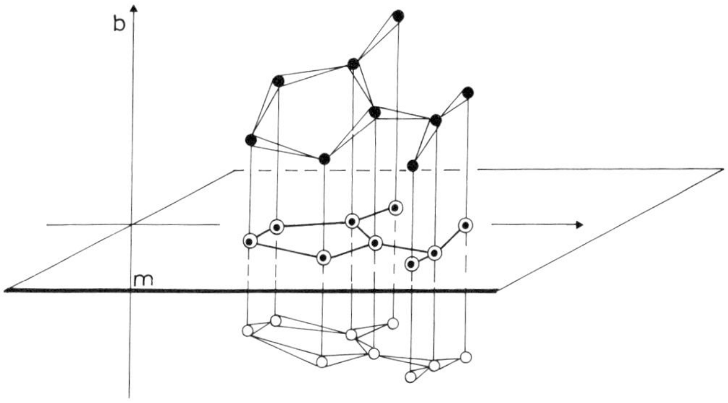

Fig. 5.A.1. Space group Pm: in the projection on the plane m the two related molecules superpose.

other symmetry elements besides the inversion centre. Let us, for instance, consider the effects due to the presence of a mirror plane, by considering the space group Pm (m perpendicular to the b axis). In Fig. 5.A.1 we can see that on the projection along the b axis, all atoms related by the m plane are superimposed; the projection of the electron density on the ac plane will only contain half of the peaks, each with double weight. The Fourier coefficients F_{h0l} relative to this projection will have a distribution corresponding to a structure of $N/2$ atoms with scattering power $2f_j$ and then

$$\langle |F_{h0l}|^2 \rangle = \sum_{j=1}^{N/2} (2f_j)^2 = 2\sum_{j=1}^{N} f_j^2 = 2\Sigma.$$

The mean intensity of the $h0l$ reflections will be twice that of the general reflections hkl. In order to take into account the symmetry element effects, (5.A.9) is generalized to

$$\langle |F|^2 \rangle = \varepsilon \sum_{j=1}^{N} f_j^2 = \varepsilon\Sigma \qquad (5.A.15)$$

where ε will be equal to 1 for the general reflections and greater than 1 for certain classes of reflections which are influenced by the presence of symmetry elements; thus in Pm $\varepsilon = 2$ for the $h0l$ reflections. The values of ε and the relative reflection classes for the different symmetry elements are tabulated in the *International Tables*.[99]

We note that the calculation of the ratio of the average observed intensity for specific classes of reflections over that of the general reflections, can be used as an indicator of the presence of some symmetry elements. For instance we can distinguish between the space groups Pm and P2 by computing the ratio $\langle |F_{h0l}|^2 \rangle / \langle |F_{hkl}|^2 \rangle$, which will be close to 2 for Pm and close to 1 for P2.

5.B. Patterson vector methods

We will consider here the so-called **vector methods**[100] which allow a systematic use of the information contained in the Patterson function also when no heavy atoms are present.

The Patterson function can be considered as formed by the superposition of N images of the crystal structure displaced one with respect to the other by the amount required to bring to the origin each atom in the unit cell.

Each image represents the ends of all the vectors from the atom at the origin to all the others. Figure 5.B.1(b) illustrates how the Patterson map of Fig. 5.4(c) can be interpreted in terms of five displaced images of the five-atom structure of Fig. 5.4(a), repeated in Fig. 5.B.1(a). It can be easily seen that the same map can also be interpreted in terms of five displaced images of the enantiomeric structure.

It is in theory rather simple[101] to extract one image of the structure from the overlapping set, but, as we shall see, the proposed methods may be difficult to apply in practice; these are the so-called **superposition methods.**

Let us first consider the **vector superposition** in which one Patterson map is superposed on an other in such a way that the origin of the second coincides with a given vector of the first; the two maps are thus separated by an interatomic vector. The set of coincident peaks will reveal one or more

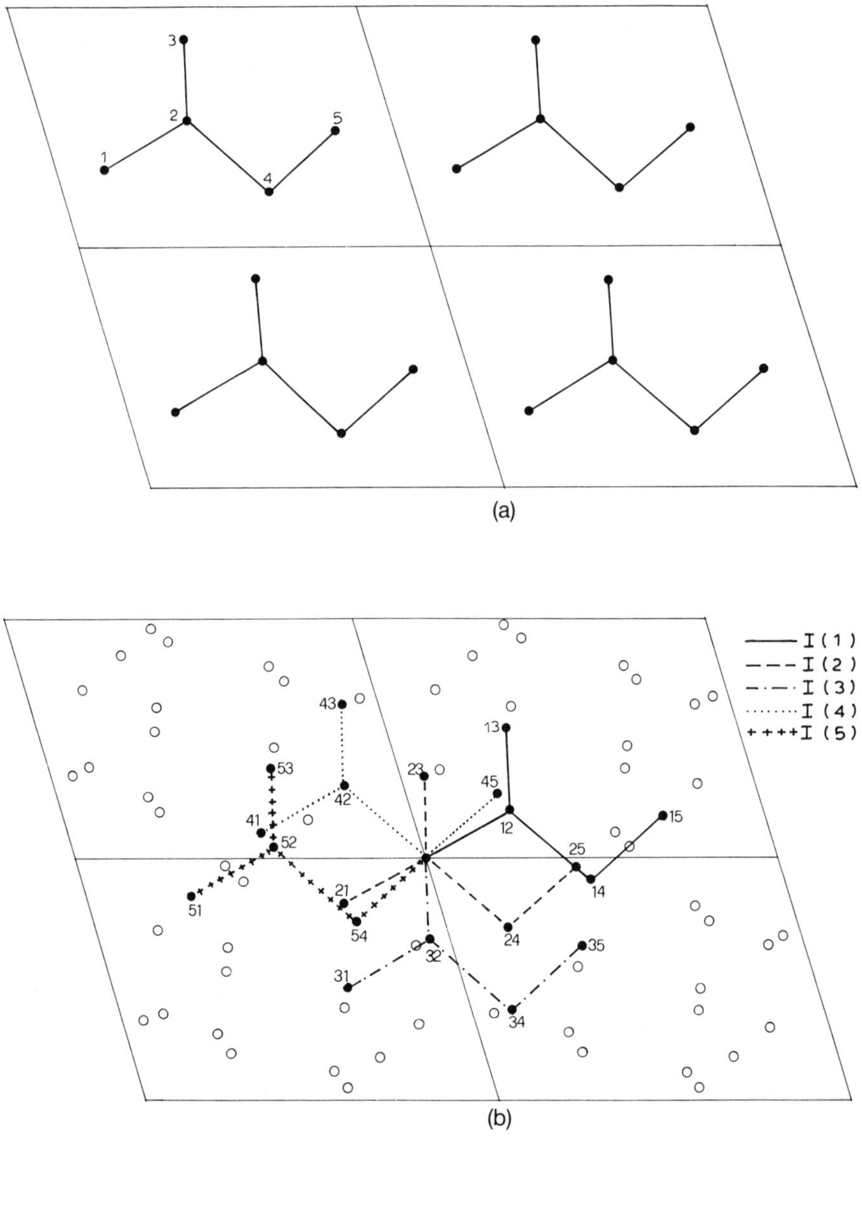

Fig. 5.B.1. (a) Four unit cells of the five-atom structure of Fig. 5.4(a); (b) four cells of the Patterson map in which the peaks indicated by ● refer to the central origin and those indicated by ○ to the other origins; the peaks referred to the central origin are interpreted as the superposition of five images and the peak numbers refer to the atoms linked by the corresponding vector.

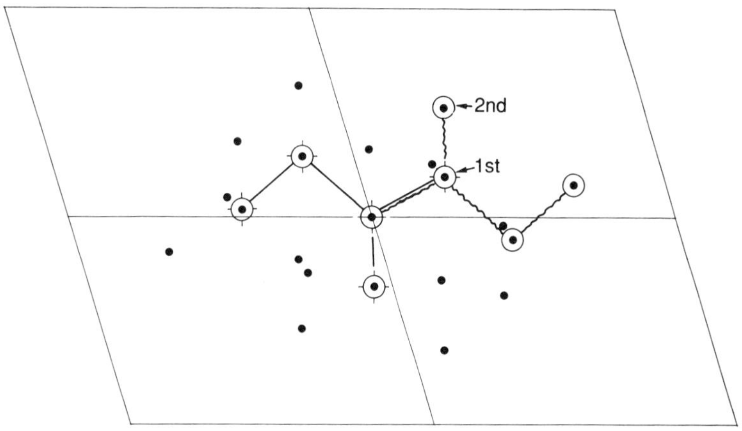

Fig. 5.B.2. Patterson superpositions: ● are the Patterson peaks as in Fig. 5.B.1; ⊙ are the coinciding peaks when the origin of the identical map of Fig. 5.B.1 is superposed on peak 12 of the present map, indicated by the first arrow (two enantiomorphic images of the molecule are shown); -⊕- are the coinciding peaks after a second superposition on peak 13, indicated by the second arrow (only one image is shown). The reader is advised to reproduce on transparent sheets three copies of Fig. 5.B.1 and then perform the above superpositions.

images of the structure. In particular, if the superposition is on a single peak, the structure plus its inverse are obtained for a non-centrosymmetric structure (Fig. 5.B.2); for a centrosymmetric structure only one image is obtained. If the superposition is on a multiple peak the number of images obtained is multiplied by the multiplicity of the peak (Fig. 5.B.4(c)). When more than one image is revealed, iteration of the superposition process on different peaks will eventually yield a single image. This is illustrated in Fig. 5.B.2, where superposition of the double image on a new peak reveals one image only.

The basic principle of the method may be understood by considering that, if the structure contains atoms $1, 2, 3, \ldots, N$ in the unit cell, the Patterson map P will be the superposition of the images $\mathrm{I}(1)$, $\mathrm{I}(2)$, $\mathrm{I}(3), \ldots, \mathrm{I}(N)$ obtained by placing atoms $1, 2, 3, \ldots, N$ in turn at the origin, i.e.

$$P = \mathrm{I}(1) + \mathrm{I}(2) + \mathrm{I}(3) + \ldots + \mathrm{I}(N).$$

After translation by a vector 1–2 the new function will be

$$P' = \mathrm{I}'(1) + \mathrm{I}'(2) + \mathrm{I}'(3) + \ldots + \mathrm{I}'(N)$$

where the different images are obtained by placing atoms $1, 2, 3, \ldots, N$ in turn, not at the origin, but at the end of the vector 1–2. In general $\mathrm{I}'(n)$ will not coincide with any image $\mathrm{I}(1)$, $\mathrm{I}(2)$, $\mathrm{I}(3), \ldots, \mathrm{I}(N)$ unless $n = 1$ or 2. In fact, because of the choice of the translation 1–2, $\mathrm{I}'(1) = \mathrm{I}(2)$ and $-\mathrm{I}'(2) = -\mathrm{I}(1)$ ($-\mathrm{I}(n)$ is the enantiomer of $\mathrm{I}(n)$) as illustrated in Fig. 5.B.3; the two images left in Fig. 5.B.2 after the first superposition are indeed $-\mathrm{I}(1)$ and $\mathrm{I}(2)$.

Unfortunately, this simple method is only applicable in the ideal case of a Patterson function corresponding to a point atom structure. In practice we have a continuous Patterson function in which the peaks of finite width overlap; for reasonably sized molecules with no heavy atoms, also in a sharpened map, only the peaks with high multiplicity appear, while single peaks are not resolved from the background. Thus superposition on any recognizable peak will give multiple images and several superpositions will be necessary to reveal a single image of the structure. A further problem arising with vector superpositions is that the obtained image is not referred to the true origin of the cell as it is apparent from Fig. 5.B.2; the correct

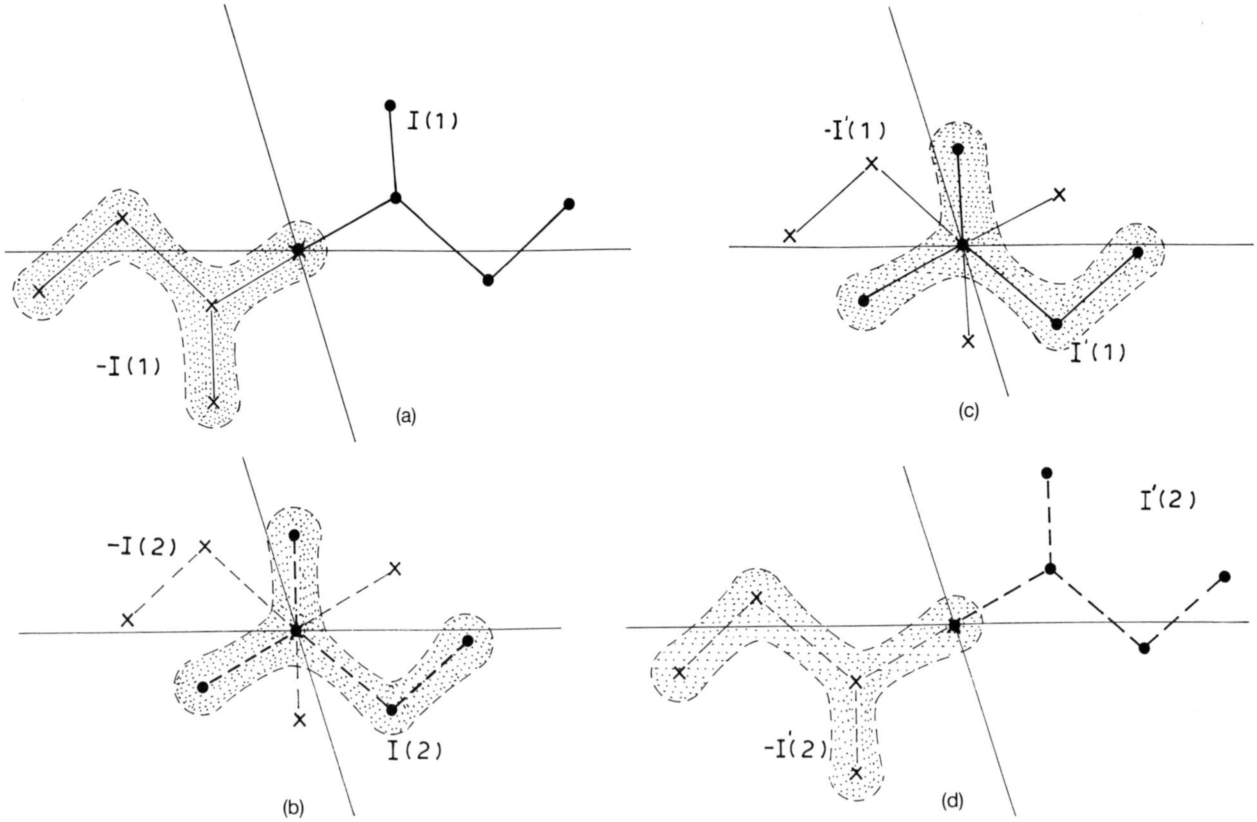

Fig. 5.B.3. (a) Image I(1) and its enantiomorph −I(1) with their atom 1 at the origin; (b) image I(2) and its enantiomorph −I(2) with their atom 2 at the origin; (c) image −I′(1) with its atom 1 on peak 12 (Fig. 5.B.1(b)), plus the image I′(1) with its atom 1 on peak 21 (enantiomorph); (d) image I′(2) with its atom 2 on peak 12, plus the image −I′(2) with its atom 2 on peak 21 (enantiomorph). The shadowed images −I(1) in (a) and I(2) in (b) are found in the superposition on peak 12 of Fig. 5.B.2; they are identical to −I′(2) in (d) and I′(1) in (c) respectively.

location of the structure with respect to the symmetry elements can be obtained by the methods described later.

Despite these problems, superposition methods can still be applied in practice, especially when some structural information is available. This can be of three types:

(1) the stereochemistry of one or more rigid molecular fragments is known, but not their orientation and position;

(2) the orientation is also known, but not the position;

(3) both orientation and position are known.

Information of the first type is often available from known bond lengths and angles or from the Cambridge Structural Database,[102] and we shall see how it can be exploited.

In order to carry out any superposition process it is necessary to evaluate the degree of coincidence between the peaks of the superposing maps. This is done by means of the so-called **image-seeking functions**;[100] three such functions have been proposed:

1. The **product function**

$$\prod(\boldsymbol{r}) = P(\boldsymbol{r}) \times P(\boldsymbol{r} - \boldsymbol{u}) \qquad (5.B.1)$$

which takes very high values when coincidences occur, but is subject to considerable background noise and is very sensitive to small errors in the value of the vector $\boldsymbol{u}$ used for the superposition.

2. The **sum function**

$$\Sigma\,(\boldsymbol{r}) = P(\boldsymbol{r}) + P(\boldsymbol{r} - \boldsymbol{u}) \qquad (5.B.2)$$

which is less sensitive to the errors on $\boldsymbol{u}$, but can often produce spurious peaks and in general a high background noise.

3. The **minimum function**

$$M(\boldsymbol{r}) = \min\,\{P(\boldsymbol{r}),\, P(\boldsymbol{r} - \boldsymbol{u})\} \qquad (5.B.3)$$

which is also sensitive to the errors on $\boldsymbol{u}$ but has the advantage of a low background noise.

Let us first consider how the superposition method can be conveniently applied when the location of one or more atoms (both independent and related by symmetry) in the cell is known. In this case Patterson maps with their origin at the sites of the known atoms are superposed. An example of this **atomic superposition** is shown in Fig. 5.B.4(d) in which two maps, with their origin at the position of atoms 1 and 1′ (related by an inversion centre) respectively, are superposed. Images I(1) and I(1′) coincide and reveal the structure properly placed with respect to the origin. Atomic superposition can be used when the crystal contains a heavy atom, the position of which has been determined by the method described on p. 328. Atomic superposition in this case is facilitated by the fact that the coinciding images are made up of vectors between heavy and light atoms which are well resolved over the background of vectors among light atoms. If the ratio (5.20) is very low, the straightforward application of the Fourier recycling procedure may be inefficient and atomic superposition may be a convenient alternative.

We can now describe one of the procedures[103–106] proposed to solve the Patterson function when the stereochemistry of part of the molecule is known. It is convenient to divide the process of locating a fragment of known stereochemistry into two steps: determination of the orientation and determination of the position with respect to the symmetry elements.

1. *Determination of the orientation.* Starting from the known molecular fragment (atomic model) it is possible to build up the corresponding vector map (vector model). This model map is superposed on the Patterson function with its origin coinciding with that of the Patterson function and then rotated in all possible orientations. For each orientation the match between the ends of the m vectors of the model and the Patterson peaks is evaluated by means of functions similar to those considered above. Their form is now:

(a) product function

$$\prod = \prod_{j=1}^{m} P(\boldsymbol{u}_j)\} \qquad (5.B.4)$$

where $P(\boldsymbol{u}_j)$ are the values of the Patterson function at the ends of the vectors;

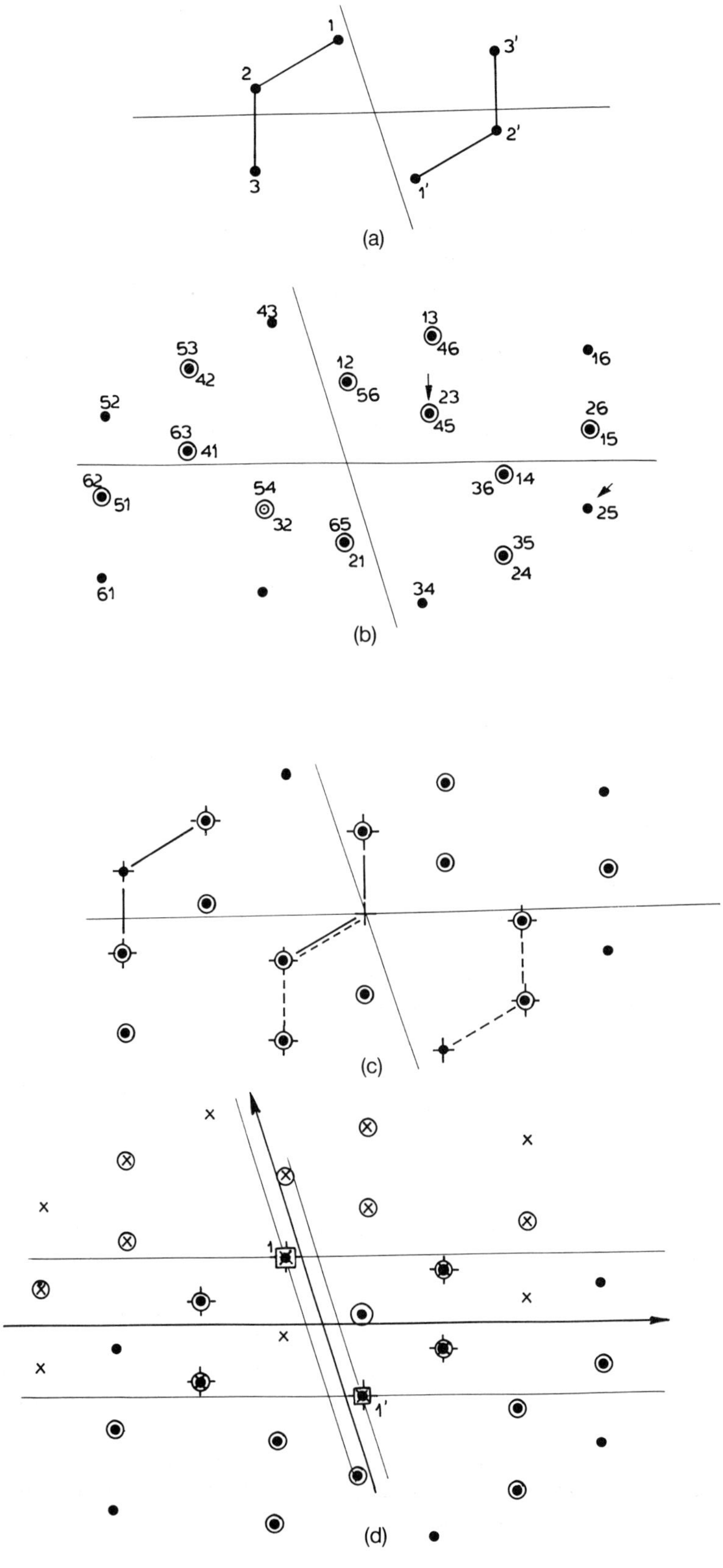

Fig. 5.B.4. (a) Centrosymmetric three-atom structure; (b) corresponding Patterson map; (c) vector superposition on the double peak 23 + 45 gives the two images in full and broken lines, while superposition on the single peak 25 gives the full-line image only, shifted from the correct origin at the inversion centre; (d) Atomic superposition: ⊡ are the known positions of the related atoms 1 and 1′; ×, ⊗ are single and double peaks of the Patterson function with its origin on atom 1; ●, ⊙ are single and double peaks of the Patterson function with its origin on atom 1′; -⌖- are the overlapping peaks showing the structure referred to the correct origin.

(b) sum function

$$\Sigma = \sum_{j=1}^{m} P(\boldsymbol{u}_j); \quad (5.B.5)$$

(c) minimum function

$$M = \min\,\{P(\boldsymbol{u}_1),\, P(\boldsymbol{u}_2)\ldots P(\boldsymbol{u}_m)\}. \quad (5.B.6)$$

This function is the most sensitive and is usually preferred, but, when the absolute minimum is assumed, small errors in the model may produce small values of M even for a correct orientation. To overcame this problem it is preferable to use the mean of the n smallest values $\langle M \rangle_n$, where n is usually equal to 10–20 per cent of the total number of vectors.[107]

The orientation of the model can be varied following the same Eulerian scheme (a detailed description is given by Stout and Jensen[108]) used in most diffractometers to define the position of the reciprocal lattice vectors in terms of the three angles θ, φ, and χ (cf. pp. 69 and 276). The number of orientations to be examined can be quite large and the symmetry of the Patterson function and that of the vector model can reduce the range of the rotational search.

The correct orientation will correspond to large values of the functions (5.B.4), (5.B.5), and (5.B.6).

2. *Determination of the position with respect to the symmetry elements*. Once a likely orientation has been defined, the location of the fragment can be determined by systematically translating it in all possible positions within the asymmetric part of the unit cell and then by generating, for each position, all the vectors between the atoms of the fragment and those of the symmetry-related fragments. Each vector set obtained in this way is compared with the Patterson function and the fit is evaluated by means of one of the functions used in the orientational search. Normally the search of the position of a fragment with respect to the different symmetry elements is carried out separately for each symmetry element, considering that:

(a) when the symmetry element is a plane the search is performed only along the direction orthogonal to the plane;

(b) when the symmetry element is an axis the search is a two-dimensional one on the plane orthogonal to the axis;

(c) when the symmetry element is an inversion centre the search is three dimensional.

Because of the errors in the model the orientation and position are determined by the above method only in an approximate way. They can be further improved by shifting the atoms of the model in such a way that a better fit of the set of interatomic vectors on the maxima in the Patterson function is obtained.

The described procedure operates in the Patterson space, but other methods, usually operating in the reciprocal space, such as the **rotation function**,[109] have been proposed[110–113] and successfully applied, especially in protein crystallography (cf. Chapter 8, p. 553). They are at the basis of the so-called **molecular replacement** methods.[114] For each orientation or position of the model fragment, instead of fitting the vector model, a fit is searched for between the structure amplitudes calculated for the model and the observed ones. A procedure combining direct and reciprocal space searches has been recently described by Rius and Miravitlles.[115]

Recently, Patterson orientation search routines have been incorporated into direct methods programs. A highly automated routine has been built into the DIRDIF system.[70] Another example is the PATSEE routine[116] incorporated into the SHELXS-86 system.[10] In both these cases the correct position of the oriented fragment is found by direct methods.

5.C Two examples of deriving phase information from positivity

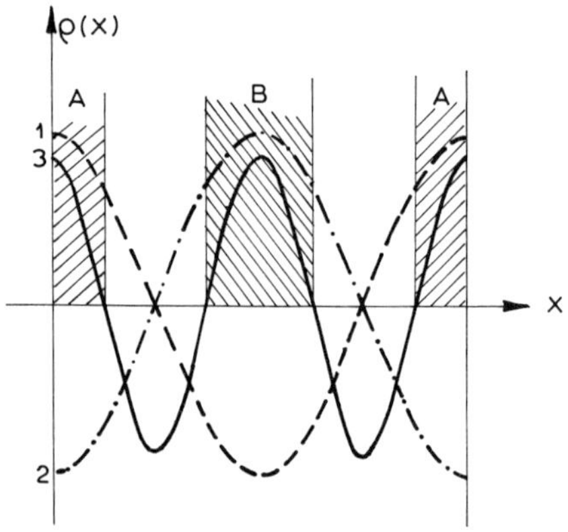

Fig. 5.C.1. One-dimensional centrosymmetric structure with $|F_{\boldsymbol{h}}|$ and $|F_{2\boldsymbol{h}}|$ large: curve 1 shows the trend of $|F_{\boldsymbol{h}}| \cos 2\pi hx$, curve 2 that of $-|F_{\boldsymbol{h}}| \cos 2\pi hx$, and curve 3 that of $|F_{2\boldsymbol{h}}| \cos 2\pi 2hx$; the hatched regions A and B are those in which a high value of the electron density is indicated.

Let us first consider a centrosymmetric one-dimensional structure for which the amplitudes $|F_{\boldsymbol{h}}|$ and $|F_{2\boldsymbol{h}}|$ are large. Their contributions to the electron density function

$$\rho(x) = (1/a) \sum_{h} F_{\boldsymbol{h}} \cos 2\pi hx$$

will be large and are illustrated in Fig. 5.C.1, where curve 1 shows the term $+|F_{\boldsymbol{h}}| \cos 2\pi hx$ ($F_{\boldsymbol{h}}$ positive), curve 2 the term $-|F_{\boldsymbol{h}}| \cos 2\pi hx$ ($F_{\boldsymbol{h}}$ negative), and curve 3 the term $+|F_{2\boldsymbol{h}}| \cos 2\pi 2hx$ ($F_{2\boldsymbol{h}}$ positive). $F_{\boldsymbol{h}}$ would then contribute to the electron density in the region indicated by the letter A, if positive, and in the region B, if negative; in both cases $F_{2\boldsymbol{h}}$ will contribute to both regions only with a positive sign. It follows that, with $\rho(x)$ everywhere positive, if $|F_{\boldsymbol{h}}|$ and $|F_{2\boldsymbol{h}}|$ are large, whatever the sign of $|F_{\boldsymbol{h}}|$, the sign of $|F_{2\boldsymbol{h}}|$ is more likely positive. This argument can be generalized to three dimensions.

As a second example we shall consider the projection of a centrosymmetric structure for which the amplitudes $|F_{\boldsymbol{h}}|$, $|F_{\boldsymbol{k}}|$, and $|F_{\boldsymbol{h}-\boldsymbol{k}}|$ are large. In Fig. 5.C.2 the traces of the three families of planes $\boldsymbol{h}$, $\boldsymbol{k}$, and $\boldsymbol{h}-\boldsymbol{k}$ are shown as full lines, while the dotted lines are located half way. If we associate to the full lines the maxima of the terms $|F_{\boldsymbol{h}}| \cos 2\pi\boldsymbol{h}\cdot\boldsymbol{r}$, $|F_{\boldsymbol{k}}| \cos 2\pi\boldsymbol{k}\cdot\boldsymbol{r}$, $|F_{\boldsymbol{h}-\boldsymbol{k}}| \cos 2\pi(\boldsymbol{h}-\boldsymbol{k})\cdot\boldsymbol{r}$ contributing to the electron density function $\rho(\boldsymbol{r})$, then the dotted lines will represent the corresponding minima. If $|F_{\boldsymbol{h}}|$, $|F_{\boldsymbol{k}}|$, $|F_{\boldsymbol{h}-\boldsymbol{k}}|$ are large, the corresponding terms must add up to give a preponderant contribution to $\rho(\boldsymbol{r})$. This will happen when $\cos 2\pi\boldsymbol{h}\cdot\boldsymbol{r} = +1$, if $F_{\boldsymbol{h}}$ is positive, and $\cos 2\pi\boldsymbol{h}\cdot\boldsymbol{r} = -1$ if it is negative; the same applies to the other two terms. It follows that the possible maxima of the electron density will be restricted to the regions indicated by A, B, C, and D in Fig. 5.C.2, where the above conditions are satisfied for all three terms at the same time. For each region the signs $S(\boldsymbol{h})$, $S(\boldsymbol{k})$, $S(\boldsymbol{h}-\boldsymbol{k})$ of the three structure factors are

Region	$S(\boldsymbol{h})$	$S(\boldsymbol{k})$	$S(\boldsymbol{h}-\boldsymbol{k})$
A	+	−	−
B	+	+	+
C	−	−	+
D	−	+	−

In all cases the product $S(\boldsymbol{h})S(\boldsymbol{k})S(\boldsymbol{h}-\boldsymbol{k})$ is always positive.

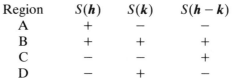

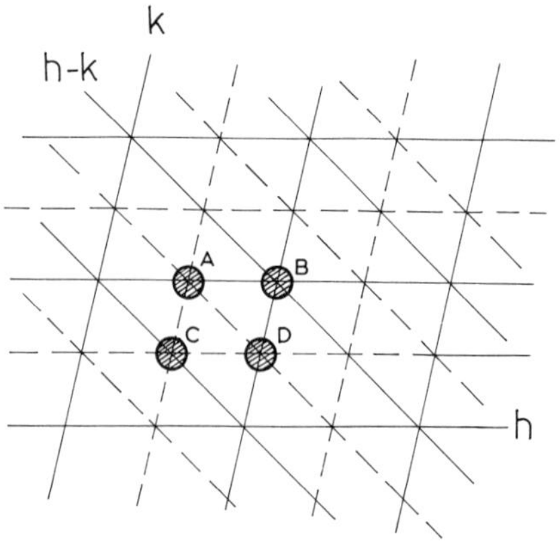

Fig. 5.C.2. Three-dimensional centrosymmetric structure with $|F_{\boldsymbol{h}}|$, $|F_{\boldsymbol{k}}|$, and $|F_{\boldsymbol{h}-\boldsymbol{k}}|$ large: the full lines represent the maxima of $|F_{\boldsymbol{h}}| \cos 2\pi\boldsymbol{h}\cdot\boldsymbol{r}$, $|F_{\boldsymbol{k}}| \cos 2\pi\boldsymbol{k}\cdot\boldsymbol{r}$ and $|F_{\boldsymbol{h}-\boldsymbol{k}}| \cos 2\pi(\boldsymbol{h}-\boldsymbol{k})\cdot\boldsymbol{r}$, the broken lines the corresponding minima; the hatched regions A, B, C, D are those in which high values of the electron density are more likely.

5.D Probability formulae for triplet invariants

For a structure with N equal atoms in a P1 unit cell, the normalized structure factor of index $\boldsymbol{h}$ can be written as the sum of N independent terms

$$E_{\boldsymbol{h}} = \frac{1}{\sqrt{N}} \sum_{j=1}^{N} \exp(2\pi i \boldsymbol{h} \cdot \boldsymbol{r}_j) = \frac{1}{\sqrt{N}} \sum_{j=1}^{N} \exp(2\pi i \boldsymbol{k} \cdot \boldsymbol{r}_j) \exp(2\pi i (\boldsymbol{h} - \boldsymbol{k}) \cdot \boldsymbol{r}_j)$$
$$= \frac{1}{\sqrt{N}} \sum_{j=1}^{N} \xi_j. \quad (5.D.1)$$

Under the assumption that all points in the unit cell have equal probability of housing an atom, the atomic position vectors $\boldsymbol{r}_j$ will form a set of independent random variables. Thus also the terms ξ_j, which are functions of $\boldsymbol{r}_j$, will be random variables. Then, because of the central limit theorem, $E_{\boldsymbol{h}}$ will have a normal distribution with mean

$$\langle E_{\boldsymbol{h}} \rangle = \frac{1}{\sqrt{N}} \sum_{j=1}^{N} \langle \xi_j \rangle \quad (5D.2)$$

and variance

$$\sigma_{\boldsymbol{h}}^2 = \frac{1}{N} \sum_{j=1}^{N} \left(\langle \xi_j^2 \rangle - \langle \xi_j \rangle^2 \right). \quad (5.D.3)$$

Supposing that the two exponential terms in the middle formula in (5.D.1) are independent, we have

$$\langle \xi_j \rangle = \langle \exp(2\pi i \boldsymbol{k} \cdot \boldsymbol{r}_j) \rangle \langle \exp(2\pi i (\boldsymbol{h} - \boldsymbol{k}) \cdot \boldsymbol{r}_j) \rangle. \quad (5.D.4)$$

Under the assumption that $E_{\boldsymbol{k}}$ has a fixed value (i.e. both its modulus and phase are known), we can write[117]

$$E_{\boldsymbol{k}} = \frac{1}{\sqrt{N}} \sum_{j=1}^{N} \exp(2\pi i \boldsymbol{k} \cdot \boldsymbol{r}_j) = \frac{1}{\sqrt{N}} N \langle \exp(2\pi i \boldsymbol{k} \cdot \boldsymbol{r}) \rangle. \quad (5.D.5)$$

With the same assumption for $E_{\boldsymbol{h}-\boldsymbol{k}}$, (5.D.4) becomes

$$\langle \xi_j \rangle = \frac{E_{\boldsymbol{k}} E_{\boldsymbol{h}-\boldsymbol{k}}}{N}, \quad (5.D.6)$$

and substituting in (5.D.2)

$$\langle E_{\boldsymbol{h}} \rangle = \frac{1}{\sqrt{N}} \sum_{j=1}^{N} \frac{E_{\boldsymbol{k}} E_{\boldsymbol{h}-\boldsymbol{k}}}{N} = \frac{1}{\sqrt{N}} E_{\boldsymbol{k}} E_{\boldsymbol{h}-\boldsymbol{k}}. \quad (5.D.7)$$

For the real and imaginary components of $E_{\boldsymbol{h}} = A_{\boldsymbol{h}} + iB_{\boldsymbol{h}}$ we have

$$\langle A_{\boldsymbol{h}} \rangle = \frac{1}{\sqrt{N}} |E_{\boldsymbol{k}} E_{\boldsymbol{h}-\boldsymbol{k}}| \cos(\varphi_{\boldsymbol{k}} + \varphi_{\boldsymbol{h}-\boldsymbol{k}})$$
$$\langle B_{\boldsymbol{h}} \rangle = \frac{1}{\sqrt{N}} |E_{\boldsymbol{k}} E_{\boldsymbol{h}-\boldsymbol{k}}| \sin(\varphi_{\boldsymbol{k}} + \varphi_{\boldsymbol{h}-\boldsymbol{k}}). \quad (5.D.8)$$

Similarly to (5.D.1), we can derive

$$A_{\boldsymbol{h}} = \frac{1}{\sqrt{N}} \sum_{j=1}^{N} \cos 2\pi (\boldsymbol{h} - \boldsymbol{k} + \boldsymbol{k}) \cdot \boldsymbol{r}_j = \frac{1}{\sqrt{N}} \sum_{j=1}^{N} [C_j(\boldsymbol{k}) C_j(\boldsymbol{h} - \boldsymbol{k}) - S_j(\boldsymbol{k}) S_j(\boldsymbol{h} - \boldsymbol{k})]$$
$$= \frac{1}{\sqrt{N}} \sum_{j=1}^{N} \alpha_j \quad (5.D.9)$$

where $C_j(\boldsymbol{k}) = \cos 2\pi\boldsymbol{k}\cdot\boldsymbol{r}_j$ and $S_j(\boldsymbol{k}) = \sin 2\pi\boldsymbol{k}\cdot\boldsymbol{r}_j$. Similarly

$$B_{\boldsymbol{h}} = \frac{1}{\sqrt{N}}\sum_{j=1}^{N}[S_j(\boldsymbol{k})C_j(\boldsymbol{h}-\boldsymbol{k}) + C_j(\boldsymbol{k})S_j(\boldsymbol{h}-\boldsymbol{k})] = \frac{1}{\sqrt{N}}\sum_{j=1}^{N}\beta_j. \quad (5.D.10)$$

Considering that

$$\langle\alpha_j^2\rangle = \langle C_j^2(\boldsymbol{k})\rangle\langle C_j^2(\boldsymbol{h}-\boldsymbol{k})\rangle - 2\langle C_j(\boldsymbol{k})\rangle\langle C_j(\boldsymbol{h}-\boldsymbol{k})\rangle\langle S_j(\boldsymbol{k})\rangle\langle S_j(\boldsymbol{h}-\boldsymbol{k})\rangle$$
$$+ \langle S_j^2(\boldsymbol{k})\rangle\langle S_j^2(\boldsymbol{h}-\boldsymbol{k})\rangle \approx \tfrac{1}{4} + 0 + \tfrac{1}{4} = \tfrac{1}{2} \quad (5.D.11)$$

and also $\langle\beta_j\rangle \approx \frac{1}{2}$, we obtain the variances

$$\sigma^2_{A_{\boldsymbol{h}}} = \frac{1}{N}\left(\sum_{j=1}^{N}\langle\alpha_j^2\rangle - \langle A_{\boldsymbol{h}}\rangle^2\right) \approx \tfrac{1}{2}; \qquad \sigma^2_{B_{\boldsymbol{h}}} \approx \tfrac{1}{2}; \quad (5.D.12)$$

where the terms $\langle A_{\boldsymbol{h}}\rangle^2/N$ and $\langle B_{\boldsymbol{h}}\rangle^2/N$ have been neglected, as they are of order $1/N^2$ [cf. eqns (5.D.8)] and become very small for any reasonable N.

The conditional probability density function $P(A_{\boldsymbol{h}}, B_{\boldsymbol{h}}|\, E_{\boldsymbol{k}}, E_{\boldsymbol{h}-\boldsymbol{k}})$ can be expressed as the product of the separate distributions of $A_{\boldsymbol{h}}$ and $B_{\boldsymbol{h}}$. The probability that $A_{\boldsymbol{h}}$ lies in the interval between $A_{\boldsymbol{h}}$ and $A_{\boldsymbol{h}} + \mathrm{d}A_{\boldsymbol{h}}$ and at the same time $B_{\boldsymbol{h}}$ lies in the interval between $B_{\boldsymbol{h}}$ and $B_{\boldsymbol{h}} + \mathrm{d}B_{\boldsymbol{h}}$, is therefore given by

$$P(A_{\boldsymbol{h}}, B_{\boldsymbol{h}} \mid E_{\boldsymbol{h}}, E_{\boldsymbol{h}-\boldsymbol{k}}) = \frac{1}{\pi}\exp\{-[(A_{\boldsymbol{h}} - \langle A_{\boldsymbol{h}}\rangle)^2 + (B_{\boldsymbol{h}} - \langle B_{\boldsymbol{h}}\rangle)^2]\} \quad (5.D.13)$$

which gives the probability that $E_{\boldsymbol{h}}$ lies within an area of surface $\mathrm{d}S = \mathrm{d}A_{\boldsymbol{h}}\,\mathrm{d}B_{\boldsymbol{h}}$ in the complex plane. When expressing $E_{\boldsymbol{h}}$ in polar coordinates (modulus and phase), then $\mathrm{d}S = |E_{\boldsymbol{h}}|\,\mathrm{d}|E_{\boldsymbol{h}}|\,\mathrm{d}\varphi_{\boldsymbol{h}}$ and

$$P(|E_{\boldsymbol{h}}|, \varphi_{\boldsymbol{h}} \mid E_{\boldsymbol{k}}, E_{\boldsymbol{h}-\boldsymbol{k}}) = \frac{|E_{\boldsymbol{h}}|}{\pi}\exp\{-[|E_{\boldsymbol{h}}|\cos\varphi_{\boldsymbol{h}}$$
$$- (1/\sqrt{N})\,|E_{\boldsymbol{k}}E_{\boldsymbol{h}-\boldsymbol{k}}|\cos(\varphi_{\boldsymbol{k}} + \varphi_{\boldsymbol{h}-\boldsymbol{k}})]^2$$
$$- [|E_{\boldsymbol{h}}|\sin\varphi_{\boldsymbol{h}} - (1/\sqrt{N})\,|E_{\boldsymbol{k}}E_{\boldsymbol{h}-\boldsymbol{k}}|\sin(\varphi_{\boldsymbol{k}} + \varphi_{\boldsymbol{h}-\boldsymbol{k}})]^2\}$$
$$= \frac{|E_{\boldsymbol{h}}|}{\pi}\exp\{-[|E_{\boldsymbol{h}}|^2 + (1/N)\,|E_{\boldsymbol{k}}E_{\boldsymbol{h}-\boldsymbol{k}}|^2]\}$$
$$\times \exp[(2/\sqrt{N})\,|E_{\boldsymbol{h}}E_{\boldsymbol{k}}E_{\boldsymbol{h}-\boldsymbol{k}}|\cos(\varphi_{\boldsymbol{h}} - \varphi_{\boldsymbol{k}} - \varphi_{\boldsymbol{h}-\boldsymbol{k}})]. \quad (5.D.14)$$

As $|E_{\boldsymbol{h}}|$ is known, we are interested in the conditional probability $P(\varphi_{\boldsymbol{h}} \mid |E_{\boldsymbol{h}}|, E_{\boldsymbol{k}}, E_{\boldsymbol{h}-\boldsymbol{k}})$, simply indicated as $P(\varphi_{\boldsymbol{h}})$, which can be obtained by dividing (5.D.14) by its integral with respect to $\varphi_{\boldsymbol{h}}$[118]

$$P(\varphi_{\boldsymbol{h}}) = \frac{\exp(G_{\boldsymbol{hk}}\cos\Phi_{\boldsymbol{hk}})}{\int_{-\pi}^{\pi}\exp(G_{\boldsymbol{hk}}\cos\Phi_{\boldsymbol{hk}})\,\mathrm{d}\varphi_{\boldsymbol{h}}} \quad (5.D.15)$$

where

$$G_{\boldsymbol{hk}} = \frac{2}{\sqrt{N}}|E_{\boldsymbol{h}}E_{\boldsymbol{k}}E_{\boldsymbol{h}-\boldsymbol{k}}| \quad (5.D.16)$$

and

$$\Phi_{\boldsymbol{hk}} = \varphi_{\boldsymbol{h}} - \varphi_{\boldsymbol{k}} - \varphi_{\boldsymbol{h}-\boldsymbol{k}}.$$

Since $\varphi_{\boldsymbol{k}}$ and $\varphi_{\boldsymbol{h}-\boldsymbol{k}}$ are known $\mathrm{d}\Phi_{\boldsymbol{hk}} = \mathrm{d}\varphi_{\boldsymbol{h}}$ and the distribution of $\varphi_{\boldsymbol{h}}$ is equivalent to that of Φ_{hk}. The value of the integral at the denominator of (5.D.15) is[58] $L = 2\pi I_0(G_{hk})$, where $I_0(x)$ is the modified Bessel function of zero order. We can finally write[24]

$$P(\varphi_h) = \frac{1}{2\pi I_0(G_{hk})} \exp(G_{hk} \cos \Phi_{hk}) \qquad (5.D.17)$$

which is equal to the Cochran distribution (5.38).

In a similar way it is possible to derive the probability formulae for centrosymmetric structures.[28] By applying the central-limit theorem to

$$E_h = \frac{1}{\sqrt{N}} \sum_{j=1}^{N/2} 2 \cos 2\pi[\boldsymbol{k} + (\boldsymbol{h} - \boldsymbol{k})] \cdot \boldsymbol{r}_j = \frac{1}{\sqrt{N}} \sum_{j=1}^{N} \xi_j \qquad (5.D.18)$$

we obtain

$$P(E_h) = \frac{1}{\sqrt{2\pi}} \exp\left[-\frac{1}{2}\left(E_h - \frac{E_k E_{h-k}}{\sqrt{N}}\right)^2\right]. \qquad (5.D.19)$$

Indicating by P^+ or P^- respectively, the probability that E_h has a positive or negative sign, we have

$$\frac{P^-}{P^+} = \exp\left\{-\frac{2}{\sqrt{N}} |E_h|\, E_k E_{h-k}\right\} = \exp\{-2x\} = \frac{1 - P^+}{P^+} \qquad (5.D.20)$$

and

$$\begin{aligned} P^+ &= \frac{1}{1 + \exp(-2x)} = \frac{\exp x}{\exp x + \exp(-x)} = \tanh x + \frac{\exp(-x)}{\exp x + \exp(-x)} \\ &= \tanh(x) + 1 - P^+ \end{aligned} \qquad (5.D.21)$$

and finally

$$P^+ = \tfrac{1}{2} + \tfrac{1}{2} \tanh\left(\frac{1}{\sqrt{N}} |E_h|\, E_k E_{h-k}\right). \qquad (5.D.22)$$

The function (5.D.22) is illustrated in Fig. 5.16; $P^+ > \frac{1}{2}$ if $S(\boldsymbol{k})S(\boldsymbol{h} - \boldsymbol{k}) = +$ and $P^+ < \frac{1}{2}$ (and $P^- > \frac{1}{2}$) if $S(\boldsymbol{k})S(\boldsymbol{h} - \boldsymbol{k}) = -$. If in (5.D.19) we consider the two cases $S(\boldsymbol{h}) = S(\boldsymbol{k})S(\boldsymbol{h} - \boldsymbol{k})$ (with probability P^+) and $S(\boldsymbol{h}) = -S(\boldsymbol{k})S(\boldsymbol{h} - \boldsymbol{k})$ (with probability P^-) we obtain eqn (5.46), where P^+ is always greater than $\frac{1}{2}$.

5.E Pseudotranslational symmetry

We will say that a structure possesses **pseudosymmetry** when a non-negligible part of the atoms approximately satisfy a higher symmetry than that of the whole structure. It can be noted that, in the special case in which these atoms exactly satisfy a higher symmetry, they will identify a supergroup (see Appendix 1.E) of the space group to which the whole structure belongs.

An important type of pseudosymmetry is the **pseudotranslational symmetry** occurring when a non-negligible amount of the electron density, say $\rho_p(\boldsymbol{r})$ is repeated by a pseudotranslation (in the sense that it does not necessarily correspond to a crystallographic translation) $\boldsymbol{u}$, i.e. $\rho_p(\boldsymbol{r}) \approx \rho_p(\boldsymbol{r} + \boldsymbol{u})$. Buerger[100,119] suggested that such structures should be treated as

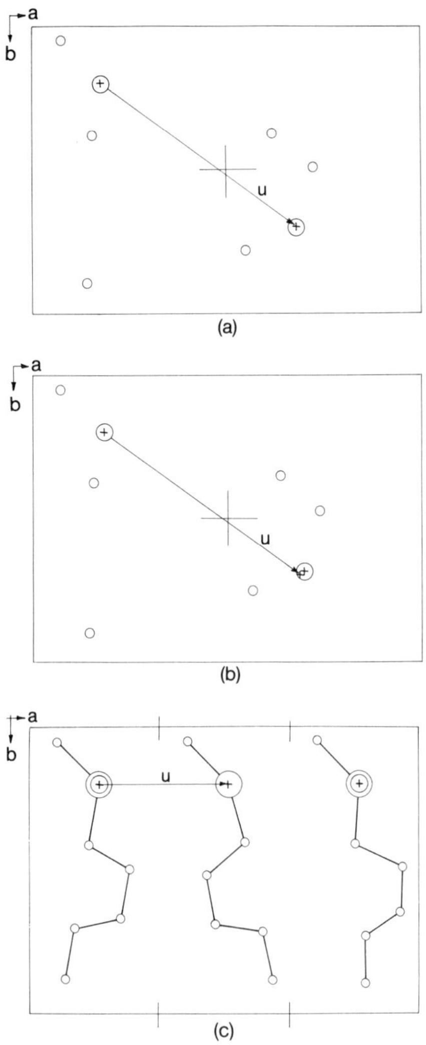

Fig. 5.E.1. (a) Heavy atoms exactly related by a pseudotranslation vector $\boldsymbol{u} = \boldsymbol{a}/2 + \boldsymbol{b}/2$; (b) heavy atoms approximately related by a pseudotranslation vector $\boldsymbol{u} = \boldsymbol{a}/2 + \boldsymbol{b}/2$; (c) slightly different heavy atoms related by a pseudotranslation vector $\boldsymbol{u} = \boldsymbol{a}/3$.

the sum of a **substructure**, $\rho_p(\boldsymbol{r})$ and the **complement structure**, $\rho_c(\boldsymbol{r}) = \rho(\boldsymbol{r}) - \rho_p(\boldsymbol{r})$.

A two-dimensional example of pseudotranslational symmetry is shown in Fig. 5.E.1(a), where the two heavy atoms 1 and 1′ are related by a centring translation, $\boldsymbol{u} = \boldsymbol{a}/2 + \boldsymbol{b}/2$, while the light atoms occupy general unrelated positions: the heavy atoms form the substructure and the light atoms the complement structure. If the structure only contained the heavy atoms, the unit cell would be centred and, according to p. 159, the reflections with $h + k = 2n + 1$ would be systematically absent. For the structure of Fig. 5.E.1(a), only the light atoms will contribute to these reflections (**superstructure reflections**), which will therefore be systematically weaker than the reflections with $h + k = 2n$ (**substructure reflections**) to which both heavy and light atoms contribute.

Because of the presence of pseudotranslational symmetry, imposing certain relations between subsets of atomic positions, the assumption that the atomic coordinates are independent random variables is violated. In the reciprocal space this is revealed by a non-uniform value of the averaged normalized intensity, $\langle |E_{\boldsymbol{h}}|^2 \rangle$, when the reflections are grouped in different ways. As we have seen, the reflections have to be divided into the set of substructure reflections with high mean intensity and the set of superstructure reflections with low mean intensity.

In Fig. 5.E.1(b) the case in which the pseudotranslation is only approximately obeyed, is shown. Finally, in Fig. 5.E.1(c) another example of pseudotranslational symmetry is shown, in which two similar but not equal atoms are related by a vector $\boldsymbol{u} = \boldsymbol{a}/3$; the substructure reflections will have $h = 3n$.

Statistical methods for identifying the type of pseudotranslational symmetry and the related sub- and superstructure reflections have been recently proposed.[49,51,120]

5.F Magic integers

Magic integers were first introduced by White and Woolfson[121] and allow one to represent several phases in terms of only one variable, through relationships of the type

$$\varphi_i = m_i x \tag{5.F.1}$$

where the m_i are the magic integers. Let us consider the example given by Woolfson.[122] Three phases, expressed for convenience in cycles ($0 \leqslant \varphi < 1$) instead of degrees, are considered; by using the three integers 3, 4, 5 in (5.F.1), we will have

$$\varphi_1 = 3x \bmod (1), \qquad \varphi_2 = 4x \bmod (1), \qquad \varphi_3 = 5x \bmod (1), \tag{5.F.2}$$

where mod(1) indicates that the phases are always taken within the interval 0–1 cycles (i.e. 0–360°). We will now see that, whatever the value of the three phases, it is always possible to find a value of x ($0 < x < 1$) for which the three equations (5.F.2) are approximately satisfied. For instance, if $\varphi_1 = 0.3$, $\varphi_2 = 0.2$, $\varphi_3 = 0.7$, the value $x = 0.766$ (Fig. 5.F.1) will give:

$\varphi_1 = 2.298 \bmod (1) = 0.298$ with an error of 0.002 cycles (1°),

$\varphi_2 = 3.064 \bmod (1) = 0.064$ with an error of 0.136 cycles (49°),

$\varphi_3 = 3.830 \bmod (1) = 0.830$ with an error of 0.130 cycles (47°).

The reader may assign any three values to the phases and find by trial and error, with the help of Fig. 5.F.1, the value of x which best satisfies (5.F.2).

So far we have used the sequence of three magic integers 3, 4, 5, but other longer or shorter sequences of different integers may be used. The accuracy with which phases can be represented by magic integers depends on the values of the integers and on the length of the sequence. Main[123,124] described the theory of magic integers and gave some rules for selecting those sequences which minimize the mean square deviations $\langle \Delta\varphi^2 \rangle$ of the represented phases. It turns out that the optimal sequences $m_1, m_2, \ldots, m_n$ are those for which $2m_1 = m_n + 1$ and the differences $m_n - m_{n-1}$, $m_{n-1} - m_{n-2}, \ldots, m_3 - m_2$, $m_2 - m_1$ form a geometric progression of integer numbers with common ratio $r \geqslant 1$. For $r = 2$ the progression 1, 2, 4, 8, 16, ... will give rise to the magic integers listed in the top part of Table 5.F.1. An integer progression with common ratio less than 2 is the so-called **Fibonacci series** 1, 1, 2, 3, 5, 8, 13, ..., where each term is defined as the sum of the two preceding ones ($F_n = F_{n-1} + F_{n-2}$) and r tends to the golden number 1.618. The corresponding magic-integer sequences are listed in the bottom part of Table 5.F.1, where the column headed $(\Delta\varphi)_{\text{r.m.s.}}$ gives the root mean square error of the phases represented with the different magic-integer sequences. The optimal sequences also have the advantage that the errors are equally distributed among the represented phases. From Table 5.F.1 it may be seen that $(\Delta\varphi)_{\text{r.m.s.}}$ increases as n increases and, for a given n it is lower when the integers are larger ($r = 2$).

In several multisolution programs the general phases in the starting set are expressed in the form (5.F.1) and x is explored at regular intervals in the range 0–1 (if the enantiomorph has to be fixed, only the interval 0–0.5 should be explored). The value of the interval Δx is chosen in such a way that the mean variation of the phases from one value of x to the other is of the same order of magnitude of the corresponding $(\Delta\varphi)_{\text{r.m.s.}}$. Δx is therefore a function of n and of the m_i values and it is smaller the larger are the integers. In order to keep the number of explored x values (each value corresponding to a trial starting set of phases) within reasonable limits, it is preferable to use magic integers based on Fibonacci series, at the cost of slightly higher $(\Delta\varphi)_{\text{r.m.s.}}$. A comparison between the use of quadrant permutations and that of magic-integer permutations is shown in Table 5.F.2. This comparison may be better illustrated by considering the case $n = 2$, for which it is possible to draw in the plane (φ_1, φ_2), the lines $\varphi_1 = 2x \bmod (1)$ and $\varphi_2 = 3x \bmod (1)$ (Fig. 5.F.2(a)). The choice of regular

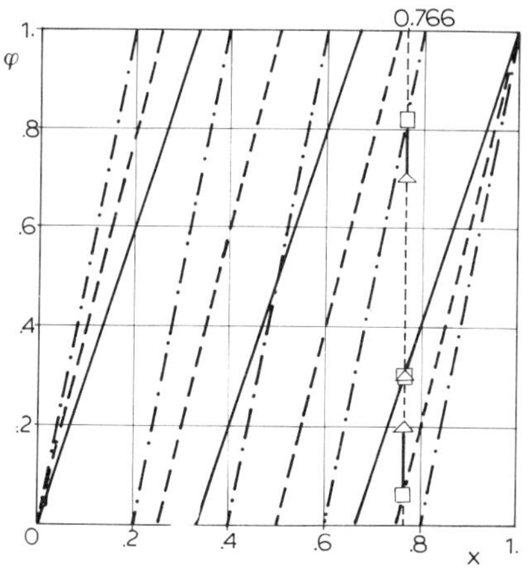

Fig. 5.F.1. Representation of the three phases in eqns (5.F.2) as a function of x.

Table 5.F.1. Magic integer sequences based on the geometric progression with common ratio $r = 2$ (top) and on the Fibonacci series (bottom); $(\Delta\varphi)_{\text{r.m.s.}}$ is the root mean square error of the phases represented by each sequence

	n	Sequence						$(\Delta\varphi)_{\text{r.m.s.}}$
$r = 2$	2	2	3					20.4
	3	4	6	7				26.6
	4	8	12	14	15			29.9
	5	16	24	28	30	31		32.5
$r - 1.618$	2	2	3					20.4
	3	3	4	5				32.0
	4	5	7	8	9			36.1
	5	8	11	13	14	15		39.2

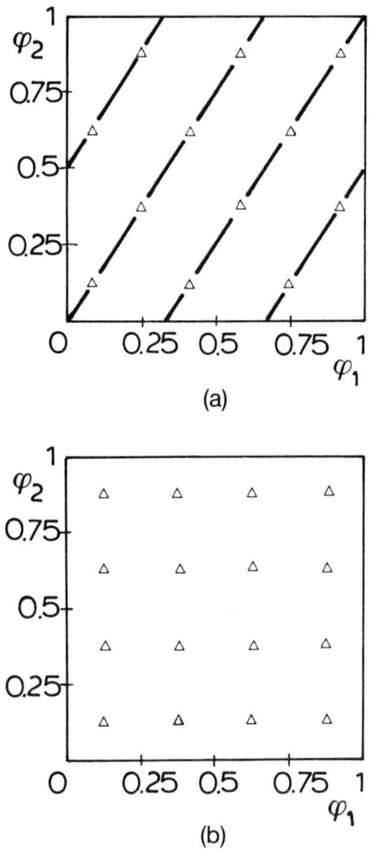

Fig. 5.F.2. Illustration of two phases in the plane (φ_1, φ_2): (a) Lines $\varphi_1 = 2x$ and $\varphi_2 = 3x$: the triangles indicate the 12 regularly spaced points, with intervals of the order of $(\Delta\varphi)_{\text{r.m.s.}}$, giving rise to the same number of phase permutations; (b) The triangles indicate the 16 points corresponding to quadrant permutation.

Table 5.F.2. ($\pm\pi/4, \pm3\pi/4$) and the permutations. Comparison between quadrant permutations of the phases obtained with magic integers based on the Fibonacci series

n	Quadrant		Magic integers	
	No. permutations	$(\Delta\varphi)_{\text{r.m.s.}}$	No. permutations	$(\Delta\varphi)_{\text{r.m.s.}}$
1	4	26.0	4	26.0
2	16	26.0	12	29.3
3	64	26.0	20	36.8
4	256	26.0	32	41.5
5	1024	26.0	50	44.7
6	4096	26.0	80	46.6
7	16384	26.0	128	48.0

intervals along x, i.e. along these lines, gives rise to a close-packed hexagonal disposition of points in the phase space. This corresponds to a more efficient way of sampling the phase space than that shown in Fig. 5.F.2(b), obtained by quadrant permutation.

This is also true for $n > 2$ and it may be concluded that the use of magic integers allows a considerable reduction in the number of trial permutations, while the accuracy of phase representation is kept within acceptable limits.

5.G New multisolution techniques

In the multisolution process described on p. 360 there are two main limitations: the restricted number of starting-set phases and the 'chain' character of the phase expansion process by means of the tangent formula. The first factor will make the initial steps of phase expansion very critical, as the first phases will be determined by few relationships, while the second factor will amplify the first effect in the following steps.

As we have seen, the use of magic integers allows an increase in the starting set of phases without a great decrease in accuracy. At the basis of all recent developments of the multisolution techniques lies the empirical observation that it is possible to solve structures, starting with a fairly large set of phases (more than 40–50 phases), even though these might be very inaccurate. The different methods proposed differ in the way in which the large initial set of phases is set up.

Those using magic integers are at the basis of the computer programs MAGIC[125] and MAGEX.[126] A limited number of phases (normally 10–15), called **primary phases**, are expressed as $\varphi_{\text{p}_i} = m_i x$. If two of these phases enter in a triplet with sufficiently large G, it will be possible to express also the phase of the third reflection as a combination of magic integers. For instance, if

$$\varphi_{\text{s}} = \varphi_{\text{p}_1} + \varphi_{\text{p}_3} + \pi \tag{5.G.1}$$

then $\varphi_{\text{s}} \approx (m_1 + m_3)x + \pi$. The new phases obtained in this way are called **secondary phases.** The starting set of phases will therefore include the origin-fixing phases, the primary and secondary phases, and their number can be up to 60–70.

The selection of the x values to be tried is done by means of the **ψ-map.** This is set up using all triplets relating the phases in the starting set. With a procedure similar to that used in deriving relation (5.64), a triplet may be written in terms of independent reflections as

$$\Phi = \varphi_{\boldsymbol{h}_1} \pm \varphi_{\boldsymbol{h}_2} \pm \varphi_{\boldsymbol{h}_3} + b \simeq 0 \bmod (2\pi) \qquad (5.G.2)$$

where b is a constant angle depending only on the symmetry operators of the space group. If in (5.G.2) we substitute the primary and secondary phases expressed in terms of magic integers, we will obtain

$$\Phi = Mx + B \simeq 0 \qquad (5.G.3)$$

where M is a combination of integers and $B = b +$ other constant terms appearing in the definition of secondary phases. The condition that all NR relationships of type (5.G.3) among the starting-set phases should be as close to zero as possible (or $\cos\Phi \approx +1$) may be expressed as

$$\psi(x) = \sum_{t=1}^{NR} G_t \cos(M_t x + B_t) = \max. \qquad (5.G.4)$$

$\psi(x)$ is a one-dimensional Fourier series and can be easily calculated at regular intervals of x of the order of $1/(4M_{\max})$. The ψ-map is thus obtained and its maxima indicate the best values of x. Some 100 x values, corresponding to the highest values of $\psi(x)$, are selected to obtain the same number of sets of numerical values of the phases, which are then further refined and expanded.

An extreme radicalization of the basic idea of enlarging the starting set by allowing larger phase errors, is the **random approach to the phase problem.**[127] The starting phases are now assigned random values between 0 and 360° and then refined with different techniques.

The first proposed random procedure is at the basis of the YZARC[127,128] program. Some 100 best interrelated phases (selected by means of the convergence procedure) form the starting set and are assigned several sets of random values. Each set is then refined by a least-squares procedure assuming as observational equations all triplet relations of type (5.G.2). The use of a least-squares procedure in the case of cyclic variables defined modulo 2π, needs an important warning. To obtain the correct phase values it is necessary to know the values of the right-hand terms of relations such as (5.G.2) **not reduced** modulo 2π; or, in another way, we must know n in the equation $\Phi = 2\pi n$ (n positive, zero, or negative integer). It may be interesting to note that this is another way in which the 'phase problem' shows up: it is generally possible to solve an overdetermined system of some thousands linear equations in terms of some hundreds of unknowns, but in our case, the cyclic nature of the variables introduces the new unknowns n, the number of which is equal to the number of observational equations. We must have some prior estimate of the n values, and this may be obtained from the starting random values of the phases; the values obtained in this way will in general not be integer and will be approximated to the nearest integer. The least-squares procedure may now be carried out and new values of the phases obtained; these in turn will yield new estimates of the ns and the process is repeated until convergence. The advantage of this refinement procedure, with respect to the tangent formula, is that all phases are refined at the same time; the cost is a longer computing time. Each phase set, after this refinement, is expanded by the tangent formula and tested by some figures of merit. Application of this method have shown that it is possible to hit the correct solution starting with a limited number (around 100) of random phase sets; the probability of getting the correct solution decreases as the structural complexity increases.

The success of this and other random methods implies that, if N ($\approx$100) is the number of starting phases, then the N-dimensional space, in which the refinement is carried out, only shows a limited number of minima; it will be sufficient to start from a limited number of random points in this space to converge to the minimum corresponding to the correct solution.

A more straightforward procedure is applied in the RANTAN[129] program in which the random starting sets of phases, including all reflections with large $|E|$ values, are directly refined by a carefully weighted tangent formula. The initial random phases are assigned a weight $w = 0.25$ and at each cycle only the phases determined with a sufficiently high $\alpha_{\mathbf{h}}$ value, so that $w_{\mathbf{h}} = \min(0.2\alpha_{\mathbf{h}}, 1) > 0.25$, are accepted. Also in this case a limited number of trials (50–200, depending on the structural complexity) is capable of yielding the correct structure. A similar random start procedure, but with a different strategy, has been implemented within the SHELXS-86[10] program.

More recently two new random procedures have been incorporated into the computer programs XMY[130,131] and SAYTAN.[132]

In the first the initial random phases are refined by **parameter shift.**[133] By this method all parameters are supposed independent so that each of them can be varied systematically from $u_j^0 - n\Delta u_j$ to $u_j^0 + n\Delta u_j$ in $2n$ steps of Δu_j; the value which maximizes (or minimizes) a selected function of the parameters is taken as a better value for this parameter and is used in the subsequent calculations with the next parameters. The process is continued until self-consistency. The advantage of the parameter-shift method is its ability to jump away from a local extremum. In the XMY procedure the function which is maximized is

$$F = \sum_{\mathbf{h}} (X_{\mathbf{h}} - Y_{\mathbf{h}}) \tag{5.G.5}$$

where

$$X_{\mathbf{h}} = \sum_{\mathbf{k}} |E_{\mathbf{h}} E_{\mathbf{k}} E_{\mathbf{h}-\mathbf{k}}| \cos(\varphi_{-\mathbf{h}} + \varphi_{\mathbf{k}} + \varphi_{\mathbf{h}-\mathbf{k}})$$

and (5.G.6)

$$Y_{\mathbf{h}} = \sum_{\mathbf{k}} |E_{\mathbf{h}} E_{\mathbf{k}} E_{\mathbf{h}-\mathbf{k}}| \sin(\varphi_{-\mathbf{h}} + \varphi_{\mathbf{k}} + \varphi_{\mathbf{h}-\mathbf{k}}).$$

A large value of F is in agreement with the most probable value of the triplets.

In SAYTAN random phases are refined by means of a modified tangent formula derived from Sayre's equation (5.24) expressed in terms of normalized structure factors. Its derivation is too long to be reported here and we will just mention that it also includes quartet invariants.

All the described random procedures have proved very successful in solving rather complex crystal structures.

Finally we mention an alternative random procedure[134] which has recently been incorporated in the SIR program[67] and is indicated with the name MESS. When phases are generated at random they are in general inconsistent with positivity and atomicity of the electron density, and only the refinement process will make them comply with these conditions; on the other hand, we have seen that these two conditions are at the basis of the probability formulae for triplets or other i.s. or s.s.. In the MESS approach, the starting phases are derived from triplets perturbed by random shifts $\Delta\Phi$ obeying a von Mises distribution; these phases will then comply with positivity and atomicity. Also in this case the number of trials necessary to get the solution is comparatively small and the procedure proved successful in solving several complex structures.

5.H Procedures for completing a partial model

Weights for Fourier syntheses

Given the positions of some atoms, what weights should be given to the coefficients in order to obtain the 'best' Fourier map revealing the remainder of the structure? This problem was first tackled by Sim[71,72]. With reference to Fig. 5.H.1, let

$$F_N = F_P + F_Q \tag{5.H.1}$$

where F_N is the structure factor of the complete structure, F_P is the contribution of the P known atoms, and F_Q is the contribution of the Q unknown atoms.

Ideally we should compute a Fourier synthesis with F_Q as coefficients, i.e.

$$\begin{aligned} F_Q = F_N - F_P &= |F_N| \exp(\mathrm{i}\varphi_N) - |F_P| \exp(i\varphi_P) \\ &= [|F_N| \exp(\mathrm{i}\alpha) - |F_P|] \exp(\mathrm{i}\varphi_P) \end{aligned} \tag{5.H.2}$$

where $\alpha = \varphi_N - \varphi_P$; but in the right-hand side of (5.H.2) α is unknown and the best estimate of F_Q is obtained by replacing $\exp(\mathrm{i}\alpha)$ by its expected value

$$\langle \exp(\mathrm{i}\alpha) \rangle = \langle \cos\alpha \rangle + \mathrm{i}\langle \sin\alpha \rangle = \langle \cos\alpha \rangle;$$

since α can equally have positive and negative values, then $\langle \sin\alpha \rangle = 0$ and (5.H.2) becomes

$$F_Q = [|F_N| \langle \cos\alpha \rangle - |F_P|] \exp(\mathrm{i}\varphi_P) \tag{5.H.3}$$

indicating that the 'best' Fourier synthesis is obtained by weighting the $|F_N|$ factors by $w = \langle \cos\alpha \rangle$. In order to derive $\langle \cos\alpha \rangle$ we must first know the probability distribution of α.

For $F_Q = A_Q + \mathrm{i}B_Q$ the probability distribution (5.A.13) becomes

$$P(A_Q, B_Q) = \frac{1}{\pi \Sigma_Q} \exp[-(A_Q^2 + B_Q^2)/\Sigma_Q] \tag{5.H.4}$$

where $\Sigma_Q = \Sigma_q f_q^2$ is a summation over the unknown atoms. From the triangle in Fig. 5.H.1 we have

$$A_Q^2 + B_Q^2 = |F_Q|^2 = |F_N|^2 + |F_P|^2 - 2\,|F_N|\,|F_P| \cos\alpha \tag{5.H.5}$$

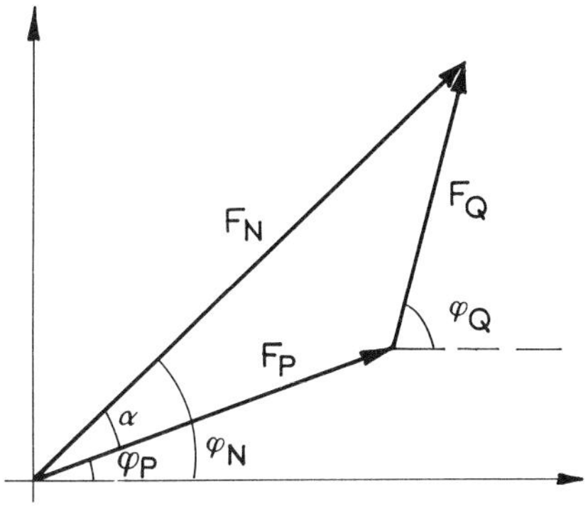

Fig. 5.H.1. Vector representation in the complex plane of the structure factor as a sum of two components.

and

$$P(A_Q, B_Q)$$

$$= \frac{1}{\pi \sum_Q} \exp\left[-(|F_N|^2 + |F_P|^2)/\textstyle\sum_Q\right] \exp\left[2\,|F_N|\,|F_P| \cos\alpha/\textstyle\sum_Q\right]. \quad (5.H.6)$$

The only unknown term is α and (5.H.6) is essentially a probability distribution of α, i.e.

$$P(\alpha) = (1/K)\exp(X\cos\alpha), \quad (5.H.7)$$

where $X = 2\,|F_N|\,|F_P|/\sum_Q$ and K is a normalizing constant given by (see eqns (5.D.15) and (5.D.17))

$$K = \int_0^{2\pi} \exp(X\cos\alpha)\,\mathrm{d}\alpha = 2\pi I_0(X). \quad (5.H.8)$$

Finally

$$P(\alpha) = \frac{\exp(X\cos\alpha)}{2\pi I_0(X)}. \quad (5.H.9)$$

Recalling eqn (5.68), we may now derive

$$w = \langle\cos\alpha\rangle = \int_0^{2\pi} \frac{\cos\alpha\exp(X\cos\alpha)}{2\pi I_0(X)}\,\mathrm{d}\alpha = \frac{I_1(X)}{I_0(X)}, \quad (5.H.10)$$

and (5.H.3) becomes

$$\langle F_Q\rangle = \left(|F_N|\frac{I_1(X)}{I_0(X)} - |F_P|\right)\exp(i\varphi_P). \quad (5.H.11)$$

When this weight is used in Fourier recycling, then $|F_N| = |F_{\mathrm{obs}}|$ and the employed coefficients are

$$F_P + \langle F_Q\rangle = |F_{\mathrm{obs}}|\frac{I_1(X)}{I_0(X)}\exp(\mathrm{i}\varphi_P). \quad (5.H.12)$$

Similarly Woolfson[135] derived that for centrosymmetric structures

$$F_P + \langle F_Q\rangle = |F_{\mathrm{obs}}|\tanh(X/2)S_P \quad (5.H.13)$$

where S_P is the sign of F_P.

Syntheses for completing a partial model

We have seen that often only part of the structure (e.g. positions of heavy atoms) can be initially determined; this information can then be combined with the Patterson function to define the rest of the structure. For this purpose Ramachandran and Raman[136] proposed several types of Fourier syntheses[137,138] with different coefficients. Among them we will recall the α and the β syntheses, with coefficients

α synthesis	$\|F_N\|^2 F_P$
β synthesis	$\|F_N\|^2/F_P^*$ (F_P^* complex conjugate of F_P)

where the terms have the same meaning as in (5.H.1).

Considering (5.H.1), the Patterson coefficients $|F_N|^2$ become

$$|F_N|^2 = |F_P|^2 + F_P^* F_Q + F_P F_Q^* + |F_Q|^2, \qquad (5.H.14)$$

and those of the α synthesis

$$\begin{aligned}|F_N|^2 F_P &= |F_P|^2 F_P + |F_P|^2 F_Q + F_P^2 F_Q^* + |F_Q|^2 F_P \\ &= \quad F_1 \quad + \quad F_2 \quad + \quad F_3 \quad + \quad F_4\end{aligned} \qquad (5.H.15)$$

where F_1 and F_4 contribute to the background and the P peaks, F_2 contributes to the background and the desired Q peaks, and F_3 contributes mainly to the background.

A similar development for the coefficients of the β synthesis yields to

$$|F_N|^2/F_P^* = F_P + F_Q + F_Q^* \exp 2i\varphi_P + |F_Q|^2/F_P^* \qquad (5.H.16)$$

where the first two terms contribute, with equal weight, to reveal the known P atoms and the unknown Q atoms respectively, while the last two terms contribute mainly to the background. An advantage of the β over the α synthesis is that it gives the electron density map when $F_P = F_N$.

Main[139] compared the β synthesis with the synthesis computed with coefficients (5.H.12) (also called weighted γ' synthesis). From (5.H.5) the expected value

$$\langle |F_Q|^2 \rangle = |F_N|^2 + |F_P|^2 - 2\,|F_N|\,|F_P| \langle \cos\alpha \rangle \qquad (5.H.17)$$

may be derived and, considering that

$$|F_N|^2 = (F_P + F_Q)(F_P^* + F_Q^*) \qquad (5.H.18)$$

the coefficients (5.H.12) are obtained as

$$\begin{aligned}|F_N| \langle \cos\alpha \rangle \exp(i\varphi_P) &= F_P + F_Q/2 \\ &+ F_Q^* \exp(2i\varphi_P)/2 + \{|F_Q|^2 - \langle |F_Q|^2 \rangle\}/2F_P^*,\end{aligned} \qquad (5.H.19)$$

where the last two terms mainly contribute to the background, the first term contributes to the known part of the structure, and the second to the desired unknown part, but only with half weight. In order to give the P and the Q peaks the same weight, as in the case of the β synthesis, the coefficients of the γ' synthesis should be altered to

$$\begin{aligned}\{2\,|F_N| \langle \cos\alpha \rangle - |F_P|\} \exp(i\varphi_P) &= F_P + F_Q + F_Q^* \exp(2i\varphi_P) \\ &+ \{|F_Q|^2 - \langle |F_Q|^2 \rangle\}/F_P^*.\end{aligned} \qquad (5.H.20)$$

The new synthesis is referred to as the $2F_o - F_c$ synthesis; it has been shown that, for non-centrosymmetric structures, it contains fewer background peaks than the β synthesis and is more effective in suppressing the peaks corresponding to wrong atoms in the initial model.

Let us conclude this paragraph with a comparison, proposed by Tollin *et al.*,[140] between the α synthesis and the optical image synthesis by holography (see any optics textbook such as Meyer-Arendt[141]). The principle of holography is illustrated in Fig. 5.H.2. The waves scattered by the object Q interfere with those from a coherent point source P and the fringe pattern is recorded on a photographic plate to form a hologram (H). The intensity pattern on the hologram will depend on the phase relationships between the interfering waves from P and Q. By illuminating the

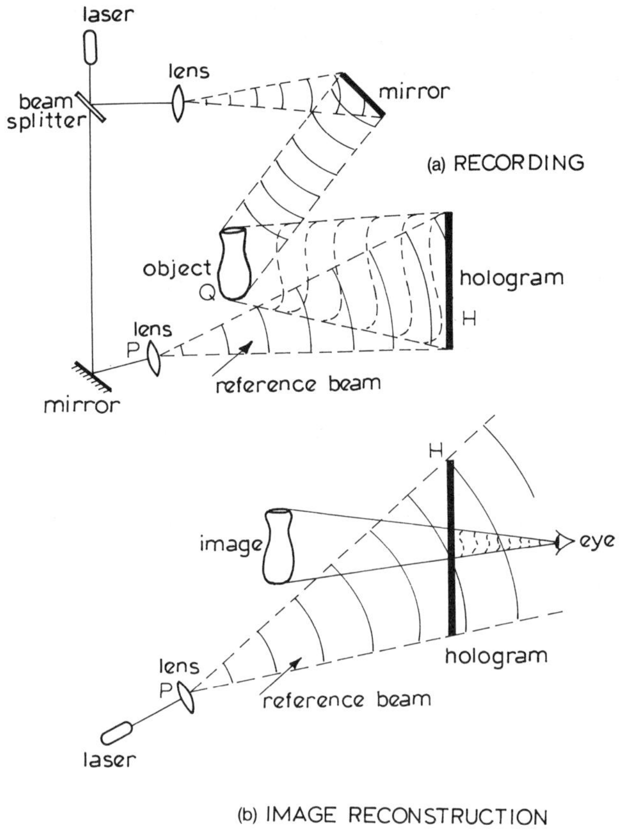

Fig. 5.H.2. Scheme of the formation of a hologram.

hologram with light identical to that from P, the image of the object Q may be reconstructed. If $t_P(x)$ and $t_Q(x)$ are the functions describing the source P and the object Q, the corresponding scattered amplitudes will be their Fourier transforms $T_P(s)$ and $T_Q(s)$. The amplitude at a point s on the hologram will be

$$A(s) = T_P(s) + T_Q(s) \exp(2\pi i a s) \qquad (5.H.21)$$

where a is the separation between $t_P(x)$ and $t_Q(x)$. The intensity at s on H will then be

$$I = AA^* = |T_P|^2 + |T_Q|^2 + T_P T_Q^* \exp\{-2\pi i a s\} + T_P^* T_Q \exp(2\pi i a s) \qquad (5.H.22)$$

where the s dependence has been omitted. When the hologram is illuminated by a light from the source P, the resulting modulated function will be

$$IT_P = T_P|T_P|^2 + T_P|T_Q|^2 + T_P^2 T_Q^* \exp(-2\pi i a s) + |T_P|^2 T_Q \exp(2\pi i a s). \qquad (5.H.23)$$

Its Fourier transform will be the combination of the transforms of the four different terms and will show the image of the object Q. Comparison with (5.H.15) shows that the four terms of the α synthesis coefficients are formally similar to those in (5.H.23). In the case of a partially known structure we observe the interference effects between the known (source) and the unknown (object) parts of the structure. In 1950 Bragg[142] had already qualitatively indicated the analogy between the heavy-atom technique and holography.

References

1. Patterson, A. L. (1944). *Physics Review,* **65,** 195.
2. Wilson, A. J. C. (1950). *Acta Crystallographica,* **3,** 397.
3. Lipson, H. and Cochran, W. (1953). *The crystalline state. vol. III: The determination of crystal structures.* G. Bell, London.
4. Wilson, A. J. C. (1949). *Acta Crystallographica,* **2,** 318.
5. Wilson, A. J. C. (1942). *Nature,* **150,** 151.
6. Viterbo, D., Gasco, A., Serafino, A. and Mortarini, V. (1975). *Acta Crystallographica,* **B31,** 2151.
7. Harker, D. (1936). *Journal of Chemical Physics,* **4,** 381.
8. Calleri, M., Chiari, G., Chiesi Villa, A., Gaetani Manfredotti, A., Guastini, C. and Viterbo, D. (1976). *Acta Crystallographica,* **B32,** 1032.
9. Gervasio, G., Rossetti, R. and Stanghellini, P. L. (1979). *Journal of Chemical Research* (S) 334, (M)3943.
10. Sheldrick, G. M. (1985). In *Crystallographic computing 3* (ed. G. M. Sheldrick, C. Krüger and R. Goddard), pp. 184–9. Oxford University Press; Robinson, W. and Sheldrick, G. M. (1988). In *Crystallographic Computing 4* (ed. N. W. Isaacs and R. M. Taylor), pp. 366–77. Oxford University Press.
11. Luger, P. and Fuchs, J. (1986). *Acta Crystallographica,* **A42,** 380.
12. Lenstra, A. T. H. and Schoone, J. C. (1973). *Acta Crystallographica,* **A29,** 419.
13. Pavelčik, F. (1989). *Journal of Applied Crystallography,* **22,** 181.
14. Argos, P. and Rossmann, M. G. (1976). *Acta Crystallographica,* **B32,** 2975.
15. Rossmann, M. G., Arnold, E. and Vriend, G. (1986). *Acta Crystallographica,* **A42,** 325.
16. Terwilliger, T. C., Sung-Hou Kim and Eisenberg, D. (1987). *Acta Crystallographica,* **A43,** 1.
17. Harker, D. and Kasper, J. S. (1948). *Acta Crystallographica,* **1,** 70.
18. Karle, J. and Hauptman, H. (1950). *Acta Crystallographica,* **3,** 181.
19. Giacovazzo, C. (1980). *Direct methods in crystallography.* Academic, London.
20. Hauptman, H. and Karle, J. (1953). *The solution of the phase problem. I. The centrosymmetric crystal,* ACA Monograph, No. 3. Polycrystal Book Service, New York.
21. Sayre, D. (1952). *Acta Crystallographica,* **5,** 60.
22. Hauptman, H. and Karle, J. (1956). *Acta Crystallographica,* **9,** 45.
23. Cascarano, G. and Giacovazzo, C. (1983). *Zeitschrift für Kristallographie,* **165,** 169.
24. Cochran, W. (1955). *Acta Crystallographica,* **8,** 473.
25. von Mises, R. (1918). *Physikalisches Zeitschrift,* **19,** 490.
26. Karle, J. and Karle, I. L. (1966). *Acta Crystallographica,* **21,** 849.
27. Karle, J. and Hauptman, H. (1956). *Acta Crystallographica,* **9,** 635.
28. Cochran, W. and Woolfson, M. M. (1955). *Acta Crystallographica,* **8,** 1.
29. Schenk, H. (1973). *Acta Crystallographica,* **A29,** 77.
30. Schenk, H. (1973). *Acta Crystallographica,* **A29,** 480.
31. Simerska, M. (1965). *Czechoslovakian Journal of Physics,* **6,** 1.
32. Hauptman, H. (1975). *Acta Crystallographica,* **A31,** 671.
33. Hauptman, H. (1975). *Acta Crystallographica,* **A31,** 680.
34. Hauptman, H. (1976). *Acta Crystallographica,* **A32,** 934.

35. Giacovazzo, C. (1977). *Acta Crystallographica,* **A33,** 933.
36. Giacovazzo, C. (1980). *Acta Crystallographica,* **A36,** 362.
37. Giacovazzo, C. (1974). *Acta Crystallographica,* **A30,** 626, 631.
38. Giacovazzo, C. (1978). *Acta Crystallographica,* **A34,** 562.
39. Burla, M. C., Nunzi, A., Polidori, G., Busetta, B. and Giacovazzo, C. (1980). *Acta Crystallographica,* **A36,** 573.
40. Burla, M. C., Nunzi, A., Giacovazzo, C. and Polidori, G. (1981). *Acta Crystallographica,* **A37,** 677.
41. Cascarano, G., Giacovazzo, C., Calabrese, G., Burla, M. C., Nunzi, A., Polidori, G. and Viterbo, D. (1984). *Zeitschrift für Kristallographie,* **167,** 34.
42. Giacovazzo, C., Spagna, R., Vicković, I. and Viterbo, D. (1979). *Acta Crystallographica,* **A35,** 401.
43. Cascarano, G., Giacovazzo, C., Polidori, G., Spagna, R., and Viterbo, D. (1982). *Acta Crystallographica,* **A38,** 663.
44. Burla, M. C., Giacovazzo, C., and Polidori, G. (1989). *Acta Crystallographica,* **A45,** 99.
45. Cascarano, G., Giacovazzo, C., Camalli, M., Spagna, R., Burla, M. C., Nunzi, A., and Polidori, G. (1984). *Acta Crystallographica,* **A40,** 278.
46. Giacovazzo, C. (1976). *Acta Crystallographica,* **A32,** 958.
47. Busetta, B., Giacovazzo, C., Burla, M. C., Nunzi, A., Polidori, G., and Viterbo, D. (1980). *Acta Crystallographica,* **A36,** 68.
48. Ladd, M. F. C. and Palmer, R. A. (ed.) (1980). *Theory and practice of direct methods in crystallography.* Plenum, New York.
49. Cascarano, G., Giacovazzo, C., and Luić, M. (1985). *Acta Crystallographica,* **A41,** 544.
50. Cascarano, G., Giacovazzo, C., and Luić, M. (1987). *Acta Crystallographica,* **A43,** 14.
51. Cascarano, G., Giacovazzo, C., and Luić, M. (1988). *Acta Crystallographica,* **A44,** 176.
52. Cascarano, G., Giacovazzo, C., and Luić, M. (1988). *Acta Crystallographica,* **A44,** 183.
53. Camalli, M., Giacovazzo, C., and Spagna, R. (1985). *Acta Crystallographica,* **A41,** 605.
54. Burla, M. C., Cascarano, G., Fares, E., Giacovazzo, C., Polidori, G., and Spagna, R. (1989). *Acta Crystallographica,* **A45,** 781.
55. Zachariasen, W. H. (1952). *Acta Crystallographica,* **5,** 68.
56. Germain, G. and Woolfson, M. M. (1968). *Acta Crystallographica,* **B24,** 91.
57. Germain, G., Main, P., and Woolfson, M. M. (1970). *Acta Crystallographica,* **B26,** 274.
58. Abramowitz, M. and Stegun, I. A. (1964). *Handbook of mathematical functions.* National Bureau of Standards, Washington, D.C.
59. Burla, M. C., Cascarano, G., Giacovazzo, C., Nunzi, A., and Polidori, G. (1987). *Acta Crystallographica,* **A43,** 370.
60. Cascarano, G., Giacovazzo, C., and Viterbo, D. (1987). *Acta Crystallographica,* **A43,** 22.
61. Cascarano, G., Giacovazzo, C., Burla, M. C., Nunzi, A., and Polidori, G. (1984). *Acta Crystallographica,* **A40,** 389.
62. Cascarano, G. (1990). Doctorate thesis. University of Bari; Cascarano, G., Giacovazzo, C., Camalli, M., Spagna, R., and Watkin J. D. (1991). *Acta Crystallographica,* **A47,** 373. (And references therein.)
63. Schenk, H. and Kiers, C. T. (1985). In *Crystallographic computing 3* (ed. G. M. Sheldrick, C. Krüger, and R. Goddard), pp. 200–5. Oxford University Press.
64. Peschar, R. and Schenk, H. (1987). *Acta Crystallographica,* **A43,** 751.

65. Karle, I. L., Karle, J., Mastropaolo, D., Camerman, A., and Camerman, N. (1983). *Acta Crystallographica,* **B39,** 625.
66. Main, P., Fiske, S. J., Hull, S. E., Lessinger, L., Germain, G., Declercq, J. P., and Woolfson, M. M. (1980). *MULTAN80*: *A system of computer programs for the automatic solution of crystal structures from x-ray diffraction data.* Universities of York and Louvain.
67. Burla, M. C., Camalli, M., Cascarano, G., Giacovazzo, C., Polidori, G., Spagna, R., and Viterbo, D. (1989). *Journal of Applied Crystallography,* **22,** 389.
68. Cerrini, S., Lamba, D., Burla, M. C., Polidori, G., and Nunzi, A. (1988). *Acta Crystallographica,* **C44,** 489.
69. Karle, J. (1968). *Acta Crystallographica,* **B24,** 182.
70. Beurskens, P. T., Bosman, W. P., Doesburg, H. M., Gould, R. O., Van den Hark, Th. E. M., Prick, P. A. J., Noordik, J. H., Beurskens, G., Parthasarathi, V., Bruins Slott, H. J., and Haltiwanger, R. C. (1985). *Program system DIRDIF,* Technical Report, Crystallography Laboratory, University of Nijmegen.
71. Sim, G. A. (1959). *Acta Crystallographica,* **12,** 813.
72. Sim, G. A. (1960). *Acta Crystallographica,* **13,** 511.
73. Stewart, J. M. and Hall, S. R. *XTAL user's manual,* Technical Report TR1364.2. Computer Science Center, University of Maryland; Skelton, B., Hall, S. R., and Stewart, J. M. (1988). In *Crystallographic computing 4,* (ed. N. W. Isaacs and M. R. Taylor), pp. 325–53. Oxford University Press.
74. Gabe, E. J., Le Page, Y., Charland, J. P., Lee, F. L., and White, P. S. (1989). *Journal of Applied Crystallography,* **22,** 384.
75. Betteridge, P. V., Prout, C. K., and Watkin, D. J. (1984). *CRYSTALS.* University of Oxford; Watkin, D. J. (1988). In *Crystallographic computing 4,* (ed. N. W. Isaacs and M. R. Taylor), pp. 354–65. Oxford University Press.
76. Viterbo, D., Calleri, M., Calvino, R., and Fruttero, R. (1984). *Acta Crystallographica,* **C40,** 1728.
77. Hughes, E. W. (1941). *Journal of the American Chemical Society,* **63,** 1737.
78. Cruickshank, D. W. J. (1970). In *Crystallographic computing* (ed. F. R. Ahmed), pp. 187–97. Munksgaard, Copenhagen.
79. Hirshfeld, F. L. and Rabinovich, D. (1973). *Acta Crystallographica,* **A29,** 510.
80. French, S. and Wilson, K. (1978). *Acta Crystallographica,* **A34,** 517.
81. Schwarzenbach, D. (1989). *Acta Crystallographica,* **A45,** 63.
82. Hamilton, W. C. (1964). *Statistics in physical science,* pp. 157–162. Ronald, New York; (1965). *Acta Crystallographica,* **18,** 502.
83. Bijvoet, J. M. (1949). *Proceedings of the Koninklijke Nederlandse Akademie van Wetenschap,* **B52,** 313.
84. Rogers, D. (1981). *Acta Crystallographica,* **A37,** 734.
85. Marsh, R. E. and Herbstein, F. H. (1983). *Acta Crystallographica,* **B39,** 280.
86. Marsh, R. E. and Herbstein, F. H. (1988). *Acta Crystallographica,* **B44,** 77.
87. Santoro, A., Mighell, A. D., and Rodgers, J. R. (1980). *Acta Crystallographica,* **A36,** 796.
88. Le Page, Y. (1982). *Journal of Applied Crystallography,* **15,** 255.
89. Mugnoli, A. (1985). *Journal of Applied Crystallography,* **18,** 183.
90. Spek, A. L. (1988). *Journal of Applied Crystallography,* **21,** 578.
91. Jones, P. G. (1984). *Chemical Society Reviews,* **13,** 157.
92. Braga, D., Gradella, C. and Grepioni, F. (1989). *Journal of the Chemical Society, Dalton Transactions,* 1721.
93. Aime, S., Botta, M., Gobetto, R., and Orlandi, A. (1990). *Magnetic Resonance in Chemistry,* **28,** 552.
94. Bijvoet, J. M., Peerdeman, A. F., and van Bommel, A. J. (1951). *Nature,* **168,** 271.

95. Jones, P. G. (1984). *Acta Crystallographica,* **A40,** 660.
96. Glazer, A. M. and Stadnicka, K. (1989). *Acta Crystallographica,* **A45,** 234.
97. Flack, H. D. (1983). *Acta Crystallographica,* **A39,** 876.
98. Bernardinelli, G. and Flack, H. D. (1985). *Acta Crystallographica,* **A41,** 500; (1987). *Acta Crystallographica,* **A43,** 75.
99. (1959). *International tables for x-ray crystallography,* Vol. II, pp. 355–7. Kynoch, Birmingham.
100. Buerger, M. J. (1959). *Vector space and its application in crystal structure investigation.* Wiley, New York.
101. Wrinch, D. M. (1939). *Philosophical Magazine,* **27,** 98.
102. Allen, F. H., Bellard, S. A., Brice, M. D., Cartwright, B. A., Doubleday, A., Higgs, H., Hummelink, T., Hummelink-Peters, B. G., Kennard, O., Motherwell, W. D. S., Rodgers, J. R., and Watson, D. G. (1979). *Acta Crystallographica,* **B35,** 2331.
103. Nordman, C. E. and Nakatsu, K. (1963). *Journal of the American Chemical Society,* **85,** 353.
104. Nordman, C. E. and Kumra, S. K. (1965). *Journal of the American Chemical Society,* **87,** 2059.
105. Nordman, C. E. (1985). In *Crystallographic computing 3* (ed. G. M. Sheldrick, C. Krüger, and R. Goddard), pp. 232–9. Oxford University Press.
106. Nordman, C. E. and Hsu, L. Y. R. (1982). In *Computational crystallography* (ed. D. Sayre), pp. 141–9. Oxford University Press.
107. High, D. H. and Krout, J. (1966). *Acta Crystallographica,* **21,** 88.
108. Stout, G. H. and Jensen, L. H. (1989). *X-ray structure determination,* pp. 312–5. Wiley, New York.
109. Rossmann, M. G. and Blow, D. M. (1962). *Acta Crystallographica,* **15,** 24.
110. Tollin, P. (1966). *Acta Crystallographica,* **21,** 613.
111. Tollin, P. (1976). In *Crystallographic computing techniques* (ed. F. R. Ahmed), pp. 212–21. Munksgaard, Copenhagen.
112. Wilson, C. C. and Tollin, P. (1986). *Journal of Applied Crystallography,* **19,** 411.
113. Wilson, C. C. (1989). *Acta Crystallographica,* **A45,** 833.
114. Rossmann, M. G. (ed.) (1972). *The molecular replacement method.* Gordon and Breach, New York.
115. Rius, J. and Miravitlles, C. (1987). *Journal of Applied Crystallography,* **20,** 261.
116. Egert, E. and Sheldrick, G. M. (1985). *Acta Crystallographica,* **A41,** 262.
117. Woolfson, M. M. (1954). *Acta Crystallographica,* **7,** 61.
118. Hamilton, W. C. (1964). *Statistics in physical science,* pp. 17–18. Ronald, New York.
119. Buerger, M. J. (1956), *Proceedings of the National Academy of Sciences of the USA,* **42,** 776.
120. Fan Hai-Fu, Yao Jia-Xing, and Qian Jin-Zi, (1988). *Acta Crystallographica,* **A44,** 688.
121. White, P. S. and Woolfson, M. M. (1975). *Acta Crystallographica,* **A31,** 53.
122. Woolfson, M. M. (1976). In *Crystallographic computing techniques* (ed. F. R. Ahmed), pp. 106–14. Munksgaard, Copenhagen.
123. Main, P. (1977). *Acta Crystallographica,* **A33,** 750.
124. Main, P. (1978). *Acta Crystallographica,* **A34,** 31.
125. Declercq, J. P., Germain, G., and Woolfson, M. M. (1975). *Acta Crystallographica,* **A31,** 367.
126. Hull, S. E., Viterbo, D., Woolfson, M. M., and Zhang Shao-Hui, (1981). *Acta Crystallographica,* **A37,** 566.
127. Baggio, R., Woolfson, M. M., Declercq, J. P., and Germain, G. (1978). *Acta Crystallographica,* **A34,** 883.

128. Declercq, J. P., Germain, G., and Woolfson, M. M. (1979). *Acta Crystallographica,* **A35,** 622.
129. Yao Jia-Xing, (1981). *Acta Crystallographica,* **A37,** 642.
130. Debaerdemaeker, T. and Woolfson, M. M. (1983). *Acta Crystallographica,* **A39,** 193.
131. Debaerdemaeker, T. and Woolfson, M. M. (1989). *Acta Crystallographica,* **A45,** 349.
132. Debaerdemaeker, T., Tate, C., and Woolfson, M. M. (1985). *Acta Crystallographica,* **A41,** 286.
133. Bhuiya, A. K. and Stanley, E. (1963). *Acta Crystallographica,* **16,** 981.
134. Burla, M. C., Giacovazzo, C., and Polidori, G. (1987). *Acta Crystallographica,* **A43,** 797.
135. Woolfson, M. M. (1956). *Acta Crystallographica,* **9,** 804.
136. Ramachandran, G. N. and Raman, S. (1959). *Acta Crystallographica,* **12,** 957.
137. Srinivasan, R. (1961). *Acta Crystallographica,* **14,** 607.
138. Ramachandran, G. N. and Srinivasan, R. (1970). *Fourier methods in crystallography.* Wiley-Interscience, London.
139. Main, P. (1979). *Acta Crystallographica,* **A35,** 779.
140. Tollin, P., Main, P., Rossmann, M. G., Stroke, G. W. and Restrick, R. C. (1966). *Nature,* **209,** 603.
141. Meyer-Arendt, J. R. (1972). *Introduction to classical and modern optics,* pp. 413–27, Prentice-Hall, Englewood Cliffs, N.J.
142. Bragg, W. L. (1950). *Nature,* **166,** 399.

Ionic crystals 6

FERNANDO SCORDARI

Introduction

This chapter deals with a class of natural and synthetic compounds that, at least to a first approximation, can be treated as being composed of oppositely charged spheres. We start by illustrating some important chemical and physical concepts that are later on applied to the organization of ionic crystals. For instance, we shall see how it is possible to rationalize some atomic structures in terms of the energetics of crystal formation. To this end, the ionic model, attractive for its simplicity, will be used even if departure from the predictions are common particularly when bonds have quite prevalent covalent character. The close-packing model of spheres is examined closely and used as a helpful tool to describe many structures. Finally we see how the concepts and principles illustrated previously are very useful for the systemization of several ionic structures with different degrees of complexity.

The structure of the atom

Atoms with a single electron

We know from quantum mechanics[1–3] that the motion of the electron (e^-) in a hydrogen atom can be described by means of the wave function $\Psi(x, y, z)$, or orbital, and that $|\Psi(x, y, z)|^2 \, dv$ expresses the probability of finding the electron in a given element of volume, dv.

The electron can assume different states of energy, each distinguished by four quantum numbers n, l, m, s:

(1) n, the **principal quantum number**, is related to the energy of the electron, and can have any positive integral value from 1 to ∞;

(2) l, the **azimuthal quantum number**, expresses the angular momentum module ($=\sqrt{l(l+1)}\,h/2\pi$) (where h is Plank's constant) of the path of the electron and determines the form of its orbital: orbitals are indicated by the letters s, p, d, f, g . . . . and correspond respectively to $l = 0, 1, 2, 3, 4 \ldots n-1$;

(3) m, the **magnetic quantum number**, $(-l \leqslant m \leqslant +l)$ represents the angular momentum component in the direction of the magnetic field and characterizes the orientation of the orbital;

(4) s, $(s = \pm\frac{1}{2})$, the **spin quantum number**, determines the way in which the electron spins.

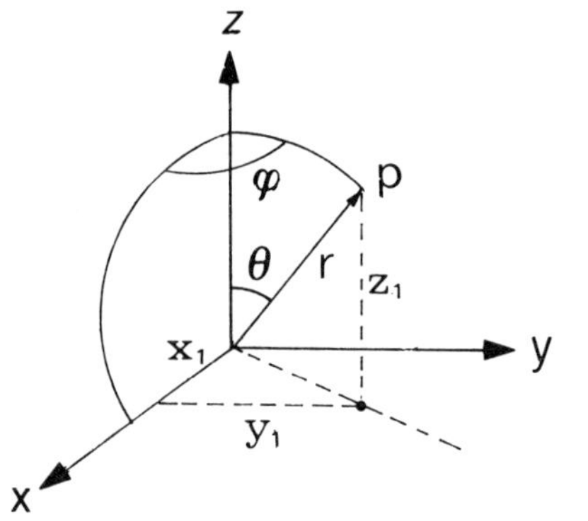

Fig. 6.1. Relationship between the orthogonal coordinates *x, y, z* and the spherical coordinates *r, ϑ, φ*.

Using the spherical coordinates r, ϑ, φ (see Fig. 6.1) $\Psi(x, y, z)$ can be resolved into the products of three functions with separate variables:

$$\Psi(r, \vartheta, \varphi) = R_{n,l}(r)\theta_{l,m}(\vartheta)\Phi_m(\varphi). \tag{6.1}$$

While $R^2_{n,l}(r)$ gives the probability of finding the electron at a certain distance from the nucleus, the form and the orientation of its orbital are determined by θ and Φ.

The functions R, θ and Φ take different algebraic forms according to the quantum numbers with which they are associated.

The probability of finding the electron in a spherical shell of volume $4\pi r^2\,dr$, at a distance of r from the nucleus, is $4\pi r^2 R^2_{n,l}(r)\,dr$. The function $D_{n,l}(r) = 4\pi r^2 R^2_{n,l}(r)$ is known as the **radial probability function**. It is represented in Fig. 6.2 for orbitals 1s, 2s, 2p, 3s, 3p, and 3d.

The presence of secondary maxima, occurring before the main maximum, produces some important effects. For example when the electron is located in the 3s orbital it is more penetrating than that when it is in the 3p, (as shown in Fig. 6.2) whereby we can deduce that the ionization energy required by the 3s electron is greater than that required by the 3p electron.

Let us examine now the angular part of $\Psi(r, \vartheta, \varphi)$. The spatial shapes of the angular probability functions are illustrated in Fig. 6.3. These configurations allow us to estimate the magnitude of the $\boldsymbol{v}$ vector that is proportional to the probability (summed over all distances) that the electron will be found in that direction.

Atoms with more than one electron

The basic configuration of the electrons of an atom can easily be deduced by means of the **aufbau principle,** or gradual building up of atoms. This exploits the **Pauli principle** and a scale of the relative energy of the orbitals. The Pauli principle can be synthesized as follows: in an atom no two electrons exist which have all four quantum numbers, n, l, m, and s, the same. Therefore two electrons belonging to the same orbital, as defined by the first three quantum numbers, must spin in opposite ways.

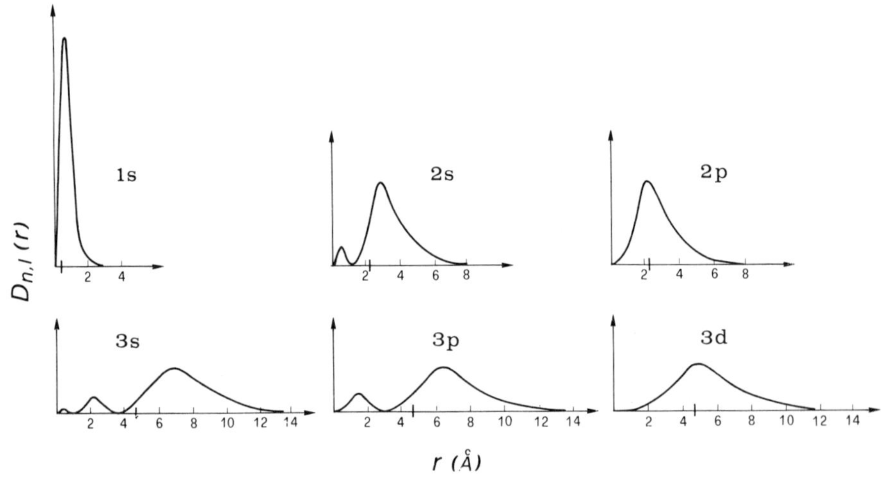

Fig. 6.2. Radial probability functions for orbitals 1s, 2s, 2p, 3s, 3p, and 3d for the hydrogen atom.

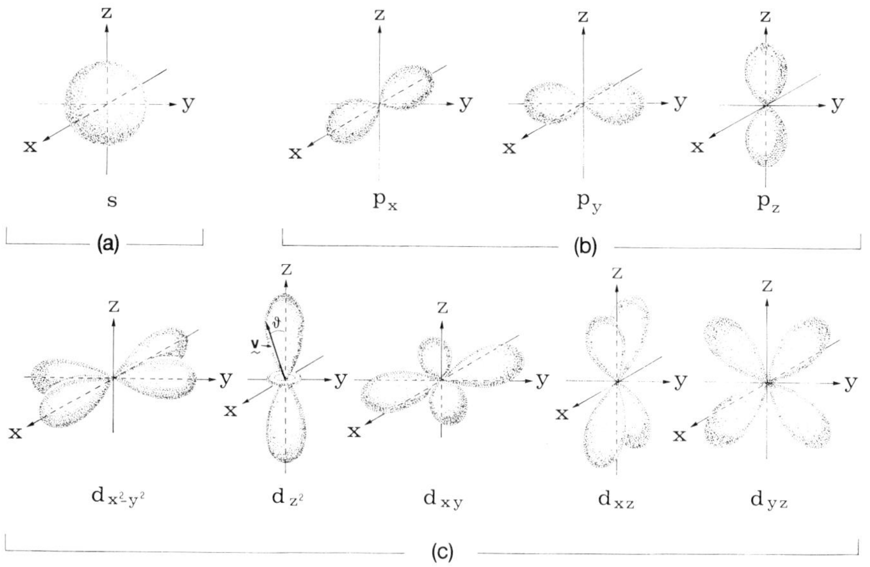

Fig. 6.3. Angular probability functions for: (a) *s*; (b) p_x, p_y, p_z; and (c) $d_{x^2-y^2}$, d_{z^2}, d_{xy}, d_{xz}, d_{yz} orbitals.

The energy of the orbitals for various neutral atoms increases in the following order:

$$1s < 2s < 2p < 3s < 3p < 4s < 3d < 4p < 5s < 4d < 5p < 6s < 4f \simeq 5d < 6p < 7s < 5f \simeq 6d.$$

This experimental scale, though not always reliable (see Fig. 6.4),

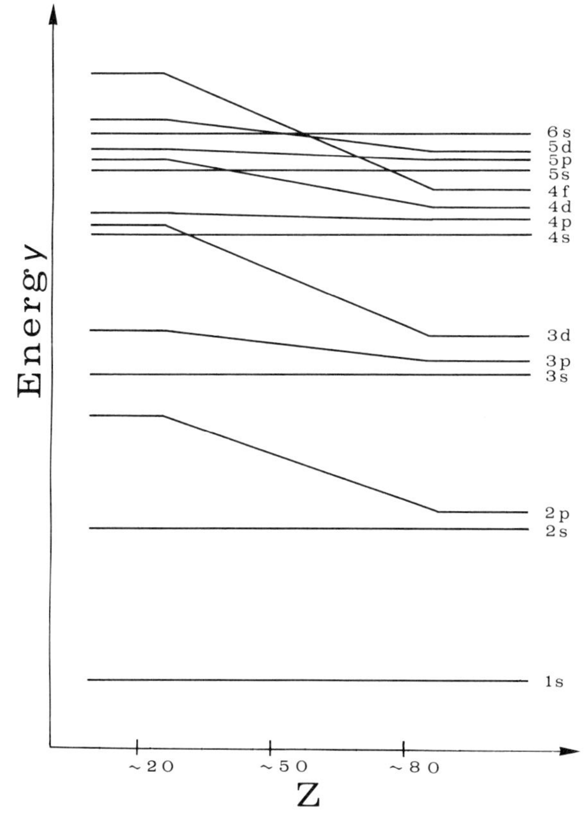

Fig. 6.4. Relative energy of various orbitals versus the atomic number *Z*.

nevertheless constitutes a good guide for the building up of atomic structures and especially for the location of valence electrons. For example the outermost electrons of Rb and Sr occupy the 5s orbital rather than the 4d, exactly as predicted by the energy scale. The relative energy of the orbitals versus Z is shown in Fig. 6.4.

Following the aufbau principle the orbitals are filled gradually as shown below:

Chemical element	Orbital	n	l	m	s	Electrons
H	1s	1	0	0	↑	1
He	1s	1	0	0	↑↓	2
Li	2s	2	0	0	↑	3
Be	2s	2	0	0	↑↓	4
B	2p	2	1	1	↑	5

On the same principle, the entire periodic table can be obtained in the same way, as illustrated in Table 6.1.

Elements whose valence electrons are in the s orbitals belong to groups IA or IIA: alkaline metals have only one valence electron in their s orbital, while alkaline-earth metals have two. The elements in groups from IB to VIIIB, known as transition metals, have their external electrons in orbitals $(n-1)$d which immediately precede the last orbital ns.

Those elements whose valence electrons are in the p orbitals belong to groups III to VIIIA. Such elements are known as post-transition metals and non-metals, and are separated roughly in the Table 6.1 by a broken line.

Helium (in the hatched square) can be placed either in group IIA, taking into account its electron configuration, or in group VIIIA, taking into account its chemical behaviour similar to that of the other noble gases. Finally there are the lanthanides and actinides which are characterized by a gradual filling of the f orbitals preceding by two places $(n-2)$ the last orbital s(n).

Interactions between ions

Notes on chemical bonds

A knowledge of the structure of an atom allows reasonable predictions to be made about interactions between adjacent atoms. At this point it will be useful to recall certain general concepts of chemistry.

The influence which the nuclear charge has on the external electrons varies according to which orbitals they occupy and to what degree they are shielded by the electrons closer to the nucleus (see p. 419). Such variations in electrons attraction can be evaluated by means of: **ionization energy** (IE); **electronic affinity** (EA); **electronegativity** (EN).

IE is the energy needed to remove an electron from the orbital of an atom to an infinite distance. The closer the electron is to the nucleus the greater the energy required. First, second energy of ionization, and so on, indicate the increasing quantities of energy necessary to free electrons 1, 2, etc. In accordance with recent thermodynamic conventions endothermic

Table 1. The periodic table of elements and configurations of internal electrons (first column) and valence electrons (first row). The oxidation states in compounds are given in the upper part of the cases, with the most important underlined

Inner electrons																			
	ns	1	2	2	2	2	1	2	2	2	2	1	2	2	2	2	2	2	2
	np	—	—	—	—	—	—	—	—	—	—	—	—	1	2	3	4	5	6
	$(n-1)$d	—	—	1	2	3	5	5	6	7	8	10	10	10	10	10	10	10	10
0	$n=1$	$^{1,-1}_{1}$H	$_{2}$He																
			a[a]	b[b]										c[c]					d[d]
		IA	IIA	IIIB	IVB	VB	VIB	VIIB		VIIIB		IB	IIB	IIIA	IVA	VA	VIA	VIIA	VIIIA
2	2	$^{1}_{3}$Li	$^{2}_{4}$Be											$^{3}_{5}$B	$^{4,2,-4}_{6}$C	$^{5,4,3,2,-3}_{7}$N	$^{-2,-1}_{8}$O		$_{10}$Ne
10	3	$^{1}_{11}$Na	$^{2}_{12}$Mg											$^{3}_{13}$Al	$^{4,-4}_{14}$Si	$^{5,3,-3}_{15}$P	$^{6,4,2,-2}_{16}$S	$^{7,5,3,1,-1}_{17}$Cl	$_{18}$Ar
18	4	$^{1}_{19}$K	$^{2}_{20}$Ca	$^{3}_{21}$Sc	$^{4,3}_{22}$Ti	$^{5,4,3,2,0}_{23}$V	$^{6,3,2,0}_{24}$Cr	$^{7,6,4,3,2,0,-1}_{25}$Mn	$^{6,3,2,0,-2}_{26}$Fe	$^{3,2,0,-1}_{27}$Co	$^{3,2,0}_{28}$Ni	$^{2,1}_{29}$Cu	$^{2}_{30}$Zn	$^{3}_{31}$Ga	$^{4}_{32}$Ge	$^{5,3,-3}_{33}$As	$^{6,4,-2}_{34}$Se	$^{7,5,3,1,-1}_{35}$Br	$^{2}_{36}$Kr
36	5	$^{1}_{37}$Rb	$^{2}_{38}$Sr	$^{3}_{39}$Y	$^{4}_{40}$Zr	$^{5,3}_{41}$Nb	$^{6,5,4,3,2,0}_{42}$Mo	$^{7}_{43}$Tc	$^{8,6,4,3,2,0,-2}_{44}$Ru	$^{5,4,3,1,2,0}_{45}$Rh	$^{4,2,0}_{46}$Pd	$^{2,1}_{47}$Ag	$^{2}_{48}$Cd	$^{3}_{49}$In	$^{4,2}_{50}$Sn	$^{5,3,-3}_{51}$Sb	$^{6,4,-2}_{52}$Te	$^{7,5,1,-1}_{53}$I	$^{2,4,6}_{54}$Xe
54	6	$^{1}_{55}$Cs	$^{2}_{56}$Ba	$_{57}$La*	$^{4}_{72}$Hf	$^{5}_{73}$Ta	$^{6,5,4,3,2,0}_{74}$W	$^{7,6,4,2,-1}_{75}$Re	$^{8,6,4,3,2,0,-2}_{76}$Os	$^{6,4,3,2,1,0,-1}_{77}$Ir	$^{4,2,0}_{78}$Pt	$^{3,1}_{79}$Au	$^{2,1}_{80}$Hg	$^{3,1}_{81}$Tl	$^{4,2}_{82}$Pb	$^{5,3}_{83}$Bi	$^{6,4,2}_{84}$Po	$^{7,5,3,1,-1}_{85}$At	$^{2}_{86}$Rn
86	7	$^{1}_{87}$Fr	$^{2}_{88}$Ra	$_{89}$Ac†	$_{104}$Rf	$_{105}$Ha													
		ns		$(n-1)$d										*np*					
	*	$^{3}_{57}$La	$^{4,3}_{58}$Ce	$^{4,3}_{59}$Pr	$^{3}_{60}$Nd	$^{3}_{61}$Pm	$^{3,2}_{62}$Sm	$^{3,2}_{63}$Eu	$^{3}_{64}$Gd	$^{4,3}_{65}$Tb	$^{3}_{66}$Dy	$^{3}_{67}$Ho	$^{3}_{68}$Er	$^{3,2}_{69}$Tm	$^{3,2}_{70}$Yb	$^{3}_{71}$Lu	Lanthanides		
	†	$^{3}_{89}$Ac	$^{4}_{90}$Th	$^{5,4}_{91}$Pa	$^{6,5,4,3}_{92}$U	$^{6,5,4,3}_{93}$Np	$^{6,5,4,3}_{94}$Pu	$^{6,5,4,3}_{95}$Am	$^{4,3}_{96}$Cm	$^{4,3}_{97}$Bk	$^{4,3}_{98}$Cf	$^{3}_{99}$Es	$^{3}_{100}$Fm	$^{3}_{101}$Md	$^{3,2}_{102}$No	$^{3}_{103}$Lr	Actinides		
		$(n-2)$f																	

[a] Light metals.
[b] Transition metals.
[c] The stair step running from boron to astatine, divides non-metals from post-transition metals.
[d] Nobel gases.

Table 6.2. Partial series of the electronegativity of elements in accordance with Sanderson[2]

Element	EN	Element	EN	Element	EN	Element	EN
H	2.31	V^{2+}	1.24	Rh	1.47	Dy	0.94
He	—	Cr^{2+}	1.35	Pb	1.57	Ho	0.96
Li	0.86	Mn^{2+}	1.44	Ag	1.72	Er	0.96
Be	1.61	Fe	1.47	Cd	1.73	Tm	0.96
B	1.88	Co^{2+}	1.70	In	1.88	Yb	0.96
C	2.47	Ni^{2+}	1.75	Sn^{2+}	1.58	Lu	0.96
N	2.93	Cu	1.14	Sn^{4+}	2.02	Hf	0.98
O	3.46	Zn^{2+}	1.86	Sb	2.19	Ta	1.04
F	3.92	Ga^{3+}	2.10	Te	2.34	W	1.13
Ne	4.38	Ge^{4+}	2.31	I	2.50	Re	1.19
Na	0.85	As^{3+}	2.53	Xe	2.63	Os	1.26
Mg	1.42	Se	2.76	Cs	0.69	Ir	1.33
Al	1.54	Br	2.96	Ba	0.93	Pt	1.36
Si	1.74	Kr	3.17	La	0.92	Au	1.72
P	2.16	Rb	0.70	Ce	0.92	Hg	1.92
S	2.66	Sr	0.96	Pr	0.92	Tl^{+}	1.36
Cl	3.28	Y	0.98	Nd	0.93	Tl^{3+}	1.96
Ar	3.92	Zr^{2+}	1.00	Pm	0.94	Pb^{2+}	1.61
K	0.74	Nb	1.12	Sm	0.94	Pb^{4+}	2.01
Ca	1.06	Mo	1.24	Eu	0.94		
Sc	1.09	Tc	1.33	Gd	0.94		
Ti^{2+}	1.13	Ru	1.40	Tb	0.94		

energy (absorbed by the atom) is marked positive, while exothermic energy (given off by the atom) is marked negative.

EA is the energy given off when an electron is added to the valence shell of an atom. The acquisition of a first electron by non-metallic atoms involves the giving off of energy. However, the acquisition of a further electron is an endothermic process, as the addition of the first electron has brought the atom to a sort of 'saturation' level.

EN expresses the property of an atom to attract electrons to itself when it combines. It is difficult to evaluate this property, since, unlike IE and EA, it depends also on the nature and number of the atoms with which it bonds. Since the vast majority of elements can form large numbers of combinations, the EN cannot be an invariable value for each element. The various methods used[2,3] to define EN have led to various electronegativity series. The semi-empirical series adopted here is that of Sanderson,[2] illustrated in Table 6.2.

Two atoms or ions are influenced by two forces: attraction (F_a) and repulsion (F_r). The algebraic sum of their relative energies, E_a and E_r, is indicated by E. The minimum of this function, E_e, in correspondence to $R = R_e$, represents the energy gain of a couple of ions in 'contact', as opposed to the energy they would possess at an infinite distance. R_e is the bond length in correspondence with which the forces of attraction and repulsion are balanced.

The bonds can be divided orientationally into two categories:

(1) strong bonds with energy $E_e > 90$ kcal mol^{-1};

(2) weak bonds with energy $E_e < 35$ kcal mol^{-1}.

Type (1) bonds are due to direct interaction between the outermost orbital electrons. If the bond electrons tend to be concentrated in the region between the atoms, then the bond is termed **covalent** (e.g. an H_2 molecule). If, instead, the outermost electrons tend to be transferred from one atom to the other, giving origin to high concentrations and impoverishments in comparison with the original electronic densities, then the bond is referred to as an **ionic bond** (e.g. an NaCl crystal). Finally, if the bond electrons belong to the whole crystal, then the bond is known as **metallic** (e.g. crystalline Na). Most compounds are characterized by bonds of an intermediate nature which are referred to as **heteropolar covalent bonds**. The **degree of ionicity** of a bond (ID) can be evaluated if the difference in electronegativity between two atoms involves (X and Y) is known, ($\Delta EN_{XY} = EN_X - EN_Y$). If the bond between them is simple, then, in accordance with Pauling,[4] ID can be estimated by means of the following equation:

$$ID = 1 - \exp[-0.25(\Delta EN_{XY})^2]. \tag{6.2}$$

Using Table 6.2 for NaCl, KCl, and LiF the following values for ID can be obtained: 0.77, 0.80, 0.90.

Type (2) bonds are due to weaker, essentially electrostatic, forces like van der Waals forces or hydrogen bond.[5] Van der Waals forces are present in all compounds: together with hydrogen bonds they play a primary role in molecular crystal structure (see Chapter 7). Hydrogen bonds are fundamental for the cohesion of many compounds like proteins, molecular crystals, hydrated salts, etc.

Ionic crystals

The periodic table (Table 6.1) and Table 6.2 show that the EN of elements varies with a certain regularity. In fact it increases much more from left to right (along the periods) than from bottom to top (along the groups). The further apart two elements are in a given period, the greater is the difference in electronegativity between them, and the more markedly ionic are the bonds of the compounds they form together. The various types of bonds which form between the atoms play an important role in determining geometrically the structural pattern and the physico-chemical properties which characterize a crystalline substance.

In **molecular crystals**, more or less complex, finite, atomic aggregates can be found, i.e. molecules, within which considerably stronger covalent bonds are formed than the van der Waals bonds which are formed between the molecules (see Chapter 7). In **non-molecular crystals** the protagonists are atoms or ions (simple or complex) which are held together by more or less localized bonds, and in limited cases of maximum polarization, by Coulomb interactions.

This chapter will deal with non-molecular crystals, in particular **ionic crystals**. The following questions will be considered:

1. What conditions must be satisfied in order that a combination of ions may be considered stable?
2. What are the configurations that such combinations give rise to?

The energy associated with a given ions' disposition will be analysed briefly in order to answer the first question, while the atomic building of some typical structures will be analysed in order to answer the second.

Gibb's free energy, $G = E - TS + PV$, indicates the stability of a crystalline structure. The term PV (pressure × volume) can be omitted if the pressure is not high, thus free energy takes the form:

$$F = E - TS \tag{6.3}$$

where F is **Helmholtz's free energy**, E is the **internal energy** of the crystal, T is the absolute temperature, and S is the entropy. E, known also as **cohesion energy**, is the sum of two terms:

$$E = U_L + E_V. \tag{6.4}$$

U_L is the static term, also known as **lattice energy**, (see p. 411), and expresses the potential energy of the immobile ions in the equilibrium positions. E_V is the dynamic term which expresses the **vibration energy** of the mobile ions oscillating around the equilibrium positions. If the temperature is not very high and the structure is well ordered, i.e. if for E_V and $TS \ll U_L$, $F \simeq U_L$ results, then the stability of a structure can be calculated by means of U_L, which, as we shall see, can be estimated experimentally using the Born–Haber cycle.

U_L can be calculated not only for ionic crystals but also for partially ionic crystals (crystals in which there are prevalently covalent bonds as well as prevalently ionic bonds), provided that the groups of atoms associated mainly by covalent bonds are considered as single ions, having a charge equal to that of the whole group.

U_L is the sum of various values: electrostatic attraction energy E_a, due to the prevailing long radius forces; repulsion energy, E_r, due to the prevailing short radius forces (such forces depend both on the repulsion of the charges (e^-–e^-, p^+–p^+) and on the repulsion between fully occupied orbitals at any attempt to superimpose them (Pauli's principle)); energy due to crystal field stabilization, E_s; energy due to covalent bonds; energy due to van der Waals forces, etc.

E_a is definitely the dominant term, being about 80 per cent of the entire potential energy, while E_r is about 10–15 per cent. E_s is not negligible when transition metals are present in the structure. The lower the ΔEN between the ions combined, the more substantial is the percentage of energy due to covalent bonds. Finally, the percentage of energy due to van der Waals forces depends on momentary interactions between induced dipoles. This type of force is always present, since the barycentre of the positive nucleus does not correspond to the barycentre of the negative charges, due to thermal agitation. Thus momentary dipoles are formed which interact electrostatically with other induced dipoles. This section will deal only with the first three contributions: E_a, E_r, E_s. Chapter 7 should be consulted for an analysis of the other terms of particular importance for molecular crystals.

Lattice energy: the contributions of attractive and repulsive terms

With reference to Table 6.1, compounds with prevalently ionic bonds are to be found amongst combinations of alkaline metals, alkaline-earth metals (groups I, IIA), part of group IIIA, transition metals with low oxidation numbers, and non-metals (groups VI and VIIA). **Lattice energy**, or rather the U_L **energy of a crystal**, can be defined as the energy necessary at absolute zero to break down a mol of crystal into its ionic components, carrying them to an infinite distance. The formation of an ionic bond can be considered as having two distinct phases in origin:

(1) a positive ion or cation forms in the following way:

$$\mathrm{M + IE \rightarrow M^+ + e^-};$$

(2) a negative ion or anion forms in the following way:

$$\mathrm{X + e^- \rightarrow X^- + EA}.$$

If ionization takes place by means of the taking off of more than one electron, then the total IE is the sum of the partial IEs.[2,3]

For example, let us consider two opposite charges supposedly concentrated in a single point, $+z_1e$, $-z_2e$ (e being the charge or an electron). At a distance of R from each other, they are mutually attracted by a force

$$F_a = z_1 z_2 e^2/R^2 \tag{6.5}$$

from which the potential electrostatic energy of the charges can be derived:

$$E_a = \int_{\infty}^{R} F_a \, dR = -z_1 z_2 e^2/R \tag{6.6}$$

In accordance with Born, the repulsion energy is given by:

$$E_r = B/R^n \tag{6.7}$$

where B is a constant of proportionality; n, Born's exponent, is a value linked to the type of ions which form the crystal and varies[4] from 5 to 12.

Alternatively the repulsion energy can be expressed as follows:

$$E_r = b \exp(-R/\rho) \tag{6.8}$$

where b and ρ are empirical parameters, which can be determined, like n, by means of compressibility measures.

The total potential energy E for a couple of ions is given by:

$$E_c = -\frac{z_1 z_2 e^2}{R} + \frac{B}{R^n}. \tag{6.9}$$

The minimum of the function E_c is obtained as follows:

$$\frac{dE_c}{dR} = \frac{z_1 z_2 e^2}{R_e^2} - \frac{nB}{R_e^{n+1}} = 0 \tag{6.10}$$

hence:

$$B = \frac{z_1 z_2 e^2 R_e^{n-1}}{\mathrm{n}}. \tag{6.11}$$

R_e is the equilibrium distance between two ions. Substituting (6.11) in (6.9) gives:

$$E_e = -\frac{z_1 z_2 e^2}{R_e}\left(1 - \frac{1}{n}\right). \tag{6.12}$$

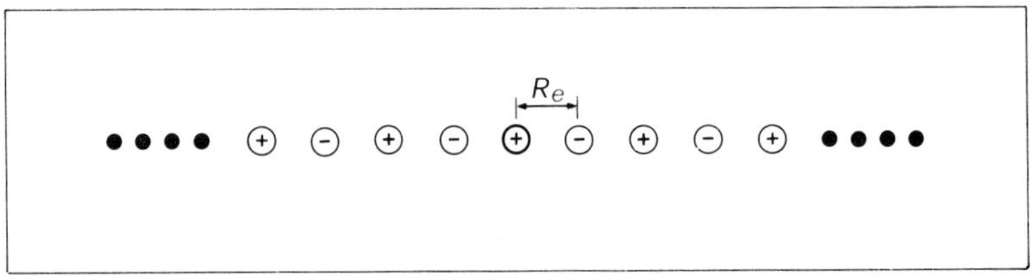

Fig. 6.5. 'Touching' cations and anions alternating in a row.

Let us calculate the potential energy in the case of a row of N couples of ions of opposite charges, positioned as illustrated in Fig. 6.5.

For the two anions on either side of the reference cation (heavily marked circle) and for the two cations immediately outside them, the potential energy will be respectively:

$$E_a = -2z_1z_2e^2/R \quad \text{and} \quad E_a = +2z_1z_2e^2/(2R).$$

Taking into account all the ions around the marked cation the total E_a is given by:

$$E_a = \frac{-2z_1z_2e^2}{R}(1 - \tfrac{1}{2} + \tfrac{1}{3} - \tfrac{1}{4} + \ldots)$$

$$= \frac{-2z_1z_2e^2}{R} \ln 2 = -1.39z_1z_2e^2/R. \tag{6.13}$$

Using the same procedure we will also obtain:

$$E_r = 2B/R^n + 2B/(2R)^n + 2B/(3R)^n + \ldots \simeq 2B/R^n. \tag{6.14}$$

Thence in the case of a row of ions (6.9) takes the form:

$$E_{ri} \simeq -1.39z_1z_2e^{-2}/R + 2B/R^n \tag{6.15}$$

Following the same procedure used in (6.10) we found $B = 1.39\, z_1z_2e^2R^{n-1}/2n$. Replacing B in (6.15) and observing that the electrostatic interactions between each ion and all the others are N, just equal to the couples of opposite charged ions, (6.15) will be:

$$E_e \simeq -\frac{1.39Nz_1z_2e^2}{R_e}\left(1 - \frac{1}{n}\right). \tag{6.16}$$

Equation (6.16) differs from (6.12) not only in the number of the interactions N, but also in the constant 1.39. This constant derives from the positions of the ions considered in Fig. 6.5: more generally it depends on the type of three-dimensional lattice which the ions form. Madelung's[2,6] constant expresses the geometric characteristics of the lattice, and will be indicated here as A_M. If N_0 (Avogadro's number) is the number of couples of ions contained in a mol, (6.16) becomes:

$$U_L \simeq -\frac{A_MN_0z_1z_2e^2}{R_e}\left(1 - \frac{1}{n}\right). \tag{6.17}$$

A_M values for some important lattice types are given in Table 6.3.

Since a crystal is the content, in terms of atoms, of a cell which is repeated many times in space, its potential energy can be calculated as

Table 6.3. Madelung constants, A_M, for some compounds

Structure	Coordination number	A_M
CsCl	8:8	1.76267
Halite (NaCl)	6:6	1.74756
Zinc blende (ZnS)	4:4	1.63806
Wurtzite (ZnS)	4:4	1.64132
Fluorite (CaF_2)	8:4	2.51939
Cuprite (Cu_2O)	4:8	2.22124
Rutile (TiO_2)	6:3	2.408
Anatase (TiO_2)	6:3	2.400

follows:

$$E_L = \tfrac{1}{2}e^2 \sum_{i,j=1}^{N} \sum_{l=-\infty}^{+\infty} z_i z_j \, |\boldsymbol{x}_{ij} + \boldsymbol{l}|^{-1} + \tfrac{1}{2} \sum_{i,j=1}^{N} \sum_{l=-\infty}^{+\infty} b_{ij} \exp\left(-|\boldsymbol{x}_{ij} + \boldsymbol{l}|/\rho\right) \quad (\text{if } i = j,\ \boldsymbol{l} \neq 0). \tag{6.18}$$

E_L is the potential energy of an elementary cell, the coefficient $\frac{1}{2}$ derives from the fact that every couple of ions is included twice in the calculation of (6.18), N is the number of ions contained in the cell, z_1 and z_2 are the number of their charges expressed in terms of electrons, $\boldsymbol{x}_{ij} = \boldsymbol{x}_i - \boldsymbol{x}_j$ are the interatomic vectors between the i and j ions of the cell, $\boldsymbol{l}$ is a generic lattice vector, b_{ij} is the repulsion coefficient between the i and j ions, and ρ is the hardness parameter due to its repulsive forces. The electrostatic potential, represented by the first term of (6.18), diminishes gradually over long distances. This term is in practice evaluated by means of particular series, such as an optimized Ewald's series,[7] which accelerate their convergence.

The **Born–Haber cycle** allows the **enthalpy** or **heat of formation** ΔH_f of an ionic crystal to be measured, starting from the elements. A scheme of the cycle is shown in Fig. 6.6.

Born and Haber applied the Hess' law to ionic crystals. Interpreting the principle of conservation of energy in thermodynamic terms, the Hess' law states that ΔH_f depends on the point of arrival and of departure, but not on the number or nature of the intermediate stages of the reaction, that is to say:

$$\Delta H_f = \Delta H_M + \Delta H_X + \Delta H_{IE} + \Delta H_{EA} + U_L \tag{6.19}$$

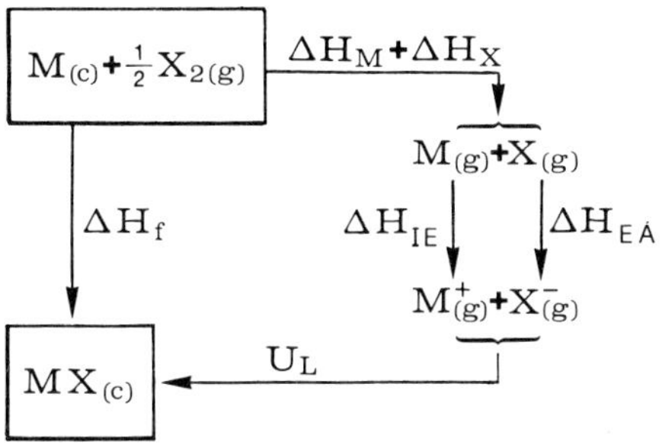

Fig. 6.6. Sketch of the Born–Haber cycle.

where ΔH_M and ΔH_X are respectively the sublimation energy of a metallic crystal ($M_{(c)}$) and the dissociation energy of the molecule X_2 in its gaseous state ($X_{2(g)}$). ΔH_{IE} and ΔH_{EA} are respectively the ionization energies of the metal in its gaseous state and the electronic affinity. For example, for NaCl an enthalpy equal to $\Delta H_f^{298} = -98.2$ kcal mol^{-1} has been determined, using the Born–Haber cycle. Knowing that $\Delta H_M = 24$, $\Delta H_X = 29$, $\Delta H_{IE} = 118.5$, $\Delta H_{EA} = -83.3$ kcal mol^{-1} then $U_L = -186.4$ kcal mol^{-1}. If in (6.17) we insert the values $A_M = 1.74745$, $R_e = 2.814 \times 10^{-8}$ cm, $n = 9$, and introduce the conversion factor ergs → kcal, we obtain:

$$U_L = \frac{1.74756 \times 6.023 \times 10^{23} \times (4.803 \times 10^{-10})^2}{2.81 \times 10^{-8} \times 4.184 \times 10^{10}} (1 - \tfrac{1}{9})$$
$$= -183.3 \text{ kcal mol}^{-1},$$

which is very close to the value obtained from experimental data.

The agreement between experimental data and the results of the calculations is particularly good in this case, although the method tends generally to underestimate the value of U_L: in particular the estimate is at least 10 kcal mol^{-1} below the experimental value for halogen compounds.

Lattice energy: CFSE contribution

The Born theory often gives U_L values for transition metals, and in particular those having incomplete d orbitals, which are too low compared with observed values. In order to explain this discrepancy, Bethe and Vleck formulated a theory known as the crystal field theory (CFT).[8–12]

This theory manages to interpret qualitatively, with conceptual simplicity, a series of experimental data. The limitations of the CFT are in its presuppositions, since it presupposes that between metal ions and ligands interactions of an electrostatic nature only develop, while the ligands are considered as point charges. To understand the type of action that the ligands have on the d orbitals of the metal ions, their spatial geometry must be borne in mind (see Fig. 6.3).

The five d orbitals can be divided into two distinct subgroups on the basis of their orientation. The first encompasses the three orbitals, d_{xy}, d_{xz}, and d_{yz}, whose lobes are orientated along the diagonals of their Cartesian axes. The second group encompasses the two orbitals, $d_{x^2-y^2}$ and d_{z^2}, whose lobes are orientated along their Cartesian axes.

Let us analyse two cases:

(1) a transition metal in an octahedral coordination;

(2) a transition metal in a tetrahedral coordination.

For case (1) we will consider a cation which has all five d orbitals empty. If the cation is isolated, the d orbitals all have the same energy (degenerate orbitals). Let us now suppose that the cation is surrounded by point-like negative charges q^-, distributed symmetrically at equal distances from the d orbitals (spherical field conditions). They will have the effect of increasing the energy of the d orbitals, which will remain degenerate (Fig. 6.7(b)).

If, however, we imagine that six q^- charges surround the cation, positioned at the vertices of an octahedron and orientated in relation to the orbitals as shown in Fig. 6.8(a), then the two subgroups of the d orbitals, $e_g(d_{x^2-y^2}, d_{z^2})$ and $t_{2g}(d_{xy}, d_{xz}, d_{yz})$, will have different energy values.

The e_g orbitals (shaded) point towards the ligands and are nearer to them than are the t_{2g} orbitals (white), which are in an intermediate position. As a result, the e_g orbitals have a higher energy value than the t_{2g}, and for this reason the electrons prefer the latter. The difference in energy between the e_g and t_{2g} orbitals, known as **crystal field splitting**, is indicated by Δ_0 (Fig. 6.7(c)). It follows the **barycentre principle**: the algebraic sum of the energies of the d orbital electrons, when all the d orbitals are filled, is equal

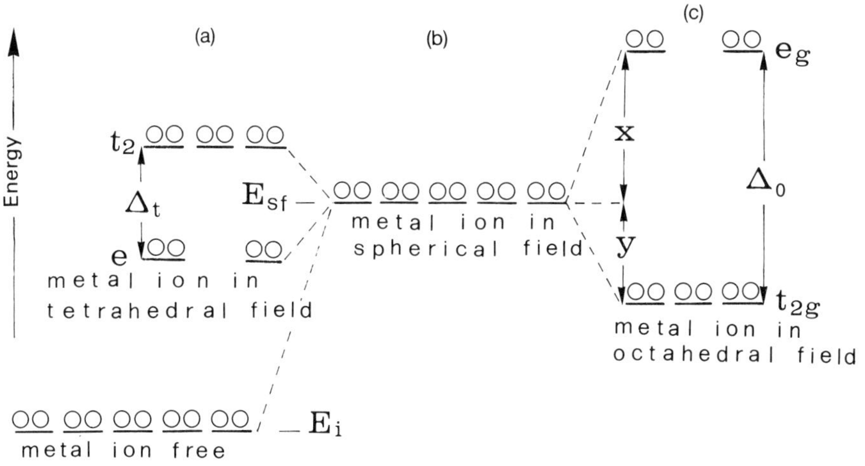

Fig. 6.7. Energy levels relative to the d orbitals in spherical, tetrahedral, and octahedral fields.

to that obtained when the cation is in spherical field conditions. Let us indicate E_i the absolute energy of a d orbital of the cation when isolated and E_{sf} the absolute energy of a d orbital of the cation in spherical field condition. When the cation is in an octahedral field the energies are indicated respectively by X and Y for a t_{2g} and an e_g orbital.

In order that the d configuration in an octahedral field should have the same energy as that in a spherical field, it is necessary that:

$$6(E_{sf} - Y) + 4(E_{sf} + X) = 10E_{sf} \quad \text{and} \quad X + Y = \Delta_0. \tag{6.20}$$

The solution which satisfies (6.20) is: $X = 0.6\Delta_0$, $Y = 0.4\Delta_0$. It might seem, judging by Fig. 6.7, that the free cation is energetically more stable than the complex, since $E_{sf} > E_i$. In reality the bond energy (disassociation energy of a pair of touching ions separated into two atoms) is more than enough to compensate for the increase in the energy of the d orbitals. Moreover, an electron which occupies a t_{2g} orbital has an energy value which is $0.4\Delta_0$ less than E_{sf}. Thus, every electron which occupies one of the t_{2g} orbitals stabilizes the metal by $0.4\Delta_0$, while every one that fills an e_g orbital reduces the metal's stability by $0.6\Delta_0$.

The value obtained from the algebraic sum $\sum_i (\pm x_i)\Delta_0$ of all the electrons of d orbitals ($x = -0.4\Delta_0$ if the electron belongs to a t_{2g} orbital, or $+0.6\Delta_0$ if it belongs to an e_g orbital) is known as the **crystal field stabilization energy** (CFSE).

Δ_0 ranges between 20 and 90 kcal mol^{-1} and depends:

(1) on the cation charge (it increases as the charge does);

(2) on the type of d orbital involved (it increases substantially from the 3d orbitals to the 5d orbitals);

(3) on the symmetry of the metal ion–ligands;

(4) on the distance (R) between the metal ion and ligands, or to be more precise, on $1/R$;[5]

(5) on the nature of the ligands (some of which in order of increasing stabilization are: $I^- < Br^- < S^{-2} < Cl^- < NO_3^- < F^- < OH^- < H_2O < NO_2^-$, **spectrochemical series**).

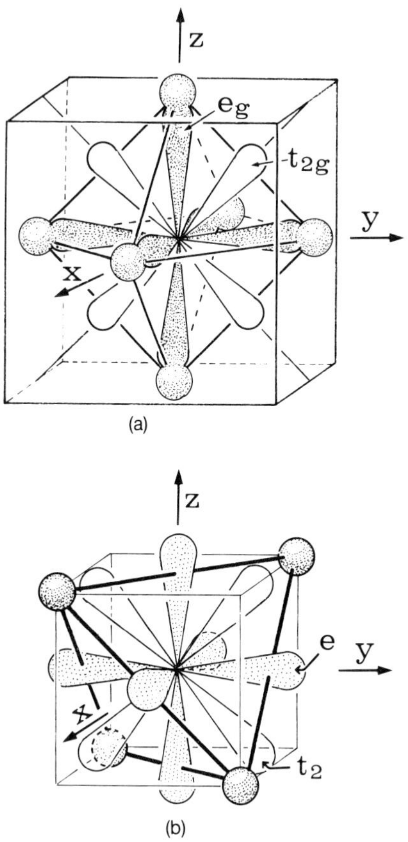

Fig. 6.8. Orientation of the orbitals e_g and t_{2g} of a transition metal in (a) an octahedral crystal field, and (b) a tetrahedral crystal field. For clarity's sake two of the three orbitals, t_{2g} and t_2, in the planes *xy* and *xz* respectively, are not shown. They are oriented at an angle of 45° to the *x* axis.

Let us now analyse how the electrons are positioned in the d orbitals. The first three electrons occupy the three t_{2g} orbitals.

The fourth electron has two possibilities: it may occupy an e_g orbital, in accordance with the Hund's first rule, or **maximum multiplicity principle** (an atom, in its fundamental state, follows the maximum multiplicity principle: in a partially filled orbital the number of electrons having the same spin orbital is maximum), or it may associate with another electron already in a t_{2g} orbital, which spins in the opposite direction, thus further stabilizing the complex.

Two factors influence the fourth electron and decide which tendency will prevail: the Δ_0 and the **pairing energy**, *P*. The latter can be defined as the energy necessary:

(1) to overcome the repulsion which exists between two electrons which occupy the same orbital;

(2) to compensate the exchange energy which is lost when Hund's first rule is violated.

If the $\Delta_0 > P$, then the fourth electron will go to fill a t_{2g} orbital, forming a low-spin complex (two electrons which spin in opposite direction plus two electrons which spin in the same way give a compound which is not very paramagnetic). If the $\Delta_0 < P$, then the fourth electron will go to fill an e_g orbital, forming a high-spin complex (four electrons which spin in the same way give a very paramagnetic compound).

Table 6.4 shows the CFSE relative to high- and low-spin configurations, for octahedral sites.

We will now examine the case of a transition metal in a tetrahedral configuration. Figure 6.8(b) shows four ligands at the vertices of a tetrahedron. The three d orbitals, which we shall call t_2, are nearer to the ligand than are the other two e orbitals. This implies that, unlike an octahedral coordination, in a tetrahedral coordination the three t_2 orbitals have more energy than the two e orbitals (Fig. 6.7(a)).

The crystal field separation value is indicated by Δ_t and is inferior to the $\Delta_0(\Delta_t/\Delta_0 = 4/9)$ for two reasons:

(1) because there is no direct interaction between the ligands and the d

Table 6.4. CFSE for transition metals of the first series in octahedral configurations

Number of d electrons	Ion	High-spin configuration			Low-spin configuration		
		t_{2g}	e_g	CFSE (Δ_0)	t_{2g}	e_g	CFSE (Δ_0)
0	$Ca^{2+}Sc^{3+}Ti^{4+}$	0		0.0	0		0.0
1	Ti^{3+}	1		−0.4	1		−0.4
2	$Ti^{2+}V^{3+}$	2		−0.8	2		−0.8
3	$V^{2+}Cr^{3+}Mn^{4+}$	3		−1.2	3		−1.2
4	$Cr^{2+}Mn^{3+}$	3	1	−0.6	4		$-1.6+P$
5	$Mn^{2+}Fe^{3+}$	3	2	0.0	5		$-2.0+2P$
6	$Fe^{2+}Co^{3+}Ni^{4+}$	4	2	$-0.4+P$	6		$-2.4+3P$
7	$Co^{2+}Ni^{3+}$	5	2	$-0.8+2P$	6	1	$-1.8+3P$
8	Ni^{2+}	6	2	$-1.2+3P$	6	2	$-1.2+3P$
9	Cu^{2+}	6	3	$-0.6+4P$	6	3	$-0.6+4P$
10	$Zn^{2+}Ga^{3+}Ge^{4+}$	6	4	$0.0+5P$	6	4	$0.0+5P$

orbitals of the metal ion;

(2) because there are four ligands instead of six.

This explains why, as experience has demonstrated, tetrahedral complexes have only high-spin configurations.

Table 6.5 shows the CFSE relative to high-spin configurations for tetrahedral sites.

At this point it will be useful to list the factors, which are at times mutually opposed, which govern cation coordination:

(1) voluminous ligands favour tetrahedral coordination (see p. 425);

(2) the position of the ligand in the spectrochemical series (see p. 415) favours tetrahedral coordination if on the left, octahedral if on the right:

(3) a high ligand charge favours octahedral coordination, because it increases Δ_0, which in turn favours the formation of low-spin complexes, increasing the number of electrons in t_{2g} orbitals;

(4) the CFSE favours octahedral complexes, both because $\Delta_0 = (9/4)\Delta_t$, and because the CFSE of tetrahedral complexes is generally less than those of octahedral complexes (cf. Tables 6.4 and 6.5).

If the CFSEs of Table 6.5 are multiplied by the scale factor 4/9, they can be compared with the CFSE of Table 6.4, and the **octahedral site stabilization energy**, OSSE (i.e. the difference between the CFSEs concerning octahedral and tetrahedral sites), can be calculated in Δ_0 units (see Table 6.6).

Table 6.5. CFSE for transition metals of the first series in tetrahedral configurations. For the ions involved see Table 6.4

Number of d electrons	High-spin configuration		
	e	t_2	CFSE (Δ_t)
0	0		0.0
1	1		−0.6
2	2		−1.2
3	2	1	−0.8
4	2	2	−0.4
5	2	3	0.0
6	3	3	$-0.6+P$
7	4	3	$-1.2+2P$
8	4	4	$-0.8+3P$
9	4	5	$-0.4+4P$
10	4	6	$0.0+5P$

Applications of lattice energy calculations

Lattice energy calculations make it possible not only to rationalize a series of observations arising from structural analyses, but also to resolve certain problems without recourse to X-ray techniques. Amongst the numerous applications of lattice energy calculations, easily computed with the aid of

Table 6.6. OSSE for transition metals of the first series. For the ions involved see Table 6.4

Number of d electrons	$CFSE_t$ Δ_0 unit	$CFSE_0$ high spin	OSSE high spin	$CFSE_0$ low spin	OSSE low spin
0	0.0	0.0	0.0	0.0	0.0
1	−0.27	−0.4	−0.13	−0.4	−0.13
2	−0.53	−0.8	−0.27	−0.8	−0.27
3	−0.35	−1.2	−0.85	−1.2	−0.85
4	−0.18	−0.6	−0.42	$-1.6+P$	$-1.42+P$
5	0.0	0.0	0.0	$-2.0+2P$	$-2.00+2P$
6	$-0.27+P$	$-0.4+P$	−0.13	$-2.4+3P$	$-2.13+2P$
7	$-0.53+2P$	$-0.8+2P$	−0.27	$-1.8+3P$	$-1.27+P$
8	$-0.35+3P$	$-1.2+3P$	−0.85	$-1.2+3P$	−0.85
9	$-0.18+4P$	$-0.6+4P$	−0.42	$-0.6+4P$	−0.42
10	$0.0+5P$	$0.0+5P$	0.0	$0.0+5P$	0.0

various automatic programmes,[7,13,14] are the following:

(1) the distribution of cations in different sites;

(2) the effect of the distribution of certain cations in certain sites on the position of others;

(3) structural predictions;

(4) solutions to structural problems;

(5) the determination of charge distribution in complex ions;

(6) the estimation of the elastic constants of stable minerals subjected to very high pressures;

(7) the effects of temperature and pressure on structures.

Let us examine some examples in more detail. U_L was calculated for 16 isostructural compounds with olivine, including certain berylites, borates, phosphates, and silicates.[15]

The structure of olivine, space group (s.g.) P*bn*m, can be described as a distorted hexagonal closest-packed system of oxygen atoms (see p. 444). The six coordinated cations occupy two sites which are not crystallographically equivalent, M(1) and M(2) respectively, at the centre (0.0, 0.0, 0.0) and on the mirror plane. In silicates with an olivine-type structure these cations may be monovalent, bivalent, or trivalent (Li^+, Na^+, Mg^{2+}, Fe^{2+}, Ca^{2+}, Fe^{3+}, etc.).

Potential energy calculations indicate that if the ionic radii of the cations are comparable, then the cations with smaller charges will prefer the M(1) site. If, instead, the ionic radii are of quite different sizes, then the cation with the greater volume will prefer the M(2) site. In accordance with the calculation, where $r_e(Li^+) = 0.76 \simeq r_e(Sc^{3+}) = 0{,}745$ Å, the compound $Li^+Sc^{3+}[SiO_4]$ is found to have Li^+ in M(1) and Sc^{3+} in M(2). Moreover, where $r_e(Mg^{2+}) = 0.72 \neq r_e(Ca^{2+}) = 1.00$ Å, the compound $MgCa[SiO_4]$, **monticellite**, has Mg^{2+} in M(1) and Ca^{2+} in M(2).

Another example is the influence of the Al/Si distribution on the sites occupied by Na^+ in albite ($AlSi_3O_8$) at high temperatures.[16] Here too the calculations agree with the experimental data, that is to say, the concentration of Al^{3+} in a particular tetrahedral site determines the site which the Na^+ will occupy.

The determination of the structure of an inorganic crystal by means only of calculations of potential energy has serious practical drawbacks if suitable crystallo-chemical information is lacking initially. It would be necessary to do an enormous number of trials, moving the atoms in all possible directions, if their precise positions are not known initially. Recently,[17] starting from the ideal structure of olivine, it has been noticed that the compact regular layers of oxygen atoms used as input in the calculations are deformed in the process of potential energy minimization, just as the actual structure shows them to be.

Ionic radius

An important factor which influences the geometry of the structure of ionic compounds is the so-called **ionic radius**. It expresses the 'dimensions' of an ion considered as a rigid sphere. Since a finite value of r, when $D_{n,l}(r) = 0$ (Fig. 6.2), does not exist, 'atomic radius', and thus also 'ionic radius', are merely conventional terms. Nevertheless, data derived from structural determination by means of X-ray techniques suggest that d_{MX} (the interionic distance between a cation M and an anion X) is the sum of the two 'radii' characteristic of the two adjacent ions.

The problem which needs solving is how to divide d_{MX} between the anion and the cation. With this aim, Pauling[4] considered certain alkaline halides, characterized by pairs of isoelectronic ions such as NaF, KCl, RbBr, CsI, and calculated their **effective charge** $Z_{eff} = Z - S$. Z expresses the formal charge, while S indicates a **screening effect** which depends on the distance of the electrons from the nucleus and on which orbital they occupy. This effect can be explained on the basis of the radial distribution curves (Fig. 6.2), which show how s orbital electrons provide better shielding for electrons of other orbitals than they, in turn, receive from them. For example, 2p orbital electrons are better shielded than the 2s orbital electrons, because the latter are nearer to the nucleus. In order to evaluate S, certain principles suggested by Slater are followed:

1. Group the electrons as follows:

$$(1s)(2s, 2p)(3s, 3p)(3d)(4s, 4p)(4d)(4f)(5s, 5p)(5d)(5f)\ldots.$$

2. Use the multiplication coefficients 0.35 and 0.85 for electrons belonging to the groups (ns, np) and $((n-1)s, (n-1)p)$ respectively and 1 for those belonging to the group $((n-2)s, (n-2)p)$ or to any preceding group. It is preferable to use 0.30 to 0.35 in the case of groups (1s).

3. If the electrons belong to groups (nd) or (nf), use the coefficients 0.35 and 1, respectively, for electrons which belong to the same group, or for those which belong to the next group to the left.

For example, for the compound NaF:

$$Z_{eff}(Na^+) = 11 - (2 \times 0.85) - (7 \times 0.35) = 6.85$$
$$Z_{eff}(F^-) = 9 - (2 \times 0.85) - (7 \times 0.35) = 4.85$$

and, in accordance with Pauling,[4] since the ionic radius is inversely proportional to Z_{eff}, and it is known that $d_{MX} = 2.32$ Å, we can obtain:

$$\begin{aligned} r_e(Na^+):r_e(F^-) &= 4.85:6.85 \\ r_e(Na^+) + r_e(F^-) &= 2.32. \end{aligned} \tag{6.21}$$

By resolving the system (6.21), $r_e(Na^+) = 0.95$ and $r_e(F^-) = 1.36$ Å can be obtained, which agree perfectly with the ionic radii quoted by Pauling.[4] This procedure can be extended to analogous isoelectronic compounds. Various ionic radius systems have been proposed, amongst the most accurate, thanks to the large number of structures examined, are those which have appeared recently.[18,19] Effective ionic radii extracted from Shannon's original work[19] are given in Table 6.7. These systems show that:

1. Cations are generally smaller than anions. The average radius of the outermost occupied orbital varies considerably from the least electronegative atom to the cation, but very little from the most electronegative atom to the anion, e.g. $r_0(Li) = 1.586$, $r_0(Li^+) = 0.186$ Å and $r_0(K) = 2.162$, $r_0(K^+) =$

Table 6.7. Effective ionic radii (r_e) versus coordination number (CN) according to Shannon[19]

Ion	CN	Spin	r_e
Ac^{3+}	6		1.12
Ag^{+}	2		0.67
	4		1.00
	4s		1.02
	5		1.09
	6		1.15
	7		1.22
	8		1.28
Ag^{2+}	4s		0.79
	6		0.94
Ag^{3+}	4s		0.67
	6		0.75
Al^{3+}	4		0.39
	5		0.48
	6		0.535
Am^{2+}	7		1.21
	8		1.26
	9		1.31
Am^{3+}	6		0.975
	8		1.09
Am^{4+}	6		0.85
	8		0.95
As^{3+}	6		0.58
As^{5+}	4		0.335
	6		0.46
At^{7+}	6		0.62
Au^{+}	6		1.37
Au^{3+}	4s		0.68
	6		0.85
Au^{5+}	6		0.57
B^{3+}	3		0.01
	4		0.11
	6		0.27
Ba^{2+}	6		1.35
	7		1.38
	8		1.42
	9		1.47
	10		1.52
	11		1.57
	12		1.61
Be^{2+}	3		0.16
Be^{2+}	4		0.27
	6		0.45
Bi^{3+}	5		0.96
	6		1.03
	8		1.17
Bi^{5+}	6		0.76
Bi^{3+}	6		0.96
Bi^{4+}	6		0.83
	8		0.93
Br^{-}	6		1.96
Br^{3+}	4s		0.59
Br^{5+}	3p		0.31
Br^{7+}	4		0.25
	6		0.39
C^{4+}	3		−0.08
	4		0.15
	6		0.16
Ca^{2+}	6		1.00
	7		1.06
	8		1.12
	9		1.18
	10		1.23
	12		1.34
Cd^{2+}	4		0.78
	5		0.87
	6		0.95
	7		1.03
	8		1.10
	12		1.31
Ce^{3+}	6		1.01
	7		1.07
	8		1.143
	9		1.196
	10		1.25
	12		1.34
Ce^{4+}	6		0.87
	8		0.97
	10		1.07
	12		1.14
Cf^{3+}	6		0.95
Cf^{4+}	6		0.821
Cf^{4+}	8		0.92
Cl^{-}	6		1.81
Cl^{5+}	3p		0.12
Cl^{7+}	4		0.08
	6		0.27
Cm^{3+}	6		0.97
Cm^{4+}	6		0.85
	8		0.95
Co^{2+}	4	h	0.58
	5		0.67
	6	l	0.65
		h	0.745
	8		0.90
Co^{3+}	6	l	0.545
		h	0.61
Co^{4+}	4		0.40
	6	h	0.53
Cr^{2+}	6	l	0.73
		h	0.80
Cr^{3+}	6		0.615
Cr^{4+}	4		0.41
	6		0.55
Cr^{5+}	4		0.345
	6		0.49
	8		0.57
Cr^{6+}	4		0.26
	6		0.44
Cs^{+}	6		1.67
	8		1.74
	9		1.78
	10		1.81
	11		1.85
	12		1.88
Cu^{1+}	2		0.46
	4		0.60
	6		0.77
Cu^{2+}	4		0.57
	4s		0.57
	5		0.65
	6		0.73
Cu^{3+}	6		0.54
D^{+}	2		−0.10
Dy^{2+}	6		1.07
	7		1.13
	8		1.19
Dy^{3+}	6		0.912
	7		0.97
	8		1.027
	9		1.083
Er^{3+}	6		0.890
	7		0.945
	8		1.004
	9		1.062
Eu^{2+}	6		1.17
	7		1.20
	8		1.25
	9		1.30
	10		1.35
Eu^{3+}	6		0.947
	7		1.01
	8		1.066
	9		1.120
F^{-}	2		1.285
	3		1.30
	4		1.31
	6		1.33
F^{7+}	6		0.08
Fe^{2+}	4	h	0.63
	4s	h	0.64
	6	l	0.61
		h	0.780
	8	h	0.92
Fe^{3+}	4	h	0.49
	5		0.58
	6	l	0.55
		h	0.645
	8	h	0.78
Fe^{4+}	6		0.585
Fe^{6+}	4		0.25
Fr^{+}	6		1.80
Ga^{3+}	4		0.47
	5		0.55
	6		0.620
Gd^{3+}	6		0.938
	7		1.00
Gd^{3+}	8		1.053
	9		1.107
Ge^{2+}	6		0.73
Ge^{4+}	4		0.390
	6		0.530
H^{+}	1		−0.38
	2		−0.18
Hf^{4+}	4		0.58
	6		0.71
	7		0.76
	8		0.83
Hg^{+}	3		0.97
	6		1.19
Hg^{2+}	2		0.69
	4		0.96
	6		1.02
	8		1.14
Ho^{3+}	6		0.901
	8		1.015
	9		1.072
	10		1.12
I^{-}	6		2.20
I^{5+}	3p		0.44
	6		0.95
I^{7+}	4		0.42
	6		0.53
In^{3+}	4		0.62
	6		0.800
	8		0.92
Ir^{3+}	6		0.68
Ir^{4+}	6		0.625
Ir^{5+}	6		0.57
K^{+}	4		1.37
	6		1.38
	7		1.46
	8		1.51
	9		1.55
	10		1.59
	12		1.64
La^{3+}	6		1.032
	7		1.10
	8		1.160
	9		1.216
	10		1.27
La^{3+}	12		1.36
Li^{+}	4		0.590
	6		0.76
	8		0.92
Lu^{3+}	6		0.861
	8		0.977
	9		1.032
Mg^{2+}	4		0.57
	5		0.66
	6		0.720
	8		0.89
Mn^{2+}	4	h	0.66
	5	h	0.75
	6	l	0.67
		h	0.830
	7	h	0.90
	8		0.96
Mn^{3+}	5		0.58
	6	l	0.58
		h	0.645
Mn^{4+}	4		0.39
	6		0.530
Mn^{5+}	4		0.33
Mn^{6+}	4		0.255
Mn^{7+}	4		0.25
	6		0.46
Mo^{3+}	4		0.69
Mo^{4+}	6		0.650
Mo^{5+}	4		0.46
	6		0.61
Mo^{6+}	4		0.41
	5		0.50
	6		0.59
	7		0.73
N^{3-}	4		1.46
N^{3+}	6		0.16
N^{5+}	3		−0.104

Table 6.7. (*Continued*)

Ion	CN	Spin	r_e
	6		0.13
Na^{+}	4		0.99
	5		1.00
	6		1.02
	7		1.12
	8		1.18
	9		1.24
Na^{+}	12		1.39
Nb^{3+}	6		0.72
Nb^{4+}	6		0.68
	8		0.79
Nb^{5+}	4		0.48
	6		0.64
	7		0.69
	8		0.74
Nd^{2+}	8		1.29
	9		1.35
Nd^{3+}	6		0.983
	8		1.109
	9		1.163
	12		1.27
Ni^{2+}	4		0.55
	4s		0.49
	5		0.63
	6		0.690
Ni^{3+}	6	l	0.56
		h	0.60
Ni^{4+}	6	l	0.48
No^{2+}	6		1.1
Np^{2+}	6		1.10
Np^{3+}	6		1.01
Np^{4+}	6		0.87
	8		0.98
Np^{5+}	6		0.75
Np^{6+}	6		0.72
Np^{7+}	6		0.71
O^{2-}	2		1.35
	3		1.36
	4		1.38
	6		1.40
	8		1.42
OH^{-}	2		1.32
	3		1.34
	4		1.35
	6		1.37
Os^{4+}	6		0.630
Os^{5+}	6		0.575
Os^{6+}	5		0.49
Os^{6+}	6		0.545
Os^{7+}	6		0.525
Os^{8+}	4		0.39
P^{3+}	6		0.44
P^{5+}	4		0.17
	5		0.29
	6		0.38
Pa^{3+}	6		1.04
Pa^{4+}	6		0.90
	8		1.01
Pa^{5+}	6		0.78
	8		0.91
	9		0.95
Pb^{2+}	4p		0.98
	6		1.19

Ion	CN	Spin	r_e
	7		1.23
	8		1.29
	9		1.35
	10		1.40
	11		1.45
	12		1.49
Pb^{4+}	4		0.65
	5		0.73
	6		0.775
	8		0.94
Pd^{+}	2		0.59
Pd^{2+}	4s		0.64
	6		0.86
Pd^{3+}	6		0.76
Pd^{4+}	6		0.615
Pm^{3+}	6		0.97
	8		1.093
	9		1.144
Po^{4+}	6		0.94
	8		1.08
Po^{6+}	6		0.67
Pr^{3+}	6		0.99
	8		1.126
	9		1.179
Pr^{4+}	6		0.85
	8		0.96
Pt^{2+}	4s		0.60
	6		0.80
Pt^{4+}	6		0.625
Pt^{5+}	6		0.57
Pu^{3+}	6		1.00
Pu^{4+}	6		0.86
	8		0.96
Pu^{5+}	6		0.74
Pu^{6+}	6		0.71
Ra^{2+}	8		1.48
	12		1.70
Rb^{+}	6		1.52
	7		1.56
	8		1.61
	9		1.63
	10		1.66
	11		1.69
	12		1.72
	14		1.83
Re^{4+}	6		0.63
Re^{5+}	6		0.58
Re^{6+}	6		0.55
Re^{7+}	4		0.38
	6		0.53
Rh^{3+}	6		0.665
Rh^{4+}	6		0.60
Rh^{5+}	6		0.55
Ru^{3+}	6		0.68
Ru^{4+}	6		0.620
Ru^{5+}	6		0.565
Ru^{7+}	4		0.38
Ru^{8+}	4		0.36
S^{2-}	6		1.84
S^{4+}	6		0.37
S^{6+}	4		0.12
	6		0.29
Sb^{3+}	4p		0.76

Ion	CN	Spin	r_e
	5		0.80
	6		0.76
Sb^{5+}	6		0.60
Sc^{3+}	6		0.745
Sc^{3+}	8		0.870
Se^{2-}	6		1.98
Se^{4+}	6		0.50
Se^{6+}	4		0.28
	6		0.42
Si^{4+}	4		0.26
	6		0.400
Sm^{2+}	7		1.22
	8		1.27
	9		1.32
Sm^{3+}	6		0.958
	7		1.02
	8		1.079
	9		1.132
	12		1.24
Sn^{4+}	4		0.55
	5		0.62
	6		0.690
	7		0.75
	8		0.81
Sr^{2+}	6		1.18
	7		1.21
	8		1.26
	9		1.31
	10		1.36
	12		1.44
Ta^{3+}	6		0.72
Ta^{4+}	6		0.68
Ta^{5+}	6		0.64
	7		0.69
	8		0.74
Tb^{3+}	6		0.923
	7		0.98
	8		1.040
	9		1.095
Tb^{4+}	6		0.76
	8		0.88
Tc^{4+}	6		0.645
Tc^{5+}	6		0.60
Tc^{7+}	4		0.37
	6		0.56
Te^{2-}	6		2.21
Te^{4+}	3		0.52
	4		0.66
	6		0.97
Te^{6+}	4		0.43
	6		0.56
Th^{4+}	6		0.94
	8		1.05
	9		1.09
	10		1.13
	11		1.18
	12		1.21
Ti^{2+}	6		0.86
Ti^{3+}	6		0.670
Ti^{4+}	4		0.42
	5		0.51
	6		0.605
	8		0.74

Ion	CN	Spin	r_e
Tl^{+}	6		1.50
	8		1.59
	12		1.70
Tl^{+}	4		0.75
	6		0.885
	8		0.98
Tm^{2+}	6		1.03
	7		1.09
Tm^{2+}	6		0.880
	8		0.994
	9		1.052
U^{3+}	6		1.025
U^{4+}	6		0.89
	7		0.95
	8		1.00
	9		1.05
	12		1.17
U^{5+}	6		0.76
	7		0.84
U^{6+}	2		0.45
	4		0.52
	6		0.73
	7		0.81
	8		0.86
V^{2+}	6		0.79
V^{3+}	6		0.640
V^{4+}	5		0.53
	6		0.58
	8		0.72
V^{5+}	4		0.355
	5		0.46
	6		0.54
W^{4+}	6		0.66
W^{5+}	6		0.62
W^{6+}	4		0.42
	5		0.51
	6		0.60
Xe^{8+}	4		0.40
	6		0.48
Y^{3+}	6		0.900
	7		0.96
	8		1.019
	9		1.075
Yb^{2+}	6		1.02
	7		1.08
	8		1.14
Yb^{2+}	6		0.868
	7		0.925
	8		0.985
	9		1.042
Zn^{2+}	4		0.60
	5		0.68
	6		0.740
	8		0.90
Zr^{4+}	4		0.59
	5		0.66
	6		0.72
	7		0.78
	8		0.84
	9		0.89

0.592 Å. In the anions the added electron increases the radius by a few per cent, e.g. $r_0(Cl) = 0.725$ Å, $r_0(Cl^-) = 0.742$ Å and $r_0(Br) = 0.851$, $r_0(Br^-) = 0.869$ Å.

2. The dimensions of the atoms decrease along the period (with n a constant), the alkaline metals having the largest dimensions, and increases sharply along the group from top to bottom (with n a variable). This occurs because the dimensions of an atom depend both on n and on Z_{eff}, which act on the atomic radius producing contrasting effects. Thus an increase in n tends to increase the atomic volume (the most probable radius increases with n, see Fig. 6.2), while an increase in Z_{eff} tends to contract orbitals. For example, the Z_{eff} calculated with Slater's principles for Na and Cl are respectively 2.20 and 6.10, while for Li and Cs they are 1.3 and 2.2. In the first case the sharp increase in Z_{eff}, which is not countered by n (which is constant), induces a decrease in the dimensions from Na to Cl. In the second case the regular increase in n, hardly countered at all by the effects of Z_{eff}, induces a notable increase in the atomic volumes from Li to Cs. Obviously this also follows for the ionic radius.

3. The ionic radius varies with the coordination number (CN). Note that the repulsion forces increases with the CN, which in turn increases the effective ionic radius r_e of the ion of the opposite charge (the cation). The lattice energy allows this trend to be quantified approximately. (6.15) can be rewritten as:

$$E = -\frac{A_M z_1 z_2 e^2}{R} + \frac{\text{CN}B}{R^n} \tag{6.22}$$

where A_M is Madelung's constant (which has taken the place of 1.39) and CN, which is defined below, is the coordination number (which has taken the place of the number 2, that is the two anions coordinated by the cation, see Fig. 6.5 and eqn (6.13)).

For $R = R_e$, $dE/dR = 0$, from which we can work out:

$$R_e = \left(\frac{\text{CN}}{A_M}\right)^{1/n-1} \left(\frac{nB}{z_1 z_2 e^2}\right)^{1/n-1}. \tag{6.23}$$

Let us suppose that ions of opposite charges can crystallize just as well into a CsCl type structure as an NaCl (halite) or ZnS (wurtzite) one. Assuming that the intermediate value $n = 9$, we obtain:

$$\frac{R_1(\text{CsCl})}{R_2(\text{NaCl})} = \left(\frac{(A_M)_2}{(A_M)_1} \times \frac{\text{CN}_1}{\text{CN}_2}\right)^{1/8} = \left(\frac{1.748}{1.763} \times \frac{8}{6}\right)^{1/8} = 1.035 \tag{6.24}$$

$$\frac{R_3(\text{ZnS})}{R_2(\text{NaCl})} = \left(\frac{1.748}{1.641} \times \frac{4}{6}\right)^{1/8} = 0.958. \tag{6.25}$$

The result of (6.24) indicates that the inter-ionic distances of a CsCl type lattice (R_1) are ≃3 per cent greater than those of an NaCl type lattice (R_2), while (6.25) shows that those of a ZnS type lattice (R_3) are smaller than it by ≃4 per cent. Since the anion radii remain more or less the same, the variations occur almost entirely in the cation radius.

4. The ionic radius depends on the electronic spin state. There is a close relationship between ionic radius and spin state in transition metals. In fact,

as explained on p. 415, Δ_0 and Δ_t depend on the type of ligands, and the coordination polyhedra which the ligands form around the transition metal. If, for example, Δ_0 is high, then the first six electrons tend to occupy the lower energy level orbitals, while if Δ_0 is low they occupy both energy levels (Fig. 6.7). These differences in turn affect the ionic radii of the transition metals which have from four to seven electrons in their d orbitals. To be more precise, electrons with high spin have a larger radius than those with low spin (Fig. 6.9).

5. The ionic radius is influenced by **polarization**. Two ions of opposite charges, if they are far enough apart, have coinciding barycentres of their positive and negative charges. When the ions are brought together a distortion of the electronic charge known as polarization, takes place, which is greater in the anion. Combination requires a sort of participation on the part of the electronic charge, and therefore a certain degree of polarization, or covalent bonding.

In accordance with Fajan,[20] the polarization depends on:

(a) the dimensions of the cation and its charge;

(b) the dimensions of the anion and its charge;

(c) the deviation of the electronic configuration from that of a noble gas.

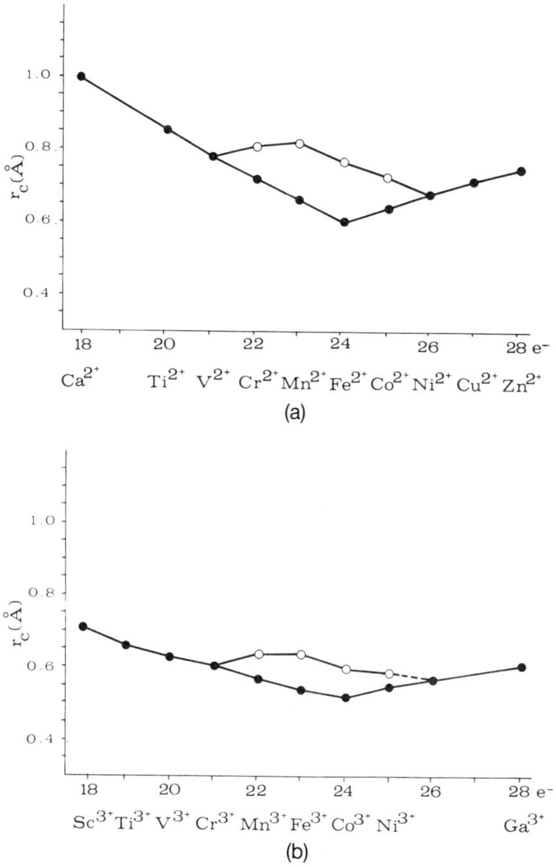

Fig. 6.9. Dependence of the effective ionic radii on the spin state in transition metals of the 1st series: (a) = bivalent; (b) = trivalent. The full circles refer to low-spin configuration, the open circles refer to high-spin configuration.

This means, for example, that for alkaline metals the polarization power increases according to the following scale: $Li^+ > Na^+ > K^+ > Rb^+ > Cs^+$ (ionic potential, $\varnothing = Z/r$, increases from right to left).

Regarding the anions, the greater the dimensions of their outermost electrons, the weaker are their bonds, due both to the greater distance from the nucleus and to the increasing shield effect. It follows that the polarizability of halogens varies according to the scale: $I^- > Br^- > Cl^- > F^-$, analogously with that of halcogens. Anions with a high negative charge, such as As^{3-} and P^{3-}, are particularly easily polarized. Moreover, it has been shown above how the different peripheral electrons are shielded by the s, p, and d type electrons. Therefore, although Ni^{2+} has approximately the same ionic radius as $Mg,^{2+}$ it polarizes more than Mg^{2+} one extra contiguous anion: this effect is even more noticeable between Cu^+ and Na^+.

Maximum filling principle

The lattice energy of N_0 pairs of ions can be expressed as the sum of the potential energy contributions (u_{ij}) relative to all the possible pairs whose components are distant $\boldsymbol{r}_{ij}$:

$$U_L = \frac{1}{2} \sum_{i,j=1}^{2N_0} u_{ij}(\boldsymbol{r}_{ij}) \quad (\text{with } i \neq j). \tag{6.26}$$

The lattice energy is minimal if the total potential energy is minimal, this occurs when:

(1) the interatomic distances are as close as possible to the equilibrium distances (R_e);

(2) the number of ions at, or very nearly at, the equilibrium distance is as high as possible.

These conditions are general and apply to crystals having simple structural units, or complex units connected to similar units by central or near central forces of attraction.

The **maximum filling principle** can be expressed as follows: **'Simple** (atoms or ions) **or complex** (groups of atoms or ions) **structural units on which central or near central attractive forces act tend to increase contact with each other to the maximum, whilst reducing their distances to a minimum'**.

By supposing that the ions behave as if they were rigid spheres on which central, or almost central, forces are acting, we can hope to:

(1) predict the most probable arrangement of the anions which surround the cation (coordination polyhedra), by means of a simple geometric rule (radius ratio rule);

(2) describe certain atomic building by means of dense mechanical models based on rigid spheres of equal radius (closest packings).

Coordination polyhedra

It is sometimes convenient to illustrate crystal structures by means of coordination polyhedra, in order to underline the geometrical relationships which characterize structural frameworks. Often such polyhedra continue to exist beyond the crystal state as definite physico-chemical realities, e.g. $[SiO_4]^{4-}$ tetrahedra in silicate melts, or $Fe(H_2O)_{6-n}(SO_4)_n$ octahedra where $1 < n < 1.3$, in aqueous solution of iron sulphate.[21]

The term coordination is used to refer to the atoms or ions which surround a central atom or ion. If it is not otherwise specified below, it may be assumed that the central ion is the cation (M) surrounded by the anions (X).

The Xs at a tangent to M are known as the first nearest neighbours of M and constitute the **first coordination shell**. The **coordination number** (CN) is equal to the number of the first nearest neighbours, i.e. to the number of bonds formed by an atom. The next Xs, at a tangent to the first nearest neighbours, are known as the second nearest neighbours of M, and form the **second coordination shell**, and so on. Sometimes the first coordination shell is not easily distinguished from the second. In such cases the uncertainty is expressed by attributing two numbers to the CN, e.g. 6 + 2. The distances d_{MX} are all equal when M occupies certain symmetrical positions, e.g. M in $\bar{3}$, with CN = 6; otherwise the d_{MX} vary, if only a little, from their average.

The **coordination polyhedron** of M can be obtained by 'joining the dots', i.e. all the centres of X. Some of the possible coordination polyhedra are shown in Figs. 6.10 and 6.11. They can be found in structures as isolated polyhedra, and/or connected to polyhedra of the same or another type.

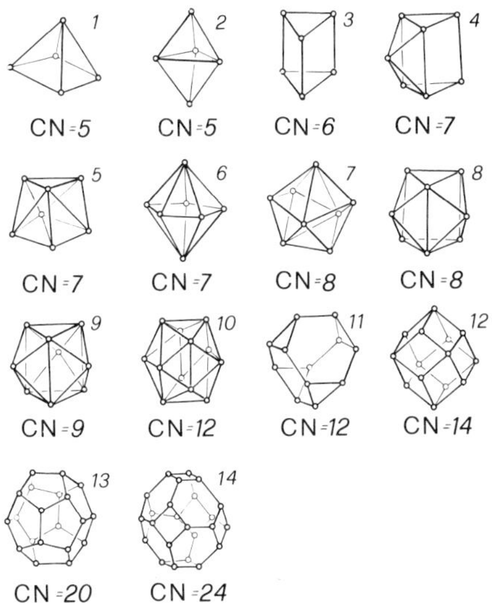

Fig. 6.10. Some coordination polyhedra. (1) tetragonal pyramid, (2) trigonal bipyramid, (3) trigonal prism, (4) one-cap trigonal prism, (5) seven-vertex polyhedron, (6) pentagonal bipyramid, (7) trigonal dodecahedron, (8) two-cap trigonal prism, (9) three-cap trigonal prism, (10) icosahedron, (11) truncated tetrahedron, (12) rhombododecahedron, (13) pentagonododecahedron, (14) truncated octahedron. The coordination number of each polyhedron is indicated below it.

Radius ratio rule

If the ions are considered as behaving like rigid spheres, an attempt can be made to predict by geometrical means which coordinations will be preferred. The parameter used for this purpose is $\rho_e = (r_M/r_X)$.

Figure 6.11 shows some of the most common regular geometrical dispositions and their relative ρ_e ratios. The ρ_e values for a given anion increases as r_M, and thus the number of contacts also increase. From Fig. 6.11 it can be deduced, for example, that within the range $0.155 \leq \rho_e < 0.225$ a triangular coordination is stable, since for $\rho_e < 0.225$ cations and anions are no longer adjacent in tetrahedral coordination; thus it is no longer possible. However, the results obtained by geometrical methods must be integrated with further energetic considerations.

Figure 6.12 shows the trend of U_L versus ρ_e for CsCl, NaCl, and ZnS type structures having the following ratios:

$$Cs^+:Cl^- = 8:8, \qquad Na^+:Cl = 6:6, \qquad Zn^{2+}:S^{2-} = 4:4.$$

This formalism not only indicates the coordination number around each ion, but also allows their stoichiometry to be deduced, which for the three

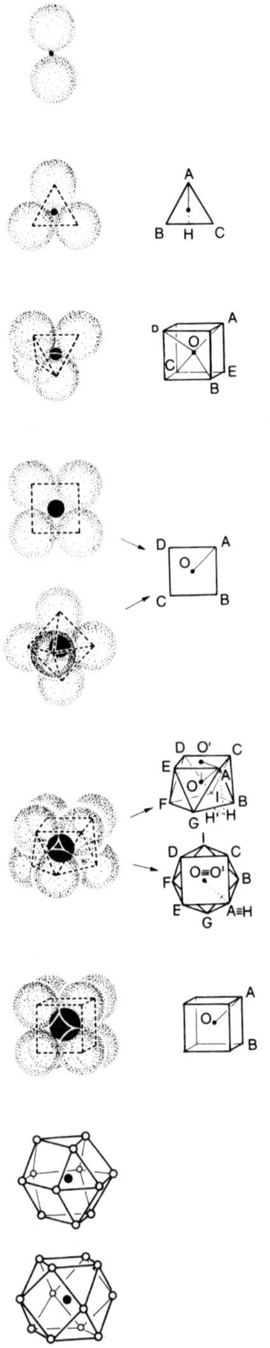

Anions	CN	Polyhedron	Relationships between r_M and r_X	ρ_e
	2	Dumb-bell	—	—
	3	Triangle	$\overline{AB} = 2r_X, \overline{AO} = r_M + r_X, \overline{AH} = r_X\sqrt{3}$; $r_M + r_X = 2/3(r_X\sqrt{3})$	0.155
	4	Tetrahedron	$\overline{AB} = 2r_X, \overline{AO} = r_M + r_X$; $\overline{AB} = \overline{AE}\sqrt{2}, \overline{AO} = \overline{AE}\sqrt{3}/2$; $\overline{AO}/\overline{AB} = \sqrt{3}/2\sqrt{2}$	0.225
	4	Square	$\overline{AB} = 2r_X, \overline{AO} = r_M + r_X$	0.414
	6	Octahedron	$2\overline{AO} = \overline{AB}\sqrt{2}$	0.414
	8	Square antiprism or Thomson's cube	$\overline{AB} = 2r_X, \overline{AO} = r_M + r_X, \overline{HH'} = (AD - FG)/2$; $\overline{AH'} = r_X\sqrt{3}, \overline{AH} = r_X\sqrt{2(2^{1/2})}, \overline{OO'} = \overline{AH}/2, \overline{O'C} = r_X\sqrt{2}$; $(\overline{OO'})^2 + (\overline{O'C})^2 = (\overline{AO})^2, (r_M/r_X)^2 + 2r_M/r_X - (\sqrt{2}+2)/2 = 0$	0.645
	8	Cube	$\overline{AB} = 2r_X, \overline{AO} = r_M + r_X$; $\overline{AO} = 2r_X\sqrt{3}/2$	0.732
	12	Hexagonal cuboc.	$r_M = r_X$	1.000
		Cuboctahedron		

Fig. 6.11. Other very common coordination polyhedra. At the side of each polyhedron there is its coordination number, name, and the procedure followed to calculate and ρ_e.

compounds in question is M:X = 1:1. From Fig: 6.12 it results that:

(1) the U_L trends are characterized by three discontinuity points in correspondence to the values $\rho_e = 0.732$, 0.414, 0.225;

(2) the U_L trends favour a CsCl type structure when $\rho_e > 0.71$, and an NaCl type when $0.71 > \rho_e > 0.32$;

(3) the U_L(NaCl) curve is closer to that of the U_L(CsCl) than to that of the U_L(ZnS).

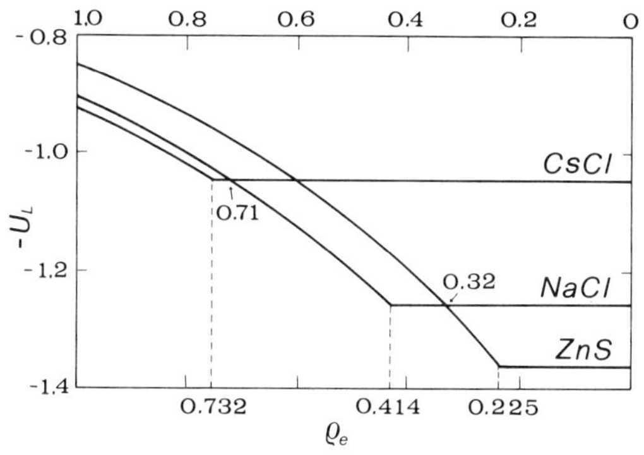

Fig. 6.12. U_L pattern versus ρ_e. From the point of view of the energy, the minimum ρ_e values for CsCl, NaCl, and ZnS type structures are lower than those predicted by geometrical means. The difference is most accentuated in the range between an NaCl and a ZnS type structure (0.32 instead of 0.41).

The reasons why these discontinuities exist become clearer on observing Fig. 6.13. In (b) an octahedron of $M(X)_6$ composition is shown, while in (a) and in (c) the same polyhedron for a larger cation ($M' > M$) and for a smaller one ($M'' < M$) are illustrated respectively. In accordance with (6.17), the value of U_L(NaCl) decreases gradually from configuration (a) to configuration (b), since $R_e < R'_e$. However, U_L does not vary in the range $0.32 < \rho_e < 0.414$, that is to say, between configurations (b) and (c), since the anions already adjacent in (b) cannot come any closer together. Thus R_e has the same value in both configurations (b) and (c). Configuration (d) is preferred to (c) if, in accordance with the U_L(NaCl) trend of Fig. 6.12, there is another cation $M''' < M''$ so that $\rho_e < 0.32$. In the same way, an NaCl type structure is likely to be preferred to a CsCl type when $\rho_e < 0.71$, i.e., for ρ_e values a little below the discontinuity point ($\rho_e = 0.732$).

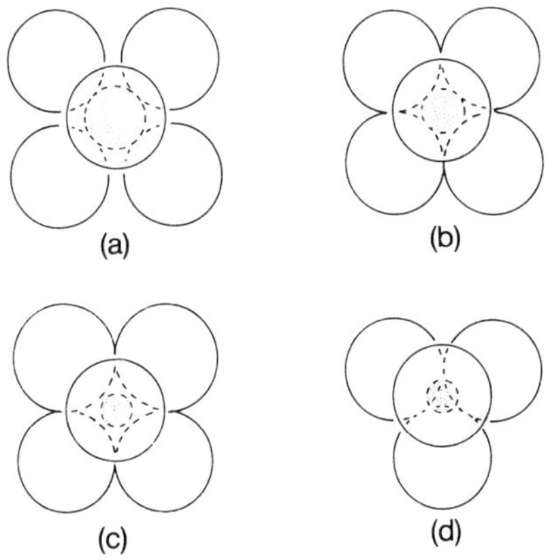

Fig. 6.13. (a), (b), and (c) illustrate the changing relationships between equal anions (white circles) in an octahedron, as the cations vary in size (broken circles). Configuration (c) becomes unstable and could evolve into configuration (d) if $\rho_e < 0.32$.

Applications of the concept of ionic radius

To assume that an atom is rigid, and also spherical, is obviously a gross simplification of the physical reality. Accurate studies[22] to determine $\rho(r)$ have shown, for example, that in MgO the electronic cloud of the oxygen only partially wraps the cation: therefore the terms sphere and ionic radius lose any meaning. Despite this, however, the concept of ionic radius often produces satisfactory results. Alkaline halides and, to a lesser extent, alkaline-earth metal oxides are the compounds to which a mechanical atomic model is most easily applied. Such compounds with an MX stoichiometry should have structures with a CN = 8, if the ionic radii r_e(M)

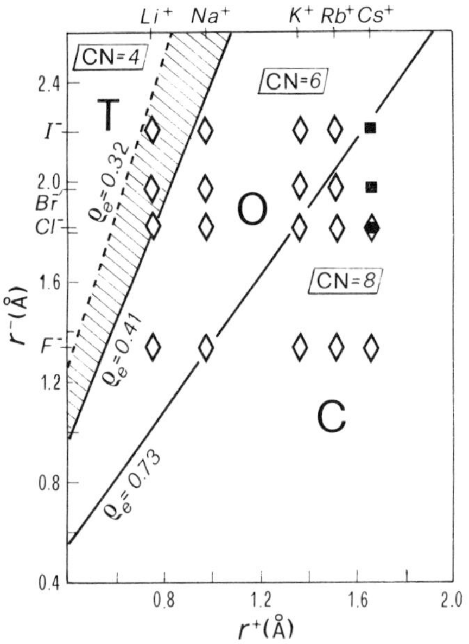

Fig. 6.14. Three fields are shown (T, O, C), delimited by two unbroken lines (ρ_e = 0.73, 0.41). In accordance with geometrical criteria, they represent stability areas for the alkaline halides. The broken line (ρ_e = 0.32) shows the reduced T field according to the energy indications of Fig. 6.12. The diamonds and squares show the type of coordination polyhedron on which the real halide structures are based: octahedra and cubes respectively.

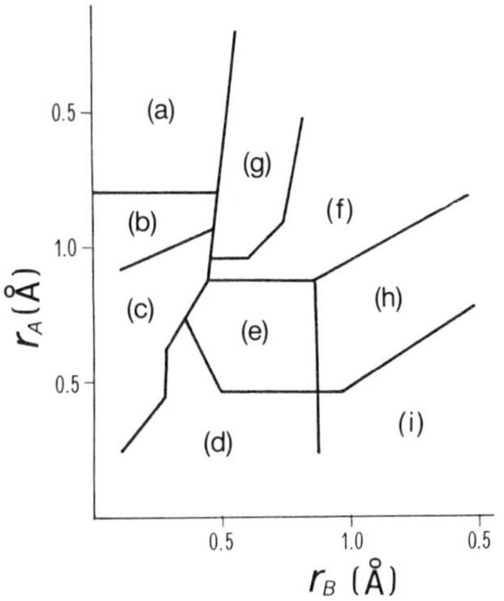

Fig. 6.15. Taking into account ionic radii r_A(A) and r_B(B) the stability fields for A_2BO_4 structures are shown. These fields conform to the following structure types: (a) β-K_2SO_4, (b) Na_2SO_4 **(thenardite)**, (c) (Mg, Fe)$_2$ SiO_4 **(olivine)**, (d) Be_2SiO_4 **(phenakite)**, (e) Al_2MgO_4 **(spinel)**, (f) Sr_2PbO_4, (g) K_2NiF_4, (h) Fe_2CaO_4, (i) Al_2BaO_4.

and r_e(X) are more or less the same. Otherwise, such compounds are governed by the cation coordination number (since the cation is generally smaller than the anion) and the stoichiometry. The dimensions of the cation depend on the CN (see p. 422). Therefore ambiguity remains as to the ionic radius dimensions when the structure (and thus the CN) is not known. Let us compare the theoretical data with the experimental data for the halides. 'Average' ionic radii for a CN = 6 are used, for example, CsBr (r_M = 1.67, r_X = 1.96 Å) forms a compound Cs^+: Br^- = 8:8, in accordance with both the radii ratio: (ρ_e = 0.85) and the stoichiometry (M:X = 1:1). Also for NaCl both the ρ_e = 0.53 and the stoichiometry indicate a compound Na^+:Cl^- = 6:6, in accordance with the results of structural analysis.

The rule of the radius ratio predicts M:X = 4:4 structures in the higher field T ($\rho_e < 0.41$), M:X = 6:6 in the intermediate field O ($0.41 \leqslant \rho_e < 0.73$), and M:X = 8:8 in the lower field C ($\rho_e > 0.73$). Experimental data, on the other hand, shows that almost all the structures are of M:X = 6:6 type, two are of M:X = 8:8 type (CsI, CsBr), and one, CsCl, can crystallize in either type of coordination.

In Fig. 6.14 a broken line is shown for ρ_e = 0.32, which corresponds to the point of intersection of the NaCl and ZnS curves in Fig. 6.12. The first anomaly involves the T field. The incongruence between the observed structure M:X = 6:6 and the predicted structure M:X = 4:4 by the radius ratio rule is only apparent. Considering Fig. 6.12, NaCl type configurations are stable for $0.32 < \rho_e < 0.41$, i.e. the values observed. The other anomaly involving the lower field, C, is constituted by M:X = 6:6 structures and can be explained by considering not only the interactions between ions of opposite charge, but also those between ions of similar charge in the CN function, and the geometric relationships between the coordination polyhedra (see p. 435).

However, if $r_e(^{IV}Li^+)$ were used for LiI, ρ_e = 0.27 would be found to be the right value. The configuration Li^+:I^- = 4:4 would gain potential energy and would therefore seem likely to be preferred, which is contrary to the experimental data.

Another example can be illustrated by means of Fig. 6.15, as deduced by Phillips.[23] The 'fields of stability' for the most important structures of A_2BO_4 composition are shown. The diagram was obtained empirically on the basis of about 130 structures using the effective radii of the cations in octahedral (r_A) and tetrahedral (r_B) coordinations. Such a diagram makes it possible to work out the field in which a certain A_2BO_4 structure prevails, even if some borderline cases may be contrary to the general trend. Improvements in structural predictions can be obtained by considering not only geometrical factors but also chemical factors, such as polarization.

Graphs for compositions of M_nX_m type have been obtained by plotting the average quantum number n (which indicates roughly the dimensions of the ions) on the y axis and the electronegativity difference, ΔEN (which indicates the degree of polarization of the bond),[24] on the x axis. The prediction is improved if $\rho_e\Delta$EN is used instead of ΔEN.

Finally using effective mean ionic radii, several authors[25–28] have found quantitative relations between crystallographic and compositional parameters in garnets. So according to Basso,[25] it is possible to calculate the oxygen fractional coordinates, the metal–oxygen distances, and the cell edge of oxide garnets and hydrogarnets as follows:

$$\begin{aligned}
X &= 0.0258(19)r(X) + 0.0093(26)r(Y) - 0.0462(15)r(Z) \\
&\quad - 0.0008(1)(OH) + 0.0171, \\
Y &= -0.0261(25)r(X) + 0.0261(34)r(Y) + 0.0310(20)r(Z) \\
&\quad + 0.0009(1)(OH) + 0.0514, \\
Z &= -0.0085(22)r(X) + 0.0323(30)r(Y) - 0.0237(17)r(Z) \\
&\quad - 0.0011(1)(OH) + 0.6501, \\
X(1)\text{–}O &= 0.558(38)r(X) + 0.298(52)r(Y) + 0.244(30)r(Z) + 0.013(1)(OH) \\
&\quad + 1.477, \\
X(2)\text{–}O &= 0.702(28)r(X) + 0.083(32)r(Y) - 0.002(1)(OH) + 1.673, \\
Y\text{–}O &= 0.154(20)r(X) + 0.754(38)r(Y) + 0.132(22)r(Z) - 0.002(1)(OH) \\
&\quad + 1.316, \\
Z\text{–}O &= 1.026(17)r(Z) + 0.025(1)(OH) + 1.366, \\
a_0 &= 1.695(98)r(X) + 1.670(93)r(Y) + 2.389(77)r(Z) \\
&\quad + 0.057(3)(OH) + 8.428,
\end{aligned}$$

where $r(X)$, $r(Y)$, and $r(Z)$ are the mean ionic radii respectively of cations 8, 6, and 4 coordinated, and (OH) are the number of OH groups present in the formula unit (f.u.) of garnets:

$$^{8}X_3{}^{6}Y_2(SiO_4)_{3-p}(H_4O_4)_p (0 \leq p < 3)$$

Closest packings

Most metals, some compounds of the type MX, MX_2 etc., and others with a more complex stoichiometry, take on structures which can be described by means of closest packings of rigid spheres of equal radius. Let us denote by **packing coefficient** (c_i) the ratio:

$$c_i = \sum (V_a/V_c) \tag{6.27}$$

where $\sum V_a$ is the volume of the atoms or anions contained in the elementary cell of volume V_c. For discussion of the concept of c_i see Appendix 6.A. According to the maximum filling principle, $c_i \rightarrow$ max. **Hexagonal closest packing** (HCP) and **cubic closest packing** (CCP) satisfy this requirement. When analysing their characteristics, it should be borne in mind that in the common structures, ions, atoms, or groups of either, replace rigid spheres forming the so-called closest-packing structures. The basis of the HCP and CCP dispositions is a layer of spheres (layer A) packed as closely as possible similar to that illustrated in Fig. 6.16.

This layer has the following characteristics: every sphere is in contact with another six spheres; between every three spheres there is a space, known as a hole, every sphere is surrounded by six holes, one third of each of which belongs to it, so a total of two holes can be associated with each sphere. Such holes, which are all alike from a physical point of view, are indicated by b (white) and c (black). Spheres belonging to the next layer may correspond to the b or the c holes of layer A, and are thus said to belong to

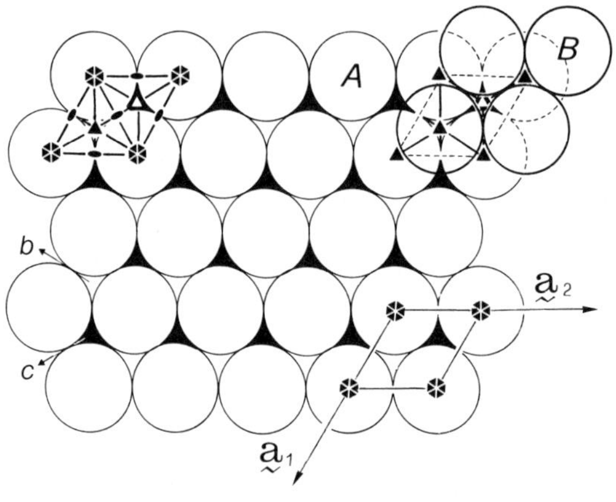

Fig. 6.16. A compact layer of spheres characterized by triangular cavities indicated by white spaces (b) and black spaces (c). At the top of the sketch on the left is shown the symmetry of the layer A (6m), while on the right is shown the symmetry of a pair of layers AB (3m).

layer B or layer C. With reference to a hexagonal mesh of sides a_1 and a_2, in general a sphere may correspond to any of three possible holes: a (for layer A) belonging to site $(0\boldsymbol{a}_1 + 0\boldsymbol{a}_2 + n\boldsymbol{h})$ b to site $((2/3)\boldsymbol{a}_1 + (1/3)\boldsymbol{a}_2 + n\boldsymbol{h})$, and c to site $((1/3)\boldsymbol{a}_1 + (2/3)\boldsymbol{a}_2 + n\boldsymbol{h})$, where n is a whole positive number including zero and $\boldsymbol{h}$ is a vector perpendicular to the compact layer of a module h (which is equal to the distance between two successive compact layers). HCP is characterized by an ABAB... type sequence, while CCP is characterized by an ABCABC... type sequence.

The two characteristic sequences which represent HCP and CCP and their orientation with respect to the compact layers, are illustrated in Fig. 6.17.

The HCP cell is projected in the direction [001] $\parallel \boldsymbol{h}$, while the CCP cell is projected in the direction [111] $\parallel \boldsymbol{h}$. An HCP cell contans two spheres (that of a simple hexagonal cell contains one); a CCP cell contains four, as does a face-centred cubic (FCC) lattice (for this reason it is possible to use FCC or CCP with the same meaning). It is possible to identify four types of polyhedra in both packings: cuboctahedra, octahedra, tetrahedra, and triangles. The first, illustrated in Fig. 6.11, can be traced by joining the centres of the 12 spheres around a central sphere of equal radius. The other three can be traced by joining the centres of the spheres which correspond to the three possible holes of the closest packings: octahedral, tetrahedral, and triangular.

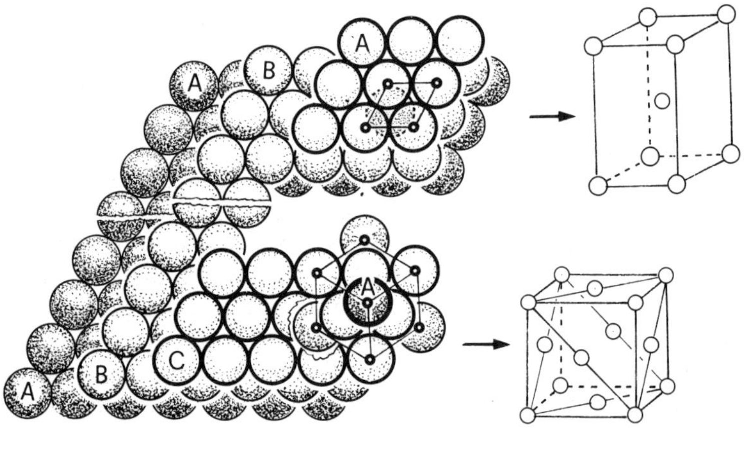

Fig. 6.17. An ABAB... sequence of layers determines a compact hexagonal cell (top), while an ABCABC... sequence determines a compact cubic cell (bottom).

Many cations have a radius of just the right size, compared with that of the oxygen atoms, that they can occupy tetrahedral or octahedral holes. Let us examine, therefore, these two types of holes more closely in order to find out what is their quantitative ratio to the spheres. In Fig. 6.18 a layer A (black spheres) and a layer B (white spheres) are shown.

Supposing that the sequence is an ABAB . . . type, a sphere marked S (belonging to layer A and not visible in the figure) is surrounded by a total of 12 spheres (six of A and six of two B layers, above and below A). It contributes to the formation of:

(1) six octahedral holes (O) three of which are shown in the figure; the other three can be obtained by a symmetry plane m which passes through A;

(2) eight tetrahedral holes, three of which, shown in the figure, are in correspondence with the three spheres marked with a thick T and one of which is in correspondence with the sphere marked S and three of the spheres marked with the same T; the other four can be obtained by a symmetry plane m which passes through A.

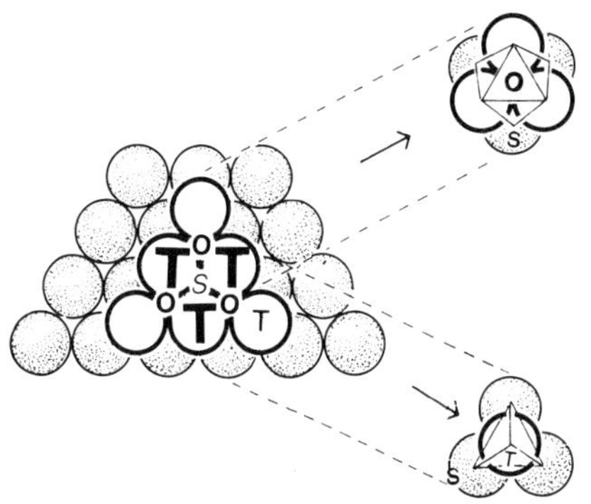

Fig. 6.18. Octahedral, O, (top right) and tetrahedral, T, (bottom right) holes with relative coordination polyhedra drawn out from two following compact layers.

Since six spheres are needed to form an octahedral hole and every sphere helps to form six holes, obviously there is a ratio of one hole to each sphere. Four spheres are needed to form each tetrahedral hole, while each sphere helps to form eight tetrahedral holes, so there is a ratio of two tetrahedral holes to each sphere.

A layer of octahedral holes (OL) is enclosed by two layers of spheres, as are two layers of tetrahedral holes (TL), through they are orientated in different directions (see Fig. 6.32). This situation is common both to HCP and CCP, consequently a double layer of spheres has the following characteristics in both HCP and CCP:

1. Octahedra which share edges.
2. Tetrahedra which, differently orientated (TL_1 and TL_2), share edges.
3. Octahedra and tetrahedra which share faces.

The differences between HCP and CCP are in the geometric elements shared by the polyhedra in the direction perpendicular to the layers; these are the differences which concern us most here. HCP has all the octahedra and half of the tetrahedra sharing faces with polyhedra of the same type, the other half of the tetrahedra sharing vertices, and finally the octahedra and tetrahedra sharing vertices and edges. Instead, CCP has (Fig. 6.19) all the octahedra sharing faces with half of the tetrahedra, the other half of the tetrahedra sharing vertices, and finally the octahedra which share edges, as do half of the tetrahedra.

It follows that, for example, while in HCP the octahedra share faces in the direction perpendicular to the OL and edges parallel to the OL, in CCP the octahedra share edges only. Such differences are very important for certain stoichiometries[29] as we shall see below, since they affect the stability of the structure and thus which type of packing is preferred.

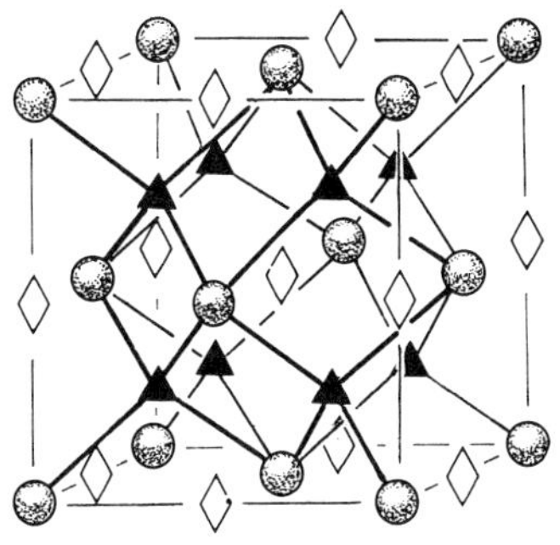

Fig. 6.19. An FCC cell. The tetrahedral (black triangles) and octahedral (white diamonds) holes of a CCP are shown. Note that the number of spheres (hatched circles) belonging to the cell is equal to that of the octahedral holes and half of that of the tetrahedral holes.

There are two coefficients which characterize closest packings. The first is c_i (6.27), which has a value of 0.74 both for HCP and CCP. It is easily found, remembering that an elementary HCP cell contains two spheres

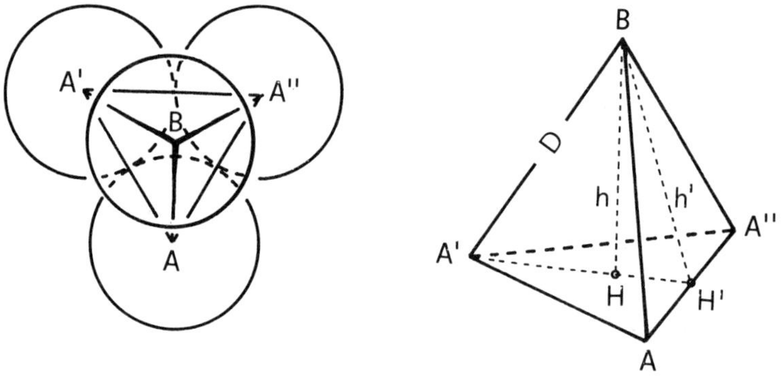

Fig. 6.20. The centres of the three spheres marked A and the centre of the sphere marked B together form a trigonal pyramid. It can be deduced from this that: $\overline{BH'} = D\sqrt{3}/2$, $\overline{HH'} = 1/3(\overline{BH'}) = D\sqrt{3}/6$, $h = D\sqrt{6}/3$, $h/D = \sqrt{6}/3 = 0.8165$.

while a CCP one contains four (Fig. 6.17). For example, the side of an elementary CCP cell will be $a_0 = D\sqrt{2}$, where D is the diameter of a sphere. The volume of a sphere is $V_a = 4.189(D/2)^3$ from which $c_i = 4V_a/a^3 = 0.74$.

The other parameter is the ratio h/D. Since $h = [(D\sqrt{3}/2)^2 - (D\sqrt{3}/6)^2]^{1/2}$ (see Fig. 6.20) it follows that $h/D = 0.8165$. For example, in HCP, $c_0 = 2h$, therefore $c_0/D = 1.63$.

Combinations of HCP and CCP can create an indefinite number of closest-packing structures. The repetition period, nh, will be in the range $2h \leqslant nh \leqslant p_c$, where p_c is the maximum dimension of the crystal perpendicular to the closest-packed layers which the sequence can reach when the structure is completely disordered.

As n increases, the number of possible sequence combinations grows very rapidly. For example, there is only one possible packing for $n = 4$: ÅBACÅ; for $n = 6$ there are two: ÅBCACBÅ and ÅBABACÅ; but for $n = 20$ the number of possible packing rises to 4625. One of the following ways can be chosen to derive the possible packings: the first is to count all the possible sequences once having generated them;[30] the second is to obtain them by means of combinatorial analysis.[31] Figure 6.16 shows the symmetry of a closest-packed layer (6mm) and of a pair of such layers (3m). The minimum symmetry content common to all closest packings is P3m1. Other symmetries are represented by all those space groups which can have as a subgroup P3m1 (see also Fig. 1.E.1), i.e. P$\bar{3}$m1, P$\bar{6}$m2, P6_3m*c*, P6_3/mm*c*, R3m, R$\bar{3}$m, Fm$\bar{3}$m. For HCP and CCP the space groups are respectively: P6_3/mm*c* and Fm$\bar{3}$m.

When real structures are considered the symmetry may be reduced, due not only to the arrangement of the cations which occupy the various holes, but also to the distortions which the cations produce in the lattice. Therefore, some structures belonging to systems different from those just examined can be described as closest packings, distorted to a greater or lesser degree.

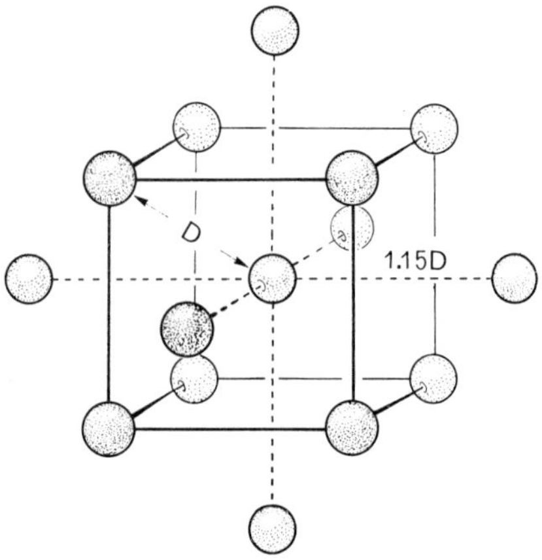

Fig. 6.21. A BCC cell. The central sphere is surrounded by eight spheres at a distance of D and six more distant spheres at $1.15D$ ($2D = l\sqrt{3}$, where l is the side of the cell).

There are other close packings characterized by spheres with fewer next neighbours. These are the PTP (primitive tetragonal packed), the BCT (body-centred tetragonal), and the BCC (body-centred cubic). The first has $c_i = 0.72$ and CN = 11, the second (not found in the known structures) has $c_i = 0.70$ and CN = 10, the third (taken on by several structures) has $c_i = 0.68$ and 8 + 6 spheres of the same type coordinated around a central sphere, eight at a distance of D and six at a distance of $1.15D$ (Fig. 6.21). HCP and CCP, on the other hand, have 12 spheres all at the same distance from the central sphere.

Pauling's rules

Ionic bonds play an important role in the building up of structures for the vast majority of minerals. In fact, even the groups $(SiO_4)^{4-}$, $(SO_4)^{2-}$, $(CO_3)^{2-}$, which have prevalently covalent bonds internally, can be seen as large negative ions bonded to the cations by essentially ionic interactions. The general principle of the minimization of potential energy can be applied to such compounds. The Pauling rules are derived from this principle.

Pauling's first rule

This rule synthesises the geometric relationships which tend to be established between cations and anions in a stable structure, introducing the concept of the ionic radius.

A coordination polyhedron is formed around each cation. The distance between the cation and the anion is determined by the sum of the radii, while the coordination number of the cation is determined by the radius ratio.

The maximum number of anions, compatible with the equilibrium conditions imposed by the minimum potential energy, surround the cation. Too many anions would increase the cation–anion distance, thus making the structure less stable than if the distances are very close to R_e. The optimum distance is the sum of the ionic radii: $r_M + r_X$. However, the cation–anion distances do not depend only on the ionic radii of the cation and anion involved, but also on the average radius of each of the other cations.[32] In structures of a certain complexity, it can happen that not all the cations have optimum CN and d_{MX} distances. In such cases some cations, above all the largest ones with the lowest valence (formal charge), are the first to take on higher-energy coordinations, followed by smaller higher-valence cations which tend to form more rigid coordinations.

It follows that the lowest coordination expected for Na^+ (which taking into account the oxygen has $\rho_e = 0.71$) is 6, but in **sodalite**, $Na_8Cl_2[Si_6Al_6O_{24}]$, it has CN = 4 ($3O^{2-} + 1Cl^-$), while Si^{4+}, half of which is substituted by Al^{3+}, has its usual CN, i.e. 4.

Pauling's second rule

Known as the **valence sum rule**, this rule analyses the electrostatic balance:

The valence of an anion in a stable ionic structure tends to compensate the strengths of the electrostatic bonds which reach it from the cations situated at the centres of the polyhedra of which the anion is a vertice, and vice versa.

In practice, this rule means that the electrostatic balance is satisfied by the cations belonging to the primary coordination.

If $-V$ is the valence of an anion, v_i and $(CN_c)_i$ are respectively the valence and the coordination number of the ith cation, then we can define the **electrostatic bond strength** as $s_i = v_i/(CN_c)_i$. The rule can be summarized as:

$$V = \sum_i s_i \qquad (6.28)$$

where i is extended to all the cations coordinated by the anion in question.

For high-symmetry, simple ionic structures it is possible to give a single value to the bond strengths and to find the valence of a given ion by multiplying the bond strength i times. For example, for NaCl (space group Fm$\bar{3}$m), the bond strength which reaches a given Cl^- from a given Na^+ is 1/6. Since there are 6 Na^+ surrounding each Cl^-, all equidistant from it follows:

$$V = \sum_{i=1}^{6} (1/6)_i = 1.$$

The same procedure can be applied as a rough guide to **rutile**, TiO_2 (space group P4_2*mnm*) whose structure is based on slightly distorted TiO_6 octahedra. Each Ti^{4+} cation coordinates six O^{2-} and around each oxygen there are three Ti^{4+}, This means that each oxygen receives 4/6 of the valence of titanium. Since very O^{2-} is surrounded by three Ti^{4+}, then:

$$V = \sum_{i=1}^{3} (2/3)_i = 2$$

which is exactly the opposite of the oxygen valence.

In structures with irregular polyhedra, like **brookite** (a polymorph of the compound TiO_2) which is characterized by strongly distorted TiO_6 octahedra, this simple model needs to be modified slightly. In order to calculate the strength of each electrostatic bond, or more correctly the **bond valence** (s), Brown and Shannon[33] proposed the following equation:

$$s = s_0(R/R_0)^{-N} \tag{6.29}$$

where s_0, R_0, and N are empirical constants. For $s_0 = 1$, $R_0 = R_1$, the parameter R_1 represents the expected bond length when the valence is assumed to be one. In such as a case (6.29) becomes:

$$s = (R/R_1)^{-N}. \tag{6.30}$$

By inserting the correct values for R_1 and N for each cation[34] and the bond distance R, which can be deduced from the structure, s can be determined. For example Na^+, Mg^{2+}, Al^{3+}, Si^{4+}, P^{5+}, and S^{6+} have values of $R_1 = 1.622$ Å and $N = 4.290$. The experimental bond valence can be obtained by applying (6.30). The agreement between the experimental and the theoretical valence is very good, generally the difference does not exceed 5 per cent. If the agreement is not satisfactory for an ionic structure it can be ascribed to the following:

(1) the structure has not been correctly determined;

(2) the structure has not been interpreted properly, or certain bonds have been overlooked, or the wrong atoms have been ascribed to certain sites (e.g. Si^{4+} in place of Al^{3+} or vice versa).

Amongst the various applications of this method[35] it is often used by crystallographers to distinguish O^{2-} from $(OH)^-$ and H_2O in some complex mineral structures.[36] Calculated bond valences can be used through (6.28) to calculate experimental atomic valence and so the oxidation state of an atom.[37]

Pauling's third rule

This summarizes, from a qualitative point of view, the influence that the fact that the coordination polyhedra share geometric elements has on the potential energy and thus on the stability of the structure:

The sharing of sides and, in particular, of faces between coordination polyhedra reduces the stability of a structure. This effect is accentuated where there are cations with high valence and low coordination number.

In order for the potential energy to be minimal it is necessary that:

(1) the distances between the cations should be as great as possible, compatible with the existence of the structure itself,

(2) the cations should be shielded as much as possible from each other.

Figure 6.22 illustrates the distances between cations for certain regular polyhedra. As can be seen, the fulfilment of conditions (1) and (2) are less likely, for any given pair of polyhedra, when they share an edge than when they share only a vertice, and least likely when a face is shared. The least probable of all the configurations are two tetrahedra which share an edge, and above all two tetrahedra which share a face. This explains why in silicates the tetrahedra $[SiO_4]^{4-}$ are found either isolated or sharing vertices; they hardly ever share edges and never share faces. For tetrahedra containing cations with a higher charge, like $[PO_4]^{3-}$, even the likelihood

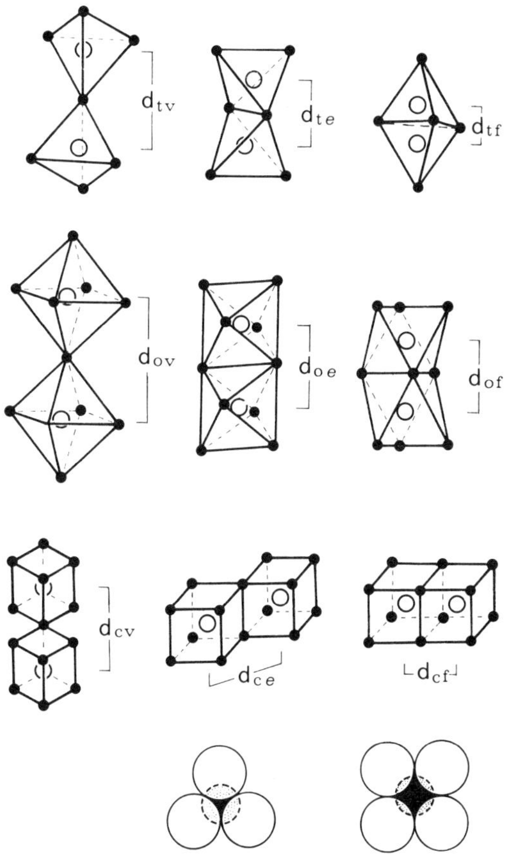

Fig. 6.22. Geometrical relationships between some regular polyhedra. Considering the oxygens as touching rigid spheres with a radius of $r_e = 1.40$ Å, the following inter-cation distances can be calculated: $d_{tv} = 3.43$, $d_{te} = 1.98$, $d_{tf} = 1.14$, $d_{ov} = 3.96$, $d_{oe} = 2.80$, $d_{of} = 2.29$, $d_{cv} = 4.85$, $d_{ce} = 3.96$, $d_{cf} = 2.80$ Å. The sketch at the bottom shows the different shielding effects produced by three or four anions together on the underlying cation.

that they will share vertices is reduced, so much so that in the case of group having yet higher cation charges, like $[SO_4]^{2-}$, that it is almost nil.

Figure 6.22 shows that $d_{oe} = d_{cf}$. However, in octahedral coordinations the shielding of the two cations is more accentuated. This explains why numerous halites tend to keep the NaCl structure rather than that of the CsCl, favoured by ρ_e. Structures like CsCl are only possible if the anions are large enough to keep the cations far enough apart, as can be observed in the case of CsCl, CsBr, CsI (in the NaCl structure type, the distance between two cations for CsF is 4.24 Å and 4.92 Å for CsCl). It should be remembered here that, in accordance with the indications of Fig. 6.22, ionic compounds with an MX stoichiometry (see p. 437) crystallize not with an HCP lattice, but with a CCP lattice (e.g. NaCl, MgO, CaO, etc.). This is because the energy levels involved in the formation of the two packings seen along ***h*** is different: in the case of CCP the octahedra MX_6 share edges, while in the case of HCP they share faces.

On the basis of what has been said, the question of how to realize the relative stability of certain TiO_2[38] polymorphs can be understood. This compound has three polymorphic modifications: **rutile, brookite,** and **anatase.** The structural differences concern mainly the way in which the TiO_2 octahedra are linked to one another. In rutile, brookite, and anatase, each octahedron shares respectively two, three, and four edges with a similar number of other octahedra. The structural stability of brookite is less than that of rutile, while anatase is the least stable, as predicted by the third Pauling rule.

Pauling's fourth rule

The fourth rule is an obvious corollary of the third, and so no further comment is necessary:

In a crystal containing various cations, those with high valence and a low coordination number tend not to allow the coordination polyhedra which they form to share any features.

Pauling's fifth rule

The fifth rule, known also as the parsimony principle, states that:

The number of essentially different constituents in a crystal tends to be small.

This rule summarizes the common observation that most of the non-fluid substances that surround us are crystalline and not amorphous. This means that the number of structural units, having considerable differences for instance with regard to volume and chemical properties, tends to be low for any given substances.

Source applications of these rules together with symmetry and other crystallochemical parameters are given in Appendix 6.B.

Ideal and defect structures

The preceding paragraphs have dealt mainly with the principles which underlie the structural building of ionic compounds.

Some simple, or at least not very complex, atomic configurations, characterized at most by few dozen atoms per elementary cell, will now be examined. Particular attention will be given to **ideal structures**, i.e. those structures in which it is possible to single out an elementary cell 100 per cent representative of all the cells that make up the crystal. In such cases all the crystallographic sites which are symmetrically equivalent (including the translation) must be occupied by the same chemical species and the resulting compound is defined as an **ideal chemical formula.**

In **defect structures** the relative positions of the atoms are not regular throughout the crystal. Thus the representative cell is a statistical cell which expresses the probability that a certain site will be occupied by a given atom. In such cases the compound is better defined by a **crystallochemical formula**, which, apart from taking into account the stoichiometry, considers also the percentage of presences (**occupancy**) of a certain chemical element in a given crystallographic site. For example, the compound $MgAl_2O_4$ (**spinel**) parent of the spinels, is a **normal spinel** (see p. 442) and adopts the following ideal structure: Mg^{2+} occupies only the tetrahedral sites, while Al^{3+} occupies the octahedral ones. The compound Fe_3O_4 (**magnetite**) is instead an **inverse spinel** (see p. 442) in which Fe^{2+} occupies only octahedral sites, while Fe^{3+} occupies half tetrahedral sites and half octahedral ones.

Therefore the structure of magnetite can be represented by a particular cell of statistical significance having its tetrahedral sites occupied by Fe^{3+} and its octahedral sites occupied by 0.5 Fe^{2+} and 0.5 Fe^{3+}. This structural peculiarity is evident from the crystallochemical formula Fe^{3+} $(Fe^{2+}_{0.5}, Fe^{3+}_{0.5})_2O_4$ which substitutes the less explanatory stoichiometric formula Fe_3O_4. Moreover, the composition of a non-stoichiometric magnetite $Fe_{2.9}O_4$, can be rationalized from its crystallochemical formula in the following way: $Fe^{3+}(Fe^{2+}_{0.35}, Fe^{3+}_{0.6}, \square_{0.05})_2O_4$. In fact the reason why it is not stoichiometric becomes clear immediately: it is due to the substitution of Fe^{2+} by Fe^{3+}, with the necessary abandoning of 5 per cent of the octahedral sites and the formation of gaps (represented by the square).

From the point of view of symmetry, the space group $Fd\bar{3}m$ adopted both by the spinel and by the magnetite, represents in the first case true symmetry and in the second statistical symmetry.

In the next section some classes of simple structures, defined on the basis of their stoichiometry, will be analysed briefly. The silicates, included amongst the ionic compounds by tradition, will be considered briefly from the point of view of their classification.

MX structures

As predicted by the third Pauling rule, the packing definitely preferred by MX compounds is CCP, while BCC is rare, and HCP is never adopted. It can be seen from the stoichiometry of these compounds that all the octahedra are occupied and that therefore an HCP-based structure is decidedly less probable than a CCP-based structure. In the first case, in fact, the octahedra share edges and faces, while in the second the octahedra only share edges. The non-existence of hexaognal packing in the structures of MX compounds can be best explained with help of Fig. 6.23.

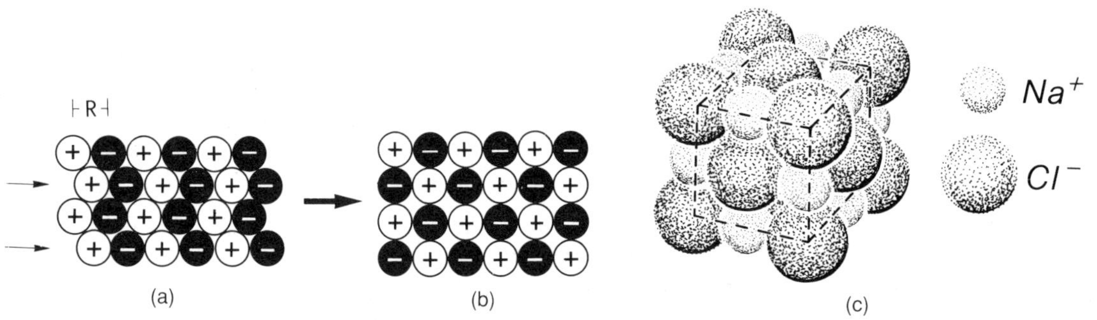

Fig. 6.23. In (a) an unstable layer of cations (white spheres) and anions (black spheres) of equal radius $R/2$, is illustrated. By moving alternate rows in the direction of the arrows a distance of $R/2$, the stable layer illustrated in (b) is obtained. In (c) the structure of NaCl, or **halite** is shown.

A compact layer like that shown in Fig. 6.23(a) is not stable due to the strength with which like charges repel each other. Such effects are considerably attenuated if alternate rows are translated a distance of $R/2$ in the direction of the arrows (Fig. 6.23(a)) to give a regular one positive–one negative ion pattern (Fig. 6.23(b)). In order that a succession of layers of the same type (along the perpendicular to the layer examined) should be stable it is necessary that a second layer be translated by R. The successive translation (equal to R) of a third layer brings us back to the initial configuration. Such a succession of layers gives rise to a cubic structure like that of NaCl, space group $Fm\bar{3}m$, or in other words, a CCP of Cl^- ions with all the octahedral sites filled by Na^+.

This tendency is strengthened if the cations are bivalent, which is why MX compounds almost always adopt CCP type packings. Apart from the halides previously mentioned (see p. 427), other examples are: **periclase** (MgO), **wüstite** (FeO), **bunsenite** (NiO), **manganosite** (MnO), and **lime** (CaO). Some alkaline halides also crystallize naturally: **villiaumite** (NaF), **halite** (NaCl), **sylvine** (KCl), and **carobbite** (KF).

MX_2 and M_2X structures

Such compounds present a wide range of structures, of which only a few are illustated here.

Fluorite (CaF_2), cubic, space group $Fm\bar{3}m$, is shown in Fig. 6.24. Each cation Ca^{2+} is surrounded by eight F^- positioned at the vertices of a cube, while each anion F^- is surrounded by four Ca^{2+} positioned at the vertices of a tetrahedron (Ca^{2+}:F^- = 8:4). The structure adopts a quasi-closest packing of F^- similar to BCC, with half of the sites at the centre of the cube empty and half occupied by Ca^{2+} alternately, in substitution of F^- ($\rho_e > 0.73$) at 1/2, 1/2, 1/2. Due to the alternate filling of these sites the cubes share edges instead of faces, and consequently the cations are further apart and better shielded (see Fig. 6.22). Various compounds such as $SrCl_2$, $BaCl_2$, and others with more complex formulae in which some cations are substituted by other cations e.g. $2Ca^{2+} \rightarrow K^+ + La^{3+}$, $3Ca^{2+} \rightarrow 2K^+ + U^{4+}$ (for $\rightarrow$ read 'are substituted by'), etc. have a CaF_2 type structure.

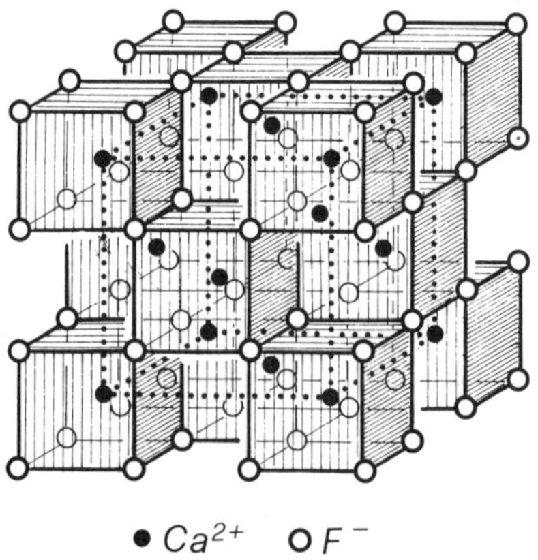

Fig. 6.24. A representation of the structure of **fluorite**, CaF_2. In the figure the cubic coordination polyhedra of the Ca^{2+} (black circles) are outlined, together with the geometrical relationships between them. The elementary cell (dotted lines) consists of four such polyhedra occupied by calcium and four similar empty polyhedra.

The ratio M:X = 6:3 is favoured when $\rho_e < 0.73$ allowing the formation of structures similar to that of **rutile** (TiO_2). The tetragonal structure of rutile, space group $P4_2/mnm$, can be described as HCP. Figure 6.25 illustrates a layer of octahedra (OL) seen (a) perpendicular to and (b) parallel to its extension, while Fig. 6.26(a) shows a packing of OL layers of the type shown in Fig. 6.25(b). The black octahedral are occupied, the white are empty. Figure 6.26(b) shows a structure of rutile obtained by the distortion of the ideal HCP as seen in Fig. 6.26(a). This is in agreemwnt with ρ_e and with the stoichiometry of rutile which predicts that half of the octahedral holes will be occupied. The occupation of the holes occurs in such a way that the octahedra share edges and vertices, never faces, as is illustrated in Fig. 6.26(c).

This feature means that HCP is preferred to CCP, in which the octahedra share edges only. Minerals with a structure like that of rutile include: **sellaite** (MgF_2), **pyrolusite** (MnO_2), **cassiterite** (SnO_2), **stishovite** (SiO_2) a high-pressure compound, with Si^{4+} hexacoordinated. A TiO_2 type structure is certainly one of the most common among strongly ionic binary compounds such as fluorides and oxides. It is also common amongst more complex compounds in which the Ti^{4+} is substituted by cations with a similar radius: e.g. if $3Ti^{4+} \rightarrow Fe^{2+} + 2Ta^{5+}$ then **tapiolite** ($FeTa_2O_6$) is formed, or if instead $3Ti^{4+} \rightarrow Mg^{2+} + 2Sb^{5+}$ **byströmite** ($MgSb_2O_6$) is formed.

Another structure typical of compounds with a ratio of M:X = 6:3 is that of **brucite** ($Mg(OH)_2$): trigonal with space group $P\bar{3}m1$. It has a layered structure which can be described by means of HCP. There are two possible ways for half of the octahedral HCP holes to be occupied without the octahedra sharing faces: the first has already been illustrated for rutile; the second involves all the octahedra of alternate octahedral layers being occupied (e.g. OLA occupied entirely and OLB left completely free in Fig. 6.26(a)). Brucite is an example of the second type. It is more advantageous from the point of view of energy to fill every other octahedron of each row, as happens in rutile, rather than to occupy all the octahedra of OLA and none of OLB. This is because in the first case the octahedra are linked by (six) vertices and (two) edges, but in the second by (six) edges only. The layered structure of rutile is due to the permanent OH^- dipoles which link together the fully occupied octahedral layer;[6] they are known as '**brucitic layers**'. Minerals with a brucite type structure include: **portlandite** ($Ca(OH)_2$), **pyrochroite** ($Mn(OH)_2$), **nibrucite** ($Mg_4NiO(OH)_9$). Another group of minerals with brucitic layers, space group $R\bar{3}m$, which conform to CCP is: **chlormagnesite** ($MgCl_2$), **lawrencite** ($FeCl_2$), **scacchite** ($MnCl_2$).

If the cation–anion positions of fluorite are inverted, an anti-fluorite

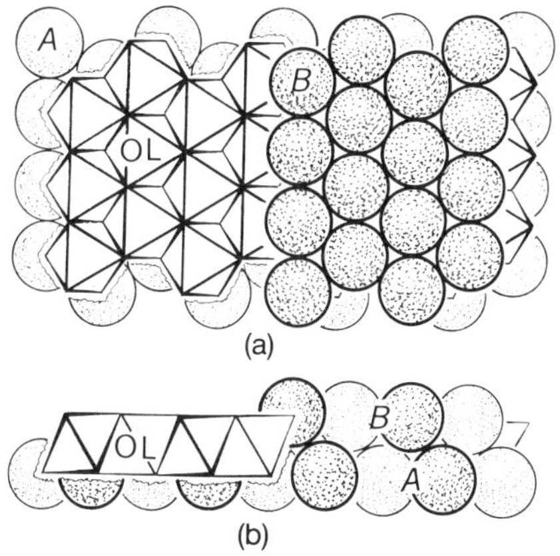

Fig. 6.25. An octahedral layer (OL) obtained by combining two compact layers of spheres of equal radius. In (a) it is seen perpendicular to its extension, in (b) parallel to its extension.

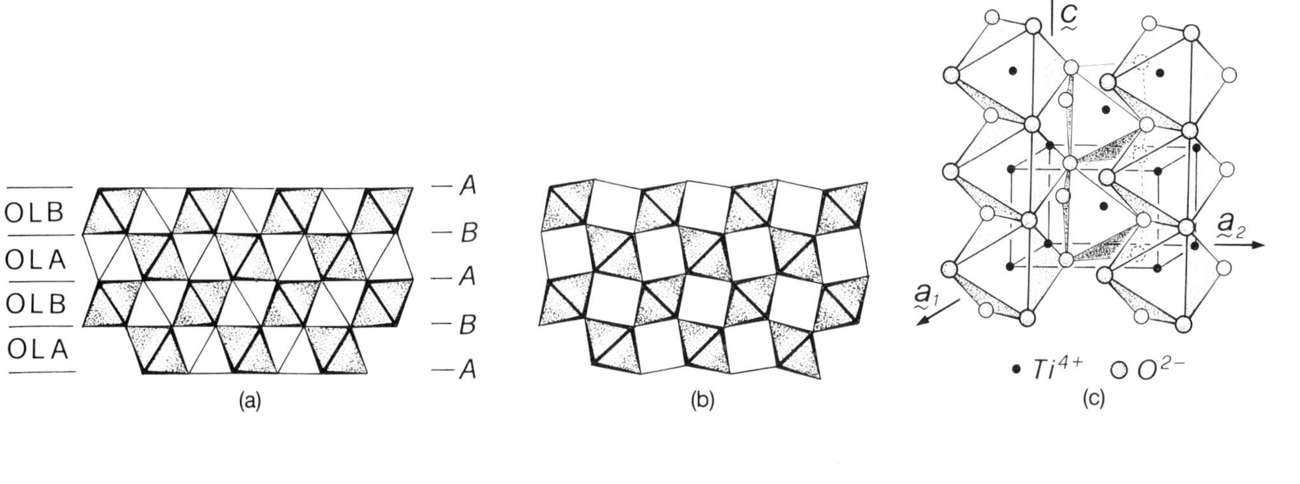

Fig. 6.26. In (a) (on the right) a sequence of ideal compact layers ABAB . . . with the corresponding sequence of octahedral layers OLA-OLB-OLA-OLB . . . (on the left), seen parallel to their extension. The rows of black octahedra (perpendicular to the sheet) are occupied, while the white ones are empty. In (b) the distorted closest packing is shown, while in (c) the rows of occupied octahedra are outlined, and their extension along the axis *c*, in the structure of rutile TiO_2.

structure is formed with a ratio of M:X = 4:8, as seen in the compounds Li_2O, Na_2O, K_2O, Rb_2O. Such a structure can be described as a CCP having all its tetrahedral holes occupied by monovalent cations. Note that in this case an HCP would be impossible, since the tretahedra would have to share faces, thus bringing the alkaline cations into close contact.

MX_3 and M_2X_3 structures

Bayerite and **gibbsite**, two polymorphic modifications of the compound $Al(OH)_3$, have MX_3 stoichiometry: they are both monoclinic with space groups C2/m and $P2_1/n$ respectively. Their structure (M:X = 6:2) is based on a layer similar to that of brucite. As can be seen from their stoichiometry, only one third of their octahedral holes are occupied, while the presence of OH^- groups indicates that their structure is layered like that of brucite. It is useful to double the stoichiometric formula: $Al_2(OH)_6$, and then to divide it into $(OH)_3^-$ and $Al_2(OH)_3^{3+}$. This expedient makes it easy to illustrate that there is a layer of unoccupied octahedra, which overlies another layer of octahedra two-thirds occupied (Fig. 6.27(a)). An HCP type packing, with partially occupied, similarly orientated octahedral layers, alternating with empty octahedral layers results[39] (bayerite, Fig. 6.27(b)).

Leaving the relative positions of the cations out of consideration, in gibbsite, two occupied octahedral layers, like that shown in Fig. 6.27(a), are related by a plane of symmetry m which passes through the empty layer, so the same orientation of the layers recurs every other full layer (Fig. 6.27(c)). In actual fact the arrangement of Al^{3+} cations does not allow for the plane m. Moreover, the substitution of $3Mg^{2+} \rightarrow 2Al^{3+} + \square$, reduces the symmetry from trigonal (brucite) to monoclinic (bayerite and gibbsite).

Both **corundum** (α-Al_2O_3) and **haematite** (α-Fe_2O_3) have an M:X = 6:4 ratio, are rhombohedral with space group $R\bar{3}c$. Their structures are characterized by a bayerite-type layer (Fig. 6.27(a)). It can be deduced from their stoichiometry that the octahedral holes are two thirds occupied. The structure of corundum can be obtained translating the bayerite layers by 1/3 $(-\boldsymbol{a}_1 + \boldsymbol{a}_2)$.

Perpendicularly to the bayerite-type layers the octahedral holes are ordered so that one empty octahedron is followed by two occupied octahedra which share a face. This explains not only the type of lattice, R, but also why the structure is stable. In fact the translation has the same effect as a rotation of the bayerite type layer through 120° would have with the consequent turning of the empty octahedral holes around a threefold

Fig. 6.27. (a) The layer characteristic of **bayerite**, $Al(OH)_3$; 1 indicates the mesh of bayerite, 2 that of **corundum**, α-Al_2O_3 whose structure is based on the same type of layer; (b) shows the structure of bayerite seen in the direction (010) as a sequence of alternately empty and occupied OL; (c) shows the sequence of OL typical of **gibbsite**, $Al(OH)_3$.

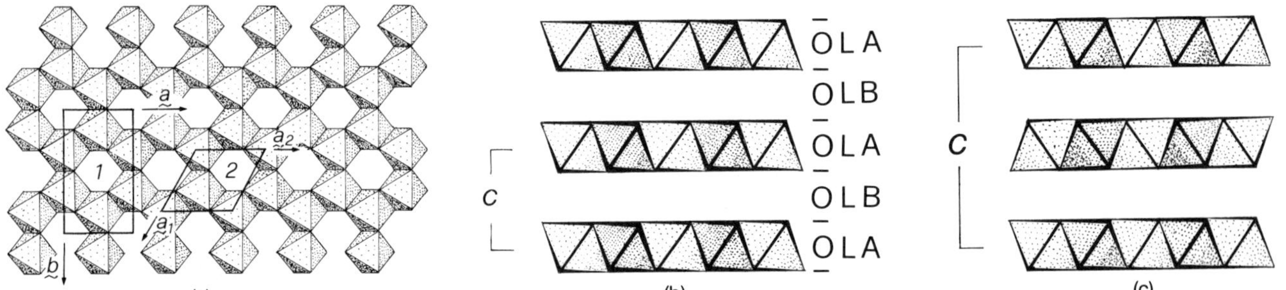

(a) (b) (c)

screw axis. As far as stability is concerned, it is indeed the presence of the two empty octahedra, above and below the two occupied octahedra, that allows the two cations which populate them to move as far apart as possible. This improves the shielding, since the three oxygens shared are brought closer together, and thus reduces the electrostatic repulsion. The octahedra occupied by Al^{3+} and Fe^{3+} are, in fact, rather distorted, having three longer distances and three shorter ones (Al–O = 1.86(×3) and 1.97 Å(×3); Fe–O = 1.95(×3) and 2.21 Å(×3)).

$A_mB_nX_p$ structures

A very large number of compounds of great geological, crystallochemical, and applicative interest belong to this class. Some structures are illustrated below in which A and B represent monovalent to pentavalent metals and X is generally oxygen.

Ilmenite ($FeTiO_3$) and **perovskite** ($CaTiO_3$) have ABX_3 stoichiometry. Ilmenite is isostructural with corundum, in which the cations are substituted according to the following scheme: $2Al^{3+} \rightarrow (Fe^{2+}, Mg^{2+}) + Ti^{4+}$. Such cations are ordered so that the octahedra of one layer are occupied by Fe^{2+} and those of the next layer are occupied by Ti^{4+}. It follows that two contiguous octahedra belonging to two adjacent layers are occupied by Fe^{2+} and by Ti^{4+}, so that the average valence (+3) of the two cations remains the same as that of the two corresponding cations in corundum. The specialization of the octahedral sites, half occupied only by Fe^{2+} and half only by Ti^{4+}, determines the disappearance of the glide *c*. Thus the space group $R\bar{3}c$ (corundum) becomes $R\bar{3}$. A number of structures with various valences (shown in brackets) have a similar structure to ilmenite, including $LiNbO_3(+1, +5)$; $MgTiO_3$, $FeTiO_3(+2, +4)$; $Mn(Fe, Sb)O_3$, α-Al_2O_3, α-Fe_2O_3 (the last two can be considered as particular cases of the structure of ilmenite) and Ti_2O_3, $V_2O_3(+3, +3)$. In these compounds A and B have similar ionic radii ($\langle r_e \rangle \simeq 0.65$ Å), therefore they occupy two octahedral holes of a CCP type packing.

If, however, the cation is too large to fit into the octahedral holes, another type of structure with the same stoichiometry is formed (perovskite type). Thus, as Na^+, K^+, Ca^{2+}, Sr^{2+}, Ba^{2+}, Pb^{2+}, etc., cations have an ionic radius equal to that of the anions when have CN = 12, they can substitute an X (O^{2-}, F^-) anion. If one quarter of the X anions are substituted by such cations, then a particular type of CCP is obtained, a representative cell of which is illustrated in Fig. 6.28.

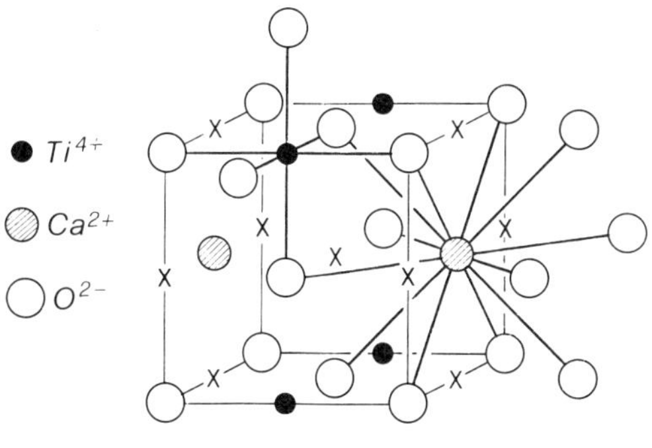

Fig. 6.28. Contents of an elementary cell of **perovskite**, $CaTiO_3$. Note the coordination 12 of the Ca^{2+} (hatched circles). The octahedral holes not occupied by Ca^{2+} are marked X.

Ti^{4+} occupies one quarter of the octahedral holes available, i.e. only of those which have no direct contact with the larger cations. Compounds with a similar structure to that of perovskite include: $NaNbO_3(+1, +5)$; $BaZrO_3(+2, +4)$; $LaAlO_3$, $LaFeO_3(+3, +3)$.

The term **spinels** is used to indicate a considerable group of compounds with a similar structure, having the general formula AB_2X_4. This term is the extension to the group of the name of **spinel** ($MgAl_2O_4$), a mineral characterized by a slightly distorted CCP type structure. In **normal spinels** the cations marked A occupy tetrahedral holes (t), while those marked B occupy octahedral holes (o). The range of their radii depends not only on the tolerance of the structure, but also on the type of anion: $X = O^{2-}$, S^{2-}, Se^{2-}, Te^{2-}, F^-, CN^-.

According to the stoichiometry of spinels one eighth of the tetrahedral and half of the octahedral holes of an FCC cell should be occupied. In actual fact, such an arrangement would result in an unstable structure because of the need for occupied tetrahedral and octahedral holes to share faces. However, the structure becomes stable if two FCC subcells share, so to speak, their work: essentially one gives space to cations in coordination of 4(T), while the other only gives space to those in coordinations of 6(O) (see Fig. 6.29).

In this case, in fact, the occupied tetrahedra and octahedra share vertices. This leads to the succession of cells TOT . . . along $\boldsymbol{a}_1$, $\boldsymbol{a}_2$, and $\boldsymbol{a}_3$, which in turn leads to a cell consisting of eight subcells with a total atomic content of $A_8B_{16}X_{32}$. In this section only spinels with AB_2O_4 will be considered. Type B cations can have a charge of +1, +2, or +3, while type A cations can have a corresponding charge of +6, +4, or +2 v.u. (valence unit). Thus from the point of view of valence, there are three possible types of spinel: (+6, +1), (+4, +2), (+2, +3).

In **inverse** or **anti-spinels**, the type A cations occupy octahedral sites, whereas the type B cations are equally divided between octahedral and

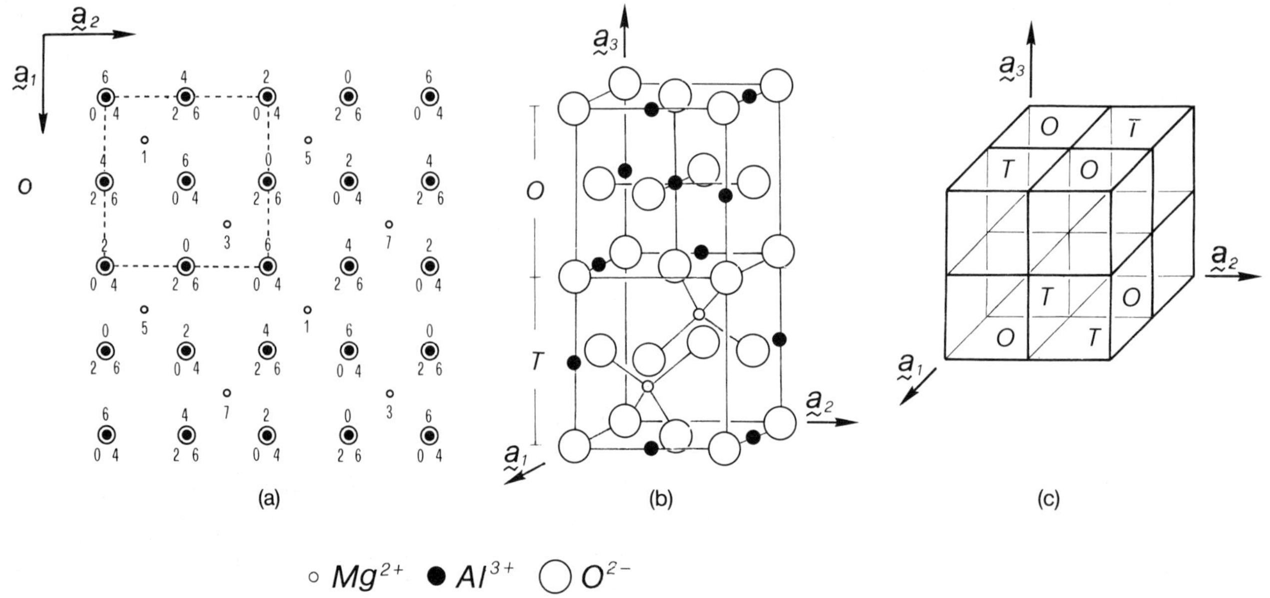

Fig. 6.29. (a) A projection, parallel to $\boldsymbol{a}_3$, of the elementary cell of **spinel** ($MgAl_2O_4$). The number next to the circles indicate the heights (×1/8) respectively of Mg^{2+} (odd) Al^{3+} (even, above the circles) and O^{2-} (even pairs, below the circles). In (b) the contents of two subcells T and O, which alternate regularly along, $\boldsymbol{a}_1$, $\boldsymbol{a}_2$, and $\boldsymbol{a}_3$ giving rise to the spinel cell (illustrated in (c)), are outlined.

tetrahedral sites. Consequently, the formula that describes **normal spinels** is $(A)_t(B_2)_oO_4$ while that of **inverse spinels** is $(B)_t(AB)_oO_4$.

The two formulae represent extreme cases, since spinels of various degrees of inversion commonly exist. The following general formula for spinels can be given $(A_{1-i}B_i)_t(A_iB_{2-i})_oO_4$, where i (**inversion parameter**) represents the fraction of A cations which are hexacoordinated.

The choice of configuration, inverse or normal, and the degree of inversion depend on a delicate network of factors[6] which can be summarized as follows: (1) temperature; (2) Madelung's constant and u parameter; (3) order–disorder phenomena; (4) ionic radius; (5) charge; (6) CFSE; (7) polarization.

If we establish the origin of the cell of spinels (space group $Fd\bar{3}m$) at $\bar{4}3m$, the equivalent 32 positions of the anions in an ideal packing, can be generated, starting with the atomic coordinates u, u, u ($u = 3/8$). Therefore the distortion of the actual packing can be measured by means of the u parameter. In order to illustrate better the structure of spinels, Fig. 6.29 was obtained by translating the origin of the cell by 1/8, 1/8, 1/8. Only (+2, +3) spinels will be analysed here.

Factor (1). Recently, by means of neutron and X-ray diffraction, the range (750–850 °C according to some authors,[40] 600–700 °C according to others[41]) within which transition of the order–disorder type occurs in the spinel $MgAl_2O_4$ has been specified. Figure 6.30(a) shows the effect of temperature[41] on the u parameter. u falls rapidly in the area within the broken lines, which corresponds to the transition normal → inverse spinel. However, it would seem that this transition is not complete, since it stops at 1200 °C, when $i = 0.30$.

Factors (2) and (3) can be examined together. It has already been shown the Madelung's constant, A_M, depends on the lattice geometry. The distortion of the lattice, which can be measured by means of u, influences the value of A_M; moreover, since A_M and U_L are in direct proportion (see eqn (6.17)), the configuration with the highest A_M value is favoured. The trend of A_M plotted against u is illustrated in Fig. 6.30(b) for normal spinels (ns), ordered inverse spinels (ois), and disordered inverse spinels (dis). From the figure it can be seen that for $u > 0.379$, $A_M(\text{ns}) > A_m(\text{dis})$ and that for $u > 0.381$, $A_M(\text{ns}) > A_M(\text{ois})$. This is the main reason why almost all (+2, +3) spinels with $u > 0.381$ are normal.

Factors (4) and (5) counteract one another. Factor (4) indicates that the smallest cations ought to occupy the tetrahedral sites, while the largest ought to occupy the octahedral sites. Since B^{3+} type cations are very often smaller than A^{2+} type cations, this means that (+2, +3) spinels should be prevalently inverse. Experience, however, shows that exactly the opposite is true. Factor (5) takes into account the Pauling's second rule, according to which the charge of the cation is more or less neutralized by the anions of the first coordination shell. Consequently, the cations with the highest charges (B^{3+}) would tend to occupy the sites with the highest coordination number (in this case octahedral sites), thus favouring normal (+2, +3) spinels.

Factor (6) considers the effects of the crystal field on transition metals. The latter, due to the weak crystal field of the oxygens, almost all adopt high-spin configurations in spinels. It should be remembered that cations with a d^o or d^5 electronic configuration (spherosymmetrical) include: Mn^{2+},

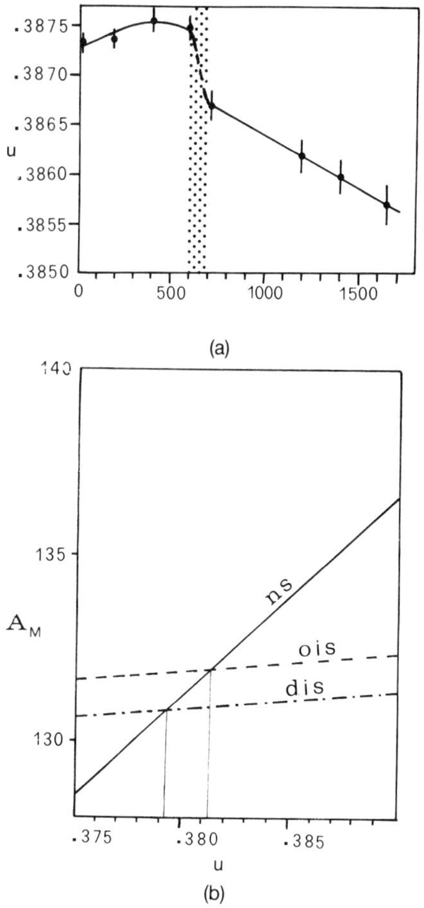

Fig. 6.30. Shows the trend of the distortion parameter, u, versus the temperature (°C) for $MgAl_2O_4$. Note the rapid fall of u (at ≃ 650 °C) which indicates the rapid inversion of the spinel. (b) Illustrates the trend of the Madelung constant A_M versus u. The abbreviations ns, ois, and dis indicate normal spinel, ordered inverse spinel, and disordered inverse spinel.

Fe^{3+}, Zn^{2+}, Ga^{3+}. It is possible to predict accurately which type of spinel will occur, if the A_M and the OSSE indications agree. If, on the other hand, they do not agree, one or other type of spinel may prevail, depending on the balance of all the factors involved. For example, (+2, +3) type spinels such as MgV_2O_4, $MgCr_2O_4$, $MgMn_2O_4$, are normal, since they have $u = 0.385$ (and thus A_M favours ns); moreover, Mg^{2+} is spherosymmetrical and the OSSE definitely suggests that octahedral sites will be preferred by V^{3+}, Cr^{3+}, Mn^{3+}.

Now let us consider the spinels $NiMn_2O_4$, $NiFe_2O_4$. They have respectively $u = 0.383$, 0.381, therefore there is a slight A_M bias towards ns. On the other hand, the OSSE favours Ni^{2+} rather than Mn^{3+} while for Fe^{3+} it is nil. Hence spinels with highest degree of inversion are the result.

Finally the minerals **magnesioferrite** ($MgFe_2O_4$), **magnetite** ($FeFe_2O_4$), and **trevorite** ($NiFe_2O_4$) have respectively $i = 0.90$, 1.0, 1.0. In the first compound the OSSE of Fe^{3+} is zero and $A_M = 0.382$, so there is no clear indication. However, Fe^{3+} tends to prefer tetrahedral sites, owing not only to the polarization (factor (7)), but also to its relatively small ionic radius ($r_e(Fe^{3+}) = 0.645$, $r_e(Mg^{2+}) = 0.72$ Å). In the other two compounds A_M does not give clear indications, while the OSSE of Fe^{2+} and Ni^{2+} decidedly favour these two cations for octahedral sites, therefore the degree of inversion is maximum.†

Factor (7) can play a decisive role when the other factors counteract each other. Therefore, if A_M does not clearly favour either of the two configurations, polarization can play an important part either strengthening, or even overturning the A_M indications.

The effects of polarization are particularly noticeable where Fe^{3+}, Ga^{3+}, and In^{3+} are concerned. For example, in **jakobsite** ($MnFe_2O_4$) the A_M favours a normal configuration ($u = 0.385$), the OSSE = 0 both for Mn^{2+} and Fe^{3+}. The effect of polarization is to partially invert the A_M indications, giving rise to a spinel with $i = 0.20$. For other interesting aspects concerning the question of spinels see, for example, Greenwood,[6] Burns,[12] Evans,[38] and Ottonello.[39]

Olivines are minerals with a general formula $M_2[SiO_4]$ where M represents various cations, which are essentially bivalent and hexacoordinated. Belonging to the olivine group are: **forsterite** $Mg_2[SiO_4]$, **fayalite** $Fe_2[SiO_4]$, **tephroite** $Mn_2[SiO_4]$, **monticellite** $CaMg[SiO_4]$, **kirschsteinite** $CaFe[SiO_4]$, **glaucochroite** $CaMn[SiO_4]$, **knebelite** $MnFe[SiO_4]$. In particular, olivine is a solid solution of forsterite and fayalite (Mg_x, Fe_{1-x}) [SiO_4], while in the Ca-olivines pairs of cations such as Ca^{2+} and Mg^{2+}, Ca^{2+} and Fe^{2+}, Ca^{2+} and Mn^{2+} occupy quite distinct sites. Referring to the two minerals forsterite and fayalite, the composition given above can also be given as Fo_xFa_{1-x}, or simply Fo_x. Olivine is found in large quantities in the earth's lower crust, and it is generally supposed to be particularly abundant in the earth's upper mantle.

The ideal structure of α-olivine can be described as HCP with (according

† Incidentally, the inversion and the complete disorder between Fe^{2+} and Fe^{3+} in octahedral sites make magnetite an excellent conductor of electricity. This is due to the ease with which Fe^{2+} and Fe^{3+} exchange electrons. Below −153 °C the Fe^{2+} cations become ordered, reducing noticeably the capacity of magnetite to conduct electricity.

to its stoichiometry) half of its octahedral holes occupied by Fe^{2+} or Mg^{2+} and one eighth of its tetrahedral sites occupied by Si^{4+}. The closest packed layers are parallel to (100) and the octahedral holes are occupied in such a way that chains of occupied and empty octahedra form zig-zags along the c axis (Fig. 6.31).

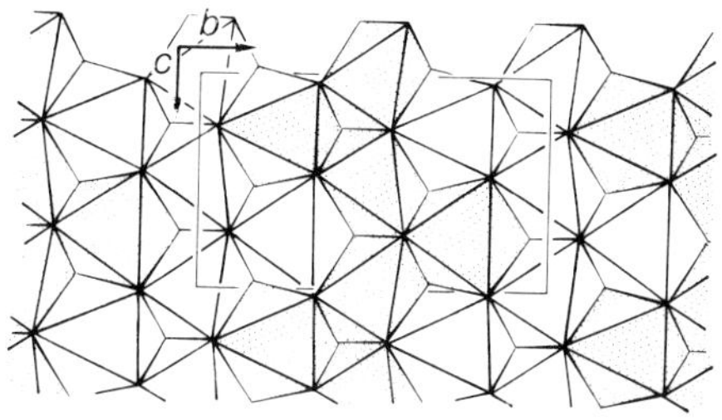

Fig. 6.31. A layer of distorted octahedra typical of the structure of **olivine**, $(Mg, Fe)_2[SiO_4]$. The dotted octahedra are populated by Mg^{2+} or Fe^{2+} and the white ones are empty.

Two layers similar to that shown in Fig. 6.31, at a distance of $a/2$ are translated by $b/2$, so that occupied octahedra overlie empty octahedra. In this way, in fact, as the Pauling's third rule predicts, it is not possible for the occupied octahedra to share faces. As pressure increases forsterite[42] becomes unstable and transforms itself giving rise first to a spinel type structure (β and γ), then to an ilmenite type structure and finally, when $P > 200$ kbar, to two structures: a perovskite type structure, and periclase as follows: $2Mg_2SiO_4 \rightarrow 2MgSiO_3 + 2MgO$.

On the classification of silicates

First of all, can the structures of silicates be described by means of closest packings? The answer to this is not always affirmative: let us examine why this is so. If $\Delta EN(Si\text{–}O) = 1.72$, then eqn (6.2) predicts that between the silicon and the oxygen interactions will be established, with a percentage of covalent bonds more or less equal to that of the ionic bonds. As far as the coordination of the Si^{4+} is concerned, the covalent bond necessitates the formation of four hybrid sp^3 orbitals, or in other words, the angular probability of finding an electron is greatest in the direction which connects the centre of a regular tetrahedron to its vertices (these directions form angles of 109°28′ between each other). On the other hand, the rule of radius ratio predicts a CN = 4 for the ionic bond. Experimentally Si^{4+} is nearly always found to be tetracoordinated, but occasionally hexacoordinated Si^{4+} has also been observed.

The cations which most frequently occur in the structures of silicates are: Al^{3+}, Fe^{3+}, Fe^{2+}, Mg^{2+}, Mn^{2+}, Ca^{2+}, Na^+, K^+. Some of these, positioned around the theoretical limit of two possible coordinations, can form various polyhedra with the anions: e.g. for Al^{3+} $\rho_e = 0.38$, therefore CN = 4, 6. Other cations are of a size to prefer coordinations of 4, 6, 8, or even 12.

It has already been mentioned that the structure of olivine can be described as HCP, even if a rather distorted one ($c_i \simeq 0.60$). The distortion is due to the fact that octahedra and tetrahedra share edges. In fact the distortion depends both on the difference in length between the O–O distances of the tetrahedron $[SiO_4]^{4-}$ ($\simeq 2.6$ Å) compared to that of the octahedron($\simeq 2.8$ Å), and on the electrostatic repulsion.

Other structures with characteristics less similar to those of closest packings are also possible in silicates. Figure 6.32 illustrates an octahedral layer (OL) and two diversely orientated tetrahedral layers (TL_1 and TL_2), which are fitted one into another to form a double layer of spheres (anions), typical of closest packing. This complex layer will be referred to as OTL.

There are several ways of filling the holes of a close-packed structure. Let us consider two border-line events. Firstly, both the tetrahedral and octahedral holes of the layer OTL shown in Fig. 6.32 could be filled, and of each successive layer. Secondly, the octahedra of one layer (OTL_1) and the

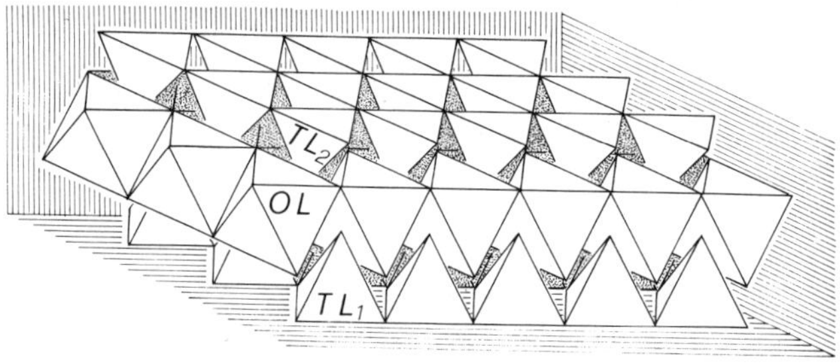

Fig. 6.32. A layer of octahedra (OL) and two diversely orientated layers of tetrahedra (TL_1 and TL_2), whose polyhedra form the holes characteristic of a double compact layer of equal spheres.

tetrahedra of the overlying layer (OTL_2) could be filled alternately, in accordance with the stoichiometric constraints.

The configurations relative to the first hypothesis imply structures of higher potential energy generally, due either to the more frequent sharing of geometric elements, or to other factors. For example, if there is not sufficient oxygen to form isolated $[SiO_4]^{4-}$ groups the tetrahedra would tend to condense. The ideal closest packing would require an Si–Ô–Si angle of 109° 28′, but, owing to the considerable percentage of ionic bond between silicon and oxygen, the two Si^{4+} atoms tend to move away from each other (thus improving the shielding) forming angles which on average are of 140°. If, however, one of the two Si^{4+} is substituted by Al^{3+}, then Al–Ô–Si angles well below 140° (≃118°) can be formed.

Intermediate figures (e.g. fully filled octahedral and partly filled tetrahedral holes) lead to structures with potential energy half way between those of the two configurations hypothesized above. Bearing in mind what has been said, the structure of **phlogopite**, $KMg_3[AlSi_3O_{10}](OH)_2$ will now be analysed. Supposing that it conformed to closest packing ($c_i = 0.56$) its stoichiometry would indicate that one quarter of the octahedral holes are occupied by Mg^{2+} and one sixth of the tetrahedral holes are occupied by Al^{3+} or Si^{4+}. In actual fact, of four successive OTL, one is completely filled by Mg^{2+} (octahedral holes), two immediately above and below each have one third of the tetrahedral holes filled, while the last layer contains K^+ cations (CN = 12) equal in number to the Al^{3+} which substitute Si^{4+}. To summarize, in phlogopite a layer of tetrahedra TL_1 is followed by a layer of octahedra OL, which is followed by another tetrahedral layer TL_2, and finally there is a layer of K^+ ions which connects two tetrahedral layers TL_1 and TL_2.

Figure 6.33(a) represents one of the two tetrahedral layers (TL_1). Every

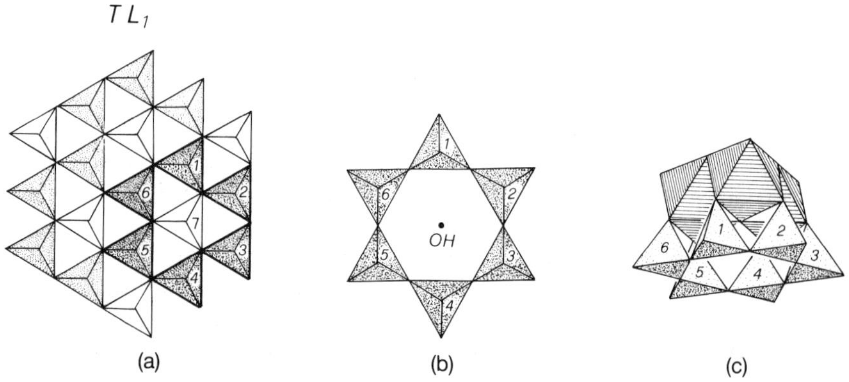

Fig. 6.33. A layer of tetrahedra is represented in (a) (TL_1). The black tetrahedra are occupied by Si^{4+}, while the white ones are empty. This determines, as can be seen in (b), the reorganization of the layer in rings with an OH^- group at the centre, and also, as shown in (c), the relationship between the ring of tetrahedra and a group of octahedra of an OL typical of **phlogopite**, $KM_{f3}[AlSi_3O_{10}](OH)_2$.

tetrahedron shares three vertices with similar tetrahedra, as shown in the figure, so that every vertex is shared by three tetrahedra. If three of the contiguous tetrahedra (e.g. 5, 6, 7 of Fig. 6.33(a)) were occupied by cations with average valence >3, the Pauling's second rule would be infringed. In fact, supposing that they are occupied by either $3Si^{4+}$, or $2Si^{4+} + Al^{3+}$, or even $Si^{4+} + 2Al^{3+}$, the shared oxygen still received respectively 3, 2.75, 2.50 v.u., far higher than the necessary value of 2.

Therefore, with respect to closest packing, there are two important differences. The first is that the group of seven tetrahedra, shown in Fig. 6.33(a), gives rise to a ring of six tetrahedra (Fig. 6.33(b)), leaving an OH^- group in the centre instead of the oxygen which acted as the vertex of the empty tetrahedron (7 of Fig. 6.33(a)).

This pseudo-hexagonal ring fits well into the overlying octahedral layer (Fig. 6.33(c)), and allows a Si–Ô–Si angle of about 140°. The layer of oxygens which constitutes the base of the ring-forming tetrahedra has $c_i = 0.76$, while the layer of oxygens to which the vertices of the tetrahedra and the OH^- group contribute has $c_i = 0.88$. The latter of these two values is nearer to the theoretical value, $c_i = 0.91$.

The other difference is represented by the anions around K^+. This cation, unlike the Ca^{2+} of perovskite, is not surrounded by six anions in the plane (001), which is characterized by K^+ cations; instead the twelve coordination is achieved by six oxygens belonging to the ring of tetrahedra of the overlying layer (TL_1) and six belonging to that of the underlying layer (TL_2).

The examples examined above refer to silicates with hexacoordinated cations which are relatively small, such as Al^{3+}, Mg^{2+}, Fe^{2+}, Fe^{3+}.

For such cations the edge of the octahedron (≃2.8 Å) is similar in length to that of the tetrahedron (≃2.6 Å), as the structure of olivine demonstrates, in which they are shared. If larger cations are present, such as Na^+ and Ca^{2+}, they adopt coordinations with CN ≥ 6. If CN > 6, then the holes made available by closest packing do not fit the requirements of the cation. If CN = 6, octahedra with edges >3.8 Å are formed, edges considerably longer than the edges of the tetrahedra. In this case a pair of tetrahedra inscribing a trigonal prism of side 4.1 Å is often necessary. Such a prism is incongruous with the polyhedra of a closest packed OTL.

As a result, in order to describe in rational terms the structures of silicates, it is necessary to abandon the mechanical model used so far. It is helpful to make use of particular coordination polyhedra, the type of links which are established between them, and other parameters.

Liebau's crystallochemical classification

The simplicity of a classification system depends basically on the number of parameters on which it is based. For the silicates this number was originally low (Bragg 1930) and has gradually increased as our knowledge of atomic structures has grown (Zoltai,[43] Liebau[44,45]).

The crystallochemical classification proposed by Liebau[44,45] takes into account the close relationships which exist between structure, chemical composition, thermodynamic stability, and reactivity in silicates. Given the strong prevalence of structures with Si^{4+} in tetrahedral coordinations over

Si^{4+} in octahedral coordinations, attention will be given here only to the former, while Liebau considers both. Since the silicon can be substituted by other cations such as Al^{3+} and to a lesser extent by P^{5+}, Ge^{4+}, Fe^{3+}, Be^{2+}, the tetrahedron $[SiO_4]^{-4}$ will be represented more generally as $[TO_4]$ below.

The first problem to be solved is that of establishing which tetrahedra can reasonably be considered as part of the silicate anion. Liebau proposes the following criterion: 'If, under certain thermodynamic conditions, an element replaces some of the silicon atoms in a given silicate crystal structure so that the crystallographic positions are statistically occupied by silicon and the other cation, then this cation is recognized as part of the silicate ion'.

For example, two phases of potassium feldspar, $KAlSi_3O_8$, known as **sanidine** hT (high temperature) and **microcline** lT (low temperature) are similar since they have Al^{3+} and Si^{4+} in tetrahedral coordination, but differ in that the two cations show different degrees of disorder. This disorder is maximum in the first phase, $K[(AlSi_3)O_8]$ while it is minimum in the second, $KAl[Si_3O_8]$, since here the tetrahedral sites tend to be completely occupied by either Al^{3+} or Si^{4+}. These two phases are the end members of a series characterized by different degrees of Al^{3+}/Si^{4+} disorder and they are an example of the phenomenon known in thermodynamics as **second-order phase transition**. As the temperature rises the feldspar passes from one phase to the other as the degree of Al/Si disorder varies, without the first phase being destroyed (in this case it would be a **first-order phase transition**). For this reason, in accordance with Liebau, the formula for fully ordered microcline lT is not $KAl[Si_3O_8]$, but is similar to that of sanidine hT, i.e. $K[AlSi_3O_8]$. It follows that, assuming $[AlSi_3O_8]$ is the building unit, both phases belong to so-called framework silicates.

Noting that Si–O bonds are generally stronger than M–O bonds (where M is a cation other than silicon) the crystallochemical classification of silicates takes into account the various possible modes of condensation of the tetrahedra $[TO_4]$.

First of all, the number of polyhedra which are linked to a given polyhedron by vertices, edges, or faces is defined as s (connectedness). When these polyhedra are tetrahedra $[TO_4]$ vertices are shared, and since each tetrahedron has four vertices, $s = 0, 1, 2, 3, 4$. If the tetrahedron is isolated it is called singular (Q^0), if it shares one vertice it is called primary (Q^1), two secondary (Q^2), three tertiary (Q^3), and four quaternary (Q^4); concisely $Q^s = Q^0, Q^1, Q^2, Q^3, Q^4$.

The parameters on which the crystallochemical classification of silicates is based are:

1. N_{an} = **number of anion types** in a silicate. $N_{an} = 1$, uniform-anion silicates; $N_{an} > 1$, mixed-anion silicates.

2. CN = **coordination number**. In naturally occurring compounds only tetrahedra $[TO_4]$ and exceptionally octahedra $[TO_6]$, are present.

3. L = **linkedness**, type of link. $L = 0, 1, 2, 3$: for $L = 0$ the tetrahedron is isolated, for $L = 1, 2, 3$ the tetrahedron shares respectively one vertex, one edge, one face with a similar tetrahedron. In Fig. 6.22 (first row) the relationships between two tetrahedra for $L = 1, 2, 3$ are shown.

4. B = **branchedness**, type of ramification. The tetrahedra $[TO_4]$ can be condensed linearly (see Fig. 6.34), giving rise to limited groups, single

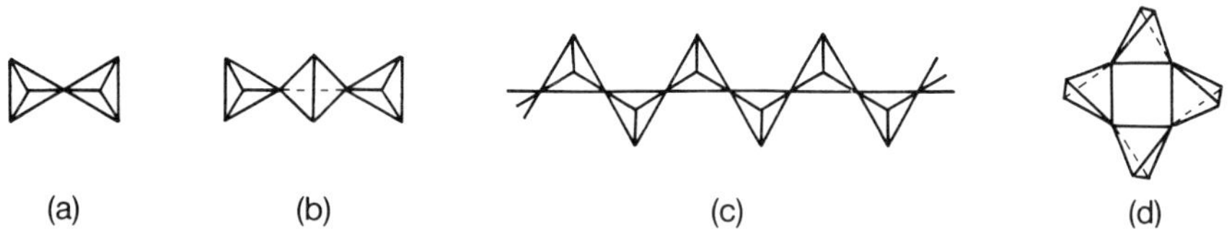

Fig. 6.34. Some unbranched fundamental anions (*uB* f.a.); (a) a pair of tetrahedra, (b) an unbranched group of three tetrahedra, (c) a single unbranched chain, (d) a single unbranched ring.

chanins and rings. They are characterized by tetrahedra with $s \leqslant 2$ and constitute the so-called **unbranched fundamental anions** (*uB*f.a.).

Complex anions of the same type (*uB*c.a.) can result from the condensation of *uB*f.a. **Branched fundamental anions** (*br*f.a.) can also be obtained by condensing tetrahedra. They include tetrahedra with $1 \leqslant s \leqslant 4$ (Fig. 6.35). By condensing *br*f.a., *br*c.a. can be obtained. Branched anions can be divided into **open-branched** (*oB*) or **loop-branched** (*lB*) ones. In the first case, one or more tetrahedra constitute a sort of 'branch' which is joined to the non-branched anions by means of a single geometric element (vertex, edge, or face). In the second case, more than one element is shared, e.g. two vertices (see Fig. 6.35). The condensation of more than one fundamental anion generates the following types of anions (see Fig. 6.36):

uB f.a. + *uB* f.a. = *uB* c.a.

uB f.a. + *oB* f.a. = *hB* c.a. (*h* = **hybrid branched** c.a.)

oB f.a. + *oB* f.a. = *oB* c.a.

oB f.a. + *lB* f.a. = *olB* c.a. (**mixed-branched** c.a.
combination of open + loop-branched anions)

lB f.a. + *lB* f.a. = *lB* c.a.

In conclusion the branchedness (*B*) of silicates can be *uB* or *br* (unbranched or branched); while branching can be of various type: *oB*, *lB*, *olB*, *hB*.

5. M = **multiplicity**, i.e. the finite number of tetrahedra that are linearly linked in a 'short chain' of tetrahedra, or the number of rings, single

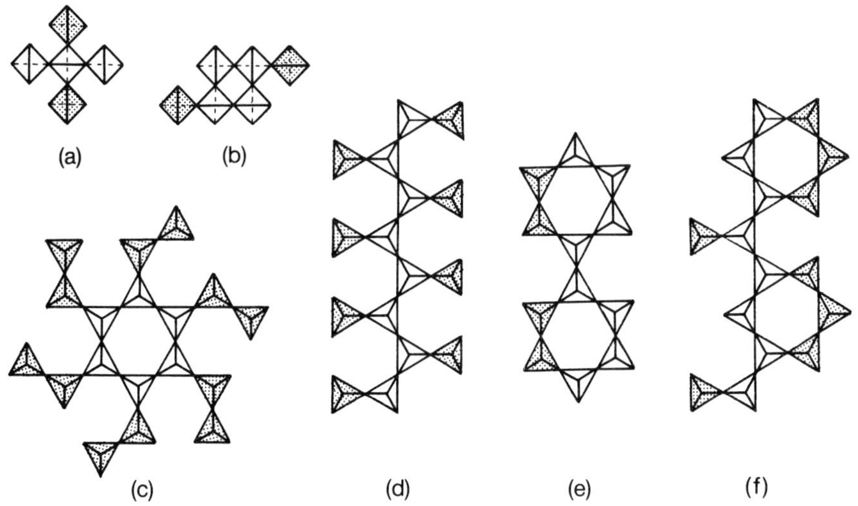

Fig. 6.35. Some branched fundamental anions (*br*f.a.). The branches can consist of one or more tetrahedra (dotted). The figure illustrates: (a) *oB* triple tetahedron; (b) and (c) *oB* single rings; (d) *oB* single chain; (e) *lB* single chain; (f) *olB* single chain.

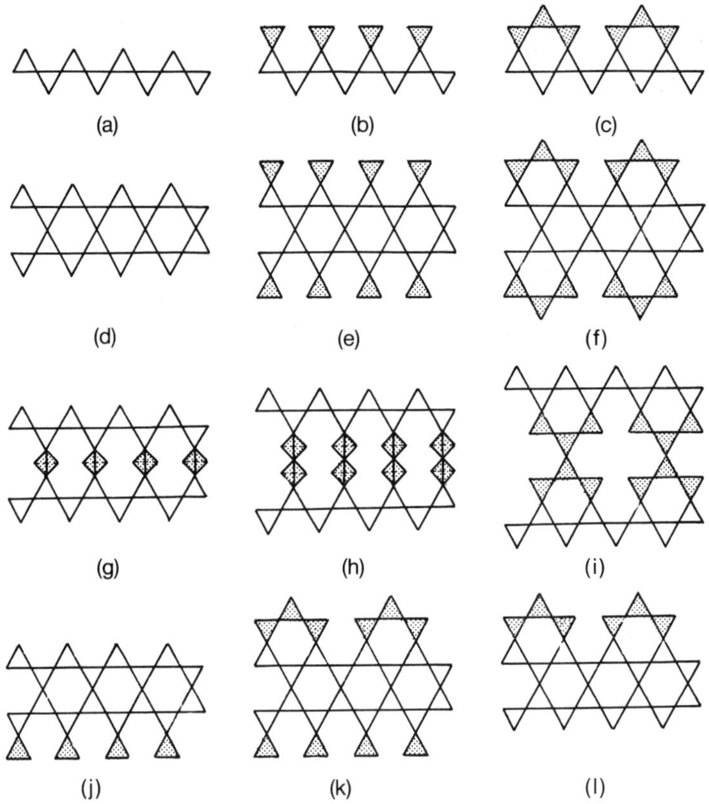

Fig. 6.36. Fundamental anions (f.a.) and complex anions (c.a.): (a) *uB* f.a., (b) *oB* f.a., (c) *lB* f.a., (d) *uB* c.a., (e) *oB* c.a., (f) *lB* c.a., (g) *hB* c.a., (h) *oB* c.a., (i) *lB* c.a., (j) *hB* c.a., (k) *olB* c.a., (l) *hB* c.a.

chains, or single layers linked to a complex ring, chain, or layer respectively; rarely is $M \geqslant 5$. The condensation of groups of connected tetrahedra (i.e. multiple tetrehedra), rings, single chains, or single layers does not alter the dimensions of the fundamental silicate anions which are involved in the condensation.

6. D = **dimensionality**. The silicate anions can have the following extensions: finite ($D = 0$), infinite in one ($D = 1$), two ($D = 2$), or three ($D = 3$) directions.

   When $D = 0$, the symbols t and r are used respectively to indicate groups of tetrahedra (**terminated anions**), and cyclic anions (**rings**).

7. P or P^r = **periodicity** of the chain (P) or ring (P^r). Silicate anions with $D = 1$, 2, 3 are based on fundamental chains of tetrahedra. The parameter P expresses the number of tetrahedra which determines the repetition period of an unbranched chain (if a branched chain is considered the tetrahedra which constitute the branches should be left out). Figure 6.37 illustrates some fundamental chains of different periodicity. If the anions are cyclic the symbol P^r, which expresses the periodicity of a ring (i.e. the number of tetrahedra within a single ring) is used, excluding, therefore any branches. In order to avoid confusion, when choosing which fundamental chain is characteristic to a silicate with $D = 1$, 2, 3, it is necessary to take into account certain rules:

   (a) Fundamental chains with lowest periodicity running parallel to the

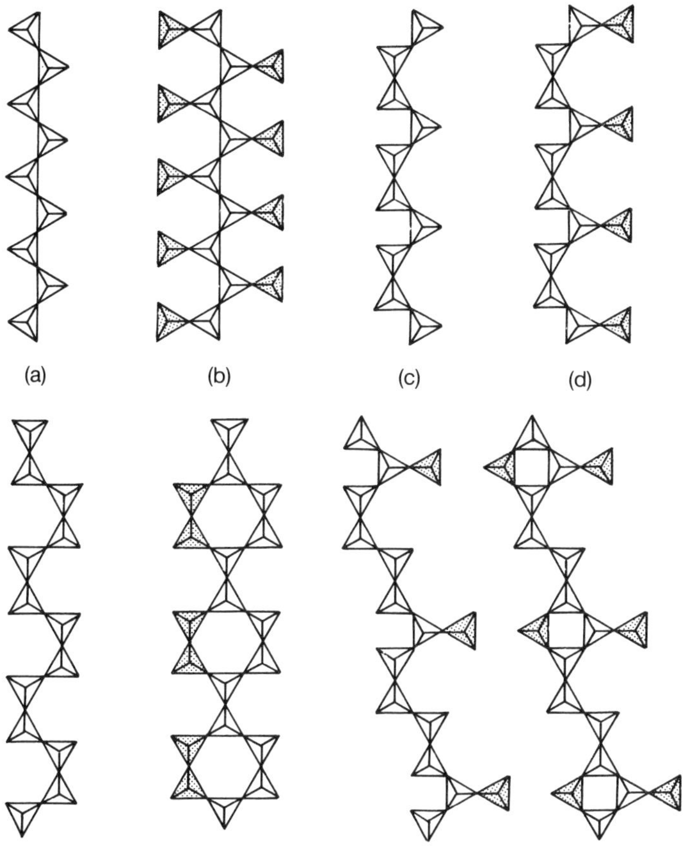

Fig. 6.37. (a), (c), and (e) illustrate unbranched chains, while (b), (d), (f), (g), and (h) illustrate branched chains. The following pairs have the same periodicity: for (a) and (b) $P = 2$, for (c) and (d) $P = 3$, for (e) and (f) $P = 4$, for (g) and (h) $P = 5$.

direction of the shortest identity period within the silicate anion are chosen.

(b) If rule (a) is satisfied by more than one chain, the fundamental chains are chosen such that their number is lowest.

(c) Once rules (a) and (b) have been satisfied, the third rule indicates the following order of preference (for fundamental chains): unbranched > loop-branched > open-branched > mixed-branched > hybrid (for > read 'is preferred to').

The order in which the various parameters have been presented represents the classification hierarchy: superclass (N_{an}), **classes** (CN), **subclasses** (L), **branches** (B), **orders** (M), **groups** (D), **sub-groups** (r or t), **families** (P, P^r).

If there are a number of silicates with the same type of silicate anion, i.e. silicates belonging the same family further subdivisions can be made. The following criteria are particularly important:

(1) the Si:O atomic ratio of the silicate anions;

(2) the degree of stretching of a chain.

As far as criterion (1) is concerned, it should be noted that two or more rings of two or more single chains having $P > 1$ can join together via all or only part of their tetrahedra. As a result, complex anions with various Si:O ratios are generated, depending on the portion of the tetrahedra involved.

As far as criterion (2) is concerned, Liebau proposes that the degree of stretching should be measured by means of the **stretching factor** (f_s), which can be obtained as follows:

$$f_s = \frac{I_c}{l_T P} \tag{6.31}$$

where I_c is the identity period of the chain, l_T is the length of the edge of the tetrahedron, both in Å, and P is the periodicity of the chain (the number of tetrahedra needed to identify the period).

Since amongst all the silicates so far discovered **shattuckite**, $Cu_5[Si_2O_6](OH)_2$, has the most stretched chain, it is taken here as a point of reference. From it the value $l_T = 2.7$ Å can be obtained (equal to half of the repetition period of the chain, which is two tetrahedra $[SiO_4]^{-4}$). Thus for shattuckite $f_s = 5.40$ Å$/2 \times 2.70$ Å $= 1.00$ while for **enstatite**, $Mg_2[Si_2O_6]$, ($I_c = 5.21$ Å, $P = 2$) $f_s = 0.965$ for **alamosite**, $Pb_{12}[Si_{12}O_{36}]$, ($I_c = 19.63$ Å, $P = 12$) $f_s = 0.606$.

The crystallochemical classification presents the same periodicity characteristics as the periodic system of the elements, so that, for example, by gradual condensation, of $[TO_4]$ tetrahedra in a linear way, *uB* anions can be generated that are linked only by means of vertices in the following way:

(1) for $D = 0$ the number of tetrahedra that can be condensed linearly increases as M increases, so that for $M \to \infty$, $D \to 1$; in which case a single *uB* chain will result;

(2) for $D = 1$ the number of chains increases with M, and for $M \to \infty$, $D \to 2$, thus forming a single *uB* layer;

(3) when single layers are condensed and for $M \to \infty$, $D \to 3$, a three-dimensional building of tetrahedra, i.e. *uB* framework is obtained.

By varying M and D the whole of Table 6.8 can be obtained, which lists

Table 6.8. Chemical and mineralogical (in round brackets) nomenclature of silicates

		$M = 1$	$M = 2$	$M = 3$	...
$D = 0$	Oligosilicates	Monosilicates	Disilicates	Trisilicates	...
	(–)	(Nesolicates)	(Sorosilicates)		
$D = 0$	Cyclosilicates	Monocyclosilicates	Dicyclosilicates	Tricyclosilicates	...
	(–)		(Cyclosilicates)		
$D = 1$	Polysilicates	Monopolysilicates	Dipolysilicates	Tripolysilicates	...
	(–)		(Inosilicates)		
$D = 2$	Phyllosilicates	Monophyllosilicates	Diphyllosilicates	Triphyllosilicates	...
	(–)		(Phyllosilicates)		
$D = 3$	Tectosilicates (–)	Tectosilicates (Tectosilicates)			

the chemical and mineralogical (in round brackets) nomenclature of silicates.

Structural formulae

A structural formula should contain the largest possible amount of information and therefore should include the parameters used in crystallochemical classification. Some of the parameters CN and P are best used only when there is some perplexity: the first is written in square brackets as a right-handed superscript to the cation, while the second is written without brackets as left-handed supercrript to the cation. Other parameters, N_{an} and L, can be deduced directly from the structural formula. Therefore a structural formula containing the essential parameters is as follows:

$$M_r\{B, M\ {}^{D}_{\infty}\}[T_xO_y]$$

M_r indicates cations which do not belong to the silicate anion; the meaning of the other symbols has already been given. For $D = 0$ the interpretation of the symbol ${}^{0}_{\infty}$ is not unequivocal and can therefore be substituted by r or t, depending on whether the silicate anion is cyclic or not. For $D = 1$, 2, 3 the symbols ${}^{1}_{\infty}$, ${}^{2}_{\infty}$, ${}^{3}_{\infty}$ indicate clearly that the silicate anions are respectively chains, layers, and three-dimensional framework of tetrahedra. Sometimes it is useful to complement the formula by the following suffixes: (lT), (mT), (hT), (lP), (mP), (hP). They indicate that the phase described by the formula is of low (l), medium (m), or high (h) temperature (T) and/or pressure (P).

Some examples:

$$Al_{12}^{[6]}Al^{[4]}\{oB, 3t\}[Si_5O_{16}](OH, F)_{18}O_4Cl \quad \textbf{zunyite}$$

The bracketed right-hand superscripts to Al^{3+}, indicate the two coordinations of the Al^{3+} in the structure of zunyite.

Since there are no indications for Si^{4+}, it is implicit that CN = 4. The symbols in curly brackets indicate that the silicate anion consists of a group of open-branched tetrahedra $\{oB\}$, three of which determine the multiplicity $M = 3$ (see Fig. 6.35(a)). The content of the square brackets indicates that the group consists of five tetrahedra and that therefore the 'branches' of the silicate anion are the remaining two tetrahedra $[SiO_4]^{4-}$.

$$Ca_2Al_2Sn\{oB, 1r\}[{}^4Si_6O_{18}](OH)_2{\cdot}2H_2O \quad \textbf{eakerite}$$

The structural formula informs us that the silicate anion is open-branched $\{oB\}$ and that it is a single ring $\{1r\}$. The content of the square brackets indicates that the unbranched part consists of four tetrahedra $[SiO_4]^{4-}$ and that the 'branches' consist of two tetrahedra $[SiO_4]^{4-}$ (see Fig. 6.35(b)).

For other structural formulae and for illustrations of the silicate anions described therein, see Figs. 6.38–6.42.

Relationship between classification parameters and properties of the cations

The properties of non-tetrahedral cations, in particular, electronegativity and valence, have a strong influence on the structure of silicate anions.[45] Here, the direct influence of the non-tetrahedral cations on the parameters which govern crystallochemical classification will be examined.

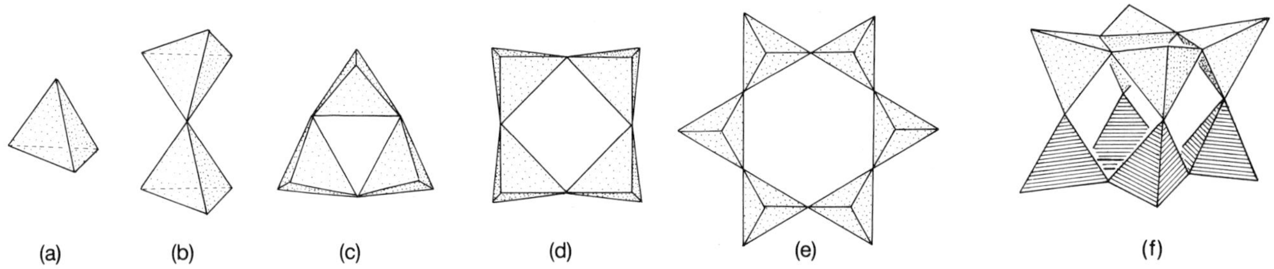

Fig. 6.38. Ideal configurations of certain silicate anions and structural formulae: (a) **olivine** $(Mg, Fe)_2\ \{uB, 1t\}[SiO_4]$; (b) **ilvaite** $CaFe_2(Fe, Mn)\{uB, 2t\}[Si_2O_7]O(OH)$; (c) **benitoite** $BaTi\{uB, 3t\}[Si_3O_9]$; (d) **taramellite** $Ba_4(Fe^{3+}, Ti)_4B_2\{uB, 4t\}[Si_4O_{12}]_2O_5Cl_x$; (e) **tourmaline** $XY_3Z_6B_3\{uB, 6t\}[Si_6O_{18}]O_9(O, OH, F)_4$ where $X = Na^+, Ca^{2+}$; $Y = Li^+, Mg^{2+}, Fe^{2+}, Mn^{2+}, Fe^{3+}, Al^{3+}$; $Z = Al^{3+}, Mg^{2+}$; (f) **steacyite** $K_{1-x}(Na, Ca)_{2-y}Th_{1-z}\{uB, 2r\}[{}^4Si_8O_{20}]$, the fundamental anion is dotted.

1. CN: the coordination number of silicon depends in particular on the EN of the anions (X) of the first coordination shell. If the anions are strongly electronegative, as is fluorine (see Table 6.2), the tendency $CN > 4$ is increased (SiF_6), if weakly electronegative tetrahedral groups are formed.

In the case of oxygen, the EN is lower than that of fluorine, but is still high. Therefore the silicon of both natural and artificial compounds is found both in tetrahedral or octahedral (a few cases) coordinations. The Si–O type of bond is influenced by the cations (M) belonging to the second coordination shell of the silicon: the highly electronegative M cations give a more ionic character to the Si–O bond. This, obviously, affects the CN ($CN > 4$ is slightly favoured), the effective charge of the silicon (positive), and, when $CN = 4$ the overall charge of the tetrahedra (negative).

The effects of repulsion between the silicate anions can be evaluated by means of the distortions which can be seen above all in the Si–Ô–Si bond angles, but also to a lesser extent in other parameters.

2. L: the stability of a silicate anion decreases as L increases, an inverse tendency would violate Pauling's third rule. This explains why no silicate

Fig. 6.39. Single chains with different periodicity (P), structural formulae and stretching factors (f_s) for the following minerals: (a) a hypothetical mineral, $Mr\{uB, 1\,{}^{1}_{\infty}\}[SiO_3]$; (b) **enstatite**, $Mg_2\{uB, 1\,{}^{1}_{\infty}\}[Si_2O_6]$, $f_s = 0.965$; (c) **wollastonite**, $Ca_3\{uB, 1\,{}^{1}_{\infty}\}[Si_3O_9]$, $f_s = 0.904$; (d) **krauskopfite**, $H_4Ba_2\{uB, {}^{1}_{\infty}\}[Si_4O_{12}]\cdot 4H_2O$, $f_s = 0.783$; (e) **rhodonite**, $(Mn, Ca)_5\{uB, 1\,{}^{1}_{\infty}\}[Si_5O_{15}]$, $f_s = 0.906$; (f) **stokesite**, $Ca_2Sn_2\{uB, 1\,{}^{1}_{\infty}\}[Si_6O_{18}]\cdot 4H_2O$, $f_s = 0.718$; (g) **pyroxmangite**, $(Fe, Ca, Mn)_7\{uB, 1\,{}^{1}_{\infty}\}[Si_7O_{21}]$, $f_s = 0.923$; (h) **alamosite**, $Pb_{12}\{uB, 1\,{}^{1}_{\infty}\}[Si_{12}O_{36}]$, $f_s = 0.606$. The periodicity (P) is as the silicon stoichiometric coefficient.

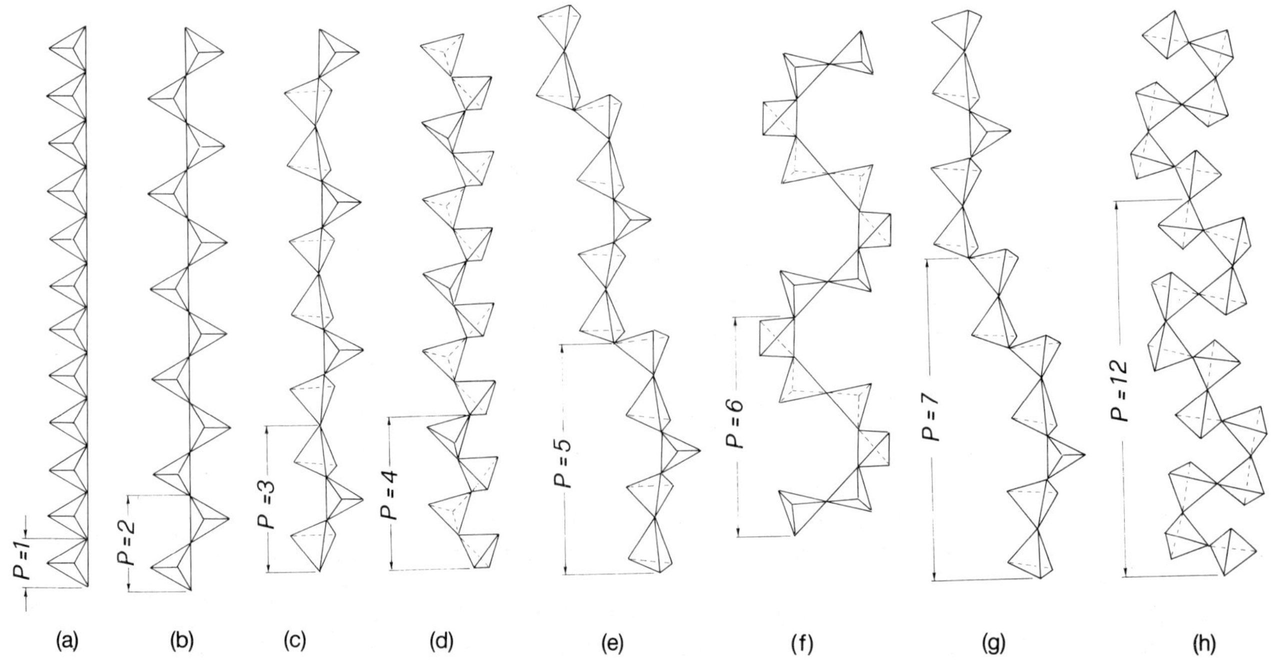

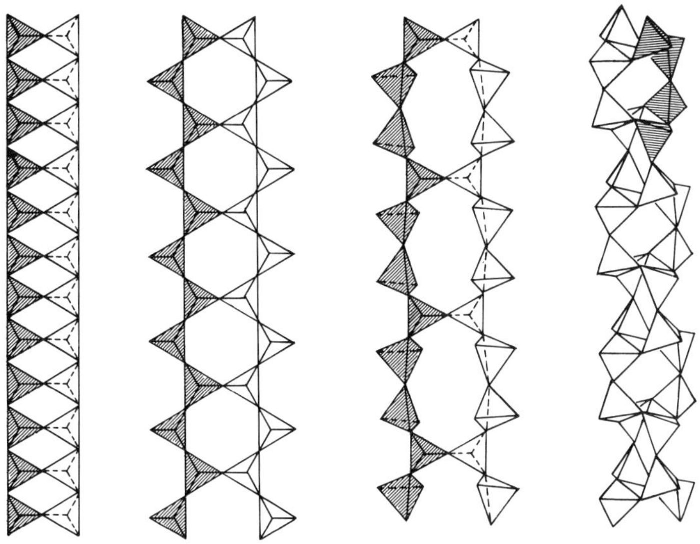

(a) (b) (c) (d)

Fig. 6.40. Double chains in which the fundamental anion is hatched, and the structural formulae of: (a) **sillimanite**, $Al^{[6]}\{uB, 2\,{}^{1}_{\infty}\}[{}^{1}(AlSiO_5)](hT)$; (b) **tremolite**, $Mg_5Ca_2\{uB, 2\,{}^{1}_{\infty}\}[{}^{2}Si_4O_{11}]_2(OH)_2$; (c) **xonotlite**, $Ca_6\{uB, 2\,{}^{1}_{\infty}\}[{}^{3}Si_6O_{17}](OH)_2$; (d) **narsarsukite**, $Na_4Ti_2\{uB, 2\,{}^{1}_{\infty}\}[{}^{4}Si_8O_{20}]O_2$; (d) can be obtained by condensing the unbranched rings $\{uB, 1r\}[Si_4O_{12}]$ (hatched in figure). In this case the structural formula of narsarsukite is: $Na_4Ti_2\{uB, 1r\}[{}^{4}Si_8O_{20}]O_2$.

anions with CN = 4 are found with $L > 1$ (at present only one case has been found with $L = 2$). Highly electronegative cations favour a higher degree of linkedness because the $[SiO_4]$ effective charge tends to be reduced.

3. B: branched silicate anions, in particular branched ring anions, are less stable than unbranched varieties, due to the shorter average Si–Si distances. If, however, cations with high EN are present, then the stability of such silicate anions is increased, since they attenuate the effective negative charge thus reducing the repulsion between the tetrahedra. The stabilizing effect of such cations decreases as D increases, because in such cases the effective charge of the tetrahedra $[TO_4]$ also decreases.

4. M: the number of structures with increasing M falls drastically both in the oligosilicates and in the cyclosilicates, and in polysilicates and phyllosilicates. This is due to the decided increase in the potential energy of structures as M increases. To examine this phenomenon the differences in energy between linear groups of tetrahedra will be analysed. It should be borne in mind that the greater their distance from the tetrahedron which terminates the group, the less two contiguous tetrahedra will vary from the point of view of energy. For example two Q^1 tetrahedra are similar from the point of view of energy because they each start and terminate a group. Two Q^1 tetrahedra and one Q^2, belonging to a group of three tetrahedra are very different from the point of view of energy, because the distance that separates Q^2 from the external tetrahedron Q^1 is only that of one tetrahedron. Finally, in a single unbranched chain every tetrahedron has practically the same energy as that next to it, since both are more or less at the same distance from the end tetrahedron. In contrast a linear group of tetrahedra could, from an energetic point of view, be regarded as being constituted by different structural units.

To this the principle of parsimony is applied (see p. 436, Pauling's fifth rule), according to which the smaller the number of structural units a

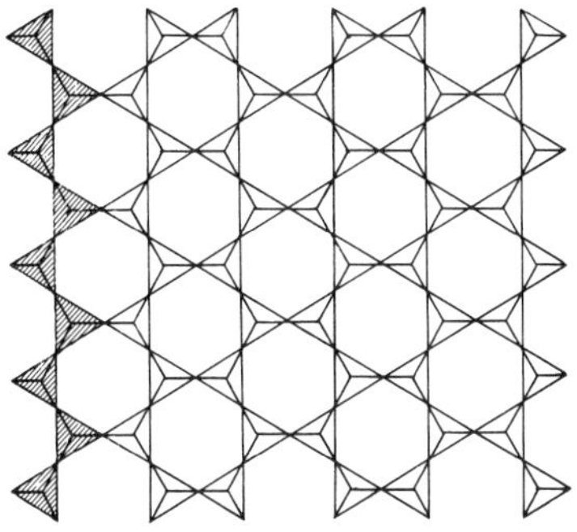

(a)

Fig. 6.41. Single and double layers (the fundamental chain is hatched) and the structural formulae of: (a) **muscovite**, $KAl_2\{uB, 1\,{}^{2}_{\infty}\}[{}^{2}(AlSi_3)O_{10}](OH)_2$ and **talc**, $Mg_3\{uB, 1\,{}^{2}_{\infty}\}[{}^{2}Si_4O_{10}](OH)_2$; (b) **apophyllite**, $KCa_4\{uB, 1\,{}^{2}_{\infty}\}[{}^{4}Si_4O_{10}]_2(F, OH)\cdot 8H_2O$; (c) **hexacelsian**, $Ba\{uB, 2\,{}^{2}_{\infty}\}[{}^{2}(AlSi)O_4]_2(hT)$.

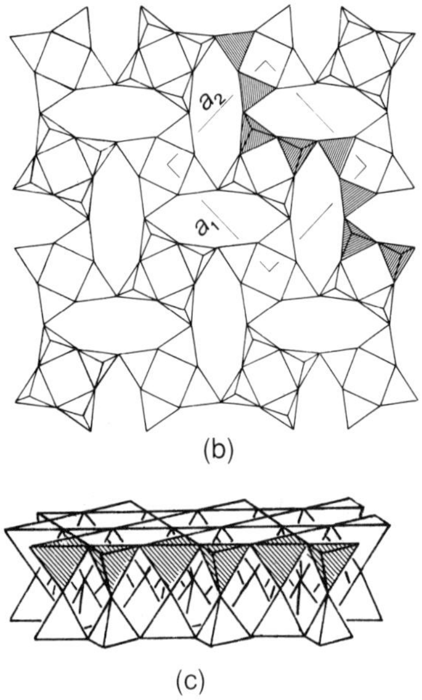

Fig. 6.41. (*Continued*)

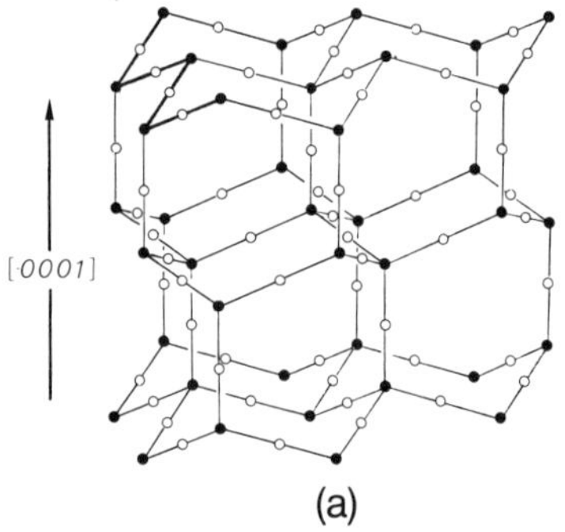

Fig. 6.42. Three-dimensional buildings of $[TO_4]$ tetrahedra, and the structural formulae of: (a) **tridymite**, $\{uB, {}^{3}_{\infty}\}[{}^{2}Si_2O_4]$; (b) **cristobalite**, $\{uB, {}^{3}_{\infty}\}[{}^{2}Si_2O_4]$ (note the thickened fundamental chain and how in cristobalite the layers perpendicular to [111] are all orientated the same way, while in tridymite the same layers perpendicular to [0001] are turned through 180° compared to one another); (c) **orthoclase**, $K\{lB\ {}^{3}_{\infty}\}[{}^{3}(AlSi_3)O_8](mT)$. The structure of orthoclase is represented schematically. The letters U and D indicate the position of the $[TO_4]$ tetrahedra with one vertex pointing either up (U) or down (D); (d) the fundamental chain *lB*, on which the structure of orthoclase is based (the 'branch' consists of the dotted tetrahedra).

structure is composed of, the more stable it is. This explains the uniform number reduction of the groups as *M* increases.

What has been said above regarding linear groups can be extended to chains or layers, by considering the whole chain or the whole layer as a structural unit, instead of a tetrahedron.

5. *D*: the general rule is that where there is a fixed Si–O ratio, the silicate anions tend to join together, in accordance with $D \rightarrow \max$. Though there are exceptions to this rule, it can be justified by the fact that condensation of the $[SiO_4]^{4-}$ tetrahedron results in a better local electrostatic valence balance, in accordance with Pauling's second rule.

6: t, r: multiple tetrahedron silicate anions are more stable than cyclic ones. In fact, since the average Si^{4+}–Si^{4+} distances are greater for the former, the force of repulsion between Si^{4+} and Si^{4+} is less.

7: *P*: the periodicity of a chain can be evaluated, in general, by means of the stretching factor f_s. The higher this factor, the more the chain is stretched, and the lower is *P*. There are, however, important exceptions where *P* increases with f_s (pyroxenoids and pyroxenes).

8. f_s: this factor depends, in particular, on EN and on the average valence $\langle v \rangle$ of the cations. Strongly electropositive cations tend to increase the negative charge of the $[TO_4]$ groups; therefore the repulsive forces which act between these groups are greater and as a result $f_s \rightarrow 1$. Vice versa, strongly electronegative cations, by reducing the charge of the $[TO_4]$ groups, reduce f_s and so *P* tends to increase.

An increase in the $\langle v \rangle$ of the cations has a similar effect. In fact, when $\langle v \rangle$ is high, a greater number of oxygens of the chain must be involved in cation–oxygen bonds. As a consequence f_s is reduced and *P* increases.

The ionic radius of the cation has more effect on the distortion of the polyhedra than on the periodicity *P*. The greater the difference between the silicon radius and that of the other cations, the greater this effect will be.

9. N_{an}: bearing in mind still the principle of parsimony, it can be stated that the number of the structural units, distinguished either chemically or geometrically, must be as small as possible. This explains why the vast majority of silicates have $N_{an} = 1$; only a few structures have $N_{an} = 2$ (e.g. okenite, $Ca_{10}\{uB, 2\,{}^{1}_{\infty}\}[{}^{3}Si_6O_{16}]\{uB, 1\,{}^{2}_{\infty}\}[{}^{3}Si_6O_{15}]_2 \cdot 18H_2O$) and there are no known structures at present with $N_{an} > 2$. Silicates with mixed anions are favoured by cations with high EN.

10. *s*: for a given Si:O ratio of the silicate anion, high EN cations favour a high *s* value.

Appendices

6.A Application of the concept of the packing coefficient (c_i)

In general, calculated packing coefficients almost never match the expected value (0.74). It is sometimes found that structures with lower c_i (≈ 0.60)

conform to closest packings model whereas others with c_i (≈ 0.70) depart from them. The main reason is that in fact ions are not hard spheres but charged, elastic, and polarizable entities. In addition the atoms are sometimes too bulky to be included in the octahedral holes of a closest packing scheme; in such cases the structure would not have the greatest economy of space, i.e. it would not be the closest-packed structure. Thus the ideal conditions demanded by the mechanical atomic model are in practice never achieved.

In spite of this, the packing coefficient provides useful information, mostly for structures with a high degree of ionic bonding. Assuming, for oxygen, $r_e = 1.40$ Å, packing coefficients are calculated for some compounds belonging to several classes of minerals with different ionic bond-character. The abbreviation c.p. shows those compounds which adopt a close-packed arrangement of anions, while δ represents density.

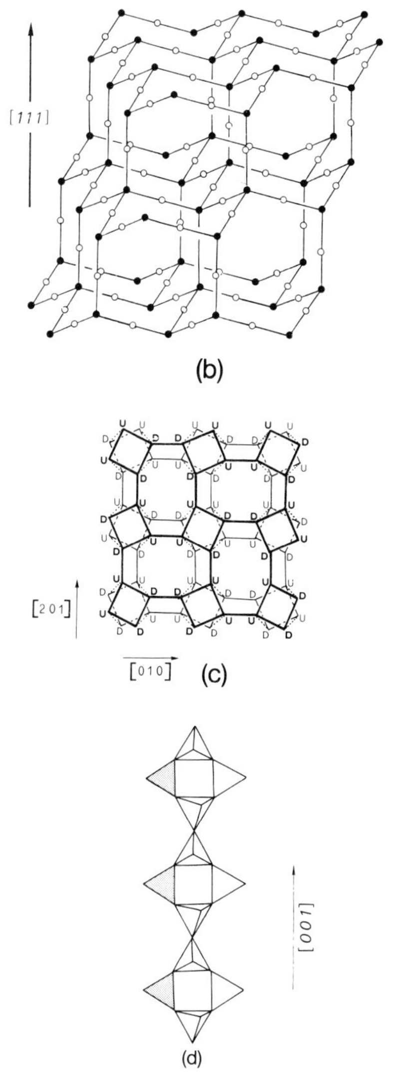

Fig. 6.42. (*Continued*)

Oxides

Periclase [MgO]	$c_i = 0.62$, c.p.
Rutile [TiO_2]	$c_i = 0.74$, c.p.
β-Tridymite [SiO_2]	$c_i = 0.50$ ($\delta = 2.22$)
Coesite [SiO_2]	$c_i = 0.67$ ($\delta = 2.91$)
Stishovite [SiO_2]	$c_i = 0.99$, c.p. ($\delta = 4.29$)
Brucite [$Mg(OH)_2$]	$c_i = 0.57$, c.p.
Gibbsite [$Al(OH)_3$]	$c_i = 0.65$, c.p.
Ilmenite [$FeTiO_3$]	$c_i = 0.66$, c.p.
Spinel [$MgAl_2O_4$]	$c_i = 0.70$, c.p.

Carbonates

Aragonite [$CaCO_3$]	$c_i = 0.61$ ($\delta = 2.93$)
Calcite [$CaCO_3$]	$c_i = 0.56$, c.p. ($\delta = 2.71$)
Alstonite [$BaCa(CO_3)_2$]	$c_i = 0.51$
Ewaldite [$Ba_3Ca_2(CO_3)_5$]	$c_i = 0.55$
Shortite [$Na_2Ca_2(CO_3)_3$]	$c_i = 0.53$
Tychite [$Na_2Mg_2(CO_3)_4(SO_4)$]	$c_i = 0.55$
Gaylussite [$Na_2Ca(H_2O)_5(CO_3)_2$]	$c_i = 0.51$

Borates

Kotoite [$Mg_3(BO_3)_2$]	$c_i = 0.67$, c.p.
Tincalconite [$Na_2B_4O_5(OH)_4 \cdot 3H_2O$]	$c_i = 0.55$
Kernite [$Na_2B_4O_6(OH)_2 \cdot 3H_2O$]	$c_i = 0.53$
Borax [$Na_2B_4O_5(OH)_4 \cdot 8H_2O$]	$c_i = 0.53$
Sinhalite [$MgAl(BO_4)$]	$c_i = 0.76$, c.p.
Aksaite [$Mg(B_3O_4(OH)_2)_2 \cdot 3H_2O$]	$c_i = 0.44$
Gowerite [$Ca(B_3O_4(OH)_2)_2 \cdot 3H_2O$]	$c_i = 0.42$

Sulphates

Baryte [$BaSO_4$]	$c_i = 0.53$
Chlorothionite [$K_2Cu(SO_4)Cl_2$]	$c_i = 0.50$
Fibroferrite [$Fe(OH)SO_4 \cdot 5H_2O$]	$c_i = 0.53$
Parabutlerite [$Fe(OH)SO_4 \cdot 2H_2O$]	$c_i = 0.60$
Hohmannite [$Fe(H_2O)_4[(SO_4)_2O] \cdot 4H_2O$]	$c_i = 0.58$
Coquimbite [$Fe_2(SO_4)] \cdot 9H_2O$	$c_i = 0.55$

Phosphates

Triphylite [$Li(Fe^{2+}, Mn^{2+})PO_4$]	$c_i = 0.63$, c.p.
Heterosite [$(Mn, Fe)PO_4$]	$c_i = 0.68$, c.p.
Hydroxylapatite [$Ca_5(OH)(PO_4)_3$]	$c_i = 0.56$
Brushite [$CaHPO_4{\cdot}2H_2O$]	$c_i = 0.55$
Moraesite [$Be_2(OH)PO_4{\cdot}4H_2O$]	$c_i = 0.56$
Struvite [$NH_4MgPO_4{\cdot}6H_2O$]	$c_i = 0.48$

Silicates

Forsterite [Mg_2SiO_4]	$c_i = 0.64$, c.p.
Larsenite [$PbZnSiO_4$]	$c_i = 0.47$
Monticellite [$CaMgSiO_4$]	$c_i = 0.54$, c.p.
Humite [$Mg_7(OH, F)_2(SiO_4)_3$]	$c_i = 0.59$, c.p.
Zircon [$ZrSiO_4$]	$c_i = 0.70$
Grossular [$Ca_3Al_2Si_3O_{12}$]	$c_i = 0.66$
Enstatite [$Mg_2Si_2O_6$]	$c_i = 0.66$
Anthophyllite [$Mg_7Si_8O_{22}(OH)_2$]	$c_i = 0.62$
Pyrophyllite [$Al_2Si_4O_{10}(OH)_2$]	$c_i = 0.65$
Orthoclase [$KAlSi_3O_8$]	$c_i = 0.51$

Some remarks concerning the relationships between the anion arrangement and the packing coefficient are helpful.

The compound $CaCO_3$ crystallizes in two structures (polymorphs): calcite ($c_i = 0.56$) and aragonite ($c_i = 0.61$). Calcite can be described by means of a close-packed oxygen arrangement with $\frac{1}{12}$ triangular and $\frac{1}{3}$ octahedral sites filled respectively by C^{4+} and Ca^{2+}. This packing is not the closest possible for $CaCO_3$ because Ca^{2+} is surrounded by six oxygens, i.e. in this structure, calcium shows the lowest CN among those adopted (see Table 6.7). If pressure increases, a more compact structure (aragonite in which Ca^{2+} has CN = 9) originates; this structure no longer conforms to the closest packing model.

The compound SiO_2 gives rise to several phases. Some of them are considered here, they are: (1) β-tridymite, stable between 870 and 1470 °C; (2) coesite, roughly stable from 30 to 100 kbar; and (3) stishovite, a very high-pressure phase, stable above 100 kbar. β-tridymite structure ($c_i = 0.50$, $\delta = 2.22$) is an infinite three-dimensional framework of $[SiO_4]^{4-}$ tetrahedra. It can be sliced into sheets of tetrahedra like muscovite (Fig. 6.41(a)) but arranged in such way that the tetrahedra vertices alternate up and down. The structure of coesite ($c_i = 0.67$, $\delta = 2.91$) is quite dense and somewhat more complex compared with that of β-tridymite. The main difference between them concerns the second and following Si^{4+} coordinations that, of course, are greater for coesite. This explains the different packing coefficients, though none of them (see p. 429 and following) conforms to the closest packing. Stishovite has an exceptionally high packing coefficient ($c_i = 0.99$) and a density ($\delta = 4.29$) much greater than that of β-tridymite or coesite. The structure is rutile type, TiO_2, (Fig. 6.26, (c)), so it conforms to the closest packing model. The very high pressure under which stishovite crystallizes forces Si^{4+} to renounce its habitual four for an unusual six coordination.

According to the mechanical closest packing model it is impossible to have $c_i \approx 1$, unless we suppose that the available space of the structure is almost completely filled by oxygens having an approximate volume of a ball with $r_e = 1.40$ Å but geometrical form by no means spherical.

The compound grossular, $Ca_3Al_2Si_3O_{12}$, belongs to an important group of minerals known as garnets. In spite of the high packing coefficient ($c_i = 0.66$), grossular does not have a close packed structure (see Appendix 6.B). Here we note that Ca^{2+} cannot replace O^{2-} as occurs in perovskite. Let us consider the structure of perovskite (Fig. 6.28) and its chemical formula in the form $Ca_4Ti_4 \boxslash_{24}O_{12}$ ($\boxslash$ represents the empty tetrahedral site).

Figures 6.19 and 6.28 show that if grossular were to adopt a perovskite-type structure in half of the cells shown in Fig. 6.28, Si^{4+} in tetrahedral holes would be in contact with Ca^{2+}. After all, according to the examples quoted, the packing coefficient is trustworthy only in a few cases, i.e. for those structures having suitable cations and a high degree of ionic bonding (as happens in several oxides). Its value can give little information on the real atomic packing, which can sometimes be enhanced only by further crystallochemical parameters (see Appendix 6.B).

6.B Structural inferences from crystallochemical parameters

The effective ionic radius r_e, the packing coefficient (c_i) and the valence electrostatic rule ($V = \Sigma_s$) can sometimes be profitably used to foresee some important structural features particularly when simple crystal structures are considered. First of all one should extract all the possible information from the chemical formula. On p. 437 we have seen that for the compounds of MX type, the coordination numbers of M and X can vary provided that the ratio 1:1 is preserved; for instance in NaCl, M:X = 6:6 and in CsBr M:X = 8:8. Again, in the MX_2 compounds the coordination number of M and X can change, the ratio M:X = 2:1 being equal. So, salts like SiO_2, TiO_2 and CaF_2 have the ratio M:X = 2:1 and respectively the following coordination numbers: 4:2, 6:3 and 8:4.

From these remarks the following rule can be derived: *in a given structure, if no bond between identical atoms forms,* M *and* X *coordination numbers of compounds* M_mX_p, *are in the ratio* $p:m$.

This rule can also be applied to more complex structures like $A_mB_nX_p$.

First let us define the average cation coordination number ($\overline{CN}_c$):

$$\overline{CN}_c = (mCN_A + nCN_B)/(m + n) \qquad (A.1)$$

where CN_A and CN_B are respectively the coordination numbers of cations A and B.

The relationship between $\overline{CN}_c$ and the average anion coordination number ($\overline{CN}_x$) is expressed by

$$\overline{CN}_c/\overline{CN}_x = p/(m + n). \qquad (A.2)$$

Replacing A.1 in A.2, the latter can be written as

$$mCN_A + nCN_B = p\overline{CN}_x. \qquad (A.3)$$

Some examples are given, in order of increasing structural complication, to understand the relationships between chemical formula, crystallochemical parameters, and symmetry, showing how they can sometimes have predictive value.

At first, we will assume tentatively that the ionic structures having $c_i > 0.60$ can be described in terms of a close-packed anion arrangement with cations filling the holes, even if in some cases this hypothesis does not hold (as some of the following examples will prove).

1. **Periclase**, MgO, $Fm\bar{3}m$, and $Z = 4$, has a packing coefficient $c_i = 0.62$. According to the accredited Mg^{2+} CN_s 4, 5, 6, 8 (see Table 6.7) this cation can fill the tetrahedral (half) or the octahedral site of a structure based on a close-packing model. Moreover the space group $Fm\bar{3}m$ and the number of formula units $Z = 4$ informs us that Mg^{2+} and O^{2-} lie at $\frac{1}{2}, \frac{1}{2}, \frac{1}{2}$ and O, O, O or vice versa, from which $\overline{CN}_c = CN_c$ follows. Using (A.3) in the more simple form $mCN_A = pCN_x$ we reach an obvious conclusion, i.e. Mg^{2+} and O^{2-} have the same CN_s (4 or 6) and of course for both coordinations the electrostatic-valence principle is satisfied. To solve the CN dilemma, we observe that periclase is isotype with lime, CaO, and that accredited CN_s for Ca^{2+} are 6, 7, 8, 9, 10, 12 (Table 6.7). Only 6 and 8 belong both to Mg^{2+} and Ca^{2+} CN_s sets, consequently only the octahedral sites of a closed-packed anion arrangement can be filled.

2. **Rutile**, TiO_2, $P4_2/mnm$, and $Z = 2$, has a packing coefficient $c_i = 0.74$. The accredited CN_s of Ti^{4+} are 4, 5, 6, 8 (Table 6.7). Therefore, according to the close packing and valence requirements, Ti^{4+} can fill tetrahedral or octahedral sites. The octahedral site stabilization energy (OSSE = 0) cannot remove the uncertainty regarding the site, even if the high electrical charge informs us that Ti^{4+} prefers the octahedral site. It can be observed that, according to the space group, Ti^{4+} can be placed at 0, 0, $\frac{1}{2}$ or 0, 0, 0 and that O^{2-} can lie on 2/m or 4 or mm (from 4e to 4g according to *International Tables of X-ray Crystallography* Vol. A) in this case $\overline{CN}_x = CN_x$. No profitable information can be drawn from isotype structures, but we know that Nb^{4+} and Ta^{4+} can replace Ti^{4+} in very high percentage (≈ 0.40). The common CN for these two cations, i.e. 6, indicates that very probably, Ti^{4+} prefers the octahedral site.

From (A.3) the calculated oxygen coordination number is $\overline{CN}_x = 3$, i.e. three Ti^{4+} surround one O^{2-} whereas, as shown above, Ti^{4+} seems linked to six oxygens.

3. **Ilmenite**, $FeTiO_3$, $R\bar{3}$, and $Z = 2$, has a packing coefficient $c_i = 0.66$. According to the space group, two Fe^{2+} and two Ti^{4+} can lie on a three-fold axis or one of them on a three-fold axis and the other on the two independent $\bar{3}$ (1a, 1b). As for the oxygen atoms, they can be situated on the two independent $\bar{1}$ (3e, 3d) or in the general position 1 (6f); in both cases ($\bar{1}$ or 1) CN_x must be an integer. The accredited CN_s for Fe^{2+} and Ti^{4+} are respectively 4, 6, 8 and 4, 5, 6, 8 (Table 6.7), so either Fe^{2+} or Ti^{4+} may fill tetrahedral and/or octahedral sites. From (A.3) three different $\overline{CN}_x$ can be calculated for ilmenite:

$$1{\cdot}4 + 1{\cdot}4 = 3{\cdot}\overline{CN}_x;\ \overline{CN}_x = 8/3$$
$$1{\cdot}4 + 1{\cdot}6 = 3{\cdot}\overline{CN}_x;\ \overline{CN}_x = 10/3$$
$$1{\cdot}6 + 1{\cdot}6 = 3{\cdot}\overline{CN}_x;\ \overline{CN}_x = 4.$$

Only the third result agrees with the statement that $\overline{CN}_x$ must be an integer and precisely it tells us that four cations are linked to each oxygen without specifying their nature. Applying the electrostatic-valence rule to the three possible combinations $1Fe^{2+} + 3Ti^{4+}$, $2Fe^{2+} + 2Ti^{2+}$, $3Fe^{2+} + 1Ti^{4+}$, we found respectively 2.33, 2, 1.67 v.u. (valence units). So $2Fe^{2+}$ and $2Ti^{4+}$ are the cations coordinated by one oxygen atom, in agreement with the stoichiometry.

4. **Perovskite**, $CaTiO_3$, $Pm\bar{3}m$, and $Z = 1$, has a packing coefficient $c_i = 0.62$. The space group and Z inform us that Ca^{2+} and Ti^{4+} are in $m\bar{3}m$ (1a, 1b) and that oxygen lies on 4/mmm (3c or 3d), i.e. the number of cations around it is an integer and, what is more, even. The accredited CN_s for Ca^{2+} are 6, 7, 8, 9, 10, 12 and for Ti^{4+} they are 4, 5, 6, 8. So according to c_i and CN_s Ca^{2+} and Ti^{4+} should fill, like ilmenite, octahedral sites. This eventuality would require that $\frac{2}{3}$ of the octahedral sites be filled by Ca^{2+} and Ti^{4+} and that consequently several edges (three are in the sheet) be shared between Ca^{2+} and Ti^{4+} octahedra. Now we observe that for $CN = 6$, different from ilmenite ($r_e = 0.61$ (Fe^{2+} low spin) and $r_e = 0.605$ Å (Ti^{4+}), the Ca^{2+} and Ti^{4+} effective ionic radii are quite different: $r_e = 1.00$ (Ca^{2+}) and $r_e = 0.605$ Å (Ti^{4+}). This strong difference is reflected on the octahedral edges that, if shared, would involve a strong strain in the structure and consequently the increase of its potential energy. Moreover for $CN = 12$, the Ca^{2+} effective ionic radius ($r_e = 1.34$ Å) matches well with the O^{2-} one, so we have a special close packing form in which $\frac{1}{4}$ of oxygens are replaced by Ca^{2+}. For Ti^{4+} in tetrahedral or octahedral sites and Ca^{2+} 12-coordinated, (A.3) gives respectively $\overline{CN}_x = \frac{16}{3}$ or $\overline{CN}_x = 6$, supporting the hypothesis that Ti^{4+} is six-coordinated like the oxygen. As far as the nature of the cations around the oxygen is concerned, the valence sum principle indicates that around the oxygen atom there are two Ti^{4+} and four Ca^{2+}.

5. **Zircon**, $ZrSiO_4$, $I4_1/amd$, and $Z = 4$, has a packing coefficient $c_i = 0.70$. In accordance with the space group and the number of formula unit Z, zirconium and silicium are located at $\bar{4}2m$ (4a and 4b). In theory, oxygen can be placed at 2/m (8c and 8d) or at 2mm (8e × 2), but it can easily be shown that starting from the unit cell parameters, $a = 6.60$ and $c = 5.98$ Å, the calculated interatomic distances Zr^{4+}–O_2^{2-} and Si^{4+}–O^{2-} do not match well with those obtained from the effective ionic radii (Table 6.7). For instance, supposing the cation (M^{4+}), i.e. Zr^{4+} or Si^{4+}, at 0, 0, 0 (4a) and O^{2-} at 0, $\frac{1}{4}$, $\frac{1}{8}$ (8c), then the M^{4+}–O^{2-} calculated distance is 1.81 Å $\neq$ 1.66 (Si^{4+}–O^{2-}) and 2.12 (Zr^{4+}–O^{2-}). On the other hand the situation does not improve if O^{2-} is located at 0, 0, z because the calculated Zr^{4+}–O and Si^{4+}–O distance sums are: 2.99 Å (5.98/2 Å) $\neq$ 3.78 Å (1.66 + 2.12 Å). These kinds of argument support the hypothesis that O^{2-} is situated on 2 or m (16f or 16g or 16h), so $\overline{CN}_x = CN_x$.

The packing coefficient would suggest that zircon adopts a close-packing structure. However, if Si^{4+} cations are located at the habitual tetrahedral holes, $\overline{CN}_x$ assumes integer values (A.3) when Zr^{4+} fill tetrahedral sites or when it is 8- or 12-coordinated. Let us now consider, besides $ZrSiO_4$, the zircon isotype structures: **coffinite**, $USiO_4$; **thorite**, $ThSiO_4$; **Hafnon**, $HfSiO_4$. The accredited CN_s (Table 6.7) of the bigger cations are: 4, 5, 6, 7, 8, 9 (Zr^{4+}); 6, 7, 8, 9, 12 (U^{4+}); 6, 8, 9, 10, 11, 12 (Th^{4+}); 4, 6, 7, 8 (Hf^{4+}). They show that some CN_s (6 and 8) are present in all the four cations. This peculiarity together with the results of (A.3) suggest that a packing, different from the close packing, in which Zr^{4+} is 8-coordinated, characterizes the crystal structure of zircon. In addition we note that the electrostatic valence balance is not compatible with a structure having Zr^{4+} in octahedral sites. In fact, since the oxygen atom is shared by Si^{4+} and Zr^{4+}, the bond strengths coming from Si^{4+} (+1 v.u.) and from Zr^{4+} (+4/6 or +8/6 v.u.) do not balance O^{2-} valence. By means of (A.3), we can conclude that around O^{2-} there are three cations, the nature of which can be specified by means of the electrostatic-valence rule: two Zr^{4+} and one Si^{4+}.

6. **Spinel**, $MgAl_2O_4$, $Fd\bar{3}m$, $Z = 8$ has a packing coefficient $c_i = 0.70$. In the unit cell there are $8Mg^{2+}$, $16Al^{3+}$, and $32O^{2-}$. Magnesium lie on $\bar{4}3m$ (8a or 8b), aluminium on $\bar{3}m$ (16c or 16d), and consequently oxygen is on 3m (32e), so $\overline{CN}_x = CN_x$ is an integer. If the c_i value and the accredited CN_s for Mg^{2+} (4, 5, 6, 8) and for Al^{3+} (4, 5, 6) are taken into account, then the spinel structure would seem to conform to a close-packing one with tetrahedral and/or octahedral sites partially filled. On the other hand, (A.3) leads to three possible configurations for $\overline{CN}_x$ integer: (1) Mg^{2+} and Al^{3+} in tetrahedral; (2) Mg^{2+} in tetrahedral and Al^{3+} in octahedral sites; and (3) half Al^{3+} in tetrahedral and half in octahedral together with Mg^{2+}. The calculated $\overline{CN}_x$ of oxygen is three for the first and four for the other two configurations. As concerns other factors influencing crystal structures, the effective ionic radius seems to favour Al^{3+} in tetrahedral and Mg^{2+} in octahedral sites, whereas the formal charge indicates that cations follow the opposite behaviour pattern. The quoted ambiguity allows us to understand why direct and inverse spinel structure exists, as is better explained in a more general way at p. 443.

Finally, for cations located in tetrahedral sites, both the chemical formula and the valence balance require two Al^{3+} and one Mg^{2+} around O^{2-} whereas when Mg^{2+} and Al^{3+} are respectively in tetrahedral and octahedral holes the chemical formula indicates that oxygen is surrounded by one Mg^{2+} and three Al^{3+}.

7. **Pyrope**, $Mg_3Al_2Si_3O_{12}$, $Ia\bar{3}d$, $Z = 8$, has packing coefficient 0.73. The number of formula units Z and the space group suggest that Al^{3+} lies on 32 or $\bar{3}$ (16b or 16a) and that Mg^{2+} and Si^{4+} are on 222 and $\bar{4}$ or vice versa. As regards the oxygen atoms these can be located on 3 (32e), 2 (48f and 48g), or 1 (96h) axes. The shortest inter-ionic distances between cations, lying on 222 (24c) and $\bar{4}$ (24d), and oxygen lying on 3 (32e) or on 2 axes (48f and 48g) have the following values: 3.09 and 2.865 Å, far from those inferred from Table 6.7, i.e. Mg^{2+}–O = 2.29 Å or Si^{4+}–O = 1.66 Å. Besides symmetry, valence balance and chemical formula requirements would seem to exclude the location of an oxygen atom on the quoted special positions, so the most reliable point symmetry for oxygen atom seems 1. The CN_s used for Mg^{2+} in the latter analysis were limited to 6 and 8 because pyrope is an isotype with grossular, $Ca_3Al_2Si_3O_{12}$ and the common CN_s between Mg^{2+} and Ca^{2+} are 6 and 8.

The modified form of equation (A.3) applicable to pyrope is

$$mCN_A + nCN_B + qCN_C = \overline{CN}_x. \tag{A.4}$$

It can be easy verified that only when Ca^{2+} has CN = 8, $\overline{CN}_x$ become a whole number, i.e. $\overline{CN}_x = 4$.

Finally by means of the chemical formula and the application of the electrostatic-valence rule it is possible to characterize the oxygen coordination, i.e. two Mg^{2+}, one Al^{3+}, and one Si^{4+}.

Generally speaking it is difficult to imagine making a correct inference on the atomic organization of a complex structure without having some preliminary information or without any assumption concerning it. For instance, coquimbite has $(Fe^{3+}, Al)_2(SO_4)_3{\cdot}9H_2O$, $P\bar{3}1c$, $Z = 4$, $V = 1760.5$ Å^3 and $c_i = 0.55$. If the usual CN_s 6 and 4 for this kind of structures are respectively assumed for Fe^{3+} and S^{6+} and if it is supposed that H^+ forms no bifurcated hydrogen bonds in coquimbite, then the $\overline{CN}_x$ of oxygen can be inferred. In fact using (A.4) we have $2{\cdot}6 + 3{\cdot}4 + 2{\cdot}18 = 21CN_x$, from which $\overline{CN}_x = 2.86$ can be estimated. The structure informs us that: O(1) and O(2) are shared by one Fe^{3+} (or Al^{3+}) and one S^{6+}; O(3) and O(4) are shared by one S^{6+} and two H^+; $O_w(1)$ and $O_w(2)$ are shared by one Fe^{3+} (or Al^{3+}) and two H^+; $O_w(3)$ is shared by four H^+. This configuration leads to the quoted $\overline{CN}_x = 2.86$.

References

1. Cartemel, E. and Fowles, G. W. A. (1966). *Valency and molecular structure.* Butterworth, London.
2. Huheey, J. E. (1983). *Inorganic chemistry. Principles of structure and reactivity,* (3rd edn). Harper, New York.
3. Vainshtein, B. K., Fridkin, V. M., and Indenbom, V. L. (1982). *Modern crystallography II.* Springer. Berlin.
4. Pauling, L. (1959). *The nature of the chemical bond.* Cornell University Press, Ithaca.
5. Hamilton, W. C. and Ibers, J. A. (1968). *Hydrogen bonding in solids.* Benjamin, New York.
6. Greenwood, N. N. (1970). *Ionic-crystals, lattice defects and non-stoichioimetry.* Butterworths, London.
7. Catti, M. (1978). *Acta Crystallographica,* **A34,** 974.
8. Cotton, F. A. and Wilkinson, G. (1980). *Advanced inorganic chemistry,* (4th edn). Wiley, New York.
9. Orgel, L. E. (1960). *An introduction to transition-metal chemistry ligand field theory.* John Wiley, New York.
10. Basolo, F. and Johnson, R. (1964). *Coordination chemistry.* Benjamin, New York.
11. Ballhausen, C. J. (1962). *Introduction to ligand field theory.* McGraw-Hill, New York.
12. Burns, R. G. (1970). *Mineralogical applications of crystal field theory.* Cambridge University Press.
13. Busing, W. R. (1981). *WMIN, computer program to model molecules and crystals in terms of potential energy functions,* U.S. National Technical Information Service, ORNL-5747.
14. Catlow, C. R. A. and Cormack, A. N. (1984). *Acta Crystallographica,* **B40,** 195.
15. Alberti, A. and Vezzalini, G. (1978). *Zeitschrift für Kristallographie,* **147,** 167.
16. Brown, G. E. and Fenn, P. M. (1979). *Physics and Chemistry of Minerals,* **4,** 83.
17. Parker, S. C., Catlow, C. R., and Cormack, A. N. (1984). *Acta Crystallographica,* **B40,** 200.
18. Shannon, R. D. and Prewitt, C. T. (1969). *Acta Crystallographica,* **B25,** 925.
19. Shannon, R. D. (1976). *Acta Crystallographica,* **A32,** 751.
20. Fajans, K. (1923). *Naturwissenschaften,* **11,** 165.

21. Magini, M. (1979). *Journal of Chemical Physics,* **70,** 317.
22. Redinger, J. and Schwartz, K. (1981). *Zeitschrift für Physik,* **B40,** 269.
23. Phillips, J. C. (ed.) (1974). *Treatise on solid state chemistry,* Vol. 1. Plenum, New York.
24. Pearson, W. B. (ed.) (1972). *Crystal chemistry and physics of metals and alloys.* Wiley, New York.
25. Basso, R. (1985). *Neues Jahrbuch für Mineralogie.* **3,** 108.
26. Hawthorne, F. C. (1981). *Journal of Solid State Chemistry,* **37,** 157.
27. Lamgley, R. H. and Sturgeon, G. D. (1979). *Journal of Solid State Chemistry,* **34,** 79.
28. Novak, G. A. and Gibbs, G. V. (1971). *American Mineralogist,* **56,** 791.
29. Moore, P. B. and Smith, J. V. (1970). *Physics of the Earth and Planetary Interiors,* **3,** 166.
30. *International tables for x-ray crystallography,* Vol. II, (1959). Birmingham: Kynock Press.
31. Iglesias, J. E. (1981). *Zeitschrift für Kristallographie,* **155,** 121.
32. Basso, R. (1985). *Neues Jahrbuch für Mineralogie,* **3,** 108.
33. Brown, I. D. and Shannon, R. D. (1973). *Acta Crystallographica,* **A29,** 266.
34. Brown, I. D. and Wu, K. K. (1976). *Acta Crystallographica,* **B32,** 1957.
35. Brown, I. D. (1978). *Chemical Society Reviews,* Vol. 7, No. 3. The Chemical Society, London.
36. Scordari, F. and Stasi, F. (1990). *Neues Jahrbuch für Mineralogie,* **161,** 241.
37. Süsse, P. and Tilmann, B. (1987). *Zeitschrift für Kristallographie,* **179,** 323.
38. Evans, R. C. (1976). *An introduction to crystal chemistry* (2nd edn). Cambridge University Press.
39. Ottonello, G. (1986). *Physics and Chemistry of Minerals,* **13,** 79.
40. Baumgartner, O., Preisinger, A., Heger, G., and Guth, H. (1981). *Acta Crystallographica,* **A37,** C187.
41. Yamanaka, T. and Takeuchi, Y. (1983). *Zeitschrift für Kristallographie,* **165,** 65.
42. Liu, L. G. (1976). *Physics of the Earth and Planetary Interiors,* **11,** 289.
43. Zoltai, T. (1960). *American Mineralogist,* **45,** 960.
44. Ribbe, P. H. (1982). *Orthosilicates,* (2nd edn), Reviews in Mineralogy No. 5. Mineralogical Society of America, Washington.
45. Liebau, F. (1985). *Structural chemistry of silicates.* Springer, Berlin.

Molecules and molecular crystals

7

GASTONE GILLI

Chemistry and X-ray crystallography

Crystal and molecular structure

X-ray diffraction from crystals was discovered at the beginning of this century and it soon became evident that it was suited to the investigation of the structure of matter at atomic resolution level. Diffraction applications of X-rays and, later on, of other radiations of comparable wave length, such as electrons or thermal neutrons, developed rapidly allowing the scientific world to obtain, for the first time, a detailed knowledge of the intimate atomic distribution in elements, organic and inorganic substances, natural and synthetic fibres, and even liquids. It became clear, however, over the years, that the level of image resolution achievable was strongly dependent on the aggregation state of matter. Currently we know that a resolution of a few thousandths of an angstrom is actually obtainable *only* by diffraction experiments carried out on relatively perfect single crystals, both organic and inorganic, while macromolecular crystals of proteins (haemoglobin, myoglobin, and enzymes) cannot give an accuracy better than, say, 0.5 Å and fibres or liquids produce poorly resolved images, which can be considered barely giving much more than a hint of the true molecular structure.

The experimental resolution achievable is to be compared with the natural resolution needed by classical chemistry. For example, carbon–carbon distances are typically 1.54 for single bond, 1.39 for aromatic compounds, and 1.33 and 1.20 Å for double and triple bonds, respectively; thus a total range of 0.034 Å covers the fantastic change of properties and reactivities between alkynes and alkanes and the much smaller range of 0.006 Å separates aromatic from alkene chemistry. Taking into account that two bond distances cannot be reasonably compared at less than four e.s.d. (estimated standard deviations), we must conclude that no valuable chemical information is contained in C–C distances known with an e.s.d. greater than 0.01 Å and, therefore, in crystal structures where the e.s.d.s on atomic positional coordinates are greater than about 0.007 Å. This value sets a clear limit on diffraction experiments. Only elements, minerals, and small molecules easily develop into long-range ordered three-dimensional arrays of perfect or ideally imperfect crystals, and only diffraction experiments on such crystals can provide independent experimental evidence of direct relevance to chemistry. Conversely, synthetic or natural macromolecules, which crystallize as highly defective hydrated monocrystals in the most favourable case of globular proteins, but often as fibrous materials having simple one-dimensional order, cannot give much more than the overall shape of the diffracting molecule and their study is to be considered more a branch of high-resolution microscopy than a part of the structural chemistry in a strict sense.

The expression **structural crystallography** is often used for identifying the branch of sciences producing direct knowledge of the structure of matter by diffraction methods. Within structural crystallography there are divisions which have become traditional: crystal chemistry of metals and alloys, minerals, molecular and macromolecular crystals are usually taught in different courses, and their scientific achievements published in different journals. This is a kind of functional partitioning, different parts of crystallography being dealt with according to their usefulness to other disciplines.

There are, however, more scientifically founded reasons for a subdivision, which originate from the differences that the knowledge of the crystal structure may have for different classes of chemical substances. This knowledge is of primary value when the atomic binding forces within the crystal can be classified as true chemical bonds, as is the case for **covalent, metallic, or ionic crystals**, where the nature of the bond can be understood only on the basis of the knowledge of the crystal itself. For instance Na_2, C_2, and Si_2 molecules are well known in the gas phase but their properties cannot explain why metallic sodium, diamond or graphite, and silicon may be electronic conductors or insulators or semiconductors. In the same way the reason for the stability of ionic substances is found in the ordered arrangement of alternating anions and cations within crystals where any simple relationship between chemical formula and crystal structure is completely lost. A good example, taken from Wells,[1] may be that of the series of AX iodides: HI, AuI, CuI, NaI, and CsI; only the first term gives biatomic molecules at room temperature, while the others are solids where A is surrounded, in order, by 2, 4, 6, and 8 atoms of X.

The other extreme is represented by the class of **molecular crystals** where chemical substances, typically organic and organometallic compounds, exist as discrete molecular entities packed together by weak dispersion or dipolar forces. Interatomic distances within the crystal can be generally grouped in two clearly distinct distributions, a first of bond distances within the molecule and a second one of the much longer **contact** or **van der Waals distances** among atoms of different molecules. Non-bonded atoms on different molecules tend to behave much like non-interacting hard spheres that cannot approach at a distance shorter than the sum of their encumbrance radii, called **van der Waals radii**, r_{vdW}, and reported in Table 7.1. If the crystal is fused or vapourized by heating or dissolved by a solvent the crystalline order is destroyed, but the geometry and pattern of interatomic connections within the molecules, apart from possible partial rotations around single bonds, remain unchanged.

Table 7.1. Atomic van der Waals radii (Å). H in aromatic rings 1.00 Å; C perpendicularly to the benzene ring 1.77 Å. Data according to Bondi[2]

			H 1.20
C 1.70	N 1.55	O 1.52	F 1.47
Si 2.10	P 1.80	S 1.80	Cl 1.75
	As 1.85	Se 1.90	Br 1.85
			I 1.98

In this case the **crystal structure** (intended as crystal packing) is less important, the chemical information being mainly concentrated in the **molecular structure** which, in principle, could be determined in any other state of aggregation of matter. In fact, methods for accurate determination of structure in liquids or in solutions are not actually available and gas phase methods (microwave spectroscopy and gas electron diffraction) allow us to study only small and preferably symmetrical molecules, so that crystal structure determination often remains the only practicable way in which to obtain detailed information on the molecular structure of chemical compounds. Taking into account the importance of the knowledge of the molecular structure of both synthetic and natural substances, it is not surprising that X-ray structure determination of molecular crystals has so far been used as a simple tool for determining molecular configurations and conformations and that crystal packing has mostly been seen either as an irrelevant factor or as an unwanted cause of perturbation of the free-molecule geometry. Only recently has the idea that crystal packing could be important in itself in the study of molecular interactions been widely, if not generally, accepted.

The growth of structural information

X-ray crystallography of molecular crystals has been so successful in chemistry that the last three decades have registered an almost exponential growth of the number of structures published per year. Structural data had to be collected in computerized databases, and the Cambridge Structural Database (CSD),[3] as at April 1990, contained bibliographic, chemical, and numerical results for some 84 464 organic and organometallic compounds. Figure 7.1 show a histogram of the new entries in the CSD in the period 1965–86.[3] The number of complete structural analyses published up to 1960 was only 579 (45 from 1935–45, 212 from 1945–55, and 322 from 1955–60) and this number is now exceeded tenfold by the structures determined *per year*.

The factors which have determined such a rapid evolution in the field are scientific (development of direct methods theory), technological (improvement in automatic single-crystal diffractometers), and organizational (distribution of complete systems of crystallographic computer programs). All these factors must also be considered with the parallel evolution of fast electronic computers, without which X-ray crystallography would have remained that slow and patient skilled work well known to the scientific world until the end of the 1950s.

The effects produced in chemistry by this extraordinary wealth of structural information are enormous, and for a vivid analysis of this

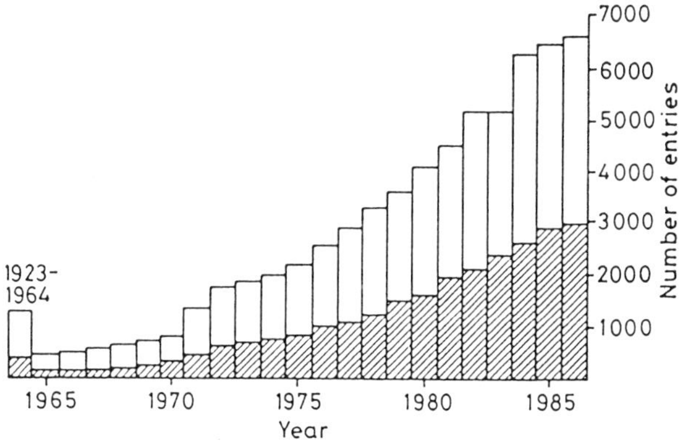

Fig. 7.1. Number of new crystal structures registered each year in the Cambridge Structural Database (CSD)[3] from 1965 to 1986. Structures containing d- or f-block metals are indicated by shading. (Reproduced by permission from Orpen *et al.*[110])

phenomenon the reader is referred to the second part of the Dunitz's book *X-ray analysis and the structure of organic molecules.*[4] The attempts aimed to interpret, rationalize, and understand this continuously growing mass of data give rise to problems far beyond the limits of a single chapter, being dealt with by the scientific discipline of **structural chemistry**, initiated more than 50 years ago by Linus Pauling[5] in his famous book *The nature of the chemical bond and the structure of molecules and crystals. An introduction to modern structural chemistry.*

The present chapter is intended to summarize the structural chemistry of molecular crystals. As regards its content, it has been necessary to be severely selective of topics. The discussion is strictly confined to the nature of molecular crystals and to the stereochemical aspects of molecules which are directly derived from the results of structural investigation. Final applications, such as structure–property studies in chemistry and molecular biology or pharmacology have been completely excluded. The topics have been divided in two parts, according to whether they refer to single molecules or molecular crystals as a whole and can be so summarized:

(1) *molecular crystals*: molecular interactions, their nature and effect on the crystal packing; elements of crystal thermodynamics and polymorphism;

(2) *single molecules*: classical stereochemistry; molecular geometry and chemical bond; molecular mechanics; interpretation of molecular structures and the structure correlation method.

The nature of molecular crystals

Generalities

Molecular crystals are ordered packings of discrete physical entities, i.e. molecules. Most of them consist of neutral molecules, and it is those which are mainly discussed here, though some may sometimes contain ions (e.g. organic carboxylates, alkylammonium halides) as well. The energy of formation of the crystal from the isolated (gas phase) molecules is called **lattice energy**, U; typical values of $-U$ for molecular crystals are 0.5–4 kcal mol^{-1} for noble gases, 10–20 kcal mol^{-1} for most neutral organic compounds and up to 40–50 kcal mol^{-1} for neutral molecules of relevant complexity. By comparison with some typical $-U$ values for ionic (e.g. 188 in NaCl, 629 in CaF_2, 3804 in Al_2O_3 and 3137 kcal mol^{-1} in SiO_2) or covalent crystals (some 171 and 283 kcal mol^{-1} for diamond and SiC, respectively) it can be seen that the forces determining the packing in molecular crystals are much weaker than those of ionic or covalent chemical bonds. Such weaker intermolecular forces can be approximately classified as follows.[6–8]

Repulsion or exchange forces

Most stable molecules have closed electron shells, that is all their molecular orbitals are doubly occupied and cannot accept other electrons without violating the Pauli principle. Accordingly, they repel other molecules trying to come too close. Repulsive forces exert their effects at very short range, decrease with the ninth to twelfth power of the interatomic distance and grow quite rapidly only when two non-bonded atoms try to get closer than

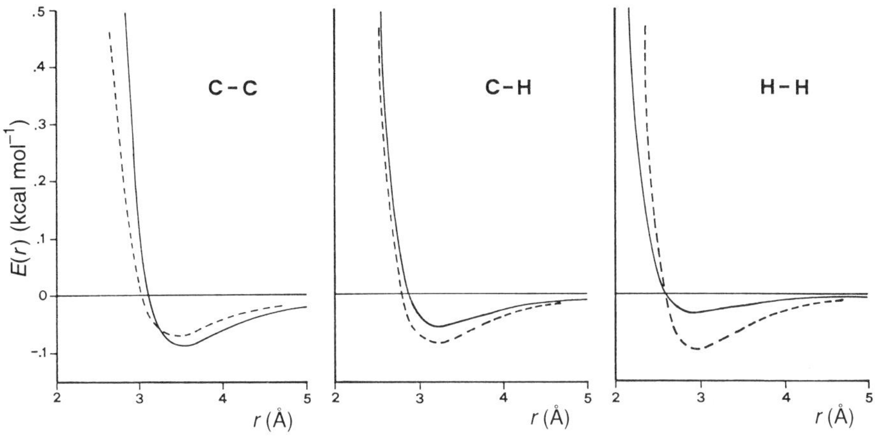

Fig. 7.2. Curves of non-bonded energy as a function of internuclear distance, $E_{nb}(r)$ versus r, for the C–C, C–H, and H–H interactions calculated by using the atom–atom potentials given by Giglio[12] (full curves) and Allinger[13] (broken curves).

their proper contact distance, given by the sum of their van der Waals radii (Table 7.1).

Dispersion or London forces[9–11]

These are weak short-range attractive forces which decrease with the sixth power of the intermolecular distance and are caused by the mutual attraction of the small transient dipoles that molecules can induce in each other. They give the greatest contribution to the lattice energy in crystals of neutral molecules even if the energy of any single interaction is very small. Dispersion energy between two molecules is usually expressed as the sum of the crossed interactions between all pairs of atoms on the two molecules and the sum of repulsion and dispersion energies is named **non-bonded energy**, E_{nb}. When plotted versus the interatomic distance, r, the non-bonded energy displays a typical minimum at a distance nearly equal to the sum of van der Waals radii of the two atoms (Fig. 7.2). Such a minimum is quite shallow, being of the order of a few tenths of a kcal mol^{-1}.

Dipolar forces

Molecules having permanent dipole moments experience electrostatic attraction when properly oriented (orientation forces). According to Kitaigorodski,[6] such attraction forces must cancel out in crystals having only translational symmetry operations (space group P1); for other space groups it has been estimated that the dipole–dipole interactions may contribute one tenth of the lattice energy for molecules having dipole moment of 3–4 D, while their contribution becomes negligible when the dipole moment is less than 1 D.

Monopolar forces

Monopoles are associated with ions, a situation which is not very common in molecular crystals and is not discussed here in detail. Two main points, however, deserve particular attention: ionic interactions are known to cause a quite relevant increase in lattice energies as can be shown, for example, by the comparison of these energies for acetic acid and sodium acetate, respectively 17.4 and 182 kcal mol^{-1}; monopole–monopole interactions are long-range forces whose energy decrease only with the first power of the interionic distance (Coulomb law) and in this respect differ from all other short- or very-short-range intermolecular forces.

Hydrogen bonding[14–18]

With the exception of monopolar forces, H-bonds are the highest-energy interactions in molecular crystals. They greatly affect the way in which molecules are packed, in the sense that the observed packing is almost inevitably that allowing the maximum number of such bonds to be made. Moreover, H-bonding is, by itself, the most relevant non-bonded interaction in nature, being the main factor determining the structure of water, the folding of proteins, and the pairing of bases in DNA. For this reason most crystal packing studies are essentially attempts to understand the laws governing the intermolecular H-bonding in an easily reproducible experimental environment, that given by the molecular crystal.

H-bonding occurs when a hydrogen atom is bonded to two (or sometimes more) other atoms. This situation may be depicted schematically as D–H––A, where D is the H-bonding donor and A the acceptor. In principle all atoms more electronegative then hydrogen (C, N, O, F, S, Cl, Se, Br, I) can play the role of A and D, though stronger hydrogen bonds are necessarily associated with the most electronegative ones (N, O, F, Cl).

Several theoretical studies have been devoted to clarifying the nature of the bond in H-bonding complexes and in particular the relative contributions of different terms to its total energy. Probably the most popular and quoted partitioning scheme is that developed by Umeyama and Morokuma[19] for the treatment of $(H_2O)_2$ and $(HF)_2$ dimers. It makes use of the energy decomposition analysis developed within the *ab initio* SCF-MO theory[20,21] where the total H-bonding energy is partitioned in four terms: the repulsion or exchange energy and the electrostatic, polarization, and charge transfer attraction energies. The authors were able to conclude that the main attractive term is electrostatic and that the contribution of charge transfer is small, so that H-bonding can be qualitatively defined as an *electrostatic more than charge transfer* or simply *electrostatic* interaction.

H-bonds can be classified (Fig. 7.3) according to their topology as **intramolecular**, **intermolecular**, or **bifurcated** and, according to their energy, as going **from weak to very strong**.

1. **Weak H-bonding** can be observed for any couple of donor and acceptor atoms whenever the two groups cannot achieve the correct approach for some sterical reason. The main factor is usually the D–H––A angle which, for maximizing the electrostatic interaction between the D–H dipole and the negatively charged acceptor, must be in the range of some 160–180°. A good example comes from the intramolecular H-bonds closing five-, six-, or seven-membered rings; the H-bond closing a five-membered ring is always weak (and so weak that the hydrogen of its D–H group forms, whenever possible, a second bifurcated hydrogen bond with another acceptor (Fig. 7.3(g′)). A second reason why a H-bond is to be classified as weak comes from the small intrinsic electronegativities of the H-bonded partners, the most classical case being that of the C–H––A interactions.[22]

2. **Medium H-bonding** is typical of water, alcohols, amines, amides, and carboxylic acids. Its geometry is rather well defined: the O–H––O group

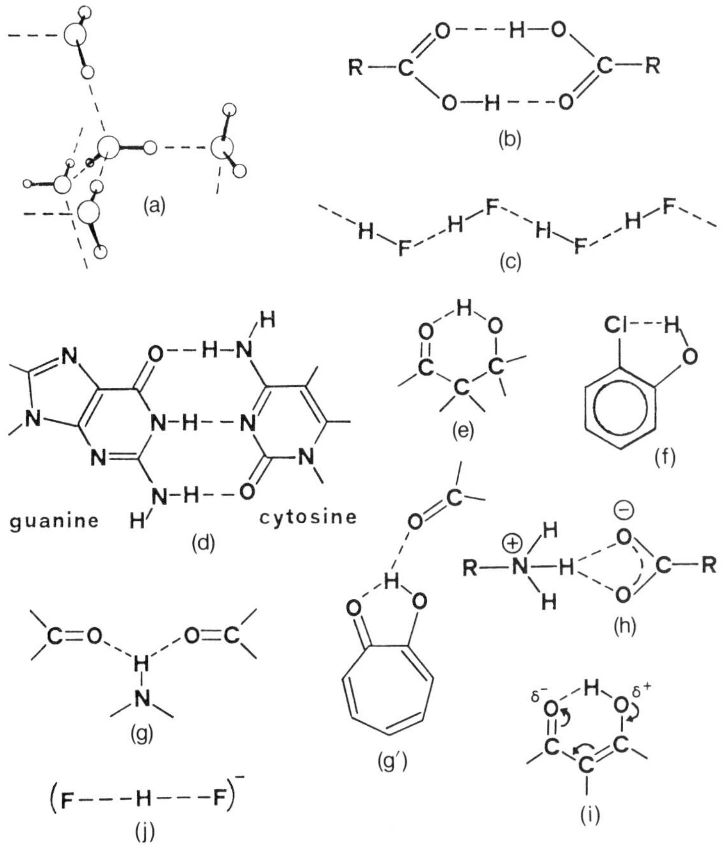

Fig. 7.3. Different types of H-bonds. **Intermolecular** H-bonds forming three-dimensional nets in water (a), dimers in carboxylic acids (b), chains in hydrofluoric acid (c), and bicyclic dimers between guanine and cytosine in DNA (d). **Intramolecular** H-bonds closing six-membered rings (e) are stronger than those closing five-membered rings (f). Two examples of intermolecular **bifurcated** H-bonds (g, g′). **Charge assisted** H-bonds (CAHB) (h) and **resonance assisted** H-bonds (RAHB) (i) are stronger than the usual H-bonds between neutral groups. An example of a **very strong** H-bond, the difluoride ion (j).

tends to be linear; the D–H distance is not significantly lengthened with respect to that observed in the absence of H-bond; the D---A contact distance is practically identical to the sum of the van der Waals radii of A and D, that is the van der Waals radius of the interleaving H atom is almost zero. A wealth of thermodynamical data[15] shows that its formation enthalpy is in the range 2–8 kcal mol^{-1}. In more detail, it is some 4.8 kcal mol^{-1} for the water dimer and 4.7, 5.4, 5.8, 8.6 kcal mol^{-1}, respectively, for the H-bonds formed by carbonyl, amide, ether, and amine acceptors with the phenol as donor.[15]

3. **Medium-strong H-bonding** can be thought to derive from medium H-bonding strengthened by the presence of positive and/or negative charges. A possible range of energies is 6–20 kcal mol^{-1}. It seems reasonable to distinguish two different cases: (a) charge assisted and (b) resonance assisted H-bonding.

(a) **Charge assisted H-bonding** (CAHB) arises from net ionic charges on the donor and/or the acceptor groups. The effect of charge in strengthening the H-bond is documented by the data collected[23] on the H-bonds formed between the intrachain or terminal groups of oligopeptides and aminoacids: the average N---O contact distance is 2.840, 2.908, 2.912, and 2.929 Å, respectively, for the H-bonds $-COO^{-}$---$H_3N^{+}-$, $\rangle C{=}O$---$H_3N^{+}-$, $-COO^{-}$---H–N$\langle$ and $\rangle C{=}O$---H–N$\langle$.

(b) **Resonance assisted H-bonding** (RAHB) is similar to CAHB but with the difference that the strengthening charges arise from polar resonance forms;[24,25] the only system extensively studied is that of β-diketone enols where the resonance O=C–C=C–OH⟨---⟩$^-$O–C=C–C=OH$^+$ strengthens the intramolecular O–H---O bond in such a way that the O---O distance can become 2.417 Å and the calculated H-bond energy 12.7–19.7 kcal mol^{-1} (values to be compared with those for the water dimer: $d_{O--O} = 2.74$ Å, $-\Delta H = 4.8$ kcal mol^{-1}). In these compounds an appreciable lengthening of the O–H bond with consequent shift of the proton towards the central interoxygen position is also observed.

4. **Very strong H-bonding** appears to be inevitably associated with ions. It is characterized by relevant lengthening of the D–H distance, nearly central position of the proton, D---A distances definitely shorter than the sum of van der Waals radii, and bond energies intermediate between true H-bonding and covalent bonding. Complexes like $[F{-}{-}H{-}{-}F]^-$ or $[H_2O{-}{-}H{-}{-}H_2O]^+$ have H-bond enthalpies of 150–250 and 130–150 kcal mol^{-1} and must be considered true chemical species where the charge transfer (covalent bonding) is relevant and we may speak of H-bonding only in a formal sense.

Charge transfer[26–29]

Intermolecular charge transfer or donor–acceptor interactions occur between electron donors (Lewis' bases) and acceptors (Lewis' acids). They establish an at least partially covalent bond between highly polarizable groups, which is often described as the formation of a molecular orbital by electron donation from the highest occupied molecular orbital (HOMO) of the donor to the lowest unoccupied molecular orbital (LUMO) of the acceptor. Classical examples are the molecular complexes $NH_3 + BF_3 = H_3N{-}BF_3$, $I_2 + I^- = [I{-}I{-}I]^-$, or the molecular crystal of iodine, where the I_2 molecule is both an acceptor along the interatomic axis and a donor perpendicular to it. Such a type of interaction occurs in crystals only in the presence of large and easily polarizable (soft) atoms. Another type of donor–acceptor interaction which has been actively studied in the recent past is that present in metallic or semiconducting organic crystals, a class of mixed crystals which contain planar donor and acceptor molecules packed in separated (segregated) infinite stacks and have given rise to great interest for their potential applications in electronics. For a recent review on their structural aspects see references.[30]

Intermolecular forces and crystal packing

We may wonder whether a generic knowledge of intermolecular forces can give us some general insight into the way in which molecular crystals are packed. It is not difficult to show that the relative molecular positions are mainly determined by short-range forces. Calling r the interatomic distance, the repulsion energy drops nearly with r^{-12}, the attractive dispersion energy tends to zero with r^{-6}, and the attractive electrostatic energy decreases with r^{-1}. These facts have important implications, because the electrostatic energy changes so slowly within the lattice that it is practically unable to displace the molecules from their sharp minima, already determined by the short-range balance of repulsion and van der Waals interactions of the outer atoms of the molecule, so that these latter remain the true controlling factor of the packing arrangement. The only exception to this rule is represented by the hydrogen bond, which is electrostatic in nature but requires a quite specific geometry in the donor–acceptor D–H--A interaction and, moreover, involves energies much greater than dispersion interactions do.

The relevant role played by short-range van der Waals forces in the crystal packing helps us to understand the apparently paradoxical fact that the large majority of molecular crystals belong to a few space groups having second-order symmetry elements. Since van der Waals forces are both adirectional and additive, the energy of any single atom is lower the higher the number of atoms of other molecules surrounding it at contact distances is or, in other words, the most stable crystal is that in which molecules pack themselves with the highest coordination number. A simple geometrical analysis of the problem has been carried out by Kitaigorodsky[6] who has shown that rows of molecules staggered by a glide lattice operation can produce a very efficient close packing (coordination of twelve) by repeating a molecule of arbitrary form in the space groups P1, $P2_1$, $P2_1/c$, $Pca2_1$, $Pna2_1$ and $P2_12_12_1$ or a centrosymmetric molecule in the space groups $P\bar{1}$, $P2_1/c$, $C2/c$ or *Pbca*. These space groups are actually those most frequently observed for molecular crystals.

A more detailed analysis of intermolecular forces

Dispersion energy

The first theoretical treatment of **dipersion forces** is due to London[9] (so that they are often called **London forces**), who used the perturbation theory to obtain the following simplified equation for the dispersion energy $E(r)$ of two molecules whose centres are separated by the distance r

$$E(r) = -(3/2)\alpha'_1\alpha'_2 I_1 I_2 r^{-6}/(I_1 + I_2) \qquad (7.1a)$$

where α'_1 and α'_2 are the molecular polarizability volumes and I_1 and I_2 their first ionization potentials. For identical molecules it becomes

$$E(r) = -(3/4)\alpha'^2 I r^{-6} \qquad (7.1b)$$

showing that the attraction energy increases for molecules having high polarizability and ionization potential values and is of the general form $-cr^{-6}$, where c is a constant mainly determined by the value of α'. $E(r)$ is the only attractive energy among neutral molecules without permanent dipole moment and determines all phase transition temperatures of the substance. Polarizabilities are known to increase with the molecular volume and with the number of π bonds (particularly extended systems of conjugated bonds) and are essentially a measure of how much electron clouds can be displaced from their equilibrium positions around the nuclei; higher polarizabilities imply stronger intermolecular attractions and this is the reason why, at room temperature, larger molecules give crystals, smaller ones give liquids, and only very small ones are gaseous. It may be said that dispersion forces (or more basically polarizabilities) are the true reason why molecular crystals can exist.

Units for dipole moments and polarizabilities

Magnitudes of dipole moments are generally quoted in debyes (D), where $1\,D = 3.336 \times 10^{-30}$ Cm. A dipole consists of two charges $+q$ and $-q$ separated by a distance R is calculated as $\boldsymbol{\mu} = q\boldsymbol{R}$ and is directed from the negative to the positive charge. Taking as unit of charge the electron ($1e = 1.602 \times 10^{-19}$C) and of length the angstrom ($1\,Å = 10^{-10}$ m) or the bohr ($1\,\text{bohr} = 0.529177\,Å = a° =$ radius of the first hydrogen orbit) the following equivalences hold: $1e\,Å = 4.800$ D and $1e$ Bohr $= 2.542$ D.

A moderate electric field $\boldsymbol{E}$ induces in a molecule a dipole moment $\boldsymbol{\mu}$ according to $\boldsymbol{\mu} = \boldsymbol{\alpha} \cdot \boldsymbol{E}$, where $\boldsymbol{\alpha}$ is called molecular **polarizability**; $\boldsymbol{\mu}$ and $\boldsymbol{E}$ being vectors, $\boldsymbol{\alpha}$ is a second rank tensor but usually only its scalar value $\alpha = (\Sigma_i \boldsymbol{\alpha}_{ii}^2)^{1/2}/3$ is given as **polarizability volume** $\alpha' = \alpha/4\pi\varepsilon°$, whose units are simply $Å^3 = 10^{-24}\,\text{cm}^3$. Typical values of α' are 0.82, 2.63, 1.48, and 10.5 $Å^3$ for H_2, Cl_2, H_2O, and CCl_4, respectively.

Repulsion (or exchange) energy and total non-bonded energy

The repulsion forces two molecules experience when their outer atoms come into contact are of quantum nature and derive from the Pauli principle; they register the impossibility for closed-shell molecules to accept other electrons in their doubly occupied molecular orbitals. Unfortunately no theoretical treatment for the calculations of repulsion energies is known, so that they must be evaluated empirically and this is done according to two different approximations. In the first, the repulsion energy is expressed as $a \exp(-br)$, where a and b are constants to be determined; the total non-bonded energy assumes the so-called **6:exp** or **Buckingham's** form[11]

$$E_{\text{nb}}(r) = a \exp(-br) - cr^{-6}. \tag{7.2}$$

In the second approximation the repulsion energy is of the type ar^{-n} where a is a constant to be determined and n an integer in the range 8–15 (usually 12); in such a case the non-bonded energy is in the so-called **6:*n*** or **Lennard-Jones'** form[10]

$$E_{\text{nb}}(r) = ar^{-n} - cr^{-6}. \tag{7.3}$$

Atom–atom potentials

When the interacting bodies are atoms (e.g. noble gases) the distance r can be taken as the internuclear distance. In the case of molecules some complications arise and the atom–atom approximation is most commonly used: the total interaction energy between two molecules is expressed as the sum of those among the constituent atoms. The two-atom potentials in the generalized form

$$E_{\text{nb}}(r) = A \exp(-Br)r^{-D} - Cr^{-6} \tag{7.4}$$

are often named **atom–atom potentials**. Values for the A, B, C, and D parameters are not derived from theory but determined by experiment, such as gas deviations from ideality, compressibility of liquids and solids, and neutron scattering by liquids. Otherwise, the parameters minimizing the difference between observed and calculated values of some physical quantity (typically lattice energies and cell parameters for crystals; bond distances, bond angles, and torsion angles for molecules) are used.

Equation (7.4) requires four constants for any pair of atoms, which can be a problem when the number of different atoms increases. For this reason other forms of (7.4) are used, such as

$$E_{nb}(r) = \varepsilon^*[(r^*/r)^{12} - 2(r^*/r)^6] \quad (7.5a)$$

which corresponds to (7.4) with $B = 0$, $D = 12$, $A = \varepsilon^* r^{*12}$, and $C = 2\varepsilon^* r^{*6}$ or

$$E_{nb}(r) = \varepsilon^*[k_1 \exp(-k_2 r/r^*) - k_3(r^*/r)^6] \quad (7.5b)$$

which is identical to (7.4) by putting $D = 0$, $A = k_1\varepsilon^*$, $B = k_2/r^*$, and $C = \varepsilon^* k_3 r^{*6}$ and where k_1, k_2, and k_3 are simple rescaling constants. For any atom i only two values are tabulated: its van der Waals radius r_i and an energy parameter ε_i, which are used to calculate the parameters of any i–j interaction as $r^* = r_i + r_j$ and $\varepsilon^* = (\varepsilon_i \varepsilon_j)^{1/2}$. The quantities r^* and ε^* assume the meaning of position and depth of the potential well of the curve E_{nb} versus r (Fig. 7.4). This formula greatly simplifies data tabulation because it requires only two values per atom instead of three or four for each pair of atoms.

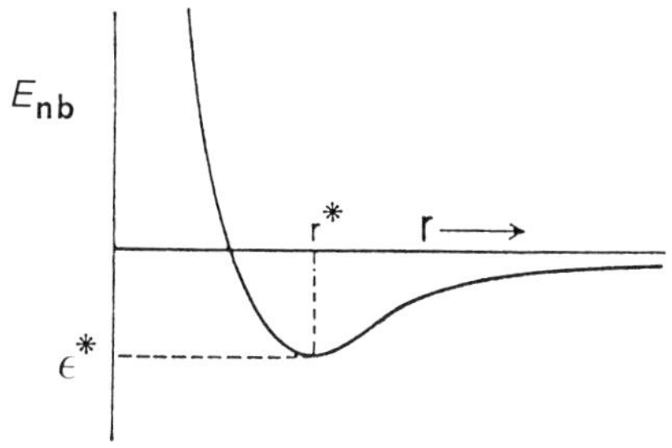

Fig. 7.4. Meaning of the parameters ε^* and r^* in eqns (7.5a, b).

Several tabulations of atom–atom potentials are available and an exhaustive literature is reported in the recent book by Pertsin and Kitaigorodski.[8] Generally speaking, potentials are not **transferable** from one application to another, being more suited to fit phenomena they are derived from. Tables 7.2 and 7.3 report typical potentials used for crystal packing studies[12] and for molecular mechanics and conformational analysis calculations.[13] As an example, Fig. 7.2 compares the curves calculated for the C–C, C–H, and H–H interactions by means of the two different potentials; although the general shape and the position of the minima are similar, the agreement is rather poor. This is typical of semi-empirical quantities since they do not derive from a unique theoretical fundamental but are just a set of parameters which only need to be internally consistent in such a way as to reproduce a given physical quantity; what really matters is that they work, which they usually do if used within their own range of applicability.

Electrostatic energies

The study of interactions among permanent molecular multipoles is simplified by the fact that the energy involved rapidly decreases with the order of the multipole itself and it is generally admitted that the role played by quadrupoles is negligible. Even considering only the first two terms of multipolar expansion the number of terms remains relevant, that is monopole with monopole, dipole, and induced dipole together with dipole with dipole and induced dipole. However, a great simplification can be reached in some cases because interactions with induced dipoles can be usually neglected in lattice energy calculations and crystals of neutral molecules (practically the only case so far studied) do not need monopoles to be taken into account.

A point of general interest concerns the value of the **electric permittivity** ε (normally expressed as the product of the **dielectric constant** ε_r and of the **vacuum permittivity** ε_0, so that $\varepsilon = \varepsilon_r \varepsilon_0$) which is included in all of the following equations. In interactions decreasing rapidly (e.g. with r^{-6} in the dispersive or dipole–dipole interactions) ε_r is assumed to be unitary as in a vacuum because the space around the atom considered can be taken as empty within the short distance (usually 15 Å) to which the calculations are

Table 7.2. Atom–atom potential parameters according to Giglio[12] to be used in crystal packing calculations. Units of kcal mol^{-1}. Q represents Cl or S and Me the methyl group. Potentials are of the form given in eqn (7.4)

Pair	*A*	$B \times 10^{-3}$	*C*	*D*
H···H	49.2	6.6	4.08	0
H···C	125.0	44.8	2.04	6
H···N	132.0	52.1	2.04	6
H···O	132.7	42.0	2.04	6
H···Me	380.5	49.1	3.705	0
H···Q	265.2	40.5	3.851	0
H···Br	366.7	25.6	3.557	0
H···I	527.8	43.4	3.540	0
C···C	327.2	301.2	0	12
C···N	340.0	340.0	0	12
C···O	342.3	278.7	0	12
C···Me	981.1	291.1	1.665	6
C···Q	684.0	255.4	1.811	6
C···Br	945.8	149.0	1.517	6
C···I	1361.0	246.0	1.500	6
N···N	354.0	387.0	0	12
N···O	356.0	316.2	0	12
N···Me	1020.5	325.9	1.665	6
N···Q	711.5	288.6	1.811	6
N···Br	983.6	165.2	1.517	6
N···I	1415.6	272.7	1.500	6
O···O	358.0	259.0	0	12
O···Me	1026.3	272.7	1.665	6
O···Q	715.5	239.2	1.811	6
O···Br	989.3	140.9	1.517	6
O···I	1423.6	232.6	1.500	6
Me···Me	2942.0	273.9	3.329	0
Me···Q	2051.1	251.6	3.475	0
Me···Br	2836.0	133.5	3.181	0
Me···I	4081.0	219.9	3.164	0
Q···Q	1430.0	220.8	3.621	0
Q···Br	1977.2	123.0	3.327	0
Q···I	2845.2	206.8	3.310	0
Br···Br	2733.8	65.8	3.033	0
Br···I	3934.0	107.6	3.016	0
I···I	5661.0	174.2	2.999	0

Table 7.3. Atom–atom potential parameters according to Allinger[13] for molecular mechanics calculations in kcal mol^{-1}. Potentials are of the form given in eqn (7.5b) with $k_1 = 8.28 \times 10^5$, $k_2 = 13.59$, and $k_3 = 2.25$

Atom	r^*	ε^*
$C(sp^3)$	1.75	0.041
$C(sp^2)$	1.85	0.030
C(sp)	1.85	0.030
H	1.50	0.063
O	1.65	0.046
N	1.70	0.039
F	1.60	0.056
Cl	1.95	0.214
Br	2.10	0.299
I	2.25	0.400
S	2.00	0.184
Si	2.10	0.137
H(OH, NH)	1.20	0.040

extended. When monopoles are involved the energy decreases only with the first or second power of distance; calculations must be extended to a wide range and the value of ε_r to be used becomes a complex and barely known function of the distance in consequence of the shielding effects produced around the ion by the interleaving atoms.

1. **Monopole–monopole** interactions occur in crystals containing ions. The interaction energy of ions i and j having formal charges q_i and q_j located at a distance r_{ij} is

$$E_{\mathrm{MM}}(r_{ij}) = -(1/4\pi\varepsilon)q_i q_j r_{ij}^{-1} \qquad (7.6)$$

where $\varepsilon = \varepsilon_r\varepsilon_0$ is the electric permittivity. The constant $(1/4\pi\varepsilon_0)$ depends on the units used and is 332.145 kcal mol^{-1} Å e^{-2} or 1389.70 kJ mol^{-1} Å e^{-2}; it has a value of one when atomic units are used, that is energies are in hartrees (1 hartree = 627.7 kcal mol^{-1}) and distances in bohr (1 bohr = 0.529177 Å).

2. **Monopole–dipole** interaction energy is calculated according to

$$E_{\mathrm{MD}}(r, \phi) = -(1/4\pi\varepsilon)q\mu \cos\phi\, r^{-2} \qquad (7.7)$$

where r is the distance between the charge q and the central point of the dipole μ, and ϕ the angle between r and the direction of the dipole.

3. The case of **dipole–dipole** interactions is of great interest as most neutral molecules have small dipole moments associated with chemical bonds between atoms of different electronegativities, whose vector sum produces, unless it does not vanish because of symmetry, the overall molecular dipole moment. The interaction energy is calculated according to two main models.

In the **dipolar model** the bond dipole moments are tabulated for all bonds of interest or directly calculated from the known values of atom electronegativities. The interaction energy between two dipoles $\boldsymbol{\mu}_i$ and $\boldsymbol{\mu}_j$ whose centres are separated by the vector $\boldsymbol{r}_{ij}$ is

$$E_{\mathrm{DD},ij} = (1/4\pi\varepsilon)(\boldsymbol{\mu}_i \cdot \boldsymbol{\mu}_j)r_{ij}^{-3} - 3(\boldsymbol{\mu}_i \cdot \boldsymbol{r}_{ij})(\boldsymbol{\mu}_j \cdot \boldsymbol{r}_{ij})r_{ij}^{-5} \qquad (7.8)$$

and the total dipole–dipole energy, E_{DD}, in the crystal is one half of the sum of all the intermolecular terms $E_{\mathrm{DD},ij}$. A slightly different way could be that of summing up all the small bond dipoles into the overall molecular dipole moment and of computing the total energy over all molecules.

In the **monopolar model** the total energy is calculated as the sum of all atomic partial charges q_i according to

$$E_{\mathrm{DD}} = -(1/8\pi\varepsilon)\sum_i \sum_j q_i q_j r_{ij}^{-1} \qquad (7.9)$$

where atoms i and j are on different molecules. Partial charges can be obtained from standard semi-empirical or *ab initio* quantum mechanical calculations, approximate semiempirical models[31,32] or, more recently, as deformation densities[33] experimentally determined by X-ray diffraction.

4. **Hydrogen bonding** has already been treated and only potential fields for the calculation of its energy remain to be discussed. Since the H-bond is essentially ionic in nature, the total interaction energy of the fragment

D–H--A (D = donor, A = acceptor) can be divided in two terms: the usual non-bonded energy, E_{nb}, and an electrostatic (or dipolar) term that takes into account the partial charges on the donor and acceptor groups. Starting from the empirical observation that the D...A contact distance is practically equal to the sum of the van der Waals radii of D and A as though the H atom does not exist, it seems clear that any potential assuming a van der Waals radius of zero for the hydrogen (see eqn (7.5)) and localizing suitable partial charges (or dipoles) on D–H and A will describe reasonably well the H-bonding energy, the preference for a linear D–H--A arrangement included, as this optimizes both the van der Waals repulsion and electrostatic attraction between A and D. This is essentially the strategy used by the potential energy fields more recently proposed for describing the H-bond,[33–36] though other approaches are also known.[37,38]

Thermodynamics of molecular crystals

It is of relevant interest to obtain even an approximate evaluation of the different terms contributing to the total free energy of a molecular crystal. The discussion essentially follows the lines of the treatment given by Kitaigorodski[6] and is restricted to the simple case of a crystal constituted by rigid molecules at atmospheric pressure. These conditions allow some simplifications. As regards the effect of pressure, the Gibbs free energy function $G = H - TS$ differs from the Helmholtz function $F = U - TS$ for the term $pV = G - F = H - U$, which is the work done by the system against the external pressure. As the coefficient of thermal expansion is very small in crystals, the expansion work is irrelevant at atmospheric pressure and we can put $G = F$. Moreover, if the molecule is conformationally rigid, intermolecular forces are unable to modify the internal molecular geometry and the following partitioning of the free energy is acceptable

$$F = F_{cr} + F_{mol} \tag{7.10}$$

where F_{cr} and F_{mol} are the inter- and intramolecular contributions to the total crystal free energy. In this approximation F_{mol} could be separately calculated at any temperature from the frequencies of the internal vibrational modes of the rigid molecule by the usual methods of statistical thermodynamics; the same methods allow us to calculate another molecular quantity which will be shown to be necessary, the contribution of internal vibrational modes to the molar heat capacity, C_{mol}.

Within our approximations the molar free energy of the crystal can be written as

$$F_{cr} = U + K_0 + F_{vib} = U_0 + \Delta U + K_0 + E_{vib} - TS_{vib}. \tag{7.11}$$

U is the **lattice energy** or total potential energy of the intermolecular interactions. Even if the interatomic forces are temperature independent, U is temperature dependent as a consequence of the slight increase of the cell parameters with temperature, so that it must be written $U(T) = U_0 + \Delta U(T)$, where U_0 is the lattice energy at absolute zero (a term large and negative) and ΔU a small positive term increasing with the temperature and equal to zero for $T = 0$, representing the expansion work against the internal cohesion forces. It causes a difference between the molar heat capacities at constant pressure and volume and from this difference can be actually calculated as $\Delta U(T) = \int_0^T (C_p - C_V)\,dT$.

K_0 is the **zero-point energy**, that is the crystal vibrational energy at zero kelvin; it is known that it may have some relevance only for molecules of extremely small moment of inertia (N_2, O_2, CO) and strongly directional bonds (H_2O). In common crystals it is very small ($<0.2\,\mathrm{kcal\,mol^{-1}}$) and can be neglected. It cannot be dissociated from the lattice energy and the global term $U + K_0$ can easily be obtained from the experimental **molar sublimation enthalpy** at the temperature T according to

$$-\Delta H^{\mathrm{sub}} = U + K_0 + 2RT \tag{7.12}$$

where $2RT$ is a small correction term (some $1.2\,\mathrm{kcal\,mol^{-1}}$ at room temperature) due to the differences of molar heat capacities between solid ($6R$) and gas ($3R$) and to the expansion work of a mole of gas ($pV = RT$ for an ideal gas). Of particular interest is the molar sublimation enthalpy extrapolated to absolute zero, which has the meaning of lattice energy at this temperature, that is

$$-\Delta H_0 = U_0 + K_0. \tag{7.13}$$

E_{vib} is the **vibrational part of the internal energy**. At not too low temperatures an oscillator gives a contribution of kT to the internal energy (equipartition principle). The Avogadro number N of molecules having six degrees of freedom (three rotational and three translational) accumulate, at the temperature T in gaseous phase, a kinetic energy of $6NkT/2 = 3RT$; in the solid state translations and rotations are hindered and become oscillations and librations around the equilibrium positions which, having both kinetic and potential components, can accumulate an energy $E_{\mathrm{vib}} = 6NkT = 6RT$ and the crystal should have a molar heat capacity $C_V = 6R$. In practice C_V tends to zero for T tending to zero because an ever decreasing number of vibrational levels is accessible and $E_{\mathrm{vib}} = \int_0^T C_V\,dT$ becomes increasingly smaller than $6R$.

S_{vib} is the **vibrational entropy**. Clearly $\lim_{T\to 0} S_{\mathrm{vib}} = S_0$ for any crystal of a pure substance without static disorder, where S_0 is the residual entropy at absolute zero. S_0, a quantity strictly related to K_0, is very small and it is assumed to be zero when the entropies are measured according to the third law of thermodynamics. This can be calculated as $S_{\mathrm{vib}} = S_{\mathrm{cr}} = \int_0^T (C_p - C_{\mathrm{mol}})\,d(\ln T)$.

The thermodynamic equations needed to evaluate the different terms in (7.11) and an actual calculation concerning naphthalene at two different temperatures are summarized in Table 7.4. The experimental data necessary are not easily found, being the molar heat capacity at constant pressure, C_p, the isobaric thermal expansivity, $\alpha = (1/V)(\delta V/\delta T)_p$, the isothermal compressibility, $\beta = -(1/V)(\delta V/\delta p)_T$, and the volume of the unit cell, V, from zero to the temperature of interest. Two other quantities are needed, U_0, obtained from the molar sublimation enthalpy according to (7.12), and the contribution of internal vibration modes to the molar heat capacity, C_{mol}, as a function of T, which has already been shown to be obtainable without great difficulty from spectroscopic data in the case that there is no mixing of internal modes (intramolecular vibrations) and external modes (vibrations and librations of the molecules as a rigid body). Figure 7.5 reports the plot

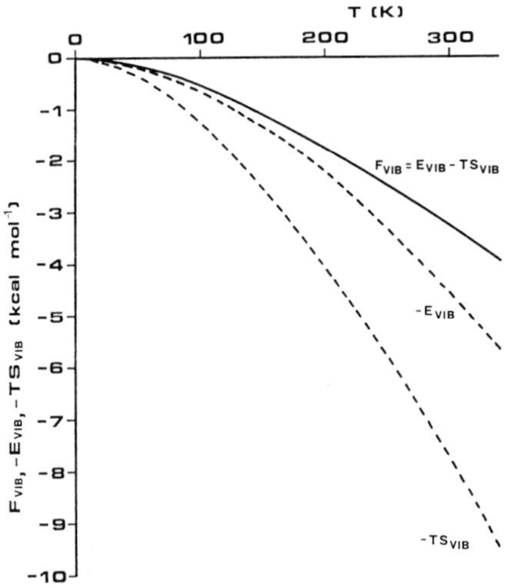

Fig. 7.5. Plot of the molar vibrational free energy, F_{vib}, and of its components as a function of the temperature for the crystal of naphthalene. Data from Kitaigorodski.[6]

Table 7.4. Quantitative evaluation of the free energy of the crystal of naphthalene at two different temperatures according to Kitaigorodski[6] (the zero-point energy K_0 has been neglected)

Isobaric thermal expansivity $\alpha = (1/V)(\delta V/\delta T)_p$		**Isothermal compressibility** $\beta = -(1/V)(\delta V/\delta p)_T$ **Calculated values (cal mol^{-1})**	
Quantity	**Way of calculus**	$T = 140$ K	$T = 300$ K
U_0	$\lim_{T\to 0} (-\Delta H^{sub})$	−16 700	−16 700
ΔU	$C_p - C_V = \alpha^2 TV/\beta$		
	$\Delta U = \int_0^T (C_p - C_V)\, dT$	53	500
E_{vib}	$C_V = C_p - (C_p - C_V)$		
	$E_{vib} = \int_0^T (C_V - C_{mol})\, dT$	1086	3195
S_{vib}	$\int_0^T (C_p - C_{mol})\, d(\ln T)$	16	26
$-TS_{vib}$		−2254	−7758
F_{vib}	$E_{vib} - TS_{vib}$	−1168	−4563
F_{cr}	$F_{cr} = U_0 + \Delta U + E_{vib} - TS_{vib}$	−17 815	−20 763

of the calculated values of molar vibrational free energy for the crystal of naphthalene over a much wider range of temperatures.

From these data it is possible to obtain some rules regarding crystals in general and molecular crystals in particular. The lattice energy constitutes by far the greatest part of crystal free energy, F_{cr}; at low temperature and even at higher temperatures it is the prevailing part. The general effect of the vibrational part is that of causing a continuous decrease of F_{cr} because $|TS_{vib}|$ increases more rapidly than E_{vib} with temperature. In other words the crystal is increasingly stabilized by higher temperatures; this, of course, is not accidental but the expression of a fundamental physical law as differentiation of the free energy expression $dG = V\,dp - S\,dT$ implies that $(\delta G/\delta T)_p = -S$ and that, S being always positive, the free energy decreases with increasing T. A last point can be that the work done against the internal cohesion forces of the crystal, ΔU, is small up to room temperature and can be neglected in a first approximation.

Free and lattice energy of a crystal from atom–atom potentials

Knowing all the individual intermolecular energy terms, the lattice energy of a crystal built up of rigid molecules can be calculated according to

$$U = (1/2)\sum_{ij}\sum E_{nb}(r_{ij}) + (1/8\pi\varepsilon)\sum_{ij}\sum q_i q_j / r_{ij} + E_{HB} \qquad (7.14)$$

where the first term is the non-bonded energy (eqn (7.4) or (7.5)), the second the electrostatic energy (7.9) and the last one the H-bonding energy. The index i runs over all atoms of a reference molecule and the index j over those of all the surrounding molecules. The summation can be truncated at some 15 Å in computing E_{nb} because dispersive forces decay rapidly with distance. This is not true for electrostatic forces and special mathematical techniques have to be used to increase series convergence.

In the crystal a rigid molecule has six degrees of freedom, three translational and three rotational (or librational); if x, y, and z are the coordinates of its centre of gravity and θ, φ and ψ the Eulerian angles describing its orientation (Fig. 7.6), U is a function of these six variables and of the cell parameters, $\boldsymbol{a}_i$ ($i = 1, 3$), that is

$$U = U(\boldsymbol{a}_i, \theta, \varphi, \psi, x, y, z). \quad (7.15)$$

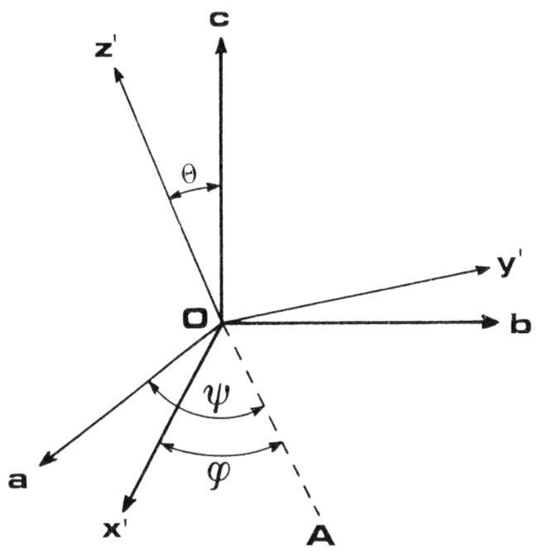

Fig. 7.6. Eulerian angles θ, φ, and ψ defining the orientation of the (x', y', z') orthogonal base of the rigid molecule with respect to the orthogonal base of the crystal, (a, b, c). The broken line OA is the intersection of planes $(x'y')$ and (ab).

The number of parameters for the unit cell goes from one for the cubic system to six for the triclinic system. Moreover, the molecule may have fewer degrees of freedom as it is in a special position; for example it has only three rotational degrees of freedom if located on a symmetry centre. So, the lattice energy of naphthalene ($P2_1/a$, $Z = 2$) depends on only seven degrees of freedom (a, b, c, β, θ, φ, ψ).

If n is the total number of parameters, it is useful to think of U as a hypersurface in an $(n + 1)$-dimensional space where the geometrical parameters are the abscissae and the energy the ordinate: the different minima on this surface correspond to all possible crystal structures in the space group chosen. The relevant number of calculations done on different crystals indicate that the experimental structure usually corresponds to the deepest minimum or, at least, to one of the deepest minima of the potential surface and that the calculated and experimental values of U compare within a few kcal mol^{-1}. This seems to indicate that the structure can be predicted from the simple evaluation of lattice energy, independently of the vibrational part of the free energy.

The reasons for this fact will be discussed in the next section. What is important here is that it allows us to obtain the best potential energy parameters (globally called a **force field**) to be used in (7.14) by a least-squares procedure on a minimum number of known crystal structures where the quantities to be reproduced are the lattice energy, the unit cell parameters, and the positional parameters of the molecule. This method has been applied to several classes of chemical compounds and is reported to give far better results than the use of potentials derived from other molecular properties. Unfortunately it has been impossible to find a force field able to give a very accurate reproduction of the crystal properties that is valid for all chemical compounds. For instance, the calculations can be very good for a class (e.g. hydrocarbons) but any attempt to extend the force field to molecules containing heteroatoms causes a worsening of the final results. Moreover, very few cases of intermolecular H-bonding have been studied so far[8,33–35] and no general evaluation of the term E_{HB} in (7.14) is therefore possible.

Another point concerns molecular flexibility. In theory, lattice energy calculations can be extended to the case in which the molecule has internal degrees of freedom caused by rotation around single bonds. If τ_j are the j torsion angles of interest, (7.15) can be rewritten as

$$U = U(\boldsymbol{a}_i, \theta, \varphi, \psi, x, y, x, \tau_j) \quad (7.16)$$

where θ, φ, ψ, x, y, z concern a reference fragment of the molecule and τ_j are the torsion angles defining the orientation of the other fragments with respect to the first one. To a first approximation the conformational energy can be calculated as the non-bonded interaction energy among the atoms belonging to the different molecular fragments by means of the same potentials used for intermolecular interactions. To obtain more accurate results, however, it will be necessary to use a specific intramolecular force field such as that described on p. 509.

The calculation of the crystal free energy turns out to be much more complicated problem because it requires evaluating, besides the lattice energy, the term $F_{\text{vib}} = E_{\text{vib}} - TS_{\text{vib}}$, that is all the possible vibrational modes of the molecules around their equilibrium positions within the crystal. This can be done by the methods of lattice dynamics within the Born–von Karman formalism[6,8,39–43] but the topic is too long and difficult to be treated here. Several general treatments of the method and a recent review on the results of the calculations carried out on molecular crystals[8] are available and the interested reader is referred to them.

Polymorphism and the prediction of crystal structures

Polymorphism is an important aspect of the thermodynamics of molecular crystals, even if little attention has been devoted to this specific problem so far. However, it frequently happens that different polymorphic forms are obtained by crystallization from different solvents (in particular when the molecule has different possible conformers) and the number of polymorphs would certainly increase steadily if a larger number of solvents and a wider range of tempertures and pressures were investigated.

The basic reason for the phenomenon lies in the fact that the hypersurface of the lattice energy U has many minima in the twelve-dimensional space of (7.15) (six cell parameters and six positional values defining the molecule position) and their number is even larger in the $(12+n)$-dimensional space of (7.16), where n is the number of possible torsions. The deepest minimum corresponds to the structure with the closest packing, which has the minimum enthalpy because it can establish the most profitable pattern of intermolecular interactions. It represents the most stable structure in the absence of the vibrational contributions, i.e. at absolute zero. When the crystal is heated, however, the total free energy changes from U_0 to $F_{\text{cr}} = U_0 + \Delta U + E_{\text{vib}} - TS_{\text{vib}}$, which is smaller than U_0 (it has already been shown that $F_{\text{vib}} = E_{\text{vib}} - TS_{\text{vib}}$ is always negative and ΔU is negligible). Comparing two hypothetical polymorphic forms of a same substance having energies of $(U_0)_1$ and $(U_0)_2$ at absolute zero (Fig. 7.7), the higher-energy form is less efficiently packed and will have a greater vibrational entropy and therefore a faster decrease of the free energy when the temperature increases. From this point of view, the hypothetical molecular arrangement having the average structure of the liquid at zero kelvin can be considered as the polymorph of highest enthalpy and having the steepest possible decrease of free energy with T. With reference to Fig. 7.7, the polymorphic transition $1 \rightarrow 2$ is possible whenever the corresponding energy curves cross at temperature T_t lower than the melting point. At the transformation point $(U_0 + \Delta U + E_{\text{vib}} - TS_{\text{vib}})_1$ is equal to $(U_0 + \Delta U + E_{\text{vib}} - TS_{\text{vib}})_2$ and a very small increase of temperature will cause a sudden phase transition by absorbing the quantity of heat $\Delta H_t = \Delta(U_0 + \Delta U + E_{\text{vib}})_{2-1}$ which is easily

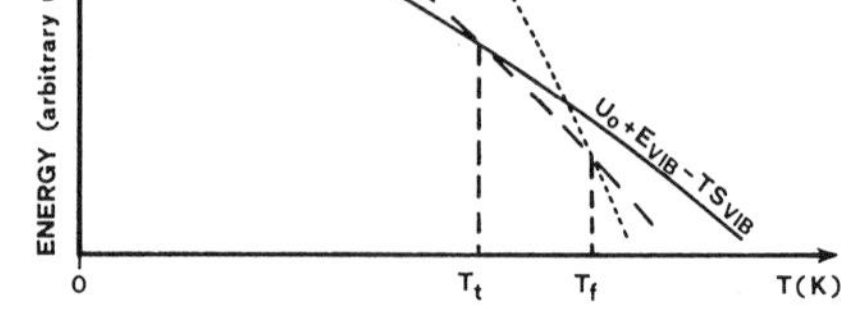

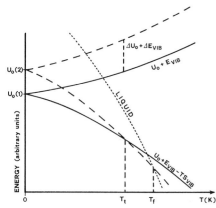

Fig. 7.7. Free energies of two hypothetical polymorphic phases of the same substance as a function of temperature. The crossing of the two curves before the melting point defines the transition from form 1 to form 2 at temperature T_t.

measured experimentally and must be numerically equal to $\Delta(TS_{\text{vib}})_{2-1} = T\Delta S_{\text{vib}}$.

We may wonder what is the energy range of the minima on the hypersurface of U_0 which could give rise, at higher temperatures, to phase transitions or, in other words, whether polymorphism could be predicted from simple lattice energy calculations. This seems unlikely, at first sight, since it has been already shown that the vibrational part of free energy may amount to a large percentage (22 per cent for naphthalene at room temperature; Table 7.4) of the total free energy. Let us reconsider, however, the expression for the transition enthalpy in the form $\Delta H_t = \Delta U_0 + \Delta(\Delta U) + \Delta E_{\text{vib}}$; ΔH_t is known from experience to rarely exceed 1 kcal mol^{-1} in molecular crystals, $\Delta(\Delta U)$ is a small variation in an already small term and can be neglected and $\Delta E_{\text{vib}} = T(\Delta C_V)_{2-1}$ (see p. 479) is a small term which is necessarily positive because the heat capacity of a less packed crystal must be slightly larger. Therefore, the total lattice energy variation energy variation, $\Delta U_0 \approx 1$ kcal mol$^{-1} - \Delta E_{\text{vib}}$, observable in a polymorphic transition should be less than 1 kcal mol^{-1}, which seems to imply that the number of possible polymorphs can actually be determined from lattice energy calculations because no polymorphism will be observed at ordinary pressures when the second deepest minimum differs from the absolute one by more, say, than 1–2 kcal mol^{-1}.

Effect of crystal forces on molecular geometry

A subject often discussed in structural literature is whether the crystal field of van der Waals crystals can modify molecular geometries. Following Kitaigorodsky,[6,44] two cases can be distinguished: deviations from the minimum-energy (or equilibrium) geometry and different choices among possible geometries.

As regards the first case and calling ΔE the hypothetical molecular energy variation caused by a displacement from the equilibrium geometry and ΔU the concomitant change of lattice energy, the deformation will be possible only when $|\Delta E| < |\Delta U|$. For dispersive interactions, $|\Delta U|$ has been estimated[44] from the changes of lattice energy consequent upon small displacements of C and H atoms in the structure of benzene: it amounts to less than 0.01 kcal mol^{-1} for an atom shift of 0.03 Å. Values for $|\Delta E|$ can be easily calculated by the methods of molecular mechanics (see p. 506). A shift of 0.03 Å in the stretching or compression of a C–C or C–H single bond requires an energy of some 0.30 kcal mol^{-1} which is much larger than $|\Delta U|$: therefore, crystal forces cannot affect bond distances. Moreover, a shift of 0.03 Å corresponds to a bond angle change of 1.2–1.6° in common organic compounds. The energy necessary to change by this amount a C–C–C or C–C–H angle in benzene is about 0.03 kcal mol^{-1} which is still three times larger than $|\Delta U|$; accordingly, it seems unlikely that relevant (say, greater than 1–2°) changes of bond angles can be produced by the crystal field.

More complex is the case of rotations around single bonds, where the same shift of 0.03 Å can be produced by rotation around a C–C–C–C torsion angle by expending only 0.0012 kcal mol^{-1} as far as the simple torsional potential is concerned. Here the controlling factor is represented by the intramolecular contacts of the atoms, or groups of atoms, which are moved by the rotation around the single bond. Let us assume that, when the molecule is in its conformational minimum, all contact distances of interest are not far from the minimum of their atom–atom potentials (Fig. 7.2): the global conformational energy minimum will inevitably be shallow and the crystal field might even produce relevant changes of the torsion angle. A classical example is diphenyl, which is not planar by itself but becomes planar under the slight compression of the crystal field. Conversely, crystal forces will weakly affect the value of the torsion angle when the walls of the potential well are steep, a situation occurring when the molecule is in tension (or rigid) because, in the conformational minimum, the atomic contact distances are in part shorter and in part longer than the optimal contact distances. From this point of view it can be stated that overcrowded molecules are the least affected by the weak crystal forces.

It remains for us to consider the case of molecules having different possible conformations which crystallize in a conformation which is not that of minimum energy for the free molecule. This is quite possible and, in fact, many conformationaly flexible molecules are found to crystallize in different space groups with different conformations (conformational polymorphism). It has already been remarked that the enthalpy of polymorphic transitions seldom exceeds 1 kcal mol^{-1} in molecular crystals and this makes it possible that a molecule, whose second most stable conformation differs from the first one by not more than 1–2 kcal mol^{-1}, may gain from a more efficient crystal packing that energy (or even more) which has been lost in consequence of the choice of an unpreferred conformation.

Elements of classical stereochemistry

Structure: constitution, configuration, and conformation

X-ray crystallography produces a quite detailed description of the spatial arrangement of atoms in molecules, a description indicated by the generic term of **structure**. However, a rigorous definition of what is structure is not a trivial problem and modern stereochemistry[45–54] tries to give an answer by means of the more basic concepts of constitution, configuration, and conformation.

Molecular **constitution** indicates the way in which atoms are interconnected, making due distinction between single and multiple bonds. It is therefore a topological concept which can be reconnected to that of graph.[55]

The **configurations** of a molecule of given constitution indicate all possible spatial arrangements of its atoms, ignoring those derived from rotation around single bonds. In carbon chemistry different configurations are mainly caused by the presence of double bonds or asymmetric carbon atoms. A change of configuration requires, as a rule, the breaking and remaking of a chemical bond and is a high-energy process.

The word **conformation** refers to the different spatial arrangements of a molecule of given configuration produced by rotation (torsion) around single bonds. The concept of conformation and the relative nomenclature are more easily illustrated by an example. Figure 7.8 reports the potential energy curve of 1, 2-dichloroethane as a function of the Cl–C–C–Cl torsion angle as calculated by the methods of molecular mechanics (p. 506–11). The

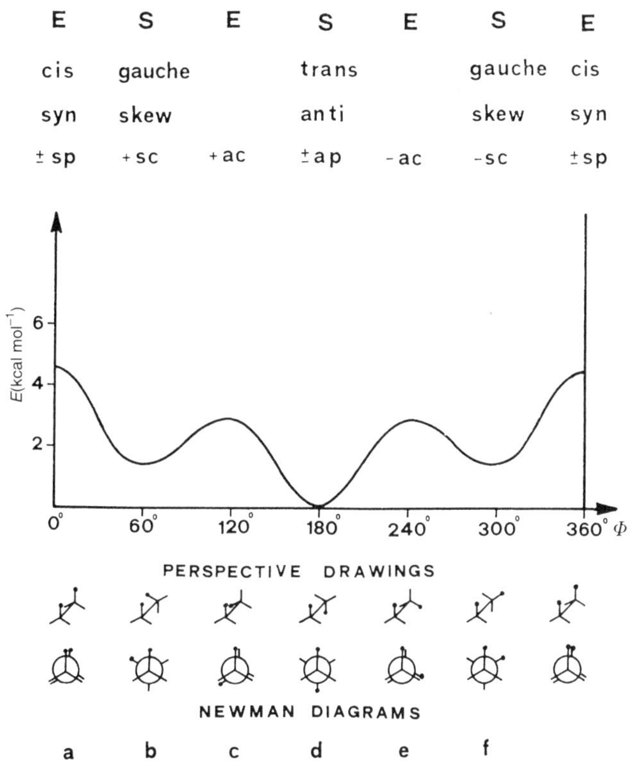

Fig. 7.8. Potential energy curve expressed as a function of the Cl–C–C–Cl torsion angle for the molecule of 1,2-dichloroethane. In the upper part of the figure the most common nomenclature systems used for describing conformations are reported (E = eclipsed, S = staggered). The fourth line reports the nomenclature proposed by Klyne and Prelog:[47] s = syn, a = anti, p = periplanar, c = clinal. At the bottom are shown two of the standard graphical representations, the perspective and Newman diagrams.

shape of the curve is clearly determined by the repulsions between C–H and C–Cl groups on the opposite sides of the molecule, repulsions decreasing in the order CCl–CCl, CH–CCl, and CH–CH. In theory there are as many possible conformations as there are values of the torsion angles. In practice the free molecules (gas, liquid, or solution) will be distributed among the possible minimum-energy conformations with the absolute minimum more populated than the others and the mean thermal energy of the environment ($3RT/2 \approx 0.89$ kcal mol^{-1} or 3.7 kJ mol^{-1} at room temperature) will cause a continuous jumping of the molecules among different minima by crossing the interleaving maxima, which are conformational transition states; the equilibrium situation will correspond to fixed populations of molecules occupying *only* the minima of the curve while all other conformations have nearly zero populations, being assumed by the molecules only instantaneously during the jumps. The rate of jumping is determined by the ordinary laws of chemical kinetics and is proportional to $\exp(-E_a/RT)$, where E_a is the activation energy required to reach the transition state. This is the reason why only conformations corresponding to minima or maxima of the energy curve are usually given names, according to the three nomenclature systems reported at the top of Fig. 7.8. For example, the most stable conformation of 1,2-dichloroethane can be indicated as **staggered(S)**, **trans**, **anti**, or ±**antiperiplanar**. In crystals the conformation is frozen at a point which will or will not correspond to the most probable conformation depending on the depth and steepness of the potential wells (as extensively discussed on p. 483).

The term **structure** includes the constitutional, configurational, and conformational aspects of the molecule. The same three aspects do not always need to be considered. For CH_4, $CHCl_3$, and CH_2Cl_2 the structure is defined by the simple constitution, for CHClBrI (asymmetric carbon) and ClHC=CHCl (*cis* or *trans* configurations) by constitution and configuration, and for $H_3C–CH_3$, $H_2BrC–CBrH_2$ and cyclohexane by constitution and conformation; finally, the structure of HRC=CH–CH=CHR′ needs the definition of constitution, configuration, and conformation, having four possible *cis–trans* configurations around the two double bonds besides the conformations produced by rotation around the central single bond.

It is important to remark that, while the crystal structure is always univocally defined by atomic coordinates, cell parameters, and symmetry operations, molecular structure is a concept which can be given a precise meaning only within the frame of the valence bond theory (p. 502–6) since all classical stereochemistry is centred on the idea of simple and multiple localized bond.

Isomerism

Molecules having the same formula but different structures are called isomers and are said to display isomerism. There are three different types of **isomerism: constitutional**, **configurational**, and **conformational**, and the two last are grouped under the common term of **stereoisomerism**.

Constitutional (or bond or topological) isomerism

Constitutional isomers represent the different modes of connecting the same set of atoms. Accordingly, 1-butene, 2-butene, and *iso*-butene (1a, 1b, 1c), the three *o*-, *m*- and *p*-disubstituted benzene derivatives (2a, 2b, 2c), and the two nitro- or nitrito-complexes (3a, 3b) are constitutional isomers. Usually isomers of this sort cannot interconvert and are to be considered as distinct chemical compounds. In case an interconversion equilibrium can be established the expressions **tautomers**, **tautomerism**, and **tautomeric equilibrium** are employed; tautomerism is normally associated with the migration of a hydrogen atom and a typical example is that of the acetylacetone keto–enol tautomerism (4a, 4b). It should be remarked that less recently constitutional isomers were termed structural isomers; it is clear that such an expression has to be avoided.

Configurational isomerism

Molecules having the same constitution but different configurations are called configurational isomers. Two main cases are to be distinguished: when the two stereoisomers can be related by a symmetry operation of reflection they are **enantiomers** and when this is not possible they are called **diastereoisomers** or, sometimes, **diastereomers**. The distinction has a precise physical meaning. Both enantiomers and diastereoisomers are identical as far as their chemical bonds are concerned. Only the former, however, have indistinguishable non-bonded interactions; they have the same physico-chemical properties in all respects except for their optical properties and reactivity towards other enantiomeric species. In particular, enantiomers rotate the plane of polarized light by the same angle but in

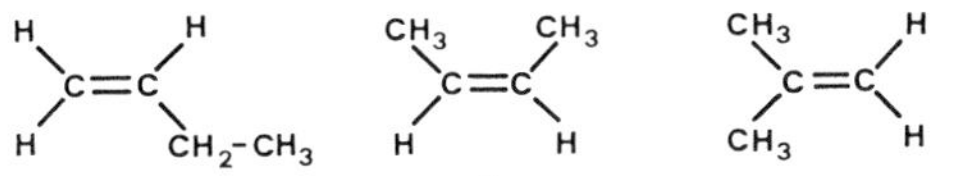

1 a, b, c

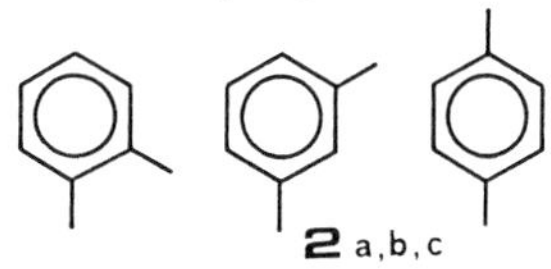

2 a, b, c

$[Co(NH_3)_5ONO]^{2+}$

$[Co(NH_3)_5NO_2]^{2+}$

3 a, b

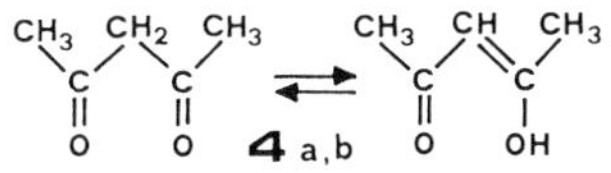

4 a, b

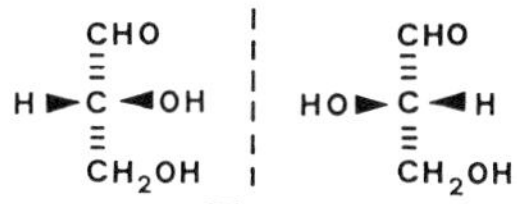

5 a, b

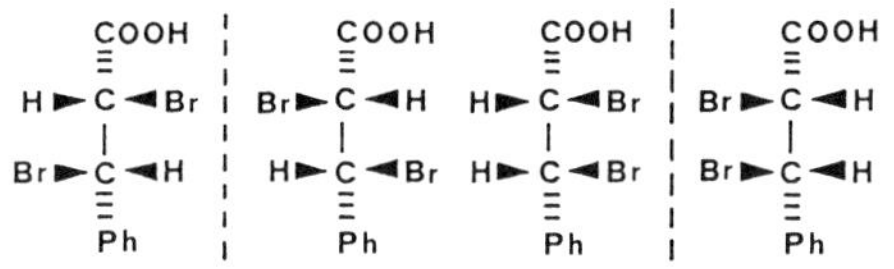

6 a, b, c, d

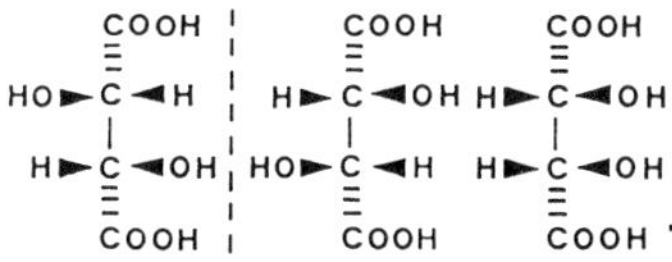

7 a, b, c

8 a, b

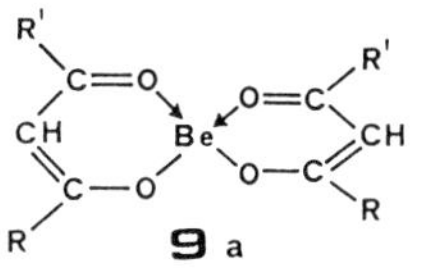

9 a

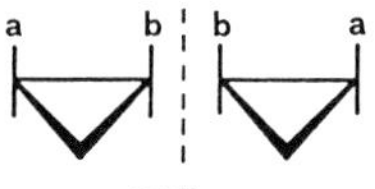

10 a, b

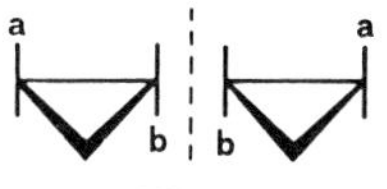

10 c, d

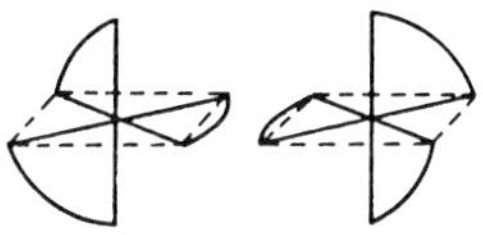

11 a, b

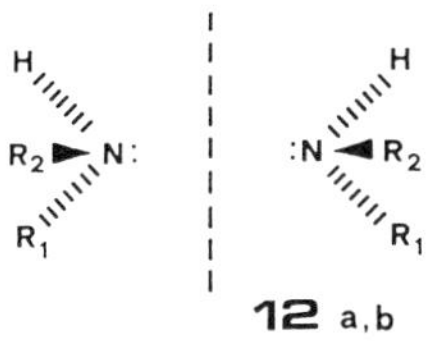

12 a, b

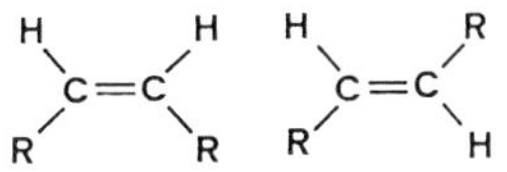

13 a, b

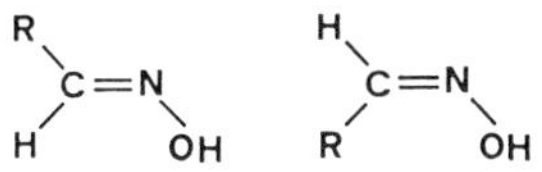

14 a, b

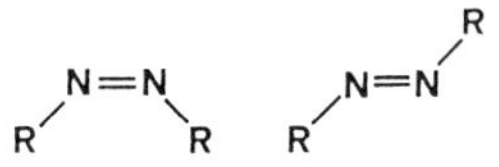

15 a, b

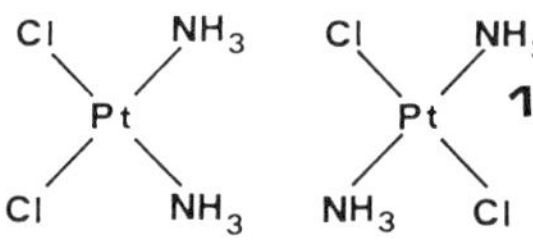

16 a, b

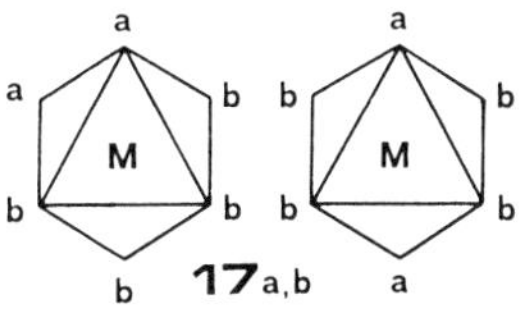

17 a, b

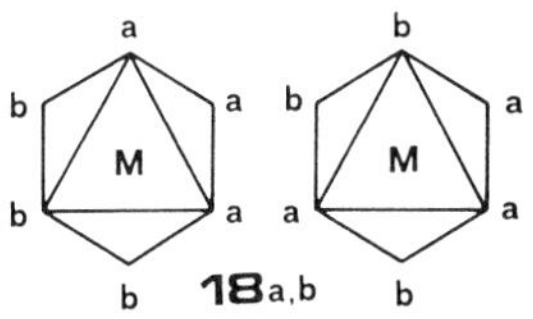

18 a, b

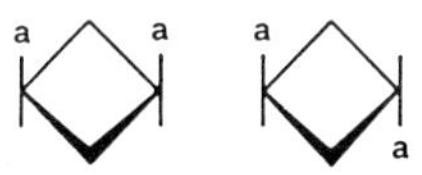

19 a, b

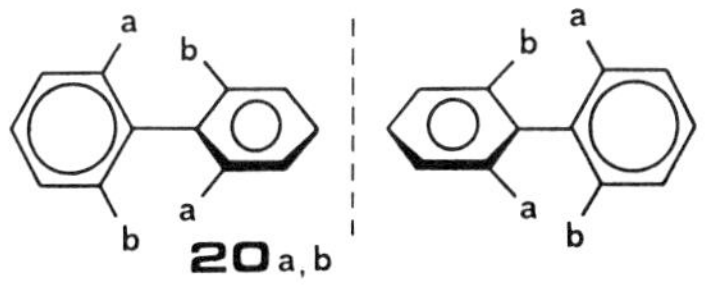

20 a, b

opposite directions, a property by which they can be identified and which explains their older name of **optical isomers**.

Molecules having an internal symmetry plane cannot have enantiomers because they are their own enantiomer. In such molecules the possible diastereoisomers are often and traditionally called **geometrical isomers**. Common cases are planar molecules (ethylene derivatives, planar rings, and square–planar complexes). However, geometrical isomerism is not another type of isomerism but just a particular name given to diastereoisomerism in a specific situation.

The condition for a molecule to show enantiomerism (i.e. to exist as a couple of enantiomers) is that it cannot be superimposed on to its mirror image in any of its possible conformations. Such a molecule is said to be **dissymmetrical** or **chiral**. Since rotations are allowed operations in trying to superimpose the molecule with its mirror image, the most general condition for chirality is the *lack of molecular rotary–reflection axes,* a condition which excludes the presence of both internal symmetry centres ($S_2 \equiv \bar{1}$) and planes ($S_1 \equiv$ m).

The most frequent cause of chirality is the presence of an asymmetric carbon atom, that is a tetrahedral carbon bonded to four different atoms or atomic groups as, for instance, in glyceraldehyde (5a, 5b). The number of possible isomers increases with the number of asymmetric atoms. The α,β-dibromocinnamic acid, having two asymmetric carbons, produces four different stereoisomers (6a–6d); the pairs (6a, 6b) and (6c, 6d) are enantiomers while all other pairs are diastereoisomers. Tartaric acid (7a–7c) illustrates the case where one of the two enantiomeric pairs vanishes because the parent diastereoisomer has an internal symmetry plane (7c: a *meso* form). This last example elucidates another interesting aspect of enantiomerism. The *meso* form depicted in (7c) cannot exist in practice because it is a high-energy eclipsed conformation. However, any pair of right- and left-handed torsions of the same entity will generate a pair of enantiomers which are isoenergetic, independently of the shape of the potential energy curve; being isoenergetic, they will be present in the pure liquid or in solution with exactly the same populations causing the compound to be optically inactive.

Chirality can be associated with a variety of other bond situations which can produce tetrahedral arrangements, as in allenes (8a, 8b), *spiro*-compounds, quaternary ammonium salts, or *bis*(benzoylacetonato)-beryllium (9a: one enantiomer shown). It can be also associated with non-aromatic cyclic systems: for example, all rings with an odd number of terms and two different non-geminal substituents have two asymmetric carbons producing two diastereoisomeric pairs of enantiomers, as shown by the cyclopropane derivative (10a–10d). An example of chirality in coordination chemistry is given by the chelate octahedral complexes of the type $[Co(en)_3]^{3+}$ (11a, 11b) or $[cis\text{-}CoCl_2(en)_2]^+$, where *en* stands for ethylenediamine.

Of particular interest are pyramidal atoms, i.e. those having tetrahedral hybridization but a hybrid orbital occupied by a non-bonding pair. It was suggested many years ago that molecules similar to the amines NHR_1R_2 and $NR_1R_2R_3$ should exist in two enantiomeric forms which, however, have never been observed. The reason is that the two enantiomers (12a) and (12b) exchange rapidly in solution by pyramidal inversion through a planar transition state having an energy higher only by a few kcal mol^{-1}. Further investigations have shown than other potentially chiral compounds of trivalent phosphorus and arsenic have higher interconversion barriers and that their enantiomers may be actually resolvable, at least at low temperature.

The **absolute configuration** of an enantiomer (i.e. its actual atomic disposition in space) is a problem which cannot be tackled by chemical methods. All attempts to obtain it from the sign of optical activity in solution, the property which is more strictly related to enantiomerism, have failed. The first absolute configuration of an enantiomer was only accomplished in 1951 by Bijvoet and co-workers[56] on (+)-tartrate of sodium and rubidium by anomalous scattering methods, as discussed in Chapter 5. Since then absolute configuration determination has become routine.

The nomenclature for identifying enantiomers has changed over the years and reflects the discovery of new techniques for their study. The oldest one simply reports the sign of the optical activity (+ or −); later on Fisher, in his fundamental studies on carbohydrates, discovered the stereochemical series and enantiomers were named after the series they belonged to (D or L). More recently Cahn, Ingold, and Prelog[46] introduced the R–S nomenclature which describes exactly the spatial arrangement of groups or atoms as obtainable from X-ray diffraction studies.

In this context it is worthwhile to remark that chemical nomenclature has been developed for solutions and not for the crystal state, so that some misunderstandings can sometimes occur. As far as enantiomers are concerned, it is known that the usual chemical synthesis can only produce a 1:1 mixture of the two enantiomers which is called **racemic mixture**; it can be separated (**resolved**) by reacting it with other chiral molecules by which the two enantiomers are transformed into diastereoisomers; only the more recent **asymmetric synthesis** can produce single R or S enantiomers. When a pure enantiomer crystallizes, it can only adopt a **polar space group** (i.e. without S_n rotary-reflection axes), otherwise the symmetry element would generate the other enantiomer; this is the case, for instance, for all natural aminoacids or proteins. The crystallization of the racemic mixture may be a more complex problem. Usually the mixture is thermodynamically more stable and crystallizes as a homogeneous solid containing equimolecular amounts of both enantiomeric molecules which is termed **racemate**. The crystal obtained is usually centrosymmetric with the two enantiomers related by a symmetry centre, though sometimes it happens to be polar with a double (or of higher even multiplicity) asymmetric unit allocated to both enantiomers. In the rare case that the racemate is less stable than its components, the racemic mixture crystallizes in a polar space group with spontaneous resolution of the enantiomers, that is it produces two types of enantiomeric crystals, each one containing just one enantiomer (which is, incidentally, the way Louis Pasteur discovered enantiomers for the first time). This fact is the origin of a not uncommon error in structural determination, that of determining the absolute molecular configuration without taking into account the fact that the bottle from which the crystal was taken contains one half of crystals of the opposite configuration. The subject of chiral crystals can become quite difficult to understand in its generality because there are also space groups which are chiral by themselves and not because they have to allocate chiral objects (e.g. $P3_1$

and $P3_2$). Moreover, what has been said for configurational enantiomerism has to be extended to conformational enantiomerism because conformations become fixed within the crystals. For more details the reader is referred to a more specialized treatment.[57]

Going back to the other isomers, it has already been remarked that **geometrical isomerism** is the traditional way of indicating diastereoisomerism in molecules having an internal S_n. Classical examples are *cis–trans* isomers in ethylene derivatives (13a, 13b), *syn–anti* isomers in oximes (14a, 14b) or azocompounds (15a, 15b), *cis–trans* isomers in square planar (16a, 16b) or octahedral (17a, 17b) complexes, and *mer–fac* (meridional–facial) octahedral complexes (18a, 18b). Geometrical isomerism is also observed in saturated cyclic compounds where the substituents can be over or under the ring plane. For instance, four-membered rings substituted as in (19) can only show geometrical isomerism in consequence of their internal mirror plane; accordingly (19a) and (19b) are *cis–trans* isomers. Complete nomenclature rules for geometrical isomers are available.[47,53]

Conformational isomerism

Conformational isomers (**rotational isomers, rotamers, conformers**) are the molecular states corresponding to minima of the potential energy curve expressed as a function of the torsion angle around a single bond. With reference to Fig. 7.8, states b, d, and f are conformers while a, c, and e are transition states; all are possible conformations. As is the case for all stereoisomers, conformers can have mutual relationships of enantiomerism or diastereoisomerism; so conformers b and f are enantiomers while b and c (or c and f) are diastereoisomers.

The energy barrier protecting conformers is usually rather small (5–15 kcal mol^{-1}) and much smaller than that protecting configurational isomers. For reference, the thermal activation barrier for *cis–trans* isomerization around a double C–C bond (13a, 13b) is some 40 kcal mol^{-1}. However, it is not the height of the barrier that can distinguish between conformational and configurational isomers but the fact that conformers differ in a rotation around what is a single bond (even if with partial double-bond character) in the ground state of the molecule. This rule allows us to classify as conformers the enantiomers produced by hindered rotation in *o, o'*-disubstituted diphenyls (20a, 20b), although the interconversion barrier is so high that they can be resolved at room temperature. Such resolvable conformers are sometimes reported as **atropoisomers**.

Ring conformations

It seems reasonable to assume that a ring containing single bonds, e.g. cyclohexane, could have different conformations describable by a potential energy curve of the type shown in Fig. 7.8, having minima and maxima corresponding to ring conformers and transition states. The difficulties arise when trying to define the independent variables that the potential energy is a function of, and this for the evident reason that the torsion angles fixing the conformation are not mutually independent but related by the condition of ring closure. The problem is not easy to solve and only some basic concepts will be discussed here; the mathematical complications will be discussed in the next section.

Let us focus our attention on the origins of the energy barrier which separates the different conformers. As it will be discussed in more detail later (p. 507), the deformation of a molecule can be conceived of in terms of displacements of its bond distances, bond angles, and torsion angles from their optimal equilibrium values. The deformation energy needed rapidly decreases in the order bond stretching or compression, bond angle bending, and torsion around single bonds, while the torsion around double bonds requires energies in between stretching and bending. Since energies of single bond torsions are very small, transitions between conformations which can occur without angle bending will have an almost null activation barrier. In this case the ring is **flexible** because many different combinations of the two isoenergetic (**degenerate**) conformations become possible. The pathway interconverting isoenergetic conformations is called, for reasons to be discussed later, a **pseudorotation path**. Classical examples of degenerate conformations are the **envelope (E)** and **twisted (T)** conformations of cyclopenthane (Fig. 7.9) and the **boat (B)** and **twist-boat or twisted (T)** conformations of cyclohexane (Fig. 7.10). The opposite occurs when it is impossible for the molecule to change conformation without bond angle deformation (or, in general, bond stretching or rotation around double bonds). In this case the molecule is **rigid** and its actual conformation is protected by a barrier which can be of the order of magnitude of 10 kcal mol^{-1} for saturated rings and somewhat greater if there are double bonds stiffening the ring. Figure 7.10 shows the somewhat idealized shape of the potential energy curve of cyclohexane as a function of a generalized coordinate of conformational interconversion as could be obtained by the molecular mechanics methods. The rigid conformation of cyclohexane is the **chair (C)**, which is the most stable and is transformed into the T form through a transition state called **half-chair (H)** which is some 11 kcal mol^{-1} higher in energy. T and B are almost isoenergetic and can be interconverted practically without any activation barrier (pseudorotation). C, B, and T are the three low-energy conformations of cyclohexane and therefore its three conformers according to the definitions given in the previous section.

The number of possible independent conformers of a ring can be determined by the use of simple considerations. A planar ring of N atoms has $2N$ degrees of freedom, out of which two are of translation, one of rotation, and $2N-3$ of in-plane vibration. Allowing the atoms to vibrate out of plane, that is the ring to become puckered, the number of vibrational

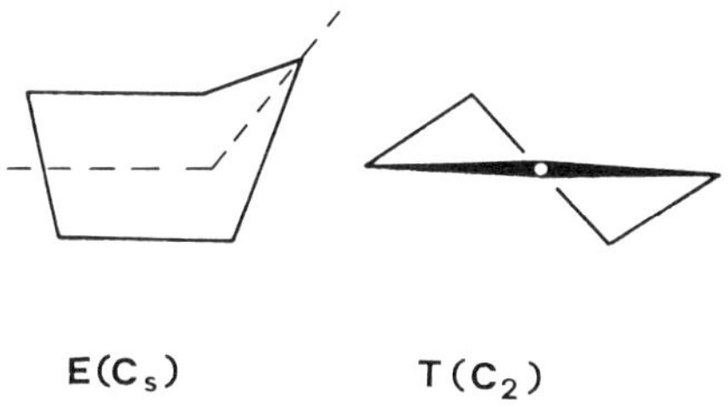

Fig. 7.9. The two degenerate conformers of cyclopenthane (E = envelope, T = twisted) and their symmetries. The ring conformation is flexible because the two conformers can interconvert with zero activation energy along a pseudorotation path. Any observed conformation can be described as a linear combination of E and T.

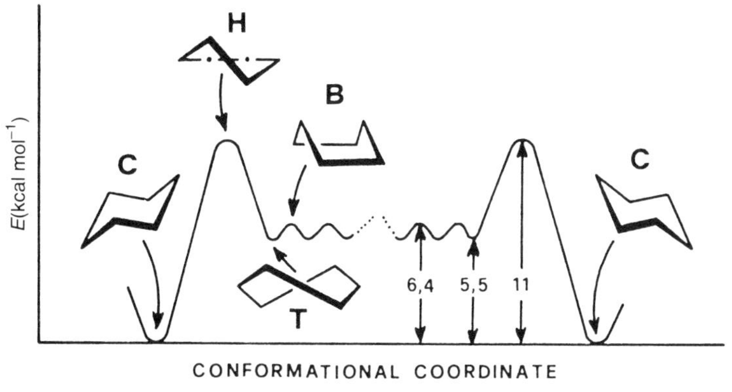

Fig. 7.10. Idealized potential energy profile of cyclohexane as a function of a generalized coordinate of conformational interconversion as could be obtained by the methods of molecular mechanics. C = chair, B = boat, T = twist-boat or twisted, H = half-chair. B and T are degenerate and produce flexible ring conformations; C is the rigid low-energy conformer.

degrees of freedom becomes $3N-6$ with an increment of $(3N-6)-(2N-3)=N-3$, which is just the number of out-of-plane independent vibrations or independent puckering conformers. So three- four- five- six- and seven-membered rings should have, respectively, zero, one, two, three, and four puckering conformers. This kind of reasoning allows us to know the total number of possible conformers but not their nature, a much more difficult task to be discussed in the next section. Another qualitative consideration can be derived from the examples already given. For cyclopentane there are $5-3=2$ conformers (E and T), which are isoenergetic or degenerate and flexible because they can interconvert along a pseudorotation path. For cyclohexane the conformers are $6-3=3$, out of which two (B and T) are degenerate and flexible while the other one (C) is single and gives the lowest-energy rigid conformation. By generalization: a N-membered ring has $N-3$ possible conformers which are arranged two by two in flexible mutually interconverting degenerate pairs; if $N-3$ is odd, the last single conformer is rigid and has the lowest possible energy. Accordingly, cycloheptane will have four possible conformers arranged in two flexible pairs while cyclooctane will have one more rigid low-energy conformation, which is the ground state of the ring.

Ring conformation and group theory

The problem of determining the $N-3$ independent low-energy conformations (conformers) of a N-membered ring is reducible to that of finding the normal out-of-plane vibrational modes of the ring, which can be easily solved by applying group theory[58] to equilateral polygons. The method will be illustrated for five- and six-membered rings and its understanding requires some previous knowledge of group theory. From a formal point of view the problem is the same as that of finding the molecular orbitals of planar rings within the Hückel theory.

Let us consider the regular hexagon of Fig. 7.11, where the out-of-plane displacements are represented by the small vertical arrows. The hexagon belongs to the point group D_{6h}, whose character table is reported in Table 7.5. The reducible representation for out-of-plane vibrations is obtained by

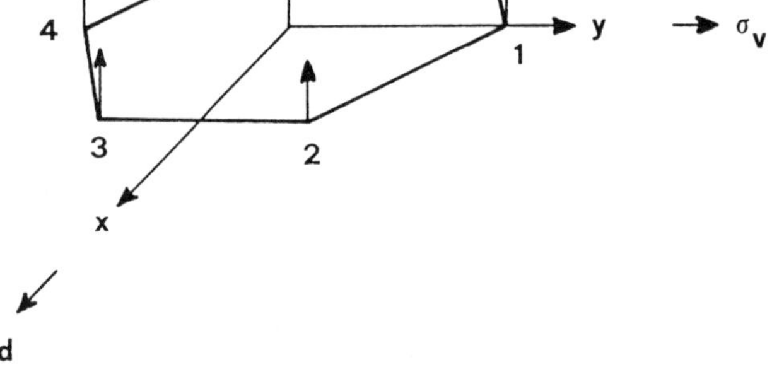

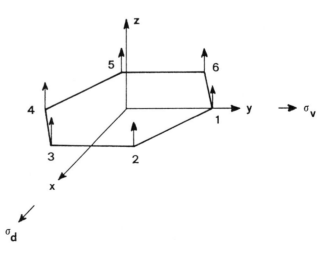

Fig. 7.11. A system of coordinates for describing the out-of-plane vibrations of a six-membered ring.

Table 7.5. Character table for the point group D_{6h}

D_{6h}	E	$2C_6$	$2C_3$	C_2	$3C_2'$	$3C_2''$	i	$2S_3$	$2S_6$	σ_h	$3\sigma_d$	$3\sigma_v$		
A_{1g}	1	1	1	1	1	1	1	1	1	1	1	1		x^2+y^2, z^2
A_{2g}	1	1	1	1	−1	−1	1	1	1	1	−1	−1	R_z	
B_{1g}	1	−1	1	−1	1	−1	1	−1	1	−1	1	−1		
B_{2g}	1	−1	1	−1	−1	1	1	−1	1	−1	−1	1		
E_{1g}	2	1	−1	−2	0	0	2	1	−1	−2	0	0	(R_x, R_y)	(xz, yz)
E_{2g}	2	−1	−1	2	0	0	2	−1	−1	2	0	0		(x^2-y^2, xy)
A_{1u}	1	1	1	1	1	1	−1	−1	−1	−1	−1	−1		
A_{2u}	1	1	1	1	−1	−1	−1	−1	−1	−1	1	1	z	
B_{1u}	1	−1	1	−1	1	−1	−1	1	−1	1	−1	1		
B_{2u}	1	−1	1	−1	−1	1	−1	1	−1	1	1	−1		
E_{1u}	2	1	−1	−2	0	0	−2	−1	1	2	0	0	(x, y)	
E_{2u}	2	−1	−1	2	0	0	−2	1	1	−2	0	0		

applying all the group symmetry operations to the six small vertical vectors of Fig. 7.11; the character for any class of symmetry operations (or column of the character table) is given by the number of small vectors not moved by the symmetry operation. It is easily seen that the characters of such representation, Γ_6, are

D_{6h}	E	$2C_6$	$2C_3$	C_2	$3C_2'$	$3C_2''$	i	$2S_3$	$2S_6$	σ_h	$3\sigma_d$	$3\sigma_v$
Γ_6	6	0	0	0	−2	0	0	0	0	−6	0	2

which can be resolved by standard methods† in a combination of the four irreducible representations (i.r.)

A_{2u}	1	1	1	1	−1	−1	−1	−1	−1	−1	1	1
E_{1g}	2	1	−1	−2	0	0	2	1	−1	−2	0	0
B_{2g}	1	−1	1	−1	−1	1	1	−1	1	−1	−1	1
E_{2u}	2	−1	−1	2	0	0	−2	1	1	−2	0	0

according to the equation $\Gamma_6 = A_{2u} + E_{1g} + B_{2g} + E_{2u}$.

The normal modes of out-of-plane vibration (or **modes of puckering**) are now to be found and, for the sake of clarity, let us make reference to the final results shown graphically in Fig. 7.12. The modes transforming as A_{2u} and E_{1g} are immediately found from Table 7.5 to correspond to rigid body motions (respectively translation along z and rotation around x and y).

A base for B_{2g} (i.e. a representation transforming under the group operations as B_{2g}) is easily found because it has to be symmetrical with respect to C_3 and antisymmetrical with respect to C_6 and therefore corresponds to alternate positive and negative out-of-plane displacements (conformation C in Fig. 7.10); calling z_i one of such displacements, the vibration is described by the function

$$\phi_C = 6^{-1/2}(z_1 - z_2 + z_3 - z_4 + z_5 - z_6) \qquad (7.17)$$

where $6^{-1/2}$ is a simple normalization factor granting that $\sum_i z_i^2 = 1$. A base

† The number of times n_i that the i.r. Γ_i appears in the reducible representation Γ' is given by

$$n_i = (1/h)\sum_R \chi_i(R)\chi'(R)$$

where h is the order of the group (in the present case 24) and $\chi_i(R)$ and $\chi'(R)$ are the characters of the symmetry operations Γ_i and Γ', respectively.

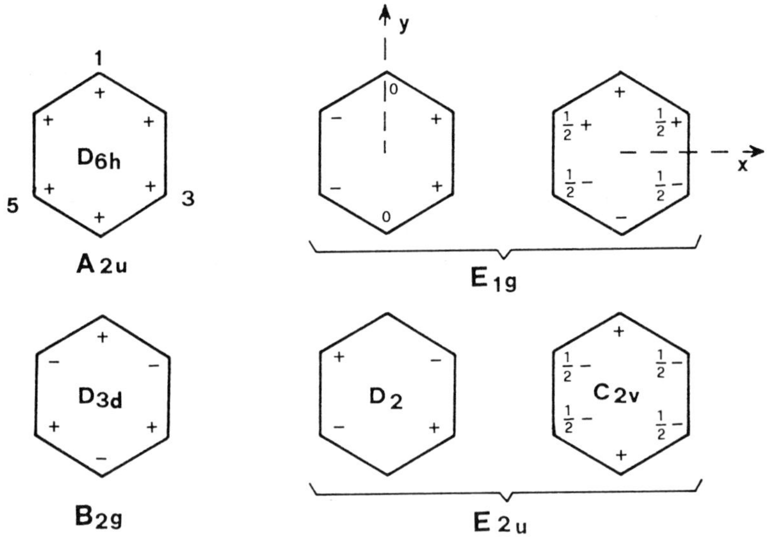

Fig. 7.12. Normal modes of out-of-plane vibration (ring puckering) of a six-membered ring and their symmetries. Any mode belongs to an irreducible representation of the point group D_{6h} and is therefore orthogonal to the others.

for E_{2u} is not intuitive and must be found by ordinary group theory methods. A first base can be obtained by associating the out-of-plane displacement to atom 1 of the ring and operating on it by all symmetry operations multiplied by their characters. The required base is the sum of all the terms obtained, that is

$$\begin{aligned}\phi &= 2Ez_1 - 1C_6^+z_1 - 1C_6^-z_1 - 1C_3^+z_1 - 1C_3^-z_1 + 2C_2z_1 \\ &\quad - 2iz_1 + 1S_3^+z_1 + 1S_3^-z_1 + 1S_6^+z_1 + 1S_6^-z_1 - 2\sigma_h z_1 \\ &= 2z_1 - z_2 - z_6 - z_3 - z_5 + 2z_4 + 2z_4 - z_3 - z_5 - z_2 - z_6 + 2z_1 \\ &= 2(2z_1 - z_2 - z_3 + 2z_4 - z_5 - z_6),\end{aligned}$$

which is a base for (or is transformed as) the boat conformation B and can be written in normalized form as

$$\phi_B = 12^{-1/2}(2z_1 - z_2 - z_3 + 2z_4 - z_5 - z_6). \qquad (7.18)$$

As E_{2u} is two-dimensional, the degenerate partner of ϕ_B remains to be found; it can be done by the same technique operating on z_2 instead of z_1 and obtaining the normalized function

$$\phi = 12^{-1/2}(2z_2 - z_3 - z_4 + 2z_5 - z_6 - z_1)$$

which, however, is not orthogonal to ϕ_B because $S = \int \phi_B\phi \, dv = -1/2$. To make the two functions orthogonal the Schmidt orthogonalization procedure is used,† obtaining the orthogonal function required in normalized form

$$\phi_T = 4^{-1/2}(-z_2 + z_3 - z_5 + z_6) \qquad (7.19)$$

which is the base for describing the puckering of the twisted conformation T. ϕ_B and ϕ_T belong to the same i.r. and are degenerate, at least for infinitesimal z_i displacements, and all their linear combinations are allowed

† If the two already normalized functions ϕ_a and ϕ_b are not orthogonal, i.e. $\int \phi_b\phi_a \, dv = S \neq 0$, the new function $\phi' = \phi_a - S\phi_b$ is orthogonal to ϕ_b, as can be shown by simple substitution.

solutions; this means that any linear combination of T and B conformers is an allowed conformation of the ring or, in other words, that the two conformers T and B can interconvert with zero activation energy along a pseudorotation path.

It is now useful to find an analytical expression for describing the displacements z_i of a ring whose puckering amplitude is $(\Sigma_i z_i^2)^{1/2} = q$. It is convenient to define the z_i values with reference to the mean ring plane, which is fixed by the condition of non-translation ($\Sigma_i z_i = 0$) and by those of non-rotation around the x and y axes; these last conditions can be obtained by equating to zero the sum of the vertical displacements weighted by the functions representing the two components of the rotational i.r. E_{1g} of Fig. 7.12, that is

$$\begin{aligned} &\sum_i z_i \cos[2\pi(i-1)/6] = 0 \\ &\sum_i z_i \sin[2\pi(i-1)/6] = 0 \end{aligned} \tag{7.20}$$

where i can assume the integer values from one to six. Moreover, it can be verified by simple computation that the following functions give the same displacements as ϕ_C in (7.17) and ϕ_B and ϕ_T in (7.18) and (7.19), and therefore are transformed by the operations of the point group D_{6h} as the i.r. B_{2g} and E_{2u} of Fig. 7.12

$$B_{2g}\colon \phi_C(i) = z_i = 6^{-1/2} q_3 \cos[3\pi(i-1)/6] = 6^{-1/2} q_3 (-1)^{i-1} \tag{7.21}$$

$$E'_{2u}\colon \phi_B(i) = z_i = 3^{-1/2} q_2 \cos[4\pi(i-1)/6] \tag{7.22}$$

$$E''_{2u}\colon \phi_T(i) = z_i = 3^{-1/2} q_2 \sin[4\pi(i-1)/6] \tag{7.23}$$

where $6^{-1/2}$ and $3^{-1/2}$ are normalization factors and q_3 and q_2 the mean puckering amplitudes of the two vibration modes.

The function ϕ_C describes the out-of-plane displacements of the C conformer while any linear combination $k_1\phi_B + k_2\phi_T$ describes those of the degenerate B and T conformers. For the properties of trigonometric functions, such combinations can be written as

$$E_{2u}\colon \phi_{B,T}(i) = z_i = 3^{-1/2} q_2 \cos[4\pi(i-1)/6 + \phi_2] \tag{7.24}$$

where ϕ_2 is a phase angle having values of 0, 60, 120, 180, 240, or 300° for the pure B conformations and 30, 90, 150, 210, 270, or 330° for the pure T conformations.

Equations (7.21) and (7.24) describe analytically the conformations of a six-membered ring in terms of the three parameters q_3, q_2, and ϕ_2 and the $N-3$ possible conformations remain characterized as the three C, B, and T conformers. The characterizing parameters (q_3, q_2, ϕ_2) are more generally identified as the $N-3$ **generalized puckering coordinates**, that is the $N-3$ variables necessary and sufficient to define any out-of-plane deformation of the ring. q_3 and q_2 are called *puckering amplitudes* and ϕ_2 the *phase angle* and any possible conformation of the six-membered ring corresponds to a point in the three-dimensional space spanned by the orthogonal base (q_2, ϕ_2, q_3).

For example $^1C_4$† corresponds to $q_3 = q$, $q_2 = 0$, and ϕ_2 undefined, while $^4C_1$ differs only for having $q_3 = -q$. Moreover, $^{1,4}B$, $B_{2,5}$, $^{3,6}B$, $B_{4,1}$, $^{2,5}B$, and $B_{3,6}$ have all $q_3 = 0$, $q_2 = q$, and $\phi_2 = 0, 60, \ldots, 300°$, respectively, while $^4T_2$, $^6T_2$, $^3T_1$, $^2T_4$, $^2T_6$, and $^1T_3$ still have $q_3 = 0$ and $q_2 = q$ but $\phi_2 = 30, 90, \ldots, 330°$, respectively. The actual value of $q = (\sum_i z_i^2)^{1/2}$ is determined by the equilibrium value of the bond angle, α, defined by the ring atoms. It is self-evident that q will be zero for an equilateral hexagonal ring having $\alpha = 120°$ and it can be shown that q_3 is equal to $6^{-1/2}R$ for an equilateral C ring of side R and tetrahedral internal ring angles (e.g. cyclohexane).

It has already been remarked that any possible ring conformation is defined by the triplet of values (q_2, ϕ_2, q_3) in a Cartesian system of axes; otherwise these coordinates can be expressed as polar coordinates (Q, θ, ϕ), where Q is the total puckering amplitude defined as

$$\sum_i z_i^2 = q_2^2 + q_3^2 = Q^2 \qquad (7.25)$$

and $\phi = \phi_2$, while θ is an angle in the range $0 - \pi$ such that

$$q_2 = Q \sin\theta \quad \text{and} \quad q_3 = Q \cos\theta. \qquad (7.26)$$

In this way any conformation is represented by a point on a sphere of radius Q, $^1C_4$ being at the north pole ($\theta = 0°$, ϕ undefined) and its enantiomer $^4C_1$ at the south pole, while B and T interconvert along a pseudorotation path which is on the equator ($\theta = 90°$) in correspondence to all possible values of ϕ.

The previous discussion was mainly concerned with the low-energy conformers C, B, and T. The ring, however, also has high-energy transition states and there is general agreement[59] that these are the **envelope (E)**, **half-chair (H)**, and **screw-boat (S)** conformations of C_s, C_2, and C_2 symmetry, respectively, as shown in Fig. 7.13. Cremer and Pople[60] have shown that the E and H forms are located at $\tan\theta = \pm\sqrt{(3/2)}$ and $\pm\sqrt{2}$, respectively, and Boeyens[61] that the S form occurs for $\tan\theta = \pm(1 + \sqrt{2})$. This last author has reported a complete map of the six conformations, usually called **canonical**, in polar coordinates, which is reported in Fig. 7.14 with some modifications. The figure has the further meaning that all maximum circles indicated are possible interconversion pathways among conformations. The T–B pseudorotation circle is the equator, while along the meridians the C conformation is converted to B through E or to T through H and S.

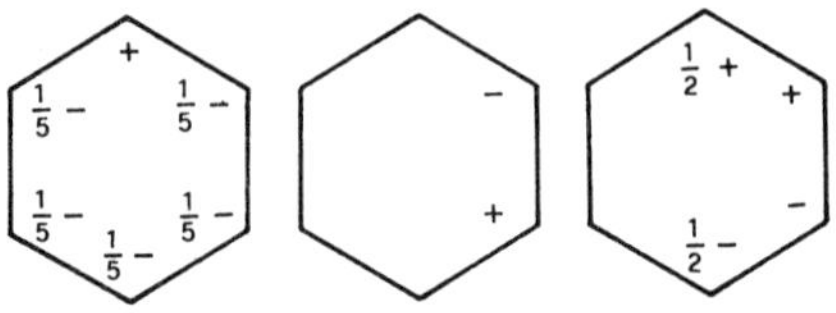

Fig. 7.13. The three activated states of puckering of a six-membered ring and their symmetries (E = envelope, H = half-chair, S = screw-boat).

The problem of the conformations of a five-membered ring can be solved by the same methods. The ring has only two puckering coordinates because $N - 3 = 2$; it belongs to the point group D_{5h}, whose character table can be obtained from any group theory text. The reducible representation for the out-of-plane vibrations is

D_{5h}	E	$2C_5$	$2C_5^2$	$5C_2$	σ_h	$2S_5$	$2S_5^3$	$5\sigma_v$
Γ_5	5	0	0	−1	−5	0	0	1

† According to the nomenclature usually employed, the conformations chair, boat, and twisted (or twist-boat) are called C, B, and T and the atoms which are over or under the mean plane of the ring are indicated by upper left or lower right indices. So, $^{2,5}B$ is a boat ring having the atoms number two and five directed upwards and $B_{2,5}$ its enantiomer while $^2T_5$ is a twisted ring with the second atom pointing up and the fifth one pointing down.

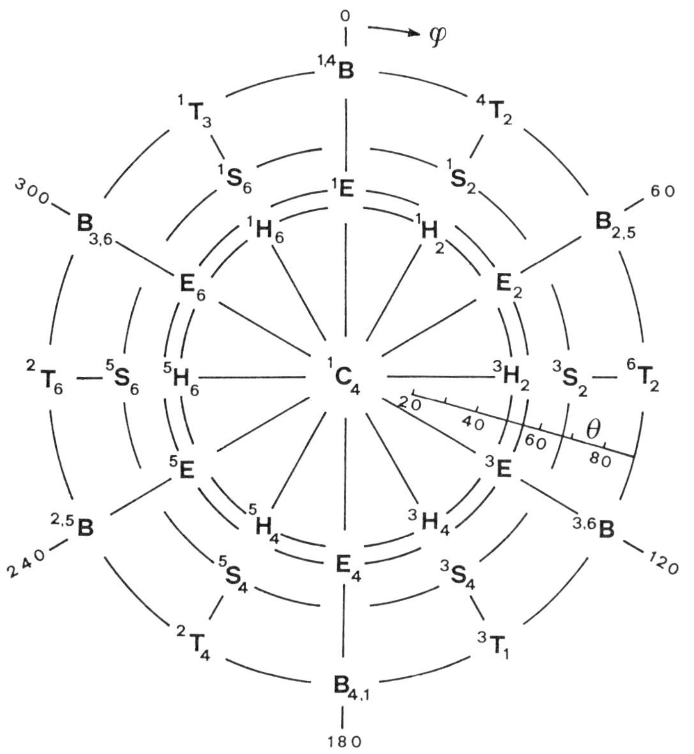

Fig. 7.14. Polar projection of the northern part of the sphere reporting, for an arbitrary value of Q, the positions θ and ϕ of the canonical conformations of a six-membered ring. The southern hemisphere contains the enantiomeric conformations in centrosymmetric positions.

which can be decomposed as $\Gamma_5 = A_2'' + E_1'' + E_2''$, where A_2'' and E_1'' correspond to translation and rigid body rotations, respectively. The puckering modes belong to the two-dimensional i.r. E_2'' and a base for them, found by the usual methods, is shown in Fig. 7.15. The two conformers found, **envelope (E)** and **twisted (T)**, are degenerate and all their linear combinations are possible conformations on a pseudorotation pathway. By analogy with (7.20–7.24) the non-translation and non-rotation conditions are

$$\sum_i z_i = 0,$$

$$\sum_i z_i \cos\left[2\pi(i-1)/5\right] = 0, \qquad (7.27)$$

$$\sum_i z_i \sin\left[2\pi(i-1)/5\right] = 0,$$

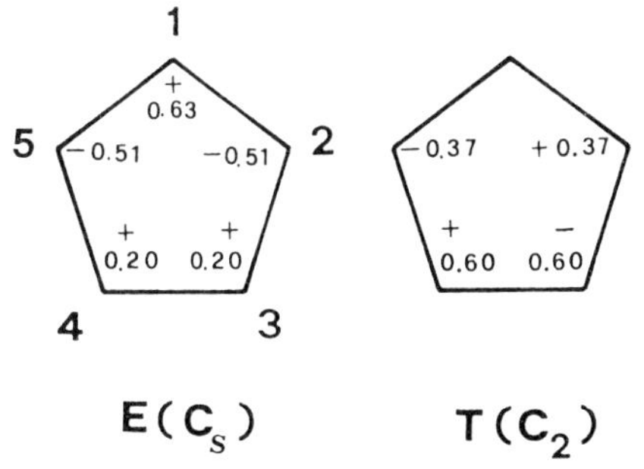

Fig. 7.15. Normal modes of out-of-plane vibration (ring puckering) of a five-membered ring and their symmetries. E = envelope, T = twisted.

and the functions describing the puckering displacements are

$$\phi_E(i) = z_i = (2/5)^{1/2} q_2 \cos\left[4\pi(i-1)/5\right], \qquad (7.28)$$

$$\phi_T(i) = z_i = (2/5)^{1/2} q_2 \sin\left[4\pi(i-1)/5\right], \qquad (7.29)$$

where i can assume all integer values from one to five and $(2/5)^{1/2}$ is a normalization factor providing that $\sum_i z_i^2 = q_2^2$, q_2 being the puckering amplitude. Since E_2'' is two-dimensional, any linear combination $k_1\phi_E + k_2\phi_T$ belongs to E_2'' and can be written as

$$\phi_{E,T}(i) = z_i = (2/5)^{1/2} q_2 \cos\left[4\pi(i-1)/5 + \phi_2\right] \qquad (7.30)$$

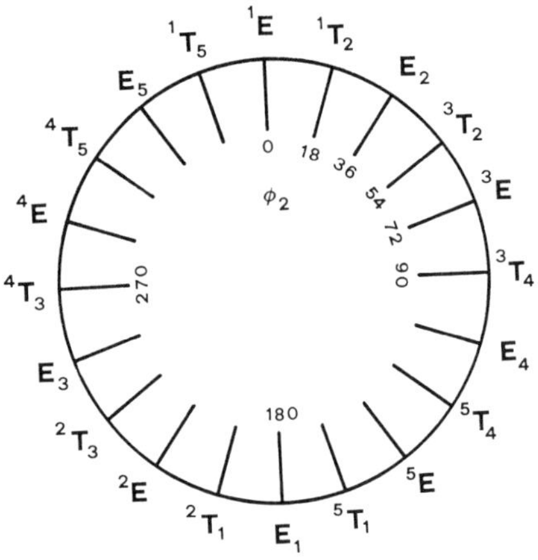

Fig. 7.16. Pseudorotation wheel describing the conformations which can be assumed by a five-membered ring during the changes of the phase angle ϕ_2.

where ϕ_2 is a phase angle having values of 0, 36, 72, 108, 144, 180, 216, 252, 288, and 324° for the ten E conformations and 18, 54, 90, 126, 162, 198, 234, 270, 306, and 342° for the other ten T conformations, intermediate values of ϕ_2 corresponding to mixtures of the two along the pseudorotation path. The complete pathway is shown in Fig. 7.16.

Computation of puckering coordinates

The theory given in the previous section concerns equilateral and isogonal rings. However, the equations obtained can also be used for irregular rings without loss of generality, at least as regards the Cremer and Pople treatment,[60] which makes use of the displacements z_i from the ring mean plane. Other methods using the values of the endocyclic torsion angles (e.g. that proposed by Altona *et al.*[62] for five-membered rings) may not have the same degree of generality.

Several computer programs[63,64] are available for the computation of the puckering parameters according to Cremer and Pople.[60] The procedure usually includes: transformation from crystal to orthogonal coordinates; passage to the system of coordinates defined by (7.20) or (7.27), having their origin at the centre of gravity of the ring, the z axis perpendicular to the mean ring plane and the y axis passing through the first ring atom; determination of the parameters q and ϕ (or Q, θ, and ϕ) satisfying eqns (7.21) and (7.24) or (7.30).

An example of application to the fructose and glucose rings of the molecule of sucrose is reported in Table 7.6. The five-membered fructose ring has $q_2 = 0.353$ Å and $\phi_2 = 265.1°$. By comparing the calculated value of ϕ_2 with those of Fig. 7.16, it is found that the ring conformation is very near $^4T_3$ with a small component of E_3. The six-membered ring has a Q value of 0.556 Å, i.e. it is more puckered than the five-membered ring. The calculated values of $\theta = 5.2°$ and $\phi = 183.7°$ indicate, when compared with

Table 7.6. Atomic coordinates and puckering parameters for fructose and glucose rings of the molecule of sucrose (coordinates from Brown and Levy, *Acta Crystallographica*, **B39**, 790 (1973))

Atom	N	Orthogonal coordinates			Puckering coordinates		
O′(2)	1	2.2445	0.8217	1.9334	0.0	1.2111	−0.0189
C′(2)	2	1.3181	1.6765	2.5591	1.1622	0.4349	0.1461
C′(3)	3	0.0762	1.6600	1.6490	0.7425	−1.0012	−0.2174
C′(4)	4	0.6860	1.4494	0.2675	−0.7221	−1.0309	0.2057
C′(5)	5	1.8665	0.5336	0.5687	−1.1826	0.3861	−0.1154

Puckering parameters: $q_2 = 0.353$ Å, $\phi_2 = 265.1°$

Atom	N	Orthogonal coordinates			Puckering coordinates		
O(5)	1	3.9935	3.4716	1.9418	0.0	1.3839	0.1976
C(1)	2	3.1719	3.1155	3.0330	1.1997	0.7624	−0.2106
C(2)	3	3.3085	4.1323	4.1739	1.2356	−0.7040	0.2393
C(3)	4	3.0216	5.5425	3.6851	0.0110	−1.4564	−0.2550
C(4)	5	3.9596	5.8402	2.5191	−1.2300	−0.7208	0.2420
C(5)	6	3.8029	4.7973	1.4170	−1.2164	0.7350	−0.2133

Puckering parameters: $q_2 = 0.050$ Å, $\phi_2 = 183.7°$
$q_3 = 0.554$ Å
$Q = 0.556$ Å, $\theta = 5.2°$, $\phi = 183.7°$

those of Fig. 7.14, that the ring conformation is an almost perfect $^1C_4$ with a very small distortion towards $^5H_4$ and E_4.

Molecular geometry and the chemical bond

An overview of bond theories

One of the central topics of structural chemistry concerns the interpretation of the observed molecular geometries in terms of the existing theories of chemical bonding. Theories derived from the fundamentals of quantum chemistry include ***ab initio***[65,66] and **semiempirical**[67–69] methods. In principle both classes of methods (in particular the first ones) are able to reproduce all the details of the experimental geometries with the great additional advantage of giving them a theoretical justification in terms of molecular orbital energies, shape, and symmetry. The main difficulties arise from the great complexity of the calculations involved even for molecules of relatively small dimensions. The situation is not so different from that discussed in connection with the problem of molecular structure determination by microwave spectroscopy or gas electron diffraction. Great accuracy can be achieved at a reasonable cost of both human effort and computer time only for molecules of very small dimensions in comparison with those X-ray crystallography is used to deal with. Another important aspect concerns the objective difficulties many crystallographers may have in understanding the fundamentals of quantum chemistry needed for a not completely acritical use even of the most known and generally available compter programs (e.g. GAUSSIAN[70] for the *ab initio* programs or MINDO/3, MNDO[71] and AM1[72] for the semiempirical ones).

It is then not surprising that, in spite of the fact that quantum-mechanical calculation programs have been implemented in most molecular graphics systems commercially available, the most popular bond theories used in interpreting molecular geometries remain the **qualitative** or **semiquantitative theories**, that is those always based on quantum chemistry but with the different aim to produce concepts of easier application for practical scientists. It is these theories which will be briefly reviewed here.

1. **Valence bond (VB) theory.**[5,73,74] In VB theory the molecular geometry is interpreted in terms of the concepts of hybridization and separability of localized σ bonds and localized or delocalized π bonds. The π bond delocalization is treated by means of the concept of resonance between canonical forms, through, more recently, the **Hückel molecular orbital** method limited to the π electrons (Pi HMO)[75] has been frequently employed. The idea of localized bonds is the logical foundation of **molecular mechanics** (see p. 506–11), which gives an interpretation of the molecular geometry in terms of balls (atoms) connected by springs (the chemical bonds), and, as previously discussed (see p. 486), is the basis of all definitions of classical stereochemistry.

2. **Valence shell electron pair repulsion (VSEPR) theory.**[76,77] This is probably the most used qualitative tool for the interpretation of the shapes of molecules containing main-group elements, while its applicability to compounds of transition elements is rather limited. The VSEPR model has

been considered destitute of theoretical background for a long time and only recently has it been interpreted in terms of basic quantum chemistry.[78]

3. **Transition metal–ligand bond theories**. The **VB** and **crystal field** (CFT)[79–81] **theories** were the first ones to try an interpretation of structure and bonding in coordination compounds. The introduction of molecular orbital methods in coordination and organometallic chemistry has given origin to the **ligand field theory** (LFT),[82,83] which is of great value for the understanding of the energetics and geometry of these compounds. A specific semiquantitative tool developed in the frame of LFT for calculations concerning electronic spectra, magnetic properties, and equilibrium geometries has been the so called **angular overlap model** (AOM),[84–87] whose application has been successively extended to the study of the molecular geometry of compounds of main-group elements.[87]

4. **Qualitative molecular orbital theories (QMOT)**. The first attempts to interpret geometries of simple compounds of the AB_2 type in terms of molecular orbitals were made by Walsh.[88] His method was successively extended to a series of small molecular systems, such as AB_3, AB_4, A_2B_4, A_2B_6, functional groups, etc.[86,87,89,90] The development of QMOTs in interpreting molecular shapes has been paralleled by the great impact these theories have had in the field of organic chemical kinetics (**frontier orbitals theory**[91,92]). Moreover, QMOTs are having increasing applications in the interpretation of the geometry of organometalic compounds by the methods developed by Hoffmann,[93–95] which are based on the partitioning of the molecule into fragments and which are of semiquantitative nature as molecular orbitals are evaluated by EHT (**extended Hückel theory**),[95] an extension of the classical Hückel method taking into account also σ electrons.

5. **Perturbative methods**. These are based on the **first-order** (FOJTE) and **second-order** (SOJTE) **Jahn–Teller effects**. FOJTE[96] is of great importance in understanding the geometrical distortions which lower the molecular energy by removing the orbital degeneration[97] and has found its most typical applications in the study of the coordination polyhedra of d^9 complexes.[98,99] The SOJTE has similar applications but in closed-shell molecules. It has been mainly developed by Pearson[100] extending a first paper by Bartell;[101] it has been extensively used for predicting geometries of compounds of main-group elements.[86,87]

6. Other qualitative theories. A rather simple model which has provided remarkable agreement between experimental and predicted structures is the **hard sphere** model proposed by Bartell[101] and developed by Glidewell.[102] It makes the hypothesis that molecular geometry is mainly determined by the repulsions among first non-bonded neighbours, each having an individual non-bonded radius. Other theories having a more limited range of application have been proposed, such as the **three centres–two electrons bond** in boron chemistry, the ***trans* influence** in square planar complexes or the theories based on the **acid–base** concept, which have proved to be of great utility in interpreting charge transfer effects in molecular crystals and have been reviewed by several authors.[29,103–105]

In view of the concise character of the present exposition, it has been decided to limit the discussion to the VSEPR and VB theories, which are certainly the most popular models used in structural crystallography. Moreover, only compounds of main-group elements will be taken into account.

The VSEPR theory[76,77]

The VSEPR theory assumes that the relative ligand arrangement around a central atom is determined by the mutual repulsions of bonding or non-bonding (lone) pairs of electrons. Bonding and non-bonding pairs are identified from the Lewis-type electronic structure of the molecule. The method can be summed up in the following few rules: **Rule 1**. Electron pairs repel each other and adopt the geometrical disposition which minimizes mutual repulsions; **Rule 2**. Lone pairs take more room than bonding pairs; **Rule 3**: The space occupied by a bonding pair decreases with the increasing electronegativity of the substituent; **Rule 4**. The two bonding pairs of a double bond (and, more so, the three bonding pairs of a triple bond) occupy more space than a single bond does.

Calling A the central atom, X the bonding pair (or the generic ligand connected by it) and E the non-bonding pair, Rule 1 can be used to predict the geometry of a large number of compounds (Fig. 7.17). $BeCl_2$, BF_3, and CF_4 are AX_2, AX_3, and AX_4 molecules and will have linear, trigonal, and tetrahedral geometries, respectively. Molecules of type AX_3E (NH_3, PH_3, NF_3, PF_3, or $AsCl_3$) are pyramidal and those of type AX_2E_2 (H_2O, H_2S, H_2Se, or OF_2) are bent (angular). Molecules of the type AX_5 may be either trigonal bipyramidal, or square pyramidal. According to VSEPR calculations, the first geometry is slightly more stable and, in fact, AX_5 molecules (PF_5, AsF_5, and PCl_5) are usually trigonal bipyramidal. Finally, molecules belonging to the types AX_6 (e.g. SF_6) and AX_5E (e.g. IF_5) display the expected octahedral and square pyramidal geometries.

For AX_4E ($TeCl_4$, SF_4), AX_3E_2 (ClF_3, BrF_3), and AX_2E_3 (XeF_2, ICl_2^-) it is necessary to establish whether the non-bonding pairs are in the axial or equatorial positions of the trigonal bipyramid. According to rule 2 the non-bonding pair takes more room and should prefer the equatorial position where it makes angles of 120, 120, and 90° with the vicinal electron pairs (the three angles would be 90° for the axial position). Similar considerations predict that AX_4E_2 molecules (e.g. $[ICl_4^-]$) are square planar with the non-bonding pairs in *trans* positions.

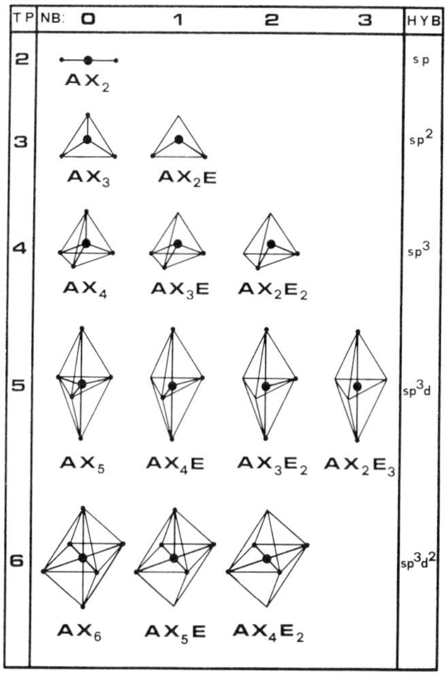

Fig. 7.17. Prediction of the shape of simple molecules according to the VSEPR and VB theories. NB = number of non-bonding electron pairs (lone pairs), TP = total number of electron pairs (bonding and non-bonding); HYB = type of VB hybridization.

An interesting feature of VSEPR is its ability to interpret fine structural details. For example, Rule 2 predicts that increasing substitution of X by E will cause a narrowing of the remaining X–A–X angles because of the greater space taken by the lone pairs. The effect is generally observed, as exemplified by the series CH_4 (AX_4), NH_3 (AX_3E), and H_2O (AX_2E_2) where the X–A–X angle decreases in the order 109.5, 107.3, and 104.5°.

Rule 2 can be considered a particular case of Rule 3 since the non-existent substituent E can be thought of as the least electronegative of all the possible X ligands. This last rule is thus the most suited for understanding the angular changes caused by substitution and predicts that the X–A–X angle decreases while the electronegativity of X increases. Although a few exceptions are known, the effect is correctly observed in a variety of cases. For example, the X–A–X angle is 103.8 in OF_2 and 104.9 in H_2O, 102.1 in NF_3 and 106.6° in NH_3. In the halides of the main-group elements of the type AX_2E_2 or AX_3E the bond angles are known to increase in the order $F < Cl < Br \approx I$. Another well documented[106] series of electronegativity-dependent angular variations concerns the monosubstituted benzenes. The endocyclic C–C–C angles in *ipso* (that is that carrying the substituent) changes with the substituent itself, being 117.2 for $N(Me)_2$, 118.1 for CH_3, 120.00 for H (unsubstituted benzene), 121.4 for Cl, 121.8 for –CN, and 123.4° for F. It is evident that the angle is wider the greater is the electronegativity of the substituent group.

Rule 4 explains why the bond angles *trans* to a multiple bond are smaller than usual, as can be illustrated by some examples in the $X_2A{=}Y$ system. The X–A–X angle should be 120° in the absence of perturbing effects but becomes respectively 108 and 111° in $F_2C{=}O$ and $Cl_2C{=}O$, where both the double-bond repulsion and the high electronegativity of the substituents contribute to the squeezing of the angle. The effect of the double bond is better seen when the X substituent is less electronegative, for example in $H_2C{=}O$ and $(NH_2)_2C{=}O$, where the X–C–X angle is 108° in both compounds.

Valence bond (VB) theory[5,73,74]

The VB theory is too well known to require a detailed exposition and only some aspects concerning the interpretation of molecular geometries will be discussed here.

The mutual orientation of the bonds departing from a central atom is interpreted in terms of **hybrid orbitals** (HO), which are linear combinations of the **atomic orbitals** (AO) of such an atom; typical sp, sp^2, sp^3, sp^3d, and sp^3d^2 HOs assume relative orientations which are respectively linear, trigonal, tetrahedral, trigonal, bipyramidal, and octahedral (Fig. 7.17) and the overlapping of the HOs with the correct AOs of the bonded atoms generates the frame of the σ bonds within the molecule.

After the formation of all the σ bonds between the central atom and its surrounding ligands, the generic set $s^lp^md^n$ of hybrid orbitals can allocate only $(l + m + n)$ bonding or non-bonding (lone) pairs, which require twice as many valence electrons. The hybridization state of the central atom remains determined by the total number of valance pairs available, while its connectivity is also affected by the number of non-bonding pairs which occupy an HO without adding any bonded atom. Moreover, electron pairs implied in π bonding are irrelevant for determining the hybridization state. So, CH_4, NH_3, and H_2O have all eight valence electrons (sp^3 hybridization) but only CH_4 is tetrahedral while NH_3 is pyramidal (one non-bonding pair) and H_2O is bent (two non-bonding pairs). Application of these rules to the examples of the previous section shows that VB gives exactly the same results obtainable by the use of VSEPR theory (see Fig. 7.17).

In VB theory the double bonds do not affect the shape of the molecule, which is strictly determined by the net of its σ bonds. Double and triple bonds are conceived of as σ bonds to which one or two further π bonds are added and the only effect of such an addition is to shorten the bond distances. The idea of single or multiple localized bonds between two given atoms is clearly an oversimplification of the complex many-body interactions between nuclei and electrons within the molecule. However, it has proved to be accurate enough to allow tabulations of characteristic bond

Table 7.7. A selection of characteristic bond distances in organic molecules (in Å); R = alkyl, Ar = aryl, ar = aromatic

C–C	sp^3–sp^3	1.54	C–P	C(sp^3)–P	1.84
	sp^3–sp^2	1.51	C–H	C(sp^3)–H	1.10
	sp^3–sp	1.46		C(ar)–H	1.08
C–C	aromatic	1.39	C–F	C(sp^3)–F	1.38
C=C	⟩C=C⟨	1.33		C(sp^2)–F	1.33
	⟩C=C=C⟨	1.31	C–Cl	C(sp^3)–Cl	1.77
	⟩C=C=C=C⟨	1.28		C(sp^2)–Cl	1.71
C≡C		1.20		C(sp)–Cl	1.64
C–N	sp^3–sp^3	1.47	C–Br	C(sp^3)–Br	1.94
C=N		1.27		C(sp^2)–Br	1.87
C≡N		1.16		C(sp)–Br	1.80
C–O	R–O–R	1.43	C–I	C(sp^3)–I	2.14
	RCO–OR	1.34		C(sp^2)–I	2.07
	RCO–OH	1.31	N–H		1.01
	RCOO–R	1.44	O–H		0.95
	C(ar)–OR	1.36	S–S	RS–SR	2.05
C–O	$RCOO^-$	1.25	O–O	RO–OR	1.48
C=O	⟩C=O	1.20	N–N	⟩N–N⟨	1.45
	=C=O	1.16	N=N	–N=N–	1.25
C–S	C(sp^3)–S–	1.82	N–O	⟩N–O–	1.36
	C(sp^2)–S–	1.76		⟩N→O	1.30
C=S		1.71		$-NO_2$	1.22
C–Si	C(sp^3)–Si–	1.87	N=O	–N=O	1.20

distances[107–110] which are reasonbly reproducible from molecule to molecule. Table 7.7 reports a selection of characteristic bond distances observed in organic molecules.

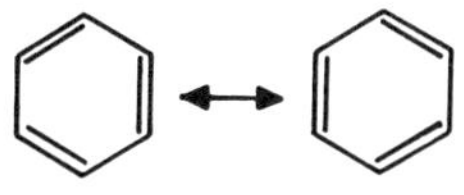

7.I

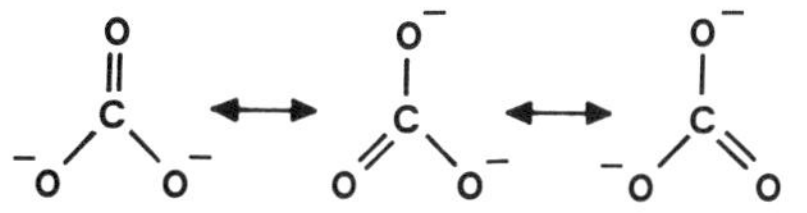

7.II

Bonds of intermediate multiplicity are interpreted within the VB theory in terms of **resonance**, as shown for the simple cases of benzene (7.I) and carbonate ion (7.II). The structure of the molecule is represented as a **resonance hybrid** between different limit forms, called **canonical**. The hybrid justifies the fractional bond order, for instance 1.5 in benzene (7.I) and 1.33 in the C–O bond of the carbonate ion (7.II). Such a fractional number has been called **bond number**, n, by Pauling,[5] who has proposed different formulae for its evaluation. The most used is

$$\Delta d = d(n) - d(1) = -c \log_{10} n \tag{7.31}$$

where $d(1)$ and $d(n)$ are the distances of a single and n-ple bond, respectively, and c is a constant to be determined. Some applications of this equation will be discussed in the last part of this chapter.

VB is particularly suited for rationalizing small changes of bond distances and angles in terms of substituent electronegativities. The discussion follows the treatment given by Bent[111] and starts from the elementary consideration that sp, sp^2, and sp^3 hybrids have decreasing s character (50, 33, and 25 per cent, respectively). This leads to important consequences:

1. Since the s AO is at a lower energy than the p orbital, the

electronegativity of an atomic valence increases with the s character of the valence itself, that is an atom binding by one of its sp HO is more electronegative than when it binds by an sp^2 HO or by an sp^3 HO. If the HO in question allocates a non-bonding pair (NB), the basicity of the atom will be affected according to the rule that *the greater is the s character of the HO donating the lone pair, the worse will be the atom as a donor*. Accordingly ketones, where the lone pair is on an sp^2 HO, are weaker bases than ethers, where it is on an sp^3 HO. Likewise nitriles R–C≡N are weaker bases than pyridine C_5H_5N, that, in turn, is weaker than ammonia NH_3, in agreement with the fact that nitrogen hybridization changes from sp to sp^2 to sp^3.

2. The best superposition of two *n*s AOs occurs at a smaller internuclear distance than that between *n*p AOs and then bond distances become shorter with the increasing s character of the HOs involved. In fact the C–H bond is 1.09 in methane, 1.07 in ethylene, and 1.06 Å in acetylene, the single bonds $C(sp^3)$–$C(sp^3)$, $C(sp^3)$–$C(sp^2)$, and $C(sp^3)$–C(sp) have decreasing bond lengths of 1.54, 1.51, and 1.46 Å and, finally, double-bond distances decrease along the series ⟩C=C⟨, ⟩C=C=C⟨, and ⟩C=C=C=C⟨ in the order 1.33, 1.31, and 1.28 Å. It may be remarked that this contraction is not a constant for all the elements but seems to increase with the electronegativity of the atom, as shown by the series carbon–halogen in Table 7.7.

3. The bond angles steadily increase with the increasing s character of the HOs, the angle between sp^3, sp^2, and sp HOs being respectively 109.47, 120, and 180°. Smaller changes caused by electronegativity differences are rationalized by the so-called **Bent's rule**:[111] *The p character of an atom tends to concentrate in its HOs pointing towards more electronegative substituents, the effect being stronger the less electronegative the atom itself is.* All the substituents can be arranged, starting from the lone pair, in order of increasing group electronegativity and the narrowest bond angles will be those implying the most electronegative ligands. The effect is well documented. The X–O–X angle is 111, 109, 104.9, and 103.8° in $(CH_3)_2O$, CH_3OH, H_2O, and F_2O, but only 100 and 92° in CH_3SH and H_2S in view of the smaller electronegativity of sulphur. In the AX_3 series the X–A–X angle is 109, 106.6, and 102.1° in $N(CH_3)_3$, NH_3, and NF_3, 100 and 93.3° in $P(CH_3)_3$ and PH_3, 96 and 91.5° in $As(CH_3)_3$ and AsH_3. Bent's rule is the VB equivalent of Rules 2 and 3 in the VSEPR theory and it can be shown to include even Rule 4. In the Y=AX_2 fragment the π bond contributes to shorten the X=Y bond length, causing a parallel increase of s character in the σ bond and, in turn, an increase of the p character of the HOs pointing to the X substituents so that the X–A–X bond angle is narrowed.

Hybridization. The machinery

A generic HO is a linear combination of AOs, ϕ,

$$\mathrm{HO}_i = \sum_j a_{ij}\phi_j \tag{7.32}$$

and the problem consists of finding the set of a_{ij} values matching the required molecular fragment geometry. A few rules allow us to set up a system of linear equations:

1. The coefficients must be such that any HO is transformed into one of its equivalent orbitals by all the symmetry operations of the fragment.
2. The HO_i must be orthonormal, that is

$$\sum_j a_{ij}a_{kj} = 1 \quad \text{if } i = k$$
$$= 0 \quad \text{if } i \neq k. \tag{7.33}$$

3. Since the atomic orbitals ϕ_j are orthonormal the sum over all HOs of the squares of the coefficients of any ϕ must be unitary, that is

$$\sum_i a_{ij}^2 = 1 \quad \text{for any } j. \tag{7.34}$$

As an example, let us build up the three trigonal HOs in the plane (xy). The axes are chosen in such a way that HO_1 points along x and HO_2 and HO_3 are in the $(-x, y)$ and $(-x, -y)$ quadrants, respectively. Let us write the HOs as

$$\begin{aligned} HO_1 &= a_1 s + b_1 p_x + c_1 p_y, \\ HO_2 &= a_2 s + b_2 p_x + c_2 p_y, \\ HO_3 &= a_3 s + b_3 p_x + c_3 p_y. \end{aligned} \tag{7.35}$$

The s orbital has spherical symmetry and must be equally shared by the three HOs; since $\sum_i a_i^2 = 1$ for condition (3), it must be that $a_1 = a_2 = a_3 = 1/\sqrt{3}$. c_1 is zero because p_y is orthogonal to HO_1 in view of the axes chosen and therefore $a_1^2 + b_1^2 = 1$ for (2), so that $b_1 = 2/\sqrt{6}$. Since HO_2 and HO_3 are symmetrical with respect to the σ_v containing x and $c_2^2 + c_3^2 = 1$, it results that $c_2 = -c_3 = 1/\sqrt{2}$. Likewise $b_2 = b_3 = -1/\sqrt{6}$ because of condition (2). The final HOs are

$$\begin{aligned} HO_1 &= 1/\sqrt{3}\, s + 2/\sqrt{6}\, p_x, \\ HO_2 &= 1/\sqrt{3}\, s - 1/\sqrt{6}\, p_x + 1/\sqrt{2}\, p_y, \\ HO_3 &= 1/\sqrt{3}\, s - 1/\sqrt{6}\, p_x - 1/\sqrt{2}\, p_y. \end{aligned} \tag{7.36}$$

It is easy to check that condition (1) is fulfilled for all the symmetry operations of the point group D_{3h} that the trigonal AX_3 fragment belongs to. The **hybridization index** n of each HO is defined as the ratio between the sum of the squares of the p AO coefficients and the square of the s coefficient. In this case $n_1 = n_2 = n_3 = 2$ and the hybridization is termed sp^2. Alternatively, the **s and p characters** (S and P, respectively) of the HO can be used, as defined by the conditions $n = P/S = P/(1 - P) = (1 - S)/S$ and $S + P = 1$ and by the inverse relationships $S = 1/(n + 1)$, $P = n/(n + 1)$. In the present case $S = 0.33_3$ and $P = 0.66_6$.

By the same methods it can be calculated that the two sp HOs ($n_1 = n_2 = 1$) oriented along x have equations $HO_1 = (s + p_x)/\sqrt{2}$ and $HO_2 = (s - p_x)/\sqrt{2}$ and that four equivalent sp^3 HOs ($n_1 = n_2 = n_3 = n_4 = 3$), oriented in such a way that each HO makes the same angles with the

Cartesian axes, have equations

$$\begin{aligned} HO_1 &= (s + p_x - p_y + p_z)/2, \\ HO_2 &= (s - p_x + p_y + p_z)/2, \\ HO_3 &= (s - p_x - p_y - p_z)/2, \\ HO_4 &= (s + p_x + p_y - p_z)/2. \end{aligned}$$

Both HO coefficients a_{ij} and hybridization indices n have been worked out starting from symmetry considerations. If the HO geometry is not exactly linear, trigonal, or tetrahedral the n values are no longer integer but become fractional numbers and not necessarily equal. Equations for the calculus of the a_{ij} and n values have been reported[112,113] for some particular geometries, that is the quasi-tetrahedral systems AX_2Y_2 of symmetry C_{2v} and AX_2YZ of symmetry C_s and the planar quasi-trigonal AX_2Y system of symmetry C_{2v}.

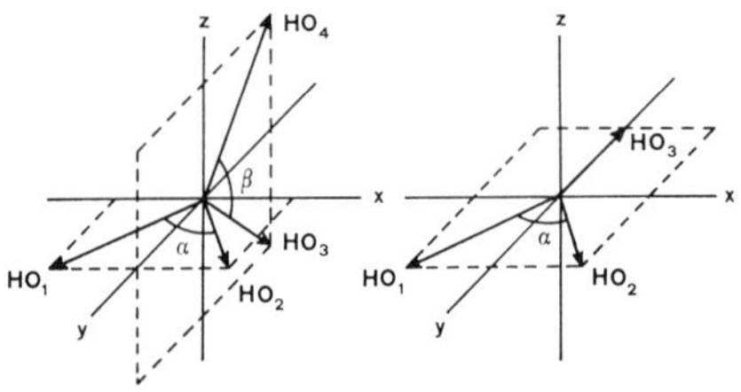

Fig. 7.18. Definition of the coordinate system and labelling of the relevant angles between hybrid orbitals in the AX_2Y_2 (left) and AX_2Y (right) systems discussed in the text.

For the AX_2Y_2 system of C_{2v} symmetry (left side of Fig. 7.18) it has been shown that $n_1 = n_2 = -\sec\alpha$, $n_3 = n_4 = -\sec\beta$ and that the angles α and β are related by the equation $\cot^2(\alpha/2) + \cot^2(\beta/2) = 1$. By putting $\alpha = 109.47°$ we obtain $n_1 = n_2 = n_3 = n_4 = 3$, $\beta = 109.47°$, $S = 0.25$, and $P = 0.75$ for any HO, in agreement with the theory. In the water molecule ($\alpha = 104.9°$) the two HOs pointing to the hydrogens have $n = 3.89$, $S = 0.20$, and $P = 0.80$ and the angle between the lone pairs is calculated to be $\beta = 114.8°$, in agreement with Bent's rule because the p character is found to concentrate on the HOs pointing to the more electronegative substituents, that is the hydrogens. The opposite case occurs in $(CH_3)_2O$, where the methyl is an electron donating group; the observed angle α is 111°, allowing us to calculate that $n = 2.79$, $S = 0.26$, and $P = 0.74$ for the HOs directed towards the CH_3 groups and $\beta = 108°$.

In the planar system AX_2Y (Fig. 7.18, right side) the important relationships become $n_1 = n_2 = -\sec\alpha$ and $n_3 = \tan^2(\alpha/2) - 1$. Only if $\alpha = 120°$ it will be that $n_1 = n_2 = n_3 = 2$, while $n_1 = n_2 > n_3$ for $\alpha < 120°$ and $n_1 = n_2 < n_3$ for $\alpha > 120°$. Since α is smaller or greater than 120° when HO_3 is directed towards substituents less or more electronegative, Bent's rule is obeyed.

These few examples can help to illustrate some interesting and not always well understood aspects of VB theory that can be summarized as follows:

(1) The hybridization indices are, in general, fractional numbers which become integers only for a few very special geometries;

(2) the hybridization indices and the p and s characters of the HOs are straightforwardly calculable from the bond angles and vice versa;

(3) Bent's rule is not a simple qualitative rule but can be quantified, at least in a number of simple cases.

Molecular mechanics

Molecular mechanics is a method for calculating the equilibrium geometries and other properties of ground state molecules on the basis of a purely classical mechanical model. The molecule is considered to be a set of atoms connected by elastic springs, and to any internal molecular coordinate (bond distances and angles, torsion angles, distances between non-bonded atoms, etc.) is assigned a natural equilibrium value. To any displacement Δx from the natural value is associated a restoring force proportional to the displacement itself, $F = -k\,\Delta x$, and an energy proportional to its square, $E = k(\Delta x)^2/2$. Molecular mechanics is essentially an attempt to assign parametric values to all **natural equilibrium values** and **force constants** k, whose complete set is called a **force field** and whose knowledge will allow us to calculate the potential energy hypersurface for the movement of the atoms within the molecule. The energy calculated assumes the meaning of the **deformation or steric energy** of a particular geometry of the molecule with respect to its equilibrium geometry.

The values of the parameters defining the force field are obtained empirically by comparison of observed and calculated molecular geometries. Other properties can be optimized as well, most frequently the formation enthalpies and the vibrational frequencies. In particular a force field able to calculate both geometries and frequencies is called a **consistent field**. The central point remains that of being able to calculate the properties of the maximum number of different molecules using the minimum number of constants transferable from one molecule to another.

The present discussion is necessarily concise and more details can be obtained from reviews[114–122] and books[123,124] available on the subject. Several molecular mechanics computer programs are distributed by QCPE (Quantum Chemistry Program Exchange).[125]

The total energy, E, is considered to be the sum of different contributions

$$E = E_s + E_b + E_{sb} + E_{oop} + E_t + E_{nb} + E_e \tag{7.37}$$

which are, in the order, the stretching, bending, stretching–bending, out-of-plane bending, torsion, non-bonded, and electrostatic energies and have the following meaning:

1. **Bond stretching.** The total energy of stretching or compression of all chemical bonds of the molecule is of the form

$$E_s = (1/2) \sum_{i=1}^{N} k_{s,i}(\Delta l_i)^2 \tag{7.38}$$

where the sum is extended to all the N bonds within the molecule, $k_{s,i}$ are their stretching force constants and $\Delta l_i = l_i - l_{0,i}$ the differences between the actual and equilibrium bond lengths. This model tends to overestimate the energy needed to produce very large lengthenings because, in such a case, the bond becomes increasingly yielding in consequence of the decreased overlapping of orbitals. It is common practice to add a small cubic term of the type $k_{sc,i}(\Delta l_i)^3$, where $k_{sc,i}$ is a small negative constant taking into account this last effect.

2. **Bond angle bending.** Its total energy is expressed as

$$E_b = (1/2) \sum_{i=1}^{M} k_{b,i}(\Delta \theta_i)^2 \tag{7.39}$$

where the sum is extended to all the M bond angles, $\Delta\theta_i = \theta_i - \theta_{0,i}$ are the differences between the actual and equilibrium bond angles, and $k_{b,i}$ the bending force constants. Bending energies are smaller than stretching energies by an order of magnitude because angles are much more yielding than distances. Also in this case a cubic term can be added, which is of the type $k_{bc,i}(\Delta\theta_i)^3$ and where $k_{bc,i}$ is a small negative constant.

3. **Stretching–bending terms.** An improved force field is obtained if proper allowance is made for the fact that the narrowing of a bond angle is paralleled by the legthening of the two encompassing bonds and *vice versa*. If $\theta_{i,ABC}$ is the angle delimited by the bonds $l_{i,AB}$ and $l_{i,BC}$ connecting the three atoms A, B, and C, the stretching–bending energy is defined as

$$E_{sb} = \sum_{i=1}^{M} k_{sb,i}(\Delta\theta_{i,ABC})(\Delta l_{i,AB} + \Delta l_{i,BC}) \tag{7.40}$$

where $k_{sb,i}$ are the corresponding force constants and Δl_i and $\Delta\theta_i$ have the usual meaning. The term E_{sb} is small in comparison with the previous terms and is neglected in some force fields. A field containing it is called a **valence force field**; this differs from the **Urey–Bradley force field** which presents a different treatment of the A . . . C geminal interactions and is not discussed here.

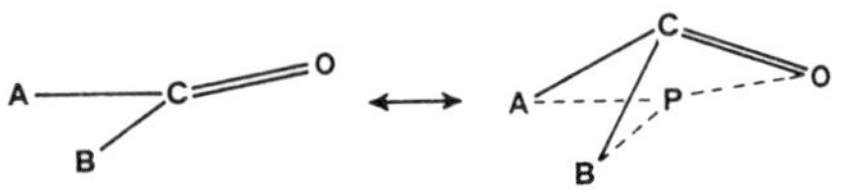

Fig. 7.19. In-plane and out-of-plane deformations of a trigonal sp^2 atom.

4. **Out-of-plane bending.** In the deformations concerning trigonal sp^2 atoms it is useful to distinguish between in-plane and out-of-plane deformations (Fig. 7.19). The in-plane deformations concern the angles A–P–B, A–P–O, and B–P–O and are dealt with by eqn (7.39), while the out-of-plane deformations are the displacements from zero ($\Delta\theta$) of the angles C–A–P, C–B–P, and C–O–P. The associated energy is

$$E_{oop} = (1/2)\sum_{i=1}^{3} k_{oop}(\Delta\theta_i)^2 \tag{7.41}$$

for any single sp^2 atom of the molecule.

5. **Torsions.** All force fields include energy terms accounting for the rotations around bonds. The corresponding force constant is relatively large for rotations around double bonds but very small for rotations around single bonds. A common expression for the torsional energy is

$$E_t = \sum [V_1(1+\cos\phi)/2 + V_2(1-\cos 2\phi)/2 + V_3(1+\cos 3\phi)/2] \tag{7.42}$$

where ϕ is the torsion angle ($-180° \leq \phi \leq 180°$), V_1, V_2, and V_3 are the force constants, and the sum is extended to all sequences 1–2–3–4 of bonded atoms in the molecule. The second term has the value zero for $\phi = 0$ and $\pm 180°$ and maxima for $\phi = \pm 90$; it is used to describe the rotation around double bonds and in this case $V_1 = V_3 = 0$. The third term is equal to zero for $\phi = \pm 60°$, $\pm 180°$ and has maxima for $\phi = 0$, $\pm 120°$; it describes the rotation around single bonds connecting sp^3 atoms. The first term is zero for $\phi = 180°$, has a maximum for $\phi = 0°$, is used as a small corrective term in special cases, and is often neglected.

6. **Non-bonded interactions.** The corresponding energy, E_{nb}, can be expressed by the atom–atom potentials discussed on p. 474. The interactions among all the atoms of the molecule are taken into account, with the exclusion of those between first neighbours (1–2 or bonded atoms) and second neighbours (1–3 or vicinal atoms). Third-neighbour (or 1–4) interactions are considered in both the E_{nb} and E_t terms. Special treatment

has sometimes been reserved for the lone pairs on etheric oxygens or aminic nitrogens by considering them as pseudo-atoms localized at about 1 Å from the atom and having their own atom–atom potentials.

7. **Electrostatic interactions.** It has already been shown on p. 477 that electrostatic interactions can be evaluated as interactions among partial charges localized on the atoms or among small dipoles associated with each chemical bond. Both methods are used in molecular mechanics.

A relevant number of different force fields proposed by different authors[125] are currently available and Table 7.8 reports, as an example, the

Table 7.8. Force field parameters for alkanes and non-conjugated alkenes according to MM1/MMP1.[13,124,125] Force constants in kcal mol^{-1} Å^{-2} or in kcal mol^{-1} deg^{-2}; bond moments (b.m.) in D. In the symbol $(\theta_0)_n$, n indicates the number of additional hydrogen atoms bonded to the central atom. 1 = C(sp^3), 2 = C(sp^2), 3 = H. Final energies in kcal mol^{-1}

'Stretching' Atoms	$(k_{sc} = -k_s)$ $k_s/2$	l_0	b.m.
1–1	316.5	1.514	0.0
1–2	316.5	1.496	0.300
1–3	330.9	1.095	0.0
2–2	690.6	1.334	0.0
2–3	330.9	1.090	0.0

'Bending' Atoms	$(k_{bc} = -0.006\ k_b/2)$ $k_b/2$	$(\theta_0)_0$	$(\theta_0)_1$	$(\theta_0)_2$
1–1–1	0.0083	109.47	110.51	110.20
1–1–3	0.0053	108.50	108.51	107.90
3–1–3	0.0042	112.80	111.10	109.47
1–1–2	0.0083	109.47	110.51	110.20
2–1–2	0.0083	109.47	110.51	110.20
2–1–3	0.0053	108.50	108.51	107.90
1–2–1	0.0083	116.60		
1–2–2	0.0083	121.70	120.50	
2–2–2	0.0131	120.00		
1–2–3	0.0053	119.50		
2–2–3	0.0053	120.00	118.80	
3–2–3	0.0042	122.40		

'Stretching–bending' Atoms	k_{sb}
A–B–C	0.301
A–B–3	0.100

'Out-of-plane bending' Atoms	$k_{oop}/2$
$2{=}2\langle^{A}_{B}$	0.0011

'Torsions' Atoms	V_1	V_2	V_3
A–1–1–A	0	0	0.53
A–1–1–B	0	0	0.53
A–1–2–A	0	0	2.34
A–1–2–B	0	0	2.34
A–1–2–2	0	0	0.30
1–2–2–1	−1.14	16.25	0
1–2–2–3	0	16.25	0
3–2–2–3	0	16.25	0

'Non-bonded interactions' Atom	r_i	ε_i
1	1.75	0.041
2	1.85	0.030
3	1.50	0.063

parameters used by the program MM1/MMP1[13] for hydrocarbons. The parameters of any force field are optimized to reproduce the experimental geometries of some reference molecules and used for predicting those of other chemically similar molecules. A force field developed for the alkanes will reproduce their observed geometries with a precision comparable with the experimental e.s.d. For other classes of molecules, such as alkenes, alcohols, ethers, thioles, esters, carboxylic acids, etc., the agreement between observed and calculated quantities is definitely worse in view of complications arising from the treatment of π-delocalized systems, partial double bonds, lone pairs, intramolecular hydrogen bonding, or highly polar groups. The discrepancies increase again when attempting to develop a force field able to deal with several functional groups at the same time and its seems unlikely that a universal field of sufficient accuracy will ever be obtained.

Let us now consider a few practical aspects of molecular mechanics calculations. Conformations undergo by far the greatest changes in a molecule, bond distances and angles being inevitably constrained to the proximity of their equilibrium values. It is therefore useful to conceive of the total potential energy as a hypersurface in a space where it is the ordinate, while the torsion angles are the abscissae. On this hypersurface there are as many minima as there are possible conformers (i.e. conformations of relative minimum energy; see pp. 484 and 490–2) of the molecule. Such minima are separated by regions of higher energy, where it will be possible to identify passages of relative minimum mapping the pathways of interconversion from one conformer to another. A typical molecular mechanics calculation starts from a given geometry, either obtained from the crystal structure or built up from standard values of distances and angles, and optimizes all the parameters in such a way as to reach, in a minimum number of program cycles, the molecular geometry corresponding to the nearest energy minimum. This is not necessarily the absolute minimum; that can be identified with certainty only by repeating the calculations starting from as many initial geometries as are necessary to systematically map the energy hypersurface. When the main minima have been found, the surrounding region can be scanned to evaluate the shape and the height of the energy barriers which separate one minimum from another. The energy profiles for 1,2-dichloroethane of Fig. 7.8 and for cyclohexane of Fig. 7.10 are typical results of molecular mechanics calculations.

A great number of organic molecules have been studied by the methods of molecular mechanics, whose success and popularity are due to the fact that it is a theoretical method which is very fast and extremely easy to understand. It does not require experimentation because the initial model can be obtained from tabulations of characteristic distances and angles. The computing time is very short with respect to the time needed by quantum-mechanical calculations. A benchmark carried out on propane[65] has shown that the full geometry optimization requires times in the ratios 1:12:13:662:5665 when accomplished, in order, by MM2, MINDO/3, MNDO, and GAUSSIAN82 with 3-21G and 6.31G* basis sets. Since molecular mechanics is so fast that it can optimize the geometry of molecules having the complexity of cholesterol in a few minutes of computer time, it has the additional advantage that large parts of a potential hypersurface can be calculated at a reasonable cost, giving information not only on the geometry of the different conformers but also a general view of the molecular motions and interconversion pathways.

This is not to say that molecular mechanics can give the best solution to all problems of conformational analysis. It presents several pitfalls, some of which have already been discussed, and all of which are essentially related to the lack of transferability of the force field parameters among different chemical classes. However, it cannot be denied that its use gives to the crystallographer the possibility of understanding aspects of the molecular geometry (particularly its dynamics) which are not accessible through crystal structure determination.

Molecular hermeneutics: the interpretation of molecular structures

Correlative methods in structural analysis

The mass of structural data obtained by X-ray crystallography of molecular crystals has been the object of many attempts at systematization and rationalization. Altogether they define a wide borderline area of interest which does not really belong to any traditional chemical discipline and is developing by itself along often not well defined tracks and with purposes which can appear somewhat confused, though all are essentially aimed at understanding the relationships between the experimentally determined molecular structure and the properties the molecule will display in gases, liquids, solutions, and in chemical and biochemical reaction environments. Though we may suppose it is a branch of structural chemistry, such an area, for the moment, does not even have a precise name, so that we are left free to use the name of our choice. *Structure–property relationships* has often been used though *interpretation of molecular structures* is certainly much more correct. To pay homage to the scholarly tradition of translating scientific sentences into ancient Greek, I would suggest the name **molecular hermeneutics** which has the meaning of the science, or the art, of molecular interpretation.

Irrespective of the name we choose, all attempts at systematic interpretation of molecular structures have one point in common: they make use of **correlative methods**, being based on the comparison of extended sets of structural parameters (distances, angles, geometries of specific molecular fragments, inter- or intramolecular contact distances, etc.) which are correlated among themselves or with other physical properties. The simplest of these attempts are intended to produce tables of standard bond distances and angles, but most of them are devoted to clarify the systematic aspects of a variety of chemical situations such as the characteristics of the inter- and intramolecular hydrogen bonding in crystals and molecules, the typical geometries of organic functional groups, the *trans* influence in coordination compounds, the role of π back-bonding in organometallic compounds, the structural requirements which organic metals or semiconductors must fulfill, or, as extensively discussed in connection with the VSEPR and VB theories, the effects that electronegativity and hybridization changes produce on the geometries of simple molecules.

As it is impossible to give even a summary of these systematic studies, it has been chosen to discuss in some detail a single method of systematic investigation of molecular geometries, the so-called **structure correlation method**[4,126–129], which can be considered to be an attempt to obtain information on the dynamic behaviour of molecules from the inevitably rather static crystal data. The basic idea of the method is as follows. Usually, internuclear distances found in molecular crystals fall into the two distinct fields of bonding and non-bonding distances, which are separated by a forbidden gap going from the longest bond to the shortest contact distance. For instance, this gap is extended from 1.48 (single-bond distance) to 3.25 Å (sum of van der Waals radii) in the case of the C–N distances. In real crystals, however, several interatomic C–N distances have been found to fall in this interval, a fact that seems interesting because any situation intermediate between bonding and non-bonding is typical of chemical reactions.

As early as 1931 Eyring and Polanyi[130] suggested that any chemical reaction could be described as a low-energy path on a **potential energy hypersurface** (or surface) of the reaction itself, which is the function giving (within the Born–Oppenheimer approximation) the total potential energy of the system as a function of the relative positions of all the atomic nuclei. The **reaction pathway** is that trajectory (**reaction coordinate**) which connects the points of relative minimum on the surface and leads from the reagents to the products through the higher energy **transition** or **activated state**. For the scope of this account it is sufficient to make reference to the classical examples reported in all books on physical chemistry, such as the reactions $H + H_2 \rightleftharpoons H_2 + H$ or $H_2 + D \rightleftharpoons H + HD$, or to the elementary accounts on the subject which are available.[131–133]

If it were possible to take snapshots of the molecules taking part in the reaction at different points of the reaction pathway, the relative positions of the atomic nuclei along the course of the reaction would become known. The basic idea of the structure correlation method consists in assuming that the instantaneous displacements of the nuclei during a reaction can be represented by the images of the reaction centre obtainable from a great number of crystal structures where the centre itself happens to be deformed by local inter- and intramolecular perturbations; in this hypothesis, the different images need only to be ordered in a rational sequence to give the progressive deformations of the reaction centre along the reaction coordinate.

This picture of a chemical reaction is directly and even more naturally applicable to a particular class of chemical transformations, the **conformational interconversions,** for which it has already been shown that the relative conformers' populations are determined by the depth of the minima on the potential energy hypersurface (the enthalpic term) and by their shape (the entropic term), while activation barriers for interconversion of conformers are straightforwardly interpretable as heights of the saddle points on the lowest-energy routes (conformational interconversion pathways) connecting different conformers on the hypersurface itself.

Some three-centre–four-electron linear systems[126,128,134]

It is well known that the solubility of iodine in water is increased by the presence of iodide ions, a fact explained by the formation of polyiodide I_3^-

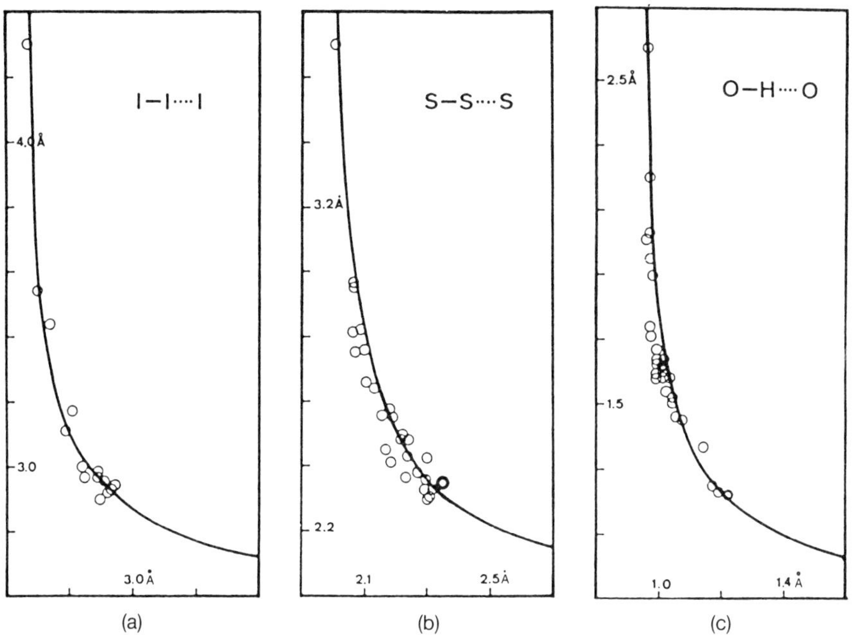

Fig. 7.20. Correlation diagrams of the interatomic distances (in Å) for three linear three-centre–four-electron systems: (a) tri-iodide anions, (b) thiathiophthenes, and (c) linear O–H ··· O hydrogen bonds. (Reproduced by permission from Bürgi.[126])

ions by a charge transfer reaction where the Lewis base I^- donates to the empty π^* orbital of I_2. Several crystal structures containing the I_3^- ion are known: the anion is nearly linear with an I_1–$I_2 \cdots I_3$ angle of 175–180°. The bond distance in the I_2 molecule is 2.67 Å and the sum of the van der Waals radii is 4.30 Å. Actual structures show many interleaving distances, the shortening of the $I_2 \cdots I_3$ distance being always associated with a lengthening of the I_1–I_2 one, and *vice versa*. In general the anion is strongly asymmetric in presence of small cations and tends to become symmetrical with larger cations. A plot of the I_1–I_2 versus $I_2 \cdots I_3$ distances is shown in Fig. 7.20(a). The experimental points are clearly arranged on an equilateral hyperboloid whose analytic form can be derived from Pauling's formula (7.31)

$$\Delta d = d(n) - d(1) = -c \log_{10} n$$

where n is the bond number, $d(n)$ and $d(1)$ the bond distances for the bond numbers n and 1, respectively, and c a constant easily determined as $c = \Delta d'/\log 2$, being $\Delta d'$ the bond length increment for the symmetrical anion having $n = 1/2$ (currently $c = 0.85$ for a $\Delta d'$ of 0.26 Å). Assuming that the single bond is shared among the three atoms, it can be written

$$n_1 + n_2 = 10^{-\Delta d_1/c} + 10^{-\Delta d_2/c} = 1 \tag{7.43}$$

and the resulting function for $c = 0.85$ is drawn as a continuous curve in Fig. 7.20(a); only one branch of the hyperboloid is shown since the second one, obtainable by interchanging I_1 with I_3, is identical because of ion symmetry. The points on the curve may be supposed to represent the geometrical changes occurring along the reaction coordinate of the process

$$I_1\text{–}I_2 + I_3^- \rightleftarrows [I_1 \cdots I_2 \cdots I_3]^- \rightleftarrows I_1^- + I_2\text{–}I_3.$$

The shape of the upper part of the curve indicates that the approach of the I_3^- ion to the iodine molecule is much more rapid than the lengthening of

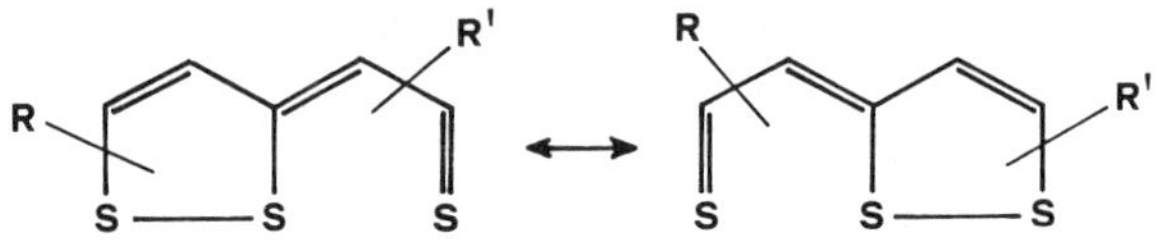

7.III

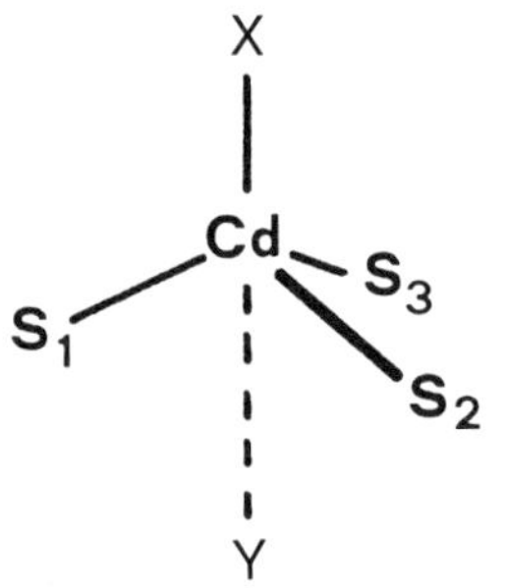

Fig. 7.21. Coordinate system used for cadmium complexes. Δx and Δy are the differences between the actual Cd–X and Cd–Y distances and the corresponding single bond standard values. Δz is the vertical displacement of the Cd atom from the $S_1S_2S_3$ plane.

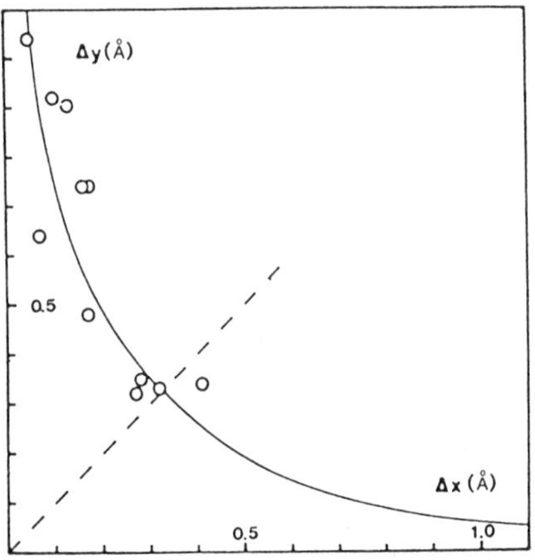

Fig. 7.22. Correlation diagram for the bond length increments Δx and Δy in cadmium complexes. (Reproduced by permission from Bürgi.[135])

the I_1–I_2 distance up to the point of formation of the activated complex, whose dissociation will quickly occur along the lower branch of the curve. From a slightly different point of view, all the curve can be considered to map the deformations along the antisymmetric stretching vibration of the activated complex up to its complete dissociation.

Similar correlations have been reported for the groups S–S $\cdots$ S (Fig. 7.20(b)) and O–H $\cdots$ O (Fig. 7.20(c)), respectively found in thiathiophthenes (7.III) and in different structures containing nearly linear hydrogen bonds. The **rule of bond number conservation** expressed by (7.43) is found to hold also in these two cases with parameters $d(1) = 2.02$, $\Delta d' = 0.31$, and $c = 1.03$ Å for (7.III) and 0.96, 0.26, and 0.85 Å for the hydrogen bond. The curves calculated in this assumption are indicated by the continuous lines and are seen to approximate well the experimental data.

Nucleophilic addition to organometallic compounds

Cadmium complexes have been studied by Bürgi in 1973 in the earliest application of the structure correlation method.[135] These compounds are normally tetrahedral but can assume quasi-trigonal bipyramidal five coordination by adding a further ligand at a distance greater than that of the other four ligands (Fig. 7.21). Eleven crystal structures were found having three sulphur atoms as equatorial ligands and a fifth ligand Y at different approaching distances on the normal to the triangular face of the tetrahedron and lying on the same line containing the Cd atom and the *trans* ligand X. X and Y happen to be iodine, sulphur, or oxygen atoms. The system of coordinates used is described in the caption of Fig. 7.21.

By plotting the experimental values of Δx versus Δy (Fig. 7.22) it is found that the points are arranged on a curve which can be interpreted in terms of bond number conservation (eqn (7.43)) with $\Delta d' = 0.32$ and $c = 1.05$ Å. Another correlation is obtained by plotting Δz as a function of Δy (upper part of Fig. 7.23) and Δx (lower part). Each structure is represented by two points having coordinates $(\Delta x, \Delta z)$ and $(\Delta y, \Delta z)$; $\Delta x = 0$ corresponds to the fragment CdS_3X of tetrahedral geometry, which is represented at the same time by the point in the left lower corner and by another point at $\Delta y = \infty$ outside the upper right corner. The leaving of X caused by the approaching of Y moves the two representative points towards the centre, where they collide at the point $\Delta x = \Delta y = 0.32$ and $\Delta z = 0$ corresponding to the trigonal bipyramidal structure.

Figures 7.22 and 7.23 may be interpreted as the geometrical changes occurring along the reaction of nucleophilic substitution S_N2

$$Y + CdS_3X \rightleftarrows [Y \cdots CdS_3 \cdots X] \rightleftarrows YCdS_3 + X.$$

The reaction starts with CdS_3X tetrahedral and Y at infinite distance. The approach of Y causes an increasing lengthening of the *trans* Cd–X bond and a flattening of the CdS_3 pyramid until trigonal bipyramidal transition state formation. Afterwards X starts leaving and finally the new tetrahedral species $YCdS_3$ is formed.

The continuous curve of Fig. 7.23 is calculated by the equations

$$\Delta x = -c \log [(0.84 + \Delta z)/1.68], \qquad (7.44)$$
$$\Delta y = -c \log [(0.84 - \Delta z)/1.68],$$

making use of the already known value of $c = 1.05$ Å. The two equations are derived from the usual bond number conservation rule $n_X + n_Y = 1$, where n_X is the bond number of Cd–X and n_Y that of Cd–Y, but expressed as a function of the angle θ, the average X–Cd–S angle. The two relations $n_X = (1 - 3\cos\theta)/2$ and $n_Y = (1 + 3\cos\theta)/2$ satisfy the rule because for $\theta = 90°$ (trigonal bipyramidal geometry) $n_X = n_Y = 0.5$ and for $\theta = 109.47°$ (tetrahedral geometry) $n_X = 1$ and $n_Y = 0$. Pauling's formula (7.31) becomes

$$\Delta x = -c \log n_X = -c \log [(1 - 3\cos\theta)/2], \qquad (7.45)$$
$$\Delta y = -c \log n_Y = -c \log [(1 + 3\cos\theta)/2],$$

and, since the equatorial Cd–S bond has a nearly constant value of 2.52 Å, we can put $\Delta z = -2.52 \cos\theta$, which gives eqns (7.44) when substituted in (7.45).

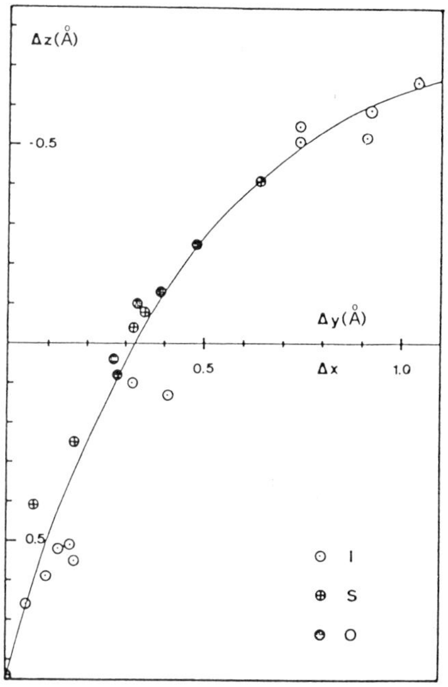

Fig. 7.23. Correlation diagram for the bond length increments Δx and Δy against the vertical displacement Δz. (Reproduced by permission from Bürgi.[135])

Another case studied is that of the organometallic compounds of Sn(IV).[136] Similar methods give a convincing mapping of what could be the reaction pathway for the S_N2 nucleophilic reaction with configuration inversion

$$Y + SnR_3X \rightleftarrows [Y \cdots SnR_3 \cdots X] \rightleftarrows YSnR_3 + X.$$

It is interesting to remark, however, that the same paper reports the mapping of the reaction pathway of the S_N3 process

$$SnR_2X_2 + 2Y \rightleftarrows [SnR_2X_2Y_2] \rightleftarrows SnR_2Y_2 + 2X,$$

a trimolecular reaction which cannot realistically occur according to the principles of chemical kinetics. This seems to suggest that structure correlations do not simply map possible reaction pathways but, more generally, the lower regions of the potential energy hypersurface for the relative movements of the nuclei irrespective of the fact that they correspond or not to real chemical reactions or conformational rearrangements.

Nucleophilic addition to the carbonyl group

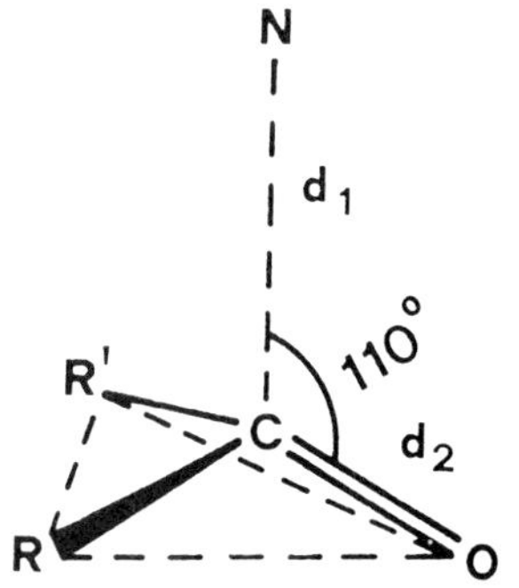

Fig. 7.24. Coordinate system used for describing the approaching of the aminic nitrogen to the carbonyl. d_1 and d_2 are the N···C and C=O distances, respectively: the distance of the carbon atom from the (RR'O) plane is a measure of carbonyl pyramidalization and is termed Δ.

Nucleophilic addition of aminic nitrogen to carbonyl is one of the first and more studied examples of structure correlations,[137,138] though the analogous addition of oxygen has been also reported.[139] Molecules where the nitrogen is forced near the carbonyl carbon at a distance shorter than the sum of the van der Waals radii are mainly rings of the type (7.IV) and (7.V) or 1,8-disubstituted naphthalenes (7.VI). In the original paper 14 structures having the required characteristic were reported and the coordinate system for their analysis is shown in Fig. 7.24. Final results are summarized in Fig. 7.25, which describes the course of the geometrical deformations of the carbonyl group caused by the approaching of the nucleophile. As this

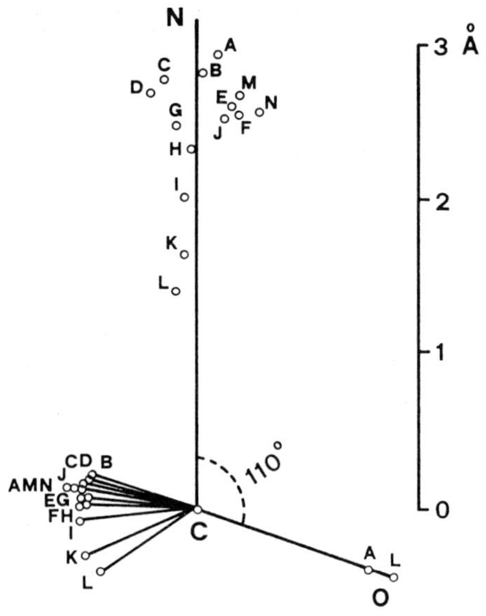

Fig. 7.25. The coordinate system is that of Fig. 7.24 projected on the (NCO) plane. The open circles indicate the positions of the nitrogen atoms (top), of the bisector of RCR′ (bottom left) and of the carbonyl oxygen (bottom right) in the 14 (A–L) structures considered. (Adapted from Bürgi and Dunitz.[128])

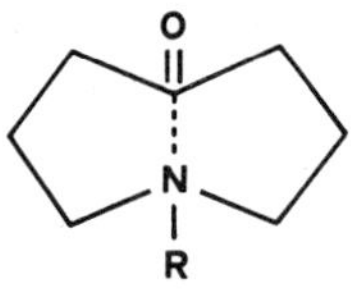
7.IV

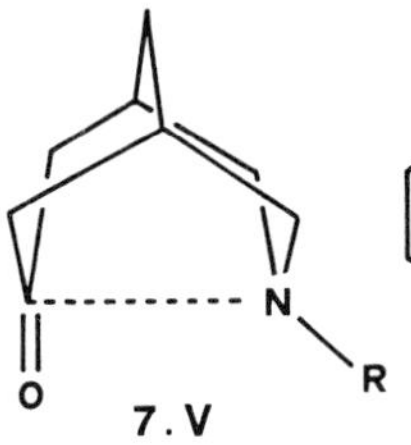
7.V

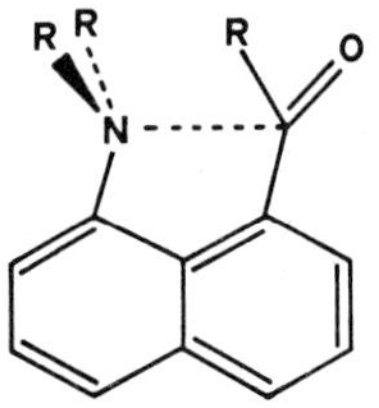
7.VI

approaches the carbon (i.e. d_1 decreases), the plane containing the carbon and the two R and R′ substituents bends away causing an increasing pyramidalization of the carbon which rehybridizes from sp^2 to sp^3, while the C=O distance is slightly increased. It seems of interest that the nucleophile does not approach the carbonyl plane perpendicularly but makes an almost constant N · · · C=O angle of 110°, which has been interpreted in terms of elementary *banana bond* considerations but also shown to match the results of *ab initio* quantum-mechanical calculations.[4] The constraint arising from the fixed 110° angle is more clearly seen in (7.VI). The two $-NR_2$ and –COR groups would both be expected to be splayed outwards in order to reduce their van der Waals repulsions but the C–N bond is actually found to be splayed inwards by the need to maintain the correct approach angle to the carbonyl C=O bond.

When looking at the dependence (Fig. 7.24) between Δ (carbonyl group pyramidality) and d_1 (N · · · C distance) it is found that they are correlated according to the regression line

$$d_1 = -1.701 \log \Delta + 0.867 \qquad (7.46)$$

a function which gives the maximum pyramidal character of the carbon atom, $\Delta_{max} = 0.437$ Å, for $d_1 = 1.479$ Å (taken as standard distance for the C–N single bond). By assuming that the C–N bond number can be expressed as $n = (\Delta/\Delta_{max})^2$ the above equation changes into

$$d_1 = -0.85 \log n + 1.479 \qquad (7.47)$$

which is nothing more than the usual Pauling equation (7.31).

A case of conformational rearrangement[140,141]

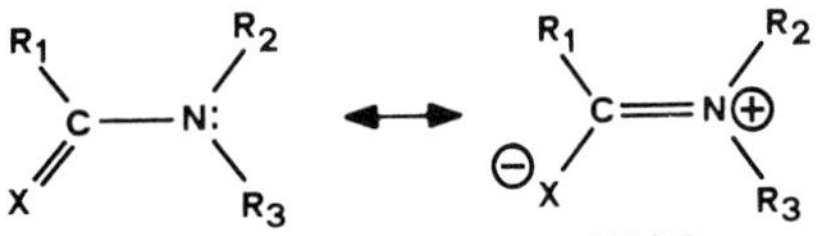
7.VII 7.VIIa

The fragment $C(sp^2)$–$N(sp^3)$ (7.VII) is contained in a variety of organic molecules, such as anilines (7.VIII A) and naphthylamines (7.VIII B), amides (7.VIII O) and thioamides (7.VIII S) (X = O, S), amidines (7.VIII N) (X = NR), and enamines (7.VIII D) ($X = CR_2$). It displays a definite tendency to be planar, which can be accounted for by the

7.VIII

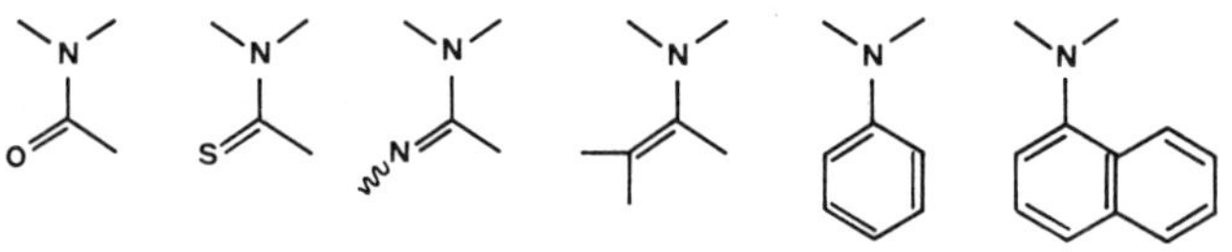

O S N D A B

contribution of the polar form (7.VIIa) to its ground state. The *cis–trans* isomerization barrier has been measured by a number of dynamic NMR experiments to be in the range 5–22 kcal mol^{-1} (on average, 20.7, 18.1, 12.8, 9.0, and 5.1 kcal mol^{-1} for thioamides, amides, amidines, enamines, and anilines respectively, in agreement with the fact that the contribution of the polar form decreases with the decrease of electronegativity of X). In spite of the general tendency to planarity, there are many crystal structures where the group is deformed by rotation around the C–N bond or by nitrogen pyramidalization, or by both, mainly in consequence of intramolecular steric hindrance. The structure correlation method was used with the aim of mapping the coordinate of *cis–trans* isomerization of the group.

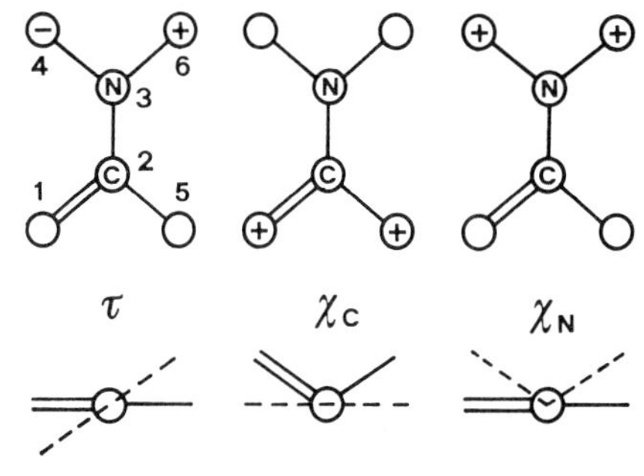

Fig. 7.26. Internal coordinate system describing the out-of-plane deformation of the $C(sp^2)–N(sp^3)$ fragment. (Reproduced by permission from Gilli *et al.*[141])

A coordinate system for the out-of-plane deformation can be obtained by the usual methods of group theory; the resulting orthogonal coordinates are the angle of twisting around the C–N bond, τ ($\tau = 0$ and 180° for the planar *cis* and *trans* geometries, $\tau = \pm 90°$ when the two halves of the group are perpendicular) and the two out-of-plane bendings of nitrogen, χ_N, and carbon, χ_C, ranging from zero (sp^2 atom) to 60° (sp^3 atom). Calling ϕ_1, ϕ_2, ϕ_3, and ϕ_4 (Fig. 7.26) the four torsion angles 1–2–3–4, 5–2–3–6, 5–2–3–4, and 1–2–3–6, it is found that[142,143]

$$\begin{aligned} \chi_C &= \phi_1 - \phi_3 + \pi = -\phi_2 + \phi_4 + \pi \pmod{2\pi}, \\ \chi_N &= \phi_2 - \phi_3 + \pi = -\phi_1 + \phi_4 + \pi \pmod{2\pi}, \qquad (7.48) \\ \tau &= (\phi_1 + \phi_2)/2. \end{aligned}$$

A total of 90 fragments belonging to 68 crystal structures were found from CSD.[3] The χ_C values are always very small (on average 1.7°), indicating that the carbon sp^2 has greater resistance to out-of-plane bending and that this motion can be neglected. Conversely χ_N shows a wide variation assuming values up to ±55°. Figure 7.27 shows the correlation diagrams of

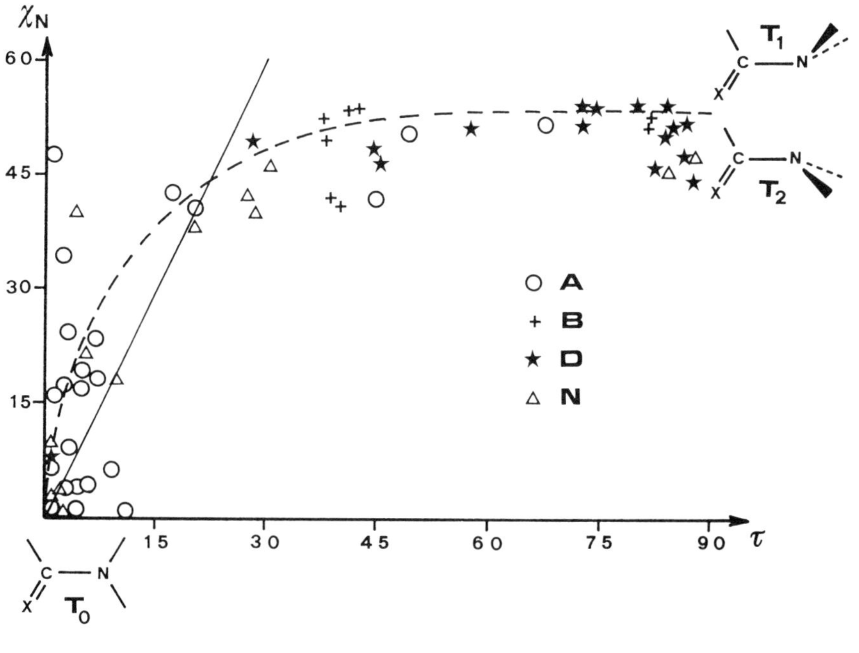

Fig. 7.27. Correlation diagram of χ_N versus τ for anilines (A), naphthylamines (B), enamines (D), and amidines (N). Data for amides and thioamides (not shown in the figure) would cluster in the lower left corner with $\tau \leq 15°$ and $\chi_N \leq 27°$. (Reproduced by permission from Gilli *et al.*[141])

χ_N against τ for all the chemical classes except amides and thioamides. These latter cluster in the lower left corner with $\tau \leqslant 15°$ and $\chi_N \leqslant 27°$, showing the greater resistance to out-of-plane deformation of these compounds in agreement with the larger electronegativities of the oxygen and sulphur atoms.

Compounds of the other classes undergo more severe distortions which appear to be of two different types. The first, producing simple nitrogen pyramidalization, occurs for τ nearly equal to zero and for increasing values of χ_N and can be called a **butterfly** deformation. The second, which could be called **combined**, associates nitrogen pyramidalization (increasing χ_N) with a rotation around the C–N bond (increasing τ). The diagonal straight line of Fig. 7.27 ideally separates mostly butterfly (on the left) from mostly combined (on the right) distortions.

The combined motion admits a simple interpretation. In the planar fragment the nitrogen is sp_2 hybridized and its p_z AO is implies in a π bond with the p_z AO on the carbon. The rotation around the C–N bond causes a decoupling of the π system while the nitrogen rehybridizes engaging its p_z AO into an sp^3 HO carrying the lone pair. The opposite mechanism is also possible: the planar nitrogen undergoes an out-of-plane vibration (butterfly motion) which causes the rehybridization of the atom from sp^2 to sp^3. The sp^3 HO on nitrogen is essentially decoupled from the p_z AO on carbon, the double bond fades, and the rotation around C–N becomes possible. This second mechanism seems to be preferred because Fig. 7.28 shows that χ_N (a measure of the nitrogen rehybridization) is strongly correlated with a lengthening of the C–N distance which almost encompasses the full variation range of d_{C-N} (from 1.27 to 1.44 Å going from double to single $C(sp^2)$–$N(sp^3)$ bond), while no similar correlation can be established with τ, the rotation angle around the C–N bond.[141]

From the point of view of the structure correlation method the butterfly motion corresponds to a simple out-of-plane vibration and does not produce

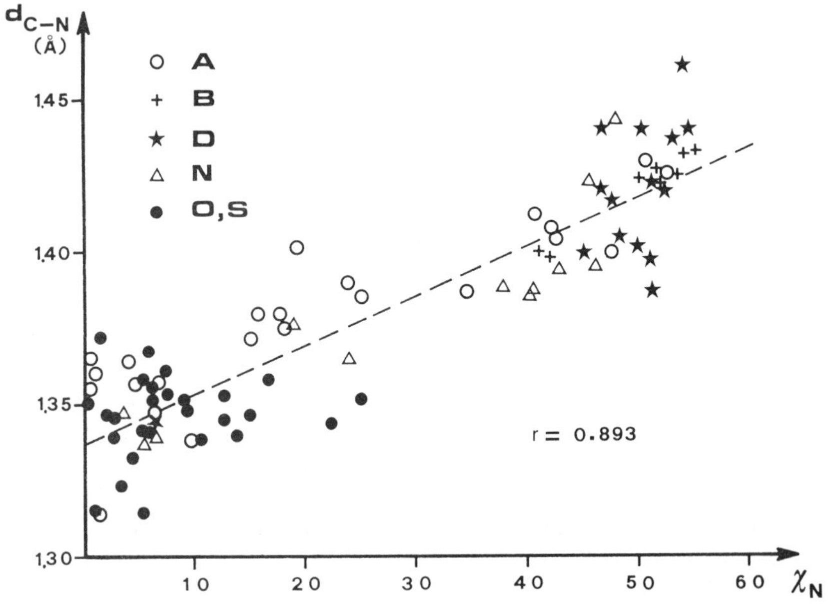

Fig. 7.28. Correlation diagram between the d_{C-N} bond distances and the χ_N values for all fragments studied (A = anilines, B = naphthylamines, D = enamines, N = amidines, O = amides, and S = thioamides). (Reproduced by permission from Gilli *et al.*[141])

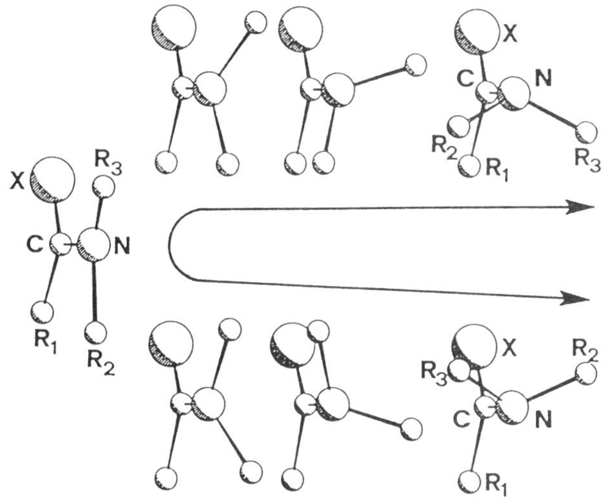

Fig. 7.29. The relative movements of the fragment atoms along the reaction pathway identified by the broken curve of Fig. 7.27, showing the steps leading from the planar ground state T_0 to the two enantiomeric transition states T_1 and T_2. Further rotation around the C–N bond would transform the *cis* isomer on the left into the *trans* isomer on the right (not shown). (Reproduced by permission from Gilli *et al.*[141])

any chemical change. Conversely, the combined motion can be considered to be representative of the geometrical changes experienced by the fragment along the pathway of a *cis–trans* isomerization reaction. The broken curve of Fig. 7.27 maps the most probable pathway leading from the planar ground state T_0, having $\tau = \chi_N = 0°$, to the transiton state located at $(\tau, \chi_N) = (90, \approx 50°)$; since the rotation can occur in two opposite directions, two enantiomeric T_1 and T_2 transition states will be equally possible and Fig. 7.29 shows the geometrical changes the fragment is undergoing along the two enantiomeric pathways indicated by the broken curve. It is seen that the atom movements start with an out-of-plane deformation of the nitrogen and are followed by the rotation around the C–N bond.

It seems of great interest to show whether the proposed reaction pathway does actually map a relative minimum path on the potential energy hypersurface. What is needed is an energy model for the out-of-plane motion of the $C(sp^2)–N(sp^3)$ fragment. A rather simple molecular mechanical model has been proposed[141] by modification of a first model given by Winkler and Dunitz,[142] that is

$$V(\tau', \chi_N) = (\mathrm{CTIB} + \mathrm{IB})(1 - \cos\tau')/2 + \mathrm{QP}(1 + \cos\tau')\chi_N^2/2 + \mathrm{IB}(1 - \cos\tau')(\cos 3\chi_N - 1)/4, \quad (7.49)$$

where the total potential energy expressed as a function of $\tau' = 2\tau$ and χ_N is assumed to depend on three predetermined parameters, CTIB = *cis–trans* isomerization barrier, IB = inversion barrier of the pyramidal sp^3 nitrogen, and QP = force constant for the out-of-plane bending of the sp^2 nitrogen. The energy is zero at the origin in correspondence to the planar geometry and increases with χ_N^2 for $\tau' = 0°$ along the out–of-plane vibration of the planar nitrogen; for $\chi_N = 0$ and $\tau' = 180°$ the function has the value CTIB + IB, differing from CTIB for the energy needed to make planar the sp^3 nitrogen, while it assumes the exact value of CTIB only for the hypothetical transition state geometry ($\tau' = 180°$ and $\chi_N = 60°$). The two expressions $(1 + \cos\tau')$ and $(1 - \cos\tau')$ are introduced in order to progressively cancel the second term of (7.49) replacing it by the third term, while

τ' changes from zero to 180° and, accordingly, the nitrogen hybridization from sp^2 to sp^3.

The values of the constants have been chosen as average values from those of the different classes of compounds. The CTIB value is in the range 5–22 kcal mol^{-1}, QP is evaluated from vibrational spectroscopy and molecular mechanics to be some 5–10 kcal mol^{-1} rad^{-2}, and IB can be assimilated to the inversion barrier of ammonia, usually reported as ≈6 kcal mol^{-1}. Values chosen were CTIB = 14, IB = 8 kcal mol^{-1}, and QP = 8 kcal mol^{-1} rad^{-2}. Equation (7.49), however, evaluates the position of the transition state at $\tau' = 180°$ and $\chi_N = 60°$, which does not correspond to that found from Fig. 7.27 as far as χ_N is concerned. The reason is that some allowance must also be made for the 1–4 non-bonded interactions, that is

$$V'(\tau', \chi_N) = V(\tau', \chi_N) + E_{nb}(\tau', \chi_N) \qquad (7.50)$$

where E_{nb} can be obtained by the methods of molecular mechanics. Here it has been calculated assuming $R_1 = R_2 = R_3$ = methyl and X equal to a nonexistent atom intermediate between oxygen and methyl group. The final V' function calculated in such a way is shown in Fig. 7.30 and represents the potential energy hypersurface for an imaginary molecule which is the average of all different compounds of all chemical classes investigated. In spite of the approximations made, it has the correct shape and the agreement with experiment becomes more convincing when the experimental points are plotted on it. They appear to be nicely located along the energy valley leading from the planar ground state to the transition state, which is now more correctly located at $\tau' = 180°$ and $\chi_N \approx 50°$.

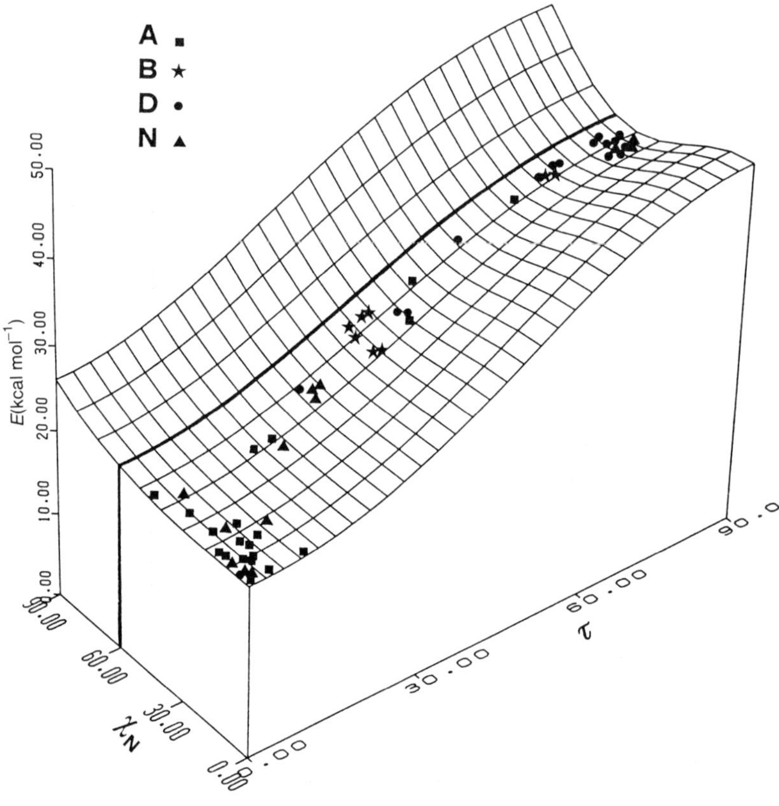

Fig. 7.30. Total potential energy (kcal mol^{-1}) surface as calculated from eqn (7.50) as a function of $\tau = \tau'/2$ and χ_N. Experimental points include all compounds except amides and thioamides. (Reproduced by permission from Gilli *et al.*[141])

Resonance assisted hydrogen bonding (RAHB)

β-diketones are known to undergo easy enolization and in two recent papers[24,25] structure correlation methods have been applied to the β-diketone fragment in its enol form (7.IXa,b) in order to understand what happens to the fragment geometry when perturbed by intramolecular, (7.X), or intermolecular, (7.XI), hydrogen bonding. The reason why this

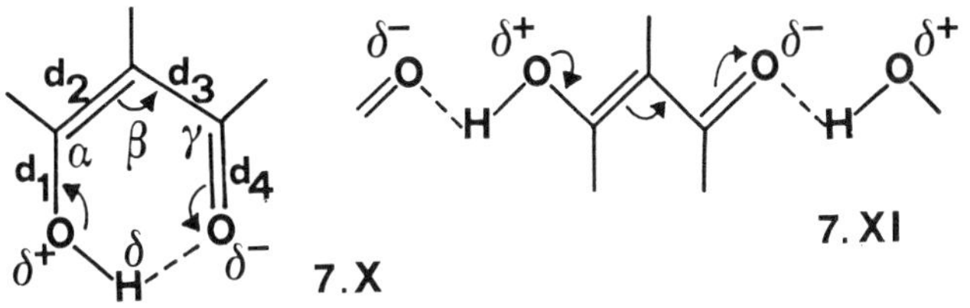

7.X

particular system was studied originates from the empirical observation that the π-conjugated system present in the β-diketone enol HOCR=CR–CR=O fragment undergoes greater delocalization when it forms either intra-molecular or infinite chain intermolecular hydrogen bonding. The nature and the entity of this effect is shown in Table 7.9. Here the *standard* distances are those tabulated[109] for pure single and double bonds, the *unperturbed* ones are an average of nine fragments that cannot make H-bonds (having OR instead of OH). The unperturbed geometry is a 87:13 mixture of the resonance forms (7.IXa) and (7.IXb) according to Pauling's formula (7.31). However, when the fragments form H-bonds all distances undergo changes consistent with an increased contribution of the polar form (7.IXb), which becomes respectively 29 per cent for the intermolecular and 48 per cent for the intramolecular cases reported in the table; at the same time the contract O--O distance, which is typically 2.74 Å in the O–H · · O bond in ice, becomes much shorter, being respectively 2.575 and 2.485 Å in

Table 7.9. Selected bond distances (Å) for the β-diketone enol fragment. %(7.IXb) = per cent contribution of the polar form (7.IXb) according to Pauling's formula (7.31); the standard O--O contact distance in parentheses is that observed in ice

	d_1(C–O)	d_2(C=C)	d_3(C–C)	d_4(C=O)	d_{O-O}	%(7.IXb)
Standard[109]	1.37	1.33	1.48	1.20	(2.74)	0
Unperturbed[24]	1.353	1.344	1.454	1.225		13
Perturbed by:						
Intermolecular H-bond (7.XI)[a]	1.316	1.372	1.431	1.238	2.575	29
Intramolecular H-bond (7.X)[b]	1.281	1.398	1.410	1.279	2.485	48

[a] Neutron data from Jones, R. D. G. (1976). *Acta Crystallographica, B*, **32**, 2133.
[b] Semmingsen, D. (1977). *Acta Chem. Scand. Sect. B*, **31**, 114.

the two cases, while the O–H bond distance (0.95 Å in the absence of H-bonding and according to Table 7.7) is lengthened up to 1.24 Å in the last compound of Table 7.9.

In a first paper[24] 25 X-ray or neutron structures of β-diketones and β-ketoesters were studied, 22 of type (7.X) and 3 of type (7.XI), and the discussion reported here concerns this first analysis. Later on[25,149] the investigation was extended to a much larger set of fragments (81 and 37 for (7X) and (7.XI), respectively) but without obtaining significant differences in the final results. In summary, all the following phenomena are observed to occur *together*:

(1) increased delocalization of the π-conjugated system;

(2) strengthening of the O–H $\cdots$ O bond as indicated by the shortening of d_{O--O} and the lengthening of $d_{O\text{–}H}$ distances;

(3) shift of the proton position towards the middle point of the O $\cdots$ O contact.

The three effects may occur with different intensities but maintain the same intercorrelation for both intramolecularly (7.X) or intermolecularly (7.XI) bonded fragments (the intramolecular H-bond is usually stronger and causes greater π delocalization) and even for different classes of compounds (e.g. β-diketone enols form stronger H-bonds and are more delocalized than the enols of β-ketoesters or β-ketoamides).

While the H-bond strengthening is easily measured by both d_{O--O} and $d_{O\text{–}H}$ changes, a specific geometric parameter is needed in order to describe the π system delocalization. The simplest way consists in using **symmetry coordinates** for the in-plane antisymmetric vibration of the group (7.X) or (7.XI), that is $q_1 = d_4 - d_1$ and $q_2 = d_2 - d_3$. Clearly $q_1 = q_2 = 0$ for the totally delocalized structure, while the greatest values will occur for the standard bond distances of Table 7.9. Figure 7.31 reports the scatter plot of

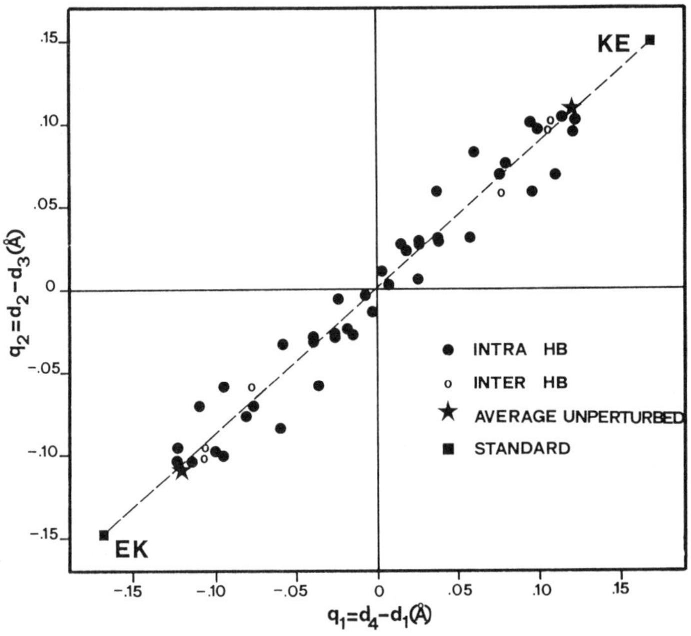

Fig. 7.31. Scatter plot of q_2 versus q_1 (in Å) for the 25 fragments studied. The plot is centrosymmetric as any point can be plotted twice in view of the equivalence of the enol–ketonic (EK) and keto–enolic (KE) isomers. The full squares and stars refer to the **standard** and **unperturbed** geometries of Table 7.9, respectively. (Reproduced by permission from Gilli *et al.*[24])

q_2 versus q_1 for the 25 compounds studied. The plot is centrosymmetric as any point can be plotted twice in view of the equivalence of the enol-ketonic (EK) and keto-enolic (KE) isomers, and its centre identifies the fully delocalized structure. The full squares and stars refer to the standard and unperturbed geometries of Table 7.9, respectively. The broken line connecting EK and KE is calculated from the condition of bond number conservation (p. 513). Since q_1 and q_2 are linearly dependent, a single coordinate $Q = q_1 + q_2$ can be used, which has values of 0.0, −0.320, and +0.320 Å for the fully π-delocalized and the fully π-localized EK and KE structures, respectively. Alternatively the fragment geometry can be described by a **coupling parameter**, λ, according to which the state of any fragment is a mixture $\lambda(\text{EK}) + (1 - \lambda)(\text{KE})$; it assumes the values of 1.0 for EK, 0.0 for KE, and 0.5 for the fully delocalized form.

The scatterplot of d_{O--O} against $Q = q_1 + q_2$ for the 25 fragments considered is reported in Fig. 7.32. The plot displays m symmetry with respect to the delocalized structure having $Q = 0$ or $\lambda = 0.5$ owing to the equivalence of the EK and KE structures. Very short O · · · O distances are seen to be associated with small values of $|Q|$, that is with relevant delocalization of the π-conjugated system. The plot of the variations of the $d_{O—H}$ and of the contact d_{OH--O} distances against d_{O--O} is shown in Fig. 7.33. Although data are necessarily more disperse owing to the known difficulties in locating protons by X-ray diffraction (only three were neutron structures), the plot clearly indicates that the shortening of the O · · · O distance below 2.5 Å causes the two bond (O–H) and contact (OH · · · O) distances to become progressively equal, though a perfectly centred positioning of the proton is never achieved.

The observed correlations are in agreement with the model which, in its qualitative aspects, is illustrated in Fig. 7.34. Let us assume firstly that the double C=C is substituted by a single C–C bond; the O–H · · · O bond established will be the balance between two different factors: the energy of

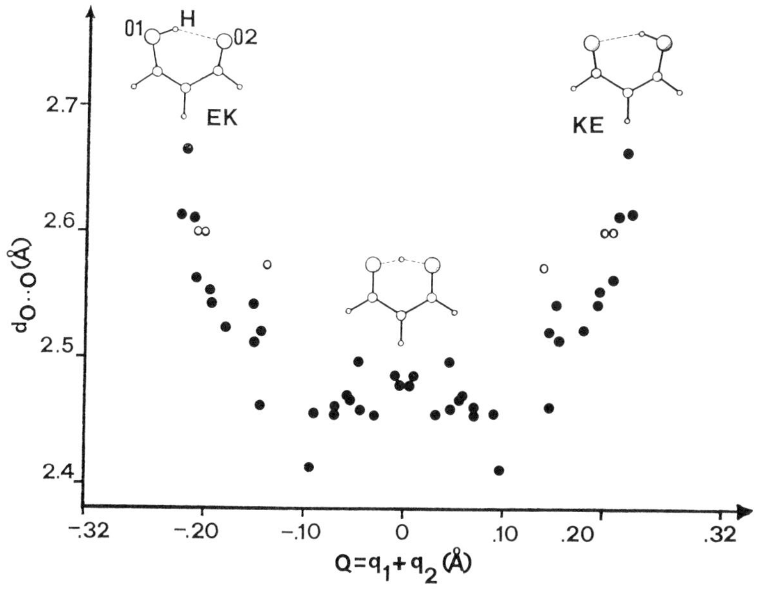

Fig. 7.32. Scatter plot of d_{O--O} versus Q. The plot has m symmetry because of the equivalence of EK and KE structures. Full and open circles indicate intra- and intermolecular H-bonds, respectively. (Reproduced by permission from Gilli *et al.*[24])

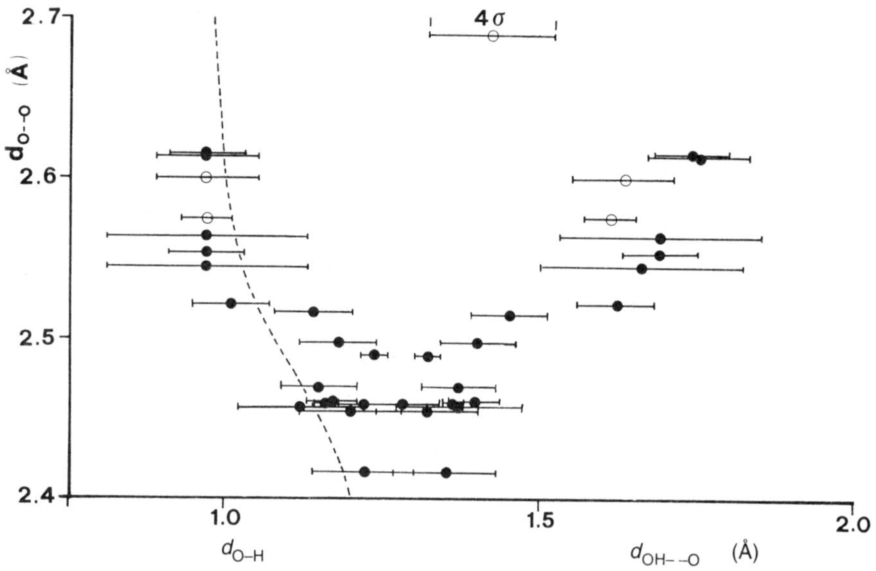

Fig. 7.33. Scatter plot of d_{O-H} and d_{OH--O} distances against the contact d_{O--O} distance.

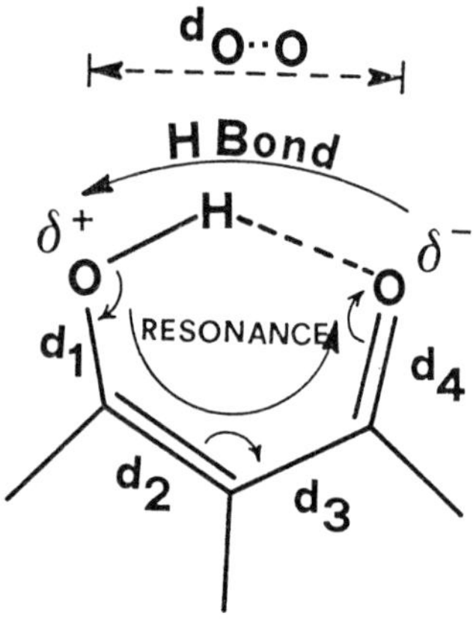

Fig. 7.34. A graphical representation of the resonance assisted hydrogen bonding (RAHB) model. (Reproduced by permission from Gilli *et al.*[24])

the H-bond itself (E_{HB}) and the van der Waals energy originated from the attractive and repulsive terms of the 1–4 interactions (E_{vdW}). Re-establishing now the double C=C bond, the resonance (7.IXa)↔(7.IXb) will cause a neat shift of electrons from left to right (Fig. 7.34), which will stop when the minimum of $E_{RES} + E_{BP}$ is attained, where the first term is the resonance energy gain and the second the energy needed to dissociate the opposite partial charges on the terminal oxygens. The charges have the correct sign for strengthening the H-bond with consequent shortening of d_{O--O} and lengthening of d_{O-H}. The movement of the proton to the right is equivalent to a negative charge going to the left and the global effect is the annihilation of the partial charges initially set up by resonance, so allowing an increased contribution of the polar form (7.IXb) and a further strengthening of the H-bond, an imaginary process going on until the minimum of the function

$$E = E_{HB} + E_{vdW} + E_{RES} + E_{BP} \qquad (7.51)$$

is attained. The H-bond formed has quite specific features and, in view of the strict interplay of π delocalization and H-bond strengthening which causes it, has been called[24] **resonance assisted hydrogen bonding** (RAHB). It is essentially a charge assisted H-bond where, however, the charges do not arise from the presence of ions but from the resonance in a heteroconjugated system.

The RAHB model is supported by a wealth of spectroscopic data and theoretical calculations for which the reader is addressed to the original paper. What is more important here is to discuss whether the correlation among geometrical parameters can be interpreted in terms of the potential energy hypersurface of the fragment. The approximate energy partitioning used in (7.51) can be evaluated quantitatively since semi-empirical equations are available for all the four terms. E_{HB} is the total energy, including both attraction and repulsion terms, of the O–H ··· O bond as a function of $d_{O-H} = r$ and $d_{O--O} = R$. It can be written as $E_{HB}(r, R)$ and has been calculated by the equation proposed by Lippincott and Schroeder[37,38] for all the desired values of r at the R value for which the energy was a relative minimum. In such a way R turns out to be a function of r and is known for any value of r. E_{RES} can be calculated by the method proposed by Krigowski *et al.*[144] which is known to give the resonance energy of a π-conjugated system from its bond distances with fairly good accuracy. E_{BP} is the energy required to create the opposite fractional charges $\pm q$ on the two terminal oxygens and can be easily evaluated by means of the coefficients of the atomic ionization energy versus electron affinity curves tabulated by Hinze and Jaffè[145,146] for the main elements. Finally, E_{vdW} can be calculated by the methods and equations already discussed in the section on molecular mechanics.

The final potential energy map calculated according to (7.51) for acetylacetone ($R_1 = R_3 = CH_3$, $R_2 = H$) is shown in Fig. 7.35 as a function of two coordinates: the bond number of the O–H bond, n(O–H) (which is 1.0 in the absence of H-bonding when $d_{O\text{—}H}$ assumes the value of 0.97 Å, and 0.5 when the hydrogen is equally shared by the two oxygens and $d_{O\text{—}H} = d_{OH\text{--}O} = 1.25$ Å), and the coupling parameter λ (which assumes the values of $\lambda = 0$, 0.5, and 1 for π-localized EK, fully π-delocalized, and π-localized KE fragment structures, respectively) or its related quantity Q. Figure 7.36 illustrates the physical meaning that can be given to the four corners and the central point of the plot of Fig. 7.35 while Fig. 7.37 shows a three-dimensional representation of the potential energy map in the same coordinate system.

The potential energy map is centrosymmetric in consequence of the fragment symmetry and displays a diagonal valley of lower energy which can

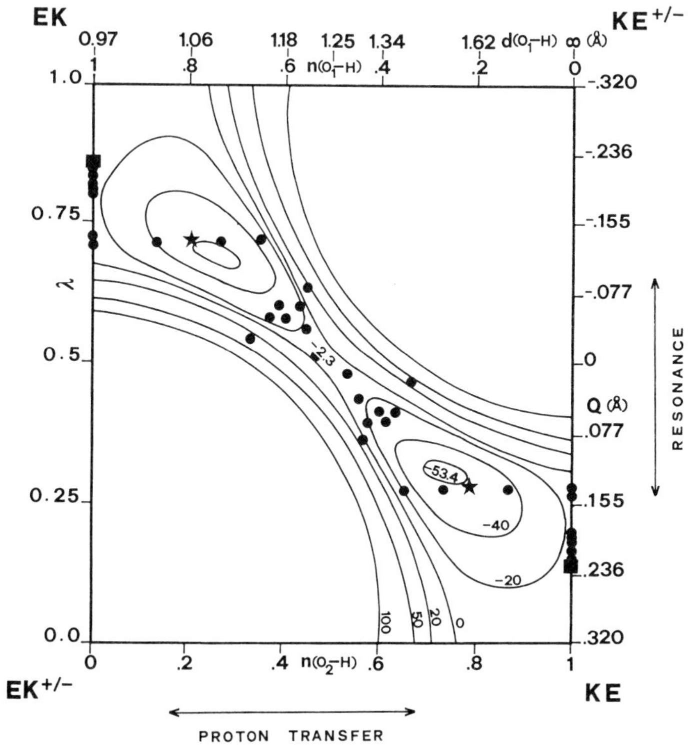

Fig. 7.35. Total potential energy map (kJ mol^{-1}) calculated according to eqn (7.51) for acetylacetone ($R_1 = R_3 = CH_3$, $R_2 = H$) as a function of the coupling parameter λ and the n(O–H) bond number. The physical meaning of the four corners and of the cental point of the plot are illustrated in Fig. 7.36. The star indicates the gas electron diffraction structure of acetylacetone[147] and the full points the X-ray or neutron structures analysed. (Reproduced by permission from Gilli *et al.*[24])

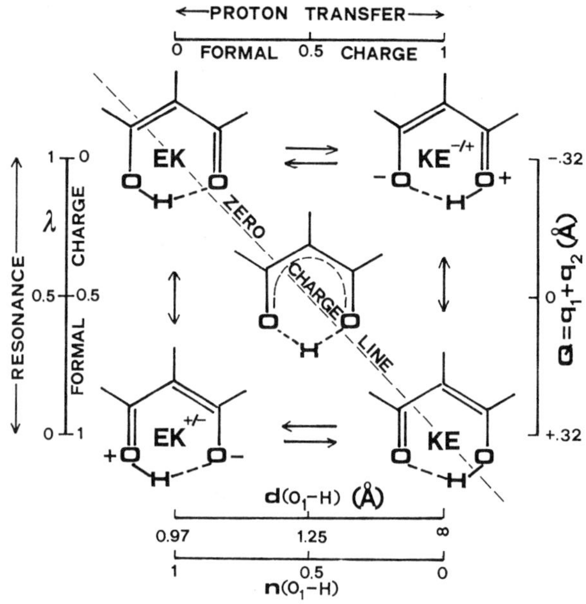

Fig. 7.36. Geometries assumed by the fragment in the five positions of the energy map of Fig. 7.35 corresponding to the four corners and to the central point. The diagonal line is a zero-charge line for the fragment because it is the *locus* of points for which the partial charges generated by resonance are exactly compensated by the shift of the proton.

be considered as the path of the reaction of proton transfer transforming the EK form into its KE isomer. Figure 7.38 shows the geometrical changes which the fragment undergoes along the reaction pathway and Fig. 7.39 reports the energy profiles along the same path. The curve A′ has been calculated without taking into account the two terms related to resonance (E_{RES} and E_{BP}); it is essentially representative of H-bonding in water and, in fact, it reproduces quite well the water distance values ($r = 1.01$ Å, $R = 2.74$ Å) and of the proton transfer barrier (some 40 kJ mol^{-1}). The curve A is calculated for acetylacetone by the use of all terms of (7.51) (Figs. 7.35 and 7.37); the introduction of the resonance term lowers the total energy by 42.5 kJ mol^{-1}, increasing the H-bond energy (already 20 kJ mol^{-1} in water) to the relevant value of 62.5 kJ mol^{-1}; at the same time the proton transfer barrier is reduced to 34.8 kJ mol^{-1} and the O–H distance is lengthened until $r = 1.08$ Å (associated $R = 2.50$ Å).

The structure of acetylacetone has been recently determined by gas electron diffraction[147] and the corresponding geometry, marked by a star in Fig. 7.35, is seen to be very close to the calculated energy minimum. In the same figure are reported as full points the geometries of all the 25 fragments investigated. It seems interesting that all of them are correctly located inside the diagonal energy valley though, of course, not in the minimum, which has been specifically calculated for acetylacetone. As far as their exact position is concerned two main factors can be discerned. The first is electronic, and we have already remarked that β-diketone enols form much stronger H-bonds than the enols of β-ketoesters or β-ketoamides. The second is steric in nature, the strongest H-bonds being associated with bulky R_1 and R_3 substituents. For example, strictly similar calculations carried out on hexamethylacetylacetone by the use of (7.51) show that the energy minimum is shifted to $r = 1.11$ and $R = 2.48$ Å, while the H-bond energy increases to 82.5 kJ mol^{-1} and the proton transfer barrier decreases to 27.5 kJ mol^{-1}.

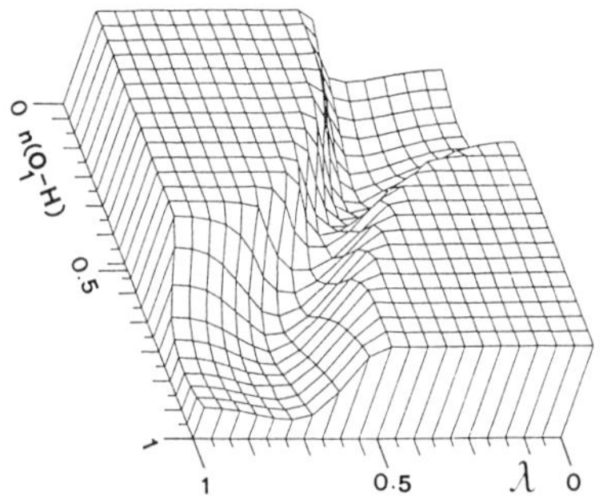

Fig. 7.37. Three-dimensional representation of the potential energy map of Fig. 7.35 in the same coordinate system. The path for the reaction of proton transfer transforming the EK into the KE isomer is clearly shown as an energy valley whose saddle point corresponds to the fully π-delocalized form having the proton in the middle of the two oxygens.

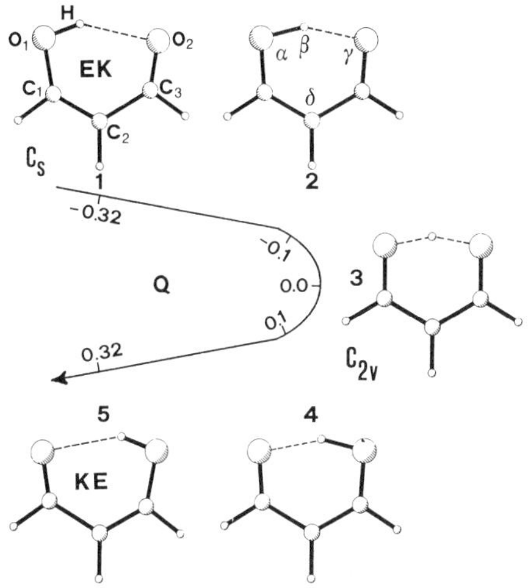

Fig. 7.38. Geometrical deformations experienced by the fragment along the pathway of the proton transfer transforming the EK into the KE isomer.

Resonance assisted hydrogen bonding (RAHB) is therefore a mechanism of synergistic interplay of resonance and H-bond formation. It can be generalized by calling it *a mechanism of interplay between H-bond and resonance in heterodienes or, more generally, heteroconjugated systems*:

$$\mathrm{H{-}X{-}C{=}C{-}C{=}Y \leftrightarrow H{-}X^{+}{=}C{-}C{=}C{-}Y^{-}} \tag{7.52}$$

$$\mathrm{H{-}X{-}C{=}C{-}C{=}C{-}C{=}Y \leftrightarrow H{-}X^{+}{=}C{-}C{=}C{-}C{=}C{-}Y^{-}} \tag{7.53}$$

$$\mathrm{H{-}X{-}C{=}C{-}C{=}C{-}C{=}C{-}C{=}Y \leftrightarrow H{-}X^{+}{=}C{-}C{=}C{-}C{=}C{-}C{=}C{-}Y^{-}} \tag{7.54}$$

$$\mathrm{H{-}X{-}C{=}Y \leftrightarrow H{-}X^{+}{=}C{-}Y^{-}}. \tag{7.55}$$

X = CH, N

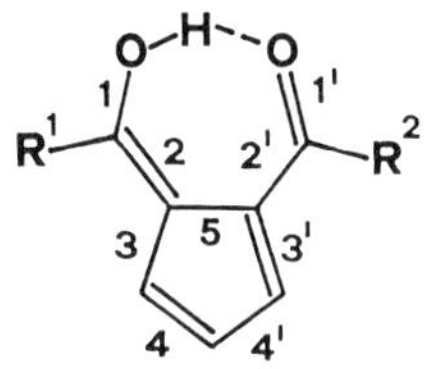

7.XII

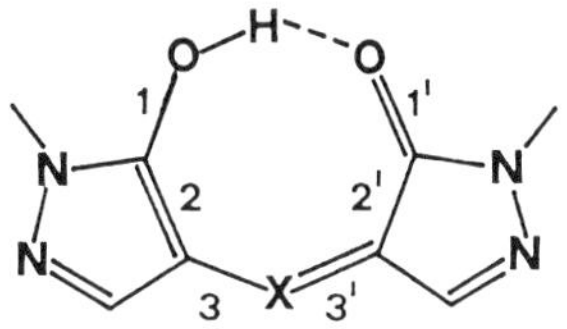

7.XIII

From this point of view, the β-diketone enols so far discussed correspond to the heterodienic scheme (7.52) (X = Y = O), but the RAHB mechanism is supposed to operate independently of the number of interleaving carbon atoms provided the conjugation between the lone pair on X and the C=Y double bond is maintained. In the X = Y = O series the scheme (7.55) represents carboxylic acids which, in fact, are well known to give strongly H-bonded dimers (Fig. 7.3, scheme b), though a more convincing proof of RAHB could come from the observation of strongly delocalized and H-bonded heterotrienes (7.53) and heterotetraenes (7.54). Structures corresponding to these fragments have been recently analysed[25] and found to display the expected behaviour. Compounds (7.XII) behave as totally

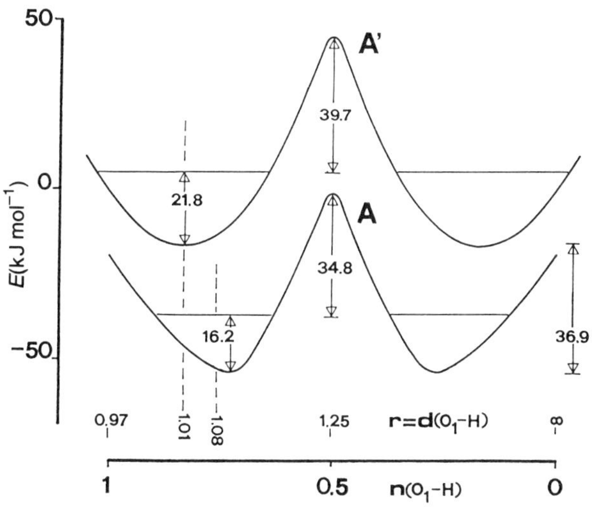

Fig. 7.39. Energy profile along the reaction coordinate of proton transfer transforming EK into KE as mapped along the mainly diagonal pathway of lowest relative energy of the map of Fig. 7.35 (curve A). Curve A′ is the same curve but calculated omitting the E_{RES} and E_{BP} terms in eqn (7.51); it can be considered to represent the proton transfer in water. The horizontal lines mark the zero-point energies which are slightly different in water and acetylacetone.

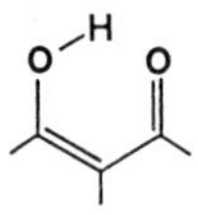
7.XIV

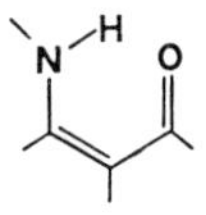
7.XV

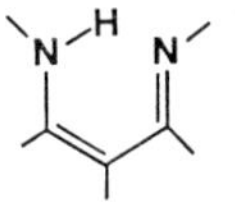
7.XVI

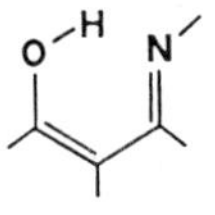
7.XVII

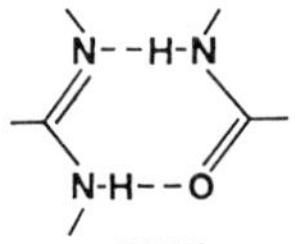
7.XVIII

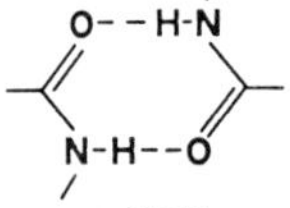
7.XIX

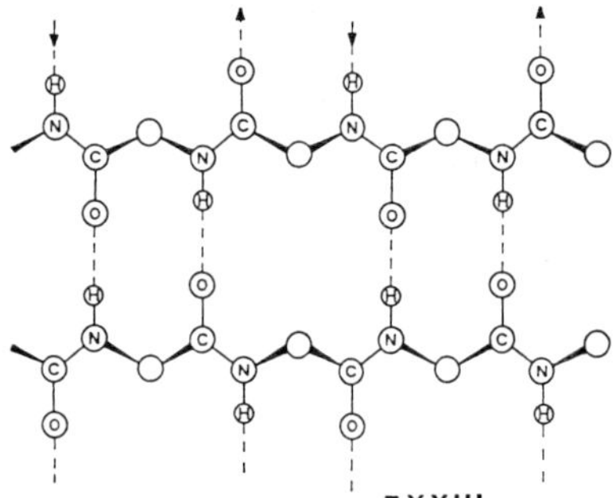
7.XXIII

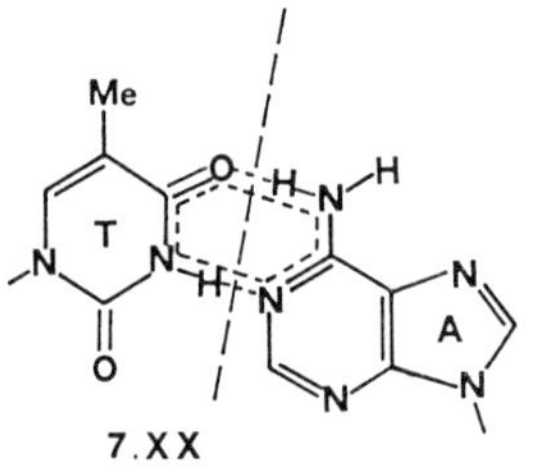
7.XX

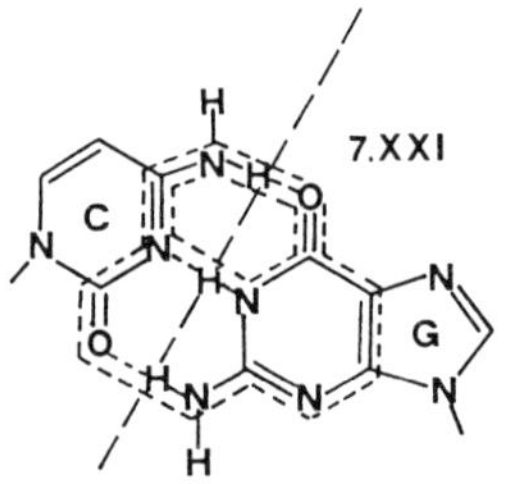
7.XXI

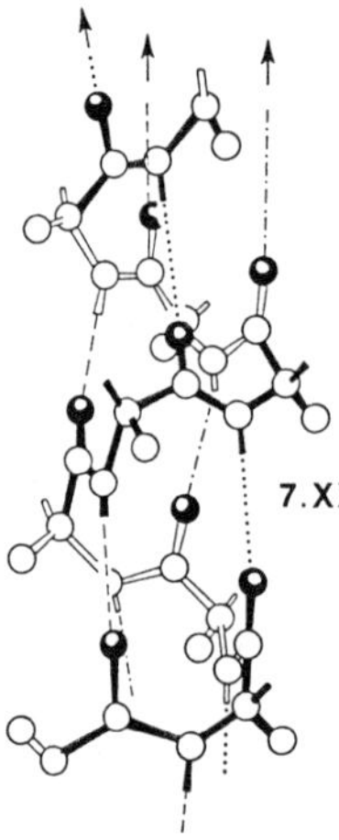
7.XXII

π-delocalized heterotetraenes with bond 5 not participating to the conjugation; the opposite bonds n and n' ($n = 1, 2, 3$, and 4) are nearly equal and the O–H$\cdots$O is extremely short (on average, $d_{O--O} = 2.47$ and $d_{O--H} = 1.18$ Å). In a strictly similar way compounds (7.XIII) show total delocalization of the heterotrienic system and display one of the shortest d_{O--O} distances ever observed, i.e. 2.425 Å.

Up to now we have proved that RAHB is active in the enolone fragment (7.XIV) (or in other diketone enols with longer conjugated systems) forming intramolecular or, sometimes, intermolecular H-bonds. All these compounds have the oxygen as the only heteroatom. However, other heteroconjugated systems having different heteroatoms are putatively implied in this phenomenon via intramolecular or intermolecular (including dimerization) H-bond formation, such as enaminones (7.XV), enaminoimines (7.XVI), enolimines (7.XVII), amide–amidine complexes (7.XVIII) and amide dimers (7.XIX). Preliminary data confirming the strong π delocalization of H-bonded enaminones (7.XV) and enolimines (7.XVII) have already been reported,[148] and the formation of amide dimers (7.XIX) has been discussed by several authors.[34,35]

Possible biochemical and biological implications originate from the fact that the amide–amidine complex (7.XVIII) is present in both thymine–adenine (7.XX) and cytosine–guanine (7.XXI) H-bonded couples determining the double-helix structure of DNA; in addition, cytosine–guanine coupling (7.XXI) implies another much wider cycle of H-bonds and π-conjugated double bonds. Strictly similar considerations can be made in connection with the H-bonds determining the secondary structure of proteins: both α-helices (7.XXII) and β-pleated sheets (7.XXIII) contain long chains of π-heteroconjugated systems connected by H-bonds and differ only because each α-helix contains three of such chains which are nearly parallel and isoriented, while in β-pleated sheets the arrangement of the chains is antiparallel.

References

1. Wells, A. F. (1975). *Structural inorganic chemistry* (4th edn). Clarendon, Oxford.
2. Bondi, A. (1964). *Journal of Physical Chemistry,* **68,** 441.
3. Allen, F. H., Bellard, S., Brice, M. D., Cartwright, B. A., Doubleday, A., Higgs, H., Hummelink, T., Hummelink-Peters, B. G., Kennard, O., Motherwell, W. D. S., Rodgers, J. R., and Watson, D. G. (1979). *Acta Crystallographica,* **B35,** 2331.
4. Dunitz, J. D. (1979). *X-ray analysis and the structure of organic molecules.* Cornell University Press, Ithaca, N.Y.
5. Pauling, L. (1960). *The nature of the chemical bond and the structure of molecules and crystals. An introduction to modern structural chemistry* (3rd edn). Cornell University Press, Ithaca, N.Y.
6. Kitaigorodski, A. I. (1973). *Molecular crystals and molecules.* Academic, New York.
7. Kaplan, I. G. (1986). *Theory of molecular interactions.* Elsevier, Amsterdam.
8. Pertsin, A. J., and Kitaigorodski, A. I. (1987). *The atom–atom potential method.* Springer, Berlin.
9. London, F. (1930). *Zeitschrift für Physikalische Chemie,* **B11,** 222, 236; (1930). *Zeitschrift für Physik,* **63,** 245.
10. Lennard-Jones, J. E. (1931). *Proceedings of the Physical Society,* **43,** 461.
11. Buckingham, R. A. (1938). *Proceedings of the Royal Society of London,* **A168,** 264.
12. Giglio, E. (1969). *Nature,* **222,** 339.
13. Allinger, N. L. (1975). *MM1/MMP1,* QCPE No. 318. Indiana University.
14. Hamilton, W. C., and Ibers, J. A. (1968). *Hydrogen bonding in solids.* Benjamin, New York.
15. Pimentel, G. C., and McClellan, A. L. (1960). *The hydrogen bond.* Freeman, San Francisco; Pimentel, G. C., and McClellan, A. L. (1971). *Annual Reviews of Physical Chemistry,* **22,** 347.
16. Kollman, P. A., and Allen, L. C. (1972). *Chemical Reviews,* **72,** 283; Kollman, P. A. and Joesten, M. D. (1982). *Journal of Chemical Education,* **59,** 362.
17. Schuster, P., Zundel, G., and Sandorfy, C. (ed.) (1976). *The hydrogen bond,* Vols. I, II, and III. North-Holland, Amsterdam.
18. Emsley, J. (1980). *Chemical Society Review,* **9,** 91.
19. Umeyama, H., and Morokuma, K. (1977). *Journal of the American Chemical Society,* **99,** 1316.
20. Morokuma, K. (1971). *Journal of Chemical Physics,* **55,** 1236.

21. Kitaura, K., and Morokuma, K. (1976). *International Journal of Quantum Chemistry*, **10,** 325.
22. Taylor, R., and Kennard, O. (1982). *Journal of the American Chemical Society*, **104,** 5063.
23. Görbitz, C. H. (1989). *Acta Crystallographica*, **B45,** 390.
24. Gilli, G., Bellucci, F., Ferretti, V., and Bertolasi, V. (1989). *Journal of the American Chemical Society*, **111,** 203.
25. Gilli, G., and Bertolasi, V. (1990). In *The chemistry of enols* (ed. Z. Rappoport). Wiley, New York. p. 713.
26. Bent, H. A. (1968). *Chemical Reviews*, **68,** 587.
27. Gutmann, V. (1978). *The donor–acceptor approach to molecular interactions*. Plenum, New York.
28. Kitaigorodski, A. I. (1984). *Mixed crystals*. Springer, Berlin.
29. Huheey, J. E. (1983). *Inorganic chemistry* (3rd edn). Harper and Row, New York.
30. Ibers, J. A., Pace, L. J., Martinsen, J., and Hoffman, B. M. (1982). *Structure and Bonding*, **50,** 1.
31. Del Re, G. (1958). *Journal of the Chemical Society*, 4031.
32. Gasteiger, J., and Marsili, M. (1980). *Tetrahedron*, **36,** 3219.
33. Berkovitch-Yellin, Z., and Leiserowitz, L. (1982). *Journal of the American Chemical Society*, **104,** 4052.
34. Lifson, S., Hagler, A. T., and Dauber, P. (1979). *Journal of the American Chemical Society*, **101,** 5111.
35. Dauber, P., and Hagler, A. T. (1980). *Accounts of Chemical Research*, **13,** 105.
36. Spackman, M. A. (1986). *Journal of Chemical Physics*, **85,** 6579, 6587.
37. Lippincott, E. R., and Schroeder, R. (1955). *Journal of Chemical Physics*, **23,** 1099.
38. Schroeder, R., and Lippincott, E. R. (1957). *Journal of Physical Chemistry*, **61,** 921.
39. Born, M., and von Karman, T. (1912). *Physikalische Zeitschrift*, **13,** 297.
40. Born, M., and Huang, K. (1954). *Dynamical theory of crystal lattices*. Clarendon, Oxford.
41. Willis, B. T. M., and Pryor, A. W. (1975). *Thermal vibrations in crystallography*. Cambridge University Press.
42. Califano, S., Schettino, V., and Neto, N. (1981). *Lattice dynamics of molecular crystals*, Lecture Notes in Chemistry, Vol. 26. Springer, Berlin.
43. Pawley, G. S. (1970). In *Crystallographic computing* (ed. F. R. Ahmed). Munksgaard, Copenhagen. p. 243.
44. Kitaigorodski, A. I. (1970). In *Advances in structure research by diffraction methods*, Vol. 3, (ed. R. Brill and R. Mason). Pergamon, Oxford. p. 173.
45. Cahn, R. S., and Ingold, C. K., (1951). *Journal of the Chemical Society*, 612.
46. Cahn, R. S., Ingold, C. K., and Prelog, V. (1956). *Experientia*, **12,** 81.
47. Klyne, W., and Prelog, V. (1960). *Experientia*, **16,** 521.
48. Mislow, K. (1965). *Introduction to stereochemistry*. Benjamin, New York.
49. Hallas, G. (1965). *Organic stereochemistry*. McGraw-Hill, New York.
50. Cahn, R. S., Ingold, C. K., and Prelog, V. (1966). *Angewandte Chemie International Edition in English*, **5,** 385.
51. Eliel, E. L., Allinger, N. L., Angyal, S. J., and Morrison, G. A. (1966). *Conformational analysis*. Wiley, New York.
52. Natta, G., and Farina, M. (1968). *Stereochimica*. Mondadori, Milan.
53. *IUPAC 1968 Tentative rules, section E, fundamental stereochemistry*, Information Bulletin No. 35; (1970). *Journal of Organic Chemistry*, **35,** 2849.
54. Stoddard, J. F. (1971). *Stereochemistry of carbohydrates*. Wiley, New York.
55. Balaban, A. T. (ed.) (1976). *Chemical applications of graph theory*. Academic, New York.

56. Bijvoet, J. M., Peerdeman, A. F., and van Bommel, A. J. (1951). *Nature*, **168,** 271.
57. Rogers, D. (1975). In *Anomalous scattering* (ed. S. Ramaseshan and S. C. Abrahams). Munksgaard, Copenhagen. p. 231.
58. Cotton, F. A. (1971). *Chemical applications of group theory* (2nd edn). Wiley, New York.
59. Cano, F. H., Foces–Foces, C., and Garcìa-Blanco, S. (1977). *Tetrahedron,* **33,** 797.
60. Cremer, D. and Pople, J. A. (1975). *Journal of the American Chemical Society,* **97,** 1354.
61. Boeyens, J. C. A. (1978). *Journal of Crystal and Molecular Structure,* **8,** 317.
62. Altona, C., Geise, H. J., and Romers, C. (1968). *Tetrahedron,* **24,** 13.
63. Parkanyi, L. (1980). *RING*. Hungarian Academy of Sciences, Budapest.
64. Nardelli, M. (1982). PARST. University of Parma, Italy.
65. Clark, T. (1985). *A handbook of computational chemistry*. Wiley, New York.
66. Hehre, W. J., Radom, L., Schleyer, P. v. R., and Pople, J. A., (1985). *Ab initio molecular orbital theory*. Wiley, New York.
67. Pople, J. A., and Beveridge, D. L. (1970). *Approximate molecular orbital theory*. McGraw-Hill, New York.
68. Dewar, M. J. S. (1969). *The molecular orbital theory of organic chemistry*. McGraw-Hill, New York.
69. Murrell, J. N., and Harget, A. J. (1972). *Semiempirical self-consistent-field molecular orbital theory of molecules*. Wiley, London.
70. Frisch, M. J., Head-Gordon, M., Schlegel, H. B., Raghavachari, K., Binkley, J. S., Gonzalez, C., *et al.* (1988). *GAUSSIAN 88*. Gaussian Inc., Pittsburgh, PA.
71. Bingham, R. C., Dewar, M. J. S., and Lo, D. H. (1975). *Journal of the American Chemical Society,* **97,** 1285, 1294, 1302, 1307, 1311; Dewar, M. J. S., and Thiel, W. (1977). *Journal of the American Chemical Society,* **99,** 4899, 4907.
72. Dewar, M. J. S., Zoebisch, E. G., Healy, E. F., and Stewart, J. J. P. (1985). *Journal of the American Chemical Society,* **107,** 3902.
73. Coulson, C. A., (1961). *Valence*. Oxford University Press, London.
74. McWeeny, R. (1979). *Coulson's valence*. Oxford University Press, London.
75. Streitwieser, A. Jr. (1961). *Molecular orbital theory for organic chemists*. Wiley, New York.
76. Gillespie, R. J. (1972). *Molecular geometry*. Van Nostrand-Reinhold, London.
77. Gillespie, R. J. (1963). *Journal of Chemical Education,* **40,** 295; (1970). **47,** 18.
78. Bader, R. F. W., Gillespie, R. J., and MacDougall, P. J. (1988). *Journal of the American Chemical Society,* **110,** 7329.
79. Orgel, L. E., (1966). *An introduction to transition-metal chemistry* (2nd edn). Wiley, New York.
80. Bethe, H. (1929). *Annalen der Physik,* **3,** 135.
81. Van Vleck, J. H. (1932). *Physical Review,* **41,** 208; (1935). *Journal of Chemical Physics,* **3,** 803, 807.
82. Ballhausen, C. J., (1962). *Introduction to ligand field theory*. McGraw-Hill, New York.
83. Dunn, T. M., McClure, D. S., and Pearson, R. G. (1965). *Some aspects of crystal field theory*. Harper and Row, New York.
84. Schäffer, C. E., and Jørgensen, C. K. (1965). *Molecular Physics,* **9,** 401.
85. Schäffer, C. E. (1973). *Structure and Bonding,* **14,** 69.
86. Burdett, J. K. (1978). *Chemical Society Review,* **7,** 507.
87. Burdett, J. K., (1980). *Molecular shapes. Theoretical models of inorganic stereochemistry,* p. 23. Wiley, New York.
88. Walsh, A. D. (1953). *Journal of the Chemical Society,* 2260, 2266, 2288, 2296, 2301, 2306.

89. Gimark, B. M. (1979). *Molecular structure and bonding*. Academic, New York.
90. Lowe, J. P. (1978). *Quantum chemistry*. Academic, New York.
91. Woodward, R. B., and Hoffmann, R. (1970). *The conservation of orbital symmetry*. Academic, New York.
92. Fujimoto, H., and Fukui, K. (1974). In *Chemical reactivity and reaction paths* (ed. G. K. Klopman). Wiley, New York.
93. Elian, M., and Hoffmann, R. (1975). *Inorganic Chemistry*, **14**, 1058.
94. Hoffmann, R. (1981). *Science*, **211**, 995.
95. Hoffmann, R. (1963). *Journal of Chemical Physics*, **39**, 1397.
96. Jahn, H. A., and Teller, E. (1937). *Proceedings of the Royal Society of London*, **A161**, 220.
97. Jotham, R. W., and Kettle, S. F. A. (1971). *Inorganic Chimica Acta*, **5**, 183.
98. Gažo, J., Bersuker, I. B., Garay, J., Kabesova, M., Kohout, J., Langfelderova, H., Melnik, M., Serator, M., and Valach, F. (1976). *Coordination Chemistry Reviews*, **19**, 253.
99. Hathaway, B., Duggan, M., Murphy, A., Mullane, J., Power, C., Walsh, A., and Walsh, B. (1981). *Coordination Chemistry Reviews*, **36**, 267.
100. Pearson, R. G. (1969). *Journal of the American Chemical Society*, **91**, 1252, 4947; (1970) *Journal of Chemical Physics*, **52**, 2167; **53**, 2986.
101. Bartell, L. S. (1960). *Journal of Chemical Physics*, **32**, 827; (1968). *Journal of Chemical Education*, **45**, 754.
102. Glidewell, C. (1975). *Inorganic Chimica Acta*, **12**, 219; (1976). *Inorganic Chimica Acta*, **20**, 113.
103. Cotton, F. A., and Wilkinson, G. (1980). *Advanced inorganic chemistry* (4th edn). Wiley, New York.
104. Pimentel, G. C., and Spratley, R. D. (1969). *Chemical bonding clarified through quantum mechanics*, Holden-Day, San Francisco.
105. Purcell, K. F., and Kotz, J. C. (1977). *Inorganic chemistry*. Saunders, Philadelphia.
106. Domenicano, A., Mazzeo, P., and Vaciago, A. (1976). *Tetrahedron Letters*, **13**, 1029.
107. Sutton, L. E. (1958). *Tables of interatomic distances and configuration in molecules and ions*, Special Publication, No. 11. The Chemical Society, London.
108. Sutton, L. E. (1965). *Tables of interatomic distances and configuration in molecules and ions*, Special Publication, No. 18. The Chemical Society, London.
109. Allen, F. H., Kennard, O., Watson, D. G., Brammer, L., Orpen, A. G., and Taylor, R. (1987). *Journal of the Chemical Society Perkin Transactions* II, S1.
110. Orpen, A. G., Brammer, L., Allen, F. H., Kennard, O., Watson, D. G., and Taylor, R. (1989). *Journal of the Chemical Society Dalton Transaction*, S1.
111. Bent, H. A. (1960). *Journal of Chemical Education*, **37**, 616; (1961). *Chemical Review*, **61**, 275.
112. Dewar, J. S., Kollmar, H., and Li, W. K. (1975). *Journal of Chemical Education*, **52**, 305.
113. Gilli, G. and Bertolasi, V. (1983). *Journal of Chemical Education*, **60**, 638.
114. Allinger, N. L. (1976). *Advances in Physical Organic Chemistry*, **13**, 1.
115. Altona, C., and Faber, D. H. (1974). *Topics in Current Chemistry*, **45**, 1.
116. Engler, E. M., Andose, J. D., and Schleyer, P. v. R. (1973). *Journal of the American Chemical Society*, **95**, 8005.
117. Ermer, O. (1976). *Structure and Bonding*, **27**, 161.
118. Hursthouse, M. B., Moss, G. P., and Sales, K. D. (1978). *Annual Reports of The Chemical Society. Section B: Organic Chemistry*, **75**, 23.

119. Kitaigorodski, A. I., (1978). *Chemical Society Review*, **7**, 133.
120. White, D. N. J. (1978). In *Molecular structure by diffraction methods*, Vol. 6, (ed. L. E. Sutton and M. R. Truter). The Chemical Society, London. p. 38.
121. Osawa, E., and Musso, H. (1982). *Topics in Stereochemistry*, **13**, 117.
122. Boyd, D. B., and Lipkowitz, K. B. (1982). *Journal of Chemical Education*, **59**, 269.
123. Ermer, O. (1981). *Aspekte von Kraftfeldrechnungen*. Volfgang Bauer, Munich.
124. Burkert, U. and Allinger, N. L. (1982). *Molecular mechanics*. American Chemical Society, Washington, D.C.
125. Allinger, N. L., *et al.*, *MM1/MMP1*, QCPE 318; Allinger, N. L., and Yu, Y. H. *MM2*, QCPE 325; Warshel, A. and Levitt, M. *QCFF/PI*, QCPE 247; Huler, E., Sharon, R., and Warshel, A. *MCA*, QCPE 325; Andose, J. D., Engler, E. M., Collins, H. B., Hummel, J. P., Mislow, K., and Schleyer, P. v. R. *BIGSTRN*, QCPE 348; Iverson, J. D., and Mislow, K. *BIGSTRN2*, QCPE 410; Browman, M. J., Carruthers, L. M., Kashuba, K. L., Momany, F. A., Pottle, M. S., Posen, S. P., *et al.* *UNICEPP*, QCPE 361.
126. Bürgi, H. B. (1975). *Angewandte Chemie, International Edition in English*, **14**, 460.
127. Murray-Rust, P. (1978). In *Molecular structure by diffraction methods*, Vol. 6, (ed. L. E. Sutton and M. R. Truter). The Chemical Society, London. p. 154.
128. Bürgi, H. B., and Dunitz, J. D. (1983). *Accounts of Chemical Research*, **16**, 153.
129. Bürgi, H. B., (1985). In *Static and dynamic implications of precise structural information*, Reports of the 11th International School of Crystallography, Erice, Italy. p. 245.
130. Eyring, H., and Polanyi, M. (1931). *Zeitschrift für Physikalische Chemie*, **B12**, 279.
131. Atkins, P. W. (1986). *Physical chemistry* (3rd edn). Oxford University Press.
132. Sathyamurthy, N. and Joseph, T. (1984). *Journal of Chemical Education*, **11**, 968.
133. Müller, K. (1980). *Angewandte Chemie, International Edition in English*, **19**, 1.
134. Bürgi, H. B. (1975). *Angewandte Chemie, International Edition in English*, **87**, 461.
135. Bürgi, H. B. (1973). *Inorganic Chemistry*, **12**, 2321.
136. Britton, D., and Dunitz, J. D. (1981). *Journal of the American Chemical Society*, **103**, 2971.
137. Bürgi, H. B., Dunitz, J. D., and Schefter, E. (1973). *Journal of the American Chemical Society*, **95**, 5065.
138. Bürgi, H. B., Dunitz, J. D., Lehn, J. M., and Wipff, G. (1974). *Tetrahedron*, **30**, 1563.
139. Bürgi, H. B., Dunitz, J. D., and Shefter, E. (1974). *Acta Crystallographica*, **B30**, 1517.
140. Gilli, G., and Bertolasi, V. (1979). *Journal of the American Chemical Soceity*, **101**, 7704.
141. Gilli, G., Bertolasi, V., Bellucci, F., and Ferretti, V. (1986). *Journal of the American Chemical Society*, **108**, 2420.
142. Winkler, F. K., and Dunitz, J. D. (1971). *Journal of Molecular Biology*, **59**, 169.
143. Bürgi, H. B., and Shefter, E., (1975). *Tetrahedron*, **31**, 2976.
144. Krygowski, T. M., Anulewicz, R., and Kruszewski, J. (1983). *Acta Crystallographica*, **B39**, 732.
145. Hinze, J. and Jaffè, H. H. (1962). *Journal of the American Chemical Society*, **84**, 540; (1963). *Journal of Physical Chemistry*, **67**, 1501.

146. Hinze, J., Whitehead, M. A., and Jaffè, H. H., (1963). *Journal of the American Chemical Soceity,* **85,** 148.
147. Iijima, K., Ohnogi, A., and Shibata, S. (1987). *Journal of Molecular Structure,* **156,** 111.
148. Ferretti, V., Bellucci, F., Bertolasi, V., and Gilli, G. (1985). *Proceedings of the IX European Crystallographic Meeting, Torino,* p. 337.
149. Bertolasi, V., Gilli, P., Ferretti, V., and Gilli, G. (1991) *Journal of the American Chemical Society,* **113,** 4917.

Protein crystallography 8

GIUSEPPE ZANOTTI

Introduction

Protein crystallography, the subject of this chapter, is a specialized branch of crystallography that investigates, by using diffraction techniques on single crystals, the three-dimensional structure of biological macromolecules.[1–5] For a long time, good quality single crystals were obtainable only for globular proteins. Nowadays, other biological macromolecules, like *t*-RNA, polysaccharides, and polynucleotides, give crystals suitable for X-ray analysis. Therefore, the techniques for structure solution and refinement described in this chapter apply to all cases where crystals with a high portion of unordered solvent can be grown.

Despite the fact that crystals of proteins have been known since the beginning of the century, only around 1960 were the first structures, myoglobin and haemoglobin, elucidated.[6] Their resolution was made possible mainly by the development of the isomorphous replacement technique. Since then, a lot of theoretical and technical advances have been made: among them, the use of anomalous dispersion, molecular replacement, and the development of oscillation techniques in data collection of crystals with large cell dimensions. At the end of the 1970s the growth of computing power made possible the refinement in reciprocal space. In the meantime, the introduction of powerful and affordable graphic stations brought the development of interactive graphic software, allowing faster interpretation of electron density maps and model building. Only in very recent years has the availability of synchrotron radiation and two-dimensional detectors made data collection a routine procedure.

An idea of the 'state of the art' of protein crystallography is given by the Brookhaven Protein Data Bank (thereafter called PDB),[7] which collects coordinates of the macromolecular structures solved, mainly by X-rays. Looking at the January 1989 release, one finds, perhaps with some surprise, that only 413 sets of coordinates are available (plus 86 bibliographic entries, that is structures solved but not yet deposited).† They include 10 polysaccharides, 26 DNA fragments, 6 *t*-RNA, and 18 model structures. The remaining 353 are protein structures, including 14 virus coat proteins: taking into account the fact that most of them are related proteins or variants of the same molecule (there are for example 34 lysozyme variants or mutants, 12 haemoglobins, and 8 myoglobins), this number must be considered quite small and demonstrates that solving the structure of a new macromolecule

† It must be remembered that the Protein Data Bank cannot be considered fully representative of the macromolecular structures so far solved, since not all of them are quickly deposited.

still remains a long and demanding job. But possibly the reason for the relatively low number of new structures published every year is the crystallization process, a field in which technical (and theoretical) advances have proceeded very slowly and which remains the more uncertain step in protein structure determination.

Protein crystals

It is the solvent content that makes the difference between a classical molecular crystal and a protein crystal: in the former, all the atoms can be described in terms of a regular lattice, whilst in the latter a crystalline array coexists with a high portion of material in the liquid state. The mother liquid, whose content can range approximately from 30 to 75 per cent or more, has a strong influence on the behaviour of this kind of crystal, making them very peculiar and creating some advantages along with some obvious disadvantages. Among the latter, the major one is that protein crystals are much less ordered than classical crystals, not only for the large amount of unordered material present in the crystal itself, but also because surface groups of the macromolecule in contact with the solvent can show a great mobility. As a consequence, diffraction data cannot be measured to the resolution normally attainable with 'small' molecules. On the other hand, among the advantages, the environment of the macromolecule in the crystal is not too different from that of the solution from which the crystal was obtained (the influence of the solvent on the conformation of the protein cannot be underestimated) and we can profit from the solvent in the preparation of heavy-atom derivatives, as will be described in the next paragraph.

A last important point must be remembered about protein crystals: since inversion symmetry elements are not allowed, the number of possible space groups is reduced from the original number of 230 to 65 (see Chapter 1, p. 24).

Principles of protein crystallization

The process of crystallization of a macromolecule is very complex and poorly understood. A theoretical treatment of it, despite recent progress,[8] is at present impossible. Nevertheless, a lot of experience has been accumulated on the crystallization of water-soluble proteins,[2,9,10,11] and a crystallographer starting with enough pure material has a reasonable chance of succeeding in obtaining X-ray quality crystals. The same does not yet apply to membrane proteins, the crystals of which have been very seldom obtained.[12,13]

Growth of a protein crystal starts from a supersaturated solution of the macromolecule, and evolves towards a thermodynamically stable state in which the protein is partitioned between a solid phase and the solution. The time necessary before the equilibrium is reached has a great influence on the final result, that can go from an amorphous or microcrystalline precipitate to large single crystals. The supersaturation conditions can be obtained by the addition of precipitating agents and/or by modifying some of the internal parameters of the solution, like pH and temperature. Since a

Table 8.1. Precipitating agents commonly used for growing protein crystals suitable for X-ray analysis. The list is not intended to be exhaustive

Class	Examples
Inorganic salts	$(NH_4)_2SO_4$, Na_2SO_4, NaCl, KCl, NH_4Cl, $MgSO_4$, $CaCl_2$, NH_4NO_3, LiCl
Organic salts	Citrate, acetate, formate cetyltrimethyl ammonium salts
Organic solvents	Ethanol, isopropanol, acetone, dioxane, 2-methyl-2,4-pentanediol(MPD)

protein is a labile molecule, all the extreme conditions of precipitation, pH, and temperature should be avoided. And indeed, the understanding of the biological and physiological properties of the macromolecule via the determination of the three-dimensional structure is much easier if crystals are grown in conditions not too far from the physiological environment in which the molecule operates.

The precipitants frequently used can be divided into three categories: salts, organic solvents, and polyethylene glycols (PEG). A list of the more common among them is reported in Table 8.1.

1. *pH.* The pH value has a strong influence on the solubility, which has a minimum at a pH close to the isoelectric point of the macromolecule.

2. *Salt concentration.* Ionic strength can have opposite effects on protein solubility, i.e. solubility decreases exponentially with increasing ionic strength, a phenomenon known as **salting-out**, but it also has a minimum at very low ionic strength, the **salting-in** effect. In practice, precipitation can be achieved by increasing the salt concentration or by dialysing the solution of protein against water.

3. *Organic solvents.* Precipitation properties of organic solvents can be ascribed to the double effect of subtracting water molecules from the solution and to decreasing the dielectric constant of the medium. They can sometimes have effects on the protein conformation too, and for this reason they should be used with caution.

4. *PEG.* PEG is a precipitating agent with peculiar properties: it is a polymer, available in molecular weights ranging from about 200 to 20 000 Da;† its effect on solubility, along with some properties in common with salts and organic solvents, is due to volume exclusion property: the solvent is restructured and phase separation is consequently promoted. PEG is of wide applicability, and has been demonstrated to work with proteins that were crystallized with other precipitating agents.[10]

Several others parameters and factors can influence the crystallization process: among them, the protein concentration, the temperature, the presence of cations that sometimes stabilize the conformation of the protein, and the purity of the sample. The presence of contaminants can be extremely important in preventing the formation of crystals suitable for X-ray analysis.

† The molecular weight is expressed in Dalton, $1\ \text{Da} = 1.6604 \times 10^{-24}$ g.

The solvent content of protein crystals

The amount of the solvent contained in a protein crystal has been discussed by Matthews,[14] who analysed data from 116 distinct crystal forms. He defined the quantity V_M, given by the ratio of the volume of the unit cell divided by the total molecular weight of the protein in the cell. Therefore, V_M represents the crystal volume per unit of protein molecular mass: it is practically independent from the volume of the asymmetric unit and mostly depends on the solvent content. In the crystals examined, V_M has a value ranging from 1.6 Å^3 Da^{-1} to about 3.5 Å^3 Da^{-1}, but more extreme values have been found. A special trend can be observed if the molecular weight of the proteins is taken into account: small proteins tend to have small V_M and big proteins a large one, which is understandable, since a low V_M means a low solvent content. Calling V_{prot} the crystal volume occupied by the protein, V'_p its fraction with respect to the total crystal volume V and M_{prot} the mass of protein in the cell:

$$V'_p = V_{prot}/V = (V_{prot}/M_{prot})(M_{prot}/V) \tag{8.1}$$

since the first term in parentheses represents the specific volume of the protein and the second the reciprocal of V_M, and remembering that molecular weight is expressed in Da:

$$V'_p = 1.6604/(d_{prot} V_M). \tag{8.2}$$

If we assume a value of about 1.35 g cm^{-3} for the protein density, as a first approximation:

$$V'_p \approx 1.23/V_M \tag{8.3}$$

$$V'_{solv} = 1 - V'_p. \tag{8.4}$$

The most important information contained in V_M is not the solvent content, but the molecular mass of macromolecule in the crystal cell. Very often the number of molecules per asymmetric unit can be determined unambiguously, when the molecular weight of the protein is known, at least approximately. For monomeric proteins, or for those without the possibility of internal symmetry, the number of molecules in the cell is relevant only for the subsequent steps in structure determination. For multimeric proteins, composed of identical subunits, this number may be of relevant biological consequence: in fact, if some of the internal symmetry elements are coincident with the crystallographic ones, the relative arrangement of such subunits in the molecule is immediately available.[15]

The experimental determination of the density of the protein crystal, along with that of the dried content of the crystal itself, can allow a quite accurate determination of the molecular mass of the protein contained in the asymmetric unit, and consequently the molecular weight of the protein. These techniques have been reviewed by Matthews,[16] but the reader must be warned that the experimental measurement of the density of such kinds of crystals is difficult and the results very often uncertain.

Preparation of isomorphous heavy-atom derivatives

Isomorphous heavy-atom derivatives are of paramount importance for the solution of the phase problem in protein crystallography.[1–3] Their prepara-

Table 8.2. Representatives of heavy-atom compounds used in protein crystallography for preparation of derivatives[2,3] (Ac = acetate)

Classes	Examples
Platinum compounds	K_2PtCl_4, $K_2Pt(CN)_4$, $K_2Pt(NO_2)_4$, $Pt(NH_3)_4Cl_2$ or $-Br_2$, $PtCl_4$, $PtCl_2$, Pt(ethylenediamine)Cl_2
Mercury compounds	$HgCl_2$, $HgAc_2$, mersalyl acid, *p*-chloromercuribenzoic acid, methylmercury acetate, ethylmercury chloride
Uranyl salts	$UO_2(NO_3)_2$
Rare earths	$SmAc_3$, $EuCl_3$
Others	$KAuCl_4$, $KAu(CN)_2$, $PbAc_2$, $PbCl_2$, $(CH_3)_3PbAc$ $AgNO_3$ K_3IrCl_6

tion is made possible by their quite high solvent content: the presence of channels of disordered or only partially ordered liquid allows the diffusion of relatively small compounds into the crystal. Reactions among the diffused compounds and eventual accessible reactive sites of the protein can take place. In some special cases, protein molecules are derivatized in solution and subsequently crystallized, but the former procedure is simpler and it is advisable to try it first. In practice, the crystal is soaked in a solution of its mother liquid in which the heavy-atom compound has been dissolved. A list of the more commonly used heavy-atom derivatives is given in Table 8.2. Concentration of heavy atoms and time of soaking are the most relevant variables: they can range from 0.001 M to 0.1 M and from a few hours to several days (these numbers must be considered only a rough indication, since extreme cases have been reported). These conditions strongly depend on the protein under study, on the precipitating agents, on the pH used, and on the temperature.

1. *pH.* At high pH (greater than 9) the acidic groups of the protein are mostly negatively charged, and potentially they can react with cations. At the same time, many heavy atoms form insoluble hydroxides. At low pH, on the other hand, potentially reactive groups will be protonated and prevented from reacting.

2. *Precipitating agents.* High salt concentration is not the ideal medium for heavy-atom reaction with proteins, not only because the solubilization of the compound can be prevented by a very high salt concentration, but also because ions in the solution will compete with the protein. Polyethylene glycol solutions on the contrary provide favourable conditions for heavy-atom reactions, since PEG does not react with most of the compounds commonly used.

3. *Temperature.* There is an obvious kinetic effect of temperature on heavy-atom reactions, which are slower at 4 °C than at room temperature.

Very often preparation of heavy-atom derivatives is a trial and error technique, unless some special features of the macromolecule are known a priori. When the three-dimensional structure has been solved, a rationale for the behaviour of the specific system can be inferred: unfortunately, at that time it becomes useless.

How isomorphous are isomorphous derivatives?

The term 'isomorphous derivative' ideally indicates a crystal where some solvent molecules have been replaced by some group of atoms with more electrons, without any alteration of the structure of the protein or of the crystal lattice itself. In practice, this never happens: the introduction of a bulky compound which interacts with some of the atoms on the surface of the protein will give rise to local movements and displacements of atomic groups, at least in the close vicinity of the binding site.

To check if some reaction has taken place after the soaking of a crystal in the heavy-atom solution, diffraction data have to be collected. To avoid wasting time, a visual inspection of a single precession photograph, compared with the same zone of the native protein, is often enough to judge about the quality of the derivative. A more quantitative test is given by the comparison of structure factor data: an agreement index, R (see pp. 309 and 312), between native and derivative data can be calculated and the entity of the substitution can be roughly evaluated from it; experience indicates that a value from 0.15 to 0.25 can be considered a reasonable agreement factor. A smaller value is an index of little substitution, too high a one must be regarded suspiciously, as a clue to non-isomorphism.

Lack of isomorphism can be confirmed by other parameters: a change in cell length of less than about 1.5 per cent can be tolerated to a resolution of about 3 Å, but it could be deleterious at 2 Å. Furthermore, the mean difference among structure factors amplitudes must be practically constant over the entire range of θ used: quite large differences at low resolution and very low at high resolution indicate disordered heavy-atom binding, whilst an increase of the differences with d spacing is an index of non-isomorphism.[1]

The solution of the phase problem

The following section is entirely concerned with primary phasing, that is the determination of approximate phase angles of protein structure factors. First the two main methods used to solve the phase problem in protein crystallography, isomorphous replacement and anomalous scattering, are described, without taking into account the experimental errors. Their treatment is introduced later, along with the techniques of phase refinement. Methods to improve the electron density maps and rotation and translation functions, that in some cases can help to solve the phase problem, have been left for the end of the chapter. A brief mention of the application of direct methods to macromolecular crystallography is given in the last subsection.

The isomorphous replacement method

The method of isomorphous replacement is a corner-stone in protein crystallography: not only has it been used to solve the phase problem for the first protein structures, but it remains, with few exceptions, the only procedure available to solve the structure of completely new proteins.[17]

Let $\boldsymbol{F}_{\mathrm{P}}$ be the structure factor of the native protein, F_{P} its magnitude and φ_{P} its phase. $\boldsymbol{F}_{\mathrm{PH}}$, F_{PH}, and φ_{PH} are the corresponding quantities of the heavy-atom derivative. If we assume a perfect, 'ideal' isomorphism, the relationships between $\boldsymbol{F}_{\mathrm{P}}$ and $\boldsymbol{F}_{\mathrm{PH}}$ is illustrated in Fig. 8.1 and can be written:

$$\boldsymbol{F}_{\mathrm{PH}} = \boldsymbol{F}_{\mathrm{P}} + \boldsymbol{F}_{\mathrm{H}} \tag{8.5}$$

where $\boldsymbol{F}_{\mathrm{H}}$ is a vector representing the contribution of the heavy atom 'alone'.

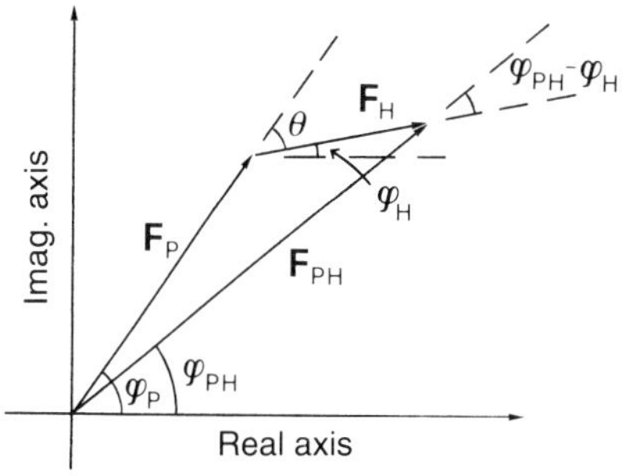

Fig. 8.1. Representation in the Argand plane of the relationship among the structure factor of native protein ($\boldsymbol{F}_{\mathrm{P}}$), heavy atom ($\boldsymbol{F}_{\mathrm{H}}$), and heavy-atom derivative ($\boldsymbol{F}_{\mathrm{PH}}$), assuming a perfect isomorphism of the two crystals.

Our final goal is to derive the value of φ_{P}, but the only quantities in the figure that can be measured are F_{P} and F_{PH}. On the contrary, $\boldsymbol{F}_{\mathrm{H}}$ cannot be measured, but since it represents the contribution of few atoms, it can be calculated if their atomic coordinates and thermal parameters are known. Let us first describe a way to determine the positions of heavy atoms and, from them, to calculate $\boldsymbol{F}_{\mathrm{H}}$ and derive an estimate for φ_{P}.

The determination of heavy-atom positions

For a small molecule, the heavy-atom positions can be determined via a Patterson map calculated using measured amplitudes (see p. 328). The same map is not useful for a macromolecular crystal: even for a small protein, the ratio among the number of electrons of the heavy atoms and those of the macromolecule is so low, that a Patterson map calculated with the coefficients of the protein derivative will look absolutely meaningless. But if the structure factor amplitudes of two isomorphous crystals are known, various kinds of **difference-Patterson** syntheses can be calculated.

The 'true' difference-Patterson[18] synthesis would be that calculated using as coefficients $|F_{\mathrm{PH}}^2 - F_{\mathrm{P}}^2|$. This map represents the difference between the Patterson of the derivative minus the Patterson of the native protein. In fact, remembering (8.5):

$$\begin{aligned} F_{\mathrm{PH}}^2 - F_{\mathrm{P}}^2 &= (\boldsymbol{F}_{\mathrm{P}} + \boldsymbol{F}_{\mathrm{H}})(\boldsymbol{F}_{\mathrm{P}}^* + \boldsymbol{F}_{\mathrm{H}}^*) - \boldsymbol{F}_{\mathrm{P}}\boldsymbol{F}_{\mathrm{P}}^* \\ &= F_{\mathrm{H}}^2 + \boldsymbol{F}_{\mathrm{H}}\boldsymbol{F}_{\mathrm{P}}^* + \boldsymbol{F}_{\mathrm{P}}\boldsymbol{F}_{\mathrm{H}}^*. \end{aligned} \tag{8.6}$$

The right-hand side of eqn (8.6) shows that the Patterson map calculated with these coefficients will contain peaks corresponding to heavy-atom heavy-atom vectors, F_{H}^2, and to heavy-atom protein-atom vectors, the mixed terms.

A more common choice among protein crystallographers is to calculate a Patterson synthesis using coefficients $(F_{\mathrm{PH}} - F_{\mathrm{P}})^2$. This was originally called **modulus difference squared synthesis**, but is commonly known as **isomorphous difference-Patterson.**[18] The resultant map has several interesting features, but the main reason for its success is that it is more representative of the heavy-atom situation. In particular, in the centrosymmetric case (that is for centric† projections in a non-centrosymmetric space group), since $\boldsymbol{F}_{\mathrm{PH}}$ and $\boldsymbol{F}_{\mathrm{P}}$ are collinear, these coefficients represent a true estimate of the value of $\boldsymbol{F}_{\mathrm{H}}$. In fact:

$$F_{\mathrm{H}} = |F_{\mathrm{PH}} \pm F_{\mathrm{P}}|. \tag{8.7}$$

† The word centric is used here in a general sense, indicating that phases are restricted to two values, not necessarily 0 or π.

Since F_H is usually small if compared with the modulus of the structure factors of protein and derivative, in general the relationship

$$F_H = |F_{PH} - F_P| \tag{8.8}$$

will be satisfied. In few cases, when F_{PH} and F_P are both very small, it may happen that $F_H = F_{PH} + F_P$. This is called a **cross-over**, and it can be removed from the calculation. For a centric projection a Patterson synthesis calculated with coefficients (8.8) will correspond to the Patterson map of the heavy atoms alone, that is it will contain peaks corresponding only to the vectors relating heavy-atom positions. Methods to derive the atomic coordinates from a Patterson map are described on pp. 328 to 335. In favourable cases, i.e. when more than one centric projection is available, the three coordinates of the heavy atoms can be obtained.

For general reflections, $|F_{PH} - F_P|$ will not be a correct estimate for F_H. From Fig. 8.1:

$$F_H^2 = F_P^2 + F_{PH}^2 - 2F_{PH}\, F_P \cos(\varphi_P - \varphi_{PH}). \tag{8.9}$$

If the phase angles of protein and derivative structure factors are very similar, that is if the two vectors $\boldsymbol{F}_{PH}$ and $\boldsymbol{F}_P$ are nearly collinear, the cosine term is close to 1 and:[17]

$$F_H^2 \approx F_P^2 + F_{PH}^2 - 2F_{PH}F_P = (F_{PH} - F_P)^2. \tag{8.10}$$

The probability distribution of the values of $\varphi_P - \varphi_{PH}$ has been analysed by Sim[19,20] (see also p. 393): the larger the value of the product $F_{PH}F_P$, the more likely the difference $|\varphi_P - \varphi_{PH}|$ is to be small, an vice versa. Equation (8.10) does not represent a good approximation for small structure factors, but the latter will give a small contribution to the Fourier transform. In practice, a Patterson map calculated with coefficients (8.8) will tend to the Patterson-difference of the heavy atoms alone in the non-centrosymmetric case too (for a general discussion of the features of difference-Patterson maps, see Ramachandran and Srinivasan[18]).

The single isomorphous replacement (SIR) method

Once the position of heavy atoms is more or less accurately known, the calculated value of $\boldsymbol{F}_H$ can be used to estimate the phase angle of the structure factor of the native protein. From Fig. 8.1, using simple trigonometric considerations,† we can write:

$$F_{PH}^2 = F_P^2 + F_H^2 + 2F_PF_H \cos\theta, \tag{8.11}$$

$$\cos\theta = (F_{PH}^2 - F_P^2 - F_H^2)/2F_PF_H; \tag{8.12}$$

θ corresponds to $\varphi_P - \varphi_H$, and since φ_H is known, φ_P, the phase angle of $\boldsymbol{F}_P$, can be derived:

$$\varphi_P = \varphi_H + \cos^{-1}[(F_{PH}^2 - F_P^2 - F_H^2)/2F_PF_H]. \tag{8.13}$$

Equation (8.13) contains a cosine term, consequently two solutions exist for φ_P. This ambiguity is illustrated in Fig. 8.2, where two circles of radius F_P and F_{PH} are drawn with centre respectively in O and at the end of vector $-\boldsymbol{F}_H$: the intersections of the two circles represent the two possible solutions

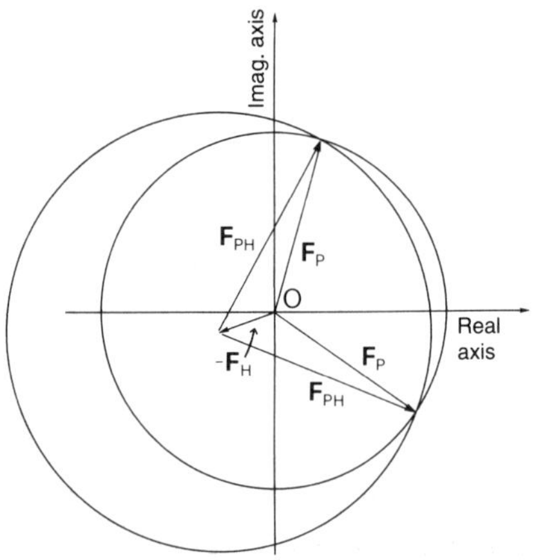

Fig. 8.2. Graphic construction of S.I.R. method, showing the two possible values for the phase angle of the native structure factor $\boldsymbol{F}_P$. The structure factor of the protein is a vector lying on a circle of radius F_P and centre O. If a circle of radius F_{PH}, the modulus of the structure factor of the heavy-atom derivative, is drawn with centre in the end of the vector $-\boldsymbol{F}_H$, the intersection of the two circles represents the two possible solutions of (8.13).

† An alternative way of deriving eqn (8.11) is given in Appendix 8.A.

for φ_P. It is important to notice that this problem is immediately solved for restricted phase reflections, for which eqn (8.12) becomes:

$$(F_{PH}^2 - F_P^2 - F_H^2)/F_P F_H = \pm 1 \quad (8.14)$$

or

$$F_{PH}^2 = F_P^2 + F_H^2 \pm F_P F_H = (F_H \pm F_P)^2. \quad (8.15)$$

Since the value of $\boldsymbol{F}_H$ is known from calculation and the values of F_{PH} and F_P from observation, the relative sign of F_P with respect to F_H can be determined unambiguously from the comparison of the magnitude of the three vectors (Fig. 8.3). For general reflections this is unfortunately not true, and the ambiguity between the two possible values for the phase angle must be solved with other techniques.

In principle, a three-dimensional electron density map could be calculated, using the two possible phase values: this kind of synthesis is called an SIR map and it will contain information about the true structure plus noise.[21] In fact, if we call φ_{PT} the correct phase value of φ_P and φ_{PW} the wrong one, looking at Fig. 8.4 we can write:

$$2\boldsymbol{F}_{SIR} = F_P \exp(i\varphi_{PT}) + F_P \exp(i\varphi_{PW}). \quad (8.16)$$

Since $\varphi_H = 1/2(\varphi_{PT} + \varphi_{PW})$, then:

$$2\boldsymbol{F}_{SIR} = F_P \exp(i\varphi_{PT}) + F_P \exp(2i\varphi_H) \exp(-i\varphi_{PT}). \quad (8.17)$$

From (8.17) it is evident that $\boldsymbol{F}_{SIR}$ will approach the correct value of $\boldsymbol{F}_P$ if $\varphi_H \approx \varphi_{PT}$. But since we have no reason to believe the numerical values of φ_H and φ_{PT} are correlated, the first term of (8.17) will give the correct electron density, while the second one will only contribute to noise. In practice the level of noise in this kind of map is so high that its interpretation is extremely difficult.

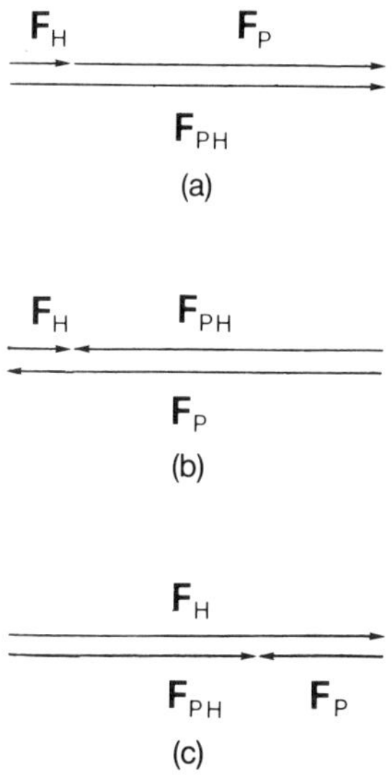

Fig. 8.3. The sign of reflections of centric zones can be fully determined using only one heavy-atom derivative (after Blundell and Johnson[1]). When F_H is small compared with F_P and F_{PH}, case (a) and (b) apply: if $F_P < F_{PH}$, the sign of F_P is equal to that of F_H; if $F_P > F_{PH}$ the sign of F_P is reversed. The so-called 'cross over', represented in (c), may occur when F_H is large and F_P is small: despite the condition $F_P < F_{PH}$, the sign of F_P is reversed with respect to F_H.

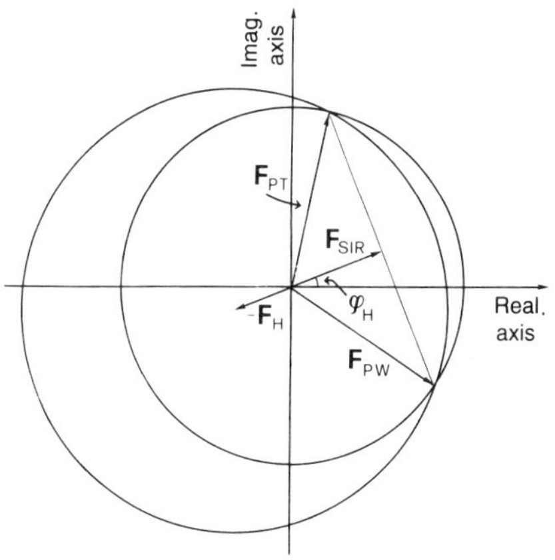

Fig. 8.4. The same diagram shown in Fig. 8.2, except that now the 'true' and the 'wrong' value for the native structure factors are explicitly indicated as $\boldsymbol{F}_{PT}$ and $\boldsymbol{F}_{PW}$ respectively. It can be seen that the further the two vectors are from each other, the more the S.I.R. phase (corresponding to φ_H) is different from φ_P.

The classical solution of the problem of phase ambiguity: the MIR technique

The multiple isomorphous replacement technique, MIR, has been the first method used to solve this phase ambiguity, and it remains the more common and is widely adopted. It is based on the preparation of a second heavy-atom derivative having heavy atoms bound to the protein in position(s) different from those of the first one. Let us assume we have measured coefficients from two derivatives, called F_{PH1} and F_{PH2}, and in some way we have calculated $\boldsymbol{F}_{H1}$ and $\boldsymbol{F}_{H2}$ (that means we know the positions of the heavy atoms of the two derivatives, relative to the same origin). The complete solution of the phase problem can be illustrated using a diagram due to Harker,[22] illustrated in Fig. 8.5: a circle of radius F_P centred on the origin O of an Argand plane is drawn, and from O the two vectors $-\boldsymbol{F}_{H1}$ and $-\boldsymbol{F}_{H2}$; they are in turn the centres of two circles of radii F_{PH1} and F_{PH2} respectively. Since the two derivatives are assumed perfectly isomorphous, the three circles must intersect in one point (B in Fig. 8.5): the vector OB will coincide with the structure factor of the protein. The graphical construction just described is equivalent to solving a pair of

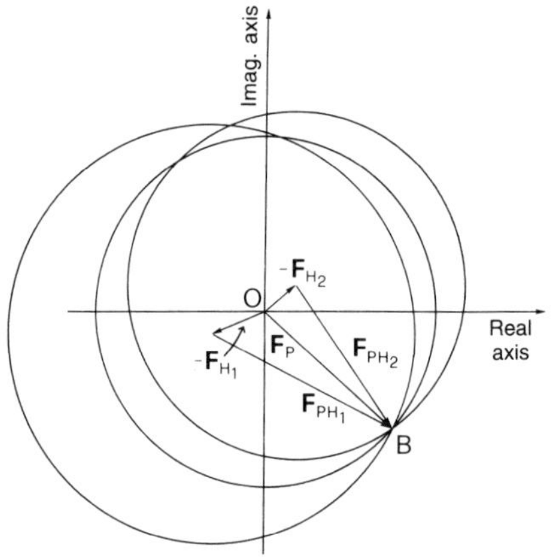

Fig. 8.5. Harker construction illustrating the M.I.R. method. A second circle of radius F_{PH2} and centred at the end of vector $-\boldsymbol{F}_{H2}$ has been added to the construction of Fig. 8.2. The three circles have only one common intersection, representing the solution of the two equations (8.18).

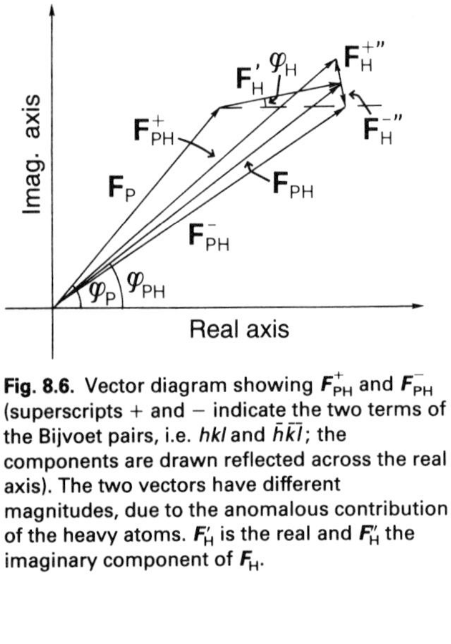

Fig. 8.6. Vector diagram showing $\boldsymbol{F}_{PH}^+$ and $\boldsymbol{F}_{PH}^-$ (superscripts + and − indicate the two terms of the Bijvoet pairs, i.e. *hkl* and $\bar{h}\bar{k}\bar{l}$; the components are drawn reflected across the real axis). The two vectors have different magnitudes, due to the anomalous contribution of the heavy atoms. $\boldsymbol{F}'_H$ is the real and $\boldsymbol{F}''_H$ the imaginary component of $\boldsymbol{F}_H$.

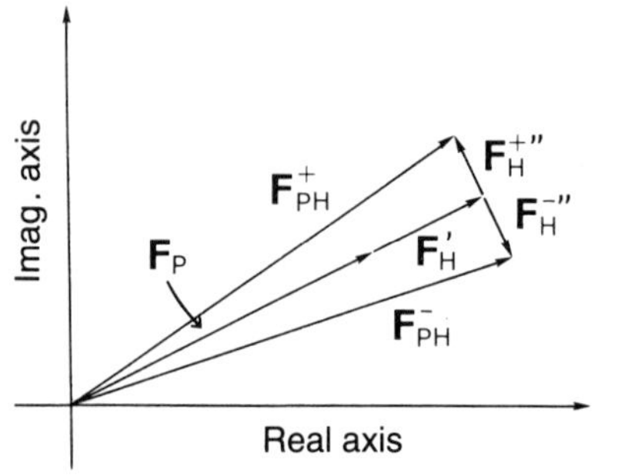

Fig. 8.7. For restricted-phase reflections, if only one type of anomalous scatterer is present, $\boldsymbol{F}_{PH}^+$ and $\boldsymbol{F}_{PH}^-$ have the same magnitude, independent of the size of the imaginary component.

simultaneous equations like (8.13):

$$\varphi_P = \varphi_{H1} + \cos^{-1}[(F_{PH1}^2 - F_P^2 - F_{H1}^2)/2F_PF_{H1}],$$
$$\varphi_P = \varphi_{H2} + \cos^{-1}[(F_{PH2}^2 - F_P^2 - F_{H2}^2)/2F_PF_{H2}]. \quad (8.18)$$

In principle, two different heavy-atom derivatives supply a unique, unambiguous value for the phase angle of the native protein. In practice, errors from various sources prevent system (8.18) from having an exact solution.

Anomalous scattering: a complementary (or alternative) approach to the solution of the phase problem

According to Chapter 3 (p. 167), the structure factor may be rewritten as the sum of two parts:

$$\boldsymbol{F}(hkl) = \boldsymbol{F}'(hkl) + i\boldsymbol{F}''(hkl) \quad (8.19)$$

where the real part of anomalous scattering (the so-called anomalous dispersion) is included in $\boldsymbol{F}'(hkl)$. If $\boldsymbol{F}''(hkl)$ is not zero, the practical result is the breakdown of Friedel's law: $\boldsymbol{F}(hkl)$ is no longer equal to $\boldsymbol{F}(\bar{h}\bar{k}\bar{l})$ and the Friedel pairs are more correctly called Bijvoet pairs. Anomalous scattering is negligible for light atoms like carbon, nitrogen, or oxygen (that is, nearly all the atoms of a protein) and can be of practical use for a native protein molecule only in special cases, when heavy ions are bound to it. Even if for a very small protein the sulphur's anomalous scattering has been used to solve a structure *ab initio*,[23] in general the anomalous scattering phenomenon becomes relevant for heavy-atom derivatives. Equation (8.5) must be in fact modified, as shown in Fig. 3.15 and Fig. 8.6:

$$\boldsymbol{F}_{PH}^+ = \boldsymbol{F}_P^+ + \boldsymbol{F}_H^{+\prime} + \boldsymbol{F}_H^{+\prime\prime} = \boldsymbol{F}_P^+ + \boldsymbol{F}_H^+$$
$$\boldsymbol{F}_{PH}^- = \boldsymbol{F}_P^- + \boldsymbol{F}_H^{-\prime} + \boldsymbol{F}_H^{-\prime\prime} = \boldsymbol{F}_P^- + \boldsymbol{F}_H^- \quad (8.20)$$

where superscripts + and − stands for (hkl) and $(\bar{h}\bar{k}\bar{l})$, respectively. If the heavy-atom derivative contains only one kind of anomalous scatterer, vectors $\boldsymbol{F}'_H$ and $\boldsymbol{F}''_H$ are mutually perpendicular, and if we call φ_H the phase angle of vector $\boldsymbol{F}'_H$, $\varphi_H \pm \pi/2$ will be that of $\boldsymbol{F}''_H$. This fact can be considered practically always true for protein derivatives, and has the important consequence that anomalous effect cannot be observed if the vectors $\boldsymbol{F}_P$ and $\boldsymbol{F}_H$ are collinear, for example for all the centric reflections (see Fig. 8.7).

An accurate measurement of the Bijvoet pairs allows the calculation of an anomalous-difference Patterson synthesis, with coefficients:

$$\Delta F_{ano} = (F_{PH}^+ - F_{PH}^-)^2. \quad (8.21)$$

Notice that we do not need native coefficients for the calculation, so (8.21) is also useful when an anomalous scatterer is present in the native protein crystal. A map calculated with coefficients (8.21), supposing the anomalous contribution F''_H is small compared with F_P, which is very often true in our case, will contain maxima corresponding to vectors relating the positions of anomalous scatterers. This synthesis was for example used by Rosmann[24] to determine the Hg position in two derivatives of horse haemoglobin. Other kind of anomalous Patterson maps with coefficients different from

(8.21) can be produced, like the α-anomalous or the β-anomalous synthesis.[18]

The most important application of anomalous scattering in protein crystallography remains the possibility of solving the phase ambiguity discussed in the previous paragraph.[25–27] Let us assume we have only one heavy-atom derivative, and we have measured F^{+}_{PH}, F^{-}_{PH}, and F_P.† Moreover, the position of the heavy atom(s) are known. Looking at Fig. 8.6, an expression for the phase angle of the derivative, whose complete derivation is given in Appendix 8.A, can be written:

$$F^{+}_{PH} - F^{-}_{PH} = 2F''_{H} \sin(\varphi_{PH} - \varphi_{H}) \tag{8.22}$$

$$\varphi_{PH} = \varphi_{H} \pm \sin^{-1}[(F^{+}_{PH} - F^{-}_{PH})/2F''_{H}]. \tag{8.23}$$

Equation (8.23) gives again two possible solutions for φ_{PH}, but since it contains a sine term, its information is complementary to that of (8.13): the phase angle can now be fully determined. Information from both sources, isomorphous replacement and anomalous scattering, is used in Fig. 8.8, where the role of the second derivative of Fig. 8.5 is played by anomalous scattering.

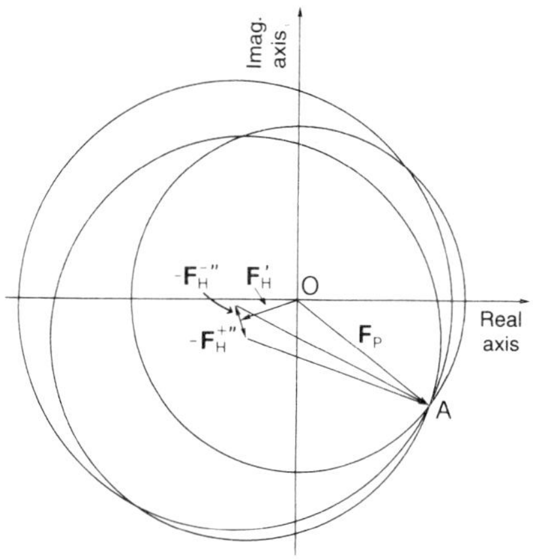

Fig. 8.8. A graphical representation of the solution of the phase problem using one heavy-atom derivative and its anomalous absorption component. Two circles of radius F^{+}_{PH} and F^{-}_{PH} are drawn at the end of vectors $-\boldsymbol{F}^{+}_{H}$ and $-\boldsymbol{F}^{-}_{H}$ respectively. OA is the native protein structure factor, whose phase angle is uniquely determined by the intersection of the three circles.

The use of anomalous scattering in the determination of the absolute configuration of the macromolecule

Two isomorphous derivatives are, in principle, enough to determine the correct value of the phase angle φ_P of the structure factor, but the M.I.R. method alone does not allow us to select the correct enantiomorph for the protein. Once a right-handed system has been chosen in indexing the reciprocal lattice, the initial arrangement of the heavy atoms for the first derivative is arbitrary. In other words, the phase of $\boldsymbol{F}_H$ can be φ_H, as in Fig. 8.2, or $-\varphi_H$. Once a hand has been arbitrarily selected for a heavy-atom derivative, all the others must simply be consistent with it, but they can all be relative to the wrong enantiomorph. If this is the case, the electron density map will represent the mirror image of the correct structure, that is all the amino acids will have D configuration. When a medium (or high) resolution electron density map can be produced, some overall features may indicate that the wrong enantiomorph has been chosen (for example, α-helices will appear left-handed instead of right-handed—the amino acid configuration and α-helices are described on p. 574). Then the problem can be immediately corrected by recalculating the phases with the heavy atoms arranged in the other way.

However, anomalous scattering differences can be used to identify from the beginning the correct enantiomorph, using an extension of the method proposed for small molecules by Bijvoet.[28] We have seen in the previous subsection that, at least in principle, structure factors for one single heavy-atom derivative along with its anomalous differences can solve the phase ambiguity of the protein. This is no longer valid if two possibilities are taken into account for the value of φ_H, since now there are two possible

† Readers must be warned that measurement of Bijvoet pairs requires very accurate data collection, since anomalous differences are very often of the same order of magnitude as the experimental errors on intensities. Synchrotron radiation is extremely helpful in this respect: the appropriate wavelengths that optimize the anomalous effect can be selected (see p. 239).

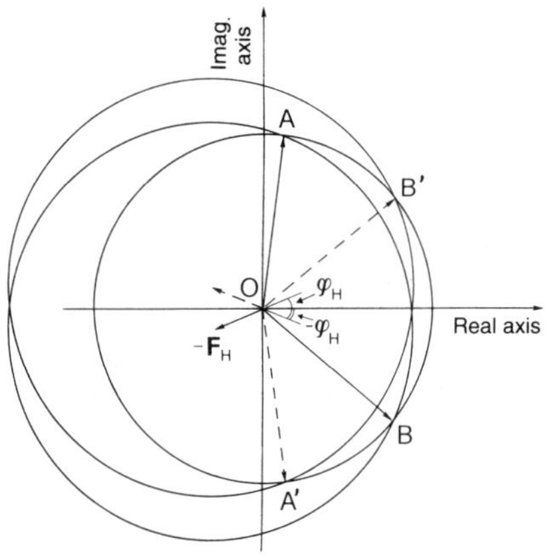

Fig. 8.9. Graphical illustration of the ambiguity in the choice of enantiomorph. This figure is equivalent to Fig. 8.2, except that now two possible arrangements of the heavy atoms for the same derivative are shown. Vectors OA and OB are the two possible structure factor values for the native protein if φ_H is selected as the phase of the heavy-atom derivative; vectors OA′ and OB′ those corresponding to $-\varphi_H$, that is to the enantiomorph.

value of φ_P associated with every φ_H. Let us assume the solution for the phase angle of the derivative structure factor, from (8.A.7) and (8.A.15), is $\varphi_{PH} = \varphi_H + \theta$. We are left with another ambiguity, since the original choice of φ_H was arbitrary: the two possibilities are in fact $\varphi_{PH} = \varphi_H + \theta$ or $\varphi_{PH} = -\varphi_H - \theta$, and they determine the selection of the enantiomorph (Fig. 8.9).

At least two heavy-atom derivatives and anomalous differences for one of them are necessary to select the correct hand of the molecule: using phases from one derivative along with anomalous contribution and the heavy atoms in the two possible arrangements, two difference-Fourier of a second one are calculated. Peaks in the difference map calculated with atoms in the correct hand will be reinforced, those with the wrong one lowered.[1]

The treatment of errors

In the previous paragraphs, we assumed that all the measurements were error-free and that a perfect isomorphism existed among crystals of the native protein and derivatives. This is obviously not realistic, since F_P and F_{PH} are experimentally measured quantities and $\boldsymbol{F}_H$ is calculated from approximate heavy-atom positions. Experimental errors can be considered to arise from three different sources:

(1) errors in intensities, mainly attributable to inaccuracy in measurements, scaling etc.;

(2) errors in position, occupancy, and temperature factor of heavy-atoms;

(3) lack of isomorphism, due to displacement of protein atoms and of solvent molecules close to the heavy-atom binding site, etc.

As a consequence, the circles of Fig. 8.5 are not expected to intersect at the same point and the system of equations (8.18) will not have an exact solution (Fig. 8.10). Following Terwilliger and Eisenberg,[29] let us denote the different contributions to the total error deriving from points (1), (2), and (3) as σ, η, and μ, respectively. It must be noticed that, whereas σ_P and σ_{PH}, the standard deviations for F_P and F_{PH}, are routinely evaluated (see pp. 271 and 280) an a priori estimate of η and μ is impossible.

The systematic treatment of errors in the isomorphous replacement was pioneered by Blow and Crick.[30] They made two basic assumptions: all the errors reside in the calculated structure factor of the derivative, $F_{PH(calc)}$; errors follow a Gaussian distribution. Consequently, a phase probability distribution for derivative j is given by:

$$P_{j(iso)}(\varphi_P) = n_j \exp\left[-(F_{PH(obs)} - F_{PH(calc)})^2/2E_j^2\right]. \qquad (8.24)$$

The difference $F_{PH(obs)} - F_{PH(calc)}$ in (8.24) is called **lack of closure** and is usually denoted by ε_j; its physical meaning is illustrated in Fig. 8.11, which is a realistic picture of what was idealized in Fig. 8.1: now the triangle formed by the three vectors $\boldsymbol{F}_P$, $\boldsymbol{F}_{PH}$, and $\boldsymbol{F}_H$ does not close. The lack of closure depends from the phase of the protein, since:

$$\varepsilon_j = F_{PH(obs)} - F_{PH(calc)} = F_{PH(obs)} - |F_{P(obs)} \exp(i\varphi_P) + \boldsymbol{F}_H|. \qquad (8.25)$$

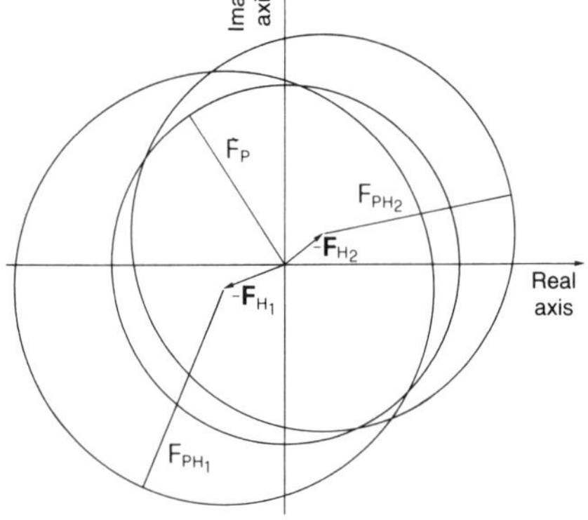

Fig. 8.10. Harker construction analogous to Fig. 8.5. where, due to lack of isomorphism and errors in the data, the three circles of radius F_P, F_{PH1}, and F_{PH2} do not intersect at the same point.

n_j in (8.24) is a normalizing factor and E_j is a measure of the total error. The choice of E_j is crucial and not obvious. In practice, the value commonly

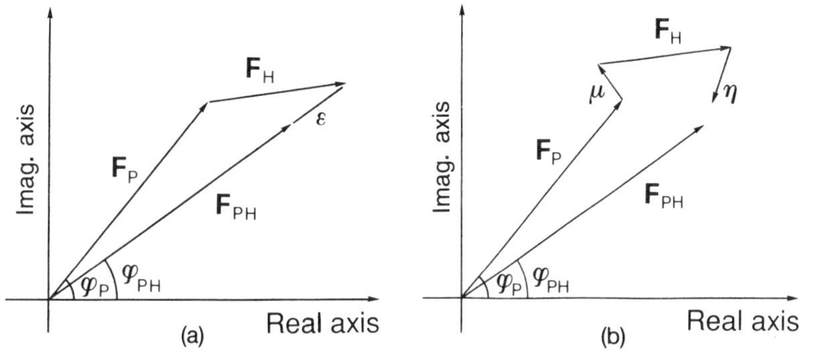

Fig. 8.11. (a) Representation of the lack of closure ε, following the Blow and Crick assumption that all the errors reside in the F_{PH} amplitude. As a consequence of it, $F_{PH(calc)}$ is equal to $F_{PH(obs)} + \varepsilon$. (b) components of the error described in (a) are now explicitly shown: μ is the lack of isomorphism and η the difference between the true and the estimated value of $\boldsymbol{F}_H$. The third source of error, the inaccuracy in the measurement of F_P and F_{PH}, prevents the closure of the polygon.

used is:[31]

$$E_j = \langle (F_{PH(obs)} - F_{PH(calc)})^2 \rangle. \tag{8.26}$$

The right-hand side of (8.26) represents the mean square value of the lack of closure residual. Since it depends from the resolution, it is evaluated in ranges of $\sin \theta/\lambda$.

An a priori derivation of (8.24) along with that of E_j^2 for centric and acentric reflections has been obtained by Terwilliger and Eisenberg[29] as a function of all the different sources of errors:

$$E^2_{j(\text{cent})} = \langle \eta^2 \rangle /2 + \langle \mu^2 \rangle /2 + \sigma_P^2 + \sigma_{PH}^2, \tag{8.27a}$$

$$E^2_{j(\text{acent})} = \langle \eta \rangle^2 + \langle \mu^2 \rangle + \sigma_p^2 + \sigma_{PH}^2. \tag{8.27b}$$

Equations (8.27a) and (8.27b) are not of direct use, since we lack a value for η and μ, but from them a theoretical justification of (8.26) can be obtained.[29]

If more than one derivative is present, (8.24) can be modified to give the total probability distribution of the phase of a reflection:

$$P(\varphi_P) = n \exp\left(-\sum_j (\varepsilon_j^2/2E_j^2) \right). \tag{8.28}$$

The sum is extended to all the j derivatives used in calculating the phases. A diagram showing $P_j(\varphi_P)$ for the hypothetical derivatives of Fig. 8.10 is reported in Figs. 8.12(a) and (b), and the total probability in Fig. 8.12(c).

The lack of closure for anomalous scattering measurements assumes a different form:[25,26,29]

$$\varepsilon_{\text{ano}} = \varepsilon^+ - \varepsilon^- = -(F^+_{PH} - F^-_{PH})(1 + F_P/F_{PH}), \tag{8.29}$$

but the probability for the phase of the structure factor due to the anomalous scattering is analogous to (8.24):

$$P_{\text{ano}}(\varphi_P) = n \exp(-\varepsilon^2_{\text{ano}}/2E^2_{\text{ano}}). \tag{8.30}$$

E_{ano} is usually smaller than E_{iso} and can be estimated for example from the agreement among centric reflections.[25] If isomorphous and anomalous information are available for the derivative j, they can be combined to give a total probability:

$$P_j(\varphi_P) = (P_{j(\text{iso})}(\varphi_P))(P_{j(\text{ano})}(\varphi_P)). \tag{8.31}$$

On the basis of a different formulation of the error model, Hendrickson and

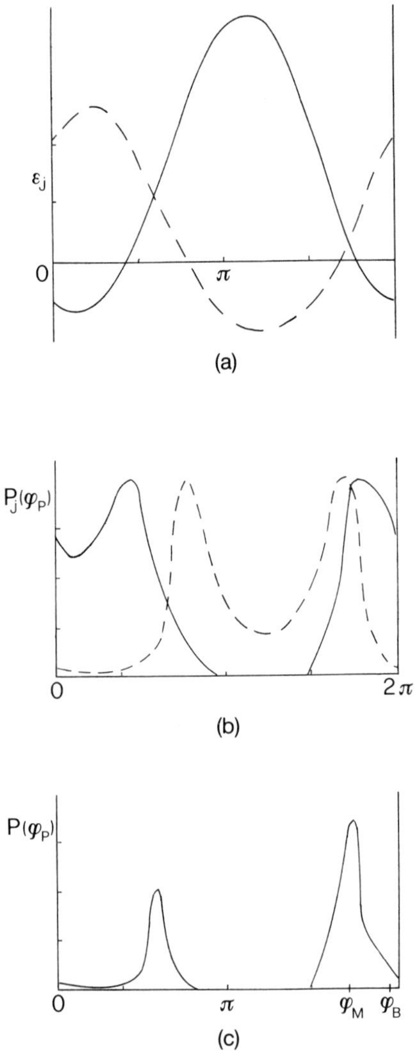

Fig. 8.12. (a) Lack of closure, ε_j, for the two derivatives shown in Fig. 8.10. ε_j assumes the value 0 at the intersections of the circles of radius F_P and F_{PHj}. Since the three circles do not intersect at the same point, this happens for values of ϕ_P different for each derivative. (b) $P_j(\varphi_P)$, calculated on an arbitrary scale using (8.24), for the two derivatives: $P_j(\varphi_P)$ has a maximum every time $\varepsilon_j = 0$. (c) Combined probability for the two derivatives. φ_M represents the most probable phase, φ_B the Best phase, that is the centroid of the probability (see text for more details).

Latmann[32] proposed a different representation of the phase probability distribution:

$$P(\varphi_P) = \exp\,(K + A\cos\varphi_P + B\sin\varphi_P + C\cos 2\varphi_P + D\sin 2\varphi_P). \quad (8.32)$$

Equation (8.32) has the advantage of being completely general and that new information can easily be included. Moreover, the old probability can be expressed using coefficients A, B, C, and D through a least-squares fitting.[33] Since the overall probability distribution is given by the product of individual probabilities:

$$P_{tot}(\varphi_P) = \left(\prod P_{iso}(\varphi_P)\right)\left(\prod P_{ano}(\varphi_P)\right), \quad (8.33)$$

if every one is cast in the form (8.32), $P_{tot}(\varphi_P)$ can be obtained by the sum of the individual coefficients:

$$P_{tot}(\varphi_P) = \exp\Big(\sum_i K_i + \sum_i A_i\cos\varphi_P + \sum_i B_i\sin\varphi_P + \sum_i C_i\cos 2\varphi_P + \sum_i D_i\sin 2\varphi_P\Big). \quad (8.34)$$

From (8.33) or (8.34) we can calculate the probability distribution of the phase angle for every structure factor and use the most probable phase in the calculation of an electron density map. This is called the **most probable Fourier**, and in principle it would be a good approximation. Unfortunately, $P(\varphi_P)$ tends to be bimodal, and the use of the most probable phase can introduce large errors (see Fig. 8.12(c)). According to Blow and Crick,[30] the Fourier which is expected to have the minimum mean square difference from the 'true' Fourier is that obtained using the centroid of the probability distribution of $\boldsymbol{F}_P$, that is:

$$\boldsymbol{F}_{P(best)} = NF_P\int_0^{2\pi} P(\varphi_P)\exp\,(i\varphi_P)\,d\varphi_P \quad (8.35)$$

where N is a normalization factor:

$$N^{-1} = \int_0^{2\pi} P(\varphi_P)\,d\varphi_P. \quad (8.36)$$

Equation (8.35) is equivalent to:

$$\boldsymbol{F}_{P(best)} = mF_P\exp(i\varphi_{P(best)}) \quad (8.37)$$

where m is a weighting function called the **figure of merit:**

$$m = N\int_0^{2\pi} P(\varphi_P)\exp\,(i\varphi_P)\,d\varphi_P. \quad (8.38)$$

$\varphi_{P(best)}$ can be calculated easily by developing (8.38) and observing that $P(\varphi_P)$ can be sampled and the integral substituted by a sum:

$$N^{-1} \approx \sum_i P(\varphi_P), \quad (8.39)$$

$$m\cos\,(\varphi_{best}) \approx N\sum_i P(\varphi_P)\cos\,(\varphi_P), \quad (8.40a)$$

$$m \sin(\varphi_{\text{best}}) \approx N \sum_i P(\varphi_P) \sin(\varphi_P). \qquad (8.40b)$$

The sum is estimated for the i values of φ_P at which the probability is evaluated, usually at intervals of 10 or 20 degrees. The figure of merit is a value ranging from 0 to 1 and is a measure of the reliability of the phase φ_P: in principle, $m = 1$ means an error of 0° in the phase angle, $m = 0.5$ means an error of 60°.

The refinement of heavy-atom parameters

Once an approximate position of one or more heavy atoms has been found, the problem immediately arising is the refinement of their coordinates, thermal parameters, and occupancy. Refinement is made difficult by the lack of a correct value of $F_{H(obs)}$: we can only have estimates of it.

Historically, the first method of refinement of heavy-atom parameters was proposed by Rossmann,[17] who suggested minimization, using classical least-squares methods:

$$S = \sum w(F_{H(obs)} - F_{H(calc)})^2. \qquad (8.41)\dagger$$

$F_{H(calc)}$ can be calculated in the usual way from the known heavy-atom parameters, but a correct estimate of $F_{H(obs)}$ is available only for centric reflections (see p. 541), so (8.41) should in principle be employed to refine that class of reflections and the refined parameters used to calculate all the phases. Nevertheless, since (8.8) can be considered an approximate estimate of $F_{H(obs)}$, expression (8.41) has sometimes been used to refine acentric data.‡

Possibly the most used and most popular refinement scheme among protein crystallographers is the so-called **phase refinement**, introduced for the first time in the refinement of myoglobin.[34] The quantity minimized is:

$$S = \sum w[F_{PH(obs)} - F_{PH(calc)}]^2 \qquad (8.42)$$

which represents the least-squares minimization of the lack of closure, where:

$$F_{PH(calc)} = |F_{P(obs)} \exp(i\varphi_P) + \boldsymbol{F}_{H(calc)}|. \qquad (8.43)$$

In general w is chosen as the reciprocal of the total error E, calculated from (8.26) for the derivative being refined.§

† In the rest of the chapter, unless otherwise specified, the sums are extended to the appropriate set of reflections.

‡ If anomalous scattering data are available, a more accurate approximation of F_H (see Blundell and Johnson[1]) is represented by

$$F_H^2 = F_{PH}^2 + F_P^2 \pm 2F_P F_{PH}\{1 - [K(F_{PH}^+ - F_{PH}^-)/2F_P]^2\}^{1/2}$$

The − and + values are called lower (F_{HLE}) and upper (F_{HUE}) estimates, respectively. It has been shown that the lower estimate would represent the correct value of F_H for the large majority of the reflections. It must be considered, in any case, that when anomalous differences are low or the measurements are affected by large errors, coefficients (8.8) can be more safely used.

§ A modification of the phase refinement scheme, called MVFC (minimum variance Fourier coefficients) has been proposed by Sygusch.[35] The expression minimized is again (8.42), but the sine and cosine of the phase are considered as variables. The values of the trigonometric functions estimated in the refinement are used directly for the electron density calculation, without the need to postulate a probability distribution for the phase error.

Whatever refinement scheme is selected, different procedures can be followed: refining only centric reflections is quite safe, if reflections with restricted phase represent a good portion of the total. The number of variables is also important in this respect: in general, three positional parameters, the occupancy, and an isotropic temperature factor are varied for every atom, but since the last two parameters are highly correlated, they must be refined separately. If general reflections are used, in order to avoid bias it is advisable to refine the parameters of one derivative omitting it from the phase calculation, that is refine one derivative and calculate the phases with all the others.[36] This can obviously be done if several different derivatives are available.

One of the difficulties in heavy-atom refinement is the check of the correctness of the parameters used: this is particularly true at the beginning of the refinement, where it is sometimes difficult even to decide if the solution found is correct or not. Many quantities can be used to monitor the progress of refinement: none of them taken alone gives an absolute indication, but together they can be considered a reliable check. Different kinds of R factors have been defined:

$$R_C = \left(\sum |F_{PH} \pm F_P| - F_{H(calc)}|\right)\left(\sum |F_{PH} - F_H|\right)^{-1}, \tag{8.44}$$

$$R_K = \left(\sum |F_{PH(obs)} - F_{PH(calc)}|\right)\left(\sum F_{PH(obs)}\right)^{-1}, \tag{8.45}$$

$$R(\text{modulus}) = \left(\sum |F_{PH(obs)} - F_{PH(calc)}|\right)\left(\sum F_H\right)^{-1}. \tag{8.46}$$

The Cullis R factor, R_C (8.44), is calculated only for centric reflections. A value between 0.40 and 0.60 is generally considered reasonable, the only drawback being the low number of reflections used in the calculation. For general reflections, the Kraut R factor (8.45) is used instead. Since its numerator is the quantity being minimized in phase refinement, R_K can be considered a good parameter to monitor the course of refinement. Unfortunately, a high degree of substitution of the derivative implies a statistically large R_K, and a very low substitution a small one, so the Kraut R factor cannot be considered an indication of the correctness of the variables being refined.†

An idea of the relative importance of a single derivative in the phase calculation is given by the R(modulus), defined by (8.46), which is the lack of closure divided by the modulus of the heavy-atom structure factor. If its value is greater than one, the circles of Fig. 8.8 will never intersect and (8.18) will not have any solution at all. R(modulus), if calculated in shells of resolution, gives an index of the utility of that derivative at different resolutions. Often the reciprocal of R(modulus), which is called the **phasing power**, is quoted. Its value must be bigger than one.

The last important quantity used in refinement is the figure of merit, defined in (8.38). The figure of merit gives a direct estimate of the errors in

† In the first refinement scheme, R_{FHLE} can also be used: $R_{FHLE} = \sum (|F_{FHLE} - F_{H(calc)}|)/\sum F_{FHLE}$, the disadvantage of it being that it is possible to obtain a completely incorrect structure with a value of R_{FHLE} ranging from 0.40 to 0.60.

phases, but it is strongly dependent on the value of the total error E: an overestimate of E will give an over-optimistic figure of merit.

Picking up minor heavy-atom sites: the difference-Fourier synthesis

When the positions of heavy-atoms have been identified and some refinement has been carried out, the next step is to check the correctness of the heavy atoms identified and the localization of other minor sites that may be present. This can be accomplished in principle by using difference-Fourier techniques (see p. 366): the map that better represents the differences between observed and calculated positions of heavy atoms should be that with coefficients:[37]

$$\delta \boldsymbol{F} = \boldsymbol{F}_{\mathrm{H(obs)}} - \boldsymbol{F}_{\mathrm{H(calc)}}. \tag{8.47}$$

Since $\boldsymbol{F}_{\mathrm{H(obs)}}$ is not directly measurable, (8.47) can be approximated only if anomalous differences are available.

Often a map with coefficients:

$$\Delta F = m(F_{\mathrm{PH}} - F_{\mathrm{P}}) \exp(\mathrm{i}\varphi_{\mathrm{P}}) \tag{8.48}$$

is calculated. Unfortunately, $F_{\mathrm{PH}} - F_{\mathrm{P}}$ does not correspond to F_{H}, and moreover φ_{P} is not the phase of F_{H}. Furthermore, since Fourier maps are strongly dominated by phases, a difference-Fourier map calculated with coefficients (8.48) will tend to reproduce the peaks of the heavy atoms used in calculating the phases. A difference-Fourier map of a derivative calculated with phases obtained from that derivative is normally useless, and the technique of cross-difference Fourier, that is the Fourier difference of a derivative calculated using phases for the protein obtained from all the others, is highly recommended. Unfortunately this is not always possible, since it requires at least three different derivatives.

A synthesis, proven useful in picking up minor sites, is the **derivative difference-map**, calculated using coefficients:

$$\delta F = m(F_{\mathrm{PH(obs)}} - F_{\mathrm{PH(calc)}}) \exp(\mathrm{i}\varphi_{\mathrm{PH(calc)}}). \tag{8.49}$$

If the position of all the major heavy atom peaks has been determined correctly, coefficients (8.49) allow the localization of low-occupancy sites. Figure 8.13 shows the coefficients used in (8.48) and (8.49) compared with the 'true' ones.[38]

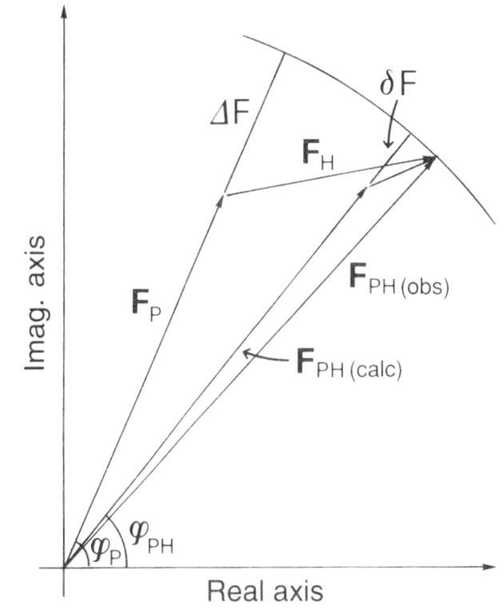

Fig. 8.13. Argand diagram of the quantities used in a Fourier-difference calculation. ΔF is the magnitude of the coefficient used instead of F_{H} if coefficients (8.48) are used. δF is used in the $F_{\mathrm{PH(obs)}} - F_{\mathrm{PH(calc)}}$ synthesis. (Adapted from Henderson and Moffatt.[38])

A third approach to the resolution of the phase ambiguity: real-space filtering

We have already seen in Chapter 4 that the positivity of the electron density can be used to obtain phase information. The same principle may be used to improve electron density maps of proteins.[39–43]

This method is called **real-space filtering** and can produce correct phases when good SIR data are available. Phase information can be extracted by an electron density map by a simple back-Fourier transform: if a suitable modification has been introduced in the map, the phases obtained by the back-transformation will be more correct than those used in calculating the map and so the new phases can be used to modify the original ones. The procedure described here, suggested by Wang,[3] is based on the same basic principle.

We have seen on p. 543 that an electron density map calculated using SIR phases will contain the correct density of the molecule, covered by an enormous amount of noise. Information about the correct phases can be obtained via the following automated procedure.

1. After calculating an electron density SIR (or single anomalous scattering, SAS) map, errors are removed in direct space by evaluating for every point an average value ρ_c from the formula:

$$(\rho_c + \langle \rho_{solv} \rangle)/(\rho_c + \rho_{max}) = S, \tag{8.50}$$

   where S is a constant estimated from previously solved protein structures.

2. The value of ρ_c estimated from (8.50) is added to any point in the map. Any negative density remaining is considered an error and set to zero. The density in solvent region is substituted with the value $\rho_c + \langle \rho_{solv} \rangle$.

3. Structure factors are calculated by Fourier inversion. Now new phases are available, and they can be combined with the old ones, using one of the well known phase combination procedures.[19,20]

4. Combined phases will produce new maps, that will undergo the procedure again, until convergence is reached.

An important feature of the Wang procedure is that an automatic method has been devised to decide what is solvent and what is protein, based on the practical assumption that solvent regions, like protein regions, are contiguous to some extent. The value of electron density at point j is substituted by a new value which depends from the points surrounding it, within a sphere of radius R,† according to the criteria:

$$\rho_j' = K \sum_i w_i \rho_i \tag{8.51}$$

where

$$\begin{aligned} w_i &= 1 - r_{ij}/R \qquad \text{if } \rho_i > 0 \text{ and } R > r_{ij} \\ w_i &= 0 \qquad \text{if } \rho_i < 0 \text{ or } R < r_{ij}. \end{aligned} \tag{8.52}$$

Since the resulting map has large homogeneous connected regions, the molecular boundary can be revealed by a threshold density level appropriately selected. Real-space filtering methods can also be used to extend the resolution of the electron density map.

Rotation and translation functions and the molecular replacement method

Sometimes it may happen that we know the three-dimensional structure of a molecule and we are interested in solving the same structure in a different

† The optimum value of radius R depends mainly on the resolution: a value between 8 and 10 Å is often used to average a 3 Å resolution isomorphous replacement map.

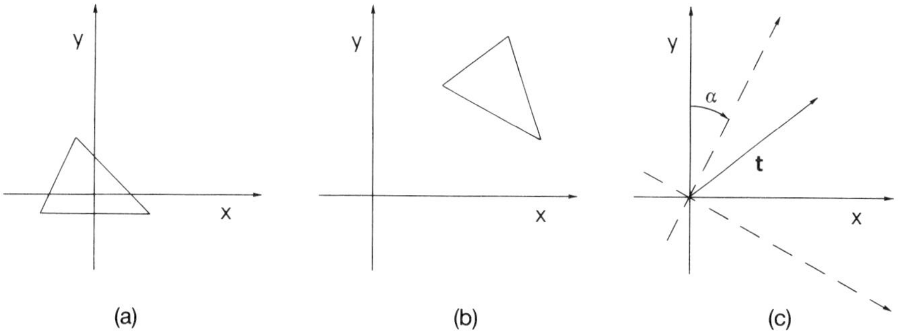

Fig. 8.14. A simplified two-dimensional illustration of (8.53). In (a) and (b) the same 'molecule' is represented in different positions with respect to the same reference system. By a rotation of an angle α and a translation of $\boldsymbol{t}$, the object in (a) can be superimposed to that in (b).

space group. At other times we have reasons to believe that the conformation of a protein is quite similar to that of another that has been previously solved, which is often the case for the same protein from different species. In all the cases mentioned above, six variables, three rotational and three translational, will approximately describe the transformation from one set of coordinates to the other. In fact, if we call $\boldsymbol{X}$ the set of vectors† representing the atoms of the original molecule and $\boldsymbol{X}'$ the transformed ones, the transformation is simply described by:

$$\boldsymbol{X}' = [\mathbf{C}]\boldsymbol{X} + \boldsymbol{t} \qquad (8.53)$$

where $[\mathbf{C}]$ is a matrix that rotates the coordinates $\boldsymbol{X}$ into the new orientation and $\boldsymbol{t}$ is a translation vector. Equation (8.53) is illustrated for a two-dimensional situation in Fig. 8.14, where a 'molecule' formed by three point atoms can be superimposed to an identical molecule in a different orientation by the translation of a vector $\boldsymbol{t}$, after the rotation of an angle α.

As mentioned in Appendix 5.B (p. 000), the technique of positioning a molecule or a fragment of known structure in a crystal cell is called **molecular replacement**. In principle it is possible to simultaneously search for the six variables which minimize the difference between F_{obs} and F_{calc}, but in practice this is a very hard task, even for the fastest computer.‡ The solution of the problem was pioneered by Rossmann and Blow,[45] who explored the possibility of finding the orientation of similar subunits in a crystal cell without any knowledge about the translation $\boldsymbol{t}$, making use of the Patterson function. After the correct orientation has been found, a search for the translation vector can be carried out (a collection of papers on molecular replacement is found in the book by Rossmann).[4] Let us first describe the methodology and the problems connected with the rotation function.

The first step in molecular replacement: the rotation function

The idea of the rotation function can be easily understood by a simple two-dimensional example. In Fig. 8.15(a) a 'molecule' of three idealized

† In the following discussion all the rotations will be performed in a Cartesian reference system. It is assumed that, if required, an appropriate orthogonalization is applied before a rotation is performed.

‡ This statement is becoming untrue, due to growing availability of computing power. Subbiah and Harrison[44] have shown, in the test case of the human histocompatibility antigen, that an exhaustive three-dimensional search at low resolution can be performed. The correct solution can also be obtained, starting from a random position, using the simulated annealing approach (see p. 569).

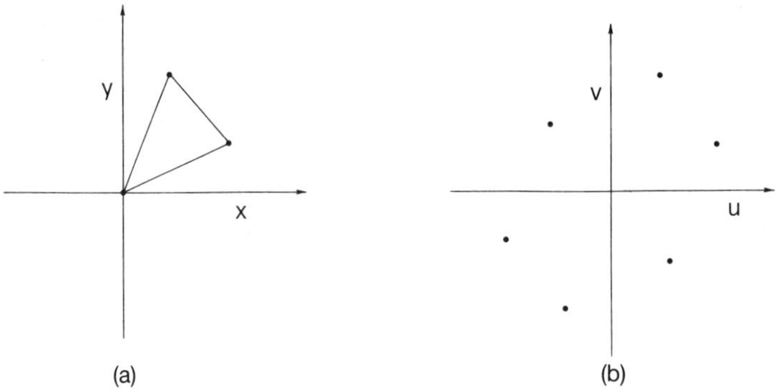

Fig. 8.15. An isolated, simplified 'molecule' of three atoms. Its self-convolution is shown in (b): since atoms are considered as points, it is everywhere 0, except when two points superimpose exactly.

point atoms is represented isolated, in a orthogonal reference system. Let us imagine a two-dimensional lattice of similar 'molecules' in a different orientation. In the lattice of Fig. 8.16(a) the unit cell is made by two of such molecules, related by a twofold axis, denoted by 1 and 2. Maxima of its idealized Patterson function, shown in Fig. 8.16(b), can be divided in two categories: those arising from intramolecular vectors, or self-vectors, and those from intermolecular or cross-vectors. Maxima belonging to the first class are indicated in the figure by circled points and are confined to a short distance from the origin. It is easy to see that by a simple rotation of 112° anti-clockwise the isolated molecule can be superimposed to molecule 1 of Fig. 8.16(a), after an appropriate translation, or, by a rotation of 292°, to molecule 2. The self-convolution function of the isolated molecule (Fig. 8.15(b)) can also be superimposed to the Patterson of the crystal if we perform the same rotation. Let us define a function $R(\mathbf{C})$:

$$R(\mathbf{C}) = \int_V P_{\text{cryst}}(\boldsymbol{u}) P_{\text{mol}}(\mathbf{C}\boldsymbol{u})\, \mathrm{d}\boldsymbol{u} \tag{8.54}$$

where $\mathbf{C}$ is a matrix that rotates the coordinates of the model molecule with respect to the reference system of the crystal, $P_{\text{cryst}}(\boldsymbol{u})$ is the Patterson

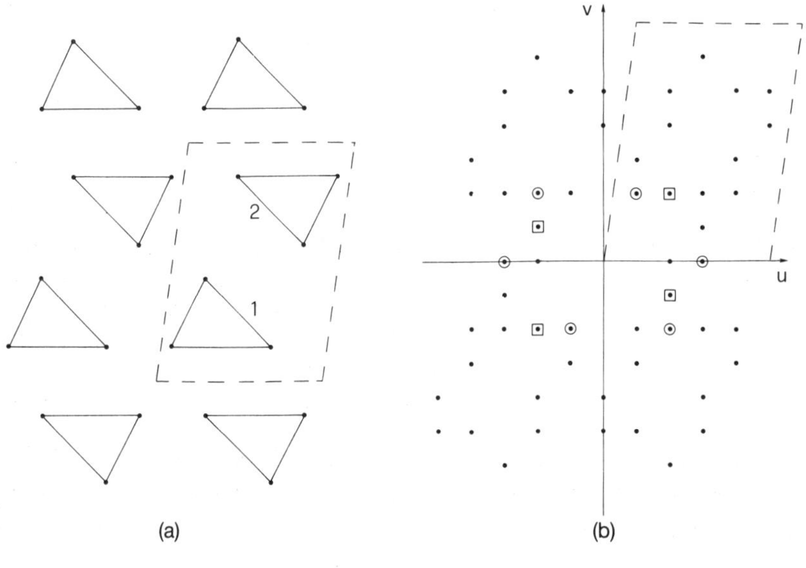

Fig. 8.16. (a) A portion of a two-dimensional lattice of a molecule identical to that of Fig. 8.15 (the unit cell is dashed). (b) Its corresponding Patterson map. Circled points indicate self-vectors. Squared points are cross-vectors close to the origin: some of the points of Fig. 8.15(b) accidentally superimpose to them during rotation, giving rise to false maxima in the rotation function.

function of the crystal and $P_{mol}(\mathbf{C}\boldsymbol{u})$ is the self-convolution function of the isolated molecule, rotated by $\mathbf{C}$. The function $R(\mathbf{C})$ will have a maximum when the peaks of the two functions superimpose, at least partially. The calculation of function $R(\mathbf{C})$ for all the possible values of the rotational variables will allow us to determine the orientation of the known molecule in the reference system of our crystal.

The right-hand side of (8.54) can be Fourier transformed[45,46] and reduced, neglecting a constant, to:

$$R(\mathbf{C}) = \sum_{p} \sum_{h} F_{mol}(\boldsymbol{p})^2 F(\boldsymbol{h})^2 G_{\boldsymbol{h},\boldsymbol{h}'}. \tag{8.55}$$

$F(\boldsymbol{h})$ are the Fourier coefficients of the crystal and $F_{mol}(\boldsymbol{p})$ the coefficients of the Fourier transform of the isolated molecule, rotated by $\mathbf{C}$ ($\boldsymbol{h}$, $\boldsymbol{h}'$ are used here to indicate different terms of (hkl) values, $\boldsymbol{p}$ represents a point in reciprocal space of a continuous transform). $G_{\boldsymbol{h},\boldsymbol{h}'}$ is an interference function whose magnitude depends on $\boldsymbol{h}$, $\boldsymbol{h}'$, and the volume used in the integration of (8.54).

The function $R(\mathbf{C})$ can be evaluated in real space using (8.54) or in reciprocal space, using (8.55). In both cases the computing time is strongly dependent on the sampling chosen, which in turn is related to resolution. In real space P_{cryst} and P_{mol} must be sampled finely enough for the resolution selected (generally this means a value around 1/2 or 1/3 the d spacing). The volume of integration is a sphere whose radius depends on the size of the isolated molecule, and this value determines the steps of the angular variables used in evaluating $R(\mathbf{C})$. In reciprocal space problems are quite similar, since $F_{mol}(\boldsymbol{p})$ is a continuous function, defined over all the reciprocal space. The isolated molecule can be put in an artificial cell, generally a cube whose edges can be about two to three times the size of the molecule, and the continuous function evaluated with a sampling appropriate to the resolution used. In practice, since (8.55) is dominated by large Fourier coefficients, it is possible to limit the numbers of $F(\boldsymbol{p})$ used.

A faster but more complex approach in evaluating the rotation function, the so-called fast-rotation function, has been devised by Crowther.[47] If we express the Patterson function in terms of spherical polar coordinates, (r, θ, φ), for a rotation $\mathbf{C}$, corresponding to the three angles α_1, α_2, α_3, the rotation function can be written:

$$R(\mathbf{C}) = \int P_{cryst}(r, \theta, \varphi) \mathrm{R}P_{mol}(r, \theta, \varphi) r^2 \sin\theta \, \mathrm{d}r \, \mathrm{d}\theta \, \mathrm{d}\varphi \tag{8.56}$$

where $\mathrm{R}P_{mol}$ is P_{mol} after a rotation $\mathbf{C}$. Equation (8.56) can be expanded using Bessel functions, more appropriate to a rotation group than a Fourier series, well suited for translation operations. The use of Bessel functions requires a lot of difficult mathematics, outside the scope of this book, but the final result is that $R(\mathbf{C})$ can be evaluated as a summation of two terms, one of them independent of the rotation itself. The computation time is consequently greatly reduced with respect to the use of (8.54) or (8.55).

The rotation matrix C and the choice of variables

Rotation is usually performed with respect to an orthogonal system, making use of different rotational variables. Quite common are the Eulerian rotation angles θ_1, θ_2, and θ_3, illustrated in Fig. 2.3(a): θ_1 is the rotation angle about the z axis and is positive when the rotation is clockwise looking from the origin; θ_2 is a rotation about the new x axis and θ_3 a rotation about the new z axis. The matrix **C** describing such a rotation is given in (2.32b).† An appropriate rotation for the three angles will cover all the space (see p. 72), but if the Patterson map presents some rotational symmetry, the rotation function will also have symmetry and a partial rotation will be sufficient.

The symmetry of the rotation function is a combination of the symmetries of the two Patterson functions, P_{cryst} and P_{mol}.[48] The Eulerian angles make easy to describe the symmetry of the rotation function.[49] Any triplets of angles θ_1, θ_2, and θ_3 can be considered as a point of a three-dimensional system, whose unit cell has dimensions 2π in all directions: a rotation α is in fact equivalent to $\alpha + 2\pi$. The resulting rotation space groups are some of those described in the *International tables for x-ray crystallography*.‡

A disadvantage in using θ angles is that when θ_2 is small, θ_1 and θ_3 represent a rotation about nearly the same axis, and maxima will resemble strips rather than maxima. The distortion effect can be avoided if a combination of Eulerian angles is used instead:[50]

$$\theta_+ = \theta_1 + \theta_3, \qquad \theta_- = \theta_1 - \theta_3 \qquad \theta_2 = \theta_2. \tag{8.57}$$

A different possibility is the use of spherical polar angles, φ, ψ, and χ (Fig. 2.3(b)). Angles φ and ψ define a spin axis, and a rotation of χ around this axis is performed. Polar angles are very useful when a particular direction has to be exploited or when a defined rotation has to take place, as is sometimes the case for self-rotation (see p. 558).

Translation functions

Once the orientation of a known molecule in an unknown cell has been found, the next step is the determination of its absolute position. Only when one molecule is present in space group P1 is this problem non-existent, since in this case the origin of the crystal cell can be chosen arbitrarily with respect to all three axes. In all the other cases, when the reference molecule, exactly oriented, is translated in the unknown cell, symmetry-related molecules move accordingly and all the intermolecular vectors change: only when all the molecules in the crystal cell are in the correct position, the calculated Patterson cross-vectors superimpose to those of the observed Patterson (intramolecular vectors are insensitive to translation). Figure 8.17 illustrates the method. Molecule 1 is positioned in the crystal cell of the unknown structure in the correct orientation: $\boldsymbol{s}_1$ is the vector defining its position with respect to the origin. Since we do not know yet the correct position of the molecule in the cell, $\boldsymbol{s}_1$ is arbitrarily chosen. Molecule 2 is generated by the twofold axis, and its position is defined by vector $\boldsymbol{s}_2$. The correct solution is shown in Fig. 8.17(a), where the correct origin of molecule 1 is indicated by $\boldsymbol{s}_1^0$. As vector $\boldsymbol{s}_1$ varies, all the intermolecular vectors among symmetry-related atoms will change: they will coincide with

† In Chapter 2 the rotation matrix **C** is called $\mathbf{R}_{Eu}$ for Eulerian angles and $\mathbf{R}_{sp}$ for spherical polar angles.

‡ The reader must be warned that the Eulerian rotation matrix is not Hermitian, that is reversing the order of the Patterson functions does not produce the same rotation-equivalent positions.

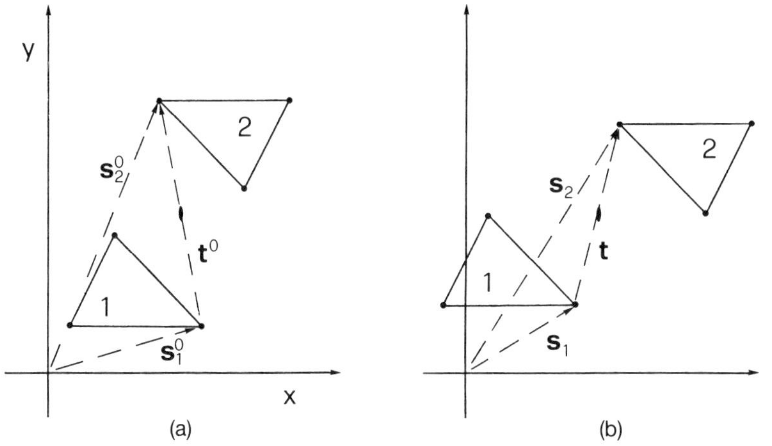

Fig. 8.17. (a) 'Molecules' 1 and 2 correctly oriented and positioned with respect to a symmetry element. $\boldsymbol{s}_1^0$ is the vector from the origin to an arbitrary point of molecule 1, and $\boldsymbol{s}_2^0$ the corresponding one for molecule 2, which is generated from 1 by twofold rotation. $\boldsymbol{t}^0 = \boldsymbol{s}_2^0 - \boldsymbol{s}_1^0$ is the translation vector from molecule 1 to 2. (b) Molecule 1 has been translated (but not rotated, so that orientation is unchanged). Molecule 2 has moved accordingly, and now all vectors defining the molecular positions in the cell are changed. Only when $\boldsymbol{t} = \boldsymbol{t}^0$ the two models, and consequently cross-vectors of Patterson maps, superimpose. Determination of $\boldsymbol{s}_1^0$ from $\boldsymbol{t}^0$ is straightforward.

those of Fig. 8.17(a) only when $\boldsymbol{s}_1 = \boldsymbol{s}_1^0$, that is when the local origin is correctly defined with respect to the symmetry element. The determination of the translation vector is performed by comparing two Patterson maps, just as before in the rotation case, except that now we are interested in maximizing the superposition of a different class of peaks. The problem described above is in practice quite difficult, since the function representing the superposition is generally very noisy and with many small maxima. Several translation functions have been proposed, and some of them are briefly summarized in Appendix 8.B. To illustrate the principles of translational search, only the T function of Crowther and Blow[51] will be described here, following the treatment of Latman.[52] In the case illustrated in Fig. 8.17, the set of cross-vectors of the calculated Patterson from molecule 1 to molecule 2 can be written as:

$$P_{12}(\boldsymbol{u}) = \int_V \rho_1(\boldsymbol{x})\rho_2(\boldsymbol{x} + \boldsymbol{u})\,\mathrm{d}\boldsymbol{x} \tag{8.58}$$

where ρ_1 and ρ_2 represent the electron density of the two molecules. If molecule 1 is now translated, a new vector $\boldsymbol{s}_1$ will define its origin. At the same time molecule 2 will move into the cell, and a new function P_{12} can be calculated for every value of $\boldsymbol{s}_1$. Since we are looking at intermolecular vectors, it is more useful to define the translation as a function of vector $\boldsymbol{t} = \boldsymbol{s}_2 - \boldsymbol{s}_1$, which defines a local origin with respect to a symmetry element. If $P_{\mathrm{obs}}(\boldsymbol{u})$ is the value of the observed Patterson at point $\boldsymbol{u}$, the T translation function is defined as:

$$T(\boldsymbol{t}) = \int_V P_{\mathrm{obs}}(\boldsymbol{u})P_{12}(\boldsymbol{u}, \boldsymbol{t})\,\mathrm{d}\boldsymbol{u}. \tag{8.59}$$

Function T will have a maximum when the two Pattersons superimpose. In reciprocal space, (8.59) can be written:[51]

$$T(\boldsymbol{t}) = \sum_{\mathrm{h}} I_{\mathrm{obs}}(\boldsymbol{h})\boldsymbol{F}_1(\boldsymbol{h})\boldsymbol{F}_1^*(\boldsymbol{h}\mathbf{A}) \exp(-2\pi \mathrm{i}\boldsymbol{h}\boldsymbol{t}) \tag{8.60}$$

where $\boldsymbol{F}_1(\boldsymbol{h})$ is the Fourier transform of the model molecule 1 and $\boldsymbol{F}_1(\boldsymbol{h}\mathbf{A})$ the calculated structure factor of molecule 1 after application of symmetry operation **A**. Looking at Fig. 8.17, since $\boldsymbol{t} = \boldsymbol{s}_2 - \boldsymbol{s}_1$:

$$T(\boldsymbol{t}) = \sum_h I_{obs}(\boldsymbol{h})\boldsymbol{F}_1(\boldsymbol{h})\boldsymbol{F}_1^*(\boldsymbol{h}\mathbf{A}) \exp\left[-2\pi i \boldsymbol{h}(\boldsymbol{s}_2 - \boldsymbol{s}_1)\right]. \tag{8.61}$$

The T function will have a peak at position $\boldsymbol{s}_2 - \boldsymbol{s}_1 = \boldsymbol{t}_0 \equiv \boldsymbol{s}_2^0 - \boldsymbol{s}_1^0$ (or $\boldsymbol{s}_1 - \boldsymbol{s}_2 = -\boldsymbol{t}_0$). The determination of $\boldsymbol{s}_1^0$ from function T is equivalent to solving a Patterson map with only two atoms in the crystal cell. In general the T function does not need to be evaluated for the entire cell: if for example two molecules are related by a screw axis along z, the maximum $\boldsymbol{t}_0$ will have coordinates $(2x, 2y, 1/2)$, that is it will be confined only to section $z = 1/2$.

Other types of translation functions have been developed, and some of the more commonly used in protein crystallography are summarized in Appendix 8.B. A general review on translation functions is reported by Beurskens *et al.*[53]

Self-rotation and self-translation functions: improving the electron density maps

Sometimes more than one molecule is present in the crystallographic asymmetric unit. If we assume that they are identical, or at least very similar, we can take advantage of the independent information present in the structure factors. In fact, if we are able to identify the non-crystallographic symmetry elements relating the independent molecules, the electron density map can be averaged and substantially improved. The presence of three molecules in the asymmetric unit has allowed the solution of the structure of the haemaglutinin of the influenza virus using only SIR phases.[54]

The self-rotation function is very similar to the general rotation function defined in (8.54):

$$R(\mathbf{C}) = \int_V P_1(\boldsymbol{u})U(\boldsymbol{u})P_1(\mathbf{C}\boldsymbol{u})\, \mathrm{d}\boldsymbol{u} \tag{8.62}$$

where $P_1(\mathbf{C}\boldsymbol{u})$ is the $P_1(\boldsymbol{u})$ Patterson function rotated by matrix **C**, and $U(\boldsymbol{u})$ is a function which is 1 inside a sphere and 0 elsewhere. The function U is necessary since both Patterson maps extend to all space, but we are interested only in the superposition of self-vectors, confined to a region around the origin of the cell. The sphere defined by U generally has a diameter slightly larger than the maximum supposed molecular dimension.

The choice of polar rotation angles is quite common for self-rotation and deserves a brief comment. Quite often the non-crystallographic symmetry is represented by a rotation axis, in a direction different from the crystallographic ones. In that event, the use of polar angles reduces the search for the position of the axis from a three-dimensional problem to a two-dimensional one: a twofold axis, for example, will correspond to a rotation of 180° around the polar axis χ, and a two-dimensional map (calculated for φ from 0° to 360°, ψ from 0° to 360°, and $\chi = 180°$) will show the presence of the axis. A clear example of that is presented by Evans *et al.*[55]

The definition of the translational component of the non-crystallographic symmetry represents the last and possibly the more difficult step. Let us for example assume that the direction of a twofold non-crystallographic axis is known. It has been shown[56] that only the component of the translation vector $\boldsymbol{t}$ in the direction of the axis can be determined precisely. The other component of vector $\boldsymbol{t}$, that is that perpendicular to the axis, is intrinsically an imprecise parameter, unless the molecular structure is perfectly known, which is not the case. In general, the self-translation function (analogous to the T function previously described) is used to detect the existence of a translational component of a rotational symmetry.

The steps and the possible different pathways described in the previous paragraphs for the solution of a crystal structure of a macromolecule are summarized in the scheme reported in Fig. 8.18.

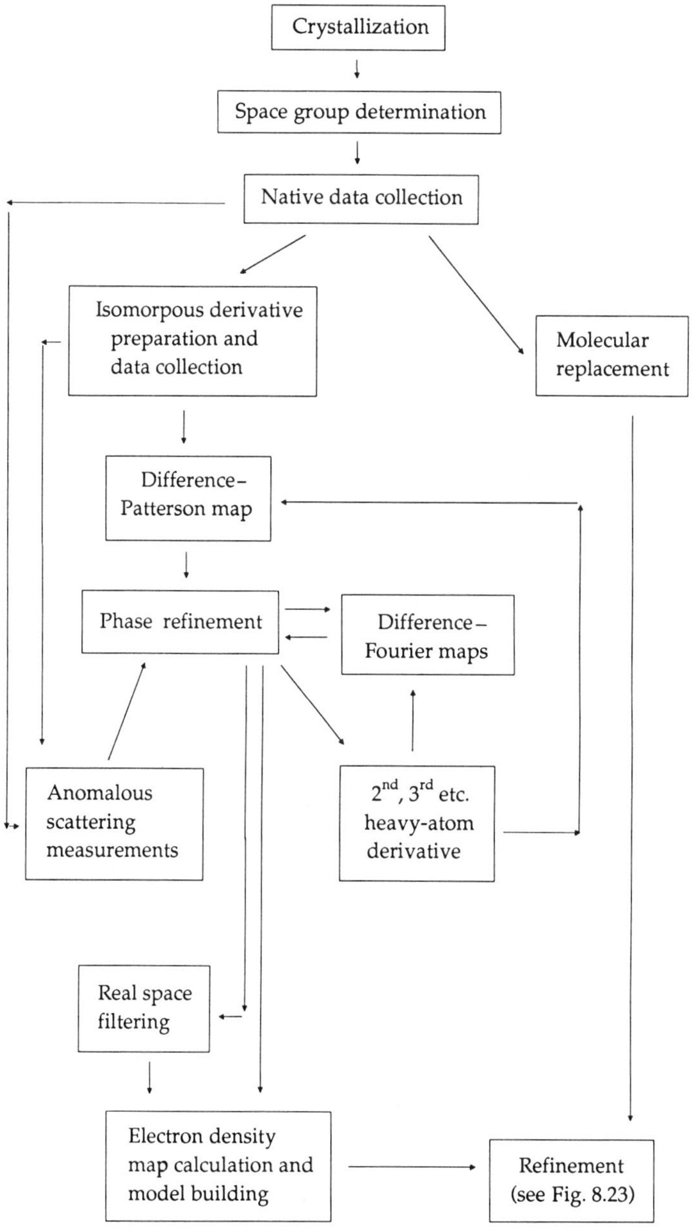

Fig. 8.18. Simplified scheme summarizing some of the possible steps in the determination of the structure of a macromolecule. Some of them can be used alternatively or combined, depending on the specific problem, that is size of the protein, previous knowledge of the structure, number of molecules in the asymmetric unit, and so on.

Direct methods and the maximum-entropy principle in macromolecular crystallography

Direct methods, extensively described in Chapter 5, are not yet applicable in the present form to the *ab initio* structural determination of a macromolecule. The main reason for that resides in the very large number of atoms to locate: additional information is needed in order to increase the reliability of the phase indication. Isomorphous replacement and anomalous-dispersion data can be classified in this category, and indeed they have been often used to locate the positions of heavy atoms. For example coefficients (8.8) can be used with conventional direct methods programs and they can be very helpful in the case of a very large number of sites.[57] Furthermore, classical direct methods have allowed, using coefficients $(F^+_{\mathrm{PH}} - F^-_{\mathrm{PH}})$, the localization of the positions of anomalous scatterers in metalloproteins.[58]

Quite promising, but not yet fully exploited, is the use of direct methods for the determination of the phase of the protein structure factor taking advantage of isomorphous replacement and anomalous-dispersion data (for a review see Giacovazzo)[59]. Let us call $\phi_{\mathrm{P}}(\boldsymbol{h})$ and $\phi_{\mathrm{P}}(\boldsymbol{k})$ the phases of reflections $\boldsymbol{h}$ and $\boldsymbol{k}$, respectively, for the native data set. $\phi_{\mathrm{PH}}(\boldsymbol{h})$ and $\phi_{\mathrm{PH}}(\boldsymbol{k})$ are the phases of the corresponding reflections of the derivative. In addition to the classical triplet invariants introduced on p. 338:

$$\begin{aligned}\Phi_1 &= \varphi_{\mathrm{P}}(-\boldsymbol{h}) + \varphi_{\mathrm{P}}(\boldsymbol{k}) + \varphi_{\mathrm{P}}(\boldsymbol{h}-\boldsymbol{k}),\\ \Phi_2 &= \varphi_{\mathrm{PH}}(-\boldsymbol{h}) + \varphi_{\mathrm{PH}}(\boldsymbol{k}) + \varphi_{\mathrm{PH}}(\boldsymbol{h}-\boldsymbol{k}),\end{aligned} \tag{8.62a}$$

we can make use of triplets containing mixed terms, that is relationships between phases of the two crystals:

$$\begin{aligned}\Phi_3 &= \varphi_{\mathrm{P}}(-\boldsymbol{h}) + \varphi_{\mathrm{PH}}(\boldsymbol{k}) + \varphi_{\mathrm{P}}(\boldsymbol{h}-\boldsymbol{k}),\\ \Phi_4 &= \varphi_{\mathrm{PH}}(-\boldsymbol{h}) + \varphi_{\mathrm{P}}(\boldsymbol{k}) + \varphi_{\mathrm{P}}(\boldsymbol{h}-\boldsymbol{k}),\\ \Phi_5 &= \varphi_{\mathrm{P}}(-\boldsymbol{h}) + \varphi_{\mathrm{PH}}(\boldsymbol{k}) + \varphi_{\mathrm{PH}}(\boldsymbol{h}-\boldsymbol{k}),\\ \Phi_6 &= \varphi_{\mathrm{PH}}(-\boldsymbol{h}) + \varphi_{\mathrm{P}}(\boldsymbol{k}) + \varphi_{\mathrm{PH}}(\boldsymbol{h}-\boldsymbol{k}),\\ \Phi_7 &= \varphi_{\mathrm{PH}}(-\boldsymbol{h}) + \varphi_{\mathrm{PH}}(\boldsymbol{k}) + \varphi_{\mathrm{P}}(\boldsymbol{h}-\boldsymbol{k}),\\ \Phi_8 &= \varphi_{\mathrm{P}}(-\boldsymbol{h}) + \varphi_{\mathrm{P}}(\boldsymbol{k}) + \varphi_{\mathrm{PH}}(\boldsymbol{h}-\boldsymbol{k}).\end{aligned} \tag{8.62b}$$

The eight different combinations of the six phases give rise to eight conditional probability distribution for Φ_j, whose formulation is not discussed here. An important point, however, must be stressed: the probability distributions do not depend only on the total number of atoms present in the crystal cell (see Appendix 5.C), but also on the difference between native and derivative data. The reliability factor for a triplet phase, G, defined in (5.37) and (5.C.16), in the presence of isomorphous data must be rewritten:[60]

$$G = (2/N_{\mathrm{P}}^{1/2})\,|E_{-\boldsymbol{h}}E_{\boldsymbol{k}}E_{\boldsymbol{h}-\boldsymbol{k}}| + (2/N_{\mathrm{H}}^{1/2})\Delta_{-\boldsymbol{h}}\Delta_{\boldsymbol{k}}\Delta_{\boldsymbol{h}-\boldsymbol{k}} \tag{8.63}$$

where N_{P} and N_{H} are the number of protein atoms and heavy atoms in the unit cell, respectively, and $\Delta_{\boldsymbol{h}}$ represents the normalized difference for reflection $\boldsymbol{h}$. The second term of (8.63) may substantially increase the value of the reliability factor G. A relevant application of (8.62) is that one single derivative allows in principle the determination of the phase of the protein structure factor.

A similar approach can also be used in presence of anomalous dispersion data, where in addition to the classical triplet invariants:

$$\begin{aligned}\Phi_1 &= \varphi_P(-\boldsymbol{h}) + \varphi_P(\boldsymbol{k}) + \varphi_P(\boldsymbol{h}-\boldsymbol{k})\\ \Phi_2 &= \varphi_P(\boldsymbol{h}) + \varphi_P(-\boldsymbol{k}) - \varphi_P(\boldsymbol{h}-\boldsymbol{k}),\end{aligned} \tag{8.64}$$

six other equations containing mixed terms, analogous to (8.62b), can be introduced. The only limitation in the previous treatment is represented by the very high degree of accuracy necessary in the measurement of the intensities, particularly in the case of anomalous scattering data. Nevertheless, anomalous dispersion techniques are becoming more relevant, due to the possibility of performing multi-wavelength experiments with synchrotron radiation.

An apparently different approach to the solution of the phase problem is based on the **maximum-entropy** principle, the application of which to crystallography was proposed by Collins[61] and that has been shown to be virtually equivalent to direct methods by Bricogne.[62] If $\rho(\boldsymbol{r})$ is a density distribution, its **entropy** is defined as:

$$H = -\int_V \rho(\boldsymbol{r}) \ln \rho(\boldsymbol{r})\, \mathrm{d}\boldsymbol{r}. \tag{8.65}$$

The concept of entropy is borrowed from information theory, where H measures the expected value of the total information. The 'better' distribution is that which maximizes H and which is at the same time consistent with other observations: in our case, obvious conditions are that the electron density over the entire cell is equal to $F(000)$ and the Fourier transform of $\rho(\boldsymbol{r})$ is in agreement with the diffraction data. Equation (8.65) can consequently be modified to include constraints. If some sort of previous density distribution is available, the relative entropy can be defined:[63]

$$H_c = -\int_V \rho(\boldsymbol{r}) \ln (\rho(\boldsymbol{r})/q(\boldsymbol{r}))\, \mathrm{d}\boldsymbol{r} \tag{8.66}$$

where $q(\boldsymbol{r})$ is a distribution inferred by a priori information: in the absence of other information, that is if q is constant, (8.66) reduces to (8.65). The maximum-entropy method has been used to expand a set of initial phases, in conjunction with other techniques.[64,65] Besides, the structure of 15 base pair of DNA (a structure of about 600 atoms) has been recently solved *ab initio*:[63] starting phases were determined by conventional direct methods and expanded using the maximum-entropy algorithm. Although this approach has been demonstrated to fail in other simpler test cases, promising applications of direct methods to macromolecular crystallography are expected in the near future.

The interpretation of electron density maps and the refinement of the model

The interpretation of electron density maps

Once a phase angle estimate for the protein structure factors is available, the calculation of an electron density map is straightforward, using (see p. 170):

$$\rho(\boldsymbol{r}) = \sum_h F_P(\boldsymbol{h}) \exp(-i\varphi_P) \exp[-2\pi i(\boldsymbol{h} \cdot \boldsymbol{r})] \qquad (8.67)$$

where the Fourier coefficients are usually weighted by their figure of merit. The initial interpretation of the map is in general not easy, unless very good phases at high resolution are available, which is seldom the case. Therefore, the strategy generally adopted is to calculate first an electron density map at low resolution, say 5–6 Å: these maps allow us to identify the contours of the molecules in the crystal cell, and to distinguish between solvent regions and protein. Eventually, some elements of secondary structure can be identified: α-helices will appear as cylindrical rods of diameter of about 4–6 Å. β-sheets are more difficult to distinguish, and in any case single β-strands are not visible.

When the position of the molecule has been located in the unit cell, a map at medium resolution, say 3.5–2.5 Å resolution, is calculated and an attempt to trace the polypeptide chain is made. Chain trace at this resolution is made easier, and sometimes possible, if the amino acid sequence is known. Mistakes are quite common in the interpretation at medium resolution: the connections among secondary structure elements are often difficult to recognize and amino acids can be positioned along the chain shifted from their correct position by one or more residues.

Higher-resolution phases, say 2 Å or more, allow us to correct for this kind of mistake and to locate more accurately the amino acid side chains. Unfortunately, MIR phases very seldom extend to that resolution, and high-resolution maps can be obtained using calculated or combined phases, as will be discussed later.

Interactive computer graphics and model building

Low-resolution maps are traditionally drawn on a small scale on transparent sheets and are known among protein crystallographers as 'minimaps': they are very useful in giving a global view of the electron density of the unit cell. Such maps can also be used at medium resolution, since they allow a preliminary, approximate tracing of the polypeptide chain.

The complete building of the molecular model in the old days was performed using an apparatus, generally home-made, called an optical comparator or 'Richard's box', from the name of its inventor.[66] Nowadays, the interpretation of the electron density can be entirely performed on a graphic display: the map is shown on a video and the operator is allowed to fit a piece of chain into the density. For this purpose, the modelling software most popular among protein crystallographers is FRODO.[67,68] The principle of the program is that objects displayed on the screen are divided in

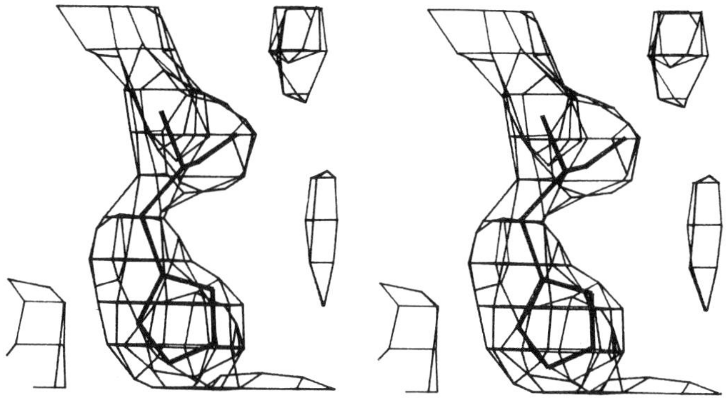

Fig. 8.19. 'Chicken wire' representation of the electron density of a side chain of a histidine residue. Maps are calculated at a resolution of 2.5 Å, with coefficients $2F_{obs} - F_{calc}$ and calculated phases. Drawing produced by an Evans and Sutherland PS330 graphic station, using program FRODO.[67]

two categories: those that can be manipulated, called 'foreground objects', and those that cannot be modified interactively, defined 'background objects'. Background and foreground models can be seen superimposed on each other. Program FRODO contains a dictionary of stereochemical information on natural amino acids and groups often occurring in protein molecules or nucleic acids. From this dictionary, building of a part of a polypeptide chain or a portion of a macromolecule in the preferred conformation is straightforward. The electron density, which does not need to be modified, is displayed as a background in a 'chicken wire' representation or similar, and the atomic model, or a part of it, as a foreground object. Figure 8.19 illustrates a simple model as it appears on a calligraphic screen. Atoms can be easily manipulated, by a simple rotation of a dial or movement of a light pen: they can be moved, alone or in groups, rotation around dihedral angles performed, or bonds stretched. The fitting of the built model into the electron density can be fast, if the starting phases are good enough; otherwise, a lot of time can be spent in trials. A convenient method to build the initial model of a protein has been devised by Jones and Thirup,[69] based on the idea that short elements of secondary structure can be taken from a data base consisting of coordinates of a restricted set of well refined protein structures: once some α-carbons (say 10–15) are roughly positioned into the density, the piece of model from the data base that better fits these atoms is searched for and used.

The only disadvantage in the use of a graphic system during electron density map interpretation is that a global view of the molecular model is practically impossible, since a drawing of all the atoms of a protein gives results which are quite confusing. Nevertheless, people who built protein models using a 'Richard's box' are strongly in favour of interactive graphic systems.

The refinement of the structure

The refinement of protein structures, with very few exceptions,[70] cannot be carried out using the classical least-squares methods. This is not due to the size of the problem, since nowadays computers are powerful enough to handle systems of equations containing thousands of variables, but to the limited number of X-ray data. It has been shown in fact on p. 98 and 104 that, for an accurate definition of the parameters, the system must be largely overdetermined, that is the ratio of observations to variables (atomic

Table 8.3. Number of theoretical independent reflections at different resolutions for a protein crystal with one molecule of 182 amino acids in the asymmetric unit. The solvent content is 40 per cent. The number of parameters is 4408 (1469 atoms times 3) if an overall *B* factor is considered, or 5876 if an individual isotropic *B* thermal parameter is assigned to each atom

Resolution range (Å)	Independent reflections	Ratio obs./var. (*x*, *y*, *z*)	Ratio obs./var. (*x*, *y*, *z*, *B*)
40–3.0	3500	0.8	—
40–2.5	6800	1.6	1.2
40–1.9	13 500	3.1	2.3
40–1.5	29 800	6.8	5.1
40–1.2	58 800	13.3	10.0
40–1.0	81 300	18.5	13.8

coordinates, thermal factors, and sometimes occupancy) must be of the order of 10 or so, and this is indeed the case for small molecules, where diffraction data can be collected to a spacing of 0.7 Å or even more. Protein crystals are intrinsically less ordered: diffraction data are often measured to a resolution of 3.0–2.5 Å, sometimes 2.0 Å. A resolution of 1.5 Å can be considered quite good, and only in exceptional cases have 1.0 Å data been collected.[23] A typical situation is illustrated in Table 8.3, where the number of independent reflections for a medium-size protein (182 amino acids) with a solvent content of 40 per cent are calculated at different ranges of resolution. For a ratio of observations to parameters of 10 it would be necessary to collect all possible diffraction data to a spacing of 1.2 Å, a resolution never attainable for such a type of crystal. For the above-mentioned reasons, the first attempts to refine a protein structure were performed in real space.[71] The method seeks to minimize the difference between the observed electron density, ρ_{obs}, computed by (8.67), and a calculated model density, ρ_{calc}, obtained by assuming a Gaussian distribution of the electron densities centred at the atomic positions of the current model:

$$S = (\rho_{\text{obs}} - \rho_{\text{calc}})^2. \tag{8.68}$$

This technique suffers from all the drawbacks of a real-space refinement procedure, the most relevant being that if poor phases are available, ρ_{obs} will be quite incorrect and the convergence of the method very slow, or it will not converge at all. There are many reasons to favour the reciprocal space refinement methods and different solutions have been proposed to overcome the problem of the underdetermination of the system: in fact, improvement of the ratio of observations to parameters can be achieved by decreasing the number of variables or by artificially increasing the number of observations. The former is called **constrained** least squares and the latter **restrained** least squares.

Constrained versus restrained least squares

Constrained or rigid-body refinement† is a well-known and widely used technique in crystallography (see p. 105): when the geometry of a group of

† Despite the distinction between them described on p. 105, **rigid-body** and **constrained** refinement are taken as synonyms in this chapter.

atoms is accurately known and there are reasons to believe that it will not be significantly modified by the environment, the entire group can be treated as a rigid entity. In the classical case of a phenyl ring, the eighteen positional variables can be reduced to only three translational and three rotational.

Bond length and valence angles in amino acids are very well known from the structures of hundreds of small peptides. In a protein, they can be held fixed to their theoretical values and only torsion angles around single bonds allowed to vary. This approach was used by Diamond[71] in real-space refinement, but it can be used in reciprocal space as well. Taking into account the fact that the peptide bond can be considered planar, only two torsion angles, called φ and ψ (see Appendix 8.D), need to be varied for the backbone chain of every amino acid: for a protein of n residues, the parameters are reduced to about $2n$ for backbone plus the torsion angles of side chains. An illustration of a possible choice of constrained parameters is reported in Fig. 8.20 for a simple dipeptide. This solves the problem of underdetermination, but the model becomes in some way too rigid and the radius of convergence, that is the maximum displacement allowed for an atom in a wrong position to be corrected, becomes quite small.

Constrained least squares can be applied to very different extent: the definition of rigid body can be applied to only some group of atoms or to the entire molecule. If, for example, an approximate solution of the structure has been found using the molecular replacement technique, a first refinement can be performed by considering the entire protein (or a subunit) as a rigid group and the best position in the new crystal cell can be searched for using only three translational and three rotational variables. In that event, there is the supplementary advantage that, since the number of variables is very limited, only low-resolution data need to be included in refinement, greatly increasing the radius of convergence of the method.

Increasing the number of observations is another possible solution of the underdetermination problem in macromolecular refinement (see p. 107). Information from other sources can, in fact, be introduced and treated in a way similar to that used for observations coming from X-ray diffraction. The use of geometrical restraints has been proposed by Konnert and Hendrickson,[72,73] following a procedure devised by Waser[74] for small

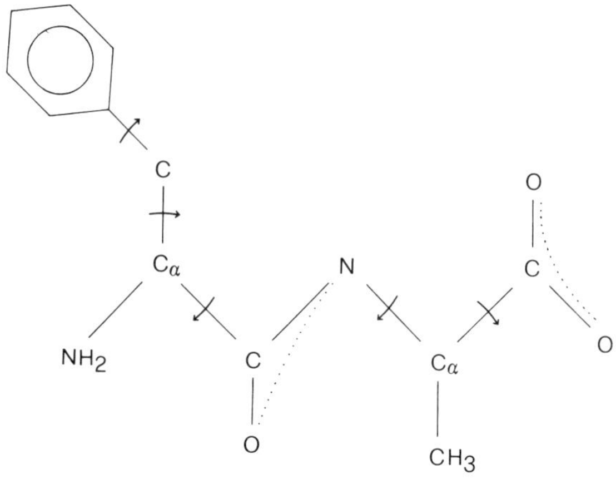

Fig. 8.20. Schematic drawing of the dipeptide phenylalanine-alanine, used to illustrate the **constrained** least-squares technique. Arrows indicate free rotation about the bond. All the bond lengths and valence angles are held fixed, and the peptide group and the phenyl ring planar. The total number of variables amount to eleven: three rotation and three translations (not indicated in figure) plus five internal torsion angles. (Some of the hydrogen atoms are indicated in figure only for clarity, but they are usually not taken into account in the refinement.)

molecules. In addition to the classical quantity minimized in crystallographic least squares:

$$S_1 = \sum_i w_i(F_{i(\text{obs})} - F_{i(\text{calc})})^2, \tag{8.69}$$

where the summation is extended to all the i reflections, other observational functions can be added. Since distances and valence angles of amino acids are well known and they are not expected to deviate significantly from the ideal value, instead of considering them as fixed, we can also minimize:

$$S_2 = \sum_j w_j(d_{j(\text{ideal})} - d_{j(\text{calc})})^2. \tag{8.70}$$

$d_{j(\text{ideal})}$ is the ideal value for the specific distance we are considering, $d_{j(\text{calc})}$ is that calculated from our present model and w_j is usually chosen as the reciprocal of the standard deviation of the distribution expected for the distances of type j. Notice that since $d_{j(\text{calc})}$ is a function of the atomic coordinates, (8.70) does not increases the number of variables. The total number of equations like (8.70) is equal to the distances that are restrained: bond lengths, the distances between one atom and the next-nearest-neighbour (which is equivalent to restraining valence angles), and the first-to-fourth atom distances, where the dihedral angle described by the four atoms is in some way fixed (this is for example the case of the planar peptide bonds). An example of the number of distances that can be restrained for a simple dipeptide is illustrated in Fig. 8.21. Other possible restraints in the Hendrickson and Konnert formulation are:

$$S_3 = \sum_k \sum_i w_{\text{m}}(\boldsymbol{m}_k \boldsymbol{r}_{i,k} - d_k)^2, \tag{8.71}$$

$$S_4 = \sum_l w_l(V_{l(\text{ideal})} - V_{l(\text{calc})})^2, \tag{8.72}$$

$$S_5 = \sum_n w_n(d_{n(\text{ideal})} - d_{n(\text{calc})})^4. \tag{8.73}$$

S_3 represents the sum of the deviations of the atoms i from the plane k, which is defined by its unit normal $\boldsymbol{m}_k$ and by the origin to plane distance

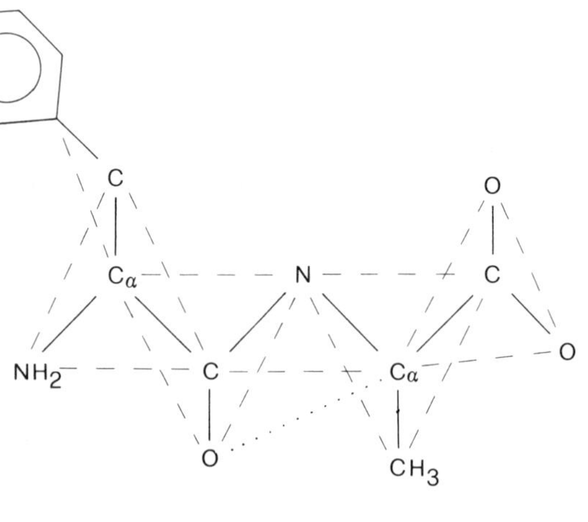

Fig. 8.21. The same dipeptide as in Fig. 8.20 illustrates the **restrained** least-squares technique. The coordinates of any atom are allowed to vary, but the stereochemistry is preserved by applying restraints on bond distances (full lines), bond angles (broken lines), torsion angles (dotted lines), and planarity. Non-bonded contacts are not shown in the figure. (Adapted from Sussman, ref. [3], Vol. 115, p. 274).

d_k; $\boldsymbol{r}_i$ is the vector that defines a point i whose distance from plane k has to be minimized.[75] S_4 restrains the volume of chiral atoms, defined for an α-carbon by the product of the interatomic vectors of the three atoms bound to it:

$$V_{C\alpha} = (\boldsymbol{r}_N - \boldsymbol{r}_{C\alpha})[(\boldsymbol{r}_C - \boldsymbol{r}_{C\alpha}) \times (\boldsymbol{r}_{C\beta} - \boldsymbol{r}_{C\alpha})]. \quad (8.74)$$

Since the sign of $V_{C\alpha}$ depends upon the handedness, S_4 restrains chiral centres to their correct configuration (Fig. 8.22). S_5 is applied to all non-bonded atoms (except those taken into account in S_2) and avoids too close contacts. Other kinds of restraints can be considered, i.e. on isotropic thermal parameters, occupancy, and non-crystallographic symmetry. It may sometimes happen, particularly during the first stages of refinement, that some part of the structure is poorly determined and the model 'blows up'. In that event, a restrain on the excessive shifts can be applied:

$$S_6 = \sum_k w_k(\boldsymbol{r}_k - \boldsymbol{r}_0)^2, \quad (8.75)$$

where $\boldsymbol{r}_k$ and $\boldsymbol{r}_0$ are the atomic vectors of the target and the initial model respectively. Using eqns (8.70)–(8.74), the number of observational functions is now greatly increased from the original number, represented by eqn (8.69). Equation (8.75) has effect only on the diagonal terms of the normal matrix. The number of restrained parameters for the example described in Table 8.3 is shown in Table 8.4.

It must be remembered that in protein crystallography 'experimental' phases are very often available. They can be included in least-squares as an additional information that imposes another restraint:[76,77]

$$S_7 = \sum_i w_p(\varphi_{i(\text{obs})} - \varphi_{i(\text{calc})})^2. \quad (8.76)$$

φ_{obs} is the estimate of phase angle from isomorphous and anomalous data and φ_{calc} is the phase calculated from the model. Weights for (8.76) must take into account the cyclic nature of phase angles.

Phase information is also used by Lunin and Urzhumtsev.[78] They suggest that only differences among crystallographic quantities be minimized, that is structure factor amplitudes and phases. Since phase probability distribution may be represented by (8.34), they assume an analogous probabilistic distribution for the module of the structure factor F for reflection i of the form:

$$P(F_i) \approx \exp[-(F_i^2 - F_{i(\text{obs})}^2)^2/2\sigma^2], \quad (8.77)$$

and if structure factors moduli and phases are assumed to be mutually independent, the joint probability distribution will be given by the product of (8.34) and (8.77). The most probable model will consequently be that which minimizes:

$$S = \sum_i \{(1/2\sigma^2)[F_i^2 - F_{i(\text{obs})}^2]^2 - [A\cos\varphi_i + B\sin\varphi_i + C\cos 2\varphi_i + D\sin 2\varphi_i]\}. \quad (8.78)$$

Using (8.78) the multimodality of the phase distribution is taken into account.

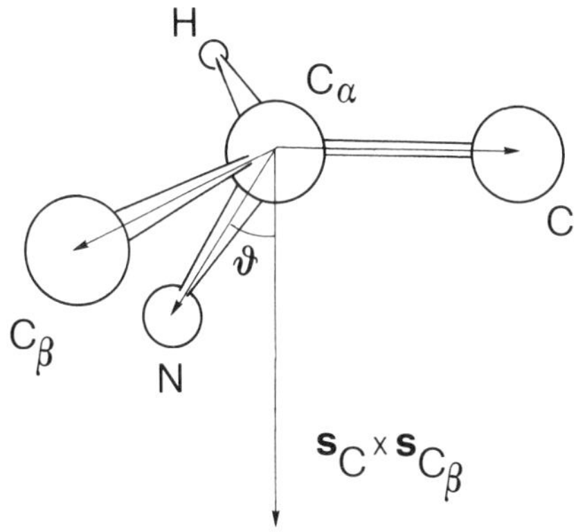

Fig. 8.22. The chiral volume for an α-carbon atom. The central atom is chosen as the origin of the coordinate system, and vectors $(\boldsymbol{r}_C - \boldsymbol{r}_{C\alpha})$, $(\boldsymbol{r}_{C\beta} - \boldsymbol{r}_{C\alpha})$, and $(\boldsymbol{r}_N - \boldsymbol{r}_{C\alpha})$ (8.74) are denoted $\boldsymbol{s}_C$, $\boldsymbol{s}_{C\beta}$, and $\boldsymbol{s}_N$ respectively. The cross product $\boldsymbol{s}_C \times \boldsymbol{s}_{C\beta}$ is a vector perpendicular to the plane $CC_\alpha C_\beta$. If it is on the same side of vector $\boldsymbol{s}_N$, as in the figure, that is if angle θ is less than 90°, the dot product between the two vectors is positive. If $\boldsymbol{s}_C$ and $\boldsymbol{s}_{C\beta}$ are reversed, that is if the wrong configuration is chosen for the α-carbon, the vector $\boldsymbol{s}_C \times \boldsymbol{s}_{C\beta}$ points in the opposite direction and the value of $V_{C\alpha}$ becomes negative.

Table 8.4. Number of restraints, following Hendrickson and Konnert,[73] for a protein molecule of 1469 atoms (excluding H) in the asymmetric unit. The example is relative to the case of Table 8.3

Number of distances:	
bond distances	1460
angle distances	1974
planar 1–4 distances	523
Planes	256
Chiral centres	201
Torsion angles	892
Possible contacts:	
contacts due to single torsion	496
contacts due to multiple torsion	967
possible H-bonds	102

A different approach to using restraints has been proposed by Jack and Levitt:[79] instead of restraining stereochemistry, they minimize:

$$S = E + D \tag{8.79}$$

where D represents the difference among observed and calculated structure factor amplitudes given by (8.69) and E is a potential-energy function:[80]

$$E = \sum k_b(b_{j(\text{calc})} - b_j^0)^2 + \sum k_\tau(\tau_{j(\text{calc})} - \tau_j^0)^2$$

$$+ \sum k_\theta\{1 + \cos(m\theta_k + \delta)\} + \sum (Ar^{-12} + Br^{-6}). \tag{8.80}$$

The four terms on the right-hand side describe bond, valence angle, dihedral torsion angle, and non-bonded interactions, respectively. K_b is the bond stretching constant and K_τ the bond angle bending force constant; k_θ is the torsional barrier and m and δ the periodicity and the phase of the barrier. A and B the repulsive and the long-range non-bonded parameters. The summation extends to the j bonds, the l valence angles, the θ torsion angles, and the n non-bonded interactions between all pairs of atoms separated by at least three bonds. Despite the apparently very different approach, the energy minimization and the geometrically restrained least-squares are not too different in practice, since the final effect of (8.80) is to impose restraints on the model.

Whatever method is used, special care is needed about the weights applied to the different functions. We are in fact dealing with non-homogeneous quantities, like structure factor amplitudes and interatomic distances, so the weights of the relative observational functions must be chosen in such a way that everything is put on the same scale: an overestimate of geometric restraints will in fact produce a stereochemically perfect model associated with a very high crystallographic R factor; on the contrary, an underestimate of the same weight will result in a good R factor with unreasonable bond lengths and angles.

Restrained and constrained least squares

The two methods described above, **restrained** and **constrained** least squares, can be combined:[81] the molecule(s) is(are) considered as made up of rigid groups, and restraints are applied to distances among such groups. The quantity minimized, S, is the sum of three terms:

$$S = w_F DF + w_D DD + w_T DT \tag{8.81}$$

where DF is (8.69), DD restrains the stereochemistry, analogously to eqns (8.70)–(8.73), and DT restrains the structure from moving away from the starting set of coordinates, (8.75). All of the terms of (8.81) are functions of the atomic coordinates, generally referred to an orthogonal reference system. If a subset of these atoms is considered as a rigid group, S can be expressed, for that particular group of atoms, as a function of six rigid-body parameters, three rotational and three translational, and an arbitrary number of torsion angles, that is:

$$S = S(\boldsymbol{t}_i, \mathbf{R}_i, \Psi_{i1}, \ldots, \Psi_{im}, B_{i1}, \ldots, B_{in}) \tag{8.82}$$

where $\boldsymbol{t}_i$ and $\mathbf{R}_i$ are the translation vector and the rotation matrix of the entire group i, $\Psi_{i1}, \ldots, \Psi_{im}$ are the m torsion angles and $B_{i1}, \ldots, B_{in}$ the n temperature factors of group i.

Since the definition of rigid group is left to the user, the entire molecule or a portion of it can be constrained, or eventually some subunits. The restrained–constrained approach was originally devised for nucleic-acid refinement, but it has been successfully used in refinement of protein structures too.[82,83] A computationally quite efficient method of combining sterical restraints and rigid-body refinement has also been recently proposed.[84]

Crystallographic refinement by molecular dynamics

The development of vectorial and parallel computers offers nowadays the possibility of performing molecular dynamics calculations on complex systems, including proteins in the crystal state. The application of molecular dynamics calculations to macromolecules is a quite widespread technique (for a review, see Karplus and McCammon[85]), but its introduction in the crystallographic refinement of protein structures had been proposed only recently.[86,87]

Molecular dynamics of free atoms consists in solving the classical Newton equation of motion:

$$m_i \, \mathrm{d}^2 x_i(t)/\mathrm{d}t^2 = -\mathrm{grad}_x \, E_{\mathrm{tot}}. \tag{8.83}$$

To take into account the effect of the medium and the approximations used to calculate the total energy, dynamical effects can be better represented by a set of Langevin equations:

$$m_i \, \mathrm{d}^2 x_i(t)/\mathrm{d}t^2 = -\mathrm{grad}_x \, E_{\mathrm{tot}} + f_i(t) - m_i b_i \, \mathrm{d}x_i(t)/\mathrm{d}t \tag{8.84}$$

where b_i is a frictional coefficient, used to prevent atoms from moving away too much from their original positions, k_{B} is Boltzmann's constant, T_0 the temperature, and $f_i(t)$ a random force with Gaussian distribution and properties:

$$\begin{aligned} \langle f_i(t) \rangle &= 0, \\ \langle f_i(t) f_i(0) \rangle &= 2k_{\mathrm{B}} T_0 b_i m i \delta(t). \end{aligned} \tag{8.85}$$

The simulation starts from an initial set of coordinates. To each atom is assigned a velocity, usually at random from a Maxwellian distribution corresponding to the temperature selected, and eqn (8.83) or (8.84) is integrated at a given temperature for a given time.† New velocities are then assigned, eventually at a new temperature, and the calculation continued. The simulation is normally performed for a short period of time, usually of the order of few picoseconds.

In the crystallographic refinement of macromolecules by molecular dynamics, the X-ray information is used to restrain the energy of the system. The total potential energy is, in fact, considered as the sum of two terms:[88]

$$E_{\mathrm{tot}} = E_{\mathrm{emp}} + E_{\mathrm{eff}}. \tag{8.86}$$

E_{emp} represents an empirical potential energy, analogous to that defined by (8.80), E_{eff} is a sort of 'experimental' potential energy, and is considered as

† Numerical integration can be performed using for example the Verlet algorithm.

the sum of three terms:

$$E_{\text{eff}} = E_{\text{xray}} + E_{\text{P}} + E_{\text{NB}}. \tag{8.87}$$

E_{xray} in (8.87) describes the difference between observed and calculated structure factor amplitudes:

$$E_{\text{xray}} = (w_{\text{A}}/N_{\text{A}}) \sum_h w_h (F_{(\text{obs})} - F_{(\text{calc})})^2, \tag{8.88}$$

w_{A} is a factor which puts E_{xray} on the same scale as the empirical potential energy term and N_{A} is given by $\sum w_h (F_{\text{obs}})^2$, to ensure that w_{A} is independent of the resolution range used. The terms E_{P} and E_{NB} can be included to take into account experimental information about MIR phases and crystal packing respectively.

Molecular dynamics simulations can be performed at ambient temperature,[87] or at higher temperature, as in the version called **simulated annealing**.[86,88] The latter consists in starting the simulation at room temperature, say 300 K, and heating up the system (for example at 2000–4000 K) and subsequently cooling down at the initial value. The advantage of going to high temperatures, unreasonable from the biological point of view, is that model can come out of local minima, and the ratio of convergence of the method is increased with respect to classical least squares.†

The result of molecular dynamics calculations is a family of conformations, but the constraints imposed by X-ray data restrict these conformations to all those with the lower crystallographic R factor.

The strategy of the refinement of protein structures

The initial model of a protein structure is, very often, not good enough to allow for a fully automated refinement. Indeed, if some serious errors are present in the model, for example the polypeptide chain is positioned more or less correctly but the amino acids are shifted one residue or more along the chain, an automatic procedure will hardly recover from that error. Besides, the radius of convergence of constrained or restrained least-squares methods can be evaluated to be around a half the resolution of the data used, that is not more than 1–1.5 Å. At the beginning of the refinement, to speed up convergence, medium-resolution data (3.0–2.5 Å) can be employed. Since the number of observations at that resolution is quite low, an overall temperature factor for all the atoms is used. Afterwards, the resolution is gradually extended, solvent molecules included, and isotropic individual B factors applied.

The same seems not to be true using the simulated annealing technique, which allows a more rapid and automatic convergence: the heating makes it easier to get out of the false minima without manual intervention. Some more experience nevertheless must be accumulated. A fully automated refinement was possible in the test case of the enzyme aspartate aminotransferase, refined with data at 2.8 Å resolution starting from MIR

† The term temperature must be regarded cautiously here: it does not indicate a physical temperature, but rather a parameter controlling the refinement. The simulated annealing is in fact virtually equivalent to the Metropolis algorithm (Metropolis, N., Rosenbluth, M., Rosenbluth, A., Teller, A. and Teller, E. (1953). *Journal of Chemical Physics,* **21,** 1087–92).

coordinates.[89] A careful comparison between the model of myohaemerytrin refined, starting from the same model, in one case with several cycles of restrained least-squares and manual rebuilding and in the other case with the 'simulated annealing' technique without manual intervention has been reported.[90] The two structures compare quite well, but molecular dynamics procedure could not bring the refinement to completion in a fully automated way: manual intervention is still necessary to correct for gross errors (say more than 3–5 Å in the main chain) and to include solvent molecules. Nevertheless, simulated annealing can save a lot of human effort, at the expense of quite a long computational time.

For the above mentioned reasons, some cycles of automatic minimization are usually followed by recalculation of the electron density maps and manual adjustment or rebuilding of the model.

In recalculating electron density maps, a major problem is the choice of the phases and the coefficients to be used. MIR phases suffer from all the errors described on p. 546, and the isomorphism of heavy-atom derivatives does not extend generally beyond 3 Å or so: high-resolution electron density maps are seldom achieved with MIR phases. On the other hand, calculated phases tend to reproduce the model used in calculating them, and an electron density map obtained wih calculated phases may be strongly biased. For these reasons, phases coming from independent sources, e.g. the phases from isomorphous derivatives data and those calculated from the model, can be combined to produce an improved electron density map.[91,92] The probability distribution of calculated phases, $P_{calc}(\varphi)$, can be evaluated by using a procedure due to Sim[20] (see Appendix 5.H, p. 393) and can be used, along with the 'experimental' probability $P_{tot}(\varphi)$, to obtain a combined probability distribution:

$$P_{comb}(\varphi) = P_{tot}(\varphi)P_{calc}(\varphi). \tag{8.89}$$

The new figure of merit, m_{comb}, obtained by (8.38) can be used to calculate a best combined electron density map. If only calculated phases are available, the Sim formula can be used to weight the Fourier coefficients. A scheme of the possible refinement procedure is illustrated in Fig. 8.23.

The coefficients and phases more commonly used for Fourier syntheses are listed below, but other combinations of them are possible:

$$mF_{obs} \exp(i\varphi_{MIR}), \tag{8.90}$$

$$w_{comb}F_{obs} \exp(i\varphi_{comb}), \tag{8.91}$$

$$w_{Sim}\,|F_{obs} - F_{calc}| \exp(i\varphi_{calc}), \tag{8.92}$$

$$w_{comb}(2F_{obs} - F_{calc}) \exp(i\varphi_{comb}), \tag{8.93}$$

$$w_{Sim}(2F_{obs} - F_{calc}) \exp(i\varphi_{calc}). \tag{8.94}$$

Equation (8.90) gives the coefficients of a classical observed Fourier synthesis. In principle, they could be used during all the stages of the refinement, but MIR phases can be improved by phase combination. Furthermore, very often they do not extend to high resolution, and calculated phases must be used instead, when a reasonable atomic model is available. To reduce bias, a Fourier map with combined phases can be calculated. Coefficients (8.92) correspond to Fourier-difference maps with calculated phases. If they are calculated from a partial model, they are

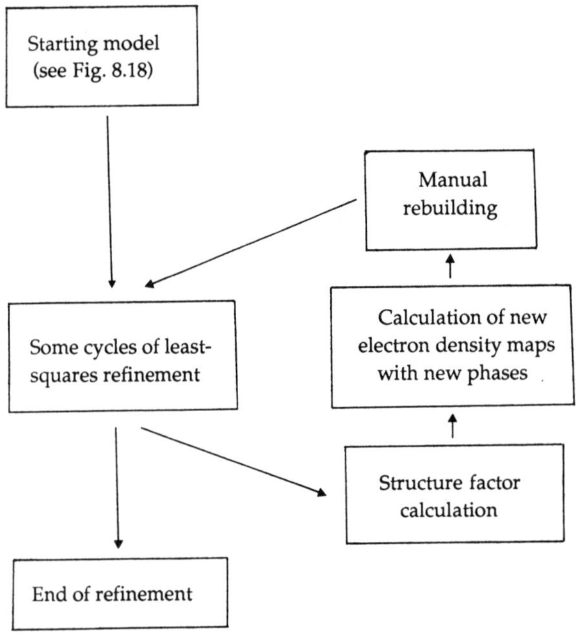

Fig. 8.23. Block diagram summarizing the phases of structural refinement. The starting model is obtained by one of the procedures schematized in Fig. 8.18. The numbers of iterating cycles necessary to reach convergence can be quite variable, depending on the quality of the initial model. If M.I.R. phases are not available (i.e. the structure has been solved by molecular replacement techniques), only maps with calculated phases can be used. Some of the coefficients used in electron density map calculation are described on p. 571.

known as **omit maps** and can be useful in positioning portions of the molecule that did not appear clearly in the M.I.R. maps. Coefficients (8.93) and (8.94) correspond to a combination of a Fourier electron density map and a difference-Fourier: if F_{obs} and F_{calc} are very similar, the magnitude of the coefficients approaches to (8.91); otherwise, terms with F_{obs} greater than F_{calc} will have a higher weight. The practical result is an enhancement of the regions of density where severe errors are present in the model. The weight used for map calculation can be a modification of the Sim scheme or the weight obtained by the combination procedure.

It must be finally remembered that the crystallographic R factor of a refined protein structure is not comparable with that of a small-molecule crystal. Besides, owing to the low ratio between observations and variables, the R factor alone does not represent an index of the reliability of the model: molecular stereochemistry must be correct also. For example, deviations from the ideality of bond lengths, valence angles, and restricted torsion angles of about 0.01 Å, 2°, and 1°, respectively, are considered reasonable. The crystallographic R factor of a crystal structure refined with these restraints at 2 Å resolution can be reduced to a value between 0.15 and 0.20.

Protein structure

The three-dimensional structure of a globular protein is the resultant of a very large number of interactions. This makes it at the same time stable and flexible: a modification in a specific site can, in fact, be assimilated by small local adjustments, without altering the overall conformation. In other cases, modifications taking place in a region of the macromolecule can be transmitted and influence the conformation in zones far apart from where the movement originated, a phenomenon called cooperative or allosteric effect.† It must also be remembered that the medium may have a relevant influence on the conformation of the macromolecule, that *in vivo* operates in a non-homogeneous environment, like for example the cell cytoplasm or the cellular membrane. Flexibility can consequently be useful, or even necessary, to the protein to fulfil its biological functions. On the contrary, the picture of a molecular model resulting from the X-ray structure determination is, at least apparently, a static one, the only 'dynamic' information residing in the atomic temperature factors. Nevertheless, this must not be considered in contrast with reality, since the X-ray structure represents an average, over time and space, of single structures around a conformational minimum, which is representative of the conformation of the protein molecule in a solution of the same medium. On the other hand, a description of a dynamic situation is quite difficult, particularly for a complex molecule: the sort of static image that will turn out from the following discusion must be considered a necessary simplification as well as the starting point for the comprehension, at the molecular level, of the behaviour of biological systems.

General aspects

Proteins are polymers of the 20 natural aminoacids (some symbols and conventions relative to amino acids are reported in Appendix 8.D). They may contain other groups, often relevant for the conformation of the molecule, like haems, prosthetic groups, carbohydrates, and so on, or they can coordinate cations (see p. 582). Nevertheless, the **building blocks** of proteins are the amino acids, and the protein fold‡ is the first and perhaps more general aspect of its structure.

A polypetide chain of a given sequence, in the appropriate conditions, can refold spontaneously into its final three-dimensional structure:[93] all the necessary structural information must be contained in the amino acid sequences. It should therefore be possible, at least in principle, to predict from it the structure of a protein. Unfortunately, despite several efforts, this remains a hope for the future. Moreover, some simple, schematic rules relative to the forces that contribute to stabilize a structure can be summarized: they are, however, only indications and exceptions are quite common.[5,94]

1. Groups potentially able to form hydrogen bonds will be positioned accordingly: or at the surface of the molecule, where they can be hydrogen bonded with the solvent, or, when in the interior, a hydrogen-bond donor will be close to an acceptor. Not necessarily all the possible interactions of this kind will be realized, but most of them usually are.

2. Polar residues in a water-soluble protein will preferably be located on the surface of the macromolecule and the hydrophobic ones in the interior (the contrary apply for lipid-interacting proteins).

3. Charged groups will be close to the surface, where they can be

† Cooperativity and allostery are not exactly synonymous, but in this context no distinction is made about them.

‡ The term 'fold' is not used here to indicate a process, but simply the way in which the main chain is wrapped on itself.

neutralized by solvent ions. When forced to be inside, they will be stabilized by a proper counter-ion.

As a corollary of the previous rules, particularly rule (2), the three-dimensional structure of the proteins will be as compact as possible. For example, the packing density of ribonuclease S, calculated approximating the atoms to spheres of van der Waals radius, is about 0.75,[95] approximately the same value found for close-packed spheres.

Levels of organization of proteins: secondary structure

The concepts illustrated before, despite generalities, refer to single interactions and in practice they are not well suited to describe overall features or general aspects of the structure. A different, and perhaps more useful, approach to understanding the structure of a protein is the identification, under the apparent complexity, of regular schemes and motifs. Looking at the structure under this aspect, a hierarchical organization becomes apparent and several levels of organization, of different complexity, can be distinguished. The first local order is imposed by the hydrogen bonds formed between the N–H and the carbonyl oxygen of the polypeptide chain: the regular structure generated by these quite strong interactions is called **secondary structure**† and results in a limited number of possible conformations, briefly described below.

1. Helices. The more common among the repetitive motifs found in polypeptide chains is the α-helix. The theoretical α-helix, predicted by Pauling *et al.* in 1951, is characterized by 3.6 residues per turn and a pitch of 5.4 Å. A hydrogen bond is formed between the carbonyl oxygen of residue n and the N–H of residue $n + 4$. The number of atoms covalently bound in the ring closed by the hydrogen bond is 13 (Fig. 8.24). Along with the number of residues per turn, this number is sufficient to characterize the secondary structure: the α-helix is consequently called a 3.6_{13}-helix. Other less common helical structures are the 3_{10}-helix and the π- or 4_{16}-helix, characterized by an exact repetition after three and four residues respectively and a hydrogen bond between residues n and $n + 3$ or $n + 5$. A special case is represented by the α_{II}-helix, which is a sort of mixture between the α- and the 3_{10}-helix.

Every type of helix can also be defined by its φ and ψ torsion angles: their indicative values for the most common repetitive structures are reported in Table 8.5. Of course, real helices present backbone torsion angles that deviate from the ideal ones and, in addition, bending and local irregularities are observed.[96] Nevertheless, they are the most regular structures found in proteins. Since natural amino acids, except glycine, are chirals, right-handed and left-handed helices are energetically not equivalent: helices found in proteins are in practice always right handed. Amino acids assuming the torsion angles characteristics of a left-handed helix, when found, are confined to very short pieces of chain (one or two residues), often at the end of a right-handed helix connecting a β-strand.

† For historical reasons, the amino acid sequence is called **primary** structure.

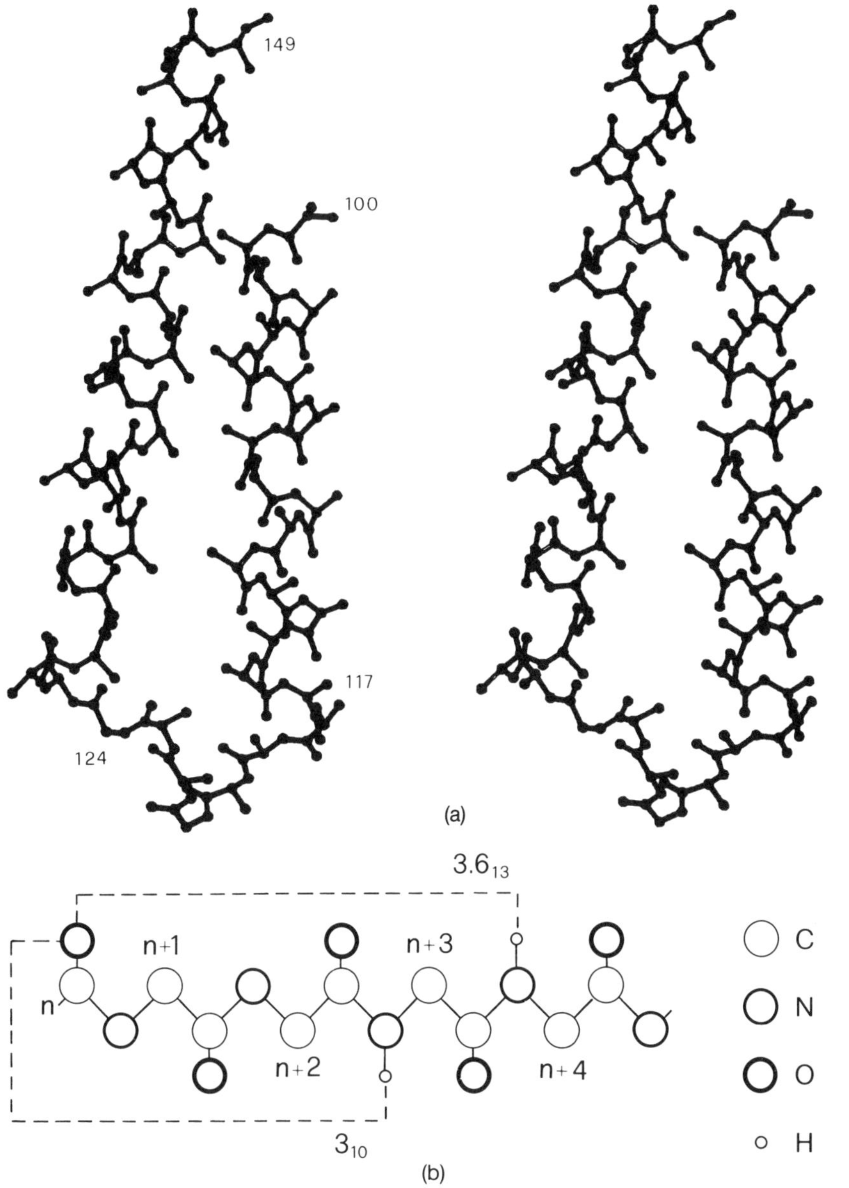

Fig. 8.24. (a) Stereo view of two α-helices of myoglobin[99] (data from PDB), connected by a short piece of chain (to simplify the representation, for every side chain only the β-carbon is drawn). Helix 7 goes from residue 100 to 118 and helix 8 from 125 to 148. Residues from 116 to 119 participate also in a type I β-turn. From 119 to 124 the structure is less regular. The two helices are not exactly aligned, but their axes are slightly bent, a classical situation of the packing of helices, (b) Scheme of the hydrogen bond pattern for the most common types of helices: in the α-helix, 13 atoms are included in the ideal ring from the carbonil oxygen to the N–H. Only 10 atoms in the 3_{10}-helix.

Table 8.5. Approximate torsion angles for some common regular structures (from IUPAC-IUB commission on biochemical nomenclature, *Journal of Molecular Biology,* **52,** 1–17 (1970)). (References about the sources of the reported values can be found in the reference; ω is assumed to be 180°)

	φ(deg)	**ψ(deg)**
Right-handed α-helix	−57	−47
Left-handed α-helix	+57	+47
3_{10}-helix	−20	−54
Parallel-chain plated sheet	−119	+113
Anti-parallel-chain pleated sheet	−139	+135
Polyglycine	−80	+150
Polyproline II	−78	+149

One of the reasons that helices are quite common is the possibility to accommodate in a very short length a large number of residues: all side chains in fact point away from the helical axis, giving rise to quite an efficient packing of bulky groups, one on the top of the other.[97] The only residue that cannot be accommodated in a α-helix is proline, due to the steric hindrance of its side chain and to the fact that its nitrogen is lacking the hydrogen necessary to form the hydrogen bond. Proline can be fitted properly only in the first turn of a helix, but will interrupt it, or introduce a bend, if present in any other place. A special kind of secondary structure, called polyproline helix, has sometimes been observed.

Another important peculiarity of helical structures is represented by their electrostatic properties: since all the dipoles of the peptides are pointing in the same direction, the result is a net total dipole for the helix[98] that may contribute, for example, in stabilizing the binding of charged species.

2. β-structures. A β-strand is a portion of polypeptide chain in a nearly extended, pleated conformation. It is less regular than helices and its torsion angle values reported in Table 8.5 must be considered only indicative, since large variations are observed in practice. A β-strand is normally not stable in itself, but it is stabilized by the contact with another β-strand, with which it forms hydrogen bonds. According to the relative direction of the two chains† β-strands can be parallel or antiparallel (Fig. 8.25): they differ in the way the hydrogen bonds are formed, making the latter slightly more stable[101] and more commonly observed in proteins. Since β-chains are quite flexible, this kind of structure can present irregularities and defects: a jump of one residue in a strand is for example called a β-bulge.

Several parallel or antiparallel β-strands form a so called β-sheet. Since in a β-strand amino acid side chains point alternatively in opposite directions, a β-sheet has the possibility of having a polar surface and an apolar one. Finally, strands in a sheet are not exactly aligned, but are twisted with respect to each other.[102]

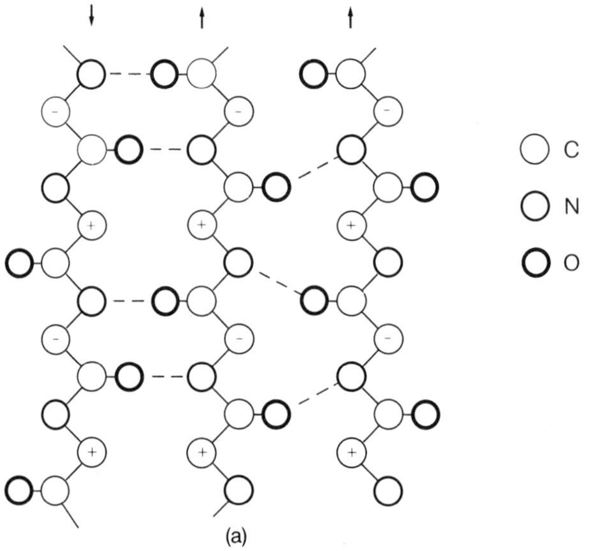

Fig. 8.25. (a) Schematic drawing of the hydrogen bond pattern for different types of β-sheets. The first two β-strands run antiparallel, the third is parallel to the second one (directions of strands are indicated by arrows). Hydrogen bonds are differently oriented in the two cases, the antiparallel being more favoured and more commonly found in proteins. A plus sign on the α-carbon indicates the side chain coming up from the sheet, and a minus a side chain going away from the reader. (b) Stereo view of three β-strands of ribonuclease[100] (data from PDB, only C_β is drawn for every side chain). The three strands 71–75, 105–110, and 120–124 run antiparallel each other. Notice that residues 121 and 122 have the side chains pointing in the same direction, an irregularity called a β-bulge.

† The chain is considered to run from the N to the C terminus.

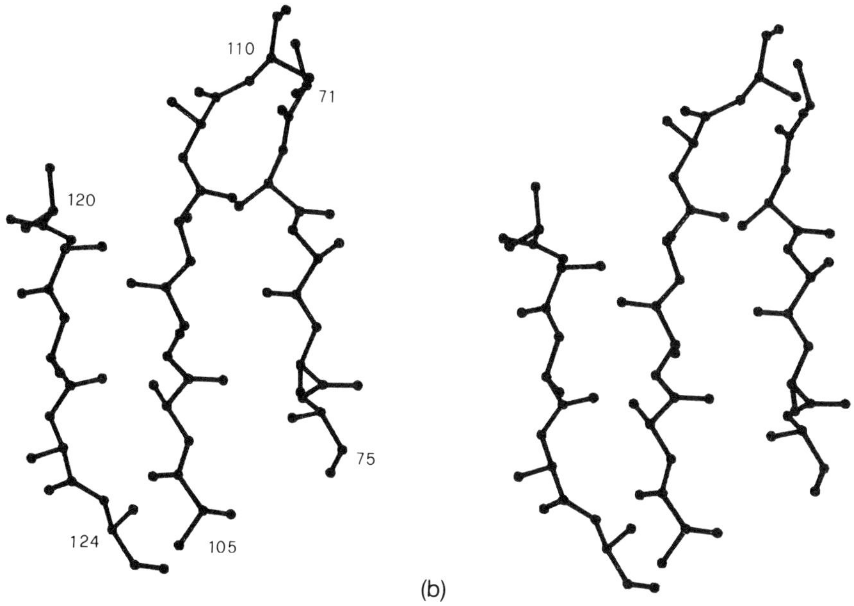

Fig. 8.25. (*Continued*).

3. Others. Short tight loops connecting two strands are called β-turns, or hairpins, or reverse turns. Different kinds of classifications have been attempted: one of them, based on the hydrogen bond patterns, is that of Milner-White and Poet.[103] The more classical β-turns are called type I, II, and III (and type I′, II′, and III′ those with φ and ψ torsion angles reversed). A schematic pattern of two turns is illustrated in Fig. 8.26.

In real structures, short pieces of polypetide chains not falling in one of the previous categories are normally found. In general their φ and ψ torsion angles correspond to those of one of the structures discussed above, but nevertheless that portion of the molecule cannot be classified as one of the previous categories.

A last important feature strongly influencing the polypeptide chain conformation is the presence of a disulphide bridge: the sulphur atoms of two cysteines form a covalent bond, connecting two pieces of polypeptide chain or even two different chains. This results in a strong stabilization of the three-dimensional structure, at the expense of some local irregularities.

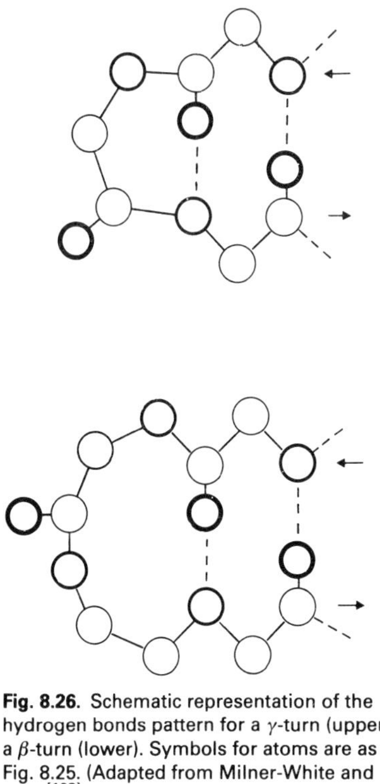

Fig. 8.26. Schematic representation of the hydrogen bonds pattern for a γ-turn (upper) and a β-turn (lower). Symbols for atoms are as in Fig. 8.25. (Adapted from Milner-White and Poet.[103])

Polypeptide chain description

In describing a polypeptide chain, the only relevant variables are in practice the two dihedral torsion angles around the N–C_α and C_α–C bonds, called φ and ψ: the torsion angle ω is very close to 180°, owing to its partial double-bond character. Therefore, only $2n-1$ numbers are enough to describe the fold of a protein composed by n amino acids.

One convenient way of representing the main chain dihedral angles is the so-called **Ramachandran plot**,[104] where every amino acid is represented by a point of coordinates φ and ψ. In such a plot, regions forbidden for sterical reasons can be individuated and they represent a large portion of the total area: the backbone torsion angles of all the amino acids, with the exception

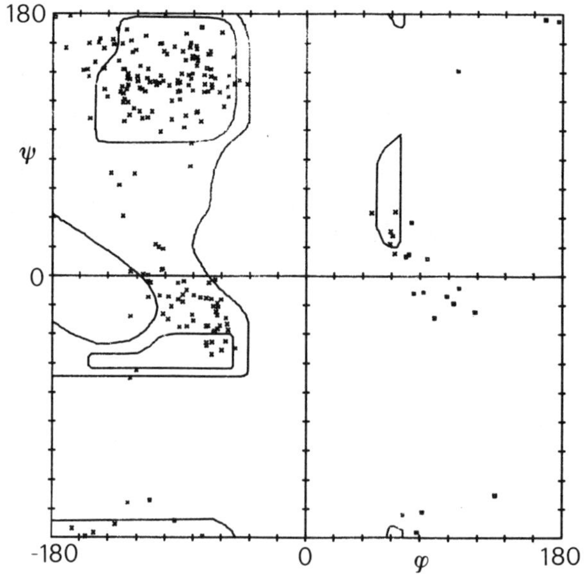

Fig. 8.27. Ramachandran plot for the model of bovine β-trypsin, orthorhombic form, refined to a crystallographic R factor of 0.18 up to a resolution of 1.8 Å[105] (data from PDB). Small squares represent glycine residues, crosses all the others. The origin of the diagram is in the centre. Contour lines correspond roughly to fully allowed and outer limit regions.

of glycine which has much more conformational freedom owing to the absence of substituents at the α-carbon, should not fall into those areas. Ramachandran plots can be used during the refinement to check for the quality of the structure: amino acids falling into forbidden regions must be corrected, and values at the border or in unfavourable zones considered suspiciously. An example of such a plot is given in Fig. 8.27. Since every type of secondary structure described in the previous subsection is characterized by its own pair of theoretical φ and ψ values, it corresponds to a point on the plot. From the diagram it is consequently possible to individuate immediately the amount of α-helix, β-sheet, and so on, present in the structure. Any information about the connectivity among residues is, on the contrary, lost.

A possibly more informative, but less used, method of mapping the folding of the main chain is the diagonal plot.[106] It is a matrix with the amino acid numbers along both axes: if the distance between two residues is less than a predefined amount, the corresponding matrix element is darkened. In this kind of graphic, the major structural elements are easily visualized: an α-helix will produce a thickening along the diagonal and two antiparallel β-strands a narrow line perpendicular to it; domains can be recognized too, and structural homologies among proteins individuated. Furthermore, the information of a diagonal plot is in a form that can easily be stored in a computer. An example of a diagonal plot is reported in Fig. 8.28.

Higher levels of organization: tertiary and quaternary structure, domains, and subunits

We have previously seen that, despite the apparently very large degree of freedom of a polypeptide chain, proteins can be considered as made up by a very limited number of secondary structure elements, that is α-helices,

β-strands and turns, plus some pieces of chain in conformations more difficult to classify. The three-dimensional relationships among secondary structure elements is called **tertiary structure** or, preferably, **super-secondary structure**. A careful examination of the structures solved to the present has allowed us to identify some macroscopic concepts that simplify the description of the way in which the secondary elements are held together. Of course these rules are not necessarily general, since they are based on quite a limited set of structures.

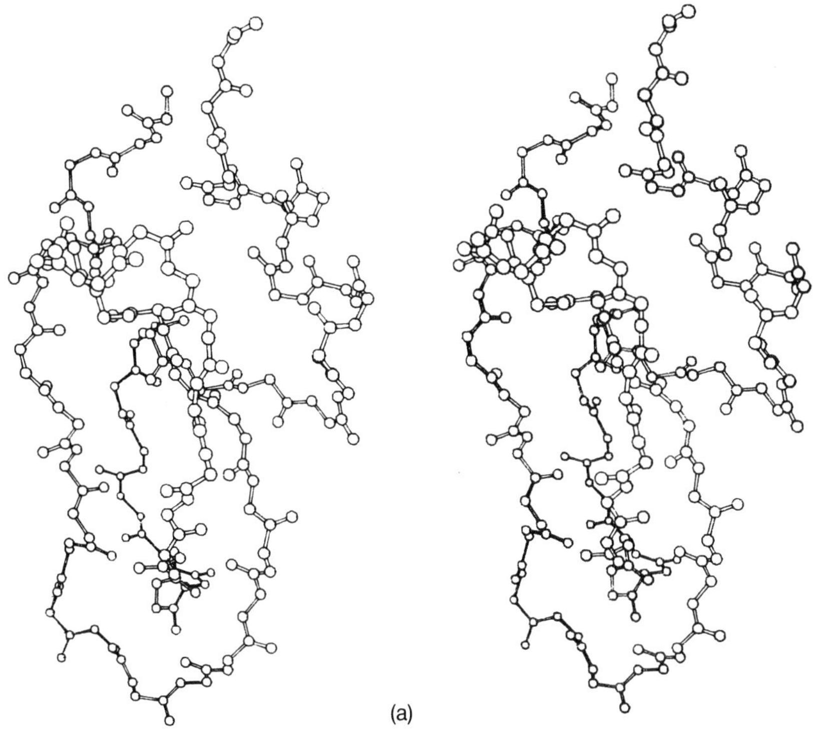

(a)

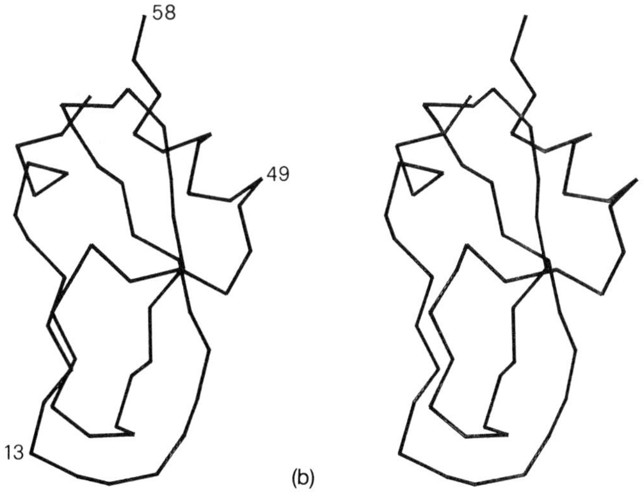

(b)

Fig. 8.28. (a) The bovine pancreatic trypsin Inhibitor, a small protein (58 amino acids) refined to an R factor of 0.16 at 1.5 Å resolution[107] (data from PDB). The backbone only, that is the atoms C_α, C, N, and O of the main chain, are shown. (b) α-carbon chain trace of the same structure. It is composed by one turn of helix, two long bent β-strands and a final α-helix from residues 47 to 56. All these elements are disposed in a quite complex way, difficult to describe. (c) Diagonal plot for the same protein. The plot was produced using the program DGPLOT (Swanson and Bernstein, distributed with PDB). Amino acid numbers are reported on both axes, and numbers inside the diagram correspond to the distance in Å, rounded to the next integer, between the two α-carbons. Distances above 9 Å are represented by letters. In the diagram the α-helix close to the C terminus of the molecule is represented by the thickening along the diagonal (A in figure) and the strands from 29 to 43 and from 6 to 25, close to each other, by a line perpendicular to the diagonal (B in figure). Other spatial relationships are immediately evident from the diagram: region 43–48 is close in space to 20–33 (marked C in figure).

```
                 1234567890123456789012345678901234567890123456789012345678901234567888

ARG    1    046879C              ACA    BC            CB          BA975757+
PRO    2     04568A              B B                  AA          BBB8798A+
ASP    3      04568B             BCA                  CA9           B9CB  +
PHE    4       04559AC           A9BA                C866AC     C  A9CC  +
CYS    5        04689C         B8687BCAAAC           A859B     C9AB8699  +
LEU    6        0479C          97769BABB             BA8B     C   C9BB   +
GLU    7        0468C          9888B    C            C987A          B    +
PRO    8        046AC         B8879BC  BBCB C  /     C9A8A                +
PRO    9         0479        CA9688B  AA9798B     C98967A      C          +
TYR   10         0469BC    CCAB9CC          C9979AA976988C                +
THR   11          04789AA9AABB            B864588879CB9C                  +
GLY   12          0468ABB C               BA7676669CCB    B               +
PRO   13          0479C                   C978779C                        +
CYS   14           0469A                  B85668A                         +
LYS   15           0479C                  98579C                          +
ALA   16            046AC                 C87569CC                        +
ARG   17             0469                 C96667B                         +
ILE   18             046A                 A85566A  B     BB9C             +
ILE   19              047A                CA665699       A989B     C      +
ARG   20              047A                A86666AAC   BCCA66578BB9         +
TYR   21               0479             A654699     CCB8656669979CCBC      +
PHE   22                0469CBB85667AA            BA9556998BA79CA9B        +
TYR   23                0469876469A               CA689CB9BA67A968B        +
ASN   24                 04656667AB                9BC   B    AA C9AC      +
ALA   25                  045578A                  B          CB C99A      +
LYS   26                  0468AB                                   CC      +
ALA   27                   04689                             C    CB       +
GLY   28                   047A                            C   B9CC979C+
LEU   29                   047A               C  C C9AB87AB87A       +
CYS   30                   047A               AA9B969A67AA88C        +
GLN   31                   047AC              BAABA7BC9A    CC       +
THR   32                    0479              B99A98C AC              +
PHE   33                    046AC  C          B9ABBB                 +
VAL   34                     0469B B    ABB                          +
TYR   35                     046898ACB8AA                            +
GLY   36                      04699C   B                             +
GLY   37                      0477B   BCC                            +
CYS   38                      0469C B                                +
ARG   39                      047ACB                                 +
ALA   40                      04787AC                                +
LYS   41                       0466A                                 +
ARG   42                       0458B    CA   BB                      +
ASN   43                       047ACC B8BC99C                        +
ASN   44                        0479ACA8B  BB                        +
PHE   45                         04679659A9AC                        +
LYS   46                         047978BCC                           +
SER   47                          045569ABC                          +
ALA   48                          046568AAA                          +
GLU   49                           045569A9                          +
ASP   50                           0455689C                          +
CYS   51                            045567A                          +
MET   52                            045548B+
ARG   53                            04668A+
THR   54                            0467A+
CYS   55                            0459+
GLY   56                            047+
GLY   57                            04+
ALA   58                            0+
TER   58                                                             +
```

B   C   A

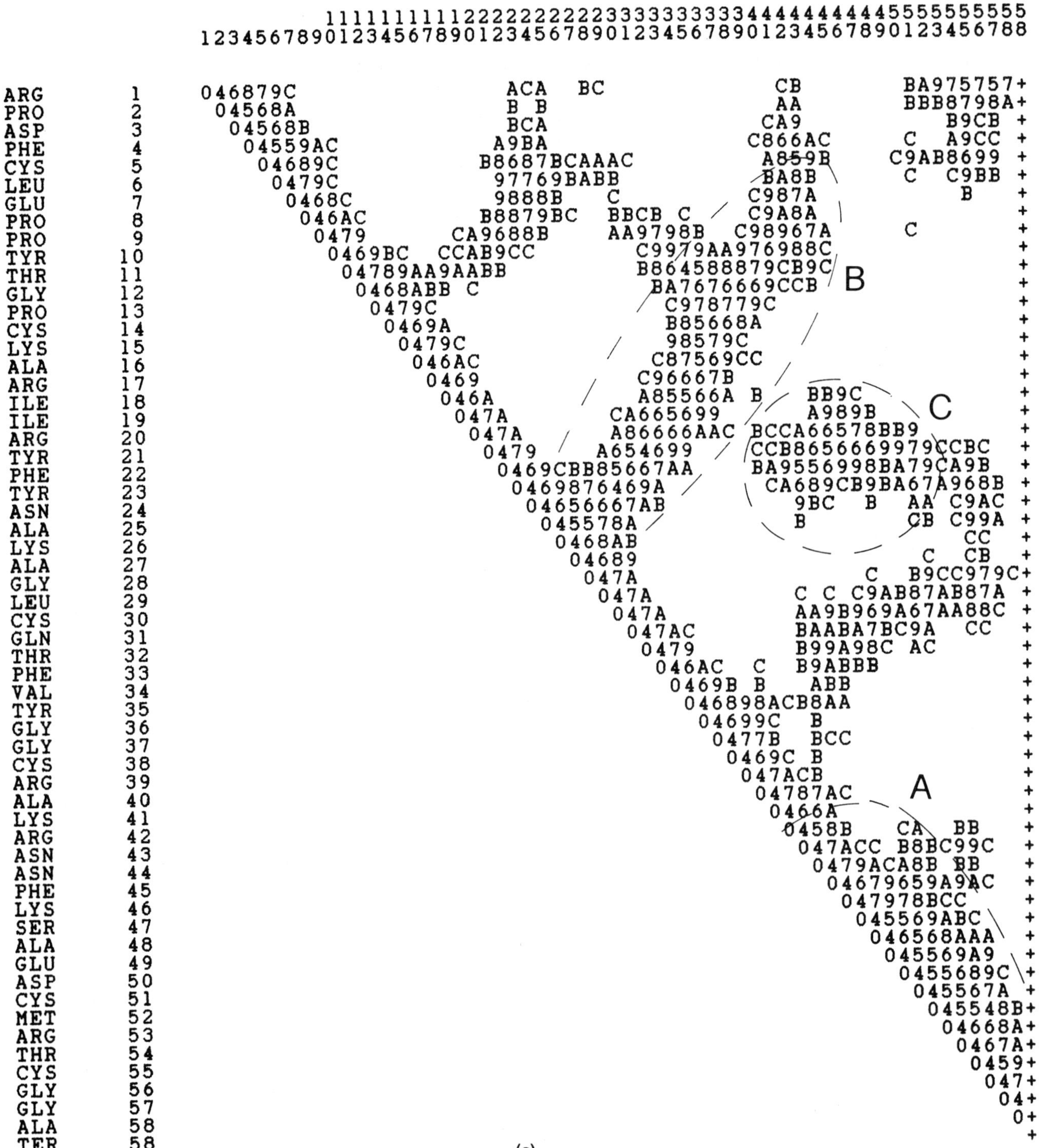

(c)

Fig. 8.28. (*Continued*).

One example is given by packing of α-helices: it has been shown that helices tend to pack not parallel to each other, but with the axes forming specific angles: according to the way they are disposed, they can be grouped in only three classes, each of them characterized by a specific angle.[108]

Connections between strands in a sheet too follow quite strict rules, thoroughly described by Richardson.[109] The topology of the sheet can be illustrated by simplified diagrams, like those reported in Fig. 8.29. The physical meaning of such a diagram is discussed on p. 585.

Every polypeptide chain does not necessarily fold into one single unit: a portion of it can be structurally independent and is called a **domain**. Although the concept is not unequivocal, its existence is widely recognized. The most simple definition of domain is that given by Richardson:[109] a domain is a contiguous portion of polypeptide chain folded into a compact, local, independent unit. Domains are sometimes easy to recognize, as in the case of immunoglobulins, or obvious, as for calmodulin (Fig. 8.30). Sometimes they are not simple to individuate, or are even questionable. Whether a domain must be thought of as a potentially independent, stable unit or only as a compact globular region contiguous in space, is matter of discussion. The domain concept is in any case justified by the classification given on p. 585.

Proteins may eventually be made up by more than one polypeptide chain: often several of them, equivalent or different, interact to form a protein molecule. The assembly of such subunits is called **quaternary structure** and is in general essential for function: dissociation of a multimeric enzyme may result for example in a change or loss of activity. Nevertheless, subunits can be considered as independent structural elements: they can often be separated and crystallized and their overall conformation is in general not too different from that of the subunit in the macromolecular complex from which they originate. A relative reorientation of subunits has been associated with allosteric properties. A classical example of it is mammalian haemoglobin, a tetramer formed by two distinct chains, denoted α and β. Two $\alpha\beta$ dimers are related by an exact twofold axis, which in some crystal form corresponds to a crystallographic one: from the oxygenated to deoxygenated state the molecule undergoes a substantial change in the quaternary (and also tertiary) structure, paired with a variation of affinity for oxygen.[111]

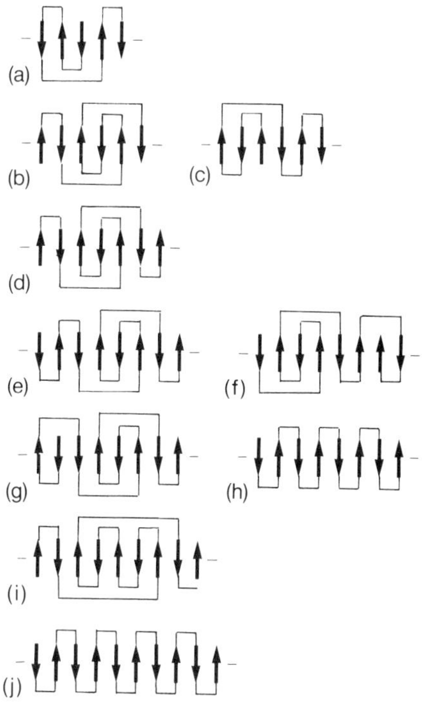

Fig. 8.29. Topology of the proteins classified as 'Greek key β-barrels'[109] (see p. 585 for more details about the classification). Barrels are unrolled and seen from the exterior. Arrows represent β-strands and thin lines connections among them. Dashes indicate that the first strand is in contact with the last one. Molecules represented are: (a) staphylococcal nuclease; (b) pyruvate kinase domain 2; (c) trypsin domain 1 and 2; (d) immunoglobulin C; (e) Cu, Zn superoxide dimutase; (f) prealbumin; (g) plastocyanin; (h) plasma retinol-binding protein; (i) immunoglobulin V; (l) fatty acid-binding proteins. Some patterns represent a family of proteins: for example, (h) corresponds also to β-lactoglobulin and bilin-binding protein.

Fig. 8.30. C_α tracing of calmodulin, a small protein that binds calcium in four different binding sites[110] (coordinates from PDB). Two 'globular' domains, composed of about 60–70 amino acids each, are completely separated in space, connected only by a long α-helix.

When a macromolecule is made up by equivalent subunits, a relevant aspect is represented by its internal symmetry. The simplest case is represented by a rotation axis: very often, in a dimeric protein the two subunits are related by a twofold axis. Multimeric proteins may present threefold and also five- or sixfold axes. A combination of rotation axes can give rise to higher point symmetries, like 222 or 32. Possibly the most complex internal symmetry is represented by the coat of spherical viruses, composed by hundreds of copies of the same protein arranged in a icosahedral symmetry.[112] It must be remembered that some or all of the internal symmetry elements may or may not correspond to those of the crystal (see p. 538).

Groups other than amino acids

The product of biosynthesis is a molecule containing only the 20 natural amino acids. To fulfil their function, proteins may need the presence of other groups: therefore, they can undergo post-translational modifications or interact with other molecules. These modifications may have a strong influence on their structure.

A post-translational modification is a chemical reaction generally involving the side chain of some amino acid.† The functional groups bound are called prosthetic groups and may have the function to assist enzymes in chemical reactions, or to add to natural amino acids properties they lack, for example photochemical ones. A quite common modification is phosphorylation, largely used in nature to control enzymatic activity.[113]

Glycoproteins are the result of binding of polysaccharides. They may be bound through the oxygen of hydroxylated residues, like serine or threonine, or through the nitrogen of asparagine. Carbohydrates are quite hydrophilic and their content can reach a high percentage of the total mass of the protein molecule: glycoproteins consequently present peculiar properties and their function is matter of discussion.[114] Very few of them have been crystallized, and in general the polysaccharide moiety has been found disordered in the crystal.

Cations bound to a protein can be divided in two categories: those which simply neutralize charged groups at the surface of the molecule and depend on the ionization conditions of these groups (and ultimately on pH), and those strongly coordinated, required for the stability of the conformation. The latter can influence protein activity by stabilizing one particular conformational state. The more common bound cation is Ca^{2+}, which is very abundant in cytoplasm. Among proteins that bind it, the most peculiar is calmodulin (Fig. 8.30), which has four calcium binding sites: the level of calcium bound to the protein modulates its function. Other bivalent cations commonly interacting are Zn^{2+}, Cu^{2+}, and Fe^{2+}, but many other metals present in nature can be found, in different oxidation states. As an example, the iron–sulphur binding site in ferredoxin is shown in Fig. 8.31.

A last important case is represented by the small, generally organic, compounds non-covalently bound to proteins. The macromolecule in this case may have the function of carrier or of storage site.

† Other possibilities are: a modification of the N or the C terminal part of the molecule, or a proteolytic cleavage of a peptide bond.

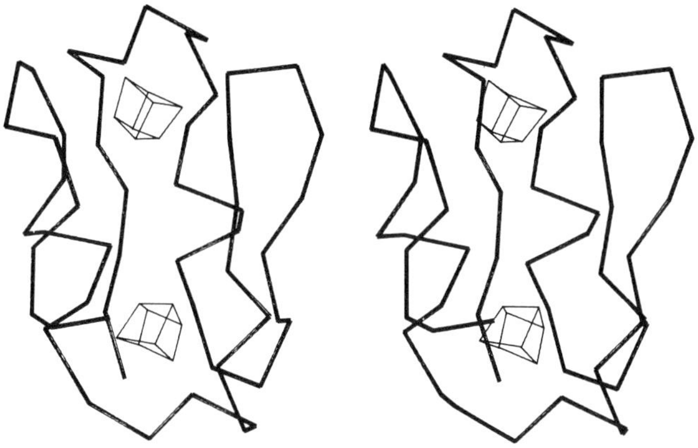

Fig. 8.31. The two sulphur–iron clusters in *Peptococcus aerogenes* ferredoxin are shown, along with the polypeptide chain trace of the protein[119] (coordinates from PDB). Fe and S atoms are disposed alternated at the vertices of a distorted cube: each Fe atom is surrounded by four S atoms, the fourth being that of a cysteine residue.

Thermal parameters and disordered structures

Special care must be used in evaluating temperature factors obtained from the refinement of a protein crystal structure. Experimental B factors include, besides thermal vibrations, the static disorder, which in such kind of crystal is particularly relevant. Owing to the high content of solvent, superficial groups of the macromolecule have the effect of partially ordering the solvent, but at the same time, as a consequence of this contact, they become very mobile. It has been shown[115] that it is possible, for very well refined structures, to distinguish the contribution to B factors of real thermal vibration from the static disorder. Moreover, during the refinement procedure thermal parameters are usually restrained, that is their variation is in some way smoothed down. In any case, the comparison of temperature factors for the same structure, ribonuclease A, independently refined by two different groups using different data[116] shows a quantitative agreement in the trends, that is regions with an high B factor roughly correspond in both structures. As a general rule, in well refined structures main chain atoms present lower thermal motion than side chains, or in any case less disorder.

In looking at crystallographic results, it must be kept in mind that a very high B factor for some residues could be either due to an intrinsic disorder of that part of the molecule, or an indication of a misinterpretation of the electron density. In some cases, small parts of the structure, often at the beginning or at the end of the polypeptide chain, or some loops protruding towards the solvent regions, are very mobile and cannot be seen at all in the map. In exceptional situations, two conformations could be differentiated for some residues.[23]

The disorder can also have some functional role, as is sometimes the case for allosteric proteins, where two conformational states are present, one of them characterized by a portion of a disordered chain that becomes ordered in the other state.[117,118] In any case, as has been pointed out by Richardson,[109] this is a case in which the crystallization process introduces a bias in the results, since less ordered or disordered proteins are likely to be more difficult or even impossible to crystallize.

Solvent structure

A high portion of the solvent contained into a crystal cell can be considered not to be relevant for the macromolecule; it is simply there to fill the channels produced by molecular contacts in the crystal. And indeed, this unordered solvent cannot be seen in an X-ray diffraction experiment. Water molecules closely bound to the protein, on the contrary, can be considered as part of the structure of the macromolecule itself: a protein cannot be completely dehydrated without a complete crash of the architecture of its three-dimensional structure. Tightly bound solvent molecules in the crystal are identified during the process of refinement, and can be distinguished in three groups:

1. Water molecules making hydrogen bonds with hydrophilic side chains on the surface of the protein, where often they take part in a tetrahedral or trigonal network of hydrogen bonds. Ordered waters are substantially on the first shell of coordination around the protein, or eventually in the second shell, bound to the water molecules of the first one.

2. Water molecules that serve as a bridge among parts of the main chain or other structural elements that are too far apart to form hydrogen bonds: for example, if two strands of a β-sheet diverge slightly, a water molecule can make an H-bond in the middle, filling the gap. This kind of solvent molecule is essential in stabilizing the protein structure.

3. Solvent molecules located in the internal cavities of the protein, where they sometimes do not form very stable interactions but simply fill a vacuum.

It should be noted that the arrangement of the solvent structure around a protein determined by X-ray analysis is strongly influenced not only by the crystal packing, but also by the pH and the solvent used in the crystallization, and it cannot be considered fully representative of the situation *in vivo*.

The influence of crystal packing

The question possibly most often asked since the beginning of protein crystallography can be summarized as follows: how is the structure in the crystal representative of the 'real' *in vivo* structure? Proteins are quite stable but largely flexible molecules: comparison of the same protein obtained in different crystal forms, and consequently subjected to completely different packing forces, present in general the same fold, with some differences, usually small, in the regions of contact among molecules in the crystal. Figure 8.32 shows the α-carbon chain trace of TRP aporepressor

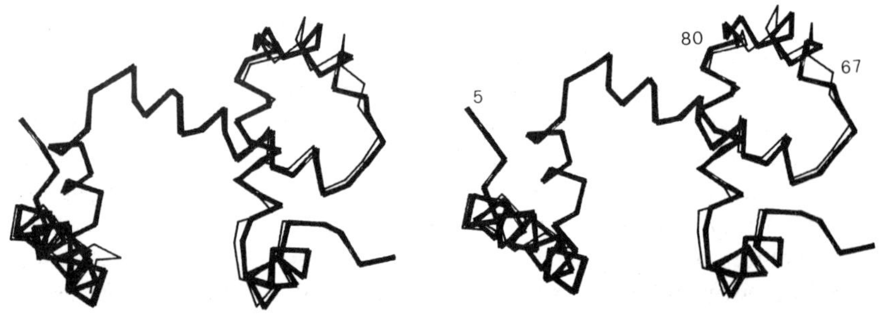

Fig. 8.32. α-carbon chain trace of TRP-aporepressor[120] (data from PDB). Thick and light lines stands for the orthorhombic and the trigonal crystal forms respectively. The two models superimpose quite well, except for the first five residues and for the region around 67–80.

from two crystal forms, trigonal and orthorhombic.[120] The fact that molecules crystallized in different conditions of pH and precipitants keep the same overall conformation is indirect evidence of the stability of protein conformation, and of the validity of the structure obtained using the crystallographic technique. On the other hand, the local variations easily observed suggest that great care has to be taken in drawing specific conclusions on functional aspects from details of the structure.

Protein classification

The secondary structural elements of a protein could be combined, at least in principle, in a nearly infinite number of ways to produce the three-dimensional structure. An examination of the structures solved till now has on the contrary shown that these elements are put together in quite a limited number of ways, giving rise to a small set of possible patterns. The previous statement must nevertheless be regarded cautiously: it could be partially ascribed to the limited size of the data base available (as explained in the introduction, in the Brookhaven Protein Data Bank only 353 sets of protein coordinates are reported, most of them relative to parent molecules); in addition, a bias could be introduced by the fact that only some classes of molecules have so far been crystallized. In any case, regularities and similarities in tertiary structures can be observed: a distribution of proteins or domains into classes and subclasses has been attempted. According to Richardson,[109] protein molecules solved up to the present belong to one of the following classes:

(1) antiparallel α structures
  - up-and-down helix bundle
  - Greek key helix bundle
  - miscellaneous

(2) antiparallel β structures
  - up-and-down β barrels
  - Greek key β barrels
  - other β barrels
  - open-face β sandwiches
  - miscellaneous

(3) parallel α/β structures
  - singly wound parallel β-barrels
  - doubly wound parallel β-sheets
  - miscellaneous

(4) small irregular proteins.

Set (4) is a very small one, and nearly all the domains of solved structures belong to one of the first three sets. A first classification in three main groups is quite obvious: domains may be composed mainly by helices, by β-strands differently arranged, or by a combination of both elements. What is more puzzling is that, inside each main group, strict rules are observed: for example, all the α-helices belonging to class (1) are arranged in an antiparallel way, that is nearby helices are pointing in opposite directions. Two structures belonging to the first two subclasses, the **up-and-down** and **Greek key antiparallel α**, are shown in Figs 8.33(a) and (b), respectively.

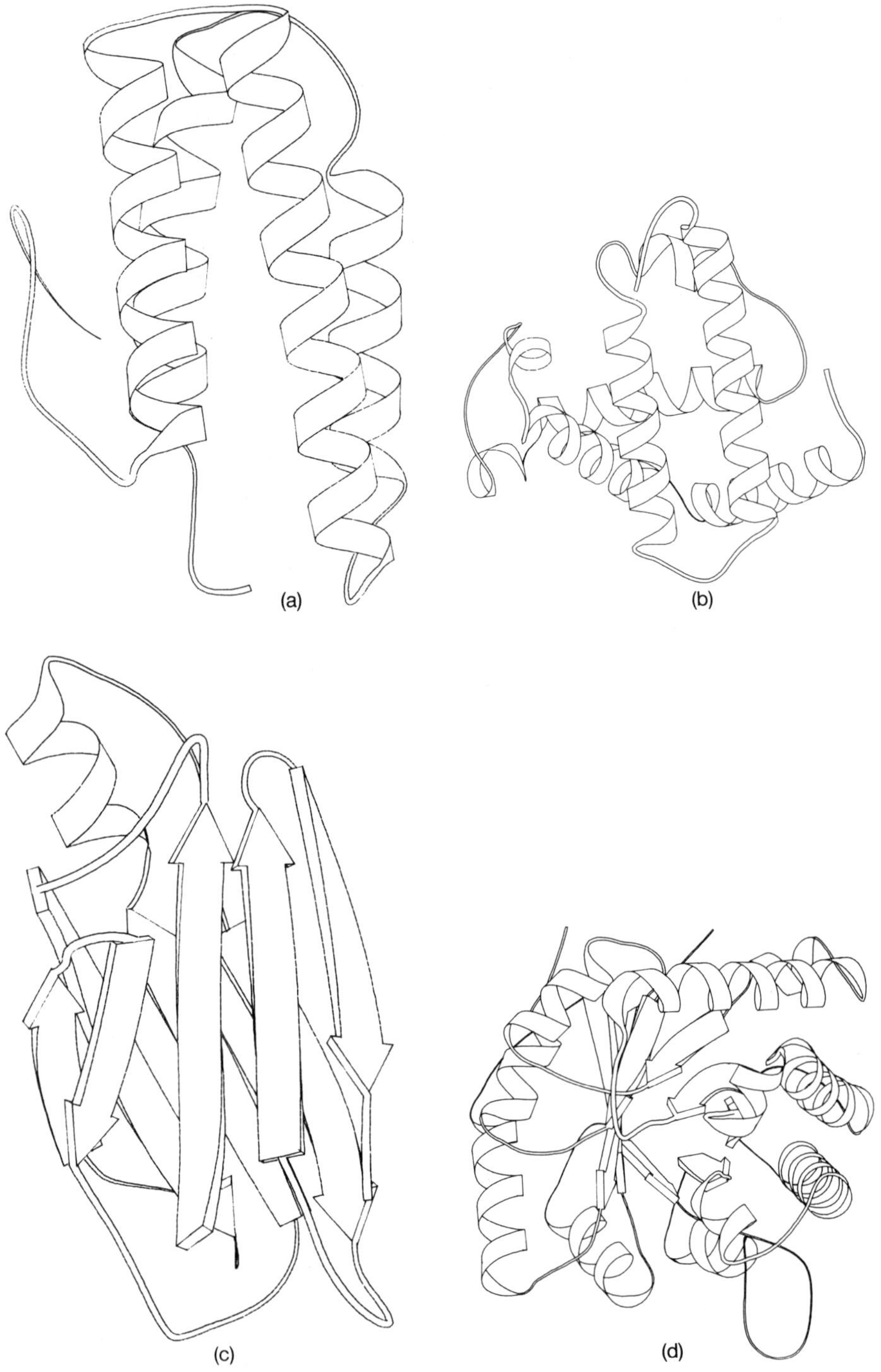

Fig. 8.33. Sketches of protein models produced following the directives of Richardson, using program RIBBON.[121] (a) myohemerythrin[122] and (b) myoglobin,[99] representatives of the antiparallel α-structures; (c) prealbumin,[123] a Greek key β-barrel, a subclass of antiparallel β structures; (d) triose phosphate isomerase (TIM),[124] one of the more elegant examples of a parallel α/β protein. All coordinates were taken from PDB.

The same holds for class (2), the all-β domains: the strands in this group are antiparallel to each other (only if more than seven strands are present, can a parallel chain be observed). Figure 8.33(c) shows a picture of prealbumin, an example of a protein of the subclass called **greek key β barrel**, and Fig. 8.29 illustrates all the topologies found in that class, which does not mean all the possible, but simply those found in the structures solved till now.

In proteins containing both structural elements, classified α/β, α-helices are arranged parallel to each other and so are β-strands. There is not a rational explanation for that, since they could be antiparallel as well. An example of this last class, the enzyme triose phosphate isomerase, classified in the subclass called **singly wound parallel β-barrel**, is shown in Fig. 8.33(d).

It is difficult to judge the real biological meaning of the previous classification, since oversimplification could have a part in it. Nevertheless, it remains surprising that molecules which are so complex give rise to such a limited number of three-dimensional patterns.

Appendices

Appendix 8.A Some formulae for isomorphous replacement and anomalous dispersion

In this appendix, formulae (8.13) and (8.23), relative to isomorphous replacement and anomalous scattering, will be derived in a general way. Equation (8.13) can also be obtained from Fig. 8.1, using simple trigonometric considerations. It will be derived to illustrate the formalism used in the rest of the appendix. Let us call:

$$\boldsymbol{F}_{\mathrm{P}} = F_{\mathrm{P}} \exp(i\varphi_{\mathrm{P}}),$$
$$\boldsymbol{F}_{\mathrm{PH}} = F_{\mathrm{PH}} \exp(\mathrm{i}\varphi_{\mathrm{PH}})$$
$$\boldsymbol{F}_{\mathrm{H}} = F_{\mathrm{H}} \exp(\mathrm{i}\varphi_{\mathrm{H}}).$$

From (8.5):

$$\boldsymbol{F}_{\mathrm{PH}}\boldsymbol{F}^*_{\mathrm{PH}} = (\boldsymbol{F}_{\mathrm{P}} + \boldsymbol{F}_{\mathrm{H}})(\boldsymbol{F}^*_{\mathrm{P}} + \boldsymbol{F}^*_{\mathrm{H}}). \tag{8.A.1}$$

By substituting in (8.A.1):

$$F_{\mathrm{PH}} \exp(\mathrm{i}\varphi_{\mathrm{PH}}) F_{\mathrm{PH}} \exp(-\mathrm{i}\varphi_{\mathrm{PH}})$$
$$= \{F_{\mathrm{P}} \exp(\mathrm{i}\varphi_{\mathrm{P}}) + F_{\mathrm{H}} \exp(\mathrm{i}\varphi_{\mathrm{H}})\}\{F_{\mathrm{P}} \exp(-\mathrm{i}\varphi_{\mathrm{P}}) + F_{\mathrm{H}} \exp(-\mathrm{i}\varphi_{\mathrm{H}})\}, \tag{8.A.2}$$

$$F^2_{\mathrm{PH}} = F^2_{\mathrm{P}} + F^2_{\mathrm{H}} + F_{\mathrm{H}}F_{\mathrm{P}} \exp[\mathrm{i}(\varphi_{\mathrm{P}} - \varphi_{\mathrm{H}})] + F_{\mathrm{P}}F_{\mathrm{H}} \exp[-\mathrm{i}(\varphi_{\mathrm{P}} - \varphi_{\mathrm{H}})], \tag{8.A.3}$$

$$F^2_{\mathrm{PH}} = F^2_{\mathrm{P}} + F^2_{\mathrm{H}} + 2F_{\mathrm{P}}F_{\mathrm{H}} \cos(\varphi_{\mathrm{P}} - \varphi_{\mathrm{H}}). \tag{8.A.4}$$

Equation (8.A.4) is equivalent to (8.11). An analogous expression can be derived for φ_{PH}. Since, from (8.5):

$$\boldsymbol{F}_{\mathrm{P}} = \boldsymbol{F}_{\mathrm{PH}} - \boldsymbol{F}_{\mathrm{H}}$$

we can derive as before:

$$F^2_{\mathrm{P}} = F^2_{\mathrm{PH}} + F^2_{\mathrm{H}} - F_{\mathrm{PH}}F_{\mathrm{H}} \exp[-\mathrm{i}(\varphi_{\mathrm{PH}} - \varphi_{\mathrm{H}})]$$
$$+ F_{\mathrm{PH}}F_{\mathrm{H}} \exp[\mathrm{i}(\varphi_{\mathrm{PH}} - \varphi_{\mathrm{H}})], \tag{8.A.5}$$

$$F^2_{\mathrm{P}} = F^2_{\mathrm{PH}} + F^2_{\mathrm{H}} - 2F_{\mathrm{PH}}F_{\mathrm{H}} \cos(\varphi_{\mathrm{PH}} - \varphi_{\mathrm{H}}), \tag{8.A.6}$$

hence:

$$\varphi_{\mathrm{PH}} = \varphi_{\mathrm{H}} + \cos^{-1}[(F^2_{\mathrm{PH}} + F^2_{\mathrm{H}} - F^2_{\mathrm{P}})/2F_{\mathrm{PH}}F_{\mathrm{H}}]. \tag{8.A.7}$$

Equations (8.A.4) and (8.A.6) are, in fact, the same equation and both give two possible solutions for the phase of the protein (or of the derivative).

For anomalous dispersion measurements, let us define:

$$\begin{aligned}
\boldsymbol{F}_{\mathrm{PH}}^{+} &= F_{\mathrm{PH}}^{+} \exp(\mathrm{i}\varphi_{\mathrm{PH}}),\\
\boldsymbol{F}_{\mathrm{PH}}^{-} &= F_{\mathrm{PH}}^{-} \exp(-\mathrm{i}\varphi_{\mathrm{PH}}),\\
\boldsymbol{F}'_{\mathrm{H}} &= F'_{\mathrm{H}} \exp(\mathrm{i}\varphi_{\mathrm{H}})\\
\boldsymbol{F}_{\mathrm{PH}} &= \boldsymbol{F}_{\mathrm{P}} + \boldsymbol{F}'_{\mathrm{H}} = F_{\mathrm{PH}} \exp(\mathrm{i}\varphi_{\mathrm{PH}}),\\
\boldsymbol{F}_{\mathrm{H}}^{+\prime\prime} &= F''_{\mathrm{H}} \exp[\mathrm{i}(\varphi_{\mathrm{H}} + \pi/2)],\\
\boldsymbol{F}_{\mathrm{H}}^{-\prime\prime} &= F''_{\mathrm{H}} \exp[-\mathrm{i}(\varphi_{\mathrm{H}} - \pi/2)].
\end{aligned}$$

From Fig. 8.7

$$\boldsymbol{F}_{\mathrm{PH}}^{+} = \boldsymbol{F}_{\mathrm{P}} + \boldsymbol{F}'_{\mathrm{H}} + \boldsymbol{F}_{\mathrm{H}}^{+\prime\prime} = \boldsymbol{F}_{\mathrm{PH}} + \boldsymbol{F}_{\mathrm{H}}^{+\prime\prime}, \tag{8.A.8}$$

$$\begin{aligned}
F_{\mathrm{PH}}^{+2} &= F_{\mathrm{PH}}^{2} + F_{\mathrm{H}}^{\prime\prime 2} + 2F_{\mathrm{PH}}F''_{\mathrm{H}} \cos(\varphi_{\mathrm{PH}} - \varphi_{\mathrm{H}} - \pi/2),\\
F_{\mathrm{PH}}^{-2} &= F_{\mathrm{PH}}^{2} + F_{\mathrm{H}}^{\prime\prime 2} - 2F_{\mathrm{PH}}F''_{\mathrm{H}} \cos(\varphi_{\mathrm{PH}} - \varphi_{\mathrm{H}} - \pi/2).
\end{aligned} \tag{8.A.9}$$

Subtracting the two:

$$F_{\mathrm{PH}}^{+2} - F_{\mathrm{PH}}^{-2} = 4F_{\mathrm{PH}}F''_{\mathrm{H}} \cos(\varphi_{\mathrm{PH}} - \varphi_{\mathrm{H}} - \pi/2) = 4F_{\mathrm{PH}}F''_{\mathrm{H}} \sin(\varphi_{\mathrm{PH}} - \varphi_{\mathrm{H}}). \tag{8.A.10}$$

We can also write:

$$F_{\mathrm{PH}}^{+2} - F_{\mathrm{PH}}^{-2} = (F_{\mathrm{PH}}^{+} + F_{\mathrm{PH}}^{-})(F_{\mathrm{PH}}^{+} - F_{\mathrm{PH}}^{-}) \tag{8.A.11}$$

and, since the anomalous contribution to the structure factor of a protein is small if compared with the non-anomalous part:

$$F_{\mathrm{PH}}^{+} + F_{\mathrm{PH}}^{-} \approx 2F_{\mathrm{PH}}. \tag{8.A.12}$$

Substitution of (8.A.12) in (8.A.11) gives:

$$F_{\mathrm{PH}}^{+2} - F_{\mathrm{PH}}^{-2} \approx 2F_{\mathrm{PH}}(F_{\mathrm{PH}}^{+} - F_{\mathrm{PH}}^{-}), \tag{8.A.13}$$

and comparison of (8.A.13) and (8.A.10):

$$F_{\mathrm{PH}}^{+} - F_{\mathrm{PH}}^{-} = 2F''_{\mathrm{H}} \sin(\varphi_{\mathrm{PH}} - \varphi_{\mathrm{H}}) \tag{8.A.14}$$

Finally:

$$\varphi_{\mathrm{PH}} = \varphi_{\mathrm{H}} + \sin^{-1}[(F_{\mathrm{PH}}^{+} - F_{\mathrm{PH}}^{-})/2F''_{\mathrm{H}}]. \tag{8.A.15}$$

The combination of (8.A.15) and (8.A.7) allows a unique determination of φ_{PH}.

Appendix 8.B Translation functions

Several types of translation functions have been defined and most of them are quite similar and sometimes virtually equivalent (for a general review, see Beurskens[53]). In the following, the more common among those used in protein crystallography are described.

1. The Q function (Tollin[125]). The Q function was devised by Tollin to determine the position $\boldsymbol{r}_j$ of a known group of atoms in a crystallographic cell, with respect to an arbitrarily chosen origin. Let us consider only two molecules, related by the symmetry operator **A**. If one of them (that we will call reference molecule) is correctly oriented in the cell of the unknown crystal, the problem of positioning it with respect to translation reduces to

finding the correct origin with respect to the symmetry operator $\mathbf{A}$. In the case of a superposition of m maps translated by vectors $\boldsymbol{r}_j$, the sum function defined in (5.B.5) (Buerger;[126] see Appendix 5.B) becomes:

$$S(\boldsymbol{u}) = \sum_h I(\boldsymbol{h}) \sum_j \cos[2\pi \boldsymbol{h}(\boldsymbol{u} - \boldsymbol{r}_j)]. \tag{8.B.1}$$

Equation (8.B.1) will present a maximum in position $\boldsymbol{r}_j = \mathbf{A}(\boldsymbol{X} + \boldsymbol{t}) - \boldsymbol{t}$, that is when the translation $\boldsymbol{t}$ will be the vector defining the position of the absolute origin with respect to the symmetry operator under consideration. Substituting in (8.B.1):

$$Q(\boldsymbol{t}) = \sum_h I(\boldsymbol{h}) \sum_{j,j'} \cos 2\pi \boldsymbol{h}[\boldsymbol{X}_j + \boldsymbol{t} - \mathbf{A}(\boldsymbol{X}_{j'} + \boldsymbol{t})]. \tag{8.B.2}$$

From the maximum of (8.B.2) it is straightforward to obtain $\boldsymbol{X}_j$.

The Q function will show several false peaks, whenever two atoms have coordinates $\boldsymbol{X}_1$ and $\boldsymbol{X}_2$ such that the relationships:

$$\mathbf{A}(\boldsymbol{X}_1 + \boldsymbol{t}) - (\boldsymbol{X}_2 - \boldsymbol{t}) = 0 \tag{8.B.3}$$

is satisfied, and these peaks will have the same magnitude as those defining the correct origin. However, the position of false peaks can be predicted, or they can be removed by modifying the $I(\boldsymbol{h})$ value.

2. The T function (Crowther and Blow[51]). The function for two molecules related by symmetry operator $\mathbf{A}$ is given by (8.60). It has been shown[127] that the T function is virtually equivalent to the Q function, except that the origin is looked at in a different way.

Modified T functions have also been defined. If more than one symmetry operation is considered, $(\mathbf{A}_1, \ldots, \mathbf{A}_n)$, the T function becomes:

$$T_1(\boldsymbol{t}) = \sum_h \Big(I_{\text{obs}}(\boldsymbol{h}) - \sum_i F_1(\boldsymbol{h}\mathbf{A}_i)\Big) F_1(\boldsymbol{h}) F_1^*(\boldsymbol{h}\mathbf{A}_j) \exp(-2\pi i \boldsymbol{h} \cdot \boldsymbol{t}) \tag{8.B.4}$$

$$T_2(\boldsymbol{t}) = \sum_h I_{\text{obs}}(\boldsymbol{h}) \Big(\sum_i \sum_j F_1(\boldsymbol{h}\mathbf{A}_i) F_1^*(\boldsymbol{h}\mathbf{A}_j)\Big) \exp(-2\pi i \boldsymbol{h} \cdot \boldsymbol{t}) \tag{8.B.5}$$

3. The T_{H} function (Harada *et al.*[128]). T_{H} represents an improvement of the T function, since it takes into account the interpenetration of molecules in the crystal cell. T_{H} is defined as the ratio of two functions:

$$T_{\text{H}} = T_{\text{O}}(\boldsymbol{t})/\text{O}(\boldsymbol{t}) \tag{8.B.6}$$

where:

$$T_{\text{O}}(\boldsymbol{t}) = \Big(\sum_h I_{\text{obs}}(\boldsymbol{h})[F_{\text{calc}}(\boldsymbol{h}, \boldsymbol{t})]^2\Big)\Big(\sum_h [I_{\text{obs}}(\boldsymbol{h})]^4\Big)^{-1}, \tag{8.B.7}$$

$$\text{O}(\boldsymbol{t}) = \Big(\sum_h [F_{\text{calc}}(\boldsymbol{h}, \boldsymbol{t})]^2\Big)\Big(N \sum_h [F(\boldsymbol{h})]^2\Big)^{-1}. \tag{8.B.8}$$

Vector $\boldsymbol{t}$ here does not have the same meaning as in Fig. 8.17: function T_{H} gives in fact the absolute value of the translation that has to be applied to the model in order to position it correctly in the crystal cell. Equation (8.B.7) is similar to the Crowther T function, but $\text{O}(\boldsymbol{t})$ is a new term, taking into account the interpenetration among molecules. T_{H} will present a maximum when cross-vectors of Patterson maps superimpose and molecules do not overlap each other.

Appendix 8.C Macromolecular least-squares refinement and the conjugate-gradient algorithm

The normal matrix used in restrained least-squares deserves a special comment. If the molecule under refinement is composed of N atoms, and we are refining atomic coordinates and individual isotropic thermal parameters, the normal matrix will be a square matrix of dimension $4N$. Even employing rigid-body constraints whenever possible, the use of classical full-matrix methods may be still prohibitive. To speed up calculations, different algorithms have been devised. The fast calculation of gradients,[129] and the use of the fast Fourier transform (FFT, see p. 131) in calculating structure factors.[130]

For matrix inversion, any algorithm able to handle system of equations with thousands of variables can be used, like the Gauss–Siedel.[77] Since in most of the restrained least-squares programs a 'sparse matrix' is used, the algorithm of conjugate gradients appears well suited for solving a system of linear equations of this kind. The only drawback of it is that it does not directly provide the diagonal elements of the inverted matrix, from which the standard deviations of the parameters can be obtained. However, approximate values for them can be calculated. A brief description of the conjugate gradients algorithm is given here, taken from Rollet[132] and Konnert.[72]

Let us have a system of n simultaneous equations in n unknowns:

$$\mathbf{A}\boldsymbol{X} = \boldsymbol{b}. \tag{8.C.1}$$

In the case of crystallographic least squares, the vector of unknowns $\boldsymbol{X}$ represents the shifts to be applied to the parameters. Classical inversion of matrix $\mathbf{A}$ for a very large number of parameters, as is the case of macromolecules, can be very inefficient or even impossible. Gradient methods are based on the iterative process that consists in obtaining, starting from an approximate solution $\boldsymbol{X}_k$, a new one from:

$$\boldsymbol{X}_{kh} = \boldsymbol{X}_k + \alpha_k \boldsymbol{r}_k \tag{8.C.2}$$

where $\boldsymbol{r}_k$ is the vector of residuals, defined by:

$$\boldsymbol{r}_k = \boldsymbol{b} - \mathbf{A}\boldsymbol{X}_k \tag{8.C.3}$$

and α_k is an appropriate constant.

The conjugate gradients method, discovered by Hasteness and Stiefel,[131] is an iterative process that gives the solution for an n-dimensional problem in approximately n steps (but generally much fewer than that). The method starts with an arbitrary value for vector $\boldsymbol{X}$: let us call it $\boldsymbol{X}_0$ (usually zero). The vector of residuals, $\boldsymbol{r}_0$, is given by:

$$\boldsymbol{r}_0 = \boldsymbol{p}_0 = \boldsymbol{b} - \mathbf{A}\boldsymbol{X}_0. \tag{8.C.4}$$

Let us also define:

$$\alpha_i = (\boldsymbol{r}_i^{\mathrm{T}}\boldsymbol{p}_i)^2/(\boldsymbol{p}_i^{\mathrm{T}}\mathbf{A}\boldsymbol{p}_i) \tag{8.C.5}$$

$$\beta_i = -(\boldsymbol{r}_{i+1}^{\mathrm{T}}\mathbf{A}\boldsymbol{p}_i)^2/(\boldsymbol{p}_i^{\mathrm{T}}\mathbf{A}\boldsymbol{p}_i). \tag{8.C.6}$$

Using (8.C.4) and (8.C.5), a new value of $\boldsymbol{X}$, called $\boldsymbol{X}_1$, will be given by:

$$\boldsymbol{X}_1 = \boldsymbol{X}_0 + \alpha_0 \boldsymbol{p}_0. \qquad (8.C.7)$$

Now a new residual and a new value for $\boldsymbol{p}_i$ can be calculated:

$$\boldsymbol{r}_1 = \boldsymbol{r}_0 - \alpha_0 \mathbf{A} \boldsymbol{p}_0, \qquad (8.C.8)$$

$$\boldsymbol{p}_1 = \boldsymbol{r}_1 + \beta_0 \boldsymbol{p}_0. \qquad (8.C.9)$$

From $\boldsymbol{p}_1$ we can now evaluate a new vector $\boldsymbol{X}$ and a new residual vector:

$$\boldsymbol{X}_2 = \boldsymbol{X}_1 + \alpha_1 \boldsymbol{p}_1, \qquad (8.C.10)$$

$$\boldsymbol{r}_2 = \boldsymbol{r}_1 - \alpha_1 \mathbf{A} \boldsymbol{p}_1. \qquad (8.C.11)$$

By analogy with (8.C.9) a new value of $\boldsymbol{p}$ can be calculated, and so on. The method can proceed until convergence, that is when the vector of residuals $\boldsymbol{r}_i$ is less than a predefined amount. The advantages of the method are that every estimate of $\boldsymbol{X}$ improves the previous one and that the elements of the matrix **A** are not moved from their positions: the non-zero elements of **A** can be stored contiguously, taking advantage of a scarcely populated matrix.

Appendix 8.D Conventions and symbols for amino acids and peptides

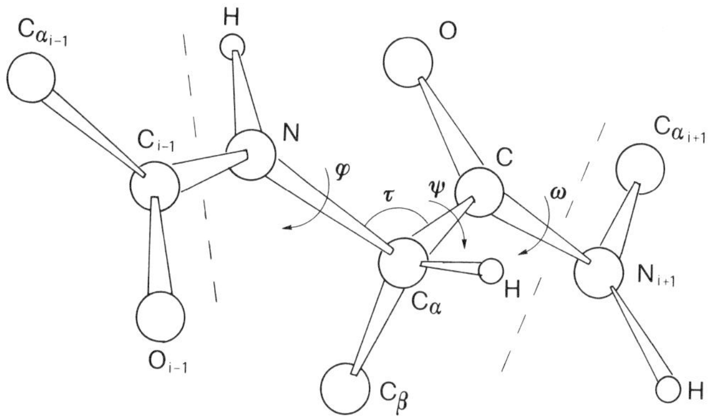

Fig. 8.D.1. Schematic representation of a peptide unit, with symbols according the recomendations of the IUPAC-IUB Commission on Biochemical Nomenclature, *Journal of Molecular Biology*, **52,** 1–17 (1970). Each repetitive unit, delimited in the figure by broken lines, has general formula –NH–CHR–CO–, where R is one of the 20 side chains illustrated in Fig. 8.D.3. Amino acids are connected by a peptide bond CO–NH. In all the protein structures known, the peptide torsion angle, ω, assumes a value of about 180°, that is the carbonyl oxygen and the H of nitrogen assume a *trans* planar conformation. The only exception is represented by proline, that in few cases has been found to be *cis*. Amino acids in a polypeptide chain are numbered from the N to the carboxyl terminal.

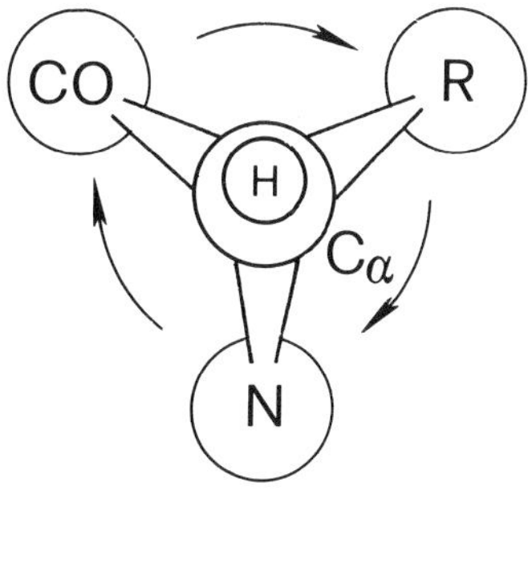

Fig. 8.D.2. A mnemonic rule for remembering the L- configuration of an amino acid: looking down from the H atom toward the C_α, the substituents (CO, R, and N) must form the word CORN if read in clockwise sense.

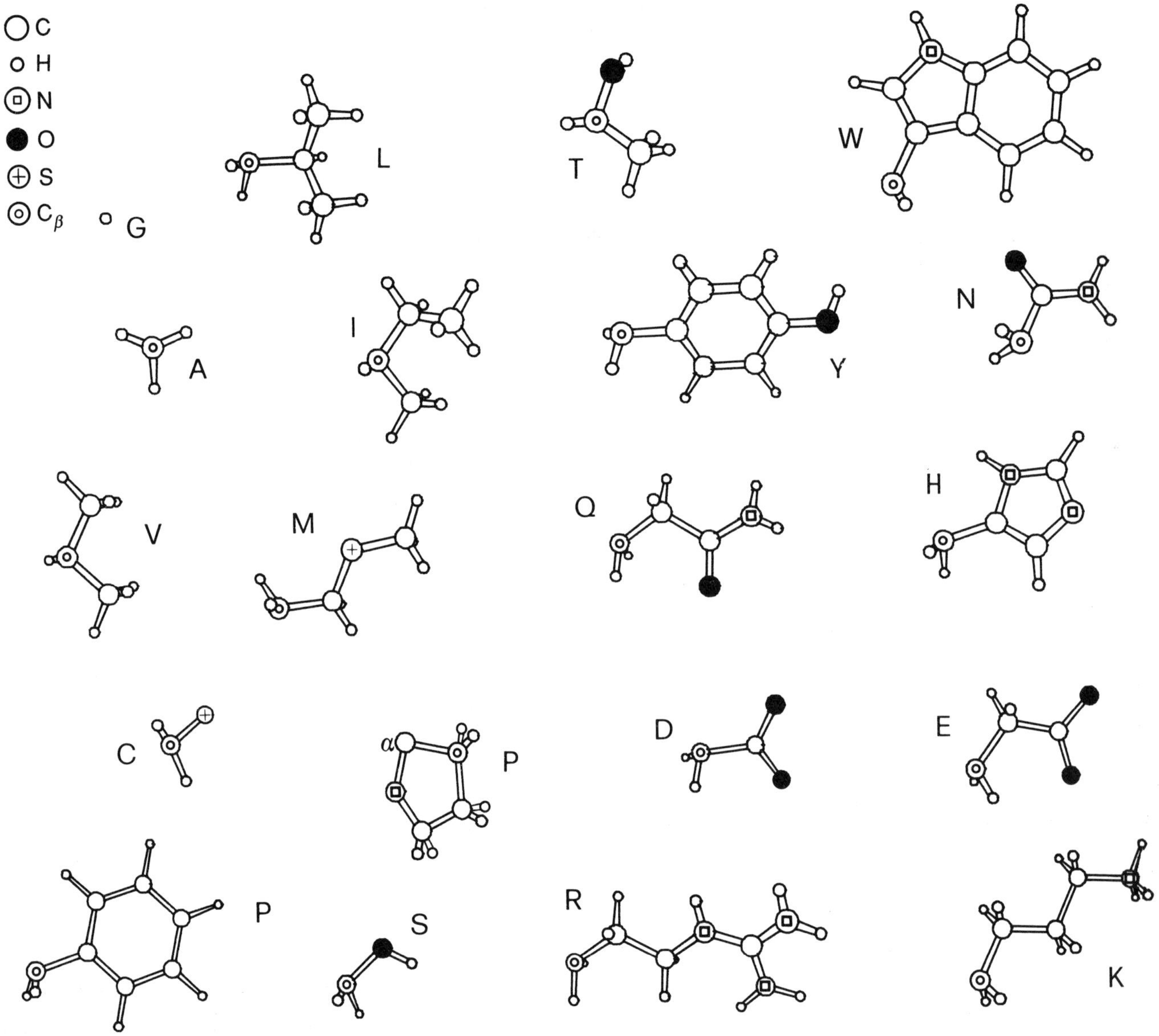

Fig. 8.D.3. The side chains of the 20 natural amino acids.

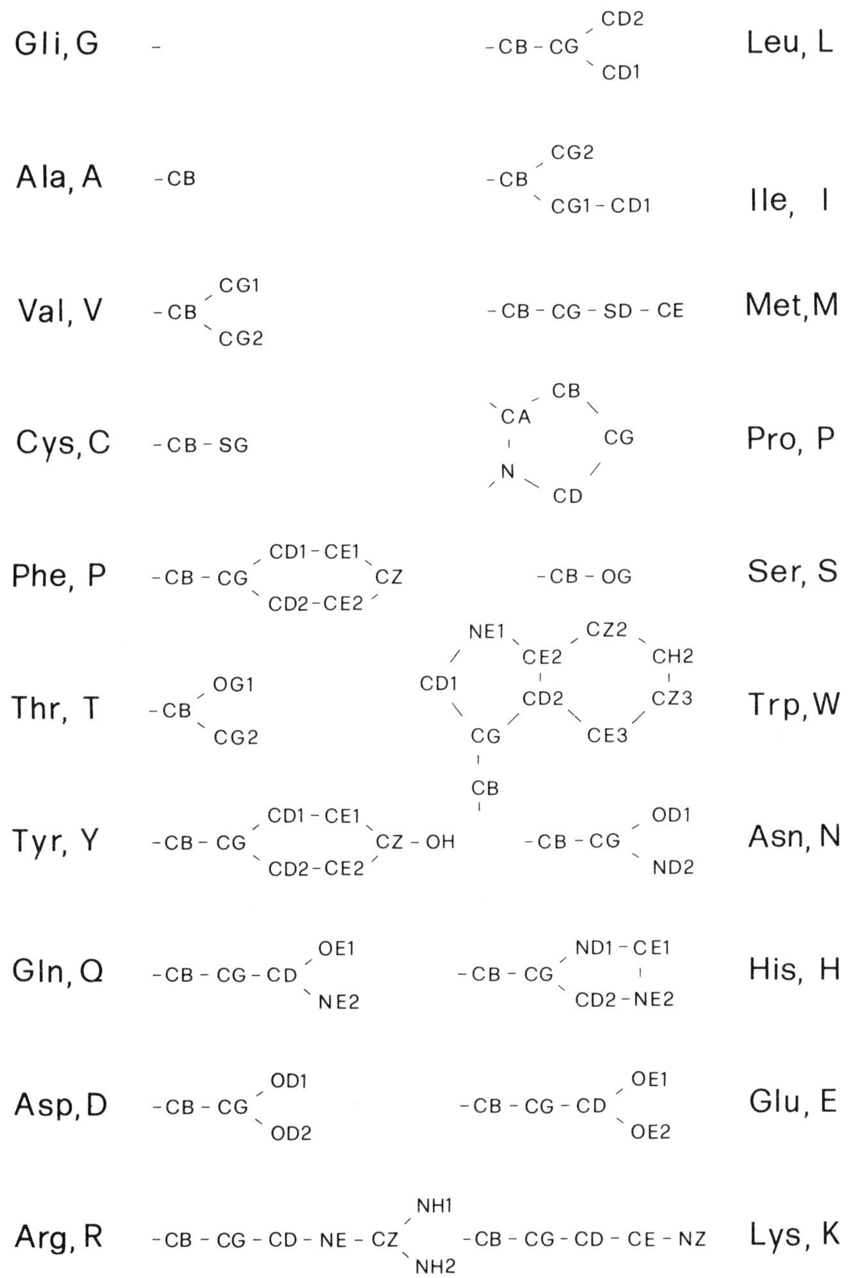

Fig. 8.D.4. Symbols used in Protein Data Bank for the atoms of the 20 side chains of Fig. 8.D.3. The same nomenclature is also used, with some small modifications, in most of the refinement programs. The three-letter and the one-letter code for the amino acids are used.

References and further reading

Further reading

1. Blundell, T. L. and Johnson, L. N. (1976). *Protein crystallography*. Academic, New York.
2. McPherson, A. (1982). *Preparation and analysis of protein crystals*. Wiley, New York.
3. Wyckoff, H. W., Hirs, C. H. W., and Timasheff, S. N. (ed.) (1985). Diffraction methods for biological macromolecules. In *Methods in enzymology,* Vols 114–15. Academic, Orlando.
4. Rossmann, M. G. (ed.) (1972). *The molecular replacement method: a collection of papers on the use of non-crystallographic symmetry*. Gordon and Breach, New York.
5. Cantor, C. and Schimmel, P. R. (1980). *Biophysical chemistry*. Freeman, San Francisco.

References

6. Kendrew, J. C., Dickerson, R. E., Strandberg, B. E., Hart, R. G., Phillips, D. C., and Shore, V. C. (1960). *Nature,* **185,** 422–7; Cullis, A. F., Muirhead, H., Perutz, M., Rossmann, M. G., and North, A. C. T. (1961). *Proceedings of the Royal Society of London,* **A265,** 161–87; Perutz, M. (1985). In *Methods in Enzymology,* Vol. 114 (ed. H. W. Wyckoff, C. H. W. Hirs, and S. Timasheff), pp. 3–46. Academic, Orlando.
7. Bernstein, F. C., Koetzle, T. F., Williams, G. J. B., Meyer, E. F. Jr., Brice, M. D., Rodgers, J. R., Kennard, O., Shimanouchi, T., and Tasumi, M. (1977). *Journal of Molecular Biology,* **112,** 535–42.
8. Pusey, M. L., Snyder, R. S., and Naumann, R. (1986). *Journal of Biological Chemistry,* **261,** 6524–9.
9. Zeppezauer, M. (1971). *Methods in Enzymology,* **22,** 253–69.
10. McPherson, A. (1976). *Journal of Biological Chemistry,* **251,** 6300–3.
11. McPherson, A. (1976). *Methods in Biochemical Analysis,* **23,** 249–345.
12. Garavito, R. M., Jenkins, J., Jansonius, J. N., Karlsson, R., and Rosenbusch, J. P. (1983). *Journal of Molecular Biology,* **164,** 313–27.
13. Michel, H. (1983). *Trends in Biological Science,* 56–9.
14. Matthews, B. W. (1968). *Journal of Molecular Biology,* **33,** 491–7.
15. Matthews, B. W., and Bernhard, S. A. (1973). *Annual Reviews in Biophysics and Bioengineering,* **2,** 257–317.
16. Matthews, B. W. (1974). *Journal of Molecular Biology,* **82,** 513–26.
17. Green, D. W., Ingram, V. M., and Perutz, M. F. (1954). *Proceedings of the Royal Society of London,* **A225,** 287–307; Rossmann, M. G. (1960). *Acta Crystallographica,* **13,** 221–6.
18. Ramachandran, G. N. and Srinivasan, R. (1970). *Fourier methods in crystallography* (ed. M. J. Buerger) pp. 96–235. Whiley, New York.
19. Sim, G. A. (1959). *Acta Crystallographica,* **12,** 813–15.
20. Sim, G. A. (1960). *Acta Crystallographica,* **13,** 511–12.
21. Blow, D. M., and Rossmann, M. G. (1961). *Acta Crystallographica,* **14,** 1195–202.
22. Harker, D. (1956). *Acta Crystallographica,* **9,** 1–9.
23. Hendrickson, W. A. and Teeter, M. M. (1981). *Nature,* **290,** 107–13.
24. Rossmann, M. G. (1961). *Acta Crystallographica*, **14,** 383–8.
25. North, A. C. T. (1965). *Acta Crystallographica,* **18,** 212–16.
26. Matthews, B. W. (1966). *Acta Crystallographica,* **20,** 82–6.
27. Parthasarathy, S. (1970). *Acta Crystallographica,* **A27,** 45–7.
28. Bijvoet, M. (1954). *Nature,* **173,** 888–91.

29. Terwilliger, T. C. and Eisenberg, D. (1987). *Acta Crystallographica,* **A43,** 6–13.
30. Blow, D. M. and Crick, F. H. C. (1959). *Acta Crystallographica,* **12,** 794–802.
31. TenEyck, L. F. and Arnone, A. (1976). *Journal of Molecular Biology,* **100,** 3–11.
32. Hendrickson, W. A. and Latman, E. E. (1970). *Acta Crystallographica,* **B26,** 136–43.
33. Hendrickson, W. A. (1971). *Acta Crystallographica,* **B27,** 1472–5.
34. Dickerson, R. E., Kendrew, J. C., and Strandberg, B. E. (1961). *Acta Crystallographica,* **14,** 1188–95.
35. Sygusch, J. (1977). *Acta Crystallographica,* **A33,** 512–28.
36. Blow, D. M. and Matthews, B. W. (1973). *Acta Crystallographica,* **A29,** 56–62.
37. Petsko, G. A. (1976). *Acta Crystallographica,* **A32,** 473–6.
38. Henderson, R. and Moffat, J. K. (1971). *Acta Crystallographica,* **B27,** 1414–20.
39. Davies, A. R. and Rollett, J. S. (1976). *Acta Crystallographica,* **A32,** 17–23.
40. Schevitz, R. W., Podjarny, A. D., Zwick, M., Hughes, J. J., and Sigler, P. B. (1981). *Acta Crystallographica,* **A37,** 669–77.
41. Bath, T. N. and Blow, D. M. (1982). *Acta Crystallographica,* **A38,** 21–9.
42. Cannillo, E., Oberti, R., and Ungaretti, L. (1983). *Acta Crystallographica,* **A39,** 68–74.
43. Bryan, R. K. and Banner, D. W. (1987). *Acta Crystallographica,* **A43,** 556–64.
44. Subbiah, S. and Harrison, S. C. (1989). *Acta Crystallographica,* **A45,** 337–42.
45. Rossmann, M. G. and Blow, D. M. (1962). *Acta Crystallographica,* **15,** 24–31.
46. Tollin, P. and Rossmann, M. G. (1966). *Acta Crystallographica,* **21,** 872–6.
47. Crowther, R. A. (1972). In *The molecular replacement method: a collection of papers on the use of non-crystallographic symmetry* (ed. M. G. Rossmann), pp. 173–5. Gordon and Breach, New York.
48. Tollin, P., Main, P., and Rossmann, M. G. (1966). *Acta Crystallographica,* **20,** 404–7.
49. Rao, S. N., Jyh-Hwang, J., and Hartsuck, J. A. (1980). *Acta Crystallographica,* **A36,** 878–84.
50. Latman, E. E. (1972). *Acta Crystallographica,* **B28,** 1065.
51. Crowther, R. A. and Blow, D. M. (1967). *Acta Crystallographica,* **23,** 544–8.
52. Latman, E. E. (1985). In *Methods in enzymology,* Vol. 115, (ed. H. W. Wyckoff, C. H. W. Hirs, and S. N. Timasheff), pp. 55–7. Academic, Orlando.
53. Beurskens, P. T., Gould, R. O., Bruins Slot, H. J., and Bosman, W. P. (1987). *Zeitschrift für Kristallographie,* **179,** 127–59.
54. Wilson, I. A., Skehel, J. J., and Wiley, D. C. (1981). *Nature,* **289,** 366–73.
55. Evans, P. R., Farrants, G. W., and Lawrence, M. C. (1986). *Journal of Molecular Biology,* **191,** 713–20.
56. Rossmann, M. G., Blow, D. M., Harding, M. M., and Coller, E. (1964). *Acta Crystallographica,* **17,** 338–42.
57. Wilson, K. (1978). *Acta Crystallographica,* **B34,** 1599–608.
58. Mukherjee, A. K., Helliwell, J. R., and Main, P. (1989). *Acta Crystallographica,* **A45,** 715–18.
59. Giacovazzo, C. (1990). In *International tables for x-ray crystallography,* vol. B. Reidel, Dordrecht.
60. Giacovazzo, C., Cascarano, G., and Chao-De, Z. (1988). *Acta Crystallographica,* **A44,** 45–57.
61. Jaynes, E. T. (1957). *Physical Review,* **106,** 620–30; Collins, D. M. (1982). *Nature,* **298,** 49–51.
62. Bricogne, G. (1984). *Acta Crystallographica,* **A40,** 410–55; (1988). *Acta Crystallographica,* **A44,** 517–45.

63. Harrison, R. W. (1989). *Acta Crystallographica,* **A45,** 4–10.
64. Podjarny, A. D., Moras, D., Navaza, J., and Alzari, P. M. (1988). *Acta Crystallographica,* **A44,** 545–51.
65. Prince, E., Sjolin, L., and Alenljung, R. (1988). *Acta Crystallographica,* **A44,** 216–22.
66. Richards, F. M. (1968). *Journal of Molecular Biology,* **37,** 225–30.
67. Jones, T. A. (1978). *Journal of Applied Crystallography,* **11,** 268.
68. Pflugrath, J. W., Saper, M. A., and Quiocho, F. A. (1984). In *Methods and applications in crystallographic computing* (ed. S. Hall and J. Ashida), p. 404. Oxford University Press.
69. Jones, A. T. and Thirup, S. (1986). *EMBO Journal,* **3,** 819–22.
70. Watenpaugh, K. D., Sieker, L. C., Herriot, J. R., and Jensen, L. H. (1973). *Acta Crystallographica*, **B29,** 943–956.
71. Diamond, R. (1971). *Acta Crystallographica,* **A27,** 436–52.
72. Konnnert, J. H. (1976). *Acta Crystallographica,* **A32,** 614–17.
73. Hendrickson, W. A. and Konnert, J. H. (1980). In *Computing in crystallography* (ed. R. Diamond, R. Ramaseshan, and K. Venkatesan) pp. 13.01–13.23. The Indian Academy of Sciences, Bangalore.
74. Waser, J. (1963). *Acta Crystallographica,* **16,** 1091–4.
75. Schomaker, V., Waser, J., Marsh, R. E., and Bergman, G. (1959). *Acta Crystallographica,* **12,** 600–4.
76. Rees, D. C. and Lewis, M. (1983). *Acta Crystallographica,* **A39,** 94–7.
77. Haneef, I., Moss, D. S., Stanford, M. J., and Borkakoti, N. (1985), *Acta Crystallographica,* **A41,** 426–33.
78. Lunin, V. Yu. and Urzhumtsev, A. G. (1985). *Acta Crystallographica,* **A41,** 327–33.
79. Jack, A. and Levitt, M. (1978). *Acta Crystallographica,* **A34,** 931–5.
80. Levitt, M. (1974). *Journal of Molecular Biology,* **82,** 393–420.
81. Sussman, J. L., Holbrook, S. R., Church, G. M., and Kim, S. H. (1977). *Acta Crystallographica,* **A33,** 800–4.
82. Sussman, J. L. and Podjarny, A. D. (1983). *Acta Crystallographica,* **B39,** 495.
83. Shoham, M., Yonath, A., Sussman, J. L., Moult, J., Traub, W., and Kalb, A. J. (1979). *Journal of Molecular Biology,* **131,** 137–55.
84. Tronrud, D. E., Ten Eyck, L. F., and Matthews, B. W. (1987). *Acta Crystallographica,* **A43,** 489–501.
85. Karplus, M. and McCammon, J. A. (1983). *Annual Reviews in Biochemistry,* **53,** 263–300.
86. Brünger, A. T., Kuriyan, J., and Karplus, M. (1987). *Science,* **235,** 458–60.
87. Fujinaga, M., Gros, P., and van Gunsteren, W. F. (1989). *Journal of Applied Crystallography,* **22,** 1–8.
88. Brünger, A. T. (1988). In *Crystallographic computing* (ed. N. W. Isaacs and M. R. Taylor), pp. 127–40. Oxford University Press.
89. Brünger, A. T. (1988). *Journal of Molecular Biology,* **203,** 803–16.
90. Kuriyan, J., Brünger, A. T., Karplus, M., and Hendrickson, W. A. (1989). *Acta Crystallographica,* **A45,** 396–409.
91. Rice, D. W. (1981). *Acta Crystallographica,* **A37,** 491–500.
92. Stuart, D. and Artymiuk, P. (1984). *Acta Crystallographica,* **A40,** 713–16.
93. Anfinsen, C. B., Haber, E., Sela, M., and White, F. H. Jr (1961). *Proceedings of the National Academy of Sciences of the USA,* **47,** 1309.
94. Schulz, G. E. and Schirmer, R. H. (1979). *Principles of protein structure.* Springer, Berlin.
95. Richards, F. M. (1974). *Journal of Molecular Biology,* **82,** 1–14.
96. Blundell, T., Barlow, D., Borkakoti, N., and Thornton, J. (1983). *Nature,* **306,** 281–3.
97. Piela, L., Nemethy, G. E., and Scheraga, H. A. (1987). *Biopolymers,* **26,** 1273–86.

98. Hol, W. G. J., van Duijnen, P. T., and Berendsen, H. J. C. (1978). *Nature,* **273,** 443–6.
99. Takano, T. (1977). *Journal of Molecular Biology,* **110,** 569–84.
100. Borkakoti, N., Moss, D. S., and Palmer, R. A. (1982). *Acta Crystallographica,* **B38,** 2210–17.
101. Richardson, J. S. (1977). *Nature,* **268,** 495–500.
102. Chothia, C. (1973). *Journal of Molecular Biology,* **75,** 295–302.
103. Milner-White, E. J. and Poet, R. (1987). *Trends in Biological Science,* 189–92.
104. Ramachandran, G. N., Ramakrishnan, C., and Sasisekharan, V. (1963). *Journal of Molecular Biology* **7,** 95–9.
105. Bode, W. and Schwager, P. (1975). *Journal of Molecular Biology*, **98,** 693–717.
106. Phillips, D. C. (1970). *British biochemistry past and present* (ed. T. W. Goodwin), pp. 11–28. Academic, London.
107. Wlodawer, A., Deisenhofer, J., and Huber, R. (1987). *Journal of Molecular Biology,* **193,** 145–56.
108. Chotia, C., Levitt, M., and Richardson, D. C. (1971). *Proceedings of the National Academy of Sciences of the USA,* **74,** 4130–4.
109. Richardson, J. S. (1981). *Advances in Protein Chemistry,* **34,** 167–339.
110. Babu, Y. S., Sack, J. S., Greenhough, T. J., Bugg, C. E., Means, A. R., and Cook, W. J. (1985). *Nature,* **315,** 37–40.
111. Dickerson, R. E. and Geiss, I. (1983). *Hemoglobin: structure, function, evolution and pathology.* Benjamin Cummings, Menlo Park.
112. Jurnak, F. A. and McPherson, A. (ed.) (1984). *Biological macromolecules and assemblies.* Wiley, New York.
113. Krebs, E. G. and Beavo, J. A. (1979). *Annual Reviews in Biochemistry,* **48,** 923–59.
114. Paulson, J. C. (1989). *Trends in Biological Science,* **14,** 272–6.
115. Petsko, G. A. and Ringe, D. (1984). *Annual Reviews in Biophysics and Bioengineering,* **13,** 331–71.
116. Wlodawer, A., Borkakoti, N., Moss, D. S., and Howlin, B. (1986). *Acta Crystallographica,* **B42,** 379–87.
117. Perutz, M. F. (1970). *Nature,* **228,** 726–39.
118. Barford, D. J. and Johnson, L. N. (1989). *Nature,* **340,** 609–16.
119. Adman, E. T., Sieker, L. C., and Jensen, L. H. (1976). *Journal of Biological Chemistry,* **251,** 3801–6.
120. Zhang, R.-g., Joachimiak, A., Lawson, C. L., Schevitz, R. W., Otwinowski, Z., and Sigler, P. B. (1987). *Nature,* **327,** 591–7.
121. Priestle, J. P. (1988). *Journal of Applied Crystallography,* **21,** 572–6.
122. Sheriff, S., Hendrickson, W. A., and Smith, J. L. (1987). *Journal of Molecular Biology,* **197,** 273–96.
123. Blacke, C. C. F., Geisow, M. J., Oatley, S. J., Rerat, B., and Rerat, C. (1978). *Journal of Molecular Biology,* **121,** 339–56.
124. Banner, D. W., Bloomer, A. C., Petsko, G. A., Phillips, D. C., and Wilson, I. A. (1976). *Biochemistry and Biophysics Research Communications,* **72,** 146.
125. Tollin, P. (1966). *Acta Crystallographica,* **21,** 613–14.
126. Buerger, M. J. (1967). *Vector space.* Wiley, New York.
127. Tollin, P. (1969). *Acta Crystallographica,* **A25,** 376–7.
128. Harada, Y., Lifchitz, A., Berthou, J., and Jolles, P. (1981). *Acta Crystallographica,* **A37,** 398–406.
129. Agarwal, R. C. (1978). *Acta Crystallographica,* **A34,** 791–809.
130. Ten Eyck, L. F. (1973). *Acta Crystallographica,* **A29,** 183–91.
131. Hestenes, M. R. and Stiefel, E. (1952). *Journal of Research of the National Bureau of Standards,* **49,** 409–36.
132. Rollet, J. S. (ed.) (1965). *Computing methods in crystallography.* Pergamon, Paris.

Physical properties of crystals

9

MICHELE CATTI

Introduction

The discipline named crystal physics has been associated traditionally with two topics:

1. The study of phenomenological and macroscopic properties of crystals, concerning mainly the effects of crystal anisotropy and symmetry on the physical properties of matter.
2. The microscopic investigation of crystal defects, i.e. deviations from the ideal periodicity of the crystal structure, and of their influence on the macroscopic physical properties.

Both subjects are closely related to crystallographic results: in particular, the first one relies upon consideration of symmetry, while the second one is founded on the structural analysis of crystals.

Crystal physics is not only a discipline of great fundamental interest, but it also has important technological applications. Crystals are used in industry because of their useful physical (optical, electrical, magnetic, etc.) properties, which are studied by crystal physics. It suffices to mention piezoelectric transducers, magnetic oxides for tapes for recorders and computers, crystals for non-linear optics of laser technology, and many other examples. Analogously, defects have large effects on the crystal properties, such as mechanical strength, electrical and thermal conductivity, etc., so as to play an outstanding role in several branches of technology. It is also worthwhile to point out the close relations between crystal physics and solid state physics. The latter discipline, in a restricted meaning, emphasizes such topics as the quantum study of electronic and vibrational energy levels, the interaction of radiation with matter, and related items, both in ideally periodic and in defective solids. All these subjects try to explain on a microscopic basis the phenomenological properties which are dealt with by crystal physics; on the other hand, their treatment relies heavily upon the geometrical features of solids studied by crystallography. Thus crystal physics can be considered to be a bridging discipline between crystallography and solid state physics, partially overlapping in some areas with both of them.

Crystal anisotropy and tensors

The behaviour of crystals is characterized typically by anisotropy: a physical agent associated with a given space orientation usually causes effects with different orientations with respect to the crystal lattice. A central role is thus played by the geometrical aspect of the problem, particularly in relation to the choice and transformations of reference frames, and taking into account the need to comply with symmetry constraints of the crystal.

As shown in Chapter 2, most crystallographic issues are treated in a lattice reference frame, where the basis vectors are three non-coplanar translations associated with the periodic symmetry of the crystal structure. However, orthonormal (Cartesian) reference frames were also introduced (p. 68), and the transformation properties between lattice **A** and Cartesian **E** bases were analysed, according to the relation $\mathbf{E} = \mathbf{MA}$ involving the orthonormalization matrix **M**. Both types of reference frames can be used in the study of physical properties of crystals; yet treatments relying upon lattice bases need a particular formalism (covariant–contravariant subscripts and superscripts) which is often awkward to use. So orthonormal frames are generally preferred, and the relations of transformations turn out to be much simpler. This method will be followed in the present book.

By working with Cartesian frames, a conventional orientation with respect to the lattice frame must be defined. In the general triclinic case, the most widely accepted convention involves: $\boldsymbol{e}_3 = \boldsymbol{c}/c$, $\boldsymbol{e}_2 = \boldsymbol{b}^*/b^*$, $\boldsymbol{e}_1 = \boldsymbol{e}_2 \times \boldsymbol{e}_3$. This implies that, for the monoclinic system, $\boldsymbol{e}_2 = \boldsymbol{b}/b$; for the trigonal and hexagonal cases, $\boldsymbol{e}_1 = \boldsymbol{a}/a$. In the latter systems, the crystallographic basis vectors $\boldsymbol{a}$ and $\boldsymbol{b}$ are chosen parallel to the twofold axes (when they are present) which are normal to the three- or sixfold unique direction. For orthorhombic, tetragonal, and cubic systems, the Cartesian basis vectors are parallel to the corresponding lattice basis vectors. The orthonormalization matrices can be found by the methods of p. 68. It should be stressed that other types of conventions are possible (for instance: $\boldsymbol{e}_3 = \boldsymbol{c}/c$, $\boldsymbol{e}_1 = \boldsymbol{a}^*/a^*$, $\boldsymbol{e}_2 = \boldsymbol{e}_3 \times \boldsymbol{e}_1$, leading to $\boldsymbol{e}_2 = \boldsymbol{b}/b$ in the trigonal and hexagonal systems). Since numerical values of the physical properties are strictly dependent on the choice of reference frame, it is necessary to know whether a different convention from the standard one is associated with tabulated physical data.

Tensorial quantities

Let us consider a functional relationship between two vectorial physical quantities, $\boldsymbol{X}$ and $\boldsymbol{Y}$. If they have a common direction, the functional relation between their moduli is dealt with by writing a Taylor expansion about $X = 0$:

$$\begin{aligned} Y &= Y_0 + \left(\frac{\mathrm{d}Y}{\mathrm{d}X}\right)_0 X + \frac{1}{2}\left(\frac{\mathrm{d}^2Y}{\mathrm{d}X^2}\right)_0 X^2 + \ldots \\ &= Y_0 + y'X + \tfrac{1}{2}y''X^2 + \ldots . \end{aligned} \tag{9.1}$$

In a crystal, where anisotropy is present, the $\boldsymbol{Y}$ vector generally has a different direction from $\boldsymbol{X}$, so that each of the three Cartesian components Y_i of $\boldsymbol{Y}$ has to be considered as a function of all three Cartesian components

X_1, X_2, X_3 of $\boldsymbol{X}$:

$$Y_i = Y_{0,i} + \sum_{h=1}^{3} \left(\frac{\partial Y_i}{\partial X_h}\right)_0 X_h + \tfrac{1}{2}\sum_{1}^{3}{}_{h,k}\left(\frac{\partial^2 Y_i}{\partial X_h\, \partial X_k}\right)_0 X_h X_k + \ldots$$

$$= Y_{0,i} + \sum_{h=1}^{3} y_{ih} X_h + \sum_{1}^{3}{}_{h,k} y_{ihk} X_h X_k + \ldots. \tag{9.2}$$

In the Taylor expansion (9.1) a single coefficient (Y_0, y', y'', etc.) is required for each term. On the other hand, according to (9.2) the constant part of the $\boldsymbol{Y}(\boldsymbol{X})$ dependence is expressed by 3 quantities $Y_{0,i}$, the linear part by 9 coefficients y_{ih}, the quadratic part by 27 coefficients y_{ihk}, and so on. In general, for a term of nth order in the Taylor expansion a number of 3^{n+1} coefficients is necessary to define the $\boldsymbol{Y}(\boldsymbol{X})$ dependence.

We now want to examine how the different coefficients appearing in the expansion (9.2) are transformed, when the orthonormal reference basis $\mathbf{E}$ is changed into another one $\mathbf{E}'$. Let $\mathbf{T}$ be the transformation matrix (cf. p. 65) relating the two Cartesian bases: $\mathbf{E}' = \mathbf{TE}$. The metric matrices of both bases are equal to the identity matrix $\mathbf{I}$, so that $\mathbf{G}' = \mathbf{G} = \mathbf{I}$; by substituting into $\mathbf{G}' = \mathbf{TG\bar{T}}$, equivalent to (2.21), the result $\mathbf{T\bar{T}} = \mathbf{I}$ is obtained. This condition is equivalent to

$$\mathbf{T}^{-1} = \mathbf{\bar{T}} \quad (\text{or } \mathbf{\bar{T}}^{-1} = \mathbf{T}) \tag{9.3}$$

and characterizes $\mathbf{T}$ as an orthogonal matrix, representing an isometric rotation in a Cartesian basis (cf. p. 36). Taking into account (2.20) and (9.3), the $\boldsymbol{X}$, $\boldsymbol{Y}$, and $\boldsymbol{Y}_0$ vector representations whose components appear in (9.2) are transformed as $\boldsymbol{X}' = \mathbf{T}\boldsymbol{X}$, $\boldsymbol{Y}' = \mathbf{T}\boldsymbol{Y}$ and $\boldsymbol{Y}_0 = \mathbf{T}\boldsymbol{Y}_0$ when the $\mathbf{E}$ basis is changed into $\mathbf{E}'$. The explicit relation for the $Y_{0,i}$ components is:

$$Y'_{0,i} = \sum_{h=1}^{3} T_{ih} Y_{0,h}. \tag{9.4}$$

As for the nine coefficients y_{ih} expressing the linear part of the $\boldsymbol{Y}(\boldsymbol{X})$ dependence, they can be considered to be components of a 3×3 square matrix $\mathbf{y}$, so that the first-order term in (9.2) is written in matrix form as:

$$\boldsymbol{Y} = \mathbf{y}\boldsymbol{X}. \tag{9.5}$$

By substituting $\boldsymbol{Y} = \mathbf{T}^{-1}\boldsymbol{Y}'$ and $\boldsymbol{X} = \mathbf{T}^{-1}\boldsymbol{X}'$ into (9.5), and taking into account (9.3), one obtains: $\boldsymbol{Y}' = \mathbf{Ty\bar{T}}\boldsymbol{X}'$. Thus the transformation law for the $\mathbf{y}$ matrix is found to be: $\mathbf{y}' = \mathbf{Ty\bar{T}}$. The explicit expression involving components is:

$$y'_{ih} = \sum_{1}^{3}{}_{k,l} T_{ik} T_{hl} y_{kl}. \tag{9.6}$$

An entity represented by nine components y_{ih} with respect to a given Cartesian basis, which obey the (9.6) law of transformation, is defined to be a tensor of second rank. Analogously, a vector whose three components follow the transformation rule (9.4) is a tensor of first rank, while a scalar is a zero-rank tensor. It should be emphasized that a tensor as such, being a generalization of the vector concept, is independent of the reference basis; its components, instead, are transformed when the basis changes.

The transformation properties of coefficients y_{ihk} in the expression (9.2) can be found by similar methods, and turn out to be:

$$y'_{ihk} = \sum_{l,p,q}^{3} T_{il} T_{hp} T_{kq} y_{lpq}. \tag{9.7}$$

An entity represented by 27 quantities y_{ihk} as components with respect to a given Cartesian basis, which obey the transformation law (9.7), is a third-rank tensor. In a general way, a tensor of rank n is defined as a set of 3^n coefficients with n subscripts, associated with a given Cartesian basis, which transform according to the formula:

$$y'_{ihkl\ldots} = \sum_{p,q,r,s\ldots}^{3} T_{ip} T_{hq} T_{kr} T_{ls} \ldots y_{pqrs\ldots}, \tag{9.8}$$

where **T** is the matrix relating the new basis to the old one. It should be noticed that the general rule (9.8) is equivalent to rule (9.4) multiplied by n times: thus we can also say that a tensor of rank n transforms in the same way as a product of n coordinates (or vector components). It is important to point out that any set of 3^n coefficients with n subscripts does not necessarily obey the (9.8) transformation rule, and so need not represent a tensor of rank n. Take for instance the 3^2 components of the matrix **T**, or those of the orthonormalization matrix **M** relating the Cartesian basis **E** to the lattice basis **A**: in no way can an expression of type (9.6) be applied to them, so that they are not components of second-rank tensors. Tensors of rank higher than two could be represented by matrices with more than two dimensions, but this is usually avoided for simplicity and the matrix formalism is limited to tensors of first and second rank. Further, in tensor calculus the Einstein convention is commonly adopted, according to which summation symbols are omitted and understood; however, for the sake of clarity this convention is not followed in this text. Because of the particular importance of second-rank tensors in representing the physical properties of crystals, their features are analysed in detail in Appendix 9.A.

Tensors of rank n have been introduced as coefficients of terms of order $n-1$ in the Taylor expansion (9.2) expressing a functional dependence between two vectors. However, tensors can represent a physical property relating not only vectors, but also other tensors. Let us consider a linear dependence only, for the sake of simplicity. Then the coefficients expressing such a dependence between vector components and second-rank tensor components are clearly themselves components of a third-rank tensor:

$$y_{ij} = \sum_{h=1}^{3} t_{ijh} v_h, \quad \text{or} \quad v_i = \sum_{h,k}^{3} q_{ihk} y_{hk}. \tag{9.9}$$

Analogously, a linear dependence between two second-rank tensors is represented by the components of a fourth-rank tensor:

$$y_{ij} = \sum_{h,k}^{3} t_{ijhk} z_{hk}. \tag{9.10}$$

The following general rule can be formulated: the coefficients of a linear dependence of the components of an nth-rank tensor on the products of the components of $n_1, \ldots, n_m$-rank tensors are themselves the components of a tensor of rank $n + n_1 + \ldots + n_m$.

From the physical point of view, tensors can represent either an intrinsic property of the crystalline medium, or an external field applied to the crystal with an arbitrary orientation. In the first case they are called matter tensors, in the second one field tensors. Several examples of both kinds of tensors will be given in the following sections.

Symmetry of tensorial properties

The symmetry constraints which must be satisfied by the physical properties of crystals are expressed in a very general form by Neumann's principle (see p. 14), which states that a crystal point group must be either the same or a subgroup of the symmetry group inherent to the physical property considered.

A useful way of applying Neumann's principle is to constrain the tensor representing the required physical property to be invariant with respect to any symmetry operation of the crystal point group. Of course it suffices to check the invariance with respect to the group generators only. For instance, in the case of a physical quantity expressed by a second-rank tensor **y**, the invariance relationship is

$$\mathbf{y} = \mathbf{R}\mathbf{y}\bar{\mathbf{R}};$$

this equation has to be satisfied for all symmetry matrices **R** corresponding to point-group generators. By solving the equation for all independent symmetry operations of a given crystal point group, the corresponding symmetry constraints on the y_{ih} components can be derived. In particular, all y_{ih} components would turn out to be zero for a point group more symmetrical than the tensor itself, thus violating Neumann's principle. This algebraic procedure has been already applied in Appendix 3.B (p. 185) for deriving restrictions on the thermal factor, and will be used in subsequent sections to obtain the symmetry conditions for a number of tensorial physical properties in all 32 crystal point groups.

Neumann's principle can also be interpreted from a geometrical point of view, by considering the symmetry of the geometrical representation of the tensorial property and comparing it straightforwardly with the crystal point symmetry. This is certainly feasible for first- and second-rank tensors, while for higher-rank tensors the method would become quite complicated. A first-rank tensorial property is represented by a vector, which may be polar (segment with arrow) or axial (segment with direction of rotation) (Fig. 9.1). For instance: velocity, force, electric field intensity are represented by polar vectors, while angular velocity, moment of force, magnetic field intensity are represented by axial vectors.

Polar and axial vectors differ by their groups of symmetry, as they belong to the ∞m and ∞/m limit groups, respectively. Limit groups of symmetry (also called Curie groups) are point groups including the infinite symmetry axes, i.e. rotations of any angles. The ∞m group of polar vectors (for instance, a moment of electric dipole) is not centrosymmetric, whereas the ∞/m group of axial vectors (e.g. a moment of magnetic dipole) is. Thus Neumann's principle states that a spontaneous electric polarization, typical of pyroelectric and ferroelectric crystals, can be observed only for the so-called polar point groups, which are subgroups of the limit group ∞m: 1, 2, 3, 4, 6, m, mm2, 3m, 4mm, 6m. The polar vector must be parallel to the symmetry axis (polar axis), when this is present; in point group m, the

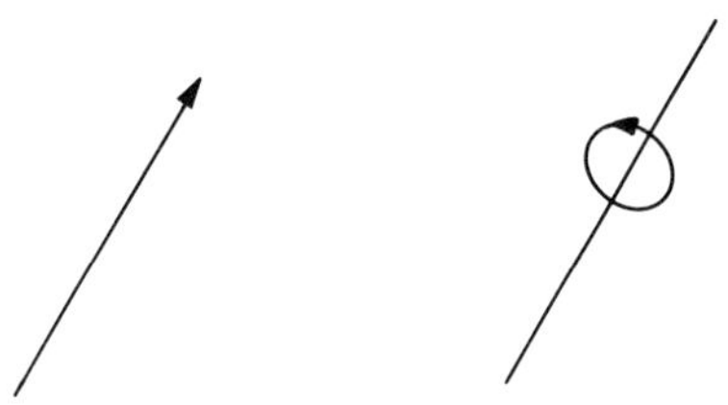

Fig. 9.1. Polar and axial vectors.

vector may have any direction lying in the mirror plane, while in group 1 its orientation is unrestricted. On the other hand, a spontaneous moment of magnetic dipole (ferromagnetism) is consistent only with crystal symmetry represented by subgroups of the limit group ∞/m: 1, 2, 3, 4, 6, m, 2/m, $\bar{6}$, 4/m, 6/m, $\bar{1}$, $\bar{3}$, $\bar{4}$. The axial vector may have any orientation in point groups 1 and $\bar{1}$, it must be normal to the mirror plane in group m, and its direction has to be parallel to the symmetry axis in all other cases.

A symmetric second-rank tensorial property is represented geometrically by a second-order surface (quadric), which may be an ellipsoid or a hyperboloid of one or two sheets (cf. Appendix 9.A). In general, such a surface has mmm symmetry, but in special cases higher symmetries can be displayed. When two tensor eigenvalues are equal but different from the third one, then the quadric becomes a revolution ellipsoid or hyperboloid with symmetry ∞/mm. Because of Neumann's principle, this must occur for all crystal point groups belonging to the tetragonal, trigonal, and hexagonal systems, which are subgroups of the limit group ∞/mm, but not of the group mmm. Besides, the principal direction of the tensor corresponding to the unique eigenvalue must be parallel to the symmetry axis 4, 3, or 6. When all three eigenvalues are equal, the quadric is a sphere with symmetry ∞∞/m; such a situation is required for all cubic point groups (subgroups of ∞∞/m), as both mmm and ∞/mm symmetries are lower than those consistent with the cubic system. For the orthorhombic, monoclinic, and triclinic systems, the second-rank tensor can be represented by a general mmm quadric with all three eigenvalues different, but with a constrained orientation in the orthorhombic and monoclinic cases. In the former instance, the principal directions are to be parallel to the crystallographic axes, while in the latter case one of the principal directions must be parallel to the unique monoclinic axis.

The forms shown by first- and symmetric second-rank tensors as imposed by symmetry are given in Table 9.1. Of course if the symmetric second-rank

Table 9.1. Forms displayed by first- and symmetric second-rank tensors referred to Cartesian axes

(a) Polar vector ***p***		(b) Axial vector ***a***	
1	$(p_1\ \ p_2\ \ p_3)$	1, $\bar{1}$	$(a_1\ \ a_2\ \ a_3)$
m	$(p_1\ \ 0\ \ p_3)$	2, m, 2/m	$(0\ \ a_2\ \ 0)$
2	$(0\ \ p_2\ \ 0)$	3, $\bar{3}$, 4, $\bar{4}$, 6, $\bar{6}$, 4/m, 6/m	$(0\ \ 0\ \ a_3)$
3, 4, 6, mm2, 3m, 4mm, 6mm	$(0\ \ 0\ \ p_3)$		

(c) Symmetric second-rank tensor **y**

Triclinic	$\begin{pmatrix} y_{11} & y_{12} & y_{13} \\ & y_{22} & y_{23} \\ & & y_{33} \end{pmatrix}$	Monoclinic	$\begin{pmatrix} y_{11} & 0 & y_{13} \\ & y_{22} & 0 \\ & & y_{33} \end{pmatrix}$
Orthorhombic	$\begin{pmatrix} y_{11} & 0 & 0 \\ & y_{22} & 0 \\ & & y_{33} \end{pmatrix}$	Tetragonal, hexagonal	$\begin{pmatrix} y_{11} & 0 & 0 \\ & y_{11} & 0 \\ & & y_{33} \end{pmatrix}$
Cubic	$\begin{pmatrix} y_{11} & 0 & 0 \\ & y_{11} & 0 \\ & & y_{11} \end{pmatrix}$		

tensor is referred to a lattice basis **A** rather than to the Cartesian basis **E**, unlike the convention followed until now, then the symmetry constraints on its components may be different from those of Table 9.1. This is true in particular for the hexagonal system, where non-zero off-diagonal components may appear (see also Tables 3.B.1 and 3.B.2).

Overview of physical properties

A number of important tensorial properties of crystals will now be examined, using the results of tensor theory previously discussed. The whole subject would include all topics of classical phenomenology applied to the behaviour of matter: mechanical, thermal, electrical, magnetic, optical properties. However, only some of them are considered here, because of limited space. First, the dielectric behaviour of crystals will be illustrated, as it is particularly suitable for a simple application of the tensorial concepts just developed. This topic is closely linked to that of optical properties, because of the relation between refraction index n and relative permittivity ε_r ($n = \sqrt{\varepsilon_r}$ in isotropic bodies); yet only a short overview will be given of that subject. Then mechanical properties (stress–strain relations, crystal elasticity), which require use of the highest (fourth) rank tensors presented in this book, will be considered in some detail. Thermal behaviour is included in that section through examination of the thermal expansion phenomenon. Eventually the piezoelectric properties, bridging electrical and mechanical behaviour of crystals, and providing an example of third-rank tensors, are analysed.

Electrical properties of crystals

Dielectric media are insulators, they ideally do not allow the transport of free charges, but only the presence of static polarization charges. Polarization of atoms is due to electrostatic induction operated by an applied electric field $\boldsymbol{E}$. The electronic component of polarization is a separation of the centre of negative from the centre of positive atomic charges, giving rise to a moment of atomic electric dipole. If atoms are ionized, then they are slightly displaced by the electric field from their equilibrium positions, and contribute to polarization by an additional ionic component. In Fig. 9.2 an unpolarized pair of ions is shown, where the positions of nuclei A and B coincide with the centres of negative (electronic) charge distributions X and Y, respectively. Under the action of an electric field E, A′ separates from X′ and B′ separates from Y′, so as to give rise to an electric dipole within each ion (electronic polarization). Besides, the relative position A′B′ changes as well with respect to AB, modifying the natural electric dipole of the ionic pair (ionic polarization).

An example of the order of magnitude of the effect can be given for NaCl crystals. The Cl^- ion displays an electronic polarizability (= ratio between induced electric dipole moment and electric field) of 3.3×10^{-40} C m^2 V^{-1}, while the Na^+ contribution is an order of magnitude smaller. This means that a potential difference of 1000 V applied to a NaCl crystal plate 1 mm thick would induce a dipole moment of about 3×10^{-34} C m on each Cl^-

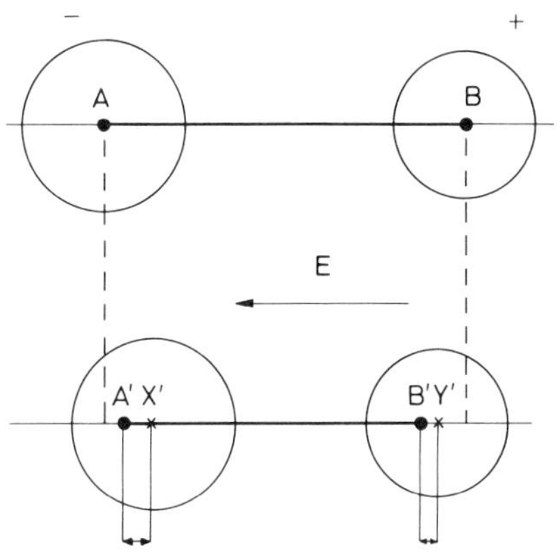

Fig. 9.2. Above: unpolarized pair of ions. Below: electronic and ionic polarization induced by the electric field $\boldsymbol{E}$.

ion, corresponding to a separation of centres of positive (nucleus–core electrons) and negative (valence electrons) charges of $\sim 3\times 10^{-6}$ Å. The contribution of ionic polarizability ($=6\times 10^{-40}\,\mathrm{C\,m^2\,V^{-1}}$ for the pair Na^+ Cl^-) would cause the Na^+–Cl^- separation to change by about 4×10^{-5} Å.

The overall effect is measured by the $\boldsymbol{P}$ vector, the intensity of polarization, which is equal to the induced moment of electric dipole per unit volume. For an isotropic body, the first-order approximation gives a relationship of proportionality:

$$\boldsymbol{P} = \varepsilon_0 \chi \boldsymbol{E}. \tag{9.11}$$

ε_0 is the vacuum permittivity (or dielectric constant), and χ is a dimensionless proportionality constant, called dielectric susceptibility, which actually expresses the magnitude of the polarization phenomenon. The material response is fully incorporated in Maxwell equations, which for a static problem take the form: div $\boldsymbol{D}=\rho$, curl $\boldsymbol{E}=0$, where ρ is the volume density of free electric charges. The electric displacement (or electric induction) vector $\boldsymbol{D}$ is related to the intensity of polarization $\boldsymbol{P}$ according to: $\boldsymbol{D}=\varepsilon_0\boldsymbol{E}+\boldsymbol{P}$. By substituting (9.11) for $\boldsymbol{P}$, the dependence of $\boldsymbol{D}$ on $\boldsymbol{E}$ for an isotropic body is found:

$$\boldsymbol{D} = (1+\chi)\varepsilon_0\boldsymbol{E} = \varepsilon_{\mathrm{r}}\varepsilon_0\boldsymbol{E} = \varepsilon\boldsymbol{E}, \tag{9.12}$$

where the relative permittivity ε_{r} is defined as $\varepsilon_{\mathrm{r}}=1+\chi$, and the product $\varepsilon_{\mathrm{r}}\varepsilon_0=\varepsilon$ is the permittivity (or dielectric constant) of the isotropic dielectric medium. When the electric field depends sinusoidally on time (e.g. in an electromagnetic wave), then ε and the related quantities are functions of the corresponding frequency.

Equation (9.12) relating $\boldsymbol{D}$ to $\boldsymbol{E}$ holds for an isotropic body and within a linear approximation is valid when the electric field $\boldsymbol{E}$ is not very large. A generalized equation extended to anisotropic media and to any field intensity takes the form of a Taylor expansion of type (cf. eqn (9.2)):

$$D_i = D_{0,i} + \sum_{h=1}^{3} \varepsilon_{ih}E_h + \sum_{1}^{3}{}_{h,k}\varepsilon_{ihk}E_hE_k + \ldots. \tag{9.13}$$

Of course a similar expansion could also be written to express the $\boldsymbol{P}(\boldsymbol{E})$ dependence, replacing equation (9.11). The coefficients ε_{ih} appearing in (9.13) are components of the second-rank permittivity tensor $\boldsymbol{\varepsilon}$; on the basis of thermodynamic arguments, this can be proved to be symmetrical ($\varepsilon_{hi}=\varepsilon_{ih}$) with positive eigenvalues. Related tensorial quantities are the relative permittivity $\boldsymbol{\varepsilon}_{\mathrm{r}}=\boldsymbol{\varepsilon}/\varepsilon_0$ and the dielectric susceptibility $\boldsymbol{\chi}=\boldsymbol{\varepsilon}_{\mathrm{r}}-\mathbf{I}$.

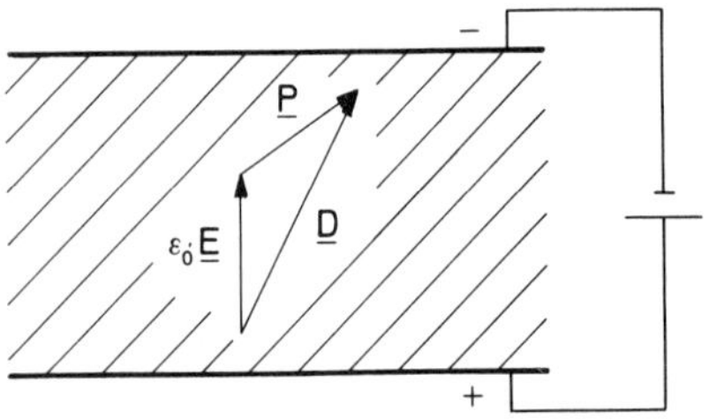

Fig. 9.3. Relations between electric field, electric displacement, and electric polarization vectors in a plane condenser with crystalline dielectric.

According to (9.13), now $\boldsymbol{D}$ and $\boldsymbol{E}$ are no longer parallel vectors; within a linear approximation corresponding to (9.12), they are related by the tensorial equation $\boldsymbol{D}=\boldsymbol{\varepsilon}\boldsymbol{E}$. However, the Maxwell equations given above and the relation $\boldsymbol{D}=\varepsilon_0\boldsymbol{E}+\boldsymbol{P}$ still hold also for anisotropic crystals. A picture of the three vectors $\boldsymbol{E}$, $\boldsymbol{D}$, $\boldsymbol{P}$ in a crystalline slab between the plates of a condenser is given in Fig. 9.3.

Pyroelectricity

The presence of the zeroth-order term $D_{0,i}$ in the expansion (9.13) implies that, in an anisotropic crystal, a spontaneous polarization given by the vector $\boldsymbol{P}_0=\boldsymbol{D}_0$ is possible when $\boldsymbol{E}=0$. This adds to the normal induced polarization depending on $\boldsymbol{E}$ and expressed by higher-order terms. When a crystal shows a spontaneous electric dipole moment $\boldsymbol{P}_0$, it is said to be pyroelectric; thus pyroelectricity is a property associated with the first-rank tensor $P_{0,i}$ or $D_{0,i}$. As $\boldsymbol{P}_0$ and $\boldsymbol{D}_0$ are polar vectors, all the symmetry considerations of p. 603 relative to that case hold: therefore pyroelectric crystals can only belong to point groups which are subgroups of ∞m (cf. Table 9.1). Of course an isotropic medium has symmetry higher than ∞m, so by Neumann's principle it can not display pyroelectricity.

It should be noticed that, from a practical point of view, the spontaneous polarization of a pyroelectric crystal has a very short life. Indeed, the combined effects of a very small crystal conductivity due to defects and of ionized particles in the air contribute to 'discharge' the condenser cancelling the electric dipole moment of the crystal. However, a quick temperature change causes a variation of $\boldsymbol{P}_0$ which makes the spontaneous polarization detectable as an electric signal. Thus pyroelectricity is properly revealed as a change of $\boldsymbol{P}_0$ with temperature, and the three pyroelectric coefficients are the derivatives $\partial P_{0,i}/\partial T$.

A particular type of pyroelectricity is that of ferroelectric crystals; for these, the vector $\boldsymbol{P}_0$ of spontaneous polarization can be inverted by applying an electric field in the opposite direction.

An example of a pyroelectric and ferroelectric crystal of technological importance (used for infrared detectors) is that of $LiTaO_3$ (symmetry group 3m). Its spontaneous electric dipole moment is aligned with the threefold axis, and the corresponding pyroelectric coefficient has value of $1.9 \times 10^{-4}\,\mathrm{C\,m^{-2}\,K^{-1}}$. Thus a temperature change of 1 K, brought about for instance by infrared radiation, causes a dipole moment per unit volume of $1.9 \times 10^{-4}\,\mathrm{C\,m^{-2}}$ to arise. Taking into account that the principal value of relative permittivity along the 3 axis is $\varepsilon_{r,3} = 46$, we can compute the electric field which would be necessary to produce the same polarization: $E = P/\varepsilon_0(\varepsilon_r - 1) = 4.77 \times 10^5\,\mathrm{V\,m^{-1}}$. This is a large field, showing that the magnitude of the pyroelectric effect for $LiTaO_3$ is noteworthy.

Dielectric impermeability and optical properties

When the first-order term in the expansion (9.13) is sufficient to account for optical behaviour of the crystal, we speak of linear optical properties. The dielectric impermeability tensor is defined as $\mathbf{b} = \boldsymbol{\varepsilon}^{-1}$, so that $\boldsymbol{E} = \mathbf{b}\boldsymbol{D}$; the relative impermeability is $\mathbf{B} = \boldsymbol{\varepsilon}_r^{-1} = \varepsilon_0\mathbf{b}$, and its positive principal values are $B_i = (\varepsilon_r^{-1})_i$. The representation surface of tensor $\mathbf{B}$ (cf. Appendix 9.A) is an ellipsoid of equations

$$\sum_{1}^{3}{}_{i,j} B_{ij}x_i x_j = 1 \quad \text{or} \quad \sum_{i=1}^{3} B_i x_i^2 = 1,$$

according to whether the reference is a general Cartesian basis or the basis of eigenvectors. This ellipsoid is called the optical indicatrix, and has the important property that its semi-axes have lengths equal to $B_i^{-1/2} = \varepsilon_{r,i}^{1/2} = n_i$, i.e. to the principal refraction indices of the crystal. Thus the optical indicatrix is simply the representation quadric of the tensor of relative dielectric impermeability.

The peculiar optical properties of crystals can be derived by use of this ellipsoid. Let us consider a propagation direction OP of an electromagnetic wave, drawn through the centre of the optical indicatrix (Fig. 9.4). The central section of the ellipsoid normal to OP is an ellipse with semi-axes OM and ON. It can be shown that not just one but two wave fronts propagate through the crystal normally to OP with different velocities $v_1 = c/n_1$ and $v_2 = c/n_2$ (c is the velocity of light in a vacuum); the corresponding refractive indices n_1 and n_2 are equal to the lengths of semi-axes OM and ON. The two waves are plane polarized with directions of vibration of the $\boldsymbol{D}$ vectors parallel to OM and ON, respectively. The whole subject of optical crystallography, which will not be dealt with here in detail, is based on the study of properties of the optical indicatrix and other related surfaces. We would just point out that the orientation and symmetry behaviour of the optical indicatrix in the different crystal point groups follows all results derived from Neumann's principle for symmetrical second-rank tensors on p. 604 and in Table 9.1.

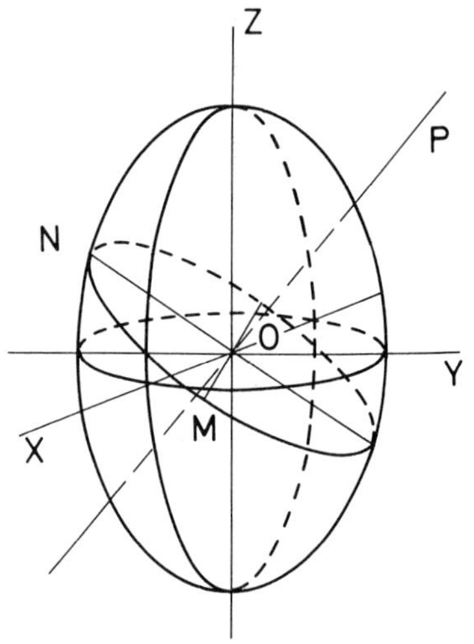

Fig. 9.4. The optical indicatrix.

Thus in the cubic system all three principal refraction indices are equal, the optical indicatrix is a sphere, and the crystal is optically isotropic in all directions. In the tetragonal, trigonal, and hexagonal systems two principal indices are equal but different from the third one, whose corresponding principal axis (called optic axis) is parallel to the high-symmetry direction. The indicatrix is an ellipsoid of revolution about the optic axis. The two wave normals propagating along that direction have equal refractive indices, and therefore coincide; no double refraction is observed, and the optic axis is a direction of optical isotropy. Crystals belonging to these symmetry systems are called uniaxial. For the orthorhombic, monoclinic, and triclinic systems, the indicatrix is a general ellipsoid with two circular central sections; the corresponding normal directions, lying on the plane of the longest and shortest semi-axes, are the optic axes and the crystals are said to be biaxial. The two optic axes are directions of optical isotropy, just as in the uniaxial case.

In some crystalline materials the optical behaviour depends significantly on terms of order higher than one (typically, second-order terms) in the $\boldsymbol{D}(\boldsymbol{E})$ expansion (9.13). This subject is called non-linear optics, and is associated with important technological applications. Optical non-linearity is also known as the electro-optical effect, and the electro-optical coefficients ε_{ihk} are evidently the components of a third-rank tensor. The (9.13) expansion truncated to second order can be rewritten as:

$$D_i = \sum_{h=1}^{3} \left(\varepsilon_{ih} + \sum_{k=1}^{3} \varepsilon_{ihk} E_k \right) E_h = \sum_{h=1}^{3} \varepsilon'_{ih} E_h, \qquad (9.14)$$

so that the ε_{ihk} coefficients may be interpreted as derivatives of permittivity components ε'_{ih} dependent on the field $\boldsymbol{E}$ with respect to the field components themselves; from this comes the name of electro-optical effect.

Elastic properties of crystals

The mechanical behaviour of solid bodies is characterized by a change of geometrical shape (deformation) as a response to applied forces, apart from rigid translations and rotations. When the mechanical stress is small enough, then the corresponding strain is reversible and follows Hooke's law, being simply proportional to the stress itself: this is the regime of elastic deformation, well known for isotropic bodies from texts of elementary mechanics. In the case of crystals, the force field acting on the body and the ensuing deformation field have to be specified in three dimensions with respect to the crystal Cartesian reference basis. Hooke's law is generalized by linear relationships between all stress and strain components, leading in a natural way to the tensor formalism for anisotropic elastic behaviour.

The study of the mechanical properties of crystals is a valuable subject for at least two reasons. First, because of technological applications: whatever use is made of a crystal in applied science or in industry, its response to mechanical agents is of primary importance and has to be known in detail. Second, there are important fundamental implications of the elastic behaviour of crystals, which concern the nature of interatomic forces giving rise to cohesion and geometrical features of the crystal structure.

Crystal strain

In a general way, the state of strain of the crystal is defined by the vector field $\boldsymbol{u} = \boldsymbol{x}' - \boldsymbol{x}$, which gives for every point the change between equilibrium $\boldsymbol{x}$ and strained $\boldsymbol{x}'$ position vectors. As usual, the dependence $\boldsymbol{u}(\boldsymbol{x})$ of the vector field can be expanded in a power series of type (9.2) (we assume that the displacement of the point at the origin is vanishing):

$$u_i = \sum_{h=1}^{3} \frac{\partial u_i}{\partial x_h} x_h + \tfrac{1}{2}\sum_{h,k\,1}^{3} \frac{\partial^2 u_i}{\partial x_h\, \partial x_k} x_h x_k + \ldots. \tag{9.15}$$

Elastic phenomena involve small deformations, for which terms of order higher than one may be neglected. In this case the strain is said to be homogeneous, and the u_i components are linear transformations of the position vector components x_h of a general crystal point:

$$u_i = \sum_{h=1}^{3} e_{ih} x_h. \tag{9.16}$$

The coefficients e_{ih} are dimensionless components of a second-rank tensor $\mathbf{e}$ which is generally non-symmetrical.

To understand the geometrical meaning of e_{ih} quantities, a two-dimensional case is illustrated in Fig. 9.5. A square is drawn with sides OA and OB parallel to the Cartesian x_1 and x_2 directions; then points A and B have coordinates $[x_1(\mathrm{A}), 0]$ and $[0, x_2(\mathrm{B})]$, respectively. After deformation, A and B change into A′ and B′; according to (9.16), the displacement vectors $\boldsymbol{u}(\mathrm{A}) = \mathrm{AA}'$ and $\boldsymbol{u}(\mathrm{B}) = \mathrm{BB}'$ have the following components:

$$u_1(\mathrm{A}) = e_{11}x_1(\mathrm{A}), \quad u_2(\mathrm{A}) = e_{21}x_1(\mathrm{A}),$$
$$u_1(\mathrm{B}) = e_{12}x_2(\mathrm{B}), \quad u_2(\mathrm{B}) = e_{22}x_2(\mathrm{B}).$$

The meaning of e_{11} and e_{22} components follows immediately: $e_{11} = u_1(\mathrm{A})/x_1(\mathrm{A})$, $e_{22} = u_2(\mathrm{B})/x_2(\mathrm{B})$. Further, the sides OA′ and OB′ are rotated by the angles φ_1 and φ_2 with respect to the original directions OA and OB, and evidently:

$$\tan\varphi_1 = \frac{u_2(\mathrm{A})}{x_1(\mathrm{A}) + u_1(\mathrm{A})} = \frac{e_{21}}{1+e_{11}}, \quad \tan\varphi_2 = \frac{u_1(\mathrm{B})}{x_2(\mathrm{B}) + u_2(\mathrm{B})} = \frac{e_{12}}{1+e_{22}}.$$

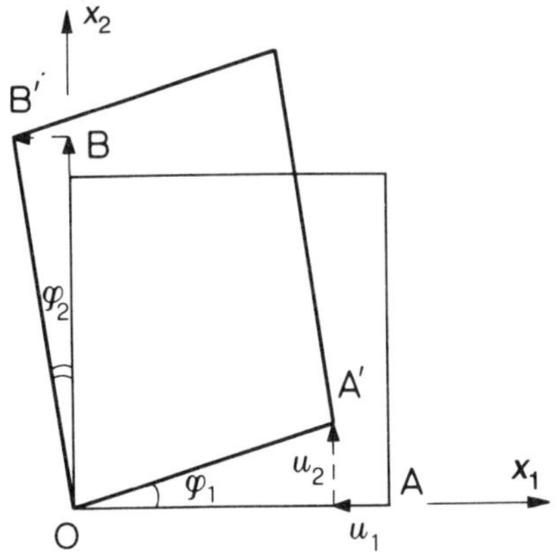

Fig. 9.5. Homogeneous deformation of a square into a parallelogram.

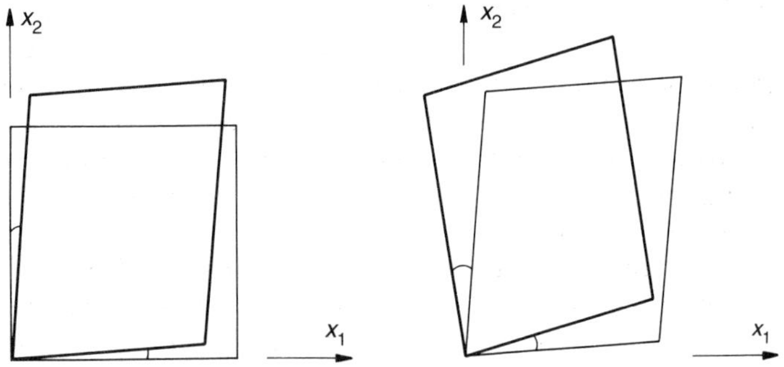

Fig. 9.6. Decomposition of the strain of Fig. 9.5 into a symmetrical strain plus a rigid rotation.

Taking into account that, for small strains, $e_{11} \ll 1$, $e_{22} \ll 1$, $\tan \varphi_1 \approx \varphi_1$, $\tan \varphi_2 \approx \varphi_2$, we obtain: $\varphi_1 \approx e_{21}$, $\varphi_2 \approx e_{12}$.

If $e_{11} = e_{22} = 0$ and $e_{12} = -e_{21}$, then the strain reduces to an anti-clockwise rigid rotation by the angle $\varphi = e_{21}$; in this case $\mathbf{e}$ is antisymmetrical ($e_{ij} = -e_{ji}$). Generally, the strain tensor $\mathbf{e}$ can always be written as the sum of a symmetrical $\boldsymbol{\varepsilon} = \frac{1}{2}(\mathbf{e} + \bar{\mathbf{e}})$ plus an antisymmetrical $\boldsymbol{\omega} = \frac{1}{2}(\mathbf{e} - \bar{\mathbf{e}})$ component: $\varepsilon_{ij} + \omega_{ij} = \frac{1}{2}(e_{ij} + e_{ji}) + \frac{1}{2}(e_{ij} - e_{ji}) = e_{ij}$. As $\boldsymbol{\omega}$ is antisymmetrical, it represents a rigid rotation; therefore $\boldsymbol{\varepsilon} = \mathbf{e} - \boldsymbol{\omega}$ corresponds to the physically relevant part of the strain. A geometrical picture of the decomposition $\mathbf{e} = \boldsymbol{\varepsilon} + \boldsymbol{\omega}$ for the planar deformation of Fig. 9.5 is shown in Fig. 9.6.

The Lagrangian strain tensor $\boldsymbol{\varepsilon}$ is called infinitesimal, because it is suitable to represent small deformations only; the use of the same symbol as that for dielectric permittivity may sometimes be confusing. Another tensor which is more convenient to express larger deformations is the finite Lagrangian strain tensor:

$$\boldsymbol{\eta} = \tfrac{1}{2}(\mathbf{e} + \bar{\mathbf{e}} + \mathbf{e}\bar{\mathbf{e}}). \tag{9.17}$$

For small strains, the difference $\boldsymbol{\eta} - \boldsymbol{\varepsilon} = 1/2\mathbf{e}\bar{\mathbf{e}}$ is vanishing, so that use of $\boldsymbol{\eta}$ or $\boldsymbol{\varepsilon}$ tensors is quite equivalent. Let us look now for a relation between the previous macroscopic representation of strain and the lattice microscopic nature of the crystal. $\mathbf{M}$ is the orthonormalization matrix of the undeformed lattice basis ($\mathbf{E} = \mathbf{MA}$), and $\mathbf{M}'$ that of the deformed basis ($\mathbf{E} = \mathbf{M}'\mathbf{A}'$); as fractional coordinates of points are not changed by a homogeneous deformation, then $\bar{\mathbf{M}}'\boldsymbol{x}' = \bar{\mathbf{M}}\boldsymbol{x}$. From (9.16), $\boldsymbol{x}' - \boldsymbol{x} = \mathbf{e}\boldsymbol{x}$; by substituting $\boldsymbol{x}' = \bar{\mathbf{M}}'^{-1}\bar{\mathbf{M}}\boldsymbol{x}$, one obtains:

$$\mathbf{e} = \bar{\mathbf{M}}'^{-1}\bar{\mathbf{M}} - \mathbf{I}. \tag{9.18}$$

Substitution into $\boldsymbol{\varepsilon} = \frac{1}{2}(\mathbf{e} + \bar{\mathbf{e}})$ and into (9.17) yields the required expressions for the $\boldsymbol{\varepsilon}$ and $\boldsymbol{\eta}$ tensors:

$$\boldsymbol{\varepsilon} = \tfrac{1}{2}(\bar{\mathbf{M}}'^{-1}\bar{\mathbf{M}} + \mathbf{M}\mathbf{M}'^{-1}) - \mathbf{I}; \quad \boldsymbol{\eta} = \tfrac{1}{2}(\mathbf{M}\mathbf{M}'^{-1}\bar{\mathbf{M}}'^{-1}\bar{\mathbf{M}} - \mathbf{I}). \tag{9.19}$$

Using the property of orthonormalization matrices ($\mathbf{M}^{-1}\bar{\mathbf{M}}^{-1} = \mathbf{G}$, cf. p. 69), one has that $\mathbf{M}'^{-1}\bar{\mathbf{M}}'^{-1} = \mathbf{G}'$ and $\mathbf{MG}\bar{\mathbf{M}} = \mathbf{I}$, so that:

$$\boldsymbol{\eta} = \tfrac{1}{2}(\mathbf{M}\mathbf{G}'\bar{\mathbf{M}} - \mathbf{I}) = \tfrac{1}{2}\mathbf{M}(\mathbf{G}' - \mathbf{G})\bar{\mathbf{M}}. \tag{9.20}$$

By comparing (9.20) with (9.19), an important difference appears between the two strain tensors $\boldsymbol{\varepsilon}$ and $\boldsymbol{\eta}$: the former depends on $\mathbf{M}'$ and then on the relative orientation of the deformed and undeformed lattices, while the latter does not. The $\boldsymbol{\eta}$ tensor, unlike $\boldsymbol{\varepsilon}$, depends only on the metrics of the deformed lattice, and is simply the transformation in a Cartesian basis of the change of metric tensor $\mathbf{G}' - \mathbf{G}$ induced by the crystal deformation.

The $\boldsymbol{\eta}$ tensor is symmetrical, and can be diagonalized by determining its principal strains η_1, η_2, η_3 and principal directions; these are the only directions in the crystal which are not changed by the deformation. The relative change of volume due to strain is given by the expression $\Delta V/V = \eta_1 + \eta_2 + \eta_3 = \mathrm{tr}\,\boldsymbol{\eta}$.

The crystal deformation is usually caused by an applied stress arbitrarily chosen, so that the tensor is a field tensor: it does not depend on intrinsic properties of the crystalline medium, but rather on an external field, being thus unconstrained by crystal symmetry. The symmetry, as well as the crystal system, may be altered by deformation. The same holds, for instance, for the vector of induced polarization $\boldsymbol{P}$, which is a field tensor of first-rank as it depends on the applied electric field $\boldsymbol{E}$. On the other hand, another important kind of crystal strain is that caused by temperature changes ΔT; the effect is represented by the tensor of thermal expansion, defined as:

$$\alpha_{ij} = \frac{\partial \eta_{ij}}{\partial T}. \qquad (9.21)$$

Its eigenvalues α_i are always positive, so that the representation quadric of equation

$$\sum_{i=1}^{3} \alpha_i x_i^2 = 1$$

is an ellipsoid, whose radius vector has a length equal to the inverse square root of the coefficient of thermal expansion along its direction. The corresponding volume thermal expansion is

$$\frac{1}{V}\frac{\partial V}{\partial T} = \sum_{i=1}^{3} \alpha_{ii} = \mathrm{tr}\,\boldsymbol{\alpha}.$$

The tensor of thermal expansion $\boldsymbol{\alpha}$ is independent of any external vector field, but is related to the inner properties of a given crystal. It is thus a matter tensor, just as tensors of dielectric permittivity or impermeability; its symmetry properties and orientation must be consistent with the point group of the crystal, according to Neumann's principle and the discussion of p. 604. An example of first-rank matter tensor is the vector of spontaneous polarization $\boldsymbol{P}_0$ of pyroelectric and ferroelectric crystals.

Inner deformation

The strain tensor $\boldsymbol{\eta}$ was related by (9.20) to a change of metric tensor $\mathbf{G}' - \mathbf{G}$, i.e. of unit-cell geometry. This corresponds to a purely homogeneous deformation of the crystal structure, leaving the atomic fractional coordinates constant (lattice strain). If, on the other hand, such coordinates vary, then in addition to the lattice deformation, an inner strain arises, which is just defined by the coordinate changes $\Delta \boldsymbol{x}_i$ for all atoms in the asymmetric unit. The inner strain generally occurs as a relaxation of atomic positions to minimize the energy of the deformed lattice, so it is a function of the lattice strain. The overall deformation of the atomic arrangement is the sum of these two effects. Changes of interatomic distances due to the total strain can be decomposed into the separate effects, according to

$$(d'_{ij})^2 - d_{ij}^2 = (\boldsymbol{x}_j - \boldsymbol{x}_i)\Delta\mathbf{G}(\boldsymbol{x}_j - \boldsymbol{x}_i) + [(\boldsymbol{x}_j - \boldsymbol{x}_i)\mathbf{G}'(\Delta\boldsymbol{x}_j - \Delta\boldsymbol{x}_i) + (\Delta\boldsymbol{x}_j - \Delta\boldsymbol{x}_i)\mathbf{G}'(\boldsymbol{x}_j - \boldsymbol{x}_i) + (\Delta\boldsymbol{x}_j - \Delta\boldsymbol{x}_i)\mathbf{G}'(\Delta\boldsymbol{x}_j - \Delta\boldsymbol{x}_i)],$$

where d_{ij} and d'_{ij} are the distances between atoms i and j before and after deformation, respectively. The first term is due to lattice strain only, while the second one (in square brackets) is mainly ascribed to inner strain contribution.

Let us consider an application of these concepts to the strain induced by thermal expansion in mica muscovite. Two crystal-structure refinements at 25 and 700 °C allowed us to determine the changes of unit-cell constants and of atomic fractional coordinates for the corresponding temperature range (Table 9.2). By calculating the metric tensor change $\mathbf{G}' - \mathbf{G}$ and applying equation (9.20), the strain tensor $\boldsymbol{\eta}$ can be obtained; dividing it by the temperature difference ΔT yields the tensor of thermal expansion $\boldsymbol{\alpha}$, which represents the lattice component of thermal strain: $\alpha_{11} = 1.12$, $\alpha_{22} = 1.18$, $\alpha_{33} = 1.89 \times 10^{-5}\,\mathrm{K}^{-1}$. The inner component of thermal strain is given by the changes of atomic fractional coordinates divided by ΔT. It appears that the largest contribution to inner deformation is related to basal oxygen atoms O(1), O(2), O(3), while the effect is very small in all other cases. This corresponds to a substantial decrease of the ditrigonal distortion of the (001) layer of (Si, Al)O_4 tetrahedra sharing corners, which approaches a more symmetrical hexagonal configuration (cf. Fig. 6.41). The contributions of lattice and internal deformations to changes of interatomic distances can be analysed in the case of K–O bond lengths; in Table 9.3 the two components

Table 9.2. Lattice constants and atomic fractional coordinates of muscovite $KAl_2[Si_3AlO_{10}](OH)_2$ (space group C2/c), at 25 °C (above) and 700 °C (below) (Catti *et al.* 1989)

$$a = \begin{cases}5.191\,\text{Å}\\5.229\,\text{Å}\end{cases},\ b = \begin{cases}9.006\,\text{Å}\\9.076\,\text{Å}\end{cases},\ c = \begin{cases}20.068\,\text{Å}\\20.322\,\text{Å}\end{cases},\ \beta = \begin{cases}95.77^\circ\\95.74^\circ\end{cases}\ V = \begin{cases}932.9\,\text{Å}^3\\959.6\,\text{Å}^3\end{cases}$$

	x	y	z
(Si, Al)(1)	0.4650	0.9298	0.1354
	0.4642	0.9306	0.1344
(Si, Al)(2)	0.4513	0.2585	0.1355
	0.4510	0.2590	0.1344
Al	0.2502	0.0834	0.0001
	0.2517	0.0836	0.0001
K	0	0.0989	1/4
	0	0.0983	1/4
O(1)	0.4146	0.0932	0.1685
	0.4297	0.0937	0.1670
O(2)	0.2519	0.8096	0.1573
	0.2402	0.8208	0.1566
O(3)	0.2515	0.3716	0.1689
	0.2397	0.3629	0.1671
O(4)	0.4623	0.9442	0.0535
	0.4631	0.9451	0.0534
O(5)	0.3835	0.2515	0.0534
	0.3795	0.2514	0.0532
O(6)	0.4577	0.5614	0.0500
	0.4576	0.5622	0.0492

Table 9.3. K–O distances (in Å) in muscovite at 25 °C (d_{ij}) and 700 °C (d'_{ij}), and differences between squared distances and corresponding lattice and inner components (in $Å^2$)

	d_{ij}	d'_{ij}	$\Delta(d^2)$	$\Delta(d^2)_l$	$\Delta(d^2)_i$
K–O(1) × 2	2.832	2.942	0.64	0.15	0.49
K–O(2) × 2	2.870	3.002	0.78	0.16	0.62
K–O(3) × 2	2.843	2.966	0.71	0.15	0.56
K–O(1)′ × 2	3.303	3.276	−0.18	0.19	−0.37
K–O(2)′ × 2	3.528	3.464	−0.43	0.23	−0.66
K–O(3)′ × 2	3.288	3.256	−0.21	0.19	−0.40

of quantities $(d'_{ij})^2 - d^2_{ij}$ are reported. A positive inner strain contribution indicates that the bond expands more than expected from unit-cell dilatation alone, whereas the opposite is implied by a negative value. Thus, in the K coordination sphere a very large effect of inner thermal strain is observed, with a positive sign for the six short bonds and a negative (even the absolute bond-distance changes are negative) for the six long contacts.

Stress tensor

The other field tensor which is necessary to define the elastic properties of crystals is the stress tensor. The mechanical forces applied externally to the crystal are 'contact forces' or 'surface forces' acting on the external surface of every volume element, and not 'body forces' acting on every point of the body (like the force of gravity). Therefore the force field is represented by the vector $\boldsymbol{p}$ (force per unit area) as a function of the unit vector $\boldsymbol{n}$ normal to the surface element dS. For a homogeneous stress, $\boldsymbol{p}$ depends not on the absolute position of dS but only on its orientation, so that $\boldsymbol{p} = \boldsymbol{p}(\boldsymbol{n})$; furthermore, the dependence is simply linear, being expressed by a tensorial relationship:

$$p_i = \sum_{h=1}^{3} \tau_{ih} n_h. \tag{9.22}$$

The τ_{ih} coefficients are components of the second-rank tensor of stress $\boldsymbol{\tau}$. A general τ_{ij} component has the physical meaning of oriented pressure along the ith direction acting on to the dS surface normal to the jth Cartesian direction. The simple case of surface forces acting onto the three faces of a cube is sketched in Fig. 9.7. The unit vector $\boldsymbol{n}$ normal to the (100) face has components [1 0 0], so that using (9.22) the components of the force per unit area acting on to (1 0 0) turn out to be $p_1 = \tau_{11}$, $p_2 = \tau_{21}$, $p_3 = \tau_{31}$; similarly for the (0 1 0) and (0 0 1) faces. The diagonal values τ_{ii} are called normal components, and the off-diagonal ones τ_{ij} $(i \neq j)$ are the shear components of stress. Analogously to what is true for the strain tensor $\boldsymbol{e}$, it can be easily proved that if no rigid rotation of the volume element is produced by the applied stress, then the $\boldsymbol{\tau}$ tensor is symmetrical: $\tau_{ji} = \tau_{ij}$. Therefore, according to the analysis in Appendix 9.A, the stress tensor has real eigenvalues and can be diagonalized; along its principal directions, the applied pressure is normal to the surface element. Important particular cases of stress are that of isotropic (hydrostatic) pressure, occurring when all three eigenvalues are equal: $\tau_{ij} = -p\,\delta_{ij}$ (the minus sign corresponds to the

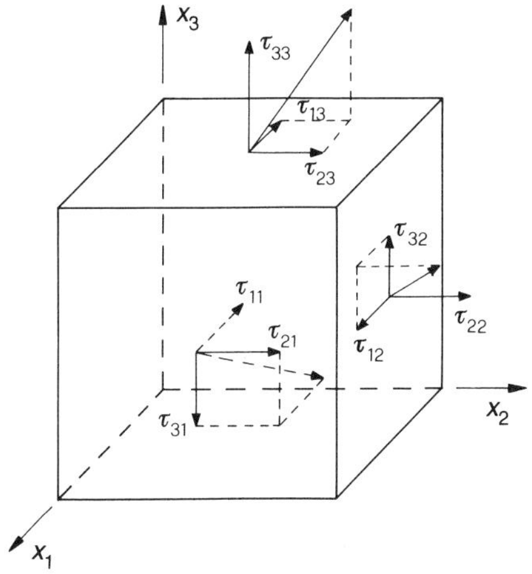

Fig. 9.7. Stress field on the (1 0 0), (0 1 0), and (0 0 1) faces of a crystal cube.

convention of considering negative a compressive stress, and positive a tensile one); and that of uniaxial pressure, when two eigenvalues are zero (or, in a more general case, different from zero but equal).

Elasticity tensor

A solid is said to be elastic when the dependence of strain on the applied stress is linear (Hooke's law). If the solid is anisotropic, as is the case for crystals, then Hooke's law is expressed in tensorial form. The coefficients of the linear dependence must be characterized by four subscripts, as they relate the η_{ij} and τ_{hk} components of two second-rank tensors (cf. eqn (9.10)):

$$\tau_{ij} = \sum_{1}^{3}{}_{h,k} c_{ijhk} \eta_{hk}, \qquad (9.23)$$

and, inversely:

$$\eta_{ij} = \sum_{1}^{3}{}_{h,k} s_{ijhk} \tau_{hk}. \qquad (9.24)$$

The quantity c_{ijhk} has the physical meaning of stress component τ_{ij} which must be applied to the crystal so as to achieve a deformation state characterized by a η_{hk} component of unit value. Similarly, s_{ijhk} is the η_{ij} strain component resulting from application of a unit stress τ_{hk} to the crystal. As the $\boldsymbol{\eta}$ components are dimensionless, and those of $\boldsymbol{\tau}$ have the dimension of pressure, it follows that the quantities c_{ijhk} and s_{ijhk} have the dimensions of pressure and of pressure^{-1}, respectively.

The coefficients c_{ijhk} and s_{ijhk} obey the transformation rule (9.8), and then are components of the fourth-rank tensors **c** and **s**, respectively; **c** is called tensor of elastic constants or stiffness coefficients, while **s** is the tensor of elastic moduli or compliance coefficients. The two tensors are related by the generalized inversion relationship:

$$\sum_{1}^{3}{}_{m,n} c_{ijmn} s_{mnhk} = \tfrac{1}{2}(\delta_{ih}\delta_{jk} + \delta_{ik}\delta_{jh}). \qquad (9.25)$$

Unlike $\boldsymbol{\eta}$ and $\boldsymbol{\tau}$, which are field tensors, **c** and **s** are matter tensors, that is they characterize an intrinsic property of the crystalline medium and are independent of the applied force field. On the basis of the symmetry relations $\eta_{ij} = \eta_{ji}$ and $\tau_{ij} = \tau_{ji}$ for the strain and stress tensors, and of the condition that the total moment of force applied to the crystal be zero (otherwise the whole crystal would rotate rigidly), a number of symmetry relations can be derived for the subscripts of c_{ijhk} and s_{ijhk} components. These are:

$$c_{ijhk} = c_{jihk} = c_{ijkh} = c_{jikh} = c_{hkij} = c_{khij} = c_{hkji} = c_{khji}, \qquad (9.26)$$

and similarly for s_{ijhk}. Thus of the 81 components of each tensor only 21 are actually independent.

The mechanical work per unit volume of an infinitesimal elastic deformation of the crystal is given by

$$\mathrm{d}W = \sum_{1}^{3}{}_{i,j} \tau_{ij}\, \mathrm{d}\eta_{ij};$$

by substituting the expression (9.23) for τ_{ij}, one obtains:

$$\mathrm{d}W = \sum_{1}^{3}{}_{i,j,h,k} c_{ijhk}\eta_{hk}\,\mathrm{d}\eta_{ij},$$

and integrating with respect to $\mathrm{d}\eta_{ij}$ the work per unit volume necessary to produce the finite strain η is derived:

$$W = \tfrac{1}{2}\sum_{1}^{3}{}_{i,j,h,k} c_{ijhk}\eta_{ij}\eta_{hk}. \tag{9.27}$$

The linear compressibility $\beta_{\boldsymbol{l}}$ of the crystal is the relative change of length along a given direction $\boldsymbol{l}$ when the crystal is subjected to a unit isotropic pressure. By the general properties of second-rank tensors (cf. Appendix 9.A and eqn (9.A.8)), the linear strain along the unit vector $\boldsymbol{l}$ is

$$\sum_{1}^{3}{}_{i,j}\eta_{ij}l_i l_j;$$

by substitution of (9.24) for η_{ij}, and remembering that $\tau_{hk} = -p\delta_{hk}$ for isotropic pressure, one obtains:

$$\beta_{\boldsymbol{l}} = \sum_{1}^{3}{}_{i,j,h} s_{ijhh} l_i l_j. \tag{9.28}$$

Analogously, for the volume compressibility $\beta = -(\partial V/\partial p)/V$; the deformation corresponding to a volume change is $\Delta V/V = \sum_{i=1}^{3}\eta_{ii}$, and substituting (9.24) for η_{ii} gives:

$$\beta = \sum_{1}^{3}{}_{i,h} s_{iihh}. \tag{9.29}$$

The expression for the reciprocal of volume compressibility, the elastic bulk modulus K, is obtained accordingly:

$$K = 1/\beta. \tag{9.30}$$

A simplified convention, based on a condensation of subscripts, is often used to express components of stress, strain, and elasticity tensors (Voigt's notation). The symmetrical pair of indices ij ($i, j = 1, 2, 3$) is substituted by a single subscript p ($p = 1, 2, 3, 4, 5, 6$), according to the rule: $11 \to 1$, $22 \to 2$, $33 \to 3$, $23 \to 4$, $13 \to 5$, $12 \to 6$. Then $\tau_p = \tau_{ij}$ and $c_{pq} = c_{ijhk}$, with the above correspondence law for subscripts implied. The tensor $\boldsymbol{\tau}$ is now represented by a 6×1 linear matrix, instead of a 3×3 symmetrical square matrix:

$$\begin{pmatrix} \tau_{11} & \tau_{12} & \tau_{13} \\ \tau_{12} & \tau_{22} & \tau_{23} \\ \tau_{13} & \tau_{23} & \tau_{33} \end{pmatrix} \to \begin{pmatrix} \tau_1 \\ \tau_2 \\ \tau_3 \\ \tau_4 \\ \tau_5 \\ \tau_6 \end{pmatrix}$$

Similarly, in Voigt's notation the elastic stiffness tensor $\mathbf{c}$ is represented by a 6×6 symmetrical square matrix ($c_{pq} = c_{qp}$). Things are slightly more complicated for the $\boldsymbol{\eta}$ and **s** tensors. In fact, in order to have relations (9.23) and (9.24) transformed into the corresponding ones

$$\tau_p = \sum_{q=1}^{6} c_{pq}\eta_q, \tag{9.31}$$

$$\eta_p = \sum_{q=1}^{6} s_{pq}\tau_q, \tag{9.32}$$

the η_p and s_{pq} components have to be defined in the following way: $\eta_p = \eta_{ii}$ $(p = 1, 2, 3)$, $\eta_p = 2\eta_{ij}$ $(p = 4, 5, 6)$; $s_{pq} = s_{iihh}$ $(p, q = 1, 2, 3)$, $s_{pq} = 2s_{iihk}$ $(p = 1, 2, 3;\ q = 4, 5, 6)$, $s_{pq} = 4s_{ijhk}$ $(p, q = 4, 5, 6)$. For example, $\eta_1 = \eta_{11}$, $\eta_4 = 2\eta_{23}$, $s_{13} = s_{1133}$, $s_{26} = 2s_{2212}$, $s_{45} = 4s_{2313}$. If the coefficients 2 and 4 were omitted in the definition of quantities η_p and s_{pq}, they would appear explicitly in the linear forms (9.31) and (9.32). Furthermore, with the chosen convention the relation of matrix inversion

$$\mathbf{s} = \mathbf{c}^{-1} \tag{9.33}$$

holds for the two 6×6 square matrices representing the elasticity tensors. Thus the relations (9.31) and (9.32) can be rewritten in matrix form as:

$$\boldsymbol{\tau} = \mathbf{c}\boldsymbol{\eta}, \tag{9.34}$$

$$\boldsymbol{\eta} = \mathbf{c}^{-1}\boldsymbol{\tau} = \mathbf{s}\boldsymbol{\tau}. \tag{9.35}$$

By use of Voigt's notation the expressions (9.27), (9.29), and (9.30) for the energy of elastic deformation, volume compressibility, and elasticity bulk modulus become, respectively:

$$W = \tfrac{1}{2}\sum_{1}^{6}{}_{p,q} c_{pq}\eta_p\eta_q, \tag{9.36}$$

$$\beta = \sum_{1}^{3}{}_{p,q} s_{pq}, \tag{9.37}$$

$$K = 1/\beta. \tag{9.38}$$

As matter tensors, the stiffness and compliance tensors **c** and **s** have to comply with the requirements of the crystal point symmetry, according to Neumann's principle. On the basis of p. 603, the simplest method to derive the symmetry constraints on c_{pq} and s_{pq} components is as follows. The symmetry operations **R** which are generators of the crystal point group are considered. For each of them, the Cartesian basis **E** is transformed into **E**′ = **RE**, and correspondingly the transformation of c_{pq} components into c'_{pq} is obtained; by symmetry, the conditions $c'_{pq} = c_{pq}$ must be satisfied leading to the wanted relations. This procedure will be illustrated by some simple examples.

A twofold symmetry axis parallel to the $\boldsymbol{e}_2$ Cartesian vector transforms the indices of Cartesian coordinates as follows: $1 \to -1$, $2 \to 2$, $3 \to -3$. As the components of a second-rank tensor transform as products of two coordinates, the corresponding transformation rule for the pair ij of indices is: $11 \to 11$, $22 \to 22$, $33 \to 33$, $23 \to -23$, $13 \to 13$, $12 \to -12$; in Voigt's notation: $1 \to 1$, $2 \to 2$, $3 \to 3$, $4 \to -4$, $5 \to 5$, $6 \to -6$. Therefore, a rotation by 180° of the crystal about $\boldsymbol{e}_2$ causes the elastic constants with only one subscript equal to 4 or 6 to change sign: but, because of symmetry invariance, such components can only have zero values. Thus in the monoclinic system, where a twofold axis parallel to $\boldsymbol{e}_2$ is present, the elastic constants c_{14}, c_{24}, c_{34}, c_{45}, c_{16}, c_{26}, c_{36}, c_{56} are always equal to zero. It is easy to show that no further constraints are imposed on the c_{pq} components by the other symmetry operations in monoclinic point groups. In fact, by the same procedure as above the **c** tensor can be proved to be invariant to action of the inversion centre. Thus elasticity is a centrosymmetrical property, and in order to derive the symmetry constraints on the c_{pq} components only the generators of the point group not containing the inversion centre need be taken into account. For instance, a mirror plane normal to $\boldsymbol{e}_2$ is the product of a twofold axis parallel to $\boldsymbol{e}_2$ and of the inversion centre, so that it is completely equivalent to the twofold axis as far as the elastic behaviour is concerned.

In the orthorhombic system, two twofold axes parallel to $\boldsymbol{e}_2$ and $\boldsymbol{e}_3$ can be considered as generators for the 222 and mmm point groups (excluding the inversion centre). Thus in this case the symmetry constraints on the elastic constants are the sum of those already found for the monoclinic system, plus those due to the twofold axis parallel to $\boldsymbol{e}_3$. Such a rotation transforms the indices of vectors $\boldsymbol{e}_i$ according to: $1 \to -1$, $2 \to -2$, $3 \to 3$, and the Voigt condensed subscripts according to: $1 \to 1$, $2 \to 2$, $3 \to 3$, $4 \to -4$, $5 \to -5$, $6 \to 6$. Following the same reasoning as before, this implies that all c_{pq} components with only one index equal to either 4 or 5 must have zero value. By combining that result with the conditions found previously for invariance to a rotation parallel to $\boldsymbol{e}_2$, one obtains that the elastic constants which may differ from zero are only c_{11}, c_{22}, c_{33}, c_{12}, c_{13}, c_{23}, c_{44}, c_{55}, c_{66}. This can be easily shown to be true for point group mm2 as well, and then holds for the whole orthorhombic system.

The symmetry restrictions on the components of the elasticity tensor **c** (Voigt's notation) are summarized in Table 9.4 for all crystal point groups.

Examples and applications

A list of experimental values of elastic constants is given for some inorganic and organic crystals in Table 9.5. As an example of numerical application, let us consider the case of anhydrite, $CaSO_4$. For orthorhombic symmetry the orthonormalization matrix has the diagonal form

$$\begin{pmatrix} 1/a & 0 & 0 \\ 0 & 1/b & 0 \\ 0 & 0 & 1/c \end{pmatrix}.$$

Then by means of (9.20), and using the Voigt notation, the following expressions are obtained for the strain components related to changes of lattice constants:

$$\eta_1 = \tfrac{1}{2}[(a'/a)^2 - 1], \quad \eta_2 = \tfrac{1}{2}[(b'/b)^2 - 1], \quad \eta_3 = \tfrac{1}{2}[(c'/c)^2 - 1],$$
$$\eta_4 = \tfrac{1}{2}(b'/b)(c'/c)\cos\alpha', \quad \eta_5 = \tfrac{1}{2}(a'/a)(c'/c)\cos\beta',$$
$$\eta_6 = \tfrac{1}{2}(a'/a)(b'/b)\cos\gamma'.$$

We want to apply, for instance, a uniaxial compression of 1 GPa ($= 10^9\,\mathrm{N\,m^{-2}}$) to a crystal of $CaSO_4$ along the x crystallographic direction, and to determine the corresponding deformation. The stress tensor takes

Table 9.4. Elastic constants and crystal symmetry

$1, \bar{1}$

$$\begin{pmatrix} c_{11} & c_{12} & c_{13} & c_{14} & c_{15} & c_{16} \\ & c_{22} & c_{23} & c_{24} & c_{25} & c_{26} \\ & & c_{33} & c_{34} & c_{35} & c_{36} \\ & & & c_{44} & c_{45} & c_{46} \\ & & & & c_{55} & c_{56} \\ & & & & & c_{66} \end{pmatrix}$$

$2, m, 2/m$

$$\begin{pmatrix} c_{11} & c_{12} & c_{13} & 0 & c_{15} & 0 \\ & c_{22} & c_{23} & 0 & c_{25} & 0 \\ & & c_{33} & 0 & c_{35} & 0 \\ & & & c_{44} & 0 & c_{46} \\ & & & & c_{55} & 0 \\ & & & & & c_{66} \end{pmatrix}$$

$222, mm2, mmm$

$$\begin{pmatrix} c_{11} & c_{12} & c_{13} & 0 & 0 & 0 \\ & c_{22} & c_{23} & 0 & 0 & 0 \\ & & c_{33} & 0 & 0 & 0 \\ & & & c_{44} & 0 & 0 \\ & & & & c_{55} & 0 \\ & & & & & c_{66} \end{pmatrix}$$

$3, \bar{3}$

$$\begin{pmatrix} c_{11} & c_{12} & c_{13} & c_{14} & c_{15} & 0 \\ & c_{11} & c_{13} & -c_{14} & -c_{15} & 0 \\ & & c_{33} & 0 & 0 & 0 \\ & & & c_{44} & 0 & -c_{15} \\ & & & & c_{44} & c_{14} \\ & & & & & c_{66} \end{pmatrix} \quad c_{66} = \tfrac{1}{2}(c_{11} - c_{12})$$

$32, 3m, \bar{3}m$

$$\begin{pmatrix} c_{11} & c_{12} & c_{13} & c_{14} & 0 & 0 \\ & c_{11} & c_{13} & -c_{14} & 0 & 0 \\ & & c_{33} & 0 & 0 & 0 \\ & & & c_{44} & 0 & 0 \\ & & & & c_{44} & c_{14} \\ & & & & & c_{66} \end{pmatrix} \quad c_{66} = \tfrac{1}{2}(c_{11} - c_{12})$$

$6, \bar{6}, 6/m, 622, 6mm, \bar{6}2m, 6/mmm$

$$\begin{pmatrix} c_{11} & c_{12} & c_{13} & 0 & 0 & 0 \\ & c_{11} & c_{13} & 0 & 0 & 0 \\ & & c_{33} & 0 & 0 & 0 \\ & & & c_{44} & 0 & 0 \\ & & & & c_{44} & 0 \\ & & & & & c_{66} \end{pmatrix} \quad c_{66} = \tfrac{1}{2}(c_{11} - c_{12})$$

$4, \bar{4}, 4/m$

$$\begin{pmatrix} c_{11} & c_{12} & c_{13} & 0 & 0 & c_{16} \\ & c_{11} & c_{13} & 0 & 0 & -c_{16} \\ & & c_{33} & 0 & 0 & 0 \\ & & & c_{44} & 0 & 0 \\ & & & & c_{44} & 0 \\ & & & & & c_{66} \end{pmatrix}$$

$422, 4mm, \bar{4}2m, 4/mmm$

$$\begin{pmatrix} c_{11} & c_{12} & c_{13} & 0 & 0 & 0 \\ & c_{11} & c_{13} & 0 & 0 & 0 \\ & & c_{33} & 0 & 0 & 0 \\ & & & c_{44} & 0 & 0 \\ & & & & c_{44} & 0 \\ & & & & & c_{66} \end{pmatrix}$$

$23, m3, 432, \bar{4}3m, m3m$

$$\begin{pmatrix} c_{11} & c_{12} & c_{12} & 0 & 0 & 0 \\ & c_{11} & c_{12} & 0 & 0 & 0 \\ & & c_{11} & 0 & 0 & 0 \\ & & & c_{44} & 0 & 0 \\ & & & & c_{44} & 0 \\ & & & & & c_{44} \end{pmatrix}$$

Table 9.5. Independent values of elastic stiffnesses c_{pq} (GPa) and compliances s_{pq} (TPa^{-1}) of some crystals (Landolt-Bornstein Tables, 1983)

	MgO periclase cubic		**$CaCO_3$ calcite trigonal**		**$CaSO_4$ anhydrite orthorhom.**		**$(C_6H_5)_2CO$ benzophenone orthorhom.**		**$CaSO_4{\cdot}2H_2O$ gypsum monoclinic**		**$C_{10}H_8$ naphahalene monoclinic**	
pq	*c*	*s*	*c*	*s*	*c*	*s*	*c*	*s*	*c*	*s*	*c*	*s*
11	294	4.01	144	11.4	93.8	11.0	10.7	130	94.5	15.4	8.0	292
22					185	5.72	10.0	157	65.2	29.5	10.0	872
33			84.0	17.4	112	9.55	7.1	165	50.2	32.8	12.2	559
44	155	6.47	33.5	41.4	32.5	30.8	2.03	493	8.6	117	3.38	302
55					26.5	37.7	1.55	645	32.4	38.2	2.21	4840
66					9.3	108	3.53	283	10.8	93.5	4.28	239
12	93	−0.96	53.9	−4.0	16.5	−0.76	5.50	−72	37.9	−8.6	4.85	−208
13			51.1	−4.5	15.2	−1.28	1.69	2	28.2	−2.2	3.38	−8
23					31.7	−1.52	3.21	−54	32.0	−15.9	2.72	−555
14			−20.5	9.5								
15									−11.0	6.6	−0.5	−181
25									6.9	−12.8	−2.5	1830
35									−7.5	10.2	3.0	−1483
46									−1.1	12.0	−0.1	−8

the form $\boldsymbol{\tau} = [-1\ 0\ 0\ 0\ 0\ 0]$ GPa, and, by using (9.32) or (9.35), the strain components are obtained: $\boldsymbol{\eta} = [-0.011\ 0.00076\ 0.00128\ 0\ 0\ 0]$. Taking into account the above relations between $\boldsymbol{\eta}$ components and lattice constants, the unit-cell edges of anhydrite undergo changes of -1.10 per cent (a), $+0.08$ per cent (b), $+0.13$ per cent (c). The mechanical work per unit volume required to perform this deformation can be calculated by (9.36), yielding a value of 5.5 MJ m^{-3}. Let us consider now an isotropic compression of 1 GPa of the same crystal, corresponding to the stress $\boldsymbol{\tau} = [-1\ -1\ -1\ 0\ 0\ 0]$ GPa. Again, by (9.35) we obtain the resulting deformation: $\eta = [-0.00896\ -0.00344\ -0.00675\ 0\ 0\ 0]$, corresponding to relative decreases of the a, b, c cell edges by -0.90 per cent, -0.34 per cent, and -0.68 per cent, respectively. The energy per unit volume amounts to 9.6 MJ m^{-3}, and the relative volume decrease is -1.9 per cent ($\sum_{q=1}^{3} \eta_q = -0.0192$). This value could also have been computed by deriving the volume compressibility $\beta = 0.01915\ (\mathrm{GPa})^{-1}$ by (9.37), which gives for a pressure of 1 GPa the same volume contraction as that calculated previously.

Piezoelectricity

Some physical properties of crystals are expressed by a relation between a vector and a second-rank tensor, rather than between two vectors or between two second-rank tensors, as have been considered until now. If the dependence between the components is linear, the relative coefficients represent a third-rank tensor. This type of third-rank tensor is different from that examined previously (cf. the electro-optical effect), which expresses a second-order dependence between the components of two vectors.

In piezoelectric crystals, by applying a mechanical stress $\boldsymbol{\tau}$ an electric dipole moment (per unit volume) $\boldsymbol{P}$ arises, whose components P_i are related linearly to the stress components:

$$P_i = \sum_{1}^{3} {}_{h,k} d_{ihk} \tau_{hk}. \tag{9.39}$$

The d_{ihk} coefficients are called piezoelectric constants (or moduli), and, as they obey the transformation rule (9.7), represent the third-rank tensor of piezoelectricity **d**. On the basis of simple thermodynamic arguments, and taking into account the symmetry of the $\boldsymbol{\tau}$ tensor, the piezoelectric constants d_{ihk} can be proved to be invariant with respect to an exchange of h and k subscripts: $d_{ihk} = d_{ikh}$. Thus only 18 out of 27 components of tensor **d** are independent.

From a microscopic point of view, piezoelectricity is a polarization of the crystalline medium due to displacements of ions from their equilibrium positions, by action of an external stress field. In this case the $\boldsymbol{P}$ vector is associated with an ionic polarization produced by the second-rank tensor $\boldsymbol{\tau}$, while in pyroelectric or ferroelectric crystals $\boldsymbol{P}$ is due to a quite spontaneous ionic polarization, and in normal dielectrics to a mainly electronic polarization induced by the electric field vector $\boldsymbol{E}$. Besides the direct piezoelectric effect (9.39), also the inverse effect is observed: this arises when an electric field $\boldsymbol{E}$ applied to the crystal produces a deformation represented by the second-rank strain tensor $\boldsymbol{\eta}$. Thermodynamics demonstrates that the inverse piezoelectric effect is a necessary consequence of the direct effect, and that the coefficients relating linearly the $\boldsymbol{\eta}$ and $\boldsymbol{E}$ components are the same piezoelectric moduli d_{ihk} which appear in (9.39):

$$\eta_{hk} = \sum_{i=1}^{3} d_{ihk} E_i. \tag{9.40}$$

As in the cases of second- ($\boldsymbol{\tau}, \boldsymbol{\eta}$) and fourth- ($\mathbf{c}, \mathbf{s}$) rank tensors, also for the third-rank tensor $\mathbf{d}$ the number of subscripts can be reduced using the contracted notation of Voigt. Of course only the second and third subscripts of d_{ihk} components, i.e. the pair of indices referring to the $\boldsymbol{\tau}$ or $\boldsymbol{\eta}$ components, are affected by contraction; the first subscript, relative to components of vector $\boldsymbol{P}$ or $\boldsymbol{E}$, is not involved. Similarly to what is done for the strain tensor $\boldsymbol{\eta}$, coefficients equal to 2 have to be introduced in the shear components in order to have the relations

$$P_i = \sum_{p=1}^{6} d_{ip} \tau_p \quad \text{and} \quad \eta_p = \sum_{i=1}^{3} d_{ip} E_i$$

satisfied: $d_{i1} = d_{i11}$, $d_{i2} = d_{i22}$, $d_{i3} = d_{i33}$, $d_{i4} = 2d_{i23}$, $d_{i5} = 2d_{i13}$, $d_{i6} = 2d_{i12}$. According to Voigt's notation, the piezoelectric tensor $\mathbf{d}$ is represented simply by a 3×6 rectangular matrix with two-subscript components d_{ip} ($i = 1, \ldots, 3; p = 1, \ldots, 6$).

Symmetry properties of the piezoelectric tensor

As the piezoelectric tensor is a matter tensor, it must be invariant with respect to symmetry operations of the crystal point group. Restrictions on the values of d_{ip} components according to the crystal symmetry can be derived by a method similar to that used in the case of elastic constants. First, it can be shown that centrosymmetric crystals are never piezoelectric. The action of the inversion centre operation leaves the centrosymmetric crystals and the stress tensor $\boldsymbol{\tau}$ unchanged: on the other hand, the vector $\boldsymbol{P}$ is reversed, changing its sign. Of course the same inversion operation can produce either a vector $\boldsymbol{P}$ or a vector $-\boldsymbol{P}$ only if $\boldsymbol{P} = 0$. The same situation was found for any property relating a polar vector to an even-rank tensor, which can thus be observed only in non-centrosymmetric crystals.

Let us now consider a twofold axis parallel to the $\boldsymbol{e}_2$ vector of the Cartesian basis: this corresponds to the point group 2 of the monoclinic system, in the standard orientation. The symmetry operation causes the indices of vectors $\boldsymbol{e}_i$, and also of Cartesian coordinates, to transform according to: $1 \to -1$, $2 \to 2$, $3 \to -3$. In Voigt's notation, the contracted subscript changes as: $1 \to 1$, $2 \to 2$, $3 \to 3$, $4 \to -4$, $5 \to 5$, $6 \to -6$. As the components of the third-rank tensor $\mathbf{d}$ transform as products of three coordinates with corresponding indices, using Voigt's convention we obtain that the quantities d_{ip} change sign for a rotation of 180° about $\boldsymbol{e}_2$ in the case of d_{11}, d_{12}, d_{13}, d_{15}, d_{24}, d_{26}, d_{31}, d_{32}, d_{33}, d_{35}. But, since by symmetry the same components cannot change, they must necessarily be equal to zero. The form taken by tensor $\mathbf{d}$ in point group m is different from that shown in point group 2, otherwise than the tensor $\mathbf{c}$ of elasticity which had a unique form for all monoclinic point groups. A mirror plane perpendicular to $\boldsymbol{e}_2$

transforms the i index of d_{ip} as $1 \rightarrow 1$, $2 \rightarrow -2$, $3 \rightarrow 3$, and the contracted p index as $1 \rightarrow 1$, $2 \rightarrow 2$, $3 \rightarrow 3$, $4 \rightarrow -4$, $5 \rightarrow 5$, $6 \rightarrow -6$. Therefore, repeating the previous reasoning shows that the components d_{14}, d_{16}, d_{21}, d_{22}, d_{23}, d_{25}, d_{34}, d_{36} must be zero.

When the generators of the point group are more than one, the constraints on the d_{ip} components imposed by all the corresponding symmetry operations must be satisfied at the same time. In the orthorhombic system, the non-centrosymmetric point groups are 222 and mm2. The former group has two twofold axes, parallel to $\boldsymbol{e}_2$ and to $\boldsymbol{e}_3$ for instance, as generators; by the previous methods the axis parallel to $\boldsymbol{e}_3$ can be shown to require that the components d_{11}, d_{12}, d_{13}, d_{16}, d_{21}, d_{22}, d_{23}, d_{26}, d_{34}, d_{35} be zero. Thus by combining these restrictions with those imposed by the symmetry axis parallel to $\boldsymbol{e}_2$, we find that in point group 222 only the piezoelectric components d_{14}, d_{25}, and d_{36} can differ from zero. Analogously, in point group mm2 the symmetry constraints of the twofold axis parallel to $\boldsymbol{e}_3$ can be combined with those of the mirror plane normal to $\boldsymbol{e}_2$,

Table 9.6. Piezoelectric tensor and crystal symmetry

1

$$\begin{pmatrix} d_{11} & d_{12} & d_{13} & d_{14} & d_{15} & d_{16} \\ d_{21} & d_{22} & d_{23} & d_{24} & d_{25} & d_{26} \\ d_{31} & d_{32} & d_{33} & d_{34} & d_{35} & d_{36} \end{pmatrix}$$

2

$$\begin{pmatrix} 0 & 0 & 0 & d_{14} & 0 & d_{16} \\ d_{21} & d_{22} & d_{23} & 0 & d_{25} & 0 \\ 0 & 0 & 0 & d_{34} & 0 & d_{36} \end{pmatrix}$$

m

$$\begin{pmatrix} d_{11} & d_{12} & d_{13} & 0 & d_{15} & 0 \\ 0 & 0 & 0 & d_{24} & 0 & d_{26} \\ d_{31} & d_{32} & d_{33} & 0 & d_{35} & 0 \end{pmatrix}$$

222

$$\begin{pmatrix} 0 & 0 & 0 & d_{14} & 0 & 0 \\ 0 & 0 & 0 & 0 & d_{25} & 0 \\ 0 & 0 & 0 & 0 & 0 & d_{36} \end{pmatrix}$$

mm2

$$\begin{pmatrix} 0 & 0 & 0 & 0 & d_{15} & 0 \\ 0 & 0 & 0 & d_{24} & 0 & 0 \\ d_{31} & d_{32} & d_{33} & 0 & 0 & 0 \end{pmatrix}$$

3

$$\begin{pmatrix} d_{11} & -d_{11} & 0 & d_{14} & d_{15} & -2d_{22} \\ -d_{22} & d_{22} & 0 & d_{15} & -d_{14} & -2d_{11} \\ d_{31} & d_{31} & d_{33} & 0 & 0 & 0 \end{pmatrix}$$

32

$$\begin{pmatrix} d_{11} & -d_{11} & 0 & d_{14} & 0 & 0 \\ 0 & 0 & 0 & 0 & -d_{14} & -2d_{11} \\ 0 & 0 & 0 & 0 & 0 & 0 \end{pmatrix}$$

3m

$$\begin{pmatrix} 0 & 0 & 0 & 0 & d_{15} & -2d_{22} \\ -d_{22} & d_{22} & 0 & d_{15} & 0 & 0 \\ d_{31} & d_{31} & d_{33} & 0 & 0 & 0 \end{pmatrix}$$

6, 4

$$\begin{pmatrix} 0 & 0 & 0 & d_{14} & d_{15} & 0 \\ 0 & 0 & 0 & d_{15} & -d_{14} & 0 \\ d_{31} & d_{31} & d_{33} & 0 & 0 & 0 \end{pmatrix}$$

622, 422

$$\begin{pmatrix} 0 & 0 & 0 & d_{14} & 0 & 0 \\ 0 & 0 & 0 & 0 & -d_{14} & 0 \\ 0 & 0 & 0 & 0 & 0 & 0 \end{pmatrix}$$

6mm, 4mm

$$\begin{pmatrix} 0 & 0 & 0 & 0 & d_{15} & 0 \\ 0 & 0 & 0 & d_{15} & 0 & 0 \\ d_{31} & d_{31} & d_{33} & 0 & 0 & 0 \end{pmatrix}$$

$\bar{6}$

$$\begin{pmatrix} d_{11} & -d_{11} & 0 & 0 & 0 & -2d_{22} \\ -d_{22} & d_{22} & 0 & 0 & 0 & -2d_{11} \\ 0 & 0 & 0 & 0 & 0 & 0 \end{pmatrix}$$

$\bar{6}$m2

$$\begin{pmatrix} 0 & 0 & 0 & 0 & 0 & -2d_{22} \\ -d_{22} & d_{22} & 0 & 0 & 0 & 0 \\ 0 & 0 & 0 & 0 & 0 & 0 \end{pmatrix}$$

$\bar{4}$

$$\begin{pmatrix} 0 & 0 & 0 & d_{14} & d_{15} & 0 \\ 0 & 0 & 0 & -d_{15} & d_{14} & 0 \\ d_{31} & -d_{31} & 0 & 0 & 0 & d_{36} \end{pmatrix}$$

$\bar{4}$2m

$$\begin{pmatrix} 0 & 0 & 0 & d_{14} & 0 & 0 \\ 0 & 0 & 0 & 0 & d_{14} & 0 \\ 0 & 0 & 0 & 0 & 0 & d_{36} \end{pmatrix}$$

23, $\bar{4}$3m

$$\begin{pmatrix} 0 & 0 & 0 & d_{14} & 0 & 0 \\ 0 & 0 & 0 & 0 & d_{14} & 0 \\ 0 & 0 & 0 & 0 & 0 & d_{14} \end{pmatrix}$$

yielding the result that the only components of tensor **d** which have non-zero values are d_{15}, d_{24}, d_{31}, d_{32}, d_{33}. By similar reasoning the conditions imposed by symmetry on the piezoelectric moduli can be derived for all other non-centrosymmetric point groups. In particular, it should be remarked that in the cubic point group 432 all d_{ip} values turn out to be zero: so this crystal symmetry, though lacking the inversion centre, is forbidden for piezoelectric crystals. The forms taken by the **d** tensor in all permitted point groups are shown in Table 9.6.

Crystal defects

In previous chapters, some properties of the crystalline state were presented and discussed on the basis of a simplified model, the ideal crystal, relying upon the fundamental assumption of an atomic structure with perfect translational periodicity. A large number of very important crystal properties can be accounted for in such a way: for example, the diffraction of X-rays, electrons, and neutrons, the main aspects of dielectric and optical behaviour, and most features of electronic and thermodynamic properties.

However, some crucial phenomena of crystalline solids can not be explained at all by the perfect crystal model. First, the mechanical behaviour in non-elastic conditions. The force per unit surface required to break brittle crystals (break modulus), and the energy necessary to produce a plastic deformation in ductile crystals have experimental values lower by several orders of magnitude with respect to theoretical values calculated on the basis of the perfect crystal model. The maximum shear stress which an ideal crystal can withstand in elastic conditions is estimated to be about $K/100$, where K is the elastic bulk modulus: on the other hand, in real crystals stresses of $K/10^5$ are sufficient to start plastic slip processes. Second, all transport phenomena in crystals can not be understood by a perfectly ordered lattice model, where all atomic sites are occupied. This includes very important processes such as diffusion, which is involved in all chemical or phase transformations in crystalline solids, and the electrical conductivity due to ionic motion within the crystal structure. Furthermore, several particular features of electronic, spectroscopic, and thermal behaviour of real crystals are not accounted for by the perfect crystal model.

Thus the deviations from ideality, or defects, have to be examined in detail in order to build up a more general model: the **real** or **defective** crystal. Two classes can be distinguished, **point defects** and **extended defects**. The former ones are a violation of translational symmetry in a single lattice site: for instance, the absence of an atom from its expected position (vacancy), or the presence of an atom in an unexpected position (interstitial). Extended defects, on the other hand, break the lattice periodicity by relating two portions of the crystal, each of which is perfect in its interior, in a 'wrong' way. The misfit may involve an extended part of a plane, and in this case we have a planar defect (e.g. stacking faults in layer structures); or it may involve just a region closely surrounding a line, so that we speak of a linear defect or dislocation. Extended defects are weak points of the crystal, from a mechanical point of view, and are then responsible for plastic behaviour and for the lower observed strength with respect to theoretical values. Point defects, on the other hand, provide vacant lattice sites which make atomic motion possible in crystals, and consequently diffusion and ionic conduction phenomena. The two kinds of defects are not quite independent of each other: it will be shown below that extended and point defects may interact strongly, so that the defective behaviour of crystals is actually a single field of investigation.

Experimental methods

Methods for direct observation (e.g. microscopy techniques) can be applied to extended but generally not to point defects, as these usually have dimensions smaller than the limit of resolution of the best instruments. Transmission electron microscopy (TEM) is the experimental technique best suited to observation of extended defects; in the case of high-resolution instruments, the resolution power can reach 2–3 Å (HRTEM). An electron beam accelerated by a potential of 100 kV is associated with a de Broglie wave of wavelength $\lambda = h/p = 0.037$ Å, where h is the Planck constant and p is the linear momentum of electrons (cf. p. 185). Such a small wavelength is consistent with a much higher resolution than can be attained by conventional optical microscopy (not better than 10^4 Å for monochromatic green light). A system of electromagnetic lenses allows the electron beam to be focused on to the sample and on to the fluorescent observation screen, analogously to what happens with light in an optical microscope. The regions of the crystal which contain extended defects give rise to changes of intensity of the transmitted beam (effect of 'contrast'), so as to visualize the defects themselves.

A scheme of the optical path of a TEM is shown in Fig. 9.8. The beams which are scattered by the sample at small angles (1 to 2°) to the transmitted beam are focused by the objective lens to form a diffraction pattern at its back focal plane. By properly focusing the intermediate and projector lens system, a magnified image of the back focal plane of the objective lens is projected on the viewing screen. The intermediate aperture can limit the sample area from which the diffraction pattern is obtained, so as to explore very small portions of the specimen (selected area diffraction mode). In the imaging mode, the aperture at the back focal plane of the objective lens is inserted in order to block all diffracted beams and let out the transmitted beam only (bright-field technique); alternatively, all but a single diffracted beam are blocked by the aperture (dark-field technique). Intermediate methods are based on selection of a number of diffracted beams, which recombine at the image plane giving a contrast contributed only by the corresponding diffracting directions.

Important sources of errors in operating conditions are due to the spherical aberration of the objective lens and to an incorrect focusing of the objective lens. The spherical aberration error causes a slight displacement $\Delta = C_s\alpha^3$ of points on the image plane. α (rad) is the maximum angle of electron scattering which can pass through the objective lens (effective aperture of the lens); C_s is the coefficient of spherical aberration ($\approx$ focal length, of order 1–3 mm). Taking into account the Rayleigh formula for the resolution of the instrument (R is the size of the resolved object):

$$R = \frac{0.61\lambda}{\alpha},$$

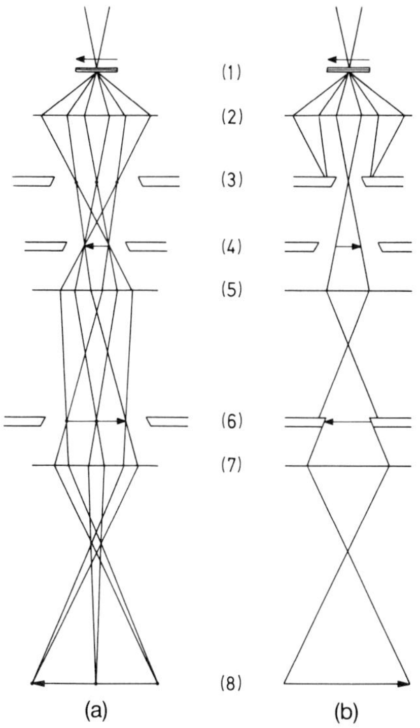

Fig. 9.8. Optical path of a transmission electron microscope: (a) selected area diffraction mode; (b) imaging mode. (1) Specimen; (2) objective lens; (3) back focal plane of objective lens; (4) first intermediate image plane; (5) intermediate lens; (6) second intermediate image plane; (7) projector lens; (8) viewing screen.

it appears that as α increases the resolution improves, but the aberration error Δ becomes larger. Thus an optimum intermediate lens aperture α would minimize both R and Δ, and this corresponds to $\alpha_{opt} \propto (\lambda/C_s)^{1/4}$, $\Delta_{opt} \propto \lambda^{3/4} C_s^{1/4}$.

The only drawback of the very powerful TEM method is the need for a very thin sample (not thicker than about 2000 Å), because of the low penetration of the electron beam into matter. Clearly such thin crystals or regions of crystals may only have a very small area, so that defects can not be observed on a large scale. A different technique which is not affected by this disadvantage is **X-ray topography**. The diffracted intensity of a very strong reflection varies locally in crystal regions where extended defects are present; thus a direct image of the defect structure of the crystal can be obtained, similarly to what occurs in electron microscopy. However, in X-ray topography the resolution is much smaller, so that the density of defects can not be very high. On the other hand, the sample is a large single crystal and then an extended area of the sample can be investigated. Thus the two techniques of X-ray topography and electron microscopy appear to be quite complementary.

The most widely used topographic technique is the **transmission Lang method** (Fig. 9.9). A collimated X-ray beam is allowed to pass through a portion of the crystal, which is oriented for Bragg reflection from a particular set of planes. The directly transmitted beam is blocked by a shield, and the diffracted beam is recorded on a photographic plate. The whole crystal is either immobile (section topograph), or is scanned by translating it slowly together with the plate (traverse topograph). Very thick samples should be avoided, to prevent too large an absorption; otherwise, a back-reflection instead of transmission technique can be used, so that only the surface crystal area contributes to diffraction. Dislocations and other extended defects are revealed by diffraction contrast, which is caused by differences of intensity of the diffracted beam from perfect and from distorted portions of the crystal. A more detailed discussion of applications of the Lang method to the study of crystal defects is presented after introducing dislocations on p. 630.

An older experimental method, which, however, can be very useful and informative, applies particularly to the study of line defects and is based on **etching** of the crystal surface by chemical or electrolytic methods. At the site where a dislocation meets the surface an etch pit is formed, revealing a number of details about the nature of the defect.

Indirect experimental techniques have usually to be used to investigate point defects. For instance, simple (but accurate) measurements of the

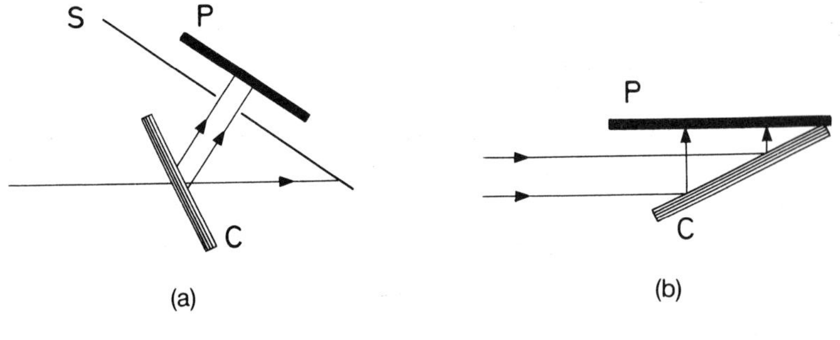

Fig. 9.9. Lang method for X-ray topography: (a) transmission mode; (b) back-reflection mode. S, shield; P, photographic plate; C, crystal.

unit-cell volume and of the crystal density may give important indications about the number and nature of point defects present in the crystal. Another physical quantity which is closely related to such defects is the electrical resistivity, particularly in ionic crystals where the very small conductivity observed is entirely due to the thermally activated motion of ions through the crystal. Ionic transport is made possible by vacancies, so that resistivity measurements as a function of temperature are able to characterize many features of point defects in the crystal. Other very important techniques are based on resonance phenomena (e.g. electron spin resonance or ESR) which may be particularly sensitive to impurities present at the concentration level of point defects. These topics will be considered in more detail in the following sections.

Planar defects

The most important planar defects are observed in layer structures and are called stacking faults. Let us examine the simplest case of layer structures, already mentioned on p. 429: the cubic and hexagonal close-packed arrangements. A single close-packed layer is shown in Figs. 9.10 and 9.11; it corresponds to a (111) or a (001) crystallographic plane in the cubic or hexagonal case, respectively. Vectors are shown along the important lattice directions lying on the plane; for example, in the cubic case (Fig. 9.10) the arrow along $[11\bar{2}]$ means the vector $\boldsymbol{a}+\boldsymbol{b}-2\boldsymbol{c}$, and is denoted by the conventional notation $a[11\bar{2}]$. The stacking sequences of layers ABCABC... (cubic) and ABAB... (hexagonal) are projected on to the plane $(1\bar{1}0)$ (cubic) or $(1\bar{2}0)$ (hexagonal) normal to the close-packed layers and containing the $[11\bar{2}]$ (cubic) or [210] (hexagonal) direction (Fig. 9.12). The three positions A, B, C of a general layer are related to one another by translation vectors equal to $a[11\bar{2}]/6$ (cubic) or $a[210]/3$ (hexagonal). Full circles represent atoms belonging to different layers, linked by thick lines to emphasize the stacking sequence.

A stacking fault occurs when a single layer takes a different position with respect to that required by the periodic sequence. This corresponds to a

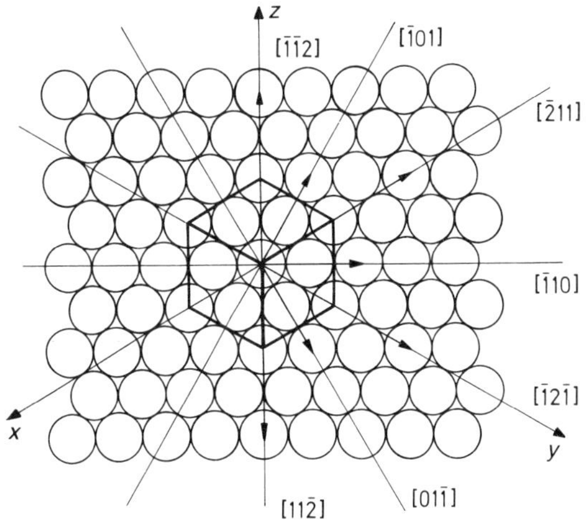

Fig. 9.10. Closed-packed (111) atomic layer of an FCC structure. The cubic cell and the significant lattice directions on the plane are outlined.

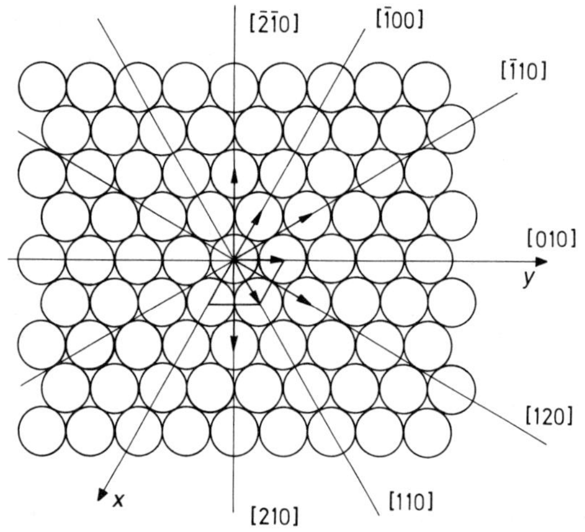

Fig. 9.11. Closed-packed (001) atomic layer of an HCP structure. The hexagonal cell and the significant lattice directions are outlined.

rigid translation of a crystal portion by the $a[11\bar{2}]/6$ vector (FCC case) on the closed-packed plane of the fault (Fig. 9.13(a)): the stacking sequence becomes ABCA/CABC. . . . A physical process which can produce such a defect is a plastic glide caused by a shear stress applied to the crystal; hence the name of deformation stacking fault given in this case. Another process, however, can give rise to the same type of fault: the removal of a single layer from the stacking sequence (a layer of type B in Fig. 9.13(a)), caused by a condensation of vacancies on the corresponding plane. This is a first example of correlation between extended and point defects. An example of a deformation stacking fault for the hexagonal close-packed structure is shown in Fig. 9.13(b). In this case, by removing, say, an A-type layer the sequence ABAB/BABAB. . . would be obtained, with an unstable configuration due to two equal B planes facing each other. Then the upper B layer shifts by the vector $a[210]/3$ so as to take the C position, attaining the stacking sequence ABAB/CBCB. . . .

It may also happen that layers following the fault plane are stacked in a symmetry-related way to those preceding the fault: we then have the sequence ABCAB/ACBA. . . (Fig. 9.14(a)), and the fault (corresponding to a twin boundary) is called a growth stacking fault, because it occurs typically during the process of crystal growth.

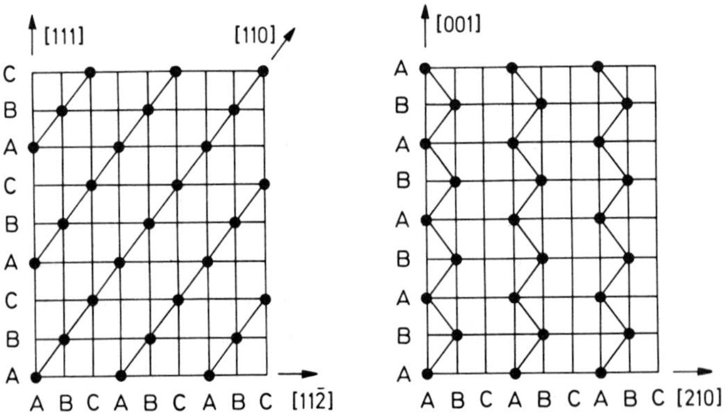

Fig. 9.12. Stacking sequences of perfect FCC (left) and HCP (right) close-packed structures shown on the $(1\bar{1}0)$ and $(1\bar{2}0)$ planes, respectively.

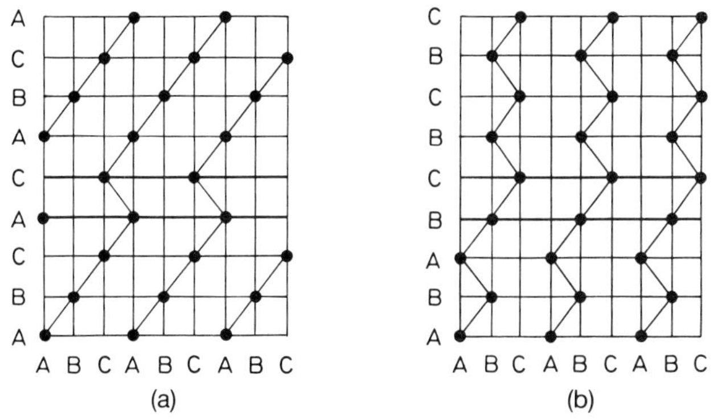

Fig. 9.13. Deformation stacking faults in FCC (a) and HCP (b) close-packed structures.

Other kinds of stacking faults in closed-packed structures are generated by a condensation of interstitial point defects, rather than vacancies as considered before. This corresponds to inserting an additional layer in the stack. An example for the FCC lattice is shown in Fig. 9.14(b), with the stacking sequence ABCA/C/BCAB. In this case the fault is referred to as an extrinsic fault, characterized by two breaks in the stacking sequence and by an extra plane not belonging to either of the lattice patterns on both sides of the fault. On the other hand, the deformation stacking fault (Figs. 9.13(a) and (b)) is also called an intrinsic fault, with just a single break in the continuing pattern of the stacking sequence.

All the examples of stacking faults discussed above refer to the simplest cases of layer structures, the FCC and HCP lattices. However, similar defects are also observed in more complicated structures built up by parallel layers related by a finite number of possible translations or rotations (cf. micas, clays, etc.). When the periodic order of the stacking sequence is violated by just a single layer, then a planar defect occurs. The basic processes are always the following: removal of a layer (condensation of vacancies), insertion of a layer (condensation of interstitials), plastic deformation by glide of atomic planes by a fraction of lattice vector. In Figs. 9.15 and 9.16 the images of extrinsic stacking faults in muscovite mica $KAl_2(Si_3AlO_{10})(OH)_2$ are shown, obtained by high-resolution transmission electron microscopy. The structure of micas (cf. p. 455) displays 10 Å thick (001) composite layers stacked in a variety of ways. A single layer is built up

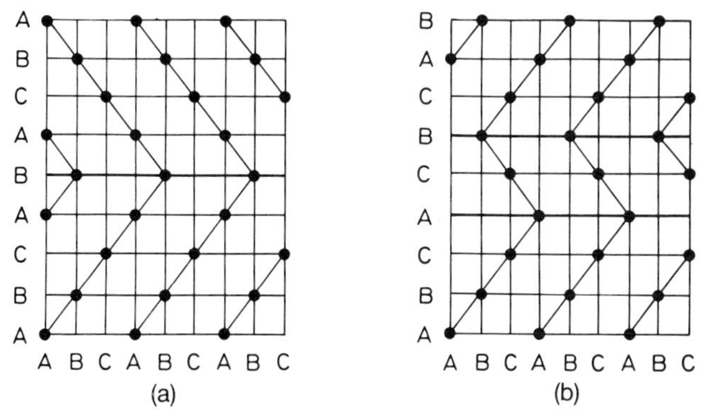

Fig. 9.14. Growth stacking fault (twin boundary) (a) and extrinsic stacking fault (b) in an FCC close-packed structure.

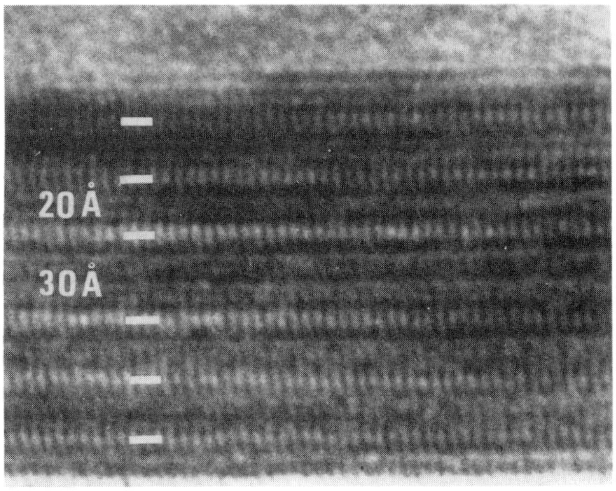

Fig. 9.15. An image of an extrinsic stacking fault in mica muscovite, by high-resolution electron microscopy. (Courtesy of M. Amouric.)

by two (Si, Al)O_4 tetrahedral sheets sandwiching cations (typically Al^{3+} or Mg^{2+}) in octahedral coordination. Adjacent layers are separated by sheets of alkali cations. The unit cell of ordinary monoclinic muscovite contains two layers, so that the $d_{(001)}$ spacing amounts to 20 Å. However, varieties of muscovite are known with one ($d_{(001)} = 10$ Å) and three ($d_{(001)} = 30$ Å) layers per unit cell. Figure 9.15 shows a single three-layer slab in a matrix of two-layer structure, giving rise to a fault in the regular stacking sequence. In Fig. 9.16 all slabs are one-layer (10 Å), but the layer corresponding to F is rotated by 120° around $\boldsymbol{c}^*$ with respect to the normal orientation, so that a fault again occurs.

Line defects: dislocations

A slip process, induced by shear stress, can give rise not only to stacking faults in layer structures, but also to line defects in any type of crystal. We want to consider a particular crystallographic plane, usually characterized by high atomic density, as a slip plane. Now let a portion of the crystal glide over the other one by a lattice vector, so that the two portions superpose perfectly after slipping. This differs from the process generating a planar

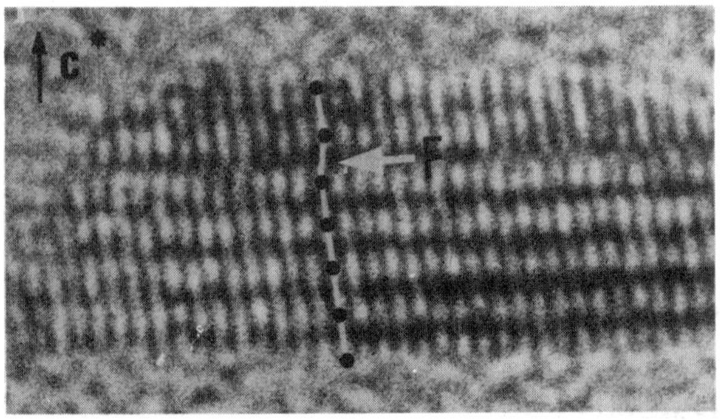

Fig. 9.16. HRTEM image of stacking fault in 10 Å muscovite. (Courtesy of M. Amouric.)

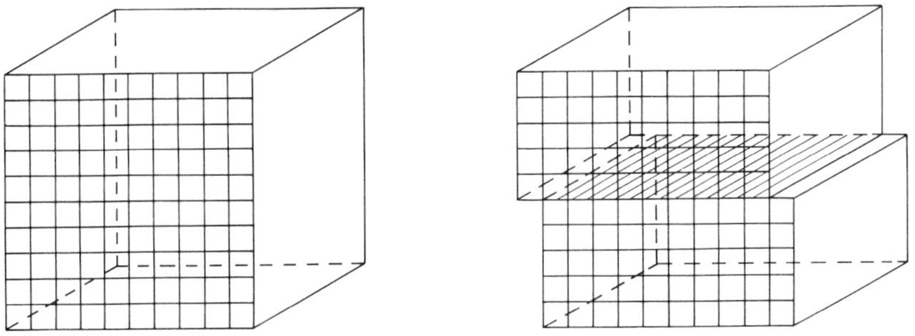

Fig. 9.17. Perfect crystal (left) and crystal with a slip plane but with perfect lattice continuity (right).

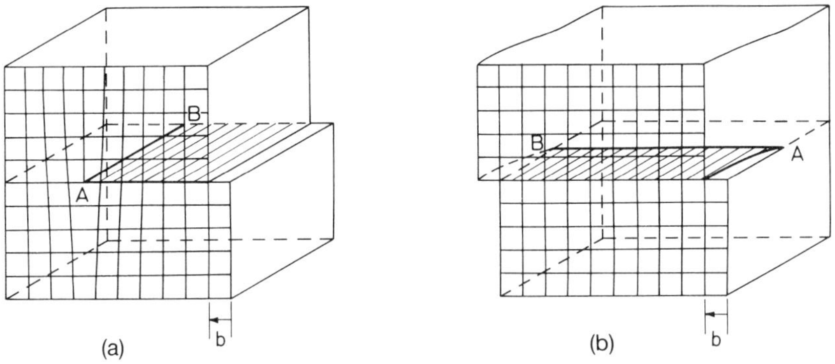

Fig. 9.18. Edge dislocation (left) and screw dislocation (right). The Burgers vector $\boldsymbol{b}$ and the dislocation line AB of length l are emphasized.

fault, where glide occurs by a fraction of lattice vector breaking the lattice continuity over the whole slip plane. In the present instance if the slip plane cuts the whole crystal, no defect is created (Fig. 9.17). On the other hand, if only a part of the slip plane, limited by a boundary line, is involved in the glide, then that line separating the slipped from the unslipped portion of the crystal is a defective region: a dislocation.

Thus a dislocation is a linear defect defined completely by the corresponding curve (which is a planar line of length l and then determines the slip plane as well), and by the lattice vector coplanar with it which measures the magnitude and direction of slip (Burgers vector $\boldsymbol{b}$). A perfect superposition of lattice points of the two slipped portions of the crystal is observed far from the dislocation line, while close to it the lattice periodicity fails. It is important to examine the orientation of the Burgers vector with respect to the dislocation line. Two limit cases are observed for a straight line: the edge dislocation, with $\boldsymbol{b}$ normal to the line, and the screw dislocation, with $\boldsymbol{b}$ parallel to the line (Fig. 9.18). If the dislocation line is a general curve (Fig. 9.19), then the dislocation character (edge or screw) changes from point to point, according to whether the unit vector $\boldsymbol{l}$ tangent to the line in a given point is normal or parallel to the Burgers vector; for an intermediate angle, the dislocation is said to have a mixed character.

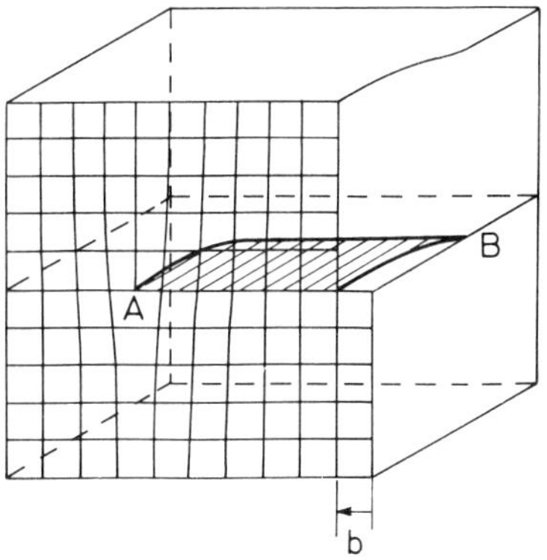

Fig. 9.19. General dislocation with mixed edge–screw character.

The Burgers circuit

Let us now try to give a more precise definition of the Burgers vector, which, as was shown before, is unique and constant for a given dislocation,

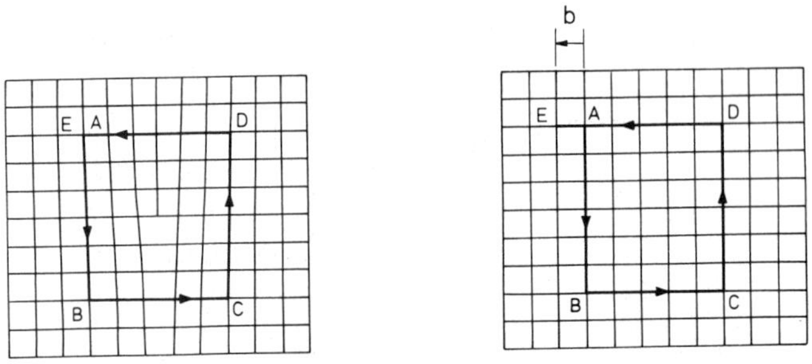

Fig. 9.20. Burgers circuits for an edge dislocation (left), and for the corresponding perfect crystal (right).

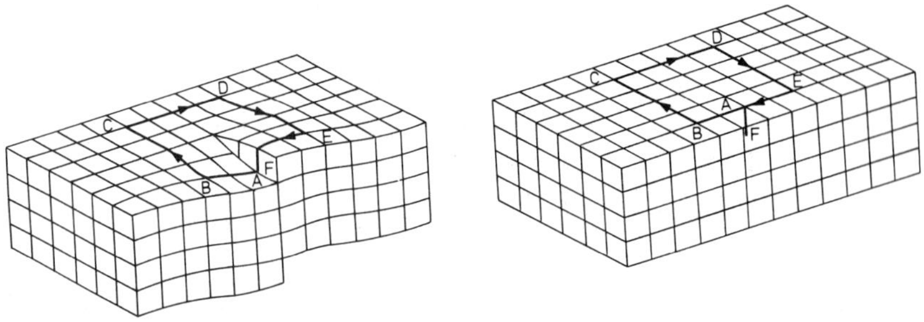

Fig. 9.21. Burgers circuits for a screw dislocation (left), and for the corresponding perfect crystal (right).

independent of the position along the dislocation line. A Burgers circuit is defined as a closed path that starts from a lattice point and comes back to the same point by encircling the dislocation line without crossing it. Now if the defect is removed, the previous Burgers circuit is modified so that it is no longer a closed line: the part exceeding or missing with respect to a loop is a lattice translation called the Burgers vector ***b*** of the dislocation. In Figs. 9.20 and 9.21 the Burgers circuits relative to an edge and a screw dislocation, respectively, are shown.

It was stressed on p. 626 that planar defects can arise not only by a slip process, but also by a condensation of point defects (vacancies or interstitials). This is true for dislocations as well. A condensation of vacancies corresponds to removing a lattice half-plane, while in the case of interstitials a new half-plane is inserted. Both processes lead to the formation of an edge dislocation.

X-ray topography of dislocations

An example of dislocations detected by X-ray topography in quartz is shown in Fig. 9.22. The nature of contrast caused by line defects in X-ray topographs is generally quite complex, as can be seen in the magnified images of two dislocations in a silicon crystal (Fig. 9.23). The contrast is contributed by three different superposing effects:

1. A direct image, due to diffraction of the X-ray beam by the defective crystal region close to the dislocation line; it is accounted for by the simple kinematical (or geometrical) theory of diffraction.

2. A dynamic image, related to the coupling of the reflected and incident waves into wave fields with directions comprised between those of the primary and diffracted beams (Borrmann fan), as required by the dynamical theory of diffraction. The dynamic image is the shadow cast by the dislocation line in the fan of these wavefields.

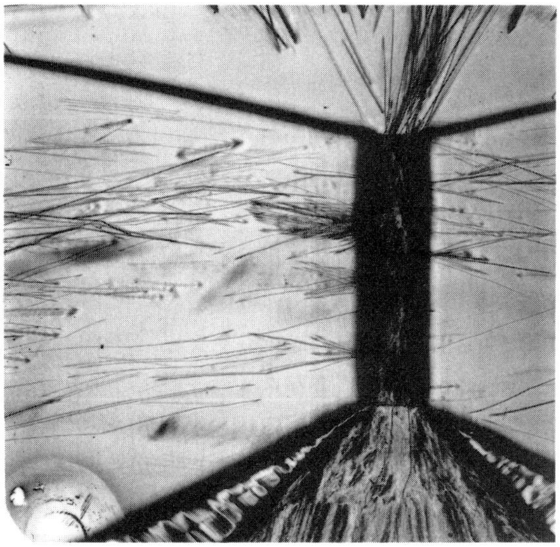

Fig. 9.22. X-ray topograph of quartz (1010 reflection), showing a great number of dislocations. (Courtesy of A. Zarka.)

3. An intermediary image, consisting of fringes due to new wave fields which are created by interaction of the dynamical wave fields with the defective region. When the crystal and photographic plate are translated (traverse topograph), the direct images form projections of the dislocations in the reflected direction (black lines).

In the section topograph of Fig. 9.23, the direct images are simply black points; the dynamic images are thick white lines, and the intermediary images a series of black fringes. Knowing both the direct and dynamic images, it is possible in most cases to reconstruct the position of the dislocation line within the crystal.

Moreover, X-ray topography is very useful for determining the direction of the Burgers vector of the line defect. Generally speaking, any crystal defect is invisible in a given Bragg reflection if its displacement field has no component parallel to the Bragg vector $\boldsymbol{r}_H$. Let $\boldsymbol{u}$ be the vector field of displacement from regular lattice periodicity caused by a dislocation of the Burgers vector $\boldsymbol{b}$ and line vector $\boldsymbol{l}$; the invisibility condition is then $\boldsymbol{u} \cdot \boldsymbol{r}_H = 0$. It can be shown that, in a general case and under the approximation of isotropic elasticity, there are non-zero displacement components of the defect parallel to $\boldsymbol{b}$, to $\boldsymbol{l} \times \boldsymbol{b}$ and to $\boldsymbol{l} \times \boldsymbol{b} \times \boldsymbol{l}$, respectively. For a pure screw dislocation, where $\boldsymbol{b}$ is parallel to $\boldsymbol{l}$, we have that $\boldsymbol{l} \times \boldsymbol{b} = 0$ and $\boldsymbol{l} \times \boldsymbol{b} \times \boldsymbol{l} = 0$, and thus there is a single invisibility condition $\boldsymbol{b} \cdot \boldsymbol{r}_H = 0$; this corresponds to complete invisibility of the dislocation for the zone of all Bragg planes which contain $\boldsymbol{l}$. In the case of a pure edge dislocation, $\boldsymbol{b}$ is normal to $\boldsymbol{l}$ and then is parallel to $\boldsymbol{l} \times \boldsymbol{b} \times \boldsymbol{l}$: therefore we have two invisibility conditions, $\boldsymbol{b} \cdot \boldsymbol{r}_H = 0$ and $\boldsymbol{l} \times \boldsymbol{b} \cdot \boldsymbol{r}_H = 0$, corresponding to the single Bragg plane orthogonal to the line vector $\boldsymbol{l}$.

In principle, mixed dislocations always have a $\boldsymbol{u}$ component parallel to $\boldsymbol{r}_H$ and then are never invisible. However, the displacement component parallel to $\boldsymbol{b}$ is dominant in causing diffraction contrast, at least when the absorption is low, so that in these conditions all dislocations are hardly visible if $\boldsymbol{b} \cdot \boldsymbol{r}_H = 0$. Besides, it can be shown that the width and the integrated excess density of the dislocation image on the topograph are approximately

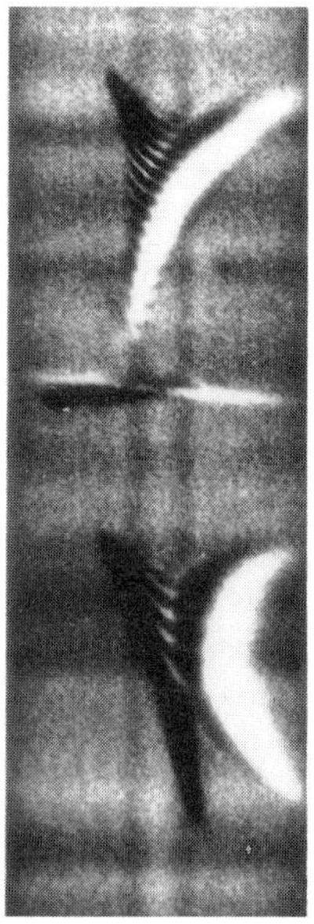

Fig. 9.23. X-ray section topograph of dislocations in silicon. Reflection 2$\bar{2}$0, crystal face (111). (Courtesy of A. Authier.)

proportional to $\boldsymbol{b} \cdot \boldsymbol{r}_H$ and $(\boldsymbol{b} \cdot \boldsymbol{r}_H)^2$, respectively. The Burgers vector direction of a given dislocation can thus be determined unambiguously by choosing properly the diffraction conditions in the X-ray topograph. By using a number of Bragg reflections, the differences in diffraction contrast of the dislocation image allow us to detect the Bragg vectors $\boldsymbol{r}_H$ for which $\boldsymbol{b} \cdot \boldsymbol{r}_H = 0$, and then the direction of $\boldsymbol{b}$. From measurements of the image width it is also possible to derive or to estimate the magnitude, b, of the Burgers vector, obtaining a complete characterization of the line defect.

Energy of a dislocation

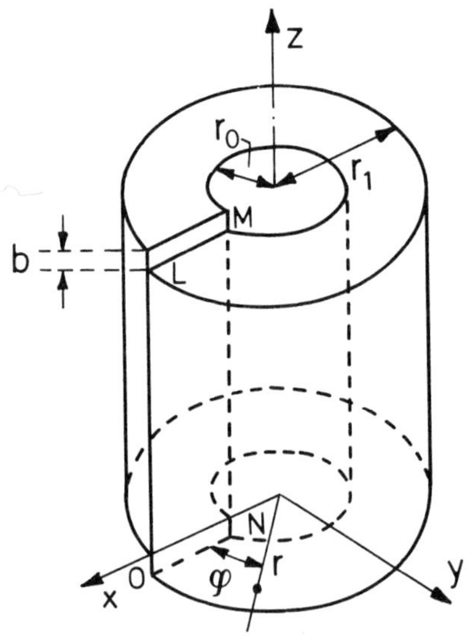

Fig. 9.24. Elastic strain of a cylindrical ring simulating the lattice distortion of a screw dislocation.

Dislocations form, move, and interact with one another in a crystal according to the energy of elastic distortion of the lattice associated with them. It is thus important to analyse the elastic strain around line defects using the basic ideas of crystal elasticity developed on p. 614, and to find out which factors control the dislocation energy. The simplest case is that of a pure screw dislocation, which can be represented for convenience as a shear deformation of a cylindrical ring of isotropic material (Fig. 9.24). A radial slit LMNO is cut parallel to the z axis, and the free surfaces are displaced rigidly with respect to each other by the distance b in the z direction (cf. Fig. 9.18(b)). Owing to the symmetry of the problem, cylindrical r, φ, z rather than Cartesian coordinates are best suited to represent the strain and stress field in the ring. The only non-zero strain component turns out to be the shear term $\varepsilon_{\varphi z} = b/2\pi r$. The corresponding stress is $\tau_{\varphi z} = G\varepsilon_{\varphi z} = Gb/2\pi r$, where G is the shear modulus (the equivalent of a shear component of the elasticity tensor).

It is easy to calculate the strain energy per unit volume W of the deformed ring as half the product of stress × strain:

$$W = \tfrac{1}{2}\tau_{\varphi z}\varepsilon_{\varphi z} = \tfrac{1}{2}G(b/2\pi r)^2. \tag{9.41}$$

As the radius r of the cylindrical ring varies between the inner r_0 and the outer r_1 values, it is necessary to integrate the above expression in order to obtain the total deformation energy per unit length of the ring. For a section of thickness dr and unit length, the volume is $2\pi r\,\mathrm{d}r$; thus the overall energy is given by:

$$E = \frac{1}{2}\int_{r_0}^{r_1} G(b/2\pi r)^2 2\pi r\,\mathrm{d}r = \frac{1}{4\pi}Gb^2 \log\frac{r_1}{r_0}. \tag{9.42}$$

This formula represents the elastic energy per unit length of a screw dislocation, neglecting the contribution of the dislocation core ($r < r_0$) where strains are very large and the elasticity theory fails. However, estimates suggest that the core energy is only a small fraction of the elastic energy, owing to the much smaller crystal volume involved in the core distortion. The r_0 length is of the order of 10^{-9} m, r_1 is of the order of the crystal dimensions, and a typical value of the shear modulus G is 4×10^{10} N m^{-2}; for a Burgers vector of, say, 2.5×10^{-10} m, and a crystal of 0.01 m the energy of a single screw dislocation amounts to 3.2×10^{-9} J m^{-1}.

The energy per unit length of an edge dislocation is given by expression (9.42) divided by $1 - \nu$, where $\nu = (3K - 2G)/[2(3K + G)]$ is Poisson's ratio (≈ 0.3). This is related to the fact that not only shear but also compressive stresses are involved in edge dislocations; thus the bulk elastic modulus K in addition to G appears in the corresponding energy expression.

An important aspect of eqn (9.42) is that the dislocation energy is proportional to the square of Burgers vector magnitude b. As a consequence, for a dislocation with a large value of b it is energetically more favourable to separate into several dislocations characterized by b values as small as possible (i.e. equivalent to a lattice spacing). Partial dislocations (cf. p. 634 below) arise from splitting of a normal dislocation so as to attain Burgers vectors shorter than a lattice spacing, in association with a stacking fault.

Motion and interaction of dislocations

An important feature of dislocations is that they can move through the crystal, as a consequence of applied stress, of thermal effects, and of change of point defect distribution. The simplest kind of motion is the conservative glide, which is equivalent to a shift of the dislocation line in a direction lying within the slip plane. This corresponds to displacing the strained part of the crystal by a glide on that plane.

In a pure edge or screw dislocation, the straight dislocation line (of length l) moves in a direction perpendicular to itself on the glide plane, covering a distance x. Thus the final result of the process would be that the line emerges on the crystal surface with disappearance of the defect (Fig. 9.17), and with an overall slip of a crystal portion by the length b of the Burgers vector. Therefore, the glide process can be considered to be driven either by a force F' applied to the moving portion of the crystal through the distance b, or by a force F applied to the moving dislocation through the distance x. The former one is evidently equal to τlx, the product of stress (= force per unit surface) by the area; the latter can be derived by equating the work performed by the two forces:

$$W = \tau lxb = Fx,$$

whence:

$$F = \tau bl. \tag{9.43}$$

So the force per unit length experienced by a dislocation when an external stress τ is applied to the crystal is simply the product of stress by the length of Burgers vector.

If two screw dislocations are close enough to each other, the stress $\tau_z = Gb/2\pi r$ associated with one of them will act on the other as if it were an external stress. Thus r in this case is equal to the distance d between the two dislocations, and substituting the expression for τ into (9.43) a formula is derived for the force per unit length between the two line defects:

$$\frac{F}{l} = \frac{Gb^2}{2\pi d}. \tag{9.44}$$

Using the G and b values assumed in the numerical example on p. 632, the interaction force per unit length between two dislocations at a distance of 100 Å turns out to be 0.04 N m^{-1}. It should be noticed that the force (9.44) decays very slowly with distance, according to the $1/d$ dependence, so that the interaction between dislocations is quite long range.

A kind of non-conservative motion of dislocations, which must be thermally activated, is the **climb**. This is typical of edge dislocations, when the line moves in a direction normal to the slip plane. Such a process is due to a condensation of vacancies or interstitials along the dislocation line. In the first case the extra lattice half-plane is eroded, with climbing of the line on the same side with respect to the slip plane. In the second case the half-plane is extended beyond the slip plane, and the dislocation line climbs on the other side of that plane.

Evidently, a dislocation line cannot end within the crystal: it either forms a closed loop, or joins another dislocation, or ends on the crystal surface. Thus in general a network of dislocations (Frank net) is present in the crystal, with its extremities ending on the surface. When three or more dislocations meet in a point (a node of the network), then the sum of their Burgers vectors can be easily proved to be zero; this result is formally analogous to Kirchhoff's law of electrical networks. A global property characterizing the network of linear defects in the crystal is the **dislocation density** N, defined either as number of dislocations intersecting a unit area, or as total length of dislocations per unit volume; in both cases the quantity N is measured in units of cm^{-2}.

Partial dislocations

A peculiar type of linear defect is observed in crystals showing layer structures, and is closely connected with stacking faults (p. 625). Let us consider the close-packed cubic structure: atomic layers (Fig. 9.10) are represented by $\{111\}$ lattice planes (i.e. the system of symmetry equivalent planes (111), $(11\bar{1})$, $(1\bar{1}1)$, $(\bar{1}11)$), and clearly most processes of plastic glide related to formation of dislocations occur on these layers. The shortest lattice vector on the (111) plane, $a[1\bar{1}0]/2$, is the most probable Burgers vector for dislocations in this case. Similarly, in a general layer structure the layer plane is preferred as slip plane for formation of dislocations. However, we have learnt (p. 626) that a slip on these layers by a suitable vector, different from a lattice vector, may give rise to a stacking fault. In the FCC structure such a vector is typically $a[11\bar{2}]/6$, translating for instance a B layer into a C position, while the whole stack above the slip plane follows it (cf. Fig. 9.13(a)). Now if the glide process occurs not on the whole plane area, but just on a part of it, then at the border between slipped and unslipped portions a linear defect arises. This is called **partial dislocation**, or **Shockley partial**: it is quite equal to a normal dislocation, except for its Burgers vector not being a lattice vector. Thus a stacking fault not extending through the whole crystal is bounded by a partial dislocation. If the fault occupies a ribbon on the slip plane, then its boundary is formed by two partial dislocations, one for each side. In this case the Burgers vectors of the two partials have as sum a lattice vector, which is the Burgers vector of the unit dislocation sum of the two partials themselves. Let us consider an ordinary dislocation with Burgers vector $a[1\bar{1}0]/2$ on the slip plane (111) of an FCC structure (cf. Fig. 9.10); it can be decomposed into two partials separated by a stacking fault according to the following relation between the corresponding Burgers vectors:

$$a[1\bar{1}0]/2 = a[2\bar{1}\bar{1}]/6 + a[1\bar{2}1]/6.$$

It should be stressed that the vector on the left-hand side of the equality is a latttice vector, while those on the right-hand side are not; this is consistent with the characters of unit and partial dislocations, respectively. The driving force of the above dislocation reaction is the minimization of the dislocation energy (9.42), brought about by a decrease of the Burgers vector $\boldsymbol{b}$.

Small-angle grain boundaries

In a densely pressed or sintered polycrystalline sample, the different crystallites (also called grains) usually have different orientations which may be quite random in some cases. Their interfaces are known as grain boundaries, and can be considered to be extended surface defects of the solid specimen. However, even a single crystal is normally divided into a mosaic of crystallites, which are only slightly misoriented with respect to one another. So in this case we speak of small-angle grain boundaries for the surfaces separating the crystallites. Two parts of a crystal with slightly different orientations, separated by a small-angle grain boundary, are sketched in Fig. 9.25. The wedge-shaped gap in between tends to be filled by portions of lattice planes, giving rise to an array of edge dislocations (points P, Q, R) on the surface of the grain boundary. Let ψ be the small angle between the surface lattice planes of two adjacent crystallites. The points P and R are separated by a distance $b/\sin(\psi/2) \approx 2b/\psi$, so that the spacing between adjacent dislocations P and Q is simply b/ψ. For a very small ψ value of 0.1° (=0.0017 rad) and a Burgers vector of 2.5×10^{-10} m, the distance between edge dislocations at the grain boundary turns out to be 1430 Å.

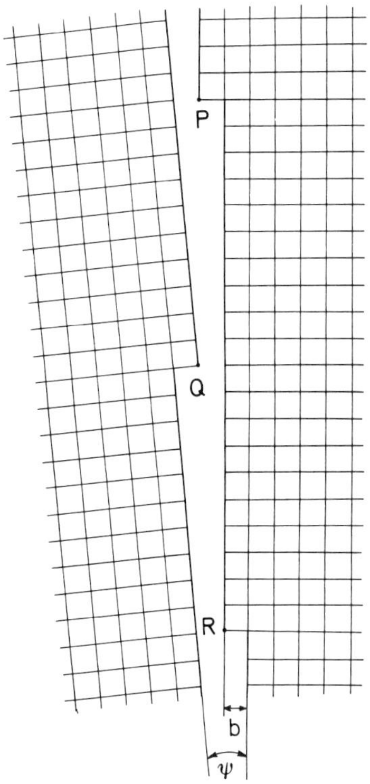

Fig. 9.25. Small-angle grain boundary, showing the array of edge dislocations at the interface.

Point defects

Besides vacancies and interstitials (intrinsic defects), which have already been mentioned on p. 622, a third kind of point defects should be considered. These are impurities, and correspond to substitution of an atom at a regular lattice site by another atom of a different chemical species (extrinsic defect). All such types of point defects are usually observed in ionic, covalent, and metal crystals. If the crystal is not monoatomic, it is necessary to distinguish between stoichiometric and non-stoichiometric compounds. In the first case the defect concentrations of different atomic species are related by the constraint of constant ratios between numbers of atoms, while this does not occur for non-stoichiometric crystals. Furthermore, in ionic solids the numbers of point defects concerning anions and cations are always constrained by the need for electroneutrality. We shall examine stoichiometric ionic crystals in more detail.

A single vacancy, either cationic or anionic, would violate the crystal electroneutrality; thus it must be associated with another vacancy of opposite sign, or to an interstitial of the same sign. In the former case we have a pair of vacancies of different signs called **Schottky defect**: in the latter one, a vacancy + interstitial pair (**Frenkel defect**) is observed. Also an ionic impurity with different charge with respect to the substituted ion introduces an electric unbalance and has then to be compensated by a vacancy or an interstitial of the same sign. For instance, a M^{2+} impurity in a

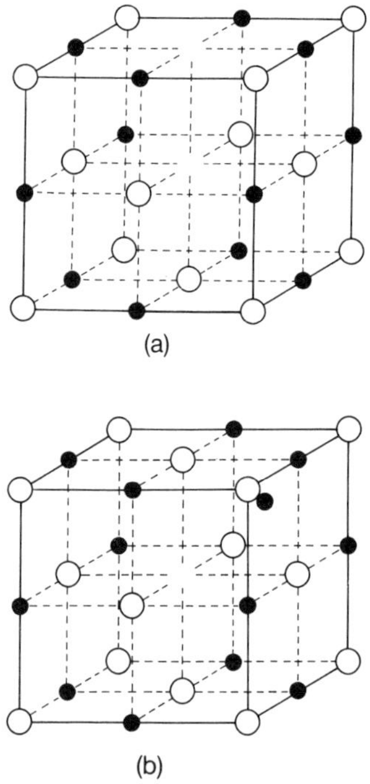

Fig. 9.26. A Schottky defect (pair of cationic and anionic vacancies) (a) and a Frenkel defect (vacancy + interstitial) (b) in the NaCl type structure.

sublattice of A^+ cations would be accompanied by a vacancy of A^+, while a A^+ impurity in a matrix of M^{2+} would require an additional A^+ interstitial.

Let us consider the class of alkali and silver halides AX with the NaCl type structure. In alkali halides the predominant defects are Schottky defects: pairs of A^+ and X^- vacancies, corresponding to empty sites in the two FCC sublattices forming the structure (Fig. 9.26(a)). The missing ions have migrated to the external surface of the crystal, or to internal surfaces (e.g. grain boundaries in polycrystals) or to extended defects (dislocations, stacking faults). Every vacancy is characterized by a relative electric charge (called effective charge), which is evidently equal in absolute value and opposite in sign to the charge of the missing ion. Thus a vacancy of Na^+ bears an effective charge of -1, and a vacancy of Cl^- a charge of $+1$; the symbols V_A^- and V_X^+ are used, respectively, so that a Schottky defect is denoted by the pair $V_A^- + V_X^+$. The space distribution of positive and negative vacancies should not be random, in principle, but is conditioned by the mutual electrostatic interaction. Vacancies of opposite signs attract each other, favouring the formation of pairs in adjacent positions, or of clusters of vacancies with peculiar configurations. However, this tendency increases with the concentration of defects, so that for very high dilution of vacancies in the crystal their interaction can be often neglected to a first approximation.

In silver halides AgX, Frenkel defects involving cations are dominant. An Ag^+ ion leaves its regular lattice position, corresponding to an octahedral hole in the FCC sublattice of X^- anions, and occupies an interstitial position in a tetrahedral hole (Fig. 9.26(b)). In a sense, the presence of Frenkel defects in this case suggests an incipient transformation of the NaCl type structure into that of ZnS type, where coordination environments of both cations and anions are tetrahedral instead of octahedral. Similarly to vacancies, interstitial defects are characterized by an effective charge, relative to the regular structure: this is simply equal to the charge of the interstitial ion itself, and the defect is denoted by the atomic symbol with the i subscript (i.e. Ag_i^+). A Frenkel defect in the AgX halide is symbolized by the pair $V_{Ag}^- + Ag_i^+$.

Thermal distribution of defects

An important question which should be asked is: why do intrinsic point defects form, and which physical variables control their concentration? To give an answer, the problem has to be tackled from a thermodynamic point of view. The formation at constant pressure of a single vacancy in a monoatomic crystal requires a change of enthalpy ΔH_v necessary to remove the atom from its lattice site and to bring it to the crystal surface. On the other hand, a disorder arises in the crystal due to the random position of the vacancy in the lattice. Let N be the number of atomic sites per unit volume, and n_v the number of vacancies per unit volume in the crystal. As the number of ways of distributing the n_v defects over the N sites is $w = N!/[(N - n_v)!n_v!]$, the corresponding increase of configurational entropy brought about by this disorder is

$$\Delta S = k \log w \approx k[N \log N - (N - n_v) \log (N - n_v) - n_v \log n_v].$$

The Stirling approximation for the natural logarithm of the factorial ($\log N! \approx N \log N - N$) was used; k is Boltzmann's constant. The change of Gibbs free energy of the crystal due to the formation of n_v vacancies includes the enthalpic and entropic effects, according to $\Delta G = \Delta H - T\Delta S = n_v \Delta H_v - T\Delta S$. At a given temperature T, equilibrium is attained when n_v is such to make ΔG a minimum:

$$[\partial(\Delta G)/\partial n_v]_T = 0.$$

By solving this equation with respect to n_v, we obtain the result $\log[n_v/(N - n_v)] = -\Delta H_v/kT$; remembering that $n_v \ll N$ (so that $N - n_v \approx N$), the formula relating the equilibrium concentration of vacancies to temperature is derived:

$$n_v \approx N \exp(-\Delta H_v/kT). \qquad (9.45)$$

If the crystal is ionic, the number of cationic vacancies must be equal to the number of the anionic ones. Thus it can be easily shown that the concentration of Schottky defects as a function of temperature is the following:

$$n_s \approx N \exp(-\Delta H_s/2kT). \qquad (9.46)$$

In the case of Frenkel defects, the corresponding distribution is:

$$n_F \approx \sqrt{NN_i} \exp(-\Delta H_F/2kT), \qquad (9.47)$$

where N_i is the number of available interstitial sites per unit volume.

It should be stressed that the relationships (9.45), (9.46) and (9.47) were derived under some simplifying assumptions, the most important of which are:

1. Defects do not interact with one another, and then are distributed in a perfectly random way. This holds to a first approximation only when the concentration of electrically charged defects is very small.
2. The contribution to entropy due to atomic vibrations and the dependence of defect enthalpy on temperature can be neglected. Such an approximation is valid for temperatures which are not very high.

The key significance of the above equations is that formation of intrinsic point defects is brought about by an entropy increase, and that their concentration changes with temperature according to an exponential law.

Diffusion

One of the fundamental phenomena made possible by point defects in crystals is the mobility of atoms and ions in the solid state. The process concerning mass transport by purely thermal activation, with absence of electric fields, is called diffusion. It is governed by Fick's laws, thc second of which is the following:

$$\frac{\partial C}{\partial t} = D\frac{\partial^2 c}{\partial x^2}, \qquad (9.48)$$

where c is the concentration of the diffusing atomic species at time t and at position x along the diffusion direction in the crystal. By integrating the differential equation (9.48) the theoretical concentration profile $c(x, t)$ with respect to space and time is obtained. Measurements based on radioactive tracers or electron microprobe can give an experimental concentration profile: by comparison with the theoretical one, the diffusion coefficient D is derived. If the whole procedure is repeated at different temperatures, the quantity D can be shown to depend on temperature according to the Arrhenius law:

$$D = D_0 \exp(-H_a/kT), \tag{9.49}$$

where D_0 is a constant factor and H_a is the activation enthalpy of the process.

From an atomistic point of view, diffusion consists of thermally activated jumps of atoms or ions from regular lattice positions into empty neighbouring sites (vacancies). The quantity H_a represents the enthalpy needed by the atom to overcome the potential barrier between starting and ending sites. Let jumps occur in a random way, n_v be the total number of vacancies per unit volume, and a the distance between regular site and vacancy; then the following expression can be proved to hold for the quantity D_0 of equation (9.49):

$$D_0 = g\nu a^2 n_v. \tag{9.50}$$

ν is the frequency of vibration of the atom attempting to hop, which is proportional to the probability of jumping into a neighbouring vacancy; g is a geometrical factor depending on the detailed crystal structure (for a primitive cubic lattice, $g = 1/6$). Therefore, D_0 and the diffusion coefficient D are proportional to the concentration of vacancies n_v.

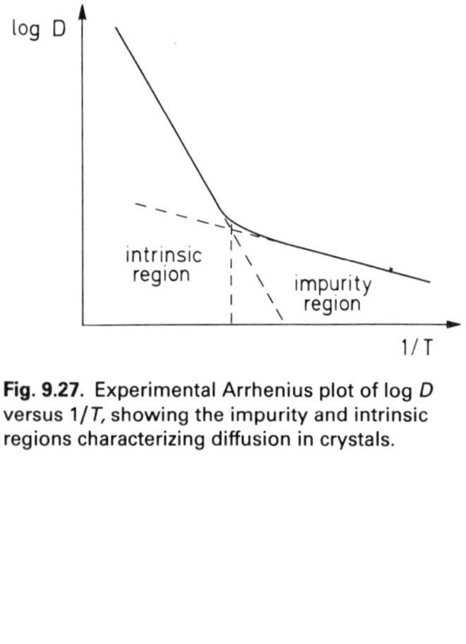

Fig. 9.27. Experimental Arrhenius plot of log D versus $1/T$, showing the impurity and intrinsic regions characterizing diffusion in crystals.

However, the diffusion coefficient D as a function of temperature shows a more complex experimental behaviour (Fig. 9.27). Two regions, of high and low temperature respectively, can be distinguished: these are characterized by quite different slopes of $\log D$ versus $1/T$. Moreover, in the low-temperature region the observed straight line may be shifted up or down by changing the crystal sample of the same substance. The explanation is related to the presence of impurities of aliovalent ions, which, as mentioned on p. 635, give rise to vacancies distinct from those due to thermal disorder. For instance, crystals of NaCl can contain small, different quantities of divalent impurities (e.g. Mn^{2+} ions) substituting Na^+ cations: then for each Mn^{2+} impurity a vacancy V_{Na}^- is created in order to keep the electroneutrality. The number of these vacancies is independent of temperature, changes from sample to sample, and at low temperature is much larger than that of thermal (intrinsic) vacancies. Thus at low temperature (impurity region) the quantity n_v of eqn (9.50) represents essentially the number of vacancies due to impurities and is a constant; the slope of the straight line $\log D$ versus $1/T$ is simply H_a, according to (9.49). At high temperature, on the other hand, intrinsic vacancies outnumber extrinsic ones (intrinsic region), and expression (9.46) for the thermal distribution of Schottky defects must replace n_v in equation (9.50) for D_0. The result obtained for the diffusion coefficient D is:

$$D = g\nu a^2 N \exp\left(-\frac{H_a + \Delta H_s/2}{kT}\right). \tag{9.51}$$

Therefore, in the intrinsic region the slope of the line $\log D$ versus $1/T$ equals $(H_a + \Delta H_s/2)/k$ instead of H_a/k, and is then larger than the corresponding one in the impurity region.

Ionic conductivity

Alternatively to the thermally activated process (diffusion), ionic transport in crystals is driven by an applied electric field. In this case electric conduction based on migration of ions rather than electrons occurs; however, the atomistic mechanism always relies upon the presence of point defects, just as for diffusion. Generally, the relevant contribution to conductivity is given either by cations (e.g. alkali or silver ions) or by anions (halide or oxide ions). In either case the conductivity σ is proportional to the number of charge carriers per unit volume, n, and to the mobility of the migrating ion, μ (e is the electron charge); for cations:

$$\sigma_+ = en_+\mu_+. \tag{9.52}$$

By considering the equilibrium condition between the field-induced drift current and the opposing diffusion current due to the concentration gradient, a very important relationship between mobility and diffusion coefficient (Nernst–Einstein equation) can be derived:

$$\mu = \frac{e}{kT}D. \tag{9.53}$$

This equation provides a link between the phenomena of diffusion and of ionic conductivity. By substitution into (9.52) one obtains:

$$\sigma_+ = \frac{n_+e^2}{kT}D = \frac{n_+e^2}{kT}D_0 \exp\left(-\frac{H_a}{kT}\right). \tag{9.54}$$

If the process occurs in the intrinsic (high-temperature) region of Fig. 9.27, where the number of Schottky defects follows Boltzmann's thermal distribution (9.46), then eqn (9.51) holds for D and can be substituted into (9.54), yielding:

$$\sigma_+ = \frac{n_+e^2}{kT}gva^2N \exp\left(-\frac{H_a + \Delta H_s/2}{kT}\right). \tag{9.55}$$

On the other hand, eqn (9.54) holds in the low-temperature impurity region, where the number of vacancies in the expression (9.50) for D_0 is independent of temperature.

From an experimental point of view, it is much easier to perform measurements of electrical conductivity than of diffusion coefficients. Hence the importance of this technique for studying the energetics of point defects appears clearly: by plotting $\log \sigma T$ against $1/T$ for an ionic crystal, one obtains diagrams quite similar to that of Fig. 9.27, showing two straight lines with different slopes for the intrinsic and impurity regions. On the basis of eqns (9.55) and (9.54), these slopes are equal to $(H_a + \Delta H_s/2)/k$ and to H_a/k, respectively. Thus it is possible to derive from these measurements both the activation enthalpy H_a for hopping of ions into vacancies, and the formation enthalpy of Schottky defects ΔH_s. Results of experiments on NaCl give values of 0.68(1) eV for H_a, and 2.30(1) eV for ΔH_s.

Appendix

9.A Properties of second-rank tensors

Second-rank tensors represent physical laws of linear type in crystalline media. A very important class is that of symmetrical second-rank tensors, which are characterized by the property $y_{ji} = y_{ij}$ for all their components in any reference basis. Antisymmetrical tensors, on the other hand, follow the rule $y_{ji} = -y_{ij}$, implying that diagonal components are zero ($y_{ii} = 0$). It may be shown easily that any tensor can be written as the sum of a symmetrical plus an antisymmetrical tensor. Symmetrical tensors are associated to the quadratic form:

$$\tilde{\mathbf{x}}\mathbf{y}\mathbf{x} = \sum_{1}^{3}{}_{i,j} y_{ij} x_i x_j = 1, \tag{9.A.1}$$

which is also the equation of a general second-order surface (a quadric); this surface is obviously invariant with respect to a change of reference basis, and is thus a correct geometrical representation of the symmetrical tensor $\mathbf{y}$.

Eigenvalues and eigenvectors

In order to study the features of the representation quadric (9.A.1), the equation $\boldsymbol{Y} = \lambda \boldsymbol{X}$, or

$$\mathbf{y}\boldsymbol{X} = \lambda \boldsymbol{X}, \tag{9.A.2}$$

should be considered. The solutions of this equation are $\boldsymbol{X}$ vectors which are transformed by action of the tensor $\mathbf{y}$ into parallel vectors $\lambda \boldsymbol{X}$; so the corresponding directions define axes of isotropy in the crystal. Such solutions are called eigenvectors, and the corresponding λ values are the eigenvalues of the tensor. It is important to stress that eigenvalues (scalars) and eigenvectors (vectors) are invariant quantities which do not depend on the reference basis; only their components do. The matrix equation (9.A.2) is equivalent to a system of three homogeneous linear equations in the unknowns X_1, X_2, X_3 which admits non-zero solutions if and only if the determinant of coefficients is zero:

$$\det(\mathbf{y} - \lambda \mathbf{I}) = 0. \tag{9.A.3}$$

By calculating the determinant explicitly, an equation of third degree in the unknown λ is obtained; it can be proved that all its three roots are real because $\mathbf{y}$ is symmetrical. By substitution of each real eigenvalue λ_1, λ_2, λ_3 into (9.A.2) the corresponding eigenvectors $\boldsymbol{X}_1$, $\boldsymbol{X}_2$, $\boldsymbol{X}_3$ can be found. One of the three linear equations implicit in (9.A.2) has to be rejected, because it is linearly dependent on the other two (cf. eqn (9.A.3)); therefore only directions and relative (not absolute) moduli of eigenvectors can be determined. It can be demonstrated that if all three eigenvalues are different, then the eigenvectors are unique and orthogonal to one another; otherwise, they may not be unique but can always be chosen to be orthogonal. It is also convenient to normalize the eigenvectors, so that the normalized components Q_{ih} are dimensionless quantities which obey the orthonormality condition:

$$\sum_{h=1}^{3} Q_{ih} Q_{jh} = \delta_{ij}, \tag{9.A.4}$$

where the first subscript is the identity number of the eigenvector, and the second one specifies the Cartesian component.

It is very useful to change the reference basis from the original one to that formed by the three normalized eigenvectors $\boldsymbol{Q}_1$, $\boldsymbol{Q}_2$, $\boldsymbol{Q}_3$, and to find out the representation of tensor $\mathbf{y}$ in the new basis. The transformation matrix $\mathbf{Q}$ has rows given by the components Q_{ij} of each eigenvector:

$$\mathbf{Q} = \begin{pmatrix} Q_{11} & Q_{12} & Q_{13} \\ Q_{21} & Q_{22} & Q_{23} \\ Q_{31} & Q_{32} & Q_{33} \end{pmatrix}.$$

By applying the tensorial transformation rule (9.6), we obtain:

$$y'_{ih} = \sum_{1}^{3}{}_{k,l} Q_{ik} Q_{hl} y_{kl} = \sum_{l=1}^{3} \left(\sum_{k=1}^{3} Q_{ik} y_{kl} \right) Q_{hl} = \sum_{l=1}^{3} \lambda_i Q_{il} Q_{hl} = \lambda_i \delta_{ih},$$

taking into account (9.A.2) and (9.A.3). This means that the representation of tensor $\mathbf{y}$ in the basis of its eigenvectors is a diagonal matrix whose non-zero components are just the eigenvalues λ_1, λ_2, λ_3. The eigenvalues take the names of principal components ($y_1 = \lambda_1$, $y_2 = \lambda_2$, $y_3 = \lambda_3$), and the directions of the eigenvectors are called principal axes (or principal directions) of tensor $\mathbf{y}$; this is represented with respect to the basis of its eigenvectors by the matrix:

$$\begin{pmatrix} y_1 & 0 & 0 \\ 0 & y_2 & 0 \\ 0 & 0 & y_3 \end{pmatrix}.$$

Along the principal directions the crystal behaves isotropically, i.e. each eigenvector $\boldsymbol{X}_i$ is transformed by tensor $\mathbf{y}$ into a parallel vector $\boldsymbol{Y}_i$ according to $\boldsymbol{Y}_i = y_i \boldsymbol{X}_i$, which is the equivalent of (9.A.2).

Representation surfaces and their properties

With respect to a general basis, the action of tensor $\mathbf{y}$ is expressed as:

$$Y_i = \sum_{h=1}^{3} y_{ih} X_h, \tag{9.A.5}$$

while in the basis of eigenvectors the correspondent relation is simply:

$$Y_i = y_i X_i. \tag{9.A.6}$$

In the latter basis, the equation of the representation quadric (9.A.1) becomes:

$$\sum_{i=1}^{3} y_i x_i^2 = 1. \tag{9.A.7}$$

From the general features of second-order surfaces and their equations, it turns out that if all eigenvalues are positive then (9.A.7) is the equation of an ellipsoid; if only one is negative, the surface is a hyperboloid of one sheet; if two eigenvalues are negative, we have a hyperboloid of two sheets, and if all of them are negative, an imaginary ellipsoid is obtained (Fig. 9.A.1). Of course in the latter case one usually considers as repre-

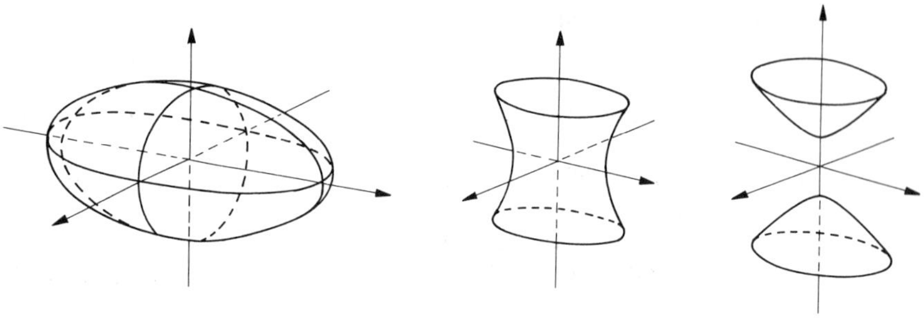

Fig. 9.A.1. Representation surfaces of second-rank tensors: ellipsoid, hyperboloid of one and two sheets.

sentative surface the real ellipsoid of equation

$$\sum_{i=1}^{3} y_i x_i^2 = -1.$$

Let us consider now the case of all eigenvalues positive. Then by writing equation (9.A.7) as $\sum_{i=1}^{3} (x_i/a_i)^2 = 1$, it turns out that the semi-axes of the representation ellipsoid have lengths given by $a_i = 1/\sqrt{y_i}$, i.e. they are equal to the inverse square root of principal tensor components. The symmetry axes of the quadric correspond to the principal directions of the tensor.

We want now to derive an expression for the component y_l of tensor $\mathbf{y}$ along a given direction characterized by direction cosines l_1, l_2, l_3 (then $\boldsymbol{l}$ is the unit vector parallel to that direction). Such a component is defined as the projection onto $\boldsymbol{l}$ of the $\boldsymbol{Y}$ vector corresponding to a $\boldsymbol{X}$ vector parallel to $\boldsymbol{l}$, divided by the modulus X: $y_l = \boldsymbol{Y} \cdot \boldsymbol{l}/X$. With respect to a general Cartesian basis, by applying the usual formula of scalar product and taking into account (9.A.5) one obtains:

$$y_l = \sum_{i=1}^{3} Y_i l_i / X = \sum_{i=1}^{3} \sum_{j=1}^{3} y_{ij}(X l_j) l_i / X = \sum_{1}^{3} {}_{i,j} y_{ij} l_i l_j. \tag{9.A.8}$$

With reference to the basis of eigenvectors of $\mathbf{y}$, a simpler expression is derived:

$$y_l = \sum_{i=1}^{3} y_i (X l_i) l_i / X = \sum_{i=1}^{3} y_i l_i^2. \tag{9.A.9}$$

A geometrical interpretation of the quantity y_l can be obtained by considering the equation of the representation surface of $\mathbf{y}$ (9.A.1), and by substituting $x_i = x l_i$. The equation becomes:

$$\sum_{1}^{3} {}_{i,j} y_{ij} l_i l_j = 1/x^2,$$

where of course x represents the length of a radius vector joining the origin to a general point on the quadric surface. Because of (9.A.8), it follows that $x = 1/\sqrt{y_l}$, i.e. the radius vector of the representation quadric has a length equal to the inverse square root of the tensor component along the same direction.

Further reading

Amelinckx, S., Gevers, R., and Van Landuyt, J. (ed.) (1978). *Diffraction and imaging techniques in material science.* North-Holland, Amsterdam.

Amouric, M. (1987). *Acta Crystallographica,* **B43,** 57–63.
Amouric, M. and Baronnet, A. (1983). *Physics and Chemistry of Minerals,* **9,** 146–59.
Authier, A. (1967). *Advances in x-ray analysis,* vol. 10, pp. 9–31. Plenum Press, New York.
Bollmann, W. (1970). *Crystal defects and crystalline interfaces.* Springer, Berlin.
Catti, M. (1985). *Acta Crystallographica,* **A41,** 494–500.
Catti, M. (1989). *Acta Crystallographica,* **A45,** 20–5.
Catti, M., Ferraris, G., and Ivaldi, G. (1989). *European Journal of Mineralogy,* **1,** 625–32.
Henderson, B. (1972). *Defects in crystalline solids.* Arnold, London.
Hull, D. (1968). *Introduction to dislocations.* Pergamon, Oxford.
Nye, F. (1985). *Physical properties of crystals.* Clarendon, Oxford.
Rosenberg, H. M. (1975). *The solid state.* Clarendon, Oxford.
Sands, D. E. (1982). *Vectors and tensors in crystallography.* Addison-Wesley, Reading, MA.
Sirotin, Yu. I. and Shaskolskaya, M. P. (1982). *Fundamentals of crystal physics.* Mir, Moscow.
Tilley, R. J. D. (1987). *Defect crystal chemistry.* Blackie, Glasgow.
Vainshtein, B. K., Fridkin, V. M., and Indenbom, V. L. (1982). *Modern crystallography,* Vol. II. Springer, Berlin.
Vainshtein, B. K. (ed.) (1988). *Modern crystallography,* Vol. IV. Springer, Berlin.
Zarka, A., Lin, L., and Sauvage, M. (1983). *Journal of Crystal Growth,* **62,** 409–24.

Subject index